Alexander Weinmann

Uncertain Models and Robust Control

Springer-Verlag Wien GmbH

Univ.-Prof. Dr. Alexander Weinmann
Department of Electrical Engineering
Technical University Vienna
Austria

With 123 Figures

ISBN 978-3-211-82299-9 ISBN 978-3-7091-6711-3 (eBook)
DOI 10.1007/978-3-7091-6711-3

To Monika

Preface

Control system analysis and design are important fields in the engineering sciences. Process identification and optimum design have developed rapidly over the past decades and resulted in great engineering progress. However, many results have suffered from considerable dependence on uncertainties, mainly incorporated in the plant, in actuators and sensors, and released by other internal disturbances.

More and more, methods tackling uncertainties form a central and important issue in designing feedback control systems. Choosing appropriate design methods, the influence of uncertainties on the closed-loop behaviour can be reduced to a large extent. Control systems particularly designed to manage uncertainties are called robust control systems. Robust control theory provides a design philosophy with respect to perturbed or uncertain parts of the system.

This monograph is devoted to plants and their approximate models and to the discussion of their uncertainties. The thrust of the book is on systematic representation of methods for robust control.

Most of the important areas of robust control are covered. The aim is to provide an introduction to the theory and methods of robust control system design, to present a coherent body of knowledge, to clarify and unify presentation, and to streamline derivations and proofs in the field of uncertainties and robust control. The book contains a thorough treatment of important material which is scattered throughout the literature.

The primary goal of the text is to present significant derivations and proofs. As far as less significant proofs or lengthy derivations are concerned, only the results are outlined and the reader is referred to the literature relevant to this subject. In a few cases, some topics are set forth only as a suggestion for further reading.

Some important problems treated in this book are:

- How is uncertainty described and bounded when applying methods of differential equations, transfer functions or transfer matrices, state-space algorithms or some approximating calculus?

- Which uncertainty or which perturbation in some special system description forces the system to instability? This question arises both for open-loop and for closed-loop systems.

- Which kind of controller is able to tolerate maximum uncertainty?

In most cases *stability robustness* is applied but *performance robustness* is also a very important subject, e.g., determining the regions in the parameter space of controllers which place all the closed-loop eigenvalues into a desired region, thus satisfying the design specifications in the face of uncertain but bounded plant parameters.

The book is intended specifically for practicing but mathematically inclined engineers, for postgraduate students and engineers on master's level. Moreover, the book is intended for those readers who wish to apply robust design principles and theories to real-life applications and problems.

Emphasis is put on practical considerations and on applicability at the expense of extensive proofs and mathematical rigor.

The book is organized in the following way:

- Part I (chapters 1 through 3) concentrates on preliminaries and on useful properties of matrix norms and measures, function norms, Kronecker calculus and analysis using vector and matrix notation.

- Part II (chapters 4 through 10) is devoted to differential sensitivity. It is advantageous to evaluate the magnitude of differential sensitivity of some system performance or property with respect to uncertainties in the parameters and to minimize this sensitivity in order to achieve small-scale robust control or desensitized control.

- In great detail Part III (chapters 11 through 20) covers the main results in the time domain such as differential or difference equations and Lyapunov theory for state-space representations.

- Part IV (chapters 21 through 28) comprehensively treats results in the frequency domain and their transfer matrix representation, stability margin, stability radius, resolvent matrix, structured singular value, performance robustness and spectral radius techniques.

- Part V (chapters 29 through 31) emphasizes the internal model principle, parametrization and coprime factorization of systems and minimax frequency domain optimization techniques in the Hardy space.

- Part VI (chapters 32 through 38) deals with sliding mode robust control, singularly perturbed systems and several decomposition techniques at a general and abstract level. Models are applied, such as orthogonal or power series expansions. This is a field where several new methods can be developed and where the uncertainty is replaced by neglected orthogonal components or is determined by the degree of expansion.

- In the Appendix, matrix computation and matrix inversion, eigenvalues and more sophisticated algebraic methods and linear regression calculus are presented.

General knowledge is preassumed, covered, e.g., by the books of *Brogan, W.L., 1985; Chen, C.T., 1970; Franklin, G.F., et al. 1988; Hsu, J.C., and Meyer, A.U., 1968; Kwakernaak, H., and Sivan, R., 1972; Maciejowski, J.M., 1989; Ogata, K., 1970; 1977; Reid, J.G., 1983; Weinmann, A., 1988.*

There are many people who gave me encouragement while I was preparing this book. I express my appreciation to Professors Manfred Deistler, Richard Eier, Peter Jörgl, Inge Troch and Robert Noisser for discussing a number of problems.

My gratitude is due to Mrs. Johanna Heinrich who capably typed the main part of the text and of the formulas in various revisions and who did the computer handling and some layout work. I thank her for her patience and exactness.

For discussing specific topics, for comments and for their help in solving numerical problems I am indebted to a number of persons: Christian Augesky, Hans-Peter Bauböck, Wolfgang Ehrendorfer, Johannes Goldynia, Wilhelm Haager, Michael Heiss, Karl Helm, Helmut Homole, Thomas Kirnbauer, Heinrich Lederer, Leopold Moosbrugger, Gerhard Silnusek, Karl Schiftner and Andreas Trenner.

My appreciation goes to Mrs. Susan Siegle for attentive language screening.

I am indebted to Wolfgang Fuchs for several software adaptations, for the back-up system and for solving problems of computer compatibility. I particularly thank Bernhard Simon and Johannes Goldynia for solving special problems concerning the computer support and the computer network. Franz Babler supported me by drawing most of the figures. Mrs. Friederike Vasina assisted in typing the author index and supplying journals and books.

I extend my thanks to many diligent and hard-working students graduating at the Technical University Vienna. I am indebted to the audience of several symposia and courses on the topic of robust control. Various interesting questions inspired me and several discussions did much to improve the structure of the book.

Finally, I thank my wife for helping me in correcting the text. More than thanks goes to my family for their kind understanding when I was writing the book.

Last but not least, I would like to express my thanks to Raimund Petri-Wieder and Frank Christian May of Springer Verlag, Vienna, for excellent cooperation.

Vienna, June 1991 Alexander Weinmann

Contents

IV Robustness in the Frequency Domain 321

V Coprime Factorization and Minimax Frequency Optimization 457

Part I

Introduction

Chapter 1

Introductory Survey

The introductory survey is intended to provide the reader with the aims of uncertain models and structures of uncertainties. An overview covering several robust control concepts is embedded into general techniques of how to overcome uncertain systems. Furthermore, concepts are included that already exist and that are likely to be applicable in the future.

A primary objective of feedback is to reject disturbances and to provide good tracking facilities. Additional requirements for an efficient control system are the properties of stability (rapidly decaying modes), noise filtering (suppression of sensor noise) and moderate effort or actuator signal magnitude. In Fig. 1.1, a block diagram briefly sketches the signals and components of concern.

Inherent in real-world systems is the fact of uncertainty. Investigating the properties of systems with uncertain structure or uncertain parameters is the aim of perturbation theory. Hence, an additional facility of controllers is to enable a controlled system to cope with uncertainties caused by varying plant structure or parameters. For *all constant values* of plant parameters $\mathbf{p}$ in a given set of parameters, the aforementioned properties must hold. Robustness is the property of dynamic systems to tolerate variations of parts of the system without exceeding predetermined tolerance bounds in the vicinity of some nominal dynamic behaviour.

Not only constant but also time-varying parameter perturbations are investigated in robust controller design.

It is worth noting that to a large extent even linear *optimal* systems are able to tolerate time-dependent or nonlinear behaviour, without affecting stability when introduced additionally to the original linear-optimal control law, see e.g. *Anderson, B.D.O., 1969*. On the other hand, dynamic systems may turn out very sensitive to perturbations in some constellation and are not able to cope with uncertainties. These circumstances require a detailed investigation.

1.1 Uncertainty and Control. An Overview

Important representatives for system uncertainties are given in what follows: Poor plant knowledge, uncertain or slowly varying parameters, nonlinearities such as hysteresis or friction, errors or unknown initial conditions, reduced-order models, additive unknown internal or external noise, environmental influence.

Uncertainties are given by individual parameter variations, bounded by inequalities or by singular values or norms of matrices containing the parameters. Uncertainties may also be represented by interdependent parameter relations. There is a great variety of mathematical descriptions.

The engineering field is not the only case of application. Uncertainties also occur in ecological or economic processes. In these topics, there may be even less insight into the model structure and parameters, as far as the matter of uncertainty is concerned.

1.1.1 Uncertainty and its Bounds

Usually, with regard to physical and engineering insight or by executing identification algorithms, the dynamical behaviour of the process is determined, that is, the structure, order, linearity and multivariability. Considering unperturbed behaviour, the system is denoted nominal. In spite of these activities, several properties remain unknown and one has to analyze the following subjects:

(i) Uncertainty of structure and parameters, *including* some information about the bounds of the uncertainty.

(ii) Uncertainty *without being informed* about the prospective variation and magnitude of the uncertainty.

Treating unperturbed systems, one remembers to maximize the distance of the nominal system to instability. Owing to uncertainties, maximization of the distance has to be performed in their presence, in either aforementioned case. Stability boundaries may be replaced by limits of the index of performance.

1.1.2 Parameter Uncertainty

In many cases of applications, the structure, linearity and order of the system are considered fixed and the parameters of the systems are allowed to change to a certain extent. Changes in parameters of parts of the system in consequence determine changes in the coefficients of the closed-loop system dynamics, that is, for instance, characteristic polynomial or state-space description. Only very simple cases are described by coefficient variations free of interdependencies. Complex processes are characterized by rather complex relations between changes in the parameters and changes in the coefficients of the closed-loop system. Nevertheless, even in those cases, sufficient conditions are required for

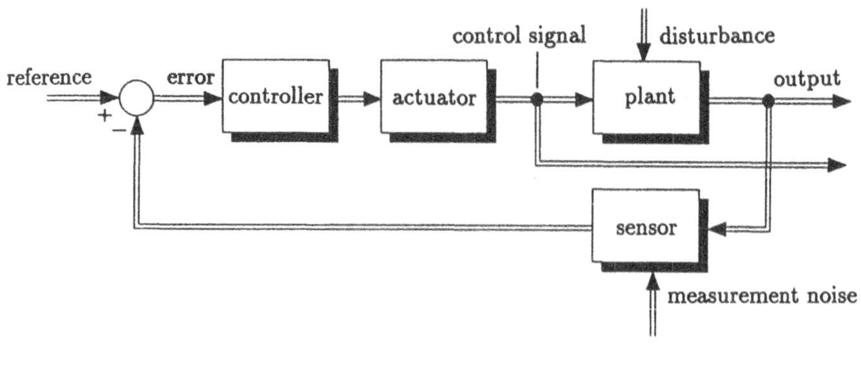

Figure 1.1: Multivariable control system

allowable parameter changes without causing inadmissible change in the dynamic system behaviour.

Usually, ***sufficient*** robustness results are required. Whether or not a robust design procedure is perhaps too cautious or too conservative must be investigated by means of several theoretical considerations or simulation tests. The standard objective is to find a criterion for a controller as least conservative as possible. An important problem in many fields of applications is to assess and discuss and reduce the region between necessary and sufficient conditions for robustness.

1.1.3 Robustness and Differential Sensitivity

Uncertainties with some additional bounds seem to be an adequate form of system description, especially from the practioner's point of view. But how to proceed if the only information is some presumable nominal behaviour and if uncertainty bounds are not available? There is a practicable way to investigate system robustness with respect to any kind of *small* parameter variation. That is, to examine small changes in system properties, such as performance and stability, with respect to small parameter variations. This procedure yields the *differential sensitivity* or *small-scale robustness* of a certain system property whenever small but unknown variations are taken into account.

Parameter differential sensitivity is a local characteristic defined by partial derivatives of first order with respect to a parameter p . This derivative only characterizes the system behaviour Ω in an infinitesimal vicinity of the nominal point p_o, that is, as briefly sketched in Fig. 1.2,

$$\frac{\partial\Omega}{\partial p} \mid_{p=p_o} \quad \leadsto \quad \Delta\Omega \doteq \frac{\partial\Omega}{\partial p} \mid_{p=p_o} \Delta p \; . \tag{1.1}$$

There is no guarantee to which extent the *environment* of p_o is sufficiently described by the local sensitivity at $p = p_o$. For small and even unknown Δp, the change of performance is small if $\partial\Omega/\partial p$ is small. The system behaviour Ω and the parameter p can be extended to vector-valued or matrix-valued variables by using Kronecker calculus and/or an index of performance with appropriate weighting matrices.

Frequently, interest focusses on examining the change of some closed-loop system behaviour caused by small open-loop perturbations. Examples are given by closed-loop eigenvalues, state variables, transition matrix, transfer function or matrix transfer function, step response, stability radius and some index of performance (*Sage, A.P., and White, C.C. III, 1977; Tomović, R., and Vukobratović, M., 1972; Kofahl, R., 1988*).

From a practical point of view, finite parameter deviations Δp with predetermined bounds seem to be the better definition. If the system behaviour Ω remains inside a certain bound $\Delta\Omega$, although p varies within a predescribed range Δp, the system is called robust with respect to the chosen system behaviour Ω (*Frank, P.M., 1985; Ficklscherer, P., and Müller, P.C., 1985*). Occasionally, differential sensitivity is only considered for comparison to robustness measures and uncertainty assumptions.

Robustness is defined as the facility of a control loop to have *one or more* properties Ω within predetermined bounds, in the presence of unknown parts or uncertain parameters of the plant. Robust analysis directly aims at the assessment whether Ω remains within predetermined bounds for the complete set of plant parameter variations Δp. The indirect sensitivity-based method uses Eq.(1.1). Without knowledge of Δp the change $\Delta\Omega$ in the property Ω will be small if $\partial\Omega/\partial p$ is sufficiently small. Hence, differential sensitivity analysis may be considered as the local property of robustness, only.

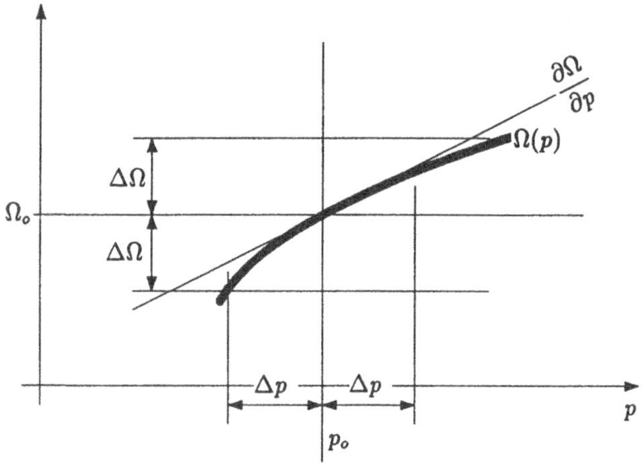

Figure 1.2: Differential sensitivity

1.1.4 Robustness and Sensitivity

Sensitivity as given by the matrix transfer function $(\mathbf{I} + \mathbf{GK})^{-1}$ is a measure of the frequency-domain property of the output variable of a control system, generated by unwanted *disturbance* input to the control system, for instance, produced by the *environment* of the control system.

Robustness is a measure of assessment to which extent the system behaviour or performance is violated by *system perturbations*.

1.1.5 Sensor and Actuator Failures

Sensor failures or actuator failures must often be taken into consideration in the process of robust control design. Robustness can be extended even to abrupt changes in the dynamic or static behaviour of sensors or actuators in control systems. Additionally, high-performance systems cause expensive back-up measurement to be implemented, in order to tolerate abrupt changes such as failures. Some statistical detection also seems appropriate (*Willsky, A.S., 1976*).

In many cases, low-performance multi-loop control systems are able to robustly manage the control function, even in the presence of failures. High-performance systems, however, require additional effort to achieve stability and performance robustness.

1.1.6 Deterministic and Stochastic Models

It is a common practice that robust control systems are characterized by deterministic unknown perturbations where bounds or norms of their magnitude are available.

Uncertainty can also be described by stochastic models, e.g., by adding noise to the signals either with fixed or uncertain stochastic parameters.

Robustness can also be defined from a stochastic point of view, estimating the probability of instability by Monte Carlo evaluation of the eigenvalues of the system (*Stengel, R.F., 1980; Stengel, R.F., and Ray, L.R., 1991*). However, the following chapters in this book deal with stochastic models of robustness in the margin, only.

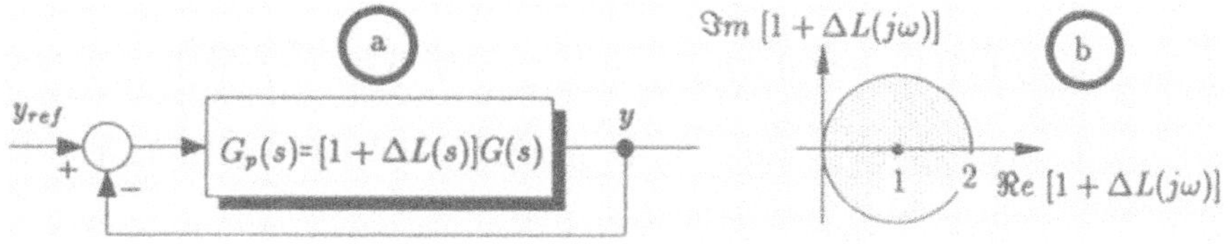

Figure 1.3: Scalar system and uncertainty location

1.1.7 Robustness and Performance I

In many cases, robustness is considered as a system property of ***stability*** being unaffected by perturbations, as far as possible. However, stability is not the only property of control systems. Stability is only a necessary requirement for robustness. On top of that, system *performance* also has to be taken into account in the field of engineering problems. In almost every case of application, increasing robustness may reduce the nominal quality of performance considerably. Roughly speaking, a trade-off between robustness and performance has to be solved.

A simple example is selected to demonstrate the problem. Let an open-loop scalar system contain perturbed dead-time delay $T_p = T + \Delta T$. Then, the open-loop transfer function is $G_p(s) = G_o(s.)e^{-s(T+\Delta T)}$. The phase margin φ_{mp} at the crossover frequency ω_r is

$$\varphi_{mp} = \pi + \arg G_p(j\omega_r) = \pi + \arg G_o(j\omega_r) + \arg e^{-j\omega_r(T+\Delta T)} \triangleq \varphi_m + \Delta\varphi_m \qquad (1.2)$$

where $\varphi_m = \pi - \omega_r T + \arg G_o(j\omega_r)$ and $|G_o(j\omega_r)| = 1$. The relation $\Delta\varphi_m$ versus ΔT is given by $\Delta\varphi_m = -\omega_r \Delta T$. The greater the delay perturbation ΔT the smaller the crossover frequency ω_r must be chosen, in particular via design of the controller part in $G_o(s)$, in order to bound the change in phase margin $\Delta\varphi_m$.

The term *resilience* is used to describe the ability of the process to move both fast and smoothly from one operating condition to another (*Morari, M., 1983a*). Dynamic resilience is significantly determined by constraints on the control signal and by robustness with respect to plant perturbations or model plant mismatch.

The problem of controlling processes in an uncertain environment may be considered as some kind of designing an intelligent machine (*Saridis, G.N., 1985*). *Knowledge* is a function of *removing* uncertainty in the operation of an intelligent machine. The *rate of knowledge* is the ability to *reduce* uncertainty, and defines a measure of intelligence. The control performance measure is given by an aggregated entropy of the system. The optimal intelligent machine is achieved by minimizing the overall entropy of the system.

1.1.8 Robustness and Performance II

Constant-gain single-input feedback controllers provide an introductory matter to limitations between robustness and performance. Consider a scalar system with multiplicative uncertainty $\Delta L(s)$, see Fig. 1.3a. The transfer functions $G(s)$ and $G_p(s)$ denote the nominal and perturbed open-loop transfer functions, respectively. The closed-loop poles are given by the zeros of $1 + G_p(s) = 1 + G(s) + \Delta L(s)G(s)$. Provided that $G(s)$ is stable, the closed-loop poles cannot be located at the imaginary axis if

$$1 + G(j\omega) + \Delta L(j\omega)G(j\omega) \neq 0 \quad \rightsquigarrow \quad |\Delta L(j\omega)| < |\frac{1 + G(j\omega)}{G(j\omega)}| . \tag{1.3}$$

Implicitly, the tracking error e_T caused by the uncertainty of the system is defined by the following equation

$$\frac{G_p(s)}{1+G_p(s)} - \frac{G(s)}{1+G(s)} = \frac{G(s)(1+\Delta L)}{1+G(s)(1+\Delta L)} - \frac{G(s)}{1+G(s)} \tag{1.4}$$

$$= \frac{G(s)}{1+G(s)} \underbrace{\frac{\Delta L(s)}{1+G(s)+G(s)\Delta L(s)}}_{e_T} \triangleq \frac{G(s)}{1+G(s)} e_T . \tag{1.5}$$

The error e_T can be kept low by small uncertainty $\Delta L(s)$ or by high loop gain $|G(s)|$. Furthermore, note the following properties.

- **High Loop Gain**: Referring to Eq.(1.3), high loop gain $|G(j\omega)|$ requires $|\Delta L(j\omega)| < 1$, i.e., the complex number $1 + \Delta L(j\omega)$ must be located inside a certain unit disc, as depicted in Fig. 1.3b. Remember that $1 + \Delta L(j\omega)$ is the ratio between perturbed and nominal transfer function at $s = j\omega$. It follows from Fig. 1.3b that the argument of $1 + \Delta L(j\omega)$ must satisfy the condition $-\pi/2 \leq \arg[1 + \Delta L(j\omega)] \leq \pi/2$. Small values e_T via high gain $|G(j\omega)|$ can be achieved only at those frequencies where the uncertainty ratio argument does not exceed $\pi/2$ in modulus.

- **Loop Gain at Crossover Frequency**: At crossover frequency ω_r , the transfer function becomes $G(j\omega_r) = 1 \times e^{j\varphi_r}$ where the quantity φ_r is the open-loop phase and $\varphi_m = \pi + \varphi_r$ is the phase margin. Then, from Eq.(1.3)

$$|\Delta L(j\omega_r)| \leq |\frac{1 + e^{j\varphi_r}}{e^{j\varphi_r}}| = \sqrt{2(1 + \cos\varphi_r)} = \sqrt{2(1 - \cos\varphi_m)} . \tag{1.6}$$

  The phase margin must exceed a minimum amount in order to permit sufficient uncertainty. In detail, for a fixed $|e_T|$, given by the aforementioned definition in Eq.(1.5), $|e_T| = |\Delta L \ / \ (1 + e^{j\varphi_r} + e^{j\varphi_r}\Delta L)|$. This, finally, yields a relation between $\Delta L = L_{\max}$ and φ_m (*Åström, K.J., 1985*).

- **Low Loop Gain**: At frequencies $\omega \gg \omega_r$, the loop gain becomes small and from Eq.(1.3) it results $|\Delta L(j\omega)| < |G^{-1}(j\omega)|$. Hence, large uncertainties require small nominal high-frequency gain.

1.2 Uncertain and Approximate Models

Uncertain and approximate models are a very different subject. Uncertain means an exact nominal matrix $\mathbf{A}$ and an uncertainty $\Delta\mathbf{A}$ bounded by a certain tolerance $\mathbf{E}$ which is also preassumed precisely known. Approximate is a description for a model $\mathbf{G}_M$ to be as close to the real plant $\mathbf{G}$ as possible. In several cases, one does not know the true plant since identification in the nominal or in selected perturbed cases is not available. Casually, one only pretends to know a nominal point and bounds for the uncertainties.

The approximate description is related to both system models and signals.

Usually, approximate linear plants and low-order plant models, including some parameter variations or structural variations are assumed. Based on these assumptions, the controller is designed. Combining this controller with the real-life plant subsequently, the resulting closed-loop system is checked with respect to some overall system performance, thus yielding a result whether the plant is sufficiently modelled, during some period of practical motion and including significant perturbations. Designing the system on the basis of an appropriate model and checking whether practical tests under worst-case conditions yield adequate overall transients, is the only practicable way, in most cases.

1.3 Categories of Uncertainties

1.3.1 Uncertainty of Some Component and Uncertainty of the System

Increasingly, uncertain system parameters became a well-known fact in the field of automatic control design, as indicated in the preceeding section. If parameters of the plant or other parts of a system are uncertain, then the coefficients of the associated differential equation and its characteristic polynomial are also uncertain. For cascading dynamic elements it is commonly known that transfer functions and associated denominator polynomials are multiplied. In consequence, system tolerances are widely scattered. For second-order polynomials, for instance,

$$\begin{aligned} p_a(s)p_b(s) &= (a_o + a_1 s + a_2 s^2)(b_o + b_1 s + b_2 s^2) \qquad (1.7) \\ &= a_o b_o + (a_o b_1 + a_1 b_o)s + (a_o b_2 + a_1 b_1 + a_2 b_o)s^2 + (a_1 b_2 + a_2 b_1)s^3 + a_2 b_2 s^4 \,. \end{aligned}$$

Each parameter of $p_a(s)$ and $p_b(s)$ and its uncertainty influences three coefficients of the product polynomial $p_a(s)p_b(s)$.

Theory of polynomials is a useful tool to examine control systems with uncertain parameters, particularly in the case where the coefficients are linearly dependent on system parameters. Including an additional term, the closed-loop behaviour is obtained, that is, $1+p_a(s)p_b(s)$. Considering multivariable control systems, the calculations in the associated polynomials are more complex.

Theory of interval polynomials provides tools for the stability check, even in the case of difficult relations between the coefficients of $p_a(s)$ and $p_b(s)$ and the coefficients of the product expression $1 + p_a(s)p_b(s) \triangleq p_c(s)$.

Bounded uncertainties which are closely related to certain coefficients c_i , that is, $l_i \le c_i \le h_i$, where l_i and h_i denote the limits low and high, respectively, are referred to as structured or even highly structured. This leads to the following generalizing statement: Shape and orientation of the sets of parameter variations in uncertain systems manifest the ***structure*** of the uncertainty (*Bartlett, A.C., et al. 1987*).

Similar relations can be achieved when the state-space description is employed. Highly structured information is given if the entries of the coefficient matrix uncertainty$\Delta\mathbf{A}$ are bounded by $\Delta a_{ij} \le E_{ij} = \max_t |\Delta a_{ij}|$ or by the relation $|\Delta\mathbf{A}| \le_e \mathbf{E}$ which is the corresponding element by element inequality relation termed by $\le_e$. The perturbation $\Delta\mathbf{A}$ can be rewritten to an interval matrix $\mathbf{E}_I$. Hence, the perturbed matrix is termed $\mathbf{A}_p = \mathbf{A} + \Delta\mathbf{A} = \mathbf{A} + \mathbf{E}_I = \mathbf{A} + [\mathbf{L} \quad \mathbf{H}]$. The matrix $\Delta\mathbf{A}$ replaces all the uncertainties. The matrices $\mathbf{E}$ and $\mathbf{E}_I$ represent the only knowledge of the uncertainty.

Another representation of uncertainty is constructed by the unidirectional matrix perturbation $\mathbf{A}_p = \mathbf{A} + \varepsilon\mathbf{E}$, as briefly sketched in Fig. 1.4b .

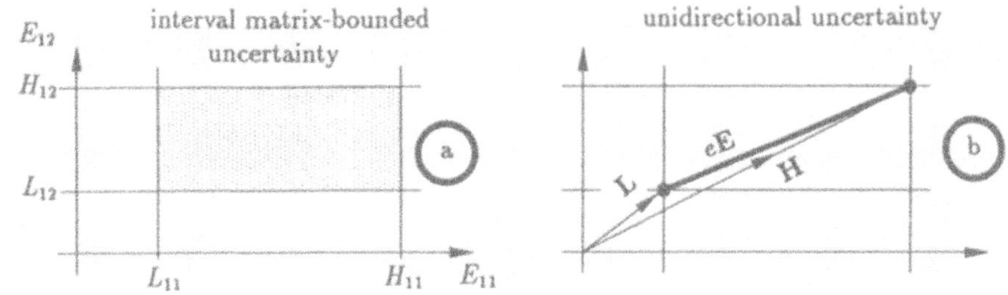

Figure 1.4: Uncertainty given by an interval matrix (a), unidirectional uncertainty (b)

If a detailed uncertainty information is not available, some other bounds are taken into account, as given by a *sum* of weighted squares of the polynomial coefficient uncertainties, by parameter space discs or some other kind of norms of the state-space coefficient uncertainty matrix. Such kinds of models are termed *unstructured* models. Since matrix norms omit structural information, various uncertainty distributions yield the same unstructured uncertainty information. In some cases, it is desirable to use unstructured uncertainty information to provide results which are independent from unknown structure of uncertainty. The perturbation matrix bound **E**, known from structured uncertainty, corresponds to a single hyperball or hyperellipsoid around the nominal system matrix in the present case of unstructured uncertainty.

A key property is: Consider the identity matrix **I** being perturbed by $\mathbf{M\Delta}$ where $\mathbf{\Delta}$ is bounded by its spectral norm $\|\mathbf{\Delta}\|_s \leq 1$. A necessary and sufficient condition for the nonsingularity of the matrix $\mathbf{I} + \mathbf{M\Delta}$ is $\sigma_{\max}[\mathbf{M}] < 1$. This is an appropriate result if the matrix $\mathbf{\Delta}$ only is subject to $\|\mathbf{\Delta}\|_s \leq 1$. The spectral-norm-bounded uncertainty is a rough bound. In many cases, additional structural information about the uncertainty constraint is available. If there exists some structural constraint, the singular-value-based stability robustness condition cited above may turn out too conservative.

Singularly perturbed systems are characterized by some additional scalar parameter. The nominal model is achieved by setting this model parameter equal to extreme values, in particular to zero. With the aid of this parameter, high-frequency parasitics of the uncertain plant can be described globally.

1.3.2 Frequency Domain

There are simple but very effective approaches to robust control and performance robustness, e.g., preserving slope and crossover frequency of the Bode plot (*Chen, K., 1961*). This simple performance-preserving controller is obtained from the Bode plot by choosing the controller part such that, in addition to the plant part, the Bode plot of the open-loop transfer functions possesses invariant crossover frequency and invariant crossover slope, even if the plant parameters vary. Then, the transient behaviour (rise time, overshoot) will be unaffected by plant parameter variations, and the control system has good robustness property.

In order to incorporate uncertainty information in the frequency domain, the transfer function is deteriorated by additive, multiplicative or feedback-oriented uncertainty terms. In the case of multivariable systems, premultiplied or postmultiplied terms are used, depending whether the uncertainty is associated with the output or input. Then, the perturbed transfer matrix is $\mathbf{G}+\Delta\mathbf{G}$, $(\mathbf{I}+\Delta\mathbf{L}_o)\mathbf{G}$ or $\mathbf{G}(\mathbf{I}+\Delta\mathbf{L}_i)$. Fractional representation is given by $(\mathbf{N}+\Delta\mathbf{N})(\mathbf{M}+\Delta\mathbf{M})^{-1}$.

Additive and multiplicative perturbation models suffer from the fact that the number of right-half-plane poles of the perturbed and unperturbed system must be equal. Uncertainty incorporating unstable poles must be left out of consideration. Hence, it is also difficult to choose some uncertainty when the unperturbed system contains poles located close to the imaginary axis. Using fractional (coprime factor) representation, the perturbations are *not* limited by these restrictions (*Glover, K., et al. 1990*).

Comparison of different non-diagonal closed-loop transfer matrices with the same diagonal elements shows that diagonal transfer matrices have best robustness properties (*Dickman, A., and Sivan, R., 1985*).

Derived from the index of performance $I = \int_0^\infty (\mathbf{x}^T\mathbf{Q}\mathbf{x} + \mathbf{u}^T\mathbf{R}\mathbf{u})dt$, the so-called performance robustness measure $\mathbf{x}_o^T(\mathbf{P}+\delta\mathbf{P})\mathbf{x}_o/\mathbf{x}_o^T\mathbf{P}\mathbf{x}_o$ is defined where $\mathbf{P}$ and $\mathbf{P}+\delta\mathbf{P}$ is the result of the algebraic Riccati equation in the nominal and in the actual (perturbed) case, respectively (*Abdul-Wahab, A.A., and Zohdy, M.A., 1988*).

Two major classes of uncertainties and corresponding robustness descriptions can be defined:

(i) Systems with unknown but constant parameters, bounded by certain limits. These robustness considerations are mainly executed in the frequency domain.

(ii) Systems with unknown and bounded parameters that may be time-varying.

It depends on the chosen derivation (frequency domain, time domain, Lyapunov method) whether or not the uncertainty is allowed for time variability.

A general mathematical view of perturbation techniques is given by *Bellman, R., 1964; Kato, T., 1976*. There are a great many of purely mathematical viewpoints when robust dynamical systems are considered.

1.4 Robustness Measure Using Decompositions

Control system applications are characterized by uncertain structure and parameters. In spite of this fact, differential equations and transfer functions with precisely predetermined or precisely bounded coefficients are used and several engineering problems (such as control design, or problems using variational calculus) are based on exact data. Since the exact solution of several problems requires great effort, there is need for approximate procedures and algorithms, which allow for replacing differential equation problems by state-space and algebraic methods, e.g., given by Taylor expansion or orthogonal decomposition. Accuracy and uncertainty bounds are implemented by the number and magnitude of decomposing elements.

Modelling the open-loop process and designing a closed-loop control system may not be considered independently. The structure and the desired performance of the control scheme influence the model variety of the process (*Skelton, R.E., and Owens, D.H., 1985*). Hence, it seems desirable to consider a great variety of models of perturbed processes.

Using time series decompositions or orthogonal function decompositions, a model $\mathbf{M}$ is defined. A measure of robustness can be stated as follows:

(i) Paralleling the process and the model, the performance I_2 is minimized and the model $\mathbf{M}(\mathbf{G},\mathbf{C},k)$ is obtained depending on the degree k of decomposition (see Fig. 1.5).

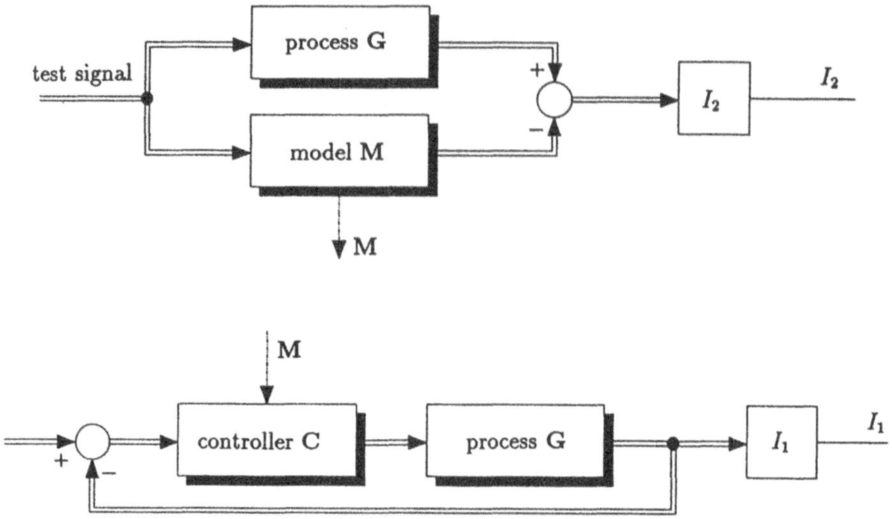

Figure 1.5: Paralleling process **G** and model **M** and determining the performance of the control system via model-based controller **C**

Decreasing number k of elements of decomposition causes an increasing measure of performance. The number k is given by the chosen upper bound in the expressions

$$x(t) = \sum_{i=0}^{k-1} \tilde{x}_i h_i(t) = \tilde{\mathbf{x}}^T \mathbf{h}(t) \quad \text{or} \quad x(t) = \sum_{i=0}^{k-1} x_i \, t^i \tag{1.8}$$

where $h_i(t)$ and $\mathbf{h}(t)$ are orthogonal functions and x_i are expansion coefficients.

(ii) Using some design method for optimal controllers **C** based on **M** yields the performance I_1 of the controlled system. The smaller I_1 in the face of larger I_2 (smaller k) the better the robustness. A measure of robustness is given by $\rho_R = I_2/I_1$.

Alternatively, as depicted in Fig. 1.6, the system is more robust the smaller I_3 where **G** is the process or a high-order approximation of the process, and **M** is a low-order model or a model with bounded modes, that is, a model which is intentively perturbed. In both paths the same controller **C** is implemented.

Some other fields also require orthogonal decomposition or some kind of series decomposition, such as nonlinear control, time-varying processes, varying-order systems, reduced-order model considerations.

1.5 Robustness by Structural Change

Sliding mode robust control systems utilize the structural change in order to overcome uncertainties. The structural change is generated by permanent switching operations.

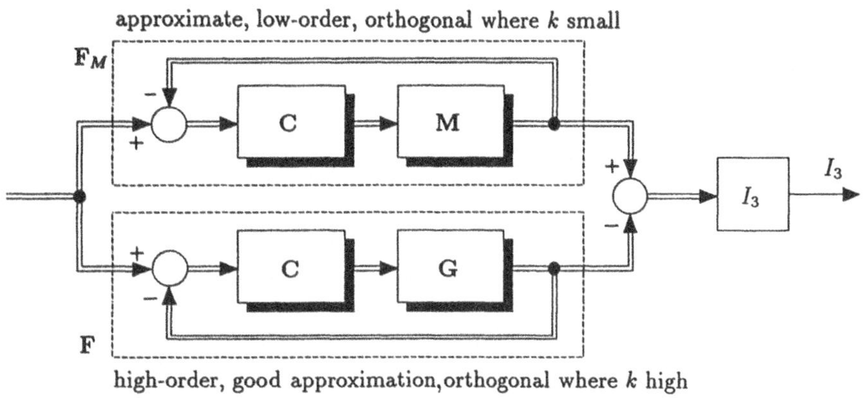

Figure 1.6: Comparing control systems, using a low-order model **M** and a high-quality identification result **G** of the real plant

1.6 Degree of Robustness

The objective of controller design often aims at the maximum perturbation which can be tolerated by the control system. Another objective is to find a controller with optimal stability margin in the face of predetermined uncertainties. Another formulation is: Minimize the index of performance with respect to a controller, in the face of uncertainties maximally deteriorating the performance. This design procedure is referred to as robustification. Performance robustness is accompanied by the following question. How are various kinds of performance representation incorporated into the investigation of robustness? Both in continuous-time and discrete-time representation, there are possibilities as follows:

- Performance I as a functional dependence of the system state $\mathbf{x}(t)$ and the control variable $\mathbf{u}(t)$, i.e., $I = \int_0^\infty (\mathbf{x}^T \mathbf{Q} \mathbf{x} + \mathbf{u}^T \mathbf{R} \mathbf{u}) dt$.
- Frequency dependent error vector norm and its maximum value versus frequency $I = \sup_\omega \|\mathbf{e}(j\omega)\|_s$.
- Integral of squared error norm versus frequency $I = \int_{-\infty}^{\infty} \|\mathbf{e}(j\omega)\|_s^2 d\omega$.
- Minimum weighted sensitivity $\|\mathbf{WS}\|_\infty$ where **W** and **S** are the weighting matrix and sensitivity matrix of the system, respectively.

1.7 Robustness, Optimality and Adaptation

The general optimal control design problem can be stated

$$\mathrm{C(G)} = \arg\min_{\mathbf{C}} I_1(\mathrm{C,G}) \tag{1.9}$$

where **C** is the multivariable controller and **G** the multivariable plant. The procedure is executed by an off-line method.

Identification is a particular approach to find a model $\mathbf{M}$ of the varying and perturbed plant $\mathbf{G}_p$

$$\mathbf{M}(\mathbf{G}_p) = \arg\min_{\mathbf{M}} I_2(\mathbf{M}, \mathbf{G}_p) \tag{1.10}$$

where $\mathbf{G}_p$ must not exceed a certain area and the execution is on-line. Design criterion I_1 and identification criterion I_2 need not be the same.

Adaptation can be formulated as

$$\mathbf{C}(\mathbf{M}) = \arg\min_{\mathbf{C}} I_1(\mathbf{C}, \mathbf{M}) \tag{1.11}$$

where the implementation of the on-line identification $\min_{\mathbf{M}} I_2$ is preassumed. Note that the adaptation is based on the model $\mathbf{M}$ not on the process $\mathbf{G}_p$.

Robustness aims at sufficient conditions of finding $\mathbf{C}$ such that

$$\mathbf{C} = \arg\min_{\mathbf{C}} \max_{\mathbf{G}_p \in \{\mathbf{G}\}} I(\mathbf{C}, \mathbf{G}_p) \tag{1.12}$$

where the uncertainty in $\mathbf{G}_p$ is limited by some given set $\{\mathbf{G}\}$ of the real plant.

In order to obtain good performance, that is, to reduce conservatism when designing the robust controller, a small set $\{\mathbf{G}\}$ proves helpful. This will be achieved if adequate identification is implemented, additionally. Adequate means that the identification algorithm should operate *fairly quickly* but need not be very accurate. The inaccuracy and identification lag is tolerated by the robust controller to some extent (*Ljung, L., 1985*).

The inclusion of robust procedures, particularly those designed by H_∞-optimization techniques, permits a larger class of perturbations in adaptive control application (*Iglesias, P.A., 1990*).

Expressed by formulae, the robust and adaptive controller are characterized by

$$\mathbf{u}(t) = \mathbf{f}[\mathbf{y}(t)] \quad \forall\, \mathbf{p} \in \mathcal{P} \quad \text{(robust controller)} \tag{1.13}$$

$$\mathbf{u}(t) = \mathbf{f}[\mathbf{y}(t), \mathbf{p}] \quad \text{(gain scheduling adaptive controller).} \tag{1.14}$$

1.8 Adaptive Controllers

Robust controllers are constant gain feedback controllers. In some cases, they are not able to overcome uncertainty. An integrator with unknown sign of the integrating operation is an illustrative example which cannot be treated by constant gain feedback. Unlike robust controllers, adaptive controllers can handle this problem. Adaptive control is an alternative and very important way to deal with uncertainty.

Adaptive control is a method to continuously increase the information about the instantaneous parameters of the uncertain process and to utilize them, by simultaneously calculating the optimal controller parameters (*Åström, K.J., 1985*). Optimality criteria of adaptation algorithms are outlined in *Tsypkin, Y.Z., 1985*. Although adaptive controllers can produce excellent control performance it may be difficult to develop general design principles which can be applied to various industrial processes. Moreover, there may arise problems concerning (i) the stability of the system, (ii) the differentiation between system parameter deviations and disturbance reactions and (iii) the supply of sufficient amplitude and frequency richness of the process signals under identification.

1.9 Methods of Robust Design

Investigating both robustness and performance leads to a great variety of procedures and design methods, depending on the assumptions and basic requirements. Some of them are suitable for designing the controller directly. Others are appropriate for subsequently checking the results of conventionally designed controllers, in various ranges of operating conditions.

1.10 Effort in Robust Design

An important objective is to reduce the computational burden which is needed to guarantee the stability and performance of the control system, including uncertain elements. There are simple formulae yielding sufficient conditions for robust controllers, suffering from suitable performance. There are better formulae providing tighter bounds and less conservativeness but accompanied by higher computational effort. If robust control can be recast to minimum search problems, the effectiveness of some robust method also depends on the convexity or nonconvexity of the risk function, or it considerably depends on the fact in which direction a bound is approximated.

1.11 Results on Experience

In some cases, it could be found out by experience that some elementary action may produce robust property of control systems. For instance, omitting penalties on the control errors within the first fake sampling interval when carrying out LQG control design yields sufficiently robust results (*Peterka, V., 1989*). Some nonlinear feedback has certain intrinsic robustness properties, in particular, infinite gain margin (*Glad, S.T., 1987*).

With regard to rather complex plant models, direct robust design methods are frequently substituted by related methods, enabling the designer to achieve satisfactory robustness, without applying fully developed and perhaps complicated design algorithms for robust systems. Applying additional noise or intentionally using numerically inaccurate calculations are examples referring to this, although they suffer from the fact that one is unaware of the perturbation bounds.

There is a great deal of uncertainty on the part of human operators when deciding between various operating modes and facilities. It is desirable that systems be designed which tolerate human peculiarities, in order to obtain unaffected overall system behaviour.

Chapter 2

Vector Norm. Matrix Norm. Matrix Measure

Norms are widespread in the techniques of uncertain modelling and robust control. Multivariable systems and the state-space representation require valuating vectors and matrices by a single number. Most of the important relations for norms are listed or derived in this chapter. This material may be considered as a kind of reference, and the chapter may be skipped when the reader is already familiar with this subject.

2.1 Rayleigh Quotient. Numerical Radius

Given a matrix $\mathbf{A} \in \mathcal{C}^{n\times n}$, the following scalar quantity is referred to as

$$\text{Rayleigh quotient} \quad R[\mathbf{x}] \triangleq \frac{\mathbf{x}^H\mathbf{A}\mathbf{x}}{\mathbf{x}^H\mathbf{x}} \quad \mathbf{x} \in \mathcal{C}^n \quad \forall \mathbf{x} \neq \mathbf{0} \,. \tag{2.1}$$

The set of Rayleigh quotients is denoted *numerical range*. The maximum modulus of the Rayleigh quotient for any $\mathbf{A}$ and for any $\mathbf{x}$ is known as

$$\text{numerical radius} \quad \rho_R \triangleq \max_{\mathbf{x}} |\frac{\mathbf{x}^H\mathbf{A}\mathbf{x}}{\mathbf{x}^H\mathbf{x}}| \,. \tag{2.2}$$

Note that usually both $\lambda_i[\mathbf{A}]$ and $R[\mathbf{x}] \in \mathcal{C}$ but $\mathbf{x}^H\mathbf{x} \in \mathcal{R}$. The Rayleigh quotient applies to any matrix $\mathbf{A}$ and any vector $\mathbf{x}$ ($\mathbf{x} \neq \mathbf{0}$)

$$\frac{\langle \mathbf{A}\mathbf{x}, \mathbf{x}\rangle}{\langle \mathbf{x}, \mathbf{x}\rangle} = \frac{\langle \mathbf{A}\mathbf{x}, \mathbf{x}\rangle}{\|\mathbf{x}\|_F^2} = \langle \mathbf{A}\frac{\mathbf{x}}{\|\mathbf{x}\|_F}, \frac{\mathbf{x}}{\|\mathbf{x}\|_F}\rangle = \frac{\mathbf{x}^T\mathbf{A}^T\mathbf{x}^*}{\mathbf{x}^T\mathbf{x}^*} = \frac{\mathbf{x}^H\mathbf{A}\mathbf{x}}{\mathbf{x}^H\mathbf{x}} \,. \tag{2.3}$$

If $\mathbf{x}$ is replaced by an eigenvector $\mathbf{a}_i$ of $\mathbf{A}$ associated with $\lambda_i[\mathbf{A}]$ then $\mathbf{A}\mathbf{a}_i = \lambda_i\mathbf{a}_i$ and the Rayleigh quotient turns out the eigenvalue

$$R[\mathbf{a}_i] = \frac{\langle \mathbf{A}\mathbf{a}_i, \mathbf{a}_i\rangle}{\langle \mathbf{a}_i, \mathbf{a}_i\rangle} = \frac{\mathbf{a}_i^H\mathbf{A}\mathbf{a}_i}{\mathbf{a}_i^H\mathbf{a}_i} = \frac{\mathbf{a}_i^H\lambda_i\mathbf{a}_i}{\mathbf{a}_i^H\mathbf{a}_i} = \lambda_i[\mathbf{A}] \,. \tag{2.4}$$

2.2 Rayleigh's Principle

From Eq.(2.3), searching a *maximum* over any vector $\mathbf{x}$,

$$\max_{\mathbf{x}\neq\mathbf{0}} \frac{\mathbf{x}^H\mathbf{A}\mathbf{x}}{\mathbf{x}^H\mathbf{x}}\Big|_{\mathbf{x}\text{ any vector and }\mathbf{A}=\mathbf{A}^H} = \lambda_{\max}[\mathbf{A}] = \max_i \lambda_i[\mathbf{A}] \tag{2.5}$$

is only true for Hermite matrices. The vector $\mathbf{x}$ turns out the eigenvector $\mathbf{a}$ associated with $\lambda_{max}[\mathbf{A}]$.

Proof: Considering the Rayleigh quotient for Hermite $\mathbf{A}$, for any $\mathbf{x}$ represented on the basis of the orthogonal and normalized eigenvectors $\mathbf{a}_i$ of $\mathbf{A} \in \mathcal{C}^{n\times n}$

$$\mathbf{x} = \sum_{i=1}^{n} \alpha_i \mathbf{a}_i \qquad \text{where } (\mathbf{a}_i, \mathbf{a}_j) = \delta_{ij}, \qquad \mathbf{a}_i, \mathbf{x} \in \mathcal{C}^n . \tag{2.6}$$

$$\langle \mathbf{A}\mathbf{x}, \mathbf{x}\rangle = (\mathbf{A}\sum_{i=1}^{n} \alpha_i \mathbf{a}_i, \sum_{i=1}^{n} \alpha_i \mathbf{a}_i) = (\sum_{i=1}^{n} \alpha_i \lambda_i \mathbf{a}_i, \sum_{i=1}^{n} \alpha_i \mathbf{a}_i) = \sum_{i=1}^{n} |\alpha_i|^2 \lambda_i . \tag{2.7}$$

The norm $\|\mathbf{x}\|_F = 1$ requires $\sum_{i=1}^{n} |\alpha_i|^2 = 1$. Thus,

$$\langle \mathbf{A}\mathbf{x}, \mathbf{x}\rangle = \sum_{i=1}^{n} |\alpha_i|^2 \lambda_i \leq \lambda_{max} \sum_{i=1}^{n} |\alpha_i^2| = \lambda_{max} . \tag{2.8}$$

Selecting $\alpha_1 = 1$ and $\alpha_j|_{j\neq 1} = 0$, i.e. $\mathbf{x} = \mathbf{a}_1$, the inequality above turns over into the equality . The inner product $\langle \mathbf{A}\mathbf{x}, \mathbf{x}\rangle$ becomes identical to the maximum. Hence, $\max_i \langle \mathbf{A}\mathbf{a}_i, \mathbf{a}_i\rangle = \lambda_{max}[\mathbf{A}]$ which equals Eq.(2.5) in the case of $\mathbf{x} = \mathbf{a}_1$.

End of the Proof

The eigenvalue λ_2 (where $\lambda_1 \geq \lambda_2 \geq \lambda_3 \ldots$) is obtained from the same formula

$$\lambda_2 = \max_{\mathbf{x}} \langle \mathbf{A}\mathbf{x}, \mathbf{x}\rangle \qquad \text{where } \|\mathbf{x}\|_F = 1 \text{ and } \langle \mathbf{x}, \mathbf{a}_1\rangle = 0, \tag{2.9}$$

i.e., $\mathbf{x}$ may only be selected orthogonal to $\mathbf{a}_1$. The eigenvalue λ_k is achieved by choosing the unit vector $\mathbf{x}$ orthogonal to $\mathbf{a}_1$ through $\mathbf{a}_{k-1}$.

2.3 Properties of the Rayleigh Quotient

For real and symmetric matrices $\mathbf{A}$ and $\mathbf{a}_i$, i.e., the eigenvector associated with the eigenvalue $\lambda_i[\mathbf{A}]$, there are two important results. The Rayleigh quotient equals the eigenvalue if $\mathbf{x}$ is an eigenvector $\mathbf{a}_i$, i.e. $R[\mathbf{a}_i] = \lambda_i$, just as before. Additionally, the stationary value of the Rayleigh quotient with respect to $\mathbf{x}$ is zero if $\mathbf{x}$ equals the eigenvector $\mathbf{a}_i$

$$\frac{\partial R[\mathbf{x}]}{\partial \mathbf{x}}\Big|_{\mathbf{x}=\mathbf{a}_i} = \frac{\partial}{\partial \mathbf{x}} \frac{\mathbf{x}^H \mathbf{A}\mathbf{x}}{\mathbf{x}^H \mathbf{x}}\Big|_{\mathbf{x}=\mathbf{a}_i} = \mathbf{0} \tag{2.10}$$

(*Lancaster, P., and Tismenetsky, M., 1985; Zurmühl, R., and Falk, S., 1984*). If an approximate $\mathbf{a}_i$, say $\mathbf{a}'_i$, is known then $R[\mathbf{a}'_i]$ is a very accurate approximation of $\lambda_i[\mathbf{A}]$, to be explained by the property of continuity and by the characteristic of the close vicinity of a maximum.

2.4 Vector Norms

For any vector in complex space $\mathbf{x} \in \mathcal{C}^n$, note the following definitions of norms

$$\|\mathbf{x}\|_\infty \triangleq \max_j |x_j| \qquad \text{infinity norm or max norm} \tag{2.11}$$

$$\|\mathbf{x}\|_1 \triangleq \sum_j |x_j| \qquad \text{sum norm} \tag{2.12}$$

Table 2.1: Rayleigh quotient $R[\mathbf{x}]$, eigenvalues $\lambda_i[\mathbf{A}]$ and eigenvectors $\mathbf{a}_i$ (Note the abbreviations: real or c.c. ≡ real or complex-conjugate; ‡ ≡ unitary eigenvectors, i.e. $\mathbf{a}_i^H \mathbf{a}_k = \delta_{ik}$; ♮ ≡ with multiplicity $n-r$)

$\mathbf{A}$	rank$\mathbf{A}$	$\mathbf{x}$	$\lambda_i[\mathbf{A}]$	$R[\mathbf{x}]$	$\mathbf{a}_i$
$\in \mathcal{C}^{n\times n}$	n	$\in \mathcal{C}^n$	$\in \mathcal{C}$	$\in \mathcal{C}$	$\in \mathcal{C}^n$
$\in \mathcal{C}^{n\times n}$	n	$= \mathbf{a}_i \in \mathcal{C}^n$	$\in \mathcal{C}$	$\lambda_i[\mathbf{A}] \in \mathcal{C}$	$\in \mathcal{C}^n$
$\in \mathcal{R}^{n\times n}$	n	$\in \mathcal{C}^n$	real or c.c.	$\in \mathcal{C}$	real or c.c.
$\in \mathcal{C}^{n\times n}$, $\mathbf{A} = \mathbf{A}^H$	n	$\in \mathcal{C}^n$	$\in \mathcal{R}$	$\in \mathcal{R}$	$\in \mathcal{C}^n$ ‡
$\in \mathcal{C}^{n\times n}$, $\mathbf{A} = \mathbf{A}^H > 0$	n	$\in \mathcal{C}^n$	> 0	> 0	$\in \mathcal{C}^n$ ‡
$\in \mathcal{C}^{n\times n}$, $\mathbf{A} = \mathbf{A}^H \geq 0$	n	$\in \mathcal{C}^n$	≥ 0	≥ 0	$\in \mathcal{C}^n$ ‡
$\in \mathcal{C}^{n\times n}$, $\mathbf{A} = \mathbf{A}^H \geq 0$	$r < n$	$\in \mathcal{C}^n$	0 ♮, ≥ 0	≥ 0	$\in \mathcal{C}^n$ ‡
$\in \mathcal{R}^{n\times n}$, $\mathbf{A} = \mathbf{A}^T$	n	$\in \mathcal{C}^n$	$\in \mathcal{R}$	$\in \mathcal{R}$	$\in \mathcal{R}^n$ ‡

$$\|\mathbf{x}\|_F \triangleq \|\mathbf{x}\|_2 \triangleq +\sqrt{\sum_j |x_j|^2} = +\sqrt{\mathbf{x}^H\mathbf{x}} = +\sqrt{\mathbf{x}^{*T}\mathbf{x}} \quad \text{Frobenius vector norm}\ . \tag{2.13}$$

In order to avoid confusion with the norm of functions on L_2 , the Frobenius (or Euclidian) vector norm is indicated $\|\mathbf{x}\|_F$, only.

In general, one has

$$\|\mathbf{x}\|_p \triangleq +(\sum_{j=1}^{n} |x_j|^p)^{1/p} \qquad \text{Hölder norms } \forall p \in [1,\infty)\ . \tag{2.14}$$

In the two-dimensional space, the locus $\|\mathbf{x}\|_\infty$ = constant is a square faced parallel to the x_1- and x_2-axes, the extensions thus reaching the axes at infinity. The locus $\|\mathbf{x}\|_F$ = constant = 1 is a unit circle centered in the origin. The locus $\|\mathbf{x}\|_1$ = constant is given by a tilted square with corners on the x_1- and x_2-axis.

Elliptic norms are defined by $\|\mathbf{x}\|_e = (\mathbf{x}^T\mathbf{W}\mathbf{x})^{1/2}$ where $\mathbf{W}$ is a symmetric positive-definite weighting matrix.

Several norms are depicted in Fig. 2.1, for the specific case of magnitude equal to unity.

Using the definition of the infinity vector norm and sum vector norm, the relations hold

$$|\mathbf{y}^T\mathbf{x}| \leq \sum_{j=1}^{n} |y_j|\,|x_j| \;\leq\; (\sum_{j=1}^{n} |y_j|) \max_j |x_j| = \|\mathbf{y}\|_1 \|\mathbf{x}\|_\infty \tag{2.15}$$

$$\text{or} \quad \leq\; (\max_j |y_j|)\,(\sum_{j=1}^{n} |x_j|) = \|\mathbf{y}\|_\infty \|\mathbf{x}\|_1\ . \tag{2.16}$$

Additionally, $|\mathbf{y}^T\mathbf{x}| \leq \|\mathbf{y}\|_F \|\mathbf{x}\|_F$.

2.5 Common Properties of Vector Norms and Matrix Norms

Vector norms and matrix norms must satisfy the following conditions listed up for the case of matrices, only. The positive scalar $\|\mathbf{A}\|_p$ is a matrix norm on $\mathcal{C}^{m\times n}$ of the matrix $\mathbf{A} \in \mathcal{C}^{m\times n}$

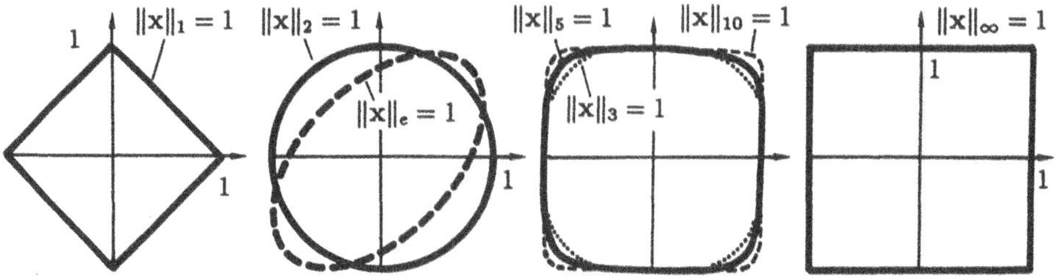

Figure 2.1: Unit ball ($\|\mathbf{x}\| \le 1$) and unit sphere ($\|\mathbf{x}\| = 1$) for several norms of $\mathbf{x} \in \mathcal{R}^2$

- if the norm is positive, i.e. $\|\mathbf{A}\|_p > 0$ if and only if $\mathbf{A} \neq \mathbf{0}$ and $\|\mathbf{0}\|_p = 0$,
- if the norm is homogenous, i.e., $\|\alpha\mathbf{A}\|_p = |\alpha| \; \|\mathbf{A}\|_p \qquad \alpha \in \mathcal{R}$,
- if the triangular (triangle) inequality is satisfied, i.e., $\|\mathbf{A}+\mathbf{B}\|_p \le \|\mathbf{A}\|_p + \|\mathbf{B}\|_p$.

A matrix norm is *consistent* if $\|\mathbf{AC}\|_p \le \|\mathbf{A}\|_p \; \|\mathbf{C}\|_p$. Induced matrix norms, see e.g. Eq.(2.26), possess this special property. For any consistent matrix norm

$$\|\mathbf{I}\| \ge 1, \qquad \|\mathbf{A}^n\| \le \|\mathbf{A}\|^n, \qquad \|\mathbf{A}^{-1}\| \ge \|\mathbf{A}\|^{-1} \quad \text{if} \quad \det\mathbf{A} \neq 0 \,. \tag{2.17}$$

The Hölder matrix norm $\|\mathbf{A}\|_p = (\sum_{i,j=1}^{n} |a_{ij}|^p)^{1/p}$ is only a consistent matrix norm for $1 \le p \le 2$, real or integer. In most cases of matrix norm applications, consistent norms are employed.

2.6 Direct Matrix Norms

For any (n,n)-matrix $\mathbf{G} = \text{matrix}[G_{ij}]$ with complex elements G_{ij} the number $\|\mathbf{G}\|$ denotes any matrix norm. The absolute norm is given by

$$\|\mathbf{G}\|_a \triangleq \max_{i,j} |G_{ij}| \quad \text{or alternatively e.g.} \quad \|\mathbf{G}\|_a \triangleq n\max_{i,j} |G_{ij}| \,, \quad \|\mathbf{G}\|_b \triangleq \sum_{i=1}^{n}\sum_{j=1}^{n} |G_{ij}| \,. \;\square \tag{2.18}$$

The Frobenius or Euclidian or Schur matrix norm $\|\mathbf{G}\|_F = +\sqrt{\sum_{i=1}^{n}\sum_{j=1}^{n} |G_{ij}|^2}$ (2.19)

is equal to

$$\|\mathbf{G}\|_F = +\sqrt{(\text{col}\mathbf{G})^H \,\text{col}\mathbf{G}} = +\sqrt{(\text{col}\mathbf{G})^{*T} \,\text{col}\mathbf{G}} = +\sqrt{\text{tr}\;\mathbf{G}^H\mathbf{G}} = +\sqrt{\sum_i \lambda_i[\mathbf{G}^H\mathbf{G}]} \,. \tag{2.20}$$

Note the distinction between the Frobenius or Euclidian norm as given above and the spectral norm $\|\mathbf{G}\|_s$ where $\lambda_{\max}$ is taken into consideration, only, and a specific matrix product is involved. The Frobenius norm is computationally cheaper to check than the spectral norm.

Hölder matrix norms $\|\mathbf{A}\|_p = (\sum_{i,j=1}^{n} |a_{ij}|^p)^{1/p}$ can be defined similarly to vector norms. Since the Hölder norm, as far as robust control is concerned, will only be applied in the case of $p = 2$ with the notation $\|\mathbf{G}\|_F$, there is no subscript confusion with induced matrix norms.

2.6.1 Matrix Infinity Norm $\|\cdot\|_\infty$

Considering the sum of the moduli of the elements of the ith row and then selecting the largest sum, the matrix infinity norm is defined

$$\|\mathbf{G}\|_\infty \triangleq \max_i \sum_{j=1}^{n} |G_{ij}| \qquad \text{largest absolute "row sum"}. \tag{2.21}$$

Remark: The same notation $\|\mathbf{G}\|_\infty$ also denotes the H_∞-norm, see Eq.(31.31). To avoid confusion, the appropriate definitions are enclosed.

2.6.2 Matrix Norm $\|\cdot\|_1$ (Matrix 1-Norm)

Calculating the sum of the jth column and selecting the largest column sum yields the norm $\|\cdot\|_1$

$$\|\mathbf{G}\|_1 \triangleq \max_j \sum_{i=1}^{n} |G_{ij}| \qquad \text{largest absolute "column sum"}. \tag{2.22}$$

2.7 Spectral Radius. Spectral Abscissa

The set of eigenvalues is denoted spectrum of a system. The eigenvalues are always located within the ***numerical range*** of a matrix as given by the set of Rayleigh quotients. The largest modulus (absolute value) of the eigenvalues of any matrix $\mathbf{G}$ is named spectral radius

$$\rho_s[\mathbf{G}] \triangleq \max_i \mid \lambda_i[\mathbf{G}] \mid . \tag{2.23}$$

A modified spectral radius $q_s[\mathbf{G}]$ is introduced when robust controllers are designed based on the spectral radius, see Eq.(28.17).

The maximum real part of the set of eigenvalues associated with a matrix $\mathbf{G}$ is denoted spectral abscissa $\alpha_s[\mathbf{G}]$.

2.8 Induced Matrix Norms

Based on the definition of vector norms, there is the following definition of the induced matrix norms (or bound matrix norms). The matrix norms are usually labelled by a subscript number $_p$ (*Lancaster, P., and Tismenetsky, M., 1985; Desoer, C.A., and Vidyasagar, M., 1975; Vidyasagar, M., 1978; Lin, S.H., et al. 1988*). As $\mathbf{x}$ varies over C^n, the induced matrix norm $\|\mathbf{G}\|_p$ in general is a least upper bound

$$\|\mathbf{G}\|_p = \sup_{\|\mathbf{x}\|_p=1} \|\mathbf{G}\mathbf{x}\|_p = \sup_{\mathbf{x} \neq 0} \frac{\|\mathbf{G}\mathbf{x}\|_p}{\|\mathbf{x}\|_p} \tag{2.24}$$

where $\mathbf{G}$ is any matrix over $\mathcal{C}^{n\times n}$, $\mathbf{G} = \text{matrix}[G_{ij}]$. The relation above is derived from the condition

$$\|\mathbf{G}\mathbf{x}\| \leq \|\mathbf{G}\|\ \|\mathbf{x}\| \quad \leadsto \quad \|\mathbf{G}\| \geq \frac{\|\mathbf{G}\mathbf{x}\|}{\|\mathbf{x}\|} \quad \leadsto \quad \|\mathbf{G}\| = \sup \frac{\|\mathbf{G}\mathbf{x}\|}{\|\mathbf{x}\|} \ . \tag{2.25}$$

The bound above holds for Hölder vector norms $p = 1, \infty$ and F. For the latter see Eq.(2.26).

A matrix norm is an induced matrix norm only if $\|\mathbf{I}\|_p = 1$.

If $\|\mathbf{G}\mathbf{x}\|_p \leq \|\mathbf{G}\|\ \|\mathbf{x}\|_p$ holds for all $\mathbf{x} \in \mathcal{C}^n$ and all $\mathbf{G} \in \mathcal{C}^{n\times n}$ then the vector norm $\|\cdot\|_p$ and the matrix norm $\|\cdot\|$ are referred to as *compatible*. A least upper bound is defined by a condition $\leq$ where the equality really holds for a certain vector $\mathbf{x}$. This property is satisfied for induced norms.

An interpretation of the induced matrix norm $\|\mathbf{G}\|_p$ defined in Eq.(2.24) is given as follows: Consider a vector $\mathbf{x}$ characterized by the norm $\|\mathbf{x}\|_p = 1$, see Fig. 2.1. After transformation with the matrix $\mathbf{G}$ the vector $\mathbf{G}\mathbf{x}$ results. The vector with maximum norm $\|\mathbf{G}\mathbf{x}\|_p$ equals the induced matrix norm $\|\mathbf{G}\|_p$. In the case $p = 2$, see Eq.(2.26), a unity vector $\mathbf{x}$ pointing to the unity circle centered at the origin is transformed to a vector $\mathbf{G}\mathbf{x}$ pointing to an ellipse. The length of the major axis of the ellipse equals the spectral norm $\|\mathbf{G}\|_s$ (*Ortega, J.M., 1972*).

2.8.1 Spectral Norm $\|\cdot\|_s$

The spectral matrix norm $\|\mathbf{G}\|_s$ is a matrix norm induced by the Euclidian (Frobenius) vector norm where $\mathbf{x} \in \mathcal{C}^n$ and $\mathbf{G}$ is any matrix $\in \mathcal{C}^{n\times n}$

$$\|\mathbf{G}\|_s \triangleq \sup_{\|\mathbf{x}\|_F=1} \|\mathbf{G}\mathbf{x}\|_F = \sup_{\mathbf{x}} \frac{\|\mathbf{G}\mathbf{x}\|_F}{\|\mathbf{x}\|_F} = \sup_{\mathbf{x}} \sqrt{\frac{\mathbf{x}^H\mathbf{G}^H\mathbf{G}\mathbf{x}}{\mathbf{x}^H\mathbf{x}}} = \sqrt{\rho_R[\mathbf{G}^H\mathbf{G}]} \ . \tag{2.26}$$

It can be proved (see below) that the same norm is attained by the following definitions

$$\|\mathbf{G}\|_s \triangleq +\sqrt{\lambda_{\max}[\mathbf{G}^H\mathbf{G}]} = \sigma_{\max}[\mathbf{G}] = \max_i \ +\sqrt{\lambda_i[\mathbf{G}^H\mathbf{G}]} = +\sqrt{\rho_s[\mathbf{G}^H\mathbf{G}]} \ , \tag{2.27}$$

i.e., the spectral norm equals the largest singular value $\sigma_{\max}$ of $\mathbf{G}$, moreover equals the root of the spectral radius ρ_s taken from $\mathbf{G}^H\mathbf{G}$. The spectral norm is also known as Hilbert norm. Note that the numerical radius ρ_R applied to the matrix $\mathbf{G}^H\mathbf{G}$ equals the spectral radius ρ_s applied to $\mathbf{G}^H\mathbf{G}$ (*Wilkinson, J.H., 1965*), i.e.,

$$\rho_R[\mathbf{G}^H\mathbf{G}] = \rho_s[\mathbf{G}^H\mathbf{G}] \ . \tag{2.28}$$

Proof that the numerical radius ρ_R equals the spectral radius ρ_s , i.e., $\rho_R[\mathbf{G}^H\mathbf{G}] = \rho_s[\mathbf{G}^H\mathbf{G}]$:

Referring to Eq.(2.26) in squared form, the vector $\mathbf{x}$ is substituted by a decomposition of $\mathbf{x} = \sum_i \alpha_i\mathbf{q}_i$ into orthogonal right-eigenvectors $\mathbf{q}_i$ associated with $\mathbf{Q} \triangleq \mathbf{G}^H\mathbf{G}$. The symbol $\mathbf{Q}$ is chosen in order to avoid any confusion or interrelation with $\mathbf{A}$. The eigenvalues of $\mathbf{Q}$ are denoted $\rho_i^2 = \lambda_i[\mathbf{Q}]$, see the proof following Eq.(22.53). For $\mathbf{x}$ and α_i arbitrary, maximizing Eq.(2.26) and assuming that the maximum actually is approached, yields

$$\begin{aligned}
\rho_R[\mathbf{G}^H\mathbf{G}] &= \max_{\alpha_i} \frac{(\sum_i \alpha_i^*\mathbf{q}_i^H)\mathbf{G}^H\mathbf{G}(\sum_i \alpha_i\mathbf{q}_i)}{(\sum_i \alpha_i^*\mathbf{q}_i^H)(\sum_i \alpha_i\mathbf{q}_i)} = \max_{\alpha_i} \frac{(\sum_i \alpha_i^*\mathbf{q}_i^H\mathbf{G}^H\mathbf{G})(\sum_i \alpha_i\mathbf{q}_i)}{\sum_i |\alpha_i|^2} && (2.29)\\
&= \max_{\alpha_i} \frac{(\sum_i \alpha_i^*\sigma_i^2\mathbf{q}_i^H)(\sum_i \alpha_i\mathbf{q}_i)}{\sum_i |\alpha_i|^2} = \max_{\alpha_i} \frac{\sum_i |\alpha_i|^2\sigma_i^2}{\sum_i |\alpha_i|^2} \ . && (2.30)
\end{aligned}$$

Choosing the order $\sigma_1^2 \geq \sigma_2^2 \geq \ldots \geq \sigma_n^2$ and replacing σ_i^2 by σ_1^2, the condition

$$\frac{\sum_i |\alpha_i|^2 \, \sigma_i^2}{\sum_i |\alpha_i|^2} \leq \frac{\sigma_1^2 \, \sum_i |\alpha_i|^2}{\sum_i |\alpha_i|^2} = \sigma_1^2 \tag{2.31}$$

results. The replacing operation can be achieved by the special choice $\alpha_1 = 1, \alpha_i = 0 \;\; \forall i = [2, n]$. Hence, σ_1^2 in the aforementioned condition is an upper bound and

$$\rho_R[\mathbf{G}^H\mathbf{G}] = \sigma_1^2 = \max_i \lambda_i[\mathbf{Q}] = \lambda_{\max}[\mathbf{G}^H\mathbf{G}] = \rho_s[\mathbf{G}^H\mathbf{G}] \tag{2.32}$$

(since $\lambda_i[\mathbf{G}^H\mathbf{G}]$ are real and non-negative in this case). Invoking the equivalence of the spectral norm and the maximum singular value, it is obvious that the spectral norm can also be applied to nonsquare matrices. (Even the spectral norm of a vector exists and is equal to the Frobenius norm.)

End of the Proof

Example:

$$\mathbf{G} = \begin{pmatrix} 0.5 & 0.5 \\ 0 & 0.5 \end{pmatrix}, \quad \mathbf{x} = \begin{pmatrix} \sqrt{1-a^2} \\ a \end{pmatrix}, \quad ||\mathbf{x}||_F = 1 \,. \tag{2.33}$$

a) Induced norm $||\mathbf{G}||_s$:

$$||\mathbf{G}||_s = \sup_{||\mathbf{x}||_F=1} ||\mathbf{Gx}||_F = \sup_{||\mathbf{x}||_F=1} || \begin{pmatrix} 0.5 & 0.5 \\ 0 & 0.5 \end{pmatrix} \begin{pmatrix} \sqrt{1-a^2} \\ a \end{pmatrix} ||_F = \max_a \; 0.5 \sqrt{1 + a^2 + 2a \sqrt{1-a^2}} \tag{2.34}$$

$$\frac{\partial}{\partial a}(1 + a^2 + 2a \sqrt{1-a^2}\,) = 0 \quad \leadsto \quad a = \sqrt{0.5 + \sqrt{0.05}} = 0.850650 \tag{2.35}$$

$$||\mathbf{G}||_s = || \begin{pmatrix} 0.5 & 0.5 \\ 0 & 0.5 \end{pmatrix} \begin{pmatrix} \sqrt{1-a^2} \\ a \end{pmatrix} || \;\Big|_{a=\sqrt{0.5+\sqrt{0.05}}} = 0.809017 \,. \tag{2.36}$$

b) Maximum singular value:

$$\sigma_{\max}\,[\mathbf{G}] = +\sqrt{\lambda_{\max}[\mathbf{G}^H\mathbf{G}]} = 0.809017 \,. \tag{2.37}$$

c) Frobenius matrix norm:

$$||\mathbf{G}||_F = +\sqrt{0.25 + 0.25 + 0.25} = 0.866025 \,. \tag{2.38}$$

End of Example

The spectral norm of the Kronecker product is given by

$$||\mathbf{A} \otimes \mathbf{B}||_s = ||\mathbf{A}||_s ||\mathbf{B}||_s \,. \; \square \tag{2.39}$$

2.8.2 Matrix Infinity Norm and Column Sum Norm (1-Norm)

Decomposing the matrix $\mathbf{G}$ into rows and columns

$$\mathbf{G} = \text{matrix}[G_{jk}] = \begin{pmatrix} \mathbf{G}_{1\cdot} \\ \mathbf{G}_{2\cdot} \\ \vdots \end{pmatrix} = (\mathbf{G}_{\cdot 1} \vdots \mathbf{G}_{\cdot 2} \; \ldots\ldots) \tag{2.40}$$

and considering the map $\mathbf{y} = \mathbf{Gx}$ yields

$$y_j = \mathbf{G}_{j\cdot}\, \mathbf{x} \quad \leadsto \quad |y_j| = |\mathbf{G}_{j\cdot}\mathbf{x}| \leq ||\mathbf{G}_{j\cdot}||_1 ||\mathbf{x}||_\infty \quad \text{and} \tag{2.41}$$

$$||\mathbf{y}||_\infty = \max_j |y_j| = \max_j ||\mathbf{G}_{j\cdot}||_1 ||\mathbf{x}||_\infty \triangleq ||\mathbf{G}||_\infty ||\mathbf{x}||_\infty \tag{2.42}$$

$$\text{where}\quad \|\mathbf{G}\|_\infty \triangleq \max_j \|\mathbf{G}_{j\cdot}\|_1 = \max_j \sum_{k=1}^{n} |G_{jk}| \,. \tag{2.43}$$

Analogously, $\|\mathbf{y}\|_1 \le \|\mathbf{G}\|_1\|\mathbf{x}\|_1$ where $\|\mathbf{G}\|_1 \triangleq \max_j \|\mathbf{G}_{\cdot j}\|_1$. Finally, from Eq.(2.26) $\|\mathbf{y}\|_F \le \|\mathbf{G}\|_s\|\mathbf{x}\|_F$. Hence, the matrix infinity norm and matrix 1-norm as given by Eqs.(2.21) and (2.22), respectively, also possess the property of an induced norm.

The relation in magnitude is briefly sketched as follows. The 1-vector norm (sum vector norm) is the largest norm. The infinity vector norm $\|\mathbf{x}\|_\infty$ is the smallest one, hence, its unity locus in Fig. 2.1 is the largest. The vector entries must attain values largest among all norms in order to match a certain amount of the norm. The Frobenius vector norm is medium size. The spectral *matrix* norm is the smallest matrix norm and, thus, yields the best estimate of the vector map. However, the computational effort is maximum (*Zurmühl, R., and Falk, S., 1984*). The spectral radius $\rho_s[\mathbf{G}]$ is always smaller than any norm.

2.9 Bounding the Gain of a Multivariable System

Consider a multivariable system with transfer matrix $\mathbf{G}(s)$ and input $\mathbf{x}(s)$. Then, from Eq.(2.26)

$$\sigma_{\max}[\mathbf{G}(s)] = \sup_{\mathbf{x}} \frac{\|\mathbf{G}(s)\mathbf{x}(s)\|_F}{\|\mathbf{x}(s)\|_F} \ge \frac{\|\mathbf{G}(s)\mathbf{x}(s)\|_F}{\|\mathbf{x}(s)\|_F} \,. \tag{2.44}$$

With regard to $\sigma_{\max}[\mathbf{G}^{-1}(s)] = 1/\sigma_{\min}[\mathbf{G}(s)]$ (if the inverse exists)

$$\frac{1}{\sigma_{\min}[\mathbf{G}(s)]} = \sup_{\mathbf{y}} \frac{\|\mathbf{G}^{-1}(s)\mathbf{y}(s)\|_F}{\|\mathbf{y}(s)\|_F} \ge \frac{\|\mathbf{G}^{-1}(s)\mathbf{y}(s)\|_F}{\|\mathbf{y}(s)\|_F} \,. \tag{2.45}$$

Substituting $\mathbf{y}(s) = \mathbf{G}(s)\mathbf{x}(s)$ and taking the reciprocal value,

$$\sigma_{\min}[\mathbf{G}(s)] \le \frac{\|\mathbf{G}(s)\mathbf{x}(s)\|_F}{\|\mathbf{x}(s)\|_F} \,. \tag{2.46}$$

Combination with $\sigma_{\max}[\mathbf{G}(s)]$ yields

$$\sigma_{\max}[\mathbf{G}(s)] = \|\mathbf{G}(s)\|_s \ge \frac{\|\mathbf{G}(s)\mathbf{x}(s)\|_F}{\|\mathbf{x}(s)\|_F} \ge \sigma_{\min}[\mathbf{G}(s)] \,, \tag{2.47}$$

i.e., the gain of a multivariable system $\|\mathbf{G}(s)\mathbf{x}(s)\|_F/\|\mathbf{x}(s)\|_F$ is bounded between the smallest and largest singular values. Referring to Eq.(20.56), the infinity norm of the transfer function matrix $\mathbf{G}(j\omega)$, i.e. $\|\mathbf{G}(j\omega)\|_\infty$, is an upper bound of the system gain.

2.10 Output Norm Estimates of a Multivariable System

Consider a multivariable system in the frequency domain with input[1] $\hat{\mathbf{u}} = \hat{\mathbf{u}}(j\omega)$, transfer matrix $\mathbf{G}(j\omega)$ and output $\hat{\mathbf{y}} = \hat{\mathbf{y}}(j\omega)$. Then, the vector norm

$$\begin{aligned}
(\|\hat{\mathbf{y}}\|_F)^2 &= (\|\mathbf{G}(j\omega)\hat{\mathbf{u}}\|_F)^2 = \hat{\mathbf{u}}^H\mathbf{G}^H\mathbf{G}\hat{\mathbf{u}} &(2.48)\\
&= \frac{\hat{\mathbf{u}}^H\mathbf{G}^H\mathbf{G}\hat{\mathbf{u}}}{\hat{\mathbf{u}}^H\hat{\mathbf{u}}}\|\hat{\mathbf{u}}\|_F^2 \le \Big(\sup_{\hat{\mathbf{u}}} \frac{\hat{\mathbf{u}}^H\mathbf{G}^H\mathbf{G}\hat{\mathbf{u}}}{\hat{\mathbf{u}}^H\hat{\mathbf{u}}}\Big)\|\hat{\mathbf{u}}\|_F^2 = \|\mathbf{G}\|_s^2\,\|\hat{\mathbf{u}}\|_F^2 &(2.49)\\
\|\hat{\mathbf{y}}(j\omega)\|_F &\le \|\mathbf{G}(j\omega)\|_s\,\|\hat{\mathbf{u}}(j\omega)\|_F \,. &(2.50)
\end{aligned}$$

[1]The hat symbol is used to make a distinction, only, between the signal in the time domain $\mathbf{u}(t)$ and in the frequency domain $\hat{\mathbf{u}}(j\omega)$ or $\hat{\mathbf{u}}(s)$.

For the sake of comparison, the definition of the Frobenius (Euclidian) vector norm, Eq.(2.13)

$$||\mathbf{x}||_F = \sqrt{|x_1|^2 + |x_2|^2 + \dots \qquad + |x_n|^2} \tag{2.51}$$

and the norm of a function on L_2^n, Eq.(3.3),

$$||\mathbf{x}||_2 = \sqrt{\int_R |x_1|^2 dt + \int_R |x_2|^2 dt + \dots \qquad + \int_R |x_n|^2 dt}\ , \tag{2.52}$$

are repeated in detail. Although both formulas have the same algebraic structure there are essential differences. If $\mathbf{x}$ is a real or complex function of t then the Frobenius vector norm as processing instantaneous values of $\mathbf{x}(t)$ is a real scalar-valued function of t, i.e. $||\mathbf{x}||_F(t)$, but the norm on L_2^n is a real, positive and constant number representing a measure of the entire vector-valued functon. Considering a constant $\mathbf{x}$ then $||\mathbf{x}||_F$ is also a constant but since a constant $\mathbf{x}$ does not belong to L_2^n the norm $||\mathbf{x}||_2$ does *not* exist.

2.11 Further Properties

An important property on induced norms (*Lancaster, P., and Tismenetsky, M., 1985 p. 359*) is given by the following. Any eigenvalue $|\ \lambda[\mathrm{G}]\ |$ of a matrix $\mathrm{G} \in \mathcal{C}^{n\times n}$ in modulus is smaller than any norm $||\mathrm{G}||_p$, i.e., $|\ \lambda[\mathrm{G}]\ |\ \leq\ ||\mathrm{G}||_p$, since

$$\lambda[\mathrm{G}]\ \mathbf{g} = \mathrm{G}\mathbf{g} \quad \leadsto \quad |\ \lambda[\mathrm{G}]\ |\ ||\mathbf{g}||_p = ||\mathrm{G}\mathbf{g}||_p \leq ||\mathrm{G}||_p\ ||\mathbf{g}||_p \quad \leadsto \quad |\ \lambda[\mathrm{G}]\ | \leq ||\mathrm{G}||_p\ . \tag{2.53}$$

In other words: The spectral radius ρ_s does not exceed any matrix norm:

$$\rho_s[\mathrm{G}]\ \leq\ ||\mathrm{G}||_p\ .\ \square \tag{2.54}$$

From Eq.(2.53) $|\ \lambda[\mathbf{F}]\ | \leq \sigma_{\max}[\mathbf{F}]$, it follows with $\mathrm{G} = \mathbf{F}^{-1}$

$$\frac{1}{|\ \lambda[\mathbf{F}]\ |} \geq \frac{1}{\sigma_{\max}[\mathbf{F}]} \quad \text{or} \quad |\ \lambda[\mathbf{F}^{-1}]\ | \geq \sigma_{\min}[\mathbf{F}^{-1}] \quad \text{or} \quad |\ \lambda[\mathrm{G}]\ | \geq \sigma_{\min}[\mathrm{G}]\ .\ \square \tag{2.55}$$

Applying Eq.(2.53) to the Hermite matrix term $\mathrm{G}^H\mathrm{G}$ for any p,

$$\lambda_{\max}[\mathrm{G}^H\mathrm{G}] \leq ||\mathrm{G}^H\mathrm{G}||_p\ , \tag{2.56}$$

$$\text{if } p = 1 \quad \lambda_{\max}[\mathrm{G}^H\mathrm{G}] = ||\mathrm{G}||_s^2 \leq ||\mathrm{G}^H\mathrm{G}||_{p=1} \leq ||\mathrm{G}^H||_1\ ||\mathrm{G}||_1\ . \tag{2.57}$$

Since $||\mathrm{G}||_\infty = ||\mathrm{G}^T||_1 = ||\mathrm{G}^H||_1$, finally,

$$||\mathrm{G}||_s^2\ \leq\ ||\mathrm{G}||_\infty\ ||\mathrm{G}||_1\ .\ \square \tag{2.58}$$

Spectral norm assessment by the Frobenius norm: Recalling the spectral norm $||\mathrm{G}||_s$ and Frobenius norm $||\mathrm{G}||_F$

$$||\mathrm{G}||_s = \sigma_{\max}[\mathrm{G}] = +\sqrt{\lambda_{\max}[\mathrm{G}^H\mathrm{G}]} = +\sqrt{\rho_s[\mathrm{G}^H\mathrm{G}]} \tag{2.59}$$

$$||\mathrm{G}||_F = +\sqrt{\sum_{i,j=1}^{n} |G_{ij}|^2} = +\sqrt{\mathrm{tr}\ \mathrm{G}^H\mathrm{G}} = +\sqrt{\sum_{i=1}^{n} \lambda_i[\mathrm{G}^H\mathrm{G}]}\ . \tag{2.60}$$

Note the fact that the eigenvalues $\lambda_i[\mathbf{G}^H\mathbf{G}]$ are throughout non-negative. Comparing the right-hand side of both equations above, the following relations are achieved:

$$\lambda_{\max} \leq \sum_{i=1}^{n} \lambda_i \quad \rightsquigarrow \quad \|\mathbf{G}\|_s \leq \|\mathbf{G}\|_F \, . \tag{2.61}$$

$$\lambda_{\max} \geq \frac{1}{n}\sum_{i=1}^{n} \lambda_i \quad \rightsquigarrow \quad \|\mathbf{G}\|_s = +\sqrt{\lambda_{\max}[\mathbf{G}^H\mathbf{G}]} \geq +\sqrt{\frac{1}{n}\sum_{i=1}^{n} \lambda_i[\mathbf{G}^H\mathbf{G}]} = \frac{1}{\sqrt{n}}\|\mathbf{G}\|_F \, . \tag{2.62}$$

Hence,

$$\frac{1}{\sqrt{n}}\,\|\mathbf{G}\|_F \leq \|\mathbf{G}\|_s \leq \|\mathbf{G}\|_F \, . \, \Box \tag{2.63}$$

If $\mathbf{G},\mathbf{H} \in \mathcal{C}^{n\times n}$ and $|\mathbf{G}| \leq_e |\mathbf{H}|$ then $\rho_s[\mathbf{G}] \leq \rho_s[\,|\mathbf{H}|\,]$ (*Ortega, J.M., 1972, p. 120*). $\Box$

Spectral radius $\rho_s[\mathbf{G}]$, numerical radius $\rho_R[\mathbf{G}]$ and spectral norm $\|\mathbf{G}\|_s$:

$$\rho_s[\mathbf{G}] \leq \rho_R[\mathbf{G}] \leq \|\mathbf{G}\|_s \tag{2.64}$$

where $\rho_s[\mathbf{G}] = \rho_R[\mathbf{G}]$ if $\mathbf{G}$ is Hermitian, see Eq.(2.28). $\Box$

Spectral norm of a block diagonal matrix:

$$\|\text{ block diag } \{\mathbf{G},\mathbf{H}\}\ \|_s = \max\ \{\|\mathbf{G}\|_s,\ \|\mathbf{H}\|_s\}\ . \, \Box \tag{2.65}$$

If $\mathbf{G} = \text{diag}\ \{g_i\}$ is a diagonal matrix then for $p = 1,\ \infty,\ s$

$$\|\mathbf{G}\|_p = |g_i|_{\max}\ . \, \Box \tag{2.66}$$

If $\mathbf{G}$ is a normal matrix, i.e. $\mathbf{G}^H\mathbf{G} = \mathbf{G}\mathbf{G}^H$, then, referring to Eq.(22.25),

$$\rho_s[\mathbf{G}] = \rho_R[\mathbf{G}] = \|\mathbf{G}\|_s \quad \text{and} \quad \lambda_i[\mathbf{G}^H\mathbf{G}] = |\ \lambda_i[\mathbf{G}]\ |^2\ . \, \Box \tag{2.67}$$

(Caution: Confusion with Eq.(2.26).)

If $\mathbf{A}$ and $\mathbf{B}$ are normal (*Taussky, O., 1964*)

$$\|\mathbf{A}-\mathbf{B}\|_F \geq \sum_i |\ \lambda_i[\mathbf{A}] - \lambda_i[\mathbf{B}]\ |\ . \, \Box \tag{2.68}$$

A matrix $\mathbf{G}$ of rank one is characterized by $\sigma_{\max}[\mathbf{G}] = \|\mathbf{G}\|_F$ since (i) for matrices with rank one the trace equals the single nonzero eigenvalue, (ii) rank$\mathbf{G} = 1$ implies rank$(\mathbf{G}^T\mathbf{G}) = 1$ and, invoking Eqs.(2.59) and (2.60),

$$\text{(iii)}\quad \sigma_{\max}[\mathbf{G}] = \sqrt{\lambda_{\max}[\mathbf{G}^T\mathbf{G}]} = \sqrt{\text{tr}[\mathbf{G}^T\mathbf{G}]} = \|\mathbf{G}\|_F \quad \text{where} \quad \text{rank}\mathbf{G} = 1\ . \, \Box \tag{2.69}$$

For any $\mathbf{H}$ and any $\mathbf{x}$, using Eqs.(2.13) and (2.63),

$$\mathbf{x}^T\mathbf{H}\mathbf{x} \leq \|\mathbf{x}^T\|_s\|\mathbf{H}\|_s\|\mathbf{x}\|_s = \|\mathbf{H}\|_s\|\mathbf{x}\|_F\|\mathbf{x}\|_F = \|\mathbf{H}\|_s\mathbf{x}^T\mathbf{x}\, . \tag{2.70}$$

If, additionally, $\mathbf{H} = \mathbf{P}\mathbf{R}$ where $\mathbf{P}$ symmetric but $\mathbf{R} \neq \mathbf{R}^T$ then, referring to Eq.(2.67),

$$\mathbf{x}^T\mathbf{P}\mathbf{R}\mathbf{x} \leq \|\mathbf{P}\|_s\|\mathbf{R}\|_s\mathbf{x}^T\mathbf{x} = \lambda_{\max}[\mathbf{P}]\ \|\mathbf{R}\|_s\mathbf{x}^T\mathbf{x}\ . \, \Box \tag{2.71}$$

If $\mathbf{P} = \mathbf{P}^T$ (since $\mathbf{P}$ Hermite and normal) then

$$\sigma[\mathbf{P}] = |\ \lambda[\mathbf{P}]\ | \quad \text{and} \quad \sigma_{\max}[\mathbf{P}] = \|\mathbf{P}\|_s = \max_i |\ \lambda_i[\mathbf{P}]\ |\ . \, \Box \tag{2.72}$$

If $\mathbf{P}$ is symmetric and positive definite ($\mathbf{P} \triangleq \mathbf{G}^T\mathbf{G}$) then $\lambda_i[\mathbf{P}] > 0$ and for any $\mathbf{x}$

$$\mathbf{x}^T\mathbf{P}\mathbf{x} = \mathbf{x}^T\mathbf{G}^T\mathbf{G}\mathbf{x}^T = \lambda_{\max}[\mathbf{G}^T\mathbf{G}]\ \mathbf{x}^T\mathbf{x} \tag{2.73}$$

$$\lambda_{\max}[\mathbf{G}^T\mathbf{G}] = \sigma^2_{\max}[\mathbf{G}] = \sigma_{\max}[\mathbf{G}^T\mathbf{G}] = \sigma_{\max}[\mathbf{P}]\ .\ \square \tag{2.74}$$

If $\mathbf{F}$ is a stable matrix and if $\rho_s[\{\mathbf{E}_s(\mathbf{F}_s)^{-1}\}_s] < 1$ then $(\mathbf{F}+\mathbf{E})$ is stable where the subscript $_s$ denotes the symmetric part (*Yedavalli, R.K., 1985a*).

Proof: By definition of ρ_s , in this specific case

$$\max_i |\ \lambda_i[\{\mathbf{E}_s(\mathbf{F}_s)^{-1}\}_s]\ | < 1 \quad \text{or} \quad |\ \lambda_i[\{\mathbf{E}_s(\mathbf{F}_s)^{-1}\}_s]\ | < 1 \quad \forall i \tag{2.75}$$

$$1 - |\ \lambda_i[\{\mathbf{E}_s(\mathbf{F}_s)^{-1}\}_s]\ | > 0 \ \leadsto\ 1 + \lambda_i[\{\mathbf{E}_s(\mathbf{F}_s)^{-1}\}_s] > 0 \quad \forall i \tag{2.76}$$

$$\lambda_i[\mathbf{I} + \{\mathbf{E}_s(\mathbf{F}_s)^{-1}\}_s] > 0 \ \leadsto\ \lambda_i[\{\mathbf{I} + \mathbf{E}_s(\mathbf{F}_s)^{-1}\}_s] > 0 \quad \forall i\ . \tag{2.77}$$

Now, the following statements are used: (i) If $\mathbf{A}$ is Hermitian then $\mathbf{A}$ is positive [negative] definite if its eigenvalues are positive [negative].(ii) If $\mathbf{A}_s$ is negative definite then $\mathbf{A}_s$ and $\mathbf{A}$ are stable. If $-\mathbf{F}_s = -\mathbf{A}_s$ is positive definite then $\mathbf{A}_s$ is stable. (iii) The product $\mathbf{C}_s\mathbf{B}_s$ has positive real part eigenvalues if and only if $\mathbf{C}_s$ and $\mathbf{B}_s$ are positive definite.

Assume $\mathbf{B}_s$ positive definite then $\lambda_i[\{1+\mathbf{E}_s(\mathbf{F}_s)^{-1}\}_s\mathbf{B}_s] > 0$. If $\mathbf{B}_s := -\mathbf{F}_s = -\mathbf{A}_s$ and, finally, $\mathbf{C}_s := [1+\mathbf{E}_s(\mathbf{F}_s)^{-1}]$

$$\{1+\mathbf{E}_s(\mathbf{F}_s)^{-1}\}_s(-\mathbf{F}_s) > 0 \ \text{(positive definite)} \quad \leadsto \quad -(\mathbf{F}_s + \mathbf{E}_s) > 0 \tag{2.78}$$

$$\mathbf{F}_s + \mathbf{E}_s < 0 \quad \leadsto \quad (\mathbf{F}+\mathbf{E})_s < 0\ . \tag{2.79}$$

Hence, $\mathbf{F}+\mathbf{E}$ has negative real-part eigenvalues and $(\mathbf{F}+\mathbf{E})$ is stable.

End of the proof

Applicable to stiff dynamic systems, the eigenvalue difference can be estimated (*Bellman, R., 1970*)

$$\max_{i,j}\ |\ \lambda_i[\mathbf{A}] - \lambda_j[\mathbf{A}]\ | \le \sqrt{2\|\mathbf{A}\|_F^2 - \frac{2}{n}|\mathrm{tr}[\mathbf{A}]|^2}\ \le \sqrt{2}\ \|\mathbf{A}\|_F\ .\ \square \tag{2.80}$$

If a given matrix $\mathbf{G} \in \mathcal{C}^{n \times n}$ with entries G_{ij} satisfies the n conditions

$$\sum_{j=1}^{n}\ \max\ \{|G_{ij}|,\ |G_{ji}|\}\ \le a \qquad \forall i = [1,n] \tag{2.81}$$

where a is a non-negative number, then the spectral norm can be estimated by

$$\|\mathbf{G}\|_s\ \le\ a\ . \tag{2.82}$$

If, additionally, the entries G_{ij} obey the n inequalities

$$|G_{ij}| - \sum_{j=1,\ j\neq i}^{n} \frac{n_i}{n_j}\frac{|G_{ij}| + |G_{ji}|}{2}\ >\ a \qquad \forall i = [1,n] \tag{2.83}$$

where a, n_i and n_j are arbitrary positive constants, then the inverse $\mathbf{G}^{-1}$ exists and the spectral norm of the inverse can be valuated by

$$\|\mathbf{G}^{-1}\|_s\ <\ \frac{1}{a} \tag{2.84}$$

(*Böcker, J., et.al. 1986; Cook, P.A., 1972*). □

Finally, for any norm $$(\mathbf{I}+\mathbf{A})^{-1} = \mathbf{I} - \mathbf{A} + \mathbf{A}^2 - \mathbf{A}^3 + \ldots \quad \text{if } \|\mathbf{A}\| < 1 \tag{2.85}$$

$$\|(\mathbf{I}+\mathbf{A})^{-1}\| \leq 1 + \|\mathbf{A}\| + \|\mathbf{A}\|^2 + \ldots \tag{2.86}$$

$$(1+\|\mathbf{A}\|)^{-1} \leq \|(\mathbf{I}+\mathbf{A})^{-1}\| \leq (1-\|\mathbf{A}\|)^{-1} \quad \text{if } \|\mathbf{A}\| < 1 . \tag{2.87}$$

Consider the matrices $\mathbf{A}, \mathbf{B} \in \mathcal{C}^{n\times m}$, $\mathbf{A}$ nonsingular. Then, for the norms $1, s, \infty$ if $\alpha = \|\mathbf{A}^{-1}\mathbf{B}\| < 1$ (or $\alpha = \|\mathbf{B}\mathbf{A}^{-1}\| < 1$) the matrix $\mathbf{A}+\mathbf{B}$ is nonsingular (*Noble, B., and Daniel, J.W., 1977*)

$$\|(\mathbf{A}+\mathbf{B})^{-1}\| \leq \|\mathbf{A}^{-1}\|/(1-\alpha) . \; \square \tag{2.88}$$

Properties of $\|\mathbf{G}\|_1, \|\mathbf{G}\|_s$ and $\|\mathbf{G}\|_\infty$: Consider the matrix $\mathbf{G} \in \mathcal{C}^{m\times n}$ and the vector $\mathbf{x} \in \mathcal{C}^n$ then

$$\|\mathbf{G}\mathbf{x}\|_1 \leq m\|\mathbf{G}\mathbf{x}\|_\infty \leq m\|\mathbf{G}\|_\infty\|\mathbf{x}\|_\infty \leq m\|\mathbf{G}\|_\infty\|\mathbf{x}\|_1 \tag{2.89}$$

$$\frac{1}{\sqrt{m}}\|\mathbf{G}\|_s \leq \|\mathbf{G}\|_\infty \leq \sqrt{n}\|\mathbf{G}\|_s \tag{2.90}$$

$$\frac{1}{\sqrt{n}}\|\mathbf{G}\|_s \leq \|\mathbf{G}\|_1 \leq \sqrt{m}\|\mathbf{G}\|_s \tag{2.91}$$

$$\frac{1}{n}\|\mathbf{G}\|_\infty \leq \|\mathbf{G}\|_1 \leq m\|\mathbf{G}\|_\infty . \tag{2.92}$$

With regard to the preceding properties, if any one of the right-hand norms is zero then the other norms also vanish. Considering an infinite sequence of matrices $\mathbf{G}_i$, if any norm tends to zero then the sequence does so in all norms (*Noble, B., and Daniel, J.W., 1977*).

2.12 Composite Form

A modified "norm" definition, regarding individual optima of matrix elements in modulus, is given by (*Chen, K.H., and Chen, W.L., 1989*)

$$\sup_\omega |\mathbf{M}(\omega)| \triangleq \text{matrix}[\sup_\omega |M_{ij}(j\omega)|\,] . \tag{2.93}$$

2.13 Matrix Measures

The matrix measure $\mu_p[\mathbf{G}]$ or directional derivative is defined as the derivative of the induced matrix norm at point $\mathbf{I}$ in the direction of $\mathbf{G}$

$$\mu_p[\mathbf{G}] = \lim_{\varepsilon\to 0^+} \frac{\|\mathbf{I}+\varepsilon\mathbf{G}\|_p - \|\mathbf{I}\|_p}{\varepsilon} = \lim_{\varepsilon\to 0^+} \frac{\|\mathbf{I}+\varepsilon\mathbf{G}\|_p - 1}{\varepsilon} \tag{2.94}$$

where the last part of the above expression requires that $p = 1$, ∞ or s .

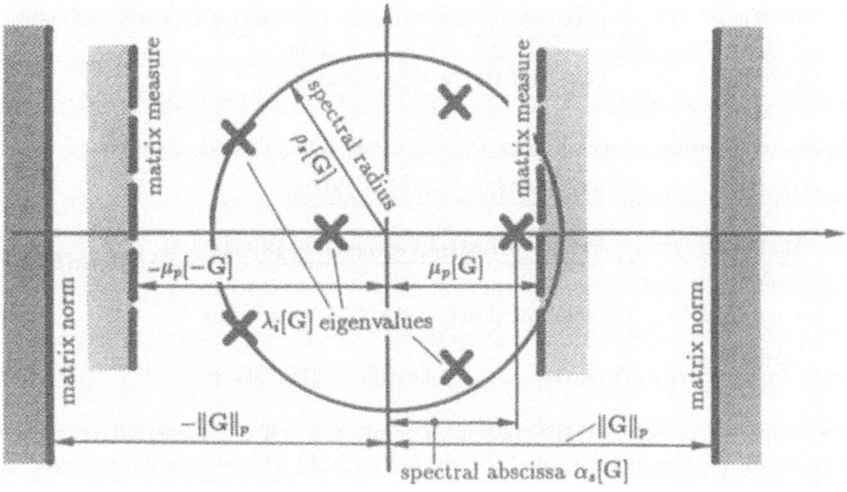

Figure 2.2: Spectral radius, eigenvalues, matrix measure

2.13.1 Matrix Measure Definitions

Considering real matrices **A** the following definitions are used

$$\mu_\infty[\mathbf{A}] = \max_i \, [A_{ii} + \sum_{j \neq i} |A_{ij}| \,] \quad \Box \tag{2.95}$$

$$\mu_1[\mathbf{A}] = \max_j [A_{jj} + \sum_{i \neq j} |A_{ij}| \,] \quad \Box \tag{2.96}$$

$$\mu_s[\mathbf{A}] = \lim_{\varepsilon \to 0} \frac{\|\mathbf{I} + \varepsilon\mathbf{A}\|_s - 1}{\varepsilon} = \lim_{\varepsilon \to 0} \frac{\sqrt{\lambda_{\max}[(\mathbf{I} + \varepsilon\mathbf{A})^T(\mathbf{I} + \varepsilon\mathbf{A})]} - 1}{\varepsilon} \tag{2.97}$$

$$= \lim_{\varepsilon \to 0} \frac{\sqrt{1 + \lambda_{\max}[\varepsilon\mathbf{A}^T + \varepsilon\mathbf{A}]} - 1}{\varepsilon} = \lim_{\varepsilon \to 0} \frac{1 + \frac{1}{2}\varepsilon\lambda_{\max}[\mathbf{A}^T + \mathbf{A}] - 1}{\varepsilon} \tag{2.98}$$

$$\mu_s[\mathbf{A}] = \frac{1}{2}\lambda_{\max}[\mathbf{A}^T + \mathbf{A}] = \lambda_{\max}\,[\frac{\mathbf{A}^T + \mathbf{A}}{2}] \, . \, \Box \tag{2.99}$$

$$-\mu_s[-\mathbf{A}] = \lambda_{\min}[\frac{\mathbf{A}^T + \mathbf{A}}{2}] \quad \Box \tag{2.100}$$

(*Jiang, C.L., 1987*) since for symmetric matrices **G** one has $\lambda_{\max}[-\mathbf{G}] = -\lambda_{\min}[\mathbf{G}]$. The subscript $_s$ often is replaced by $_2$. If the subscript is omitted for abbreviation, then $\mu \triangleq \mu_s$ is denoted.

2.13.2 Bounding the Solution of a Differential Equation

Consider the system of differential equations of first order $\dot{\mathbf{x}}(t) = \mathbf{A}(t)\mathbf{x}(t)$. Emphasis is put onto the change in norm of the solution $\|\mathbf{x}(t)\|$ when Δt is an infinitesimal and positive increment in time

$$\frac{\Delta\|\mathbf{x}(t)\|_F}{\Delta t} = \lim_{\Delta t\to 0}\frac{\|\mathbf{x}(t+\Delta t)\|_F - \|\mathbf{x}(t)\|_F}{\Delta t} \doteq \lim_{\Delta t\to 0}\frac{\|\mathbf{x}(t)+\dot{\mathbf{x}}(t)\Delta t\|_F - \|\mathbf{x}(t)\|_F}{\Delta t}$$

$$\frac{\Delta\|\mathbf{x}(t)\|_F}{\Delta t} \le \lim_{\Delta t\to 0}\frac{\|\mathbf{I}+\mathbf{A}(t)\Delta t\|_s\|\mathbf{x}(t)\|_F - \|\mathbf{x}(t)\|_F}{\Delta t} = \mu_s[\mathbf{A}(t)]\ \|\mathbf{x}(t)\|_F\,. \quad (2.101)$$

Alternatively, using a general norm property yields a more conservative bound

$$\frac{\Delta\|\mathbf{x}(t)\|_F}{\Delta t} = \lim_{\Delta t\to 0}\frac{\|\mathbf{x}(t)+\mathbf{A}(t)\mathbf{x}(t)\Delta t\|_F - \|\mathbf{x}(t)\|_F}{\Delta t} \quad (2.102)$$

$$\frac{\Delta\|\mathbf{x}(t)\|_F}{\Delta t} \le \lim_{\Delta t\to 0}\frac{\|\mathbf{x}(t)\|_F + \|\mathbf{A}(t)\|_s\|\mathbf{x}(t)\|_F\Delta t - \|\mathbf{x}(t)\|_F}{\Delta t} = \|\mathbf{A}(t)\|_s\|\mathbf{x}(t)\|_F\,.$$

Following Eq.(2.101), the solution of $\dot{\mathbf{x}}(t) = \mathbf{A}\mathbf{x}(t)$ is bounded by

$$\frac{\Delta\|\mathbf{x}(t)\|_F}{\|\mathbf{x}(t)\|_F} \le \mu_s[\mathbf{A}(t)]\ \Delta t \quad \rightsquigarrow \quad \|\mathbf{x}(t)\|_F \le \|\mathbf{x}(0^+)\|_F\ e^{\int_0^t \mu_s[\mathbf{A}(t)]dt}\,. \quad (2.103)$$

For comparison, observe (i) the state-space result $\mathbf{x}(t) = \mathbf{\Phi}\mathbf{x}(0^+)$ where $\mathbf{\Phi}(t)$ is given by Eq.(B.103), (ii) the significance of the eigenvalue $\lambda_i[\mathbf{A}]$ and (iii) the upper bound given by Eq.(2.109).

2.13.3 Matrix Measure Properties

For $\mathbf{G},\mathbf{H} \in \mathcal{C}^{n\times n}$ one has the following matrix measure properties: The matrix measure μ_p is sign-sensitive, the matrix norm is not. Furthermore,

$$\mu_p[\mathbf{G}+\mathbf{H}] \le \mu_p[\mathbf{G}] + \mu_p[\mathbf{H}] \quad \square \quad (2.104)$$

$$-\|\mathbf{G}\|_p \le -\mu_p[-\mathbf{G}] \le \mu_p[\mathbf{G}] \le \|\mathbf{G}\|_p \quad \square \quad (2.105)$$

$$\mu_p[\mathbf{G}] - \mu_p[\mathbf{H}] \le \mu_p[\mathbf{G}+\mathbf{H}] \quad \square \quad (2.106)$$

$$\max\left\{\mu_p[\mathbf{G}] - \mu_p[-\mathbf{H}],\ -\mu_p[-\mathbf{G}] + \mu_p[\mathbf{H}]\right\} \le \mu_p[\mathbf{G}+\mathbf{H}] \le \mu_p[\mathbf{G}] + \mu_p[\mathbf{H}] \quad (2.107)$$

$$\mu_p[\alpha\mathbf{G} + (1-\alpha)\mathbf{H}] \le \alpha\mu_p[\mathbf{G}] + (1-\alpha)\mu_p[\mathbf{H}] \quad \text{where} \quad \alpha \in [0,1]\,. \ \square \quad (2.108)$$

Matrix properties if $\mathbf{A} \in \mathcal{R}^{n\times n}$: Estimating the real part of $\lambda_i[\mathbf{A}]$ by use of matrix measures

$$-\mu_p[-\mathbf{A}] \le \Re e\lambda_i[\mathbf{A}] \le \mu_p[\mathbf{A}]\,. \quad (2.109)$$

For $\mathbf{A},\mathbf{B} \in \mathcal{R}^{n\times n}$ with $\mathbf{B} \ge 0$ it results (*Saniuk, J.M., and Rhodes, I.B., 1987*)

$$\text{tr}\,(\mathbf{AB}) \le |\,\text{tr}\,(\mathbf{AB})| \le \|\mathbf{A}\|_s\ \text{tr}\,\mathbf{B}\,, \quad (2.110)$$

$$\text{e.g.,}\quad \mathbf{A} = \begin{pmatrix} -3 & 1 \\ 2 & -5 \end{pmatrix},\ \mathbf{B} = \begin{pmatrix} 1 & -1 \\ -1 & 1 \end{pmatrix},\ \|\mathbf{A}\|_s = 5.834 \quad \rightsquigarrow \quad -11 \le 11 \le 11.668\,. \ \square \quad (2.111)$$

For the same assumption $\mathbf{A},\mathbf{B} \in \mathcal{R}^{n\times n}$ with $\mathbf{B} \ge 0$ (positive semidefinite), there exists an improved property (*Hmamed, A., 1989*) using the matrix measure

$$-\mu_s[-\mathbf{A}]\ \text{tr}\,\mathbf{B} \le \text{tr}\,(\mathbf{AB}) \le \mu_s(\mathbf{A})\ \text{tr}\,\mathbf{B}\,. \quad (2.112)$$

For the same numerical example as given in Eq.(2.111) it results

$$\mu_s[\mathbf{A}] = -2.197,\ \mu_s[-\mathbf{A}] = 5.802 \quad \rightsquigarrow \quad -11.605 \le -11 \le -4.394\,. \ \square \quad (2.113)$$

2.14 Diagonally Weighted Norms

For $D_i \neq 0,\ i = 1\dots n,\ \mathbf{D} = \text{diag}\ D_i,\ p \in [1,\infty)$ the diagonally weighted or D-weighted Hölder norm is

$$\|\mathbf{x}\|_{pD} \triangleq \|\mathbf{D}\mathbf{x}\|_p \triangleq \Big(\sum_{i=1}^{n} |D_i x_i|^p\Big)^{1/p} . \tag{2.114}$$

Special cases of weighted norm:

$$\|\mathbf{x}\|_{1D} \triangleq \sum_j |D_j x_j| \tag{2.115}$$

$$\|\mathbf{x}\|_{\infty D} \triangleq \|\mathbf{D}\mathbf{x}\|_\infty \triangleq \max |D_j x_j| \tag{2.116}$$

$$\|\mathbf{x}\|_F = \|\mathbf{x}\|_{2D} \ \Big|_{\ \mathbf{D}=\mathbf{I},\ p=2} . \ \square \tag{2.117}$$

The D-weighted direct matrix norms are defined as

$$\|\mathbf{A}\|_{1D} \triangleq \|\ |\mathbf{A}^T|\mathbf{d}\|_{\infty D} = \|\mathbf{D}\ |\mathbf{A}^T|\ \mathbf{d}\|_\infty = \text{max row sum of } \mathbf{D}|\mathbf{A}^T|\mathbf{D}^{-1} \tag{2.118}$$

$$\|\mathbf{A}\|_{\infty D} \triangleq \|\ |\mathbf{A}|\ \mathbf{d}\|_{\infty D} = \|\mathbf{D}\ |\mathbf{A}|\ \mathbf{d}\|_\infty = \text{max row sum of } \mathbf{D}|\mathbf{A}|\mathbf{D}^{-1} \tag{2.119}$$

where $\mathbf{d} = (\frac{1}{d_1}\ \frac{1}{d_2} \cdots \frac{1}{d_n})^T$ and $\mathbf{D}^{-1} = \text{diag}_n d_i$. □

The absolute-value sign applied to a matrix, i.e. $|\mathbf{A}|$, retains the magnitude of each element and omits the phase.

The D-weighted induced matrix norm is defined

$$\|\mathbf{A}\|_{pD} \triangleq \sup_{\mathbf{x}\neq 0} \frac{\|\mathbf{A}\mathbf{x}\|_{pD}}{\|\mathbf{x}\|_{pD}} = \sup_{\mathbf{x}\neq 0} \frac{\|\mathbf{D}\mathbf{A}\mathbf{x}\|_p}{\|\mathbf{D}\mathbf{x}\|_p} = \sup_{\mathbf{y}=\mathbf{D}\mathbf{x}} \frac{\|\mathbf{D}\mathbf{A}\mathbf{D}^{-1}\mathbf{y}\|_p}{\|\mathbf{y}\|_p} = \|\mathbf{D}\mathbf{A}\mathbf{D}^{-1}\|_p . \tag{2.120}$$

Note that for diagonal matrices $\mathbf{A}_d$ with regard to the commutativity of diagonal matrices the result is simply $\|\mathbf{A}_d\|_{pD} = \|\mathbf{A}_d\|_p$. Hence, D-weighted norms of diagonal perturbation matrices are independent of the weighting factors of the diagonally weighted norms. In the case $p = 2$, diagonally weighted norms are identical to optimal scaling operations using the maximum singular value.

For application to scaling operations, see Eq.(25.59) and (25.60).

Useful property for any $\mathbf{G} \in \mathcal{C}^{n\times n}$ and for $p \in [1,\ \infty]$ (*Safonov, M.G., 1982*)

$$\|\mathbf{G}\|_{pD} = \|\mathbf{G}^H\|_{q\mathbf{D}^{*,-1}} \qquad \frac{1}{p} + \frac{1}{q} = 1, \qquad \mathbf{D} = \text{diag}\ D_i, \qquad \det \mathbf{D} \neq 0 \ . \ \square \tag{2.121}$$

2.15 Hankel Norm $\|\mathbf{G}(s)\|_H$

Assume $\mathbf{G}(s) \in H_\infty$ and $\mathbf{G}(s) = \mathbf{C}(s\mathbf{I} - \mathbf{A})^{-1}\mathbf{B} + \mathbf{D}$ where $\mathbf{D}$ is the input-output matrix. Then, solve the Lyapunov equations

$$\mathbf{A}\mathbf{L}_C + \mathbf{L}_C\mathbf{A}^T + \mathbf{B}\mathbf{B}^T = \mathbf{0} \tag{2.122}$$

$$\mathbf{A}^T\mathbf{L}_O + \mathbf{L}_O\mathbf{A} + \mathbf{C}^T\mathbf{C} = \mathbf{0} . \tag{2.123}$$

The solutions $\mathbf{L}_C$ and $\mathbf{L}_O$ are real symmetric matrices and are denoted controllability and observability gramians, respectively. Then, the Hankel norm is given by

$$\|\mathbf{G}(s)\|_H \triangleq \sigma_{H\max}[\mathbf{G}(s)] \triangleq \sigma_{\max}[(\mathbf{L}_C\mathbf{L}_O)^{1/2}] = \|(\mathbf{L}_C\mathbf{L}_O)^{1/2}\|_s = \sqrt{\lambda_{\max}[\mathbf{L}_C\mathbf{L}_O]} \tag{2.124}$$

The matrix $\mathbf{D}$ does not influence $\|\mathbf{G}\|_H$.

If $\mathbf{A}$ is stable the Hankel singular value is given by

$$\sigma_{Hi}[\mathbf{G}(s)] \triangleq \sqrt{\lambda_i[\mathbf{L}_C\mathbf{L}_O]} \,. \tag{2.125}$$

Computing reduced-order models or reducing models the Hankel norm is an efficient tool.

2.15.1 Controllability and Observability Gramians

Notice the following properties

$$\frac{\partial}{\partial t}e^{\mathbf{A}t} = \mathbf{A}e^{\mathbf{A}t} \quad \text{and} \quad \frac{\partial}{\partial t}e^{\mathbf{A}^Tt} = \mathbf{A}^Te^{\mathbf{A}^Tt} = e^{\mathbf{A}^Tt}\mathbf{A}^T \tag{2.126}$$

which can easily be proved using Taylor expansion of $e^{\mathbf{A}t}$ and $e^{\mathbf{A}^Tt}$.

Differentiating the product $e^{\mathbf{A}t}\mathbf{B}\mathbf{B}^Te^{\mathbf{A}^Tt}$ with respect to t and using

$$\frac{\partial}{\partial t}\mathbf{F}_1(t)\mathbf{F}_2(t) = \frac{\partial \mathbf{F}_1(t)}{\partial t}\mathbf{F}_2(t) + \mathbf{F}_1(t)\frac{\partial \mathbf{F}_2(t)}{\partial t} \tag{2.127}$$

where $\mathbf{F}_1 \triangleq e^{\mathbf{A}t}\mathbf{B}$ and $\mathbf{F}_2 \triangleq \mathbf{B}^Te^{\mathbf{A}^Tt}$, one has

$$\frac{\partial}{\partial t}e^{\mathbf{A}t}\mathbf{B}\mathbf{B}^Te^{\mathbf{A}^Tt} = \mathbf{A}e^{\mathbf{A}t}\mathbf{B}\mathbf{B}^Te^{\mathbf{A}^Tt} + e^{\mathbf{A}t}\mathbf{B}\mathbf{B}^Te^{\mathbf{A}^Tt}\mathbf{A}^T \,. \tag{2.128}$$

Integrating both sides with the integration bounds 0 and ∞, defining the controllability gramian $\mathbf{L}_C$ and combining yields

$$\mathbf{L}_C \triangleq \int_0^\infty e^{\mathbf{A}t}\mathbf{B}\mathbf{B}^Te^{\mathbf{A}^Tt}dt \tag{2.129}$$

$$\int_0^\infty \frac{\partial}{\partial t}e^{\mathbf{A}t}\mathbf{B}\mathbf{B}^Te^{\mathbf{A}^Tt}dt = e^{\mathbf{A}t}\mathbf{B}\mathbf{B}^Te^{\mathbf{A}^Tt}\,\Big|_0^\infty = -\mathbf{B}\mathbf{B}^T = \mathbf{A}\mathbf{L}_C + \mathbf{L}_C\mathbf{A}^T \,. \tag{2.130}$$

Thus, the matrix $\mathbf{L}_C$ is proved to be the unique solution of the Lyapunov equation

$$\mathbf{A}\mathbf{L}_C + \mathbf{L}_C\mathbf{A}^T = -\mathbf{B}\mathbf{B}^T \,. \tag{2.131}$$

Similarly, the observability gramian $\mathbf{L}_O$ is the unique solution of a corresponding Lyapunov equation

$$\mathbf{L}_O \triangleq \int_0^\infty e^{\mathbf{A}^Tt}\mathbf{C}^T\mathbf{C}e^{\mathbf{A}t}dt \qquad \mathbf{A}^T\mathbf{L}_O + \mathbf{L}_O\mathbf{A} = -\mathbf{C}^T\mathbf{C} \,. \tag{2.132}$$

2.15.2 Hankel Operator and L_2-Gain Factor

The Hankel operator that corresponds to the matrix $\mathbf{G}$ is termed Γ_G^{op} and defined by

$$\mathbf{y}(t) = (\Gamma_G^{op}\mathbf{v})(t) \triangleq \int_0^\infty \mathbf{C}e^{\mathbf{A}(t+\tau)}\mathbf{B}\mathbf{v}(\tau)d\tau \tag{2.133}$$

$$\mathbf{v}(t) = \mathbf{u}(-t) \qquad (\Gamma_G^{op}\mathbf{u})(-t) = \mathbf{y} \qquad 0 < t < \infty \tag{2.134}$$

$$\text{or} \quad \mathbf{u}(t) = \mathbf{v}(-t) \;\; \forall t < 0 \quad \text{and} \quad \mathbf{y}(t) = (\Gamma_G^{op}\mathbf{v})(t) \quad \forall t > 0 \,. \tag{2.135}$$

The initial state variable $\mathbf{x}(0) = \mathbf{x}_o$ is considered as the result of past inputs $\mathbf{u}(-t)$ for $t > 0$, i.e.,

$$\mathbf{x}(0) = \mathbf{x}_o = \int_0^\infty e^{\mathbf{A}\tau}\mathbf{B}\mathbf{u}(-\tau)d\tau \,. \tag{2.136}$$

From the initials, the output results $\mathbf{y}(t) \triangleq \Gamma_G^{op}\mathbf{v}(t) = \mathbf{C}e^{\mathbf{A}t}\mathbf{x}_o$. Applying the complex conjugate transpose Γ_G^{opH} ,

$$(\Gamma_G^{opH}\mathbf{y})(t) = \int_0^\infty (\mathbf{C}e^{\mathbf{A}(t+\tau)}\mathbf{B})^T\mathbf{y}(\tau)d\tau = \int_0^\infty \mathbf{B}^T e^{\mathbf{A}^T(t+\tau)}\mathbf{C}^T\mathbf{y}(\tau)d\tau \,. \tag{2.137}$$

The expression above and $\mathbf{y}(t) = \mathbf{C}e^{\mathbf{A}t}\mathbf{x}_o$ is used in $\Gamma_G^{opH}\Gamma_G^{op}\mathbf{v} = \Gamma_G^{opH}\mathbf{y}$

$$\Gamma_G^{opH}\mathbf{y} = \mathbf{B}^T e^{\mathbf{A}^T t}\int_0^\infty e^{\mathbf{A}^T\tau}\mathbf{C}^T\mathbf{y}(\tau)d\tau = \mathbf{B}^T e^{\mathbf{A}^T t}\underbrace{\int_0^\infty e^{\mathbf{A}^T\tau}\mathbf{C}^T\mathbf{C}e^{\mathbf{A}\tau}d\tau}_{\mathbf{L}_O}\ \mathbf{x}_o = \mathbf{B}^T e^{\mathbf{A}^T t}\mathbf{L}_O\mathbf{x}_o \,. \tag{2.138}$$

By definition $\Gamma_G^{opH}\Gamma_G^{op}\mathbf{v} = \sigma_i^2\mathbf{v}$, i.e., $\sigma_i^2[\Gamma_G^{op}]$ is an eigenvalue corresponding to $\Gamma_G^{opH}\Gamma_G^{op}$ and σ_i is a singular value of Γ_G^{op} . Combining with Eq.(2.138) yields

$$\mathbf{B}^T e^{\mathbf{A}^T t}\mathbf{L}_O\mathbf{x}_o = \sigma_i^2\mathbf{v}(t) \quad \leadsto \quad \mathbf{v}(t) = \mathbf{B}^T e^{\mathbf{A}^T t}\mathbf{L}_O\mathbf{x}_o\sigma_i^{-2} \,. \tag{2.139}$$

Combination of Eqs.(2.139) and (2.136) yields

$$\mathbf{x}_o = \int_0^\infty e^{\mathbf{A}\tau}\mathbf{B}\mathbf{v}(\tau)d\tau = \underbrace{\int_0^\infty e^{\mathbf{A}\tau}\mathbf{B}\mathbf{B}^T e^{\mathbf{A}^T\tau}d\tau}_{\mathbf{L}_C}\,\mathbf{L}_O\mathbf{x}_o\sigma_i^{-2} = \mathbf{L}_C\mathbf{L}_O\mathbf{x}_o\sigma_i^{-2} \,. \tag{2.140}$$

Rearranging, $\sigma_i^2\mathbf{x}_o = \mathbf{L}_C\mathbf{L}_O\mathbf{x}_o$ is obtained. Thus, $\mathbf{x}_o$ is an eigenvector of the matrix $\mathbf{L}_C\mathbf{L}_O$ corresponding to the eigenvalue $\sigma_i^2 = \lambda_i[\mathbf{L}_C\mathbf{L}_O]$. By a former definition, one has $\sigma_i^2[\Gamma_G^{op}] = \lambda_i[\Gamma_G^{opH}\Gamma_G^{op}]$. Selecting the maximum singular value and taking the square root, the Hankel norm $\sigma_{i\max}[\Gamma_G^{op}] = \|\mathbf{G}(s)\|_H$ is achieved. Finally,

$$(\Gamma_G^{op}\mathbf{v})(t) \triangleq \int_0^\infty \mathbf{C}e^{\mathbf{A}(t+\tau)}\mathbf{B}\mathbf{u}(-\tau)d\tau = \int_{-\infty}^0 \mathbf{C}e^{\mathbf{A}(t-\eta)}\mathbf{B}\mathbf{u}(\eta)d\eta = \mathbf{y}(t) \,. \tag{2.141}$$

Since $\mathbf{u}(t) = \mathbf{0}$ for $t \geq 0$, see Fig. 3.3, in Eq.(2.141) the convolution of the input $\mathbf{u}(t)$ and the weighting matrix $\mathbf{C\Phi}(t)\mathbf{B}$ is the result where $\mathcal{L}^{-1}\mathbf{G}(s) = \mathbf{C\Phi}(t)\mathbf{B}$ and $\mathbf{\Phi}(s) = (s\mathbf{I}-\mathbf{A})^{-1} = \mathcal{L}\mathbf{\Phi}(t) = \mathcal{L}e^{\mathbf{A}t}$.

If the input $\mathbf{u}(t)$ and output $\mathbf{y}(t)$ are restricted to both $\mathbf{y}(t) = \mathbf{0}$ where $t < 0$ and $\mathbf{u}(t) = \mathbf{0}$ where $t > 0$ then the largest gain from input $\mathbf{u}(t)$ to output $\mathbf{y}(t)$ can be used as given by the spectral norm and the result is identical to the Hankel norm $\|\mathbf{G}(s)\|_H = \sup_{\mathbf{u}} \|\mathbf{y}(t)\|_2 \,/\, \|\mathbf{u}(t)\|_2$. This expression is the L_2-gain relating the past inputs $\mathbf{u}(t)$ $(t < 0)$ and the future outputs $\mathbf{y}(t)$ (*Glover, K., 1984*).

2.15.3 Control and Output Energy

The controllability gramian can be illustrated by determining the minimum control energy

$$\min_{\mathbf{u}} \int_{-\infty}^0 \mathbf{u}^T(t)\mathbf{u}(t)dt \quad \text{subject to} \quad \mathbf{x}(0) = \mathbf{x}_o \tag{2.142}$$

Then, see *Glover, K., 1984* ,

$$\arg\min_{\mathbf{u}} \int_{-\infty}^0 \mathbf{u}^T(t)\mathbf{u}(t)dt = \mathbf{u}^*(t) = \mathbf{B}^T e^{-\mathbf{A}^T t}\mathbf{L}_C^{-1}\mathbf{x}_o \ \text{ and } \ \min_{\mathbf{u}} \int_{-\infty}^0 \mathbf{u}^T(t)\mathbf{u}(t)dt = \mathbf{x}_o^T\mathbf{L}_C^{-1}\mathbf{x}_o \,. \tag{2.143}$$

If $\mathbf{u}(t) = \mathbf{0}$ for $t \geq 0$ and the system is released from $\mathbf{x}_o$, the output energy is $\int_0^\infty \mathbf{y}^T(t)\mathbf{y}(t)dt = \mathbf{x}_o^T\mathbf{L}_O\mathbf{x}_o$. The observability gramian relates the output energy to the initial conditions.

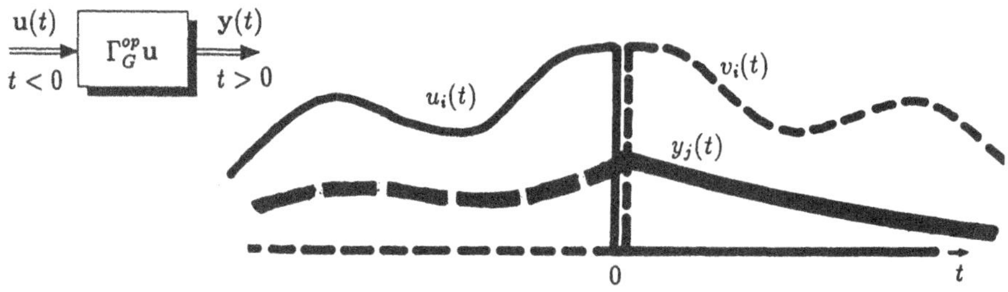

Figure 2.3: Illustrating the Hankel operator

2.15.4 Hankel Norm and Infinity Norm

The Hankel norm is linked to the H_∞-optimization by the relation

$$\inf_{\mathbf{F}} \|\mathbf{G}(s) - \mathbf{F}(s)\|_\infty = \inf_{\mathbf{F}} \sup_{\omega} \|\mathbf{G}(j\omega) - \mathbf{F}(j\omega)\|_s = \|\mathbf{G}(s)\|_H \qquad (2.144)$$

where $\mathbf{G}(s)$ is a given causal matrix-valued function, analytic in the closed right-half s-plane whereas $\mathbf{F}(s)$ is analytic in the open left-half s-plane.

Chapter 3

Functional Analysis, Function Norms and Control Signals

3.1 Function Space L_p of a Continuous Scalar-Valued Function f

Consider a function f with the value $f(t)$ at a certain instant t and assume that the integral of the pth power of the absolute value $|f(t)|$ exists within a region R. Then, this function f is said to be within the function space (or set of functions) $L_p(R)$. In most cases, R is given by an interval $[t_o, \infty)$. The function norm $\|f\|_p$ of the entire function f is defined by

$$\|f\|_p \triangleq [\int_R |f(t)|^p dt]^{1/p} \quad \forall p \in [1, \infty) \; . \tag{3.1}$$

If the integral does not converge for some power p then this function does not belong to the space L_p.

If $p = \infty$ the maximum value of $|f(t)|$ within the region R is dominant. Hence, the infinity space $L_\infty(R)$ is a space of functions bounded within R. Its norm is given by

$$\|f\|_\infty \triangleq \sup_{t \in R} |f(t)| \tag{3.2}$$

and is denoted as supremum norm or infinity norm.

3.2 Function Space L_p^n of Continuous Vector-Valued Function $\mathbf{f}$

Consider a vector-valued function $\mathbf{f} = (f_1 \;\, f_2 \;\, f_3 \ldots f_n)^T$. Then, the L_p^n space is given by the set of functions $\mathbf{f}$ with elements $f_i \in L_p$. Thus, $L_p \equiv L_p^1$. The norm on the L_p^n space is defined by

$$\|\mathbf{f}\|_p \triangleq [\sum_{i=1}^{n}(\|f_i\|_p)^p]^{1/p} = [\sum_{i=1}^{n} \int_R |f_i(t)|^p dt]^{1/p} \quad \forall p \in [1, \infty) \; . \tag{3.3}$$

In the infinity case the norm is given by

$$\|\mathbf{f}\|_\infty \triangleq \max_i \|f_i\|_\infty \quad \text{or alternatively by} \quad \|\mathbf{f}\|_\infty \triangleq \sum_{i=1}^{n} \|f_i\|_\infty \; . \tag{3.4}$$

The function spaces L_p^n are named Banach spaces if the function spaces are complete, i.e., if every Cauchy sequence $\{\mathbf{f}_n\}$ of L_p^n converges to a point $\mathbf{f}_\infty$ in L_p^n (*Martin,*

R.H., Jr., 1976; Curtain, R.F., and Pritchard, A.J., 1977; Milanese, M., and Negro, A., 1973; Böcker, J., et al. 1986)

$$\lim_{n\to\infty} \|\mathbf{f}_n - \mathbf{f}_\infty\| = 0 . \tag{3.5}$$

Using the norm $\|\mathbf{f}\|_p$ a non-negative scalar value describes the quantity of the entire function $\mathbf{f}$ within the region R . L_p-norms can be used to express convergence conditions. If the function $f(f : R^+ \to R)$ belongs to L_p (for $p = 1$ or 2 or ∞) then $\mathcal{L}\{f(t)\} = \hat{f}(s)$ exists (and is analytic) for $\Re e\ s > 0$.

3.3 Function Space of a Continuous Scalar-Valued Function

The Hardy space H_2 is defined by complex-valued functions $F(s)$ analytic in the open right half-plane $\Re e\ s > 0$ and satisfying the H_2-norm $\|F(s)\|_2$

$$\|F(s)\|_2 = \sqrt{\sup_\sigma \frac{1}{2\pi}\int_{-\infty}^{\infty} |F(\sigma + j\omega)|^2 d\omega} < \infty \quad \forall \sigma > 0 . \tag{3.6}$$

The real-rational function $F(s)$ (using the prefix R) belongs to RH_2 if $F(s)$ is stable and strictly proper. Then, the integration path can be chosen identical to the imaginary axis. Hence,

$$\|F(s)\|_2 = \sqrt{\frac{1}{2\pi}\int_{-\infty}^{\infty} |F(j\omega)|^2 d\omega} . \tag{3.7}$$

Referring to Eq.(20.56) the squared H_2-norm is equivalent to the energy of the signal $f(t) = \mathcal{L}^{-1}F(s)$ preassuming that $F(s) \in RH_2$ (*Francis, B.A., 1987*).

The Hardy space H_∞ consists of complex-valued functions $F(s)$ of a complex variable s analytic and bounded in the open right half-plane (*Duren, P.L., 1970; Garnett, J.B., 1981*)

$$|F(s)| < m_o \quad \forall\, \Re e\ s > 0, \quad \text{i.e.,} \quad F(s) \in H_\infty . \tag{3.8}$$

The least bound is the H_∞-norm given by the supremum

$$\|F(s)\|_\infty \triangleq \sup_{\Re e\ s>0} |F(s)| . \tag{3.9}$$

Considering rational functions with real coefficients, only, i.e., real-rational functions, this subset is termed RH_∞.

If the real-rational function $F(s)$ is proper (i.e., $|F(\infty)|$ is finite) and stable (i.e., $F(s)$ has no poles in the closed right half-plane $\Re e\ s \geq 0$) then $F(s) \in RH_\infty$. With regard to the maximum-modulus theorem the open right half-plane can be replaced by the imaginary axis

$$\|F(s)\|_\infty = \sup_{\omega \in \mathcal{R}} |F(j\omega)| \quad 0 \leq \omega \leq \infty \tag{3.10}$$

The right-hand side of the above equation is the maximum distance of the Nyquist plot of $F(j\omega)$ from the origin.

Suppose a single-input single-output control system with a stable controller $K(s)$ and a stable plant $G(s)$ as given e.g. in Fig. 31.3 in the case $\Delta G(s) \equiv 0$. Putting $F(s)$ equal to the open-loop transfer function $K(s)G(s)$ then

$$\|F(s)\|_\infty = \|K(s)G(s)\|_\infty < 1 \tag{3.11}$$

is sufficient for stability since the Nyquist plot cannot encircle the point $(-1, j0)$ in the $F(s)$-plane.

3.4 Function Space of a Continuous Matrix-Valued Function

The matrix-valued function $\mathbf{F}(j\omega)$ is described by its Hardy norm H_2 and H_∞ as follows

$$\|\mathbf{F}(j\omega)\|_2 \triangleq \sqrt{\mathrm{tr}\,\frac{1}{2\pi}\int_{-j\infty}^{j\infty} \mathbf{F}^H(j\omega)\mathbf{F}(j\omega)d\omega} \tag{3.12}$$

$$\|\mathbf{F}(j\omega)\|_\infty \triangleq \sup_\omega \sigma_{\max}[\mathbf{F}(j\omega)] = \sup_\omega \|\mathbf{F}(j\omega)\|_s \quad 0 \leq \omega \leq \infty \tag{3.13}$$

where $\|\mathbf{F}\|_s$ is the spectral norm.

The linear quadratic regulator problem can be recast to the general H_2 or H_∞ model-matching problem which stands for minimizing

$$\|\mathbf{T}_{11} + \mathbf{T}_{12}\mathbf{Q}\mathbf{T}_{21}\|_2^2 \quad \text{or} \quad \|\mathbf{T}_{11} + \mathbf{T}_{12}\mathbf{Q}\mathbf{T}_{21}\|_\infty^2 \tag{3.14}$$

with respect to $\mathbf{Q}$ where $\mathbf{Q} \in RH_\infty$. RH_∞ is the space of matrices with entries real-rational proper functions of s with no pole in $\Re e\ s \geq 0$. In addition, also $\mathbf{T}_{11}, \mathbf{T}_{12}, \mathbf{T}_{21} \in RH_\infty$. Defining $\mathbf{H}^R(s) = \mathbf{H}^T(-s)$ the notation

$$\|\mathbf{H}(s)\|_2^2 = \frac{1}{2\pi}\int_{-\infty}^{\infty} \mathrm{tr}[\mathbf{H}^R(j\omega)\mathbf{H}(j\omega)]d\omega \tag{3.15}$$

is also frequently used (*Francis, B.A., 1987; Wilson, D.A., 1989*).

3.5 Function Space of Discontinous Functions

Replacing the integral by a sum and the region R by a set of subscripts i the function spaces L_p and L_∞ in the case of discontinuous functions f_i are defined in a corresponding manner.

3.6 Function Space of an mth Power of a Vector-Valued Function $\mathbf{f}_m$

An ordinary power of $\mathbf{f}$, e.g., $\mathbf{f}^2$ or $\mathbf{f}^m$ makes no sense. Hence, $\mathbf{f}^m$ can be utilized to characterize another operation, namely

$$\mathbf{f}^m \triangleq (|f_1|^m, |f_2|^m, \ldots |f_n|^m)^T . \tag{3.16}$$

The norm of the mth power as defined above is given by

$$\begin{aligned}\|\mathbf{f}^m\|_p &\triangleq [\sum_{i=1}^{n}\int_R |f_i(t)|^{mp}dt]^{1/p} = \{[\sum_{i=1}^{n}\int_R |f_i(t)|^{mp}dt]^{1/(mp)}\}^m &(3.17)\\ &= (\|\mathbf{f}\|_{mp})^m = \|\mathbf{f}\|_{mp}^m \quad \forall 1 \leq p \leq \infty, \quad \forall 0 < m < \infty . &(3.18)\end{aligned}$$

3.7 Hölder Inequality

Consider two functions $\mathbf{f}$ and $\mathbf{u}$ on various spaces

$$\mathbf{f} \in L_p^n,\ \mathbf{u} \in L_q^n \quad \forall 1 \leq p, q \leq \infty \quad \text{where} \quad \frac{1}{p} + \frac{1}{q} = 1 . \tag{3.19}$$

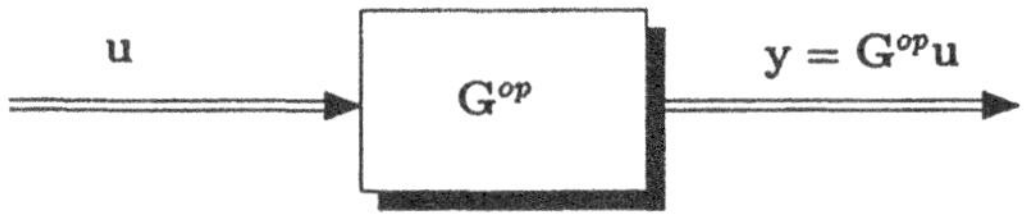

Figure 3.1: Mapping operation

Then, $\mathbf{f}^T\mathbf{u} \in L_1$ and an estimate of $\mathbf{f}^T\mathbf{u}$ is given by the Hölder inequality

$$\|\mathbf{f}^T\mathbf{u}\|_1 \leq \|\mathbf{f}\|_p \, \|\mathbf{u}\|_q \quad \text{where} \quad \frac{1}{p} + \frac{1}{q} = 1, \tag{3.20}$$

proof see *Yosida, K., 1980*. For a fixed function $\mathbf{f}$ the Hölder inequality yields the smallest possible estimate for any function $\mathbf{u}$. The special case of $p = q = 2$

$$\|\mathbf{f}^T\mathbf{u}\|_1 \leq \|\mathbf{f}\|_2 \, \|\mathbf{u}\|_2 \tag{3.21}$$

is known as Cauchy-Schwartz inequality.

3.8 Minkowski Inequality

Consider $\mathbf{f}, \mathbf{g} \in L_p^n \quad \forall p = [1, \infty]$ then

$$\|\mathbf{f} + \mathbf{g}\|_p \leq \|\mathbf{f}\|_p + \|\mathbf{g}\|_p \tag{3.22}$$

(*Ljusternik, L.A., and Sobolew, W.J., 1979; Chen, K.H., and Chen, W.L., 1989*). An interesting application of the Minkowski inequality occurs in developing the robust stability index, defined as $\|\Delta y(t)\|_2 / \|y(t)\|_2$ where Δy is the difference between the actual output and the model output $y(t)$. In view of this, a system is considered stable if the robust stability index is smaller than unity (*Wu, W.T., et al. 1991*).

3.9 Operator and Input-Output Mapping

By definition, the operator $\mathbf{G}^{op}$ is mapping the entire vector-valued input function $\mathbf{u}$ to the entire vector-valued output function $\mathbf{y}$. Any single-variable or multivariable system is described by the equation

$$\mathbf{y} = \mathbf{G}^{op}\mathbf{u} \qquad \mathbf{y}, \mathbf{u} \in \mathcal{R}^n \tag{3.23}$$

regardless if the system is continuous or discontinuous. $\mathbf{G}^{op}$ is an operator on L_p^n (*Hsu, J.C., and Meyer, A.U., 1968*).

3.9.1 Time-Invariant System

With the help of a retardation operator $\mathbf{R}_\tau^{op}$ a time delay is defined

$$(\mathbf{R}_\tau^{op}\mathbf{u})(t) \triangleq \mathbf{u}(t - \tau) \; . \tag{3.24}$$

If a system given by the operator $\mathbf{G}^{op}$ is time-invariant a delayed input $\mathbf{u}$ only causes a delayed output $\mathbf{y}$, i.e.,

$$\text{if} \quad \mathbf{y} = \mathbf{G}^{op}\mathbf{u} = \mathbf{G}^{op}(\mathbf{u}) \quad \forall \mathbf{u}, \mathbf{y}, \mathbf{G}^{op} \in L_p^n \quad \text{then} \quad \mathbf{G}^{op}(\mathbf{R}_\tau^{op}\mathbf{u}) = \mathbf{R}_\tau^{op}\mathbf{y} \tag{3.25}$$

and the operator or mapping function $\mathbf{G}^{op}(\mathbf{u})$ on L_p^n is denoted time-invariant.

3.9.2 Nonanticipative (Causal) System

The definition of a truncation operation applied to a function **u** and, equivalently, the definition of a truncation operator $\mathbf{P}_T^{op}$ are, respectively,

$$\mathbf{u}_T(t) \triangleq \begin{cases} \mathbf{u}(t) & \text{if} \quad t \leq T \\ \mathbf{0} & \text{if} \quad t > T \end{cases} \qquad \mathbf{u}_T \triangleq \mathbf{P}_T^{op}\mathbf{u} \,. \tag{3.26}$$

A function **u** is a function on the extended space L_{pe}^n if the truncated function $\mathbf{u}_T$ is on $L_p^n \ \forall T$.

Now, the causality theorem is stated: Applying a truncated input function $\mathbf{u}_T$ the corresponding output must not be influenced by any input component of the original nontruncated function **u** occurring at $t = T + \Delta(\Delta > 0)$, i.e.,

$$[\mathbf{G}^{op}(\mathbf{u}_T)]_T = [\mathbf{G}^{op}(\mathbf{u})]_T \quad \forall T, \forall \mathbf{u} \in L_p^n \quad \text{or} \quad \mathbf{P}_T^{op}\mathbf{G}^{op}\mathbf{P}_T^{op} = \mathbf{P}_T^{op}\mathbf{G}^{op} \,. \tag{3.27}$$

3.9.3 Convolution

The mapping operation from **u** to **y**, i.e. $\mathbf{y} = \mathbf{G}^{op}\mathbf{u}$, can be described by the convolution integral

$$\mathbf{y}(t) = \int_0^t \mathbf{G}(t-\tau)\mathbf{u}(\tau)d\tau = (\mathbf{G} \star \mathbf{u})(t) \tag{3.28}$$

in the case of a continuous-time system and by the convolution sum

$$\mathbf{y}(k) = \sum_{i=0}^{k} \mathbf{G}(k-i)\ \mathbf{u}(i) = (\mathbf{G} \star \mathbf{u})(k) \tag{3.29}$$

for a discontinuous-time system where $\mathbf{G}(t-\tau)$ and $\mathbf{G}(k-i)$ is the weighting function matrix $\in \mathcal{R}^{n \times n}$ of a time-invariant, nonanticipative, continuous-time and discontinuous-time system, respectively, associated with the operator $\mathbf{G}^{op}$ on L_p^n .

3.10 Operator Norm

The operator norm $\|\mathbf{G}^{op}\|_p$ of an operator $\mathbf{G}^{op}$ on any space L_p^n is the smallest scalar $r_{\min} \in \mathcal{R}$ such that

$$\|\mathbf{G}^{op}\mathbf{u}\|_p \leq r_{\min}\, \|\mathbf{u}\| = \|\mathbf{G}^{op}\|_p\, \|\mathbf{u}\|_p \quad \forall \mathbf{u} \in L_p^n \quad \rightsquigarrow \quad \|\mathbf{G}^{op}\|_p = \sup_{\mathbf{u} \neq 0} \frac{\|\mathbf{G}^{op}\mathbf{u}\|_p}{\|\mathbf{u}\|_p} \,. \tag{3.30}$$

The operator norm is also known as induced gain (*Sevaston, G.E., and Longman, R.W., 1985 and 1988*).

Using the truncation operator the L_2^n-induced norm of the matrix **G** is defined as

$$\|\mathbf{G}\|_2 \triangleq \sup_{\mathbf{u}_T \neq 0} \frac{\|(\mathbf{G}\mathbf{u})_T\|_2}{\|\mathbf{u}_T\|_2} \qquad \mathbf{u} \in L_{2e}^n,\ T = [0, \infty) \tag{3.31}$$

and the matrix **G** again is said to be L_2-bounded if $\|\mathbf{G}\|_2$ is finite.

3.11 Sobolev Space

The Sobolev space is a space of functions $\mathbf{f}(t) \in L_2^n$ which have derivatives $\dot{\mathbf{f}}(t) \in L_2^n$. Sobolev space is a subspace of L_2^n. The Sobolev norm is given by

$$\|\mathbf{f}(t)\|_{S2} = \sqrt{\|\mathbf{f}(t)\|_2^2 + \|\dot{\mathbf{f}}(t)\|_2^2} \tag{3.32}$$

preassuming that the function $\mathbf{f}(t)$, its derivative and the time scale are already transformed into a space without physical dimension. The L_2-S_2-gain of $\mathbf{G}$ is defined as

$$g_{LS}[\mathbf{G}] \triangleq \sup_{\mathbf{u}_T \neq 0} \frac{\|(\mathbf{G}\mathbf{u})_T\|_{S2}}{\|\mathbf{u}_T\|_2} = \sup_{\omega} \mid (1+j\omega)\sigma_{\max}[\mathbf{G}(j\omega)] \mid . \tag{3.33}$$

3.12 Estimates of the Convolution of Scalar-Valued Functions

An estimate of the convolution can be given as follows. Assume $g \in L_p$, $u \in L_q$ then the convolution $g \star u$ exists on L_r . The norms are related by

$$\|g \star u\|_r \leq \|g\|_p \, \|u\|_q \quad \text{where} \quad \frac{1}{r} = \frac{1}{p} + \frac{1}{q} - 1 \quad \text{and} \quad p,q,r \in [1,\infty] \, . \; \square \tag{3.34}$$

The special case where the input u and output $g \star u$ of a system belong to the same space L_r, i.e., $q = r$ is

$$p = 1 \qquad \|g \star u\|_r \leq \|g\|_1 \, \|u\|_r \quad \forall r \in [1,\infty] \, . \; \square \tag{3.35}$$

An alternative special case $r = \infty$ yields the L_∞-norm, namely the convolution infinity norm $\|g \star u\|_\infty$ for a fixed g and for any u . Since Eq.(3.20)

$$\begin{aligned} \|g \star u(t)\|_\infty &= \sup_t |g \star u| = \sup_t \left| \int_{-\infty}^{\infty} g(t-\tau)u(\tau)d\tau \right| \leq \int_{-\infty}^{\infty} |g(t-\tau)u(\tau)d\tau| \\ &= \|g(t-\tau)u(\tau)\|_1 \leq \|g(t-\tau)\|_p \|u(\tau)\|_q = \|g(t)\|_p \|u(t)\|_q \end{aligned} \tag{3.36}$$

$$\rightsquigarrow \qquad \|g \star u\|_\infty \leq \|g\|_p \, \|u\|_q \quad \text{where} \quad \frac{1}{p} + \frac{1}{q} = 1 \, . \tag{3.37}$$

The inequality above is the smallest bound. From Eq.(3.35) $\forall r \in [1,\infty]$ it results

$$\|g \star u\|_r \leq \|g\|_1 \, \|u\|_r \quad \text{or} \quad \frac{\|g \star u\|_r}{\|u\|_r} \leq \|g\|_1 \, . \tag{3.38}$$

The convolution operator norm is defined by

$$\|G_C^{op}\|_p \triangleq \sup_{u \neq 0} \frac{\|g \star u\|_p}{\|u\|_p} \, . \tag{3.39}$$

Comparison with Eq.(3.38) yields the convolution operator property

$$\|G_C^{op}\|_p \leq \|g\|_1 \qquad \forall p \in [1,\infty] \tag{3.40}$$

where in the case $p = \infty$ the equality is reached exactly

$$\|G_C^{op}\|_\infty = \|g\|_1 \, . \tag{3.41}$$

3.13 Parseval Theorem. Continuous-Time Signals

Consider continuous complex functions in time domain $f(t), f_1(t), f_2(t), g(t) \in L_2$ and the Fourier transformation

$$\mathcal{F}\{f(t)\} = \hat{f}(j\omega) \triangleq \int_{-\infty}^{\infty} f(t)e^{-j\omega t}dt \qquad f(t) = \mathcal{F}^{-1}\{\hat{f}(j\omega)\} = \frac{1}{2\pi}\int_{-\infty}^{\infty} \hat{f}(j\omega)e^{j\omega t}d\omega \,. \tag{3.42}$$

Recalling the property $\mathcal{F}\{f^*(t)\} = \hat{f}^*(-j\omega)$ and the correspondence between multiplication in time domain and convolution in frequency domain,

$$\mathcal{F}\{f_1(t)f_2(t)\} = \frac{1}{2\pi}\hat{f}_1(j\omega) \star \hat{f}_2(t) \qquad \mathcal{F}\{f_1(t)f_2^*(t)\} = \frac{1}{2\pi}\hat{f}_1(j\omega) \star \hat{f}_2^*(j\omega) \,. \tag{3.43}$$

Explicitly,

$$\int_{-\infty}^{\infty} f_1(t)f_2^*(t)e^{-j\omega t}\,dt = \frac{1}{2\pi}\int_{-\infty}^{\infty} \hat{f}_1(j\xi)\hat{f}_2^*[-(j\omega - j\xi)]d\xi \,. \tag{3.44}$$

Selecting $\omega = 0$ and, then, resubstituting the integration variable ξ by ω, it results

$$\int_{-\infty}^{\infty} f_1(t)f_2^*(t)dt = \frac{1}{2\pi}\int_{-\infty}^{\infty} \hat{f}_1(j\omega)\hat{f}_2^*(j\omega)d\omega \,. \tag{3.45}$$

With $f_1 = f_2 = f$ follows

$$\int_{-\infty}^{\infty} |f(t)|^2 = \frac{1}{2\pi}\int_{-\infty}^{\infty} |\hat{f}(j\omega)|^2 d\omega \quad \leadsto \quad \|f(t)\|_2 = \frac{1}{\sqrt{2\pi}}\|\hat{f}(j\omega)\|_2 \tag{3.46}$$

and the fact $\hat{f}_1, \hat{f}_2, \hat{f} \in L_2$, i.e., the Fourier transform maps L_2 onto L_2. The L_2-norms of f and $\hat{f}$ are equal except for the factor $1/\sqrt{2\pi}$. Substituting $f_1 := f$ and $f_2^* := g$ in Eq.(3.44),

$$\int_{-\infty}^{\infty} f(t)g(t)e^{-j\omega t}\,dt = \frac{1}{2\pi}\int_{-\infty}^{\infty} \hat{f}(j\xi)\hat{g}(j\omega - j\xi)d\xi \tag{3.47}$$

$$\int_{-\infty}^{\infty} f(t)g(t)dt = \frac{1}{2\pi}\int_{-\infty}^{\infty} \hat{f}(j\omega)\hat{g}(-j\omega)d\omega \,. \tag{3.48}$$

From the equation above the commonly used version is obtained if $f(t)$ and $g(t)$ are assumed real

$$\int_{-\infty}^{\infty} f(t)g(t)dt = \frac{1}{2\pi}\int_{-\infty}^{\infty} \hat{f}(j\omega)\hat{g}^*(j\omega)d\omega \,. \tag{3.49}$$

The "signal energy" is

$$\int_{-\infty}^{\infty} f^2(t)dt = \frac{1}{2\pi}\int_{-\infty}^{\infty} |\hat{f}(j\omega)|^2 d\omega \,. \tag{3.50}$$

For applying the Parseval theorem, tables of integrals can be used (*Eveleigh, V.W., 1967*).

3.14 Parseval Theorem. Discrete-Time Signals

Consider discrete real functions $f_{1i}, f_{2i} \in L_2$ and the corresponding z-transform

$$\mathcal{Z}\{f_{1i}\} = \mathcal{Z}\{f_1(iT)\} \triangleq \sum_{i=0}^{\infty} f_{1i}z^{-1} = \hat{f}_1, \qquad \mathcal{Z}\{f_{2i}\} \triangleq \hat{f}_2 \tag{3.51}$$

where $z = e^{sT}$ and T is the sampling period. The complex convolution in z-domain

$$\mathcal{Z}\{f_{1i}f_{2i}\} = \frac{1}{2\pi j}\oint \hat{f}_1(\xi)\hat{f}_2(\frac{z}{\xi})\xi^{-1}d\xi \tag{3.52}$$

is used with $z = 1$ in order to calculate the sum $\sum_{i=0}^{\infty} f_{1i}f_{2i}$.

$$\sum_{i=0}^{\infty} f_{1i}f_{2i} = \mathcal{Z}\{f_{1i}f_{2i}\} \mid_{z=1,\ \xi:=z} = \frac{1}{2\pi j}\oint \hat{f}_1(z)\hat{f}_2(z^{-1})z^{-1}dz\ . \tag{3.53}$$

The integration path is chosen the unit circle because all the poles of $\hat{f}_1$ and $\hat{f}_2$ are located within the unit circle of the complex z-plane. If $f_{1i} = f_{2i} = f_i$

$$\sum_{i=0}^{\infty} f_i^2 = \sum_{i=0}^{\infty} f^2(iT) = \frac{1}{2\pi}\int_{-\pi}^{\pi} |\hat{f}(e^{j\Omega})|^2 d\Omega\ . \tag{3.54}$$

Tables of integrals are given by *Günther, M., 1986*.

3.15 Generalized Parseval Theorem. Continuous-Time Signals

3.15.1 Two-Sided Laplace Transformation

In the case of two-sided Laplace transform $\hat{f}(s) = \int_{-\infty}^{\infty} f(t)e^{-st}dt$, assume that

$$\int_{-\infty}^{\infty} e^{-\alpha_i t}\ |f_i(t)|dt < \infty \quad \text{and} \quad \int_{-\infty}^{\infty} e^{-2\alpha_i t}\ |f_i(t)|^2 dt < \infty \qquad i = 1,2 \tag{3.55}$$

where the former of the conditions above can be included by the latter using additional considerations. Then, the generalized Parseval theorem

$$\int_{-\infty}^{\infty} e^{-(\alpha_1+\alpha_2)t}\ f_1(t)f_2^*(t)dt = \frac{1}{2\pi}\int_{-\infty}^{\infty} \hat{f}_1(\alpha_1 + j\omega)\hat{f}_2^*(\alpha_2 + j\omega)d\omega \tag{3.56}$$

holds (*Doetsch, G., 1967*). If $f_1 = f_2 = f$ and $\alpha_1 = \alpha_2 = \alpha$

$$\int_{-\infty}^{\infty} e^{-2\alpha t}\ |f(t)|^2 dt = \frac{1}{2\pi}\int_{-\infty}^{\infty} |\hat{f}(\alpha + j\omega)|^2 d\omega\ . \tag{3.57}$$

3.15.2 One-Sided Laplace Transformation

In the case of commonly used one-sided Laplace transform for functions $f(t)$ identically zero for $t < 0$ $(f : R^+ \rightarrow R)$

$$\hat{f}(s) \triangleq \mathcal{L}\{f(t)\} = \int_0^{\infty} f(t)e^{-st}dt\ . \tag{3.58}$$

Similarly, it follows

$$\int_0^{\infty} e^{-2\alpha t}\ |f(t)|^2 dt = \frac{1}{2\pi}\int_{-\infty}^{\infty} |\hat{f}(\alpha + j\omega)|^2 d\omega\ . \tag{3.59}$$

Part II

Differential Sensitivity. Small-Scale Perturbation

Chapter 4

Kronecker Calculus in Control Theory

4.1 Kronecker Product

In sensitivity analysis and in system analysis there is a frequent need for Kronecker products. The Kronecker product of two matrices (or direct product, tensor product) is defined by a partitioned matrix whose (i,j)-partition is $A_{ij}\mathbf{B}$

$$\mathbf{A}\otimes\mathbf{B} \triangleq \begin{pmatrix} A_{11}\mathbf{B} & A_{12}\mathbf{B} & \dots & A_{1n}\mathbf{B} \\ A_{21}\mathbf{B} & A_{22}\mathbf{B} & \dots & \vdots \\ \vdots & \vdots & \ddots & \vdots \\ A_{m1}\mathbf{B} & A_{m2}\mathbf{B} & \dots & A_{mn}\mathbf{B} \end{pmatrix} = \text{matrix}[A_{ij}\mathbf{B}]\,, \qquad \begin{array}{l} \mathbf{A} \in \mathcal{C}^{n\times m} \\ \mathbf{B} \in \mathcal{C}^{r\times s} \\ \mathbf{A}\otimes\mathbf{B} \in \mathcal{C}^{nr\times ms}\,. \end{array} \tag{4.1}$$

Some properties of the Kronecker product (*Brewer, J.W., 1978; Graham, A., 1981*) are listed below:

$$\text{Distributivity:} \qquad \mathbf{A}\otimes(\mathbf{B}+\mathbf{C}) = \mathbf{A}\otimes\mathbf{B}+\mathbf{A}\otimes\mathbf{C} \tag{4.2}$$

$$\text{Associativity:} \qquad (\mathbf{A}\otimes\mathbf{B})\otimes\mathbf{C} = \mathbf{A}\otimes(\mathbf{B}\otimes\mathbf{C})\,. \tag{4.3}$$

Permuting the Kronecker operands using Kronecker permutation matrix,

$$\mathbf{B}^{(k\times l)}\otimes\mathbf{A}^{(n\times m)} = \mathbf{U}_{kn}\left(\mathbf{A}^{(n\times m)}\otimes\mathbf{B}^{(k\times l)}\right)\mathbf{U}_{ml}\,. \tag{4.4}$$

Permutation matrix in Kronecker matrix sense: Square matrix with entries zero except one solitary digit one in each row and column

$$\mathbf{U}_{kl} \triangleq \mathbf{U}_{kl}^{(kl\times kl)} \triangleq \sum_{i=1}^{k}\sum_{j=1}^{l}\mathbf{E}_{ij}^{(k\times l)}\otimes\mathbf{E}_{ji}^{(l\times k)} = \sum_{i}^{k}\sum_{j}^{l}\mathbf{E}_{ij}^{(k\times l)}\otimes\left(\mathbf{E}_{ij}^{(k\times l)}\right)^{T}. \tag{4.5}$$

Some properties:

$$\mathbf{U}_{kl}^{-1} = \mathbf{U}_{kl}^{T} = \mathbf{U}_{lk} \ \ \text{(symmetric and orthogonal)} \qquad \mathbf{U}_{kl}^{T}\mathbf{U}_{kl} = \mathbf{I}_{(kl)}^{(kl\times kl)}\,. \ \square \tag{4.6}$$

$$(\mathbf{I}_r\otimes\mathbf{U}_{kl})^{-1} = \mathbf{I}_r\otimes\mathbf{U}_{kl}^{T} = \mathbf{I}_r\otimes\mathbf{U}_{lk}\,. \tag{4.7}$$

Self-derivative matrix (rectangular matrix $\partial\mathbf{A}/\partial\mathbf{A}$), see also Eq.(5.82), is given by

$$\bar{\mathbf{U}}_{kl} = \bar{\mathbf{U}}_{kl}^{(k^2\times l^2)} \triangleq \sum_{i=1}^{k}\sum_{j=1}^{l}\mathbf{E}_{ij}^{(k\times l)}\otimes\mathbf{E}_{ij}^{(k\times l)}\,. \tag{4.8}$$

Examples:

$$\bar{\mathbf{U}}_{22} = \begin{pmatrix} 1 & 0 & 0 & 1 \\ 0 & 0 & 0 & 0 \\ 0 & 0 & 0 & 0 \\ 1 & 0 & 0 & 1 \end{pmatrix}, \qquad \bar{\mathbf{U}}_{23} = \begin{pmatrix} 1 & 0 & 0 & 0 & 1 & 0 & 0 & 0 & 1 \\ 0 & 0 & 0 & 0 & 0 & 0 & 0 & 0 & 0 \\ 0 & 0 & 0 & 0 & 0 & 0 & 0 & 0 & 0 \\ 1 & 0 & 0 & 0 & 1 & 0 & 0 & 0 & 1 \end{pmatrix}. \tag{4.9}$$

End of Examples

Mixed product rule (only applicable if $\mathbf{A}, \mathbf{D}$ and $\mathbf{B}, \mathbf{G}$ are conformable):

$$(\mathbf{A} \otimes \mathbf{B})\ (\mathbf{D} \otimes \mathbf{G}) = (\mathbf{AD}) \otimes (\mathbf{BG})\ . \ \square \tag{4.10}$$

Transpose and inverse of the Kronecker product, respectively,

$$(\mathbf{A} \otimes \mathbf{B})^T = \mathbf{A}^T \otimes \mathbf{B}^T \qquad (\mathbf{N} \otimes \mathbf{M})^{-1} = \mathbf{N}^{-1} \otimes \mathbf{M}^{-1}\ . \tag{4.11}$$

Pseudo-inverse of the Kronecker product:

$$(\mathbf{A} \otimes \mathbf{B})^{\sharp} = (\mathbf{A}^{\sharp}) \otimes (\mathbf{B}^{\sharp})\ . \tag{4.12}$$

Defining $\mathbf{M} \in \mathcal{C}^{m \times m}\ , \quad \mathbf{N} \in \mathcal{C}^{n \times n}\ ,$

$$\det \mathbf{N}{\otimes}\mathbf{M} = (\det \mathbf{N})^m (\det \mathbf{M})^n\ , \quad \operatorname{tr} \mathbf{N}{\otimes}\mathbf{M} = (\operatorname{tr} \mathbf{N})(\operatorname{tr} \mathbf{M})\ , \quad e^{\mathbf{N}\otimes\mathbf{M}} = (e^{\mathbf{N}}){\otimes}(e^{\mathbf{M}})\ . \tag{4.13}$$

Powers of the block-diagonal matrix:

$$(\mathbf{I}_s \otimes \mathbf{M})(\mathbf{I}_s \otimes \mathbf{M}) = (\mathbf{I}_s\mathbf{I}_s) \otimes (\mathbf{MM}) = \mathbf{I}_s \otimes \mathbf{M}^2\ , \qquad (\mathbf{I}_s \otimes \mathbf{M})^k = \mathbf{I}_s \otimes \mathbf{M}^k\ . \tag{4.14}$$

Polynomial in $(\mathbf{I}_s \otimes \mathbf{M})$:

$$\operatorname{polyn}[\mathbf{I}_s \otimes \mathbf{M}] = \mathbf{I}_s \otimes \operatorname{polyn}[\mathbf{M}]\ . \tag{4.15}$$

Properties of uniform block diagonal matrices where f is any scalar analytic function

$$f(\mathbf{I}_m \otimes \mathbf{N}) = \mathbf{I}_m \otimes f(\mathbf{N}) \qquad f(\mathbf{N} \otimes \mathbf{I}_m) = f(\mathbf{N}) \otimes \mathbf{I}_m \tag{4.16}$$

$$(\mathbf{I}_m \otimes \mathbf{N})(\mathbf{M} \otimes \mathbf{I}_n) = (\mathbf{M} \otimes \mathbf{I}_n)(\mathbf{I}_m \otimes \mathbf{N}) = (\mathbf{M}\mathbf{I}_m) \otimes (\mathbf{I}_n\mathbf{N}) = \mathbf{M} \otimes \mathbf{N}\ . \ \square \tag{4.17}$$

Kronecker product between matrix and vector

$$\mathbf{A} \otimes \mathbf{z} = (\mathbf{I}_n \otimes \mathbf{z})\mathbf{A} \qquad \mathbf{A} \otimes \mathbf{z}^T = \mathbf{A}(\mathbf{I}_m \otimes \mathbf{z}^T) \qquad \mathbf{A} \in \mathcal{C}^{n \times m}\ . \tag{4.18}$$

The matrix $(\mathbf{I}_k \otimes c)$ and the (n, m)-matrix $\mathbf{A}$ "commute" if k is conformable with the dimensional assumptions of the matrix $\mathbf{A}$

$$(\mathbf{I}_n \otimes c)\mathbf{A} = (c\mathbf{I}_n)\mathbf{A} = c\mathbf{A} = c\mathbf{A}\mathbf{I}_m = \mathbf{A}\mathbf{I}_m c = \mathbf{A}(\mathbf{I}_m \otimes c) \qquad \mathbf{A} \in \mathcal{C}^{n \times m}\ . \tag{4.19}$$

4.2 Eigenvalues and Eigenvectors of the Kronecker Product

Consider the product of the matrices $\mathbf{A}\otimes\mathbf{B}$ and $\mathbf{a}_i\otimes\mathbf{b}_i$, applying the eigenvector definition $\mathbf{A}\mathbf{a}_i = \lambda_i[\mathbf{A}]\mathbf{a}_i$ and the mixed product rule

$$(\mathbf{A}\otimes\mathbf{B})(\mathbf{a}_k\otimes\mathbf{b}_i) = (\mathbf{A}\mathbf{a}_k)\otimes(\mathbf{B}\mathbf{b}_i) = \lambda_k[\mathbf{A}]\mathbf{a}_k\otimes\lambda_i[\mathbf{B}]\mathbf{b}_i = \lambda_k[\mathbf{A}]\lambda_i[\mathbf{B}](\mathbf{a}_k\otimes\mathbf{b}_i)\ . \quad (4.20)$$

From this relation the following important properties can be achieved

$$\mathbf{A}\otimes\mathbf{B} = \mathbf{C} \quad \leadsto \quad \lambda_j[\mathbf{C}] = \lambda_j[\mathbf{A}\otimes\mathbf{B}] = \lambda_k[\mathbf{A}]\lambda_i[\mathbf{B}]\ , \qquad \mathbf{c}_j^\triangleright = \mathbf{a}_k^\triangleright\otimes\mathbf{b}_i^\triangleright\ . \quad (4.21)$$

Example: Kronecker product:

$$\mathbf{A} = \begin{pmatrix} 1 & 0 \\ 0 & 2 \end{pmatrix},\quad \mathbf{B} = \begin{pmatrix} 0 & -1 \\ -5 & 4 \end{pmatrix},\quad \mathbf{C} = \begin{pmatrix} 0 & -1 & 0 & 0 \\ -5 & 4 & 0 & 0 \\ 0 & 0 & 0 & -2 \\ 0 & 0 & -10 & 8 \end{pmatrix}. \quad (4.22)$$

$$\text{Eigenvalues} \quad \lambda_k[\mathbf{A}] = 1;\ 2 \qquad \lambda_i[\mathbf{B}] = 5;\ -1 \qquad \lambda_j[\mathbf{C}] = 5;\ -1;\ 10;\ -2\ . \quad (4.23)$$

$$\text{Eigenvectors normalized:} \quad \mathbf{a}_k^\triangleright = \begin{pmatrix}1\\0\end{pmatrix};\begin{pmatrix}0\\1\end{pmatrix} \qquad \mathbf{b}_i^\triangleright = \frac{1}{\sqrt{26}}\begin{pmatrix}1\\-5\end{pmatrix};\ \frac{1}{\sqrt{2}}\begin{pmatrix}1\\1\end{pmatrix} \quad (4.24)$$

$$\mathbf{c}_j^\triangleright = \frac{1}{\sqrt{26}}\begin{pmatrix}1\\-5\\0\\0\end{pmatrix};\ \frac{1}{\sqrt{2}}\begin{pmatrix}1\\1\\0\\0\end{pmatrix};\ \frac{1}{\sqrt{26}}\begin{pmatrix}0\\0\\1\\-5\end{pmatrix};\ \frac{1}{\sqrt{2}}\begin{pmatrix}0\\0\\1\\1\end{pmatrix}. \quad (4.25)$$

Special case $k = 2, i = 1, j = 3$:

$$\lambda_3[\mathbf{C}] = 2\times 5 = 10\ , \qquad \mathbf{c}_3^\triangleright = \mathbf{a}_2^\triangleright\otimes\mathbf{b}_1^\triangleright = \begin{pmatrix}0\\1\end{pmatrix}\otimes\frac{1}{\sqrt{26}}\begin{pmatrix}1\\-5\end{pmatrix} = \frac{1}{\sqrt{26}}\begin{pmatrix}0\\0\\1\\-5\end{pmatrix}.\ \square \quad (4.26)$$

4.3 Kronecker Sum

The Kronecker sum is defined by

$$\mathbf{N}\oplus\mathbf{M} = \mathbf{N}\otimes\mathbf{I}_m + \mathbf{I}_n\otimes\mathbf{M} \qquad \mathbf{N}\in\mathcal{C}^{n\times n},\ \ \mathbf{M}\in\mathcal{C}^{m\times m},\ \ \mathbf{N}\oplus\mathbf{M}\in\mathcal{C}^{mn\times mn}\ . \quad (4.27)$$

Examples:

$$\mathbf{A}^{(n\times m)}\otimes\mathbf{0}^{(r\times s)} = \mathbf{0}_{(nr\times ms)}\ ; \qquad \mathbf{N}\oplus\mathbf{0} = \mathbf{N}\otimes\mathbf{I}_m \neq \mathbf{N}\ ; \qquad \mathbf{A}\otimes 1 = \mathbf{A}\ . \quad (4.28)$$

For $\mathbf{M}\in\mathcal{R}^{n\times n}$ and $\mathbf{M} = \text{matrix}[m_{ij}]$

$$\mathbf{M}\oplus\mathbf{M} = \text{matrix}[m_{ij}\mathbf{I}] + \text{diag}\{\mathbf{M},\mathbf{M},\ldots\mathbf{M}\} \quad \in\mathcal{R}^{n^2\times n^2}\ . \quad (4.29)$$

End of Examples

4.3.1 Exponential Algebra with Kronecker Sum

$$e^{\mathbf{N}\oplus\mathbf{M}} = e^{\mathbf{N}\otimes\mathbf{I}_m+\mathbf{I}_n\otimes\mathbf{M}} = e^{\mathbf{N}\otimes\mathbf{I}_m}e^{\mathbf{I}_n\otimes\mathbf{M}} = e^{\mathbf{I}_n\otimes\mathbf{M}+\mathbf{N}\otimes\mathbf{I}_m} = e^{\mathbf{I}_n\otimes\mathbf{M}}e^{\mathbf{N}\otimes\mathbf{I}_m}. \quad (4.30)$$

From Eq.(4.30) using mixed product rule,

$$e^{\mathbf{N}\oplus\mathbf{M}} = (e^{\mathbf{N}}\otimes\mathbf{I}_m)(\mathbf{I}_n\otimes e^{\mathbf{M}}) = (e^{\mathbf{N}}\mathbf{I}_n)\otimes(\mathbf{I}_m e^{\mathbf{M}}) = e^{\mathbf{N}}\otimes e^{\mathbf{M}}\ . \quad (4.31)$$

4.3.2 Eigenvalues and Eigenvectors of the Kronecker Sum

$$\mathbf{A} \oplus \mathbf{B} = \mathbf{C} \quad \leadsto \quad \lambda_j[\mathbf{C}] = \lambda_j[\mathbf{A} \oplus \mathbf{B}] = \lambda_k[\mathbf{A}] + \lambda_i[\mathbf{B}] \qquad \mathbf{c}_j^{\triangleright} = \mathbf{a}_k^{\triangleright} \otimes \mathbf{b}_i^{\triangleright} , \quad (4.32)$$

i.e., Kronecker sum and Kronecker product share the eigenvectors.

Example: Kronecker sum:

$$\mathbf{A} = \begin{pmatrix} 1 & 0 \\ 0 & 2 \end{pmatrix}, \quad \mathbf{B} = \begin{pmatrix} 0 & -1 \\ -5 & 4 \end{pmatrix} \tag{4.33}$$

$$\mathbf{C} = \begin{pmatrix} 1 & 0 \\ 0 & 2 \end{pmatrix} \otimes \begin{pmatrix} 1 & 0 \\ 0 & 1 \end{pmatrix} + \begin{pmatrix} 1 & 0 \\ 0 & 1 \end{pmatrix} \otimes \begin{pmatrix} 0 & -1 \\ -5 & 4 \end{pmatrix} = \begin{pmatrix} 1 & -1 & 0 & 0 \\ -5 & 5 & 0 & 0 \\ 0 & 0 & 2 & -1 \\ 0 & 0 & -5 & 6 \end{pmatrix} \tag{4.34}$$

$$\lambda_k[\mathbf{A}] = 2 \qquad \lambda_i[\mathbf{B}] = 5;\ -1 \qquad \lambda_j[\mathbf{C}] = 6;\ 0;\ 7;\ 1\,. \tag{4.35}$$

The eigenvectors $\mathbf{c}_j^{\triangleright}$ are identical to Eq. (4.25). Check with $k = 2, i = 1, j = 3$:

$$\begin{pmatrix} 1 & -1 & 0 & 0 \\ -5 & 5 & 0 & 0 \\ 0 & 0 & 2 & -1 \\ 0 & 0 & -5 & 6 \end{pmatrix} \frac{1}{\sqrt{26}} \begin{pmatrix} 0 \\ 0 \\ 1 \\ -5 \end{pmatrix} = \frac{1}{\sqrt{26}} \begin{pmatrix} 0 \\ 0 \\ 7 \\ -35 \end{pmatrix} = 7 \frac{1}{\sqrt{26}} \begin{pmatrix} 0 \\ 0 \\ 1 \\ -5 \end{pmatrix} . \tag{4.36}$$

End of Example

4.4 Exponential of the Kronecker Product

The matrix $\mathbf{N}^T \otimes \mathbf{M}^T$, evolved from $(\mathbf{N} \otimes \mathbf{M})^T$, possesses eigenvectors $\mathbf{n}_i^{\triangleleft} \otimes \mathbf{m}_k^{\triangleleft}$. If $\mathbf{M}$ has m distinct eigenvectors then $\mathbf{I}_n \otimes \mathbf{M}$ has mn eigenvectors, namely m distinct eigenvectors each with multiplicity n. From orthonomality follows

$$(\mathbf{n}_i^{\triangleleft} \otimes \mathbf{m}_k^{\triangleleft})^T (\mathbf{n}_i \otimes \mathbf{m}_k) = (\mathbf{n}_i^{\triangleleft T} \otimes \mathbf{m}_k^{\triangleleft T})(\mathbf{n}_i \otimes \mathbf{m}_k) = (\mathbf{n}_i^{\triangleleft T} \mathbf{n}_i) \otimes (\mathbf{m}_k^{\triangleleft T} \mathbf{m}_k) = 1\,. \tag{4.37}$$

The exponential of the Kronecker product can be written by analogy to the exponential of a simple matrix $e^{\mathbf{A}t} = \sum_i \mathbf{a}_i \mathbf{a}_i^{\triangleleft T} e^{\lambda_i[\mathbf{A}]t}$ (*Brewer, J.W., 1977*) as

$$e^{\mathbf{N} \otimes \mathbf{M}\, t} = \sum_{i=1}^{n} \sum_{k=1}^{m} (\mathbf{n}_i \otimes \mathbf{m}_k)(\mathbf{n}_i^{\triangleleft T} \otimes \mathbf{m}_k^{\triangleleft T}) e^{\lambda_i[\mathbf{N}]\lambda_k[\mathbf{M}]\, t} \qquad \mathbf{N} \in \mathcal{C}^{n \times n} , \quad \mathbf{M} \in \mathcal{C}^{m \times m} . \tag{4.38}$$

For a particular choice $\mathbf{N} = \mathbf{I}_n$, $\lambda_i[\mathbf{N}] = 1$ (with multiplicity n) it results $\mathbf{N} = \mathbf{N}^T \leadsto$ $\mathbf{n}_i = \mathbf{n}_i^{\triangleleft} = \mathbf{e}_i^{(n \times 1)} \leadsto \sum_{i=1}^{n} \mathbf{e}_i \mathbf{e}_i^T = \mathbf{I}_n$

$$e^{\mathbf{I}_n \otimes \mathbf{M} t} = \sum_{k=1}^{m} \sum_{i=1}^{n} \mathbf{e}_i \mathbf{e}_i^T \mathbf{m}_k \mathbf{m}_k^{\triangleleft T} e^{\lambda_k[\mathbf{M}]\, t} = \sum_{k=1}^{m} \mathbf{I}_s \otimes \mathbf{m}_k \mathbf{m}_k^{\triangleleft T} e^{\lambda_k[\mathbf{M}]\, t} . \tag{4.39}$$

4.5 Column Operator and Kronecker Matrix Product

Column operator (or stacking operator) applied to matrix product, i.e., a vector-valued function of a matrix

$$\operatorname{col} \mathbf{ACB} = (\mathbf{B}^T \otimes \mathbf{A}) \operatorname{col} \mathbf{C}\,. \tag{4.40}$$

Proof for one column of **ACB**:

$$(\mathbf{ACB})_{\cdot j} = \sum_i (\mathbf{AC})_{\cdot i} B_{ij} = \sum_i (B_{ij}\mathbf{A})\mathbf{C}_{\cdot i} \tag{4.41}$$

$$= [(\mathbf{B}_{\cdot j})^T \otimes \mathbf{A}]\text{col } \mathbf{C} = [(\mathbf{B}^T)_{j\cdot} \otimes \mathbf{A}]\text{col } \mathbf{C} \ . \tag{4.42}$$

Assuming matrices **A** and **C** with dimensions $n \times m$, $m \times q$, respectively,

$$\text{col } (\mathbf{AC}) = \text{col } (\mathbf{ACI}_q) = (\mathbf{I}_q \otimes \mathbf{A})\text{col } \mathbf{C} \tag{4.43}$$

$$= \text{col } (\mathbf{I}_n\mathbf{AC}) = (\mathbf{C}^T \otimes \mathbf{I}_n)\text{col } \mathbf{A} \tag{4.44}$$

$$= \text{col } (\mathbf{AI}_m\mathbf{C}) = (\mathbf{C}^T \otimes \mathbf{A})\text{col } \mathbf{I}_m \tag{4.45}$$

$$= \sum_{j=1}^{m} (\mathbf{C}^T)_{\cdot j} \otimes \mathbf{A}_{\cdot j} = \sum_{j=1}^{m} \mathbf{C}_{j\cdot} \otimes \mathbf{A}_{\cdot j} \ . \tag{4.46}$$

Establishing a row string *row* and transposing it to obtain row^T in order to define a (column) vector yields an expression where the ordering $\mathbf{A}, \mathbf{B}$ and $\mathbf{A}, \mathbf{B}^T$ is maintained when applied to the product **ACB** (*Barnett, S., 1973*):

$$\text{row}^T(\mathbf{ACB}) = \text{col } [(\mathbf{ACB})^T] = \text{col } (\mathbf{B}^T\mathbf{C}^T\mathbf{A}^T) \tag{4.47}$$

$$= (\mathbf{A} \otimes \mathbf{B}^T)\text{col } (\mathbf{C}^T) = (\mathbf{A} \otimes \mathbf{B}^T)\text{row}^T\mathbf{C} \ .$$

$$\text{row}^T\mathbf{A} = \text{col } \mathbf{A}^T = \mathbf{U}_{nm} \text{ col } \mathbf{A} \qquad \mathbf{A} \in \mathcal{C}^{n\times m} \ . \ \square \tag{4.48}$$

Example: $n = m = 2$, permutation matrix $\mathbf{U}_{22} = \mathbf{U}_{22}^{(4\times4)}$:

$$\mathbf{U}_{22} = \begin{pmatrix}1&0\\0&0\end{pmatrix} \otimes \begin{pmatrix}1&0\\0&0\end{pmatrix}^T + \begin{pmatrix}0&1\\0&0\end{pmatrix} \otimes \begin{pmatrix}0&1\\0&0\end{pmatrix}^T + \begin{pmatrix}0&0\\1&0\end{pmatrix} \otimes \begin{pmatrix}0&0\\1&0\end{pmatrix}^T + \begin{pmatrix}0&0\\0&1\end{pmatrix} \otimes \begin{pmatrix}0&0\\0&1\end{pmatrix}^T \tag{4.49}$$

$$= \begin{pmatrix}1&0&0&0\\0&0&0&0\\0&0&0&0\\0&0&0&0\end{pmatrix} + \begin{pmatrix}0&0&0&0\\0&0&1&0\\0&0&0&0\\0&0&0&0\end{pmatrix} + \begin{pmatrix}0&0&0&0\\0&0&0&0\\0&1&0&0\\0&0&0&0\end{pmatrix} + \begin{pmatrix}0&0&0&0\\0&0&0&0\\0&0&0&0\\0&0&0&1\end{pmatrix} = \begin{pmatrix}1&0&0&0\\0&0&1&0\\0&1&0&0\\0&0&0&1\end{pmatrix} \tag{4.50}$$

$$\mathbf{A} = \begin{pmatrix}a&b\\c&d\end{pmatrix}, \quad \text{col } \mathbf{A} = \begin{pmatrix}a\\c\\b\\d\end{pmatrix}, \quad \text{col } \mathbf{A}^T = \begin{pmatrix}a\\b\\c\\d\end{pmatrix}; \quad \text{col } \mathbf{A}^T = \mathbf{U}_{22} \text{ col } \mathbf{A} \ . \tag{4.51}$$

End of Example

Column operator and unit vector $\quad \text{col } \mathbf{A} = \sum_{j=1}^{m} (\mathbf{e}_j^{(m\times1)} \otimes \mathbf{I}_n)\mathbf{A}\mathbf{e}_j^{(m\times1)} \ . \ \square \qquad (4.52)$

4.6 Inverse of the Kronecker Sum

The matrix equation (occuring in stability theory, orthogonal function expansion etc.)

$$\mathbf{LX} + \mathbf{XN} = \mathbf{Q} \qquad \mathbf{X}, \mathbf{L}, \mathbf{N}, \mathbf{Q} \in \mathcal{C}^{n\times n} \tag{4.53}$$

is solved with respect to the unknown $\mathbf{X}$. This symmetric mixed linear matrix equation is the well-known Lyapunov equation. Applying the Kronecker product and the Kronecker sum yields

$$[(\mathbf{I}_n \otimes \mathbf{L}) + (\mathbf{N}^T \otimes \mathbf{I}_n)]\text{col } \mathbf{X} = \text{col } \mathbf{Q} \tag{4.54}$$

$$(\mathbf{N}^T \oplus \mathbf{L}) \text{ col } \mathbf{X} = \text{col } \mathbf{Q} \quad \rightsquigarrow \quad \text{col } \mathbf{X} = (\mathbf{N}^T \oplus \mathbf{L})^{-1}\text{col } \mathbf{Q} \,. \tag{4.55}$$

The dimensionality problem can be reduced by employing an important formula in what follows, namely if all $\lambda[\mathbf{N}^T \oplus \mathbf{L}]$ are distinct, if $\mathbf{N}^T \oplus \mathbf{L}$ has linearly independent eigenvectors and if the eigenvectors of $\mathbf{N}, \mathbf{N}^T$ and $\mathbf{L}, \mathbf{L}^T$ are normalized, $\mathbf{n}_k^T \mathbf{n}_k^{\triangleleft} = 1$, $\mathbf{l}_i^T \mathbf{l}_i^{\triangleleft} = 1$, i.e. the nondefective case (*Brewer, J.W., 1978*):

Following to the principle of spectral representation (dyadic decomposition) Eq.(B.101), the inverse $(\mathbf{N}^T \oplus \mathbf{L})^{-1}$ can be decomposed into a sum of dyadic products of the right-eigenvector of $(\mathbf{N}^T \oplus \mathbf{L})$ and the transposed left-eigenvectors of $(\mathbf{N}^T \oplus \mathbf{L})$ weighted with the inverse of the sum of the eigenvalues of $\mathbf{N}$ and $\mathbf{L}$

$$(\mathbf{N}^T \oplus \mathbf{L})^{-1} = \sum_{i=1}^{n}\sum_{k=1}^{n} \frac{(\mathbf{n}_k^{\triangleleft} \otimes \mathbf{l}_i)(\mathbf{n}_k^T \otimes \mathbf{l}_i^{\triangleleft T})}{\lambda_k[\mathbf{N}] + \lambda_i[\mathbf{L}]} = \sum_{i=1}^{n}\sum_{k=1}^{n} \frac{(\mathbf{n}_k^{\triangleleft}\mathbf{n}_k^T) \otimes (\mathbf{l}_i \mathbf{l}_i^{\triangleleft T})}{\lambda_k[\mathbf{N}] + \lambda_i[\mathbf{L}]} \tag{4.56}$$

$$\text{col } \mathbf{X} = \sum_{i=1}^{n}\sum_{k=1}^{n} \frac{(\mathbf{n}_k^{\triangleleft}\mathbf{n}_k^T) \otimes (\mathbf{l}_i \mathbf{l}_i^{\triangleleft T})}{\lambda_k[\mathbf{N}] + \lambda_i[\mathbf{L}]} \text{ col } \mathbf{Q} \,. \tag{4.57}$$

Applying Eq.(4.40), $\mathbf{C} = \mathbf{Q}$, $\text{col } \mathbf{X} = \text{col } \mathbf{ACB}$, $\mathbf{B}^T = \mathbf{n}_k^{\triangleleft}\mathbf{n}_k^T$ and $\mathbf{A} = \mathbf{l}_i \mathbf{l}_i^{\triangleleft T}$

$$\mathbf{X} = \sum_{i=1}^{n}\sum_{k=1}^{n} \frac{1}{\lambda_k[\mathbf{N}] + \lambda_i[\mathbf{L}]} \mathbf{l}_i \mathbf{l}_i^{\triangleleft T} \mathbf{Q} \mathbf{n}_k \mathbf{n}_k^{\triangleleft T}. \tag{4.58}$$

Hence, the Lyapunov equation can be solved in a straightforward way

$$\mathbf{A}^T\mathbf{X} + \mathbf{X}\mathbf{A} = \mathbf{Q} \qquad \mathbf{X}, \mathbf{A}, \mathbf{Q} \in \mathcal{C}^{n \times n} \tag{4.59}$$

$$\text{col } \mathbf{X} = (\mathbf{A}^T \otimes \mathbf{I}_n + \mathbf{I}_n \otimes \mathbf{A}^T)^{-1} \text{ col } \mathbf{Q} = (\mathbf{A}^T \oplus \mathbf{A}^T)^{-1} \text{ col } \mathbf{Q} \tag{4.60}$$

$$\mathbf{X} = \sum_i \sum_k \frac{1}{\lambda_k[\mathbf{A}] + \lambda_i[\mathbf{A}^T]} \; \mathbf{a}_i^{\triangleleft} \mathbf{a}_i^T \; \mathbf{Q} \; \mathbf{a}_k \mathbf{a}_k^{\triangleleft T}. \tag{4.61}$$

Chapter 5

Analysis Using Matrices and Control Theory

In control theory, employing vectors and matrices, formulas become short and simple. When studying dynamic behaviour and its parameter sensitivity, matrix calculus frequently is applied. In these cases, the differentiation of scalar-, vector- and matrix- valued functions with respect to scalars, vectors or matrices is required. An overview of the nine cases arising is given in Table 5.1. Moreover, Kronecker algebra as discussed in former chapters provides shorter and more concise calculations.

The first column in Table 5.1 shows derivative operations with respect to a scalar variable e . These differentiating operations are the basis upon which more complex differentiating operations are stated by applying elementary matrix operations. Based thereon, systematic composition and partition is obtained.

Although the derivatives with respect to a scalar seem to be trivial some complex operations exist, e.g. differentiating trace or determinant functions. Most of the formulas are derived and developed in detail, but some of them are only presented as a result (see *Vetter, W.J., 1970*). The elements of all vectors or matrices are independent variables with no interelement dependencies.

5.1 Derivative of a Scalar-Valued Function with Respect to a Scalar: $\frac{\partial c}{\partial e}$

$$c = \det \mathbf{A} \qquad \frac{\partial \det \mathbf{A}}{\partial e} = \mathrm{tr}\,[\frac{\partial \mathbf{A}}{\partial e}\ \mathrm{adj}\ \mathbf{A}] \tag{5.1}$$

Proof: Using the definition of the adjoint

$$\mathrm{adj}\ \mathbf{A} = [\mathrm{matrix}\ (\mathrm{cof}_{ik}\mathbf{A})]^T = \begin{pmatrix} \mathrm{cof}_{11}\mathbf{A} & \mathrm{cof}_{21}\mathbf{A} & \mathrm{cof}_{31}\mathbf{A} & \dots \\ \mathrm{cof}_{12}\mathbf{A} & \ddots & & \\ \vdots & & & \end{pmatrix}, \tag{5.2}$$

Table 5.1: Derivatives overview

Operand	$\frac{\partial}{\partial \mathrm{scalar}}$	$\frac{\partial}{\partial \mathrm{vector}}$	$\frac{\partial}{\partial \mathrm{matrix}}$
scalar	$\frac{\partial c}{\partial e}$	$\frac{\partial c}{\partial \mathbf{p}}$	$\frac{\partial c}{\partial \mathbf{M}}$
vector	$\frac{\partial \mathbf{f}}{\partial e}$	$\frac{\partial \mathbf{f}}{\partial \mathbf{p}}$	$\frac{\partial \mathbf{f}}{\partial \mathbf{M}}$
matrix	$\frac{\partial \mathbf{A}}{\partial e}$	$\frac{\partial \mathbf{A}}{\partial \mathbf{p}}$	$\frac{\partial \mathbf{A}}{\partial \mathbf{M}}$

postmultiplying with the matrix $\mathbf{M}$ and, finally, replacing $\mathbf{M}$ by $\partial\mathbf{A}/\partial e$ yields the ij-element

$$[(\mathrm{adj}\mathbf{A})\mathbf{M}]_{ij} = [\begin{pmatrix} \mathrm{cof}_{11}\mathbf{A} & \mathrm{cof}_{21}\mathbf{A} & \mathrm{cof}_{31}\mathbf{A} & \dots \\ \mathrm{cof}_{12}\mathbf{A} & \ddots & & \\ \vdots & & & \end{pmatrix}\begin{pmatrix} M_{11} & M_{12} & M_{13} & \dots \\ M_{21} & \ddots & & \\ \vdots & & & \end{pmatrix}]_{ij} \tag{5.3}$$

$$[(\mathrm{adj}\mathbf{A})\frac{\partial\mathbf{A}}{\partial e}]_{ij} = \sum_k(\mathrm{cof}_{ki}\mathbf{A})M_{kj} = \sum_k(\mathrm{cof}_{ki}\mathbf{A})\frac{\partial A_{kj}}{\partial e}\ . \tag{5.4}$$

Specializing for $i = j$ yields

$$\sum_j[(\mathrm{adj}\mathbf{A})\frac{\partial\mathbf{A}}{\partial e}]_{jj} = \sum_{jk}(\mathrm{cof}_{kj}\mathbf{A})\frac{\partial A_{kj}}{\partial e}\ . \tag{5.5}$$

Using the total differential

$$\frac{d}{de}\ \det\ \mathbf{A}(e) = \sum_{jk}[\frac{\partial}{\partial A_{jk}}\ \det\ \mathbf{A}(e)]\frac{dA_{jk}}{de} \tag{5.6}$$

and the definition of the cofactor,

$$\frac{d}{de}\ \det\ \mathbf{A}(e) = \sum_{jk}[\frac{\partial}{\partial A_{jk}}\sum_{k=1}^{n}A_{jk}\ \mathrm{cof}_{jk}\mathbf{A}]\frac{dA_{jk}}{de} = \sum_{jk}[\mathrm{cof}_{jk}\mathbf{A}]\frac{dA_{jk}}{de}\ . \tag{5.7}$$

Combination with Eq.(5.5) yields the result. Partial derivatives are used to emphasize that it is assumed that no interelement dependencies exist

$$\frac{\partial}{\partial e}\ \det\ \mathbf{A}(e) = \sum_j[(\mathrm{adj}\ \mathbf{A})\frac{\partial\mathbf{A}}{\partial e}]_{jj} = \mathrm{tr}\ [(\mathrm{adj}\ \mathbf{A})\frac{\partial\mathbf{A}}{\partial e}]\ . \tag{5.8}$$

End of the proof

Consider $\mathbf{A} = \mathbf{A}(e) \in \mathcal{C}^{n\times n}$ the derivative with respect to the scalar e is

$$\frac{\partial}{\partial e}\ \det\ \mathbf{A} = \sum_{i=1}^{n}\ \det\ \mathbf{A}_{red\ i} \tag{5.9}$$

where $\mathbf{A}_{red\ i}$ is a square matrix obtained from $\mathbf{A}$ by replacing the ith row by $\partial\mathbf{A}_{i.}/\partial e$, e.g.,

$$\frac{\partial}{\partial e}\det\begin{pmatrix} A_{11} & A_{12} \\ A_{21} & A_{22} \end{pmatrix} = \det\begin{pmatrix} \partial A_{11}/\partial e & \partial A_{12}/\partial e \\ A_{21} & A_{22} \end{pmatrix} + \det\begin{pmatrix} A_{11} & A_{12} \\ \partial A_{21}/\partial e & \partial A_{22}/\partial e \end{pmatrix}\ . \ \square \tag{5.10}$$

Derivation of $\det(e\mathbf{I}_n - \mathbf{A})$, $\mathbf{A} \in \mathcal{C}^{n\times n}$, e scalar, yields

$$\frac{\partial}{\partial e}\ \det\ (e\mathbf{I}_n - \mathbf{A}) = \sum_{i=1}^{n}\ \det\ (e\mathbf{I}_{n-1} - \mathbf{A}_{red\ ii}) \tag{5.11}$$

where $\mathbf{A}_{red\ ii}$ is a square matrix evolved from $\mathbf{A}$ deleting row i and column i . $\square$

Finally, $$\frac{\partial}{\partial e}\ \mathrm{tr}\mathbf{A}^T(e)\mathbf{A}(e) = \frac{\partial}{\partial e}\|\mathbf{A}(e)\|_F^2 = 2\ \mathrm{tr}[\mathbf{A}^T(e)\frac{\partial\mathbf{A}(e)}{\partial e}]\ , \tag{5.12}$$

see also Eq.(5.110).

5.2 Derivative of a Scalar-Valued Function With Respect to a Vector: $\frac{\partial c}{\partial \mathbf{p}}$

The vector derivative operator (or gradient or nabla operator) with respect to the vector $\mathbf{p} = \text{vector}[p_i]$ is termed

$$\frac{\partial}{\partial \mathbf{p}} \triangleq \begin{pmatrix} \frac{\partial}{\partial p_1} \\ \frac{\partial}{\partial p_2} \\ \vdots \\ \frac{\partial}{\partial p_n} \end{pmatrix} \triangleq \text{grad}_{\mathbf{p}} = \nabla_p \, . \tag{5.13}$$

Only applicable to scalars, the notation is used

$$\frac{\partial}{\partial \mathbf{p}^T} = \Big(\frac{\partial}{\partial p_1} \; \frac{\partial}{\partial p_2} \; \cdots \; \frac{\partial}{\partial p_n}\Big) = \Big(\frac{\partial}{\partial \mathbf{p}}\Big)^T . \tag{5.14}$$

$$\text{Gradient}: \quad \frac{\partial}{\partial \mathbf{p}} I \triangleq \nabla_{\mathbf{p}} I \triangleq \text{grad}_{\mathbf{p}}\, I = \mathbf{g} = \begin{pmatrix} g_1 \\ g_2 \\ \vdots \\ g_{n_p} \end{pmatrix} \qquad \mathbf{p} \in \mathcal{R}^{n_p} \, . \tag{5.15}$$

$$\text{Divergence}: \quad \Big(\frac{\partial}{\partial \mathbf{p}}\Big)^T \mathbf{g} \triangleq \nabla_{\mathbf{p}} \cdot \mathbf{g} = \sum_{i=1}^{n_p} \frac{\partial g_i}{\partial p_i} \triangleq \text{div}\; \mathbf{g} \, . \tag{5.16}$$

$$\text{If } \mathbf{g} = \frac{\partial I}{\partial \mathbf{p}} \text{ then } \quad \text{div}\; \mathbf{g} = \Big(\frac{\partial}{\partial \mathbf{p}}\Big)^T \frac{\partial I}{\partial \mathbf{p}} = \nabla_{\mathbf{p}}^2 I = \Delta I = \sum_{i=1}^{n_p} \frac{\partial^2 I}{\partial p_i^2} \, . \tag{5.17}$$

Derivative of an inner vector product:

$$c = \mathbf{a}^T(\mathbf{p})\, \mathbf{b}(\mathbf{p}) = \sum_{i=1}^{n} a_i(\mathbf{p})\, b_i(\mathbf{p}) \qquad \mathbf{a}, \mathbf{b} \in \mathcal{R}^n, \quad \mathbf{p}, \frac{\partial c}{\partial \mathbf{p}} \in \mathcal{R}^r \tag{5.18}$$

$$\frac{\partial c}{\partial p_1} = \sum_{i=1}^{n} \frac{\partial a_i(\mathbf{p})}{\partial p_1} b_i(\mathbf{p}) + a_i(\mathbf{p}) \frac{\partial b_i(\mathbf{p})}{\partial p_1} \tag{5.19}$$

$$\frac{\partial}{\partial \mathbf{p}}\, \mathbf{a}^T(\mathbf{p})\, \mathbf{b}(\mathbf{p}) = \text{vector}\; \Big[\Big(\frac{\partial \mathbf{a}}{\partial p_i}\Big)^T \mathbf{b} + \Big(\frac{\partial \mathbf{b}}{\partial p_i}\Big)^T \mathbf{a}\Big] = \Big(\frac{\partial \mathbf{a}^T}{\partial \mathbf{p}}\Big)\mathbf{b} + \Big(\frac{\partial \mathbf{b}^T}{\partial \mathbf{p}}\Big)\mathbf{a} \, , \tag{5.20}$$

$$\text{e.g.,} \qquad \mathbf{a}(\mathbf{p}) = \mathbf{p}, \; \mathbf{b}(\mathbf{p}) = \mathbf{b} = \text{constant}, \quad \frac{\partial}{\partial \mathbf{p}} \mathbf{a}^T(\mathbf{p})\mathbf{b}(\mathbf{p}) = \mathbf{I}\,\mathbf{b} + \mathbf{0p} = \mathbf{b} \, . \; \Box \tag{5.21}$$

As far as the derivatives of a vector $\mathbf{a}$ with respect to another vector $\mathbf{p}$ are concerned see Eq.(5.62).

Derivative of a square function:

$$\frac{\partial}{\partial \mathbf{p}} \mathbf{p}^T \mathbf{A} \mathbf{p} = (\mathbf{A} + \mathbf{A}^T)\mathbf{p} \qquad \text{if } \mathbf{A} \text{ independent of } \mathbf{p} \, . \; \Box \tag{5.22}$$

$$\text{If } \mathbf{Q} = \mathbf{Q}^T \text{independent of } \mathbf{p}: \qquad \frac{\partial}{\partial \mathbf{p}}\, \mathbf{a}^T(\mathbf{p})\, \mathbf{Q}\, \mathbf{a}(\mathbf{p}) = \frac{\partial}{\partial \mathbf{p}} \text{tr}\; \mathbf{Q}\mathbf{a}\mathbf{a}^T = 2\frac{\partial \mathbf{a}^T}{\partial \mathbf{p}} \mathbf{Q}\mathbf{a} \, . \tag{5.23}$$

(See, e.g., minimization of the Frobenius vector norm of the controller vector $\|\mathbf{k}(\mathbf{p})\|_F$ of a single-input single-output system of dimension n with respect to $\mathbf{p}$, when only m

eigenvalues are assigned and $\mathbf{p}$ is the $(n-m)$-vector of the remaining free eigenvalues (*Bühler, H., 1986*).

$$\text{If } \mathbf{a}=\mathbf{R}\mathbf{p}\ , \quad \mathbf{R} \text{ constant}: \qquad \frac{\partial}{\partial \mathbf{p}}\mathbf{p}^T\mathbf{R}^T\mathbf{Q}\mathbf{R}\mathbf{p} = 2\mathbf{R}^T\mathbf{Q}\mathbf{R}\mathbf{p}\ . \ \square \tag{5.24}$$

$$\text{If } \mathbf{h} \text{ constant}: \qquad \frac{\partial}{\partial \mathbf{p}}\mathbf{p}^T\mathbf{h} = \frac{\partial}{\partial \mathbf{p}}\mathbf{h}^T\mathbf{p} = \mathbf{h}\ . \ \square \tag{5.25}$$

5.3 Derivative of a Scalar-Valued Function with Respect to a Matrix: $\frac{\partial c}{\partial \mathbf{M}}$

Vetter, W.J., 1971 and *Brewer, J.W., 1978* presented a detailed collection of formulas.

$$\text{Common rule:} \qquad \frac{\partial c(\mathbf{M})}{\partial \mathbf{M}^T} = \left(\frac{\partial c(\mathbf{M})}{\partial \mathbf{M}}\right)^T. \tag{5.26}$$

$$\text{If } \mathbf{a} \text{ and } \mathbf{b} \text{ are independent of } \mathbf{M}: \qquad \frac{\partial}{\partial \mathbf{M}}(\mathbf{a}^T\mathbf{M}\mathbf{b}) = \frac{\partial}{\partial \mathbf{M}}(\mathbf{a}^T\mathbf{M}\mathbf{b}) = \mathbf{a}\mathbf{b}^T. \tag{5.27}$$

$$\frac{\partial}{\partial \mathbf{M}}\mathbf{e}_i^T\mathbf{M}\mathbf{e}_j = \mathbf{e}_j\mathbf{e}_i^T = \mathbf{E}^{ji} \qquad \mathbf{M} \in \mathcal{R}^{m\times m}, \quad \mathbf{e}_i \in \mathcal{R}^m\ . \ \square \tag{5.28}$$

$$\text{If } \mathbf{B} \text{ and } \mathbf{C} \text{ independent of } \mathbf{M}: \qquad \frac{\partial}{\partial \mathbf{M}}\mathrm{tr}(\mathbf{B}\mathbf{M}^T\mathbf{C}) = \mathbf{C}\mathbf{B}. \tag{5.29}$$

Remark as a mnemonic aid: If $\mathbf{B} \in \mathcal{R}^{n\times s}$, $\mathbf{M} \in \mathcal{R}^{r\times s}$ and $\mathbf{C} \in \mathcal{R}^{r\times n}$ the argument of the trace is square. Other result speculations $(\mathbf{BC}, \mathbf{BC}^T, \mathbf{B}^T\mathbf{C}, \mathbf{B}^T\mathbf{C}^T, \mathbf{CB}^T, \mathbf{C}^T\mathbf{B}, \mathbf{C}^T\mathbf{B}^T)$ have to be rejected by dimension nonconformity since the derivative of a scalar with respect to an (r,s)-matrix is of dimension $r\times s$ and the matrix multiplicands must be conformable for multiplication.

Further memorandum comment: Since the trace can only be applied to square matrix arguments and with regard to the cyclic property of the trace it follows

$$\text{if } \mathbf{B} \text{ is independent of } \mathbf{M}: \qquad \frac{\partial}{\partial \mathbf{M}}\mathrm{tr}\ (\mathbf{B}\mathbf{M}^T) = \frac{\partial}{\partial \mathbf{M}}(\mathbf{M}^T\mathbf{B}) = \mathbf{B}, \tag{5.30}$$

$$\frac{\partial}{\partial \mathbf{M}}\mathrm{tr}\ (\mathbf{B}\mathbf{M}) = \frac{\partial}{\partial \mathbf{M}}\mathrm{tr}\ (\mathbf{M}^T\mathbf{B}^T) = \frac{\partial}{\partial \mathbf{M}}\mathrm{tr}\ (\mathbf{B}^T\mathbf{M}^T) = \mathbf{B}^T. \ \square \tag{5.31}$$

Example: $\partial\lambda_i[\mathbf{A}]/\partial\mathbf{A} = ?$ Premultiplying $\mathbf{A}\mathbf{a}_i = \lambda_i\mathbf{a}_i$ with $\mathbf{a}_i^{\triangleleft H}$ as given by $\mathbf{A}^T\mathbf{a}_i^{\triangleleft *} = \lambda_i\mathbf{a}_i^{\triangleleft *}$ and invoking Eq.(B.11) yields $\mathbf{a}_i^{\triangleleft H}\mathbf{A}\mathbf{a}_i = \lambda_i$. Differentiating with respect to $\mathbf{A}$, inserting the trace operator to the scalar argument and using the cyclic property of the trace,

$$\frac{\partial\lambda_i}{\partial\mathbf{A}} = \frac{\partial(\mathbf{a}_i^{\triangleleft H}\mathbf{A}\mathbf{a}_i)}{\partial\mathbf{A}} = \frac{\partial\mathrm{tr}\ (\mathbf{A}\mathbf{a}_i\mathbf{a}_i^{\triangleleft H})}{\partial\mathbf{A}} = (\mathbf{a}_i\mathbf{a}_i^{\triangleleft H})^T = \mathbf{a}_i^{\triangleleft *}\mathbf{a}_i^T \tag{5.32}$$

$$\text{or} \quad \frac{\partial\lambda_i}{\partial\mathbf{A}} = \frac{\partial(\mathbf{a}_i^{\triangleleft H}\mathbf{A}\mathbf{a}_i)}{\partial\mathbf{A}} = \frac{\partial(\mathbf{a}_i^T\mathbf{A}^T\mathbf{a}_i^{\triangleleft *})}{\partial\mathbf{A}} = \frac{\partial\ \mathrm{tr}\ (\mathbf{a}_i^T\mathbf{A}^T\mathbf{a}_i^{\triangleleft *})}{\partial\mathbf{A}} = \frac{\partial\ \mathrm{tr}\ (\mathbf{A}^T\mathbf{a}_i^{\triangleleft *}\mathbf{a}_i^T)}{\partial\mathbf{A}} = \mathbf{a}_i^{\triangleleft *}\mathbf{a}_i^T\ . \tag{5.33}$$

With regard to the normalization of the eigenvectors, the increment of the eigenvectors caused by changes in $\mathbf{A}$ does not influence the result, see Eqs.(6.69), (6.79) and (6.84). From

$$\mathbf{A}\mathbf{a} = \lambda\mathbf{a} \quad \text{and} \quad \mathbf{A}^H\mathbf{a}^{\triangleleft} = \lambda^*\mathbf{a}^{\triangleleft} \qquad (\text{or} \quad \mathbf{a}^{\triangleleft H}\mathbf{A} = \lambda\mathbf{a}_i^{\triangleleft H}) \tag{5.34}$$

and using a scalar e, it becomes readily transparent that the derivative of the eigenvectors does not affect the derivative of the eigenvalues

$$\frac{\partial\mathbf{A}}{\partial e}\mathbf{a} + \mathbf{A}\frac{\partial\mathbf{a}}{\partial e} = \frac{\partial\lambda}{\partial e}\mathbf{a} + \lambda\frac{\partial\mathbf{a}}{\partial e}\ . \tag{5.35}$$

Premultiplication with $\mathbf{a}^{\triangleleft H}$ yields

$$\mathbf{a}^{\triangleleft H}\frac{\partial\mathbf{A}}{\partial e}\mathbf{a} + (\mathbf{a}^{\triangleleft H}\mathbf{A} - \lambda\mathbf{a}^{\triangleleft H})\frac{\partial\mathbf{a}}{\partial e} = \mathbf{a}^{\triangleleft H}\frac{\partial\lambda}{\partial e}\mathbf{a} \quad \rightsquigarrow \quad \mathbf{a}^{\triangleleft H}\frac{\partial\mathbf{A}}{\partial e}\mathbf{a} + 0\times\frac{\partial\mathbf{a}}{\partial e} = \frac{\partial\lambda}{\partial e}\mathbf{a}^{\triangleleft H}\mathbf{a} = \frac{\partial\lambda}{\partial e} \tag{5.36}$$

which is identical to Eq.(6.73) . **End of Example**

If $\mathbf{A},\mathbf{B}$ are independent of $\mathbf{M}$: $\frac{\partial}{\partial \mathbf{M}}\text{tr }(\mathbf{AMB}) = \mathbf{A}^T\mathbf{B}^T.$ (5.37)

If $\mathbf{a},\mathbf{B},\mathbf{b}$ are independent of $\mathbf{M}$: $\frac{\partial}{\partial \mathbf{M}}(\mathbf{a}^T\mathbf{MBM}^T\mathbf{b}) = \mathbf{ab}^T\mathbf{MB}^T+\mathbf{ba}^T\mathbf{MB}\ .$ (5.38)

If $\mathbf{a}$ is independent of $\mathbf{M}$: $\frac{\partial}{\partial \mathbf{M}}\mathbf{a}^T\mathbf{M}^{-1}\mathbf{a} = -(\mathbf{M}^{-1})^T\mathbf{aa}^T(\mathbf{M}^{-1})^T.$ (5.39)

Sensitivity of the eigenvalues on the generating matrix:

$$\frac{\partial\lambda_i[\mathbf{M}]}{\partial \mathbf{M}} = \frac{\text{adj}^T(\lambda_i\mathbf{I}-\mathbf{M})}{\text{tr adj }(\lambda_i\mathbf{I}-\mathbf{M})},\quad \det\,(\lambda_i[\mathbf{M}]\mathbf{I}-\mathbf{M}) = 0\ . \tag{5.40}$$

Some determinant derivative properties: $\frac{\partial}{\partial \mathbf{M}}\ \det\mathbf{M} = (\mathbf{M}^{-1})^T\ \det\ \mathbf{M} = \text{adj}^T(\mathbf{M}),$ (5.41)

$$\frac{\partial}{\partial \mathbf{M}}\ \det\,(\mathbf{M}^n) = n(\mathbf{M}^{-1})^T(\det\ \mathbf{M})^n, \tag{5.42}$$

$$\frac{\partial}{\partial \mathbf{M}}\det\ e^{\mathbf{M}} = \det\ e^{\mathbf{M}},\qquad \frac{\partial}{\partial \mathbf{M}}\ \log\det\ \mathbf{M} = (\mathbf{M}^{-1})^T\ .\ \Box \tag{5.43}$$

If $\mathbf{V}$ is independent of $\mathbf{M}$: $\frac{\partial}{\partial \mathbf{M}}\det\,(\mathbf{MVM}^T) = 2(\mathbf{MVM}^T)^{-1}\mathbf{MV}\ .\ \Box$ (5.44)

If $\mathbf{A}$ and $\mathbf{B}$ are independent of $\mathbf{M}$: $\frac{\partial}{\partial \mathbf{M}}\det\,(\mathbf{AMB}) = (\mathbf{M}^{-1})^T\det\,(\mathbf{AMB}).$ (5.45)

If $\mathbf{A}$ and $\mathbf{B}$ and $\mathbf{C}$ are independent of $\mathbf{K}$; proof see Eq.(8.12):

$$\frac{\partial}{\partial \mathbf{K}}\ \det(s\mathbf{I}_n-\mathbf{A}-\mathbf{BKC}) = -\mathbf{B}^T[\text{adj}(s\mathbf{I}_n-\mathbf{A}-\mathbf{BKC})]^T\mathbf{C}^T\ . \tag{5.46}$$

Some further trace derivative properties (*Athans, M., 1968*):

$$\frac{\partial}{\partial \mathbf{M}}\text{tr }(\mathbf{M}^n) = n(\mathbf{M}^{n-1})^T,\qquad \frac{\partial}{\partial \mathbf{M}}\text{tr }\mathbf{MM}^T = 2\mathbf{M}, \tag{5.47}$$

$$\frac{\partial}{\partial \mathbf{M}}||\mathbf{M}||_F = \frac{\partial}{\partial \mathbf{M}}\sqrt{\text{tr}(\mathbf{MM}^T)} = \frac{\mathbf{M}}{||\mathbf{M}||_F}, \tag{5.48}$$

$$\frac{\partial}{\partial \mathbf{M}}\text{tr }(e^{\mathbf{M}}) = e^{(\mathbf{M}^T)} = (e^{\mathbf{M}})^T,\qquad \frac{\partial\text{tr}\{\sin\mathbf{M}\}}{\partial \mathbf{M}} = (\cos\mathbf{M})^T, \tag{5.49}$$

if $\mathbf{A}$ independent of $\mathbf{M}$: $\frac{\partial}{\partial \mathbf{M}}\text{tr }(\mathbf{AM}^m) = [\sum_{i=0}^{n-1}\mathbf{M}^i\mathbf{AM}^{m-i-1}\ ]^T;$ (5.50)

if $\mathbf{A}$ and $\mathbf{B}$ independent of $\mathbf{A}$: $\frac{\partial}{\partial \mathbf{M}}\text{tr }(\mathbf{AMBM}) = \mathbf{A}^T\mathbf{M}^T\mathbf{B}^T+\mathbf{B}^T\mathbf{M}^T\mathbf{A}^T,$ (5.51)

$$\frac{\partial}{\partial \mathbf{M}}\text{tr }(\mathbf{AMBM}^T) = \mathbf{A}^T\mathbf{MB}^T+\mathbf{AMB}, \tag{5.52}$$

$$\frac{\partial}{\partial \mathbf{M}}\text{tr }(\mathbf{AM}^{-1}\mathbf{B}) = -(\mathbf{M}^{-1}\mathbf{BAM}^{-1})^T\ .\ \Box \tag{5.53}$$

Using the trace functions for calculating the differential of a scalar-valued function c(**M**) where $\mathbf{M} \in \mathcal{R}^{r\times s}$

$$d\,c(\mathbf{M}) \triangleq \sum_{i=1}^{r}\sum_{j=1}^{s}\Big(\frac{dc(\mathbf{M})}{dM_{ij}}\Big)dM_{ij} = \text{tr}\,[\frac{dc(\mathbf{M})}{d\mathbf{M}}d\mathbf{M}^T] = \text{tr}\,[\frac{dc(\mathbf{M})}{d\mathbf{M}^T}d\mathbf{M}] \tag{5.54}$$

$$= \text{tr}\,[\begin{pmatrix} \frac{dc}{dM_{11}} & \frac{dc}{dM_{21}} & \cdots \\ \frac{dc}{dM_{12}} & \frac{dc}{dM_{22}} & \\ \vdots & \vdots & \ddots \end{pmatrix}\begin{pmatrix} dM_{11} & dM_{12} & \cdots \\ dM_{21} & dM_{22} & \cdots \\ \vdots & \vdots & \ddots \end{pmatrix}] \tag{5.55}$$

$$\mathbf{F} \triangleq \frac{dc(\mathbf{M})}{d\mathbf{M}} = \frac{\text{tr}\,[\frac{dc(\mathbf{M})}{d\mathbf{M}^T}d\mathbf{M}]}{d\mathbf{M}} = \frac{\text{tr}\,[\mathbf{F}^T d\mathbf{M}]}{d\mathbf{M}} = \frac{d\,\text{tr}[\mathbf{F}^T\mathbf{M}]}{d\mathbf{M}} \tag{5.56}$$

$$\text{or} \quad \frac{\partial\,\text{tr}(\mathbf{F}^T\mathbf{M})}{\partial\mathbf{M}} = \mathbf{F} \quad \text{or} \quad \frac{\partial\,\text{tr}(\mathbf{G}\mathbf{M})}{\partial\mathbf{M}} = \mathbf{G}^T. \tag{5.57}$$

Kleinman lemma: If a scalar function $c(\mathbf{M})$ is a trace function of the specific structure

$$c_1(\mathbf{M} + \beta\mathbf{E}_{ij}) = c_1(\mathbf{M}) + \beta\,\text{tr}\,\{\mathbf{G}(\mathbf{M})\mathbf{E}_{ij}\} \tag{5.58}$$

then selecting the limit $\beta \to 0$, the Kleinman lemma is obtained

$$\frac{\partial}{\partial M_{ij}}c_1(\mathbf{M}) = \lim_{\beta\to 0}\frac{c_1(\mathbf{M}+\beta\mathbf{E}_{ij}) - c_1(\mathbf{M})}{\beta} = \text{tr}\,\{\mathbf{G}(\mathbf{M})\mathbf{E}_{ij}\} \tag{5.59}$$

$$\frac{\partial}{\partial M_{ij}}c_1(\mathbf{M}) = \text{tr}\{[0\;0\ldots\big(\mathbf{G}(\mathbf{M})\big)_{.i}\ldots 0]\;j\text{th column only}\} = [\mathbf{G}(\mathbf{M})]_{ji} \quad \leadsto \quad \frac{\partial c_1(\mathbf{M})}{\partial\mathbf{M}} = \mathbf{G}^T(\mathbf{M}). \tag{5.60}$$

This lemma yields the gradient matrix of a trace function with respect to its matrix argument. In control engineering the trace function is frequently represented by the performance index I and the matrix **M** by the state controller **K**. □

$$\text{Finally,} \qquad \frac{\partial}{\partial\mathbf{A}}\,\text{tr}[\mathbf{A}^T\mathbf{B}(\mathbf{A})] = \mathbf{B}(\mathbf{A}) + \text{matrix}[\text{tr}\mathbf{A}^T\frac{\partial\mathbf{B}(\mathbf{A})}{\partial A_{jk}}]\,. \tag{5.61}$$

5.4 Derivation of a Vector-Valued Function With Respect to a Scalar: $\frac{\partial\mathbf{f}}{\partial e}$

See Eqs.(5.112) and (5.113).

5.5 Derivation of a Vector-Valued Function with Respect to a Vector: $\frac{\partial\mathbf{f}}{\partial\mathbf{p}}$

The matrix $\frac{\partial\mathbf{f}^T(\mathbf{p})}{\partial\mathbf{p}}$ can be treated as an outer product of the vector operator $\frac{\partial}{\partial\mathbf{p}}$ and the row $\mathbf{f}^T(\mathbf{p})$ where $\mathbf{f}(\mathbf{p}) \in \mathcal{R}^n$, $\mathbf{p} \in \mathcal{R}^r$

$$\frac{\partial\,\mathbf{f}^T(\mathbf{p})}{\partial\,\mathbf{p}} = \frac{\partial}{\partial\mathbf{p}}\mathbf{f}^T(\mathbf{p}) = \begin{pmatrix} \frac{\partial}{\partial p_1} \\ \vdots \\ \frac{\partial}{\partial p_r} \end{pmatrix}(f_1\;f_2\ldots f_n) = \begin{pmatrix} \frac{\partial f_1}{\partial p_1} & \frac{\partial f_2}{\partial p_1} & \frac{\partial f_3}{\partial p_1} & \cdots \\ \frac{\partial f_1}{\partial p_2} & \frac{\partial f_2}{\partial p_2} & \cdots & \\ \vdots & \vdots & \ddots & \\ \frac{\partial f_1}{\partial p_r} & \cdots & \cdots & \frac{\partial f_n}{\partial p_r} \end{pmatrix} = \text{matrix}[\frac{\partial f_j}{\partial p_i}]\,. \tag{5.62}$$

Jacobian matrix:

$$\frac{\partial \mathbf{f}(\mathbf{p})}{\partial \mathbf{p}^T} = \left(\frac{\partial \mathbf{f}(\mathbf{p})}{\partial p_1} \vdots \frac{\partial \mathbf{f}(\mathbf{p})}{\partial p_2} \cdots \frac{\partial \mathbf{f}(\mathbf{p})}{\partial p_r}\right) = \left(\frac{\partial \mathbf{f}^T}{\partial \mathbf{p}}\right)^T = \begin{pmatrix} \frac{\partial f_1}{\partial p_1} & \frac{\partial f_1}{\partial p_2} & \frac{\partial f_1}{\partial p_3} & \cdots \\ \frac{\partial f_2}{\partial p_1} & \frac{\partial f_2}{\partial p_2} & \cdots & \\ \vdots & \vdots & \ddots & \\ \frac{\partial f_n}{\partial p_1} & \cdots & & \frac{\partial f_n}{\partial p_r} \end{pmatrix} . \tag{5.63}$$

Further derivative properties:

$$\frac{\partial \mathbf{f}^T(p)}{\partial \mathbf{p}^T} = \left(\frac{\partial f_1}{\partial p_1} \vdots \frac{\partial f_2}{\partial p_1} \vdots \frac{\partial f_3}{\partial p_1} \cdots \frac{\partial f_1}{\partial p_2} \vdots \frac{\partial f_2}{\partial p_2} \vdots \frac{\partial f_3}{\partial p_2} \cdots \frac{\partial f_n}{\partial p_r}\right) = \left(\frac{\partial \mathbf{f}}{\partial \mathbf{p}}\right)^T \tag{5.64}$$

$$\frac{\partial \mathbf{f}(\mathbf{p})}{\partial \mathbf{p}} = \operatorname{col} \frac{\partial \mathbf{f}(\mathbf{p})}{\partial \mathbf{p}^T} \qquad \frac{\partial \mathbf{p}}{\partial \mathbf{p}} = \operatorname{col} \mathbf{I}_r \qquad \frac{\partial \mathbf{p}^T}{\partial \mathbf{p}} = \mathbf{I}_r \tag{5.65}$$

$$\frac{\partial(\mathbf{A}\mathbf{p})}{\partial \mathbf{p}} = \operatorname{col} \mathbf{A} \qquad \frac{\partial(\mathbf{A}\mathbf{p})}{\partial \mathbf{p}^T} = \mathbf{A} \qquad (\text{if } \mathbf{A} \text{ independent of } \mathbf{p}) \tag{5.66}$$

$$\frac{\partial}{\partial \mathbf{p}}[\mathbf{a}^T(\mathbf{p})\mathbf{A}] = \frac{\partial \mathbf{a}^T(\mathbf{p})}{\partial \mathbf{p}}\mathbf{A} \qquad (\text{if } \mathbf{A} \text{ is independent of } \mathbf{p}) \tag{5.67}$$

$$\mathbf{a} \in \mathcal{R}^n, \quad \mathbf{p} \in \mathcal{R}^r, \quad \mathbf{A} \in \mathcal{R}^{n\times m}, \quad \frac{\partial \mathbf{a}^T(\mathbf{p})}{\partial \mathbf{p}} \in \mathcal{R}^{r\times n} . \ \square$$

Derivative

$$\frac{\partial \mathbf{f}(\mathbf{p})}{\partial \mathbf{p}} \quad \text{where} \quad \mathbf{f}(\mathbf{p}) = \mathbf{A}\mathbf{p}\mathbf{a}^T\mathbf{B}\mathbf{p} \quad \text{and} \quad \mathbf{a}, \mathbf{A} \text{ and } \mathbf{B} \text{ independent of } \mathbf{p} : \tag{5.68}$$

$$f_i = \sum_k A_{ik}p_k\Big(\sum_l a_l \sum_m B_{lm}p_m\Big) \tag{5.69}$$

$$\frac{\partial f_i}{\partial p_j} = A_{ij}\Big(\sum_l a_l \sum_m B_{lm}p_m\Big) + \Big(\sum_k A_{ik}p_k\Big)\sum_l a_l B_{lj} \tag{5.70}$$

$$\frac{\partial}{\partial \mathbf{p}}\mathbf{A}\mathbf{p}\mathbf{a}^T\mathbf{B}\mathbf{p} = \frac{\partial \mathbf{f}(\mathbf{p})}{\partial \mathbf{p}} = \mathbf{A}(\mathbf{a}^T\mathbf{B}\mathbf{p}) + \mathbf{A}\mathbf{p}\mathbf{a}^T\mathbf{B} . \ \square \tag{5.71}$$

5.6 Derivative of a Vector-Valued Function with Respect to a Matrix: $\frac{\partial \mathbf{f}}{\partial \mathbf{M}}$

$$\frac{\partial \mathbf{A}_{\cdot j}}{\partial \mathbf{M}} = \frac{\partial \mathbf{A}}{\partial \mathbf{M}}(\mathbf{I}_s \otimes \mathbf{e}_j^{(m\times 1)}) \qquad \mathbf{A} \in \mathcal{R}^{n\times m}, \ \mathbf{M} \in \mathcal{R}^{r\times s} \tag{5.72}$$

(see Eq.(5.79) for the definition).

5.7 Derivation of a Matrix-Valued Function with Respect to a Scalar: $\frac{\partial \mathbf{A}}{\partial e}$

$$\frac{\partial}{\partial e}\mathbf{A}(e)\mathbf{B}(e)\mathbf{C}(e) = \frac{\partial \mathbf{A}}{\partial e}\mathbf{B}\mathbf{C} + \mathbf{A}\frac{\partial \mathbf{B}}{\partial e}\mathbf{C} + \mathbf{A}\mathbf{B}\frac{\partial \mathbf{C}}{\partial e} \tag{5.73}$$

$$\text{or} \qquad d(\mathbf{A}\mathbf{B}\mathbf{C}) = (d\mathbf{A})\mathbf{B}\mathbf{C} + \mathbf{A}(d\mathbf{B})\mathbf{C} + \mathbf{A}\mathbf{B}(d\mathbf{C}) , \tag{5.74}$$

e.g., $\mathbf{A}^T\mathbf{X}+\mathbf{X}\mathbf{A} = \text{constant} \quad \rightsquigarrow \quad (d\mathbf{A}^T)\mathbf{X}+\mathbf{A}^T d\mathbf{X}+(d\mathbf{X})\mathbf{A}+\mathbf{X}d\mathbf{A} = \mathbf{0}\ .\ \Box$ (5.75)

$$\frac{\partial \mathbf{A}^m}{\partial e} = \frac{\partial \mathbf{A}}{\partial e}\mathbf{A}^{m-1} + \mathbf{A}\frac{\partial \mathbf{A}}{\partial e}\mathbf{A}^{m-2} + \ldots + \mathbf{A}^{m-1}\frac{\partial \mathbf{A}}{\partial e}\ .\ \Box \tag{5.76}$$

From

$$\mathbf{A}^{-1}\mathbf{A} = \mathbf{I} \quad \rightsquigarrow \quad \frac{d(\mathbf{A}^{-1}\mathbf{A})}{de} = 0 \quad \rightsquigarrow \tag{5.77}$$

$$\rightsquigarrow \quad \frac{d\mathbf{A}^{-1}}{de}\mathbf{A} + \mathbf{A}^{-1}\frac{d\mathbf{A}}{de} = 0 \quad \rightsquigarrow \quad \frac{\partial \mathbf{A}^{-1}}{\partial e} = -\mathbf{A}^{-1}\frac{\partial \mathbf{A}}{\partial e}\mathbf{A}^{-1}\ .\ \Box \tag{5.78}$$

See also Eq.(5.114).

5.8 Derivative of a Matrix-Valued Function With Respect to a Matrix: $\frac{\partial \mathbf{A}}{\partial \mathbf{M}}$

In this section, the derivative operations on matrix-valued functions are presented. Several theorems and rules are listed. Of course, most of these theorems also can be specialized to vector operations.

5.8.1 Definition and Main Properties

Starting with the definition

$$\frac{\partial \mathbf{A}}{\partial \mathbf{M}} \stackrel{\Delta}{=} \sum_{ij} \mathbf{E}_{ij}^{(r\times s)} \otimes \frac{\partial \mathbf{A}}{\partial M_{ij}} = \begin{pmatrix} \frac{\partial \mathbf{A}}{\partial M_{11}} & \frac{\partial \mathbf{A}}{\partial M_{12}} & \cdots & \frac{\partial \mathbf{A}}{\partial M_{1s}} \\ \frac{\partial \mathbf{A}}{\partial M_{21}} & \frac{\partial \mathbf{A}}{\partial M_{22}} & \cdots & \\ \vdots & \vdots & \ddots & \\ \frac{\partial \mathbf{A}}{\partial M_{r1}} & \frac{\partial \mathbf{A}}{\partial M_{r2}} & & \frac{\partial \mathbf{A}}{\partial M_{rs}} \end{pmatrix} = \text{matrix}[\frac{\partial \mathbf{A}}{\partial M_{ij}}] \tag{5.79}$$

$$\mathbf{A} = \text{matrix}[A_{ij}]; \quad \mathbf{M} = \text{matrix}[M_{ij}]; \quad \mathbf{A} \in \mathcal{R}^{n\times m}; \quad \mathbf{M} \in \mathcal{R}^{r\times s}; \quad \frac{\partial \mathbf{A}}{\partial \mathbf{M}} \in \mathcal{R}^{nr\times ms} \tag{5.80}$$

it must be emphasized that the matrix $\mathbf{M}$ provides the scheme in which the derivatives of $\mathbf{A}$ with respect to M_{ij} are arranged.

$$\text{Derivative and its transpose:} \quad \left(\frac{\partial \mathbf{A}}{\partial \mathbf{M}}\right)^T = \frac{\partial \mathbf{A}^T}{\partial \mathbf{M}^T} \tag{5.81}$$

$$\frac{\partial \mathbf{M}^T}{\partial \mathbf{M}} = \mathbf{U}_{rs}^{(rs\times rs)}\ , \quad \frac{\partial \mathbf{M}}{\partial \mathbf{M}} = \bar{\mathbf{U}}_{rs}^{(r^2\times s^2)} \quad \text{since} \quad \frac{\partial M_{ij}}{\partial M_{\nu\mu}} = \delta_{i\nu}\,\delta_{j\mu} \quad \mathbf{M} \in \mathcal{R}^{r\times s}\ . \tag{5.82}$$

$$\text{If} \quad \mathbf{M} = \mathbf{M}^T,\ r = s: \quad \frac{\partial \mathbf{M}}{\partial \mathbf{M}} = \bar{\mathbf{U}}_{rr} + \mathbf{U}_{rr} - \sum_{i=1}^{r} \mathbf{E}_{ii}^{(r\times r)} \otimes \mathbf{E}_{ii}^{(r\times r)}\ . \tag{5.83}$$

5.8.2 Matrix Product Rule

From the definitions in Eqs. (5.73) and (5.79)

$$\frac{\partial}{\partial M_{ik}}\mathbf{A}\mathbf{B} = \frac{\partial \mathbf{A}}{\partial M_{ik}}\mathbf{B} + \mathbf{A}\frac{\partial \mathbf{B}}{\partial M_{ik}} \qquad \mathbf{A} \in \mathcal{R}^{n\times m},\ \mathbf{B} \in \mathcal{R}^{m\times q},\ \mathbf{M} \in \mathcal{R}^{r\times s} \tag{5.84}$$

$$\begin{aligned}\frac{\partial}{\partial \mathbf{M}}\mathbf{A}(\mathbf{M})\mathbf{B}(\mathbf{M}) &= \sum_{i,k}\mathbf{E}_{ik}^{(r\times s)}\otimes\frac{\partial(\mathbf{AB})}{\partial M_{ik}} = \sum_{i,k}\mathbf{E}_{ik}^{(r\times s)}\otimes\frac{\partial \mathbf{A}}{\partial M_{ik}}\mathbf{B} + \mathbf{E}_{ik}^{(r\times s)}\otimes\mathbf{A}\frac{\partial \mathbf{B}}{\partial M_{ik}} \\ &= \sum_{i,k}\mathbf{E}_{ik}^{(r\times s)}\,\mathbf{I}_s\otimes\frac{\partial \mathbf{A}}{\partial M_{ik}}\mathbf{B} + \sum_{i,k}\mathbf{I}_r\mathbf{E}_{ik}^{(r\times s)}\otimes\mathbf{A}\frac{\partial \mathbf{B}}{\partial M_{ik}} \qquad (5.85)\end{aligned}$$

$$= \sum_{i,k}\Big(\mathbf{E}_{ik}^{(r\times s)}\otimes\frac{\partial \mathbf{A}}{\partial M_{ik}}\Big)(\mathbf{I}_s\otimes\mathbf{B}) + \sum_{i,k}(\mathbf{I}_r\otimes\mathbf{A})\Big(\mathbf{E}_{ik}^{(r\times s)}\otimes\frac{\partial \mathbf{B}}{\partial M_{ik}}\Big)\,. \qquad (5.86)$$

Hence, the matrix product rule is

$$\frac{\partial}{\partial \mathbf{M}}\mathbf{A}(\mathbf{M})\mathbf{B}(\mathbf{M}) = \frac{\partial \mathbf{A}}{\partial \mathbf{M}}(\mathbf{I}_s\otimes\mathbf{B}) + (\mathbf{I}_r\otimes\mathbf{A})\frac{\partial \mathbf{B}}{\partial \mathbf{M}} \qquad \frac{\partial(\mathbf{AB})}{\partial \mathbf{M}} \in \mathcal{R}^{nr\times qs}\,. \qquad (5.87)$$

5.8.3 Derivative of the Kronecker Product with Respect to a Matrix

The derivative with respect to a single element M_{ij} of the matrix $\mathbf{M}^{(r\times s)}$ is

$$\frac{\partial(\mathbf{A}\otimes\mathbf{B})}{\partial M_{ij}} = \frac{\partial \mathbf{A}}{\partial M_{ij}}\otimes\mathbf{B} + \mathbf{A}\otimes\frac{\partial \mathbf{B}}{\partial M_{ij}} \qquad \mathbf{A}\in\mathcal{R}^{n\times m},\quad \mathbf{B}\in\mathcal{R}^{k\times l},\quad \mathbf{M}\in\mathcal{R}^{r\times s}\,. \qquad (5.88)$$

Using Eq.(5.79) to compose the complete derivative,

$$\frac{\partial(\mathbf{A}\otimes\mathbf{B})}{\partial \mathbf{M}} = \sum_{i,j}\mathbf{E}_{ij}^{(r\times s)}\otimes\Big[\frac{\partial \mathbf{A}}{\partial M_{ij}}\otimes\mathbf{B} + \mathbf{A}\otimes\frac{\partial \mathbf{B}}{\partial M_{ij}}\Big] \qquad (5.89)$$

$$\frac{\partial(\mathbf{A}\otimes\mathbf{B})}{\partial \mathbf{M}} = \frac{\partial \mathbf{A}}{\partial \mathbf{M}}\otimes\mathbf{B} + \sum_{i,j}\mathbf{E}_{ij}^{(r\times s)}\otimes\Big\{\mathbf{A}\otimes\frac{\partial \mathbf{B}^{(k\times l)}}{\partial M_{ij}}\Big\}\,. \qquad (5.90)$$

Permuting the last term with the last but one (both within the brace),

$$\frac{\partial(\mathbf{A}\otimes\mathbf{B})}{\partial \mathbf{M}} = \frac{\partial \mathbf{A}}{\partial \mathbf{M}}\otimes\mathbf{B} + \sum_{i,j}(\mathbf{I}_r\mathbf{E}_{ij}^{(r\times s)}\,\mathbf{I}_s)\otimes\Big\{\mathbf{U}_{nk}\Big(\frac{\partial \mathbf{B}^{(k\times l)}}{\partial M_{ij}}\otimes\mathbf{A}^{(n\times m)}\Big)\,\mathbf{U}_{lm}\Big\}\,. \qquad (5.91)$$

Applying Eq.(4.10), namely

$$\mathbf{I}_r\mathbf{D}\otimes\mathbf{U}_{nk}\mathbf{G} = (\mathbf{I}_r\otimes\mathbf{U}_{nk})(\mathbf{D}\otimes\mathbf{G}) \quad \text{where} \quad \mathbf{D} = \mathbf{E}_{ij}^{(r\times s)}\,\mathbf{I}_s \quad \text{and} \quad \mathbf{G} = \frac{\partial \mathbf{B}}{\partial M_{ij}}\otimes\mathbf{A} \qquad (5.92)$$

$$\rightsquigarrow \qquad \frac{\partial(\mathbf{A}\otimes\mathbf{B})}{\partial \mathbf{M}} = \frac{\partial \mathbf{A}}{\partial \mathbf{M}}\otimes\mathbf{B} + \sum_{i,j}(\mathbf{I}_r\otimes\mathbf{U}_{nk})\Big(\mathbf{E}_{ij}^{(r\times s)}\mathbf{I}_s\otimes\frac{\partial \mathbf{B}}{\partial M_{ij}}\otimes\mathbf{A}\Big)\mathbf{U}_{lm}\,. \qquad (5.93)$$

Applying once more Eq.(4.10) and $\mathbf{A} = \mathbf{E}_{ij}$, $\mathbf{D} = \mathbf{I}_s$, $\mathbf{G} = \mathbf{U}_{lm}$ finally one has

$$\frac{\partial(\mathbf{A}\otimes\mathbf{B})}{\partial \mathbf{M}} = \frac{\partial \mathbf{A}}{\partial \mathbf{M}}\otimes\mathbf{B} + \sum_{i,j}(\mathbf{I}_r\otimes\mathbf{U}_{nk})\Big(\mathbf{E}_{ij}\otimes\frac{\partial \mathbf{B}}{\partial M_{ij}}\otimes\mathbf{A}\Big)(\mathbf{I}_s\otimes\mathbf{U}_{lm}) \qquad (5.94)$$

$$\frac{\partial(\mathbf{A}\otimes\mathbf{B})}{\partial \mathbf{M}} = \frac{\partial \mathbf{A}}{\partial \mathbf{M}}\otimes\mathbf{B} + (\mathbf{I}_r\otimes\mathbf{U}_{nk})\Big(\frac{\partial \mathbf{B}}{\partial \mathbf{M}}\otimes\mathbf{A}\Big)(\mathbf{I}_s\otimes\mathbf{U}_{lm})\,. \qquad (5.95)$$

Since the matrices $\mathbf{I}_r, \mathbf{U}_{nk}$ etc. are not conformable for matrix multiplication, this equation cannot be reduced further by applying the mixed product rule.

Examples:

$$\frac{\partial(\mathbf{p}\otimes\mathbf{p})}{\partial \mathbf{p}^T} = \mathbf{I}_r\otimes\mathbf{p} + \mathbf{p}\otimes\mathbf{I}_r \qquad \mathbf{p}\in\mathcal{R}^r\,. \qquad (5.96)$$

$$\frac{\partial}{\partial \mathbf{p}}(\mathbf{e}_r\otimes\mathbf{p}^T) = (\mathbf{I}_r\otimes\mathbf{U}_{r1})\Big(\frac{\partial \mathbf{p}^T}{\partial \mathbf{p}}\otimes\mathbf{e}_r\Big)(\mathbf{I}_1\otimes\mathbf{U}_{r1}) = (\mathbf{I}_r\otimes\mathbf{I}_r)(\mathbf{I}_r\otimes\mathbf{e}_r)(\mathbf{I}_r\otimes\mathbf{I}_r) = \mathbf{I}_r\otimes\mathbf{e}_r\,. \qquad (5.97)$$

End of Examples

5.8.4 Scalar Product Rule

$$\text{Specializing}\quad \mathbf{A}(\mathbf{M}) := a(\mathbf{M}),\quad n=1,\ m=1,\qquad a\mathbf{B} = a\otimes\mathbf{B} = \mathbf{B}\otimes a \tag{5.98}$$

yields the scalar product rule

$$\frac{\partial}{\partial\mathbf{M}}a(\mathbf{M})\mathbf{B}(\mathbf{M}) = \frac{\partial a}{\partial\mathbf{M}}\otimes\mathbf{B} + (\mathbf{I}_r\otimes\mathbf{U}_{1k})\Big(\frac{\partial\mathbf{B}}{\partial\mathbf{M}}\otimes a\Big)(\mathbf{I}_s\otimes\mathbf{U}_{l1}) \tag{5.99}$$

$$= \frac{\partial a}{\partial\mathbf{M}}\otimes\mathbf{B} + (\mathbf{I}_r\otimes\mathbf{I}_k)\Big(\frac{\partial\mathbf{B}}{\partial\mathbf{M}}\otimes a\Big)(\mathbf{I}_s\otimes\mathbf{I}_l) \tag{5.100}$$

$$= \frac{\partial a}{\partial\mathbf{M}}\otimes\mathbf{B} + \mathbf{I}_{rk}\Big(\frac{\partial\mathbf{B}}{\partial\mathbf{M}}\otimes a\Big)^{(kr\times ls)}\mathbf{I}_{ls} \tag{5.101}$$

$$\frac{\partial}{\partial\mathbf{M}}a(\mathbf{M})\mathbf{B}(\mathbf{M}) = \frac{\partial a}{\partial\mathbf{M}}\otimes\mathbf{B} + \frac{\partial\mathbf{B}}{\partial\mathbf{M}}\otimes a = \frac{\partial a(\mathbf{M})}{\partial\mathbf{M}}\otimes\mathbf{B} + a(\mathbf{M})\frac{\partial\mathbf{B}}{\partial\mathbf{M}}\,. \tag{5.102}$$

5.8.5 Chain Rule

$$\frac{\partial}{\partial\mathbf{M}}\mathbf{A}[\mathbf{B}(\mathbf{M})] = \Big(\mathbf{I}_r\otimes\frac{\partial\mathbf{A}}{\partial(\text{row }\mathbf{B})}\Big)\Big(\frac{\partial\text{col }\mathbf{B}^T}{\partial\mathbf{M}}\otimes\mathbf{I}_m\Big) \tag{5.103}$$

$$= \Big(\frac{\partial(\text{col }\mathbf{B})^T}{\partial\mathbf{M}}\otimes\mathbf{I}_n\Big)\Big(\mathbf{I}_s\otimes\frac{\partial\mathbf{A}}{\partial\text{col }\mathbf{B}}\Big) \tag{5.104}$$

$$\mathbf{A}\in\mathcal{R}^{n\times m},\qquad \mathbf{B}\in\mathcal{R}^{k\times l},\qquad \mathbf{M}\in\mathcal{R}^{r\times s},\qquad \frac{\partial}{\partial\mathbf{M}}\mathbf{A}[\mathbf{B}(\mathbf{M})]\in\mathcal{R}^{nr\times ms}\,.$$

Derivative of the inverse matrix (*Vetter, W.J., 1973*):

$$\frac{\partial\mathbf{A}^{-1}}{\partial\mathbf{M}} = -(\mathbf{I}_r\otimes\mathbf{A}^{-1})\frac{\partial\mathbf{A}}{\partial\mathbf{M}}(\mathbf{I}_s\otimes\mathbf{A}^{-1})\qquad \mathbf{A}\in\mathcal{R}^{n\times m}\,. \tag{5.105}$$

Partly specializing to scalar functions, the implicit function rule for scalar differentials on matrices ($\mathbf{A} := a,\ \mathbf{B} := b,\ n=m=k=l=1,\ \mathbf{M}\in\mathcal{R}^{r\times s}$) is

$$\frac{\partial a[b(\mathbf{M})]}{\partial\mathbf{M}} = \Big(\mathbf{I}_r\otimes\frac{\partial a}{\partial b}\Big)\Big(\frac{\partial b}{\partial\mathbf{M}}\otimes\mathbf{I}_m\Big) \tag{5.106}$$

$$= \Big(\frac{\partial a}{\partial b}\mathbf{I}_r\Big)\Big(\frac{\partial b}{\partial\mathbf{M}}\otimes 1\Big) = \frac{\partial a}{\partial b}\mathbf{I}_r\frac{\partial b}{\partial\mathbf{M}} = \frac{\partial a}{\partial b}\frac{\partial b}{\partial\mathbf{M}}\,. \tag{5.107}$$

5.9 Various Derivatives with Respect to Time

Scalar function c dependent on the vector $\mathbf{x}(t)$ as a particular case of Eq.(5.104):

$$n=m=r=s=1,\quad \mathbf{A} := c,\quad \mathbf{B} := \mathbf{x}(t),\quad \mathbf{M} := t,\quad \mathbf{x}\in\mathcal{R}^q \tag{5.108}$$

$$\frac{d}{dt}c[\mathbf{x}(t)] = \Big(\frac{d\mathbf{x}^T}{dt}\otimes\mathbf{I}_n\Big)\Big(\mathbf{I}_s\otimes\frac{\partial c}{\partial\mathbf{x}}\Big) = \frac{d\mathbf{x}^T(t)}{dt}\frac{\partial c}{\partial\mathbf{x}}\,. \tag{5.109}$$

Total derivative if c is dependent on t, additionally,

$$\frac{d}{dt}c[\mathbf{x}(t),t] = \frac{d\mathbf{x}^T(t)}{dt}\frac{\partial c}{\partial\mathbf{x}} + \frac{\partial c}{\partial t}\,. \tag{5.110}$$

Vector (column) function $\mathbf{z}$ dependent on $\mathbf{x}(t)$ and t: Taking particular cases from Eq.(5.104), $\mathbf{A} := \mathbf{z}$, $\mathbf{z} \in \mathcal{R}^m$, $\mathbf{x}(t) \in \mathcal{R}^q$, $\mathbf{M} := t$

$$\frac{d\mathbf{z}[\mathbf{x}(t)]}{dt} = \Big(\frac{d\mathbf{x}^T}{dt} \otimes \mathbf{I}_m\Big)\Big(1 \otimes \frac{\partial \mathbf{z}}{\partial \mathbf{x}}\Big) = \Big(\frac{d\mathbf{x}^T}{dt} \otimes \mathbf{I}_m\Big)\frac{\partial \mathbf{z}}{\partial \mathbf{x}} \tag{5.111}$$

$$\frac{d\mathbf{z}[\mathbf{x}(t),t]}{dt} = \Big(\frac{d\mathbf{x}^T}{dt} \otimes \mathbf{I}_m\Big)\frac{\partial \mathbf{z}}{\partial \mathbf{x}} + \frac{\partial \mathbf{z}}{\partial t} \,. \tag{5.112}$$

Row function $\mathbf{y}^T$ dependent on $\mathbf{x}(t)$ and t: From Eq.(5.104) with $\mathbf{A} := \mathbf{y}^T$; $\mathbf{A}, \mathbf{y}^T \in \mathcal{R}^{1\times n}$

$$\frac{d\mathbf{y}^T[\mathbf{x}(t),t]}{dt} = \Big(\frac{d\mathbf{x}^T}{dt} \otimes \mathbf{I}_1\Big)\Big(\mathbf{I}_1 \otimes \frac{\partial \mathbf{y}^T}{\partial \mathbf{x}}\Big) = \frac{d\mathbf{x}^T}{dt}\frac{\partial \mathbf{y}^T}{\partial \mathbf{x}} \,. \tag{5.113}$$

Matrix-valued function $\mathbf{A}$ dependent on the vector $\mathbf{x}(t)$: Applying Eq.(5.104) and specializing $\mathbf{B} := \mathbf{x}(t)$, $\mathbf{x}(t) \in \mathcal{R}^q$, $\mathbf{M} := t$, $r = s = 1$,

$$\frac{d\mathbf{A}[\mathbf{x}(t),t]}{dt} = \Big(\frac{d\mathbf{x}^T}{dt} \otimes \mathbf{I}_n\Big)\Big(\mathbf{I}_s \otimes \frac{\partial \mathbf{A}}{\partial \mathbf{x}}\Big) + \frac{\partial \mathbf{A}}{\partial t} = \Big(\frac{d\mathbf{x}^T}{dt} \otimes \mathbf{I}_n\Big)\frac{\partial \mathbf{A}}{\partial \mathbf{x}} + \frac{\partial \mathbf{A}}{\partial t} \,. \tag{5.114}$$

Example 1: For $\mathbf{y} \in \mathcal{R}^n$, $\mathbf{z} \in \mathcal{R}^m$, $\mathbf{x} \in \mathcal{R}^q$, $\mathbf{A} \in \mathcal{R}^{n\times m}$ calculate

$$\frac{d}{dt}\,\mathbf{y}^T[\mathbf{x}(t),t]\;\mathbf{A}[\mathbf{x}(t),t]\;\mathbf{z}[\mathbf{x}(t),t] = \frac{d\mathbf{y}^T}{dt}\mathbf{A}\mathbf{z} + \mathbf{y}^T\frac{d\mathbf{A}}{dt}\mathbf{z} + \mathbf{y}^T\mathbf{A}\frac{d\mathbf{z}}{dt} \,. \tag{5.115}$$

The last expression contains three derivatives. The result simply can be obtained by substituting the results of Eqs.(5.113), (5.114) and (5.112), respectively (*Vetter, W.J., 1970*).

Various further derivatives on $c = \mathbf{y}^T[\mathbf{x}(t),t]\;\mathbf{A}[\mathbf{x}(t),t]\;\mathbf{z}[\mathbf{x}(t),t]$:

$$\frac{\partial c}{\partial \mathbf{y}} = \mathbf{A}\mathbf{z} \qquad \frac{\partial c}{\partial \mathbf{z}} = (\mathbf{y}^T\mathbf{A})^T = \mathbf{A}^T\mathbf{y} \tag{5.116}$$

$$\frac{\partial c}{\partial \mathbf{A}} = \frac{\partial}{\partial \mathbf{A}}\mathrm{tr}\;(\mathbf{y}^T\mathbf{A}\mathbf{z}) = \frac{\partial}{\partial \mathbf{A}}\mathrm{tr}\;(\mathbf{z}\mathbf{y}^T\mathbf{A}) = (\mathbf{z}\mathbf{y}^T)^T = \mathbf{y}\mathbf{z}^T \,. \;\square \tag{5.117}$$

Example 2: (*Vetter, W.J., 1970*)

$$\frac{\partial}{\partial \mathbf{x}^T}\,\mathbf{x}^T\mathbf{Q}(\mathbf{x})\,\mathbf{x} = \mathbf{x}^T[\mathbf{Q}^T(\mathbf{x}) + \mathbf{Q}(\mathbf{x}) + \frac{\partial \mathbf{Q}(\mathbf{x})}{\partial \mathbf{x}^T}(\mathbf{I}_q \otimes \mathbf{x})] \qquad \mathbf{x} \in \mathcal{R}^q \,. \;\square \tag{5.118}$$

Example 3: If $\mathbf{A}$ independent of t

$$\frac{d}{dt}e^{\mathbf{A}t} = \mathbf{A}e^{\mathbf{A}t} = e^{\mathbf{A}t}\mathbf{A} \quad \text{and} \quad \int_0^t e^{\mathbf{A}\tau}d\tau = \mathbf{A}^{-1}(e^{\mathbf{A}t} - \mathbf{I}) \,. \;\square \tag{5.119}$$

End of Examples

5.10 Taylor Expansion

The Taylor expansion of a matrix $\mathbf{A}$ dependent on a vector $\mathbf{p}$ is

$$\mathbf{A}(\mathbf{p}) \doteq \mathbf{A}(\mathbf{p}_o) + \sum_{i=1}^{n}\frac{1}{i!}\Big(\frac{\partial^i}{\partial \mathbf{p}^{Ti}}\mathbf{A}(\mathbf{p})\Big)_{\mathbf{p}=\mathbf{p}_o}[(\mathbf{p}-\mathbf{p}_o)^{[i]} \otimes \mathbf{I}_m] \tag{5.120}$$

where $\mathbf{A} \in \mathcal{R}^{n\times m}$. The superscript $^{[i]}$ is the ith Kronecker power

$$(\mathbf{p}-\mathbf{p}_o)^{[i]} = (\mathbf{p}-\mathbf{p}_o) \otimes (\mathbf{p}-\mathbf{p}_o) \otimes \ldots \otimes (\mathbf{p}-\mathbf{p}_o) \,. \tag{5.121}$$

An equivalent expression is

$$\mathbf{A}(\mathbf{p}) \doteq \mathbf{A}(\mathbf{p}_o) + \sum_{i=1}^{n}\frac{1}{i!}[(\mathbf{p}-\mathbf{p}_o)^{T[i]} \otimes \mathbf{I}_n]\Big(\frac{\partial^i}{\partial \mathbf{p}^i}\mathbf{A}(\mathbf{p})\Big)_{\mathbf{p}=\mathbf{p}_o} \tag{5.122}$$

(see *Vetter, W.J., 1970*). Further matrix calculus operations such as matrix integrals, integration by parts structures are detailed by *Vetter, W.J., 1973*.

5.11 Incremental Notation

Note that the definition of the derivative of a matrix-valued function with respect to a matrix as given in Eq.(5.79) is a definition of a scheme, only. This abbreviation is suitable when cascades of calculations are taken into consideration. In this section, various examples of the incremental notation are posed which can be applied advantageously in control theory. The increment of a matrix $\mathbf{A}$ is termed $\Delta\mathbf{A}$.

5.11.1 Derivative of the Frobenius Norm

From the scalar function given by the Frobenius matrix norm

$$\|\mathbf{A}-\mathbf{KC}\|_F^2 = \text{tr}\ \{(\mathbf{A}-\mathbf{KC})^T(\mathbf{A}-\mathbf{KC})\} = \text{tr}\ \mathbf{A}^T\mathbf{A} - 2\text{tr}\ \{(\mathbf{KC})^T\mathbf{A}\} + \text{tr}\ \{(\mathbf{KC})^T\mathbf{KC}\} \tag{5.123}$$

the differential or increment is

$$\Delta\|\mathbf{A}-\mathbf{KC}\|_F^2 = \text{tr}\ \mathbf{A}^T\Delta\mathbf{A} + \text{tr}\ \Delta\mathbf{A}^T\mathbf{A} - 2\text{tr}\ \{(\mathbf{KC})^T\Delta\mathbf{A}\} = 2\text{tr}\ \{[\mathbf{A}^T-(\mathbf{KC})^T]\Delta\mathbf{A}\}\ . \tag{5.124}$$

The increment of the inverse is termed $\Delta(\mathbf{X}^{-1})$ and is given by

$$\mathbf{X}^{-1}\mathbf{X} = \mathbf{I} \ \rightsquigarrow\ [\Delta(\mathbf{X}^{-1})]\mathbf{X} + \mathbf{X}^{-1}\Delta\mathbf{X} = \mathbf{0} \ \rightsquigarrow\ \Delta(\mathbf{X}^{-1}) = \mathbf{X}^{-1}(\Delta\mathbf{X})\mathbf{X}^{-1}\ . \tag{5.125}$$

If $\mathbf{A} = \mathbf{G}\mathbf{X}^{-1}$ and both $\mathbf{G}$ and $\mathbf{X}$ are considered variable

$$\Delta\mathbf{A} = (\Delta\mathbf{G})\mathbf{X}^{-1} + \mathbf{G}\ \Delta(\mathbf{X}^{-1}) = (\Delta\mathbf{G})\mathbf{X}^{-1} - \mathbf{G}\mathbf{X}^{-1}(\Delta\mathbf{X})\mathbf{X}^{-1} \tag{5.126}$$

$$= (\Delta\mathbf{G})\mathbf{X}^{-1} - \mathbf{A}(\Delta\mathbf{X})\mathbf{X}^{-1} = (\Delta\mathbf{G} - \mathbf{A}\ \Delta\mathbf{X})\mathbf{X}^{-1}\ . \tag{5.127}$$

Finally, the result is

$$\Delta\|\mathbf{A}-\mathbf{KC}\|_F^2 = 2\text{tr}\ \{[\mathbf{A}^T - (\mathbf{KC})^T](\Delta\mathbf{G} - \mathbf{A}\ \Delta\mathbf{X})\mathbf{X}^{-1}\}\ . \tag{5.128}$$

Example: Equivalence of Eq.(5.124) and (5.47) when $\mathbf{K} \equiv \mathbf{0}$. From Eq.(5.124) the increment follows $\Delta\|\mathbf{A}\|_F^2 = 2\,\text{tr}\ (\mathbf{A}^T\Delta\mathbf{A})$. Assuming a matrix $\mathbf{A}$ with changes only in the (i,k)-position, i.e., $A_{ik} + \Delta A_{ik}$, then $\Delta\mathbf{A} = (\Delta A_{ik})\mathbf{E}_{ik}$ and

$$\Delta\|\mathbf{A}\|_F^2 = 2\ \text{tr}(\mathbf{A}^T\Delta\mathbf{A}) = 2\ A_{ik}\ \Delta A_{ik} \quad \rightsquigarrow \quad \frac{\Delta\|\mathbf{A}\|_F^2}{\Delta A_{ik}} = 2\ A_{ik}\ . \tag{5.129}$$

Arranging this result according to Eq.(5.79),

$$\frac{\partial\|\mathbf{A}\|_F^2}{\partial\mathbf{A}} = \text{matrix}\ [\frac{\Delta\|\mathbf{A}\|_F^2}{\Delta A_{ik}}] = 2\ \mathbf{A} \tag{5.130}$$

which is equivalent to Eq.(5.47).

End of Example

5.11.2 Derivative of the Spectral Condition Number

$$\text{From}\quad \kappa_s[\mathbf{X}] = \frac{\sigma_{\max}[\mathbf{X}]}{\sigma_{\min}[\mathbf{X}]} \quad \text{derive} \quad \Delta\kappa_s[\mathbf{X}] = \frac{\sigma_{\min}\Delta\sigma_{\max} - \sigma_{\max}\Delta\sigma_{\min}}{\sigma_{\min}^2}\ . \tag{5.131}$$

Using $\partial\sigma_{\max}/\partial\mathbf{X} = \mathbf{u}_a\mathbf{v}_a^T$ and $\partial\sigma_{\min}/\partial\mathbf{X} = \mathbf{u}_i\mathbf{v}_i^T$, see Eq.(22.80), where $\mathbf{v}_i, \mathbf{v}_a, \mathbf{u}_i, \mathbf{u}_a$ are the real eigenvectors of $\mathbf{X}^H\mathbf{X}$ and $\mathbf{X}\mathbf{X}^H$ associated with $\sigma_{\min}$ (subscript i) and $\sigma_{\max}$ (subscript a), respectively, it follows

$$\Delta\sigma_{\max} = \text{tr}\ (\Delta\sigma_{\max}) = \text{tr}\ (\mathbf{v}_a\mathbf{u}_a^T\Delta\mathbf{X}) \quad \text{and} \quad \Delta\sigma_{\min} = \text{tr}\ (\mathbf{v}_i\mathbf{u}_i^T\Delta\mathbf{X}) \tag{5.132}$$

$$\Delta\kappa_s[\mathbf{X}] = \frac{\text{tr}\,[(\sigma_{\min}[\mathbf{X}]\mathbf{v}_a\mathbf{u}_a^T - \sigma_{\max}[\mathbf{X}]\mathbf{v}_i\mathbf{u}_i^T)\Delta\mathbf{X}]}{\sigma_{\min}^2[\mathbf{X}]} \quad \text{or} \tag{5.133}$$

$$\frac{\partial\kappa_s[\mathbf{X}]}{\partial\mathbf{X}} = \frac{\sigma_{\min}[\mathbf{X}]\mathbf{u}_a\mathbf{v}_a^T - \sigma_{\max}[\mathbf{X}]\mathbf{u}_i\mathbf{v}_i^T}{\sigma_{\min}^2[\mathbf{X}]}\,. \tag{5.134}$$

5.12 Homogenous Matrix Differential Equations

5.12.1 First-Order. Time-Invariant Coefficients

Let a matrix differential equation with constant and square coefficient matrices **A** and **B** be

$$\dot{\mathbf{X}}(t) = \mathbf{A}\mathbf{X}(t) + \mathbf{X}(t)\mathbf{B} \qquad \mathbf{X} \in \mathcal{R}^{m\times n}, \quad \mathbf{A} \in \mathcal{R}^{m\times m}, \quad \mathbf{B} \in \mathcal{R}^{n\times n}\,. \tag{5.135}$$

The initial value of $\mathbf{X}(t)$ is $\mathbf{X}(0)$. Applying the column operator yields

$$\text{col}\,\dot{\mathbf{X}} = \text{col}\,(\mathbf{A}\mathbf{X}\mathbf{I}_n + \mathbf{I}_m\mathbf{X}\mathbf{B}) = (\mathbf{I}_n \otimes \mathbf{A} + \mathbf{B}^T \otimes \mathbf{I}_m)\text{col}\,\mathbf{X} \triangleq \mathbf{K}_C\,\text{col}\,\mathbf{X}\,. \tag{5.136}$$

In accordance with scalar diffential equations and using Eq.(4.16),

$$\begin{aligned} \text{col}\,\mathbf{X} &= e^{(\mathbf{I}_n\otimes\mathbf{A}+\mathbf{B}^T\otimes\mathbf{I}_m)t}\,\mathbf{X}(0) = e^{\mathbf{I}_n\otimes\mathbf{A}t}e^{\mathbf{B}^T\otimes\mathbf{I}_m t}\,\mathbf{X}(0) &(5.137)\\ &= (\mathbf{I}_n \otimes e^{\mathbf{A}t})(e^{\mathbf{B}^Tt} \otimes \mathbf{I}_m)\mathbf{X}(0) = (\mathbf{I}_n e^{\mathbf{B}^Tt}) \otimes (e^{\mathbf{A}t}\mathbf{I}_m)\mathbf{X}(0) &(5.138)\\ &= (e^{\mathbf{B}^Tt} \otimes e^{\mathbf{A}t})\mathbf{X}(0) = \text{col}\,(e^{\mathbf{A}t}\mathbf{X}(0)e^{\mathbf{B}t}) &(5.139)\\ \mathbf{X}(t) &= e^{\mathbf{A}t}\mathbf{X}(0)e^{\mathbf{B}t} &(5.140) \end{aligned}$$

(*Bellman, R., 1967*). If the differential equation is given in a more complex structure, e.g.,

$$\dot{\mathbf{X}}(t) = \mathbf{A}^2\mathbf{X}(t) + 2\mathbf{A}\mathbf{X}(t)\mathbf{B} + \mathbf{X}(t)\mathbf{B}^2, \tag{5.141}$$

by applying Eqs.(A.13) and (4.14), a transformation into Eq.(5.136) easily can be performed

$$\text{col}\,\dot{\mathbf{X}}(t) = (\mathbf{I}_n \otimes \mathbf{A}^2 + 2\mathbf{B}^T \otimes \mathbf{A} + \mathbf{B}^{T2} \otimes \mathbf{I}_m)\,\text{col}\,\mathbf{X}(t) \tag{5.142}$$

$$\text{col}\,\dot{\mathbf{X}}(t) = (\mathbf{I}_n \otimes \mathbf{A} + \mathbf{B}^T \otimes \mathbf{I}_m)^2\,\text{col}\,\mathbf{X}(t) = \mathbf{K}_C^2\,\text{col}\,\mathbf{X}(t)\,. \tag{5.143}$$

5.12.2 Order r. Time-Invariant Coefficients

Suppose a matrix differential equation of rth order (*Barnett, S., 1973*) structured as

$$\begin{aligned} \mathbf{X}^{(r)}(t) &+ a_{r-1}[\mathbf{A}\mathbf{X}^{(r-1)}(t) + \mathbf{X}^{(r-1)}(t)\mathbf{B}]\\ &+ a_{r-2}[\mathbf{A}^2\mathbf{X}^{(r-2)}(t) + 2\mathbf{A}\mathbf{X}^{(r-2)}(t)\mathbf{B} + \mathbf{X}^{(r-2)}\mathbf{B}^2] + \ldots = \mathbf{0} \end{aligned} \tag{5.144}$$

$$\text{col}\,\mathbf{X}^{(r)}(t) + a_{r-1}\mathbf{K}_C\,\text{col}\,\mathbf{X}^{(r-1)}(t) + a_{r-2}\mathbf{K}_C^2\,\text{col}\,\mathbf{X}^{(r-2)}(t) + \ldots = \mathbf{0}\,. \tag{5.145}$$

For the sake of comparison,

$$x(t) = \sum_{i=1}^{q}\sum_{j=1}^{v_i} e^{\lambda_i t}c_{ij}g_{ij}(t) \quad \text{where} \quad g_{ij}(t) = \frac{t^{(v_i-j)}}{(v_i-j)!} \tag{5.146}$$

is the general solution of the general scalar linear homogeneous equation

$$x^{(r)}(t) + a_{r-1}x^{(r-1)}(t) + a_{r-2}x^{(r-2)}(t) + \ldots = 0 \tag{5.147}$$

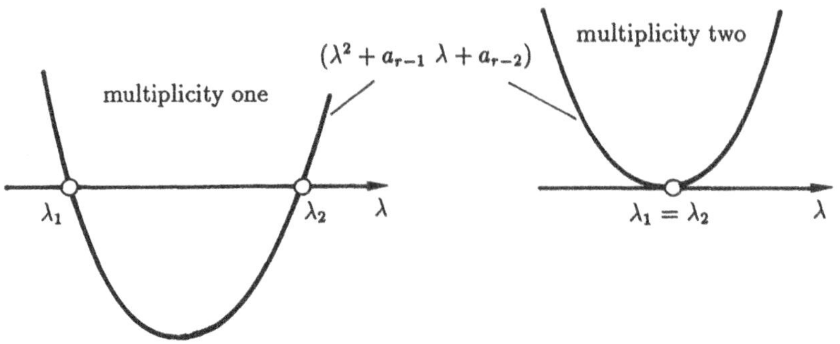

Figure 5.1: Characteristic polynomial with multiplicity one and multiplicity two

where

$$c_{ij} = \frac{1}{(j-1)!}\,\frac{d^{j-1}}{ds^{j-1}}\Big[(s-\lambda_i)^{v_i}\,\frac{1}{s^r + \sum_{k=o}^{r-1} a_k s^k}\Big]_{s=\lambda_i} \quad \text{and} \quad r = \sum_{i=1}^{q} v_i\ . \tag{5.148}$$

The differential equation of order r possesses a characteristic equation with q various distinct roots λ of multiplicity v_i

$$s^r + \sum_{k=o}^{r-1} a_k s^k = 0\ . \tag{5.149}$$

By comparison with Eq.(5.146), the solution of the matrix differential equation Eq.(5.144) using column notation is

$$\text{col}\,\mathbf{X}(t) = \sum_{i=1}^{q}\sum_{j=1}^{v_i} e^{\lambda_i \mathbf{K}_C t}\mathbf{d}_{ij} g_{ij}(t) = \sum_{i=1}^{q}\sum_{j=1}^{v_i} e^{\lambda_i(\mathbf{I}_n\otimes\mathbf{A}+\mathbf{B}^T\otimes\mathbf{I}_m)t}\,\mathbf{d}_{ij} g_{ij}(t) \tag{5.150}$$

$$= \sum_{i=1}^{q}\sum_{j=1}^{v_i} e^{\mathbf{I}_n\otimes(\lambda_i\mathbf{A}t)} e^{(\lambda_i\mathbf{B}^T t)\otimes\mathbf{I}_m}\,\mathbf{d}_{ij} g_{ij}(t) \tag{5.151}$$

$$= \sum_{i=1}^{q}\sum_{j=1}^{v_i} (\mathbf{I}_n\otimes e^{\lambda_i\mathbf{A}t})(e^{\lambda_i\mathbf{B}^T t}\otimes\mathbf{I}_m)\mathbf{d}_{ij} g_{ij}(t) \tag{5.152}$$

$$\text{col}\,\mathbf{X}(t) = \sum_{i=1}^{q}\sum_{j=1}^{v_i} (e^{\lambda_i\mathbf{B}^T t}\otimes e^{\lambda_i\mathbf{A}t})\mathbf{d}_{ij} g_{ij}(t) \tag{5.153}$$

$$\mathbf{X}(t) = \sum_{i=1}^{q}\sum_{j=1}^{v_i} e^{\lambda_i\mathbf{A}t}\mathbf{C}_{ij} e^{\lambda_i\mathbf{B}t} g_{ij}(t) \quad \text{where} \quad \mathbf{d}_{ij} = \text{col}\,\mathbf{C}_{ij}\ . \tag{5.154}$$

Example:

(i) Second-order case with multiplicity one: Suppose $r = 2$ in Eq.(5.144). The solution col $\mathbf{X}(t)$ is

$$\text{col}\,\mathbf{X}(t) = \sum_i\sum_j e^{\lambda_i\mathbf{K}_C t}\mathbf{d}_{ij}. \tag{5.155}$$

Substituting into Eq.(5.145) yields

$$(\lambda_i^2 + a_{r-1}\lambda_i + a_{r-2})\mathbf{K}_C^2 e^{\lambda_i \mathbf{K}_C t}\mathbf{d}_{ij} = \mathbf{O}. \tag{5.156}$$

(ii) Second-order case with multiplicity two: Substituting the solution

$$\text{col } \mathbf{X}(t) = \sum_i \sum_j e^{\lambda_i \mathbf{K}_C t}\mathbf{d}_{ij} g_{ij}(t) \qquad g_{ij}(t) = t \tag{5.157}$$

into Eq.(5.145) by applying product rule of differentiation,

$$(\lambda_i^2 + a_{r-1}\lambda_i + a_{r-2})\mathbf{K}_C^2 e^{\lambda_i \mathbf{K}_C t}\mathbf{d}_{ij} + (2\lambda_i + a_{r-1})\mathbf{K}_C e^{\lambda_i \mathbf{K}_C t}\mathbf{d}_{ij}\frac{dgij}{dt} = 0\,. \tag{5.158}$$

In this case $d^2 g_{ij}/dt^2$ is zero, the expression $(\lambda_i^2 + a_{r-1}\lambda_i + a_{r-2}) = 0$ has a root λ_1 of multiplicity two and $(2\lambda_i + a_{r-1}) = 0$ yields $\lambda_2 = -0.5\, a_{r-1}$. Fig. 5.1 shows the characteristic polynomial $(\lambda^2 + a_{r-1}\lambda + a_{r-2})$ in both cases. In case of multiplicity two the roots coincide with the minimum. Hence $(2\lambda + a_{r-1})$ is the derivative of $\lambda^2 + a_{r-1}\lambda + a_{r-2}$ with respect to λ .

(iii) Numerical assumptions

$$r = 2,\;\; a_{r-1} = 6,\;\; a_{r-2} = 9,\;\; \lambda_1 = \lambda_2 = -3 \;:\quad \mathbf{X}(t) = e^{-3\mathbf{A}t}\mathbf{C}_o e^{-3\mathbf{B}t} + e^{-3\mathbf{A}t}\mathbf{C}_1 e^{-3\mathbf{B}t} t \tag{5.159}$$

where $\mathbf{C}_o$ and $\mathbf{C}_1$ depend on the initial condition $\mathbf{X}(0)$. □

5.12.3 Linear Time-Varying Matrix Differential Equation

$$\dot{\mathbf{x}}(t) = \mathbf{A}(t)\mathbf{x}(t_o) \qquad \rightsquigarrow \qquad \mathbf{x}(t) = \mathbf{\Phi}(t, t_o)\mathbf{x}(t_o) \tag{5.160}$$

where $\mathbf{\Phi}(t, t_o)$ is the solution of linear matrix differential equation in the state transition matrix

$$\dot{\mathbf{\Phi}}(t, t_o) = \mathbf{A}(t)\mathbf{\Phi}(t, t_o) \qquad \mathbf{\Phi}(t_o, t_o) = \mathbf{I} \quad \text{and} \quad \mathbf{\Phi}^{-1}(t, \tau) = \mathbf{\Phi}(\tau, t) \qquad \forall t, \tau\,. \tag{5.161}$$

Note that $\mathbf{\Phi}(t, \tau)\mathbf{\Phi}(\tau, \tau_1) = \mathbf{\Phi}(t, \tau_1) \qquad \forall t, \tau, \tau_1$.

Chapter 6

Eigenvalue and Eigenvector Differential Sensitivity

Differential sensitivity of an eigenvalue is given by the quotient of an infinitesimal change of the eigenvalue $\lambda[\mathbf{F}]$ and an infinitesimal change of a matrix $\mathbf{K}$ on which the eigenvalue depends when $\mathbf{F} = \mathbf{A} + \mathbf{BKC}$. Small-scale robustness is obtained when the differential sensitivity is small. A geometric interpretation of the differential sensitivity of an eigenvalue with respect to a matrix is given in Eq.(32.91).

6.1 Closed-Loop Eigenvalue Sensitivity to Controller Matrix

There exist various applications demonstrating that an appropriate choice of the eigenvalues of a closed-loop system does not necessarily give a reliable indication of sufficient stability margin since the eigenvalue sensitivity with respect to parameter variations may be very large. This section is devoted to the eigenvalue sensitivity with respect to matrix perturbations.

A good alternative measure of the stability margin is the Frobenius norm of the smallest perturbation $\Delta\mathbf{A}$ of $\mathbf{A}$ that shifts one of the eigenvalues $\lambda[\mathbf{A}]$ to the imaginary axis. When trajectory optimization is investigated using the calculus of variations, absolute and relative sensitivity measures can also be given (*Rohrer, R.A., and Sobral, M.,Jr., 1965*).

The eigenvalues of a closed-loop control system in many cases depend on a matrix, e.g., the controller matrix $\mathbf{K}$. Let the coefficient matrix of the closed-loop system be $\mathbf{F}$ and let the eigenvalues be distinct. By definition, the right-eigenvectors of $\mathbf{F}$ are $\mathbf{f}_i$, the left-eigenvectors of $\mathbf{F}$ are $\mathbf{f}_i^{\triangleleft}$. Taking the derivative of both sides of $\mathbf{F}\mathbf{f}_i = \mathbf{f}_i\lambda_i[\mathbf{F}]$ with respect to the (m,n)-matrix $\mathbf{K}$ by using the matrix product rule Eq.(5.87) yields

$$\frac{\partial \mathbf{F}}{\partial \mathbf{K}}(\mathbf{I}_n \otimes \mathbf{f}_i) + (\mathbf{I}_m \otimes \mathbf{F})\frac{\partial \mathbf{f}_i}{\partial \mathbf{K}} = \frac{\partial \mathbf{f}_i}{\partial \mathbf{K}}(\mathbf{I}_n \otimes \lambda_i[\mathbf{F}]) + (\mathbf{I}_m \otimes \mathbf{f}_i)\frac{\partial \lambda_i[\mathbf{F}]}{\partial \mathbf{K}} \ . \tag{6.1}$$

Taking the conjugate transpose of both sides, it results (*Feliachi, A., 1986*)

$$(\mathbf{I}_n \otimes \mathbf{f}_i^{*T})(\frac{\partial \mathbf{F}}{\partial \mathbf{K}})^T + \underbrace{(\frac{\partial \mathbf{f}_i}{\partial \mathbf{K}})^T(\mathbf{I}_m \otimes \mathbf{F}^T)}_{\mathbf{L}} = \underbrace{(\mathbf{I}_n \otimes \lambda_i^*)(\frac{\partial \mathbf{f}_i}{\partial \mathbf{K}})^T}_{\mathbf{R}} + \frac{\partial \lambda_i^*}{\partial \mathbf{K}^T}(\mathbf{I}_m \otimes \mathbf{f}_i^{*T}) \ . \tag{6.2}$$

Postmultiplying the above equation by $(\mathbf{I}_m \otimes \mathbf{f}_i^{\triangleleft})$, it can be observed that the term $\mathbf{L}$ of left side (after postmultiplication) becomes

$$\begin{aligned}
\mathbf{L}(\mathbf{I}_m \otimes \mathbf{f}_i^{\triangleleft}) &= (\frac{\partial \mathbf{f}_i}{\partial \mathbf{K}})^T(\mathbf{I}_m \otimes \mathbf{F}^T)(\mathbf{I}_m \otimes \mathbf{f}_i^{\triangleleft}) = (\frac{\partial \mathbf{f}_i}{\partial \mathbf{K}})^T(\mathbf{I}_m \otimes \mathbf{F}^T\mathbf{f}_i^{\triangleleft}) \\
&= (\frac{\partial \mathbf{f}_i}{\partial \mathbf{K}})^T[\mathbf{I}_m \otimes (\mathbf{f}_i^{\triangleleft}\lambda_i^*)] = (\frac{\partial \mathbf{f}_i}{\partial \mathbf{K}})^T(\mathbf{I}_m \otimes \mathbf{f}_i^{\triangleleft})(\mathbf{I}_m \otimes \lambda_i^*) \ .
\end{aligned} \tag{6.3}$$

Next, the rule

$$(\mathbf{I}_n \otimes c)\mathbf{Q} = (c\mathbf{I}_n)\mathbf{Q} = c\mathbf{Q} = c\mathbf{Q}\mathbf{I}_m = \mathbf{Q}\mathbf{I}_m c = \mathbf{Q}(\mathbf{I}_m \otimes c) \tag{6.4}$$

for any scalar c and any (n, m)-matrix $\mathbf{Q}$ is applied. By substituting

$$\mathbf{Q} = (\frac{\partial \mathbf{f}_i}{\partial \mathbf{K}})^T (\mathbf{I}_m \otimes \mathbf{f}_i^a) \quad \text{and} \quad c = \lambda_i^* \tag{6.5}$$

into the rule Eq.(6.4),

$$(\mathbf{I}_n \otimes \lambda_i^*)(\frac{\partial \mathbf{f}_i}{\partial \mathbf{K}})^T (\mathbf{I}_m \otimes \mathbf{f}_i^a) = (\frac{\partial \mathbf{f}_i}{\partial \mathbf{K}})^T (\mathbf{I}_m \otimes \mathbf{f}_i^a)(\mathbf{I}_m \otimes \lambda_i^*) \ . \tag{6.6}$$

Now, it can be noticed, first, that the right-hand side of Eq.(6.6) is equal to the result of Eq.(6.3) and, second, that the left-hand side of Eq.(6.6) is equal to the term $\mathbf{R}$ postmultiplied by $(\mathbf{I}_m \otimes \mathbf{f}_i^a)$. Thus, $\mathbf{L}$ and $\mathbf{R}$ after postmultiplication can be cancelled and from Eq.(6.2) follows

$$(\mathbf{I}_n \otimes \mathbf{f}_i^{*T})(\frac{\partial \mathbf{F}}{\partial \mathbf{K}})^T (\mathbf{I}_m \otimes \mathbf{f}_i^a) = \frac{\partial \lambda_i^*}{\partial \mathbf{K}^T} (\mathbf{I}_m \otimes \mathbf{f}_i^{*T})(\mathbf{I}_m \otimes \mathbf{f}_i^a) = \frac{\partial \lambda_i^*}{\partial \mathbf{K}^T} \ . \tag{6.7}$$

The last simplification on the right is obtained by assumption of $\mathbf{f}_i$ and $\mathbf{f}_i^a$ being normalized, see Eq.(B.11). Finally, taking the conjugate transpose one has

$$\frac{\partial \lambda_i[\mathbf{F}]}{\partial \mathbf{K}} = (\mathbf{I}_m \otimes \mathbf{f}_i^{a*T}) \frac{\partial \mathbf{F}}{\partial \mathbf{K}} (\mathbf{I}_n \otimes \mathbf{f}_i) \tag{6.8}$$

which is the desired sensitivity of the closed-loop eigenvalues(poles) $\lambda_i[\mathbf{F}]$ with respect to any matrix $\mathbf{K}$. Eq.(6.8) documents the influence of the sensitivity of the matrix $\mathbf{F}$ with respect to $\mathbf{K}$ as well as the influence of the eigenvectors $\mathbf{f}_i$ of $\mathbf{F}$.

Example:

$$\mathbf{F} = \begin{pmatrix} 0 & \alpha \\ 2 & \beta \end{pmatrix} \qquad \mathbf{K} = (\beta \quad \alpha) \qquad m = 1,\ n = 2\ . \tag{6.9}$$

Assume the nominal values $\alpha = -1$, $\beta = -2$ in $\mathbf{K}$. Using the results of the tabular following Eq.(B.12) and substituting into Eq.(6.8) yields

$$\frac{\partial \lambda_i}{\partial \mathbf{K}} \mid_{i=1} = \begin{pmatrix} -0.5j \\ 0.25 + 0.25j \end{pmatrix}^T \begin{pmatrix} 0 & 0 & 0 & 1 \\ 0 & 1 & 0 & 0 \end{pmatrix} \left(\mathbf{I}_2 \otimes \begin{pmatrix} 1+j \\ 2 \end{pmatrix} \right) \tag{6.10}$$

$$= (-0.5j \ \vdots\ 0.25 + 0.25j) \begin{pmatrix} 0 & 0 & 0 & 1 \\ 0 & 1 & 0 & 0 \end{pmatrix} \begin{pmatrix} 1+j & 0 \\ 2 & 0 \\ 0 & 1+j \\ 0 & 2 \end{pmatrix} = (0.5 + 0.5j \ \vdots\ -j)\ . \tag{6.11}$$

Direct calculation:

$$\lambda_1[\mathbf{F}] = 0.5\beta + \sqrt{0.25\beta^2 + 2\alpha}, \qquad \lambda_1|_{\alpha=-1;\ \beta=-2} = -1 + j \tag{6.12}$$

$$(\lambda_1 + \Delta\lambda_1)|_{\alpha=\text{constant}\ =-1;\ \beta=-2} = -1 + 0.5\Delta\beta + j + 0.5j\Delta\beta = -1 + j + 0.5(1+j)\Delta\beta \tag{6.13}$$

$$\frac{\partial \lambda_1}{\partial \beta} = \frac{\Delta \lambda_1}{\Delta \beta} = 0.5(1+j)\ . \tag{6.14}$$

End of Example

6.2 Centralized Control and Closed-Loop Differential Sensitivity

Assume the plant and output controller

$$\dot{\mathbf{x}}(t) = \mathbf{A}\mathbf{x}(t) + \mathbf{B}\mathbf{u}(t), \qquad \mathbf{x}(0) = \mathbf{x}_o \quad \in \mathcal{R}^n \tag{6.15}$$

$$\mathbf{y}(t) = \mathbf{C}\mathbf{x}(t) \quad \in \mathcal{R}^r \qquad \mathbf{u}(t) = \mathbf{K}\mathbf{y}(t) \quad \in \mathcal{R}^m, \quad \mathbf{K} \in \mathcal{R}^{m\times r} \tag{6.16}$$

and $\mathbf{A}, \mathbf{B}, \mathbf{C}$ constant. Then, the closed-loop system is

$$\dot{\mathbf{x}}(t) = (\mathbf{A} + \mathbf{BKC})\mathbf{x}(t) = \mathbf{F}\mathbf{x}(t) . \tag{6.17}$$

For abbreviation, $\mathbf{H} \triangleq \mathbf{KC}$ is introduced. Calculating $\partial\mathbf{F}/\partial\mathbf{K}$ and applying Eq.(5.87),

$$\frac{\partial \mathbf{F}}{\partial \mathbf{K}} = \frac{\partial(\mathbf{BKC})}{\partial \mathbf{K}} = \frac{\partial(\mathbf{BH})}{\partial \mathbf{K}} = (\mathbf{I}_m \otimes \mathbf{B})\frac{\partial \mathbf{H}}{\partial \mathbf{K}} \tag{6.18}$$

is obtained. Note that $\partial\mathbf{A}/\partial\mathbf{K} = \mathbf{0}$. Using Eqs.(4.8) and (5.87) again yields

$$\frac{\partial \mathbf{H}}{\partial \mathbf{K}} = \frac{\partial \mathbf{KC}}{\partial \mathbf{K}} = \frac{\partial \mathbf{K}}{\partial \mathbf{K}}(\mathbf{I}_r \otimes \mathbf{C}) = \bar{\mathbf{U}}_{mr}(\mathbf{I}_r \otimes \mathbf{C}) \tag{6.19}$$

$$\frac{\partial \mathbf{F}}{\partial \mathbf{K}} = (\mathbf{I}_m \otimes \mathbf{B})\bar{\mathbf{U}}_{mr}^{m^2 \times r^2}(\mathbf{I}_r \otimes \mathbf{C}) \quad \in \mathcal{R}^{mn\times nr} . \tag{6.20}$$

6.3 Centralized Control. Component Connection Framework

Assume an interconnected large-scale system in the component connection model framework as shown in Fig. 34.2 and described in Eqs.(34.72) and (34.73). The system is completed by the state feedback controller

$$\mathbf{u}_c = \mathbf{K}\mathbf{x} \qquad \mathbf{K} \in \mathcal{R}^{m_c\times n} . \tag{6.21}$$

Combining Eqs.(34.72) and (6.21) it results

$$\dot{\mathbf{x}} = [\mathbf{A} + \mathbf{B}(\mathbf{I} - \mathbf{M}_y\mathbf{D})^{-1}\mathbf{M}_y\mathbf{C} + \mathbf{B}(\mathbf{I} - \mathbf{M}_y\mathbf{D})^{-1}\mathbf{M}_u\mathbf{K}]\mathbf{x} = \mathbf{F}\mathbf{x} \tag{6.22}$$

$$\frac{\partial \mathbf{F}}{\partial \mathbf{K}} = \left(\mathbf{I}_m \otimes [\mathbf{B}(\mathbf{I}_m - \mathbf{M}_y\mathbf{D})^{-1}\mathbf{M}_u]\right)\frac{\partial \mathbf{K}}{\partial \mathbf{K}} = \left(\mathbf{I}_m \otimes [\mathbf{B}(\mathbf{I}_m - \mathbf{M}_y\mathbf{D})^{-1}\mathbf{M}_u]\right)\bar{\mathbf{U}}_{m_cn} . \tag{6.23}$$

If the plant component connection model is controlled by an output feedback

$$\mathbf{u}_c(t) = \bar{\mathbf{K}}\mathbf{y}_c(t) \qquad \bar{\mathbf{K}} \in \mathcal{R}^{m_c\times r_c} \tag{6.24}$$

then the combination with Eqs.(34.72) and (34.73) yields (omitting some manipulations) a very complicated overall system matrix $\mathbf{F} = \mathbf{F}(\bar{\mathbf{K}})$:

$$\mathbf{F} = \mathbf{A} + \mathbf{B}(\mathbf{I}_m - \mathbf{M}_y\mathbf{D})^{-1}\mathbf{M}_y\mathbf{C} - \mathbf{B}(\mathbf{I}_m - \mathbf{M}_y\mathbf{D})^{-1}\mathbf{M}_y\bar{\mathbf{K}}\mathbf{F}_{DM}\mathbf{N}_y(\mathbf{I}_r - \mathbf{D}\mathbf{M}_y)^{-1}\mathbf{C} \tag{6.25}$$

$$\text{where} \qquad \mathbf{F}_{DM} = \{\mathbf{I}_{r_c} + [\mathbf{N}_y(\mathbf{I}_r - \mathbf{D}\mathbf{M}_y)^{-1}\mathbf{D}\mathbf{M}_u + \mathbf{N}_u]\bar{\mathbf{K}}\}^{-1} . \tag{6.26}$$

6.4 Decentralized Control. Closed-Loop Differential Sensitivity

Consider a large-scale system to be controlled by a decentralized state feedback. Naturally, a plant model has to be chosen that is prepared for the structure of a decentralized controller. Thus, the large-scale system is composed by a component connection model again. The controller is

$$\mathbf{K} = \text{block diag } \mathbf{K}_i \quad \in \mathcal{R}^{m_i \times n_i} \quad \forall i = 1 \ldots L \,. \tag{6.27}$$

All the entries $K_{\nu\mu}$ outside of this block diagonal structure are zero. Applying the notation $\partial\lambda_i/\partial\mathbf{K}$ and $\partial\mathbf{F}/\partial\mathbf{K}$ it must be kept in mind that $\partial\lambda_i/\partial K_{\nu\mu}$ and $\partial\mathbf{F}/\partial K_{\nu\mu}$ do not exist and, hence, are put to zero by definition. As a consequence, some vector and matrix definitions can be condensed to non-zero positions.

6.5 Decentralized Fixed Modes

An eigenvalue λ_i is denoted as a decentralized fixed eigenvalue (decentralized fixed mode) if its location in the s-plane cannot be influenced by the decentralized controller. The sensitivity calculations can be used to prove the ability of eigenvalue (pole) assignment. If the sensitivity matrix $\partial\boldsymbol{\lambda}/\partial\mathbf{K}$ or any of its matrix norms is zero, a decentrally fixed pole exists. All the fixed poles must be stable in order to guarantee a stable overall system. Engineers are interested in a pole allocability to a certain extent which means that the fact "fixed pole and sensitivity near zero" must be assessed appropriatedly. As shown by the rather complicated interdependencies of Eqs.(6.8), (6.20), (6.23) and (6.25), the sensitivity and the decentralized fixed modes practically depend on $\mathbf{K}$ to a large extent.

6.6 Centralized Pole Allocation

Apart from the pole assignment method solving det $(s\mathbf{I} - \mathbf{A} - \mathbf{BK}) = \prod_1^n (s - \lambda_i)$ with respect to $\mathbf{K}$ for given λ_i and its ambiguity for multivariable systems, an algorithm is studied using first-order Taylor extrapolation. The quantity $\delta\lambda_i$ is a pole placement increment of λ_i caused by altering $\mathbf{K}$ with the increments $\Delta K_{\alpha\beta}$ $\forall\, \alpha = 1 \ldots m, \beta = 1 \ldots n$:

$$\delta\lambda_i = \sum_{\alpha=1}^{m} \sum_{\beta=1}^{n} \frac{\partial\lambda_i}{\partial K_{\alpha\beta}} \, \Delta K_{\alpha\beta} = \left(\text{col } \frac{\partial\lambda_i}{\partial\mathbf{K}}\right)^T \text{col } \Delta\mathbf{K} \tag{6.28}$$

$$\delta\boldsymbol{\lambda} = \begin{pmatrix} \text{col}^T(\partial\lambda_1/\partial\mathbf{K}) \\ \text{col}^T(\partial\lambda_2/\partial\mathbf{K}) \\ \vdots \\ \text{col}^T(\partial\lambda_n/\partial\mathbf{K}) \end{pmatrix} \text{col } \Delta\mathbf{K} \;\triangleq\; \mathbf{M}_\lambda \text{ col } \Delta\mathbf{K} \,. \tag{6.29}$$

There are n entries in $\delta\boldsymbol{\lambda}$ and mn entries in the controller matrix $\mathbf{K}$. Thus, Eq.(6.29) is underdetermined. With the condition minimizing the norm of a vector containing all $\Delta K_{\alpha\beta}$, an optimal solution from Eq.(D.5) is

$$\text{col } \Delta\mathbf{K}^* = \mathbf{M}_\lambda^{\sharp R} \, \delta\boldsymbol{\lambda} \,. \tag{6.30}$$

6.7 Decentralized Pole Allocation

First, it has to be taken into account that the matrix of the decentralized controller is block diagonal and a great number of positions is filled up with zeros

$$\mathbf{K} = \begin{pmatrix} \mathbf{K}_1 & \mathbf{0} & \dots & \mathbf{0} \\ \mathbf{0} & \mathbf{K}_2 & \dots & \mathbf{0} \\ \vdots & \vdots & \ddots & \vdots \\ \mathbf{0} & \mathbf{0} & \dots & \mathbf{K}_L \end{pmatrix} \qquad \mathbf{K}_1 = \begin{pmatrix} K_{11} & K_{12} & \dots & K_{1n_1} \\ K_{21} & K_{22} & \dots & K_{2n_1} \\ \vdots & \vdots & \ddots & \vdots \\ K_{m_1 1} & \dots & \dots & K_{m_1 n_1} \end{pmatrix} . \tag{6.31}$$

Second, the sensitivity only exists in cases of non-zero entries and is defined zero in the case of zero entries. Thus, the first-order Taylor expansion can be confined to non-zero elements

$$\delta\lambda_i = \begin{pmatrix} \text{col}\,(\partial\lambda_i/\partial\mathbf{K}_1) \\ \text{col}\,(\partial\lambda_i/\partial\mathbf{K}_2) \\ \vdots \\ \text{col}\,(\partial\lambda_i/\partial\mathbf{K}_L) \end{pmatrix}^T \begin{pmatrix} \text{col}\,\Delta\mathbf{K}_1 \\ \text{col}\,\Delta\mathbf{K}_2 \\ \vdots \\ \text{col}\,\Delta\mathbf{K}_L \end{pmatrix} \quad \text{where} \quad \frac{\partial\lambda_i}{\partial\mathbf{K}_j} \in \mathcal{R}^{m_j \times n_j} \qquad \forall j = 1 \dots L . \tag{6.32}$$

$$\delta\boldsymbol{\lambda} = \begin{pmatrix} \delta\lambda_1 \\ \delta\lambda_2 \\ \vdots \\ \delta\lambda_n \end{pmatrix} = \begin{pmatrix} \text{col}^T(\partial\lambda_1/\partial\mathbf{K}_1) & \text{col}^T(\partial\lambda_1/\partial\mathbf{K}_2) & \dots & \text{col}^T(\partial\lambda_1/\partial\mathbf{K}_L) \\ \text{col}^T(\partial\lambda_2/\partial\mathbf{K}_1) & \dots & \dots & \vdots \\ \vdots & \vdots & \ddots & \vdots \\ \text{col}^T(\partial\lambda_n/\partial\mathbf{K}_1) & \dots & \dots & \text{col}^T(\partial\lambda_n/\partial\mathbf{K}_L) \end{pmatrix} \begin{pmatrix} \text{col}\,\Delta\mathbf{K}_1 \\ \text{col}\,\Delta\mathbf{K}_2 \\ \vdots \\ \text{col}\,\Delta\mathbf{K}_L \end{pmatrix} . \tag{6.33}$$

If $\delta\boldsymbol{\lambda}$ is given and all the derivatives $\partial\lambda_i/\partial\mathbf{K}_j$ have been calculated then the increments $\Delta\mathbf{K}_j$ can be optimally designed by using Eq.(D.5)

$$\begin{pmatrix} \text{col}\,\Delta\mathbf{K}_1^* \\ \text{col}\,\Delta\mathbf{K}_2^* \\ \vdots \\ \text{col}\,\Delta\mathbf{K}_L^* \end{pmatrix} = \begin{pmatrix} \text{col}^T(\partial\lambda_1/\partial\mathbf{K}_1) & \text{col}^T(\partial\lambda_1/\partial\mathbf{K}_2) & \dots & \text{col}^T(\partial\lambda_1/\partial\mathbf{K}_L) \\ \text{col}^T(\partial\lambda_2/\partial\mathbf{K}_1) & \dots & \dots & \vdots \\ \vdots & \vdots & \ddots & \vdots \\ \text{col}^T(\partial\lambda_n/\partial\mathbf{K}_1) & \dots & \dots & \text{col}^T(\partial\lambda_n/\partial\mathbf{K}_L) \end{pmatrix}^{IR} \delta\boldsymbol{\lambda} \triangleq \mathbf{M}_k^{IR}\,\delta\boldsymbol{\lambda} . \tag{6.34}$$

In the calculation shown above there is a number of $\sum_1^L n_j m_j$ unknown controller elements and only $\sum_1^L n_i = n$ known increments $\delta\lambda_i$. The result is obtained by minimizing the length of the vector of all $\Delta\mathbf{K}_j$ components. The quantity $\delta\lambda_i$ is the desired increment from the starting point $\lambda_i^{(0)}$ to the first step $\lambda_i^{(1)} = \lambda_i^{(0)} + \delta\lambda_i$ where $\lambda_i^{(0)}$ is the eigenvalue applying the controller $\mathbf{K}^{(0)}$. By means of one step (0) to (1) the desired result $\lambda_i^{(1)}$ is not arrived because only a first-order extrapolation was used. An algorithm must be established repeating the above matrix operation and changing its operation index from 0, 1 through $\mu, \mu+1$. Moreover, it seems to be adequate to alter $\lambda_i^{(\mu+1)}$ in appropriate steps in order to avoid algorithm overshoot.

The λ_i setpoint can also be assigned to stabilize a given control system.

6.8 Combining Differential Sensitivity and Eigenvalue Assignment

Based on differential sensitivity, the eigenvalue assignment method is modified. Only assign a subset $\lambda_i[\mathbf{F}]$, $\forall i \in \mathcal{M}$, of the closed-loop eigenvalues to predetermined and desired eigenvalues s_i . The degree of freedom is utilized for minimizing the distance $|\lambda_i[\mathbf{F}] - s_i|^2$ with respect to the entries of the controller matrix $\mathbf{K}$ and for performing minimum differential sensitivity $\frac{\partial \lambda_i[\mathbf{F}]}{\partial A_{jk}}$ of the eigenvalue(s) λ_i with the largest real part with respect to the uncertain system parameters A_{jk} where $\mathbf{F} = \mathbf{A} + \mathbf{BK}$ and α_{jk}, β_i are weighting factors

$$\frac{\partial}{\partial K_{pq}}[\sum_{i \notin \mathcal{M}} \beta_i |\lambda_i[\mathbf{F}] - s_i|^2 + \sum_{jk} \alpha_{jk} (\frac{\partial \lambda_i[\mathbf{F}]}{\partial A_{jk}})^2] \rightarrow \min . \tag{6.35}$$

By appropriate assumption of s_i it must be avoided that $\max_i \Re e\ \lambda_i[\mathbf{F}]$ approaches the $j\omega$-axis (*Weinmann, A., 1991*).

6.9 Open-Loop Eigenvalue Sensitivity

In Eq.(6.8) the sensitivity of $\lambda_i[\mathbf{F}]$ with respect to the controller matrix $\mathbf{K}$ is investigated where $\mathbf{F}$ is any matrix $\mathbf{F}(\mathbf{K})$, in particular $\mathbf{F} = \mathbf{A} + \mathbf{BK}$. In this section, the sensitivity of distinct eigenvalues $\lambda_i[\mathbf{A}]$ to $\mathbf{A}$ is presented. If needed, the matrix $\mathbf{A}$ of the open-loop linear system can be substituted by any other matrix.

This particular case is solved via Eq.(6.8) by substituting $\mathbf{K} := \mathbf{A}$, $\mathbf{F} := \mathbf{A} \in \mathcal{R}^{n \times n}$, $m = n$, $\mathbf{f}_i := \mathbf{a}_i$ and $\mathbf{f}_i^{\triangleleft} := \mathbf{a}_i^{\triangleleft}$ and by using Eqs.(5.82) and (4.8)

$$\begin{aligned}\frac{\partial \lambda_i[\mathbf{A}]}{\partial \mathbf{A}} &= (\mathbf{I}_n \otimes \mathbf{a}_i^{\triangleleft H})\frac{\partial \mathbf{A}}{\partial \mathbf{A}}(\mathbf{I}_n \otimes \mathbf{a}_i) = (\mathbf{I}_n \otimes \mathbf{a}_i^{\triangleleft H})[\sum_{\mu,\nu=1}^{n} (\mathbf{e}_\mu \mathbf{e}_\nu^T) \otimes (\mathbf{e}_\mu \mathbf{e}_\nu^T)](\mathbf{I}_n \otimes \mathbf{a}_i) \\ &= \sum_{\mu,\nu=1}^{n} (\mathbf{e}_\mu \mathbf{e}_\nu^T) \otimes (\mathbf{a}_i^{\triangleleft H} \mathbf{e}_\mu \mathbf{e}_\nu^T \mathbf{a}_i) \rightsquigarrow \left(\frac{\partial \lambda_i}{\partial \mathbf{A}}\right)_{\mu\nu} = \frac{\partial \lambda_i}{\partial \mathbf{A}_{\mu,\nu}} = (\mathbf{a}_i^{\triangleleft *})_\mu (\mathbf{a}_i)_\nu \qquad (6.36) \\ \frac{\partial \lambda_i[\mathbf{A}]}{\partial \mathbf{A}} &= \mathbf{a}_i^{\triangleleft *} \mathbf{a}_i^T = (\mathbf{a}_i^{\triangleleft} \mathbf{a}_i^H)^* \qquad (6.37)\end{aligned}$$

$$\text{or} \quad \delta\lambda_i[\mathbf{A}] = \sum_{\mu,\nu=1}^{n} (\mathbf{a}_i^{\triangleleft *})_\mu (\mathbf{a}_i)_\nu \delta A_{\mu,\nu} + \text{ h.o.t.} = \mathbf{a}_i^{\triangleleft H}\ \delta\mathbf{A}\ \mathbf{a}_i + \text{ h.o.t.} . \tag{6.38}$$

In Eq.(6.37) the vector $\mathbf{a}_i^{\triangleleft *}$ denotes the complex conjugate of the left-eigenvector of $\mathbf{A}$ (right-eigenvector of $\mathbf{A}^T$). With regard to a previous assumption, $\mathbf{a}_i$ and $\mathbf{a}_i^{\triangleleft}$ are normalized, i.e., $\mathbf{a}_i^{\triangleleft H} \mathbf{a}_j = \delta_{ij}$. The result in Eq. (6.37) is identical to a well-known Jacobi formula (*Pierre, C., 1989*).

6.9.1 Sensitivity of the Real and Imaginary Part

From Eq.(6.36) follows

$$\frac{\partial \Re e \lambda_i[\mathbf{A}]}{\partial A_{\mu,\nu}} = \Re e(\mathbf{a}_i^{\triangleleft *})_\mu (\mathbf{a}_i)_\nu \quad \text{or} \quad \Re e\ \delta\lambda_i[\mathbf{A}] = \sum_{\mu,\nu=1}^{n} \Re e(\mathbf{a}_i^{\triangleleft *})_\mu (\mathbf{a}_i)_\nu \delta A_{\mu\nu} \tag{6.39}$$

$$\frac{\partial \Im m\ \lambda_i[\mathbf{A}]}{\partial A_{\mu\nu}} = \Im m(\mathbf{a}_i^{\triangleleft *})_\mu (\mathbf{a}_i)_\nu \quad \text{or} \quad \Im m\ \delta\lambda_i[\mathbf{A}] = \sum_{\mu,\nu=1}^{n} \Im m(\mathbf{a}_i^{\triangleleft *})_\mu (\mathbf{a}_i)_\nu \delta A_{\mu\nu} . \tag{6.40}$$

6.9.2 Sensitivity of the Modulus

The first-order perturbation of the modulus $\delta|\lambda_i[\mathbf{A}]|$ is obtained from

$$\begin{aligned} |\lambda_i + \delta\lambda_i| &= \sqrt{(\Re e\ \lambda_i + \Re e\ \delta\lambda_i)^2 + (\Im m\ \lambda_i + \Im m\ \delta\lambda_i)^2} \qquad (6.41) \\ &= \sqrt{(\Re e\ \lambda_i)^2 + (\Im m\ \lambda_i)^2}\sqrt{1 + 2\frac{(\Re e\ \lambda_i)(\Re e\ \delta\lambda_i) + (\Im m\ \lambda_i)(\Im m\ \delta\lambda_i)}{(\Re e\ \lambda_i)^2 + (\Im m\ \lambda_i)^2}} + h.o.t. \\ |\lambda_i + \delta\lambda_i| &= |\lambda_i|(1 + \frac{\Re e\ \lambda_i \Re e\ \delta\lambda_i + \Im m\ \lambda_i \Im m\ \delta\lambda_i}{|\lambda_i|^2}) + h.o.t.\,. \qquad (6.42) \end{aligned}$$

The first-order perturbation $\delta|\lambda_i|$ becomes

$$\delta|\lambda_i| = |\lambda_i + \delta\lambda_i| - |\lambda_i| = \frac{\Re e\ \lambda_i}{|\lambda_i|}\Re e\ \delta\lambda_i + \frac{\Im m\ \lambda_i}{|\lambda_i|}\Im m\ \delta\lambda_i\ . \qquad (6.43)$$

Since $\Re e\ \delta\lambda_i = \delta\ \Re e\ \lambda_i$ it results

$$\frac{\partial|\lambda_i|}{\partial A_{\mu\nu}} = \frac{\Re e\ \lambda_i}{|\lambda_i|}\frac{\partial \Re e \lambda_i}{\partial A_{\mu\nu}} + \frac{\Im m\ \lambda_i}{|\lambda_i|}\frac{\partial \Im m \lambda_i}{\partial A_{\mu\nu}} \qquad (6.44)$$

$$\frac{\partial|\ \lambda_i[\mathbf{A}]\ |}{\partial \mathbf{A}} = \frac{\Re e\ \lambda_i}{|\lambda_i|}\frac{\partial \Re e \lambda_i}{\partial \mathbf{A}} + \frac{\Im m\ \lambda_i}{|\lambda_i|}\frac{\partial \Im m \lambda_i}{\partial \mathbf{A}}\ . \qquad (6.45)$$

6.10 Norm of the Sensitivity Matrices

The Frobenius norm $\|\mathbf{S}\|_F$ of a complex sensitivity matrix $\mathbf{S}$ is derived from the definition $\|\mathbf{S}\|_F = +\sqrt{tr\ \mathbf{S}^H\mathbf{S}}$ and Eq.(6.37) as follows

$$\begin{aligned} \|\frac{\partial\lambda_i[\mathbf{A}]}{\partial \mathbf{A}}\|_F &= +\sqrt{\mathrm{tr}\ (\mathbf{a}_i^{\triangleleft *}\mathbf{a}_i^T)^H(\mathbf{a}_i^{\triangleleft *}\mathbf{a}_i^T)} = +\sqrt{\mathrm{tr}\ \mathbf{a}_i^{*}\mathbf{a}_i^{\triangleleft T}\mathbf{a}_i^{\triangleleft *}\mathbf{a}_i^T} \qquad (6.46) \\ &= +\sqrt{\mathrm{tr}\ \mathbf{a}_i^T\mathbf{a}_i^{*}\mathbf{a}_i^{\triangleleft T}\mathbf{a}_i^{\triangleleft *}} = +\sqrt{\mathrm{tr}\ \|\mathbf{a}_i\|_F^2\|\mathbf{a}_i^{\triangleleft}\|_F^2} = \|\mathbf{a}_i\|_F\ \|\mathbf{a}_i^{\triangleleft}\|_F\ . \qquad (6.47) \end{aligned}$$

From the Frobenius norm representation using the squared modulus of the matrix elements and Eq.(6.36), an equivalent result is given

$$\|\frac{\partial\lambda_i[\mathbf{A}]}{\partial \mathbf{A}}\|_F = \sqrt{\sum_{\mu,\nu=1}^{n} (\mathbf{a}_i^{\triangleleft *})_\mu(\mathbf{a}_i)_\nu(\mathbf{a}_i^{\triangleleft})_\mu(\mathbf{a}_i^{*})_\nu} = \sqrt{\sum_{\mu,\nu=1}^{n} |(\mathbf{a}_i^{\triangleleft})_\mu|^2\ |(\mathbf{a}_i)_\nu|^2}\ . \qquad (6.48)$$

Conditions for minimum norm sensitivity (*Gilbert, E.G., 1984*), e.g., in the case of real (and distinct) eigenvalues are given by

$$\mathcal{N}(\lambda_i\mathbf{I}_n - \mathbf{A}) = \mathcal{N}(\lambda_i\mathbf{I}_n - \mathbf{A}^T) \quad \rightsquigarrow \quad \mathbf{a}_i = \mathbf{a}_i^{\triangleleft} \qquad (6.49)$$

which can easily be verified by inspection of Eq.(6.47). Therefore, symmetric matrices have a strong minimality property since minimal norm sensitivity is given for all of their real eigenvalues.

The Frobenius norm of the real-valued sensitivity matrices derived from the real and imaginary part with respect to $\mathbf{A}$ can be obtained as follows

$$\|\frac{\partial \Re e\lambda_i[\mathbf{A}]}{\partial \mathbf{A}}\|_F = \sqrt{\sum_{\mu,\nu=1}^{n} [\Re e(\mathbf{a}_i^{\triangleleft *})_\mu(\mathbf{a}_i)_\nu]^2} \qquad (6.50)$$

$$\|\frac{\partial \Im m \lambda_i[\mathbf{A}]}{\partial \mathbf{A}}\|_F = \sqrt{\sum_{\mu,\nu=1}^{n} [\Im m (\mathbf{a}_i^{\triangleleft *})_\mu (\mathbf{a}_i)_\nu]^2} \ . \tag{6.51}$$

Since

$$\frac{\partial \Re e \lambda_i}{\partial A_{\mu\nu}} = \Re e \frac{\partial \lambda_i}{\partial A_{\mu\nu}} = \Re e[(\mathbf{a}_i^{\triangleleft *})_\mu (\mathbf{a}_i)_\nu] \ , \tag{6.52}$$

the norm of the modulus sensitivity matrix is (*Pierre, C., 1989*)

$$\|\frac{\partial |\lambda_i[\mathbf{A}]|}{\partial \mathbf{A}}\|_F = \frac{1}{|\lambda_i|}\sqrt{\sum_{\mu,\nu=1}^{n} \{\Re e\ \lambda_i \Re e[(\mathbf{a}_i^{\triangleleft *})_\mu (\mathbf{a}_i)_\nu] + \Im m\ \lambda_i \Im m[(\mathbf{a}_i^{\triangleleft *})_\mu (\mathbf{a}_i)_\nu]\}^2} \ . \tag{6.53}$$

6.11 Lower Bound For the Sensitivity Norm

Applying Schwartz inequality,

$$|\mathbf{a}_i^{\triangleleft H} \mathbf{a}_i| < \|\mathbf{a}_i^{\triangleleft H}\|_F \ \|\mathbf{a}_i\|_F = \|\mathbf{a}_i^{\triangleleft}\|_F \ \|\mathbf{a}_i\|_F \ . \tag{6.54}$$

Since $\mathbf{a}_i^{\triangleleft H}\mathbf{a}_i = 1$ the relation $\|\mathbf{a}_i^{\triangleleft}\| \ \|\mathbf{a}_i\| \ \geq 1$ exists. Referring to Eq.(6.47),

$$\|\frac{\partial \lambda_i[\mathbf{A}]}{\partial \mathbf{A}}\|_F \geq 1 \quad \text{and} \quad S_{RSM} \stackrel{\Delta}{=} \sum_{1}^{n} \|\frac{\partial \lambda_i[\mathbf{A}]}{\partial \mathbf{A}}\|_F \geq n \ , \tag{6.55}$$

i.e., the sensitivity norm is bounded from below. The quantity S_{RSM} is the complete root sensitivity metric. The smallest sensitivity norm is given if $\mathbf{A}$ is normal, i.e. $\mathbf{A}\mathbf{A}^H = \mathbf{A}^H\mathbf{A}$, then the eigenvectors $\mathbf{a}_i^{\triangleleft}$ and $\mathbf{a}_i$ are colinear: $\mathbf{a}_i^{\triangleleft} = \mathbf{a}_i$ (*Skelton, R.E., and Wagie, D.A., 1984*). Near normal $\mathbf{A}$ implies minimal root sensitivity, but minimizing abnormality in combination with quadratic performance may unexpectedly cause increase of root sensitivity. This happens if two roots are close together since root sensitivity tends to infinity in the case of multiple roots.

6.12 Statistical Approach to Eigenvalue Perturbation

Assume that the plant matrix $\mathbf{A}$ depends on a parameter $\mathbf{p} \in \mathcal{R}^{n_p}$ and that $\Delta\mathbf{p}$ is a small perturbation vector of random variables. Then, the first variation is

$$\delta\mathbf{A} = \sum_{k=1}^{n_p} \frac{\partial \mathbf{A}}{\partial p_k}\Delta p_k \quad \text{or} \quad \delta A_{\mu,\nu} = \sum_{k=1}^{n_p} \frac{\partial A_{\mu\nu}}{\partial p_k}\Delta p_k \ . \tag{6.56}$$

Combination with Eq.(6.38) yields

$$\delta\lambda_i[\mathbf{A}] = \sum_{k=1}^{n_p} \sum_{\mu,\nu=1}^{n} (\mathbf{a}_i^{\triangleleft *})_\mu (\mathbf{a}_i)_\nu \frac{\partial A_{\mu\nu}}{\partial p_k}\Delta p_k \ . \tag{6.57}$$

With the assumption of zero mean random parameter, $E\{\Delta\mathbf{p}\} = \mathbf{0}$, the expectation $E\{\delta\lambda_i[\mathbf{A}]\} = 0$, i.e., the mean eigenvalue is unperturbed. For a given covariance matrix $\mathrm{cov}\ \mathbf{p} = E\{\Delta\mathbf{p}\Delta\mathbf{p}^T\}$ the variance of the eigenvalue can be evolved from Eq.(6.57)

$$\sigma_{\lambda_i}^2 = E\{|\delta\lambda|^2\} = E\{\delta\lambda\ (\delta\lambda)^*\} \tag{6.58}$$

$$\sigma^2_{\lambda_i} = \sum_{k,k'=1}^{n_p} \Big[\sum_{\mu,\nu,\mu',\nu'=1}^{n} (\mathbf{a}_i^{\triangleleft *})_\mu (\mathbf{a}_i)_\nu (\mathbf{a}_i^{\triangleleft})_{\mu'} (\mathbf{a}_i^{*})_{\nu'} \frac{\partial A_{\mu\nu}}{\partial p_k} \frac{\partial A_{\mu'\nu'}}{\partial p_{k'}} \Big] E\{\Delta p_k \Delta p_{k'}\} \, . \tag{6.59}$$

The variance of λ_i is determined by the covariance of the parameter variation. In the very special case that each element $A_{\mu\nu}$ is determined by a separate parameter p_k and that the entries $A_{\mu\nu}$ are uncorrelated and identically distributed with variance σ_A^2, the equation above can be rewritten to

$$\sigma^2_{\lambda_i} = \sum_{\mu,\nu,\mu',\nu'=1}^{n} (\mathbf{a}_i^{\triangleleft *})_\mu (\mathbf{a}_i)_\nu (\mathbf{a}_i^{\triangleleft})_{\mu'} (\mathbf{a}_i^{*})_{\nu'} \; E\{\delta A_{\mu\nu} \delta A_{\mu'\nu'}\} \tag{6.60}$$

$$= \sum_{\mu,\nu,\mu',\nu'=1}^{n} (\mathbf{a}_i^{\triangleleft *})_\mu (\mathbf{a}_i)_\nu (\mathbf{a}_i^{\triangleleft})_{\mu'} (\mathbf{a}_i^{*})_{\nu'} \; \sigma_A^2 \delta_{\mu\nu} \delta_{\mu'\nu'} \tag{6.61}$$

$$\sigma^2_{\lambda_i} = \Big[\sum_{\mu,\nu=1}^{n} (\mathbf{a}_i^{\triangleleft *})_\mu (\mathbf{a}_i)_\nu (\mathbf{a}_i^{\triangleleft})_\mu (\mathbf{a}_i^{*})_\nu \Big] \sigma_A^2 = \Big[\sum_{\mu,\nu=1}^{n} |(\mathbf{a}_i^{\triangleleft})_\mu|^2 \, |(\mathbf{a}_i)_\nu|^2 \Big] \sigma_A^2 \, . \tag{6.62}$$

Comparison with Eq.(6.48) yields

$$\frac{\sigma_{\lambda_i}}{\sigma_A} = \| \frac{\partial \lambda_i[\mathbf{A}]}{\partial \mathbf{A}} \|_F \, , \tag{6.63}$$

i.e., the standard deviation $\sigma_{\lambda_i}/\sigma_A$ is precisely equivalent to the norm of the eigenvalue sensitivity.

Similarly, *Pierre, C., 1989* presented

$$\frac{\sigma_{\Re e\lambda}}{\sigma_A} = \| \frac{\partial \Re e \; \lambda[\mathbf{A}]}{\partial \mathbf{A}} \|_F, \quad \frac{\sigma_{\Im m\lambda}}{\sigma_A} = \| \frac{\partial \Im m \; \lambda[\mathbf{A}]}{\partial \mathbf{A}} \|_F, \quad \frac{\sigma_{|\lambda|}}{\sigma_A} = \| \frac{\partial | \; \lambda[\mathbf{A}] \; |}{\partial \mathbf{A}} \|_F \, . \tag{6.64}$$

Furthermore, the norm sensitivity to parameter uncertainties $\Delta \mathbf{p}$ equals the relative standard deviation in the special case that the parameters are uncorrelated and equally distributed

$$\| \frac{\partial \lambda[\mathbf{p}]}{\partial \mathbf{p}} \|_F = \frac{\sigma_\lambda}{\sigma_p} \, . \tag{6.65}$$

The norm sensitivity analysis and the first-order statistical perturbation method yield identical results. Comparing the norm sensitivity Eq.(6.48) and the statistical perturbation results Eq.(6.59), it can be noticed that the latter is more general. The entries of the sensitivity matrix are all weighted equally whilst the statistical approach, Eq.(6.59), contains particular elements suitable for weighting purpose and for correlation among system parameters.

6.13 Eigenvector Differential Sensitivity

Let $\mathbf{A} \in \mathcal{C}^{n \times n}$ be a matrix with distinct eigenvalues $\lambda_i[\mathbf{A}]$ and with corresponding right- and left-eigenvectors $\mathbf{a}_i$ and $\mathbf{a}_i^{\triangleleft}$, respectively, defined by

$$\mathbf{A}\mathbf{a}_i \triangleq \lambda_i[\mathbf{A}] \; \mathbf{a}_i \qquad \mathbf{A}^H \mathbf{a}_i^{\triangleleft} \triangleq \lambda_i^*[\mathbf{A}] \; \mathbf{a}_i^{\triangleleft} \, . \tag{6.66}$$

If $\mathbf{A}$ changes to $\mathbf{A} + \delta\mathbf{A}$ the changes $\delta\lambda_i[\mathbf{A}]$ and $\delta\mathbf{a}_i$ are determined as follows. Considering

$$(\mathbf{A} + \delta\mathbf{A})(\mathbf{a}_i + \delta\mathbf{a}_i) = (\lambda_i + \delta\lambda_i)(\mathbf{a}_i + \delta\mathbf{a}_i) \tag{6.67}$$

it must be noticed that the special case $\delta\mathbf{A} = \mathbf{0}$ and $\delta\lambda_i = 0$ results in $\mathbf{A}\delta\mathbf{a}_i = \lambda_i\delta\mathbf{a}_i$ which is satisfied by any $\delta\mathbf{a}_i = \beta\ \mathbf{a}_j \quad (\beta \neq 0)$. Normalizing the eigenvector $\mathbf{a}_i$ by keeping $\delta\mathbf{a}_i = \mathbf{0}$ if $\delta\mathbf{A} = \mathbf{0}$ requires that the decomposition of

$$\mathbf{a}_i + \delta\mathbf{a}_i = \sum_{k=1}^{n} \alpha_{ik}\mathbf{a}_k = \sum_{k=1,\ k\neq i}^{n} \alpha_{ik}\mathbf{a}_k + \alpha_{ii}\mathbf{a}_i \tag{6.68}$$

contains the coefficient $\alpha_{ii} = 1$. This may be regarded as a normalization operation. In other words, the expansion of $\delta\mathbf{a}_i$ is

$$\delta\mathbf{a}_i = \sum_{k=1,\ k\neq i}^{n} \alpha_{ik}\mathbf{a}_k\ . \tag{6.69}$$

The eigenvector increment $\delta\mathbf{a}_i$ is a linear combination of $\mathbf{a}_k$ where $k \neq i$ *(Franklin, J.N., 1968)*. Hence, the inner product with "postmultiplied" $\mathbf{a}_i^{\triangleleft}$ is

$$\delta\mathbf{a}_i^T\mathbf{a}_i^{\triangleleft *} = \sum_{k=1,\ k\neq i}^{n} \alpha_{ik}\mathbf{a}_k^T\mathbf{a}_i^{\triangleleft *} = 0 \qquad \text{since} \qquad \mathbf{a}_k^T\mathbf{a}_i^{\triangleleft *} = \delta_{ki} \tag{6.70}$$

which is the usual normalization operation.

6.13.1 Eigenvalue Increment

From Eq.(6.67) it follows

$$\mathbf{A}\ \delta\mathbf{a}_i + \delta\mathbf{A}\ \mathbf{a}_i = \lambda_i\ \delta\mathbf{a}_i + \delta\lambda_i\ \mathbf{a}_i\ . \tag{6.71}$$

The inner product with $\mathbf{a}_i^{\triangleleft}$ yields

$$(\mathbf{A}\ \delta\mathbf{a}_i)^T\mathbf{a}_i^{\triangleleft *} + (\delta\mathbf{A}\ \mathbf{a}_i)^T\mathbf{a}_i^{\triangleleft *} = \lambda_i\ \delta\mathbf{a}_i^T\mathbf{a}_i^{\triangleleft *} + \delta\lambda_i\ \mathbf{a}_i^T\mathbf{a}_i^{\triangleleft *}\ . \tag{6.72}$$

The first term vanishes since $\delta\mathbf{a}_i^T\mathbf{A}^T\mathbf{a}_i^{\triangleleft *} = \delta\mathbf{a}_i^T\lambda_i\mathbf{a}_i^{\triangleleft *}$, the third term is zero with respect to Eq.(6.70). Thus, from Eq.(6.72), using $\mathbf{a}_k^T\mathbf{a}_i^{\triangleleft *} = \delta_{ki}$ again,

$$(\delta\mathbf{A}\ \mathbf{a}_i)^T\mathbf{a}_i^{\triangleleft *} = \mathbf{a}_i^T\delta\mathbf{A}^T\mathbf{a}_i^{\triangleleft *} = \delta\lambda_i\ \mathbf{a}_i^T\mathbf{a}_i^{\triangleleft *} \quad \rightsquigarrow \quad \delta\lambda_i = \frac{\mathbf{a}_i^T\delta\mathbf{A}^T\mathbf{a}_i^{\triangleleft *}}{\mathbf{a}_i^T\mathbf{a}_i^{\triangleleft *}} = \mathbf{a}_i^{\triangleleft * T}\delta\mathbf{A}\ \mathbf{a}_i\ . \tag{6.73}$$

6.13.2 Eigenvector Increment

The inner product of Eq.(6.71) with $\mathbf{a}_k^{\triangleleft}$ yields

$$\delta\mathbf{a}_i^T\mathbf{A}^T\mathbf{a}_k^{\triangleleft *} + (\delta\mathbf{A}\ \mathbf{a}_i)^T\mathbf{a}_k^{\triangleleft *} = (\lambda_i\ \delta\mathbf{a}_i)^T\mathbf{a}_k^{\triangleleft *} + \delta\lambda_i\ \mathbf{a}_i^T\mathbf{a}_k^{\triangleleft *} \tag{6.74}$$

$$\delta\mathbf{a}_i^T\lambda_k\mathbf{a}_k^{\triangleleft *} + \mathbf{a}_i^T\delta\mathbf{A}^T\mathbf{a}_k^{\triangleleft *} = \lambda_i\ \delta\mathbf{a}_i^T\mathbf{a}_k^{\triangleleft *} + 0 \tag{6.75}$$

$$(\lambda_i - \lambda_k)\delta\mathbf{a}_i^T\mathbf{a}_k^{\triangleleft *} = \mathbf{a}_i^T\delta\mathbf{A}^T\mathbf{a}_k^{\triangleleft *}\ . \tag{6.76}$$

The inner product of Eq.(6.69) with $\mathbf{a}_k^{\triangleleft}$ is

$$\delta\mathbf{a}_i^T\mathbf{a}_k^{\triangleleft *} = \sum_{l=1,\ l\neq j}^{n} \alpha_{il}\ \mathbf{a}_l^T\mathbf{a}_k^{\triangleleft *} = \alpha_{jk}\ \mathbf{a}_k^T\mathbf{a}_k^{\triangleleft *}\ . \tag{6.77}$$

Substituting $\delta\mathbf{a}_i^T\mathbf{a}_k^{\triangleleft *}$ from Eq.(6.77) into (6.76),

$$(\lambda_i - \lambda_k)\alpha_{ik}\mathbf{a}_k^T\mathbf{a}_k^{\triangleleft *} = \mathbf{a}_i^T\delta\mathbf{A}^T\mathbf{a}_k^{\triangleleft *} \quad \rightsquigarrow \quad \alpha_{ik} = \frac{\mathbf{a}_i^T\ \delta\mathbf{A}^T\mathbf{a}_k^{\triangleleft *}}{(\lambda_i - \lambda_k)\mathbf{a}_k^T\mathbf{a}_k^{\triangleleft *}} \quad \forall\, i \neq k \tag{6.78}$$

and substituting into Eq.(6.69) yields the result

$$\delta\mathbf{a}_i = \sum_{k=1,\ k\neq i}^{n} \frac{\mathbf{a}_i^T\ \delta\mathbf{A}^T\mathbf{a}_k^{\triangleleft *}}{(\lambda_i - \lambda_k)\mathbf{a}_k^T\mathbf{a}_k^{\triangleleft *}}\mathbf{a}_k = \sum_{k=1,\ k\neq i}^{n} \frac{\mathbf{a}_k^{\triangleleft *T}\delta\mathbf{A}\ \mathbf{a}_i}{(\lambda_i - \lambda_k)\mathbf{a}_k^{\triangleleft *T}\mathbf{a}_k}\mathbf{a}_k\ . \tag{6.79}$$

The derivative of the eigenvector $\mathbf{a}_i$ with respect to an arbitrary matrix $\mathbf{M} \in \mathcal{R}^{r\times s}$ is given by using Eq.(5.79)

$$\frac{\delta\mathbf{a}_i}{\delta M_{\nu\mu}} = \frac{\partial\mathbf{a}_i}{\partial M_{\nu\mu}} = \sum_{k=1,\ k\neq i}^{n} \frac{\mathbf{a}_k^{\triangleleft *T}\frac{\partial\mathbf{A}}{\partial M_{\nu\mu}}\mathbf{a}_i}{(\lambda_i - \lambda_k)\mathbf{a}_k^{\triangleleft *T}\mathbf{a}_k}\mathbf{a}_k \quad \rightsquigarrow \quad \frac{\partial\mathbf{a}_i}{\partial\mathbf{M}} = \sum_{\nu\mu}\mathbf{E}_{\nu\mu}^{(r\times s)}\otimes\frac{\partial\mathbf{a}_i}{\partial M_{\nu\mu}}\ . \tag{6.80}$$

6.13.3 Increment of an Outer Product

With regard to Eq.(6.69) and using the definition of a corresponding β_{ik} re $\mathbf{a}_i^{\triangleleft}$, the increment of the outer product, $\delta(\mathbf{a}_i^{\triangleleft *}\mathbf{a}_i^T)$, has the property

$$\delta(\mathbf{a}_i^{\triangleleft *}\mathbf{a}_i^T) = \mathbf{a}_i^{\triangleleft *}\delta\mathbf{a}_i^T + (\delta\mathbf{a}_i^{\triangleleft *})\mathbf{a}_i^T = \mathbf{a}_i^{\triangleleft *}\sum_{k=1,\ k\neq i}^{n}\alpha_{ik}\mathbf{a}_k^T + (\sum_{k=1,\ k\neq i}^{n}\beta_{ik}\mathbf{a}_k^{\triangleleft *})\mathbf{a}_i^T\ . \tag{6.81}$$

Premultiplication by $\mathbf{A}^T$ and taking the trace and using its cyclic property yields

$$\begin{aligned}
\text{tr}\ \mathbf{A}^T\ \delta(\mathbf{a}_i^{\triangleleft *}\mathbf{a}_i^T) &= \text{tr}\ \mathbf{A}^T\mathbf{a}_i^{\triangleleft *}\sum_{k=1,\ k\neq i}^{n}\alpha_{ik}\mathbf{a}_k^T + \text{tr}\ (\sum_{k=1,\ k\neq i}^{n}\beta_{ik}\mathbf{A}^T\mathbf{a}_k^{\triangleleft *})\mathbf{a}_i^T &(6.82)\\
&= \text{tr}\ \lambda_i\mathbf{a}_i^{\triangleleft *}\sum_{k=1,\ k\neq i}^{n}\alpha_{ik}\mathbf{a}_k^T + \text{tr}\ (\sum_{k=1,\ k\neq i}^{n}\beta_{ik}\lambda_i\mathbf{a}_k^{\triangleleft *})\mathbf{a}_i^T &(6.83)\\
&= \text{tr}\ \lambda_i\sum_{k=1,\ k\neq i}^{n}\alpha_{ik}\mathbf{a}_k^T\mathbf{a}_i^{\triangleleft *} + \text{tr}\sum_{k=1,\ k\neq i}^{n}\beta_{ik}\lambda_i\mathbf{a}_i^T\mathbf{a}_k^{\triangleleft *} = 0 + 0\ . &(6.84)
\end{aligned}$$

Abbreviating the outer product $\mathbf{B} \triangleq (\mathbf{a}_i^{\triangleleft *}\mathbf{a}_i^T)$ and referring to Eq.(6.84), it results

$$\delta\text{tr}\ [\mathbf{A}^T\mathbf{B}] = \delta\sum_{k=1}^{n}\sum_{i=1}^{n}A_{ik}B_{ik} = \text{tr}\ [(\delta\mathbf{A})^T\mathbf{B}] + \text{tr}\ [\mathbf{A}^T\ \delta\mathbf{B}] = \text{tr}\ [(\delta\mathbf{A})^T\mathbf{B}] + 0 \tag{6.85}$$

$$\frac{\partial\ \text{tr}\ [\mathbf{A}^T\mathbf{B}]}{\partial A_{\nu\mu}} = \frac{\delta\ \text{tr}\ [\mathbf{A}^T\mathbf{B}]}{\delta A_{\nu\mu}} = \text{tr}\ [\frac{(\delta\mathbf{A})^T}{\delta A_{\nu\mu}}\mathbf{B}] = B_{\nu\mu}\ . \tag{6.86}$$

Invoking Eq.(6.73) yields

$$\frac{\partial\lambda_i[\mathbf{A}]}{\partial\mathbf{A}} = \mathbf{a}_i^{\triangleleft *}\mathbf{a}_i^T\ , \tag{6.87}$$

as given in Eqs.(5.33) and (6.37). No additional term in Eq.(5.33) is required.

Example:

$$\mathbf{A} = \begin{pmatrix} 0 & 1 \\ -3 & -4 \end{pmatrix},\quad \delta\mathbf{A} = \begin{pmatrix} 0 & 0 \\ 0.1 & 0 \end{pmatrix},\quad \lambda_{1,2}[\mathbf{A}] = -1;\ -3\ . \tag{6.88}$$

$$\mathbf{a}_1 = \begin{pmatrix}1\\-1\end{pmatrix}, \quad \mathbf{a}_2 = \begin{pmatrix}1\\-3\end{pmatrix}, \quad \mathbf{a}_1^{\triangleleft} = \begin{pmatrix}1.5\\0.5\end{pmatrix}, \quad \mathbf{a}_2^{\triangleleft} = \begin{pmatrix}-0.5\\-0.5\end{pmatrix} . \tag{6.89}$$

Since tr $\delta\mathbf{A} = 0$, one may expect $\delta\lambda_1 + \delta\lambda_2 = 0$. From Eq.(6.73) it follows

$$\delta\lambda_1 = (1.5 \;\; 0.5)\begin{pmatrix}0 & 0\\0.1 & 0\end{pmatrix}\begin{pmatrix}1\\-1\end{pmatrix} = 0.05 \quad \text{and} \quad \delta\lambda_2 = -0.05 . \tag{6.90}$$

By Eq.(6.79) the eigenvector increment is

$$\delta\mathbf{a}_1 = \frac{(-0.5 \;\; -0.5)\begin{pmatrix}0 & 0\\0.1 & 0\end{pmatrix}\binom{1}{-1}}{2(-0.5 \;\; -0.5)\binom{1}{-3}}\begin{pmatrix}1\\-3\end{pmatrix} = \begin{pmatrix}-0.025\\0.075\end{pmatrix} \tag{6.91}$$

which satisfies Eq.(6.69) with $\alpha_{ik} = \alpha_{12} = -0.025$. Furthermore, $\delta\mathbf{a}_2 = (-0.025 \;\;\; 0.025)^T$.

$$\text{Perturbed matrix} \quad \mathbf{A}' \triangleq \mathbf{A} + \delta\mathbf{A} = \begin{pmatrix}0 & 1\\-2.9 & -4\end{pmatrix} \qquad \lambda_{1,2}[\mathbf{A}'] = -0.9512, \;\; -3.0488 \tag{6.92}$$

$$\mathbf{a}_1' = \begin{pmatrix}0.975\\-0.925\end{pmatrix}, \quad \mathbf{a}_2' = \begin{pmatrix}0.975\\-2.975\end{pmatrix}, \quad \mathbf{a}_1'^{\triangleleft} = \begin{pmatrix}1.4889\\0.4883\end{pmatrix}, \quad \mathbf{a}_2'^{\triangleleft} = \begin{pmatrix}-0.4645\\-0.4884\end{pmatrix} \tag{6.93}$$

$$\text{Perturbed outer product} \quad \mathbf{a}_i'^{\triangleleft *}\mathbf{a}_i'^T = \begin{pmatrix}1.4889\\0.4883\end{pmatrix}(0.975 \;\vdots\; -0.925) = \begin{pmatrix}1.4517 & -1.3772\\0.4760 & -0.4517\end{pmatrix} \tag{6.94}$$

$$\text{Increment of the outer product} \quad \mathbf{a}_i'^{\triangleleft *}\mathbf{a}_i'^T - \mathbf{a}_i^{\triangleleft *}\mathbf{a}_i^T = \delta(\mathbf{a}_i^{\triangleleft *}\mathbf{a}_i^T) = \begin{pmatrix}-0.0483 & 0.1228\\-0.0240 & 0.0483\end{pmatrix} \tag{6.95}$$

$$\text{tr}\,[\mathbf{A}^T\,\delta(\mathbf{a}_i^{\triangleleft *}\mathbf{a}_i^T)] = \text{tr}\,[\begin{pmatrix}1 & -3\\1 & -4\end{pmatrix}\begin{pmatrix}-0.0483 & 0.1228\\-0.0240 & 0.0483\end{pmatrix}] \doteq 0 . \tag{6.96}$$

Drawing a parallel, Eq.(6.37) for $i = 1$ yields

$$\frac{\partial\lambda_1}{\partial\mathbf{A}} = \mathbf{a}_1^{\triangleleft *}\mathbf{a}_1^T = \begin{pmatrix}1.5\\0.5\end{pmatrix}(1 \quad -1) = \begin{pmatrix}1.5 & -1.5\\0.5 & -0.5\end{pmatrix} \tag{6.97}$$

where the component 0.5 corresponds to $\delta\lambda_1 \,/\, \delta A_{21} = 0.05/0.1 = 0.5$.

End of Example

Chapter 7

Transition Matrix Differential Sensitivity

7.1 Continuous-Time Systems. Using Eigenvectors

Referring to Eq.(4.39) let $\mathbf{M}$ be the (n,n)-matrix $\mathbf{A}$ of the open-loop or closed-loop coefficients

$$e^{\mathbf{I}_m \otimes \mathbf{A}t} = \sum_{k=1}^{n} \mathbf{I}_m \otimes (\mathbf{a}_k \mathbf{a}_k^{aT}) e^{\lambda_k[\mathbf{A}]t} \tag{7.1}$$

where $\mathbf{I}_m$ is the identity matrix with arbitrary dimension $m \times m$, $\mathbf{a}_k$ is the kth (right) eigenvector of $\mathbf{A}$ associated with eigenvalue $\lambda_k[\mathbf{A}]$. Furthermore, $\mathbf{a}_k^a$ is the left eigenvector or the right eigenvector of $\mathbf{A}^T$, normalized with $\mathbf{a}_k$ (i.e., $\mathbf{a}_k^{aT}\mathbf{a}_i = \delta_{ik}$).

Now, consider the derivative of the exponential matrix $e^{\mathbf{A}t}$ with respect to an arbitrary (m,r)-matrix $\mathbf{K}$

$$\frac{\partial}{\partial \mathbf{K}} e^{\mathbf{A}t} = \mathbf{f}\Big(\frac{\partial \mathbf{A}}{\partial \mathbf{K}}\Big) \quad \rightsquigarrow \quad \mathbf{f} = ? \tag{7.2}$$

The aim is to find the interrelation function $\mathbf{f}(\partial\mathbf{A}/\partial\mathbf{K})$. The exponential matrix $e^{\mathbf{A}t}$ or transition matrix $\mathbf{\Phi}(t)$ obeys the equations (*Ludyk, G., 1990*)

$$\mathbf{\Phi}(t) = e^{\mathbf{A}t}, \quad \frac{d}{dt}\mathbf{\Phi}(t) = \mathbf{A}\mathbf{\Phi}(t), \quad \mathbf{\Phi}(0) = \mathbf{I}\,. \tag{7.3}$$

Differentiating with respect to $\mathbf{K}$ and using the product rule Eq.(5.87) yields

$$\frac{\partial}{\partial \mathbf{K}}\frac{d}{dt}\mathbf{\Phi}(t) = (\mathbf{I}_m \otimes \mathbf{A})\frac{\partial \mathbf{\Phi}}{\partial \mathbf{K}} + \frac{\partial \mathbf{A}}{\partial \mathbf{K}}(\mathbf{I}_r \otimes \mathbf{\Phi}) \tag{7.4}$$

$$\frac{d}{dt}\frac{\partial \mathbf{\Phi}(t)}{\partial \mathbf{K}} - (\mathbf{I}_m \otimes \mathbf{A})\frac{\partial \mathbf{\Phi}(t)}{\partial \mathbf{K}} = \frac{\partial \mathbf{A}}{\partial \mathbf{K}}\big(\mathbf{I}_r \otimes \mathbf{\Phi}(t)\big)\,. \tag{7.5}$$

For comparison, the scalar inhomogeneous differential equation and its solution are

$$\frac{dy(t)}{dt} - fy(t) = g(t), \quad f \text{ const.}, \quad \rightsquigarrow \quad y(t) = \int_0^t e^{f(t-\tau)}\, g(\tau)d\tau + e^{ft}y(o)\,. \tag{7.6}$$

This can easily be proved by substitution and by applying the differentiation rule with respect to the upper integration boundary

$$\frac{\partial}{\partial t}\int_a^t q(t,\tau)d\tau = \int_a^t \frac{\partial}{\partial t} q(t,\tau)d\tau + q(t,t)\,. \tag{7.7}$$

In view of this, the corresponding solution of the matrix differential equation is

$$\frac{d\mathbf{Y}(t)}{dt} - \mathbf{F}\mathbf{Y}(t) = \mathbf{G}(t) \quad \rightsquigarrow \quad \mathbf{Y}(t) = \int_0^t e^{\mathbf{F}(t-\tau)}\, \mathbf{G}(\tau)d\tau + e^{\mathbf{F}t}\mathbf{Y}(0)\,. \tag{7.8}$$

This can be checked in the same way using $\partial e^{\mathbf{A}t}/\partial t = \mathbf{A}e^{\mathbf{A}t} = e^{\mathbf{A}t}\mathbf{A}$. Using this result and $\partial\mathbf{\Phi}(0)/\partial\mathbf{K} = \mathbf{0}$ (from $\mathbf{\Phi}(0) = \mathbf{I}$), the solution is

$$\frac{\partial\mathbf{\Phi}(t)}{\partial\mathbf{K}} = \int_0^t e^{\mathbf{I}_m\otimes\mathbf{A}(t-\tau)} \frac{\partial\mathbf{A}}{\partial\mathbf{K}} \left(\mathbf{I}_r \otimes e^{\mathbf{A}}\right) d\tau \,. \tag{7.9}$$

The proof can easily be given by substitution into Eq.(7.5). Note that $\frac{\partial\mathbf{\Phi}}{\partial\mathbf{K}}$ may be treated as an autonomous variable and that the solution of Eq. (7.5) in the variable $\frac{\partial\mathbf{\Phi}}{\partial\mathbf{K}}$ can be treated irrespective of the fact that $\mathbf{\Phi}$ itself acts as a forcing component (see *Knobloch, H.W., and Kappel, F., 1974, chapter III.3.2 and III.4*).

Applying the formulas Eqs.(4.16) and (7.1) (*Brewer, J.W., 1977 and 1978*),

$$\frac{\partial\mathbf{\Phi}(t)}{\partial\mathbf{K}} = \int_0^t \sum_{k=1}^{n}[\mathbf{I}_m \otimes (\mathbf{a}_k\mathbf{a}_k^{\triangleleft T})]e^{\lambda_k(t-\tau)} \frac{\partial\mathbf{A}}{\partial\mathbf{K}} \sum_{i=1}^{n}[\mathbf{I}_r \otimes \mathbf{a}_i\mathbf{a}_i^{\triangleleft T}]e^{\lambda_i\tau}d\tau \tag{7.10}$$

$$\frac{\partial\mathbf{\Phi}(t)}{\partial\mathbf{K}} = \sum_{k=1}^{n}\sum_{i=1}^{n}[\mathbf{I}_m \otimes (\mathbf{a}_k\mathbf{a}_k^{\triangleleft T})]\frac{\partial\mathbf{A}}{\partial\mathbf{K}}[\mathbf{I}_r \otimes \mathbf{a}_i\mathbf{a}_i^{\triangleleft T}] \int_0^t e^{\lambda_k(t-\tau)}\, e^{\lambda_i\tau}d\tau \,. \tag{7.11}$$

The integral term turns out as an inverse Laplace transform and can be calculated in detail

$$\int_0^t e^{\lambda_k(t-\tau)}\, e^{\lambda_i\tau}d\tau = \mathcal{L}^{-1}\frac{1}{(s-\lambda_i)(s-\lambda_k)} = \begin{cases} \frac{e^{\lambda_i t}-e^{\lambda_k t}}{\lambda_i-\lambda_k} & \text{if } \lambda_i \neq \lambda_k \\ te^{\lambda_i t} & \text{if } \lambda_i = \lambda_k \,. \end{cases} \tag{7.12}$$

If $\mathbf{K}$ is specialized to a scalar $\mathbf{K} := K$ then $m = r = 1$ and

$$\frac{\partial e^{\mathbf{A}t}}{\partial K} = \sum_{k=1}^{n}\sum_{i=1}^{n} \mathbf{a}_k\mathbf{a}_k^{\triangleleft T} \frac{\partial\mathbf{A}}{\partial K} \mathbf{a}_i\mathbf{a}_i^{\triangleleft T} \int_0^t e^{\lambda_k(t-\tau)}\, e^{\lambda_i\tau}d\tau \,. \tag{7.13}$$

If both $\mathbf{A}$ and $\mathbf{K}$ are scalars $\mathbf{A} := A$, $\mathbf{K} := K$ then $n = m = r = 1$ and

$$\frac{\partial e^{At}}{\partial K} = \frac{\partial A}{\partial K} te^{At} = te^{At}\frac{\partial A}{\partial K} \,. \tag{7.14}$$

Assuming $\mathbf{K}$ as a matrix of slowly varying parameters (e.g., state feedback controller, observer coefficient matrix), the solution of Eq. (7.11) is very useful for determining the sensitivity of the system dynamics with respect to parameter variations. The dynamic behaviour incorporated by the state transition matrix $\mathbf{\Phi}(t)$ is expressed by the coefficient matrix sensitivity $\partial\mathbf{A}/\partial\mathbf{K}$.

7.2 Continuous-Time Systems. Using Matrix Components

7.2.1 Minimum Polynomial and Function of a Matrix

Minimal polynomial of $\mathbf{A}$: Define m_m the smallest integer up to which the powers of $\mathbf{A}$ are linearly dependent. Thus,

$$\mathbf{A}^{m_m} + d_{m_m-1}\mathbf{A}^{m_m-1} + \ldots + d_0\mathbf{I} = \mathbf{0} \,. \tag{7.15}$$

Then, the minimum polynomial is given by

$$\lambda^{m_m} + d_{m_m-1}\lambda^{m_m-1} + \ldots + d_0 \,. \tag{7.16}$$

The minimal polynomial is a factor of the characteristic polynomial. Minimum polynomial and characteristic polynomial are equal if all $\lambda_i[\mathbf{A}]$ are distinct or if $q_i = 1$ (simple degeneracy) holds. The characteristic polynomial $c(\lambda)$ in factored form with multiplicities m_i is

$$c(\lambda) = (\lambda - \lambda_1)^{m_1}(\lambda - \lambda_2)^{m_2} \dots (\lambda - \lambda_p)^{m_p} . \tag{7.17}$$

The minimal polynomial $m(\lambda)$ with multiplicities (indices) m_{mi} ($m_{mi} \leq m_i$) is

$$m(\lambda) = (\lambda - \lambda_1)^{m_{m1}}(\lambda - \lambda_2)^{m_{m2}} \dots (\lambda - \lambda_p)^{m_{mp}} \quad \text{where} \quad \sum_{i=1}^{p} m_{mi} = m_m . \tag{7.18}$$

Each $\lambda_i[\mathbf{A}]$ is a zero of the minimum polynomial. The multiplicity m_{mi} of the eigenvalue λ_i in the minimal polynomial is less in comparison with the multiplicity m_i in the characteristic polynomial. The multiplicity m_{mi} is in close relation with the dimension of the Jordan block, see Eq.(B.112). If the zeros of the minimum polynomial are $\lambda_1 \dots \lambda_p$ and if the multiplicities are $m_{m1} \dots m_{mp}$ it can be shown that for any analytic function $f(\cdot)$

$$f(\mathbf{A}) = \sum_{i=1}^{p} \sum_{j=0}^{m_{mi}-1} \frac{d^j f(\lambda)}{d\lambda^j} |_{\lambda=\lambda_i} \mathbf{Z}_{ij} . \tag{7.19}$$

The matrices $\mathbf{Z}_{ij}$ are named interpolating polynomials or matrix components of $\mathbf{A}$. The matrices $\mathbf{Z}_{ij}$ are expressible as a polynomial in $\mathbf{A}$: $\mathbf{Z}_{ij} = \mathbf{Z}_{ij}(\mathbf{A})$. If all eigenvalues are distinct (multiplicity m_i throughout one) then $\mathbf{A}$ has a full set of eigenvectors. The characteristic polynomial and minimal polynomial are equal. The matrices $\mathbf{Z}_{ij}$ are related to $\mathbf{a}_i$ and $\mathbf{a}_i^q$ (normalized) by $\mathbf{Z}_{ij} = \mathbf{a}_i \mathbf{a}_i^{qT}$.

7.2.2 Matrix Exponential

Specializing $f(\mathbf{A}) = e^{\mathbf{A}t}$ and using matrix components (*Zadeh, L.A., and Desoer, C.A., 1963 p.300*) and preassuming that the minimum polynomial has p distinct eigenvalues $\lambda_1, \lambda_2 \dots \lambda_i \dots \lambda_p$ of index m_{mi} and $\mathbf{A}$ corresponds to a stable matrix ($\Re e \lambda_i \leq 0$), it results

$$e^{\mathbf{A}t} = \mathbf{\Phi}(t) = \sum_{i=1}^{p} \sum_{j=0}^{m_{mi}-1} t^j e^{\lambda_i t} \mathbf{Z}_{ij} . \tag{7.20}$$

7.2.3 Differential Sensitivity

Following *Brewer, J.W., 1978a*, the derivative of the state transition matrix $\mathbf{\Phi}(t)$ or matrix exponential $\mathbf{\Phi}(t) = \exp \mathbf{A}t$ with respect to the (m,r)-matrix $\mathbf{K}$ is

$$\frac{\partial \mathbf{\Phi}}{\partial \mathbf{K}} = \sum_{i=1}^{p} \sum_{\mu=1}^{p} \sum_{j=0}^{m_{mi}-1} \sum_{\nu=0}^{m_{m\mu}-1} (\mathbf{I}_m \otimes \mathbf{Z}_{ij}) \frac{\partial \mathbf{A}}{\partial \mathbf{K}} (\mathbf{I}_r \otimes \mathbf{Z}_{\mu\nu}) f_{i\mu j\nu}(t) \tag{7.21}$$

$$\text{where} \quad f_{i\mu j\nu}(t) = \mathcal{L}^{-1} \frac{j!\nu!}{(s+\lambda_i)^{j+1}(s+\lambda_\mu)^{\nu+1}} . \tag{7.22}$$

7.3 Continuous-Time Systems. Using Interpolation Method

7.3.1 Interpolation Method

The interpolation method representation of the matrix exponential is given by

$$\mathbf{\Phi}(t) = e^{\mathbf{A}t} = \sum_{i=0}^{m_m-1} \alpha_i(t)\mathbf{A}^i \quad \text{where} \quad \alpha_i(t) = \mathcal{L}^{-1} \sum_{j=0}^{n-i-1} \frac{c_{i+j+1}\, s^j}{c(s)}\,, \tag{7.23}$$

the number m_m is the degree of the minimal polynomial of $\mathbf{A}$, $c(s)$ is the characteristic polynomial of $\mathbf{A}$ from Eq.(B.1). The coefficients c_i are calculated from Leverrier's algorithm.

7.3.2 Leverrier's Algorithm

The Leverrier's algorithm (or Faddeev method, Bôcher formula) is defined as follows. Given a square matrix $\mathbf{A}$, $\mathbf{A} \in \mathcal{R}^{n\times n}$, the inverse of the characteristic matrix is

$$\mathcal{L}e^{\mathbf{A}t} = (s\mathbf{I}-\mathbf{A})^{-1} = \frac{\text{adj }(s\mathbf{I}-\mathbf{A})}{\det\,(s\mathbf{I}-\mathbf{A})} = \frac{\mathbf{\Omega}(s)}{c(s)} = \frac{\mathbf{\Omega}_{n-1}s^{n-1} + \mathbf{\Omega}_{n-2}s^{n-2} + \ldots + \mathbf{\Omega}_0}{s^n + c_{n-1}s^{n-1} + \ldots + c_0} \tag{7.24}$$

where $c(s)$ is the characteristic polynomial, $\mathbf{\Omega}(s)$ a matrix polynomial in s of degree $n-1$ with (n,n)-matrix factors $\mathbf{\Omega}_i$ associated with s^i . The algorithm runs as follows (*Zadeh, L.A., and Desoer, C.A., 1963 p. 303*)

$$\mathbf{\Omega}_{n-1} = \mathbf{I}_n \tag{7.25}$$

$$c_{n-1} = -\text{tr }\mathbf{A}\mathbf{\Omega}_{n-1} \qquad \mathbf{\Omega}_{n-2} = \mathbf{A}\mathbf{\Omega}_{n-1} + c_{n-1}\mathbf{I} \tag{7.26}$$

$$c_{n-2} = -\frac{1}{2}\text{ tr }\mathbf{A}\mathbf{\Omega}_{n-2} \qquad \mathbf{\Omega}_{n-3} = \mathbf{A}\mathbf{\Omega}_{n-2} + c_{n-2}\mathbf{I} \tag{7.27}$$

$$\vdots \qquad\qquad \vdots$$

$$\mathbf{\Omega}_1 = \mathbf{A}\mathbf{\Omega}_2 + c_2\mathbf{I} \tag{7.28}$$

$$c_1 = -\frac{1}{n-1}\text{ tr }\mathbf{A}\mathbf{\Omega}_1 \qquad \mathbf{\Omega}_0 = \mathbf{A}\mathbf{\Omega}_1 + c_1\mathbf{I} \tag{7.29}$$

$$c_0 = -\frac{1}{n}\text{ tr }\mathbf{A}\mathbf{\Omega}_0 \qquad \mathbf{\Omega}_{-1} = \mathbf{A}\mathbf{\Omega}_0 + c_0\mathbf{I} = \mathbf{0}\,. \tag{7.30}$$

The last equation can be utilized to check the round-off errors presented by the magnitude $\mathbf{A}\mathbf{\Omega}_0$ differs from the diagonal matrix $c_0\mathbf{I}$.

7.3.3 Differential Sensitivity $\partial\mathbf{\Phi}/\partial\mathbf{K}$

Following *Brewer, J.W., 1978a*, the sensitivity with respect to the (m,r)-matrix $\mathbf{K}$ is

$$\frac{\partial\mathbf{\Phi}}{\partial\mathbf{K}} = \sum_{i=0}^{n-1}\sum_{\mu=0}^{n-1}(\mathbf{I}_m \otimes \mathbf{A}^i)\frac{\partial\mathbf{A}}{\partial\mathbf{K}}(\mathbf{I}_r \otimes \mathbf{A}^\mu)q_{i\mu}(t) \tag{7.31}$$

$$\text{where} \quad q_{i\mu}(t) = \mathcal{L}^{-1}\{\alpha_i(s)\ \alpha_\mu(s)\} \quad \text{and} \quad \alpha_i(s) = \sum_{j=0}^{n-i-1} \frac{c_{i+j+1}\, s^j}{c(s)}\,. \tag{7.32}$$

7.3.4 Differential Sensitivity $\partial\Phi^k/\partial K$

The differential sensitivity of the kth power of Φ is required with discrete systems. The sensitivity is calculated by making repeated use of the product rule for differentiation with the (m,r)-matrix $\mathbf{K}$

$$\frac{\partial\Phi^k}{\partial\mathbf{K}} = \frac{\partial\ \Phi\Phi\ldots\Phi}{\partial\mathbf{K}} = \sum_{i=1}^{k}(\mathbf{I}_m\otimes\Phi^{k-i})\frac{\partial\Phi}{\partial\mathbf{K}}(\mathbf{I}_r\otimes\Phi^{i-1})\,. \tag{7.33}$$

7.4 State Variable Differential Sensitivity of Continuous-Time Systems

Assume a linear time-invariant process with the solution

$$\dot{\mathbf{x}}(t) = \mathbf{A}\mathbf{x}(t)+\mathbf{B}\mathbf{u}(t) \qquad \mathbf{x}(0)=\mathbf{x}_o \tag{7.34}$$

$$\mathbf{x}(t) = \Phi(t)\mathbf{x}_o + \int_0^t \Phi(t-\tau)\mathbf{B}\mathbf{u}(\tau)d\tau\,. \tag{7.35}$$

Using the product rule for differentiation with respect to the (m,r)-matrix $\mathbf{K}$ from Eq.(5.87) the trajectory sensitivity vector is

$$\begin{aligned}\frac{\partial\mathbf{x}(t)}{\partial\mathbf{K}} &= \frac{\partial\Phi}{\partial\mathbf{K}}(\mathbf{I}_r\otimes\mathbf{x}_0)+(\mathbf{I}_m\otimes\Phi)\frac{\partial\mathbf{x}_o}{\partial\mathbf{K}}\\ &+\int_0^t\Big(\frac{\partial\Phi(t-\tau)}{\partial\mathbf{K}}[\mathbf{I}_r\otimes\{\mathbf{B}\mathbf{u}(\tau)\}]\ +\ [\mathbf{I}_m\otimes\{\Phi(t-\tau)\}]\frac{\partial[\mathbf{B}\mathbf{u}(\tau)]}{\partial\mathbf{K}}\Big)d\tau\,.\end{aligned} \tag{7.36}$$

The state variable sensitivity can be calculated by substituting the state transition matrix sensitivity from Eqs.(7.21) or (7.31).

For a single-input system the matrix $\mathbf{A}$ in companion form or phase variable form is

$$\mathbf{A} = \begin{pmatrix} 0 & 1 & 0 & \ldots & 0\\ 0 & 0 & 1 & \ldots & 0\\ \vdots & \vdots & \vdots & \ddots & 0\\ -p_1 & -p_2 & -p_3 & \ldots & -p_n\end{pmatrix} \qquad \text{and} \qquad \mathbf{B} := \mathbf{b} = \mathbf{e}_n\,. \tag{7.37}$$

The parameter vector $\mathbf{p}$ corresponds to the coefficients c_i of the characteristic equation Eq.(B.1) as follows

$$\mathbf{p}^T = (p_1\ \ p_2\ldots p_n) = (c_o\ \ c_1\ldots c_{n-1})\,. \tag{7.38}$$

Referring to Eq.(9.44), the parameter sensitivity is given by

$$\frac{\partial\mathbf{A}}{\partial\mathbf{p}^T} = (\mathbf{E}_{n1}\ \vdots\ \mathbf{E}_{n2}\ldots\mathbf{E}_{nn}) = (\mathrm{col}\ {}^T\mathbf{I}_n)\otimes\mathbf{e}_n\,. \tag{7.39}$$

Assume that there is no eigenvalue multiplicity: $m_i = m_{mi} = 1$, $\mathbf{Z}_{ij} = \mathbf{a}_i\mathbf{a}_j^{\triangleleft T}$. *Brewer, J.W., 1978a* showed that the Jacobian matrix is

$$\frac{\partial\mathbf{x}(t)}{\partial\mathbf{p}^T} = \sum_j\sum_l(\mathbf{a}_j^{\triangleleft})_n\ (\mathbf{a}_l^{\triangleleft})_n\ [\mathbf{a}_j\ \vdots\ \lambda_l\mathbf{a}_j\ldots\lambda_l^{n-1}a_j]\int_0^t f_{jl}(t-\tau)u(\tau)d\tau \tag{7.40}$$

$$\text{where} \qquad f_{jl} = \mathcal{L}^{-1}\frac{1}{(s+\lambda_j)(s+\lambda_l)}\,. \tag{7.41}$$

The simplification is also true

$$\Big(\frac{\partial\mathbf{x}}{\partial\mathbf{p}^T}\Big)_{ik} = \Big(\frac{\partial\mathbf{x}}{\partial\mathbf{p}^T}\Big)_{i-1,k+1} \qquad \forall i=1\ldots n,\ k=1\ldots n-1\,. \tag{7.42}$$

7.5 Differential Sensitivity of Discrete-Time Systems

Consider a discrete-time linear system with sampling interval T, $\mathbf{x}(0) = \mathbf{x}_o$, $\boldsymbol{\Phi} = \boldsymbol{\Phi}(T)$

$$\mathbf{x}(k+1) = \boldsymbol{\Phi}\mathbf{x}(k) + \boldsymbol{\Psi}\mathbf{u}(k) \qquad \mathbf{x}(k) = \boldsymbol{\Phi}^k\mathbf{x}_o + \sum_{\nu=1}^{k} \boldsymbol{\Phi}^{k-\nu}\boldsymbol{\Psi}\mathbf{u}(\nu-1) . \tag{7.43}$$

The derivative with respect to the (m,r)-matrix $\mathbf{K}$ is given by Eq.(7.33) and

$$\begin{aligned}\frac{\partial \mathbf{x}(k)}{\partial \mathbf{K}} = \frac{\partial \boldsymbol{\Phi}^k}{\partial \mathbf{K}}(\mathbf{I}_r \otimes \mathbf{x}_0) \;&+\; (\mathbf{I}_m \otimes \boldsymbol{\Phi}^k)\frac{\partial \mathbf{x}_o}{\partial \mathbf{K}} + \sum_{\nu=1}^{k} \frac{\partial \boldsymbol{\Phi}^{k-\nu}}{\partial \mathbf{K}}[\mathbf{I}_r \otimes \{\mathbf{F}\mathbf{u}(\nu-1)\}] \\ &+\; (\mathbf{I}_m \otimes \boldsymbol{\Phi}^{k-\nu})\frac{\partial[\mathbf{F}\mathbf{u}(\nu-1]}{\partial \mathbf{K}} .\end{aligned} \tag{7.44}$$

7.5.1 Eigenvalues of kth Power $\mathbf{A}^k$

The eigenvalues $\lambda[\mathbf{A}^k]$ are given by

$$\det(s\mathbf{I} - \mathbf{A}^k) \triangleq \det(s_1^k\mathbf{I} - \mathbf{A}^k) \tag{7.45}$$

$$= [\det(s_1\mathbf{I} - \mathbf{A})]\det(s_1^{k-1}\mathbf{I} + s_1^{k-2}\mathbf{A} + s_1^{k-3}\mathbf{A}^2 + \ldots \mathbf{A}^{k-1}) = 0 . \tag{7.46}$$

With regard to the factor $\det(s_1\mathbf{I} - \mathbf{A})$, in the expression above $s_1 = \lambda[\mathbf{A}]$ is a solution. Thus $s_1^k = \lambda^k$ is an eigenvalue of $\mathbf{A}^k$ where s_1 is an eigenvalue of $\mathbf{A}$: $\lambda[\mathbf{A}^k] = \lambda^k[\mathbf{A}]$. Applying spectral decomposition

$$\mathbf{A}^k = (\sum_i \mathbf{a}_i\mathbf{a}_i^{\triangleleft T}\lambda_i[\mathbf{A}])(\sum_i \mathbf{a}_i\mathbf{a}_i^{\triangleleft T}\lambda_i[\mathbf{A}])\ldots(\sum_i \mathbf{a}_i\mathbf{a}_i^{\triangleleft T}\lambda_i[\mathbf{A}])\ldots(k \text{ times}) , \tag{7.47}$$

recognize that mainly mixed products $\mathbf{a}_i^{\triangleleft T}\mathbf{a}_j$ $\forall i \neq j$ appear. All of them are zero. Hence, $\mathbf{A}^k = \sum_i \mathbf{a}_i\mathbf{a}_i^{\triangleleft T}\lambda_i^k[\mathbf{A}]$.

7.5.2 Eigenvectors and Eigenvalues of $\boldsymbol{\Phi}^k$

Assume that $\boldsymbol{\Phi}(T)$ has only distinct eigenvalues. The right eigenvectors are denoted φ_i. If the right eigenvectors of $\boldsymbol{\Phi}^T(T)$ are $\varphi_i^{\triangleleft}$ and normalized

$$\boldsymbol{\Phi}^k(T) = \sum_{i=1} \varphi_i\varphi_i^{\triangleleft T}\lambda^k[\boldsymbol{\Phi}], \quad \boldsymbol{\Phi}(T) = e^{\mathbf{A}T}, \quad \lambda[\boldsymbol{\Phi}] = e^{\lambda_i[\mathbf{A}]T}, \quad \lambda[\boldsymbol{\Phi}^k] = \lambda^k[\boldsymbol{\Phi}] . \tag{7.48}$$

7.5.3 Transition Matrix Sensitivity

From Eqs.(7.33) and (7.48)

$$\frac{\partial \boldsymbol{\Phi}^k}{\partial \mathbf{K}} = \sum_{j=1}^{k}\Big(\mathbf{I}_m \otimes \sum_{l=1}^{n} \varphi_l\varphi_l^{\triangleleft T}\lambda_l^{k-j}\Big)\frac{\partial \boldsymbol{\Phi}}{\partial \mathbf{K}}\Big(\mathbf{I}_r \otimes \sum_{i=1}^{n} \varphi_i\varphi_i^{\triangleleft T}\lambda_i^{j-1}\Big) \tag{7.49}$$

$$= \Big(\mathbf{I}_m \otimes \sum_{l=1}^{n} \varphi_l\varphi_l^{\triangleleft T}\Big)\frac{\partial \boldsymbol{\Phi}}{\partial \mathbf{K}}\Big(\mathbf{I}_r \otimes \sum_{i=1}^{n} \varphi_i\varphi_i^{\triangleleft T}\Big)\sum_{j=1}^{k}\lambda_l^{k-j}\lambda_i^{j-1} . \tag{7.50}$$

Applying the geometric progression rule $\sum_{j=1}^{k} \lambda^{j-1} = (\lambda^k - 1)/(\lambda - 1)$ to the aforementioned expression,

$$\begin{aligned}
\lambda_l \neq \lambda_i : \ d_{li}(k) &= \lambda_l^{k-1} \sum_{j=1}^{k} \lambda_l^{1-j} \lambda_i^{j-1} = \lambda_l^{k-1} \sum_{j=1}^{k} \left(\frac{\lambda_i}{\lambda_l}\right)^{j-1} = \lambda_l^{k-1} \frac{\left(\frac{\lambda_i}{\lambda_l}\right)^k - 1}{\left(\frac{\lambda_i}{\lambda_l}\right) - 1} = \frac{\lambda_i^k - \lambda_l^k}{\lambda_i - \lambda_l} \\
\lambda_l = \lambda_i : \ d_{li}(k) &= \lambda_i^{k-1} \sum_{j=1}^{k} \lambda_i^{1-j} \lambda_i^{j-1} = \lambda_i^{k-1} \sum_{j=1}^{k} 1 = k \lambda_i^{k-1} \qquad (7.51)
\end{aligned}$$

and
$$\frac{\partial \mathbf{\Phi}^k}{\partial \mathbf{K}} = \sum_{l=1}^{n} \sum_{i=1}^{n} (\mathbf{I}_m \otimes \varphi_l \varphi_l^{dT}) \frac{\partial \mathbf{\Phi}}{\partial \mathbf{K}} (\mathbf{I}_r \otimes \varphi_i \varphi_i^{dT}) d_{li}(k) \, . \qquad (7.52)$$

Chapter 8

Characteristic Polynomial Differential Sensitivity

8.1 Differential Sensitivity and Transfer Zeros

Consider the dynamic system with output feedback controller as depicted in Fig. 8.1

$$\begin{aligned} \dot{\mathbf{x}}(t) &= \mathbf{A}\mathbf{x}(t) + \mathbf{B}\mathbf{u}(t) & \mathbf{x} \in \mathcal{R}^n \quad (8.1)\\ \mathbf{y}(t) &= \mathbf{C}\mathbf{x}(t) & \mathbf{y} \in \mathcal{R}^r \quad (8.2)\\ \mathbf{u}(t) &= \mathbf{K}\mathbf{y}(t) + \mathbf{y}_{ref}(t) & \mathbf{u}, \mathbf{y}_{ref} \in \mathcal{R}^m \,. \quad (8.3) \end{aligned}$$

Applying the Laplace transform yields the system transfer matrix $\mathbf{G}(s)$ and closed-loop transfer matrix $\mathbf{F}(s)$

$$\mathbf{y}(s) = \mathbf{C}(s\mathbf{I}_n - \mathbf{A})^{-1}\mathbf{B}\mathbf{u}(s) \triangleq \mathbf{G}(s)\mathbf{u}(s) \quad (8.4)$$

$$\mathbf{y}(s) = \mathbf{C}(s\mathbf{I}_n - \mathbf{A} - \mathbf{B}\mathbf{K}\mathbf{C})^{-1}\mathbf{B}\mathbf{y}_{ref}(s) \triangleq \mathbf{F}(s)\mathbf{y}_{ref}(s) \,, \quad (8.5)$$

see Fig. 8.1. Taking the numerator of each entry of $\mathbf{F}(s)$ and establishing a new matrix $\mathbf{F}_N(s)$ with these numerators, irrespective of several alternative methods defining zeros (*Patel, R.V., 1975* and *Sinswat, V., 1976*)

$$\mathbf{F}(s) \triangleq \frac{\mathbf{F}_N(s)}{\det\,(s\mathbf{I}_n - \mathbf{A} - \mathbf{B}\mathbf{K}\mathbf{C})} = \mathbf{C}(s\mathbf{I}_n - \mathbf{A} - \mathbf{B}\mathbf{K}\mathbf{C})^{-1}\mathbf{B} \quad (8.6)$$

$$\mathbf{F}_N(s) = \mathbf{C}[\text{adj}\,(s\mathbf{I}_n - \mathbf{A} - \mathbf{B}\mathbf{K}\mathbf{C})]\mathbf{B} \,. \quad (8.7)$$

8.1.1 Differential Sensitivity

The differential sensitivity of the closed-loop characteristic polynomial with respect to the controller matrix $\mathbf{K}$ is given by $\frac{\partial}{\partial \mathbf{K}} \det\,(s\mathbf{I}_n - \mathbf{A} - \mathbf{B}\mathbf{K}\mathbf{C})$. Applying Eq.(5.1), the partial derivative of the characteristic polynomial with respect to the (i,j)-entry K_{ij} of the matrix $\mathbf{K}$ is considered first

$$\frac{\partial}{\partial K_{ij}}\det\,(s\mathbf{I}_n - \mathbf{A} - \mathbf{B}\mathbf{K}\mathbf{C}) = \text{tr}\,[\frac{\partial(s\mathbf{I}_n - \mathbf{A} - \mathbf{B}\mathbf{K}\mathbf{C})}{\partial K_{ij}}\,\text{adj}\,(s\mathbf{I}_n - \mathbf{A} - \mathbf{B}\mathbf{K}\mathbf{C})] \,. \quad (8.8)$$

With the help of Eq.(5.73)

$$\text{tr}[\mathbf{B}\frac{\partial \mathbf{K}}{\partial K_{ij}}\mathbf{C}\,\text{adj}(s\mathbf{I}_n - \mathbf{A} - \mathbf{B}\mathbf{K}\mathbf{C})] = \text{tr}\,[\mathbf{B}\mathbf{E}_{ij}\mathbf{C}\,\text{adj}(s\mathbf{I}_n - \mathbf{A} - \mathbf{B}\mathbf{K}\mathbf{C})] \quad (8.9)$$

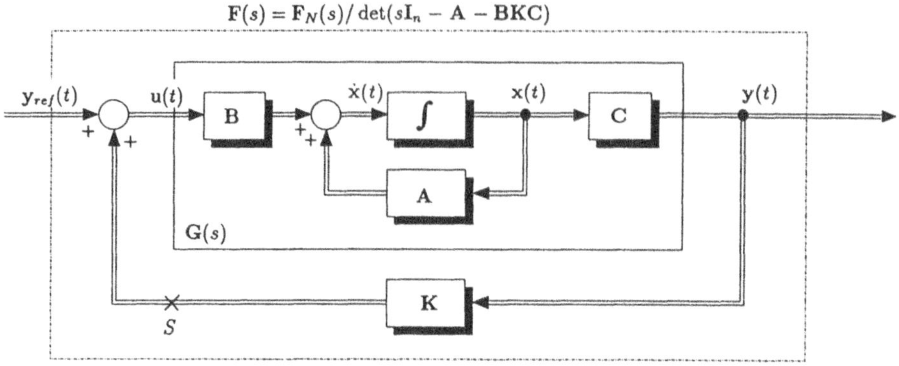

Figure 8.1: Multivariable control with output feedback

$$= \mathrm{tr}[\mathrm{adj}^T(s\mathbf{I}_n - \mathbf{A} - \mathbf{BKC})\mathbf{C}^T\mathbf{E}_{ji}\mathbf{B}^T] = \mathrm{tr}\ [\mathbf{B}^T\ \mathrm{adj}^T(s\mathbf{I}_n - \mathbf{A} - \mathbf{BKC})\mathbf{C}^T\mathbf{E}_{ji}] \quad (8.10)$$

$$= [\mathbf{B}^T\ \mathrm{adj}^T(s\mathbf{I}_n - \mathbf{A} - \mathbf{BKC})\mathbf{C}^T]_{ij}\ . \quad (8.11)$$

The last term was achieved using Eq.(5.59). With the definition of Eq.(5.79) from the derivative with respect to K_{ij} , the derivative with respect to the matrix $\mathbf{K}$ is obtained

$$\frac{\partial}{\partial \mathbf{K}}\ \det\ (s\mathbf{I}_n - \mathbf{A} - \mathbf{BKC}) = -\mathbf{B}^T[\mathrm{adj}^T(s\mathbf{I}_n - \mathbf{A} - \mathbf{BKC})]\mathbf{C}^T\ . \quad (8.12)$$

8.1.2 Differential Sensitivity and Transfer Zeros

Comparing Eqs.(8.7) and (8.12) yields

$$\mathbf{F}_N^T = -\frac{\partial}{\partial \mathbf{K}}\ \det\ (s\mathbf{I}_n - \mathbf{A} - \mathbf{BKC})\ . \quad (8.13)$$

The transpose of the numerator matrix of the closed-loop transfer matrix equals the sensitivity function of the characteristic polynomial with respect to the controller matrix $\mathbf{K}$. The numerator of the scalar transfer function from the jth position of the input $\mathbf{y}_{ref}$ to the ith position of the output $\mathbf{y}$ is given by (*Konigorski, U., 1988*)

$$F_{Nij}(s) = -\frac{\partial}{\partial K_{ji}}\ \det\ (s\mathbf{I}_n - \mathbf{A} - \mathbf{BKC})\ . \quad (8.14)$$

The transfer function is $F_{Nij}(s)/\det\ (s\mathbf{I}_n - \mathbf{A} - \mathbf{BKC})$.

8.1.3 Interdependence between Transfer Matrices

Transfer matrices mentioned above are repeated for the sake of completeness:

Open-loop return-difference matrix with respect to loop break S (see Fig. 8.1):

$$\mathbf{I}_m - \mathbf{KG} \equiv \mathbf{I}_m - \mathbf{KC}(s\mathbf{I}_n - \mathbf{A})^{-1}\mathbf{B}\ . \quad (8.15)$$

Closed-loop behaviour:

$$\mathbf{y} = \mathbf{G}[\mathbf{I}_m - \mathbf{K}\mathbf{G}]^{-1}\mathbf{y}_{ref} = \mathbf{F}\mathbf{y}_{ref} \qquad \mathbf{G}, \mathbf{F} \in \mathcal{R}^{r \times m} \tag{8.16}$$

The equivalence in Eq.(8.16) can be proved easily by using matrix inversion lemma Eq.(C.13)

$$\mathbf{F} = \mathbf{C}(s\mathbf{I}_n - \mathbf{A} - \mathbf{B}\mathbf{K}\mathbf{C})^{-1}\mathbf{B} = \mathbf{G}[\mathbf{I}_m - \mathbf{K}\mathbf{G}]^{-1} \,. \tag{8.17}$$

8.2 Zero Placement with a Second Internal Input

Eigenvalue assignment is a well-known feature when state feedback is applied to a single-input single-output system. Zeros of the entries of the transfer matrix accompanying the pole placement problem are often neglected although the zeros influence the dynamic behaviour of the closed loop. Taking the inverse Laplace transform of a variable in s-domain, the coefficients of the exponential functions exp λt, known as the residues, depend as well on the poles as on the zeros of the transfer function. Moreover, the initial conditions of each t-domain signal cannot be chosen arbitrarily in the presence of zeros, see Eq.(A.80).

8.2.1 Zero Placement

If zeros can be placed in the s-plane arbitrarily without being coupled with the placement of poles or if they are unaffected by feedback design they are termed ***invariant zeros*** (*Sinswat, V., et al. 1976*). In addition to this problem, the question arises whether it is possible to satisfy the invariant-zero placement without exact knowledge of the plant and independent of the pole placement design operations.

Assume a given system with state feedback and two inputs u_1 and u_2 and a single output y as depicted in Fig. 8.2. Let the system matrix $\mathbf{A}$ and the input vector $\mathbf{b}_1$ be given in the phase variable (companion) form

$$\dot{\mathbf{x}}(t) = \mathbf{A}\mathbf{x}(t) + \mathbf{B}\mathbf{u}(t) = \mathbf{A}\mathbf{x} + (\mathbf{b}_1 \vdots \mathbf{b}_2)\begin{pmatrix} u_1 \\ u_2 \end{pmatrix} \tag{8.18}$$

$$y(t) = \mathbf{c}^T\mathbf{x}(t) \qquad \mathbf{x} \in \mathcal{R}^n, \quad \mathbf{B} \in \mathcal{R}^{n \times 2} \tag{8.19}$$

$$\mathbf{A} = \begin{pmatrix} 0 & 1 & 0 & \dots & 0 \\ 0 & 0 & 1 & \dots & 0 \\ \vdots & \vdots & & \ddots & \\ 0 & 0 & 0 & & 1 \\ -a_o & -a_1 & \dots & & -a_{n-1} \end{pmatrix}, \; \mathbf{b}_1 = \begin{pmatrix} 0 \\ 0 \\ \vdots \\ 0 \\ 1 \end{pmatrix}, \; \mathbf{b}_2 = \begin{pmatrix} b_{12} \\ b_{22} \\ \vdots \\ b_{n-1,2} \\ b_{n2} \end{pmatrix}, \; \mathbf{c} = \begin{pmatrix} c_1 \\ c_2 \\ \vdots \\ c_{n-1} \\ c_n \end{pmatrix}. \tag{8.20}$$

If a state feedback is applied in such a way that the first input u_1 is related to a reference $y_{ref}(t)$ and the second input u_2 is only forced by an internal feedback $\mathbf{k}_2$, the following features are obtained:

- The invariant-zero placement is achieved by choosing $\mathbf{k}_2$ (in companion with c_i) being influenced neither by $\mathbf{k}_1$ nor by a_i where $\mathbf{k}_1, \mathbf{k}_2 \in \mathcal{R}^n$.

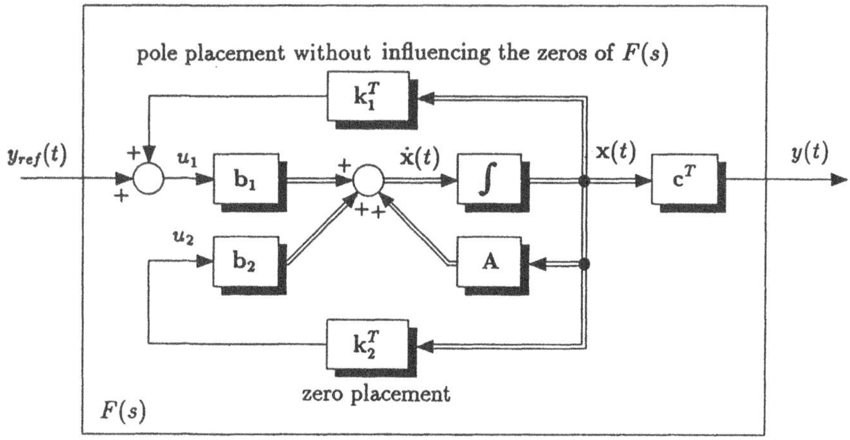

Figure 8.2: Control structure with invariant-zero placement

- The pole placement is obtained by selecting appropriate $\mathbf{k}_1$ (in companion with a_i and $\mathbf{k}_2$). If the zero placement should be redesigned by $\mathbf{k}_2$, this must be compensated for by altering $\mathbf{k}_1$.

To assign n poles and n zeros it is not sufficient to have only n entries of a single state variable feedback at one's disposal. Another input variable must be utilized to obtain $2n$ degrees of freedom. Combining Eqs.(8.18) to (8.20), the scalar overall transfer function of the resulting SISO-system of Fig. 8.2 is

$$F(s) = \mathbf{c}^T(s\mathbf{I}_n - \mathbf{A} - \mathbf{b}_1\mathbf{k}_1^T - \mathbf{b}_2\mathbf{k}_2^T)^{-1}\mathbf{b}_1 = \frac{F_N(s)}{F_D(s)} . \tag{8.21}$$

It is well known from the theory of a classical SISO-system $(\mathbf{A}, \mathbf{b}_1)$ in phase variable form of Eq. (8.20) that the zeros of the corresponding transfer function are only given by c_i and are not influenced by a_i

$$\dot{\mathbf{x}} = \mathbf{A}\mathbf{x} + \mathbf{b}_1 u \qquad y = \mathbf{c}^T\mathbf{x} \qquad \leadsto \qquad \frac{y(s)}{u(s)} = \frac{\sum_{i=1}^{n} c_i s^{i-1}}{\sum_{i=0}^{n-1} a_i s^i + s^n} . \tag{8.22}$$

With regard to the fact that the system with feedback $u = \mathbf{k}_1^T\mathbf{x}$ keeps the phase variable structure, the transfer function remains unchanged if $\mathbf{A} := \mathbf{A} + \mathbf{b}_1\mathbf{k}_1^T$ or if the parameters obey $a_i := a_i + (\mathbf{k}_1)_i$. Hence, $\mathbf{k}_1$ is of no influence on the zeros. Since the zero placement is not influenced by a_i and $\mathbf{k}_1$, the zero placement can be declared robust with respect to a_i and $\mathbf{k}_1$, i.e., robust with respect to the uncertain model (a_i) and the pole placement parameter vector $\mathbf{k}_1$. These special features, summarized in Fig. 8.3, first were presented by *Sadeghi, T., et al. 1983*. *Schmidt, J., 1988* proved the robustness of the zero placement with respect to a_i and $\mathbf{k}_1$ by means of the Rosenbrock matrix (*Rosenbrock, H.H., 1970*).

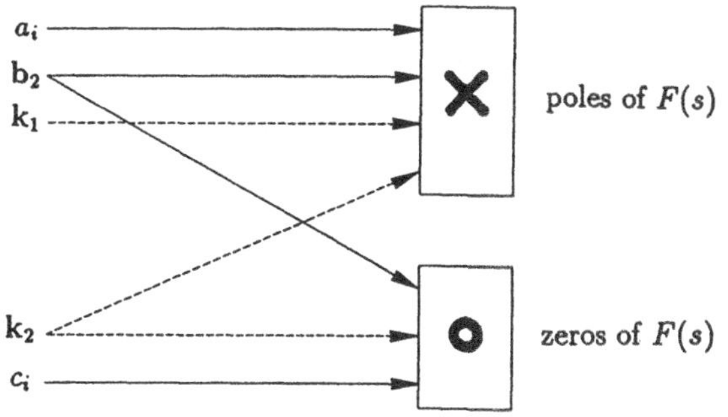

Figure 8.3: Influence of system parameters and of controller parameters on poles/zeros of $F(s)$ depicted by full and dotted lines, respectively

8.2.2 Pole Placement

The poles are given by the scalar denominator function

$$F_D(s) = \det\left(s\mathbf{I}_n - \mathbf{A} - \mathbf{b}_1\mathbf{k}_1^T - \mathbf{b}_2\mathbf{k}_2^T\right) \triangleq \det\left(s\mathbf{I}_n - \mathbf{A}' - \mathbf{b}_2\mathbf{k}_2^T\right) = 0 \ . \tag{8.23}$$

With respect to the phase variable structure of $\mathbf{A}$ and $\mathbf{b}_1$, the coefficients a_i' of the matrix $\mathbf{A}'$ are given by

$$a_i' = a_i - (\mathbf{k}_1)_{i+1} \quad \forall\, i = 0, 1 \ldots n-1 \ . \tag{8.24}$$

After having chosen $\mathbf{k}_2$ to place the zeros, then, $\mathbf{k}_1$ is utilized to assign the poles. By varying $\mathbf{k}_1$ only the poles are determined without backlash on the zeros.

8.2.3 Conclusion

Since the differential sensitivity of the closed-loop characteristic polynomial with respect to the controller matrix is given by the numerator matrix of the closed-loop transfer matrix, invariant zeros of the transfer matrix correspond to an invariant differential sensitivity function. In the special case of a single-input single-output system with an additional internal input, the poles and zeros can be placed in an independent way. Hence, robust sensitivity of the characteristic polynomial with respect to the controller matrix $\mathbf{K}$ can be achieved although the poles may be assigned arbitrarily or plant uncertainties may occur. This property holds irrespective of the fact that in Eq.(8.13) the derivative has to be taken at the objective value $\mathbf{K}$ which *does not* remain constant.

8.3 Transfer Function Differential Sensitivity

Now, the transfer function differential sensitivity with respect to the open-loop characteristic polynomial is studied. Consider a single-input single-output plant

$$\dot{\mathbf{x}}(t) = \mathbf{A}\mathbf{x}(t) + \mathbf{b}u \qquad y(t) = \mathbf{c}^T\mathbf{x}(t) \qquad \mathbf{A} \in \mathcal{R}^{n\times n} \tag{8.25}$$

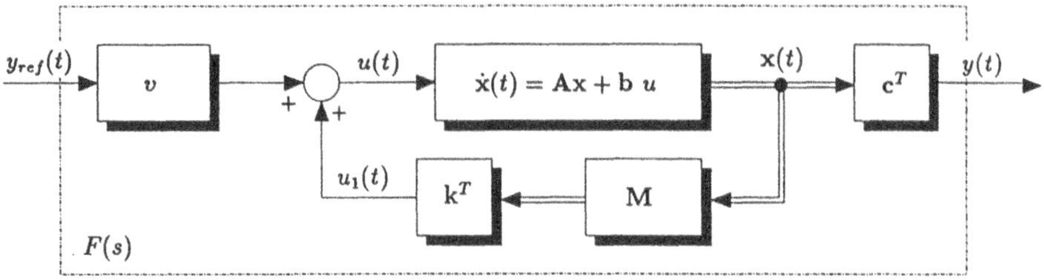

Figure 8.4: Single-input single-output system using a measurement variable

and a feedback controller $\mathbf{k}$ using a measurement variable $\mathbf{M}\mathbf{x}$ (see Fig. 8.4)

$$u(t) = v\; y_{ref}(t) + \mathbf{k}^T\mathbf{M}\mathbf{x} \qquad \mathbf{M} \in \mathcal{R}^{r\times n} \,. \tag{8.26}$$

In order to obtain $y(t)|_{t\to\infty} = y_{ref}(\infty)$

$$v = \frac{-1}{\mathbf{c}^T(\mathbf{b}\mathbf{k}^T\mathbf{M} + \mathbf{A})^{-1}\mathbf{b}} \equiv \frac{\det(\mathbf{b}\mathbf{k}^T\mathbf{M} + \mathbf{A})}{\mathbf{c}^T\mathbf{A}^{-1}\mathbf{b}\ \det\mathbf{A}} \tag{8.27}$$

must be satisfied. The polynomials

$$a(s) = \det(s\mathbf{I} - \mathbf{A}) \qquad \text{and} \qquad f(s) = \det(s\mathbf{I} - \mathbf{A} - \mathbf{b}\mathbf{k}^T\mathbf{M}) \tag{8.28}$$

are the open-loop and closed-loop characteristic polynomials, respectively.

$$f(s) = \det\{(s\mathbf{I} - \mathbf{A})[\mathbf{I} - (s\mathbf{I} - \mathbf{A})^{-1}\mathbf{b}\mathbf{k}^T\mathbf{M}]\} = a(s)\det[\mathbf{I} - (s\mathbf{I} - \mathbf{A})^{-1}\mathbf{b}\mathbf{k}^T\mathbf{M}] = f(s)\,. \tag{8.29}$$

Since $\det(\mathbf{I}+\mathbf{Y}\mathbf{Z}) = \det(\mathbf{I}+\mathbf{Z}\mathbf{Y})$ with $\mathbf{Y} := (s\mathbf{I}-\mathbf{A})^{-1}\mathbf{b}$ and $\mathbf{Z} := -\mathbf{k}^T\mathbf{M}$, from Eq.(8.29) it follows

$$f(s) = a(s)\det[\mathbf{I} - \mathbf{k}^T\mathbf{M}(s\mathbf{I} - \mathbf{A})^{-1}\mathbf{b}] = a(s)[1 - \mathbf{k}^T\mathbf{M}(s\mathbf{I} - \mathbf{A})^{-1}\mathbf{b}] \tag{8.30}$$

$$f(s) = \det(s\mathbf{I} - \mathbf{A}) - \mathbf{k}^T\mathbf{M}(s\mathbf{I} - \mathbf{A})^{-1}\det(s\mathbf{I} - \mathbf{A})\mathbf{b} = a(s) - \mathbf{k}^T\mathbf{M}\ \mathrm{adj}(s\mathbf{I} - \mathbf{A})\mathbf{b}\,. \tag{8.31}$$

The closed-loop system is characterized by the transfer functions (*Litz, L., and Preuss, H.P., 1977*)

$$u(s) = \frac{v}{1 - \mathbf{k}^T\mathbf{M}(s\mathbf{I} - \mathbf{A})^{-1}\mathbf{b}}y_{ref}(s) = \frac{v\;a(s)}{f(s)}y_{ref}(s) \tag{8.32}$$

$$y(s) = \mathbf{c}^T(s\mathbf{I} - \mathbf{A})^{-1}\mathbf{b}u(s) = \frac{\mathbf{c}^T(s\mathbf{I} - \mathbf{A})^{-1}\mathbf{b}va(s)}{f(s)}y_{ref}(s) \triangleq F(s)y_{ref}(s)\,. \tag{8.33}$$

It is well known that a system with open-loop transfer function $n(s)/a(s)$ and closed-loop transfer function $F(s) = n(s)/[n(s)+a(s)]$ is characterized by the sensitivity function S_a^F defined by *Bode, H.W., 1945; Horowitz, I., 1963*

$$\frac{\partial F(s)}{\partial a(s)} = \frac{\partial \frac{n}{n+a}}{\partial a} = \frac{-n}{(n+a)^2} \tag{8.34}$$

$$S_a^F \triangleq \frac{\partial F(s)}{\partial a(s)} \frac{\frac{1}{F(s)}}{\frac{1}{a(s)}} = \frac{-n}{(n+a)^2} \frac{\frac{n+a}{n}}{\frac{1}{a}} = \frac{-a}{n+a} = \frac{-a(s)}{f(s)} \tag{8.35}$$

since the polynomial $n(s) + a(s)$ corresponds with $f(s)$.

Considering separate plant and feedback controller transfer functions $G(s)$ and $H(s)$, respectively,

$$\frac{\partial F(s)}{\partial G(s)} = \frac{\partial \frac{G}{1+GH}}{\partial G} = \frac{+1}{(1+GH)^2} \tag{8.36}$$

$$S_G^F \triangleq \frac{\partial F(s)}{\partial G(s)} \frac{\frac{1}{F(s)}}{\frac{1}{G(s)}} = \frac{\partial \frac{G}{1+GH}}{\partial G} \frac{\frac{1+GH}{G}}{\frac{1}{G}} = \frac{1}{1+GH} \tag{8.37}$$

this expression corresponds to the sensitivity or to the dynamic control factor and bridges differential sensitivity and ordinary sensitivity of control systems. □

Apart from eigenvalue assignment the sensitivity function S_a^F can be used to design the control system. Remember the very similar relations, see Eq.(8.32),

$$\frac{u(s)}{y_{ref}(s)} = \frac{v\, a(s)}{f(s)} \quad \text{and} \quad S_a^F \triangleq \frac{-a(s)}{f(s)} . \tag{8.38}$$

In order to obtain

$$|S_a^F(j\omega)| < |S_a^F(\infty)| \qquad \forall\, \omega , \tag{8.39}$$

i.e., reduction of the differential sensitivity by employing the controller, the relation

$$|\frac{u}{y_{ref}}(j\omega)| < |\frac{u}{y_{ref}}(\infty)| \qquad \forall\, \omega \tag{8.40}$$

must hold. This requires that the absolute-value-integral theorem (see below) is *not* satisfied.

- Absolute-value-integral theorem of Bode (*Frank, P.M., 1976*): Assume, first, the transfer function $GH(s)$ of the open loop is stable and, second, the return ratio $1 + GH(s)$ does not contain zeros in the right-half s-plane and, third, the pole excess of the open-loop transfer function $GH(s)$ is at least two, then,

$$\int_0^\infty \ln |1 + GH(j\omega)|\, d\omega = 0 . \ \square \tag{8.41}$$

Since Eq.(8.30) $1 + GH(s) = 1 - \mathbf{k}^T\mathbf{M}(s\mathbf{I}_n - \mathbf{A})^{-1}\mathbf{b} = f(s)/a(s)$ and if absolute-value-integral theorem is satisfied one has

$$\int_0^\infty \ln |1 + GH(j\omega)|\, d\omega = \int_0^\infty \ln |\frac{f(j\omega)}{a(j\omega)}|\, d\omega = -\int_0^\infty \ln |S_a^F(j\omega)|\, d\omega = 0 \tag{8.42}$$

which is in opposition to Eq.(8.39).

The absolute-value-intergral theorem is *not* satisfied if the degree of the numerator is $n-1$ or $\mathbf{k}^T\mathbf{M}\mathbf{b} \neq 0$, resulting from the open-loop transfer function $u_1(s)/u(s)$ and from the expansion of $\text{adj}(s\mathbf{I} - \mathbf{A})$ into powers of s with coefficient matrices $\mathbf{H}_\nu$ $\forall \nu = 0 \ldots n-2$

$$-GH(s) \triangleq \frac{u_1(s)}{u(s)} = \mathbf{k}^T\mathbf{M}(s\mathbf{I} - \mathbf{A})^{-1}\mathbf{b} = \frac{\mathbf{k}^T\mathbf{M}\ \text{adj}(s\mathbf{I} - \mathbf{A})\mathbf{b}}{a(s)} \tag{8.43}$$

$$= \frac{\mathbf{k}^T\mathbf{M}(\mathbf{I}_n s^{n-1} + \mathbf{H}_{n-2}s^{n-2} + \ldots + \mathbf{H}_o)\mathbf{b}}{a(s)} . \tag{8.44}$$

Example:

$$\mathbf{A} = \begin{pmatrix} 0 & 1 \\ -3 & -4 \end{pmatrix}, \quad \mathbf{b} = \begin{pmatrix} 0 \\ 1 \end{pmatrix}, \quad \mathbf{M} = \mathbf{I}, \quad \mathbf{k}^T = (-22 \vdots -6), \quad \mathbf{F} = \begin{pmatrix} 0 & 1 \\ -25 & -10 \end{pmatrix} \tag{8.45}$$

$$a(s) = s^2 + 4s + 3 \qquad f(s) = s^2 + 10s + 25 \; . \tag{8.46}$$

The Nyquist plot of $S_a^F(j\omega) = -a(s)/f(s) \mid_{s=j\omega}$ is given by a curve quite similar to a semi-circle. Its modulus is obviously smaller than $a(\infty)/f(\infty)$. □

Chapter 9

Optimal Control and Performance Sensitivity

This chapter is devoted to the following topics:

- Riccati controller and Riccati matrix differential sensitivity with respect to changes in parameters, supposing that the system matrix and the input matrix depend on these parameters.
- Differential sensitivity of an index of performance with respect to parameters contained in the system and input matrix. When I should be minimized with respect to $\mathbf{p}$, the zeros of $\partial I / \partial \mathbf{p}$ are calculated. In order to apply Newton-Raphson algorithm the second derivative $\partial^2 I \, / \, \partial \mathbf{p}^T \partial \mathbf{p}$ is determined.
- Reducing performance sensitivity can be achieved via gradient methods, augmenting the index of performance by differential sensitivity expressions.
- Control actions softer than optimal control are obtained when a prespecified index of performance is approached via gradient methods.

Assume the time-invariant plant model

$$\dot{\mathbf{x}}(t) = \mathbf{A}(\mathbf{p})\mathbf{x}(t) + \mathbf{B}(\mathbf{p})\mathbf{u}(t), \qquad \mathbf{x}(0) = \mathbf{x}_o, \quad \mathbf{A} \in \mathcal{R}^{n \times n}, \quad \mathbf{B} \in \mathcal{R}^{n \times m} \tag{9.1}$$

where $\mathbf{A}(\mathbf{p})$ and $\mathbf{B}(\mathbf{p})$, respectively, are the (n, n)-system matrix and the (n, m)-control matrix depending on the n_p-vector $\mathbf{p}$ of the parameters. The index of performance is

$$I = \int_0^\infty [\mathbf{x}^T(t)\mathbf{Q}\mathbf{x}(t) + \mathbf{u}^T(t)\mathbf{R}\mathbf{u}(t)] \, dt \qquad \mathbf{Q} \in \mathcal{R}^{n \times n}, \quad \mathbf{R} \in \mathcal{R}^{m \times m} \, . \tag{9.2}$$

Minimizing I, the controller $\mathbf{K}$ is given by

$$\mathbf{u}(t) = \mathbf{K}(\mathbf{p})\mathbf{x}(t) = -\mathbf{R}^{-1}\mathbf{B}^T(\mathbf{p})\mathbf{P}(\mathbf{p})\mathbf{x}(t) \qquad \mathbf{K} \in \mathcal{R}^{m \times n} \tag{9.3}$$

where $\mathbf{P}(\mathbf{p})$ is the solution of the algebraic Riccati equation

$$\mathbf{A}^T(\mathbf{p})\mathbf{P}(\mathbf{p}) + \mathbf{P}(\mathbf{p})\mathbf{A}(\mathbf{p}) + \mathbf{Q} - \mathbf{P}(\mathbf{p})\mathbf{B}(\mathbf{p})\mathbf{R}^{-1}\mathbf{B}^T(\mathbf{p})\mathbf{P}(\mathbf{p}) = \mathbf{0} \, . \tag{9.4}$$

9.1 Riccati Controller Differential Sensitivity

Both $\mathbf{P}(p)$ and $\mathbf{K}(\mathbf{p})$ depend on the parameter $\mathbf{p}$. Using Taylor expansion and maintaining first-order truncation, the vicinity of $\mathbf{p}_o$ is described by relations to be derivated from Eqs.(9.3) and (9.4)

$$\mathbf{A}(\mathbf{p}_o + \delta\mathbf{p}) = \mathbf{A}(\mathbf{p}_o) + \delta\mathbf{A} = \mathbf{A}_o + \delta\mathbf{A}, \qquad \mathbf{B}(\mathbf{p}_o + \delta\mathbf{p}) = \mathbf{B}(\mathbf{p}_o) + \delta\mathbf{B} = \mathbf{B}_o + \delta\mathbf{B}, \tag{9.5}$$

$$\mathbf{K}(\mathbf{p}_o+\delta\mathbf{p})=\mathbf{K}(\mathbf{p}_o)+\delta\mathbf{K}=\mathbf{K}_o+\delta\mathbf{K}, \qquad \mathbf{P}(\mathbf{p}_o+\delta\mathbf{p})=\mathbf{P}(\mathbf{p}_o)+\delta\mathbf{P}=\mathbf{P}_o+\delta\mathbf{P}. \tag{9.6}$$

The objective is to find the interdependence between $\delta\mathbf{K}$ and $\delta\mathbf{A}$ and $\delta\mathbf{B}$ caused by variation $\delta\mathbf{p}$ (*Vetter, W.J., Balchen, J.G., and Pohner, F., 1972*). From Eq.(9.4) a relation $\delta\mathbf{P}$ of $\delta\mathbf{A}$ and $\delta\mathbf{B}$ can be found which has to be introduced in Eq.(9.3). Using Eq.(5.73), the Eqs.(9.3) and (9.4) are rewritten to the incremental differential forms

$$\delta\mathbf{K}=-\mathbf{R}^{-1}\delta\mathbf{B}^T\mathbf{P}_o-\mathbf{R}^{-1}\mathbf{B}_o\delta\mathbf{P} \tag{9.7}$$

$$\begin{aligned}&\delta\mathbf{A}^T\mathbf{P}_o+\mathbf{A}_o^T\delta\mathbf{P}+(\delta\mathbf{P})\mathbf{A}_o+\mathbf{P}_o\delta\mathbf{A}\\&-(\delta\mathbf{P})\mathbf{B}_o\mathbf{R}^{-1}\mathbf{B}_o^T\mathbf{P}_o-\mathbf{P}_o(\delta\mathbf{B})\mathbf{R}^{-1}\mathbf{B}_o^T\mathbf{P}_o-\mathbf{P}_o\mathbf{B}_o\mathbf{R}^{-1}\delta\mathbf{B}^T\mathbf{P}_o-\mathbf{P}_o\mathbf{B}_o\mathbf{R}^{-1}\mathbf{B}_o^T\delta\mathbf{P}=\mathbf{0}\,.\end{aligned} \tag{9.8}$$

Keeping in mind that small variations $\delta\mathbf{p}$ in the vicinity of $\mathbf{p}_o$ and $\mathbf{K}_o=-\mathbf{R}^{-1}\mathbf{B}_o^T\mathbf{P}_o$ are considered, only, the index $_o$ is omitted for convenience. With the abbreviation $\mathbf{K}=-\mathbf{R}^{-1}\mathbf{B}^T\mathbf{P}$ for the nominal value, from Eq.(9.8)

$$(\mathbf{A}^T+\mathbf{K}^T\mathbf{B}^T)\delta\mathbf{P}+(\delta\mathbf{P})(\mathbf{A}+\mathbf{B}\mathbf{K})=-\delta\mathbf{A}^T\mathbf{P}-\mathbf{P}\delta\mathbf{A}-\mathbf{P}(\delta\mathbf{B})\mathbf{K}-\mathbf{K}^T\delta\mathbf{B}^T\mathbf{P}\,. \tag{9.9}$$

Transforming this equation to the equivalent column representation as given in Eq.(4.40), the following relations can be achieved (with $\mathbf{P}=\mathbf{P}^T$)

$$[\mathbf{I}_n\otimes(\mathbf{A}^T+\mathbf{K}^T\mathbf{B}^T)+(\mathbf{A}^T+\mathbf{K}^T\mathbf{B}^T)\otimes\mathbf{I}_n]\ \mathrm{col}\ \delta\mathbf{P}$$

$$=-(\mathbf{P}\otimes\mathbf{I}_n)\ \mathrm{col}\ \delta\mathbf{A}^T-(\mathbf{I}_n\otimes\mathbf{P})\ \mathrm{col}\ \delta\mathbf{A}-(\mathbf{K}^T\otimes\mathbf{P})\ \mathrm{col}\ \delta\mathbf{B}-(\mathbf{P}\otimes\mathbf{K}^T)\ \mathrm{col}\ \delta\mathbf{B}^T \tag{9.10}$$

$$=-[(\mathbf{I}_n\otimes\mathbf{P})+(\mathbf{P}\otimes\mathbf{I}_n)\mathbf{U}_{nn}]\ \mathrm{col}\ \delta\mathbf{A}-[(\mathbf{K}^T\otimes\mathbf{P})+(\mathbf{P}\otimes\mathbf{K}^T)\mathbf{U}_{nm}]\ \mathrm{col}\ \delta\mathbf{B}\,. \tag{9.11}$$

In the preceding equation use is made of Eq.(4.48), i.e. $\mathrm{col}\ \mathbf{B}^T=\mathbf{U}_{nm}\ \mathrm{col}\ \mathbf{B}, \mathbf{B}\in\mathcal{R}^{n\times m}$. With the abbreviation

$$\mathbf{W}_g\triangleq[\mathbf{I}_n\otimes(\mathbf{A}^T+\mathbf{K}^T\mathbf{B}^T)+(\mathbf{A}^T+\mathbf{K}^T\mathbf{B}^T)\otimes\mathbf{I}_n]^{-1} \tag{9.12}$$

it results

$$\mathrm{col}\ \delta\mathbf{P}=-\mathbf{W}_g[\mathbf{I}_n\otimes\mathbf{P}+(\mathbf{P}\otimes\mathbf{I}_n)\mathbf{U}_{nn}]\ \mathrm{col}\ \delta\mathbf{A}-\mathbf{W}_g[\mathbf{K}^T\otimes\mathbf{P}+(\mathbf{P}\otimes\mathbf{K}^T)\mathbf{U}_{nm}]\ \mathrm{col}\ \delta\mathbf{B}\,. \tag{9.13}$$

Taking the column representation of Eq.(9.7)

$$\mathrm{col}\ \delta\mathbf{K}=-(\mathbf{P}\otimes\mathbf{R}^{-1})\ \mathrm{col}\ \delta\mathbf{B}^T-[\mathbf{I}_n\otimes(\mathbf{R}^{-1}\mathbf{B}^T)]\ \mathrm{col}\ \delta\mathbf{P} \tag{9.14}$$

and substituting col $\delta\mathbf{P}$ from Eq.(9.13) yields, see Fig. 9.1,

$$\begin{aligned}\mathrm{col}\ \delta\mathbf{K}&=[\mathbf{I}_n\otimes(\mathbf{R}^{-1}\mathbf{B}^T)]\mathbf{W}_g[\mathbf{I}_n\otimes\mathbf{P}+(\mathbf{P}\otimes\mathbf{I}_n)\mathbf{U}_{nm}]\ \mathrm{col}\ \delta\mathbf{A}\\&+\Big([\mathbf{I}_n\otimes(\mathbf{R}^{-1}\mathbf{B}^T)]\mathbf{W}_g[\mathbf{K}^T\otimes\mathbf{P}+(\mathbf{P}\otimes\mathbf{K}^T)\mathbf{U}_{nm}]-(\mathbf{P}\otimes\mathbf{R}^{-1})\mathbf{U}_{nm}\Big)\ \mathrm{col}\ \delta\mathbf{B}\,.\end{aligned} \tag{9.15}$$

Relating this equation in $\delta\mathbf{K}$ to the increment $\delta\mathbf{p}$, the corresponding derivative expression or sensitivity functions can be obtained

$$\begin{aligned}\frac{d\ \mathrm{col}\ \mathbf{K}}{d\mathbf{p}}&=[\mathbf{I}_n\otimes(\mathbf{R}^{-1}\mathbf{B}^T)]\mathbf{W}_g[\mathbf{I}_n\otimes\mathbf{P}+(\mathbf{P}\otimes\mathbf{I}_n)\mathbf{U}_{nm}]\frac{d\ \mathrm{col}\ \mathbf{A}}{d\ \mathbf{p}}\\&+\Big([\mathbf{I}_n\otimes(\mathbf{R}^{-1}\mathbf{B}^T)]\mathbf{W}_g[\mathbf{K}^T\otimes\mathbf{P}+(\mathbf{P}\otimes\mathbf{K}^T)\mathbf{U}_{nm}]-(\mathbf{P}\otimes\mathbf{R}^{-1})\mathbf{U}_{nm}\Big)\frac{d\ \mathrm{col}\ \mathbf{B}}{d\mathbf{p}}\,.\end{aligned} \tag{9.16}$$

The derivatives are only true in the vicinity of nominal parameter $\mathbf{p}_o$. The matrix-updating algorithm saves computation effort. Recomputations of the matrix Riccati equation can be avoided.

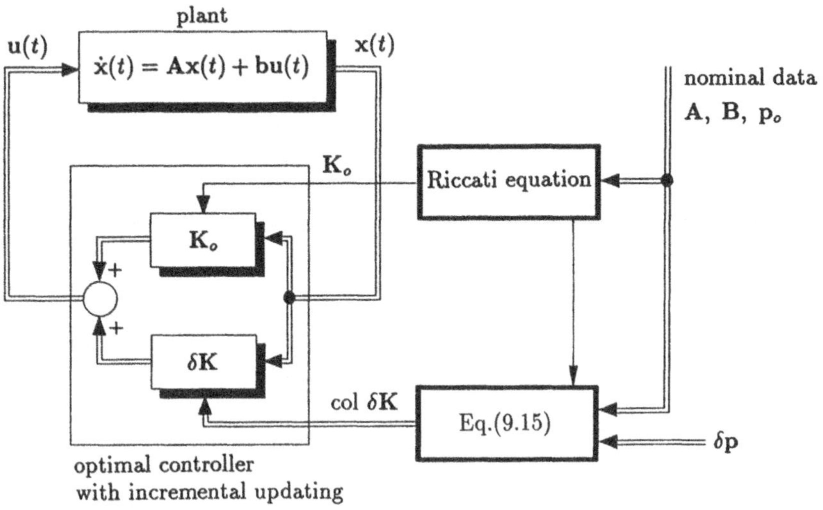

Figure 9.1: Near optimal controller incrementally updating the controller feedback gain matrix in the vicinity of the nominal Riccati equation solution

9.2 Riccati Matrix Sensitivity with Respect to any L

The Riccati matrix $\mathbf{P}(t)$ is denoted as the solution of the Riccati differential equation

$$-\dot{\mathbf{P}}(t) = \mathbf{A}^T\mathbf{P} + \mathbf{P}\mathbf{A} + \mathbf{Q} - \mathbf{P}\mathbf{B}\mathbf{R}^{-1}\mathbf{B}^T\mathbf{P} \qquad \mathbf{P}(0) = \mathbf{P}_o \,. \tag{9.17}$$

Decomposing the matrix $\mathbf{P}(t)$,

$$\mathbf{P}(t) = (\mathbf{\Theta}_{21} + \mathbf{\Theta}_{22}\mathbf{P}_o)\,(\mathbf{\Theta}_{11} + \mathbf{\Theta}_{12}\mathbf{P}_o)^{-1} \tag{9.18}$$

where $\mathbf{\Theta}_{ij}$ are matrix elements of a partitioned matrix $\mathbf{\Theta}$. The matrix $\mathbf{\Theta}$ is a matrix-valued exponentional function of the partitioned Riccati coefficient matrix $\mathbf{M}_R$

$$\mathbf{\Theta} = \begin{pmatrix} \mathbf{\Theta}_{11} & \mathbf{\Theta}_{12} \\ \mathbf{\Theta}_{21} & \mathbf{\Theta}_{22} \end{pmatrix} = e^{\mathbf{M}_R t} \qquad \mathbf{M}_R = \begin{pmatrix} \mathbf{A} & \mathbf{B}\mathbf{R}^{-1}\mathbf{B}^T \\ \mathbf{Q} & -\mathbf{A}^T \end{pmatrix} \qquad \mathbf{\Theta},\, \mathbf{M}_R \in \mathcal{R}^{2n\times 2n} \,. \tag{9.19}$$

Defining a square $(2n, 2n)$-matrix

$$\mathbf{I}^{\oslash} \triangleq \begin{pmatrix} \mathbf{0} & -\mathbf{I}_n \\ \mathbf{I}_n & \mathbf{0} \end{pmatrix} \qquad \mathbf{I}^{\oslash -1} = -\mathbf{I}^{\oslash} \,, \tag{9.20}$$

the Riccati coefficient matrix $\mathbf{M}_R$ obeys the Hamiltonian property

$$\mathbf{I}^{\oslash}\, \mathbf{M}_R^T\, \mathbf{I}^{\oslash} = \mathbf{M}_R \tag{9.21}$$

as can be proved easily. Moreover, several eigenvalue and eigenvector properties associated with $\mathbf{M}_R$ exist $\forall i = 1 \ldots n$ (*Brewer, J.W., 1977a*), namely $\lambda_{i+n}\,[\mathbf{M}_R] = -\lambda_i\,[\mathbf{M}_R]$ and

$$\mathbf{M}_R\mathbf{m}_i = \lambda_i\mathbf{m}_i, \quad \mathbf{M}_R\mathbf{m}_{i+n} = -\lambda_i\mathbf{m}_{i+n}\,, \qquad \mathbf{m}_i^{\triangleleft} = \mathbf{I}^{\oslash}\mathbf{m}_{i+n}\,, \qquad \mathbf{m}_{i+n}^{\triangleleft} = \mathbf{I}^{\oslash}\mathbf{m}_i \,. \tag{9.22}$$

Applying the product rule for differentiation with respect to any (m,r)-matrix $\mathbf{L}$, and the formula $\partial\mathbf{A}^{-1}/\partial\mathbf{M}$ of Eq.(5.105) to Eq.(9.18), it results (if $\mathbf{P}_o$ does not depend on $\mathbf{L}$)

$$\frac{\partial}{\partial\mathbf{L}}\mathbf{P}(t) = \Big(\frac{\partial\Theta_{21}}{\partial\mathbf{L}} + \frac{\partial\Theta_{22}}{\partial\mathbf{L}}(\mathbf{I}_r\otimes\mathbf{P}_o)\Big)[\mathbf{I}_r\otimes(\Theta_{11}+\Theta_{12}\mathbf{P}_o)^{-1}]$$

$$+[\mathbf{I}_m\otimes(\Theta_{21}+\Theta_{22}\mathbf{P}_o)][-\mathbf{I}_m\otimes(\Theta_{11}+\Theta_{12}\mathbf{P}_o)^{-1}]\frac{\partial(\Theta_{11}+\Theta_{12}\mathbf{P}_o)}{\partial\mathbf{L}}[\mathbf{I}_r\otimes(\Theta_{11}+\Theta_{12}\mathbf{P}_o)^{-1}] \tag{9.23}$$

$$\frac{\partial}{\partial\mathbf{L}}\mathbf{P}(t) = \{\frac{\partial\Theta_{21}}{\partial\mathbf{L}} + \frac{\partial\Theta_{22}}{\partial\mathbf{L}}(\mathbf{I}_r\otimes\mathbf{P}_o) - \Big(\mathbf{I}_m\otimes[(\Theta_{21}+\Theta_{22}\mathbf{P}_o)(\Theta_{11}+\Theta_{12}\mathbf{P}_o)^{-1}]\Big)\times$$

$$\times[\frac{\partial\Theta_{11}}{\partial\mathbf{L}} + \frac{\partial\Theta_{12}}{\partial\mathbf{L}}(\mathbf{I}_r\otimes\mathbf{P}_o)]\}[\mathbf{I}_r\otimes(\Theta_{11}+\Theta_{12}\mathbf{P}_o)^{-1}]\,. \tag{9.24}$$

The exponential matrix and its sensitivity has been studied with the derivation of Eq.(7.11). Applying this solution to the exponential $\exp(\mathbf{M}_R t)$ yields

$$\frac{\partial\Theta}{\partial\mathbf{L}} = \sum_{k=1}^{2n}\sum_{i=1}^{2n}[\mathbf{I}_m\otimes(\mathbf{m}_k\mathbf{m}_k^{aT})]\frac{\partial\mathbf{M}_R}{\partial\mathbf{L}}[\mathbf{I}_r\otimes(\mathbf{m}_i\mathbf{m}_i^{aT})]g_{ki}(t)\,. \tag{9.25}$$

Primarily needed in Eq.(9.24) are the derivatives $\partial\Theta_{11}/\partial\mathbf{L}$, $\partial\Theta_{12}/\partial\mathbf{L}$ etc. whereas Eq.(9.25) provides $\partial\Theta/\partial\mathbf{L}$ and its partitions $(\partial\Theta/\partial\mathbf{L})_{11}$ etc. (Partitioning and derivative operation do not commute.) *Brewer, J.W., 1977a* presented a formula containing pre- and postmultiplication of $\partial\Theta/\partial\mathbf{L}$ by permutation matrix expressions in order to obtain the derivatives of the partitions Θ_{ij} etc. ($ij = 1,2$)

$$\frac{\partial\Theta_{ij}}{\partial\mathbf{L}} = \Big(\mathbf{U}_{2,nm}(\mathbf{I}_m\otimes\mathbf{U}_{2,n}^T)\frac{\partial\Theta}{\partial\mathbf{L}}(\mathbf{I}_r\otimes\mathbf{U}_{n,2}^T)\mathbf{U}_{nr,2}\Big)_{\text{partition } ij}\,. \tag{9.26}$$

The permuation matrix is known from Kronecker calculus applied when Kronecker operands are permuted. Finally, a performance information can be obtained by applying the trace function to $\mathbf{P}(t)$, i.e. $I = \text{tr }\mathbf{P}(t)$. Its derivative with respect to the (m,r)-matrix $\mathbf{L}$ yields a performance sensitivity $\partial I/\partial\mathbf{L}$ and can be calculated as follows

$$I = \text{tr }\mathbf{P} = \sum_{i=1}^{n}\mathbf{e}_i^T\mathbf{P}\mathbf{e}_i \tag{9.27}$$

$$\frac{\partial\text{tr }\mathbf{P}}{\partial\mathbf{L}} = \frac{\partial I}{\partial\mathbf{L}} = \sum_{i=1}^{n}(\mathbf{I}_m\otimes\mathbf{e}_i^T)\frac{\partial\mathbf{P}\mathbf{e}_i}{\partial\mathbf{L}} = \sum_{i=1}^{n}(\mathbf{I}_m\otimes\mathbf{e}_i^T)\frac{\partial\mathbf{P}}{\partial\mathbf{L}}(\mathbf{I}_r\otimes\mathbf{e}_i)\,. \tag{9.28}$$

The order for computation is:

$$\frac{\partial\mathbf{M}_R}{\partial\mathbf{L}},\ \frac{\partial\Theta}{\partial\mathbf{L}}\ \text{Eq.(9.25)},\ \frac{\partial\Theta_{ij}}{\partial\mathbf{L}}\ \text{Eq.(9.26)},\ \frac{\partial\mathbf{P}}{\partial\mathbf{L}}\ \text{Eq.(9.24)},\ \frac{\partial I}{\partial\mathbf{L}}\ \text{Eq.(9.28)}\,. \tag{9.29}$$

9.3 Performance Sensitivity and Performance Gradient $\partial I/\partial\mathbf{p}$

Suppose a homogenous time-invariant linear system model

$$\dot{\mathbf{x}}(t) = \mathbf{A}(\mathbf{p})\mathbf{x}(t) \quad\rightsquigarrow\quad \mathbf{x} = \mathbf{x}(t;\mathbf{p}), \qquad \mathbf{x}\in\mathcal{R}^n \tag{9.30}$$

where $\mathbf{A}$ depends on an r-dimensional parameter vector $\mathbf{p}$, $\mathbf{x}(t)$ is approximated by a set of measurements[1] $\mathbf{x}_m(t_m)$ at discrete sampling instants t_m. The best estimate $\hat{\mathbf{p}}$ of the parameter vector $\mathbf{p}$ is defined that vector which minimizes a quadratic index of performance

$$\min_{\mathbf{p}} I(\mathbf{p}) = I(\hat{\mathbf{p}}) \qquad \text{or} \qquad \hat{\mathbf{p}} = \arg \min_{\mathbf{p}} I(\mathbf{p}) \tag{9.31}$$

$$\text{where} \quad I = \sum_m [\mathbf{x}(t_m) - \mathbf{x}_m(t_m)]^T [\mathbf{x}(t_m) - \mathbf{x}_m(t_m)] \ . \tag{9.32}$$

There are several methods available for minimizing a function $f(\mathbf{p})$ with respect to the parameter $\mathbf{p}$, e.g., the function minimization by conjugate gradients which proves very powerful since the minimum is exactly located at the nth iteration (*Fletcher, R., and Reeves, C.M., 1964*). An iteration method solving this estimation problem with the iteration step $\mathbf{p}_\nu$ is established as follows

$$\mathbf{p}_{\nu+1} = \mathbf{p}_\nu + k_I \frac{\partial I}{\partial \mathbf{p}} \big|_{\mathbf{p}=\mathbf{p}_\nu} \ . \tag{9.33}$$

The gradient can be calculated from Eq.(5.23)

$$\frac{\partial I}{\partial \mathbf{p}} = 2 \sum_m \frac{\partial \mathbf{x}^T(t_m)}{\partial \mathbf{p}} [\mathbf{x}(t_m) - \mathbf{x}_m(t_m)] \ . \tag{9.34}$$

It follows from state transition equation that

$$\left(\frac{\partial \mathbf{x}^T(t_m)}{\partial \mathbf{p}}\right)^T = \frac{\partial \mathbf{x}(t_m)}{\partial \mathbf{p}^T} = \frac{\partial}{\partial \mathbf{p}^T}[\mathbf{\Phi}(t_m)\mathbf{x}_0] \tag{9.35}$$

where $\mathbf{x}_0$ is an arbitrarily chosen initial value, e.g. $\mathbf{x}_0 = \mathbf{x}_m(t_0)$. From product rule Eq.(5.87)

$$\frac{\partial \mathbf{x}(t_m)}{\partial \mathbf{p}^T} = \frac{\partial \mathbf{\Phi}(t_m)}{\partial \mathbf{p}^T}(\mathbf{I}_r \otimes \mathbf{x}_0) \ . \tag{9.36}$$

Applying Eq.(7.11) to $\mathbf{\Phi}(t) = e^{\mathbf{A}(\mathbf{p})t}$, the derivative with respect to the row $\mathbf{p}^T$ of dimension r is

$$\frac{\partial \mathbf{\Phi}(t)}{\partial \mathbf{p}^T} = \sum_i \sum_j \mathbf{a}_i \mathbf{a}_i^{\triangleleft T} \frac{\partial \mathbf{A}}{\partial \mathbf{p}^T}(\mathbf{I}_r \otimes \mathbf{a}_j \mathbf{a}_j^{\triangleleft T}) g_{ij}(t) \ . \tag{9.37}$$

Combining Eqs.(9.36) and (9.37) yields

$$\frac{\partial \mathbf{x}(t_m)}{\partial \mathbf{p}^T} = \sum_i \sum_j \mathbf{a}_i \mathbf{a}_i^{\triangleleft T} \frac{\partial \mathbf{A}}{\partial \mathbf{p}^T}(\mathbf{I}_r \otimes \mathbf{a}_j \mathbf{a}_j^{\triangleleft T} \mathbf{x}_o) g_{ij}(t_m) \ . \tag{9.38}$$

The resulting gradient from Eq.(9.34) is (*Brewer, J.W., 1978a*)

$$\frac{\partial I}{\partial \mathbf{p}} = 2 \sum_m \sum_i \sum_j \mathbf{a}_i \mathbf{a}_i^{\triangleleft T} \frac{\partial \mathbf{A}}{\partial \mathbf{p}^T}(\mathbf{I}_r \otimes \mathbf{a}_j \mathbf{a}_j^{\triangleleft T} \mathbf{x}_o) g_{ij}(t_m)[\mathbf{x}(t_m) - \mathbf{x}_m(t_m)] \tag{9.39}$$

where $\mathbf{a}_i$ and $\mathbf{a}_i^{\triangleleft}$ are the (right) eigenvectors of $\mathbf{A}$ and $\mathbf{A}^T$, respectively, and $g_{ij}(t)$ is the inverse Laplace transform of $(s - \lambda_i[\mathbf{A}])^{-1} \ (s - \lambda_j[\mathbf{A}])^{-1}$.

[1]The subscript m as a symbol for measurement values may not be confused with the dimension m of multivariable controlling variable $\mathbf{u}$.

9.4 Gradient for Phase Variable Form

Using the companion form $\mathbf{A}$ to describe an unforced linear time-invariant system

$$\dot{\mathbf{x}}(t) = \mathbf{A}(\mathbf{p})\mathbf{x}(t) = \begin{pmatrix} 0 & 1 & 0 & \dots & 0 \\ 0 & 0 & 1 & \dots & 0 \\ \vdots & \vdots & \vdots & \ddots & \vdots \\ 0 & 0 & 0 & & 1 \\ -p_1 & -p_2 & -p_3 & \dots & -p_n \end{pmatrix} \mathbf{x}(t) , \tag{9.40}$$

the last row of $\mathbf{A}$ and the coefficients c_i of the characteristic polynomial $c(\lambda)$ associated with $\mathbf{A}$ are in a close relation, see Eq.(B.1). The coefficient matrix $\mathbf{A}$ in companion form can be written as

$$\mathbf{A} = \sum_{i=1;j=i+1}^{n-1} \mathbf{E}_{ij} - \mathbf{e}_n \otimes \mathbf{p}^T \qquad \in \mathcal{R}^{n \times n} \tag{9.41}$$

$$\mathbf{p} = (p_1 \;\; p_2 \dots p_n)^T = (c_0 \;\; c_1 \dots c_{n-1})^T \in \mathcal{R}^n \; , \quad r = n \; . \tag{9.42}$$

If there are no interdependencies within $\mathbf{p}$ (*Brewer, J. W., 1978*), using Eq.(5.95)

$$\frac{\partial \mathbf{A}}{\partial \mathbf{p}^T} = -(\mathbf{I}_1 \otimes \mathbf{U}_{n1})\Big(\frac{\partial \mathbf{p}^T}{\partial \mathbf{p}^T} \otimes \mathbf{e}_n\Big)(\mathbf{I}_n \otimes \mathbf{U}_{n1}), \qquad \mathbf{U}_{n1} = \mathbf{U}_{1n} = \mathbf{I}_n \tag{9.43}$$

$$= -\mathbf{I}_n[(\mathrm{col}^T \mathbf{I}_n) \otimes \mathbf{e}_n](\mathbf{I}_n \otimes \mathbf{I}_n) = (\mathrm{col}\;^T \mathbf{I}_n) \otimes \mathbf{e}_n \; . \tag{9.44}$$

From Eq.(9.38), by transposition and for $r = n$,

$$\frac{\partial \mathbf{x}^T(t_m)}{\partial \mathbf{p}} = \Big(\frac{\partial \mathbf{x}(t_m)}{\partial \mathbf{p}^T}\Big)^T = \sum_i \sum_j (\mathbf{I}_n \otimes \mathbf{x}_0^T \mathbf{a}_j^q \mathbf{a}_j^T) \frac{\partial \mathbf{A}^T}{\partial \mathbf{p}} (\mathbf{a}_i^q \mathbf{a}_i^T) g_{ij}(t) \; . \tag{9.45}$$

Combining Eqs.(9.34) and (9.44) yields

$$\frac{\partial I}{\partial \mathbf{p}} = 2 \sum_m \sum_i \sum_j (\mathbf{I}_n \otimes \mathbf{x}_0^T \mathbf{a}_j^q \mathbf{a}_j^T)[(\mathrm{col}\; \mathbf{I}_n) \otimes \mathbf{e}_n^T]\mathbf{a}_i^q \mathbf{a}_i^T [\mathbf{\Phi}(t_m)\mathbf{x}_0 - \mathbf{x}_m(t_m)] g_{ij}(t_m) \; . \tag{9.46}$$

9.5 Hessian Matrix. Second-Derivative Sensitivity

The Newton-Raphson method is based on finding the zeros of $\partial I/\partial \mathbf{p}$ by a numerical procedure. This procedure utilizes the Taylor expansion of $\partial I/\partial \mathbf{p}$ at $\mathbf{p} = \mathbf{p}_\nu$. The expansion usually is restricted to the first-order term of $\partial I/\partial \mathbf{p}$ which corresponds to the second-order term of I (*Brewer, J. W., 1978*). The matrix of second partial derivative is the Hessian matrix and defined as $\partial^2 I/[\partial \mathbf{p}^T \partial \mathbf{p}]$. This Hessian can be calculated by the use of the product rule Eq.(5.87) and

$$\frac{\partial \mathbf{M}}{\partial \mathbf{p}^T} = \Big(\frac{\partial \mathbf{M}^T}{\partial \mathbf{p}}\Big)^T, \qquad \mathbf{M} \triangleq \frac{\partial \mathbf{x}^T}{\partial \mathbf{p}}[\mathbf{x}(t_m) - \mathbf{x}_m(t_m)] \; . \tag{9.47}$$

Differentiating $\mathbf{M}$ with respect to $\mathbf{p}^T$ equals the expression obtained by differentiating $\mathbf{M}^T$ with respect to $\mathbf{p} \in \mathcal{R}^r$ and by transposing the result

$$\begin{aligned} \frac{\partial^2 I}{\partial \mathbf{p}^T \partial \mathbf{p}} &= 2\sum_m \Big(\frac{\partial \mathbf{M}^T}{\partial \mathbf{p}}\Big)^T = 2\sum_m \Big(\frac{\partial (\mathbf{x} - \mathbf{x}_m)^T}{\partial \mathbf{p}}(\mathbf{I}_1 \otimes \frac{\partial \mathbf{x}}{\partial \mathbf{p}^T}) + [\mathbf{I}_r \otimes (\mathbf{x} - \mathbf{x}_m)^T]\frac{\partial^2 \mathbf{x}}{\partial \mathbf{p} \partial \mathbf{p}^T}\Big)^T \\ &= 2\sum_m \frac{\partial \mathbf{x}^T}{\partial \mathbf{p}} \frac{\partial \mathbf{x}}{\partial \mathbf{p}^T} + \frac{\partial^2 \mathbf{x}^T}{\partial \mathbf{p}^T \partial \mathbf{p}}[\mathbf{I}_r \otimes (\mathbf{x} - \mathbf{x}_m)] \end{aligned} \tag{9.48}$$

where abbreviated $\mathbf{x} = \mathbf{x}(t_m)$ and $\mathbf{x}_m = \mathbf{x}_m(t_m)$. It follows from elementary properties that

$$\frac{\partial^2 I}{\partial \mathbf{p}^T \partial \mathbf{p}} = 2\sum_m \Big(\frac{\partial \mathbf{x}}{\partial \mathbf{p}^T}\Big)^T \Big(\frac{\partial \mathbf{x}}{\partial \mathbf{p}^T}\Big) + \Big(\frac{\partial}{\partial \mathbf{p}}\frac{\partial \mathbf{x}}{\partial \mathbf{p}^T}\Big)^T [\mathbf{I}_r \otimes (\mathbf{x} - \mathbf{x}_m)] \ . \tag{9.49}$$

The first expression $(\partial \mathbf{x}/\partial \mathbf{p}^T)$ can be substituted from Eq.(9.38). The second expression $\partial^2 \mathbf{x}/\partial \mathbf{p} \partial \mathbf{p}^T$ is detailed below. From Eq.(9.36)

$$\frac{\partial^2 \mathbf{x}}{\partial \mathbf{p} \partial \mathbf{p}^T} = \frac{\partial}{\partial \mathbf{p}} \frac{\partial \mathbf{\Phi}}{\partial \mathbf{p}^T} (\mathbf{I}_r \otimes \mathbf{x}_0) \ . \tag{9.50}$$

Applying Eq.(7.4) with $\mathbf{K} := \mathbf{p}^T$ yields the interrelation of $\mathbf{\Phi}(\mathbf{p})$ and $\mathbf{A}(\mathbf{p})$. Setting $\mathbf{K} := \mathbf{p}^T$ in Eq.(7.4) and, then, differentiating once more with respect to $\mathbf{p}$ using product rule for matrix product and Kronecker product, Eqs. (5.87) and (5.95) respectively, yields

$$\begin{aligned} \frac{d}{dt}\Big(\frac{\partial}{\partial \mathbf{p}}\frac{\partial \mathbf{\Phi}}{\partial \mathbf{p}^T}\Big) &= \mathbf{I}_r \otimes \mathbf{A}\Big(\frac{\partial}{\partial \mathbf{p}}\frac{\partial \mathbf{\Phi}}{\partial \mathbf{p}^T}\Big) + \frac{\partial \mathbf{A}}{\partial \mathbf{p}}\frac{\partial \mathbf{\Phi}}{\partial \mathbf{p}^T} \\ &+ \Big(\frac{\partial}{\partial \mathbf{p}}\frac{\partial \mathbf{A}}{\partial \mathbf{p}^T}\Big)(\mathbf{I}_r \otimes \mathbf{\Phi}) + \Big(\mathbf{I}_r \otimes \frac{\partial \mathbf{A}}{\partial \mathbf{p}^T}\mathbf{U}_{rn}\Big)\Big(\frac{\partial \mathbf{\Phi}}{\partial \mathbf{p}} \otimes \mathbf{I}_r\Big)\mathbf{U}_{nr} \ . \end{aligned} \tag{9.51}$$

This matrix differential equation in the variable $\frac{\partial}{\partial \mathbf{p}}\frac{\partial \mathbf{\Phi}}{\partial \mathbf{p}^T}$ can be solved similarly to the derivation of Eq.(7.11). The result is (*Brewer, J.W., 1978*)

$$\begin{aligned} \frac{\partial}{\partial \mathbf{p}}\frac{\partial \mathbf{\Phi}}{\partial \mathbf{p}^T} &= \sum_{i=1}^n \sum_{j=1}^n (\mathbf{I}_r \otimes \mathbf{a}_i \mathbf{a}_i^{\triangleleft T})\Big(\frac{\partial}{\partial \mathbf{p}}\frac{\partial \mathbf{A}}{\partial \mathbf{p}^T}\Big)(\mathbf{I}_r \otimes \mathbf{a}_j \mathbf{a}_j^{\triangleleft T}) g_{ij}(t) \\ &+ \sum_{i=1}^n \sum_{j=1}^n \sum_{k=1}^n \{(\mathbf{I}_r \otimes \mathbf{a}_k \mathbf{a}_k^{\triangleleft T})\Big(\frac{\partial \mathbf{A}}{\partial \mathbf{p}}(\mathbf{a}_i \mathbf{a}_i^{\triangleleft T})\frac{\partial \mathbf{A}}{\partial \mathbf{p}^T}\Big)(\mathbf{I}_r \otimes \mathbf{a}_j \mathbf{a}_j^{\triangleleft T}) \\ &+ (\mathbf{I}_r \otimes \mathbf{a}_k \mathbf{a}_k^{\triangleleft T})\frac{\partial \mathbf{A}}{\partial \mathbf{p}^T}\mathbf{U}_{rn}]\Big([(\mathbf{I}_r \otimes \mathbf{a}_i \mathbf{a}_i^{\triangleleft T})\frac{\partial \mathbf{A}}{\partial \mathbf{p}}\mathbf{a}_j \mathbf{a}_j^{\triangleleft T}] \otimes \mathbf{I}_r\Big)\mathbf{U}_{nr}\} h_{ijk}(t) \end{aligned} \tag{9.52}$$

$$\text{where} \qquad h_{ijk}(t) = \mathcal{L}^{-1}\frac{1}{(s - \lambda_i[\mathbf{A}])(s - \lambda_j[\mathbf{A}])(s - \lambda_k[\mathbf{A}])} \ . \tag{9.53}$$

The iteration procedure for finding the zeros of $\partial I/\partial \mathbf{p}$ makes use of extrapolation at the gradient rate taken from $\partial I/\partial \mathbf{p}$ resulting in $\partial^2 I/(\partial \mathbf{p}^T \partial \mathbf{p})$. Hence, the iteration step finally is given by

$$\mathbf{p}_{\nu+1} = \mathbf{p}_\nu - \Big(\frac{\partial^2 I}{\partial \mathbf{p}^T \partial \mathbf{p}}\Big)^{-1}_{\mathbf{p}=\mathbf{p}_\nu} \Big(\frac{\partial I}{\partial \mathbf{p}}\Big)_{\mathbf{p}=\mathbf{p}_\nu} \ . \tag{9.54}$$

Note that this method can only be applied in the case of distinct eigenvalues. The iteration need not converge if the initial value $\mathbf{p}_0$ is not close enough to the final point. Combinations of gradient and Newton-Raphson schemes frequently give good opportunities for convergence.

9.6 Nonlinear Regression

In Eq.(D.13) a linear model $\mathbf{M}\mathbf{p} = \mathbf{y} - \boldsymbol{\varepsilon}$ is considered. Now, the measurements $\mathbf{y}$ are assumed to be in nonlinear relation to the parameter $\mathbf{p}$. The expression $\mathbf{M}\mathbf{p}$ is replaced by a vector-valued function $\mathbf{m}(\mathbf{p})$. Thus,

$$\mathbf{y} = \mathbf{m}(\mathbf{p}) + \boldsymbol{\varepsilon} \ . \tag{9.55}$$

Selecting the parameter $\mathbf{p}$ optimally requires the definition of an index of performance using, e.g., a symmetric weighting matrix $\mathbf{W}$ and a square structure

$$C = \boldsymbol{\varepsilon}^T \mathbf{W} \boldsymbol{\varepsilon} = [\mathbf{y}-\mathbf{m}(\mathbf{p})]^T \mathbf{W} [\mathbf{y}-\mathbf{m}(\mathbf{p})] = \mathbf{y}^T \mathbf{W} \mathbf{y} - 2\mathbf{m}^T(\mathbf{p}) \mathbf{W} \mathbf{y} + \mathbf{m}^T(\mathbf{p}) \mathbf{W} \mathbf{m}(\mathbf{p}) \,. \tag{9.56}$$

Referring to Eq.(9.54), one has to calculate the gradient and the Hessian matrix. Using Eq.(5.20), the derivatives are

$$\frac{\partial \mathbf{m}^T \mathbf{W} \mathbf{y}}{\partial \mathbf{p}} = \frac{\partial \mathbf{m}^T}{\partial \mathbf{p}} \mathbf{W} \mathbf{y} \tag{9.57}$$

$$\frac{\partial \mathbf{m}^T \mathbf{W} \mathbf{m}}{\partial \mathbf{p}} = \frac{\partial \mathbf{m}^T}{\partial \mathbf{p}} \mathbf{W} \mathbf{m} + \frac{\partial (\mathbf{W}\mathbf{m})^T}{\partial \mathbf{p}} \mathbf{m} = 2\frac{\partial \mathbf{m}^T}{\partial \mathbf{p}} \mathbf{W} \mathbf{m} \,. \tag{9.58}$$

Applying these derivatives to calculate the first and second derivative of C with respect to $\mathbf{p}$ and $\mathbf{p}^T$

$$\frac{\partial C}{\partial \mathbf{p}} = -2\frac{\partial \mathbf{m}^T(\mathbf{p})}{\partial \mathbf{p}} \mathbf{W} [\mathbf{y} - \mathbf{m}(\mathbf{p})] \tag{9.59}$$

$$\frac{\partial^2 C}{\partial \mathbf{p}^T \partial \mathbf{p}} = 2\frac{\partial \mathbf{m}^T(\mathbf{p})}{\partial \mathbf{p}} \mathbf{W} \frac{\partial \mathbf{m}(\mathbf{p})}{\partial \mathbf{p}^T} + \text{small terms in } \frac{\partial^2 \mathbf{m}^T}{\partial \mathbf{p}^T \partial \mathbf{p}} \,. \tag{9.60}$$

Neglecting the small terms in the equation above and referring to Eq.(9.54), the iterative algorithm solving the nonlinear least-squares regression problem is given by

$$\mathbf{p}_{\nu+1} = \mathbf{p}_\nu + \left(\frac{\partial \mathbf{m}^T(\mathbf{p})}{\partial \mathbf{p}} \mathbf{W} \frac{\partial \mathbf{m}(\mathbf{p})}{\partial \mathbf{p}^T}\right)^{-1} \Big|_{\mathbf{p}=\mathbf{p}_\nu} \frac{\partial \mathbf{m}^T(\mathbf{p})}{\partial \mathbf{p}} \Big|_{\mathbf{p}=\mathbf{p}_\nu} \mathbf{W} \, [\mathbf{y} - \mathbf{m}(\mathbf{p})] \,. \tag{9.61}$$

9.7 Controllers With Reduced Performance Sensitivity

When parameters of the system are varying, optimal control systems must be designed with reduced differential sensitivity with respect to varying parameters. Otherwise the influence of varying parameters on the overall performance of the control system may grow to an unadmissible extent. With the help of weighting matrices the sensitivity with respect to selected parameters can be governed individually.

9.7.1 State Space Representation and Index of Performance

From the plant and the output feedback controller equations

$$\dot{\mathbf{x}}(t) = \mathbf{A}\mathbf{x}(t) + \mathbf{B}\mathbf{u}(t), \qquad \mathbf{x}(0) = \mathbf{x}_o \tag{9.62}$$

$$\mathbf{y}(t) = \mathbf{C}\mathbf{x}(t) \tag{9.63}$$

$$\mathbf{u}(t) = \mathbf{K}\mathbf{y}(t) = \mathbf{K}\mathbf{C}\mathbf{x}(t) \tag{9.64}$$

the closed-loop control system is

$$\dot{\mathbf{x}}(t) = (\mathbf{A} + \mathbf{B}\mathbf{K}\mathbf{C})\mathbf{x} \triangleq \mathbf{F}\mathbf{x} \quad \rightsquigarrow \quad \mathbf{x}(t) = e^{\mathbf{F}t}\mathbf{x}_o \tag{9.65}$$

$$I = \int_0^\infty (\mathbf{x}^T \mathbf{Q} \mathbf{x} + \mathbf{u}^T \mathbf{R} \mathbf{u}) d\tau = \int_0^\infty \mathbf{x}^T(\tau)(\mathbf{Q} + \mathbf{C}^T \mathbf{K}^T \mathbf{R} \mathbf{K} \mathbf{C}) \mathbf{x}(\tau) \, d\tau \tag{9.66}$$

$$= \int_0^\infty \mathbf{x}_o^T e^{\mathbf{F}^T \tau} (\mathbf{Q} + \mathbf{C}^T \mathbf{K}^T \mathbf{R} \mathbf{K} \mathbf{C}) e^{\mathbf{F}\tau} \, \mathbf{x}_o d\tau \,. \tag{9.67}$$

9.7.2 Constraint on Λ

Assume $I \triangleq \mathbf{x}_o^T \Lambda \mathbf{x}_o$ is an admissible structure of the performance, then, by comparison with the above expression, the matrix Λ is given by

$$\Lambda = \int_0^\infty e^{\mathbf{F}^T\tau}(\mathbf{Q} + \mathbf{C}^T\mathbf{K}^T\mathbf{R}\mathbf{K}\mathbf{C})e^{\mathbf{F}\tau}\, d\tau\ . \tag{9.68}$$

Now, consider the expression $\mathbf{F}^T\Lambda + \Lambda\mathbf{F}$. Using Eq.(9.68),

$$\begin{aligned}
\mathbf{F}^T\Lambda + \Lambda\mathbf{F} &= \int_0^\infty [\mathbf{F}^T e^{\mathbf{F}^T\tau}(\mathbf{Q} + \mathbf{C}^T\mathbf{K}^T\mathbf{R}\mathbf{K}\mathbf{C})e^{\mathbf{F}\tau} + e^{\mathbf{F}^T\tau}(\mathbf{Q} + \mathbf{C}^T\mathbf{K}^T\mathbf{R}\mathbf{K}\mathbf{C})e^{\mathbf{F}\tau}\mathbf{F}]\, d\tau \\
&= \int_0^\infty [\frac{d}{d\tau} e^{\mathbf{F}^T\tau}(\mathbf{Q} + \mathbf{C}^T\mathbf{K}^T\mathbf{R}\mathbf{K}\mathbf{C})e^{\mathbf{F}\tau}]\, d\tau \qquad (9.69) \\
&= e^{\mathbf{F}^T\tau}(\mathbf{Q} + \mathbf{C}^T\mathbf{K}^T\mathbf{R}\mathbf{K}\mathbf{C})e^{\mathbf{F}\tau}\ |_0^\infty = -(\mathbf{Q} + \mathbf{C}^T\mathbf{K}^T\mathbf{R}\mathbf{K}\mathbf{C}) \qquad (9.70)
\end{aligned}$$

where a stable matrix $\mathbf{F}$ was preassumed. From Eq.(9.70) the constraint in Λ is given as a Lyapunov equation in Λ

$$\mathbf{\Gamma}_\Lambda \triangleq \mathbf{C}^T\mathbf{K}^T\mathbf{R}\mathbf{K}\mathbf{C} + \mathbf{F}^T\Lambda + \Lambda\mathbf{F} + \mathbf{Q} = \mathbf{0}\ . \tag{9.71}$$

9.7.3 Constraint on Lagrange Multiplier V

Using a symmetric Lagrange multiplier matrix $\mathbf{V}^T$, the performance index to be minimized is

$$I_1 = I + \text{tr}\ [\mathbf{V}\mathbf{\Gamma}_\Lambda] = \text{tr}\ [\Lambda\mathbf{x}_o\mathbf{x}_o^T + \mathbf{V}\mathbf{\Gamma}_\Lambda] \rightarrow \min\ . \tag{9.72}$$

Differentiating with respect to Λ and equating to zeros yields the necessary condition $\mathbf{\Gamma}_V$,

$$\frac{\partial}{\partial\Lambda}\text{tr}\ [\Lambda\mathbf{x}_o\mathbf{x}_o^T + \mathbf{V}(\mathbf{C}^T\mathbf{K}^T\mathbf{R}\mathbf{K}\mathbf{C} + \mathbf{F}^T\Lambda + \Lambda\mathbf{F} + \mathbf{Q})] \triangleq \mathbf{\Gamma}_V = \mathbf{0} \tag{9.73}$$

$$\mathbf{\Gamma}_V = \mathbf{x}_o\mathbf{x}_o^T + \mathbf{F}\mathbf{V} + \mathbf{V}\mathbf{F}^T = \mathbf{0}\ . \tag{9.74}$$

The matrix $\mathbf{\Gamma}_V$ is regarded as an additional constraint (*Yahagi, T., 1973*).

9.7.4 Variation of the Performance Index

Varying $\mathbf{A}$ to $\mathbf{A} + \delta\mathbf{A}$, $\mathbf{F}$ changes to $\mathbf{F} + \delta\mathbf{F}$ where $\delta\mathbf{F} = \delta\mathbf{A}$. From Eq.(9.71)

$$\delta\mathbf{F}^T\Lambda + \mathbf{F}^T\delta\Lambda + \delta\Lambda\ \mathbf{F} + \Lambda\ \delta\mathbf{F} = 0 \tag{9.75}$$

$$\text{or} \qquad \mathbf{F}^T\delta\Lambda + \delta\Lambda\ \mathbf{F} = -(\delta\mathbf{F}^T\Lambda + \Lambda\ \delta\mathbf{F}) = -(\delta\mathbf{A}^T\Lambda + \Lambda\ \delta\mathbf{A})\ . \tag{9.76}$$

Assume

$$\delta\Lambda = \int_0^\infty e^{\mathbf{F}^T\tau}(\delta\mathbf{A}^T\Lambda + \Lambda\ \delta\mathbf{A})e^{\mathbf{F}\tau}\, d\tau\ . \tag{9.77}$$

Evaluating yields

$$\begin{aligned}
\mathbf{F}^T\delta\Lambda + \delta\Lambda\ \mathbf{F} &= \int_0^\infty \mathbf{F}^T e^{\mathbf{F}^T\tau}(\delta\mathbf{A}^T\Lambda + \Lambda\ \delta\mathbf{A})e^{\mathbf{F}\tau}\, d\tau + e^{\mathbf{F}^T\tau}(\delta\mathbf{A}^T\Lambda + \Lambda\ \delta\mathbf{A})e^{\mathbf{F}\tau}\mathbf{F}\, d\tau \\
&= \int_0^\infty [\frac{d}{d\tau} e^{\mathbf{F}^T\tau}(\delta\mathbf{A}^T\Lambda + \Lambda\ \delta\mathbf{A})e^{\mathbf{F}\tau}]\, d\tau = -(\delta\mathbf{A}^T\Lambda + \Lambda\ \delta\mathbf{A}) \qquad (9.78)
\end{aligned}$$

which verifies that the assumption Eq.(9.77) is a solution of Eq.(9.76). Hence, the first variation δI using Eq.(9.77) is given by

$$\delta I = \mathbf{x}_o^T \delta\Lambda\ \mathbf{x}_o = \int_0^\infty \mathbf{x}^T(\delta\mathbf{A}^T\Lambda + \Lambda\ \delta\mathbf{A})\mathbf{x}\ d\tau\ . \tag{9.79}$$

The solution of Eq.(9.74) is

$$\mathbf{V} = \int_0^\infty e^{\mathbf{F}\tau}\mathbf{x}_o\mathbf{x}_o^T e^{\mathbf{F}^T\tau}\ d\tau \tag{9.80}$$

since

$$\mathbf{F}\mathbf{V} + \mathbf{V}\mathbf{F}^T = \int_0^\infty (\mathbf{F}e^{\mathbf{F}\tau}\mathbf{x}_o\mathbf{x}_o^T e^{\mathbf{F}^T\tau} + e^{\mathbf{F}\tau}\mathbf{x}_o\mathbf{x}_o^T e^{\mathbf{F}^T\tau}\mathbf{F}^T)\ d\tau \tag{9.81}$$

$$= \int_0^\infty \frac{d}{d\tau}[e^{\mathbf{F}\tau}\mathbf{x}_o\mathbf{x}_o^T e^{\mathbf{F}^T\tau}]\ d\tau = -\mathbf{x}_o\mathbf{x}_o^T\ . \tag{9.82}$$

For comparison see Eq.(16.46).

9.7.5 Derivative of the Performance with Respect to A, B, C, K

In order to desensitize the system with respect to parameter variations, the derivatives of I with respect to $\mathbf{A}, \mathbf{B}, \mathbf{C}$ and $\mathbf{K}$ are required. From Eq.(9.79)

$$\frac{\partial I}{\partial \mathbf{A}} = \frac{\partial\ \mathrm{tr}\ I}{\partial \mathbf{A}} = \frac{\partial}{\partial \mathbf{A}}\int_0^\infty \mathrm{tr}\ [\Lambda\mathbf{x}\mathbf{x}^T\delta\mathbf{A}^T + \mathbf{x}\mathbf{x}^T\Lambda\ \delta\mathbf{A}]\ d\tau \tag{9.83}$$

$$= \int_0^\infty [\Lambda\mathbf{x}\mathbf{x}^T + \Lambda^T\mathbf{x}\mathbf{x}^T]\ d\tau = 2\Lambda\mathbf{V} \tag{9.84}$$

where $\mathbf{x}$ is the nominal solution and Λ the nominal constraint. Similarly,

$$\frac{\partial I}{\partial \mathbf{B}} = 2\Lambda\mathbf{V}\mathbf{C}^T\mathbf{K}^T \tag{9.85}$$

$$\frac{\partial I}{\partial \mathbf{C}} = 2\mathbf{K}^T(\mathbf{R}\mathbf{K}\mathbf{C} + \mathbf{B}^T\Lambda)\mathbf{V} \tag{9.86}$$

$$\frac{\partial I}{\partial \mathbf{K}} = 2(\mathbf{R}\mathbf{K}\mathbf{C} + \mathbf{B}^T\Lambda)\mathbf{V}\mathbf{C}^T\ . \tag{9.87}$$

For comparison see Eq.(16.48). From Eq.(9.87)

$$\frac{\partial I}{\partial \mathbf{K}} = 0 \ \leadsto\ \mathbf{R}\mathbf{K}\mathbf{C}\mathbf{V}\mathbf{C}^T + \mathbf{B}^T\Lambda\mathbf{V}\mathbf{C}^T = 0 \ \leadsto\ \mathbf{K} = -\mathbf{R}^{-1}\mathbf{B}^T\Lambda\mathbf{V}\mathbf{C}^T(\mathbf{C}\mathbf{V}\mathbf{C}^T)^{-1}\ . \tag{9.88}$$

9.7.6 Optimum Desensitized Controller

A reduced sensitivity of the performance index with respect to small parameter variations can be achieved when the index of performance is augmented by the partial derivatives given in the above. This leads to

$$I_2 = I + \mathrm{tr}\ [(\partial I/\partial \mathbf{A})^T\mathbf{Q}_A(\partial I/\partial \mathbf{A}) + (\partial I/\partial \mathbf{B})^T\mathbf{Q}_B(\partial I/\partial \mathbf{B})$$

$$+ (\partial I/\partial \mathbf{C})\mathbf{Q}_c(\partial I/\partial \mathbf{C})^T + (\partial I/\partial \mathbf{K})\mathbf{Q}_K(\partial I/\partial \mathbf{K})^T] \tag{9.89}$$

subject to the constraints $\mathbf{\Gamma}_\Lambda$ and $\mathbf{\Gamma}_V$. Adjoining the constraints via symmetric Lagrange multiplicators $\mathbf{L}_\Lambda$ and $\mathbf{L}_V$

$$H = I_2 + \text{tr}\,[\mathbf{L}_\Lambda \mathbf{\Gamma}_\Lambda] + \text{tr}\,[\mathbf{L}_V \mathbf{\Gamma}_V]\,. \tag{9.90}$$

With the abbreviation $\mathbf{J} \triangleq \mathbf{RKC} + \mathbf{B}^T\mathbf{\Lambda}$

$$\begin{aligned} H = \mathbf{x}_o\mathbf{\Lambda}\mathbf{x}_o &+ 4\text{tr}[\mathbf{V\Lambda}\ \mathbf{Q}_A\mathbf{\Lambda V} + \mathbf{KCV\Lambda}\ \mathbf{Q}_B\mathbf{\Lambda VC}^T\mathbf{K}^T + \mathbf{K}^T\mathbf{JV}\ \mathbf{Q}_C\mathbf{VJ}^T\mathbf{K} + \mathbf{JVC}^T\ \mathbf{Q}_K\mathbf{CVJ}^T] \\ &+ \text{tr}[\mathbf{L}_\Lambda(\mathbf{C}^T\mathbf{K}^T\mathbf{RKC} + \mathbf{F}^T\mathbf{\Lambda} + +\mathbf{\Lambda F} + \mathbf{Q}) + \mathbf{L}_V(\mathbf{xx}_o^T + \mathbf{FV} + \mathbf{VF}^T)]\,. \end{aligned} \tag{9.91}$$

The necessary condition for the optimum solution is given by taking the partial derivatives of H with respect to $\mathbf{K}, \mathbf{\Lambda}$ and $\mathbf{V}$ and equating them to zero. (Differentiating with respect to $\mathbf{L}_\Lambda$ and $\mathbf{L}_V$ yields the constraints $\mathbf{\Gamma}_\Lambda$ and $\mathbf{\Gamma}_V$.) This calculation presented by *Yahagi, T., 1977* in detail is rather straightforward but extensive:

Differentiating H with respect to $\mathbf{K}$ yields

$$\mathbf{JL}_\Lambda\mathbf{C}^T + \mathbf{B}^T\mathbf{L}_V\mathbf{VC}^T + \mathbf{KCV\Lambda}\ \mathbf{Q}_B\mathbf{\Lambda VC}^T + \mathbf{JV}\ \mathbf{Q}_C\mathbf{VJ}^T\mathbf{K} + \mathbf{RKK}^T\mathbf{JV}\ \mathbf{Q}_C\ \mathbf{VC}^T + \mathbf{RJVC}^T\ \mathbf{Q}_K\ \mathbf{CVC}^T = 0\,. \tag{9.92}$$

Differentiating H with respect to $\mathbf{\Lambda}$ yields

$$\begin{aligned} \mathbf{FL}_\Lambda &+ \mathbf{L}_\Lambda\mathbf{F}^T + \mathbf{x}_o\mathbf{x}_o^T + \mathbf{VV\Lambda Q}_A + \mathbf{Q}_A\mathbf{\Lambda VV} + \mathbf{VC}^T\mathbf{K}^T\mathbf{KCV\Lambda Q}_B + \mathbf{Q}_B\mathbf{\Lambda VC}^T\mathbf{K}^T\mathbf{KCV} \\ &+ \mathbf{BKK}^T\mathbf{JVQ}_B\mathbf{V} + \mathbf{VQ}_C\mathbf{VJ}^T\mathbf{KK}^T\mathbf{B}^T + \mathbf{BJVC}^T\mathbf{Q}_K\mathbf{CV} + \mathbf{VC}^T\mathbf{Q}_K\mathbf{CVJ}^T\mathbf{B}^T = \mathbf{0}\,. \end{aligned} \tag{9.93}$$

Differentiating H with respect to $\mathbf{V}$ yields

$$\begin{aligned} \mathbf{F}^T\mathbf{L}_V &+ \mathbf{L}_V\mathbf{F} + \mathbf{V\Lambda}\ \mathbf{Q}_A\mathbf{\Lambda} + \mathbf{\Lambda}\ \mathbf{Q}_A\mathbf{\Lambda V} + \mathbf{C}^T\mathbf{K}^T\mathbf{KCV\Lambda}\ \mathbf{Q}_B\mathbf{\Lambda} + \mathbf{\Lambda}\ \mathbf{Q}_B\mathbf{\Lambda VC}^T\mathbf{K}^T\mathbf{KC} \\ &+ \mathbf{J}^T\mathbf{KK}^T\mathbf{JV}\ \mathbf{Q}_C + \mathbf{Q}_C\mathbf{VJ}^T\mathbf{KK}^T\mathbf{J} + \mathbf{J}^T\mathbf{JVC}^T\mathbf{Q}_K\mathbf{C} + \mathbf{C}^T\mathbf{Q}_K\mathbf{CVJ}^T\mathbf{J} = \mathbf{0}\,. \end{aligned} \tag{9.94}$$

The equations above can be solved with respect to the unknown matrices $\mathbf{V}, \mathbf{\Lambda}, \mathbf{K}, \mathbf{L}_\Lambda$ and $\mathbf{L}_V$. However, with regard to the high degree only numerical computation is feasible.

9.7.7 Solution by Gradient Method

The numerical solution can be obtained by changing $\mathbf{K}$ incrementally to $\mathbf{K} + \delta\mathbf{K}$ such that δI_2 is maximized. Since $\delta\mathbf{F} = \mathbf{B}\ \delta\mathbf{K}\ \mathbf{C}$ and from Eq.(9.71) and (9.74)

$$\mathbf{F}^T\delta\mathbf{\Lambda} + \delta\mathbf{\Lambda}\ \mathbf{F} = -\delta\mathbf{F}^T\mathbf{\Lambda} - \mathbf{\Lambda}\delta\mathbf{F} - \delta(\mathbf{C}^T\mathbf{K}^T\mathbf{RKC} + \mathbf{Q}) \tag{9.95}$$

$$= -\delta\mathbf{F}^T\mathbf{\Lambda} - \mathbf{\Lambda}\delta\mathbf{F} - \mathbf{C}^T\delta\mathbf{K}^T\mathbf{RKC} - \mathbf{C}^T\mathbf{K}^T\mathbf{R}\ \delta\mathbf{K}\ \mathbf{C} \triangleq -\delta\mathbf{Z} \tag{9.96}$$

$$\mathbf{F}\delta\mathbf{V} + \delta\mathbf{V}\ \mathbf{F}^T + \delta\mathbf{F}\ \mathbf{V} + \mathbf{V}\delta\mathbf{F}^T = \mathbf{0}\,. \tag{9.97}$$

Using Eq.(4.40),

$$(\mathbf{I} \otimes \mathbf{F}^T + \mathbf{F}^T \otimes \mathbf{I})\ \text{col}\ \delta\mathbf{\Lambda} = -\text{col}\ \delta\mathbf{Z} \tag{9.98}$$

$$(\mathbf{I} \otimes \mathbf{F} + \mathbf{F} \otimes \mathbf{I})\ \text{col}\ \delta\mathbf{V} = -\text{col}\ (\delta\mathbf{F}\ \mathbf{V} + \mathbf{V}\delta\mathbf{F}^T) \tag{9.99}$$

the varation $\delta\mathbf{\Lambda}$ and $\delta\mathbf{V}$ can be calculated as a function of $\mathbf{\Lambda}$, $\mathbf{V}$, $\mathbf{F}$ and $\mathbf{K}$ at any stage of an iteration. The matrices $\mathbf{\Lambda}$ and $\mathbf{V}$ are obtained from Eqs.(9.71) and (9.74)

$$(\mathbf{I} \otimes \mathbf{F}^T + \mathbf{F}^T \otimes \mathbf{I})\ \text{col}\ \mathbf{\Lambda} = -\text{col}\ (\mathbf{Q} + \mathbf{C}^T\mathbf{K}^T\mathbf{RKC}) \tag{9.100}$$

$$(\mathbf{I} \otimes \mathbf{F} + \mathbf{F} \otimes \mathbf{I})\ \text{col}\ \mathbf{V} = -\text{col}\ (\mathbf{xx}_o)\,. \tag{9.101}$$

Finally, δI_2 is achieved from Eq.(9.89) by the same operations as applied above.

$$\begin{aligned}\delta I_2 = \mathbf{x}_o^T\delta\Lambda\mathbf{x}_o &+ 2\mathrm{tr}[(\delta\mathbf{V}\ \Lambda + \mathbf{V}\delta\Lambda)\mathbf{Q}_A\Lambda\mathbf{V} + (\delta\mathbf{K}\ \mathbf{C}\mathbf{V}\Lambda + \mathbf{K}\mathbf{C}\delta\mathbf{V}\ \Lambda + \mathbf{K}\mathbf{C}\mathbf{V}\delta\Lambda)\mathbf{Q}_B\Lambda\mathbf{V}\mathbf{C}^T\mathbf{K}^T\\ &+ \delta\mathbf{K}^T\mathbf{J}\mathbf{V}\ \mathbf{Q}_C\mathbf{V}\mathbf{J}^T\mathbf{K} + \mathbf{K}^T(\mathbf{R}\delta\mathbf{K}\ \mathbf{C} + \mathbf{B}^T\Lambda)\mathbf{V}\ \mathbf{Q}_C\mathbf{V}\mathbf{J}^T\mathbf{K} + \mathbf{K}^T\mathbf{J}\delta\mathbf{V}\ \mathbf{Q}_C\mathbf{V}\mathbf{J}^T\mathbf{K}\\ &+ (\mathbf{R}\delta\mathbf{K}\ \mathbf{C} + \mathbf{B}^T\delta\Lambda)\mathbf{V}\mathbf{C}^T\ \mathbf{Q}_K\mathbf{C}\mathbf{V}\mathbf{J}^T + \mathbf{J}\delta\mathbf{V}\ \mathbf{C}^T\mathbf{Q}_K\mathbf{C}\mathbf{V}\mathbf{J}^T]\ . \end{aligned} \tag{9.102}$$

Varying $\delta\mathbf{K}$ the gradient is obtained. Thus, the algorithm is completed in the sense of steepest descent.

Choosing multiple initial values of $\mathbf{K}$, local minima can be avoided.

9.7.8 State Feedback Controller

If all the state variables are available, $\mathbf{C}$ becomes $\mathbf{I}$. Then, from Eq.(9.88) and (9.71)

$$\mathbf{K} = -\mathbf{R}^{-1}\mathbf{B}^T\Lambda \quad \text{or} \quad \mathbf{R}\mathbf{K} = -\mathbf{B}^T\Lambda \tag{9.103}$$

$$\begin{aligned}\mathbf{K}^T\mathbf{R}\mathbf{K} + (\mathbf{A}^T + \mathbf{K}^T\mathbf{B}^T)\Lambda + \Lambda(\mathbf{A} + \mathbf{B}\mathbf{K}) + \mathbf{Q} &= \mathbf{K}^T\mathbf{R}\mathbf{K} + \mathbf{A}^T\Lambda - \mathbf{K}^T\mathbf{R}\mathbf{K} + \Lambda\mathbf{A} + \Lambda\mathbf{B}\mathbf{K} + \mathbf{Q}\\ &= \mathbf{A}^T\Lambda + \Lambda\mathbf{A} - \Lambda\mathbf{B}\mathbf{R}^{-1}\mathbf{B}^T\Lambda + \mathbf{Q} = \mathbf{0}\ . \end{aligned} \tag{9.104}$$

9.7.9 Random Initial State

Hitherto, $\mathbf{x}_o$ was treated as a fixed initial state. If $\mathbf{x}_o$ is a random variable where the mathematical expectation $E\{\mathbf{x}_o\mathbf{x}_o^T\}$ is known, Eq.(9.74) is replaced by

$$\mathbf{F}\mathbf{V} + \mathbf{V}\mathbf{F}^T + E\{\mathbf{x}_o\mathbf{x}_o^T\} = \mathbf{0}\ . \tag{9.105}$$

The performance must be substituted by $E\{I\}$.

9.8 Control with Prespecified Performance

Consider a given linear time-invariant process

$$\dot{\mathbf{x}}(t) = \mathbf{A}\mathbf{x}(t) + \mathbf{B}\mathbf{u}(t), \qquad \mathbf{x}(0) = \mathbf{x}_o \in \mathcal{R}^n \tag{9.106}$$

and a permissible measure or setpoint I_s of the performance index

$$I = \int_0^\infty [\mathbf{x}^T(t)\mathbf{Q}\mathbf{x}(t) + \mathbf{u}^T(t)\mathbf{R}\mathbf{u}(t)]dt\ . \tag{9.107}$$

The problem to be solved is the design of a linear constant state feedback controller $\mathbf{u}(t) = \mathbf{K}\mathbf{x}(t)$, $\mathbf{K} \in \mathcal{R}^{m\times n}$, in order to obtain the prespecified I_s .

Assume that this controller $\mathbf{K}$ lies in the vicinity of the *optimum* Riccati controller $\mathbf{K}^*$. Then, $\mathbf{K}^*$ is calculated from the algebraic Riccati equation

$$\mathbf{A}^T\mathbf{P} + \mathbf{P}\mathbf{A} + \mathbf{Q} - \mathbf{P}\mathbf{B}\mathbf{R}^{-1}\mathbf{B}^T\mathbf{P} = \mathbf{0} \tag{9.108}$$

$$\mathbf{K}^* = -\mathbf{R}^{-1}\mathbf{B}^T\mathbf{P} \tag{9.109}$$

and is applied for the initial matrix $\mathbf{K}_o$ in the following algroithm.

If $I_s > I(\mathbf{K}_o) = I(\mathbf{K}^*)$, increments $\Delta\mathbf{K}$ are used in order to alter $\mathbf{K}_o$ and to achieve a softer control process and to relax the performance $I(\mathbf{K}_o)$ towards I_s. The next step in the loosening process is to arrive at the performance I_1. By Taylor series expansion (*Feliachi, A., 1988*),

$$I_1 = I_o + \delta I = I_o + \left(\frac{\partial I}{\partial K_{11}} \; \vdots \; \frac{\partial I}{\partial K_{12}} \; \cdots \; \frac{\partial I}{\partial K_{mn}}\right) \begin{pmatrix} \Delta K_{11} \\ \Delta K_{12} \\ \vdots \\ \Delta K_{mn} \end{pmatrix} \tag{9.110}$$

where K_{ij} is the entry of $\mathbf{K}$ in the (i,j)-position. Defining a vector of dimension $mn \times 1$

$$\mathbf{k} = \text{ col } (\mathbf{K}^T) \tag{9.111}$$

and a row matrix $\mathbf{M} \in \mathcal{R}^{1 \times mn}$

$$\mathbf{M} = \left(\frac{\partial I}{\partial \mathbf{k}}\right)^T = \frac{\partial I}{\partial \mathbf{k}^T} , \tag{9.112}$$

the first-order increment δI of the performance can be written as

$$\delta I = \mathbf{M} \; \Delta\mathbf{k} = \frac{\partial I}{\partial \mathbf{k}^T}\Delta\mathbf{k} = \frac{\partial I}{\partial \text{ col}^T(\mathbf{K}^T)}\Delta \text{ col } (\mathbf{K}^T). \tag{9.113}$$

For a given δI and I_o the gradient $\partial I/\partial \mathbf{k}^T$ can be calculated. If the objective is to solve Eq.(9.113) with respect to $\Delta\mathbf{k}$, this task is extremely underdetermined. There is only one scalar equation and there are mn unknowns in the vector $\Delta\mathbf{k}$. However, a solution can be obtained by minimizing the norm $\|\Delta\mathbf{k}\|_F$ in the presence of the boundary condition given in Eq.(9.113). This is exactly the problem solved in Eq.(D.5) with the help of the right-pseudo-inverse

$$\Delta\mathbf{k} = \mathbf{M}^{\#R}\delta I = \mathbf{M}^T(\mathbf{M}\mathbf{M}^T)^{-1}\delta I = \left(\frac{\partial I}{\partial \mathbf{k}}\right) [\left(\frac{\partial I}{\partial \mathbf{k}^T}\right)\left(\frac{\partial I}{\partial \mathbf{k}}\right)]^{-1} \; \delta I \; . \tag{9.114}$$

The gradient $\partial I/\partial \mathbf{k}$ is easily determined from $\partial I/\partial \mathbf{K}$ by rearranging. Suppose $\mathbf{x}_o$ is independent of $\mathbf{K}$ the matricial gradient from Eq.(5.87) is

$$\frac{\partial I}{\partial \mathbf{K}} = \frac{\partial \mathbf{x}_o^T\mathbf{P}\mathbf{x}_o}{\partial \mathbf{K}} = (\mathbf{I}_m \otimes \mathbf{x}_o^T)\frac{\partial \mathbf{P}\mathbf{x}_o}{\partial \mathbf{K}} = (\mathbf{I}_m \otimes \mathbf{x}_o^T)\frac{\partial \mathbf{P}}{\partial \mathbf{K}}(\mathbf{I}_n \otimes \mathbf{x}_o) \; . \tag{9.115}$$

The sensitivity function $\partial\mathbf{P}/\partial\mathbf{K}$ results from Lyapunov equation

$$(\mathbf{A}+\mathbf{B}\mathbf{K})^T\mathbf{P} + \mathbf{P}(\mathbf{A}+\mathbf{B}\mathbf{K}) + \mathbf{Q} + \mathbf{K}^T\mathbf{R}\mathbf{K} = \mathbf{0} \tag{9.116}$$

by deriving the product terms with respect to $\mathbf{K}$. Noting that $\mathbf{A}, \mathbf{B}, \mathbf{Q}, \mathbf{R}$ do not depend on $\mathbf{K}$, some components of Eq.(9.116) are derived first:

$$\frac{\partial(\mathbf{A}+\mathbf{B}\mathbf{K})}{\partial \mathbf{K}} = \frac{\partial \mathbf{B}\mathbf{K}}{\partial \mathbf{K}} = \frac{\partial \mathbf{B}}{\partial \mathbf{K}}(\mathbf{I}_n \otimes \mathbf{K}) + (\mathbf{I}_m \otimes \mathbf{B})\frac{\partial \mathbf{K}}{\partial \mathbf{K}} = (\mathbf{I}_m \otimes \mathbf{B})\frac{\partial \mathbf{K}}{\partial \mathbf{K}} \tag{9.117}$$

$$\frac{\partial(\mathbf{A}+\mathbf{B}\mathbf{K})^T}{\partial \mathbf{K}} = \frac{\partial \mathbf{K}^T\mathbf{B}^T}{\partial \mathbf{K}} = \frac{\partial \mathbf{K}^T}{\partial \mathbf{K}}(\mathbf{I}_n \otimes \mathbf{B}^T) \tag{9.118}$$

$$\frac{\partial \mathbf{P}(\mathbf{A}+\mathbf{BK})}{\partial \mathbf{K}} = \frac{\partial \mathbf{P}}{\partial \mathbf{K}}[\mathbf{I}_n \otimes (\mathbf{A}+\mathbf{BK})] + (\mathbf{I}_m \otimes \mathbf{P})\frac{\partial(\mathbf{A}+\mathbf{BK})}{\partial \mathbf{K}} \tag{9.119}$$

$$\frac{\partial(\mathbf{A}+\mathbf{BK})^T\mathbf{P}}{\partial \mathbf{K}} = \frac{\partial(\mathbf{A}+\mathbf{BK})^T}{\partial \mathbf{K}}(\mathbf{I}_n \otimes \mathbf{P}) + [\mathbf{I}_m \otimes (\mathbf{A}+\mathbf{BK})^T]\frac{\partial \mathbf{P}}{\partial \mathbf{K}} \tag{9.120}$$

$$\begin{aligned} \frac{\partial \mathbf{K}^T\mathbf{RK}}{\partial \mathbf{K}} &= \frac{\partial \mathbf{K}^T}{\partial \mathbf{K}}(\mathbf{I}_n \otimes \mathbf{RK}) + (\mathbf{I}_m \otimes \mathbf{K}^T)\frac{\partial \mathbf{RK}}{\partial \mathbf{K}} && (9.121)\\ &= \frac{\partial \mathbf{K}^T}{\partial \mathbf{K}}[\mathbf{I}_n \otimes (\mathbf{RK})] + [\mathbf{I}_m \otimes (\mathbf{K}^T\mathbf{R})]\frac{\partial \mathbf{K}}{\partial \mathbf{K}} \,. && (9.122) \end{aligned}$$

Using these formulas, the derivative of the Lyapunov equation Eq.(9.116) can be written as

$$\begin{aligned} [\mathbf{I}_m \;\otimes\; &(\mathbf{A}+\mathbf{BK})^T]\frac{\partial \mathbf{P}}{\partial \mathbf{K}} + \frac{\partial \mathbf{K}^T}{\partial \mathbf{K}}(\mathbf{I}_n \otimes \mathbf{B}^T)(\mathbf{I}_n \otimes \mathbf{P}) + \frac{\partial \mathbf{P}}{\partial \mathbf{K}}[\mathbf{I}_n \otimes (\mathbf{A}+\mathbf{BK})] \\ + \;&(\mathbf{I}_m \otimes \mathbf{P})(\mathbf{I}_m \otimes \mathbf{B})\frac{\partial \mathbf{K}}{\partial \mathbf{K}} + \frac{\partial \mathbf{K}^T}{\partial \mathbf{K}}[\mathbf{I}_n \otimes (\mathbf{RK})] + [\mathbf{I}_m \otimes (\mathbf{K}^T\mathbf{R})]\frac{\partial \mathbf{K}}{\partial \mathbf{K}} = \mathbf{0} \,. \end{aligned} \tag{9.123}$$

Applying

$$\frac{\partial \mathbf{K}^T}{\partial \mathbf{K}} = \mathbf{U}_{mn} = \mathbf{U}_{mn}^{(mn \times mn)} \quad \text{and} \quad \frac{\partial \mathbf{K}}{\partial \mathbf{K}} = \bar{\mathbf{U}}_{mn} = \bar{\mathbf{U}}_{mn}^{(m^2 \times n^2)} \,, \tag{9.124}$$

it results

$$\begin{aligned} &[\mathbf{I}_m \otimes (\mathbf{A}+\mathbf{BK})^T]\frac{\partial \mathbf{P}}{\partial \mathbf{K}} + \frac{\partial \mathbf{P}}{\partial \mathbf{K}}[\mathbf{I}_n \otimes (\mathbf{A}+\mathbf{BK})] \\ &\qquad + \mathbf{U}_{mn}[\mathbf{I}_n \otimes (\mathbf{RK}+\mathbf{B}^T\mathbf{P})] + [\mathbf{I}_m \otimes (\mathbf{K}^T\mathbf{R}+\mathbf{PB})]\bar{\mathbf{U}}_{mn} = \mathbf{0} \,. \end{aligned} \tag{9.125}$$

Remark: If Eq.(9.109) is satisfied (Riccati optimum controller), from Eq.(9.125) one has $\partial\mathbf{P}/\partial\mathbf{K} = \mathbf{0}$ and from Eq.(9.115) $\partial I/\partial\mathbf{K} = \mathbf{0}$.

Finally, the jth step of the iteration is

$$\mathbf{k}_{j+1} = \mathbf{k}_j + [I_s - I(\mathbf{k}_j)] \left(\frac{\partial I}{\partial \mathbf{k}}\right)_j \left[\left(\frac{\partial I}{\partial \mathbf{k}^T}\right)_j \left(\frac{\partial I}{\partial \mathbf{k}}\right)_j\right]^{-1} \tag{9.126}$$

where $[\cdot]^{-1}$ above is a scalar. The sensitivity $\partial\mathbf{P}/\partial\mathbf{K}$ is obtained from Eq.(9.125) and the performance matricial gradient $\partial I/\partial\mathbf{K}$ from Eq.(9.115). Note that $\partial\mathbf{P}/\partial\mathbf{K}$ is an (mn, nn)-matrix of high dimension. The computation runs until a stopping condition is satisfied.

Chapter 10

Desensitizing Control

Consider the plant

$$\dot{\mathbf{x}}(t) = \mathbf{A}(\mathbf{p})\mathbf{x}(t) + \mathbf{B}(\mathbf{p})\mathbf{u}(t) \qquad \mathbf{B} \in \mathcal{R}^{n \times m}, \quad \mathbf{p} \in \mathcal{R}^{n_p} \tag{10.1}$$

where $\mathbf{A}(\mathbf{p})$ and $\mathbf{B}(\mathbf{p})$ are matrix-valued functions of a slowly varying parameter vector $\mathbf{p}$. Hence, $\mathbf{x}(t)$ depends on $\mathbf{p}$. The subscript $_o$ denotes the nominal values. Assume that a quadratic performance has to be minimized

$$I = \int_0^\infty [\mathbf{x}^T(t)\mathbf{Q}\mathbf{x}(t) + \mathbf{u}^T(t)\mathbf{R}\mathbf{u}(t)]dt \tag{10.2}$$

and the optimal control variable $\mathbf{u}(t)$ is

$$\mathbf{u}^*(t) = \mathbf{K}\mathbf{x}(t) = -\mathbf{R}^{-1}\mathbf{B}_o\mathbf{P}\mathbf{x}(t) . \tag{10.3}$$

The matrix $\mathbf{P}$ is given by the algebraic Riccati equation

$$\mathbf{A}_o^T\mathbf{P} + \mathbf{P}\mathbf{A}_o + \mathbf{Q} - \mathbf{P}\mathbf{B}_o\mathbf{R}^{-1}\mathbf{B}_o^T\mathbf{P} = 0 . \tag{10.4}$$

Then, $\mathbf{u}^*(t)$ also is a function of $\mathbf{p}$. Studying the differential sensitivity of the system state variable with respect to variations $\mathbf{p}$, the state variable differential sensitivity (trajectory differential sensitivity) $\boldsymbol{\sigma}_i$ is

$$\boldsymbol{\sigma}_i(t) - \frac{\partial}{\partial p_i}\mathbf{x}(t) \mid_{\mathbf{p}=\mathbf{p}_o} \qquad \frac{\partial \mathbf{x}(t)}{\partial \mathbf{p}} = [\boldsymbol{\sigma}_1^T(t) \vdots \boldsymbol{\sigma}_2^T(t) \ldots \boldsymbol{\sigma}_{n_p}^T(t)]^T = \text{vec}\, \boldsymbol{\sigma}_i(t) \tag{10.5}$$

(*Cruz, I., 1973*). Differentiating Eq.(10.1) with respect to p_i ,

$$\dot{\boldsymbol{\sigma}}_i(t) = \frac{\partial \mathbf{A}}{\partial p_i}|_{\mathbf{p}_o}\mathbf{x}_o(t) + \mathbf{A}_o\boldsymbol{\sigma}_i(t) + \frac{\partial \mathbf{B}}{\partial p_i}|_{\mathbf{p}_o}\mathbf{u}_o(t) + \mathbf{B}_o\frac{\partial \mathbf{u}(t)}{\partial p_i} . \tag{10.6}$$

This sensitivity variable can be weighted according to the index

$$I_\sigma = \sum_{i=1}^{n_p} \int_0^\infty \boldsymbol{\sigma}_i^T(t)\mathbf{Q}_\sigma\boldsymbol{\sigma}_i(t)dt . \tag{10.7}$$

Combined weighting of performance, control effort and differential sensitivity leads to minimizing of the sum

$$I_{xu\sigma} = I + I_\sigma . \tag{10.8}$$

10.1 Additional Differential Sensitivity Feedback

Let the optimal feedback $\mathbf{u}^*(t) = \mathbf{K}\mathbf{x}(t)$ be augmented by an additional feedback given by the sensitivity variables $\boldsymbol{\sigma}_j(t)$

$$\mathbf{u}(t) = \mathbf{K}\mathbf{x}(t) + \sum_{j=1}^{n_p} \mathbf{K}_j \boldsymbol{\sigma}_j(t) . \tag{10.9}$$

Equating $\mathbf{u}_o = \mathbf{u}$, $\mathbf{x}_o = \mathbf{x}$, combining with Eq.(10.6) and omitting the argument t yields

$$\dot{\boldsymbol{\sigma}}_i = \frac{\partial \mathbf{A}}{\partial p_i}\mathbf{x} + \mathbf{A}_o \boldsymbol{\sigma}_i + \frac{\partial \mathbf{B}}{\partial p_i}(\mathbf{K}\mathbf{x} + \sum_{j=1}^{n_p} \mathbf{K}_j \boldsymbol{\sigma}_j) + \mathbf{B}_o(\mathbf{K}\frac{\partial \mathbf{x}}{\partial p_i} + \sum_{j=1}^{n_p} \mathbf{K}_j \frac{\partial \boldsymbol{\sigma}_j}{\partial p_i}) . \tag{10.10}$$

It has been shown by *Newmann, M.M., 1970* that the nominal problem of Eqs.(10.1) through (10.9) has no solution. However, neglecting the last term on the right-hand side containing $\partial \boldsymbol{\sigma}_j / \partial p_i$ (*Kreindler, E., 1968; Fleming, P.J., Newmann, M.M., 1977*), defining a new sensitivity variable $\boldsymbol{\rho}_i$ and resubstituting Eq.(10.9) yields

$$\dot{\boldsymbol{\rho}}_i = (\mathbf{A}_o + \mathbf{B}_o \mathbf{K})\boldsymbol{\rho}_i + \frac{\partial \mathbf{A}}{\partial p_i}\mathbf{x} + \frac{\partial \mathbf{B}}{\partial p_i}\mathbf{u} . \tag{10.11}$$

Combining with Eq.(10.1),

$$\begin{pmatrix} \dot{\mathbf{x}} \\ \dot{\boldsymbol{\rho}}_i \end{pmatrix} = \begin{pmatrix} \mathbf{A}_o & \vdots & \mathbf{0} \\ \frac{\partial \mathbf{A}}{\partial p_i}|_{\mathbf{p}_o} & \vdots & \mathbf{A}_o + \mathbf{B}_o \mathbf{K} \end{pmatrix} \begin{pmatrix} \mathbf{x} \\ \boldsymbol{\rho}_i \end{pmatrix} + \begin{pmatrix} \mathbf{B}_o \\ \frac{\partial \mathbf{B}}{\partial p_i} \end{pmatrix} \mathbf{u} . \tag{10.12}$$

Minimization of Eq.(10.8) can be obtained by applying LQ-algorithm to the extended system of Eq.(10.12). Since its system matrix depends on the unknown controller matrix $\mathbf{K}$, the algorithm has to be iterated (*Weinmann, A., 1988, Vol.3, p. 226*). The new matrix $\mathbf{K}$ is given by the result of the previous Riccati calculus. Assuming convergence, the algorithm is repeated until a stopping condition is satisfied, e.g., $\|\Delta \mathbf{K}\| < \Delta K_o$ (*Breinl, W., 1980*).

10.2 Replacing $\partial \mathbf{u}/\partial p_i$ by x and by the Sensitivity Signal

10.2.1 Fixed Combination

Following *Elmetwally, E.E., and Rao, N.D., 1974*,

$$\partial \mathbf{u}/\partial p_i = \mathbf{N}_x \mathbf{x} + \mathbf{N}_\sigma \boldsymbol{\sigma}_i \tag{10.13}$$

is chosen where $\mathbf{N}_x$ and $\mathbf{N}_\sigma$ are fixed matrices. Substituting this assumption of $\partial \mathbf{u}/\partial p_i$ into Eq.(10.6) yields

$$\begin{pmatrix} \dot{\mathbf{x}} \\ \dot{\boldsymbol{\sigma}}_i \end{pmatrix} = \begin{pmatrix} \mathbf{A}_o & \vdots & \mathbf{0} \\ \frac{\partial \mathbf{A}}{\partial p_i}|_{\mathbf{p}_o} + \mathbf{B}_o \mathbf{N}_x & \vdots & \mathbf{A}_o + \mathbf{B}_o \mathbf{N}_\sigma \end{pmatrix} \begin{pmatrix} \mathbf{x} \\ \boldsymbol{\sigma}_i \end{pmatrix} + \begin{pmatrix} \mathbf{B}_o \\ \frac{\partial \mathbf{B}}{\partial p_i}|_{\mathbf{p}_o} \end{pmatrix} \mathbf{u} . \tag{10.14}$$

10.2.2 Using a Lyapunov Equation

Byrne, P.C., and Burke, M., 1976 introduced

$$\frac{\partial \mathbf{u}}{\partial p_i} = -\mathbf{R}^{-1}(\mathbf{B}_o^T\mathbf{P}_{p_i} + \frac{\partial \mathbf{B}}{\partial p_i}|_{\mathbf{p}_o}\mathbf{P})\mathbf{x} + \mathbf{K}_o\boldsymbol{\sigma}_i \tag{10.15}$$

where $\mathbf{K}_o = \mathbf{K}$ as given by Eq.(10.3). The matrix $\mathbf{P}$ follows from Eq.(10.4) and $\mathbf{P}_{p_i}$ is the result of a Lyapunov equation in $\mathbf{P}_{p_i}$, i.e.,

$$\begin{aligned}\mathbf{P}_{p_i}(\mathbf{A}_o + \mathbf{B}_o\mathbf{K}_o) &+ (\mathbf{A}_o + \mathbf{B}_o\mathbf{K}_o)^T\mathbf{P}_{p_i} + (\mathbf{P}\frac{\partial \mathbf{A}}{\partial p_i}|_{\mathbf{p}_o} + \frac{\partial \mathbf{A}^T}{\partial p_i}|_{\mathbf{p}_o}\mathbf{P}) \\ &- \mathbf{P}(\frac{\partial \mathbf{B}}{\partial p_i}|_{\mathbf{p}_o}\mathbf{R}^{-1}\mathbf{B}_o^T + \mathbf{B}_o\mathbf{R}^{-1}\frac{\partial \mathbf{B}^T}{\partial p_i}|_{\mathbf{p}_o})\mathbf{P} = 0 \,.\end{aligned} \tag{10.16}$$

Combining Eqs.(10.15) and (10.6) yields

$$\begin{pmatrix}\dot{\mathbf{x}} \\ \dot{\boldsymbol{\sigma}}_i\end{pmatrix} \begin{pmatrix} \mathbf{A}_o & \vdots & 0 \\ \frac{\partial \mathbf{A}}{\partial p_i}|_{\mathbf{p}_o} - \mathbf{B}_o\mathbf{R}^{-1}(\mathbf{B}_o^T\mathbf{P}_{p_i} + \frac{\partial \mathbf{B}}{\partial p_i}|_{\mathbf{p}_o})\mathbf{P} & \vdots & \mathbf{A}_o + \mathbf{B}_o\mathbf{K}_o\end{pmatrix}\begin{pmatrix}\mathbf{x} \\ \boldsymbol{\sigma}_i\end{pmatrix} + \begin{pmatrix}\mathbf{B}_o \\ \frac{\partial \mathbf{B}}{\partial p_i}|_{\mathbf{p}_o}\end{pmatrix}\mathbf{u} \,. \tag{10.17}$$

10.2.3 Minimization by One-Step Riccati Solution

A resulting criterion $I_{xu\sigma}$ is established to weight $\mathbf{x}, \boldsymbol{\sigma}_i$ and $\mathbf{u}$

$$I_{xu\sigma} = I + I_\sigma = \int_0^\infty [\begin{pmatrix}\mathbf{x} \\ \boldsymbol{\sigma}_i\end{pmatrix}^T \begin{pmatrix}\mathbf{Q} & 0 \\ 0 & \mathbf{Q}_\sigma\end{pmatrix}\begin{pmatrix}\mathbf{x} \\ \boldsymbol{\sigma}_i\end{pmatrix} + \mathbf{u}^T\mathbf{R}\mathbf{u}]\, dt \,. \tag{10.18}$$

Under the constraint of the differential equation of the extended system $\binom{\mathbf{x}}{\boldsymbol{\sigma}}$ as given in the previous subsections, the minimization of $I_{xu\sigma}$ can be carried out by a one-step Riccati scheme.

10.3 Extended State Controller

Consider the plant with the state vector $\mathbf{x}(t)$ as given by Eq.(10.1) and the sensitivity vector $\boldsymbol{\sigma} \triangleq \partial\mathbf{x}/\partial p$ with respect to one parameter p, only ($n_p = 1$, $p_i = p$). Using Eq.(10.6) yields

$$\dot{\boldsymbol{\sigma}}(t) = \mathbf{A}_p\mathbf{x}(t) + \mathbf{A}\boldsymbol{\sigma}(t) + \mathbf{B}_p\mathbf{u} + \mathbf{B}\mathbf{u}_p \quad \text{where} \quad \mathbf{A}_p = \partial\mathbf{A}/\partial p \text{ etc.}\,. \tag{10.19}$$

Assume a combined linear state and sensitivity feedback controller and the sensitivity of its signals with respect to p

$$\mathbf{u}(t) = \mathbf{K}_1\mathbf{x}(t) + \mathbf{K}_2\boldsymbol{\sigma}(t) \quad \rightsquigarrow \quad \mathbf{u}_p = \frac{\partial \mathbf{u}(t)}{\partial p} = \mathbf{K}_1\boldsymbol{\sigma} + \mathbf{K}_2\frac{\partial \boldsymbol{\sigma}}{\partial p} \,. \tag{10.20}$$

Substitution into Eq.(10.19) yields

$$\dot{\boldsymbol{\sigma}} = (\mathbf{A}_p + \mathbf{B}_p\mathbf{K}_1)\mathbf{x} + (\mathbf{A} + \mathbf{B}\mathbf{K}_1 + \mathbf{B}_p\mathbf{K}_2)\boldsymbol{\sigma} + \mathbf{B}\mathbf{K}_2\frac{\partial \boldsymbol{\sigma}}{\partial p} \,. \tag{10.21}$$

Neglecting the last expression and taking the derivatives at p_o, only, the equation for a new variable $\rho(t)$ is

$$\dot{\rho} = (\mathbf{A}_{p_o} + \mathbf{B}_{p_o}\mathbf{K}_1)\mathbf{x} + (\mathbf{A} + \mathbf{B}\mathbf{K}_1 + \mathbf{B}_{p_o}\mathbf{K}_2)\rho \tag{10.22}$$

where $\mathbf{A}_{p_o} \triangleq \partial\mathbf{A}/\partial p|_{p_o}$ etc. Reducing the controller assumption to $\mathbf{u} = \mathbf{K}_1\mathbf{x} + \mathbf{K}_2\rho$ and substituting this control algorithm into Eq.(10.6) yields

$$\dot{\mathbf{x}} = (\mathbf{A} + \mathbf{B}\mathbf{K}_1)\mathbf{x} + \mathbf{B}\mathbf{K}_2\rho \tag{10.23}$$

$$\dot{\sigma} = \frac{\partial\dot{\mathbf{x}}}{\partial p} = (\mathbf{A}_p + \mathbf{B}_p\mathbf{K}_1)\mathbf{x} + \mathbf{B}_p\mathbf{K}_2\rho + (\mathbf{A} + \mathbf{B}\mathbf{K}_1)\sigma + \mathbf{B}\mathbf{K}_2\frac{\partial\rho}{\partial p}\,. \tag{10.24}$$

From Eq.(10.22), remembering that $\mathbf{A}_o, \mathbf{A}_{p_o}, \mathbf{B}_o$ etc. are constant (*Fleming, P.J., and Newmann, M.M., 1977*)

$$\frac{\partial\dot{\rho}}{\partial p} = (\mathbf{A}_{p_o} + \mathbf{B}_{p_o}\mathbf{K}_1)\sigma + (\mathbf{A}_o + \mathbf{B}_o\mathbf{K}_1 + \mathbf{B}_{p_o}\mathbf{K}_2)\frac{\partial\rho}{\partial\alpha}\,. \tag{10.25}$$

The initial conditions $\sigma(0), \rho(0), \partial\rho/\partial p(0)$ may be regarded zero. Combination of Eqs. (10.23), (10.22), (10.21) and (10.25), in this order, yields

$$\dot{\mathbf{v}} \triangleq \begin{pmatrix} \dot{\mathbf{x}} \\ \dot{\rho} \\ \dot{\sigma} \\ \frac{\partial\dot{\rho}}{\partial\alpha} \end{pmatrix} = \begin{pmatrix} \mathbf{A}+\mathbf{B}\mathbf{K}_1 & -\mathbf{B}\mathbf{K}_2 & 0 & 0 \\ \mathbf{A}_{p_o}+\mathbf{B}_{p_o}\mathbf{K}_1 & \mathbf{A}_o+\mathbf{B}_o\mathbf{K}_1+\mathbf{B}_{p_o}\mathbf{K}_2 & 0 & 0 \\ \mathbf{A}_p+\mathbf{B}_p\mathbf{K}_1 & \mathbf{B}_p\mathbf{K}_2 & \mathbf{A}+\mathbf{B}\mathbf{K}_1 & \mathbf{B}\mathbf{K}_2 \\ 0 & 0 & \mathbf{A}_{p_o}+\mathbf{B}_{p_o}\mathbf{K}_1 & \mathbf{A}_o+\mathbf{B}_o\mathbf{K}_1+\mathbf{B}_{p_o}\mathbf{K}_2 \end{pmatrix} \begin{pmatrix} \mathbf{x} \\ \rho \\ \sigma \\ \frac{\partial\rho}{\partial p} \end{pmatrix} \tag{10.26}$$

The block-triangle matrix above can be rewritten to

$$\dot{\mathbf{v}} = \bar{\mathbf{A}}\mathbf{v} = (\mathbf{A}_v + \sum_{i=1}^{4}\mathbf{B}_i\mathbf{K}\mathbf{E}_i)\mathbf{v} \qquad \bar{\mathbf{A}} \in \mathcal{R}^{4n\times 4n} \tag{10.27}$$

where $\mathbf{K} \triangleq (\mathbf{K}_1 \vdots \mathbf{K}_2) \in \mathcal{R}^{m\times 2n}$, $\mathbf{B}_i \in \mathcal{R}^{4n\times m}$ and

$$\bar{\mathbf{A}}_v \triangleq \begin{pmatrix} \mathbf{A} & 0 & \mathbf{A}_p & 0 \\ \mathbf{A}_{p_o} & \mathbf{A}_{p_o} & 0 & 0 \\ \mathbf{A}_p & 0 & \mathbf{A} & 0 \\ 0 & 0 & \mathbf{A}_{p_o} & \mathbf{A}_{p_o} \end{pmatrix},\ \mathbf{B}_1 \triangleq \begin{pmatrix} \mathbf{B} \\ \mathbf{B}_{p_o} \\ \mathbf{B}_p \\ 0 \end{pmatrix},\ \mathbf{B}_2 \triangleq \begin{pmatrix} 0 \\ 0 \\ \mathbf{B} \\ \mathbf{B}_{p_o} \end{pmatrix},\ \mathbf{B}_3 \triangleq \begin{pmatrix} 0 \\ \mathbf{B}_{p_o} \\ 0 \\ 0 \end{pmatrix},\ \mathbf{B}_4 \triangleq \begin{pmatrix} 0 \\ 0 \\ 0 \\ \mathbf{B}_{p_o} \end{pmatrix} \tag{10.28}$$

$$\mathbf{E}_1 \triangleq \begin{pmatrix} \mathbf{I}_n & 0 & 0 & 0 \\ 0 & \mathbf{I}_n & 0 & 0 \end{pmatrix},\ \mathbf{E}_2 \triangleq \begin{pmatrix} 0 & 0 & \mathbf{I}_n & 0 \\ 0 & 0 & 0 & \mathbf{I}_n \end{pmatrix},\ \mathbf{E}_3 \triangleq \begin{pmatrix} 0 & \mathbf{I}_n & 0 & 0 \\ 0 & 0 & 0 & 0 \end{pmatrix},\ \mathbf{E}_4 \triangleq \begin{pmatrix} 0 & 0 & 0 & \mathbf{I}_n \\ 0 & 0 & 0 & 0 \end{pmatrix} \tag{10.29}$$

$$\mathbf{K}\mathbf{E}_1 \triangleq (\mathbf{K}_1\ \mathbf{K}_2\ 0\ 0),\ \mathbf{K}\mathbf{E}_2 \triangleq (0\ 0\ \mathbf{K}_1\ \mathbf{K}_2),\ \mathbf{K}\mathbf{E}_3 \triangleq (0\ \mathbf{K}_1\ 0\ 0),\ \mathbf{K}\mathbf{E}_4 \triangleq (0\ 0\ 0\ \mathbf{K}_1)\,. \tag{10.30}$$

10.3.1 Structure of the Controller

With regard to the block-triangle structure the variables $\mathbf{x}$ and ρ are decoupled from σ and $\partial\rho/\partial p$. The first and the second row in Eq.(10.26) are sufficient to realize the structure of the controller.

10.3.2 Minimization of a Performance Index

The entire system Eq.(10.26) is used to calculate the performance index

$$I = \int_0^\infty [\mathbf{x}^T(t)\mathbf{Q}\mathbf{x}(t) + \boldsymbol{\sigma}^T(t)\mathbf{Q}_\sigma\boldsymbol{\sigma}(t) + \mathbf{u}^T(t)\mathbf{R}\mathbf{u}(t)]dt \ . \tag{10.31}$$

At the expense of a high dimension, the problem of improving the approximation of $\partial\sigma/\partial p$ is solved. Before minimizing, I is rewritten to

$$I = \int_0^\infty \mathbf{v}^T(\bar{\mathbf{Q}} + \mathbf{E}_1^T\mathbf{K}^T\mathbf{R}\mathbf{K}\mathbf{E}_1)\mathbf{v}dt \quad \text{where} \quad \mathbf{Q} = \text{block diag } \{\mathbf{Q},\ \mathbf{0},\ \mathbf{Q}_\sigma,\ \mathbf{0}\} \ . \tag{10.32}$$

Invoking Eqs.(16.83) and (16.84),

$$I = \text{tr } \{\int_0^\infty \boldsymbol{\Phi}[\mathbf{Q} + \mathbf{K}^T\mathbf{R}\mathbf{K}]\boldsymbol{\Phi}\ dt\ \mathbf{x}_o\mathbf{x}_o^T\} = \text{tr } \{\boldsymbol{\Lambda}\mathbf{x}_o\mathbf{x}_o^T\} \ . \tag{10.33}$$

The symmetric matrix $\boldsymbol{\Lambda}$ is the solution of Eq.(16.50) which results subject to the constraint

$$\mathbf{L} \triangleq \boldsymbol{\Lambda}\bar{\mathbf{A}} + \bar{\mathbf{A}}^T\boldsymbol{\Lambda} + \bar{\mathbf{Q}} + \mathbf{E}_1^T\mathbf{K}^T\mathbf{R}\mathbf{K}\mathbf{E}_1 = \mathbf{0} \ . \tag{10.34}$$

Minimizing I with the constraint above leads to

$$I_1 = I + \text{tr } \mathbf{V}^T\mathbf{L} \tag{10.35}$$

where $\mathbf{V}$ is a Lagrange multiplier matrix of dimension $4n \times 4n$. Necessary conditions for the minimum are

$$\frac{\partial I_1}{\partial \mathbf{K}} = 2\sum_{i=1}^{4}\mathbf{B}_i^T\boldsymbol{\Lambda}\mathbf{V}\mathbf{E}_i^T + 2\mathbf{R}\mathbf{K}\mathbf{E}_1\mathbf{V}\mathbf{E}_1^T = \mathbf{0} \tag{10.36}$$

$$\frac{\partial I_1}{\partial \boldsymbol{\Lambda}} = \mathbf{x}_o\mathbf{x}_o^T + \mathbf{V}\bar{\mathbf{A}}^T + \bar{\mathbf{A}}\mathbf{V} = \mathbf{0} \tag{10.37}$$

and $\partial I_1/\partial\mathbf{V} = \mathbf{0}$ which is identical to Eq.(10.34).

10.4 Desensitizing Without Differential-Sensitivity Feedback

The results obtained in Eq.(10.26) can be simplified by using a state feedback, only. The controller structure

$$\mathbf{u}(t) = \mathbf{K}_1\mathbf{x}(t) \qquad (\mathbf{K}_2 \equiv \mathbf{0}) \tag{10.38}$$

leads to a system $\mathbf{w} \triangleq (\mathbf{x}^T\boldsymbol{\sigma}^T)^T$ which can be found by simply cancelling second and forth row and column both in Eq.(10.26) and in $\bar{\mathbf{A}}_v, \mathbf{B}_1, \mathbf{B}_2$. The matrices $\mathbf{B}_3 = \mathbf{B}_4 = \mathbf{0}$ are out of action. The algorithm of Eq.(10.34), thus, is simplified substantially.

Applying a state feedback $\mathbf{u}(t) = \mathbf{K}\mathbf{x}(t)$ with constant $\mathbf{K}$, only, and minimizing $I_{xu\sigma}$ yields a simple but effective control strategy, see *Rillings, J.H., and Roy, R.J., 1970* for a plant $\dot{\mathbf{x}}(t) = \mathbf{f}(\mathbf{x}, \mathbf{u}, p_i, t)$ with vector-valued parameter $\mathbf{p}$.

10.5 Desensitizing by Increasing Q

Increasing the weighing matrix $\mathbf{Q}$ to $\beta\mathbf{Q}$ $(\beta > 0)$ causes a noticable differential-sensitivity reduction (*Subbayyan, R., et al. 1978*). Increasing $\mathbf{Q}$ to $\mathbf{Q} + \beta\mathbf{S}$ yields improved results (*Verde, C., and Frank, P.M., 1987 and 1988*). The matrix $\mathbf{S}$ is found by the following procedure. Corresponding to the system equations and to the differential-sensitivity equations

$$\dot{\mathbf{x}}(t) = [\mathbf{A}(p) + \mathbf{B}(p)\mathbf{K}]\mathbf{x}(t) \qquad \mathbf{x}(0) = \mathbf{x}_o \tag{10.39}$$

$$\dot{\boldsymbol{\sigma}}_i(t) = [\frac{\partial \mathbf{A}}{\partial p_i}|_{\mathbf{p}_o} + \frac{\partial \mathbf{B}}{\partial p_i}|_{\mathbf{p}_o}\mathbf{K}]\ \mathbf{x}(t) - (\mathbf{A}_o + \mathbf{B}_o\mathbf{K})\boldsymbol{\sigma}_i(t) \qquad \boldsymbol{\sigma}_i(0) = \mathbf{0} \tag{10.40}$$

and to the performance index

$$I_\sigma = \int_0^\infty \sum_{i=1}^{n_p} \boldsymbol{\sigma}_i^T(t)\mathbf{Q}_{\sigma_i}\boldsymbol{\sigma}_i(t)\ dt \tag{10.41}$$

the Lyapunov equations below yield the matrices $\mathbf{H}_i$, namely

$$\tilde{\mathbf{A}}_i\mathbf{H}_i + \mathbf{H}_i\tilde{\mathbf{A}}_i = -\tilde{\mathbf{Q}}_{\sigma_i}, \qquad \mathbf{H}_i \in \mathcal{R}^{2n\times 2n} \qquad \text{where} \tag{10.42}$$

$$\tilde{\mathbf{A}}_i = \begin{pmatrix} \mathbf{A}_o + \mathbf{B}_o\mathbf{K} & \mathbf{0} \\ \frac{\partial \mathbf{A}}{\partial p_i}|_{\mathbf{p}_o} + \frac{\partial \mathbf{B}}{\partial p_i}|_{\mathbf{p}_o}\mathbf{K} & \vdots\ \mathbf{A}_o + \mathbf{B}_o\mathbf{K} \end{pmatrix} \qquad \tilde{\mathbf{Q}}_{\sigma i} = \begin{pmatrix} 0 & 0 \\ 0 & \mathbf{Q}_{\sigma i} \end{pmatrix} . \tag{10.43}$$

Partitioning of $\mathbf{H}_i$ and taking the sum

$$\mathbf{H} = \sum_{i=1}^{n_p} \mathbf{H}_i = \sum_{i=1}^{n_p} \begin{pmatrix} \mathbf{S}_i & \mathbf{P}_{i12} \\ \mathbf{P}_{i12}^T & \mathbf{P}_{i22} \end{pmatrix} \triangleq \begin{pmatrix} \mathbf{S} & \mathbf{P}_{12} \\ \mathbf{P}_{12}^T & \mathbf{P}_{22} \end{pmatrix} \tag{10.44}$$

yields the matrix $\mathbf{S}$. The matrix $\mathbf{S}$ is that partition of $\mathbf{H}$ which gives the interrelation between $\mathbf{x}_o$ and the resulting performance. Hence, $\mathbf{S}$ is an adequate matricial increment to change $\mathbf{Q}$. The strength of change can be varied by β.

10.6 Implicit and Virtual Model-Following Approach

Consider the plant state variable as given in Eq.(10.1), an output $\mathbf{y} = \mathbf{Cx}$ and an output feedback $\mathbf{u} = \mathbf{Ky}$. The output matrix $\mathbf{C}$ is considered independent of p . Assume $n_p = 1$. Partially differentiating with respect to p and combining with Eq.(10.1) yields

$$\begin{pmatrix} \dot{\mathbf{x}} \\ \dot{\boldsymbol{\sigma}} \end{pmatrix} = \begin{pmatrix} \mathbf{A} + \mathbf{BKC} & \mathbf{0} \\ \frac{\partial \mathbf{A}}{\partial p}|_{\mathbf{p}_o} + \frac{\partial \mathbf{B}}{\partial p}|_{\mathbf{p}_o}\mathbf{KC} & \vdots\ \mathbf{A} + \mathbf{BKC} \end{pmatrix} \begin{pmatrix} \mathbf{x} \\ \boldsymbol{\sigma} \end{pmatrix} . \tag{10.45}$$

Fleming, P.J., 1978 suggested a two-stage design method. From a first stage a closed-loop system behaviour with a system matrix $\mathbf{F}$ is obtained. In a second stage the sensitivity is minimized according to the performance index by an implicit model-following term derived from $\dot{\mathbf{x}} - \mathbf{Fx}$. Thus,

$$I = \int_0^\infty [(\dot{\mathbf{x}} - \mathbf{Fx})^T\mathbf{Q}_1(\dot{\mathbf{x}} - \mathbf{Fx}) + \boldsymbol{\sigma}^T\mathbf{Q}_\sigma\boldsymbol{\sigma} + \mathbf{u}^T\mathbf{Ru}]dt\ . \tag{10.46}$$

An alternative is given by using a virtual model-following concept

$$I = \int_0^\infty \left([\mathbf{y} - \mathbf{C}e^{\mathbf{F}t}\mathbf{x}(0)]^T\mathbf{Q}_3[\mathbf{y} - \mathbf{C}e^{\mathbf{F}t}\mathbf{x}(0)] + \boldsymbol{\sigma}^T\mathbf{Q}_\sigma\boldsymbol{\sigma} + \mathbf{u}^T\mathbf{Ru}\right)\ dt \tag{10.47}$$

based on the difference between the actual output $\mathbf{y}$ and the response $\mathbf{C}e^{\mathbf{F}t}\mathbf{x}(0)$ of the virtual model. The minimizing matrix $\mathbf{K}$ is found by a gradient search algorithm.

10.7 Conclusion

Various methods for desensitizing control systems to parameter variations are shown from an engineering point of view. It is possible to reduce the differential sensitivity via methods of linear quadratic regulator or via incremental techniques. Various methods provide a tradeoff between differential-sensitivity reduction and increase in control effort or system error. In detail, a comparison between several methods considering the example of a bearing magnet is documented by *Verde, C., and Frank, P.M., 1987.*

Part III

Robustness in the Time Domain

Chapter 11

General Stability Bounds in Perturbed Systems

11.1 Linear First-Order Time-Varying System

Solving the linear first-order time-varying differential equation

$$\dot{x}(t) = a(t)x(t) + u(t) \tag{11.1}$$

the homogeneous differential equation

$$\dot{x}(t) = a(t)x(t) \qquad \text{or} \qquad \frac{\dot{x}(t)}{x(t)} = \frac{d}{dt}\ln x(t) = a(t) \tag{11.2}$$

is considered first. Integrating yields

$$\ln x(t) = \int_{t_o}^{t} a(\tau)d\tau + \ln k \tag{11.3}$$

$$x(t) = k\exp\int_{t_o}^{t} a(\tau)d\tau \triangleq k\varphi(t,t_o)\ . \tag{11.4}$$

In accordance with Eq.(11.2), $\dot{\varphi}(t,t_o)$ obeys the homogeneous differential equation

$$\dot{\varphi}(t,t_o) = a(t)\varphi(t,t_o) \qquad \text{where} \qquad \varphi(t_o,t_o) = 1\ . \tag{11.5}$$

In order to obtain a particular solution, the method of variation of parameters is applied, i.e., the number k is considered a function of time $k(t)$. Thus,

$$x(t) = k(t)\varphi(t,t_o) \tag{11.6}$$

$$\dot{x}(t) = \dot{k}(t)\varphi(t,t_o) + k(t)\dot{\varphi}(t,t_o)\ . \tag{11.7}$$

Combining with Eq.(11.1)

$$\dot{x}(t) = a(t)x(t) + u(t) = a(t)k(t)\varphi(t,t_o) + u(t) = \dot{k}(t)\varphi(t,t_o) + k(t)\dot{\varphi}(t,t_o) \tag{11.8}$$

and with Eq.(11.5) yields

$$u(t) = \dot{k}(t)\varphi(t,t_o) \qquad \text{and by integration} \tag{11.9}$$

$$k(t) = k(t_o) + \int_{t_o}^{t} \varphi^{-1}(\tau,t_o)u(\tau)d\tau = k(t_o) + \int_{t_o}^{t} e^{-\int_{t_o}^{\tau} a(\tau')d\tau'}u(\tau)d\tau\ . \tag{11.10}$$

Multiplying with $\varphi(t,t_o)$ yields

$$x(t) = x(t_o)e^{\int_{t_o}^{t} a(\tau)d\tau} + e^{\int_{t_o}^{t} a(\tau)d\tau}\int_{t_o}^{t} e^{[-\int_{t_o}^{\tau} a(\tau')d\tau']}\,u(\tau)d\tau \tag{11.11}$$

where $k(t_o) = x(t_o)$ is obvious.

11.2 Linear nth Order Time-Varying System

Generalizing the approach to an nth order linear time-varying problem

$$\dot{\mathbf{x}}(t) = \mathbf{A}(t)\mathbf{x}(t) + \mathbf{B}(t)\mathbf{u}(t) \qquad \mathbf{x}_o = \mathbf{x}(t_o) \tag{11.12}$$

yields the solution

$$\mathbf{x}(t) = \mathbf{\Phi}(t,t_o)\mathbf{k}(t) = \mathbf{\Phi}(t,t_o)\mathbf{x}(t_o) + \int_{t_o}^{t} \mathbf{\Phi}^{-1}(\tau,t_o)\mathbf{B}(\tau)\mathbf{u}(\tau)d\tau \,. \tag{11.13}$$

11.3 Bellman-Gronwall Lemma

Assume $h(t)$ and $k(t)$ are real functions of t, c constant and let

$$h(t) \le c + \int_{t_o}^{t} k(\tau)h(\tau)d\tau \,. \tag{11.14}$$

This inequality is rewritten to an equality by introducing a positive function $z(t) \ge 0$. Substituting $h(t) \triangleq x(t) - z(t)$ into Eq.(11.14)

$$h(t) \le x(t) \triangleq c + \int_{t_o}^{t} k(\tau)h(\tau)d\tau \,. \tag{11.15}$$

Taking the time derivative

$$\dot{x}(t) = k(t)h(t) = k(t)x(t) - k(t)z(t) \,. \tag{11.16}$$

Comparison with Eq.(11.1) shows $a(t) := k(t)$ and $u(t) := -k(t)z(t)$. Substituting into Eq.(11.11) yields

$$x(t) = x(t_o)e^{\int_{t_o}^{t} k(\tau)d\tau} - e^{\int_{t_o}^{t} k(\tau)d\tau} \int_{t_o}^{t} e^{[-\int_{t_o}^{\tau} k(\tau')d\tau']} \, k(\tau)z(\tau)d\tau \,. \tag{11.17}$$

Since $z(t) > 0$, $k(t) > 0$ and $x(t_o) = c$

$$x(t) \le x(t_o)e^{\int_{t_o}^{t} k(\tau)d\tau} = c\, e^{\int_{t_o}^{t} k(\tau)d\tau} \,. \tag{11.18}$$

Using, once more, the expressions $h(t) = x(t) - z(t)$ and $h(t) \le x(t)$, the solution of Eq.(11.14) is

$$h(t) \le c\, e^{\int_{t_o}^{t} k(\tau)d\tau} \,. \tag{11.19}$$

11.4 Bellman-Gronwall Lemma. General Form

Assume (i) the functions f, g, k are real and defined for $t \ge 0$ and are locally integrable, (ii) g and k are nonnegative and locally integrable and (iii) $g \in L_{\infty e}$. If the function u satisfies the integral inequation

$$h(t) \le f(t) + g(t)\int_{t_o}^{t} k(\tau)h(\tau)d\tau \tag{11.20}$$

then $h(t)$ explicitly satisfies the inequality

$$h(t) \le f(t) + g(t)\int_{t_o}^{t} k(\tau)f(\tau)e^{\int_{\tau}^{t} k(\eta)g(\eta)d\eta}\, d\tau \tag{11.21}$$

(*Desoer, C.A., and Vidyasagar, M., 1975; de la Sen, M., 1990*).

11.5 Stability of Single-Input Time-Invariant Systems

For a system as depicted in Fig. 20.4 but single-input single-output, assume the nonlinearity and the linear dynamical part are given and bounded as follows

$$u(t) = N^{op}e(t) = N^{op}\{y_{ref}(t) - y(t)\} \quad \text{where} \quad |u(t)| \leq \gamma_N |e(t)|, \quad |y_{ref}(t)| \leq c_r e^{-\alpha t} \tag{11.22}$$

$$y(t) = \int_o^t g(t-\tau)u(\tau)d\tau \qquad \text{where} \qquad |g(t)| \leq c_g e^{-\alpha t} \tag{11.23}$$

and c_g, c_r, γ_N are positive constant numbers. The closed loop is characterized by

$$y - y_{ref} = -e = -y_{ref} + \int_o^t g(t-\tau)u(\tau)d\tau \tag{11.24}$$

$$-e(t) = -y_{ref}(t) + \int_o^t g(t-\tau)u(\tau)d\tau \tag{11.25}$$

$$|e(t)| \leq |y_{ref}(t)| + \int_o^t |g(t-\tau)|\,|u(\tau)|d\tau \tag{11.26}$$

$$|e(t)| \leq c_r e^{-\alpha t} + \int_o^t c_g e^{-\alpha t} e^{\alpha \tau} \gamma_N |e(\tau)| d\tau \tag{11.27}$$

$$|e(t)|e^{\alpha t} \leq c_r + c_g \gamma_N \int_o^t e^{\alpha \tau}|e(\tau)|d\tau \ . \tag{11.28}$$

Applying the Bellman-Gronwall lemma and comparing with Eq.(11.14) yields the relations $h(t) := |e(t)|e^{\alpha t}$, $k(t) := c_g\gamma_N$, $c := c_r$. From Eq.(11.19)

$$|e(t)|e^{\alpha t} \leq c_r e^{\int_o^t c_g \gamma_N d\tau} = c_r e^{c_g \gamma_N t} \qquad \rightsquigarrow \qquad |e(t)| \leq c_r e^{-(\alpha - c_g\gamma_N)t} \ . \tag{11.29}$$

11.6 Stability of Multivariable Time-Varying Nonlinear Systems

The linear portion of the system as given in Fig. 20.4 obeys

$$\mathbf{x}(t) = \mathbf{\Phi}(t,t_o)\mathbf{x}(t_o) + \int_{t_o}^t \mathbf{\Phi}(t,\tau)\mathbf{u}(\tau)d\tau \ . \tag{11.30}$$

Obviously, in order to hold $\|\mathbf{x}(t)\|_F < \infty$ the conditions

$$\|\mathbf{\Phi}(t,t_o)\|_F \leq b_o < \infty \quad \forall t \geq t_o \quad \text{or} \quad \sup_{t,t_o} \|\mathbf{\Phi}(t,t_o)\|_F \triangleq b_o \tag{11.31}$$

and

$$\int_{t_o}^t \|\mathbf{\Phi}(t,\tau)\|d\tau \leq c_o < \infty \quad \forall t \geq t_o \tag{11.32}$$

must be guaranteed. From Eq.(11.32) by changing the order of the arguments t, τ of $\mathbf{\Phi}$ and by replacing t by t_o ,

$$\sup_{t,t_o} \int_{t_o}^t \|\mathbf{\Phi}(t,\tau)\|d\tau = \sup_{t,t_o} \int_{t_o}^t \|\mathbf{\Phi}(\tau,t_o)\|d\tau \triangleq c_o \ . \tag{11.33}$$

Defining a sequence b_i, i.e.,

$$b_i \triangleq \sup_{t,t_o} (t-t_o)^i \|\mathbf{\Phi}(t,t_o)\|_F, \quad t \geq t_o, \forall i = [0, 1 \ldots \infty] \tag{11.34}$$

where b_i can be rewritten using an integral expression,

$$b_i = \sup_{t,t_o} i \int_{t_o}^{t} (t-\tau)^{i-1} \|\mathbf{\Phi}(t,t_o)\|_F \le \sup_{t,t_o} i \int_{t_o}^{t} (t-\tau)^{i-1} \|\mathbf{\Phi}(t,\tau)\|_F \|\mathbf{\Phi}(\tau,t_o)\|_F dt \qquad (11.35)$$

$$b_i \le i b_{i-1} c_o \quad \leadsto \quad b_i = i! b_o c_o^i \ . \qquad (11.36)$$

Combination with Eq.(11.34) yields

$$\sup_{t,t_o} (t-t_o)^i \|\mathbf{\Phi}(t,t_o)\|_F \le i! b_o c_o^i \ . \qquad (11.37)$$

Multiplying with α^i(where $\alpha < 1/c_o$) and calculating the sum,

$$\|\mathbf{\Phi}(t,t_o)\|_F \ \alpha^i \frac{(t-t_o)^i}{i!} \le b_o c_o^i \alpha^i \qquad (11.38)$$

$$\|\mathbf{\Phi}(t,t_o)\|_F \sum_{i=0}^{\infty} \alpha^i \frac{(t-t_o)^i}{i!} = \|\mathbf{\Phi}(t,t_o)\|_F e^{\alpha(t-t_o)} \le b_o \frac{1}{1-c_o\alpha} \ . \qquad (11.39)$$

Finally, the upper bounds for $\|\mathbf{\Phi}(t,t_o)\|_F$ and $\|\mathbf{x}(t)\|_F$ are

$$\|\mathbf{\Phi}(t,t_o)\|_F \le \frac{b_o}{1-c_o\alpha} e^{\alpha(t-t_o)} \triangleq d_o e^{\alpha(t-t_o)} \quad \forall t \ge t_o \qquad (11.40)$$

$$\|\mathbf{x}(t)\|_F \le \|\mathbf{\Phi}(t,t_o)\|_F \|\mathbf{x}(t_o)\|_F \le d_o e^{-\alpha(t-t_o)} \ \|\mathbf{x}(t_o)\|_F \qquad (11.41)$$

(*Perron, O., 1930; Hsu, J.C., and Meyer, A.U., 1968; Ludyk, G., 1977*). The admissible region for $\|\mathbf{\Phi}(t,t_o)\|_F$ is given both by the constant bound Eq.(11.31)and by the exponentially decaying bound Eq.(11.40).

11.7 Bounded Input-Bounded Output Stability

Consider the feedback system Fig. 20.4 with $\mathbf{y}$ replaced by $\mathbf{x}$ and $\mathbf{w} = \mathbf{0}$, $\mathbf{y}_{ref} = \mathbf{0}$, $\mathbf{e} = -\mathbf{x}$. Let the nonlinearity $\mathbf{u}(t) = \mathbf{u}[\mathbf{x}(t)] = \mathbf{N}^{op}\mathbf{x}$ be bounded, using Frobenius norm,

$$\|\mathbf{N}^{op}\mathbf{x}(t)\|_F(t) \le k_N \|\mathbf{x}(t)\|_F(t) \ . \qquad (11.42)$$

Combining Eqs.(11.30) and (11.40),

$$\|\mathbf{x}(t)\|_F \le d_o e^{-\alpha(t-t_o)} \ \|\mathbf{x}(t_o)\|_F + \int_{t_o}^{t} d_o e^{-\alpha(t-\tau)} \ k_N \|\mathbf{x}(\tau)\|_F d\tau \qquad (11.43)$$

$$\|\mathbf{x}(t)\|_F e^{\alpha(t-t_o)} \le d_o \|\mathbf{x}(t_o)\|_F + d_o k_N e^{-\alpha t_o} \int_{t_o}^{t} e^{\alpha\tau} \|\mathbf{x}(\tau)\|_F d\tau \qquad (11.44)$$

$$\|\mathbf{x}(t+t_o)\|_F e^{\alpha t} \le d_o \|\mathbf{x}(t_o)\|_F + d_o k_N \int_{o}^{t} e^{\alpha\tau} \|\mathbf{x}(\tau+t_o)\|_F d\tau \ . \qquad (11.45)$$

Applying Bellman-Gronwall lemma,

$$\|\mathbf{x}(t+\tau_o)\|_F e^{\alpha t} \le d_o \|\mathbf{x}(t_o)\|_F e^{d_o k_N t} \quad \forall t > 0 \qquad (11.46)$$

$$\|\mathbf{x}(t)\|_F \le d_o \|\mathbf{x}(t_o)\|_F e^{-(\alpha - d_o k_N)(t-t_o)} \quad \forall t > t_o \ . \qquad (11.47)$$

11.8 Estimating Linear System Solution

Consider a linear time-invariant system in state-space representation

$$\dot{\mathbf{x}}(t) = \mathbf{A}\mathbf{x}(t) + \mathbf{v}(t) \tag{11.48}$$

where $\mathbf{v}(t)$ is any excitation. Then,

$$\mathbf{x}(t) = \mathbf{\Phi}(t)\mathbf{x}(0) + \int_0^t \mathbf{\Phi}(t-\tau)\mathbf{v}(\tau)\, d\tau \qquad \mathbf{\Phi}(t) = e^{\mathbf{A}t} \tag{11.49}$$

$$\|\mathbf{x}(t)\|_F \leq \|\mathbf{\Phi}(t)\|_s \|\mathbf{x}(0)\|_F + \int_0^t \|\mathbf{\Phi}(t-\tau)\|_s \|\mathbf{v}(t)\|_F \, d\tau \, . \tag{11.50}$$

Let the norm of $\mathbf{\Phi}(t)$ be bounded by

$$\|\mathbf{\Phi}(t)\|_s = \|e^{\mathbf{A}t}\|_s < m_o e^{\lambda_n t} \, . \tag{11.51}$$

The number $\lambda_n = \alpha_s[\mathbf{A}]$ if all the eigenvalues of $\mathbf{A}$ are distinct. The number λ_n exceeds the dominant eigenvalue and is an upper bound to the descending norm, only, if $\mathbf{A}$ has multiple eigenvalues. Then, (*Doraiswami, R., and Bordry, F., 1987*)

$$\|\mathbf{x}(t)\|_F < m_o\|\mathbf{x}(0)\|_F e^{\lambda_n t} + m_o \left| \int_0^t e^{\lambda_n(t-\tau)}\, d\tau \right| \int_0^t \|\mathbf{v}(\tau)\|_F d\tau \tag{11.52}$$

$$\|\mathbf{x}(t)\|_F < m_o[\ \|\mathbf{x}(0)\|_F e^{\lambda_n t} + \frac{1-e^{\lambda_n t}}{\lambda_n}\frac{1}{2\pi}\int_{-\infty}^{\infty} \|\mathbf{v}(j\omega)\|_F \, d\omega \ ] \, . \tag{11.53}$$

11.9 Linear Systems Stability with Time Delay

Consider the linear autonomous time-delay system

$$\dot{\mathbf{x}}(t) = \mathbf{A}\mathbf{x}(t) + \mathbf{A}_d\mathbf{x}(t-T_d) \qquad t \geq 0, \quad \mathbf{A},\ \mathbf{A}_d \in \mathcal{R}^{n\times n} \tag{11.54}$$

$$\mathbf{x}(t) = \mathbf{x}_{in}(t) \qquad -T_d \leq t < 0, \quad \mathbf{x}(t) \in \mathcal{R}^n \tag{11.55}$$

with the initial function $\mathbf{x}_{in}(t)$ and time delay T_d. The solution $\mathbf{x}(t)$ of Eq.(11.54) can be written as

$$\mathbf{x}(t) = e^{\mathbf{A}t}\mathbf{x}(0) + \int_0^t e^{\mathbf{A}(t-\eta)}\mathbf{A}_d\mathbf{x}(\eta - T_d)d\eta \, . \tag{11.56}$$

Differentiating with respect to t yields the proof [1]

$$\dot{\mathbf{x}}(t) = \mathbf{A}e^{\mathbf{A}t}\mathbf{x}(0) + e^{\mathbf{A}(t-\eta)}\mathbf{A}_d\mathbf{x}(\eta - T_d)|_{\eta=t} + \int_0^t \mathbf{A}e^{\mathbf{A}(t-\eta)}\mathbf{A}_d\mathbf{x}(\eta-T_d)d\eta \, . \tag{11.57}$$

Applying Eq.(11.56) for the integral expression above,

$$\dot{\mathbf{x}}(t) = \mathbf{A}e^{\mathbf{A}t}x(0) + \mathbf{A}_d\mathbf{x}(t-T) + \mathbf{A}[\mathbf{x}(t) - e^{\mathbf{A}t}\mathbf{x}(0)] = \mathbf{A}\mathbf{x}(t) + \mathbf{A}_d\mathbf{x}(t-T_d) \, . \tag{11.58}$$

[1] Differentiation with respect to t requires, first, differentiation with respect to the upper bound of the integral yielding the integrand on $\eta = t$ and, second, differentiation with respect to the parameter t within the integrand.

Calculating the vector norm and the induced matrix norm for any p, it results from Eq.(11.56)

$$\|\mathbf{x}(t)\|_p \leq \|e^{\mathbf{A}t}\|_p\|\mathbf{x}(0)\|_p + \|\int_0^t e^{\mathbf{A}(t-\eta)}\mathbf{A}_d\mathbf{x}(\eta - T_d)d\eta\|_p \tag{11.59}$$

$$\|\mathbf{x}(t)\|_p \leq \|e^{\mathbf{A}t}\|_p \; \|\mathbf{x}(0)\|_p + \int_0^t \|e^{\mathbf{A}(t-\eta)}\|_p \; \|\mathbf{A}_d\|_p \; \|x(\eta - T_d)\|_p \; d\eta \; . \tag{11.60}$$

Applying the inequality using the matrix measure $\mu_p[\mathbf{A}]$,

$$\|e^{\mathbf{A}t}\|_p \leq e^{\mu_p[\mathbf{A}]t} \qquad t \geq 0 \tag{11.61}$$

and replacing the vector norm $\|\mathbf{x}\|_p$ by the scalar variable $v(t)$ (*Mori, T., et al. 1981*),

$$v(t) \leq e^{\mu_q[\mathbf{A}]t} \; v(0) + \|\mathbf{A}_d\|_p \int_0^t e^{\mu_p[\mathbf{A}](t-\eta)} \; v(\eta - T_d)d\eta \qquad t \geq 0 \tag{11.62}$$

$$v(t) = \|\mathbf{x}_{in}(t)\|_p \qquad -T_d \leq t < 0 \; . \tag{11.63}$$

The *inequality* expression in Eq.(11.62) is rewritten as an *equation* satisfied by a new scalar variable $z(t)$ using the identical initial condition

$$z(t) = e^{\mu_q[\mathbf{A}]t} \; v(0) + \|\mathbf{A}_d\|_p \int_0^t e^{\mu_q[\mathbf{A}](t-\eta)} \; z(\eta - T_d)d\eta \qquad t \geq 0 \tag{11.64}$$

$$z(t) = \|\mathbf{x}_{in}(t)\|_p \qquad -T_d \leq t < 0 \; . \tag{11.65}$$

These expressions may be considered as the solution of the scalar differential equation

$$\dot{z}(t) = \mu_q[\mathbf{A}] \; z(t) + \|\mathbf{A}_d\|_p \; z(t - T_d) \qquad t \geq 0, \qquad z(0) = v(0) \; . \tag{11.66}$$

Using the comparison theorem of the theory of differential inequalities in Eqs.(11.62) and (11.65) gives $v(t) \leq z(t)$. If asymptotic stability of $z(t)$ can be proved then $v(t)$ or $\|\mathbf{x}\|_p$ also is asymptotically stable.

It can be shown that

$$\dot{x}(t) = ax(t) + a_d x(t - T_d) \tag{11.67}$$

in the case $a_d \geq 0$ is asymptotically stable if and only if $a < -a_d$ irrespective of the amount of time delay T_d. (The closed-loop differential equation Eq.(11.67) corresponds to the open-loop transfer function $-a_d e^{-sT_d}/(s-a)$. The polar plots $s = j\omega$ do not encircle the critical point $(-1, j0)$ as long as $a < -a_d$.) Combining Eqs.(11.67) and (11.66) yields

$$-\mu_p[\mathbf{A}] > \|\mathbf{A}_d\|_p \;\; (\geq 0) \; . \tag{11.68}$$

Eq.(11.68) is a sufficient condition for asymptotic stability. The solution of Eq. (11.54) is bounded by the condition

$$\|\mathbf{x}(t)\|_p \leq [\sup_{\zeta \in (-T_d, 0)} \|\mathbf{x}(\zeta)\|_p \,] \; e^{-\sigma t} \qquad t > 0 \tag{11.69}$$

where σ is given by (*Mori, T., et al. 1981*)

$$\sigma + \mu_p[\mathbf{A}] + \|\mathbf{A}_d\|_p \; e^{\sigma T_d} = 0 \; . \tag{11.70}$$

The results can be extended to the set of subsystems

$$\dot{\mathbf{x}}_l(t) = \mathbf{A}_l\mathbf{x}_l(t) + \mathbf{A}_{dl} \; \mathbf{x}_l(t - T_d) + \mathbf{B}_l\mathbf{u}_l \qquad \forall \, l = 1, 2 \ldots N \tag{11.71}$$

where the controlling variable $\mathbf{u}_l$ is derived from state variables of the other subsystems by

$$\mathbf{u}_l = \sum_{k=1, \; k \neq l}^{N} \mathbf{C}_{lk}\mathbf{x}_k(t - T_d) \; . \tag{11.72}$$

11.10 Robust Stability Criteria for Time-Delay Systems

Consider the multivariable perturbed linear system with time-delay

$$\dot{\mathbf{x}} = (\mathbf{A} + \Delta\mathbf{A})\mathbf{x}(t) + (\mathbf{A}_d + \Delta\mathbf{A}_d)\mathbf{x}(t - T_d) + (\mathbf{B} + \Delta\mathbf{B})\mathbf{u}(t) \qquad (11.73)$$

$$\mathbf{y}(t) = (\mathbf{C} + \Delta\mathbf{C})\mathbf{x}(t) \qquad t > 0 \qquad (11.74)$$

$$\mathbf{x}(t) = \mathbf{x}_{in}(t) \qquad -T_d \le t \le 0\,. \qquad (11.75)$$

The initial vector $\mathbf{x}(0)$ is included in $\mathbf{x}_{in}(t)$, $\mathbf{u}(t)$ is an m-vector, $\Delta\mathbf{A}, \Delta\mathbf{A}_d, \Delta\mathbf{B}$ and $\Delta\mathbf{C}$ represent the uncertainties. Applying a linear state feedback controller $\mathbf{u}(t) = \mathbf{K}\mathbf{x}(t)$ with (m, n)-matrix $\mathbf{K}$ of constant entries, one has (*Su, T.J., et al. 1988*)

$$\dot{\mathbf{x}}(t) = [\mathbf{A} + \mathbf{B}\mathbf{K} + \Delta\mathbf{A} + (\Delta\mathbf{B})\mathbf{K}]\mathbf{x}(t) + (\mathbf{A}_d + \Delta\mathbf{A}_d)\mathbf{x}(t - T_d)\,. \qquad (11.76)$$

Applying Eq.(11.68) with $\mu_p = \mu_s = \mu$,

$$\mu[\mathbf{A} + \mathbf{B}\mathbf{K} + \Delta\mathbf{A} + (\Delta\mathbf{B})\mathbf{K}] + \|\mathbf{A}_d + \Delta\mathbf{A}_d\| \le 0\,. \qquad (11.77)$$

11.10.1 Unstructured Uncertainties

If the uncertainty has the following unstructured norm bounds

$$\|\Delta\mathbf{A}\|_p \le \alpha, \quad \|\Delta\mathbf{A}_d\|_p \le \alpha_d, \quad \|\Delta\mathbf{B}\|_p \le \beta, \quad \|\Delta\mathbf{C}\|_p \le \gamma \qquad (11.78)$$

from Eq.(11.77) using matrix measure properties it results

$$\mu[\mathbf{A} + \mathbf{B}\mathbf{K}] + \alpha + \beta\|\mathbf{K}\|_p + \|\mathbf{A}_d\|_p + \alpha_d \le 0 \qquad (11.79)$$

as a sufficient condition for asymptotic stability. Modified results are presented by *Hmamed, A., 1991*.

11.10.2 Structured Uncertainties

If the uncertainty bounds are known, element by element, the (i,j)th position is

$$|\Delta A_{ij}| \le |\Delta A_{ij}|_{\max} = \alpha_{ij} = (\Xi_\alpha)_{ij} \text{ and } \Xi_\alpha = \sum_{ij} \alpha_{ij}\mathbf{E}_{ij}\,, \qquad (11.80)$$

$$|\Delta B_{ij}| \le |\Delta B_{ij}|_{\max} = \beta_{ij} = (\Xi_\beta)_{ij} \text{ and } \Xi_\beta = \sum_{ij} \beta_{ij}\mathbf{E}_{ij}\,. \qquad (11.81)$$

The matrix $\Xi_{\alpha d}$ is defined in a similar way. By using (i) the modulus matrix $|\mathbf{P}| \triangleq$ matrix$[\,|P_{ij}|\,]$ of the matrix $\mathbf{P} = $ matrix$[P_{ij}]$, (ii) $|\mathbf{P}| >_e |\mathbf{Q}|$ if $|P_{ij}| > |Q_{ij}|\ \forall\, i,j$ and (iii)

$$|\mathbf{P}| \ge_e |\mathbf{Q}| \quad \rightsquigarrow \quad \|\,|\mathbf{P}|\,\|_p \ge \|\,|\mathbf{Q}|\,\|_p \ge \|\mathbf{Q}\|_p\,, \qquad (11.82)$$

an inequality chain is obtained as follows

$$\|\mathbf{A}_d + \Delta\mathbf{A}_d\|_p \le \|\,|\mathbf{A}_d + \Delta\mathbf{A}_d|\,\|_p \le \|\,|\mathbf{A}_d| + |\Delta\mathbf{A}_d|\,\|_p \le \|\,|\mathbf{A}_d| + \Xi_{\alpha d}\|_p \qquad (11.83)$$

$$\mu[\mathbf{A} + \Delta\mathbf{A} + (\mathbf{B} + \Delta\mathbf{B})\mathbf{K}] \le \mu[\mathbf{A} + \mathbf{B}\mathbf{K}] + \|(\Delta\mathbf{B})\mathbf{K}\|_p + \|\Delta\mathbf{A}\|_p \qquad (11.84)$$

$$\mu[\mathbf{A} + \Delta\mathbf{A} + (\mathbf{B} + \Delta\mathbf{B})\mathbf{K}] \le \mu[\mathbf{A} + \mathbf{B}\mathbf{K}] + \|\Xi_\beta\|_p\,\|\,|\mathbf{K}|\,\|_p + \|\Xi_\alpha\|_p\,. \qquad (11.85)$$

Combining Eqs.(11.68), (11.77), (11.83) and (11.85) yields a sufficient condition

$$\mu[\mathbf{A} + \mathbf{B}\mathbf{K} + \Delta\mathbf{A} + (\Delta\mathbf{B})\mathbf{K}] + \|\mathbf{A}_d + \Delta\mathbf{A}_d\|_p < 0 \qquad (11.86)$$

$$\Leftarrow \quad \mu[\mathbf{A} + \mathbf{B}\mathbf{K}] + \|\Xi_\beta\|_p\,\|\,|\mathbf{K}|\,\|_p + \|\Xi_\alpha\|_p + \|\,|\mathbf{A}_d| + \Xi_{\alpha d}\|_p < 0\,. \qquad (11.87)$$

11.10.3 Design of the Feedback Controller for Structured Uncertainty

To design the controller matrix **K** for given $\Xi_\alpha, \Xi_{\alpha d}, \Xi_\beta$ the following algorithm can be used: Compute

$$\mu[\mathbf{A}],\ \|\mathbf{A}_d\|_p,\ \|\mathbf{B}\|_p; \qquad g = \frac{\mu[\mathbf{A}] + \|\Xi_\alpha\|_p + \|\ |\mathbf{A}_d| + \Xi_{\alpha d}\|_p}{\|\mathbf{B}\|_p - \|\Xi_\beta\|_p}\ ; \tag{11.88}$$

$$h = -\|\Xi_\alpha\|_p - \|\ |\mathbf{A}_d| + \Xi_{\alpha d}\|_p - \|\Xi_\beta\|_p \max(g, 0)\ . \tag{11.89}$$

Then, find **K** by pole assignment method that $\Re e\lambda_i[\mathbf{A} + \mathbf{BK}] \leq h$. If **K** does not satisfy Eq.(11.87) repeat pole assignment method by shifting $\lambda_i[\mathbf{A} + \mathbf{BK}]$ more to the left (*Su, T.J., et al. 1988*).

11.11 Robust Linear Dynamic Controller

Consider the multivariable nonlinear time-varying plant

$$\dot{\mathbf{x}}(t) = \mathbf{Ax}(t) + \mathbf{Bu}(t) + \Delta\mathbf{A}(\mathbf{x}) + \Delta\mathbf{B}(\mathbf{u}) \qquad \mathbf{x}(0) = \mathbf{x}_o \tag{11.90}$$

$$\mathbf{y}(t) = \mathbf{Cx}(t) + \mathbf{Du}(t) + \Delta\mathbf{C}(\mathbf{x}) + \Delta\mathbf{D}(\mathbf{u}) \qquad \mathbf{B} \in \mathcal{R}^{n\times m},\ \ \mathbf{C} \in \mathcal{R}^{r\times n}\ . \tag{11.91}$$

Let the parametrical uncertainties be bounded by the sum norm and maximum column sum, respectively. Note that vector and matrix norms are applied.

$$\|\Delta\mathbf{A}(\mathbf{x})\|_1 \leq \beta_A\|\mathbf{x}\|_1 \qquad \|\Delta\mathbf{B}(\mathbf{u})\|_1 \leq \beta_B\|\mathbf{u}\|_1 \tag{11.92}$$

$$\|\Delta\mathbf{C}(\mathbf{x})\|_1 \leq \beta_C\|\mathbf{x}\|_1 \qquad \|\Delta\mathbf{D}(\mathbf{u})\|_1 \leq \beta_D\|\mathbf{u}\|_1\ . \tag{11.93}$$

The dynamic state feedback controller is given by

$$\dot{\mathbf{x}}_r = \mathbf{A}_r\mathbf{x}_r + \mathbf{B}_r\mathbf{y} \qquad \mathbf{x}_r(0) = \mathbf{x}_{ro} \qquad \mathbf{x}_r \in \mathcal{R}^{n_r} \tag{11.94}$$

$$\mathbf{u} = \mathbf{Kx} + \mathbf{K}_r\mathbf{x}_r \qquad \mathbf{K} \in \mathcal{R}^{m\times n},\ \mathbf{K}_r \in \mathcal{R}^{m\times n_r}\ , \tag{11.95}$$

see Fig. 11.1. The aim is to design the controller matrices $\mathbf{A}_r, \mathbf{B}_r, \mathbf{K}_r$ and **K** in order to obtain a stable closed-loop system. If such a controller exists in the presence of the plant uncertainties, namely if the uncertainties do not affect the stability of the closed loop, then the controller is termed robust.

11.11.1 Unified State Representation

Defining a state vector $\mathbf{x}_{pr}$ and combining the plant equation and controller equation Eq.(11.95), the closed loop is given by

$$\dot{\mathbf{x}}_{pr} \triangleq \begin{pmatrix}\dot{\mathbf{x}}\\ \dot{\mathbf{x}}_r\end{pmatrix} = \begin{pmatrix}\mathbf{A} + \mathbf{BK} & \mathbf{BK}_r\\ \mathbf{B}_r\mathbf{C} + \mathbf{B}_r\mathbf{DK}\ \vdots & \mathbf{A}_r + \mathbf{B}_r\mathbf{DK}_r\end{pmatrix}\mathbf{x}_{pr} + \begin{pmatrix}\Delta\mathbf{A}(\mathbf{x}) + \Delta\mathbf{B}(\mathbf{u})\\ \mathbf{B}_r[\Delta\mathbf{C}(\mathbf{x}) + \Delta\mathbf{D}(\mathbf{u})]\end{pmatrix}$$

$$\dot{\mathbf{x}}_{pr} \triangleq \mathbf{A}_{pr}\mathbf{x}_{pr} + \Delta\mathbf{A}_{pr}(\mathbf{x}_{pr}) \qquad \mathbf{x}_{pr}(0) = \begin{pmatrix}\mathbf{x}_o\\ \mathbf{x}_{ro}\end{pmatrix} \triangleq \mathbf{x}_{pro} \tag{11.96}$$

$$\mathbf{y} = (\mathbf{C} + \mathbf{DK}\ \vdots\ \mathbf{DK}_r)\mathbf{x}_{pr} + \Delta\mathbf{C}(\mathbf{x}) + \Delta\mathbf{D}(\mathbf{u}) \triangleq \mathbf{C}_{pr}\mathbf{x}_{pr} + \Delta\mathbf{C}_{pr}(\mathbf{x}_{pr})\ . \tag{11.97}$$

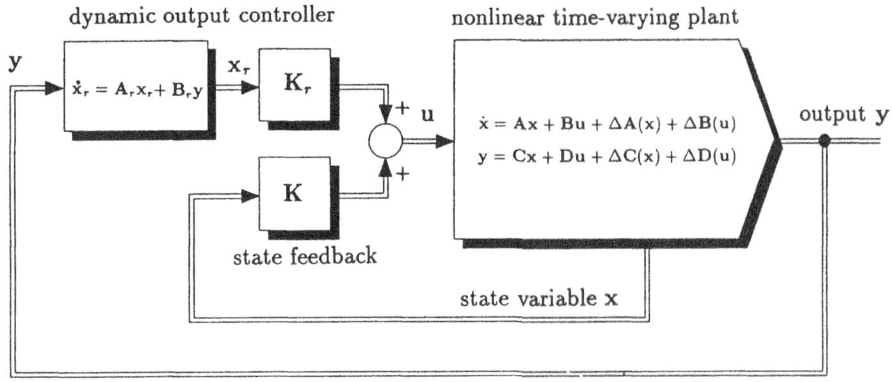

Figure 11.1: Nonlinear time-varying plant with parametrical uncertainty and dynamic controller

11.11.2 Robust Design Using Bellman-Gronwall Lemma

The transition matrix $\mathbf{\Phi}_{pr}$ of the closed loop is given by $\mathbf{\Phi}_{pr}(t) = e^{\mathbf{A}_{pr}t}$. In order to satisfy the stability requirements, the largest column sum norm $\|\cdot\|_1$ is chosen. Assume $\|\mathbf{\Phi}_{pr}(t)\|_1$ decaying quicker than an exponential function

$$\|\mathbf{\Phi}_{pr}(t)\|_1 \leq m_o e^{-\alpha t} \qquad \alpha > 0,\ \forall t \geq 0\ . \tag{11.98}$$

Referring to Eq.(11.96), $\Delta\mathbf{A}_{pr}(\mathbf{x}_{pr})$ is considered as an input function to the closed-loop system characterized by the system matrix $\mathbf{A}_{pr}$

$$\mathbf{x}_{pr}(t) = \mathbf{\Phi}_{pr}(t)\mathbf{x}_{pro} + \int_0^t \mathbf{\Phi}_{pr}(t-\tau)\Delta\mathbf{A}_{pr}[\mathbf{x}_{pr}(\tau)]d\tau\ . \tag{11.99}$$

Performing the norm $\|\cdot\|_1$ (*Chen, B.S., and Wong, C.C., 1987*) and omitting the argument t, for abbreviaton,

$$\|\mathbf{x}_{pr}\|_1 \leq \|\mathbf{\Phi}_{pr}\|_1\|\mathbf{x}_{pro}\|_1 + \int_0^t \|\mathbf{\Phi}_{pr}\|_1\|\Delta\mathbf{A}_{pr}\left(\mathbf{x}_{pr}(\tau)\right)\|_1 d\tau\ . \tag{11.100}$$

From the definition of $\Delta\mathbf{A}_{pr}(\mathbf{x}_{pr})$, omitting the argument τ , it results

$$\begin{aligned}
\|\Delta\mathbf{A}_{pr}(\mathbf{x}_{pr})\|_1 &\leq \|\Delta\mathbf{A}(\mathbf{x}) + \Delta\mathbf{B}(\mathbf{u})\|_1 + \|\mathbf{B}_r(\Delta\mathbf{C} + \Delta\mathbf{D})\|_1 && (11.101)\\
&\leq \|\Delta\mathbf{A}\|_1 + \|\Delta\mathbf{B}(\mathbf{u})\|_1 + \|\mathbf{B}_r\|_1\left(\|\Delta\mathbf{C}\|_1 + \|\Delta\mathbf{D}(\mathbf{u})\|_1\right) && (11.102)\\
&\leq \beta_A\|\mathbf{x}\|_1 + \beta_B\|\mathbf{u}\|_1 + \|\mathbf{B}_r\|_1\left(\beta_C\|\mathbf{x}\|_1 + \beta_D\|\mathbf{u}\|_1\right) && (11.103)\\
&\leq \beta_A\|\mathbf{x}\|_1 + \beta_B\|(\mathbf{K}\ \vdots\ \mathbf{K}_r)\|_1\|\mathbf{x}_{pr}\|_1 \\
&\qquad +\|\mathbf{B}_r\|_1\beta_C\|\mathbf{x}\|_1 + \|\mathbf{B}_r\|_1\beta_D\|(\mathbf{K}\ \vdots\ \mathbf{K}_r)\|_1\|\mathbf{x}_{pr}\|_1 && (11.104)\\
\|\Delta\mathbf{A}_{pr}[\mathbf{x}_{pr}(\tau)]\ \|_1 &\leq [\ \beta_A + \left(\beta_B + \beta_D\|\mathbf{B}_r\|_1\right)\|(\mathbf{K}\ \vdots\ \mathbf{K}_r)\|_1 + \beta_C\|\mathbf{B}_r\|_1\ ]\ \|\mathbf{x}_{pr}(\tau)\|_1 \\
&\triangleq \rho_A\|\mathbf{x}_{pr}(\tau)\|_1\ . && (11.105)
\end{aligned}$$

Similarly,

$$\begin{aligned}\|\Delta \mathbf{C}_{pr}[\mathbf{x}_{pr}(\tau)]\|_1 &= \|\Delta \mathbf{C}[\mathbf{x}(\tau)] + \Delta \mathbf{D}[\mathbf{u}(\tau)]\|_1 \leq [\ \beta_C + \beta_D\|(\mathbf{K} \vdots \mathbf{K}_r)\|_1\ ]\ \|\mathbf{x}_{pr}(\tau)\|_1 \\ &\triangleq \rho_C\|\mathbf{x}_{pr}(\tau)\| \ . \end{aligned} \tag{11.106}$$

Using Eqs.(11.98) and (11.105) from Eq.(11.100),

$$\|\mathbf{x}_{pr}(t)\|_1 \leq m_o e^{-\alpha t}\|\mathbf{x}_{pro}\|_1 + \int_0^t m_o e^{-\alpha(t-\tau)}\ \rho_A\|\mathbf{x}_{pr}(\tau)\|_1 d\tau \tag{11.107}$$

$$\|\mathbf{x}_{pr}(t)\|_1 e^{\alpha t} \leq m_o\|\mathbf{x}_{pro}\|_1 + \int_0^t m_o\rho_A\|\mathbf{x}_{pr}(\tau)\|_1 e^{\alpha\tau} d\tau \ . \tag{11.108}$$

Now, applying the Bellman-Gronwall lemma from Eq.(11.20), by substituting

$$h(t) := \|\mathbf{x}_{pr}(t)\|_1 e^{\alpha t}, \quad f(t) := m_o\|\mathbf{x}_{pro}\|_1, \quad k(\tau) := m_o\rho_A, \quad g(t) := 1 \ ,$$

$$\begin{aligned} h(t) = \|\mathbf{x}_{pr}(t)\|_1 e^{\alpha t} &\leq m_o\|\mathbf{x}_{pro}\|_1 + \int_0^t m_o\rho_A m_o\|\mathbf{x}_{pro}\|_1 e^{m_o\rho_A(t-\tau)}\ d\tau & (11.109) \\ &\leq m_o\|\mathbf{x}_{pro}\|_1(1 + e^{m_o\rho_A t} - 1) & (11.110) \end{aligned}$$

$$\|\mathbf{x}_{pr}(t)\|_1 \leq m_o\|\mathbf{x}_{pro}\|_1 e^{(-\alpha + m_o\rho_A)t} \ . \tag{11.111}$$

Performing the norm to Eq.(11.97) and referring to Eq.(11.106),

$$\|\mathbf{y}\|_1 < \|\mathbf{C}_{pr}\|\ \|\mathbf{x}_{pr}\| + \|\Delta \mathbf{C}_{pr}(\mathbf{x}_{pr})\| \leq (\|\mathbf{C}_{pr}\| + \rho_C)\|\mathbf{x}_{pr}\| \ . \tag{11.112}$$

Combination with Eq.(11.111) yields

$$\|\mathbf{y}\|_1 = m_o(\|\mathbf{C}_{pr}\|_1 + \rho_C)\|\mathbf{x}_{pro}\|_1 e^{(-\alpha + m_o\rho_A)t} \ . \tag{11.113}$$

A sufficient condition for stability is given by $\alpha > m_o\rho_A$.

11.11.3 Dynamic Output Feedback Controller

Now, consider a dynamic output feedback controller

$$\mathbf{u}(t) = \mathbf{K}_y\mathbf{y}(t) + \mathbf{K}_r\mathbf{x}_r(t) \ . \tag{11.114}$$

Replacing the state feedback controller of Eq.(11.95) by the output controller above, the method can be repeated to some extent, e.g. if parametrical uncertainties have to be investigated in the matrix **A**, only. In this case, the matrix $\mathbf{A}_{pr}$ turns out as

$$\begin{aligned}\mathbf{A}_{pr} &= \begin{pmatrix} \mathbf{A} + \mathbf{B}\mathbf{K}_y\mathbf{C} + \mathbf{B}\mathbf{K}_y\mathbf{D}(\mathbf{I} - \mathbf{K}_y\mathbf{D})^{-1}\mathbf{K}_y\mathbf{C} & \vdots\ \mathbf{B}\mathbf{K}_r + \mathbf{B}\mathbf{K}_y\mathbf{D}(\mathbf{I} - \mathbf{K}_y\mathbf{D})^{-1}\mathbf{K}_r \\ \mathbf{B}_r\mathbf{C} + \mathbf{B}_r\mathbf{D}(\mathbf{I} - \mathbf{K}_y\mathbf{D})^{-1}\mathbf{K}_y\mathbf{C} & \mathbf{A}_r + \mathbf{B}_r\mathbf{D}(\mathbf{I} - \mathbf{K}_y\mathbf{D})^{-1}\mathbf{K}_r \end{pmatrix} \\ &= \begin{pmatrix} \mathbf{A} + \mathbf{B}(\mathbf{I} - \mathbf{K}_y\mathbf{D})^{-1}\mathbf{K}_y\mathbf{C} & \mathbf{B}(\mathbf{I} - \mathbf{K}_y\mathbf{D})^{-1}\mathbf{K}_r \\ \mathbf{B}_r\mathbf{C} + \mathbf{B}_r\mathbf{D}(\mathbf{I} - \mathbf{K}_y\mathbf{D})^{-1}\mathbf{K}_y\mathbf{C} & \vdots\ \mathbf{A}_r + \mathbf{B}_r\mathbf{D}(\mathbf{I} - \mathbf{K}_y\mathbf{D})^{-1}\mathbf{K}_r \end{pmatrix} \end{aligned} \tag{11.115}$$

$$\mathbf{y} = \left(\mathbf{C} + \mathbf{D}(\mathbf{I} - \mathbf{K}_y\mathbf{D})^{-1}\mathbf{K}_y\mathbf{C}\ \vdots\ \mathbf{D}(\mathbf{I} - \mathbf{K}_y\mathbf{D})^{-1}\mathbf{K}_r\right)\begin{pmatrix}\mathbf{x} \\ \mathbf{x}_r\end{pmatrix} . \tag{11.116}$$

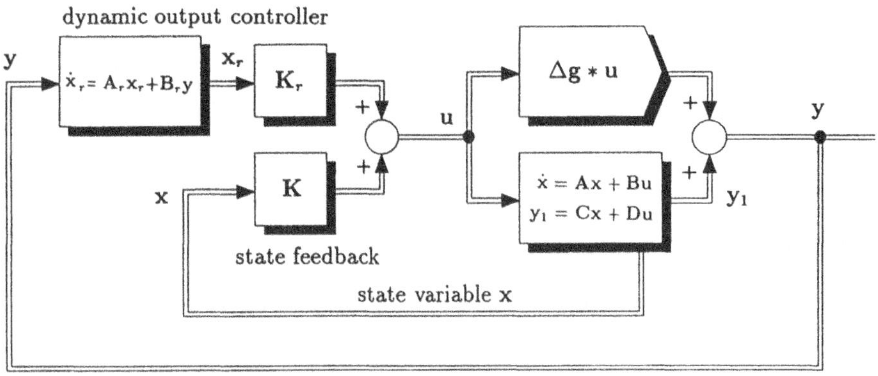

Figure 11.2: Time-varying plant with structural model error

11.11.4 Plant With a Structural Model Error

An alternative representation of the plant uncertainties is given by the structural model error (see Fig. 11.2)

$$\dot{\mathbf{x}}(t) = \mathbf{A}\mathbf{x}(t) + \mathbf{B}\mathbf{u}(t), \qquad \mathbf{x}(0) = \mathbf{x}_o \tag{11.117}$$

$$\mathbf{y}(t) = \mathbf{C}\mathbf{x}(t) + \mathbf{D}\mathbf{u}(t) + \Delta\mathbf{g}(t)\star\mathbf{u}(t) \tag{11.118}$$

where $\star$ is the convolution operator. The function $\Delta\mathbf{g}(t)$ is linear, possibly time-varying and bounded by

$$\|\Delta\mathbf{g}(t)\|_1 \leq m_g e^{-\beta t} \,. \tag{11.119}$$

Following the derivation in the previous sections, but taking Eq.(11.118) into account, the closed-loop system is

$$\dot{\mathbf{x}}_{pr} = \mathbf{A}_{pr}\mathbf{x}_{pr} + \begin{pmatrix} 0 \\ \mathbf{B}_r\Delta\mathbf{g}\star(\mathbf{K}\mathbf{x} + \mathbf{K}_r\mathbf{x}_r) \end{pmatrix} \tag{11.120}$$

$$\mathbf{y} = \mathbf{C}_{pr}\mathbf{x}_{pr} + \Delta\mathbf{g}\star(\mathbf{K}\mathbf{x} + \mathbf{K}_r\mathbf{x}_r) \,. \tag{11.121}$$

Integration of the differential equation and using the state transition matrix yields (t omitted)

$$\mathbf{x}_{pr} = \mathbf{\Phi}_{pr}\mathbf{x}_{pro} + \mathbf{\Phi}_{pr}\star\begin{pmatrix} 0 \\ \mathbf{B}_r\Delta\mathbf{g}\star(\mathbf{K}\mathbf{x} + \mathbf{K}_r\mathbf{x}_r) \end{pmatrix} \,. \tag{11.122}$$

Performing the column sum norm and integrating between the bounds 0 and ∞,

$$\|\mathbf{x}_{pr}\|_1 \leq \|\mathbf{\Phi}_{pr}\|_1\|\mathbf{x}_{pro}\|_1 + \|\mathbf{\Phi}_{pr}\star\mathbf{B}_r\Delta\mathbf{g}\star(\mathbf{K}\mathbf{x} + \mathbf{K}_r\mathbf{x}_r)\|_1 \tag{11.123}$$

$$\int_0^\infty \|\mathbf{x}_{pr}\|_1 dt \leq \int_0^\infty \|\mathbf{\Phi}_{pr}\|_1\|\mathbf{x}_{pro}\|_1 dt + \int_0^\infty \|\mathbf{\Phi}_{pr}\star\mathbf{B}_r\Delta\mathbf{g}\star(\mathbf{K}\mathbf{x} + \mathbf{K}_r\mathbf{x}_r)\|_1 dt \,. \tag{11.124}$$

Applying the property

$$\int_0^\infty \|\mathbf{a}(t)\star\mathbf{b}(t)\star\mathbf{c}(t)\|dt \leq (\int_0^\infty \|\mathbf{a}(t)\|dt)(\int_0^\infty \|\mathbf{b}(t)\|dt)(\int_0^\infty \|\mathbf{c}(t)\|dt) \tag{11.125}$$

$$\text{and} \quad \int_0^\infty \|\Delta \mathbf{g}\|_1 dt \le \int_0^\infty m_g e^{-\beta t} dt = m_g/\beta \quad \text{yields} \tag{11.126}$$

$$\begin{aligned}
\int_0^\infty \|\mathbf{x}_{pr}\|_1 dt &\le \|\mathbf{x}_{pro}\|_1 \int_0^\infty \|\mathbf{\Phi}_{pr}\|_1 dt \quad (11.127) \\
&\quad + \|\mathbf{B}_r\|_1 \Big(\int_0^\infty \|\mathbf{\Phi}_{pr}\|_1 dt\Big)\Big(\int_0^\infty \|\Delta \mathbf{g}\|_1 dt\Big)\Big(\int_0^\infty \|(\mathbf{K}\mathbf{x} + \mathbf{K}_r \mathbf{x}_r)\|_1 dt\Big) \\
\int_0^\infty \|\mathbf{x}_{pr}\|_1 dt &\le \frac{m_o}{\alpha}\|\mathbf{x}_{pro}\|_1 + \frac{m_o m_g}{\alpha\beta}\|\mathbf{B}_r\|_1 \|(\mathbf{K}_r \vdots \mathbf{K})\|_1 \int_0^\infty \|\mathbf{x}_{pr}\|_1 dt \quad (11.128) \\
\int_0^\infty \|\mathbf{x}_{pr}\|_1 dt &\le \frac{(m_o/\alpha)\|\mathbf{x}_{pro}\|_1}{1 - (m_o m_g / \lfloor \alpha\beta \rfloor)\|\mathbf{B}_r\|_1 \|(\mathbf{K}_r \vdots \mathbf{K})\|_1}\,, \quad (11.129)
\end{aligned}$$

preassuming that the denominator is positive

$$1 - \frac{m_o m_g}{\alpha\beta}\|\mathbf{B}_r\|_1 \|(\mathbf{K}_r \vdots \mathbf{K})\|_1 > 0\,. \tag{11.130}$$

Finally, it can easily be verified that $\|\mathbf{y}\|_1$ is bounded in the sense of

$$\int_0^\infty \|\mathbf{y}\|_1 dt \le \frac{(m_o/\alpha)\Big(\|\mathbf{C}_{pr}\|_1 + (m_g/\beta)\|(\mathbf{K}_r \vdots \mathbf{K})\|_1\Big)\|\mathbf{x}_{pro}\|_1}{1 - (m_o m_g/\lfloor \alpha\beta \rfloor)\|\mathbf{B}_r\|_1 \|(\mathbf{K}_r \vdots \mathbf{K})\|_1}\,. \tag{11.131}$$

Thus, Eq.(11.130) is the sufficient stability condition.

Using similar considerations, *Lin, C.L., et al. 1990; Su, T.J., et al. 1990* reported about the stabilization procedure for large structural systems under mode truncations and actuator saturations.

11.11.5 Stability Robustness Via Lyapunov Function

Simplifying Eq.(11.115) for $\mathbf{D} = \mathbf{0}$ yields

$$\mathbf{A}_{pr} = \begin{pmatrix} \mathbf{A} + \mathbf{B}\mathbf{K}_y\mathbf{C} & \mathbf{B}\mathbf{K}_r \\ \mathbf{B}_r\mathbf{C} & \mathbf{A}_r \end{pmatrix} \qquad \mathbf{x}_{pr}(t) = \begin{pmatrix} \mathbf{x}(t) \\ \mathbf{x}_r(t) \end{pmatrix}. \tag{11.132}$$

Using a positive-definite symmetric weighting matrix $\mathbf{P}$, a sufficient condition for the over-all system to be asymptotically stable (with decay rate α) is given by

$$\dot{V}(\mathbf{x}_{pr}) \le 0 \qquad \text{where} \quad V(\mathbf{x}) = \mathbf{x}_{pr}^T \mathbf{P} \mathbf{x}_{pr}\,. \tag{11.133}$$

Using the abbreviations

$$\mathbf{A}_c \triangleq \begin{pmatrix} \mathbf{K}_y & \mathbf{K}_r \\ \mathbf{B}_r & \mathbf{A}_r \end{pmatrix},\ \bar{\mathbf{A}} \triangleq \begin{pmatrix} \mathbf{A} & \mathbf{0} \\ \mathbf{0} & \mathbf{0} \end{pmatrix},\ \bar{\mathbf{B}} \triangleq \begin{pmatrix} \mathbf{B} & \mathbf{0} \\ \mathbf{0} & \mathbf{I}_{n_r} \end{pmatrix},\ \bar{\mathbf{C}} \triangleq \begin{pmatrix} \mathbf{C} & \mathbf{0} \\ \mathbf{0} & \mathbf{I}_{n_r} \end{pmatrix}, \tag{11.134}$$

from Eqs.(11.132) and (11.95)

$$\mathbf{A}_{pr} = \bar{\mathbf{A}} + \bar{\mathbf{B}}\mathbf{A}_c\bar{\mathbf{C}}\,. \tag{11.135}$$

Hence, the resulting model of the closed-loop system with dynamic controller in action is

$$\mathbf{x}_{pr} = \begin{pmatrix} \mathbf{x} \\ \mathbf{x}_r \end{pmatrix}, \qquad \dot{\mathbf{x}}_{pr} = \mathbf{A}_{pr}\mathbf{x}_{pr}, \qquad \begin{pmatrix} \mathbf{u} \\ \dot{\mathbf{x}}_r \end{pmatrix} = \mathbf{A}_c \begin{pmatrix} \mathbf{y} \\ \mathbf{x}_r \end{pmatrix} = \mathbf{A}_c\bar{\mathbf{C}}\mathbf{x}_{pr}\,. \tag{11.136}$$

The Lyapunov solution $\mathbf{P}$ is

$$(\mathbf{A}_{pr} + \alpha \mathbf{I}_{n+n_r})^T \mathbf{P} + \mathbf{P}(\mathbf{A}_{pr} + \alpha \mathbf{I}_{n+n_r}) = -\mathbf{Q}_c \quad \text{where} \quad \mathbf{Q}_c^T = \mathbf{Q}_c > 0 \tag{11.137}$$

$$(\bar{\mathbf{A}} + \alpha \mathbf{I}_{n+n_r})^T \mathbf{P} + \mathbf{P}(\bar{\mathbf{A}} + \alpha \mathbf{I}_{n+n_r}) + (\bar{\mathbf{B}}\mathbf{A}_c\bar{\mathbf{C}})^T \mathbf{P} + \mathbf{P}(\bar{\mathbf{B}}\mathbf{A}_c\bar{\mathbf{C}}) = -\mathbf{Q}_c \,. \tag{11.138}$$

The matrix $\mathbf{P}$ is the solution of the Lyapunov equation in the nominal case without controller and perturbation. The positive definite matrix $\mathbf{Q}_c$ is considered a given matrix which plays a key role in the Lyapunov equation. The matrix $\mathbf{Q}_c$ is given by the sum of a constant matrix $\mathbf{Q}$ and an additional matrix, depending on $\mathbf{A}_c$ of the controller,

$$\bar{\mathbf{Q}}_c = \mathbf{Q} + (\mathbf{A}_c\bar{\mathbf{C}})^T \mathbf{R} (\mathbf{A}_c\bar{\mathbf{C}}) \tag{11.139}$$

where $\mathbf{Q} > 0$ and $\mathbf{R} > 0$. If $\mathbf{A}_c = \mathbf{0}$ then $\mathbf{Q}_c = \mathbf{Q}$.

Nonlinear Perturbation

Adding a perturbation term to the system of Eq.(11.132), the perturbed closed-loop system is

$$\dot{\mathbf{x}}_{pr}(t) = \mathbf{A}_{pr}\mathbf{x}_{pr}(t) + \mathbf{g}_{pr}(\mathbf{x}_{pr}) \,. \tag{11.140}$$

Then, the sufficient condition for stability is that the perturbation vector $\mathbf{g}_{pr}$ satisfies (*Abdul-Wahab, A.A., 1989*)

$$\frac{\|\mathbf{g}_{pr}(\mathbf{z})\|_F}{\|\mathbf{z}\|_F} \leq \frac{\lambda_{\min}[\mathbf{Q}_c] + 2\alpha\lambda_{\min}[\mathbf{P}]}{2\lambda_{\max}[\mathbf{P}]} \qquad \forall \mathbf{z} \in \mathcal{R}^{n+n_r} \,. \tag{11.141}$$

Proof: From Eqs.(11.133) and (11.140)

$$\dot{V}(\mathbf{x}_{pr}) = \dot{\mathbf{x}}_{pr}^T \mathbf{P}\mathbf{x}_{pr} + \mathbf{x}_{pr}^T \mathbf{P}\dot{\mathbf{x}}_{pr} = \mathbf{x}_{pr}^T(\mathbf{A}_{pr}^T\mathbf{P} + \mathbf{P}\mathbf{A}_{pr})\mathbf{x}_{pr} + 2\mathbf{g}_{pr}^T\mathbf{P}\mathbf{x}_{pr} \,. \tag{11.142}$$

If Eq.(11.141) is satisfied then

$$\begin{aligned} \mathbf{g}_{pr}^T\mathbf{P}\mathbf{x}_{pr} &\leq \|\mathbf{g}_{pr}\|_F\|\mathbf{P}\mathbf{x}_{pr}\| \leq \|\mathbf{g}_{pr}\|_F\|\mathbf{P}\|_s\|\mathbf{x}_{pr}\|_F \leq \{\frac{\lambda_{\min}[\mathbf{Q}_c]}{2\lambda_{\max}[\mathbf{P}]} + \frac{\alpha\lambda_{\min}[\mathbf{P}]}{\lambda_{\max}[\mathbf{P}]}\}\lambda_{\max}[\mathbf{P}]\,\|\mathbf{x}_{pr}\|_F^2 \\ \mathbf{g}_{pr}^T\mathbf{P}\mathbf{x}_{pr} &\leq \{0.5\,\lambda_{\min}[\mathbf{Q}_c] + \alpha\lambda_{\min}[\mathbf{P}]\}\|\mathbf{x}_{pr}\|_F^2 \,. \end{aligned} \tag{11.143}$$

Combining Eqs.(11.137), (11.142) and Eq.(11.143), the time derivative $\dot{V}$ is

$$\begin{aligned} \dot{V} &= \mathbf{x}_{pr}^T\left(-2\alpha\mathbf{P} - \mathbf{Q}_c + \lambda_{\min}[\mathbf{Q}_c]\,\mathbf{I}_{n+n_r} + 2\alpha\lambda_{\min}[\mathbf{P}]\,\mathbf{I}_{n+n_r}\right)\mathbf{x}_{pr} & (11.144) \\ &= \mathbf{x}_{pr}^T\,2\alpha\left(-\mathbf{P} + \lambda_{\min}[\mathbf{P}]\,\mathbf{I}_{n+n_r}\right)\mathbf{x}_{pr} + \mathbf{x}_{pr}^T\left(-\mathbf{Q}_c + \lambda_{\min}[\mathbf{Q}_c]\,\mathbf{I}_{n+n_r}\right)\mathbf{x}_{pr} \,. & (11.145) \end{aligned}$$

For $\alpha > 0$ from Rayleigh theorem it results $\dot{V} < 0$. Even in the case of time-varying perturbation this is true.

End of the Proof

Linear Perturbation

Consider the linear perturbation

$$\dot{\mathbf{x}}(t) = (\mathbf{A} + \Delta\mathbf{A})\mathbf{x}(t) + (\mathbf{B} + \Delta\mathbf{B})\mathbf{u}(t) \tag{11.146}$$

$$\mathbf{y}(t) = (\mathbf{C} + \Delta\mathbf{C})\mathbf{x}(t) \; . \tag{11.147}$$

In order to obtain the model of Eq.(11.140) with abbreviations as given by Eq.(11.136), the perturbation term is

$$\mathbf{g}_{pr}(\mathbf{x}_{pr}) = (\Delta\bar{\mathbf{A}} + \Delta\bar{\mathbf{B}}\;\mathbf{A}_c\bar{\mathbf{C}} + \bar{\mathbf{B}}\mathbf{A}_c\Delta\bar{\mathbf{C}})\mathbf{x}_{pr} \tag{11.148}$$

where

$$\Delta\bar{\mathbf{A}} = \begin{pmatrix} \Delta\mathbf{A} & 0 \\ 0 & 0 \end{pmatrix}, \quad \Delta\bar{\mathbf{B}} = \begin{pmatrix} \Delta\mathbf{B} & 0 \\ 0 & 0 \end{pmatrix}, \quad \Delta\bar{\mathbf{C}} = \begin{pmatrix} \Delta\mathbf{C} & 0 \\ 0 & 0 \end{pmatrix} \; . \tag{11.149}$$

Using norm properties, the result of Eq.(11.141) can be employed and the sufficient stability condition is

$$\|\Delta\bar{\mathbf{A}}\|_s + \|\Delta\bar{\mathbf{B}}\|_s\|\mathbf{A}_c\|_s\|\bar{\mathbf{C}}\|_s + \|\bar{\mathbf{B}}\|_s\|\mathbf{A}_c\|_s\|\Delta\bar{\mathbf{C}}\|_s \leq \frac{\lambda_{\min}[\mathbf{Q}_c] + 2\alpha\lambda_{\min}[\mathbf{P}]}{2\lambda_{\max}[\mathbf{P}]} \; . \tag{11.150}$$

11.11.6 Dynamic Controller Reductions

The dynamic feedback controller as outlined in the previous sections can be reduced to various classical control concepts.

Optimal State Feedback Controller

If $n_r = 0$, $\mathbf{C} = \mathbf{I}_n$, the output controller $\mathbf{u} = \mathbf{K}_y\mathbf{C}\mathbf{x} = \mathbf{K}_y\mathbf{x}$ equals a state feedback controller where

$$\mathbf{A}_c = \mathbf{K}_y, \quad \bar{\mathbf{A}} = \mathbf{A}, \quad \bar{\mathbf{B}} = \mathbf{B}, \quad \bar{\mathbf{C}} = \mathbf{C} = \mathbf{I}_n, \quad \mathbf{A}_{pr} = \mathbf{A}+\mathbf{B}\mathbf{A}_c\mathbf{C} = \mathbf{A}+\mathbf{B}\mathbf{K}_y \; . \tag{11.151}$$

Its optimal result in the sense of quadratic performance is given by $\mathbf{K}_y = -\mathbf{R}^{-1}\mathbf{B}^T\mathbf{P}$ where $\mathbf{P}$ is the solution of the algebraic Riccati equation

$$(\mathbf{A} + \alpha\mathbf{I}_n)^T\mathbf{P} + \mathbf{P}(\mathbf{A} + \alpha\mathbf{I}_n) - \mathbf{P}\mathbf{B}\mathbf{R}^{-1}\mathbf{B}^T\mathbf{P} + \mathbf{Q} = 0, \qquad \mathbf{R} > 0, \; \mathbf{Q} \geq 0 \; . \tag{11.152}$$

Comparison with Eq.(11.139) yields

$$\mathbf{Q}_c = \mathbf{Q} + \mathbf{P}\mathbf{B}\mathbf{R}^{-1}\mathbf{B}^T\mathbf{P} \; . \tag{11.153}$$

From the Lyapunov equation Eq.(11.137), $\mathbf{Q}_c$ is given by

$$\mathbf{Q}_c = (\mathbf{A} + \alpha\mathbf{I}_n)^T\mathbf{P} + \mathbf{P}(\mathbf{A} + \alpha\mathbf{I}_n) + 2\mathbf{Q} \; . \tag{11.154}$$

The matrix $\mathbf{Q}_c$ is required to calculate the upper bound of $\mathbf{g}_{pr}$ in Eq.(11.141). The result $\mathbf{Q}_c$ corresponds to $\mathbf{D}$ in Eq.(13.137) and to the results given by *Patel, R.V., and Toda, M., 1980*.

Optimal Output Feedback Controller

Invoking Eqs.(16.75), (16.49) and (16.50) and substituting $\mathbf{V} := \mathbf{L}$, $\mathbf{\Lambda} := \mathbf{P}$, $\mathbf{K}^* := \mathbf{K}_y$, the optimal output feedback controller $\mathbf{K}_y$ is

$$\mathbf{K}_y = -\mathbf{R}^{-1}\mathbf{B}^T\mathbf{P}\mathbf{L}\mathbf{C}^T(\mathbf{C}\mathbf{L}\mathbf{C}^T)^{-1} \tag{11.155}$$

where $\mathbf{P}$ and $\mathbf{L}$ are the solution of the algebraic Riccati equation and of the Lyapunov equation, respectively

$$(\mathbf{A}+\mathbf{B}\mathbf{K}_y\mathbf{C}+\alpha\mathbf{I}_n)^T\mathbf{P}+\mathbf{P}(\mathbf{A}+\mathbf{B}\mathbf{K}_y\mathbf{C}+\alpha\mathbf{I}_n)+(\mathbf{K}_y\mathbf{C})^T\mathbf{R}\mathbf{K}_y\mathbf{C}+\mathbf{Q}=\mathbf{0} \quad (11.156)$$

$$(\mathbf{A}+\mathbf{B}\mathbf{K}_y\mathbf{C}+\alpha\mathbf{I}_n)\mathbf{L}+\mathbf{L}(\mathbf{A}+\mathbf{B}\mathbf{K}_y\mathbf{C}+\alpha\mathbf{I}_n)^T+\mathbf{I}_n=\mathbf{0}\,. \quad (11.157)$$

Referring to Eq.(11.139)

$$\begin{aligned}\mathbf{Q}_c &= \mathbf{Q}+\mathbf{C}^T\mathbf{K}_y^T\mathbf{R}\mathbf{K}_y\mathbf{C} \quad (11.158)\\ &= (\mathbf{A}+\mathbf{B}\mathbf{K}_y\mathbf{C}+\alpha\mathbf{I}_n)^T\mathbf{P}+\mathbf{P}(\mathbf{A}+\mathbf{B}\mathbf{K}_y\mathbf{C}+\alpha\mathbf{I}_n)+2(\mathbf{C}^T\mathbf{K}_y^T\mathbf{R}\mathbf{K}_y\mathbf{C}+\mathbf{Q})\,.\end{aligned}$$

Full-Order Observer Feedback

Applying a full-order state observer by setting $\mathbf{x}_r \triangleq \hat{\mathbf{x}},\ n_r = n$,

$$\dot{\hat{\mathbf{x}}}(t)=\mathbf{A}\hat{\mathbf{x}}(t)+\mathbf{B}\mathbf{u}(t)+\mathbf{N}[\mathbf{y}(t)-\mathbf{C}\hat{\mathbf{x}}(t)]=(\mathbf{A}-\mathbf{N}\mathbf{C})\hat{\mathbf{x}}(t)+\mathbf{B}\mathbf{u}(t)+\mathbf{N}\mathbf{y}(t) \quad (11.159)$$

$$\mathbf{u}(t)=\mathbf{K}\hat{\mathbf{x}}(t) \quad \leadsto \quad \dot{\hat{\mathbf{x}}}=(\mathbf{A}+\mathbf{B}\mathbf{K}-\mathbf{N}\mathbf{C})\hat{\mathbf{x}}+\mathbf{N}\mathbf{y}\,. \quad (11.160)$$

The full-order observer-based feedback is rewritten to the dynamic controller

$$\begin{pmatrix}\mathbf{u}\\ \dot{\mathbf{x}}_r\end{pmatrix}=\begin{pmatrix}\mathbf{u}\\ \dot{\hat{\mathbf{x}}}\end{pmatrix}=\mathbf{A}_c\begin{pmatrix}\mathbf{y}\\ \hat{\mathbf{x}}\end{pmatrix}=\begin{pmatrix}\mathbf{0} & \mathbf{K}\\ \mathbf{N} & \mathbf{A}+\mathbf{B}\mathbf{K}-\mathbf{N}\mathbf{C}\end{pmatrix}\begin{pmatrix}\mathbf{y}\\ \hat{\mathbf{x}}\end{pmatrix}. \quad (11.161)$$

Reduced-Order Observer Feedback

Denoting $\mathbf{z}$, $\hat{\mathbf{x}}$ the observer states and the plant state estimate, respectively, and defining $\mathbf{x}_r \triangleq \mathbf{z} \in \mathcal{R}^{n-r} = \mathcal{R}^{n_r}$ it follows

$$\dot{\mathbf{z}}(t)=\mathbf{F}_z\mathbf{z}+\mathbf{H}\mathbf{u}+\mathbf{N}\mathbf{y} \qquad \mathbf{y}\in\mathcal{R}^r \quad (11.162)$$

$$\hat{\mathbf{x}}=\mathbf{E}\mathbf{z}+\mathbf{D}\mathbf{y} \qquad \mathbf{u}=\mathbf{K}\hat{\mathbf{x}} \quad (11.163)$$

$$\begin{aligned}\dot{\mathbf{z}} &= \mathbf{F}_z\mathbf{z}+\mathbf{H}\mathbf{K}\hat{\mathbf{x}}+\mathbf{N}\mathbf{y}=\mathbf{F}_z\mathbf{z}+\mathbf{H}\mathbf{K}(\mathbf{E}\mathbf{z}+\mathbf{D}\mathbf{y})+\mathbf{N}\mathbf{y} \quad (11.164)\\ &= (\mathbf{F}_z+\mathbf{H}\mathbf{K}\mathbf{E})\mathbf{z}+(\mathbf{H}\mathbf{K}\mathbf{D}+\mathbf{N})\mathbf{y}\,. \quad (11.165)\end{aligned}$$

Hence, the reduced-order observer-based feedback corresponds to the general dynamic feedback controller

$$\begin{pmatrix}\mathbf{u}\\ \dot{\mathbf{z}}\end{pmatrix}=\mathbf{A}_c\begin{pmatrix}\mathbf{y}\\ \mathbf{z}\end{pmatrix}=\begin{pmatrix}\mathbf{K}\mathbf{D} & \mathbf{K}\mathbf{E}\\ \mathbf{H}\mathbf{K}\mathbf{D}+\mathbf{N} & \mathbf{F}_z+\mathbf{H}\mathbf{K}\mathbf{E}\end{pmatrix}\begin{pmatrix}\mathbf{y}\\ \mathbf{z}\end{pmatrix}. \quad (11.166)$$

11.12 Structured State-Space Uncertainty

Consider a linear system with perturbed state-space model

$$\dot{\mathbf{x}}(t)=\mathbf{A}\mathbf{x}(t)+\mathbf{B}\mathbf{u}(t)+\Delta\mathbf{A}(t)\mathbf{x}(t)+\Delta\mathbf{B}(t)\mathbf{u}(t) \quad (11.167)$$

$$\mathbf{y}(t)=\mathbf{C}\mathbf{x}(t)\,. \quad (11.168)$$

The time-varying parametric uncertainties $\Delta\mathbf{A}(t)$ and $\Delta\mathbf{B}(t)$ in the entries of $\mathbf{A}$ and $\mathbf{B}$ are assumed bounded element by element, according with

$$|\Delta\mathbf{A}(t)| \leq_e \mathbf{A}_m \quad \text{and} \quad |\Delta\mathbf{B}(t)| \leq_e \mathbf{B}_m \; . \tag{11.169}$$

Applying an output feedback controller with constant gain $\mathbf{u}(t) = \mathbf{K}\mathbf{y}(t)$, the closed-loop system is described by

$$\dot{\mathbf{x}}(t) = [\mathbf{F} + \Delta\mathbf{F}(t)]\mathbf{x}(t) \tag{11.170}$$

$$\mathbf{F} \triangleq \mathbf{A} + \mathbf{BKC} \qquad \Delta\mathbf{F} \triangleq \Delta\mathbf{A}(t) + \Delta\mathbf{B}(t)\mathbf{KC} \qquad |\Delta\mathbf{F}(t)| \leq_e \mathbf{A}_m + \mathbf{B}_m|\mathbf{KC}| \; . \tag{11.171}$$

By employing modal transformation to the closed-loop system and assuming distinct eigenvalues,

$$\mathbf{x}(t) = \mathbf{T}\mathbf{x}^{mo}(t) = \mathbf{T}[\mathbf{A} + \mathbf{BKC}]\; \mathbf{x}^{mo}(t) \tag{11.172}$$

$$\alpha_s = \alpha_s[\mathbf{A} + \mathbf{BKC}] = \max|\lambda_i[\mathbf{A} + \mathbf{BKC}]\,| \qquad \Lambda_n \triangleq \mathrm{diag}_n \lambda_i[\mathbf{A} + \mathbf{BKC}] \; . \tag{11.173}$$

Since $\|\mathbf{x}(t)\|_F \leq \|\mathbf{T}\|_s \|\mathbf{x}^{mo}(t)\|_F$ and assuming $\mathbf{T}$ nonsingular, it is sufficient to consider $\mathbf{x}^{mo}$ instead of $\mathbf{x}$

$$\dot{\mathbf{x}}^{mo}(t) = \mathbf{T}^{-1}\mathbf{F}\; \mathbf{T}\mathbf{x}^{mo}(t) + \mathbf{T}^{-1}\Delta\mathbf{F}\; \mathbf{T}\mathbf{x}^{mo}(t) \tag{11.174}$$

$$\mathbf{x}^{mo}(t) = e^{\Lambda_n t}\mathbf{x}^{mo}(0) + \int_0^t e^{\Lambda(t-\tau)}\; \mathbf{T}^{-1}\Delta\mathbf{F}(\tau)\mathbf{T}\mathbf{x}^{mo}(\tau)d\tau \tag{11.175}$$

$$\|\mathbf{x}^{mo}(t)\|_F \leq e^{\alpha_s t}\|\mathbf{x}^{mo}(0)\|_F + \int_0^t e^{\alpha_s(t-\tau)}\; \|\mathbf{T}^{-1}\Delta\mathbf{F}\;\mathbf{T}\|_s\; \|\mathbf{x}^{mo}(\tau)\|_F d\tau \tag{11.176}$$

$$\|\mathbf{x}^{mo}(t)\|_F e^{-\alpha_s t} \leq \|\mathbf{x}^{mo}(0)\|_F + \int_0^t e^{-\alpha_s \tau}\|\mathbf{T}^{-1}\Delta\mathbf{F}\;\mathbf{T}\|_s\; \|\mathbf{x}^{mo}(\tau)\|_F d\tau \; . \tag{11.177}$$

Using, now, Bellman-Gronwall lemma given by Eq.(11.19) with $c := \|\mathbf{x}^{mo}(0)\|_F$ and $k(\tau) := \|\mathbf{T}^{-1}\Delta\mathbf{F}\;\mathbf{T}\|_s$

$$\|\mathbf{x}^{mo}(t)\|_F e^{-\alpha_s t} \leq \|\mathbf{x}^{mo}(0)\|_F e^{\int_0^t \|\mathbf{T}^{-1}\Delta\mathbf{F}(\tau)\mathbf{T}\|_s d\tau} \; . \tag{11.178}$$

With regard to $\|\mathbf{M}\|_s \leq \|\;|\mathbf{M}|\;\|_s$ (*Kouvaritakis, B., and Latchman, H., 1985*)

$$\begin{aligned}
\|\mathbf{T}^{-1}\Delta\mathbf{F}(t)\mathbf{T}\|_s &= \|\mathbf{T}^{-1}[\Delta\mathbf{A}(t) + \Delta\mathbf{B}(t)\mathbf{KC}]\mathbf{T}\|_s \leq \|\;|\mathbf{T}^{-1}[\Delta\mathbf{A}(t) + \Delta\mathbf{B}(t)\mathbf{KC}]\mathbf{T}|\;\|_s \\
&= \sup_{\mathbf{z}} \frac{\|\;|\mathbf{T}^{-1}[\Delta\mathbf{A} + \Delta\mathbf{B}\;\mathbf{KC}]\mathbf{T}|\;|\mathbf{z}|\;\|_F}{\|\;|\mathbf{z}|\;\|_F} && (11.179) \\
&\leq \sup_{\mathbf{z}} \frac{\|\;|\mathbf{T}^{-1}|\left(\mathbf{A}_m + \mathbf{B}_m|\mathbf{KC}|\;\right)|\mathbf{T}|\;\|_F}{\|\;|\mathbf{z}|\;\|_F} && (11.180) \\
&\leq \|\;|\mathbf{T}^{-1}|\left(\mathbf{A}_m + \mathbf{B}_m|\mathbf{KC}|\right)|\mathbf{T}|\;\|_s \; . && (11.181)
\end{aligned}$$

The time-varying exponent in Eq.(11.178) is bounded by a constant one. Continuing the calculation with Eq.(11.178),

$$\|\mathbf{x}^{mo}(t)\|_F \leq e^{\alpha_s t}\|\mathbf{x}^{mo}(0)\|_F \exp\left\{\|\;|\mathbf{T}^{-1}|\left(\mathbf{A}_m + \mathbf{B}_m|\mathbf{KC}|\right)|\mathbf{T}|\;\|_F t\right\} \; . \tag{11.182}$$

Hence, $\|\mathbf{x}^{mo}(t)\|_F$ and $\|\mathbf{x}(t)\|_F$ decay if the spectral abscissa of the nominal closed-loop system matrix satisfies (*Sobel, K.M., et al. 1989*)

$$\alpha_s < -\|\;|\mathbf{T}^{-1}|\left(\mathbf{A}_m + \mathbf{B}_m|\mathbf{KC}|\right)|\mathbf{T}|\;\|_s \; . \tag{11.183}$$

11.12.1 Singular-Value Decomposition of B

Full-State Feedback Controller

Suppose a full-state feedback controller $\mathbf{C} = \mathbf{I}$. Applying the singular-value decomposition of $\mathbf{B}$ and using abbreviations,

$$\mathbf{B} = (\mathbf{U}_o \vdots \mathbf{U}_1)\begin{pmatrix} \boldsymbol{\Sigma}_m \mathbf{V}^T \\ \mathbf{0} \end{pmatrix} = \mathbf{U}_o \boldsymbol{\Sigma}_m \mathbf{V}^T \tag{11.184}$$

$$\boldsymbol{\Lambda}_r \triangleq \text{diag}_r \lambda_i \qquad \boldsymbol{\Sigma}_m \triangleq \text{diag}_m \sigma_i[\mathbf{B}] \qquad \boldsymbol{\Sigma}_r \triangleq \text{diag}_r \sigma_{ri}[\mathbf{T}_r] \tag{11.185}$$

$$\mathbf{B}, \mathbf{U}_o \in \mathcal{R}^{n \times m}; \qquad \mathbf{V}, \boldsymbol{\Sigma}_m \in \mathcal{R}^{m \times m}; \qquad \mathbf{U}_1 \in \mathcal{R}^{n \times (n-m)} \tag{11.186}$$

where $(\mathbf{U}_o \vdots \mathbf{U}_1)$ and $\mathbf{V}$ are the matrices of the ("left" and "right") singular vectors associated with $\mathbf{B}\mathbf{B}^H$ and $\mathbf{B}^H\mathbf{B}$, respectively. Then,

$$\mathbf{A} + \mathbf{B}\mathbf{K} = \mathbf{A} + \mathbf{U}_o \boldsymbol{\Sigma}_m \mathbf{V}^T \mathbf{K} = \mathbf{F} = \mathbf{T}\boldsymbol{\Lambda}_n \mathbf{T}^{-1} \tag{11.187}$$

$$\mathbf{K} = \mathbf{V}\boldsymbol{\Sigma}_m^{-1}\mathbf{U}_o^T[\mathbf{T}\boldsymbol{\Lambda}_n \mathbf{T}^{-1} - \mathbf{A}] \tag{11.188}$$

where the properties of unitary matrices have been used for $\mathbf{U}_o$ and $\mathbf{V}$ (see remark). Hence, $\mathbf{K}$ is replaced by data of the control matrix $\mathbf{B}$ and by data of the closed loop, i.e. λ_i and $\mathbf{T}$. Combining Eq.(11.188) and (11.183) yields

$$\alpha_s < -\| \, |\mathbf{T}^{-1}| \left(\mathbf{A}_m + \mathbf{B}_m |\mathbf{V}\boldsymbol{\Sigma}_m^{-1}\mathbf{U}_o^T[\mathbf{T}\boldsymbol{\Lambda}_n\mathbf{T}^{-1} - \mathbf{A}]\, |\right) |\mathbf{T}| \, \|_s \,. \tag{11.189}$$

Remark: When solving

$$\mathbf{B}\mathbf{K} = (\mathbf{U}_o \vdots \mathbf{U}_1)\begin{pmatrix} \boldsymbol{\Sigma}_m \\ \mathbf{0} \end{pmatrix}\mathbf{V}^T\mathbf{K} = \mathbf{F} - \mathbf{A} \tag{11.190}$$

$$\begin{pmatrix} \boldsymbol{\Sigma}_m \\ \mathbf{0} \end{pmatrix}\mathbf{V}^T\mathbf{K} = (\mathbf{U}_o \vdots \mathbf{U}_1)^{-1}(\mathbf{F} - \mathbf{A}) = \mathbf{U}^T(\mathbf{F} - \mathbf{A}) = \begin{pmatrix} \mathbf{U}_o^T \\ \mathbf{U}_1^T \end{pmatrix}(\mathbf{F} - \mathbf{A}) \,, \tag{11.191}$$

the matrix equation above can be separated into two matrix equations

$$\boldsymbol{\Sigma}_m\mathbf{V}^T\mathbf{K} = \mathbf{U}_o^T(\mathbf{F} - \mathbf{A}) \quad \text{and} \quad \mathbf{0}_{(n-m)\times m}\mathbf{V}^T\mathbf{K} = \mathbf{0}_{(n-m)\times n} = \mathbf{U}_1^T(\mathbf{F} - \mathbf{A}) \tag{11.192}$$

where the last one provides a condition for solving Eq.(11.188), only. □

Example: Singular-value decomposition of **B**:

$$n = 3, \quad m = 2, \quad \mathbf{B} = \begin{pmatrix} 0 & 0 \\ 1 & 0 \\ 0 & 2 \end{pmatrix}, \quad \mathbf{U}_o = \begin{pmatrix} 0 & 0 \\ 1 & 0 \\ 0 & 1 \end{pmatrix}, \quad \mathbf{U}_1 = \begin{pmatrix} 1 \\ 0 \\ 0 \end{pmatrix} \tag{11.193}$$

$$\sigma_i[\mathbf{B}] = 1; 2 \qquad \mathbf{V} = \begin{pmatrix} 1 & 0 \\ 0 & 1 \end{pmatrix}, \qquad \boldsymbol{\Sigma}_m = \text{diag}_m \sigma_i[\mathbf{B}] = \begin{pmatrix} 1 & 0 \\ 0 & 2 \end{pmatrix} \tag{11.194}$$

$$\mathbf{B} = \mathbf{U}\begin{pmatrix} \boldsymbol{\Sigma}_m \\ \mathbf{0} \end{pmatrix}\mathbf{V}^T = \mathbf{U}\begin{pmatrix} \boldsymbol{\Sigma}_m\mathbf{V}^T \\ \mathbf{0} \end{pmatrix} = \begin{pmatrix} 0 & 0 & 1 \\ 1 & 0 & 0 \\ 0 & 1 & 0 \end{pmatrix}\begin{pmatrix} 1 & 0 \\ 0 & 2 \\ 0 & 0 \end{pmatrix}\begin{pmatrix} 1 & 0 \\ 0 & 1 \end{pmatrix} = \begin{pmatrix} 0 & 0 & 1 \\ 1 & 0 & 0 \\ 0 & 1 & 0 \end{pmatrix}\begin{pmatrix} 1 & 0 \\ 0 & 2 \\ 0 & 0 \end{pmatrix}. \tag{11.195}$$

End of Example

Output Feedback Controller

If an output feedback controller $\mathbf{u}(t) = \mathbf{K}\mathbf{y}(t)$ is applied where $\mathbf{K} \in \mathcal{R}^{m\times r}$, r eigenvalues of the closed-loop system can be assigned. Supposing that $\mathbf{K}$ is such that $\mathbf{A}+\mathbf{BKC}$ is asymptotically stable with distinct eigenvalues λ_i, the eigenvectors $\mathbf{v}_i$ associated with λ_i satisfy

$$\mathbf{F}\mathbf{f}_i = (\mathbf{A}+\mathbf{BKC})\mathbf{f}_i = \lambda_i\mathbf{f}_i \quad \forall i = 1\ldots r \tag{11.196}$$

$$(\mathbf{A}+\mathbf{BKC})\mathbf{T}_r = \mathbf{T}_r\mathbf{\Lambda}_r \triangleq (\mathbf{f}_1 \vdots \mathbf{f}_2 \ldots \mathbf{f}_r)\,\mathrm{diag}_r\lambda_i \quad \rightsquigarrow \quad \mathbf{BKCT}_r = \mathbf{T}_r\mathbf{\Lambda}_r - \mathbf{AT}_r\,. \tag{11.197}$$

Applying singular-value decompositions for $\mathbf{B} \in \mathcal{R}^{n\times m}$ and $\mathbf{T}_r \in \mathcal{R}^{n\times r}$,

$$(\mathbf{U}_o \vdots \mathbf{U}_1)\begin{pmatrix}\mathbf{\Sigma}_m\mathbf{V}^T\\ \mathbf{0}\end{pmatrix}\mathbf{KC}(\mathbf{U}_{ro} \vdots \mathbf{U}_{r1})\begin{pmatrix}\mathbf{\Sigma}_r\mathbf{V}_r^T\\ \mathbf{0}\end{pmatrix} = \mathbf{T}_r\mathbf{\Lambda}_r - \mathbf{AT}_r \tag{11.198}$$

the matrix can be separated. Using the unitary properties of $\mathbf{U}_o, \mathbf{U}_{ro}, \mathbf{V}, \mathbf{V}_r$,

$$\mathbf{KC} = \mathbf{V}\mathbf{\Sigma}_m^{-1}\mathbf{U}_o^T(\mathbf{T}_r\mathbf{\Lambda}_r - \mathbf{AT}_r)\mathbf{V}_r\mathbf{\Sigma}_r^{-1}\mathbf{U}_{ro}^T \tag{11.199}$$

and from Eq.(11.183)

$$\alpha_s < -\|\ |\mathbf{T}^{-1}|\ \left(\mathbf{A}_m + \mathbf{B}_m\ |\mathbf{V}\mathbf{\Sigma}_m^{-1}\mathbf{U}_o^T(\mathbf{T}_r\mathbf{\Lambda}_r - \mathbf{AT}_r)\mathbf{V}_r\mathbf{\Sigma}_r^{-1}\mathbf{U}_{ro}^T|\ \right)\ |\mathbf{T}|\ \|_s\,. \tag{11.200}$$

Norm-Bounded Perturbation Matrices

Using norm bounds of unstructured perturbation matrices

$$\|\Delta\mathbf{A}(t)\|_s \le \beta_A \qquad \|\Delta\mathbf{B}(t)\|_s < \beta_B\,, \tag{11.201}$$

from Eq.(11.183) a simplified result is obtained using the spectral condition number $\kappa_s[\mathbf{T}]$

$$\alpha_s < -(\beta_A + \beta_B\|\mathbf{KC}\|_s)\kappa_s[\mathbf{T}]\,. \tag{11.202}$$

Optimal Weighting

Using diagonal weighting of the nominal closed-loop modal matrix $\mathbf{T}$, the conservatism of the sufficient stability can be reduced. Replacing $\mathbf{T}$ by $\mathbf{TD}$, $\mathbf{D} = \mathrm{diag}\ d_i$, $d_i > 0$, the matrix $\mathbf{T}$ is scaled and Eq.(11.183) becomes

$$\alpha_s < -\|\ \mathbf{D}^{-1}\ |\mathbf{T}^{-1}|\ \left(\mathbf{A}_m + \mathbf{B}_m|\mathbf{KC}|\right)\ |\mathbf{T}|\ \mathbf{D}\ \|_s \tag{11.203}$$

which corresponds to the so-called D-weighted spectral norm. Selecting the infimum with respect to $\mathbf{D}$

$$\alpha_s < -\inf_{\mathbf{D}}\|\ \mathbf{D}^{-1}\ |\mathbf{T}^{-1}|\ \left(\mathbf{A}_m + \mathbf{B}_m|\mathbf{KC}|\right)\ |\mathbf{T}|\ \mathbf{D}\ \|_s \tag{11.204}$$

the spectral abscissa is chosen as near as possible to the imaginary axis. By the way, the right-hand side of the last equation corresponds to the Perron eigenvalue.

Eigenvalue Increment

Suppose $\Delta\mathbf{A}$ and $\Delta\mathbf{B}$ time-invariant, from Eq.(11.171) and from the Bauer-Fike theorem Eq.(15.43)

$$\min_i|\lambda_i[\mathbf{F}] - \lambda_i[\mathbf{F} - \Delta\mathbf{F}]\ | \le \|\mathbf{T}^{-1}\Delta\mathbf{F}\ \mathbf{T}\|_s\,. \tag{11.205}$$

Using Eq.(11.181), the eigenvalue increment of the closed-loop system when subject to the perturbation is given by

$$\min_i|\lambda_i[\mathbf{F}] - \lambda_i[\mathbf{F} + \Delta\mathbf{F}]\ | \le \|\ |\mathbf{T}^{-1}|\ \left(\mathbf{A}_m + \mathbf{B}_m|\mathbf{KC}|\right)\ |\mathbf{T}|\ \|_s\,. \tag{11.206}$$

11.12.2 Discrete-Time System

The results given in the previous sections can easily be extended to discrete-time systems. Consider

$$\mathbf{x}(k+1) = \mathbf{\Phi x}(k) + \mathbf{\Psi u}(k) \qquad \mathbf{y}(k) = \mathbf{Cx}(k) \quad \text{and} \tag{11.207}$$

$$\mathbf{x}(k+1) = [\mathbf{\Phi} + \Delta\mathbf{\Phi}(k)]\mathbf{x}(k) + [\mathbf{\Psi} + \Delta\mathbf{\Psi}(k)]\mathbf{u}(k) \tag{11.208}$$

in the nominal and in the perturbed case, respectively. Suppose that the bounds

$$|\Delta\mathbf{\Phi}(k)| \leq_e \mathbf{\Phi}_m \qquad |\Delta\mathbf{\Psi}(k)| \leq_e \mathbf{\Psi}_m \tag{11.209}$$

exist and that a constant gain feedback is employed such that the nominal system is stable with distinct eigenvalues. Then, from a calculation very similar to the derivation of Eq.(11.183), the perturbed closed-loop discrete-time system is asymptotically stable (*Sobel, K.M., et al. 1989*) if the spectral radius of $\mathbf{\Phi}_{cl}$ satisfies

$$\rho_s[\mathbf{\Phi} + \mathbf{\Psi KC}] = \max_i |\lambda_i| < 1 - \|\, |\mathbf{T}_\Phi^{-1}| \left(\mathbf{\Phi}_m + \mathbf{\Psi}_m|\mathbf{KC}|\right) |\mathbf{T}_\Phi| \,\|_s \tag{11.210}$$

where $\mathbf{T}_\Phi$ is the modal matrix corresponding to the closed-loop matrix $\mathbf{\Phi}_{cl} = \mathbf{\Phi} + \mathbf{\Psi KC}$.

11.13 Comparison Norm-Bounded and Structured Uncertainty

Comparing the approaches given by *Chen, B.S., and Wong, C.C., 1987; Sobel, K.M., et al. 1989* yields the following results. *Sobel, K.M., et al. 1989* include details of the structure of the uncertainty, the result is less conservative, particularly if D-weighted. If the approach of *Sobel, K.M., et al. 1989* is considered only time-invariant, the uncertainty is only norm-bounded, the condition number $\kappa_s(\mathbf{T})$ is replaced by m_o and the spectral norm is selected instead of $\|\mathbf{\Phi}_{pr}(t)\|_1$ and $\|\Delta\mathbf{A}_{pr}\|_1$, then the result Eq.(11.105) of *Chen, B.S., and Wong, C.C., 1987* is achieved.

Moreover, *Sobel, K.M., et al. 1989* present a control example with a detailed comparison of the methods mentioned above.

Chapter 12

Robust Dynamic Interval Systems

Stability results in interval polynomials and interval matrices can be applied to solve the problem of robust control in spite of the fact that there is no simple relation between the tolerances or the uncertainty of the plant and the interval matrix or interval characteristic polynomial of the closed-loop control system. A property is said true with respect to the interval matrix $\mathbf{A}_I$ if the property holds for every $\mathbf{A} \in \mathbf{A}_I$.

12.1 Stability Properties of an Interval Matrix Set

An interval matrix set is defined by the set of real matrices

$$\mathbf{A}_I \triangleq [\mathbf{L} \;\; \mathbf{H}] \triangleq \{\mathbf{A} \in \mathcal{R}^{n\times n} \; : \; \mathbf{L} \leq_e \mathbf{A} \leq_e \mathbf{H}\} \tag{12.1}$$

where the meaning of $\mathbf{A} = \text{matrix}[A_{ij}] \leq_e \mathbf{M} = \text{matrix}[M_{ij}]$ is $A_{ij} \leq M_{ij}$ and $\mathbf{L}$, $\mathbf{H}$ denote low, high, respectively. The modulus matrix $|\mathbf{P}|$ is given by $|\mathbf{P}| = \text{matrix}[\,|P_{ij}|\,]$.

Using a parameter ε in an interval $[0 \;\; 1]$, the auxiliary matrix $\mathbf{A}(\varepsilon)$ can be established as

$$\mathbf{A}(\varepsilon) = \mathbf{L} + \varepsilon\mathbf{E} \quad \text{where} \quad \mathbf{A}(0) = \mathbf{L}, \;\; \mathbf{A}(1) = \mathbf{H}, \;\; \mathbf{E} = \mathbf{H} - \mathbf{L}\,. \tag{12.2}$$

A stable matrix $\mathbf{A}(\varepsilon)$ is characterized by eigenvalues $\lambda[\mathbf{A}(\varepsilon)]$ in the open left-half complex plane

$$\det\{\lambda[\mathbf{A}(\varepsilon)]\,\mathbf{I}_n - \mathbf{A}(\varepsilon)\} \;\Big|_{\,\Re e\lambda<0} = 0\,, \tag{12.3}$$

i.e., the root locus $\lambda = \lambda[\mathbf{A}(\varepsilon)]$ may not pass through or touch the imaginary axis. Eigenvalues $\lambda = j\omega \;\; \forall\omega \geq 0$ are considered unstable. Hence, if $\mathbf{A}(\varepsilon)$ is stable

$$\det[j\omega\mathbf{I}_n - \mathbf{A}(\varepsilon)] \neq 0 \qquad \forall\omega \geq 0 \qquad \varepsilon \in [0 \;\; 1]\,. \tag{12.4}$$

12.2 Stability Robustness Using Resolvent Matrix

Consider the state-space representation of a system with nominal system matrix $\mathbf{L}$. Given a non-negative tolerance (or uncertainty) matrix $\mathbf{U}_o$ with entries $U_{oij} \geq 0$, i.e. $\mathbf{U}_o \geq_e \mathbf{0}$. Then, the following statement can be easily proved: If $\mathbf{E} <_e \mathbf{U}_o$ and $\mathbf{L}$ is stable then $\mathbf{H}$ is also stable if

$$\rho_s[\,|\mathbf{N}_o(j\omega)|\,\mathbf{U}_o] < 1 \;\; \forall\omega \geq 0 \qquad \text{where} \qquad \mathbf{N}_o(j\omega) \triangleq (j\omega\mathbf{I}_n - \mathbf{L})^{-1} \tag{12.5}$$

is the resolvent matrix of $\mathbf{L}$.

Proof: Two statements are used. First,

$$|\mathbf{N}_o(j\omega)\mathbf{E}| \le_e |\mathbf{N}_o(j\omega)|\ |\mathbf{E}| \le_e |\mathbf{N}_o(j\omega)|\mathbf{U}_o \tag{12.6}$$

and, second,

$$\rho_s[\mathbf{G}] \le \rho_s[\ |\mathbf{H}|\ ] \quad \text{if} \quad |\mathbf{G}| \le_e |\mathbf{H}|\ . \tag{12.7}$$

The condition above is derived from the diagonal dominance theorem (*Ortega, J.M., 1972 p.120*). Thus, the inequality results

$$\rho_s[\mathbf{N}_o(j\omega)\mathbf{E}] \le \rho_s[\ |\mathbf{N}_o(j\omega)|\ \mathbf{U}_o]\ . \tag{12.8}$$

The eigenvalues $\lambda_{N\varepsilon}$ of the matrix $\varepsilon\mathbf{N}_o(j\omega)\mathbf{E}$ are given by the characteristic equation

$$\det\ [\lambda_{N\varepsilon}\mathbf{I}_n - \varepsilon\mathbf{N}_o(j\omega)\mathbf{E}] = 0 \tag{12.9}$$

and satisfy

$$\lambda_{N\varepsilon} = \varepsilon\lambda_{N,\varepsilon=1} \quad \text{or} \quad |\lambda_{N\varepsilon}| \le |\lambda_{N1}| = |\lambda[\mathbf{N}_o(j\omega)\mathbf{E}]\ |\ . \tag{12.10}$$

Since Eq.(12.5) and (12.8) $|\lambda_{N\varepsilon}| < 1$. Hence, replacing $\lambda_{N\varepsilon}$ in Eq.(12.9) by unity, the equality never can occur, i.e.

$$\det[\mathbf{I}_n - \varepsilon\mathbf{N}_o(j\omega)\mathbf{E}] \ne 0 \quad \text{or} \tag{12.11}$$

$$\det\ (j\omega\mathbf{I}_n - \mathbf{L})^{-1}(j\omega\mathbf{I}_n - \mathbf{L} - \varepsilon\mathbf{E}) = [\det\ (j\omega\mathbf{I}_n - \mathbf{L})^{-1}][\det\ (j\omega\mathbf{I}_n - \mathbf{L} - \varepsilon\mathbf{E})] \ne 0\ . \tag{12.12}$$

Using Eq.(12.4) with $\varepsilon = 0$ for the stable matrix $\mathbf{L}$, it results

$$\det\ (j\omega\mathbf{I}_n - \mathbf{L} - \varepsilon\mathbf{E}) \ne 0 \quad \forall\omega \ge 0 \quad \varepsilon \in [0 \quad 1] \tag{12.13}$$

and both $\mathbf{A}(\varepsilon)$ and $\mathbf{H} = \mathbf{A}(\varepsilon = 1)$ are stable.

End of Proof

An alternative formulation is given as follows: The stability of $\mathbf{A}_o$ implies stability of $\mathbf{A}_o + \mathbf{U}_o$ if $\rho_s[\ |\mathbf{N}_o|\ \mathbf{U}_o] < 1$. If the calculation is repeated with $\varepsilon \in [0 \quad -1]$ the similar result is achieved that stability of $\mathbf{A}_o$ also implies stability of $\mathbf{A}_o - \mathbf{U}_o$. Thus, the symmetric interval matrix $\mathbf{A}_I = [\mathbf{A}_o - \mathbf{U}_o \quad \mathbf{A}_o + \mathbf{U}_o] = [\mathbf{A}_1 \quad \mathbf{A}_2]$ is stable if $\mathbf{A}_o$ is stable and $\rho_s[\ |\mathbf{N}_o(j\omega)|\ \mathbf{U}_o] < 1 \ \ \forall\omega \ge 0$ where $\mathbf{N}_o$ is given by Eq.(12.5) with $\mathbf{L}$ replaced by $\mathbf{A}_o$. In other words, the interval matrix set

$$\mathbf{A}_I = [\mathbf{A}_1 \quad \mathbf{A}_2] \quad \text{is stable if} \quad \rho_s[\ |(j\omega\mathbf{I}_n - \frac{\mathbf{A}_1 + \mathbf{A}_2}{2})^{-1}|\ \frac{(\mathbf{A}_2 - \mathbf{A}_1)}{2}\ ] < 1\ . \tag{12.14}$$

The matrices $\mathbf{A}_o$ and $\mathbf{U}_o$ correspond to $(\mathbf{A}_2 + \mathbf{A}_1)/2$ and $(\mathbf{A}_2 - \mathbf{A}_1)/2$, respectively (*Juang, Y.T., et al. 1987*). Note that $\rho_s[\ |\mathbf{N}_o|\ \mathbf{U}_o] < 1$ is implied by $\|\ |\mathbf{N}_o|\ \|\ \|\mathbf{U}_o\| < 1$. In the case of output feedback control, the relation $\mathbf{N}_o(j\omega) = (j\omega\mathbf{I}_n - \mathbf{A}_o - \mathbf{BKC})^{-1}$ can be utilized to design $\mathbf{K}$ when $\mathbf{A}_o, \mathbf{B}, \mathbf{C}$ and $\mathbf{U}_o$ are given.

12.3 Unidirectional Perturbation. Perron Eigenvalue Approach

Consider a linear continuous time-invariant system with additive state-space perturbance

$$\dot{\mathbf{x}}(t) = \mathbf{A}_I\mathbf{x}(t) = (\mathbf{A} + \Delta\mathbf{A})\mathbf{x}(t) \qquad \mathbf{x}(t) \in \mathcal{R}^n\ . \tag{12.15}$$

The matrix $\mathbf{A}_I$ is an interval matrix $\mathbf{A}_I \triangleq [\mathbf{A} \quad \mathbf{A} + \Delta\mathbf{A}]$ and represents the open-loop or closed-loop control system, $\mathbf{A}$ is the nominal state matrix. The matrix $\Delta\mathbf{A}$ is the perturbation which is assumed of the structure $\Delta\mathbf{A} = \epsilon\mathbf{E}$ where $\mathbf{E}$ is a fixed known positive matrix, i.e., $e_{ij} > 0 \ \forall\, i,j$. Then, if $\mathbf{A}$ is stable the system $\mathbf{A} + \Delta\mathbf{A}$ is also stable if e is sufficiently small, namely

$$e < \{\sup_{\omega \geq 0} \pi[\ |(j\omega \mathbf{I}_n - \mathbf{A})^{-1}|\mathbf{E}]\}^{-1} \tag{12.16}$$

where $\pi[\cdot]$ is the Perron-Frobenius eigenvalue. The matrix $(j\omega\mathbf{I}_n - \mathbf{A})^{-1}$ is denoted resolvent matrix, see Eq.(23.3). Then, $[\mathbf{A} - e\mathbf{E} \quad \mathbf{A} + e\mathbf{E}]$ is a stable interval matrix.

On the other hand, if an interval matrix $\mathbf{A}_I = [\mathbf{L} \ \ \mathbf{H}]$ is given, define the average matrix (average whiting matrix) $\bar{\mathbf{A}} = (\mathbf{L}+\mathbf{H})/2$ and the positive maximal bias or deviation matrix $\mathbf{M} = \mathbf{H} - \bar{\mathbf{A}} \geq_e \mathbf{0}$. The bias matrix is defined by $\mathbf{A}_b \triangleq \mathbf{A} - \bar{\mathbf{A}}$. Then,

$$\Delta\mathbf{A} = e\mathbf{E} = \mathbf{H} - \bar{\mathbf{A}} \quad \rightsquigarrow \quad \mathbf{E} = \frac{\mathbf{M}}{e} = \frac{\mathbf{H} - \bar{\mathbf{A}}}{e} = \frac{0.5\,(\mathbf{H} - \mathbf{L})}{e} \tag{12.17}$$

and from Eq.(12.16) (*Qiu, L., and Davison, E.J., 1986; Juang, Y.T., et al. 1987; Petkovski, D.B., 1988*)

$$1 < \{\sup_{\omega \geq 0} \pi[\ |(j\omega\mathbf{I}_n - \bar{\mathbf{A}})^{-1}|(\mathbf{H} - \bar{\mathbf{A}})]\}^{-1} . \tag{12.18}$$

Example 1:

$$\mathbf{A} = \begin{pmatrix} 0 & 1 \\ -3 & -2 \end{pmatrix} \qquad \mathbf{E} = \mathbf{I} \tag{12.19}$$

From Fig. 12.1 $e < (1.25)^{-1} = 0.8$ is a sufficient solution. In this simple example the exact limit e_E can be calculated from $\det[j\omega - (\mathbf{A} + e_E\mathbf{E})] = 0 \quad \rightsquigarrow \quad e_E = 1$. □

Example 2: Similarly, for $n = 3$, $\mathbf{A}$ in companion form, Eq.(7.37), $p_1 = 6$, $p_2 = 11$, $p_3 = 6$ tolerates a maximum reduction of p_2 by $\Delta p_2 = -10$. □

12.4 Unidirectional Perturbation. Nonsingularity Approach

The considerations to follow (*Fu, M., and Barmish, B.R., 1988; Tesi, A., and Vicino, A., 1990*) are applicable to the matrix and the polynomial case. The matrix case is presented, only.

12.4.1 Transformation of Stability to Nonsingularity

A linear mapping $\mathcal{T} : \mathcal{R}^{n \times n} \to \mathcal{R}^{m \times m}$ is defined such that (i) the matrix $\mathbf{T}_A$ generated from $\mathbf{A}$ by the transformation $\mathcal{T}$ has at least one real eigenvalue and (ii) the spectral abscissae $\alpha_s[\mathbf{A}]$ and $\alpha_s[\mathbf{T}_A]$ have the same sign or both of them are zero. This transformation $\mathcal{T}$ is said to transform the stability problem into a nonsingularity problem. Usually $m > n$.

Now, consider a perturbed matrix $\mathbf{A}_p \in \mathcal{R}^{n \times n}$

$$\mathbf{A}_p = \mathbf{A} + e\mathbf{E} \qquad e \in (e_1,\ e_2) \qquad e_1 < 0,\ e_2 > 0 . \tag{12.20}$$

Since $\mathbf{E}$ is a specified matrix, the perturbation is called unidirectional. The set of all matrices $\mathbf{A}_p$ is stable if and only if for any mapping $\mathcal{T}$ defined above the matrices $\mathbf{T}_{Ap}$ are nonsingular. The set of matrices $\mathbf{T}_{Ap}$ results from applying the transformation $\mathcal{T}$ to the set $\mathbf{A}_p$. Hence,

$$\mathbf{A}_p \ \text{stable} \ \forall\, e \in (e_1, e_2) \quad \Longleftrightarrow \quad \mathbf{T}_{Ap} \ \text{nonsingular} . \tag{12.21}$$

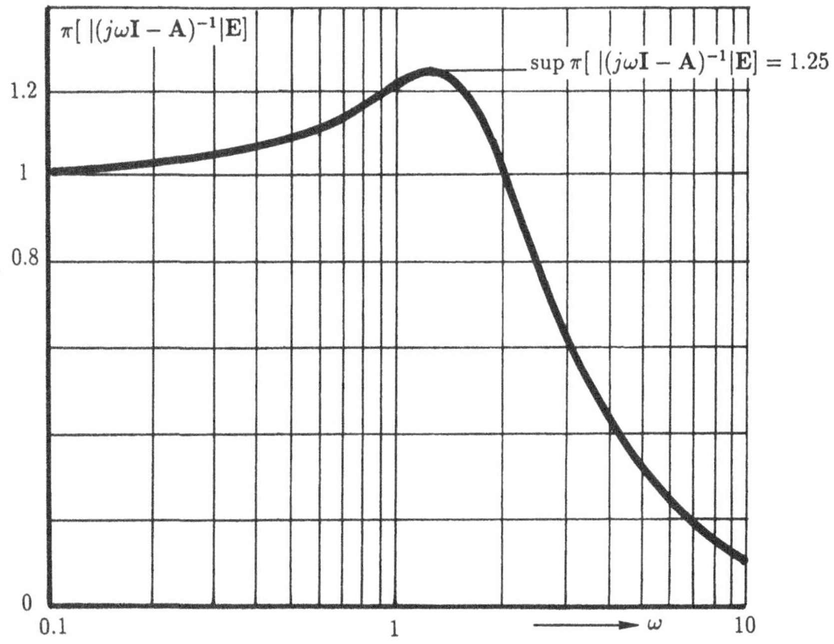

Figure 12.1: Perron eigenvalue versus ω

Proof: Since $\mathbf{A}_p$ stable implies $\Re e\ \lambda_i[\mathbf{T}_{Ap}] < 0$, stability of $\mathbf{T}_{Ap}$ is necessary for stability of $\mathbf{A}_p$. Sufficiency is proved by the properties to follow. Nonsingularity of every $\mathbf{T}_{Ap}$ implies that every $\alpha_s[\mathbf{A}_p]$ must be nonzero. Since $\Re e\ \lambda_i[\mathbf{A}] < 0$ and the eigenvalues depend continuously on e, the set of perturbed matrices $\mathbf{A}_p$ is stable.

Example: Let the transformation $\mathcal{T}$ be given by the Kronecker sum $\mathbf{A}_p \oplus \mathbf{A}_p$

$$\mathbf{T}_{Ap} = \mathbf{A}_p \oplus \mathbf{A}_p = \mathbf{A}_p \otimes \mathbf{I}_n + \mathbf{I}_n \otimes \mathbf{A}_p = \text{matrix}[a_{ij}\ \mathbf{I}_n] + \text{block diag}_{n^2}(\mathbf{A}, \mathbf{A}, ...\mathbf{A})\ . \qquad (12.22)$$

With regard to $\lambda_k[\mathbf{A}_p \oplus \mathbf{A}_p] = \lambda_i + \lambda_j \quad \forall\ i,j = 1,2...n,\ k = 1,2...n^2$ one has: If the matrix $\mathbf{A}_p$ has eigenvalues on the imaginary axis the corresponding eigenvalues $\lambda_k[\mathbf{A}_p \oplus \mathbf{A}_p]$ are zero and $\mathbf{A}_p \oplus \mathbf{A}_p$ is singular. The eigenvalues $\lambda_i[\mathbf{A}_p]$ need not be distinct. □

In a similar way, Schäflian forms, the bialternate product, guardian maps and semi-guardian maps are applied to check generalized stability (*Saydy, L., et al. 1990*).

12.4.2 Admissible Range of Perturbation Parameter e

Evaluation of the perturbed matrix

$$\mathbf{A}_p = \mathbf{A} + e\mathbf{E} = e\mathbf{A}[\frac{1}{e}\mathbf{I} - (-\mathbf{A}^{-1}\mathbf{E})] \qquad (12.23)$$

reveals that $\mathbf{A}_p$ is nonsingular, i.e., $\det \mathbf{A}_p \neq 0$, if and only if $\frac{1}{e}$ is not an eigenvalue of $-\mathbf{A}^{-1}\mathbf{E}$ $\forall\ 1/e_{\max} < 1/e < 1/e_{\min}$. Hence,

$$\frac{1}{e_{\min}} = \lambda^{-}_{\min}[-\mathbf{A}^{-1}\mathbf{E}] \quad \text{and} \quad \frac{1}{e_{\max}} = \lambda^{+}_{\max}[-\mathbf{A}^{-1}\mathbf{E}] \tag{12.24}$$

where $\lambda^{+}_{\max}$ and $\lambda^{-}_{\min}$ denote the maximum positive real eigenvalue and minimum negative real eigenvalue, respectively, preassuming that it exists. The number $\lambda^{+}_{\max}$ yields the smallest inverse among all eigenvalues. This number is required to guarantee the largest parameter $e_{\max}$ without causing an unstable matrix $\mathbf{A}_p$. If these real eigenvalues do not exist they are respectively replaced by 0^+ and 0^- in Eq.(12.24).

12.4.3 Admissible Range Using Nonsingularity Representation

Applying the transformation $\mathcal{T}$ to $\mathbf{A}_p = \mathbf{A} + e\mathbf{E}$ and denoting the transformed matrices $\mathbf{A}$ and $\mathbf{E}$ by $\mathbf{T}_A$ and $\mathbf{T}_E$, respectively,

$$\mathbf{T}_{Ap} = \mathbf{T}_A + e\mathbf{T}_E \qquad e \in (e_1, e_2)\ . \tag{12.25}$$

Since $\mathbf{A}_p$ is stable if and only if $\mathbf{T}_{Ap}$ is nonsingular, the nonsingularity property of $\mathbf{T}_{Ap}$ is checked. This is done by the same considerations which led to Eq.(12.24). Hence,

$$\frac{1}{e_{\min}} = \lambda^{-}_{\min}[-\mathbf{T}_A^{-1}\mathbf{T}_E] \quad \text{and} \quad \frac{1}{e_{\max}} = \lambda^{+}_{\max}[-\mathbf{T}_A^{-1}\mathbf{T}_E]\ . \tag{12.26}$$

12.5 Unidirectional Perturbation. Lyapunov Approach

For $\mathbf{E} \geq_e \mathbf{0}$ and stable $\mathbf{A}$ the perturbed system $\dot{\mathbf{x}}(t) = (\mathbf{A} + e\mathbf{E})\mathbf{x}(t)$ remains stable if

$$\frac{1}{\lambda_{\min}[\mathbf{Q}^{-1}(\mathbf{E}^T\mathbf{P} + \mathbf{P}\mathbf{E})]} = e_{\min} \leq e \leq e_{\max} = \frac{1}{\lambda_{\max}[\mathbf{Q}^{-1}(\mathbf{E}^T\mathbf{P} + \mathbf{P}\mathbf{E})]} \tag{12.27}$$

$$\text{where} \quad \mathbf{A}^T\mathbf{P} + \mathbf{P}\mathbf{A} + \mathbf{Q} = \mathbf{0} \quad \text{and} \quad \mathbf{Q} = \mathbf{Q}^T > 0 \ \text{arbitrary}\ . \tag{12.28}$$

Proof see Eq.(12.48) and *Petkovski, D.B., 1985*.

If the minimum eigenvalue mentioned above is not negative then $e_{\min}$ has to be replaced by $-\infty$. If the maximum eigenvalue is not positive then $e_{\max}$ has to be replaced by $+\infty$. Then, every interval matrix $[\mathbf{A} + e_{\min}\mathbf{E} \quad \mathbf{A} + e_{\max}\mathbf{E}]$ is stable.

An improved result is obtained by the following algorithm. Starting with an initial $(e_{\min})_o = e_{\min}$ and $j = 1$ consider the perturbed system as an unperturbed one, i.e.,

$$\mathbf{A} := \mathbf{A} + (e_{\min})_{j-1}\mathbf{E} \qquad j = 1, 2, 3\ldots\ , \tag{12.29}$$

and determine $(e_{\min})_j$ using Eq.(12.27). If the system at stage j is stable increase j by one and return to Eq.(12.29). If the system at stage j is not stable the improved minimum factor is given by

$$e'_{\min} = \sum_{p=0}^{j-1} (e_{\min})_p\ . \tag{12.30}$$

In the same manner proceed for $e'_{\max}$ (*Petkovski, D.B., 1988*).

12.6 Using Perron-Frobenius Eigenvalue and Scaled Perturbation Matrix

Consider $\dot{\mathbf{x}} = (\mathbf{A} + \Delta\mathbf{A})\mathbf{x}$ and assume $\Delta\mathbf{A}$ is structured using known scaling matrices $\mathbf{L}$ and $\mathbf{R}$

$$\Delta\mathbf{A} = \mathbf{L}\Delta\bar{\mathbf{A}}\,\mathbf{R} \tag{12.31}$$

$$\mathbf{A},\ \Delta\mathbf{A} \in \mathcal{R}^{n\times n},\quad \mathbf{L} \in \mathcal{R}^{n\times p},\quad \Delta\bar{\mathbf{A}} \in \mathcal{R}^{p\times q},\quad \mathbf{R} \in \mathcal{R}^{q\times n},\quad p \leq n,\quad q \leq n\,. \tag{12.32}$$

Now, let $|\Delta\bar{\mathbf{A}}| \leq_e e\mathbf{E}$. The common positive parameter e is unknown. Then, the perturbed system $\mathbf{A}_p = \mathbf{A} + \Delta\mathbf{A}$ is stable if

$$e < \{\sup_{\omega\geq 0} \pi[\ |\mathbf{R}(j\omega\mathbf{I}_n - \mathbf{A})^{-1}\mathbf{L}|\mathbf{E}]\}^{-1}\,. \tag{12.33}$$

See *Qiu, L., and Davison, E.J., 1986* for the proof. The results obtained by this generalization are less conservative than those given by Eq.(13.87).

12.7 Using Stability Radius

The perturbed system $\dot{\mathbf{x}}(t) = (\mathbf{A} + \Delta\mathbf{A})\mathbf{x}(t)$ where $\Delta\mathbf{A} = \sum_{i=1}^{r} \beta_i\mathbf{E}_i$ is Hurwitz stable if

$$|\beta_i| < \{\sup_{\omega\geq 0} \rho_s[\sum_{i=1}^{r} |(j\omega\mathbf{I} - \mathbf{A})^{-1}\mathbf{E}_i|\ ]\ \}^{-1} \tag{12.34}$$

(*Yedavalli, R.K., 1989*) since stability results for $\sup_{\omega\geq 0} \rho_s[(j\omega\mathbf{I} - \mathbf{A})^{-1}\Delta\mathbf{A}] < 1$ and the property is true that

$$\sup_{\omega\geq 0} \rho_s[(j\omega\mathbf{I} - \mathbf{A})^{-1}\sum_{i=1}^{r}\beta_i\mathbf{E}_i] \leq \max_j |\beta_j| \sup_{\omega\geq 0} \rho_s[\sum_{i=1}^{r} |(j\omega\mathbf{I} - \mathbf{A})^{-1}\mathbf{E}_i|\ ]\,. \tag{12.35}$$

12.8 Stability Robustness of Decentralized Control with Sensor Perturbations

12.8.1 Unperturbed System

Consider the system with N local subsystems

$$\dot{\mathbf{x}}(t) = \mathbf{A}\mathbf{x}(t) + \sum_{i=1}^{N}\mathbf{B}_i\mathbf{u}_i(t) \qquad \mathbf{x}(0) = \mathbf{x}_o \tag{12.36}$$

$$\mathbf{u}_i(t) = \mathbf{K}_i\mathbf{y}_i(t) \qquad \mathbf{y}_i(t) = \mathbf{C}_i\mathbf{x}(t) \tag{12.37}$$

where the local input vector $\mathbf{u}_i(t)$ is derived from the local output signal $\mathbf{y}_i(t)$. Minimizing

$$I = \int_0^{\infty} e^{2\alpha t}(\mathbf{x}^T\mathbf{Q}\mathbf{x} + \sum_{i=1}^{N}\mathbf{u}_i^T\mathbf{R}_i\mathbf{u}_i)dt \qquad \mathbf{Q} = \mathbf{Q}^T \geq 0,\quad \mathbf{R}_i = \mathbf{R}_i^T > 0,\quad \alpha > 0 \tag{12.38}$$

yields three matrix equations with unknown $\mathbf{E}_i$, $\mathbf{V}$, $\mathbf{\Lambda}$, see Eqs. (16.75), (16.50) and (16.49), respectively,

$$\mathbf{K}_i = -\mathbf{R}_i^{-1}\mathbf{B}_i^T\mathbf{\Lambda}\mathbf{V}\mathbf{C}_i^T(\mathbf{C}_i\mathbf{V}\mathbf{C}_i^T)^{-1} \qquad i = 1,2...N \tag{12.39}$$

$$(\mathbf{A} + \alpha\mathbf{I} + \sum_{i=1}^{N}\mathbf{B}_i\mathbf{K}_i\mathbf{C}_i)^T\Lambda + \Lambda(\mathbf{A} + \alpha\mathbf{I} + \sum_{i=1}^{N}\mathbf{B}_i\mathbf{K}_i\mathbf{C}_i) + \mathbf{Q}_1 = \mathbf{0} \tag{12.40}$$

$$\text{where} \quad \mathbf{Q}_1 \triangleq \sum_{i=1}^{N}\mathbf{C}_i^T\mathbf{K}_i^T\mathbf{R}_i\mathbf{K}_i\mathbf{C}_i + \mathbf{Q} \quad \text{and} \tag{12.41}$$

$$(\mathbf{A} + \alpha\mathbf{I} + \sum_{i=1}^{N}\mathbf{B}_i\mathbf{K}_i\mathbf{C}_i)\mathbf{V} + \mathbf{V}(\mathbf{A} + \alpha\mathbf{I} + \sum_{i=1}^{N}\mathbf{B}_i\mathbf{K}_i\mathbf{C}_i)^T + \mathbf{I} = \mathbf{0}\ . \tag{12.42}$$

The solution $\mathbf{x}(t)$ is obtained from

$$\dot{\mathbf{x}}(t) = (\mathbf{A} + \sum_{i=1}^{N}\mathbf{B}_i\mathbf{K}_i\mathbf{C}_i)\mathbf{x}(t) \quad \text{where} \quad \Re e\ \lambda_i[\mathbf{A} + \sum_{i=1}^{N}\mathbf{B}_i\mathbf{K}_i\mathbf{C}_i] < -\alpha\ . \tag{12.43}$$

A quantitative relationship between robustness of stability and the weighting matrices of the quadratic index of performance is presented by *Wong, P.K., et al. 1978*.

12.8.2 Sensor Perturbations

The perturbations $\mathbf{M}_i(t)$ associated with the sensor are considered time-varying

$$\mathbf{y}_i(t) = \mathbf{M}_i(t)\mathbf{C}_i\mathbf{x}(t)\ . \tag{12.44}$$

The Lyapunov function $V(\mathbf{x}) = \mathbf{x}^T(t)\mathbf{P}\mathbf{x}(t)$, $\mathbf{P} > 0$ is used. In order to formulate the stability robustness, the negative definiteness of the time derivative $\dot{V}(\mathbf{x})$ is checked

$$\dot{V}(\mathbf{x}) = \mathbf{x}^T(t)[\mathbf{P}(\mathbf{A} + \sum_{i=1}^{N}\mathbf{B}_i\mathbf{K}_i\mathbf{M}_i\mathbf{C}_i) + (\mathbf{A} + \sum_{i=1}^{N}\mathbf{B}_i\mathbf{K}_i\mathbf{M}_i\mathbf{C}_i)^T\mathbf{P}]\mathbf{x}(t) \tag{12.45}$$

where $\mathbf{P}$ satisfies

$$(\mathbf{A} + \alpha\mathbf{I} + \sum_{i=1}^{N}\mathbf{B}_i\mathbf{K}_i\mathbf{C}_i)^T\mathbf{P} + \mathbf{P}(\mathbf{A} + \alpha\mathbf{I} + \sum_{i=1}^{N}\mathbf{B}_i\mathbf{K}_i\mathbf{C}_i) + \mathbf{Q}_1 = \mathbf{0}\ . \tag{12.46}$$

Combining Eq.(12.45) and Eq.(12.46), the condition for negative definiteness is

$$\mathbf{Q}_1 + 2\alpha\mathbf{P} - \mathbf{P}\sum_{i=1}^{N}\mathbf{B}_i\mathbf{K}_i(\mathbf{M}_i - \mathbf{I})\mathbf{C}_i - \sum_{i=1}^{N}\mathbf{C}_i^T(\mathbf{M}_i - \mathbf{I})^T\mathbf{K}_i^T\mathbf{B}_i^T\mathbf{P} > \mathbf{0}\ . \tag{12.47}$$

Defining $\mathbf{Q}_2 \triangleq \mathbf{Q}_1 + 2\alpha\mathbf{P}$ and premultiplying Eq.(12.47) by $\mathbf{Q}_2^{-1}$ (*Petkovski, D.B., 1985*),

$$\mathbf{I} - \mathbf{Q}_2^{-1}\{\mathbf{P}\sum_{i=1}^{N}\mathbf{B}_i\mathbf{K}_i(\mathbf{M}_i - \mathbf{I})\mathbf{C}_i + \sum_{i=1}^{N}\mathbf{C}_i^T(\mathbf{M}_i - \mathbf{I})^T\mathbf{K}_i^T\mathbf{B}_i^T\mathbf{P}\} > \mathbf{0} \tag{12.48}$$

$$\lambda_{\max}[\mathbf{Q}_2^{-1}\{\mathbf{P}\sum_{i=1}^{N}\mathbf{B}_i\mathbf{K}_i(\mathbf{M}_i - \mathbf{I})\mathbf{C}_i + \sum_{i=1}^{N}\mathbf{C}_i^T(\mathbf{M}_i - \mathbf{I})^T\mathbf{K}_i^T\mathbf{B}_i^T\mathbf{P}\}] < 1\ . \tag{12.49}$$

Alternatively, using $\mathbf{Q}_2 \triangleq \mathbf{Q}_1 + 2\alpha\mathbf{P}$, Eq.(12.47) is implied by

$$\lambda_{\min}[\mathbf{Q}_2] > 2\lambda_{\max}[\mathbf{P}]\sum_{i=1}^{N}\|\mathbf{B}_i\|_s\|\mathbf{K}_i\|_s\|\mathbf{M}_i - \mathbf{I}\|_s\|\mathbf{C}_i\|_s\ . \tag{12.50}$$

If the perturbation matrix may be replaced by a scalar perturbation function $m(t)$, the bounds $m_{\min} < m(t) < m_{\max}$ are required. From Expr.(12.48)

$$1 - [m(t) - 1]\ \lambda_{\max}[\mathbf{Q}_2^{-1}\{\sum_{i=1}^{N} \mathbf{P}\mathbf{B}_i\mathbf{K}_i\mathbf{C}_i + \mathbf{C}_i^T\mathbf{K}_i^T\mathbf{B}_i^T\mathbf{P}\}] > 0 \tag{12.51}$$

$$m_{\max} = 1 + \lambda_{\max}^{-1}[\mathbf{Q}_2^{-1}\{\sum_{i=1}^{N} \mathbf{P}\mathbf{B}_i\mathbf{K}_i\mathbf{C}_i + \mathbf{C}_i^T\mathbf{K}_i^T\mathbf{B}_i^T\mathbf{P}\}] \quad \text{if} \quad \lambda_{\max}[\cdot] > 0\ . \tag{12.52}$$

Similarly, if $\lambda_{\min}[\cdot] < 0$

$$m_{\min} = 1 + \lambda_{\min}^{-1}[\mathbf{Q}_2^{-1}\{\sum_{i=1}^{N} \mathbf{P}\mathbf{B}_i\mathbf{K}_i\mathbf{C}_i + \mathbf{C}_i^T\mathbf{K}_i^T\mathbf{B}_i^T\mathbf{P}\}]\ . \tag{12.53}$$

If $\lambda_{\max}[\cdot] \not> 0$ and $\lambda_{\min}[\cdot] \not< 0$ then $m(t) \in [m_{\min}, \infty)$ and $m(t) \in (-\infty, m_{\max}]$, respectively.
From Expr.(12.48) another condition can be derived

$$1 - \lambda_{\max}[\mathbf{Q}_2^{-1}]\ \lambda_{\max}[\mathbf{P}]\ |1 - m(t)| \times 2 \times \|\sum_{i=1}^{N} \mathbf{B}_i\mathbf{K}_i\mathbf{C}_i\|_s > 0 \tag{12.54}$$

$$|1 - m(t)|\ \|\sum_{i=1}^{N} \mathbf{B}_i\mathbf{K}_i\mathbf{C}_i\|_s < \frac{\lambda_{\min}[\mathbf{Q}_2]}{2\lambda_{\max}[\mathbf{P}]}\ . \tag{12.55}$$

12.9 Eigenvalue Bounds of an Interval Matrix

For any (n, n)-matrix the spectral radius is smaller than any norm of the matrix, see Eq.(2.53). If $\mathbf{A}_I$ is an interval matrix, this property can be used to bound the location of the eigenvalues $\lambda_i[\mathbf{A}_I]$. Considering a real matrix $\mathbf{A}$ within the set $\mathbf{A}_I$ and substituting $\mathbf{A} - \alpha\mathbf{I}_n$ into Eq.(2.53),

$$|\ \lambda_i[\mathbf{A} - \alpha\mathbf{I}_n]\ | \le \|\mathbf{A} - \alpha\mathbf{I}_n\| \quad \forall\, i = 1 \dots n\ . \tag{12.56}$$

This property must also be satisfied by the maximum value of both sides even in the presence of $\mathbf{A}$ varying within $\mathbf{A}_I (\mathbf{A} \in \mathbf{A}_I)$

$$\max_{\mathbf{A}\in\mathbf{A}_I} |\ \lambda_i[\mathbf{A} - \alpha\mathbf{I}_n]\ | = \max_{\mathbf{A}\in\mathbf{A}_I} |\lambda_i[\mathbf{A}] - \alpha| < \max_{\mathbf{A}\in\mathbf{A}_I} \|\mathbf{A} - \alpha\mathbf{I}_n\| \triangleq a_r\ . \tag{12.57}$$

The maximum norm of $\|\mathbf{A} - \alpha\mathbf{I}_n\|$ is denoted a_r. As depicted in Fig. 12.2, the eigenvalues $\lambda_i[\mathbf{A}_I]$ are located within a disc centered at a real α with a radius a_r. This property holds for any matrix norm (*Lin, S.H., et al. 1988*). Depending on the chosen norm, the eigenvalues are located within the circular region if one of the sufficient conditions in the following subsections is satisfied. Considering realizable systems, only, the elements A_{1ii}, A_{1ij}, A_{2ii} etc. are real, positive or negative where $\mathbf{A}_I = [\mathbf{L}\ \ \mathbf{H}]$.

- Norm $\|\cdot\|_\infty$

$$\max_{1\le i\le n} [\max(|L_{ii} - \alpha|, |H_{ii} - \alpha|) + \sum_{j=1, j\ne i}^{n} \max(|L_{ij}|, |H_{ij}|)] < a_r\ . \tag{12.58}$$

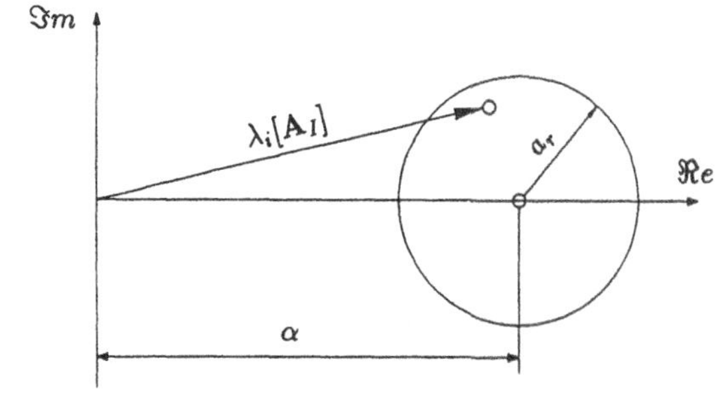

Figure 12.2: Disc bounding the eigenvalues

- Norm $\|\cdot\|_1$

$$\max_{1\leq j\leq n}\ [\max(|L_{jj}-\alpha|,|H_{jj}-\alpha|)+\sum_{i=1,i\neq j}^{n}\ \max\ (|L_{ij}|,|H_{ij}|)] < a_r\ . \tag{12.59}$$

- Frobenius Norm $\|\cdot\|_F$

$$\sqrt{\sum_{i=1}^{n}\ \max\ [(L_{ii}-\alpha)^2,\ (H_{ii}-\alpha)^2]+\sum_{j=1,j\neq i}^{n}\ \max\ (L_{ij}^2,H_{ij}^2)} < a_r\ . \tag{12.60}$$

12.10 Degree of Robust Stability

If all the eigenvalues of a stable system $\mathbf{A}_I$ are located in the left of $-h$ where $h \geq 0$, i.e.,

$$\max_{\mathbf{A}\in\mathbf{A}_I} \Re e\ \lambda_i[\mathbf{A}] < -h\ , \tag{12.61}$$

the system is said to have a degree of stability h. Using Eq.(12.56), the unknown eigenvalues can be replaced by the norm

$$|\Re e\ \lambda_i[\mathbf{A}]-\Re e\ \alpha| \leq |\lambda_i[\mathbf{A}-\alpha\mathbf{I}_n]\ | \leq \|\mathbf{A}-\alpha\mathbf{I}_n\| \qquad \alpha\in\mathcal{R} \tag{12.62}$$

$$\max_{\mathbf{A}\in\mathbf{A}_I}\ \Re e\ \lambda_i[\mathbf{A}] \leq \max_{\mathbf{A}\in\mathbf{A}_I}\ (\alpha+\|\mathbf{A}-\alpha\mathbf{I}_n\|) \leq \max_{\mathbf{A}\in\mathbf{A}_I}\ \min_{\alpha}\ (\alpha+\|\mathbf{A}-\alpha\mathbf{I}_n\|)\ . \tag{12.63}$$

Hence, the sufficient interval system stability condition (with degree of stability h) is

$$\max_{\mathbf{A}\in\mathbf{A}_I}\ \min_{\alpha}\ (\alpha+\|\mathbf{A}-\alpha\mathbf{I}_n\|) < -h\ . \tag{12.64}$$

Consider the interval matrix $\mathbf{A}_I$ given by its bounds $\mathbf{L}$ and $\mathbf{H}$, $\mathbf{A}_I = [\mathbf{L}\ \ \mathbf{H}]$. Replacing the norm $\|\cdot\|$, e.g., by $\|\cdot\|_\infty$ and choosing $\alpha = \{A_{kk}\}$ (where A_{kk} satisfies $L_{kk} \leq A_{kk} \leq H_{kk}\ \ \forall k = 1\ldots n$), the sufficient stability condition for the degree of stability h can be rewritten to

$$\max_{\mathbf{A}\in\mathbf{A}_I}\ \min_{\alpha}\ (\alpha+\|\mathbf{A}-\alpha\mathbf{I}_n\|_\infty) \leq \max_{1\leq i\leq n}\ [H_{ii}+\sum_{j=1,j\neq i}^{n}\ \max\ (|L_{ij}|,\ |H_{ij}|)] < -h\ . \tag{12.65}$$

Proof see *Lin, S.H., et al. 1988* or *Heinen, J.A., 1984*.

12.11 Hurwitz Stability Using Row or Column Sum of $\mathbf{A}^\flat$

From Gershgorin's theorem the stability of continuous-time interval systems can be derived. The system is characterized by the interval matrix $\mathbf{A}_I = [\mathbf{L} \;\; \mathbf{H}]$. Assume that the elements h_{jj} in the main diagonal of $\mathbf{H}$ are negative. Define (*Heinen, J.A., 1984*)

$$\mathbf{A}^\flat = \text{matrix}[a^\flat_{ij}] \quad \text{where} \quad a^\flat_{ij} \triangleq \begin{cases} \max\{|l_{ij}|,\ |h_{ij}|\} & i \neq j \\ h_{jj}\ , \end{cases} \tag{12.66}$$

then, Hurwitz stability is guaranteed if all the row or column sums of $\mathbf{A}^\flat$ are negative

$$\sum_{i=1}^{n} a^\flat_{ij} < 0 \;\; \forall j \qquad \text{or} \qquad \sum_{j=1}^{n} a^\flat_{ij} < 0 \;\; \forall i\ . \tag{12.67}$$

Note that the stability condition above can be modified using the diagonal scaling matrix $\mathbf{R} = \text{diag}\ \{r_i\},\ \ r_i > 0$, and checking the row and column sum of the scaled matrix $\mathbf{R}^{-1}\mathbf{A}^\flat\mathbf{R}$. The conditions Eq.(12.67) require $h_{jj} < 0$ otherwise they cannot be satisfied.

Proof: Gershgorin's theorem states that every eigenvalue $\lambda_k[\mathbf{A}]$ of $\mathbf{A} \in \mathcal{C}^{n\times n}$ is located in a closed disc with radius $\sum_{i=1,\ i\neq j}^{n} |a_{ij}|$ and with center at a_{ii} . The radius is given by the moduli of the off-diagonal elements of the jth column, the center is given by the diagonal element. Applying Gershgorin's theorem to interval matrices, a worst case matrix $\mathbf{A}^\flat$ must be found, i.e., a matrix the elements of which yield worst stability conditions with respect to the sufficient conditions given by Gershgorin's theorem. The worst case is given if the bounds on the circular regions are as far to the right in the complex s-plane as possible. This case is given if the center a_{jj} equals its upper bound h_{jj} and if the radius is maximum. That interval bound, low or high, must be selected the modulus of which is maximum. □

If the worst case is Hurwitz stable then every $\mathbf{A} \in \mathbf{A}_I$ is Hurwitz stable. The worst case is Hurwitz stable if the Gershgorin discs of $\mathbf{A}^\flat$ are located completely in the left-half s-plane. Hence,

$$h_{jj} + \sum_{i=1,\ i\neq j}^{n} \max\{|l_{ij}|,\ |h_{ij}|\} < 0\ . \tag{12.68}$$

This condition is equal to the condition $\sum_{i=1}^{n} a^\flat_{ij} < 0$ in Eq.(12.67) with the definition Eq.(12.66).

The conditions can be reformulated using the row sum rather than the column sum.

12.12 Hurwitz Stability of Two-Matrix Combination

12.12.1 Combination $\mathbf{M}_r$

For $\mathbf{M} \in \mathcal{R}^{n\times n}$

$$\mathbf{M} \oplus \mathbf{M} = \text{matrix}[m_{ij}\mathbf{I}] + \text{diag}\{\mathbf{M}, \mathbf{M}, ...\mathbf{M}\} \in \mathcal{R}^{n^2\times n^2}\ . \tag{12.69}$$

Suppose that $\mathbf{M}$ possesses only real eigenvalues. Then,

$$\lambda_{\max+}[\mathbf{M}] \triangleq \begin{cases} \lambda_{\max} & \text{if} \;\; \lambda_{\max}[\mathbf{M}] > 0 \\ 0^+ & \text{if} \;\; \lambda_{\max}[\mathbf{M}] < 0\ . \end{cases} \tag{12.70}$$

Similarly, $\lambda_{\max-}[\mathbf{M}]$ is defined. Now, define the family of matrices $\mathbf{M}_r = \mathbf{M}_o + r\mathbf{M}_1$ where $\mathbf{M}_o$ is strictly Hurwitz. Then, $\forall r = [0,1]$ $\mathbf{M}_r$ is strictly Hurwitz if and only if (*Fu, M., and Barmish, B.R., 1987*)

$$\lambda[(\mathbf{M}_o \oplus \mathbf{M}_o)^{-1}(\mathbf{M}_1 \oplus \mathbf{M}_1)] \notin (-\infty, 0] \ . \ \square \tag{12.71}$$

The maximal range $r \in (r_{\min}, r_{\max})$ for $\mathbf{M}_r$ to be strictly Hurwitz is given by

$$r_{\min} = \lambda^{-1}_{\min-}[-(\mathbf{M}_o \oplus \mathbf{M}_o)^{-1}(\mathbf{M}_1 \oplus \mathbf{M}_1)] \tag{12.72}$$

$$r_{\max} = \lambda^{-1}_{\max+}[-(\mathbf{M}_o \oplus \mathbf{M}_o)^{-1}(\mathbf{M}_1 \oplus \mathbf{M}_1)] \ . \ \square \tag{12.73}$$

If $\mathbf{M}_o$ and $\mathbf{M}_1$ are symmetric, then, stability is guaranteed in the range

$$r_{\min} = \lambda^{-1}_{\min-}[-\mathbf{M}_o^{-1}\mathbf{M}_1] \tag{12.74}$$

$$r_{\max} = \lambda^{-1}_{\max+}[-\mathbf{M}_o^{-1}\mathbf{M}_1] \ . \ \square \tag{12.75}$$

12.12.2 Combination $\mathbf{M}_R$

The matrix combination

$$\mathbf{M}_R = (1-r)\mathbf{M}_o + r\mathbf{M}_1 \qquad \mathbf{M}_o, \mathbf{M}_1 \in \mathcal{R}^{n \times n}, \ \mathbf{M}_o \ \ \text{nonsingular} \tag{12.76}$$

is stable $\forall r \in [0,1]$ if

$$\lambda[\mathbf{M}_o^{-1}\mathbf{M}_1] \notin (-\infty, 0] \ . \tag{12.77}$$

The maximal range of r for stability is determined by identical bounds as given by Eqs.(12.74) and (12.75). If $\mathbf{M}_o$ and $\mathbf{M}_1$ are symmetric then stability of $\mathbf{M}_R$ $\forall r = [0,1]$ is given if and only if $\mathbf{M}_o$ and $\mathbf{M}_1$ individually are strictly Hurwitz (*Fu, M., and Barmish, B.R., 1987*).

12.13 Discrete-Time Dynamic Interval System

A discrete-time system

$$\mathbf{x}(k+1) = \mathbf{\Phi}(T)\mathbf{x}(k) \tag{12.78}$$

with the sampling period T is stable if the eigenvalues $\lambda_i[\mathbf{\Phi}(T)]$ are located within the unit circle in the z-plane. If $\mathbf{\Phi}(T)$ is an interval matrix $\mathbf{\Phi}_I(T)$ any $\mathbf{\Phi}(T) \in \mathbf{\Phi}_I(T)$ has to guarantee $|\lambda_i| < 1$. Using Eq.(12.57) with $\alpha = 0$ and $a_r = 1$,

$$\max_{\mathbf{\Phi}(T) \in \mathbf{\Phi}_I(T)} \|\mathbf{\Phi}(T)\| < 1 \ . \tag{12.79}$$

12.13.1 Using Direct Matrix Norms

Let the interval matrix $\mathbf{\Phi}_I = [\mathbf{L} \ \ \mathbf{H}] \in \mathcal{R}^{n \times n}$ of a discrete-time system be given and $\mathbf{\Phi} \in \mathbf{\Phi}_I$. Depending on the fact which element of $\mathbf{\Phi}_I$ by modulus characterizes the boundary of the interval matrix $\mathbf{\Phi}_I$, the matrix $\mathbf{\Phi}^\partial$ is defined

$$\mathbf{\Phi}^\partial \triangleq \text{matrix}[\varphi^\partial_{ij}] : \qquad \varphi^\partial_{ij} \triangleq \begin{cases} l_{ij} & \text{if } |l_{ij}| > |h_{ij}| \\ h_{ij} & \text{if } |l_{ij}| < |h_{ij}| \\ l_{ij} \ \text{or} \ h_{ij} & \text{if } |l_{ij}| = |h_{ij}| \ . \end{cases} \tag{12.80}$$

The superscript ∂ is chosen with regard to the outermost boundary. By definition of the direct matrix norms $\|\cdot\|_1$, $\|\cdot\|_\infty$ and $\|\cdot\|_F$, it results that for given matrices $\mathbf{M}$, $\mathbf{N}$

$$\text{if } |\mathbf{M}| \geq_e |\mathbf{N}| \text{ then } \|\mathbf{M}\|_1 \geq \|\mathbf{N}\|_1, \quad \|\mathbf{M}\|_\infty \geq \|\mathbf{N}\|_\infty, \quad \|\mathbf{M}\|_F \geq \|\mathbf{N}\|_F . \tag{12.81}$$

Hence, for the norms cited above and for $\mathbf{\Phi} \in \mathbf{\Phi}_I = [\mathbf{L} \;\; \mathbf{H}]$ one has:

$$\text{If } \|\mathbf{\Phi}^\partial\|_p < 1 \quad \text{then} \quad \|\mathbf{\Phi}\|_p < 1 \tag{12.82}$$

since $\mathbf{\Phi}^\partial$ yields a higher norm than each $\mathbf{\Phi} \in \mathbf{\Phi}_I$. Regarding $\rho_s[\mathbf{\Phi}] \leq \|\mathbf{\Phi}\|_p \leq 1$ for any $\mathbf{\Phi} \in \mathbf{\Phi}_I$ and $p = 1, \infty, F$ the matrix is Schur stable.

12.13.2 Using Spectral Radius and Spectral Norm

From $|\mathbf{G}| \leq_e |\mathbf{H}| \;\; \leadsto \;\; \rho_s[\mathbf{G}] \leq \rho_s[\,|\mathbf{H}|\,]$, see *Ortega, J.M., 1972*. By substituting $\mathbf{G} := \mathbf{\Phi} \in \mathbf{\Phi}_I$ and $\mathbf{H} := \mathbf{\Phi}^\partial$ and, finally, using Eq.(12.80), it results

$$|\mathbf{\Phi}| \leq_e |\mathbf{\Phi}^\partial| \quad \leadsto \quad \rho_s[\mathbf{\Phi}] \leq \rho_s[\,|\mathbf{\Phi}^\partial|\,] . \tag{12.83}$$

Hence,

$$\text{if } \rho_s[\,|\mathbf{\Phi}^\partial|\,] < 1 \quad \text{or} \quad \|\,|\mathbf{\Phi}^\partial|\,\|_s < 1 \quad \text{then} \quad \rho_s[\mathbf{\Phi}] < 1 \tag{12.84}$$

and the system is Schur stable.

12.14 Absolute Value of the Interval Matrix

Denoting

$$a_l \leq a \leq a_h \quad a \in a_I = [a_l \;\, a_h]; \quad b_l \leq b \leq b_h \quad b \in b_I = [b_l \;\, b_h]; \quad c_l \leq c \leq c_h \quad c \in c_I = [c_l \;\, c_h], \tag{12.85}$$

the product of interval scalars $a_I b_I = c_I$ is given by

$$c_l = \min_{a \in a_I,\; b \in b_I} ab \quad \text{and} \quad c_h = \max_{a \in a_I,\; b \in b_I} ab . \tag{12.86}$$

Using the above definition of interval multiplication, the product of matrices is obtained applying the well-known calculus for point matrices. But note that $(\mathbf{A}_I\mathbf{B}_I)\mathbf{C}_I \neq \mathbf{A}_I(\mathbf{B}_I\mathbf{C}_I)$ and $\mathbf{A}_I^2\mathbf{A}_I \neq \mathbf{A}_I\mathbf{A}_I^2$. □

The matrix $|\mathbf{\Phi}^\partial|$ is also denoted the absolute value of the interval matrix $|\mathbf{\Phi}_I|$. The interval matrix is convergent if $\rho_s[\,|\mathbf{\Phi}^\partial|\,] < 1$ (*Mayer, G., 1984*).

The matrix $\mathbf{\Phi}_I$ is said right convergent if the product of interval matrices in the order

$$\mathbf{A}_I^{k-1}\mathbf{A}_I \triangleq \mathbf{A}_I^k \tag{12.87}$$

tends to $\mathbf{0}$ as k tends to infinity. (For contrast: left convergence is termed ${}^k\mathbf{A}_I$.) Convergence is given if one of the following conditions is satisfied (*Mayer, G., 1984*)

$$\text{using 1-norm} \quad \max_j \sum_{i=1}^{n} |a_{Iij}| < 1 \tag{12.88}$$

$$\text{using } \infty-\text{norm} \quad \max_i \sum_{j=1}^{n} |a_{Iij}| < 1 \tag{12.89}$$

$$\text{using Frobenius norm} \quad \sum_{i=1}^{n}\sum_{j=1}^{n} |a_{Iij}|^2 < 1 \quad \text{where } |a_{Iij}| \triangleq \max\{|l_{ij}|,\; |h_{ij}|\} . \tag{12.90}$$

These conditions are equivalent to the condition given by the theorem of Eq.(12.82) using the outermost boundary matrix $\mathbf{\Phi}^\partial$.

Eq.(12.84) can also be proved using the fact (*Mayer, G., 1984*) $|\mathbf{\Phi}_I^k| \leq |\mathbf{\Phi}_I|^k$.

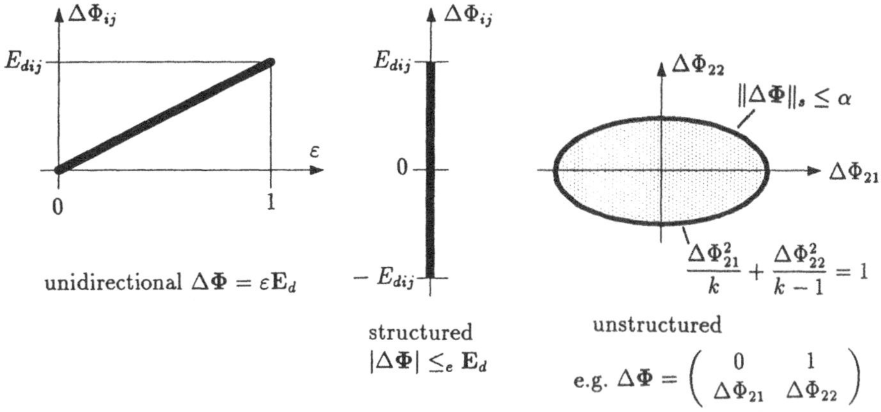

Figure 12.3: Unidirectional, structured and unstructured perturbation

12.15 Unidirectional Perturbation of Discrete-Time Systems

Consider the autonomous discrete-time system

$$\mathbf{x}_{k+1} = \mathbf{x}(k+1) = \mathbf{\Phi}_p \mathbf{x}(k) \quad \text{where} \quad \mathbf{\Phi}_p = \mathbf{\Phi} + \varepsilon \mathbf{E}_d \tag{12.91}$$

is the perturbed fundamental matrix and $\mathbf{E}_d$ denotes the maximum perturbation. The matrix $\mathbf{E}_d$ is fixed. The parameter ε varies in the range $[0\ 1]$, see Fig. 12.3. Assume that the unperturbed system is stable. When the perturbed system is stable all the eigenvalues of $\mathbf{\Phi}_p$ for any ε must lie inside the unit circle. When the perturbed system is unstable there exists a region of numbers ε such that the corresponding eigenvalues $\lambda_i[\mathbf{\Phi}_p]$ are located on or outside the unit circle. Since the eigenvalues of a matrix depend continuously on the elements of the matrix and since $\lambda_i[\mathbf{\Phi}]$ are located inside the unit circle, it must be true in the unstable case that n eigenvalues exist on the unit circle for a certain $\varepsilon \in [0\ 1]$. Replacing the eigenvalues $\lambda_i[\mathbf{\Phi}_p]$ by $|z| = 1$ or $z = e^{j\theta}$,

$$\det\left[e^{j\theta}\mathbf{I} - (\mathbf{\Phi} + \varepsilon \mathbf{E}_d)\right] = 0 \qquad \theta \in [0\ \pi], \quad \varepsilon \in [0\ 1] \tag{12.92}$$

is the condition for an unstable behaviour. Hence, the condition for robust stability is given by Eq.(12.92), the equality replaced by $\neq$ and $\forall \theta \in [0\ \pi], \varepsilon \in [0\ 1]$. This condition is both necessary and sufficient. The range $\theta \in [0\ \pi]$ suffices for the investigation due to the symmetry with respect to the real axis of the z-plane. For comparison, see the relations on continuous-time systems and Eq.(12.4).

12.15.1 Determinant $\det(\mathbf{I} \pm \varepsilon \mathbf{\Omega})$ if $|\ \lambda_i[\mathbf{\Omega}]\ | < 1$

It is well known that for any $\mathbf{\Omega} \in \mathcal{C}^{n \times n}$

$$\det\left[\mathbf{I} + \mathbf{\Omega}\right] = 0 \quad \Longleftrightarrow \quad \lambda[\mathbf{\Omega}] = -1 \tag{12.93}$$

$$\text{or} \quad |\ \lambda_i[\mathbf{\Omega}]\ | < 1 \quad \leadsto \quad \det\left(\mathbf{I} + \mathbf{\Omega}\right) \neq 0\ . \tag{12.94}$$

Considering $\mid \lambda_i[\varepsilon\mathbf{\Omega}] \mid < 1$ implies $\det\,[\mathbf{I} + \varepsilon\mathbf{\Omega}] \neq 0$. If $\mid \lambda_i[\mathbf{\Omega}] \mid < 1$ is stated for all $i = [1, 2 \ldots n]$ and if $\varepsilon \in [0\ 1]$ since $\mid \lambda_i[\varepsilon\mathbf{\Omega}] \mid = \mid \varepsilon\lambda_i[\mathbf{\Omega}] \mid$ the statement $\mid \lambda_i[\varepsilon\mathbf{\Omega}] \mid < 1$ is true, even for $-1 \leq \varepsilon \leq 0$. Hence, (*Juang, Y.T., et al. 1987*)

$$\mid \lambda_i[\mathbf{\Omega}] \mid < 1 \qquad \forall i = 1 \ldots n \quad \rightsquigarrow \quad \det\,[\mathbf{I} \pm \varepsilon\mathbf{\Omega}] \neq 0 \qquad \forall \varepsilon = [0\ 1]\,. \tag{12.95}$$

12.15.2 Asymptotic Stability Using Characteristic Resolvent Matrix

For the perturbed autonomous system $\mathbf{x}(k+1) = \mathbf{\Phi}_p \mathbf{x}(k)$ using the nominal (unperturbed) matrix $\mathbf{\Phi}$, the characteristic resolvent matrix is defined as $(e^{j\theta}\mathbf{I} - \mathbf{\Phi})^{-1}$. Referring to Eq.(12.92), if the nominal system $\mathbf{\Phi}$ is stable then

$$\det(e^{j\theta}\mathbf{I} - \mathbf{\Phi}) \neq 0 \qquad \forall \theta \in [0\ \pi]\,. \tag{12.96}$$

Referring to Eq.(12.92), the perturbed system $\mathbf{\Phi}_p$ is stable if $\det(e^{j\theta}\mathbf{I} - \mathbf{\Phi}_p) \neq 0\ \forall \theta \in [0\ \pi]$. Evaluating yields

$$\det[e^{j\theta}\mathbf{I} - \mathbf{\Phi} - \varepsilon\mathbf{E}_d] = \det(e^{j\theta}\mathbf{I} - \mathbf{\Phi})\,\det[\mathbf{I} - \varepsilon(e^{j\theta}\mathbf{I} - \mathbf{\Phi})^{-1}\mathbf{E}_d] \neq 0\,. \tag{12.97}$$

Invoking Eq.(12.96) and applying Eq.(12.95) to the right-hand determinant above yields that

$$|\lambda_i[(e^{j\theta}\mathbf{I} - \mathbf{\Phi})^{-1}\mathbf{E}_d] \mid < 1 \qquad \forall \theta \in [0\ \pi], \quad \forall i \tag{12.98}$$

$$\text{or} \quad \rho_s[(e^{j\theta}\mathbf{I} - \mathbf{\Phi})^{-1}\mathbf{E}_d] < 1 \qquad \forall \theta \in [0\ \pi] \tag{12.99}$$

are sufficient for asymptotic stability of the perturbed system. If the spectral radius associated with the matrix product of the characteristic resolvent matrix times perturbation matrix is smaller than unity, a sufficient stability condition is given. In the formula of Eq.(12.99) the parameter ε is no more involved.

12.15.3 Replacing the Spectral Radius by Any Norm

Since the spectral radius ρ_s is smaller than any matrix norm $\|\cdot\|$, Eq.(12.99) is implied by $\|(e^{j\theta}\mathbf{I} - \mathbf{\Phi})^{-1}\mathbf{E}_d\| < 1 \quad \forall \theta \in [0\ \pi]$ or by

$$\|\mathbf{R}^{-1}(e^{j\theta}\mathbf{I} - \mathbf{\Phi})^{-1}\mathbf{L}^{-1}\mathbf{L}\mathbf{E}_d\mathbf{R}\| < 1 \quad \forall \theta \in \ [0\ \ \pi] \tag{12.100}$$

$$\Leftarrow \quad \|\mathbf{R}^{-1}(e^{j\theta}\mathbf{I} - \mathbf{\Phi})^{-1}\mathbf{L}^{-1}\|\ \|\mathbf{L}\mathbf{E}_d\mathbf{R}\| < 1 \quad \forall \theta \in \ [0\ \ \pi] \tag{12.101}$$

where $\mathbf{L}$ and $\mathbf{R}$ are arbitrary scaling matrices which should be carefully chosen in order to avoid useless conservativeness, see Eq.(25.77).

12.16 Structured Perturbation of Discrete-Time Systems

Suppose the perturbation is structured such that

$$|\Delta\mathbf{\Phi}| \leq_e \mathbf{E}_d \tag{12.102}$$

where $\mathbf{E}_d$ is a positive matrix. Now, consider the system

$$\mathbf{x}_{k+1} = (\mathbf{\Phi} + \Delta\mathbf{\Phi})\mathbf{x}_k \tag{12.103}$$

with structured perturbation given by Eq.(12.102). Then, the system is asymptotically stable if

$$\rho_s[\ |(e^{j\theta}\mathbf{I} - \mathbf{\Phi})^{-1}|\ \mathbf{E}_d] < 1 \qquad \forall \theta \in [0\ \pi]\,. \tag{12.104}$$

The proof follows immediately from Eqs.(12.8) and (12.95).

12.17 Stability Robustness of Discrete-Time Closed-Loop Systems

Considering the perturbed multivariable feedback system with output feedback,

$$\mathbf{x}_{k+1} = \mathbf{\Phi}_p \mathbf{x}_k + \mathbf{\Psi}_p \mathbf{u}_k \qquad \mathbf{\Psi} \in \mathcal{R}^{n \times m} \tag{12.105}$$

$$\mathbf{y}_k = \mathbf{C}_p \mathbf{x}_k \qquad \mathbf{C}_p \in \mathcal{R}^{r \times n} \tag{12.106}$$

$$\mathbf{u}_k = \mathbf{K} \mathbf{y}_k \qquad \mathbf{K} \in \mathcal{R}^{m \times r} . \tag{12.107}$$

Reducing yields

$$\mathbf{x}_{k+1} = (\mathbf{\Phi}_p + \mathbf{\Psi}_p \mathbf{K} \mathbf{C}_p)\mathbf{x}_k \triangleq (\mathbf{\Phi} + \mathbf{\Psi}\mathbf{K}\mathbf{C} + \Delta\mathbf{\Phi}_{CL})\mathbf{x}_k \tag{12.108}$$

where additive perturbations

$$\mathbf{\Phi}_p = \mathbf{\Phi} + \Delta\mathbf{\Phi} \qquad \mathbf{\Psi}_p = \mathbf{\Psi} + \Delta\mathbf{\Psi} \qquad \mathbf{C}_p = \mathbf{C} + \Delta\mathbf{C} \tag{12.109}$$

$$\Delta\mathbf{\Phi}_{CL} \triangleq \Delta\mathbf{\Phi} - \Delta\mathbf{\Psi}\ \mathbf{K}\mathbf{C} - \mathbf{\Psi}\mathbf{K}\Delta\mathbf{C} - \Delta\mathbf{\Psi}\ \mathbf{K}\Delta\mathbf{C} \tag{12.110}$$

are applied. If the perturbations are bounded by

$$|\Delta\mathbf{\Phi}| \leq_e \mathbf{E}_d \qquad |\Delta\mathbf{\Psi}| \leq_e \mathbf{E}_\Psi \qquad |\Delta\mathbf{C}| \leq_e \mathbf{E}_C , \tag{12.111}$$

then, from Eq.(12.99), the robust stability conditions are given by

$$\rho_s[\ |(e^{j\theta}\mathbf{I} - \mathbf{\Phi} - \mathbf{\Psi}\mathbf{K}\mathbf{C})^{-1}|\ \mathbf{E}_{CL}] < 1 \qquad \theta \in [0\ \pi] \tag{12.112}$$

where

$$\mathbf{E}_{CL} = \mathbf{E}_d + \mathbf{E}_\psi|\mathbf{K}\mathbf{C}| + \big(|\mathbf{\Psi}\mathbf{K}| + \mathbf{E}_\psi|\mathbf{K}|\big)\mathbf{E}_C \tag{12.113}$$

(*Juang, Y.T., et al. 1987*).

12.18 Robust Controller Using Index of Performance

Robustly stabilizing a dynamic interval plant by a state feedback controller **K** is outlined in the following design procedure. Let the dynamic interval plant be given by

$$\dot{\mathbf{x}}(t) = \mathbf{A}_I\mathbf{x}(t) + \mathbf{B}\mathbf{u}(t), \qquad \mathbf{A}_I = [\mathbf{L}\ \ \mathbf{H}], \qquad \mathbf{B} \in \mathcal{R}^{n \times m} . \tag{12.114}$$

If a most likely plant matrix $\mathbf{A}_o$ is known a linear quadratic controller $\mathbf{K}_o$ can be achieved. The actual **K** is obtained by minimizing $\|\mathbf{K} - \mathbf{K}_o\|$ under the condition that the closed loop with state matrix $\mathbf{A}_I + \mathbf{B}\mathbf{K}$ has at least the degree of stability h, i.e.,

$$\min_{\mathbf{K}} \|\mathbf{K} - \mathbf{K}_o\|_F \ \mid\ \max_{\mathbf{F} \in \mathbf{A}_I + \mathbf{B}\mathbf{K}}\ \min_{\alpha}\ (\alpha + \|\mathbf{F} - \alpha\mathbf{I}_n\|) < -h \tag{12.115}$$

or

$$\min_{\mathbf{K}} \|\mathbf{K} - \mathbf{K}_o\|_F \ \mid\ \max_i [F_{2ii} + \sum_{j=1, j\neq i}^{n} \max(|F_{1ij}|,\ |F_{2ij}|)] < -h \tag{12.116}$$

$$\text{where } \mathbf{F}_1 = \mathbf{L} + \mathbf{B}\mathbf{K} = \text{matrix}[F_{1ij}] \text{ and } \mathbf{F}_2 = \mathbf{H} + \mathbf{B}\mathbf{K} = \text{matrix}[F_{2ij}] . \tag{12.117}$$

Solving the problem for unknown $\mathbf{K} = \text{matrix}[K_{ij}]$, one has to minimize a function of the type

$$[(K_{11} - K_{0,11})^2 + (K_{12} - K_{0,12})^2 + \ldots]^{1/2} \tag{12.118}$$

subject to several inequalities. These inequalities contain linear combinations of K_{ij} resulting from interval arithmetic. The minimization procedure can only be carried out by computational search. The results strongly depend on the matrix norm applied to the stability determination.

12.19 Robust Controller Using Pole Assignment

Using a circular pole region as given in Eq.(12.57) and Fig. 12.2 provides another algorithm to design a robust state feedback controller for an interval plant. Replacing $\mathbf{A}$ by the matrix $\mathbf{F} = \mathbf{A} + \mathbf{BK}$ in Eq.(12.57) for a given α and a_r , inequalities are obtained that have to be satisfied by selecting proper K_{ij}. See also Eq.(15.15).

12.20 Stability of Interval Matrices and Symmetric Part of Corner Matrices

12.20.1 Hurwitz Stability

The following lemma concerning the Hurwitz stability of the sum of matrices is used. Consider $\mathbf{A} = \sum_{k=1}^{N} \beta_k \mathbf{A}_k$ where $\beta_k \geq 0$. If the symmetric part $\mathbf{A}_{ks} = 0.5\ (\mathbf{A}_k + \mathbf{A}_k^T)$ is asymptotically stable then $\mathbf{A}_{ks}$ is negative definite, since $\mathbf{A}_{ks}$ is Hermite. Then, the symmetric part $\mathbf{A}_s < 0$ and $\mathbf{A}$ is asymptotically stable, i.e.,

$$\Re e\ \lambda_i[\mathbf{A}] < 0 \quad \text{if} \quad \Re e\ \lambda_i[\mathbf{A}_{ks}] < 0\ .\ \square \tag{12.119}$$

Since $\mathbf{A}_I$ is a closed and bounded convex set the matrix $\mathbf{A}_I$ is the convex hull of its corner matrices and any matrix $\mathbf{A} \in \mathbf{A}_I = [\mathbf{L}\ \ \mathbf{H}] \in \mathcal{R}^{n \times n}$ can be decomposed by a linear combination of corner matrices

$$\mathbf{A} = \sum_{k=1}^{N} \beta_k \mathbf{A}_k^{\square}, \qquad \beta_k \geq 0, \qquad \sum_{k=1}^{N} \beta_k = 1, \qquad N = 2^{(n^2)}\ . \tag{12.120}$$

Referring to Eq.(12.119), a sufficient condition for the Hurwitz stability of $\mathbf{A} \in \mathbf{A}_I$ is that $0.5\ (\mathbf{A}_\nu^{\square T} + \mathbf{A}_\nu^{\square})\ \ \forall \nu = 1, 2 ... N$ is asymptotically stable (*Jiang, C.L., 1987; Mansour, M., 1988*), i.e.,

$$\Re e\ \lambda_i[\mathbf{A}] < 0 \quad \Leftarrow \quad \Re e\ \lambda_i[\mathbf{A}_{\nu s}^{\square}] < 0 \quad \forall i = 1, 2 ... n\ \ \forall \nu = 1, 2 ... N\ . \tag{12.121}$$

12.20.2 Schur Stability

Consider the matrix $\mathbf{A} \in \mathbf{A}_I$ and its corner matrices $\mathbf{A}_\nu^{\square}\ \forall \nu = 1, 2 ... N$. If the matrices $(\mathbf{A}_\nu^{\square T} \mathbf{A}_\nu^{\square} - \mathbf{I})$ are negative definite then $\sigma_{\max}[\mathbf{A}_\nu^{\square}] = \|\mathbf{A}_\nu^{\square}\|_s < 1$. Referring to Eq.(12.136), this is a sufficient condition for Schur stability of $\mathbf{A}$

$$\rho_s[\mathbf{A}] < 1 \quad \text{if} \quad \mathbf{A}_\nu^{\square T} \mathbf{A}_\nu^{\square} - \mathbf{I} < 0 \quad \forall \nu = 1, 2 ... N\ . \tag{12.122}$$

12.20.3 Stability of Interval Matrices and Convex Hull of Corner Matrix Norms

Decomposing $\mathbf{A} \in \mathbf{A}_I$ according with Eq.(12.120), for any p

$$\max_i |\lambda_i[\mathbf{A}]\ | \ = \ \max_i |\lambda_i[\sum_{k=1}^{N} \beta_k \mathbf{A}_k^{\square}]\ | = \rho_s[\sum_{k=1}^{N} \beta_k \mathbf{A}_k^{\square}] \tag{12.123}$$

$$\leq \ \|\sum_{k=1}^{N} \beta_k \mathbf{A}_k^{\square}\|_p \leq \sum_{k=1}^{N} \beta_k \|\mathbf{A}_k^{\square}\|_p < 1 \tag{12.124}$$

where < 1 above results from $\|\mathbf{A}_k^{\square}\|_p < 1 \;\; \forall\; k = 1,2...N$ and $\sum_{k=1}^{N} \beta_k = 1$. Hence, every $\mathbf{A} \in \mathbf{A}_I$ is Schur stable (*Juang, Y.T., et al. 1989*).

The condition that the convex hull

$$\mathrm{conv}(\mathbf{A}_1^{\square}, \mathbf{A}_2^{\square}, ...\mathbf{A}_N^{\square}) = \{\mathbf{A}: \;\; \mathbf{A} = \sum_{k=1}^{N} \beta_k \mathbf{A}_k^{\square}, \;\; \sum_{k=1}^{N} \beta_k = 1, \;\; 0 \leq \beta_k \leq 1 \;\; \forall k = 1,2...N\;\} \tag{12.125}$$

is asymptotically stable does not provide a sufficient condition (*Jiang, C.L., 1988a*). The linear combination of the norms, Eq.(12.124), is correct.

12.21 Reduced Number of Corner Points

Consider an interval matrix $\mathbf{A}_I$ and the matrix $\mathbf{A} = \mathrm{matrix}[a_{ij}] \in \mathbf{A}_I \;\; l_{ij} \leq a_{ij} \leq h_{ij} \;\; \forall i,j = [1,2..n]$ where the entries a_{ij} may vary continuously between low and high bounds l_{ij} and h_{ij}, respectively (*Jiang, C.L., 1987*). Define the corner matrices

$$\mathbf{A}_\nu^{\square} = \mathrm{matrix}[s_{ij}] \qquad s_{ij} \in \{l_{ij}, h_{ij}\} \;\; \forall i,j = [1,2...n]\;, \tag{12.126}$$

i.e., the entries are identical to the lower or higher bounds. There are $2^{(n^2)}$ corner matrices $\mathbf{A}_\nu^{\square}$. Since an (n,n)-matrix contains n^2 entries and each entry possesses two bounds there are $2^{(n^2)}$ corner matrices $\mathbf{A}_\nu^{\square}$ or the subscript ν is $\nu = [1,2...2^{(n^2)}]$. A sufficient condition for asymptotic Hurwitz stability of $\mathbf{A}$ is given if all the symmetric parts $\mathbf{A}_{\nu s}^{\square} = 0.5\;(\mathbf{A}_\nu^{\square} + \mathbf{A}_\nu^{\square T})\;\; \forall \nu$ are asymptotically stable. The number of polynomials whose stability has to be checked equals the number of corner points. Since its number grows exponentially with the order of the system the computational effort is considerably large.

A strongly simplified condition (*Mansour, M., 1989*) is the asymptotic stability condition of the matrices

$$\tilde{\mathbf{A}}_{\mu s}^{\square} = 0.5\;(\tilde{\mathbf{A}}_\mu^{\square} + \tilde{\mathbf{A}}_\mu^{\square T}) \qquad \forall \mu = [1,2...2^{n(n-1)/2}\,] \tag{12.127}$$

where

$$\tilde{\mathbf{A}}_\mu^{\square} = \mathrm{matrix}[\tilde{s}_{ij}], \qquad \left\{\begin{array}{c} \tilde{s}_{ii} = h_{ii} \\ \tilde{s}_{ij} = h_{ij} \\ \tilde{s}_{ji} = h_{ji} \end{array}\right\} \quad \text{or} \quad \left\{\begin{array}{c} \tilde{s}_{ii} = h_{ii} \\ \tilde{s}_{ij} = l_{ij} \\ \tilde{s}_{ji} = l_{ji} \end{array}\right\}. \tag{12.128}$$

Owing to the symmetry of $0.5\;(\tilde{\mathbf{A}}_\mu^{\square} + \tilde{\mathbf{A}}_\mu^{\square T})$, there are only $(n^2 - n)/2 = n(n-1)/2$ remaining positions to be checked in the extreme values. In the diagonal elements only the maximum values are needed. A large number of matrices is *not required*, e.g.,

$$\left\{\begin{array}{c} \tilde{s}_{ii} = h_{ii} \\ \tilde{s}_{ij} = h_{ij} \\ \tilde{s}_{ji} = l_{ji} \end{array}\right\}, \quad \left\{\begin{array}{c} \tilde{s}_{ii} = h_{ii} \\ \tilde{s}_{ij} = l_{ij} \\ \tilde{s}_{ji} = h_{ji} \end{array}\right\}, \quad \left\{\begin{array}{c} \tilde{s}_{ii} = l_{ii} \\ \tilde{s}_{ij} = h_{ij} \\ \tilde{s}_{ji} = l_{ji} \end{array}\right\} \quad \text{etc.}\;. \tag{12.129}$$

The proof whether $\tilde{\mathbf{A}}_\mu^{\square}$ is negative definite requires the Sylvester inequalities for the symmetric parts $0.5\;(\tilde{\mathbf{A}}_\mu^{\square} + \tilde{\mathbf{A}}_\mu^{\square T})$. Evaluating these Sylvester inequalities it turns out that, e.g., the diagonal elements only appear linearly with negative coefficients for positivity and with positive coefficients for negativity. Since the diagonal elements are negative, throughout, the worst case is given only if the diagonal elements are maximum, i.e., negative with smallest absolute value.

Example 1:

$$\mathbf{L} = \text{matrix}[l_{ij}] = \begin{pmatrix} -8 & -1 \\ -2 & -10 \end{pmatrix} \qquad \mathbf{H} = \text{matrix}[h_{ij}] = \begin{pmatrix} -6 & -0.5 \\ -1 & -9 \end{pmatrix} \tag{12.130}$$

$$\tilde{\mathbf{A}}_1^\square = \begin{pmatrix} -6 & -1 \\ -2 & -9 \end{pmatrix} \qquad \tilde{\mathbf{A}}_2^\square = \begin{pmatrix} -6 & -0.5 \\ -1 & -9 \end{pmatrix} . \tag{12.131}$$

Fourteen matrices are not required, e.g.,

$$\begin{pmatrix} -6 & -1 \\ -1 & -9 \end{pmatrix} , \begin{pmatrix} -6 & -0.5 \\ -2 & -9 \end{pmatrix} . \tag{12.132}$$

Then,

$$\tilde{\mathbf{A}}_{1s}^\square = 0.5\,(\tilde{\mathbf{A}}_1^\square + \tilde{\mathbf{A}}_1^{\square T}) = \begin{pmatrix} -6 & -1.5 \\ -1.5 & -9 \end{pmatrix} , \quad \tilde{\mathbf{A}}_{2s}^\square = 0.5\,(\tilde{\mathbf{A}}_2^\square + \tilde{\mathbf{A}}_2^{\square T}) = \begin{pmatrix} -6 & -0.75 \\ -0.75 & -9 \end{pmatrix} \tag{12.133}$$

and since $\tilde{\mathbf{A}}_{1s}^\square$ and $\tilde{\mathbf{A}}_{2s}^\square$ are stable the interval matrix $\mathbf{A}$ is Hurwitz stable.

Example 2: Stable interval matrix but the sufficient condition is failed:

$$\mathbf{L} = \text{matrix}[l_{ij}] = \begin{pmatrix} 0 & -2 \\ 1 & -3 \end{pmatrix} \qquad \mathbf{H} = \text{matrix}[h_{ij}] = \begin{pmatrix} 0 & -1 \\ 1 & -2 \end{pmatrix} \tag{12.134}$$

$$\tilde{\mathbf{A}}_1^\square = \begin{pmatrix} 0 & -2 \\ 1 & -2 \end{pmatrix} , \quad \tilde{\mathbf{A}}_2^\square = \begin{pmatrix} 0 & -1 \\ 1 & -2 \end{pmatrix} , \quad \tilde{\mathbf{A}}_{1s}^\square = \begin{pmatrix} 0 & -0.5 \\ -0.5 & -2 \end{pmatrix} , \quad \tilde{\mathbf{A}}_{2s}^\square = \begin{pmatrix} 0 & 0 \\ 0 & -2 \end{pmatrix} . \tag{12.135}$$

12.22 Conditions in Norm For Corner Matrices

With regard to Kharitonov's theorem, the Hurwitz stability of polynomials can easily be proved necessarily and sufficiently by checking the stability of four (or less) polynomials. In general, there is no parallel in the Schur stability case and in the interval matrix case. A theorem given by *Mori, T., and Kokame, H., 1987* shows a certain kind of parallel if in the discrete-time case the stability radius problem is replaced by the problem of proving matrix norm. Note that in general the maximum modulus of eigenvalues is always smaller than any norm. Hence, $\|\mathbf{\Phi}\|_p$ is larger than $\rho_s[\mathbf{\Phi}]$, in general, and $\|\mathbf{\Phi}\|_p < 1$ for any p is sufficient for $\rho_s[\mathbf{\Phi}] < 1$, only. In the case of (i) symmetric matrices $\mathbf{\Phi}$ and (ii) $p = 2$, the condition to be derived in what follows is necessary and sufficient.

The corner matrices of the interval matrix $\mathbf{A}_I = [\mathbf{L}\ \ \mathbf{H}] \in \mathcal{R}^{n\times n}$ are termed $\mathbf{A}_\nu^\square$ $\forall \nu = 1, 2 \ldots 2^{(n^2)}$. The corner matrices are obtained by setting $a_{ij} = l_{ij}$ or $a_{ij} = h_{ij}$ $\forall\, i, j = 1, 2 \ldots n$. If the set of corner matrices is characterized by $\|\mathbf{A}_\nu^\square\|_p < 1$ $\forall\, \nu = 1, 2 \ldots 2^{(n^2)}$ for any p it is necessary and sufficient that every matrix $\mathbf{A} \in \mathbf{A}_I$ has $\|\mathbf{A}\|_p < 1$ (*Mori, T., and Kokame, H., 1987*) , i.e.,

$$\mathbf{A} \in N_1 \quad \text{if and only if} \quad \mathbf{A}_\nu^\square \in N_1 \tag{12.136}$$

where N_1 is the set of matrices with p-norm < 1 .

Proof: Necessity is obvious: If a_{ij} happens to equal l_{ij} or h_{ij} the matrix $\mathbf{A}$ must be stable. To prove sufficiency matrices $\mathbf{A}'(a_{ij})$ are defined. The matrix $\mathbf{A}'(a_{ij})$ is a function of a_{ij} and denotes the matrix $\mathbf{A}$ where the entries a_{kl} coincide with the corner points $\forall\, k, l = 1, 2 \ldots n$ $k \neq i$ and $l \neq j$. Then, for the parameter ρ and $0 \leq \rho \leq 1$

$$\mathbf{A}'(a_{ij}) = \rho \mathbf{A}'(l_{ij}) + (1 - \rho)\mathbf{A}'(h_{ij}) \tag{12.137}$$

is a matrix with $a_{ij} = \rho l_{ij} + (1-\rho)h_{ij}$ and $a_{kl} = l_{kl}$ or $h_{kl} \forall k, l = 1, 2...n,\ k \neq i, l \neq j$. Applying the norm operation to Eq.(12.137), by virtue of the convexity of matrix norms,

$$\|\mathbf{A}'(a_{ij})\|_p \leq \rho\|\mathbf{A}'(l_{ij})\|_p + (1-\rho)\|\mathbf{A}'(h_{ij})\|_p = \rho\|\mathbf{A}^{\square}_{\nu_1}\|_p + (1-\rho)\|\mathbf{A}^{\square}_{\nu_2}\|_p < 1 \qquad (12.138)$$

since both norms above are smaller than unity.

Next, another element a_{rs} is selected additionally to a_{ij} and the matrix $\mathbf{A}'(a_{ij}, a_{rs})$ denotes a matrix whose elements coincide with the corner points except a_{ij} and a_{pq}. Following the arguments of above, this yields $\|\mathbf{A}'(a_{ij}, a_{rs})\|_p < 1$. Hence, for any $\mathbf{A} \in \mathbf{A}'$ it results $\|\mathbf{A}\|_p < 1$.

End of Proof

In the case of symmetric matrices $\mathbf{A}$ it is true that $\sigma_{\max}[\mathbf{A}] = \|\mathbf{A}\|_s = \rho_s[\mathbf{A}]$ and, hence, $\rho_s[\mathbf{A}] < 1$ if and only if $\rho_s[\mathbf{A}^{\square}] < 1$. In the case of nonsymmetric matrices since $\|\mathbf{A}\| < 1$ the relation is only sufficient.

With regard to the one-to-one correspondence given by Eqs.(12.139) and (12.119) there is a sufficient condition for the interval polynomial. The interval polynomial corresponding to Eq.(12.140) is Schur stable if the corner matrices $\mathbf{A}^{\square}$ associated with Eq.(12.139) belong to the set N_1.

Recall the fact that any monic polynomial $a(s) = \sum_{i=0}^{n} a_i s^{n-i}$ where $a_o = 1$ can be rewritten using the matrix $\mathbf{A}$ in phase-variable canonical form (companion form), i.e. $a(s) = \det(s\mathbf{I} - \mathbf{A})$ and

$$\mathbf{A} = \begin{pmatrix} 0 & 1 & 0 & \dots & 0 \\ 0 & 0 & 1 & \dots & 0 \\ \vdots & \vdots & & \ddots & \\ 0 & 0 & 0 & & 1 \\ -a_n & -a_{n-1} & -a_{n-2} & & -a_1 \end{pmatrix} \qquad (12.139)$$

$$a(s) = s^n + a_1 s^{n-1} + \dots + a_i s^{n-i} + \dots + a_{n-1}s + a_n \ . \qquad (12.140)$$

12.23 Maximum Perturbation of Stable Matrices

12.23.1 Perturbation Given by Interval Vectors

Consider the linear time-invariant system where $\mathbf{A}$ is perturbed by the matrix $\varepsilon\mathbf{g}\mathbf{h}^T$ and $\mathbf{g}$ and $\mathbf{h} \in \mathcal{R}^n$ are interval vectors

$$\dot{\mathbf{x}}(t) = (\mathbf{A} + \varepsilon\mathbf{g}\mathbf{h}^T)\mathbf{x}(t) \qquad \mathbf{g} = \begin{pmatrix} g_1 \\ \vdots \\ g_n \end{pmatrix} \qquad \mathbf{h} = \begin{pmatrix} h_1 \\ \vdots \\ h_n \end{pmatrix} \qquad \varepsilon > 0 \qquad (12.141)$$

$$g_i \in [-c_{gi}, c_{gi}]\ , \qquad h_i \in [-c_{hi}, c_{hi}]\ , \qquad c_{gi}, c_{hi} \in [0,1] \quad \forall\, i = [1,n]\ . \qquad (12.142)$$

Assume that the nominal system and its matrix $\mathbf{A}$ are stable. Find the largest ε ensuring that the perturbed matrix $\mathbf{A} + \varepsilon\mathbf{g}\mathbf{h}^T$ remains stable. Let $a(s)$ and $b(s)$ denote the characteristic polynomials of $\mathbf{A}$ and $\mathbf{A} + \varepsilon\mathbf{g}\mathbf{h}^T$. Then,

$$\begin{aligned} b(s) &= \det(s\mathbf{I}_n - \mathbf{A} - \varepsilon\mathbf{g}\mathbf{h}^T) = \det\{(s\mathbf{I}_n - \mathbf{A})[\mathbf{I}_n - \varepsilon(s\mathbf{I}_n - \mathbf{A})^{-1}\mathbf{g}\mathbf{h}^T]\} && (12.143) \\ &= \det(s\mathbf{I}_n - \mathbf{A})\det[\mathbf{I}_n - \varepsilon(s\mathbf{I}_n - \mathbf{A})^{-1}\mathbf{g}\mathbf{h}^T] = \mathbf{a}(s)[1 - \varepsilon\mathbf{h}^T(s\mathbf{I}_n - \mathbf{A})^{-1}\mathbf{g}] && (12.144) \end{aligned}$$

where $\det(\mathbf{I}-\mathbf{XY}) = \det(\mathbf{I}-\mathbf{YX})$ has been used. Denoting the characteristic polynomials in monic and reciprocated version,

$$a(s) = s^n + a_1 s^{n-1} + \ldots + a_n \qquad b(s) = s^n + b_1 s^{n-1} + \ldots + b_n \tag{12.145}$$

and evaluating the inverse $(s\mathbf{I}_n - \mathbf{A})^{-1}$ it results

$$(s\mathbf{I}_n - \mathbf{A})^{-1} \triangleq \frac{\mathbf{N}}{a(s)} = \frac{\text{adj }(s\mathbf{I}_n - \mathbf{A})^{-1}}{a(s)} \triangleq \frac{s^{n-1}\mathbf{I}_n + s^{n-2}\mathbf{N}_1 + \ldots + \mathbf{N}_{n-1}}{a(s)} \tag{12.146}$$

where the coefficients a_i and the coefficient matrices $\mathbf{N}_i$ are given from $\mathbf{A}$ by Leverrier's algorithm, see Eq.(7.24), by

$$\mathbf{N}_i = \mathbf{\Omega}_{n-i-1}\,, \qquad \mathbf{N}_o = \mathbf{I}_n, \qquad a_i = c_{n-i}\,. \tag{12.147}$$

From Eqs.(12.144) and (12.146)

$$b(s) = a(s) - \varepsilon a(s)\mathbf{h}^T(s\mathbf{I}_n - \mathbf{A})^{-1}\mathbf{g} = a(s) - \varepsilon \mathbf{h}^T\mathbf{N}(s)\mathbf{g} \tag{12.148}$$

$$b_i = a_i - \varepsilon \mathbf{h}^T\mathbf{N}_{i-1}\mathbf{g} \quad \text{or} \quad \mathbf{b} = \begin{pmatrix} b_1 \\ \vdots \\ b_n \end{pmatrix} = \mathbf{a} - \varepsilon \begin{pmatrix} \mathbf{h}^T\mathbf{N}_o \\ \mathbf{h}^T\mathbf{N}_1 \\ \vdots \\ \mathbf{h}^T\mathbf{N}_{n-1} \end{pmatrix} \mathbf{g}\,. \tag{12.149}$$

Now, consider the Frobenius norm $\|\mathbf{b}-\mathbf{a}\|_F$ as a constraint which should be maximized

$$\max_{\mathbf{g},\mathbf{h}} \|\mathbf{b}-\mathbf{a}\|_F = \max_{\mathbf{g},\mathbf{h}} \varepsilon^2 \sum_{i=0}^{n-1} \mathbf{g}^T\mathbf{N}_i^T\mathbf{h}\mathbf{h}^T\mathbf{N}_i\mathbf{g} \le \varepsilon^2 \sum_{i=0}^{n-1} \max_{\mathbf{g},\mathbf{h}} \mathbf{g}^T\mathbf{N}_i^T\mathbf{h}\mathbf{h}^T\mathbf{N}_i\mathbf{g} \triangleq \varepsilon^2 R_{m1}^2\,. \tag{12.150}$$

The inequality above is proved by *Soh, Y.C., and Evans, R.J., 1988*. The last maximum in the above inequation is obtained if $\mathbf{g},\mathbf{h}$ are put to vectors with maximum entries, i.e.

$$\mathbf{g} = (c_{g1},\ c_{g2}, \ldots, c_{gn})^T \qquad \mathbf{h} = (c_{h1},\ c_{h2}, \ldots, c_{hn})^T\,. \tag{12.151}$$

Another upper bound is simply given by

$$\max \|\mathbf{b}-\mathbf{a}\|_F \le \varepsilon^2 (\sum_{i=1}^{n} c_{hi}^2)(\sum_{i=1}^{n} c_{gi}^2) \sum_{i=0}^{n-1} \|\mathbf{N}_i\|_s^2 \triangleq \varepsilon^2 R_{m2}^2\,. \tag{12.152}$$

12.23.2 Perturbation Matrix Bounded by Ellipsoids

If the perturbation matrix $\mathbf{gh}^T$ is bounded by ellipsoids, i.e., $\mathbf{g}^T\mathbf{Q}_g\mathbf{g} \le 1$ and $\mathbf{h}^T\mathbf{Q}_h\mathbf{h} \le 1$ where $\mathbf{Q}_g$ and $\mathbf{Q}_h > 0$ and symmetric, then, using $\mathbf{Q} = (\mathbf{Q}^T)^{1/2}\mathbf{Q}^{1/2}$, the bound can be formulated as (*Soh, Y.H., and Evans, R.J., 1988*)

$$\max_{\mathbf{g},\mathbf{h}} \|\mathbf{b}-\mathbf{a}\|_F^2 \le \varepsilon^2 \sum_{k=0}^{n-1} \lambda_{\max}[\mathbf{Q}_g^{-1/2}\mathbf{Q}_h^{-1/2}\mathbf{N}_k\mathbf{N}_k^T\mathbf{Q}_h^{1/2}\mathbf{Q}_g^{1/2}] \triangleq \varepsilon^2 R_{m3}^2\,. \tag{12.153}$$

12.23.3 Maximum of ε

It must be emphasized that in the previous subsections the maxima $\|\mathbf{b}-\mathbf{a}\|_F$ versus $\mathbf{g}$ and $\mathbf{h}$ have not been calculated in fact but have been *replaced* by a simpler algorithm and by R_{mj}^2 $\forall\, j = 1,2,3$. However, referring to Eq.(21.150),

$$\max_{\mathbf{g},\mathbf{h}} \|\mathbf{b}-\mathbf{a}\|_F^2 \le R \quad \rightsquigarrow \quad \varepsilon^2 R_{mj}^2 \le R^2 \quad \rightsquigarrow \quad \varepsilon_{\max} = R/R_{mj} \ \ \forall j = 1,2,3\,. \tag{12.154}$$

12.24 Maximum Perturbation Matrices Given by Hyperellipsoids

Suppose that the coefficient extremes of the characteristic polynomial are already given. Let the extreme coefficients be termed $b_{i\,\max}$ and $b_{i\,\min}$. Then, from Eq.(12.149)

$$b_{i\,\min} = a_i - \varepsilon \max_{\mathbf{g},\mathbf{h}} \mathbf{h}^T \mathbf{N}_{i-1}\mathbf{g} \tag{12.155}$$

$$b_{i\,\max} = a_i - \varepsilon \min_{\mathbf{g},\mathbf{h}} \mathbf{h}^T \mathbf{N}_{i-1}\mathbf{g} \,. \tag{12.156}$$

The problem is to find those $\mathbf{g}, \mathbf{h}$ and ε in order to determine the interval matrix and to satisfy Eq.(12.155) and (12.156). If the vectors $\mathbf{g}$ and $\mathbf{h}$ are specified by a hyperellipsoid

$$\begin{pmatrix}\mathbf{g}\\ \mathbf{h}\end{pmatrix}^T \mathbf{Q}_{gh} \begin{pmatrix}\mathbf{g}\\ \mathbf{h}\end{pmatrix} \le 1 \tag{12.157}$$

there is a straightforward method for the determination of the matrix interval extremes (*Soh, Y.C., and Evans, R.J., 1988*).

12.24.1 Combining g and h

The scalar $\mathbf{h}^T\mathbf{N}_{i-1}\mathbf{g}$ can be rewritten using a combined vector comprising $\mathbf{g}$ and $\mathbf{h}$

$$\mathbf{h}^T\mathbf{N}_{i-1}\mathbf{g} = \begin{pmatrix}\mathbf{g}\\ \mathbf{h}\end{pmatrix}^T \begin{pmatrix} 0 & 0.5\,\mathbf{N}_{i-1} \\ 0.5\,\mathbf{N}_{i-1}^T & 0 \end{pmatrix} \begin{pmatrix}\mathbf{g}\\ \mathbf{h}\end{pmatrix} \triangleq \mathbf{y}^T\mathbf{M}_{i-1}\mathbf{y} \,. \tag{12.158}$$

12.24.2 Scaling

Consider the problem $\max \mathbf{y}^T\mathbf{M}\mathbf{y}$ under the constraint $\mathbf{y}^T\mathbf{R}\mathbf{y} \le 1$. Substituting

$$\mathbf{x} = \mathbf{U}\mathbf{y}, \quad \mathbf{G}^T\mathbf{G} = \mathbf{M}, \quad \mathbf{U}^T\mathbf{U} = \mathbf{R} \,, \tag{12.159}$$

one has $\mathbf{y}^T\mathbf{U}^T\mathbf{U}\mathbf{y} = \mathbf{x}^T\mathbf{x} = \|\mathbf{x}\|_F^2 = 1$. Selecting the maximum, $\le$ has been substituted by the equality. Now, since $\mathbf{M}$ is symmetric

$$\max \mathbf{y}^T\mathbf{M}\mathbf{y} = \max_{\|\mathbf{x}\|=1} \mathbf{x}^T(\mathbf{U}^T)^{-1}\mathbf{G}^T\mathbf{G}\mathbf{U}^{-1}\mathbf{x} = \sigma_{\max}^2[\mathbf{G}\mathbf{U}^{-1}] \tag{12.160}$$

$$= \lambda_{\max}[(\mathbf{U}^T)^{-1}\mathbf{M}\mathbf{U}^{-1}] = \|\mathbf{G}\mathbf{U}^{-1}\|_s^2 \,. \tag{12.161}$$

12.24.3 Optimum g, h and ε

Applying the result of the previous subsections for a given shape of the hyperellipsoids, i.e. $\mathbf{Q}_{gh}$, it is required to calculate the matrices $\mathbf{U}$ and $\mathbf{G}_{i-1}$ from $\mathbf{R} = \mathbf{U}^T\mathbf{U} := \mathbf{Q}_{gh}$ and $\mathbf{G}_{i-1}^T\mathbf{G}_{i-1} := \mathbf{M}_{i-1}$. Then, the maximum and minimum values can be obtained from

$$\max_{\mathbf{g},\mathbf{h}} \mathbf{h}^T\mathbf{N}_{i-1}\mathbf{g} = \|\mathbf{G}\mathbf{U}^{-1}\|_s^2 \quad \text{and} \quad \min_{\mathbf{g},\mathbf{h}} \mathbf{h}^T\mathbf{N}_{i-1}\mathbf{g} = \sigma_{\min}^2[\mathbf{G}\mathbf{U}^{-1}] \,. \tag{12.162}$$

For a fixed $\mathbf{Q}_{gh}$, finally, the maximum allowable ε results from the worst condition among all in Eq.(12.155) and (12.156). Variable $\mathbf{Q}_{gh}$ enables the designer to include the shape of the ellipsoids into the optimization algorithm.

12.25 Stable Interval Matrices and Gershgorin's Theorem

12.25.1 Simple Stability

Consider an interval matrix $\mathbf{A}_I \in \mathcal{R}^{n\times n}$ whose elements satisfy $l_{ij} \leq a_{ij} \leq h_{ij}$. Assume that $\mathbf{L} = \text{matrix}[l_{ij}]$ and $\mathbf{H} = \text{matrix}[h_{ij}]$ are stable. The matrix $\mathbf{A}_I$ can be replaced by $\mathbf{A}_p = \mathbf{A} + \Delta\mathbf{A}$ where $|\Delta a_{ij}| \leq e_{ij}$ or $|\Delta\mathbf{A}| \leq_e \mathbf{E}$, $e_{ij} = (h_{ij} - l_{ij})/2$ and $a_{ij} = (h_{ij} + l_{ij})/2$. Using the modal matrix $\mathbf{T}$ in the case of distinct eigenvalues, only, $\text{diag}\lambda_i[\mathbf{A}] = \mathbf{T}^{-1}\mathbf{A}\mathbf{T}$. Applying modal transformation, see Eq.(B.13), to the perturbation $\mathbf{E}$ the matrix $\mathbf{D}$ is defined

$$\mathbf{D} \triangleq |\mathbf{T}^{-1}|\ \mathbf{E}\ |\mathbf{T}| \tag{12.163}$$

which usually is not diagonal. Invoking Gershgorin's theorem in a modification (*Juang, Y.T., and Shao, C.S., 1989*), the eigenvalues of the perturbed system λ_{pi} are located in the union of discs with centres at λ_i and radii $\sum_{j=1}^{n} d_{ij}$ or $\sum_{i=1}^{n} d_{ij}$ $\forall\, i \in [1, 2...n]$. Since

$$|\mathbf{T}^{-1}\Delta\mathbf{A}\ \mathbf{T}| \leq |\mathbf{T}^{-1}|\ |\Delta\mathbf{A}|\ |\mathbf{T}| \leq |\mathbf{T}^{-1}|\ \mathbf{E}\ |\mathbf{T}| \triangleq \mathbf{D} \tag{12.164}$$

the maximum row or column sum of $\mathbf{T}^{-1}\Delta\mathbf{A}\ \mathbf{T}$ is less than or equal to $\sum_{j=1}^{n} d_{ij}$. To avoid changing the location of the centres of the Gershgorin discs as the system is perturbed the discs are enlarged to include the changes in the diagonal elements. Then, the theorem exists that the eigenvalues of $\mathbf{A}_p$ must be included in the union of discs centered at $\lambda_i[\mathbf{A}]$.

If the modulus operation is omitted in Eq. (12.163) even if the radii are calculated by the modulus sum of d_{ij} , the sufficiency of the condition does not hold (*Juang, Y.T., and Shao, C.S., 1989*).

12.25.2 Stability With Stability Margin M

From the previous subsection it can be derived that

$$|\ \lambda_i[\mathbf{A}_I] - \lambda_i[\mathbf{A}]\ | \leq \sum_{j=1}^{n} d_{ij} \tag{12.165}$$

$$\Re e\ \lambda_i[\mathbf{A}_I] - \Re e\ \lambda_i[\mathbf{A}] \leq \sum_{j=1}^{n} d_{ij} \quad \rightsquigarrow \quad \Re e\ \lambda_i[\mathbf{A}_I] \leq \Re e\ \lambda_i[\mathbf{A}] + \sum_{j=1}^{n} d_{ij}\ . \tag{12.166}$$

Hence, if $\Re e\ \lambda_i[\mathbf{A}] + \sum_{j=1}^{n} d_{ij} < -M$ $\forall\, i = 1, 2...n$ then $\Re e\ \lambda_i[\mathbf{A}_I] < -M$ where $M > 0$. For discrete-time systems from Eq.(12.165)

$$|\ \lambda_i[\mathbf{A}_I]\ | - |\ \lambda_i[\mathbf{A}]\ | \leq |\ \lambda_i[\mathbf{A}_I] - \lambda_i[\mathbf{A}]\ | \leq \sum_{j=1}^{n} d_{ij} \tag{12.167}$$

$$|\ \lambda_i[\mathbf{A}_I]\ | \leq |\ \lambda_i[\mathbf{A}]\ | + \sum_{j=1}^{n} d_{ij} \leq 1 - M_d \quad \text{where} \quad M_d > 0\ . \tag{12.168}$$

Hence, the interval matrix is stable with stability margin M_d if the moduli of the eigenvalues of the average matrix $\mathbf{A}$ augmented by the row sum of the matrix $|\mathbf{T}^{-1}|\mathbf{E}_d|\mathbf{T}|$ remains smaller than $1 - M_d$. The row sum $\sum_{j=1}^{n} d_{ij}$ may be replaced by the column sum as known from Gershgorin's theorem.

Mansour, M.,1989a presented a version for Schur stability using matrix norms

$$\|\mathbf{R}^{-1}\mathbf{W}[\mathbf{A}_I]\ \mathbf{R}\|_1 < 1 \quad \text{or} \quad \|\mathbf{R}^{-1}\mathbf{W}[\mathbf{A}_I]\ \mathbf{R}\|_\infty < 1 \tag{12.169}$$

where $\mathbf{R} = \text{diag}\{r_k\}$, $r_k > 0$ (e.g. $\mathbf{R} = \mathbf{I}$), $\mathbf{W}[\mathbf{A}_I]$ =matrix$[w_{ij}]$ and $\forall\, i,\ j$

$$w_{ij} = \begin{cases} l_{ij} & \text{if } |l_{ij}| > |h_{ij}| \\ h_{ij} & \text{if } |l_{ij}| < |h_{ij}| \\ l_{ij} \text{ or } h_{ij} & \text{if } |l_{ij}| = |h_{ij}|\,. \end{cases} \tag{12.170}$$

The algorithm is modified (*Argoun, M.B., 1986a and b*) by decomposing the perturbation $\mathbf{E}$ additively into successive matrices $\mathbf{E}_k$ $\forall\ k = 1...r$ and by applying the modal transformation $\mathbf{T}$ to each $\mathbf{E}_k$. The matrix $\mathbf{E}_k$ is a matrix with elements zero except the kth perturbed element. Each perturbation contributes to the radius of each disc. For all the perturbations the radii result from a sum given by the individual contributions. The sufficient condition for stability is satisfied if the union of discs is located in the left-half s-plane or inside the unit circle in the z-plane for continuous-time and discrete-time systems, respectively.

Conservatism may occur to a high extent. By the fact that the moduli of the off-diagonal elements are used the same radii are obtained irrespective if the error occurs in a direction that relative stability increases or decreases.

12.26 Lyapunov Equation and Finite Parameter Variations

12.26.1 Continuous-Time Stable Systems

If $\mathbf{A}$ is any Hurwitz stable matrix then so is the sum $\mathbf{A} + \mathbf{E}$ where

$$\mathbf{E} = \mathbf{P}^{-1}(\mathbf{S} - \mathbf{Q}_o) \qquad \mathbf{A}^T\mathbf{P} + \mathbf{P}\mathbf{A} = -\mathbf{Q} \tag{12.171}$$

and $\mathbf{S}$ is arbitrary *skew-symmetric*, $\mathbf{Q}_o[\mathbf{Q}]$ is arbitrary positive-semidefinite [definite] symmetric.

Consider a stable monic polynomial

$$a(\lambda) = \lambda^n + a_{n-1}\lambda^{n-1} + \ldots + a_o = \lambda^n + \sum_{i=1}^{n} a_{i-1}\lambda^{i-1}\,. \tag{12.172}$$

Then, the polynomial $a(\lambda) + e(\lambda)$ is also stable if the coefficient e_i of $e(\lambda) = \sum_{i=1}^{n} e_{i-1}\lambda^{i-1}$ is given by

$$e_{i-1} = \alpha\, p_{ni} \qquad \forall i = 1\ldots n \tag{12.173}$$

where α is any non-negative scalar, p_{ni} is the component of the last row $(\mathbf{P})_{n.}$ of $\mathbf{P}$ given by Eq.(12.171) if $\mathbf{A}$ is the matrix in companion form associated with $a(\lambda)$ □. (*Shane, B.A., and Barnett, S., 1972a*).

12.26.2 Continuous-Time System. General Case

Consider the system $\dot{\mathbf{x}}(t) = \mathbf{A}\mathbf{x}(t)$ where $\mathbf{A}$ is given in the companion form in which the last row is $(-a_o \ldots - a_{n-1})$ and the characteristic polynomial is given by Eq.(12.172). Assume the distribution of the zeros of $a(\lambda)$ with respect to a given region Γ in the complex s-plane is known. No root should be located on the boundary of Γ.

The problem is to find finite changes in the parameters a_i which do not affect the root distribution. Then, $\mathbf{A}$ or $a(\lambda)$ is said robust with respect to these variations in the parameters. If $\mathbf{A}$ has r characteristic roots inside Γ and $n - r$ outside then $\mathbf{A}$ has inertia

$(n-r,r,0)$. From Lyapunov theory it is known, see Eq.(13.16), that then $\mathbf{P}$ has inertia $(r,n-r,0)$ where

$$\mathbf{A}^T\mathbf{P}+\mathbf{P}\mathbf{A}=-\mathbf{Q} \tag{12.174}$$

and $\mathbf{Q}$ is any positive definite matrix (*Barnett, S., 1971*).

Summarizing,

$$\mathbf{A} \text{ has } r \text{ eigenvalues } \lambda[\mathbf{A}] \text{ inside } \Gamma \text{ and inertia } (n-r,r,0)$$

$$\mathbf{P} \text{ has } n-r \text{ eigenvalues } \lambda[\mathbf{P}] \text{ inside } \Gamma \text{ and inertia } (r,n-r,0)\ .$$

If, e.g., $\mathbf{A}$ is stable, $\Gamma \triangleq \{s : \Re e\ s<0\}, r=n$, $\mathbf{A}$ and $\mathbf{P}$ have inertia $(0,n,0)$ and $(n,0,0)$, respectively. If a matrix $\mathbf{E}$ can be found satisfying

$$(\mathbf{A}+\mathbf{E})^T\mathbf{P}+\mathbf{P}(\mathbf{A}+\mathbf{E})=-(\mathbf{Q}+\mathbf{Q}_o) \tag{12.175}$$

where $(\mathbf{Q}+\mathbf{Q}_o)$ is positive definite then the inertia of $\mathbf{A}+\mathbf{E}$ is identical to the inertia of $\mathbf{A}$. If a matrix $\mathbf{E}$ is found obeying

$$\mathbf{E}=\mathbf{P}^{-1}(\mathbf{S}-\mathbf{Q}_o/2) \tag{12.176}$$

where $\mathbf{S}$ is any skew-Hermite matrix then $\mathbf{A}+\mathbf{E}$ and $\mathbf{A}$ have the same inertia. □

If a polynomial $e(\lambda)$

$$e(\lambda)=e_{n-1}\lambda^{n-1}+e_{n-2}\lambda^{n-2}+....+e_o \tag{12.177}$$

is added to $a(\lambda)$ where $\mathbf{P}=\text{matrix}[p_{jk}]$ is the solution of the Lyapunov equation associated with $\mathbf{A}$ in companion form, then $a(\lambda)+e(\lambda)$ and $a(\lambda)$ have the same inertia (*Shane, B.A., and Barnett, S., 1972*) if

$$e_i=qp_{ni} \qquad \forall\ i=0\ ....\ n-1 \qquad \text{and} \qquad \frac{-\lambda_{\min}[\mathbf{Q}]}{2\sum_{j=1}^n p_{nj}^2}<q\ .\ \square \tag{12.178}$$

Consider $a(\lambda)$ a (complex) polynomial with zeros inside Γ. If only just one particular coefficient a_i is changed by addition of e_i ($\forall\ i=0\ ...\ n-1$) then the roots of the perturbed polynomial are in the circle with

$$\text{centre}\ :\ \frac{\lambda_{\min}[\mathbf{Q}]}{\Re e\ p_{(i+1),n}-\Im m\ p_{(i+1),n}\sum_{j=1,j\neq i+1}^n\ |p_{jn}|^2} \tag{12.179}$$

$$\text{radius}\ :\ \lambda_{\min}[\mathbf{Q}]\sqrt{\sum_{j=1}^n|p_{jn}|^2\ /\ \sum_{j=1,j\neq i+1}^n\ |p_{jn}|^2} \tag{12.180}$$

where $\mathbf{P}=\text{matrix}[p_{jk}]$ and $\mathbf{Q}$ are given by Eq.(12.174). (Proof see *Shane, B.A., and Barnett, S., 1972.*) □

12.26.3 Discrete-Time Systems

Let a discrete-time system be $\mathbf{x}_{i+1}=\mathbf{\Phi}(T)\mathbf{x}_i$. The fundamental matrix $\mathbf{\Phi}(T)$ is considered given in companion form. Using a Lyapunov candidate $V(\mathbf{x}_i)=\mathbf{x}_i^T\mathbf{P}\mathbf{x}_j$ where $\mathbf{P}$ is positive definite, the system is stable if $V(\mathbf{x}_{i+1})-V(\mathbf{x}_i)$ is negative definite. Hence, with abbreviation $\mathbf{\Phi}=\mathbf{\Phi}(T)$ the equation

$$\mathbf{\Phi}^T\mathbf{P}\mathbf{\Phi}-\mathbf{P}=-\mathbf{Q} \qquad \mathbf{\Phi},\mathbf{P}\in\mathcal{R}^{n\times n} \tag{12.181}$$

must have a positive definite solution $\mathbf{P}$ for positive definite $\mathbf{Q}$.

Generalizing, if Γ_d is the unit disc and

$$\mathbf{\Phi} \text{ has } r \text{ eigenvalues } \lambda[\mathbf{\Phi}] \text{ inside } \Gamma_d \text{ and inertia } (n-r, r, 0)$$

$$\mathbf{P} \text{ has } n-r \text{ eigenvalues } \lambda[\mathbf{P}] \text{ inside } \Gamma_d \text{ and inertia } (r, n-r, 0)\ .$$

For a given $\mathbf{\Phi}(T)$ with r eigenvalues inside the unit disc Γ_d the addition of a matrix $\mathbf{E}_d$ does not affect the eigenvalue distribution with respect to Γ_d if

$$\mathbf{E}_d = (\mathbf{S} - \mathbf{P}/2)^{-1}(\mathbf{SY} + \mathbf{PY}/2 + \mathbf{P\Phi}) \tag{12.182}$$

where $\mathbf{Y}$ is any solution of the algebraic Riccati equation

$$\mathbf{\Phi}^T\mathbf{PY} + \mathbf{YP\Phi} + \mathbf{YPY} = -\mathbf{Q}_o\ , \tag{12.183}$$

$\mathbf{Q}_o$ is a Hermite matrix for which $\mathbf{Q} + \mathbf{Q}_o$ is positive definite and $\mathbf{S}$ is any skew-Hermite matrix. □

Given the polynomial $\varphi(\lambda)$ with known root distribution with respect to Γ_d , the distribution remains the same if another polynomial $e_d(\lambda)$ is added to $\varphi(\lambda)$, if the following properties are satisfied

$$e_{di} = \frac{q\alpha_{(i+1),n}}{2 + qp_{nn}} \quad \forall\, i = 0....n-1 \tag{12.184}$$

and

$$\text{either} \quad q > 0 \quad \text{or} \quad \frac{-2q}{(2+qp_{nn})^2} < \lambda_{\min}[\mathbf{Q}] \sum_{j=1}^{n} \alpha_{jn}^2 \tag{12.185}$$

where $\alpha_{j_n} = (\mathbf{\Phi}^H\mathbf{P})_{j_n}$ $\forall\, j = 1...n$ (*Shane, B.A., and Barnett, S., 1972 and 1973*).

12.27 Variation Preserving Stability. Lyapunov Approach

Given an asymptotically stable autonomous system $\dot{\mathbf{x}}(t) = \mathbf{Ax}(t)$. Then, for an arbitrary $\mathbf{Q} = \mathbf{Q}^T > 0$ there exists a unique $\mathbf{P} = \mathbf{P}^T > 0$ such that $\mathbf{A}^T\mathbf{P} + \mathbf{PA} + \mathbf{Q} = \mathbf{0}$. If $\mathbf{Q} = \mathbf{0}$ then $\mathbf{A}$ has purely imaginary eigenvalues. Suppose a perturbation $\Delta\mathbf{A}$ to $\mathbf{A}$, i.e., $\mathbf{A}_p = \mathbf{A} + \Delta\mathbf{A}$, then the system $\dot{\mathbf{x}}(t) = \mathbf{A}_p\mathbf{x}(t)$ remains asymptotically stable if

$$\Delta\mathbf{A}\ \mathbf{Q}^{-1}\Delta\mathbf{A}^T < \frac{1}{4}\mathbf{P}^{-1}\mathbf{Q}\mathbf{P}^{-1}\ . \tag{12.186}$$

Proof: The Lyapunov equation in the unperturbed case is premultiplied and postmultiplied by $\mathbf{P}^{-1}$

$$\mathbf{P}^{-1}\mathbf{A}^T + \mathbf{AP}^{-1} + \mathbf{P}^{-1}\mathbf{QP}^{-1} = 0\ . \tag{12.187}$$

Adding and subtracting equal terms yields

$$(\mathbf{A} + \Delta\mathbf{A})\mathbf{P}^{-1} + \mathbf{P}^{-1}(\mathbf{A} + \Delta\mathbf{A})^T \underbrace{- \Delta\mathbf{A}\ \mathbf{P}^{-1} - \mathbf{P}^{-1}\Delta\mathbf{A}^T + \mathbf{P}^{-1}\mathbf{QP}^{-1}}_{\mathbf{Q}'} = 0\ . \tag{12.188}$$

Using the above definition of $\mathbf{Q}'$, the above equation may be considered the Lyapunov equation for the perturbed system. The perturbed system remains stable if $\mathbf{Q}' > 0$. Using the relation $\P\P^T \geq 0$

$$\P\P^T \triangleq \Big(\frac{1}{\sqrt{2}}\mathbf{P}^{-1}\mathbf{Q}^{1/2} - \sqrt{2}\,\Delta\mathbf{A}\ \mathbf{Q}^{-1/2}\Big)\Big(\frac{1}{\sqrt{2}}\mathbf{P}^{-1}\mathbf{Q}^{1/2} - \sqrt{2}\,\Delta\mathbf{A}\ \mathbf{Q}^{-1/2}\Big)^T \geq 0 \tag{12.189}$$

$$\frac{1}{2}\mathbf{P}^{-1}\mathbf{Q}\mathbf{P}^{-1} - \mathbf{P}^{-1}\Delta\mathbf{A}^T - \Delta\mathbf{A}\ \mathbf{P}^{-1} + 2\ \Delta\mathbf{A}\ \mathbf{Q}^{-1}\Delta\mathbf{A}^T \geq 0 \tag{12.190}$$

$$\mathbf{P}^{-1}\Delta\mathbf{A}^T + \Delta\mathbf{A}\ \mathbf{P}^{-1} \leq \frac{1}{2}\mathbf{P}^{-1}\mathbf{Q}\mathbf{P}^{-1} + 2\ \Delta\mathbf{A}\ \mathbf{Q}^{-1}\Delta\mathbf{A}^T\ . \tag{12.191}$$

From the definition of $\mathbf{Q}'$ and using Eq.(12.191),

$$\mathbf{Q}' \triangleq -\Delta\mathbf{A}\ \mathbf{P}^{-1} - \mathbf{P}^{-1}\Delta\mathbf{A}^T + \mathbf{P}^{-1}\mathbf{Q}\mathbf{P}^{-1} \geq -\frac{1}{2}\mathbf{P}^{-1}\mathbf{Q}\mathbf{P}^{-1} - 2\ \Delta\mathbf{A}\ \mathbf{Q}^{-1}\Delta\mathbf{A}^T + \mathbf{P}^{-1}\mathbf{Q}\mathbf{P}^{-1} \tag{12.192}$$

$$\Leftarrow \quad \frac{1}{2}\mathbf{P}^{-1}\mathbf{Q}\mathbf{P}^{-1} - 2\ \Delta\mathbf{A}\ \mathbf{Q}^{-1}\Delta\mathbf{A}^T > 0 \quad \rightsquigarrow \quad \Delta\mathbf{A}\ \mathbf{Q}^{-1}\Delta\mathbf{A}^T < \frac{1}{4}\mathbf{P}^{-1}\mathbf{Q}\mathbf{P}^{-1}\ . \tag{12.193}$$

End of Proof

An equivalent result

$$\Delta\mathbf{A}^T\mathbf{P}\mathbf{Q}^{-1}\mathbf{P}\Delta\mathbf{A} < \frac{1}{4}\mathbf{Q} \tag{12.194}$$

is achieved if the Lyapunov equation $\mathbf{AP}+\mathbf{PA}^T+\mathbf{Q} = \mathbf{0}$ is used (*Eslami, M., and Russell, D.L., 1980*).

12.27.1 Single Parameter Variation

The following theorem will be used: Given the Hermitian matrices $\mathbf{M}$ and $\mathbf{N}$ of dimension $n \times n$. If $\mathbf{M} > 0$ then an (n,n)-matrix $\mathbf{R}$ exists such that

$$\mathbf{R}^H\mathbf{M}\mathbf{R} = \mathbf{I} \quad \text{and} \quad \mathbf{R}^H\mathbf{N}\mathbf{R} = \text{diag}\{\nu_i\} \tag{12.195}$$

where $\nu_i \in \mathcal{R}$. If $\mathbf{N} > 0$ then the numbers ν_i are generalized eigenvalues obeying

$$\mathbf{N}\mathbf{r}_i = \nu_i\ \mathbf{M}\mathbf{r}_i \qquad \mathbf{r}_i \neq 0 \quad \forall i = 1,2...n\ . \tag{12.196}$$

If $\mathbf{M}$ and $\mathbf{N}$ are real then $\mathbf{R}$ is real and $\mathbf{r}_i$ are columns of $\mathbf{R}$ and ν_i are given by $\det(\nu_i\mathbf{M} - \mathbf{N}) = 0$ (*Franklin, J.N., 1968*).

Consider the single parameter a_i perturbed to $a_i + \Delta a_i$. This coefficient in $a(s)$ corresponds to the parameter at the location $(n, n-i+1)$. With $j \triangleq n-i+1$ the location is (n,j). The matrix $\Delta\mathbf{A}$ turns out $\Delta a_j\mathbf{E}_{nj}$ where $\mathbf{E}_{nj}$ is a matrix with entry 1 in the (n,j)-position and zero elsewhere. Denoting the (k,j)-position of $\mathbf{Q}^{-1}$ by $c_{kj}/\det\mathbf{Q}$ where c_{kj} denotes the cofactor $\text{cof}_{kj}\mathbf{Q}$, then, omitting calculations,

$$\Delta\mathbf{A}\ \mathbf{Q}^{-1}\Delta\mathbf{A}^T = \frac{1}{\det\mathbf{Q}}c_{jj}\Delta a_j^2\text{diag}\{0,0,...1\} = \frac{1}{\det\mathbf{Q}}c_{jj}\Delta a_j^2\mathbf{E}_{nn}\ . \tag{12.197}$$

From Eq.(12.186)

$$\frac{c_{jj}\Delta a_j^2\mathbf{E}_{nn}}{\det\mathbf{Q}} < \frac{1}{4}\mathbf{M}\ . \tag{12.198}$$

Abbreviating $\mathbf{M} \triangleq \mathbf{P}^{-1}\mathbf{Q}\mathbf{P}^{-1} = \mathbf{M}^T > 0$, $\mathbf{N} \triangleq \mathbf{E}_{nn} = \mathbf{N}^T \geq 0$ and using Eq.(12.195), i.e., simultaneously diagonalizing $\mathbf{M}$ and $\mathbf{E}_{nn}$, yields

$$\frac{c_{jj}\Delta a_j^2}{\det\mathbf{Q}}\text{diag}\{\nu_i\} < \frac{1}{4}\mathbf{I}\ . \tag{12.199}$$

Selecting the largest ν_i from $\det(\nu_i\mathbf{M} - \mathbf{N}) = 0$ and renaming ν_i to $\nu_{\max}$ yields

$$\frac{c_{jj}\Delta a_j^2}{\det\mathbf{Q}}\nu_{\max} < \frac{1}{4} \quad \rightsquigarrow \quad \Delta a_j^2 < \frac{\det\mathbf{Q}}{4\nu_{\max}\text{cof}_{jj}\mathbf{Q}} \tag{12.200}$$

which is an upper bound for the single parameter variation preserving stability. Regarding the structure $\mathbf{N} = \text{diag}\{0,0,...0,1\}$ from $\det[\nu\mathbf{M} - \mathbf{N}] = 0$ finally it results $\nu_{\max} = \text{cof}_{nn}\mathbf{M}/\det\mathbf{M}$ and

$$\Delta a_j^2 < \frac{\det\mathbf{Q}\ \ \det\mathbf{M}}{4\ \text{cof}_{jj}\mathbf{Q}\ \ \text{cof}_{nn}\mathbf{M}}\ . \tag{12.201}$$

12.27.2 Variation of Several Coefficients

In most control applications a perturbation influences more than a single parameter. Now, suppose a variation $\Delta\mathbf{A}$ with several nonzero elements $-\Delta a_n$ through $-\Delta a_1$ only in the last row. Then, using Eq.(12.194),

$$\Delta\mathbf{A}^T\mathbf{P}\mathbf{Q}^{-1}\mathbf{P}\Delta\mathbf{A} = (\mathbf{P}\mathbf{Q}^{-1}\mathbf{P})_{nn}\begin{pmatrix} \Delta a_n^2 & \Delta a_n\Delta a_{n-1} & \cdots & \\ \vdots & & \ddots & \\ & & \cdots & \Delta a_1^2 \end{pmatrix} < \frac{1}{4}\mathbf{Q}\,. \tag{12.202}$$

The matrix containing the entries $\Delta a_\rho \Delta a_\sigma$ is a nonlinear function of the perturbations. Using the fact that its rank is one, its maximum eigenvalue (i.e., its unique nonzero eigenvalue) equals the trace. Following *Eslami, M., and Russell, D.L., 1980*

$$\sum_{j=1}^{n}\Delta a_j^2 < \frac{\lambda_{\min}[\mathbf{Q}]}{4(\mathbf{P}\mathbf{Q}^{-1}\mathbf{P})_{nn}}\,. \tag{12.203}$$

12.28 Robust Stability Using Infinity Matrix Norm

12.28.1 Infinity Matrix Norm of the Maximum Bias Matrix

Consider the continuous-time system

$$\dot{\mathbf{x}}(t) = \mathbf{A}\mathbf{x}(t) + \mathbf{B}\mathbf{u}(t), \qquad \mathbf{x}(0) = \mathbf{x}_o \qquad \mathbf{y}(t) = \mathbf{C}\mathbf{x}(t) \tag{12.204}$$

where $\mathbf{B}$ and $\mathbf{C}$ are known (white) matrices and $\mathbf{A}$ is an interval (grey) matrix limited by minimum and maximum whiting matrices $\mathbf{L}$ and $\mathbf{H}$, respectively, $\mathbf{L} \leq_e \mathbf{A} \leq_e \mathbf{H}$. The following definitions are used: the average whiting matrix $\bar{\mathbf{A}} \triangleq 0.5\ (\mathbf{L}+\mathbf{H})$, the maximum bias or deviation matrix $\mathbf{M} \triangleq \mathbf{H} - \bar{\mathbf{A}} = 0.5\ (\mathbf{H} - \mathbf{L})$, the bias matrix $\mathbf{A}_b \triangleq \mathbf{A} - \bar{\mathbf{A}}$ and, finally, (using $\|\mathbf{M}\|_\infty = \max_i \sum_j |M_{ij}|$) the maximum absolute row sum plus maximum absolute column sum of the maximal bias matrix termed by $\alpha_M \triangleq \|\mathbf{M}\|_\infty + \|\mathbf{M}^T\|_\infty$.

12.28.2 Open-Loop Continuous-Time System

Using the properties Eqs.(B.75) and (2.53),

$$\begin{aligned} \Re e\ \lambda_i[\mathbf{A}] &= \Re e\ \lambda_i[\bar{\mathbf{A}} + \mathbf{A}_b] \leq \lambda_{\max}[\frac{\bar{\mathbf{A}} + \mathbf{A}_b + (\bar{\mathbf{A}}^T + \mathbf{A}_b^T)}{2}] & (12.205)\\ &\leq \lambda_{\max}[\frac{\bar{\mathbf{A}} + \bar{\mathbf{A}}^T}{2}] + \lambda_{\max}[\frac{\mathbf{A}_b + \mathbf{A}_b^T}{2}] \leq 0.5\ \lambda_{\max}[\bar{\mathbf{A}} + \bar{\mathbf{A}}^T] + \|\frac{\mathbf{A}_b + \mathbf{A}_b^T}{2}\|_\infty & \\ &\leq 0.5\left(\lambda_{\max}[\bar{\mathbf{A}} + \bar{\mathbf{A}}^T] + \|\mathbf{A}_b\|_\infty + \|\mathbf{A}_b^T\|_\infty\right)\,. & (12.206) \end{aligned}$$

Since $\mathbf{A}_b \triangleq \mathbf{A} - \bar{\mathbf{A}}$ and $\mathbf{M} \triangleq \mathbf{H} - \bar{\mathbf{A}}$ it results $\mathbf{A}_b \leq_e \mathbf{M}$ and

$$\|\mathbf{A}_b\|_\infty \leq \|\mathbf{M}\|_\infty \quad \leadsto \quad \|\mathbf{A}_b\|_\infty + \|\mathbf{A}_b^T\|_\infty \leq \|\mathbf{M}\|_\infty + \|\mathbf{M}^T\|_\infty = \alpha_M\,. \tag{12.207}$$

Combining Eq.(12.206) and Eq.(12.207),

$$\Re e\ \lambda_i[\mathbf{A}] \leq 0.5\left(\lambda_{\max}[\bar{\mathbf{A}} + \bar{\mathbf{A}}^T] + \alpha_M\right) = 0.5\ \lambda_{\max}[\bar{\mathbf{A}} + \bar{\mathbf{A}}^T + \alpha_M\ \mathbf{I}\ ]\,. \tag{12.208}$$

Hence, a sufficient condition for $\Re e\ \lambda_i[\mathbf{A}] \leq 0$ is

$$\bar{\mathbf{A}} + \bar{\mathbf{A}}^T + \alpha_M\ \mathbf{I} < 0 \tag{12.209}$$

(Zhou, C.S., and Deng, J.L., 1986).

12.28.3 Open-Loop Continuous-Time System. Modification

A modification for robust Hurwitz stability is obtained by applying the Lyapunov equation

$$\mathbf{A}^T\mathbf{P}+\mathbf{P}\mathbf{A}+\mathbf{Q}=\mathbf{0} \quad \text{where} \quad \mathbf{A}_I=[\mathbf{L}\ \mathbf{H}]=\bar{\mathbf{A}}+\mathbf{E}_I=\bar{\mathbf{A}}+[-\mathbf{M}\ \mathbf{M}], \quad \mathbf{A}=\bar{\mathbf{A}}+\mathbf{E} \quad (12.210)$$

with specified $\mathbf{P}=\mathbf{I}$ and $\mathbf{Q}>0$

$$\mathbf{A}^T\mathbf{P}+\mathbf{P}\mathbf{A}+\mathbf{Q}=\bar{\mathbf{A}}^T+\bar{\mathbf{A}}+\mathbf{E}^T+\mathbf{E}+\mathbf{Q}=\mathbf{0} \quad (12.211)$$

$$\bar{\mathbf{A}}^T+\bar{\mathbf{A}}+\mathbf{E}^T+\mathbf{E}=-\mathbf{Q}<\mathbf{0} \quad \rightsquigarrow \quad \mathbf{x}^T(\bar{\mathbf{A}}^T+\bar{\mathbf{A}})\mathbf{x}+\mathbf{x}^T(\mathbf{E}^T+\mathbf{E})\mathbf{x}<0 \quad (12.212)$$

$$\Leftarrow \quad \mathbf{x}^T(\bar{\mathbf{A}}^T+\bar{\mathbf{A}})\mathbf{x}+\lambda_{\max}[\mathbf{E}^T+\mathbf{E}]\mathbf{x}^T\mathbf{x}<0 \quad (12.213)$$

$$\Leftarrow \quad \bar{\mathbf{A}}^T+\bar{\mathbf{A}}+2\sqrt{\|\mathbf{M}\|_1\|\mathbf{M}\|_\infty}\,\mathbf{I}<0 \quad (12.214)$$

since

$$\lambda_{\max}[\mathbf{E}^T+\mathbf{E}]\le\|\mathbf{E}^T+\mathbf{E}\|_s\le 2\ \|\mathbf{E}\|_s\le 2\sqrt{\|\mathbf{E}\|_1\|\mathbf{E}\|_\infty}\le 2\sqrt{\|\mathbf{M}\|_1\|\mathbf{M}\|_\infty}\,. \quad (12.215)$$

12.28.4 Closed-Loop Continuous-Time System

Consider the uncertain linear plant and state feedback controller, respectively,

$$\dot{\mathbf{x}}(t)=(\bar{\mathbf{A}}+\mathbf{A}_b)\mathbf{x}(t)+\mathbf{B}\mathbf{u}(t) \qquad \mathbf{u}(t)=-\mathbf{R}\mathbf{B}^T\mathbf{P}^{-1}\mathbf{x}(t)\,. \quad (12.216)$$

Then, the closed-loop system is stable if under $\mathbf{P},\mathbf{R},\mathbf{W}\in\mathcal{R}^{n\times n}$, $\mathbf{P}>0$, $\mathbf{W}>0$, $\mathbf{R}=\mathbf{R}^T$ the condition as follows is satisfied

$$\bar{\mathbf{A}}\mathbf{P}+\mathbf{P}\bar{\mathbf{A}}^T-2\mathbf{B}\mathbf{R}\mathbf{B}^T+\mathbf{P}\mathbf{W}\mathbf{P}+\frac{\|\mathbf{M}\|_\infty\|\mathbf{M}^T\|_\infty}{\lambda_{\min}[\mathbf{W}]}\,\mathbf{I}<0\,. \quad (12.217)$$

Proof: Using a Lyapunov function $V(\mathbf{x})=\mathbf{x}^T\mathbf{P}^{-1}\mathbf{x}$ and substituting the closed-loop system equation with system matrix $(\mathbf{A}_b+\bar{\mathbf{A}}-\mathbf{B}\mathbf{R}\mathbf{B}^T\mathbf{P}^{-1})$ yields

$$\dot{V}=\mathbf{x}_1^T[\mathbf{P}\bar{\mathbf{A}}^T+\bar{\mathbf{A}}\mathbf{P}-2\mathbf{B}\mathbf{R}\mathbf{B}^T+\mathbf{P}\mathbf{W}\mathbf{P}+\mathbf{A}_b\mathbf{W}^{-1}\mathbf{A}_b^T-(\mathbf{P}\mathbf{W}-\mathbf{A}_b)\mathbf{W}^{-1}(\mathbf{P}\mathbf{W}-\mathbf{A}_b)^T]\mathbf{x}_1 \quad (12.218)$$

where $\mathbf{x}_1=\mathbf{P}^{-1}\mathbf{x}$. In order to obtain a sufficient condition, the last term within the bracket is omitted and the properties (*Zhou, C.S., and Deng, J.L., 1986*)

$$\mathbf{x}_1^T\mathbf{A}_b\mathbf{W}^{-1}\mathbf{A}_b^T\mathbf{x}_1\le\frac{\mathbf{x}_1^T\mathbf{x}_1}{\lambda_{\min}[\mathbf{W}]}\mathbf{A}_b\mathbf{A}_b^T \quad (12.219)$$

$$\mathbf{x}_1^T\mathbf{A}_b\mathbf{A}_b^T\mathbf{x}_1\le\lambda_{\max}[\mathbf{A}_b\mathbf{A}_b^T]\mathbf{x}_1^T\mathbf{x}_1\le\|\mathbf{A}_b\mathbf{A}_b^T\|_\infty\mathbf{x}_1^T\mathbf{x}_1\le\|\mathbf{A}_b\|_\infty\|\mathbf{A}_b^T\|_\infty\mathbf{x}_1^T\mathbf{x}_1 \quad (12.220)$$

are employed. Then, simple calculations omitted, $\dot{V}(\mathbf{x}_1)$ is a negative-definite function if the condition Eq.(12.217) is satisfied.

End of the Proof

Special cases: Setting $\mathbf{R}=\mathbf{0}$, the *open-loop* system stability, Eq.(12.204), can be checked. Additionally, setting $\mathbf{P}=\mathbf{I}$ and

$$\mathbf{W}=\sqrt{\|\mathbf{M}\|_\infty\|\mathbf{M}^T\|_\infty}\,\mathbf{I}, \quad \text{it results} \quad \bar{\mathbf{A}}+\bar{\mathbf{A}}^T+2\sqrt{\|\mathbf{M}\|_\infty\|\mathbf{M}^T\|_\infty}\,\mathbf{I}<0\,. \quad (12.221)$$

With regard to different assumptions during the derivation the result is similar but not equal to Eq.(12.209).

12.28.5 Open-Loop Discrete-Time System

By similar evaluation, the discrete-time interval matrix $\mathbf{\Phi}_I = \bar{\mathbf{\Phi}} + [-\mathbf{M} \;\; \mathbf{M}]$ is stable if

$$\bar{\mathbf{\Phi}}^T\bar{\mathbf{\Phi}} - \mathbf{I} + \alpha_D\mathbf{I} < 0 \tag{12.222}$$

where

$$\alpha_D = \|\mathbf{M}\|_1\|\bar{\mathbf{\Phi}}\|_\infty + \|\mathbf{M}\|_\infty\|\bar{\mathbf{\Phi}}\|_1 + \|\mathbf{M}\|_1\|\mathbf{M}\|_\infty = 2\sqrt{\|\mathbf{M}\|_1\|\mathbf{M}\|_\infty\|\bar{\mathbf{\Phi}}\|_1\|\bar{\mathbf{\Phi}}\|_\infty} + \|\mathbf{M}\|_1\|\mathbf{M}\|_\infty \; . \tag{12.223}$$

12.29 Matrix Variation Using Kronecker Sum

Consider a perturbed system $\dot{\mathbf{x}}(t) = (\mathbf{A} + \Delta\mathbf{A})\mathbf{x}(t) = \mathbf{A}_p\mathbf{x}(t)$. The Gastinel-Kahan theorem, Eq.(23.8), states that the matrix $\Delta\mathbf{A}$, which may be added to $\mathbf{A}$ to force the matrix $\mathbf{A} + \Delta\mathbf{A}$ singular, is characterized by the induced matrix norm for any p

$$\min \| \;\Delta\mathbf{A}\|_p = \|\mathbf{A}^{-1}\|_p^{-1} \; . \tag{12.224}$$

In other words, if $\|\Delta\mathbf{A}\|_p < \|\mathbf{A}^{-1}\|_p^{-1}$ the system remains stable. Recalling the Kronecker sum, $\mathbf{A}_p \oplus \mathbf{A}_p$ is given by

$$\begin{aligned} \mathbf{A}_p \oplus \mathbf{A}_p &= (\mathbf{A} + \Delta\mathbf{A}) \oplus (\mathbf{A} + \Delta\mathbf{A}) = (\mathbf{A} + \Delta\mathbf{A}) \otimes \mathbf{I} + \mathbf{I} \otimes (\mathbf{A} + \Delta\mathbf{A}) & (12.225)\\ &= \mathbf{A} \otimes \mathbf{I} + \Delta\mathbf{A} \otimes \mathbf{I} + \mathbf{I} \otimes \mathbf{A} + \mathbf{I} \otimes \Delta\mathbf{A} = \mathbf{A} \oplus \mathbf{A} + \Delta\mathbf{A} \oplus \Delta\mathbf{A} \; . & (12.226) \end{aligned}$$

Now, apply a theorem given by *Fu, M., and Barmish, B.R., 1987* which states that the matrix $\mathbf{A}_p + \Delta\mathbf{A}$ is stable if $\mathbf{A}$ is stable and $\mathbf{A}_p \oplus \mathbf{A}_p$ is nonsingular, see Eq.(12.21). Combining this result with Eq. (12.226) and with the Gastinel-Kahan theorem as modified above yields the stability condition (*Soh, C.B., 1989*)

$$\|\Delta\mathbf{A} \oplus \Delta\mathbf{A}\|_p < \|(\mathbf{A} \oplus \mathbf{A})^{-1}\|_p^{-1} \; . \tag{12.227}$$

12.30 Robust Stability and Positivity Test

Consider structured perturbations given by the class

$$\dot{\mathbf{x}}(t) = \mathbf{A}_p(\mathbf{p})\mathbf{x}(t) \qquad \text{where} \quad \mathbf{A}_p(\mathbf{p}) = \mathbf{A}_o(\mathbf{p}) + \sum_{i=1}^{l} f_i(\mathbf{p} - \mathbf{p}_o)\mathbf{A}_i \tag{12.228}$$

and where the matrix $\mathbf{A}_o$ is the nominal coefficient matrix and f_i are scalar-valued functions, $f_i(0) = 0$, $\mathbf{A}_p, \mathbf{A}_o, \mathbf{A}_i \in \mathcal{R}^{n \times n}$. The uncertainty is bounded by a hyperrectangular $\Omega(\rho)$ such that for $\rho > 0$

$$\Omega(\rho) \triangleq \{\mathbf{p} \in \mathcal{R}^q \; : \; \|\mathbf{T}(\mathbf{p} - \mathbf{p}_o)\|_\infty \le \rho \;\} \tag{12.229}$$

where, e.g., $\mathbf{T} = \mathrm{diag}\{w_i^{-1}\}$. Assume $\mathbf{A}_o$ is Γ-stable then the class of matrices $\mathbf{A}_p(\mathbf{p})$ is Γ-stable if

$$\det\Big(\mathbf{I} + \sum_{i=1}^{l} f_i(\mathbf{p} - \mathbf{p}_o)\mathbf{L}_1[\mathbf{A}_i]\mathbf{L}_1^{-1}[\mathbf{A}_o]\Big) > 0 \quad \forall \mathbf{p} \in \Omega(\rho) \tag{12.230}$$

where

$$\mathbf{L}_1[\mathbf{M}] = \mathbf{E}(\mathbf{M} \oplus \mathbf{M})\mathbf{S}\mathbf{E}^T \qquad \mathbf{L}_2[\mathbf{M}] = \mathbf{E}(\mathbf{M} \otimes \mathbf{M})\mathbf{S}\mathbf{E}^T \tag{12.231}$$

$$\mathbf{E} = \text{diag}\{\mathbf{E}_i\}, \quad \mathbf{E}_i = (\mathbf{e}_1, \ldots\ldots, \mathbf{e}_i)^T \in \mathcal{R}^{i \times n} \quad \mathbf{S} = (\mathbf{s}_1, \mathbf{s}_2, \ldots\ldots, \mathbf{s}_{n^2}) \in \mathcal{R}^{n^2 \times n^2} \tag{12.232}$$

$$\mathbf{s}_{n(i-1)+j} = \mathbf{l}_{n(i-1)+j} + (1 - \delta_{ij})\mathbf{l}_{n(j-1)+i} \qquad i,j = 1,2\ldots\ldots n \tag{12.233}$$

and δ_{ij} is the Kronecker delta, $\mathbf{e}_i$ and $\mathbf{l}_i$ are unit vectors $\mathbf{e}_i \in \mathcal{R}^n$, $\mathbf{l}_i \in \mathcal{R}^{n^2}$

$$\lambda_i[\mathbf{L}_1] = \lambda_i[\mathbf{M}] + \lambda_j[\mathbf{M}] \qquad \lambda_i[\mathbf{L}_2] = \lambda_i[\mathbf{M}] \times \lambda_j[\mathbf{M}] \,. \tag{12.234}$$

Eq.(12.230) shows that robust stability can be reduced to a positivity test (*Vicino, A., and Tesi, A., 1989*).

12.31 Minimal Dimension of Stable Faces

The minimal dimension of stable faces to guarantee D-stability of a matrix polytope is of basic theoretical interest. Consider a polytope $\mathcal{P}$ of (n,n)-matrices $\mathbf{A}$. The problem is to determine D-stability of an interval matrix $\mathbf{A}$ by checking stability of low-dimensional faces of the polytope. D-stability is given if each point in $\mathcal{P}$ or of certain faces of $\mathcal{P}$ yields eigenvalues of $\mathbf{A}$ in a specified, open, convex, conjugate-symmetric subset D of the complex plane of the form (*Cobb, J.D., 1988; 1990*)

$$D = \{s \in \mathcal{C} : \quad \alpha < \Re e\ s < \beta, \quad -\infty \leq \alpha < \beta \leq \infty \} \,. \tag{12.235}$$

A polytope $\mathcal{P}$ in a vector space is the convex hull of any finite subset of the vector space. The smallest linear variety containing $\mathcal{P}$ is denoted dimension of $\mathcal{P}$. Given a supporting hyperplane $\mathcal{H}$ of $\mathcal{P}$, a vertex of $\mathcal{P}$ is any singleton $\mathcal{H} \cap \mathcal{P}$. A k-dimensional subpolytope of $\mathcal{P}$ is any k-dimensional polytope which equals the convex hull of some of the vertices of $\mathcal{P}$. A k-dimensional face is a k-dimensional subpolytope of $\mathcal{P}$ contained in the boundary of $\mathcal{P}$. (The 1-dimensional faces are edges.)

A polytope in $\mathcal{R}^n$ is sketched by the coefficient vector of a monic polynomial to be D-stable.

For $n \geq 3$ an integer $m_A = 2n - 4$ is presented by *Cobb, J.D., and DeMarco, C.L., 1988* with the property that checking stability of all m_A-dimensional faces of $\mathcal{P} \in \mathcal{R}^{n \times n}$ is minimal and sufficient to guarantee stability of $\mathcal{P}$. For $n = 2$ the result is $m_A = 1$.

For any other kind of conjugate-symmetric set not satisfying D, the integer m_B is given by $m_B = 2n - 2$. Hence, e.g., Schur stability of interval matrices requires $(2n-2)$-dimensional faces.

12.32 Suggestions for Further Reading

The concept of guarding maps is introduced as a unifying tool to study the generalized stability of one-parameter and two-parameter families of matrices (*Saydy, L., et al. 1988*).

Robustness without making conservative steps and a method for sampling time selection for scalar systems is presented by *Zafiriou, E., and Morari, M., 1986*.

For numerical iterations on large and composite matrices involved in robustness problems see *Qiu, L., and Davison, E.J., 1988a*.

Robustness to structural changes corresponding to coefficient cancellations, breakdown of sensors or actuators using a structured state model and a graphic characterization called cactus approach was investigated by *Rech, R., 1990*.

Noldus, E., 1982 presented a robustly stabilizing state feedback controller for uncertain nonlinear plants of the type

$$\dot{\mathbf{x}} = \mathbf{A}\mathbf{x} + \mathbf{B}\mathbf{u} + \mathbf{D}_1\mathbf{E}(\mathbf{x}, t)\mathbf{D}_2\mathbf{x} \tag{12.236}$$

where $\mathbf{D}_1$, $\mathbf{D}_2$, $\mathbf{R} = \mathbf{R}^T$ are constant matrices. The uncertainty is expressed by the matrix $\mathbf{E}(\mathbf{x}, t)$. It is assumed that $\mathbf{E}(\mathbf{x}, t)$ satisfies a so-called sector-type inequality condition, i.e.,

$$\mathbf{x}^T\mathbf{D}_2^T\mathbf{E}^T(\mathbf{x},t)[\mathbf{I} - \mathbf{R}^{-1}\mathbf{E}(\mathbf{x},t)]\mathbf{D}_2\mathbf{x} \geq 0 \quad \forall \mathbf{x}, \ t \ . \tag{12.237}$$

The approach is based on Lyapunov's direct method.

Sufficient conditions for the *positive definiteness* of symmetric interval matrices are documented by *Hmamed, A., 1991a.*

Chapter 13

Lyapunov-Based Methods for Perturbed Continuous-Time Systems

13.1 Quadratic Forms and Lyapunov Stability

The matrix $\mathbf{U}$ denotes a positive matrix if each entry $u_{ij} > 0 \ \forall i, j$. A matrix is termed positive definite if it has properties in what is to follow. A square matrix $\mathbf{Q}$ is positive definite if

$$\mathbf{x}^T\mathbf{Q}\mathbf{x} > 0 \qquad \forall \mathbf{x} \neq \mathbf{0} \tag{13.1}$$

and positive semidefinite (non-negative definite) if

$$\mathbf{x}^T\mathbf{Q}\mathbf{x} \geq 0 \qquad \forall \mathbf{x}. \tag{13.2}$$

If the relations $>\geq$ are substituted by $<\leq$, respectively, negative definite and negative semidefinite (non-positive definite) forms are established.

The quadratic matrix $\mathbf{Q} = \text{matrix}[q_{ii}]$ is positive [negative] definite if and only if $q_{ij} = q_{ji}$ and the following successive main subdeterminants are positive [negative/positive] according with the following scheme (Sylvester theorem):

$$q_{11} > 0 \ [< 0], \quad \det\begin{pmatrix} q_{11} & q_{12} \\ q_{21} & q_{22} \end{pmatrix} > 0, \quad \det\begin{pmatrix} q_{11} & q_{12} & q_{13} \\ q_{21} & q_{22} & q_{23} \\ q_{31} & q_{32} & q_{33} \end{pmatrix} > 0 \ [< 0] \text{ etc.} . \tag{13.3}$$

Necessary and sufficient conditions for $\mathbf{Q}$ to be positive (negative) semidefinite see e.g. *Strejc, V., 1981*.

The above theorems are also true in the case of Hermite forms $\mathbf{x}^H\mathbf{Q}\mathbf{x}$, where $\mathbf{x} \in \mathcal{C}^n$ and $\mathbf{Q}$ is Hermite ($\mathbf{Q}^H = \mathbf{Q}$, i.e., $q_{ij} = q_{ji}^*$).

An alternative notation for the bilinear form $\mathbf{x}^T\mathbf{Q}\mathbf{y}$ is

$$\mathbf{x}^T\mathbf{Q}\mathbf{y} = \langle \mathbf{x}, \mathbf{Q}\mathbf{y} \rangle = \langle \mathbf{Q}^T\mathbf{x}, \mathbf{y} \rangle = \mathbf{y}^T\mathbf{Q}^T\mathbf{x} = \langle \mathbf{Q}\mathbf{y}, \mathbf{x} \rangle . \tag{13.4}$$

All eigenvalues $\lambda_i[\mathbf{Q}]$ of a positive definite matrix $\mathbf{Q}$ are positive. All eigenvalues of a positive semidefinite matrix are zero or positive (i.e., non-negative), see also Eq.(B.77).

If $\mathbf{Q}$ is a positive semidefinite matrix the expression $\mathbf{B}^T\mathbf{Q}\mathbf{B}$ is positive semidefinite for $\mathbf{B}$ any matrix.

The sum of positive semidefinite matrices is positive semidefinite again.

A quadratic form associated with a matrix $\mathbf{Q}$ structured as the dyadic product of two vectors $\mathbf{Q} = \mathbf{a}\mathbf{a}^T$ always is positive semidefinite

$$\mathbf{x}^T\mathbf{Q}\mathbf{x} = \mathbf{x}^T\mathbf{a}\mathbf{a}^T\mathbf{x} = (\mathbf{a}^T\mathbf{x})^T(\mathbf{a}^T\mathbf{x}) = (\mathbf{a}^T\mathbf{x})^2 \geq 0 . \tag{13.5}$$

In this calculation $(\mathbf{a}^T\mathbf{x})$ is a scalar. With regard to the possibility of $\mathbf{a} = \mathbf{0}$ the quadratic form is only semidefinite. In this case, the quadratic form vanishes even with $\mathbf{x} \neq \mathbf{0}$. Besides this, the form mentioned above can be written as

$$\mathbf{x}^T\mathbf{Q}\mathbf{x} = \text{tr}\ (\mathbf{Q}\mathbf{x}\mathbf{x}^T) = \text{tr}\ (\mathbf{a}\mathbf{a}^T\mathbf{x}\mathbf{x}^T) \geq 0 . \tag{13.6}$$

Example: Symmetric weighting matrix: The scalar quantity γ of the quadratic form with non- symmetric weighting matrix $\mathbf{Q}$ is not altered by transposing the elements of the quadratic form:

$$\gamma = \mathbf{x}^T\mathbf{Q}\mathbf{x} = (\mathbf{x}^T\mathbf{Q}\mathbf{x})^T = \mathbf{x}^T\mathbf{Q}^T\mathbf{x}, \quad \mathbf{Q} = \text{matrix}[q_{ij}] . \tag{13.7}$$

It would be incorrect to claim the equality of the weighting matrix on the equality of the second and fourth term in Eq.(13.7). Symmetry in the quadratic form emerges from the summation procedure as can readily be seen from the arithmetic presentation $\gamma = \sum_i q_{ii}x_i^2 + \sum_{ij}(q_{ij} + q_{ji})x_i x_j$. If $\gamma = \mathbf{x}^T\mathbf{Q}\mathbf{x}$ is to be expressed with a symmetric weighting matrix although $\mathbf{Q}$ is nonsymmetric Eq.(13.7) is rewritten as follows

$$\gamma = \mathbf{x}^T\mathbf{Q}\mathbf{x} = 0.5(\mathbf{x}^T\mathbf{Q}\mathbf{x} + \mathbf{x}^T\mathbf{Q}\mathbf{x}) = 0.5(\mathbf{x}^T\mathbf{Q}\mathbf{x} + \mathbf{x}^T\mathbf{Q}^T\mathbf{x}) = 0.5\ \mathbf{x}^T(\mathbf{Q} + \mathbf{Q}^T)\mathbf{x}\ . \ \square \qquad (13.8)$$

13.2 Bounds For the Lyapunov Function Dynamics

13.2.1 Autonomous System

Assume a stable autonomous system $\dot{\mathbf{x}} = \mathbf{A}\mathbf{x}$ with $\mathbf{A} \in \mathcal{R}^{n \times n}$, first. Defining a positive definite function (Lyapunov function)

$$V = V(\mathbf{x}) = \mathbf{x}^T\mathbf{P}\mathbf{x} \qquad (13.9)$$

the derivative with respect to time is

$$\dot{V}(\mathbf{x}) = -\mathbf{x}^T\mathbf{Q}\mathbf{x} \quad \text{where} \quad -\mathbf{Q} = \mathbf{A}^T\mathbf{P} + \mathbf{P}\mathbf{A}\ . \qquad (13.10)$$

Both $\mathbf{P}$ and $\mathbf{Q}$ are positive definite matrices. If $\mathbf{P}$ and $\mathbf{Q}$ are symmetric an upper and a lower bound of the relation $\dot{V}/V$ and $V[\mathbf{x}(t)]$ can be formulated as follows

$$-\frac{\lambda_{\max}[\mathbf{Q}]}{\lambda_{\min}[\mathbf{P}]} \le \frac{\dot{V}(\mathbf{x})}{V(\mathbf{x})} = -\frac{\mathbf{x}^T\mathbf{Q}\mathbf{x}}{\mathbf{x}^T\mathbf{P}\mathbf{x}} = -\frac{\mathbf{x}^T\mathbf{Q}\mathbf{x}/\mathbf{x}^T\mathbf{x}}{\mathbf{x}^T\mathbf{P}\mathbf{x}/\mathbf{x}^T\mathbf{x}} \le -\frac{\lambda_{\min}[\mathbf{Q}]}{\lambda_{\max}[\mathbf{P}]} \qquad (13.11)$$

$$\exp\{-\frac{\lambda_{\max}[\mathbf{Q}]}{\lambda_{\min}[\mathbf{P}]}t\} \le \frac{V[\mathbf{x}(t)]}{V[\mathbf{x}_o]} \le \exp\{-\frac{\lambda_{\min}[\mathbf{Q}]}{\lambda_{\max}[\mathbf{P}]}t\}\ , \qquad (13.12)$$

i.e., the closer the (positive) eigenvalues $\lambda_{\max}$ and $\lambda_{\min}$ the better the upper and lower bound.

13.2.2 Non-Autonomous System

Considering a non-autonomous stable system $\dot{\mathbf{x}}(t) = \mathbf{A}\mathbf{x}(t) + \mathbf{B}\mathbf{u}(t)$ and the identical Lyapunov function. Then, its derivative is

$$\dot{V}(\mathbf{x}) = -\mathbf{x}^T\mathbf{Q}\mathbf{x} + 2\mathbf{x}^T\mathbf{P}\mathbf{B}\mathbf{u}\ . \qquad (13.13)$$

A control algorithm can be achieved by choosing a controlling variable $\mathbf{u}(t)$ such that the amount of $\dot{V}$ given by Eq.(13.13) is pushed below the lower bound of $\dot{V}$, resulting from Eq.(13.12)

$$\frac{-\mathbf{x}^T\mathbf{Q}\mathbf{x} + 2\mathbf{x}^T\mathbf{P}\mathbf{B}\mathbf{u}}{\mathbf{x}^T\mathbf{P}\mathbf{x}} < -\frac{\lambda_{\max}[\mathbf{Q}]}{\lambda_{\min}[\mathbf{P}]}\ . \qquad (13.14)$$

13.3 Lyapunov Equation and Lyapunov Stability

Consider the linear continuous time-invariant system $\dot{\mathbf{x}}(t) = \mathbf{A}\mathbf{x}(t)$ and the Lyapunov function $V(\mathbf{x}) = \mathbf{x}^T\mathbf{P}\mathbf{x}$ as before, then, the derivative $\dot{V}(\mathbf{x})$ with respect to time is

$$\dot{V}(\mathbf{x}) = \dot{\mathbf{x}}^T\mathbf{P}\mathbf{x} + \mathbf{x}^T\mathbf{P}\dot{\mathbf{x}} = \mathbf{x}^T(\mathbf{A}^T\mathbf{P} + \mathbf{P}\mathbf{A})\mathbf{x} \triangleq -\mathbf{x}^T\mathbf{Q}\mathbf{x}\,. \tag{13.15}$$

The matrices $\mathbf{P}, \mathbf{Q} \in \mathcal{R}^{n\times n}$ must be found such that both $\mathbf{P}$ and $\mathbf{Q}$ are positive definite. Then, the equilibrium state $\mathbf{x} = \mathbf{0}$ is globally and asymptotically stable. Lyapunov theory is already settled choosing *any* positive definite $\mathbf{Q}$ and studying $\mathbf{P}$ whether or not it is positive definite. It must be emphasized that the system is only stable if and only if the Lyapunov equation

$$\mathbf{A}^T\mathbf{P} + \mathbf{P}\mathbf{A} = -\mathbf{Q} \tag{13.16}$$

has a unique and positive definite solution for $\mathbf{P}$. In other cases (no solution, more than one solution, unique but not positive definite solution) not all eigenvalues of $\mathbf{A}$ have negative real parts and the system is not asymptotically stable (*Kalman, R.E., and Bertram, J.E., 1960; Taussky, O., 1964*).

Summarizing: Given the system $\dot{\mathbf{x}} = \mathbf{A}\mathbf{x}$ and the matrix $\mathbf{Q} = \mathbf{Q}^T > 0$ then $\mathbf{P}$ results from the equation $\mathbf{A}^T\mathbf{P} + \mathbf{P}\mathbf{A} = -\mathbf{Q}$. The system and the matrix $\mathbf{A}$ are stable if and only if $\mathbf{P}^T = \mathbf{P} > 0$. If $\mathbf{A}$ is stable the solution is unique.

Proof: Assume $\mathbf{P} > 0$ and define the Lyapunov function $V(\mathbf{x}) \triangleq \mathbf{x}^T\mathbf{P}\mathbf{x}$. Since $\dot{V}(\mathbf{x}) = -\mathbf{x}^T\mathbf{Q}\mathbf{x} < 0$ one has the result:

$$\mathbf{P} > 0 \quad \text{implies} \quad \mathbf{A} \ \text{stable}\,. \tag{13.17}$$

Viceversa, if $\mathbf{A}$ is stable then using $\mathbf{x} = e^{\mathbf{A}t}\mathbf{x}_o$ and Eq.(13.22)

$$\int_0^\infty \mathbf{x}^T\mathbf{Q}\mathbf{x}dt = \mathbf{x}_o^T \int_0^\infty (e^{\mathbf{A}\tau})^T\mathbf{Q}e^{\mathbf{A}\tau}d\tau\, \mathbf{x}_o \quad \rightsquigarrow \quad \mathbf{x}_o^T\mathbf{P}\mathbf{x}_o \tag{13.18}$$

and $\mathbf{Q} > 0$ implies $\mathbf{P} > 0$. □

The Lyapunov equation in $\mathbf{P}$ need not be solved if only an upper bound for $\|\mathbf{P}\|_s$ is required, since an inequality holds using a matrix measure (*Mori, T., 1980*)

$$\|\mathbf{P}\|_s \le \frac{\|\mathbf{Q}\|_s}{-2\mu_s[\mathbf{A}]} \quad \text{in which} \quad \mu_s[\mathbf{A}] < 0\,. \ \square \tag{13.19}$$

A discrete-time system $\mathbf{x}_{k+1} = \mathbf{\Phi}\mathbf{x}_k$ is Schur stable if and only if for $\mathbf{Q} = \mathbf{Q}^T > 0$ the (unique) solution $\mathbf{P}$ of the Lyapunov equation

$$\mathbf{\Phi}^T\mathbf{P}\mathbf{\Phi} - \mathbf{P} = -\mathbf{Q} \tag{13.20}$$

is positive definite. □

Remark: Consider an autonomous system $\dot{\mathbf{x}}(t) = \mathbf{A}\mathbf{x}(t)$. Eq.(13.16) has a unique solution for $\mathbf{P}$ if and only if $\lambda_i[\mathbf{A}] + \lambda_j^*[\mathbf{A}] \neq 0 \quad \forall i,j$ (necessary and sufficient). If $\lambda_i[\mathbf{A}] + \lambda_j^*[\mathbf{A}] \neq 0\ \forall i,j$ and $\mathbf{Q} > 0$ then the number of negative eigenvalues of $\mathbf{P}$ is identical to the number of eigenvalues $\lambda[\mathbf{A}]$ with positive real part (*Vidyasagar, M., 1978*).

Lyapunov functions can be modified to structured ones. For some given $\mathbf{A}$ the structured Lyapunov function yields a convex minimization problem for the numerical solution. The special structure may be block-diagonal similarly to structured singular values (*Boyd, S., and Yang, Q., 1989*).

13.4 Solution of the Lyapunov Equation

Consider the Lyapunov matrix equation Eq.(13.16) as the special case $t \to \infty$ of the linear matrix differential equation

$$\dot{\mathbf{P}}(t) = \mathbf{A}^T\mathbf{P}(t) + \mathbf{P}(t)\mathbf{A} + \mathbf{Q} \qquad \text{where} \qquad \mathbf{P}(0) = \mathbf{0} \,. \tag{13.21}$$

It will be shown that the unique solution of this differential equation is given by

$$\mathbf{P}(t) = \int_0^t e^{\mathbf{A}^T\tau}\, \mathbf{Q}\, e^{\mathbf{A}\,\tau}\, d\tau \,. \tag{13.22}$$

Differentiating Eq.(13.22) with respect to t and substituting into Eq.(13.21) yields the proof

$$e^{\mathbf{A}^T t}\, \mathbf{Q}\, e^{\mathbf{A}t} = \int_0^t (\mathbf{A}^T e^{\mathbf{A}^T\tau}\, \mathbf{Q}\, e^{\mathbf{A}\,\tau} + e^{\mathbf{A}^T\tau}\, \mathbf{Q}\, e^{\mathbf{A}\,\tau}\, \mathbf{A})d\tau + \mathbf{Q} \tag{13.23}$$

$$= \int_0^t \frac{d}{d\tau}(e^{\mathbf{A}^T\tau}\, \mathbf{Q}\, e^{\mathbf{A}\,\tau})d\tau + \mathbf{Q} = (e^{\mathbf{A}^T t}\, \mathbf{Q}\, e^{\mathbf{A}\,t} - \mathbf{Q}) + \mathbf{Q} = e^{\mathbf{A}^T t}\, \mathbf{Q}\, e^{\mathbf{A}\,t} \,. \tag{13.24}$$

If all eigenvalues of $\mathbf{A}$ are in the left-half s-plane then $\mathbf{P}(t) \to 0$ as $t \to \infty$ and Eq.(13.21) approaches Eq.(13.16). Hence, the solution of the algebraic Lyapunov equation Eq.(13.16) is

$$\mathbf{P} = \int_0^\infty e^{\mathbf{A}^T\tau}\, \mathbf{Q}\, e^{\mathbf{A}\,\tau}\, d\tau \,. \tag{13.25}$$

13.5 Positive Definiteness of P

The proof that $\mathbf{P}$ is positive definite for *any* positive definite $\mathbf{Q}$ can be furnished as follows: If $\mathbf{A}$ has eigenvalues throughout with negative real parts and $\mathbf{Q}$ is positive definite then $\mathbf{Q}$ can be written as $\mathbf{Q} = \mathbf{M}^T\mathbf{M}$ where $\mathbf{M}$ is nonsingular, $\mathbf{M} \in \mathcal{R}^{n\times n}$:

$$\mathbf{P} = \int_0^\infty e^{\mathbf{A}^T\tau}\, \mathbf{M}^T\mathbf{M}\, e^{\mathbf{A}\,\tau}\, d\tau \tag{13.26}$$

$$\mathbf{x}^T\mathbf{P}\mathbf{x} = \int_0^\infty \mathbf{x}^T\, e^{\mathbf{A}^T\tau}\, \mathbf{M}^T\mathbf{M}\, e^{\mathbf{A}\,\tau}\, \mathbf{x}\, d\tau = \int_0^\infty \|\, \mathbf{M}\, e^{\mathbf{A}\,\tau}\, \mathbf{x}\, \|_F^2\, d\tau \geq 0 \,. \tag{13.27}$$

The Euclidian norm is always positive, the case $= 0$ only appears if $\mathbf{M}e^{\mathbf{A}\,\tau}\mathbf{x} = \mathbf{0}$. Since $\mathbf{M}$ is nonsingular zeros are only possible if $\mathbf{x} = \mathbf{0}$ which implies that $\mathbf{P}$ is positive definite. Checking whether all the entries $P_{ij} > 0$ serves as a sufficient stability condition.

13.6 Sufficient Stability Conditions Via Lyapunov Approach

Without solving the Lyapunov equation the following sufficient conditions exist:

Let $\mathbf{P} = \mathbf{P}^T > 0$. If $\mathbf{Q} = -(\mathbf{A}^T\mathbf{P} + \mathbf{P}\mathbf{A}) > 0$ then $\mathbf{A}$ is stable.

Let $\mathbf{P} = \mathbf{P}^T \leq 0$. If $\mathbf{Q} = -(\mathbf{A}^T\mathbf{P} + \mathbf{P}\mathbf{A}) > 0$ then $\mathbf{A}$ is unstable.

13.7 Definiteness of a Sum of Matrices

Consider

$$\mathbf{A}(\mathbf{d}) = \mathbf{A}_o + \sum_{i=1}^r d_i\mathbf{A}_i \qquad \text{where} \quad \mathbf{d} = \mathbf{d}(\varepsilon) \in \mathcal{R}^r \tag{13.28}$$

and let ε vary from 0 to 1, i.e.,

$$\mathbf{d}(\varepsilon) = \mathbf{l} + \varepsilon(\mathbf{h} - \mathbf{l}) \quad \forall\, \varepsilon \in [0,1] \,. \tag{13.29}$$

Then,

$$\begin{aligned}
\mathbf{Q}(\mathbf{d}) &= -\mathbf{A}^T(\mathbf{d})\mathbf{P} - \mathbf{P}\mathbf{A}(\mathbf{d}) = -\mathbf{A}_o^T\mathbf{P} - \sum_{i=1}^{r} d_i\mathbf{A}_i^T\mathbf{P} - \mathbf{P}\mathbf{A}_o - \mathbf{P}\sum_{i=1}^{r} d_i\mathbf{A}_i && (13.30)\\
&= -\mathbf{A}_o^T\mathbf{P} - \sum_{i=1}^{r} l_i\mathbf{A}_i^T\mathbf{P} - \varepsilon\sum_{i=1}^{r}(h_i - l_i)\mathbf{A}_i^T\mathbf{P} - \mathbf{P}\mathbf{A}_o - \mathbf{P}\sum_{i=1}^{r} l_i\mathbf{A}_i - \varepsilon\sum_{i=1}^{r}(h_i - l_i)\mathbf{P}\mathbf{A}_i\\
&= \mathbf{Q}(\mathbf{l}) - \varepsilon\sum_{i=1}^{r}(h_i - l_i)(\mathbf{A}_i^T\mathbf{P} + \mathbf{P}\mathbf{A}_i) > 0 \,. && (13.31)
\end{aligned}$$

Assume $\mathbf{A}$ stable in the corners, i.e.,

$$\mathbf{Q}(\mathbf{l}) > 0 \tag{13.32}$$

$$\mathbf{Q}(\mathbf{h}) = \mathbf{Q}(\mathbf{l}) - \sum_{i=1}^{r}(h_i - l_i)(\mathbf{A}_i^T\mathbf{P} + \mathbf{P}\mathbf{A}_i) > 0 \,. \tag{13.33}$$

Since the right-hand side in Eq.(13.31) depends linearly on ε, the assumption above is sufficient for $\mathbf{Q}(\mathbf{d}) > 0 \;\; \forall\, \mathbf{d} = \mathbf{d}(\varepsilon),\; \varepsilon \in [0,1]$ (*Kiendl, H., 1985*).

13.8 Robust Design Via Definiteness of Q

Based on the above theorem, assume $\mathbf{Q} > 0$ and $\mathbf{d}$ as a polytope of parameters with "centre" $\mathbf{l}$ and corners $\mathbf{h}_k$, e.g., $\{\mathbf{d} : \; |d_i| < d_{oi}\}$ and $\mathbf{l} = \mathbf{0}$. The matrix $\mathbf{P}_c$ denotes the solution of

$$\mathbf{A}^T(\mathbf{l})\mathbf{P}_c + \mathbf{P}_c\mathbf{A}(\mathbf{l}) = -\mathbf{Q} \,. \tag{13.34}$$

Then, one has:

- (i) If by Eq.(13.34) $\mathbf{P}_c$ is calculated and

$$\mathbf{P}_c > 0 \quad \text{and if} \quad \mathbf{Q}(\mathbf{d}) = -[\mathbf{A}^T(\mathbf{h}_k)\mathbf{P}_c + \mathbf{P}_c\mathbf{A}(\mathbf{h}_k)] > 0 \qquad \forall\, k \,, \tag{13.35}$$

  i.e., for all corners of the polytope, then $\mathbf{A}(\mathbf{d})$ is stable *for all* $\mathbf{d}$ inside the polytope.

- (ii) If from Eq.(13.34) $\mathbf{P}_c \leq 0$ then $\mathbf{A}(\mathbf{l})$ is unstable. If, additionally, the matrix $\mathbf{Q}(\mathbf{d}) = -[\mathbf{A}^T(\mathbf{d})\mathbf{P}_c + \mathbf{P}_c\mathbf{A}(\mathbf{d})] > 0$ then $\mathbf{A}(\mathbf{d})$ is unstable $\forall\, \mathbf{d}$. □

Via Eq.(13.35) stability robustness can also be determined. Repeatedly applying Eq.(13.35), the stability boundary in the parameter space is approached (*Kiendl, H., 1985*).

13.9 Nonlinear Time-Varying Perturbation

The problem of stability robustness is a very important one for all kinds of control applications. Analysis in time domain is considered in this chapter. For a given nominal system matrix $\mathbf{A}$ the Lyapunov matrix equation usually must be solved. The Lyapunov solution matrix $\mathbf{P}$ as well as bounds based on $\mathbf{P}$ are used to determine upper bounds on the perturbation $\Delta\mathbf{A}$ of a linear system $\mathbf{A}$ which is asymptotically stable in the nominal (unperturbed) operation. In the case of closed-loop systems the matrix $\mathbf{A}$ might be replaced by $\mathbf{F}$.

Consider the system

$$\dot{\mathbf{x}}(t) = \mathbf{A}\mathbf{x}(t) + \mathbf{g}(\mathbf{x},t) \tag{13.36}$$

with time-varying nonlinear perturbation $\mathbf{g}(\mathbf{x},t)$ where $\mathbf{g}(\mathbf{0},t) = \mathbf{0}\ \forall t$. Let the Lyapunov function be $V(\mathbf{x}) = \mathbf{x}^T\mathbf{P}\mathbf{x}$ then its derivative $\dot{V}(\mathbf{x})$ is

$$\dot{V}(\mathbf{x}) = \mathbf{x}^T(\mathbf{A}^T\mathbf{P}+\mathbf{P}\mathbf{A})\mathbf{x} + \mathbf{g}^T(\mathbf{x},t)\mathbf{P}\mathbf{x} + \mathbf{x}^T\mathbf{P}\mathbf{g}(\mathbf{x},t) = \mathbf{x}^T(\mathbf{A}^T\mathbf{P}+\mathbf{P}\mathbf{A})\mathbf{x} + 2\mathbf{g}^T(\mathbf{x},t)\mathbf{P}\mathbf{x}\ . \tag{13.37}$$

If $\mathbf{P}$ is the positive definite solution of the Lyapunov equation

$$\mathbf{A}^T\mathbf{P}+\mathbf{P}\mathbf{A} = -2\mathbf{Q} \qquad \mathbf{Q} > 0 \tag{13.38}$$

the condition $\dot{V}(\mathbf{x}) \leq 0$ requires

$$2\mathbf{g}^T(\mathbf{x},t)\mathbf{P}\mathbf{x} \leq 2\mathbf{x}^T\mathbf{Q}\mathbf{x} \quad \Leftarrow \quad \mathbf{g}^T(\mathbf{x},t)\mathbf{P}\mathbf{x} \leq \lambda_{\min}[\mathbf{Q}]\mathbf{x}^T\mathbf{x} \tag{13.39}$$

where Rayleigh's principle has been applied. Hence,

$$\Leftarrow \quad \|\mathbf{g}(\mathbf{x},t)\|_F\|\mathbf{P}\|_s\|\mathbf{x}\|_F = \|\mathbf{g}(\mathbf{x},t)\|_F\lambda_{\max}[\mathbf{P}]\ \|\mathbf{x}\|_F \leq \lambda_{\min}[\mathbf{Q}]\ \|\mathbf{x}\|_F^2 \tag{13.40}$$

$$\frac{\|\mathbf{g}(\mathbf{x},t)\|_F}{\|\mathbf{x}(t)\|_F} \leq \frac{\lambda_{\min}[\mathbf{Q}]}{\lambda_{\max}[\mathbf{P}]} \triangleq \mu_P[\mathbf{P},\mathbf{Q}] \tag{13.41}$$

(*Patel, R.V., and Toda, M., 1980*).

13.9.1 Maximum Bound $\|\mathbf{g}(\mathbf{x},t)\|_F$

The bound as given in Eq.(13.41) is maximum when $\mathbf{Q} = \mathbf{I}_n$.

This assertion is verified by replacing $\mathbf{Q}$ by $\bar{\mathbf{Q}} = q\mathbf{Q}$ $(q > 0)$, first. Then, the solution $\bar{\mathbf{P}}$ of Eq.(13.38) becomes $\bar{\mathbf{P}} = q\mathbf{P}$, each eigenvalue of $\bar{\mathbf{Q}}$ and $\bar{\mathbf{P}}$ is enlarged by the factor q. The bound of $\|\mathbf{g}(\mathbf{x},t)\|_F$ in Eq.(13.41) remains unaffected by q , i.e.,

$$\mu_P[\bar{\mathbf{P}},\bar{\mathbf{Q}}] = \mu_P[q\mathbf{P},q\mathbf{Q}] = \mu_P[\mathbf{P},\mathbf{Q}]\ . \tag{13.42}$$

Second, for arbitrary positive definite $\mathbf{Q}$ the factor q is chosen equal to $1/\lambda_{\min}[\mathbf{Q}]$. Then, $\lambda_{\min}[\bar{\mathbf{Q}}] = 1$ and the perturbation bound is

$$\mu_P[\bar{\mathbf{P}},\bar{\mathbf{Q}}]\ \mid\ q = 1/\lambda_{\min}[\mathbf{Q}] \quad = \frac{1}{\lambda_{\max}[\bar{\mathbf{P}}]}\ . \tag{13.43}$$

Third, $\mathbf{Q} = \mathbf{I}_n$ is chosen. The corresponding solution of Eq.(13.38) is termed $\mathbf{P}_I$ and the corresponding perturbation bound is $\mu_P[\mathbf{P}_I,\mathbf{I}_n] = 1/\lambda_{\max}[\mathbf{P}_I]$. Invoking Eq.(13.22),

$$\bar{\mathbf{P}} - \mathbf{P}_I = 2\int_0^\infty e^{\mathbf{A}^T\tau}(\bar{\mathbf{Q}} - \mathbf{I}_n)e^{\mathbf{A}\tau}d\tau\ . \tag{13.44}$$

Since $\lambda_{\min}[\bar{\mathbf{Q}}] = 1$ all eigenvalues $\lambda_i[\bar{\mathbf{Q}}] \geq 1$. It follows that $\bar{\mathbf{Q}} - \mathbf{I}_n \geq 0$ which implies $\bar{\mathbf{P}} - \mathbf{P}_I \geq 0$, $\lambda_{\max}[\bar{\mathbf{P}}] \geq \lambda_{\max}[\mathbf{P}_I]$ and the perturbation bound relation

$$\mu_P[\mathbf{P}_I,\mathbf{I}_n] = \frac{1}{\lambda_{\max}[\mathbf{P}_I]} \geq \mu_P[\bar{\mathbf{P}},\bar{\mathbf{Q}}]\ \mid\ q = 1/\lambda_{\min}[\mathbf{Q}] \quad = \frac{1}{\lambda_{\max}[\bar{\mathbf{P}}]}\ . \tag{13.45}$$

Hence, the perturbation factor for any $\mathbf{Q}$ with $\lambda_{\min}[\mathbf{Q}] = 1$ is smaller than the perturbation factor for the choice $\mathbf{Q} = \mathbf{I}_n$. Noting that the perturbation factor is not affected by q the result holds for all $\bar{\mathbf{Q}} > 0$ and not just for those $\bar{\mathbf{Q}}$ with $\lambda_{\min}[\bar{\mathbf{Q}}] = 1$ (*Patel, R.V., and Toda, M., 1980*)

$$\mu_P[\mathbf{P}_I, \mathbf{I}_n] \geq \mu_P[\mathbf{P}, \mathbf{Q}] \ . \tag{13.46}$$

Analogously to Eq.(13.41), for discrete-time systems $\sup_{\mathbf{Q}} \{\lambda_{\min}[\mathbf{Q}]/\lambda_{\max}[\mathbf{P}]\}$ is achieved for $\mathbf{Q} = \mathbf{I}$ where $\mathbf{\Phi}^T\mathbf{P}\mathbf{\Phi} - \mathbf{P} = -\mathbf{Q}, \quad \mathbf{Q} > 0$.

13.9.2 Perturbation Bound and Spectral Abscissa $\alpha_s[\mathbf{A}]$

The eigenvector relation $\mathbf{A}\mathbf{a}_i = \lambda_i \mathbf{a}_i$ is rewritten to $\mathbf{a}_i^T\mathbf{A}^T = \lambda_i\mathbf{a}_i^T$ or $\mathbf{a}_i^H\mathbf{A}^T = \lambda_i^*\mathbf{a}_i^H$ where $\mathbf{a}_i$ is the right eigenvector of $\mathbf{A}$ associated with $\lambda_i[\mathbf{A}]$. Selecting $\mathbf{Q} = \mathbf{I}_n$, pre- and postmultiplying the Lyapunov equation Eq.(13.38) by $\mathbf{a}_i^H$ and $\mathbf{a}_i$, respectively,

$$\mathbf{a}_i^H(\mathbf{A}^T\mathbf{P} + \mathbf{P}\mathbf{A})\mathbf{a}_i = (\lambda_i^* + \lambda_i)\mathbf{a}_i^H\mathbf{P}\mathbf{a}_i = -2\mathbf{a}_i^H\mathbf{a}_i \tag{13.47}$$

$$2\ \Re e\ \lambda_i[\mathbf{A}]\ \mathbf{a}_i^H\mathbf{P}\mathbf{a}_i = -2\mathbf{a}_i^H\mathbf{a}_i \ . \tag{13.48}$$

Using Rayleigh's principle, Eq.(2.5),

$$\lambda_{\max}[\mathbf{P}] \geq \frac{\mathbf{a}_i^H\mathbf{P}\mathbf{a}_i}{\mathbf{a}_i^H\mathbf{a}_i} = -\frac{1}{\Re e\ \lambda_i[\mathbf{A}]} \geq -\frac{1}{\max_i \Re e\ \lambda_i[\mathbf{A}]} \ . \tag{13.49}$$

Combining with Eq.(13.45) with $\mathbf{Q} = \mathbf{I}_n$,

$$\mu_P[\mathbf{P}, \mathbf{I}_n] = \frac{1}{\lambda_{\max}[\mathbf{P}]} \leq -\max_i \Re e\ \lambda_i[\mathbf{A}] = -\alpha_s[\mathbf{A}] \ . \tag{13.50}$$

The spectral abscissa $\alpha_s[\mathbf{A}]$ with the opposite sign, positive for stable $\mathbf{A}$, is an upper limit for the optimal perturbation bound.

In the special case of normal $\mathbf{A}$ the inequality above turns out the equality. Evaluating the solution $\mathbf{P}$ of the Lyapunov equation, see Eq.(13.22),

$$\mathbf{P} = 2\int_0^\infty e^{\mathbf{A}^T\tau}e^{\mathbf{A}\ \tau}d\tau \ . \tag{13.51}$$

Invoking the property $e^{\mathbf{A}^T\tau} \equiv (e^{\mathbf{A}\ \tau})^T$ (see Taylor expansion), the matrix $e^{\mathbf{A}\ \tau}$ is normal. The singular values are

$$\sigma_i[e^{\mathbf{A}\ \tau}] = \sqrt{\lambda_i[e^{\mathbf{A}\ \tau}e^{\mathbf{A}^T\tau}]} = \sqrt{\lambda_i[e^{(\mathbf{A}+\mathbf{A}^T)\tau}]} = \sqrt{e^{\lambda_i[\mathbf{A}+\mathbf{A}^T]\ \tau}} \ . \tag{13.52}$$

Referring to Eq.(22.28),

$$e^{\mathbf{A}\ \tau} = \mathbf{U}\mathbf{\Sigma}_n\mathbf{U}^H = \mathbf{U}(\mathrm{diag}_n \sqrt{e^{\lambda_i[\mathbf{A}+\mathbf{A}^T]\tau}}\ \mathbf{U}^H \ . \tag{13.53}$$

It can be seen that $e^{\mathbf{A}^T\tau}$ yields the same decomposition since only the sum of $\mathbf{A} + \mathbf{A}^T$ is required. Calculating the product for any $\mathbf{A}$,

$$e^{\mathbf{A}\ \tau}e^{\mathbf{A}^T\tau} = \mathbf{U}(\mathrm{diag}_n e^{\lambda_i[\mathbf{A}+\mathbf{A}^T]\tau}\mathbf{U}^H \leq \mathbf{U}e^{(\mathbf{A}+\mathbf{A}^*)\tau}\mathbf{U}^H \ , \tag{13.54}$$

employing $\max_i \lambda_i[\mathbf{A} + \mathbf{A}^T] = \max_i(\lambda_i[\mathbf{A}] + \lambda_i^*[\mathbf{A}])$ from Rayleigh's theorem, Eqs.(2.5) and (B.73), for normal $\mathbf{A}$,

$$\mathbf{P} = 2\mathbf{U}\int_0^\infty e^{(\mathbf{A}^*+\mathbf{A})\tau}\ d\tau\mathbf{U}^H \ . \tag{13.55}$$

If the matrix $\mathbf{P}$ is symmetric positive definite, $\lambda_{max}[\mathbf{P}] = \|\mathbf{P}\|_s$. Since the matrix $\mathbf{U}$ is unitary $\|\mathbf{U}\|_s\|\mathbf{U}^H\|_s = 1$

$$\lambda_{max}[\mathbf{P}] = 2\| \int_0^{\infty} e^{(\mathbf{A}^* + \mathbf{A})\tau} d\tau \|_s = -\frac{1}{\max_i \Re e \ \lambda_i[\mathbf{A}]} \quad (13.56)$$

and the equality holds (*Patel, R.V., and Toda, M., 1980*)

$$\mu_P[\mathbf{P}, \mathbf{I}_n] = -\max_i \Re e \ \lambda_i[\mathbf{A}] = -\alpha_s[\mathbf{A}] \leq \min_i |\lambda_i[\mathbf{A}]| \ . \quad (13.57)$$

Normal matrices have the property of making the most of the perturbation bound provided that the structure and degree of freedom of the system enable the designer to choose a normal matrix $\mathbf{A}$.

13.9.3 Transition Matrix Approach

Suppose that $\mathbf{g}[\mathbf{x}(t), t]$ of Eq.(13.36) is bounded by

$$\frac{\|\mathbf{g}[\mathbf{x}(t), t]\|_F}{\|\mathbf{x}(t)\|_F} \leq \mu_{CN} \triangleq -\frac{\alpha_s[\mathbf{A}]}{\kappa_s[\mathbf{T}]} = -\frac{\max_i \Re e \ \lambda_i[\mathbf{A}]}{\kappa_s} \quad (13.58)$$

where $\kappa_s[\mathbf{T}]$ is the spectral condition number of the modal matrix $\mathbf{T}[\mathbf{A}]$ associated with $\mathbf{A}$ diagonalizing $\mathbf{A}$ and $\mathbf{\Phi}(t)$. It can be verified that μ_{CN} is a sufficient condition for the stability of the perturbed system.

For a given $\mathbf{x}(t_o)$ the solution of Eq.(13.36) in time-domain is

$$\mathbf{x}(t) = e^{\mathbf{A}(t-t_o)} \ \mathbf{x}(t_o) + \int_{t_o}^{t} e^{\mathbf{A}(t-\tau)} \ \mathbf{g}[\mathbf{x}(\tau), \tau] d\tau \ . \quad (13.59)$$

Employing the Frobenius norm,

$$\|\mathbf{x}(t)\|_F \leq \|e^{\mathbf{A}(t-t_o)}\|_s \|\mathbf{x}(t_o)\|_F + \int_{t_o}^{t} \|e^{\mathbf{A}(t-\tau)}\|_s \|\mathbf{g}[\mathbf{x}(\tau), \tau] \ \|_F d\tau \quad (13.60)$$

$$\|\mathbf{x}(t)\|_F \leq \|e^{\mathbf{A}(t-t_o)}\|_s \|\mathbf{x}(t_o)\|_F - \frac{\alpha_s[\mathbf{A}]}{\kappa_s[\mathbf{T}]} \int_{t_o}^{t} \|e^{\mathbf{A}(t-\tau)}\|_s \|\mathbf{x}(\tau)\|_F d\tau \ . \quad (13.61)$$

Using the modal matrix $\mathbf{T}$ to diagonalize $\mathbf{\Phi}$ (assuming $\mathbf{A}$ diagonalizable),

$$e^{\mathbf{A}t} = \mathbf{\Phi}(t) = \mathbf{T} e^{\Lambda t} \mathbf{T}^{-1} \quad \text{where} \quad \Lambda = \text{diag} \ \{\lambda_i[\mathbf{A}]\} \quad (13.62)$$

$$\|e^{\mathbf{A}t}\|_s \leq \|\mathbf{T}\|_s \|\mathbf{T}^{-1}\|_s \|e^{\Lambda t}\|_s = \kappa_s[\mathbf{T}] e^{\alpha_s[\mathbf{A}]t} \quad (13.63)$$

$$\|\mathbf{x}(t)\|_F \leq \kappa_s e^{\alpha_s(t-t_o)} \ \|\mathbf{x}(t_o)\|_F - \frac{\alpha_s}{\kappa_s} \int_{t_o}^{t} \kappa_s e^{\alpha_s(t-\tau)} \ \|\mathbf{x}(\tau)\|_F d\tau \quad (13.64)$$

$$e^{-\alpha_s t} \|\mathbf{x}(t)\|_F \leq \kappa_s e^{-\alpha_s t_o} \|\mathbf{x}(t_o)\|_F + \int_{t_o}^{t} (-\alpha_s) e^{-\alpha_s \tau} \ \|\mathbf{x}(\tau)\|_F d\tau \ . \quad (13.65)$$

Using Bellman-Gronwall Lemma, Eq.(11.19) with $k(\tau) := -\alpha_s$,

$$e^{-\alpha_s t} \ \|\mathbf{x}(t)\|_F \leq \kappa_s e^{-\alpha_s t_o} \ \|\mathbf{x}(t_o)\|_F e^{\int_{t_o}^{t} -\alpha_s d\tau} = \kappa_s e^{-\alpha_s t_o} \|\mathbf{x}(t_o)\|_F e^{-\alpha_s(t-t_o)} \quad (13.66)$$

$$\|\mathbf{x}(t)\|_F \leq \kappa_s[\mathbf{T}] \ \|\mathbf{x}(t_o)\|_F \quad \text{where} \quad \mathbf{T} = \mathbf{T}[\mathbf{\Phi}] \ . \quad (13.67)$$

If $\mathbf{A}$ is normal, $\mathbf{T}[e^{\mathbf{A}t}]$ is unitary, $\|\mathbf{T}\|_s = \|\mathbf{T}^{-1}\|_s = 1$ and $\kappa_s[\mathbf{T}] = 1$. Hence, it results $\mu_{CN} = -\alpha_s[\mathbf{A}]$ which equals μ_P in the case of normal matrices, see Eq.(13.57).

Both stabilization and observation of nonlinear and uncertain systems by means of Lyapunov theory and Bellman-Gronwall lemma is discussed by *Zak, S.H., 1990*.

13.10 Linear Time-Varying Perturbation

Consider the homogeneous equation of a linear system

$$\dot{\mathbf{x}}(t) = \mathbf{A}\mathbf{x}(t) + \Delta\mathbf{A}\mathbf{x}(t) = (\mathbf{A} + \Delta\mathbf{A})\mathbf{x}(t) \qquad \mathbf{x} \in \mathcal{R}^n \tag{13.68}$$

where the perturbation or error matrix $\Delta\mathbf{A}$ is bounded by

$$|\Delta A_{ij}| \leq E_{ij} \qquad e_{\max} \triangleq \max_{i,j} E_{ij} \ . \tag{13.69}$$

The bound notation $\Delta A_{ij} \leq \max_t |\Delta A_{ij}| \triangleq E_{ij}$ equals the formulation given by $|\Delta A_{ij}| \leq E_{ij} \triangleq |E_{ij}|_{\max}$.

13.10.1 Perturbation Matrix Bounded by $\|\mathbf{P}\|_s^{-1}$

The system of Eq.(13.68) is stable if

$$e_{\max} < \frac{1}{n\ \sigma_{\max}[\mathbf{P}]} = \frac{1}{n\|\mathbf{P}\|_s} = \frac{\|\mathbf{P}\|_s^{-1}}{n} \tag{13.70}$$

where $\mathbf{P}$ is the solution of the Lyapunov equation

$$\mathbf{A}^T\mathbf{P} + \mathbf{P}\mathbf{A} = -2\mathbf{I}_n \tag{13.71}$$

(*Patel, R.V., and Toda, M., 1980*). The perturbation is considered known but no distinction is made as to which is the origin of the perturbation. Only the overall maximum $e_{\max}$ according to Eq.(13.69) is bounded by Eq.(13.70).

The result of Eq.(13.70) is obtained in what follows. Comparing Eq.(13.68) and (13.36),

$$\mathbf{g}(\mathbf{x}, t) = \Delta\mathbf{A}(t)\mathbf{x}(t) \quad \rightsquigarrow \quad \|\mathbf{g}(\mathbf{x}, t)\|_F \leq \|\Delta\mathbf{A}(t)\|_s\|\mathbf{x}(t)\|_F \ . \tag{13.72}$$

From Eq.(13.41) and since $\mathbf{P}$ is Hermite

$$\Leftarrow \qquad \|\Delta\mathbf{A}(t)\|_s \leq \mu_P[\mathbf{P}, \mathbf{I}] = \frac{1}{\lambda_{\max}[\mathbf{P}]} = \frac{1}{\sigma_{\max}[\mathbf{P}]} = \|\mathbf{P}\|_s^{-1} \ . \tag{13.73}$$

This condition is satisfied if

$$\|\Delta\mathbf{A}(t)\|_F \leq \|\mathbf{P}\|_s^{-1} \quad \text{or if} \quad n|\Delta A_{ij}| \leq \|\mathbf{P}\|_s^{-1} \quad \text{or if} \quad n e_{\max} \leq \|\mathbf{P}\|_s^{-1} \ . \tag{13.74}$$

If $\mathbf{A}$ is diagonalizable the transition matrix approach yields the condition for robust stability

$$\|\Delta\mathbf{A}(t)\|_s \leq \mu_{CN} \quad \text{or} \quad \|\Delta\mathbf{A}(t)\|_F < \mu_{CN} \quad \text{or} \quad n|\Delta A_{ij}| < \mu_{CN} = \frac{-\alpha_s[\mathbf{A}]}{\kappa_s[\mathbf{T}]} \tag{13.75}$$

where $\mathbf{T}$ is the modal matrix associated with $\mathbf{A}$.

From Bauer-Fike theorem, Eq.(15.43), the perturbed matrix $\mathbf{\Phi}_p = \mathbf{\Phi} + \Delta\mathbf{\Phi}$, where $|\Delta\mathbf{\Phi}| \leq_e \mathbf{E}_d$, is Schur stable if

$$\|\mathbf{E}_d\|_p < \frac{1 - \rho_s[\mathbf{\Phi}]}{\kappa_p[\mathbf{T}_\Phi]} \qquad p = 1, 2, \infty, s, F \tag{13.76}$$

where $\mathbf{T}_\Phi$ is the modal matrix associated with $\mathbf{\Phi}$. For comparison see Eq.(13.75): The distance of the worst eigenvalue to the stability boundary, $-\alpha_s[\mathbf{A}]$ and $1 - \rho_s[\mathbf{\Phi}]$, and the associated condition number of nominal $\mathbf{A}$ and $\mathbf{\Phi}$ determine the norm of the maximum allowable perturbation matrix.

13.10.2 Perturbation Matrix Bounded by Globally Weighted |P|

The system Eq.(13.68) is stable if the error matrix $\Delta\mathbf{A}$ and the Lyapounov solution $\mathbf{P}$, Eqs.(13.69) and 13.71), satisfy

$$e_{\max} < \frac{1}{\sigma_{\max}[\ (\ |\mathbf{P}|\ \mathbf{U}_n)_s]} = \|(\ |\mathbf{P}|\ \mathbf{U}_n)_s\|_s^{-1} \tag{13.77}$$

where $|\mathbf{P}|$ is the modulus matrix (given by the modulus of each entry), $\mathbf{U}_n$ is the unity matrix, i.e. $(\mathbf{U}_n)_{ij} = 1\ \forall i,j$, and the subscript $_s$ to the bracket denotes the symmetric part and $\|\cdot\|_s$ is the spectral norm (*Yedavalli, R.K., 1985b*). Let $V(\mathbf{x}) = \mathbf{x}^T\mathbf{P}\mathbf{x}$ be the Lyapunov function for the system Eq.(13.68) where $\mathbf{P}$ is the symmetric and positive definite solution of Eq.(13.71), i.e., $\mathbf{A}^T\mathbf{P} + \mathbf{P}\mathbf{A} = -2\mathbf{I}_n$. Referring to Eq.(5.115), the derivative $\dot{V}(\mathbf{x})$ with respect to time is

$$\begin{aligned}\dot{V}(\mathbf{x}) &= \frac{d}{dt}\mathbf{x}^T\mathbf{P}\mathbf{x} = \frac{d\mathbf{x}^T}{dt}\mathbf{P}\mathbf{x} + \mathbf{x}^T\mathbf{P}\frac{d\mathbf{x}}{dt} = \mathbf{x}^T(\mathbf{A}^T + \Delta\mathbf{A}^T)\mathbf{P}\mathbf{x} + \mathbf{x}^T\mathbf{P}(\mathbf{A} + \Delta\mathbf{A})\mathbf{x} \\ &= \mathbf{x}^T(\mathbf{A}^T\mathbf{P} + \Delta\mathbf{A}^T\mathbf{P} + \mathbf{P}\mathbf{A} + \mathbf{P}\Delta\mathbf{A})\mathbf{x} = \mathbf{x}^T(-2\mathbf{I}_n + \Delta\mathbf{A}^T\mathbf{P} + \mathbf{P}\Delta\mathbf{A})\mathbf{x}\ .\end{aligned} \tag{13.78}$$

From Eq.(13.77) and by defining $\mathbf{\Delta} \triangleq e_{\max}\mathbf{U}_n$ and $e_{\max} \triangleq \max_{i,j}|\Delta A_{ij}|$,

$$e_{\max}\ \sigma_{\max}[\ (\ |\mathbf{P}|\ \mathbf{U}_n)_s] < 1 \tag{13.79}$$

$$\sigma_{\max}[\ (\ |\mathbf{P}|\ e_{\max}\mathbf{U}_n)_s] = \sigma_{\max}[\ (\ |\mathbf{P}|\ \mathbf{\Delta})_s] < 1\ . \tag{13.80}$$

Invoking the properties

$$\rho_s[\ |\mathbf{M}|\ ] \geq \rho_s[\mathbf{M}] \tag{13.81}$$

and $\rho_s[\mathbf{M}_1] \geq \rho_s[\mathbf{M}_2]$ if $M_{1ij} > M_{2ij}$ where $\mathbf{M}_1$ and $\mathbf{M}_2$ are square and non-negative, it results (*Yedavalli, R.K., 1985b*)

$$\sigma_{\max}[\ (\ |\mathbf{P}\ \Delta\mathbf{A}|\ )_s] < 1 \quad \rightsquigarrow \quad \sigma_{\max}[\ (\mathbf{P}\ \Delta\mathbf{A})_s] < 1\ . \tag{13.82}$$

Since the maximum modulus of a matrix eigenvalue is always smaller then any norm

$$\max|\ \lambda[\ (\mathbf{P}\ \Delta\mathbf{A})_s]\ | < 1 \quad \rightsquigarrow \quad \lambda_i[\ (\mathbf{P}\ \Delta\mathbf{A})_s - \mathbf{I}_n] < 0 \tag{13.83}$$

$$(\mathbf{P}\ \Delta\mathbf{A})_s - \mathbf{I}_n < 0\ \text{ (negative definite)} \quad \rightsquigarrow \quad \frac{\mathbf{P}\ \Delta\mathbf{A} + \Delta\mathbf{A}^T\mathbf{P}}{2} - \mathbf{I}_n < 0\ . \tag{13.84}$$

Combination with Eq.(13.78) yields

$$\dot{V}(\mathbf{x}) < 0 \quad \text{or} \quad (\mathbf{A} + \Delta\mathbf{A})\ \text{ is stable}\ . \tag{13.85}$$

13.10.3 Perturbation Matrix Bounded by Individually Weighted |P|

Lyapunov Equation Based on A

Consider an error matrix $\Delta\mathbf{A}$ where the information E_{ij} is available. A matrix $\mathbf{U}_e$ is defined with entries

$$(\mathbf{U}_e)_{ij} = E_{ij}/e_{\max} = E_{ij}/\max_{i,j} E_{ij}\ . \tag{13.86}$$

Then, a sufficient stability condition is given by

$$e_{\max} < \frac{1}{\sigma_{\max}[\ (\ |\mathbf{P}|\ \mathbf{U}_e)_s]} \tag{13.87}$$

where the subscript $_s$ denoted the symmetric part (*Yedavalli, R.K., 1985b*). Since $\sigma_{\max}[\,(\,|\mathbf{P}|\,\mathbf{U}_e)_s] < \sigma_{\max}[\,(\,|\mathbf{P}|\,\mathbf{U}_n)_s]$ the bound in Eq.(13.87) is higher (better) than the bound in Eq.(13.77). The entry $(\mathbf{U}_e)_{ij}$ is chosen 0 or 1 in order to take into account zero or maximum perturbation and the i,j-position of the perturbation. In Eq.(13.87) the matrix $|\mathbf{P}|$ serves as a weighting matrix in the product $|\mathbf{P}|\mathbf{U}_e$. This is a great advantage since the system behaviour highly depends on the ij-position. Eqs.(13.69) and (13.70) do not require the i,j-position of the perturbation and, thus, the resulting bound is lower than in Eq.(13.87).

When $\Delta\mathbf{A}$ is considered unknown and has to be maximized within the condition $|\Delta\mathbf{A}| \leq_e \mathbf{E}$ then from Eq.(13.87)

$$e_{\max} \leq \frac{1}{\sigma_{\max}[(\,|\mathbf{P}|\,\mathbf{E}/e_{\max})_s]} = \frac{e_{\max}}{\sigma_{\max}[(\,|\mathbf{P}|\,\mathbf{E})_s]} \tag{13.88}$$

the matrix $\mathbf{E}$ may be chosen such that

$$\sigma_{\max}[(\,|\mathbf{P}|\,\mathbf{E})_s] = 1\ . \tag{13.89}$$

Lyapunov Equation Based on the Bound Matrices and Average Matrices

In Eq.(13.87) the solution matrix $\mathbf{P}$ of the Lyapunov equation with the nominal matrix $\mathbf{A}$ is used. Replacing $\mathbf{A}$ in $\mathbf{A}^T\mathbf{P}+\mathbf{P}\mathbf{A} = -2\mathbf{I}$ by the lower and upper bound of an interval matrix $\mathbf{A}_I$, i.e., $\mathbf{L}$ and $\mathbf{H}$, respectively, then the matrices $\mathbf{P}_L$ and $\mathbf{P}_H$ are obtained, i.e.,

$$\mathbf{L}^T\mathbf{P}_L + \mathbf{P}_L\mathbf{L} = -2\mathbf{I} \quad \text{and} \quad \mathbf{H}^T\mathbf{P}_H + \mathbf{P}_H\mathbf{H} = -2\mathbf{I}\ . \tag{13.90}$$

From $\mathbf{A}_I = [\mathbf{L}\ \ \mathbf{H}] = \text{matrix}[a_{ij}]$ and $l_{ij} \leq a_{ij} \leq h_{ij}\ \ \forall i,j = 1,2...n$ the deviation matrix $\mathbf{D} \triangleq \mathbf{H} - \mathbf{L} = \text{matrix}[d_{ij}]$ results. Selecting $d \triangleq \max_{ij} d_{ij}$ and defining a matrix $\mathbf{U}_d$

$$\mathbf{U}_d \triangleq \text{matrix}[\frac{d_{ij}}{d}] = \frac{1}{d}\mathbf{D} \tag{13.91}$$

there exist the following alternative stability conditions

$$d_l < \frac{1}{\sigma_{\max}[(\,|\mathbf{P}_L|\,\mathbf{U}_d)_s]} \quad \text{and} \quad d_h < \frac{1}{\sigma_{\max}[(\,|\mathbf{P}_H|\,\mathbf{U}_d)_s]} \tag{13.92}$$

(*Yedavalli, R.K., 1986*). Since $\mathbf{L}$ and $\mathbf{H}$ need not be symmetric to $\mathbf{A}$ another sufficient result can be achieved (*Jiang, C.L., 1988*) by using the average matrix $(\mathbf{L}+\mathbf{H})/2$ and the Lyapunov equation

$$(\frac{\mathbf{L}+\mathbf{H}}{2})^T\mathbf{P}_M + \mathbf{P}_M(\frac{\mathbf{L}+\mathbf{H}}{2}) = -2\mathbf{I}\ . \tag{13.93}$$

The result $\mathbf{P}_M$ determines the condition

$$d_m < \frac{1}{\sigma_{\max}[(\,|\mathbf{P}_M|\,\mathbf{U}_d)_s]}\ . \tag{13.94}$$

This condition has wider application and yields improved results in comparison to the result of Eqs.(13.87) and (13.92).

13.10.4 Lyapunov Approach Using Interval Matrix Plus Additive $\mathbf{A}_o$

When the interval matrix is given by the representation $\mathbf{A}_I = \mathbf{A}_o + \mathbf{Z}_I$ where $\mathbf{A}_o$ is Hurwitz stable the Lyapunov equation

$$\mathbf{A}_o^T\mathbf{P}_o + \mathbf{P}_o\mathbf{A}_o = -2\mathbf{I} \tag{13.95}$$

is solved, first. Then, the sufficient condition for Hurwitz stability of the system $\mathbf{A}$ (where $\mathbf{A} \in \mathbf{A}_I$, $\mathbf{A} = \mathbf{A}_o + \mathbf{Z}$, $\mathbf{Z} \in \mathbf{Z}_I$) using Eq.(12.80) is given by

$$\|\mathbf{Z}^\vartheta\|_s < \frac{1}{\sigma_{\max}[\mathbf{P}_o]} \ . \tag{13.96}$$

In the case of symmetric interval matrix $\mathbf{Z}_I = [-\mathbf{D}_o \ \ \mathbf{D}_o]$ the sufficient condition can be expressed in terms of $\mathbf{D}_o$

$$\|\mathbf{D}_o\|_s < \frac{1}{\sigma_{\max}[\mathbf{P}_o]} \ . \tag{13.97}$$

13.10.5 Robustness Measure of Lee

Using the orthogonal matrix $\mathbf{U}$ and the polar decomposition of $\mathbf{A}$

$$\mathbf{A} = \mathbf{U}\mathbf{H}_R \qquad \mathbf{A} = \mathbf{H}_L\mathbf{U} \tag{13.98}$$

and assuming that the matrix $\mathbf{U}$ is stable,then

$$\sigma_{\max}[\Delta\mathbf{A}] < -\sigma_{\min}[\mathbf{A}]\cos\theta_{\min} \tag{13.99}$$

is a stability condition (*Lee, W.H., 1982*) where $\theta_{\min}$ is the smallest principal phase of $\mathbf{A}$ measured counter-clockwise from the positive real axis (*Maciejowski, J.M., 1989*).

13.10.6 Negative Definite Symmetric Part $\mathbf{A}_s$

Under the assumption that $\mathbf{A}$ is stable and $\mathbf{A}_s$ negative definite the matrix sum $\mathbf{A} + \Delta\mathbf{A}$ is stable if

$$\sigma_{\max}[\Delta\mathbf{A}] < \sigma_{\min}[\mathbf{A}_s] \ . \tag{13.100}$$

Proof: From Eq.(13.100) it follows $\sigma_{\max}[\Delta\mathbf{A}_s] < \sigma_{\min}[\mathbf{A}_s]$. The left-hand side is replaced by a smaller quantity. The maximum modulus of any eigenvalue is smaller than the maximum singular value (spectral norm). Replacing the right-hand side by a greater expression, a condition is obtained which is necessary for the relation above

$$\Rightarrow \quad |\lambda_i[\Delta\mathbf{A}_s]\,| < -\lambda_{\max}[\mathbf{A}_s] \quad \Rightarrow \quad \lambda_{\max}[\Delta\mathbf{A}_s] < -\lambda_{\max}[\mathbf{A}_s] \ . \tag{13.101}$$

Hence, $\lambda_{\max}[\Delta\mathbf{A}_s] + \lambda_{\max}[\mathbf{A}_s] < 0 \ \Rightarrow$ by Eq.(2.5) $\Delta\mathbf{A}_s + \mathbf{A}_s < 0$ negative definite and

$$\Re e\ \lambda_i[\mathbf{A} + \Delta\mathbf{A}] < 0 \quad \text{or} \quad (\mathbf{A} + \Delta\mathbf{A}) \ \text{is stable} \ . \ \square \tag{13.102}$$

13.10.7 Nonlinear Delayed Perturbation

Consider the system

$$\dot{\mathbf{x}}(t) = \mathbf{A}\mathbf{x} + \mathbf{f}[\mathbf{x}(t - T_d), t] \qquad \text{with} \qquad \| \mathbf{f}[\mathbf{x}(t - T_d), t] \|_F \leq \beta \|\mathbf{x}(t - T_d)\|_F \quad (13.103)$$

as nonlinear perturbation bound. Applying similarity transformation $\mathbf{Tz} = \mathbf{x}$ the Lyapunov equation is given by

$$\mathbf{P}(\mathbf{T}^{-1}\mathbf{A}\mathbf{T}) + (\mathbf{T}^{-1}\mathbf{A}\mathbf{T})^T\mathbf{P} = -\mathbf{Q} \qquad \mathbf{P} > 0, \;\; \mathbf{Q} > 0 \,. \quad (13.104)$$

The system is stable if (*Cheres, E., et al. 1989a*)

$$\beta < \frac{\lambda_{\min}[\mathbf{Q}]\sqrt{\lambda_{\min}[\mathbf{P}]}}{2\;\|\mathbf{T}\|_s\|\mathbf{T}^{-1}\|_s\sqrt{\lambda^3_{\max}[\mathbf{P}]}} \,. \quad (13.105)$$

13.11 Uncertainty With Common Matrix-Valued Factor B

Consider a perturbed multivariable system where the uncertainty matrices $\Delta\mathbf{A}$, $\Delta\mathbf{B}$ and $\mathbf{H}$ contain the same matrix factor $\mathbf{B}$. There exists a common time-variant factor p . The uncertainty matrices $\Delta\mathbf{A}$, $\Delta\mathbf{B}$ and $\Delta\mathbf{C}$ depend on this factor.

$$\dot{\mathbf{x}}(t) = [\mathbf{A} + \Delta\mathbf{A}(p)]\mathbf{x}(t) + [\mathbf{B} + \Delta\mathbf{B}(p)]\mathbf{u}(t) + \mathbf{H}\mathbf{v}(t) \qquad \mathbf{x}(0) = \mathbf{x}_o \quad (13.106)$$

$$\mathbf{y}(t) = [\mathbf{C} + \Delta\mathbf{C}(p)]\mathbf{x}(t) + \mathbf{w}(t) \qquad p = p(t) \qquad \mathbf{B} \in \mathcal{R}^{n\times m}, \mathbf{C} \in \mathcal{R}^{r\times n} \,. \quad (13.107)$$

Assuming that $\mathbf{A} \in \mathcal{R}^{n\times n}$ is stable and that matrix functions $\mathbf{D}(p), \mathbf{E}(p)$ and $\mathbf{H}_1$ exist such that

$$\Delta\mathbf{A}(p) = \mathbf{B}\mathbf{D}(p), \quad \Delta\mathbf{B}(p) = \mathbf{B}\mathbf{E}(p), \quad \mathbf{H} = \mathbf{B}\mathbf{H}_1 \,, \quad (13.108)$$

choose matrices $\mathbf{R} = \mathbf{R}^T \in \mathcal{R}^{n\times n}$ positive definite and $\mathbf{K} \in \mathcal{R}^{m\times r}$. Then, solve for $\mathbf{P} = \mathbf{P}^T > 0$

$$\mathbf{A}^T\mathbf{P} + \mathbf{P}\mathbf{A} + 2\mathbf{Q} = \mathbf{0} \quad (13.109)$$

for a given $\mathbf{Q} = \mathbf{Q}^T > 0$. Define a test matrix

$$\mathbf{\Psi}(p) \triangleq \mathbf{P}\mathbf{B}[\mathbf{I} + \mathbf{E}(p)]\mathbf{\Pi}(p) + \mathbf{\Pi}^T(p)[\mathbf{I} + \mathbf{E}^T(p)]\mathbf{B}^T\mathbf{P} \quad (13.110)$$

$$\text{where} \qquad \mathbf{\Pi}(p) \triangleq -\mathbf{K}[\mathbf{C} + \Delta\mathbf{C}(p)] - \mathbf{R}\mathbf{B}^T\mathbf{P} \,. \quad (13.111)$$

Calculate $\lambda_o = \min_p \lambda_{min}[\mathbf{\Psi}(p)]$. Assume $\lambda_o \geq 0$. Then, the system is denoted practically stable. Under the assumption

$$-\,2\lambda_{\min}[\mathbf{R}] > \theta \triangleq \min_p \lambda_{\min}[\mathbf{E}(p)\mathbf{R} + \mathbf{R}\mathbf{E}^T(p)] \quad (13.112)$$

the output feedback control is $\mathbf{u} = \gamma\mathbf{K}\mathbf{y}$ where

$$\gamma > \gamma_o \triangleq \frac{1}{2\lambda_{\min}[\mathbf{R}] + \theta}\left(\frac{\rho_D^2}{C_1\lambda_{\min}[\mathbf{Q}]} + \frac{\rho_v^2}{C_2}\right) \quad (13.113)$$

$$\rho_D \triangleq \max_p \|\mathbf{D}(p)\|_s, \qquad \rho_v \triangleq \max_v \|\mathbf{v}\|_F, \qquad C_1 \in (0,1) \quad C_2 \in (0,\infty) \,. \quad (13.114)$$

In the special case of state feedback ($\mathbf{C} = \mathbf{I}$) and $\Delta\mathbf{C} = \mathbf{0}$, $\mathbf{K} = -\mathbf{R}\mathbf{B}^T\mathbf{P}$ and $\mathbf{\Psi} = \mathbf{0}$ is a possible solution. For the proof see *Barmish, B.R., et al. 1983; Chen, Y.H., 1987*. An indirect design uses a state observer and a similar concept.

The concept of practical stability for nonlinear uncertain systems was presented by *Barmish, B.R., et al. 1983* and the problem was solved based on Lyapunov derivative and realized as a linear time-invariant feedback law. Robust stabilization of some other class of norm-bounded time-varying uncertainty via linear control can be solved by using a parameter-dependent Riccati equation (*Zhou, K., and Khargonekar, P., 1988*).

A similar method for state delay systems using the min-max controller approach is given by *Cheres, E., et al. 1989*.

13.12 Regions of Stability for Bilinear Systems

13.12.1 Single-Input Single-Output System

Consider the single-input bilinear system

$$\dot{\mathbf{x}}(t) = \mathbf{A}\mathbf{x}(t) + u(t)\mathbf{B}\mathbf{x}(t) + \mathbf{b}u(t)\ , \qquad y(t) = \mathbf{c}^T\mathbf{x}(t) \tag{13.115}$$

with single-output feedback $u(t) = ky(t)$. Then,

$$\dot{\mathbf{x}} = (\mathbf{A} + k\mathbf{b}\mathbf{c}^T)\mathbf{x} + k\mathbf{c}^T\mathbf{x}\mathbf{B}\mathbf{x}\ . \tag{13.116}$$

Suppose $(\mathbf{A} + k\mathbf{b}\mathbf{c}^T)$ stable and diagonizable. Introducing the similarity transformation $\mathbf{z} \triangleq \mathbf{M}^{-1}\mathbf{x}$ or $\mathbf{x} \triangleq \mathbf{M}\mathbf{z}$ and abbreviating $\mathbf{F} \triangleq \mathbf{M}^{-1}(\mathbf{A} + k\mathbf{b}\mathbf{c}^T)\mathbf{M}$,

$$\dot{\mathbf{z}}(t) = \mathbf{F} + \mathbf{M}^{-1}k\mathbf{c}^T\mathbf{M}\mathbf{z}\mathbf{B}\mathbf{M}\mathbf{z} = \mathbf{F} + k(\mathbf{c}^T\mathbf{M}\mathbf{z})\mathbf{M}^{-1}\mathbf{B}\mathbf{M}\mathbf{z} \triangleq \mathbf{A}_1[\mathbf{z}(t)]\ \mathbf{z}(t) \tag{13.117}$$

the solution $\mathbf{z}(t)$ can be bounded by using Eq.(2.103)

$$\|\mathbf{z}(t)\|_F < \|\mathbf{z}(0)\|_F e^{\int_0^t \mu_s[\mathbf{A}_1(\tau)]d\tau} \tag{13.118}$$

where $\mu_s[\cdot]$ is the induced matrix measure. Moreover, the properties

$$\mu_s[\mathbf{A}_1] = \mu_s[\mathbf{F} + k(\mathbf{c}^T\mathbf{M}\mathbf{z})\mathbf{M}^{-1}\mathbf{B}\mathbf{M}] \leq \mu_s[\mathbf{F}] + |k(\mathbf{c}^T\mathbf{M}\mathbf{z})|\bar{\mu}_s \tag{13.119}$$

$$\bar{\mu}_s \triangleq \max\{\mu_s[\mathbf{M}^{-1}\mathbf{B}\mathbf{M}], \mu_s[-\mathbf{M}^{-1}\mathbf{B}\mathbf{M}]\} \tag{13.120}$$

are used. A stable solution of Eq.(13.118) is given if

$$\mu_s[\mathbf{F}] + |k(\mathbf{c}^T\mathbf{M}\mathbf{z})|\ \bar{\mu}_s < 0 \quad \rightsquigarrow \quad k|\mathbf{c}^T\mathbf{M}\mathbf{z}| = k\|\mathbf{c}^T\mathbf{M}\mathbf{z}\|_F < \frac{-\mu_s[\mathbf{F}]}{\bar{\mu}_s} \tag{13.121}$$

$$\Leftarrow \qquad \|\mathbf{z}(t)\|_F < \frac{-\mu_s[\mathbf{F}]}{k\bar{\mu}_s\|\mathbf{c}^T\mathbf{M}\|_F}\ . \tag{13.122}$$

If the norm $\|\mathbf{z}(t)\|_F$ satisfies the condition above the point $\mathbf{z}(0)$ belongs to the domain of attraction. The steady-state $\mathbf{z} = \mathbf{0}$ will be approached. The upper bound for the time constant T_o is given by

$$T_o = -\{\max_{\mathbf{z}}\big(\mu_s[\mathbf{F}] + k(\mathbf{c}^T\mathbf{M}\mathbf{z})\bar{\mu}_s\big)\}^{-1} \tag{13.123}$$

and the transient behaviour is bounded by (*Genesio, R., and Tesi, A., 1988*)

$$\|\mathbf{z}(t)\|_F < \|\mathbf{z}(0)\|_F e^{-t/T_o}\ . \tag{13.124}$$

13.12.2 Multivariable System

Consider the bilinear system (*Mohler, R.R., 1973*)

$$\dot{\mathbf{x}}(t) = \mathbf{A}\mathbf{x}(t) + \mathbf{B}\mathbf{u}(t) + \sum_{i=1}^{m} \mathbf{B}_i u_i(t)\mathbf{x}(t) \qquad \mathbf{A}, \mathbf{B}_i \in \mathcal{R}^{n\times n}, \mathbf{B} \in \mathcal{R}^{n\times m} \tag{13.125}$$

where $u_i(t)$ is a component of the control vector $\mathbf{u}(t)$. Suppose a state controller given by $\mathbf{u}(t) = \mathbf{K}\mathbf{x}(t)$ or $u_i(t) = \mathbf{k}_i^T\mathbf{x}(t)$. Defining $g_i[\mathbf{x}(t)]$ and $\mathbf{G}_i$,

$$g_i[\mathbf{x}(t)] \triangleq \sum_{i=1}^{m} \mathbf{b}_{ij}^T u_i \mathbf{x} = \sum_{i=1}^{m} u_i \mathbf{b}_{ij}^T \mathbf{x} = \sum_{i=1}^{m} \mathbf{k}_i^T \mathbf{x} \mathbf{b}_{ij}^T \mathbf{x} = \mathbf{x}^T \sum_{i=1}^{m} \mathbf{k}_i \mathbf{b}_{ij}^T \mathbf{x} \triangleq \mathbf{x}^T \mathbf{G}_i \mathbf{x} \tag{13.126}$$

$$\mathbf{g}[\mathbf{x}(t)] \triangleq \begin{pmatrix} \mathbf{x}^T\mathbf{G}_1\mathbf{x} \\ \mathbf{x}^T\mathbf{G}_2\mathbf{x} \\ \vdots \\ \mathbf{x}^T\mathbf{G}_n\mathbf{x} \end{pmatrix} \qquad \mathbf{k}_i^T \triangleq (\mathbf{K})_{i.} \in \mathcal{R}^n, \qquad \mathbf{b}_{ij}^T \triangleq (\mathbf{B}_i)_{j.} \in \mathcal{R}^n \tag{13.127}$$

$$\dot{\mathbf{x}}(t) = (\mathbf{A} + \mathbf{B}\mathbf{K})\mathbf{x}(t) + \mathbf{g}[\mathbf{x}(t)] \qquad \mathbf{K} \in \mathcal{R}^{m\times n} . \tag{13.128}$$

Then,

$$\|\mathbf{g}(\mathbf{x})\|_F^2 = \sum_{i=1}^{n} |\mathbf{x}^T\mathbf{G}_i\mathbf{x}|^2 \le \sum_{i=1}^{n} (\|\mathbf{G}_i\|_s \|\mathbf{x}\|_F^2)^2 \ge [\sum_{i=1}^{n} \|\mathbf{G}_i\|_s^2]\ \|\mathbf{x}\|_F^4 \tag{13.129}$$

$$\frac{\|\mathbf{g}(\mathbf{x})\|_F}{\|\mathbf{x}\|_F} \le \sqrt{\sum_{i=1}^{n} \|\mathbf{G}_i\|_s^2}\ \|\mathbf{x}\|_F \le \mu_P . \tag{13.130}$$

The stability region near the origin finally is

$$\|\mathbf{x}\|_F \le \frac{\mu_P}{\sqrt{\sum_{i=1}^{n} \|\mathbf{G}_i\|_s^2}} \tag{13.131}$$

where μ_P is given by Eq.(13.50) as $\mu_P = -\alpha_s[\mathbf{A} + \mathbf{B}\mathbf{K}]$. The perturbation bound μ_P can be replaced by μ_{CN} if $\mathbf{A} + \mathbf{B}\mathbf{K}$ is diagonizable.

13.13 Robustness of Linear Quadratic State Controllers

Consider a system optimized with respect to the performance index I

$$\dot{\mathbf{x}}(t) = \mathbf{A}\mathbf{x}(t) + \mathbf{B}\mathbf{u}(t) \qquad I = \int_0^\infty e^{2\alpha t}[\mathbf{x}^T(t)\mathbf{Q}\mathbf{x}(t) + \mathbf{u}^T(t)\mathbf{R}\mathbf{u}(t)]dt \tag{13.132}$$

where $\mathbf{Q} \ge 0, \mathbf{R} > 0$ and $\alpha \ge 0$ is the degree of stability. The optimal control variable is

$$\mathbf{u}(t) = -\mathbf{R}^{-1}\mathbf{B}^T\mathbf{P}\mathbf{x}(t) \quad \text{where} \tag{13.133}$$

$$(\mathbf{A} + \alpha\mathbf{I})^T\mathbf{P} + \mathbf{P}(\mathbf{A} + \alpha\mathbf{I}) - \mathbf{P}\mathbf{B}\mathbf{R}^{-1}\mathbf{B}^T\mathbf{P} + \mathbf{Q} = 0 . \tag{13.134}$$

In the case $\mathbf{R} = \mathbf{I}$ the optimal controller can be modified to obtain robust facilities. Replacing $\mathbf{R}^{-1}$ by $1 + \gamma$, a norm bounded optimal controller can be achieved taking into account that $\mathbf{B}$ is perturbed to $\mathbf{B} + \Delta\mathbf{B}$ (*Leitmann, G., et al. 1986*).

13.13.1 Nonlinear Perturbation

The dynamic behaviour including an additional nonlinear excitation $\mathbf{g}[\mathbf{x}(t)]$ is given by

$$\dot{\mathbf{x}}(t) = (\mathbf{A} - \mathbf{B}\mathbf{R}^{-1}\mathbf{B}^T\mathbf{P})\mathbf{x}(t) + \mathbf{g}[\mathbf{x}(t)] \triangleq \mathbf{F}\mathbf{x}(t) + \mathbf{g}[\mathbf{x}(t)] \; . \tag{13.135}$$

Studying bounds for closed-loop stability, a Lyapunov function $V(\mathbf{x}) = \mathbf{x}^T\mathbf{P}\mathbf{x}$ is chosen using the solution $\mathbf{P}$ of the Riccati equation. Since $\mathbf{P} > 0, V(\mathbf{x}) > 0 \; \forall \mathbf{x} \neq \mathbf{0}$.The first derivative of $V(\mathbf{x})$ with respect to time is

$$\dot{V}(\mathbf{x}) = \dot{\mathbf{x}}^T\mathbf{P}\mathbf{x} + \mathbf{x}^T\mathbf{P}\dot{\mathbf{x}} = \mathbf{x}^T(\mathbf{F}^T\mathbf{P} + \mathbf{P}\mathbf{F})x + 2\mathbf{g}^T\mathbf{P}\mathbf{x} \; . \tag{13.136}$$

With the abbreviations

$$\mathbf{Q}_c \triangleq 2\mathbf{Q} + (\mathbf{A} + \alpha\mathbf{I})^T\mathbf{P} + \mathbf{P}(\mathbf{A} + \alpha\mathbf{I}) \tag{13.137}$$

$$\kappa_s(\mathbf{P}) \triangleq \|\mathbf{P}\|_s\|\mathbf{P}^{-1}\|_s \quad \text{(spectral condition number)} \tag{13.138}$$

the closed-loop system is stable if the nonlinear excitation function $\mathbf{g}(\mathbf{x})$ is bounded by (*Patel, R.V., et al. 1977; Abdul-Wahab, A.A., 1990b*)

$$\frac{\|\mathbf{g}(\mathbf{x})\|_F}{\|\mathbf{x}\|_F} \leq \frac{1}{2\|\mathbf{Q}_c^{-1}\|_s\|\mathbf{P}\|_s} + \frac{\alpha}{\kappa_s[\mathbf{P}]} \; . \tag{13.139}$$

Proof:

(i) Combining the definition of $\mathbf{Q}_c$ with the Riccati equation, it results

$$\mathbf{Q}_c = \mathbf{Q} + \mathbf{P}\mathbf{B}\mathbf{R}^{-1}\mathbf{B}^T\mathbf{P} > 0 \quad \text{and} \quad \lambda_{\min}[\mathbf{Q}_c] > 0 \; . \tag{13.140}$$

(ii) Since $\mathbf{Q}_c - \mathbf{Q} \geq 0, \; \lambda_{\min}[\mathbf{Q}_c] \geq \lambda_{\min}[\mathbf{Q}] > 0$.

(iii) Since $\mathbf{P} > 0$ and symmetric, $\mathbf{P}$ can be decomposed into the product $\mathbf{P} \triangleq \mathbf{W}^T\mathbf{W}$ and the quadratic form is

$$\mathbf{x}^T\mathbf{P}\mathbf{x} = \mathbf{x}^T\mathbf{W}^T\mathbf{W}\mathbf{x} = (\mathbf{W}\mathbf{x})^T(\mathbf{W}\mathbf{x}) > 0 \; , \tag{13.141}$$

i.e. is positive definite. Then, the singular value of $\mathbf{P}$ is

$$\sigma_i[\mathbf{P}] = \lambda_i[\sqrt{\mathbf{P}^T\mathbf{P}}] = \lambda_i[\sqrt{\mathbf{P}\mathbf{P}}] = \lambda_i[\mathbf{P}] \; . \tag{13.142}$$

(iv) Since $\mathbf{Q}_c > 0, \lambda[\mathbf{Q}_c^{-1}] = 1/\lambda[\mathbf{Q}_c]$. Moreover

$$\lambda_i[\mathbf{Q}_c^{-1}] = 1/\lambda_i[\mathbf{Q}_c] = 1/\sigma_i[\mathbf{Q}_c] \tag{13.143}$$

$$\lambda_{\min}[\mathbf{Q}_c^{-1}] = 1/\lambda_{\max}[\mathbf{Q}_c] = 1/\sigma_{\max}[\mathbf{Q}_c] = \frac{1}{\|\mathbf{Q}_c\|_s} \tag{13.144}$$

$$\lambda_{\max}[\mathbf{Q}_c^{-1}] = 1/\lambda_{\min}[\mathbf{Q}_c^{-1}] = \sigma_{\max}[\mathbf{Q}_c^{-1}] = \|\mathbf{Q}_c^{-1}\|_s = \sigma_{\max}[\mathbf{Q}_c^{-1}] = \lambda_{\max}[\mathbf{Q}_c^{-1}] = 1/\lambda_{\min}[\mathbf{Q}_c] \; . \tag{13.145}$$

Therefore, Eq.(13.139) is

$$\frac{\|\mathbf{g}(\mathbf{x})\|_F}{\|\mathbf{x}\|_F} \leq \frac{\lambda_{\min}[\mathbf{Q}_c]}{2\lambda_{\max}[\mathbf{P}]} + \frac{\alpha\lambda_{\min}[\mathbf{P}]}{\lambda_{\max}[\mathbf{P}]} \tag{13.146}$$

$$\text{or} \quad \|\mathbf{g}(\mathbf{x})\|_F \leq \Big(0.5\,\lambda_{\min}[\mathbf{Q}_c] + \alpha\lambda_{\min}[\mathbf{P}]\Big)\|\mathbf{x}\|_F/\lambda_{\max}[\mathbf{P}] \; . \tag{13.147}$$

Substituting $\|\mathbf{g}(\mathbf{x})\|_F$ from above into the relation

$$\mathbf{g}^T(\mathbf{x})\mathbf{P}\mathbf{x} \leq \|\mathbf{g}(\mathbf{x})\|_F\|\mathbf{P}\mathbf{x}\|_F \leq \|\mathbf{g}(\mathbf{x})\|_F\|\mathbf{P}\|_s\|\mathbf{x}\|_F = \|\mathbf{g}(\mathbf{x})\|_F\|\mathbf{x}\|_F\lambda_{\max}[\mathbf{P}] \; , \tag{13.148}$$

$$\mathbf{g}^T(\mathbf{x})\mathbf{P}\mathbf{x} < \Big(0.5\,\lambda_{\min}[\mathbf{Q}_c] + \alpha\lambda_{\min}[\mathbf{P}]\Big)\|\mathbf{x}\|_F^2 \; . \tag{13.149}$$

Substituting Eqs.(13.134) and (13.137) into Eq.(13.136) yields

$$\dot{V}(\mathbf{x}) = -\mathbf{x}^T(\mathbf{Q}_c + 2\alpha\mathbf{P})\mathbf{x} + 2\mathbf{g}^T\mathbf{P}\mathbf{x} \,. \tag{13.150}$$

Replacing $2\mathbf{g}^T\mathbf{P}\mathbf{x}$ by a greater number as given by Eq.(13.149) yields a sufficient condition for Eq.(13.150)

$$\dot{V}(\mathbf{x}) = -\mathbf{x}^T(\mathbf{Q}_c + 2\alpha\mathbf{P})\mathbf{x} + \Big(\lambda_{\min}[\mathbf{Q}_c] + 2\alpha\lambda_{\min}[\mathbf{P}]\Big)\mathbf{x}^T\mathbf{x} \tag{13.151}$$

$$\dot{V}(\mathbf{x}) = -\mathbf{x}^T\Big(\mathbf{Q}_c - \mathbf{I}\lambda_{\min}[\mathbf{Q}_c] + 2\alpha(\mathbf{P} - \mathbf{I}\lambda_{\min}[\mathbf{P}])\Big)\mathbf{x} \,. \tag{13.152}$$

Since $\mathbf{x}^T\mathbf{Q}_c\mathbf{x} > \mathbf{x}^T\lambda_{\min}[\mathbf{Q}_c]\mathbf{x}$, $\dot{V}(\mathbf{x})$ is negative definite if $\alpha > 0$ $\forall\mathbf{x}$ and the closed-loop system with nonlinearity $\mathbf{g}(\mathbf{x})$ is stable. □

13.13.2 Linear Perturbation

If the perturbation $\mathbf{g}(\mathbf{x})$ is specialized to $\Delta\mathbf{A}\mathbf{x}(t)$ and $\mathbf{B}$ is considered changed to $\mathbf{B}+\Delta\mathbf{B}$

$$\mathbf{g}(\mathbf{x}) = \Delta\mathbf{A}\mathbf{x}(t) + \Delta\mathbf{B}\mathbf{u}(t) = (\Delta\mathbf{A} + \Delta\mathbf{B}\ \mathbf{K})\mathbf{x}(t) \tag{13.153}$$

then the system equation is

$$\dot{\mathbf{x}}(t) = (\mathbf{A} + \Delta\mathbf{A})\mathbf{x}(t) + (\mathbf{B} + \Delta\mathbf{B})\mathbf{u}(t) = [\mathbf{A} + \Delta\mathbf{A} + (\mathbf{B} + \Delta\mathbf{B})\mathbf{K}]\mathbf{x}(t) \,. \tag{13.154}$$

The system matrix $\mathbf{A} + \mathbf{B}\mathbf{K}$ is perturbed with the additional matrix $\Delta\mathbf{A} + \Delta\mathbf{B}\ \mathbf{K}$. From Eq.(13.153)

$$\begin{aligned} \frac{\|\mathbf{g}(\mathbf{x})\|_F}{\|\mathbf{x}\|_F} &= \frac{\|(\Delta\mathbf{A} + \Delta\mathbf{B}\ \mathbf{K})\mathbf{x}\|_F}{\|\mathbf{x}\|_F} < \sup_{\|\mathbf{x}\|} \frac{\|(\Delta\mathbf{A} + \Delta\mathbf{B}\ \mathbf{K})\mathbf{x}\|_F}{\|\mathbf{x}\|_F} = \|\Delta\mathbf{A} + \Delta\mathbf{B}\mathbf{K}\|_s \\ &\le \|\Delta\mathbf{A}\|_s + \|\Delta\mathbf{B}\|_s\|\mathbf{K}\|_s \,. \end{aligned} \tag{13.155}$$

Invoking Eq.(13.139), a sufficient condition for stability is

$$\|\Delta\mathbf{A}\|_s + \|\Delta\mathbf{B}\|_s\|\mathbf{K}\|_s \le \frac{1}{2\|\mathbf{Q}_c^{-1}\|_s\|\mathbf{P}\|_s} + \frac{\alpha}{\kappa_s[\mathbf{P}]} \,. \tag{13.156}$$

Since $\|\mathbf{G}\|_F \ge \|\mathbf{G}\|_s$ in order to facilitate computation, the left-hand side can be replaced by Frobenius norms of the perturbation matrices $\|\Delta\mathbf{A}\|_F + \|\Delta\mathbf{B}\|_F\|\mathbf{K}\|_s$. If $\Delta\mathbf{A}$ and $\Delta\mathbf{B}$ are structured and the elements individually bounded,

$$|\Delta A_{ik}| \le e_{\max}, \qquad |\Delta B_{ij}| < e_{B\max} \quad \forall i,k = 1\ldots n, \forall j = 1\ldots m \tag{13.157}$$

then the properties of the Frobenius norm $\|\Delta\mathbf{A}\|_F < n e_{\max}$ and $\|\Delta\mathbf{B}\|_F < \sqrt{nm}\ e_{B\max}$ can be used.

13.13.3 Perturbation Bounds and Closed-Loop Behaviour

With the definition $\mathbf{F} = \mathbf{A} - \mathbf{B}\mathbf{R}^{-1}\mathbf{B}^T\mathbf{P}$ it results

$$\mathbf{F}^T\mathbf{P} + \mathbf{P}\mathbf{F} = \mathbf{A}^T\mathbf{P} + \mathbf{P}\mathbf{A} - 2\mathbf{P}\mathbf{B}\mathbf{R}^{-1}\mathbf{B}^T\mathbf{P} \,. \tag{13.158}$$

Substituting, first, $\mathbf{Q}$ from Eq.(13.137) into (13.134) and, second, $\mathbf{A}^T\mathbf{P} + \mathbf{P}\mathbf{A}$ from Eq.(13.158) into Eq.(13.134) yields (calculations omitted)

$$\mathbf{F}^T\mathbf{P} + \mathbf{P}\mathbf{F} = -\mathbf{Q}_c - 2\alpha\mathbf{P} \,. \tag{13.159}$$

Pre- and post-multiplying Eq.(13.159) by $\mathbf{f}^H$ and $\mathbf{f}$ where $\mathbf{f}$ is the eigenvector associated with the dominant eigenvalue $\lambda[\mathbf{F}]$ of the closed loop

$$\mathbf{f}^H\mathbf{F}^T\mathbf{P}\mathbf{f} + \mathbf{f}^H\mathbf{P}\mathbf{F}\mathbf{f} = -\mathbf{f}^H\mathbf{Q}_c\mathbf{f} - 2\alpha\mathbf{f}^H\mathbf{P}\mathbf{f} \ . \tag{13.160}$$

Using the eigenvector definition $\mathbf{F}\mathbf{f} = \lambda[\mathbf{F}]\mathbf{f}$, its conjugate transpose and since $\mathbf{F}$ is real

$$\lambda^*[\mathbf{F}]\mathbf{f}^H\mathbf{P}\mathbf{f} + \lambda[\mathbf{F}]\mathbf{f}^H\mathbf{P}\mathbf{f} = 2(\max \Re e\ \lambda[\mathbf{F}])\mathbf{f}^H\mathbf{P}\mathbf{f} = -\mathbf{f}^H\mathbf{Q}_c\mathbf{f} - 2\alpha\mathbf{f}^H\mathbf{P}\mathbf{f} \tag{13.161}$$

$$-\max \Re e\ \lambda[\mathbf{F}] = \frac{\mathbf{f}^H\mathbf{Q}_c\mathbf{f}}{2\mathbf{f}^H\mathbf{P}\mathbf{f}} + \alpha \ . \tag{13.162}$$

Since $\mathbf{Q}_c$ and $\mathbf{P}$ are positive definite, $\mathbf{P}$ and $\mathbf{Q}_c$ are symmetric matrices, their spectrum is real and the Rayleigh quotient is bounded according to

$$\lambda_{\min}[\mathbf{Q}_c] \le \frac{\mathbf{x}^H\mathbf{Q}_c\mathbf{x}}{\mathbf{x}^H\mathbf{x}} \le \lambda_{\max}[\mathbf{Q}_c] \ . \tag{13.163}$$

Hence, from Eq.(13.162) it follows

$$-\max \Re e\ \lambda[\mathbf{F}] > \frac{\lambda_{\min}[\mathbf{Q}_c]}{2\lambda_{\max}[\mathbf{P}]} + \alpha \ . \tag{13.164}$$

Since the condition number $\kappa_s[\mathbf{P}]$ always exceeds unity, an expression is obtained as given by the bound in the right-hand side of Eq.(13.139)

$$-\max \Re e\ \lambda[\mathbf{F}] \ge \frac{1}{2\|\mathbf{Q}_c^{-1}\|_s\|\mathbf{P}\|_s} + \frac{\alpha}{\kappa_s[\mathbf{P}]} \ . \tag{13.165}$$

The quantity in the right-hand side of Eq.(13.139) is both an upper bound for the system nonlinearity and a minimum distance of the closed-loop poles to the imaginary axis (*Patel, R.V., et al. 1977*).

13.14 Linear Perturbation Via Constant Matrices

Consider the perturbed system

$$\dot{\mathbf{x}}(t) = (\mathbf{A} + \sum_{i=1}^{n_E} p_i\mathbf{E}_i)\mathbf{x}(t) \qquad \mathbf{A}, \mathbf{E}_i \in \mathcal{R}^{n\times n} \tag{13.166}$$

where $\mathbf{E}_i$ are constant perturbation matrices and the matrix $\mathbf{A}$ denotes the stable nominal (unperturbed) system and p_i are unknown perturbation parameters. Bounds on p_i have to be chosen such that the perturbed system remains stable. The Lyapunov function is $V(\mathbf{x}) = \mathbf{x}^T\mathbf{P}\mathbf{x}$ where $\mathbf{P}$ is positive definite and the solution of the Lyapunov equation $\mathbf{A}^T\mathbf{P} + \mathbf{P}\mathbf{A} = -\mathbf{Q}$ for some symmetric $\mathbf{Q} > 0$. Defining

$$\mathbf{H}_i \triangleq \mathbf{E}_i^T\mathbf{P} + \mathbf{P}\mathbf{E}_i \ , \tag{13.167}$$

the time derivative $\dot{V}(\mathbf{x})$ is

$$\dot{V}(\mathbf{x}) = -\mathbf{x}^T\mathbf{Q}\mathbf{x} + \mathbf{x}^T(\sum_{i=1}^{n_E} p_i\mathbf{H}_i)\mathbf{x} \ . \tag{13.168}$$

Using

$$\mathbf{x}^T\mathbf{R}\mathbf{x} = \mathbf{x}^T\mathbf{Q}^{1/2}\mathbf{Q}^{-1/2}\ \mathbf{R}\ \mathbf{Q}^{-1/2}\mathbf{Q}^{1/2}\mathbf{x} = (\mathbf{Q}^{1/2}\mathbf{x})^T\mathbf{Q}^{-1/2}\ \mathbf{R}\ \mathbf{Q}^{-1/2}(\mathbf{Q}^{1/2}\mathbf{x}) \tag{13.169}$$

and $\mathbf{x}^T\mathbf{M}\mathbf{x} \le \sigma_{\max}[\mathbf{M}]\mathbf{x}^T\mathbf{x}$, it follows

$$\dot{V}(\mathbf{x}) \le -\left\{1 - \sigma_{\max}[\sum_{i=1}^{n_E} p_i\mathbf{Q}^{-1/2}\mathbf{H}_i\mathbf{Q}^{-1/2}]\right\}\ \|\mathbf{Q}^{1/2}\mathbf{x}\|_F^2 \tag{13.170}$$

and

$$\sigma_{\max}[\sum_{i=1}^{n_E} p_i\mathbf{Q}^{-1/2}\mathbf{H}_i\mathbf{Q}^{-1/2}] < 1 \tag{13.171}$$

is a sufficient stability condition.

A similar concept using the matrix measure is suggested by *Wang, W.J., et al. 1991*.

13.14.1 Choosing $\mathbf{Q} = \mathbf{I}$

Solving the Lyapunov equation for $\mathbf{Q} = \mathbf{I}$ the results are denoted $\mathbf{P}_I$ and $\mathbf{H}_{Ii}$. Then, sufficient conditions to satisfy Eq.(13.171) are

$$\sigma_{\max}[\sum_{i=1}^{n_E} p_i\mathbf{H}_{Ii}] < 1 \quad \Leftarrow \quad \sum_{i=1}^{n_E} |p_i|\sigma_{\max}[\mathbf{H}_{Ii}] < 1 \tag{13.172}$$

$$\text{or} \quad \Leftarrow \quad \max_i |p_i| \ < \ \sigma_{\max}^{-1}[\ \sum_{i=1}^{n_E} |\mathbf{H}_{Ii}|\ ] \tag{13.173}$$

$$\text{or} \quad \Leftarrow \quad \sum_{i=1}^{n_E} p_i^2 \ < \ \sigma_{\max}^{-1}[\ \sum_{i=1}^{n_E} \mathbf{H}_{Ii}^2\ ]\ . \tag{13.174}$$

13.14.2 Using Additional Singular-Value Properties

Using Eq.(22.69), alternative bounds are obtained, e.g.,

$$\sum_{i=1}^{n_E} |p_i|\sigma_{\max}[\mathbf{H}_i]\sigma_{\max}[\mathbf{Q}^{-1}] < 1 \quad \rightsquigarrow \quad \sum_{i=1}^{n_E} |p_i|\sigma_{\max}[\mathbf{H}_i] < \sigma_{\min}[\mathbf{Q}]\ . \tag{13.175}$$

13.14.3 Scaling the Uncertain Parameters

Introducing scaling factors α_i and setting

$$p_i' = p_i/\alpha_i \qquad \mathbf{E}_i' = \alpha_i\mathbf{E}_i \qquad \text{where} \qquad \alpha_i > 0 \qquad \text{and} \qquad \sum_{i=1}^{n_E} \alpha_i = 1\ , \tag{13.176}$$

the bound Eq.(13.173) becomes

$$\max_i |\frac{p_i}{\alpha_i}| = \sigma_{\max}^{-1}[\ \sum_{i=1}^{n_E} \alpha_i|\mathbf{Q}^{-1/2}\mathbf{H}_i\mathbf{Q}^{-1/2}|\ ]\ . \tag{13.177}$$

13.14.4 Scaling the Matrices Involved

Introducing a nonsingular scaling matrix $\mathbf{T}$, the result can be rewritten to

$$\sigma_{\max}[\sum_{i=1}^{n_E} p_i \mathbf{T}^T \mathbf{H}_i \mathbf{T}] < \sigma_{\min}[\mathbf{T}^T \mathbf{Q} \mathbf{T}] \ . \tag{13.178}$$

This approach considerably reduces the conservativeness of the aforementioned bounds (*Sezer, M.E., and Šiljak, D.D., 1989*). In the case of block-diagonal structure of $\mathbf{A}$, similar sufficient conditions can be derived.

13.15 Extensive Linear Perturbation. Robustification

13.15.1 Dynamic Output and State Feedback Controller

Plant:

$$\dot{\mathbf{x}}(t) = \mathbf{A}\mathbf{x}(t) + \mathbf{B}\mathbf{u}(t), \qquad \mathbf{x}(0) = \mathbf{x}_o \tag{13.179}$$

$$\mathbf{y}(t) = \mathbf{C}\mathbf{x}(t) \tag{13.180}$$

Controller:

$$\dot{\mathbf{x}}_c(t) = \mathbf{A}_c\mathbf{x}_c(t) + \mathbf{B}_c\mathbf{y}(t) \tag{13.181}$$

$$\mathbf{u}(t) = \mathbf{C}_c\mathbf{x}_c(t) + \mathbf{D}_c\mathbf{y}(t) \tag{13.182}$$

Closed-loop system:

$$\begin{aligned} \begin{pmatrix} \dot{\mathbf{x}} \\ \dot{\mathbf{x}}_c \end{pmatrix} &= \begin{pmatrix} \mathbf{A}+\mathbf{B}\mathbf{D}_c\mathbf{C} & \mathbf{B}\mathbf{C}_c \\ \mathbf{B}_c\mathbf{C} & \mathbf{A}_c \end{pmatrix} \begin{pmatrix} \mathbf{x} \\ \mathbf{x}_c \end{pmatrix} = [\begin{pmatrix} \mathbf{A} & 0 \\ 0 & 0 \end{pmatrix} + \begin{pmatrix} \mathbf{B} & 0 \\ 0 & \mathbf{I} \end{pmatrix} \begin{pmatrix} \mathbf{D}_c & \mathbf{C}_c \\ \mathbf{B}_c & \mathbf{A}_c \end{pmatrix} \begin{pmatrix} \mathbf{C} & 0 \\ 0 & \mathbf{I} \end{pmatrix}] \begin{pmatrix} \mathbf{x} \\ \mathbf{x}_c \end{pmatrix} \\ &\triangleq (\mathbf{A}_t + \mathbf{B}_t\mathbf{K}_t\mathbf{C}_t) \begin{pmatrix} \mathbf{x} \\ \mathbf{x}_c \end{pmatrix} \end{aligned} \tag{13.183}$$

$\mathbf{B} \in \mathcal{R}^{n \times m}$, $\mathbf{C} \in \mathcal{R}^{r \times n}$, $\mathbf{D}_c \in \mathcal{R}^{m \times r}$, $\mathbf{C}_c \in \mathcal{R}^{m \times n_c}$, $\mathbf{B}_c \in \mathcal{R}^{n_c \times r}$, $\mathbf{A}_t \in \mathcal{R}^{(n+n_c)\times(n+n_c)}$.

The closed-loop structure $\mathbf{A}+\mathbf{B}\mathbf{K}\mathbf{C}$ can be used in static or dynamic feedback controllers.

13.15.2 Linear Perturbation Caused by Weighted Sum of Matrices

Let the perturbation of the system matrix $\mathbf{A}$ be given by the sum of fixed matrices $\mathbf{E}_i$ weighted by (variable) coefficients p_i. Then, the closed-loop system is described by

$$\dot{\mathbf{x}}(t) = (\mathbf{A} + \sum_{i=1}^{n_E} p_i \mathbf{E}_i + \mathbf{B}\mathbf{K}\mathbf{C})\mathbf{x}(t) \ . \tag{13.184}$$

Choosing the Lyapunov function $V(\mathbf{x}) = \mathbf{x}^T\mathbf{P}\mathbf{x}$ where $\mathbf{P}$ denotes the unique positive definite symmetric solution of the Lyapunov equation

$$(\mathbf{A}+\mathbf{B}\mathbf{K}\mathbf{C})^T\mathbf{P} + \mathbf{P}(\mathbf{A}+\mathbf{B}\mathbf{K}\mathbf{C}) = -\mathbf{Q} \qquad \mathbf{Q} > 0 \ , \tag{13.185}$$

$$\begin{aligned} \dot{V}(\mathbf{x}) &= \mathbf{x}^T[(\mathbf{A}+\mathbf{B}\mathbf{K}\mathbf{C})^T\mathbf{P} + \mathbf{P}(\mathbf{A}+\mathbf{B}\mathbf{K}\mathbf{C})]\mathbf{x} + \mathbf{x}^T(\sum_{i=1}^{n_E} p_i\mathbf{E}_i^T\mathbf{P} + \sum_{i=1}^{n_E} p_i\mathbf{P}\mathbf{E}_i)\mathbf{x} \\ &= -\mathbf{x}^T\mathbf{Q}\mathbf{X} + \mathbf{x}^T(\sum_{i=1}^{n_E} p_i\mathbf{E}_i^T\mathbf{P} + \sum_{i=1}^{n_E} p_i\mathbf{P}\mathbf{E}_i)\mathbf{x} \le 0 \end{aligned} \tag{13.186}$$

$$\mathbf{x}^T(\sum_{i=1}^{n_E} p_i \mathbf{E}_i^T \mathbf{P} + \sum_{i=1}^{n_E} p_i \mathbf{P}\mathbf{E}_i)\mathbf{x} \leq \mathbf{x}^T \mathbf{Q} \mathbf{x} \tag{13.187}$$

$$|\mathbf{x}^T(\sum_{i=1}^{n_E} p_i \mathbf{E}_i^T \mathbf{P}\mathbf{x} + \sum_{i=1}^{n_E} p_i \mathbf{P}\mathbf{E}_i)\mathbf{x}| \leq \|\mathbf{x}\|_F^2 \sum_{i=1}^{n_E} |p_i| \; \|\mathbf{E}_i^T \mathbf{P} + \mathbf{P}\mathbf{E}_i\|_s \leq \sigma_{\min}[\mathbf{Q}] \; \mathbf{x}^T \mathbf{x} \tag{13.188}$$

$$|\mathbf{p}|^T \begin{pmatrix} \|\mathbf{E}_1^T \mathbf{P} + \mathbf{P}\mathbf{E}_1\|_s \\ \vdots \\ \|\mathbf{E}_{n_E}^T \mathbf{P} + \mathbf{P}\mathbf{E}_{n_E}\|_s \end{pmatrix} < \sigma_{\min}[\mathbf{Q}] \tag{13.189}$$

$$\|\mathbf{p}\|_F^2 \sum_{i=1}^{n_E} \|\mathbf{E}_i^T \mathbf{P} + \mathbf{P}\mathbf{E}_i\|_s^2 < \sigma_{\min}^2[\mathbf{Q}] \tag{13.190}$$

$$\sum_{i=1}^{n_E} |p_i|^2 \leq \frac{\sigma_{\min}^2[\mathbf{Q}]}{\sum_{i=1}^{n_E} \|\mathbf{E}_i^T \mathbf{P} + \mathbf{P}\mathbf{E}_i\|_s^2} \triangleq \rho^2(\mathbf{K}, \mathbf{Q}) \, . \tag{13.191}$$

13.15.3 Perturbation of A and B

Consider the alternate perturbation of **A** and **B**

$$\mathbf{A} = \mathbf{A}_o + \sum_{i=1}^{n_E} p_i \mathbf{A}_i \qquad\qquad \mathbf{B} = \mathbf{B}_o + \sum_{i=1}^{n_E} p_i \mathbf{B}_i \tag{13.192}$$

then

$$\mathbf{A} + \mathbf{B}\mathbf{K}\mathbf{C} = \mathbf{A}_o + \sum_{i=1}^{n_E} p_i \mathbf{A}_i + \mathbf{B}_o \mathbf{K}\mathbf{C} + \sum_{i=1}^{n_E} p_i \mathbf{B}_i \mathbf{K}\mathbf{C} \tag{13.193}$$

$$= \mathbf{A}_o + \mathbf{B}_o \mathbf{K}\mathbf{C} + \sum_{i-1}^{n_E} p_i (\mathbf{A}_i + \mathbf{B}_i \mathbf{K}\mathbf{C}) \, . \tag{13.194}$$

In view of the above calculation the following result is obtained. Replacing **A** and $\mathbf{E}_i$ according to

$$\mathbf{A} := \mathbf{A}_o + \mathbf{B}_o \mathbf{K}\mathbf{C} \quad \text{and} \quad \mathbf{E}_i := \mathbf{A}_i + \mathbf{B}_i \mathbf{K}\mathbf{C} \, , \tag{13.195}$$

the largest hypersphere $\rho^2(\mathbf{K}, \mathbf{Q})$ as given in the previous section is still of relevance in the generalized perturbation case of this section.

13.15.4 Structured and Bounded Perturbation

The system with the alternate uncertainty model

$$\mathbf{A} = \mathbf{A}_o + \sum_{i=1}^{n_o} r_i(t) \mathbf{A}_i \tag{13.196}$$

with constant matrices $\mathbf{A}_i$ and bounded parameters $|r_i(t)| \leq r_{\max} > 0$ minimizing the quadratic performance index is presented by *Kosmidou, O.I., 1990*. The control matrix is structured in a similar way. The Lyapunov type function $V = \mathbf{x}^T(0)\mathbf{P}(0)\mathbf{x}(0)$ is considered as a guaranteed cost where $\mathbf{P}(t)$ is derived from Riccati differential equation in $\mathbf{P}(t)$ under terminal condition $\mathbf{P}(t_f) = \mathbf{F}_f$. Several auxiliary matrices are defined to find appropriate bounds. The basic idea was given by *Chung, S.S.L., and Peng, T.K.C., 1972*. For numerical procedures see *Bernussou, J., et al. 1990*. A similar method for quadratic stabilizability via a two level optimization is presented by *Gu, K., et al. 1990;1991*.

13.15.5 Robustification Procedure

The problem of robustification is stated as to find both $\mathbf{Q}$ and $\mathbf{K}$ in order to minimize the radius $\rho(\mathbf{K},\mathbf{Q})$ of the hypersphere or to minimize its inverse

$$\max_{\mathbf{K}} \max_{\mathbf{Q}} \rho^2(\mathbf{K},\mathbf{Q}) \quad \text{or} \quad \min_{\mathbf{K}} \min_{\mathbf{Q}} \rho^{-2}(\mathbf{K},\mathbf{Q}) = \min_{\mathbf{K},\mathbf{Q}} \frac{\sum_{i=1}^{n_E} \|\mathbf{E}_i^T\mathbf{P}+\mathbf{P}\mathbf{E}_i\|_s^2}{\sigma_{\min}^2[\mathbf{Q}]} \,. \tag{13.197}$$

The minimization with respect to $\mathbf{K}$ and $\mathbf{Q}$ must, first, comply with the Lyapunov equation

$$(\mathbf{A}+\mathbf{BKC})^T\mathbf{P}+\mathbf{P}(\mathbf{A}+\mathbf{BKC}) = -\mathbf{Q} \tag{13.198}$$

which gives the dependence of $\mathbf{P}$ on $\mathbf{K}$ and $\mathbf{Q}$ and, second, satisfy the constraint

$$I_{\alpha_s} \triangleq \max_i \Re e\, \lambda_i[\mathbf{A}+\mathbf{BKC}] = \alpha_s[\mathbf{A}+\mathbf{BKC}] < 0 \,. \tag{13.199}$$

13.15.6 Perturbation of the Lyapunov Equation

Consider the Lyapunov equation Eq.(13.198) and its perturbation with respect to $\mathbf{L}$

$$(\mathbf{A}+\mathbf{BKC})^T\Delta\mathbf{P}+\Delta\mathbf{P}(\mathbf{A}+\mathbf{BKC}) = -(\Delta\mathbf{L}^T\mathbf{L}+\mathbf{L}^T\Delta\mathbf{L}) \,. \tag{13.200}$$

Then, from *deSouza, E., and Bhattacharyya, S.P., 1981* the increment in the Lyapunov matrix due to the perturbation is given by

$$\Delta\mathbf{P} = \sum_{j=1}^{n}\sum_{k=1}^{n} \gamma_{ik}\, (\mathbf{A}+\mathbf{BKC})^{T,(j-1)}(\Delta\mathbf{L}^T\mathbf{L}+\mathbf{L}^T\Delta\mathbf{L})(\mathbf{A}+\mathbf{BKC})^{k-1} \,. \tag{13.201}$$

13.15.7 Increments in $\rho^{-2}(\mathbf{K},\mathbf{Q})$

From Eq.(13.197) using

$$c = \frac{u}{v} \quad \rightsquigarrow \quad dc = \frac{du}{v} - \frac{u}{v^2}dv \,, \tag{13.202}$$

$$\Delta\rho^{-2}(\mathbf{K},\mathbf{Q}) = \sum_{i=1}^{n_E} \frac{2\|\mathbf{E}_i^T\mathbf{P}+\mathbf{P}\mathbf{E}_i\|_s\, \Delta\|\mathbf{E}_i^T\mathbf{P}+\mathbf{P}\mathbf{E}_i\|_s}{\sigma_{\min}^2[\mathbf{L}^T\mathbf{L}]} - \frac{\|\mathbf{E}_i^T\mathbf{P}+\mathbf{P}\mathbf{E}_i\|_s 2\sigma_{\min}[\mathbf{L}^T\mathbf{L}]\, \Delta\sigma_{\min}[\mathbf{L}^T\mathbf{L}]}{\sigma_{\min}^4[\mathbf{L}^T\mathbf{L}]} \tag{13.203}$$

where

$$\Delta\|\mathbf{E}_i^T\mathbf{P}+\mathbf{P}\mathbf{E}_i\|_s = \mathrm{tr}\, [\mathbf{v}_{ai}\mathbf{u}_{ai}^T\Delta(\mathbf{E}_i^T\mathbf{P}+\mathbf{P}\mathbf{E}_i)] = \mathrm{tr}\, [\mathbf{v}_{ai}\mathbf{u}_{ai}^T(\mathbf{E}_i^T\Delta\mathbf{P}+\Delta\mathbf{P}\,\mathbf{E}_i)] \tag{13.204}$$

$$\Delta\sigma_{\min}[\mathbf{L}^T\mathbf{L}] = \mathbf{v}_m\mathbf{u}_m^T\Delta(\mathbf{L}^T\mathbf{L}) = \mathbf{v}_m\mathbf{u}_m^T(\Delta\mathbf{L}^T\mathbf{L}+\mathbf{L}^T\Delta\mathbf{L}) \,. \tag{13.205}$$

The vectors $\mathbf{v}_{ai}$, $\mathbf{u}_{ai}$ denote the singular vectors associated with $\mathbf{E}_i^T\mathbf{P}+\mathbf{P}\mathbf{E}_i$. The singular vectors $\mathbf{v}_m, \mathbf{u}_m$ correspond with $\mathbf{L}^T\mathbf{L}$.

13.15.8 Gradient Descent Procedure

Combining Eqs.(13.201), (13.204) and (13.205) yields the derivatives of $\rho^{-2}(\mathbf{K},\mathbf{Q})$ with respect to $\mathbf{L}$ and $\mathbf{K}$ (*Bhattacharyya, S.P., 1987*)

$$\frac{\partial \rho^{-2}(\mathbf{K},\mathbf{Q})}{\partial \mathbf{L}} = \frac{2}{\sigma^2_{\min}[\mathbf{L}^T\mathbf{L}]}\mathbf{L}\,[\mathbf{V}^T - \sum_{i=1}^{n_E}\frac{\|\mathbf{E}_i^T\mathbf{P}+\mathbf{P}\mathbf{E}_i\|_s^2}{\sigma_{\min}[\mathbf{L}^T\mathbf{L}]}(\mathbf{u}_m\mathbf{v}_m^T+\mathbf{v}_m\mathbf{u}_m^T)] \quad (13.206)$$

$$\frac{\partial \rho^{-2}(\mathbf{K},\mathbf{Q})}{\partial \mathbf{K}} = \frac{2}{\sigma^2_{\min}[\mathbf{L}^T\mathbf{L}]}[\sum_{i=1}^{n_E}\|\mathbf{E}_i^T\mathbf{P}+\mathbf{P}\mathbf{E}_i\|_s\mathbf{B}_i^T\mathbf{P}\mathbf{W}_i + \mathbf{B}^T\mathbf{P}^T\mathbf{V}^T]\mathbf{C}^T \quad (13.207)$$

where $\mathbf{V}$ and $\mathbf{W}_i$ are given by

$$(\mathbf{A}_o+\mathbf{B}_o\mathbf{K}\mathbf{C})\mathbf{V}+\mathbf{V}(\mathbf{A}_o+\mathbf{B}_o\mathbf{K}\mathbf{C})^T = -\sum_{i=1}^{n_E}\|\mathbf{E}_i^T\mathbf{P}+\mathbf{P}\mathbf{E}_i\|_s[\mathbf{E}_i\mathbf{W}_i+\mathbf{W}_i\mathbf{E}_i^T] \quad (13.208)$$

$$\mathbf{W}_i = \mathbf{u}_{ai}\mathbf{v}_{ai}^T + \mathbf{v}_{ai}\mathbf{u}_{ai}^T\,. \quad (13.209)$$

A gradient-based descent procedure is given by the derivatives above satisfying the constraint Eq.(13.199). Its matricial gradients are

$$\frac{\partial I_{\alpha_s}}{\partial \mathbf{L}} = 0 \quad \text{and} \quad \frac{\partial I_{\alpha_s}}{\partial \mathbf{K}} = \frac{\partial}{\partial \mathbf{K}}\max_i \Re e\lambda_i[\mathbf{A}+\mathbf{B}\mathbf{K}\mathbf{C}]\,. \quad (13.210)$$

With $\mathbf{F} := \mathbf{A}_o+\mathbf{B}_o\mathbf{K}\mathbf{C} \triangleq \mathbf{F}_o$ as given by Eq.(13.194) and using Eq.(6.8),

$$\frac{\partial I_{\alpha_s}}{\partial \mathbf{K}} = \Re e(\mathbf{I}_m \otimes \mathbf{f}_{o,\max}^{a*T})\mathbf{B}_o\frac{\partial \mathbf{K}}{\partial \mathbf{K}}\mathbf{C}(\mathbf{I}_n\otimes\mathbf{f}_{o,\max})\,. \quad (13.211)$$

The self-derivative matrix $\frac{\partial \mathbf{K}}{\partial \mathbf{K}}$ is obtained invoking Eq.(4.8). The vectors $\mathbf{f}_{o,\max}$ denote the eigenvectors associated with $\max_i \Re e\lambda_i[\mathbf{A}_o+\mathbf{B}_o\mathbf{K}\mathbf{C}]$.

The problem of maximizing the robustness region in the parameter space based on a Lyapunov function that varies linearly with the uncertain parameters was solved by *Leal, M.A., and Gibson, J.S., 1990*.

13.16 Bounding the Lyapunov Derivative

Consider the system

$$\dot{\mathbf{x}}(t) = \mathbf{A}(t)\mathbf{x}(t)+\mathbf{B}(t)\mathbf{u}(t) \qquad \mathbf{B}(t)\in\mathcal{R}^{n\times m} \quad (13.212)$$

and a nonlinear controller

$$\mathbf{u}(t) = \mathbf{g}[\mathbf{x}(t)]\,. \quad (13.213)$$

The time-varying matrices $\mathbf{A}(t)$ and $\mathbf{B}(t)$ denote slow motions caused by the system uncertainty. The closed-loop system is said quadratically stabilizable if the Lyapunov function $V(\mathbf{x},t) = \mathbf{x}^T(t)\mathbf{P}\mathbf{x}(t)$ has a derivative $\dot{V}(\mathbf{x},t)$ satisfying the condition

$$\dot{V}(\mathbf{x},t) = 2\mathbf{x}^T\mathbf{P}\mathbf{A}\mathbf{x}+2\mathbf{x}^T\mathbf{P}\mathbf{B}\mathbf{g} \le -\alpha\,\|\mathbf{x}\|_F^2 \quad (13.214)$$

for any initial condition $\mathbf{x}_o$. Note that $\mathbf{P} = \mathbf{P}^T > 0,\ \alpha > 0$. Usually, the controller $\mathbf{g}[\mathbf{x}(t)]$ can be chosen continuously differentiable with $\mathbf{g}[0] = \mathbf{0}$. Linear control is given by $\mathbf{g}[\mathbf{x}(t)] = \mathbf{K}\mathbf{x}(t)$.

A uniform-boundedness result for the tracking error $\mathbf{e}$ of a PD-type controller using the Lyapunov approach via $V = \mathbf{e}^T\mathbf{P}\mathbf{e}$ is presented by *Dawson, D.M., et al. 1990*.

13.16.1 Linear Control and Augmented System

As shown by *Barmish, B.R., 1983*, the closed-loop control given by Eqs.(13.212) and (13.213) is quadratically stabilizable via linear control if and only if the augmented system

$$\dot{\mathbf{x}}'(t) = \begin{pmatrix} \mathbf{A}(t) & \mathbf{B}(t) \\ \mathbf{0} & \mathbf{0} \end{pmatrix} \mathbf{x}'(t) + \begin{pmatrix} \mathbf{0}_{(n-m)\times m} \\ \mathbf{I}_m \end{pmatrix} \mathbf{u}'(t) \qquad \mathbf{x}' \in \mathcal{R}^{n+m}, \quad \mathbf{u}' \in \mathcal{R}^m \tag{13.215}$$

is quadratically stabilizable via linear control. Note that the augmented system contains a constant input matrix.

13.16.2 Nonlinear Control and Linear Bound

Consider the nonlinearity and its Taylor expansion

$$\mathbf{g}[\mathbf{x}(t)] = \left(\frac{\partial \mathbf{g}}{\partial \mathbf{x}^T}\right)_o \mathbf{x}(t) + \mathbf{g}_1(\|\mathbf{x}\|_F) \qquad \mathbf{g}, \mathbf{g}_1 \in \mathcal{R}^m, \quad \mathbf{K}, \left(\frac{\partial \mathbf{g}}{\partial \mathbf{x}^T}\right)_o \in \mathcal{R}^{m\times n} \tag{13.216}$$

$$\text{where} \quad \lim_{\|\mathbf{x}\|_F \to 0} \frac{\| \ \mathbf{g}_1(\|\mathbf{x}\|_F) \ \|_F}{\|\mathbf{x}\|_F} = 0 \, . \tag{13.217}$$

Selecting $\mathbf{K} = (\partial \mathbf{g} \ / \ \partial \mathbf{x}^T)_o$ the derivative $\dot{V}$ is

$$\dot{V} = 2\mathbf{x}^T\mathbf{P}\mathbf{A}\mathbf{x} + 2\mathbf{x}^T\mathbf{P}\mathbf{B}(\mathbf{K}\mathbf{x} + \mathbf{g}_1) = 2\mathbf{x}^T\mathbf{P}(\mathbf{A} + \mathbf{B}\mathbf{K})\mathbf{x} + 2\mathbf{x}^T\mathbf{P}\mathbf{B}\mathbf{g}_1 \tag{13.218}$$

$$\dot{V} \le -\alpha_p \ \|\mathbf{x}\|_F^2 + \beta_p \ \|\mathbf{x}\|_F^2 \frac{\|\mathbf{g}_1\|_F}{\|\mathbf{x}\|_F} \qquad \text{where} \qquad \beta_p \triangleq 2\|\mathbf{P}\|_F \sup \|\mathbf{B}\|_F \, . \tag{13.219}$$

The supremum is selected for that region of parameters causing the uncertainty. Assuming bounded $\|\mathbf{g}_1\|_F \ / \ \|\mathbf{x}\|_F \triangleq c$, the number $-\alpha_p + \beta_p c$ must be kept below α in order to satisfy the stability condition as given by Eq. (13.214) (*Barmish, B.R., 1983*).

13.17 Adaptive Robust Control Systems

13.17.1 General Uncertain System

Consider the uncertain system

$$\dot{\mathbf{x}}(t) = \mathbf{A}\mathbf{x}(t) + \mathbf{B}\mathbf{u}(t) + \mathbf{e}(\mathbf{x}, \mathbf{u}, \mathbf{v}, t) \qquad \mathbf{x}(0) = \mathbf{x}_o \qquad \mathbf{B} \in \mathcal{R}^{n\times m} \tag{13.220}$$

where $\mathbf{v} = \mathbf{v}(t) \in \mathcal{R}^p$ is an uncertain vector function (or slowly time-dependent parameter). Let the control signal $\mathbf{u}$ be subdivided into a constant and a varying feedback component represented by $\mathbf{K}\mathbf{x}$ and $\mathbf{p}$, respectively,

$$\mathbf{u}(\mathbf{x}, \mathbf{b}_\beta, t) = \mathbf{K}\mathbf{x} + \mathbf{p}(\mathbf{x}, \mathbf{b}_\beta, t) \qquad \mathbf{p} \in \mathcal{R}^m \, . \tag{13.221}$$

The component $\mathbf{p}$ is regarded as the direction of control. Designing $\mathbf{p}$ the structure of physical limitation is employed as given by

$$\mathbf{p}(\mathbf{x}, \mathbf{b}_\beta, t) \triangleq \begin{cases} -\frac{\mathbf{m}_\mu(\mathbf{x}, \mathbf{b}_\beta, t)}{\|\mathbf{m}_\mu\|_F} \ \rho(\mathbf{x}, \mathbf{b}_\beta, t) & \text{if } \|\mathbf{m}_{mu}\|_F > \varepsilon \\ -\frac{\mathbf{m}_\mu(\mathbf{x}, \mathbf{b}_\beta, t)}{\varepsilon} \ \rho(\mathbf{x}, \mathbf{b}_\beta, t) & \text{if } \|\mathbf{m}_{mu}\|_F < \varepsilon \, . \end{cases} \tag{13.222}$$

The vector variable $\mathbf{m}_\mu \in \mathcal{R}^m$ results from $\mathbf{x}$ by matrix premultiplication $\mathbf{B}^T\mathbf{P}$ and regarding the scalar upper bound $\rho(\mathbf{x}, \mathbf{b}_\beta, t)$

$$\mathbf{m}_\mu(\mathbf{x}, \mathbf{b}_\beta, t) \triangleq 2\mathbf{B}^T\mathbf{P}\mathbf{x}(t)\ \rho(\mathbf{x}, \mathbf{b}_\beta, t) \tag{13.223}$$

$$\text{where}\quad \mathbf{P}\bar{\mathbf{A}} + \bar{\mathbf{A}}^T\mathbf{P} = -\mathbf{Q} \qquad \mathbf{Q} > 0,\ \bar{\mathbf{A}} = \mathbf{A} + \mathbf{B}\mathbf{K}\ . \tag{13.224}$$

The vector $\mathbf{b}_\beta \in \mathcal{R}^s$ is a constant vector to bound the uncertainty whereas the scalar $\rho(\mathbf{x}, \mathbf{b}_\beta, t) \in \mathcal{R}$ is an upper bound for $\|\hat{\mathbf{e}}\|_F$. The scalar ε obeys $0 < \varepsilon < 1$. It is assumed that $\mathbf{A} + \mathbf{B}\mathbf{K}$ is stable and $\mathbf{e}$ lies in the range space of $\mathbf{B}$, i.e., there exists a variable $\hat{\mathbf{e}}$ such that $\mathbf{e} = \mathbf{B}\hat{\mathbf{e}}$. Furthermore, $\hat{\mathbf{e}}$ is assumed bounded

$$\|\hat{\mathbf{e}}\|_F < \rho(\mathbf{x}, \mathbf{b}_\beta, t)\ . \tag{13.225}$$

The control variable is said robust stabilizing if for $\mathbf{P} > 0$

$$V(\mathbf{x}) = \mathbf{x}^T\mathbf{P}\mathbf{x} \qquad \text{and} \qquad \dot{V}(\mathbf{x}) = 2\mathbf{x}^T\mathbf{P}(\mathbf{A}\mathbf{x}+\mathbf{B}\mathbf{u}+\mathbf{e}) < r_1\|\mathbf{x}\|_F^2 + r_2\|\mathbf{x}\|_F + r_3 \tag{13.226}$$

(*Barmish, B.R., and Leitmann, G., 1982*). Finally, it is assumed that the scalar variable $h(\mathbf{x}, \mathbf{b}_\beta, t)$ exists

$$\mathbf{p}(\mathbf{x}, \mathbf{b}_\beta, t) = -\mathbf{B}^T\frac{\partial V}{\partial \mathbf{x}} h(\mathbf{x}, \mathbf{b}_\beta, t) \triangleq -\alpha(\mathbf{x})\ h(\mathbf{x}, \mathbf{b}_\beta, t)\ , \tag{13.227}$$

representing the direction of control. The scalar h is regarded as the magnitude of control and is assumed concave, i.e.,

$$h(\mathbf{x}, \mathbf{b}_{\beta 1}, t) - h(\mathbf{x}, \mathbf{b}_{\beta 2}, t) \le \frac{\partial h}{\partial \mathbf{b}_\beta^T}(\mathbf{x}, \mathbf{b}_{\beta 2}, t)\ (\mathbf{b}_{\beta 1} - \mathbf{b}_{\beta 2})\ . \tag{13.228}$$

13.17.2 Linear Uncertain System

For the special case of a linear uncertain system

$$\dot{\mathbf{x}}(t) = [\mathbf{A} + \Delta\mathbf{A}(\mathbf{v})]\mathbf{x}(t) + [\mathbf{B} + \Delta\mathbf{B}(\mathbf{v})]\mathbf{u}(t) + \mathbf{C}\mathbf{w}(\mathbf{v}) \tag{13.229}$$

the uncertainty $\mathbf{e}(t)$ results in

$$\mathbf{e}(t) = \Delta\mathbf{A}(\mathbf{v})\mathbf{x}(t) + \Delta\mathbf{B}(\mathbf{v})\mathbf{u}(t) + \mathbf{C}\mathbf{w}(\mathbf{v})\ . \tag{13.230}$$

If matrix-valued functions $\mathbf{D}, \mathbf{E}$ and $\mathbf{F}$ exist such that

$$\Delta\mathbf{A}(\mathbf{v}) = \mathbf{B}\mathbf{D}(\mathbf{v}), \qquad \Delta\mathbf{B}(\mathbf{v}) = \mathbf{B}\mathbf{E}(\mathbf{v}), \qquad \mathbf{C} = \mathbf{B}\mathbf{F} \qquad \text{then} \tag{13.231}$$

$$\mathbf{e} = \mathbf{B}\hat{\mathbf{e}} = \mathbf{B}\mathbf{D}\mathbf{x} + \mathbf{B}\mathbf{E}\mathbf{u} + \mathbf{B}\mathbf{F}\mathbf{v} \quad \rightsquigarrow \quad \hat{\mathbf{e}} = \mathbf{D}\mathbf{x} + \mathbf{E}\mathbf{u} + \mathbf{F}\mathbf{w}\ . \tag{13.232}$$

Using the scalar function ρ limiting the norm $\|\hat{\mathbf{e}}\|_F$

$$\|\hat{\mathbf{e}}\|_F < \rho = \beta_1\|\mathbf{x}\|_F + \beta_2 \tag{13.233}$$

and applying $\mathbf{u} = \mathbf{K}\mathbf{x} + \mathbf{p}$ and $\mathbf{p} = -\mathbf{m}_\mu\ \rho/\|\mathbf{m}_\mu\|_F$

$$\|\hat{\mathbf{e}}\|_F = \|(\mathbf{D} + \mathbf{E}\mathbf{K})\mathbf{x} - \frac{\mathbf{E}\mathbf{m}_\mu}{\|\mathbf{m}_\mu\|_F}\rho + \mathbf{F}\mathbf{w}\|_F < \rho = \beta_1\|\mathbf{x}\|_F + \beta_2 \tag{13.234}$$

$$(\max_{v\in\Sigma}\|\mathbf{D}\|_s)\|\mathbf{x}\|_F + (\max_{v\in\Sigma}\|\mathbf{EK}\|_s)\|\mathbf{x}\|_F + (\max_{v\in\Sigma}\|\mathbf{E}\|_s)(\beta_1\|\mathbf{x}\|_F + \beta_2) + (\max_{v\in\Sigma}\|\mathbf{Fw}\|_F) < \beta_1\|\mathbf{x}\|_F + \beta_2 \tag{13.235}$$

$$\frac{(\max_{v\in\Sigma}\|\mathbf{D}\|_s) + (\max_{v\in\Sigma}\|\mathbf{EK}\|_s)}{1-\max_{v\in\Sigma}\|\mathbf{E}\|_s}\|\mathbf{x}\|_F + \frac{\max_{v\in\Sigma}\|\mathbf{Fw}\|_F}{(1-\max_{v\in\Sigma}\|\mathbf{E}\|_s} < \beta_1\|\mathbf{x}\|_F + \beta_2 \ . \tag{13.236}$$

By the inequation above, the constants β_1 and β_2 are determined. If $r_1 = -\lambda_{\min}[\mathbf{Q}]$ and $r_2 = 0$ and $r_3 = 2\varepsilon$ (*Corless, M.J., and Leitmann, G., 1981*) with regard to the saturation-like structure of $\mathbf{p}$

$$\dot{V} < -\lambda_{\min}[\mathbf{Q}] \ \|\mathbf{x}\|_F^2 + 2\varepsilon \ . \tag{13.237}$$

Since the assumptions stated before are met, the adaptive algorithm can be implemented completely.

Adaptive schemes are applied to control systems with interval bounded plant parameters by *Kreisselmeier, G., 1986*.

Recent research activities are devoted to robustly designing an adaptive linear controller irrespective of its control law (*Giri, F., et al. 1990; Krause, J.M., et al. 1990*).

Chapter 14

Lyapunov-Based Methods for Perturbed Discrete-Time Systems

This chapter is devoted to the time-domain approach of stability robustness of discrete-time systems. Uncertainty frequently occurs associated with the parameters of the process. However, with regard to varying random sampling another type of uncertainty may arise in control systems. The stability margin and the expectation of the stability margin can be assessed by simple formulae (*Jury, E.I., and Tsypkin, Y.Z., 1971; Weinmann, A., 1981 and 1985*).

Individually, the sampling interval T does not act as an additional design parameter; it is seen from several examples that only the products of physical parameters of the plant and the sampling period appear in the z-transfer function. On the one hand, scaling with respect to T gives the opportunity to treat continuous-time and discrete-time systems with a unified parameter; on the other hand, if T is an uncertain parameter then all the parameters in the z-transfer function associated with the physical parameters are in a dependent relationship to the uncertainty of the sampling interval (*Ackermann, J., and Hu, H.Z., 1990a*).

For the sake of final design considerations, robustness and sensitivity versus sampling period is often investigated only by means of simulation (*Diduch, C.P., and Doraiswami, R., 1988*).

Now, consider a discrete-time homogeneous system with fixed sampling period with nonlinear perturbation $\mathbf{g}_k(\mathbf{x}_k)$

$$\mathbf{x}_{k+1} = \mathbf{\Phi}(T)\mathbf{x}_k + \mathbf{g}_k(\mathbf{x}_k) \qquad \mathbf{x}_k \in \mathcal{R}^n \tag{14.1}$$

where the eigenvalues of $\mathbf{\Phi}(T)$ are located inside a disc of radius $\sqrt{1+\alpha}$, $\alpha > 0$, and T is the sampling period. Preassuming stable $\mathbf{A}$ in the continuous-time background, i.e., $\Re e\, \lambda[\mathbf{A}] < \lambda_o$,

$$\lambda[\mathbf{\Phi}(T)] = \lambda[e^{\mathbf{A}T}] = e^{\lambda[\mathbf{A}T]} = e^{T\lambda[\mathbf{A}]} \tag{14.2}$$

$$|\ \lambda[\mathbf{\Phi}(T)]\ | = |\ e^{T\lambda[\mathbf{A}]}\ | = e^{T\ \Re e\ \lambda[\mathbf{A}]} < e^{T\lambda_o} = \frac{1}{\sqrt{1+\alpha}} \tag{14.3}$$

the inequality $\lambda_o \leq 0$ corresponds to $\alpha \geq 0$, see also Eq.(14.27).

14.1 Linear System

A discrete-time system is stable, i.e. $|\ \lambda[\mathbf{\Phi}(T)]\ | < 1$, if there exists a unique positive definite solution $\mathbf{P}$ to the Lyapunov matrix equation

$$\mathbf{\Phi}^T\mathbf{P}\mathbf{\Phi} - \mathbf{P} = -\mathbf{Q} \tag{14.4}$$

for $\mathbf{Q} > 0$ where abbreviated $\mathbf{\Phi}(T) = \mathbf{\Phi}$. Replacing $\mathbf{\Phi}$ by $\sqrt{1+\alpha}\ \mathbf{\Phi}$ in Eq.(14.4) and substituting $|\ \lambda[\sqrt{1+\alpha}\ \mathbf{\Phi}]\ | = \sqrt{1+\alpha}\ |\ \lambda[\mathbf{\Phi}]\ |$, the matrix equation

$$(\sqrt{1+\alpha}\ \mathbf{\Phi})^T\ \mathbf{P}\ (\sqrt{1+\alpha}\ \mathbf{\Phi}) - \mathbf{P} = (1+\alpha)\mathbf{\Phi}^T\mathbf{P}\mathbf{\Phi} - \mathbf{P} = -\mathbf{Q} \tag{14.5}$$

has a unique positive definite solution $\mathbf{P}$ if $|\ \lambda[\mathbf{\Phi}]\ | < 1/\sqrt{1+\alpha}$. (Differential sensitivity of discrete-time Lyapunov equations is analyzed by *Gahinet, P.M., et al. 1990* and provides an interrelation to the sensitivity and the stability radius of the open-loop state matrix.)

14.2 Nonlinear Perturbation

The system might be perturbed by $\mathbf{g}_k(\mathbf{x}_k)$ where a gain factor γ relates the Frobenius vector norms

$$\mathbf{x}_{k+1} = \mathbf{\Phi}\mathbf{x}_k + \mathbf{g}_k(\mathbf{x}_k) \qquad \|\mathbf{g}_k(\mathbf{x}_k)\|_F \leq \gamma\|\mathbf{x}_k\|_F\ . \tag{14.6}$$

The Lyapunov function $\mathbf{V}_k = \mathbf{x}_k^T\mathbf{P}\mathbf{x}_k$ is used where $\mathbf{P}$ is a positive definite matrix. Hence, $\mathbf{a}^T\mathbf{P}\mathbf{a} > 0$ for any $\mathbf{a}$. Selecting $\mathbf{a} = \sqrt{\alpha}\mathbf{\Phi}\mathbf{x}_k - \mathbf{g}_k/\sqrt{\alpha}$, then

$$(\sqrt{\alpha}\ \mathbf{\Phi}\mathbf{x}_k - \mathbf{g}_k/\sqrt{\alpha}\ )^T\ \mathbf{P}\ (\sqrt{\alpha}\ \mathbf{\Phi}\mathbf{x}_k - \mathbf{g}_k/\sqrt{\alpha}\ ) \geq 0 \tag{14.7}$$

$$\mathbf{x}_k^T\mathbf{\Phi}^T\mathbf{P}\mathbf{g}_k + \mathbf{g}_k^T\mathbf{P}\mathbf{\Phi}\mathbf{x}_k \leq \alpha\mathbf{x}_k^T\mathbf{\Phi}^T\mathbf{P}\mathbf{\Phi}\mathbf{x}_k + \mathbf{g}_k^T\mathbf{P}\mathbf{g}_k/\alpha\ . \tag{14.8}$$

The first difference of the Lyapunov function is

$$\begin{aligned} V_{k+1} - V_k &= \mathbf{x}_{k+1}^T\mathbf{P}\mathbf{x}_{k+1} - \mathbf{x}_k^T\mathbf{P}\mathbf{x}_k = (\mathbf{x}_k^T\mathbf{\Phi}^T + \mathbf{g}_k^T)\mathbf{P}(\mathbf{\Phi}\mathbf{x}_k + \mathbf{g}_k) - \mathbf{x}_k^T\mathbf{P}\mathbf{x}_k \\ &= \mathbf{x}_k^T\mathbf{\Phi}^T\mathbf{P}\mathbf{g}_k + \mathbf{g}_k^T\mathbf{P}\mathbf{\Phi}\mathbf{x}_k + \mathbf{x}^T\mathbf{\Phi}^T\mathbf{P}\mathbf{\Phi}\mathbf{x}_k + \mathbf{g}_k^T\mathbf{P}\mathbf{g}_k - \mathbf{x}_k^T\mathbf{P}\mathbf{x}_k\ . \end{aligned} \tag{14.9}$$

Since it is intended to satisfy the condition $V_{k+1} - V_k < 0$ for stability the right-hand side of Eq.(14.9) may be replaced by a greater quantity several times in order to obtain a sufficient condition. Substituting the left-hand side of Eq.(14.8) into the first two terms of the right-hand side of Eq.(14.9) yields

$$\begin{aligned} V_{k+1} - V_k &\leq \alpha\mathbf{x}_k^T\mathbf{\Phi}^T\mathbf{P}\mathbf{\Phi}\mathbf{x}_k + \mathbf{g}_k^T\mathbf{P}\mathbf{g}_k/\alpha + \mathbf{x}^T\mathbf{\Phi}^T\mathbf{P}\mathbf{\Phi}\mathbf{x}_k + \mathbf{g}_k^T\mathbf{P}\mathbf{g}_k - \mathbf{x}_k^T\mathbf{P}\mathbf{x}_k \\ V_{k+1} - V_k &\leq \mathbf{x}_k[(1+\alpha)\mathbf{\Phi}^T\mathbf{P}\mathbf{\Phi} - \mathbf{P}]\mathbf{x}_k + (1+1/\alpha)\mathbf{g}_k^T\mathbf{P}\mathbf{g}_k\ . \end{aligned} \tag{14.10}$$

Since $\mathbf{P}$ is symmetric

$$\mathbf{g}_k^T\mathbf{P}\mathbf{g}_k \leq \mathbf{g}_k^T\mathbf{g}_k\lambda_{\max}[\mathbf{P}] \leq \gamma^2\mathbf{x}_k^T\mathbf{x}_k\lambda_{\max}[\mathbf{P}] \tag{14.11}$$

$$V_{k+1} - V_k \leq \mathbf{x}_k^T\{-\mathbf{Q} + (1+1/\alpha)\mathbf{I}\gamma^2\lambda_{\max}[\mathbf{P}]\}\mathbf{x}_k\ . \tag{14.12}$$

In order to guarantee a negative definite matrix on the right-hand side it results (*Yaz,E. and Niu,X. 1989*)

$$\lambda_{\min}[\mathbf{Q}] \geq (1+1/\alpha)\gamma^2\lambda_{\max}[\mathbf{P}] \quad \rightsquigarrow \quad \gamma \leq \gamma_1 = \sqrt{\frac{\lambda_{\min}[\mathbf{Q}]}{(1+1/\alpha)\lambda_{\max}[\mathbf{P}]}}\ . \tag{14.13}$$

Then, the nonlinearly perturbed system is globally uniformly asymptotically stable.

14.3 Maximum Gain for Nonlinear Perturbation

14.3.1 Replacing Q by qQ

Before investigating the maximum gain γ, consider Eq.(14.5) where $\mathbf{Q}$ is replaced by $q\mathbf{Q}$ and the solution is termed $\mathbf{P}_1$.

$$(1+\alpha)\mathbf{\Phi}^T\mathbf{P}_1\mathbf{\Phi} - \mathbf{P}_1 = -\mathbf{Q}_1 = -q\mathbf{Q} \qquad q > 0 \tag{14.14}$$

$$(1+\alpha)\mathbf{\Phi}^T(q^{-1}\mathbf{P}_1)\mathbf{\Phi} - (q^{-1}\mathbf{P}_1) = -\mathbf{Q}\ . \tag{14.15}$$

Comparison with Eq.(14.5) shows $q^{-1}\mathbf{P}_1 = \mathbf{P}$. Hence, the ratio of eigenvalues

$$\frac{\lambda_{\min}[\mathbf{Q}_1]}{\lambda_{\max}[\mathbf{P}_1]} = \frac{q\ \lambda_{\min}[\mathbf{Q}]}{q\ \lambda_{\max}[\mathbf{P}]} = \frac{\lambda_{\min}[\mathbf{Q}]}{\lambda_{\max}[\mathbf{P}]} \tag{14.16}$$

as applied in Eq.(14.13) is independent of q.

14.3.2 Maximizing the Gain γ by Q = I

The solution of the Lyapunov equation Eq.(14.14) is given by the sum decomposition

$$\mathbf{P}_1 = \sum_{i=0}^{\infty}(1+\alpha)^i(\mathbf{\Phi}^T)^i\mathbf{Q}_1\mathbf{\Phi}^i \tag{14.17}$$

which can easily be proved by substitution into Eq.(14.14). Now, in order to maximize the right-hand side of Eq.(14.13) for a given $\mathbf{Q}$ (and given $\lambda_{\min}[\mathbf{Q}]$) the maximum eigenvalue $\lambda_{\max}[\mathbf{P}]$ must be minimized. Considering a pair of Lyapunov matrices $\mathbf{P}_Q$ and $\mathbf{P}_R$ associated with positive definite matrices $\mathbf{Q}$ and $\mathbf{R}$, respectively, from Eq.(14.17)

$$\mathbf{P}_Q - \mathbf{P}_R = \sum_{i=0}^{\infty}(1+\alpha)^i(\mathbf{\Phi}^T)^i(\mathbf{Q}-\mathbf{R})\mathbf{\Phi}^i\ . \tag{14.18}$$

The smaller the matrix difference $(\mathbf{Q}-\mathbf{R})$ the smaller $\mathbf{P}_Q - \mathbf{P}_I$, in any scalar weighting index. The maximum eigenvalue $\lambda_{\max}[\mathbf{P}_Q]$ is located the closer to $\lambda_{\max}[\mathbf{R}]$ the closer $\mathbf{Q}$ is to $\mathbf{R}$. In order to keep $\lambda_{\max}[\mathbf{P}_Q]$ as small as possible for a fixed $\lambda_{\min}[\mathbf{R}]$, the eigenvalues $\lambda_i[\mathbf{R}]$ must be as close together as possible. This is obtained if $\mathbf{R}$ is chosen a diagonal matrix with n identical eigenvalues. Invoking Eq.(14.16), it is not essential whether $\mathbf{R}$ is chosen $\mathbf{I}$ or $q\mathbf{I}$. Reassuming, for a fixed stability margin $\alpha > 0$ the right-hand side of Eq.(14.13) is maximized by $\mathbf{Q} = \mathbf{I}$ and $\mathbf{P}$ is the Lyapunov matrix associated with $\mathbf{Q} = \mathbf{I}$.

14.4 Upper Bound Using the Spectral Radius

14.4.1 Any Matrix $\mathbf{\Phi}$

Using, first, the eigenvector φ_i of $\mathbf{\Phi}$ associated with $\lambda_i[\mathbf{\Phi}]$, i.e. $\mathbf{\Phi}\varphi_i = \lambda_i[\mathbf{\Phi}]\ \varphi_i$, and, second, Eq.(14.5) with $\mathbf{Q} = \mathbf{I}$, the scalar vector product is

$$\begin{aligned} -\varphi_i^H\varphi_i &= -\varphi_i^H\mathbf{I}\varphi_i = \varphi_i^H[(1+\alpha)\mathbf{\Phi}^T\mathbf{P}\mathbf{\Phi} - \mathbf{P}]\varphi_i = (1+\alpha)\varphi_i^H\mathbf{\Phi}^T\mathbf{P}\mathbf{\Phi}\varphi_i - \varphi_i^H\mathbf{P}\varphi_i \\ &= (1+\alpha)\lambda_i^*\varphi_i^H\mathbf{P}\lambda_i\varphi_i - \varphi_i^H\mathbf{P}\varphi_i = [(1+\alpha)|\lambda_i|^2 - 1]\varphi_i^H\mathbf{P}\varphi_i \end{aligned} \tag{14.19}$$

$$\frac{\varphi_i^H\mathbf{P}\varphi_i}{\varphi_i^H\varphi_i} = [\,1-(1+\alpha)|\lambda_i|^2\,]^{-1}\ . \tag{14.20}$$

Applying Rayleigh's principle, for any vector φ_i

$$\lambda_{\max}[\mathbf{P}] > \frac{\varphi_i^H \mathbf{P} \varphi_i}{\varphi_i^H \varphi_i} = \{1 - (1+\alpha)\rho_s^2[\mathbf{\Phi}]\}^{-1} . \tag{14.21}$$

The inequality above usually cannot be replaced by equality since φ_i is not identical to the eigenvector of $\mathbf{P}$. If $\lambda_{\max}[\mathbf{P}]$ in Eq.(14.13) is replaced by a smaller quantity as given by Eq.(14.21), a necessary condition for Eq.(14.13) is obtained

$$\gamma_1 = \sqrt{\frac{1}{(1+1/\alpha)\lambda_{\max}[\mathbf{P}]}} \leq \sqrt{\frac{1-(1+\alpha)\rho_s^2[\mathbf{\Phi}]}{\frac{1+\alpha}{\alpha}}} = \sqrt{\frac{\alpha}{1+\alpha} - \alpha\rho_s^2[\mathbf{\Phi}]} . \tag{14.22}$$

14.4.2 Normal Matrix $\mathbf{\Phi}$

If $\mathbf{\Phi}$ is normal, i.e. $\mathbf{\Phi}\mathbf{\Phi}^H = \mathbf{\Phi}^H\mathbf{\Phi}$ where $\mathbf{\Phi}^H = \mathbf{\Phi}^T$ since $\mathbf{\Phi}$ is real, from Eq.(14.17) with $\mathbf{Q} = \mathbf{I}$

$$\mathbf{P} = \sum_{i=0}^{\infty}(1+\alpha)^i(\mathbf{\Phi}^H)^i\mathbf{\Phi}^i = \sum_{i=0}^{\infty}(1+\alpha)^i\mathbf{T}\mathbf{\Lambda}^{Hi}\mathbf{\Lambda}^i\mathbf{T}^{-1} = \mathbf{T}[\sum_{i=0}^{\infty}(1+\alpha)^i(\mathbf{\Lambda}^H\mathbf{\Lambda})^i]\mathbf{T}^{-1} . \tag{14.23}$$

Note that $\mathbf{\Lambda}^{Hi}\mathbf{\Lambda}^i = (\mathbf{\Lambda}^H\mathbf{\Lambda})^i$ because diagonal matrices commute. Since for a unitary matrix $\mathbf{T}$ - see Eq.(22.21) -

$$\det(s\mathbf{I} - \mathbf{\Omega}) = \det\mathbf{T}\det(s\mathbf{I} - \mathbf{\Omega})\det\mathbf{T}^H = \det(s\mathbf{I} - \mathbf{T}\mathbf{\Omega}\mathbf{T}^H) \tag{14.24}$$

it follows $\lambda[\mathbf{T}\mathbf{\Omega}\mathbf{T}^H] = \lambda[\mathbf{\Omega}]$ and

$$\begin{aligned}
\lambda_{\max}[\mathbf{P}] &= \lambda_{\max}[\sum_{i=0}^{\infty}(1+\alpha)^i(\mathbf{\Lambda}^H\mathbf{\Lambda})^i] = \lambda_{\max}[\sum_{i=0}^{\infty}(1+\alpha)^i \operatorname{diag}\{(\lambda_k^*\lambda_k)^i\}] && (14.25)\\
&= \lambda_{\max}[\operatorname{diag}\{\sum_{i=0}^{\infty}(1+\alpha)^i \mid \lambda_k \mid^{2i}\}] = \max_k \sum_{i=0}^{\infty}(1+\alpha)^i \mid \lambda_k \mid^{2i} && (14.26)\\
&= \sum_{i=0}^{\infty}(1+\alpha)^i\rho_s^{2i}[\mathbf{\Phi}] = \sum_{i=0}^{\infty}\beta^i = (1-\beta)^{-1} = \{1-(1+\alpha)\rho_s^2[\mathbf{\Phi}]\}^{-1} . && (14.27)
\end{aligned}$$

Note that $\beta \triangleq (1+\alpha)\rho_s^2[\mathbf{\Phi}] < 1$ with regard to the assumption that the eigenvalues of $\mathbf{\Phi}$ are located inside a disc of radius $1/\sqrt{1+\alpha}$. Hence, if $\mathbf{\Phi}$ is normal the inequality in Eq.(14.22) becomes an equality.

14.4.3 Optimum Gain Versus α When $\mathbf{\Phi}$ is Normal

Maximizing the gain with respect to α or differentiating γ^2 of Eq.(14.22) with respect to α yields the optimum $\alpha^\star$

$$\frac{\partial\gamma^2}{\partial\alpha} = -\alpha(1+\alpha)^{-2} + (1+\alpha)^{-1} - \rho_s^2[\mathbf{\Phi}] = 0 \tag{14.28}$$

$$\alpha^\star = \frac{1}{\rho_s[\mathbf{\Phi}]} - 1 \quad \rightsquigarrow \quad \gamma^\star = 1 - \rho_s[\mathbf{\Phi}] . \tag{14.29}$$

14.5 Uncertainties Structured Via Modulus Matrix

Consider the discrete-time system with given $\mathbf{\Phi}$

$$\mathbf{x}(k+1) = \mathbf{x}_{k+1} = (\mathbf{\Phi} + \Delta\mathbf{\Phi})\mathbf{x}_k \tag{14.30}$$

where $\Delta\mathbf{\Phi}$ is a perturbation matrix.

14.5.1 Uncertainty Bounded by $|\Delta\Phi_{ij}| < E_{d,ij} \quad \forall i,j$

A necessary and sufficient condition for the above system Eq.(14.30) being stable is that $\rho_s[\mathbf{\Phi} + \Delta\mathbf{\Phi}] < 1$. A sufficient condition is that $\mathbf{I} - |\mathbf{\Phi} + \Delta\mathbf{\Phi}|$ is a Metzler matrix (*Rachid, A., 1989*). Furthermore, it is sufficient if $\lambda[\ |\mathbf{\Phi} + \Delta\mathbf{\Phi}|\ ] < 1$ or $\rho_s[\ |\mathbf{\Phi} + \Delta\mathbf{\Phi}|\ ] < 1$. Since

$$|\Phi_{ij} + \Delta\Phi_{ij}| \le |\Phi_{ij}| + |\Delta\Phi_{ij}| < |\Phi_{ij}| + E_{d,ij} \tag{14.31}$$

$$\text{or} \quad |\mathbf{\Phi} + \Delta\mathbf{\Phi}| \le_e |\mathbf{\Phi}| + |\Delta\mathbf{\Phi}| <_e |\mathbf{\Phi}| + \mathbf{E}_d \tag{14.32}$$

the condition

$$\rho_s[\ |\mathbf{\Phi}| + \mathbf{E}_d] < 1 \tag{14.33}$$

is sufficient for stability of Eq.(14.30) (*Rachid, A., 1989*).

Note that

$$\rho_s[-\mathbf{\Phi}] = |-\lambda[\mathbf{\Phi}]\ |_{\max} = |\lambda[\mathbf{\Phi}]\ |_{\max} = \rho_s[\mathbf{\Phi}] = |\lambda[\mathbf{\Phi}]\ |_{\max} = e^{\{\Re e\lambda[\mathbf{A}]\ \}_{\max}\ T}\ , \tag{14.34}$$

i.e., the spectral radius of $\mathbf{\Phi}$ is insensitive to changes in the sign of $\mathbf{\Phi}$ where $\mathbf{\Phi} = e^{\mathbf{A}T}$ and T is the sampling period. The same result is obtained, no matter whether $\mathbf{\Phi}$ is a positive or negative matrix, i.e. $\Phi_{ij} > 0$ or $\Phi_{ij} < 0 \ \ \forall\, i,j$.

Using structured bounds on uncertain plant parameters where the uncertainties are continuous functions of a common variable, robust single-input nonlinear controllers are suggested by *Yang, W.C., and Tomizuka, M., 1990*.

14.5.2 Uncertainty Bounded by $|\Delta\Phi_{ij}| < \varepsilon|\Phi_{ij}|$

Considering a particular perturbation $|\mathbf{E}_d| <_e \varepsilon|\mathbf{\Phi}|$, then

$$\rho_s[\ |\mathbf{\Phi}| + \mathbf{E}_d] = \rho_s[(1+\varepsilon)|\mathbf{\Phi}|\ ] = (1+\varepsilon)\rho_s[\ |\mathbf{\Phi}|\ ] < 1 \quad \rightsquigarrow \quad \varepsilon < -1 + 1/\rho_s[\ |\mathbf{\Phi}\ |\ ]\ . \tag{14.35}$$

$$\text{e.g.} \quad \mathbf{\Phi} = \begin{pmatrix} 0.5 & 1 \\ 0.3 & 0.2 \end{pmatrix}, \qquad \lambda_1 = 0.918 = \rho_s[\mathbf{\Phi}]; \qquad \lambda_2 = -0.217 \tag{14.36}$$

$$\varepsilon < -1 + 1/\rho_s[\mathbf{\Phi}] = -1 + 1/0.918 = 0.089\ .\ \square \tag{14.37}$$

14.6 Interval Matrices

Consider $\mathbf{\Phi}_1$ as an interval matrix $\mathbf{\Phi}_1 \in [\mathbf{L} \quad \mathbf{H}]$, the stability of the system $\mathbf{x}_{k+1} = \mathbf{\Phi}_1\mathbf{x}_k$ can be studied by defining

$$\mathbf{\Phi} = \frac{\mathbf{L} + \mathbf{H}}{2} \qquad \mathbf{E}_d = \frac{\mathbf{H} - \mathbf{L}}{2} = |\mathbf{E}_d| \tag{14.38}$$

and using Eq.(14.33). Since Eq.(14.33) is not sensitive to the sign of $\mathbf{\Phi}$ and since $\mathbf{E}_d$ is positive, the discrete-time system with interval matrix $\mathbf{\Phi}_1$ is stable if the substitutions of Eq.(14.38) are made and Eq.(14.33) is satisfied.

14.7 Upper Bound Using the Spectral Norm

Using $\mathbf{P}$ of Eq.(14.5) and the Lyapunov function $V_k = \mathbf{x}_k^T\mathbf{P}\mathbf{x}_k$, the first difference from Eq.(14.9) is

$$V_{k+1} - V_k = \mathbf{x}_k^T(\mathbf{\Phi}^T\mathbf{P}\mathbf{\Phi} - \mathbf{P})\mathbf{x}_k + 2\mathbf{g}_k^T\mathbf{P}\mathbf{\Phi}\mathbf{x}_k + \mathbf{g}_k^T\mathbf{P}\mathbf{g}_k \ . \tag{14.39}$$

The following expressions are substituted to achieve a sufficient condition for global uniform asymptotical stability. In order to check $V_{k+1} - V_k < 0$, note that all terms have to be replaced by greater ones (negative terms have to be substituted by those smaller in magnitude): From Eq.(14.5)

$$\mathbf{\Phi}^T\mathbf{P}\mathbf{\Phi} = \mathbf{P}/(1+\alpha) - \mathbf{Q}/(1+\alpha) \tag{14.40}$$

$$\mathbf{g}_k^T\mathbf{P}\mathbf{g}_k < \gamma^2\mathbf{x}_k^T\mathbf{P}\mathbf{x}_k < \gamma^2\mathbf{x}_k^T\mathbf{x}_k\lambda_{\max}[\mathbf{P}] \ , \tag{14.41}$$

$$\mathbf{x}_k^T\mathbf{Q}\mathbf{x}_k > \mathbf{x}_k^T\mathbf{x}_k\lambda_{\min}[\mathbf{Q}] \tag{14.42}$$

$$2\mathbf{g}_k^T\mathbf{P}\mathbf{\Phi}\mathbf{x}_k < 2\gamma\lambda_{\max}[\mathbf{P}] \ \|\mathbf{\Phi}\|_s\mathbf{x}_k^T\mathbf{x}_k \ . \tag{14.43}$$

Hence,

$$V_{k+1} - V_k \le \mathbf{x}_k^T\{\mathbf{P}/(1+\alpha) - \mathbf{Q}/(1+\alpha) - \mathbf{P} + 2\gamma\lambda_{\max}[\mathbf{P}] \ \|\mathbf{\Phi}\|_s + \gamma^2\lambda_{\max}[\mathbf{P}]\}\mathbf{x}_k \tag{14.44}$$

$$\begin{aligned} V_{k+1} - V_k \le \mathbf{x}_k^T\{ & - \alpha/(1+\alpha)\mathbf{I} \ \lambda_{\min}[\mathbf{P}] - \mathbf{I} \ \lambda_{\min}[\mathbf{Q}]/(1+\alpha) & (14.45)\\ & + 2\gamma\mathbf{I}\lambda_{\max}[\mathbf{P}] \ \|\mathbf{\Phi}\|_s + \gamma^2\mathbf{I}\lambda_{\max}[\mathbf{P}] \ \} \ \mathbf{x}_k \ . & (14.46) \end{aligned}$$

$$V_{k+1} - V_k = 0 \quad \leadsto \quad \gamma_o^2 + 2\gamma_o \ \|\mathbf{\Phi}\|_s - \frac{\alpha\lambda_{\min}[\mathbf{P}] + \lambda_{\min}[\mathbf{Q}]}{(1+\alpha)\lambda_{\max}[\mathbf{P}]} = 0 \tag{14.47}$$

$$\text{or} \quad \gamma_o = -\|\mathbf{\Phi}\|_s + \sqrt{\|\mathbf{\Phi}\|_s^2 + \frac{\alpha\lambda_{\min}[\mathbf{P}] + \lambda_{\min}[\mathbf{Q}]}{(1+\alpha)\lambda_{\max}[\mathbf{P}]}} \ . \tag{14.48}$$

Only the positive sign of γ_o is permissible. If $\gamma < \gamma_o$ then the first difference of the Lyapunov function $V_{k+1} - V_k < 0$ and the system is stable.

In the case of $\mathbf{Q} = \mathbf{I}$, Eq.(14.48) can be rewritten to

$$\gamma_o = \frac{\alpha\lambda_{\min}[\mathbf{P}] + 1}{(1+\alpha)\lambda_{\max}[\mathbf{P}]\{\|\mathbf{\Phi}\|_s + \sqrt{\|\mathbf{\Phi}\|_s^2 + \frac{\alpha\lambda_{\min}[\mathbf{P}]+1}{(1+\alpha)\lambda_{\max}[\mathbf{P}]}} \ \}} \tag{14.49}$$

(calculations omitted). Replacing $\|\mathbf{\Phi}\|_s$ by $\rho_s[\mathbf{\Phi}]$ causes the right-hand side to increase which is admissible when a necessary condition is required. Next, in Eq.(14.48) $\lambda_{\min}[\mathbf{P}]/\lambda_{\max}[\mathbf{P}]$ is replaced by unity and - invoking Eq.(14.21) - $\lambda_{\max}^{-1}[\mathbf{P}]$ by the expression $\{1-(1+\alpha)\rho_s^2[\mathbf{\Phi}]\}$, since both substituting quantities are greater than the replaced ones. Thus, from Eq.(14.48), with $\mathbf{Q} = \mathbf{I}$

$$\gamma_o \ < \ -\rho_s[\mathbf{\Phi}] + \sqrt{\rho_s^2[\mathbf{\Phi}] + \frac{1}{1+\alpha}\Big(\alpha\frac{\lambda_{\min}[\mathbf{P}]}{\lambda_{\max}[\mathbf{P}]} + \lambda_{\max}^{-1}[\mathbf{P}]\Big)} \tag{14.50}$$

$$\gamma_o \ < \ -\rho_s[\mathbf{\Phi}] + \sqrt{\rho_s^2[\mathbf{\Phi}] + \frac{1}{1+\alpha}\Big(\alpha \times 1 + 1 - (1+\alpha)\rho_s^2[\mathbf{\Phi}]\Big)} = 1 - \rho_s[\mathbf{\Phi}] \ . \tag{14.51}$$

14.8 Upper Bound For Diagonizable Φ

The solution of Eq.(14.1) is given by

$$\mathbf{x}_k = \Phi^k \mathbf{x}_o + \sum_{i=0}^{k-1} \Phi^{k-i-1}\, \mathbf{g}_i \tag{14.52}$$

which can easily be proved by substitution

$$\mathbf{x}_{k+1} = \Phi^{k+1}\mathbf{x}_o + \sum_{i=0}^{k} \Phi^{k-i}\mathbf{g}_i = \Phi\mathbf{x}_k + \mathbf{g}_k = \Phi\Big(\Phi^k\mathbf{x}_o + \sum_{i=0}^{k-1}\Phi^{k-i-1}\mathbf{g}_i\Big) + \mathbf{g}_k\,. \tag{14.53}$$

Taking norms of both sides of Eq.(14.52) and using the gain bound γ yields

$$\|\mathbf{x}_k\|_F \le \|\Phi^k\|_s\, \|\mathbf{x}_o\|_F + \gamma \sum_{i=0}^{k-1} \|\Phi^{k-i-1}\|_s\, \|\mathbf{x}_i\|_F\,. \tag{14.54}$$

Note that any matrix can be decomposed into the sum of a diagonizable matrix and another matrix. The other matrix may be treated as a part of the perturbation (*Patel, R.V., and Toda, M., 1980*). If Φ is considered diagonizable in the above sense

$$\Phi = \mathbf{T}\Lambda\mathbf{T}^{-1} \;\rightsquigarrow\; \Phi^k = \mathbf{T}\Lambda^k\mathbf{T}^{-1} \text{ where } \Lambda = \text{diag}\,\{\lambda_i[\Phi]\}, \tag{14.55}$$

$$\|\Phi^k\|_s \le \|\mathbf{T}\|_s\, \|\mathbf{T}^{-1}\|_s\, \|\Lambda^k\|_s = \kappa_s[\mathbf{T}]\rho_s^k[\Phi] \tag{14.56}$$

where $\kappa_s[\mathbf{T}]$ is the condition number and

$$\|\Lambda^k\|_s = \sigma_{\max}[\Lambda^k] = \lambda_{\max}[\sqrt{(\Lambda^H\Lambda)^k}] = |\,\lambda_{\max}[\Phi]\,|^k = \rho_s^k[\Phi]\,. \tag{14.57}$$

From Eq.(14.54)

$$\|\mathbf{x}_k\|_F \le \kappa_s[\mathbf{T}]\;\rho_s^k[\Phi]\;\|\mathbf{x}_o\|_F + \gamma\kappa_s[\mathbf{T}]\sum_{i=0}^{k-1}\rho_s^{k-i-1}[\Phi]\;\|\mathbf{x}_i\|_F\,. \tag{14.58}$$

With regard to the assumption $|\,\lambda[\Phi]\,| < 1/\sqrt{1+\alpha}$ it results $\rho_s[\Phi] < 1/\sqrt{1+\alpha}$ and

$$\|\mathbf{x}_k\|_F \le \kappa_s[\mathbf{T}](1+\alpha)^{-k/2}\;\|\mathbf{x}_o\|_F + \gamma\kappa_s[\mathbf{T}]\sum_{i=0}^{k-1}(1+\alpha)^{-(k-i-1)/2}\;\|\mathbf{x}_i\|_F \tag{14.59}$$

$$(1+\alpha)^{k/2}\|\mathbf{x}_k\|_F \le \kappa_s[\mathbf{T}]\;\|\mathbf{x}_o\|_F + \sum_{i=0}^{k-1}(1+\alpha)^{1/2}\gamma\kappa_s[\mathbf{T}](1+\alpha)^{i/2}\|\mathbf{x}_i\|_F\,. \tag{14.60}$$

Using the Bellman-Gronwall lemma in discrete form (see below) by setting

$$h_i := (1+\alpha)^{i/2}\;\|\mathbf{x}_i\|_F\,,\quad f_i = f := \kappa_s[\mathbf{T}]\;\|\mathbf{x}_o\|_F = \text{constant and } k_i = k\; := (1+\alpha)^{1/2}\gamma\kappa_s[\mathbf{T}] = \text{constant} \tag{14.61}$$

in Eq.(14.73),

$$(1+\alpha)^{k/2}\|\mathbf{x}_k\|_F \le \kappa_s[\mathbf{T}]\;\|\mathbf{x}_o\|_F\Big(1 + (1+\alpha)^{1/2}\gamma\kappa_s[\mathbf{T}]\Big)^k \tag{14.62}$$

$$\|\mathbf{x}_k\|_F \le \kappa_s[\mathbf{T}]\;\|\mathbf{x}_o\|_F\Big(1 + (1+\alpha)^{1/2}\gamma\kappa_s[\mathbf{T}]\Big)^k(1+\alpha)^{-k/2} \tag{14.63}$$

$$\|\mathbf{x}_k\|_F \leq \kappa_s[\mathbf{T}]\ \|\mathbf{x}_o\|_F\Big(\frac{1}{\sqrt{1+\alpha}} + \gamma\kappa_s[\mathbf{T}]\Big)^k . \tag{14.64}$$

In order to guarantee stability it is sufficient that

$$\frac{1}{\sqrt{1+\alpha}} + \gamma\kappa_s[\mathbf{T}] < 1 \quad \leadsto \quad \gamma < \frac{1}{\kappa_s[\mathbf{T}]}\Big(1 - \frac{1}{\sqrt{1+\alpha}}\Big) \triangleq \gamma_3 . \tag{14.65}$$

Since $\rho_s[\mathbf{\Phi}] \leq \frac{1}{\sqrt{1+\alpha}}$ and $\kappa_s[\mathbf{T}] \geq 1$, a necessary condition for Eq.(14.65) is given by (*Yaz, E., and Niu, X., 1989*)

$$\gamma_3 < 1 - \rho_s[\mathbf{\Phi}] . \tag{14.66}$$

14.9 Bellman-Gronwall Lemma. Discrete Form

Consider the real-valued positive sequences h_l, k_l and f_l. Assume the following inequality holds

$$h_l \leq f_l + \sum_{i=0}^{l-1} k_i h_i \qquad l = 0,1,2\dots . \tag{14.67}$$

In detail,

$$\begin{aligned} h_o &\leq f_o & (14.68)\\ h_1 &\leq f_1 + k_o h_o \leq f_1 + k_o f_o & (14.69)\\ h_2 &\leq f_2 + k_1 h_1 + k_o h_o \leq f_2 + (1+k_1)k_o f_o + k_1 f_1 & (14.70)\\ h_3 &\leq f_3 + (1+k_1)(1+k_2)k_o f_o + (1+k_2)k_1 f_1 + k_2 f_2 \quad \text{etc.} & (14.71) \end{aligned}$$

By induction,

$$h_l \leq f_l + \sum_{i=0}^{l-1} \prod_{j=i+1}^{l-1} (1+k_j)k_i f_i \tag{14.72}$$

putting $\prod_j^{l-1}(1+k_j) = 1$ if $j > l-1$. If k and f are constant from Eq.(14.71) it results

$$h_l \leq f(1+k)^l . \tag{14.73}$$

14.10 Error Matrix Representation

Let the uncertain system be

$$\mathbf{x}_{k+1} = (\mathbf{\Phi} + \mathbf{E}_d)\mathbf{x}_k . \tag{14.74}$$

The error matrix $\mathbf{E}_d$ is represented by a sum of constant matrices weighted by scalar coefficients d_i and combined in a perturbation vector $\mathbf{d}$ (*Yaz, E., 1988*)

$$\mathbf{E}_d \triangleq \sum_{i=1}^{m_d} d_i \mathbf{D}_i \qquad \mathbf{d} \triangleq (d_1\ d_2 \dots d_{m_d})^T . \tag{14.75}$$

The first difference of the Lyapunov function is given by

$$\begin{aligned} V_{k+1} - V_k &= \mathbf{x}_{k+1}^T \mathbf{P}\mathbf{x}_{k+1} - \mathbf{x}_k^T \mathbf{P}\mathbf{x}_k & (14.76)\\ &= \mathbf{x}_k^T[-\frac{\alpha}{1+\alpha}\mathbf{P} - \frac{1}{1+\alpha}\mathbf{I} + \mathbf{E}_d^T\mathbf{P}\mathbf{\Phi} + \mathbf{\Phi}^T\mathbf{P}\mathbf{E}_d + \mathbf{E}_d^T\mathbf{P}\mathbf{E}_d]\mathbf{x}_k & (14.77) \end{aligned}$$

where Eq.(14.4) was used with $\mathbf{Q} = \mathbf{I}$.

14.10.1 Singular-Value Properties

If singular-value properties are applied, see Eq.(2.71),

$$V_{k+1} - V_k \le -\frac{1+\alpha\lambda_{\min}[\mathbf{P}]}{1+\alpha}\mathbf{x}_k^T\mathbf{x}_k + \{2\|\mathbf{E}_d\|_s\lambda_{\max}[\mathbf{P}]\ \|\mathbf{\Phi}\|_s\|\mathbf{E}_d\|_s^2\lambda_{\max}[\mathbf{P}]\}\mathbf{x}_k^T\mathbf{x}_k\ . \quad (14.78)$$

If the quantity on the right-hand side is negative definite a sufficient condition for stability is achieved. From

$$\|\mathbf{E}_d\|_s^2 + 2\|\mathbf{E}_d\|_s\|\mathbf{\Phi}\|_s - \frac{1+\alpha\lambda_{\min}[\mathbf{P}]}{(1+\alpha)\lambda_{\max}[\mathbf{P}]} = 0 \quad (14.79)$$

$$\|\mathbf{E}_d\|_s < -\|\mathbf{\Phi}\|_s + \sqrt{\|\mathbf{\Phi}\|_s^2 + \frac{1+\alpha\lambda_{\min}[\mathbf{P}]}{(1+\alpha)\lambda_{\max}[\mathbf{P}]}}\ . \quad (14.80)$$

14.10.2 Using Inner Product Property

Consider the simple statements

$$\|\mathbf{\Lambda}^{1/2}\mathbf{a} - \mathbf{\Lambda}^{-1/2}\mathbf{b}\|_F^2 \ge 0 \qquad \text{where} \qquad \mathbf{\Lambda} = \mathbf{\Lambda}^T > 0 \quad (14.81)$$

$$(\mathbf{\Lambda}^{1/2}\mathbf{a} - \mathbf{\Lambda}^{-1/2}\mathbf{b})^T(\mathbf{\Lambda}^{1/2}\mathbf{a} - \mathbf{\Lambda}^{-1/2}\mathbf{b}) = \mathbf{a}^T\mathbf{\Lambda}\mathbf{a} - \mathbf{b}^T\mathbf{a} - \mathbf{a}^T\mathbf{b} + \mathbf{b}^T\mathbf{\Lambda}^{-1}\mathbf{b} \ge 0 \quad (14.82)$$

$$\mathbf{a}^T\mathbf{\Lambda}\mathbf{a} + \mathbf{b}^T\mathbf{\Lambda}^{-1}\mathbf{b} \ge 2\mathbf{a}^T\mathbf{b}\ . \quad (14.83)$$

From Eq.(14.74) and (14.76) it follows

$$V_{k+1} - V_k = \mathbf{x}_k^T[\mathbf{\Phi}\mathbf{P}\mathbf{\Phi} - \mathbf{P} + \mathbf{E}_d^T\mathbf{P}\mathbf{\Phi} + \mathbf{\Phi}^T\mathbf{P}\mathbf{E}_d + \mathbf{E}_d^T\mathbf{P}\mathbf{E}_d]\mathbf{x}_k\ . \quad (14.84)$$

Selecting two terms from above and applying Eq.(14.83) yields with $\mathbf{\Lambda} = \alpha\mathbf{I}$

$$\begin{aligned}
\mathbf{x}_k^T[\mathbf{E}_d^T\mathbf{P}\mathbf{\Phi} + \mathbf{\Phi}^T\mathbf{P}\mathbf{E}_d]\mathbf{x}_k &= 2\mathbf{x}_k\mathbf{\Phi}^T\mathbf{P}\mathbf{E}_d\mathbf{x}_k = 2(\mathbf{P}^{1/2}\mathbf{\Phi}\mathbf{x}_k)^T(\mathbf{P}^{1/2}\mathbf{E}_d\mathbf{x}_k) \triangleq 2\mathbf{a}^T\mathbf{b} \\
&\le (\mathbf{P}^{1/2}\mathbf{\Phi}\mathbf{x}_k)^T\alpha(\mathbf{P}^{1/2}\mathbf{\Phi}\mathbf{x}_k) + (\mathbf{x}_k^T\mathbf{E}_d^T\mathbf{P}^{1/2})\frac{1}{\alpha}(\mathbf{P}^{1/2}\mathbf{E}_d\mathbf{x}_k) \\
&\le \alpha\mathbf{x}_k^T\mathbf{\Phi}^T\mathbf{P}\mathbf{\Phi}\mathbf{x}_k + \frac{1}{\alpha}\mathbf{x}_k^T\mathbf{E}_d^T\mathbf{P}\mathbf{E}_d\mathbf{x}_k\ . \qquad (14.85)
\end{aligned}$$

Combining with Eq.(14.84), followed by employing Eq.(14.5) with $\mathbf{Q} = \mathbf{I}$, one has

$$\begin{aligned}
V_{k+1} - V_k &= \mathbf{x}_k^T[(1+\alpha)\mathbf{\Phi}^T\mathbf{P}\mathbf{\Phi} - \mathbf{P} + (1+\frac{1}{\alpha})\mathbf{E}_d^T\mathbf{P}\mathbf{E}_d]\mathbf{x}_k \qquad (14.86) \\
&= \mathbf{x}_k^T[-\mathbf{I} + (1+\frac{1}{\alpha})\mathbf{E}_d^T\mathbf{P}\mathbf{E}_d]\mathbf{x}_k < 0 \qquad (14.87)
\end{aligned}$$

and, finally,

$$\|\mathbf{E}_d\|_s < \frac{1}{\sqrt{(1+\frac{1}{\alpha})\ \lambda_{\max}[\mathbf{P}]}}\ . \quad (14.88)$$

14.10.3 Stochastic Parameter Perturbation

Consider the matrix representation of $\mathbf{E}_d$ with stochastic parameter d_{si}

$$\mathbf{E}_d = \sum_{i=1}^{m_d} d_{si}\mathbf{D}_i \tag{14.89}$$

where $\mathbf{D}_i$ are constant matrices and d_{si} are zero mean scalar random sequences characterized by variances $v_i = E\{d_{si}^2\} = \mathrm{cov}\ d_{si}$. The first difference of the Lyapunov function is established

$$E\{V_{k+1} \mid \mathbf{x}_k, \mathbf{x}_{k-1}\ldots\} - V_k = \mathbf{x}_k^T[\boldsymbol{\Phi}^T\mathbf{P}\boldsymbol{\Phi} + \sum_{i=1}^{m_d} v_i\mathbf{D}_i\mathbf{P}\mathbf{D}_i]\mathbf{x}_k - \mathbf{x}_k^T\mathbf{P}\mathbf{x}_k \tag{14.90}$$

$$= \mathbf{x}_k^T[-\frac{1}{1+\alpha}\mathbf{I} - \frac{\alpha}{1+\alpha}\mathbf{P} + \sum_{i=1}^{m_d} v_i\bar{\mathbf{P}}_i]\mathbf{x}_k \quad \text{where } \bar{\mathbf{P}}_i = \mathbf{D}_i^T\mathbf{P}\mathbf{D}_i \tag{14.91}$$

$$\|\sum_{i=1}^{m_d} v_i\bar{\mathbf{P}}_i\|_s = \sigma_{\max}[\sum_{i=1}^{m_d} v_i\bar{\mathbf{P}}_i] < \frac{1+\alpha\lambda_{\min}[\mathbf{P}]}{1+\alpha}\,. \tag{14.92}$$

14.10.4 Norm Relations

There are three possibilities of evaluating the singular value $\|\mathbf{E}_d\|_s$ by its known components $\mathbf{D}_i$ and its unknown coefficients d_i. First,

$$\begin{aligned}\|\sum_{i=1}^{m_d} d_i\mathbf{D}_i\|_s &= \sigma_{\max}[\sum_{i=1}^{m_d} d_i\mathbf{D}_i] = \|\ (\mathbf{D}_1 \vdots \mathbf{D}_2\ldots\mathbf{D}_{m_d})(d_1\mathbf{I} \vdots d_2\mathbf{I}\ldots d_{m_d}\mathbf{I})^T\|_s \\ &\le \|\mathbf{D}_e\|_s\sqrt{\sum_{i=1}^{m_d} d_i^2} = \|\mathbf{d}\|_F\|\mathbf{D}_e\|_s \quad \text{where } \mathbf{D}_e \triangleq [\mathbf{D}_1 \vdots \mathbf{D}_2\ldots\mathbf{D}_{m_d}]\,,\end{aligned} \tag{14.93}$$

second,

$$\|\sum_{i=1}^{m_d} d_i\mathbf{D}_i\|_s \le \sum_{i=1}^{m_d} |d_i|\ \|\mathbf{D}_i\|_s \le \Big(\sum_{i=1}^{m_d} |d_i|\Big)\Big(\max_i \|\mathbf{D}_i\|_s\Big) \tag{14.94}$$

and, third,

$$\|\sum_{i=1}^{m_d} d_i\mathbf{D}_i\|_s \le \|\sum_{i=1}^{m_d} |d_i\mathbf{D}_i|\ \|_s \le \Big(\max_i |d_i|\Big)\|\sum_{i=1}^{m_d} |\mathbf{D}_i|\ \|_s = \|\mathbf{d}\|_\infty\|\sum_{i=1}^{m_d} |\mathbf{D}_i|\ \|_s\,. \tag{14.95}$$

If stochastic parameters d_{si} and their variances are used to model the uncertainty d_i, then the expressions $|d_i|, \mathbf{d}, \mathbf{D}_e$ can be replaced by $v_i, v_i, \mathbf{v} = (v_1\ v_2\ldots v_{m_d})^T$ and $\mathbf{P}_e \triangleq (\bar{\mathbf{P}}_1 \vdots \bar{\mathbf{P}}_2\ldots\bar{\mathbf{P}}_{m_d})$, respectively.

14.11 Nonlinear and Random Perturbations

Consider the linear discrete-time single-input single-output system with a feedback nonlinearity $g_k(y_k)$ subject to multiplicative noises ξ_{1i}, ξ_{2i} and ξ_{3i} (Fig. 14.1)

$$\mathbf{x}_{k+1} = (\boldsymbol{\Phi} + \sum_{i=1}^{n_1} \xi_{1ik}\boldsymbol{\Phi}_i)\mathbf{x}_k + (\boldsymbol{\Psi} + \sum_{i=1}^{n_3} \xi_{3ik}\boldsymbol{\Psi}_i)u_k - (\mathbf{d} + \sum_{i=1}^{n_2} \xi_{2ik}\mathbf{d}_i)g_k(y_k) \tag{14.96}$$

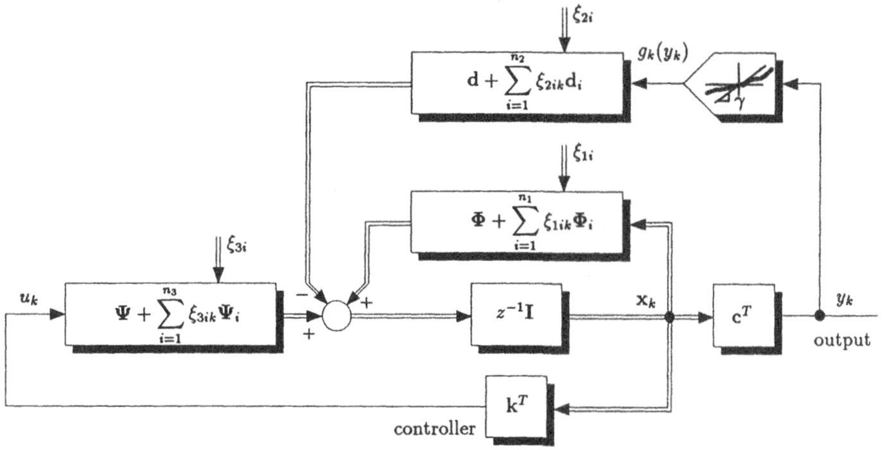

Figure 14.1: Discrete-time system with nonlinear feedback and random perturbations

$$y_k = \mathbf{c}^T \mathbf{x}_k \qquad u_k = \mathbf{k}^T \mathbf{x}_k \qquad 0 \leq |g_k(y_k)| \leq \gamma\, y_k \,. \tag{14.97}$$

The question is to which extent the sector bound γ of $g_k(y_k)$ may be increased without affecting the stability of the overall system (*Yaz, E., and Yildizbayrak, N., 1985*).

The independent sequences ξ_{1ik}, ξ_{2ik} and ξ_{3ik} are assumed scalar white Gaussian of zero mean and with variances σ_{1i}, σ_{2i} and σ_{3i}. The output signal y_k is the input of the nonlinearity represented by the $g_k(y_k)$. The matrix $\mathbf{\Phi}_c = \mathbf{\Phi} + \mathbf{\Psi}\mathbf{k}^T$ is assumed stable. A sufficient condition for the origin to be asymptotically stable is stated by

$$\Delta V_k \triangleq E\{V(\mathbf{x}_{k+1})|\mathbf{x}_k, \mathbf{x}_{k-1} \ldots \mathbf{x}_o\} - V(\mathbf{x}_k) < -\gamma_o(\|\mathbf{x}_k\|_F) < 0 \qquad \forall \mathbf{x}_o \,. \tag{14.98}$$

That is, the difference of the expectation of the Lyapunov function at $k+1$, based on the previous states up to $\mathbf{x}_k$, minus the Lyapunov function $V(\mathbf{x}_k)$ should be smaller than a negative quantity $-\gamma_o$. Choosing the Lyapunov function $\mathbf{x}_k^T\mathbf{P}\mathbf{x}_k$ with $\mathbf{P} > 0$ and evaluating ΔV_k according to the difference $\mathbf{x}_{k+1}^T\mathbf{P}\mathbf{x}_{k+1} - \mathbf{x}_k^T\mathbf{P}\mathbf{x}_k$ yields

$$\begin{aligned}\Delta V_k = \mathbf{x}_k^T\mathbf{\Phi}_c^T\mathbf{P}\mathbf{\Phi}_c\mathbf{x}_k &- \mathbf{x}_k^T\mathbf{P}\mathbf{x}_k - 2\mathbf{d}^T\mathbf{P}\mathbf{\Phi}_c\mathbf{x}_k g_k(y_k) + \mathbf{d}^T\mathbf{P}\mathbf{d}g_k^2(y_k) \\ &+ \mathbf{x}_k^T\Sigma_1\mathbf{x}_k + \mathbf{x}_k^T\mathbf{c}\Sigma_3\mathbf{c}^T\mathbf{x}_k + \Sigma_2 g_k^2(y_k)\end{aligned} \tag{14.99}$$

$$\text{where} \quad \mathbf{\Sigma}_1 = \sum_{i=1}^{n_1}\sigma_{1i}\mathbf{\Phi}_i^T\mathbf{P}\mathbf{\Phi}_i, \quad \Sigma_2 = \sum_{i=1}^{n_2}\sigma_{2i}\mathbf{d}_i^T\mathbf{P}\mathbf{d}_i, \quad \Sigma_3 = \sum_{i=1}^{n_3}\sigma_{3i}\mathbf{\Psi}_i^T\mathbf{P}\mathbf{\Psi}_i \,. \tag{14.100}$$

In the above calculation the independence and zero mean properties have been utilized. Since the nonlinearity obeys the sector condition

$$-\mathbf{d}^T\mathbf{P}\mathbf{\Phi}_c\mathbf{x}_k g_k(y_k) \leq |\mathbf{d}^T\mathbf{P}\mathbf{\Phi}_c\mathbf{x}_k g_k(y_k)| \leq |\mathbf{d}^T\mathbf{P}\mathbf{\Phi}_c\mathbf{x}_k|\;|g_k(y_k)| \leq \|\mathbf{d}^T\mathbf{P}\mathbf{\Phi}_c\|_F\|\mathbf{x}_k\|_F\gamma\|\mathbf{c}\|_F\|\mathbf{x}_k\|_F \,. \tag{14.101}$$

Using the sector inequality $\Sigma_2 g_k^2(y_k) \leq \Sigma_2\gamma^2(\mathbf{c}^T\mathbf{x}_k)^2$ once more,

$$\begin{aligned}\Delta V_k < -\mathbf{x}_k^T\{\mathbf{P} &- \mathbf{\Phi}_c^T\mathbf{P}\mathbf{\Phi}_c - \Sigma_1 - \Sigma_3\mathbf{k}\mathbf{k}^T \\ &- 2\gamma\|\mathbf{d}^T\mathbf{P}\mathbf{\Phi}_c\|_F\|\mathbf{c}\|_F\,\mathbf{I} - \gamma^2(\mathbf{d}^T\mathbf{P}\mathbf{d} + \Sigma_2)\mathbf{c}\mathbf{c}^T\}\mathbf{x}_k \,.\end{aligned} \tag{14.102}$$

A sufficient condition for ΔV_k to be negative is that the matrix-valued brace term above is positive definite which is guaranteed if

$$\lambda_{\min}[\mathbf{P} - \mathbf{\Phi}_c^T\mathbf{P}\mathbf{\Phi}_c - \Sigma_1 - \Sigma_3\mathbf{k}\mathbf{k}^T] > \lambda_{\max}[2\gamma\|\mathbf{d}^T\mathbf{P}\mathbf{\Phi}_c\|_F\|\mathbf{c}\|_F\,\mathbf{I} + \gamma^2(\mathbf{d}^T\mathbf{P}\mathbf{d} + \Sigma_2)\mathbf{c}\mathbf{c}^T] \tag{14.103}$$

$$\lambda_{\min}[\mathbf{P} - \mathbf{\Phi}_c^T\mathbf{P}\mathbf{\Phi}_c - \Sigma_1 - \Sigma_3\mathbf{k}\mathbf{k}^T] > 2\gamma\|\mathbf{d}^T\mathbf{P}\mathbf{\Phi}_c\|_F\|\mathbf{c}\|_F + \gamma^2(\mathbf{d}^T\mathbf{P}\mathbf{d} + \Sigma_2)\ \lambda_{\max}[\mathbf{c}\mathbf{c}^T]\ . \tag{14.104}$$

Since rank $[\mathbf{c}\mathbf{c}^T] = 1$ and $\lambda_{\max}[\mathbf{c}\mathbf{c}^T] = \mathrm{tr}\,(\mathbf{c}\mathbf{c}^T) = \mathbf{c}^T\mathbf{c} = \|\mathbf{c}\|_F^2$

$$\gamma^2(\mathbf{d}^T\mathbf{P}\mathbf{d}+\Sigma_2)\|\mathbf{c}\|_F^2+2\gamma\|\mathbf{d}^T\mathbf{P}\mathbf{\Phi}_c\|_F\|\mathbf{c}\|_F-\lambda_{\min}[\mathbf{P}-\mathbf{\Phi}_c^T\mathbf{P}\mathbf{\Phi}_c-\Sigma_1-\Sigma_3\mathbf{k}\mathbf{k}^T] < 0\ . \tag{14.105}$$

The result of this quadratic inequality in γ is given, if γ is smaller or equal to

$$\frac{-\|\mathbf{d}^T\mathbf{P}\mathbf{\Phi}_c\|_F\|\mathbf{c}\|_F + \sqrt{\|\mathbf{d}^T\mathbf{P}\mathbf{\Phi}_c\|_F^2\|\mathbf{c}\|_F^2 + (\mathbf{d}^T\mathbf{P}\mathbf{d} + \Sigma_2)\|\mathbf{c}\|_F^2\lambda_{\min}[\mathbf{P} - \mathbf{\Phi}_c^T\mathbf{P}\mathbf{\Phi}_c - \Sigma_1 - \Sigma_3\mathbf{k}\mathbf{k}^T]}}{(\mathbf{d}^T\mathbf{P}\mathbf{d}^T + \Sigma_2)\|\mathbf{c}\|_F^2}\ . \tag{14.106}$$

The solution only is positive and, thus, of interest if

$$\mathbf{P} - \mathbf{\Phi}_c^T\mathbf{P}\mathbf{\Phi}_c - \Sigma_1 - \Sigma_3\mathbf{k}\mathbf{k}^T > 0\ . \tag{14.107}$$

14.12 Robust Predictive Regulators With Computation Delays

Computation delays in digital control systems can be compensated using state predictors. Unstructured perturbations are considered primarily. Both the return difference matrices and the characteristic equation are used. Associated expressions utilizing singular values provide tighter bounds on the stability robustness than related approaches using Lyapunov functions.

Consider the discrete-time linear plant with sampling period T

$$\mathbf{x}(k+1) = \mathbf{\Phi}\mathbf{x}(k) + \mathbf{\Psi}\mathbf{u}(k) \qquad \mathbf{\Phi} \triangleq \mathbf{\Phi}(T) \tag{14.108}$$

$$\mathbf{y}(k) = \mathbf{C}\mathbf{x}(k) \tag{14.109}$$

and an L-step ahead predictive controller

$$\mathbf{u}(k+L) = \mathbf{K}\hat{\mathbf{x}}(k+\mathbf{L}|k)\ . \tag{14.110}$$

This optimal controller minimizes the index of performance

$$I = \sum_{k=0}^{\infty}\mathbf{y}^T(k)\mathbf{y}(k) + \mathbf{u}^T(k+L)\mathbf{R}\mathbf{u}(k+L) \qquad \mathbf{R} \geq 0 \tag{14.111}$$

where $\mathbf{K}$ is given by the optimal gain matrix in the case $L = 0$. The L-step ahead state prediction is (*Mita, T., 1985*)

$$\hat{\mathbf{x}}(k+L|k) \triangleq \mathbf{\Phi}^L\mathbf{x}(k) + \mathbf{\Phi}^{L-1}\mathbf{\Psi}\mathbf{u}(k) + \mathbf{\Phi}^{L-2}\mathbf{\Psi}\mathbf{u}(k+1) + \ldots + \mathbf{\Psi}\mathbf{u}(k+L-1)\ . \tag{14.112}$$

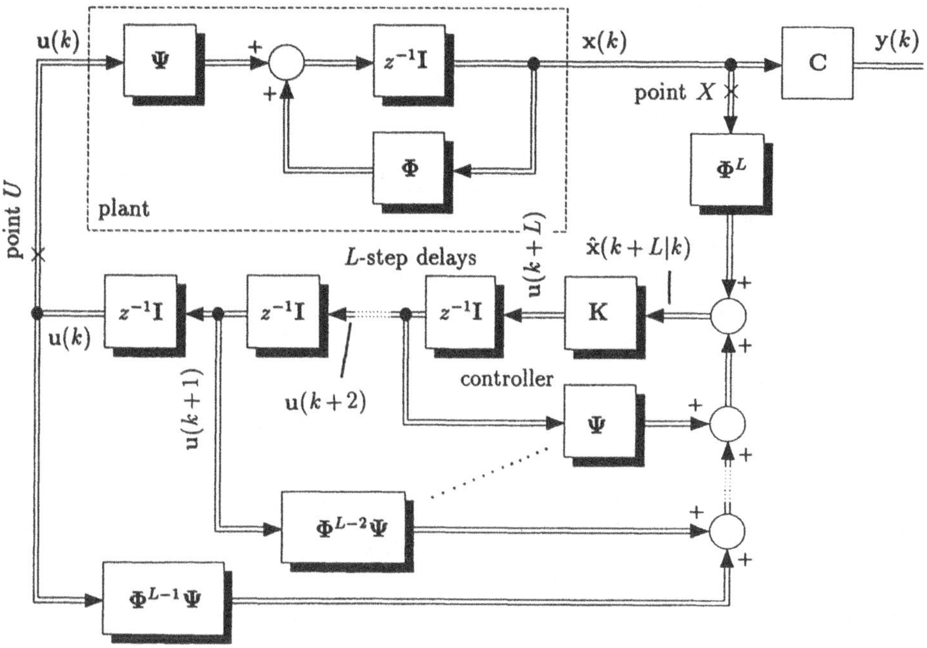

Figure 14.2: Predictive controller compensating for the computation delay

14.12.1 Transfer Matrices

The transfer matrix $\mathbf{H}_{Lux}$ from $\mathbf{x}(k)$ to $\mathbf{u}(k)$ in the L-step prediction case is given by

$$\mathbf{H}_{Lux}(z) = \mathbf{W}_L^{-1}\mathbf{K}\mathbf{\Phi}^L \qquad \text{or} \qquad = \mathbf{K}\mathbf{V}_L^{-1}\mathbf{\Phi}^L \tag{14.113}$$

where

$$\mathbf{W}_L \triangleq z^L\mathbf{I} - \mathbf{K}(z^{L-1}\mathbf{I} + z^{L-2}\mathbf{\Phi} + \ldots + \mathbf{\Phi}^{L-1})\mathbf{\Psi} \tag{14.114}$$

$$\mathbf{V}_L \triangleq z^L\mathbf{I} - (z^{L-1}\mathbf{I} + z^{L-2}\mathbf{\Phi} + \ldots + \mathbf{\Phi}^{L-1})\mathbf{\Psi}\mathbf{K}\ . \tag{14.115}$$

14.12.2 Return Difference Matrices

Note that $\mathbf{H}_{Lxu} \mid_{L=0} = \mathbf{H}_{Oxu} = \mathbf{K}$. Defining

$$\mathbf{S}_{OU} \triangleq \mathbf{I} - \mathbf{K}(z\mathbf{I} - \mathbf{\Phi})^{-1}\mathbf{\Psi}\ , \tag{14.116}$$

the following return difference matrices can be stated. Detailed calculations see *Ishihara, T., 1988*. The index L denotes the L-step ahead prediction case.

Break at point U:

$$\mathbf{S}_{LU} \triangleq \mathbf{I} - \mathbf{H}_{Lxu}(z\mathbf{I} - \mathbf{\Phi})^{-1}\mathbf{\Psi} = z^L\mathbf{W}_L^{-1}\mathbf{S}_{OU} \tag{14.117}$$

$$\mathbf{T}_{LU} \triangleq \mathbf{I} - \mathbf{S}_{LU}^{-1} = -z^{-L}\mathbf{K}(z\mathbf{I} - \mathbf{\Phi} + \mathbf{\Psi}\mathbf{K})^{-1}\mathbf{\Phi}^L\mathbf{\Psi}\ . \tag{14.118}$$

Break at point X:

$$\begin{aligned}
\mathbf{S}_{LX} &\triangleq \mathbf{I} - (z\mathbf{I} - \mathbf{\Phi})^{-1}\mathbf{\Psi}\mathbf{H}_{Lxu} = z^L \mathbf{\Phi}^L \mathbf{S}_{OX} \mathbf{V}_L^{-1} \mathbf{\Phi}^L && (14.119)\\
\mathbf{T}_{LX} &\triangleq \mathbf{I} - \mathbf{S}_{LX}^{-1} = -z^{-L}\mathbf{T}_{OX}\mathbf{\Phi}^L && (14.120)\\
\text{and for } L = 0: \quad \mathbf{S}_{OX} &= \mathbf{I} - (z\mathbf{I} - \mathbf{\Phi})^{-1}\mathbf{\Psi}\mathbf{K} && (14.121)\\
\mathbf{T}_{OX} &= -(z\mathbf{I} - \mathbf{\Phi} - \mathbf{\Psi}\mathbf{K})^{-1}\mathbf{\Psi}\mathbf{K}\,. && (14.122)
\end{aligned}$$

14.13 Perturbed Plant and Optimal Regulator Without Delay

When the system is considered without computation delay ($L = 0$) then

$$\mathbf{x}(k+1) = (\mathbf{\Phi} + \mathbf{\Psi}\mathbf{K} + \Delta\mathbf{\Phi} + \Delta\mathbf{\Psi}\ \mathbf{K})\mathbf{x}(k) \triangleq (\mathbf{F}_d + \Delta\mathbf{F}_d)\mathbf{x}(k)\,. \qquad (14.123)$$

Using the Lyapunov function $V(k) \triangleq \mathbf{x}^T(k)\mathbf{P}\mathbf{x}(k)$ and aiming at a stability robust system, the optimal $\mathbf{K}$ obeys the Riccati equation (*Shaked, U., 1986*)

$$\begin{aligned}
\mathbf{P} &= (\mathbf{\Phi} + \mathbf{\Psi}\mathbf{K})^T\mathbf{P}(\mathbf{\Phi} + \mathbf{\Psi}\mathbf{K}) + \mathbf{C}^T\mathbf{C} + \mathbf{K}^T\mathbf{R}\mathbf{K} && (14.124)\\
&= \mathbf{F}_d^T\mathbf{P}\mathbf{F}_d + \mathbf{C}^T\mathbf{C} + \mathbf{K}^T\mathbf{R}\mathbf{K} \triangleq \mathbf{F}_d^T\mathbf{P}\mathbf{F}_d + \mathbf{Q}\,. && (14.125)
\end{aligned}$$

Using Eq.(14.125), the first difference of the Lyapunov function is

$$\begin{aligned}
V(k+1) - V(k) &= \mathbf{x}^T(k+1)\mathbf{P}\mathbf{x}(k+1) - \mathbf{x}^T(k)\mathbf{P}\mathbf{x}(k) && (14.126)\\
&= \mathbf{x}^T(\mathbf{F}_d^T + \Delta\mathbf{F}_d^T)\mathbf{P}(\mathbf{F}_d + \Delta\mathbf{F}_d)\mathbf{x} - \mathbf{x}^T\mathbf{P}\mathbf{x} && (14.127)\\
&= \mathbf{x}^T[(\mathbf{F}_d^T + \Delta\mathbf{F}_d^T)\mathbf{P}(\mathbf{F}_d + \Delta\mathbf{F}_d) - \mathbf{F}_d^T\mathbf{P}\mathbf{F}_d - \mathbf{C}^T\mathbf{C} - \mathbf{K}^T\mathbf{R}\mathbf{K}]\mathbf{x}^T\\
&= \mathbf{x}^T[\Delta\mathbf{F}_d^T\mathbf{P}\Delta\mathbf{F}_d + \Delta\mathbf{F}_d^T\mathbf{P}\mathbf{F}_d + \mathbf{F}_d^T\mathbf{P}\Delta\mathbf{F}_d - \mathbf{C}^T\mathbf{C} - \mathbf{K}^T\mathbf{R}\mathbf{K}]\mathbf{x}^T\,.
\end{aligned}$$

To continue, there are three theorems for sufficient conditions and bounds on unstructured and structured time-varying perturbations (*Kolla, S.R., 1989*).

14.13.1 Unstructured Perturbation

If the inequality holds that

$$\sigma_{\min}[\mathbf{Q}] > \sigma_{\max}^2[\Delta\mathbf{F}_d]\ \sigma_{\max}[\mathbf{P}] + 2\sigma_{\max}[\Delta\mathbf{F}_d]\ \sigma_{\max}[\mathbf{F}_d]\ \sigma_{\max}[\mathbf{P}]\ , \qquad (14.128)$$

then $V(k+1) - V(k) < 0$ and the perturbed closed-loop system is stable. Solving with respect to $\sigma_{\max}[\Delta\mathbf{F}_d]$ yields

$$\sigma_{\max}[\Delta\mathbf{F}_d] < -\sigma_{\max}[\mathbf{F}_d] + \sqrt{\sigma_{\max}^2[\mathbf{F}_d] + \frac{\sigma_{\min}[\mathbf{Q}]}{\sigma_{\max}[\mathbf{P}]}}\,. \qquad (14.129)$$

As in the continuous-time case the bound is maximum when $\mathbf{C}^T\mathbf{C} + \mathbf{K}^T\mathbf{R}\mathbf{K} = \mathbf{I}$.

14.13.2 Structured Perturbation by Selecting Maximum Entries

For structured perturbations $\Delta\mathbf{\Phi}(k)$ select and define

$$\Delta\varphi_{ij}(k) \le \max_k\ |\Delta\varphi_{ij}(k)| \triangleq e_{dij} = (\mathbf{E}_d)_{ij} \qquad \text{and} \qquad e_{\max} = \max_{i,j} e_{dij}\,. \qquad (14.130)$$

Since an (n,n)-matrix $\mathbf{E}_1$ all of whose elements are e possesses the spectral matrix norm $\|\mathbf{E}_1\|_s = ne$ and since $\|\mathbf{E}_d\|_s < \|\mathbf{E}_1\|_s$, the bound is given by the right-hand side of Eq.(14.129) multiplied by $1/n$.

14.13.3 Unidirectional Perturbation

When the matrix $\mathbf{E}_d(k)$ is known exactly, define $\mathbf{U}$ with entries $u_{ij} = e_{dij}/e_{\max}$. Then, $0 \le u_{ij} \le 1$. If e_{dij} is not known for some i, j the entry can be chosen arbitrarily, if there is no perturbation at the (i, j)-position select $u_{ij} = 0$. Thus $|\Delta\mathbf{\Phi}(k)| \le_e e_{\max}\mathbf{U}$ where $\mathbf{U}$ is a specified matrix with non-negative elements. Combining Eqs.(14.127) and (14.125) yields

$$\Delta V(\mathbf{x}) = V[\mathbf{x}(k+1)] - V[\mathbf{x}(k)] = \mathbf{x}^T[-\mathbf{Q} + 2(\Delta\mathbf{\Phi}^T\mathbf{P}\mathbf{\Phi})_s + \Delta\mathbf{\Phi}^T\mathbf{P}\Delta\mathbf{\Phi}]\mathbf{x} < 0 \quad (14.131)$$

which is satisfied if (*Kolla, S.R., 1989*)

$$\sigma_{\max}[2(\Delta\mathbf{\Phi}^T\mathbf{P}\mathbf{\Phi})_s + \Delta\mathbf{\Phi}^T\mathbf{P}\Delta\mathbf{\Phi}] < \sigma_{\min}[\mathbf{Q}] \quad (14.132)$$

$$\Leftarrow \quad 2\sigma_{\max}[\,|\Delta\mathbf{\Phi}^T\mathbf{P}\mathbf{\Phi}|_s] + \sigma_{\max}[\,|\Delta\mathbf{\Phi}^T\mathbf{P}\Delta\mathbf{\Phi}|\,] < \sigma_{\min}[\mathbf{Q}] \quad (14.133)$$

$$\Leftarrow \quad 2e_{\max}\sigma_{\max}[(\mathbf{U}^T|\mathbf{P}\mathbf{\Phi}|)_s] + e_{\max}^2\sigma_{\max}[\mathbf{U}^T|\mathbf{P}|\mathbf{U}] < \sigma_{\min}[\mathbf{Q}] \quad (14.134)$$

$$e_{\max} < -\frac{\sigma_{\max}[(\mathbf{U}^T|\mathbf{P}\mathbf{\Phi}|)_s]}{\sigma_{\max}[\mathbf{U}^T|\mathbf{P}|\mathbf{U}]} + \sqrt{\left(\frac{\sigma_{\max}[(\mathbf{U}^T|\mathbf{P}\mathbf{\Phi}|)_s]}{\sigma_{\max}[\mathbf{U}^T|\mathbf{P}|\mathbf{U}]}\right)^2 + \frac{\sigma_{\min}[\mathbf{Q}]}{\sigma_{\max}[\mathbf{U}^T|\mathbf{P}|\mathbf{U}]}} \; . \quad (14.135)$$

14.13.4 Using State Transformation into a Quasi-Diagonal Form

Improved bounds can be achieved when a state transformation into a quasi-diagonal form is used. Replace the transition matrix $\mathbf{\Phi} = \mathbf{\Phi}(T)$ by $\bar{\mathbf{\Phi}} = \mathbf{M}^{-1}\mathbf{\Phi}\mathbf{M}$ and solve

$$\bar{\mathbf{\Phi}}^T\bar{\mathbf{P}}\bar{\mathbf{\Phi}} - \bar{\mathbf{P}} + \mathbf{Q} = 0 \; . \quad (14.136)$$

Then, the bound for the unstructured additive perturbation $\mathbf{E}_d$ in $\mathbf{\Phi}_p = \mathbf{\Phi} + \mathbf{E}_d$ is given by

$$\sigma_{\max}[\mathbf{E}_d] < \kappa_s^{-1}[\mathbf{M}]\;\bar{\mu}_u \quad (14.137)$$

where $\bar{\mu}_u$ equals the right-hand side of Eq.(14.129) when $\mathbf{F}_d$, $\mathbf{P}$ are replaced by $\bar{\mathbf{\Phi}}, \bar{\mathbf{P}}$, respectively.

In the case of structured perturbations better bounds are obtained by using a diagonal matrix $\mathbf{M} = \mathrm{diag}\{m_i\}$. The perturbed system is stable if

$$e_{\max} < \frac{\bar{\mu}_s}{\max_{ij}\;|m_j/m_i|\;u_{ij}} \quad (14.138)$$

where

$$e_{ij}(k) \le \max_k |e_{dij}(k)| \qquad \mathbf{E}_d(k) = \mathrm{matrix}[e_{dij}(k)] \quad (14.139)$$

$$e_{\max} \triangleq \max_{ij} \quad |m_j/m_i|\;[\,\max_k |e_{dij}(k)|\,] \quad (14.140)$$

$$\bar{\mathbf{U}} = \mathrm{matrix}[\bar{u}_{ij}] = \mathrm{matrix}[\frac{|m_j/m_i|\;\max_k |e_{dij}(k)|}{e_{\max}}] \quad (14.141)$$

and $\bar{\mu}_s$ results from the right-hand side of Eq.(14.135) when $\mathbf{U}$ and $\mathbf{P}$ are replaced by $\bar{\mathbf{U}}$ and $\bar{\mathbf{P}}$, respectively (*Kolla, S.R., and Farison, J.B., 1990a*).

14.14 Perturbed System Including Computation Delay

Associated with Fig. 14.2, a combined vector is defined comprising the vectors $\mathbf{x}(k)$ and $\mathbf{u}(k+l)$ $\forall l = [0, 1 \dots L-1]$

$$\bar{\mathbf{x}}(k) \triangleq [\mathbf{x}^T(k) \vdots \mathbf{u}^T(k) \vdots \mathbf{u}^T(k+1) \dots \mathbf{u}^T(k+L-1)]^T . \tag{14.142}$$

Then, the perturbed system with L-step control input delays is rewritten to

$$\bar{\mathbf{x}}(k+1) = (\mathbf{\Phi}_L + \Delta\mathbf{\Phi}_L)\bar{\mathbf{x}}(k) + \mathbf{\Psi}_L \mathbf{u}(k+L) \tag{14.143}$$

$$\mathbf{y}(k) = \mathbf{C}_L \bar{\mathbf{x}}(k) \tag{14.144}$$

where

$$\mathbf{\Phi}_L \triangleq \begin{pmatrix} \mathbf{\Phi} & \mathbf{\Psi} & \mathbf{0} & \dots & \mathbf{0} \\ \mathbf{0} & \mathbf{0} & \mathbf{I} & \ddots & \mathbf{0} \\ \vdots & \vdots & \ddots & \ddots & \vdots \\ \mathbf{0} & \vdots & \ddots & \ddots & \mathbf{I} \\ \mathbf{0} & \mathbf{0} & \dots & \dots & \mathbf{0} \end{pmatrix}, \quad \Delta\mathbf{\Phi}_L \triangleq \begin{pmatrix} \Delta\mathbf{\Phi} & \Delta\mathbf{\Psi} & \mathbf{0} & \dots & \mathbf{0} \\ \mathbf{0} & \mathbf{0} & \mathbf{0} & \ddots & \mathbf{0} \\ \vdots & \vdots & \ddots & \ddots & \vdots \\ \mathbf{0} & \vdots & \ddots & \ddots & \mathbf{0} \\ \mathbf{0} & \mathbf{0} & \dots & \dots & \mathbf{0} \end{pmatrix}, \quad \mathbf{\Psi}_L \triangleq \begin{pmatrix} \mathbf{0} \\ \mathbf{0} \\ \vdots \\ \mathbf{0} \\ \mathbf{I} \end{pmatrix} . \tag{14.145}$$

Referring to $\mathbf{F}_d = \mathbf{\Phi} + \mathbf{\Psi}\mathbf{K}$ and replacing $\mathbf{\Phi}, \mathbf{\Psi}, \mathbf{C}$ by $\mathbf{\Phi}_L, \mathbf{\Psi}_L, \mathbf{C}_L$, respectively, in Eq.(14.129), the robust stability condition in the case of perturbed plant and predictor-based optimal regulator is given. Using Eq.(14.135), improved results are obtained (*Kolla, S.R., and Farison, J.B., 1990*).

14.15 Singular Value of the Return Difference Matrix

On the unit circle, i.e. at $|z| = 1$,

$$\sigma_{\min}[\mathbf{S}_{LU}] \geq \frac{\sigma_{\min}[\mathbf{S}_{OU}]}{\sigma_{\max}[\mathbf{W}_L]} \tag{14.146}$$

$$\Leftarrow \quad \sigma_{\min}[\mathbf{S}_{LU}] \geq \frac{\sigma_{\min}[\mathbf{S}_{OU}]}{1 + \sigma_{\max}[\mathbf{K}]\ \sigma_{\max}[\mathbf{\Psi}]\big(1 + \sigma_{\max}[\mathbf{\Phi}] + \dots + \sigma_{\max}^{L-1}[\mathbf{\Phi}]\big)} . \tag{14.147}$$

In the relation above the effect of computation delays is expressed. The bounds on the minimum singular value of the return difference matrix $\mathbf{S}_{LU}$ is a robustness measure for the control loop when unstructured perturbations of $\mathbf{\Phi}$ and $\mathbf{\Psi}$ are introduced. Using the minimum singular value of $\mathbf{S}_{LX}$ at $|z| = 1$

$$\sigma_{\min}[\mathbf{S}_{LX}] \geq \sigma_{\max}^{-L}[\mathbf{\Phi}]\ \sigma_{\min}[\mathbf{S}_{OX}]\ \sigma_{\max}[\mathbf{V}_L]\ \sigma_{\min}^{L}[\mathbf{\Phi}] \tag{14.148}$$

$$\Leftarrow \quad \sigma_{\min}[\mathbf{S}_{LX}] \geq \frac{\kappa_s^{-L}[\mathbf{\Phi}]\ \sigma_{\min}[\mathbf{S}_{OX}]}{1 + \sigma_{\max}[\mathbf{K}]\ \sigma_{\max}[\mathbf{\Psi}]\big(1 + \sigma_{\max}[\mathbf{\Phi}] + \dots + \sigma_{\max}^{L-1}[\mathbf{\Phi}]\big)} . \tag{14.149}$$

This inequality shows the strong effect of the inverse spectral condition number on the robustness measure, especially in cases of ill-conditioned matrices $\mathbf{\Phi}$.

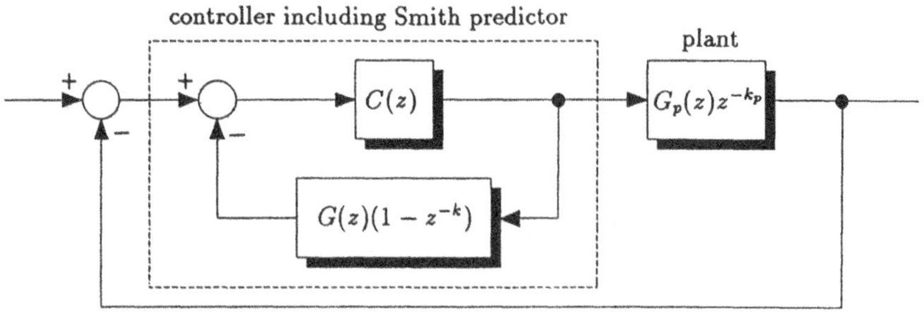

Figure 14.3: Discrete-time Smith predictor and perturbed plant

14.16 Regulator Satisfying $\sigma_{\max}[\mathbf{F}_d] < 1$

The return difference matrix at X is given by $\mathbf{S}_{LX}(z)$. Calculating the determinant, its zeros are the poles of the closed-loop system. For the perturbed plant it results

$$\det \mathbf{S}_{LX}(z) = \det\,[\mathbf{I} - (z\mathbf{I} - \mathbf{\Phi} - \Delta\mathbf{\Phi})^{-1}(\mathbf{\Psi} + \Delta\mathbf{\Psi})\mathbf{K}\mathbf{V}_L^{-1}(z)\mathbf{\Phi}^L] = 0\,. \tag{14.150}$$

If $\mathbf{K}$ is assumed to satisfy the strong closed-loop stability condition $\sigma_{\max}[\mathbf{\Phi} + \mathbf{\Psi}\mathbf{K}] < 1$ then

$$\begin{aligned}\Big(\frac{\sigma_{\max}[\mathbf{\Phi}]}{\sigma_{\min}[\mathbf{\Phi}]}\Big)^L\Big(1 \;&+\; \sigma_{\max}[\mathbf{\Psi}]\;\sigma_{\max}[\mathbf{K}]\{1 + \sigma_{\max}[\mathbf{\Phi}] + ... + \sigma_{\max}^{L-1}[\mathbf{\Phi}]\}\Big)\sigma_{\max}[\Delta\mathbf{\Phi}] \\ &+\; \sigma_{\max}^L[\mathbf{\Phi}]\;\sigma_{\max}[\mathbf{K}]\;\sigma_{\max}[\Delta\mathbf{\Psi}] \;<\; 1 - \sigma_{\max}[\mathbf{\Phi} + \mathbf{\Psi}\mathbf{K}]\end{aligned} \tag{14.151}$$

is a sufficient condition. (Proof see *Ishihara, T., 1988.*) The above result is an explicit function of the computation delays, nominal matrices $\mathbf{\Phi}, \mathbf{\Psi}$ and its perturbation $\Delta\mathbf{\Phi}$ and $\Delta\mathbf{\Psi}$. Note the interesting phenomenon that the Lth power of the condition number and $\sigma_{\max}^L[\mathbf{\Phi}]$ weights the perturbations $\sigma_{\max}[\Delta\mathbf{\Phi}]$ and $\sigma_{\max}[\Delta\mathbf{\Psi}]$, respectively.

14.17 Robust Smith Predictor

In plants delay time is often incorporated. In order to compensate for dead-time phenomena, the well-known Smith predictor makes use of a model of the plant to obtain effective control. Consider the scalar perturbed plant termed as $G_p(z)z^{-k_p}$. The z-transfer function $G(z)z^{-k}$ denotes the unperturbed plant, see Fig. 14.3. The characteristic polynomial is given by

$$1 + C(z)\;[\;G_p(z)z^{-k_p} + G(z)(1 - z^{-k})\;]\,. \tag{14.152}$$

Dividing by the characteristic polynomial of the unperturbed system, i.e. $1 + C(z)G(z)$, yields

$$\begin{aligned}1 + \frac{C(z)}{1 + C(z)G(z)}[G_p(z)z^{-k_p} - G(z)z^{-k}] \;&=\; 1 + \underbrace{\frac{C(z)G(z)}{1 + C(z)G(z)}}_{M(z)}\underbrace{[\frac{G_p(z)}{G(z)}z^{-(k_p-k)} - 1]}_{U(z)}z^{-k} \\ &\triangleq\; 1 + M(z)U(z)z^{-k}\,.\end{aligned} \tag{14.153}$$

The function $M(z)$ may be considered as the complementary sensitivity function, $U(z)$ contains all the uncertainties. By Nyquist theorem for stable $M(z)$ and $U(z)$, a sufficient stability condition is given if for all z on the unit circle

$$\sup_{|z|=1} |M(z)U(z)z^{-k}| < 1 \quad \rightsquigarrow \quad \sup_{|z|=1} |M(z)U(z)| < 1 \; . \tag{14.154}$$

Supposing that the unperturbed system is stable, a necessary condition for stability is that the intersection of the Nyquist plot with the real axis at $z = e^{j\pi} = -1$ is located in the right of the Nyquist point $(-1, j0)$ (*Palmor, Z.J., and Halevi, Y., 1990*).

Now, assume that $G_p(z) = G(z)$ and that there is an uncertainty $\Delta k = -1$, only. For $z = -1$ the necessary condition is

$$M(z)[z^{-1} - 1]z^{-k} \;|_{z=-1} > 1 \quad \rightsquigarrow \quad M(-1) < 0.5\,(-1)^k \; . \tag{14.155}$$

The same result is obtained for $\Delta k = +1$. Hence, the delay uncertainty may be $\pm T$ where T is the sampling period. This special case is very important from a practical point of view since mismatch in the delay time frequently happens and affects stability.

For

$$M(z) = \frac{z^q(\ldots + b_1 z + b_o)}{a_n z^n + \ldots + a_o} \qquad (n \geq m + q) \tag{14.156}$$

the characteristic equation is

$$1 + M(z)[z^{-1} - 1]z^{-k} = 0 \tag{14.157}$$

$$a_n z^n + \ldots + a_o + z^{q-k}[z^{-1} - 1](\ldots + b_1 z + b_o) = 0 \tag{14.158}$$

Rearranging yields

$$a_n z^{n-q+k+1} + \ldots + b_o = 0 \quad \text{or} \quad z^{n-q+k+1} + \ldots + b_o/a_n = 0 \; . \tag{14.159}$$

Since the coefficient of z^0 equals the product of all roots and no root may exceed unity a necessary condition is $|b_o/a_n| < 1$.

Robustness of the Smith predictor can be improved by a method based on an m-loop-filter prediction providing a prediction m steps ahead (*Cao, C.T., 1989*).

14.18 Lyapunov Min-Max-Controllers

Consider the uncertain discrete-time system

$$\mathbf{x}_{k+1} = \mathbf{x}(k+1) = \mathbf{\Phi x}(k) + \mathbf{\Psi u}(k) + \Delta \mathbf{g}(k) \qquad \mathbf{x} \in \mathcal{R}^n, \mathbf{u} \in \mathcal{R}^m \tag{14.160}$$

where $\Delta\mathbf{g}(k)$ represents the uncertainties. The Lyapunov min-max controllers $\mathbf{K}$ are defined as controllers for the uncertain system which satisfy (*Corless, M., 1985*)

$$\min_{\mathbf{u}} \max_{\mathbf{v}} \hat{V}(\mathbf{x}, \mathbf{u}, \mathbf{v}) = \hat{V}(\mathbf{x}, \mathbf{Kx}, \Delta\mathbf{g}) \qquad \forall k, \mathbf{x} \tag{14.161}$$

$$\text{where} \quad V(\mathbf{x}_k) = \mathbf{x}_k^T \mathbf{P} \mathbf{x}_k \qquad \mathbf{P} = \mathbf{P}^T > 0 \tag{14.162}$$

$$V(\mathbf{x}_{k+1}) = \hat{V}(\mathbf{x}_k, \mathbf{u}_k, \mathbf{v}_k) \qquad \mathbf{v}_k = \Delta\mathbf{g}(k) \; . \tag{14.163}$$

Assuming that for each $\Delta\mathbf{g}$ from the class of admissible uncertainties $-\Delta\mathbf{g}$ belongs to the same class and that a function $\mathbf{e}$ exists such that $\Delta\mathbf{g} = \mathbf{\Psi e}$ then the Lyapunov min-max controller is uniquely given by the linear control law $\mathbf{Kx}$ where

$$\mathbf{K} = -(\mathbf{\Psi}^T\mathbf{P\Psi})^{-1}\mathbf{\Psi}^T\mathbf{P\Phi} \tag{14.164}$$

independently from $\Delta\mathbf{g}$. The system is globally, uniformly asymptotically stable about the zero state. The controller above is identical to the well-known optimal controller minimizing $\Delta V = V(\mathbf{x}_{k+1}) - V(\mathbf{x}_k)$ with respect to $\mathbf{u}$, see e.g. *Sage, A.P., and White, Ch.C.III, 1977*. Using the definition of the closed-loop system

$$\mathbf{\Phi}_{CL} \triangleq \mathbf{\Phi} + \mathbf{\Psi K} = [\mathbf{I} - \mathbf{\Psi}(\mathbf{\Psi}^T\mathbf{P\Psi})^{-1}\mathbf{\Psi}^T\mathbf{P}]\mathbf{\Phi}\ , \tag{14.165}$$

in order to have $\Delta V < 0$ for the nominal closed-loop system (i.e. $\Delta\mathbf{g} \equiv \mathbf{0}$), the relation

$$\mathbf{\Phi}^T\mathbf{P\Phi} - \mathbf{P} - \mathbf{\Phi}^T\mathbf{P\Psi}(\mathbf{\Psi}^T\mathbf{P\Psi})^{-1}\mathbf{\Psi}^T\mathbf{P\Phi} < 0 \tag{14.166}$$

must be satisfied. Let $\mathbf{\Omega} \triangleq \mathbf{P}^{-1}\mathbf{\Phi}_{CL}^T\mathbf{P\Phi}_{CL}$ and apply Eq.(14.165). Then, Eq.(14.166) can be rewritten to $\mathbf{P\Omega} - \mathbf{P} < 0$ which leads to $\lambda_{\max}[\mathbf{\Omega}] < 1$.

Under the boundedness assumption

$$\|\Delta\mathbf{g}\|_{FP} \leq \gamma_o + \gamma_1\|\mathbf{x}\|_{FP} \quad \text{where} \quad \|\mathbf{x}\|_{FP} = \sqrt{\mathbf{x}^T\mathbf{Px}} \tag{14.167}$$

the uncertain system is globally uniformly asymptotically stable about the P-ball of radius d, i.e. $\|\mathbf{x}\|_{PF} \leq d$, if

$$\lambda_{\max}^2[\mathbf{\Omega}] + \gamma_1^2 < 1 \quad \text{where} \quad d = \frac{\gamma_o}{\sqrt{1 - \lambda_{\max}^2[\mathbf{\Omega}]} - \gamma_1}\ . \tag{14.168}$$

Example: Uncertain single input system with additional uncertain single input $w(k)$.

Consider an nth order system in companion form with uncertain parameters a_{ip} as given by Eq.(7.37) where $-p_i = a_{ip}$. With the notation

$$a_{il} \leq a_{ip} \leq a_{ih};\quad \bar{a}_i \triangleq 0.5(a_{il} + a_{ih});\quad \Delta\bar{a}_i \triangleq 0.5(a_{ih} - a_{il});\quad \Delta a_i \triangleq a_{ip} - \bar{a}_i;\quad |\Delta a_i| \leq \Delta\bar{a}_i \tag{14.169}$$

the perturbed system $\mathbf{\Phi}_p(k)$ with perturbed input $w(k)$

$$\mathbf{x}(k+1) = \mathbf{\Phi}_p(k)\mathbf{x}(k) + \mathbf{B}u(k) + \mathbf{B}w(k) \tag{14.170}$$

can be rewritten to a system with uncertainty $\Delta\mathbf{g}(k)$

$$\mathbf{x}(k+1) = \mathbf{\Phi}(k)\mathbf{x}(k) + \mathbf{B}\mathbf{u}(k) + \Delta\mathbf{g}(k) \tag{14.171}$$

where $\Delta\mathbf{g}(k) = \mathbf{B}[w(k) + \sum_{i=1}^n \Delta a_i(k)x_i]$.

Replacing the right-hand side of Eq.(14.166) by $-\mathbf{I}$ satisfies the stability requirements and a solution is $\mathbf{P} = \mathrm{diag}_n\{i\}$. If the uncertain input is bounded by $|w(k)| \leq \gamma_o'$ then

$$\|\Delta\mathbf{g}(k)\|_{FP} \leq \gamma_o + \gamma_1\|\mathbf{x}(k)\|_{FP} \quad \text{where} \quad \gamma_o = \sqrt{n}\,\gamma_o'\ ;\quad \gamma_1 = \sqrt{n\sum_1^n(\Delta\bar{a}_i)^2}\ . \tag{14.172}$$

The Lyapunov min-max controller is given by $-\sum_1^n a_i x_i$ and the closed-loop nominal system turns out as an n-stage integrator with eigenvalues throughout zero. Finally,

$$\lambda_{\max}[\mathbf{\Omega}] = \sqrt{1 - \frac{1}{n}} \qquad \gamma_1 < \frac{1}{\sqrt{n}} \qquad d = \frac{n\gamma_o'}{1 - \gamma_1\sqrt{n}}\ . \tag{14.173}$$

End of Example

14.19 Optimal Robust Stability of Discrete Large-Scale Systems

Quadratic or norm-like Lyapunov funtions can be chosen in order to obtain largest robustness bounds on nonlinear time-varying perturbations. The perturbations are given by the interconnections among the subsystems of a large-scale system.

14.19.1 Degree of Exponential Stability

Autonomous systems are considered first. The autonomous time-invariant discrete system

$$\mathbf{x}(k+1) = \mathbf{\Phi x}(k) \qquad \mathbf{x}_o = \mathbf{x}(0) \tag{14.174}$$

is *exponentially* stable[1] if the solution $\mathbf{x}(k)$ satisfies the condition

$$\|\mathbf{x}(k)\|_F \le \beta \; \|\mathbf{x}_o\|_F \exp\,(k \ln \rho_s) = \beta \; \|\mathbf{x}_o\|_F \rho_s^k \quad \text{where} \quad \beta > 0,\; 0 \le \rho_s < 1 \,. \tag{14.175}$$

The decay of the norm $\|\mathbf{x}(k)\|_F$ of the system motion $\mathbf{x}(k)$ must be quicker than a scalar motion with time constant $(-\ln \rho_s)$. The scalar ρ_s is the spectral radius of $\mathbf{\Phi}$ or the degree of stability.

The system is *asymptotically* stable if the following Lyapunov equation in $\mathbf{P}$

$$\mathbf{\Phi}^T\mathbf{P\Phi} - \mathbf{P} = -\mathbf{Q} \qquad \mathbf{Q} > 0 \tag{14.176}$$

has a unique solution $\mathbf{P} > 0$. The solution $\mathbf{P}$ is expanded as

$$\mathbf{P} = \sum_{i=0}^{\infty}(\mathbf{\Phi}^T)^i\mathbf{Q\Phi}^i \tag{14.177}$$

and can be proved by simple substitution

$$\mathbf{\Phi}^T \sum_{i=0}^{\infty}(\mathbf{\Phi}^T)^i\mathbf{Q\Phi}^i\mathbf{\Phi} - \sum_{j=0}^{\infty}(\mathbf{\Phi}^T)^j\mathbf{Q\Phi}^j = -(\mathbf{\Phi}^T)^0\mathbf{Q\Phi}^0 = -\mathbf{Q}\,. \tag{14.178}$$

The terms with $i = j - 1$ subsequently can be cancelled $\forall j = 1 \ldots \infty$.

14.19.2 Quadratic Lyapunov Function

A quadratic Lyapunov Function V

$$V(\mathbf{x}) = \mathbf{x}^T\mathbf{Px} \tag{14.179}$$

can be bounded by the maximum and minimum singular value of $\mathbf{P}$, see Eq.(2.26),

$$\sigma_{\min}[\mathbf{P}] \; \|\mathbf{x}\|_F^2 \le V(\mathbf{x}) \le \sigma_{\max}[\mathbf{P}] \; \|\mathbf{x}\|_F^2 \,. \tag{14.180}$$

The increase of the Lyapunov function is

$$\Delta V[\mathbf{x}(k)] = V[\mathbf{x}(k+1)] - V[\mathbf{x}(k)] = \mathbf{x}^T\mathbf{\Phi}^T\mathbf{P\Phi x} - \mathbf{x}^T\mathbf{Px} = -\mathbf{x}^T\mathbf{Qx} \tag{14.181}$$

$$\sigma_{\min}[\mathbf{Q}] \; \|\mathbf{x}\|_F^2 \le \mathbf{x}^T\mathbf{Qx}^T \tag{14.182}$$

[1] A system is called exponentially stable if it is proper and all its poles have negative real parts.

$$\Delta V[\mathbf{x}(k)] = V[\mathbf{x}(k+1)] - V[\mathbf{x}(k)] \leq -\sigma_{\min}[\mathbf{Q}]\ \|\mathbf{x}(k)\|_F^2 . \tag{14.183}$$

From Eq.(14.180)

$$V[\mathbf{x}(k)] \leq \sigma_{\max}[\mathbf{P}]\ \|\mathbf{x}(k)\|_F^2 . \tag{14.184}$$

Substituting $\|\mathbf{x}(k)\|_F^2$ from Eq.(14.184) into Eq.(14.183) and using the inequality rules

$$(a < b) \text{ and } (b < c) \quad \Rightarrow \quad a < c \tag{14.185}$$

$$(f < -e) \text{ and } (e > g) \quad \Rightarrow \quad f < -g , \tag{14.186}$$

it results

$$\Rightarrow \quad V[\mathbf{x}(k+1)] - V[\mathbf{x}(k)] \leq -\sigma_{\min}[\mathbf{Q}]\ \frac{V[\mathbf{x}(k)]}{\sigma_{\max}[\mathbf{P}]} \tag{14.187}$$

$$V[\mathbf{x}(k+1)] \leq \left(1 - \frac{\sigma_{\min}[\mathbf{Q}]}{\sigma_{\max}[\mathbf{P}]}\right) V[\mathbf{x}(k)] . \tag{14.188}$$

Repeated application of the above equation leads to

$$V[\mathbf{x}(k)] \leq \left(1 - \frac{\sigma_{\min}[\mathbf{Q}]}{\sigma_{\max}[\mathbf{P}]}\right)^k V[\mathbf{x}_o] \tag{14.189}$$

$$\sigma_{\min}[\mathbf{P}]\ \|\mathbf{x}(k)\|_F^2 \leq \left(1 - \frac{\sigma_{\min}[\mathbf{Q}]}{\sigma_{\max}[\mathbf{P}]}\right)^k \sigma_{\max}[\mathbf{P}]\ \|\mathbf{x}_o\|_F^2 \tag{14.190}$$

$$\|\mathbf{x}(k)\|_F \leq \sqrt{\frac{\sigma_{\max}[\mathbf{P}]}{\sigma_{\min}[\mathbf{P}]}} \left(\sqrt{1 - \frac{\sigma_{\min}[\mathbf{Q}]}{\sigma_{\max}[\mathbf{P}]}}\right)^k \|\mathbf{x}_o\|_F . \tag{14.191}$$

Comparing Eq.(14.191) with (14.175), an estimate of ρ_s is achieved. The spectral radius ρ_s depends on the choice of $\mathbf{Q}$ and is labelled with $_Q$ indicating the quadratic Lyapunov function

$$\hat{\rho}_{sQ}(\mathbf{Q}) = \sqrt{1 - \frac{\sigma_{\min}[\mathbf{Q}]}{\sigma_{\max}[\mathbf{P}]}} . \tag{14.192}$$

The stability is optimized in the exponential sense if ρ_s can be minimized with respect to the matrix $\mathbf{Q}$. Hence, from Eq.(14.179) using Eq.(14.177) it follows

$$\mathbf{x}^T\mathbf{P}\mathbf{x} = \mathbf{x}^T\Big(\sum_{i=0}^{\infty}(\mathbf{\Phi}^T)^i\mathbf{Q}\mathbf{\Phi}^i\Big)\mathbf{x} \geq \mathbf{x}^T\Big(\sum_{i=0}^{\infty}(\mathbf{\Phi}^T)^i\sigma_{\min}[\mathbf{Q}]\mathbf{\Phi}^i\Big)\mathbf{x} \tag{14.193}$$

$$\mathbf{x}^T\mathbf{P}\mathbf{x} \geq \sigma_{\min}[\mathbf{Q}]\ \mathbf{x}^T\Big(\sum_{i=0}^{\infty}(\mathbf{\Phi}^T)^i\mathbf{\Phi}^i\Big)\mathbf{x} . \tag{14.194}$$

The expression $\sum_{i=0}^{\infty}(\mathbf{\Phi}^T)^i\mathbf{\Phi}^i$ is considered as the solution $\mathbf{P} = \mathbf{P}_I$ if $\mathbf{Q} = \mathbf{I}$ and $\mathbf{x}^T\mathbf{P}\mathbf{x}$ is substituted by its upper bound

$$\mathbf{x}^T\sigma_{\max}[\mathbf{P}]\ \mathbf{x} \geq \sigma_{\min}[\mathbf{Q}]\ \mathbf{x}^T\mathbf{P}_I\mathbf{x} \quad \rightsquigarrow \quad \frac{\sigma_{\min}[\mathbf{Q}]}{\sigma_{\max}[\mathbf{P}]} \leq \frac{\mathbf{x}^T\mathbf{x}}{\mathbf{x}^T\mathbf{P}_I\mathbf{x}} . \tag{14.195}$$

Using

$$\hat{\rho}_{sQ}[\mathbf{I}] = \hat{\rho}_{sQ}[\mathbf{Q} = \mathbf{I}] = \sqrt{1 - \frac{\sigma_{\min}[\mathbf{I}]}{\sigma_{\max}[\mathbf{P}_I]}} = \sqrt{1 - \frac{1}{\sigma_{\max}[\mathbf{P}_I]}} \tag{14.196}$$

and comparing with Eq.(14.195) yields

$$\frac{\sigma_{\min}[\mathbf{Q}]}{\sigma_{\max}[\mathbf{P}]} \leq \frac{\mathbf{x}^T\mathbf{I}\mathbf{x}}{\mathbf{x}^T\mathbf{P}_I\mathbf{x}} \leq \frac{\mathbf{x}^T\mathbf{x}}{\sigma_{\max}[\mathbf{P}_I]\mathbf{x}^T\mathbf{x}} \tag{14.197}$$

$$1 - \frac{\sigma_{\min}[\mathbf{Q}]}{\sigma_{\max}[\mathbf{P}]} \geq 1 - \frac{1}{\sigma_{\max}[\mathbf{P}_I]} \quad \rightsquigarrow \quad \hat{\rho}_{sQ}[\mathbf{Q}] \geq \hat{\rho}_{sQ}[\mathbf{I}] \; . \tag{14.198}$$

Hence, $\mathbf{Q} = \mathbf{I}$ yields the minimum $\hat{\rho}_Q$.

14.19.3 Norm-Like Lyapunov Functions

Let the norm-like Lyapunov function be

$$V_N(\mathbf{x}) = \sqrt{\mathbf{x}^T\mathbf{P}\mathbf{x}} \; . \tag{14.199}$$

Taking the square root of Eq.(14.180) yields

$$\sqrt{\sigma_{\min}[\mathbf{P}]}\; \|\mathbf{x}\|_F \leq V_N(\mathbf{x}) \leq \sqrt{\sigma_{\max}[\mathbf{P}]}\; \|\mathbf{x}\|_F \; . \tag{14.200}$$

Repeating the derivations similarly to the preceding section (*Sezer, M.E., and Šiljak, D.D., 1988*) the estimate $\hat{\rho}_{sN}[\mathbf{Q}]$ is achieved by the following calculations

$$\Delta V_N[\mathbf{x}(k)] = \sqrt{\mathbf{x}^T(k+1)\mathbf{P}\mathbf{x}(k+1)} - \sqrt{\mathbf{x}^T(k)\mathbf{P}\mathbf{x}(k)} \tag{14.201}$$

$$= \frac{\mathbf{x}^T(k+1)\mathbf{P}\mathbf{x}(k+1) - \mathbf{x}^T(k)\mathbf{P}\mathbf{x}(k)}{\sqrt{\mathbf{x}^T(k+1)\mathbf{P}\mathbf{x}(k+1)} + \sqrt{\mathbf{x}^T(k)\mathbf{P}\mathbf{x}(k)}} \tag{14.202}$$

$$= \frac{-\mathbf{x}^T(k)\mathbf{Q}\mathbf{x}(k)}{\sqrt{\mathbf{x}^T(k+1)\mathbf{P}\mathbf{x}(k+1)} + \sqrt{\mathbf{x}^T(k)\mathbf{P}\mathbf{x}(k)}} \; . \tag{14.203}$$

Since

$$\mathbf{x}^T(k+1)\mathbf{P}\mathbf{x}(k+1) > \mathbf{x}^T(k)(\mathbf{P}-\mathbf{Q})\mathbf{x}(k) \tag{14.204}$$

$$\Delta V_N[\mathbf{x}(k)] \leq \frac{-\mathbf{x}^T(k)\mathbf{Q}\mathbf{x}(k)}{\sqrt{\mathbf{x}^T(k)(\mathbf{P}-\mathbf{Q})\mathbf{x}(k)} + \sqrt{\mathbf{x}^T(k)\mathbf{P}\mathbf{x}(k)}} \leq \frac{-\sigma_{\min}[\mathbf{Q}]}{\sqrt{\sigma_{\max}[\mathbf{P}-\mathbf{Q}]} + \sqrt{\sigma_{\max}[\mathbf{P}]}}\|\mathbf{x}(k)\|_F \; . \tag{14.205}$$

By analogy to the derivation of Eq.(14.191),

$$\hat{\rho}_{sN}(\mathbf{Q}) = 1 - \frac{\sigma_{\min}[\mathbf{Q}]}{\sigma_{\max}[\mathbf{P}] + \sqrt{\sigma_{\max}[\mathbf{P}]\sigma_{\max}[\mathbf{P}-\mathbf{Q}]}} \; . \tag{14.206}$$

The optimal estimate again is obtained if $\mathbf{Q} = \mathbf{I}$

$$\hat{\rho}_{sN}(\mathbf{Q}) \geq \hat{\rho}_{sN}(\mathbf{I}) = 1 - \frac{1}{\sigma_{\max}[\mathbf{P}_I] + \sqrt{\sigma_{\max}[\mathbf{P}_I]\sigma_{\max}[\mathbf{P}_I - \mathbf{I}]}} \; . \tag{14.207}$$

14.19.4 Comparing $\hat{\rho}_{sQ}[\mathbf{Q}]$ and $\hat{\rho}_{sN}[\mathbf{Q}]$

Comparison of $\hat{\rho}_{sQ}[\mathbf{Q}]$ and $\hat{\rho}_{sN}[\mathbf{Q}]$ yields (calculations omitted) that $\hat{\rho}_{sQ}[\mathbf{Q}] \geq \hat{\rho}_{sN}[\mathbf{Q}]$ with equality for $\mathbf{Q} = \mathbf{I}$. The norm-like Lyapunov function is superior to the quadratic form. In the continuous-time case both Lyapunov functions would yield the same result (*Šiljak, D.D., 1978*).

14.19.5 Nonlinear Perturbed System

The objective is the investigation of the norm-like Lyapunov stability. Let the nonlinear perturbed system be

$$\mathbf{x}(k+1) = \mathbf{\Phi}\mathbf{x}(k) + \mathbf{h}[k, \mathbf{x}(k)] \,. \tag{14.208}$$

The perturbation is considered bounded by

$$\|\mathbf{h}[k, \mathbf{x}(k)]\|_F \le \xi \|\mathbf{x}(k)\|_F, \qquad \xi > 0 \,. \tag{14.209}$$

The norm of $\mathbf{h}$ does not exceed the norm of $\mathbf{x}$ multiplied with a fixed gain ξ. Assuming that $\mathbf{\Phi}$ is stable and $\hat{\rho}_{sN}(\mathbf{Q})$ exists, the norm-like Lyapunov functions are taken into consideration. Now,

$$\begin{aligned} \Delta V[\mathbf{x}] &= V[\mathbf{\Phi x} + \mathbf{h}] - V[\mathbf{x}] \le V[\mathbf{\Phi x}] + \mid V[\mathbf{\Phi x} + \mathbf{h}] - V[\mathbf{\Phi x}] \mid - V[\mathbf{x}] && (14.210) \\ &\le V[\mathbf{\Phi x}] - V[\mathbf{x}] + \mid V[\mathbf{\Phi x} + \mathbf{h}] - V[\mathbf{\Phi x}] \mid \,. && (14.211) \end{aligned}$$

Using the relation

$$|\sqrt{\mathbf{a}^T\mathbf{P}\mathbf{a}} - \sqrt{\mathbf{b}^T\mathbf{P}\mathbf{b}}| \le \sqrt{\sigma_{\max}[\mathbf{P}]}\ \|\mathbf{a} - \mathbf{b}\|_F \tag{14.212}$$

and substituting $\mathbf{a} = \mathbf{\Phi x} + \mathbf{h}$ and $\mathbf{b} = \mathbf{\Phi x}$ (*Sezer, M.E., and Šiljak, D.D., 1988*), the increment $\Delta V[\mathbf{x}]$ becomes negative if ξ obeys

$$\xi < \hat{\xi}_N(\mathbf{Q}) = \frac{\sigma_{\min}[\mathbf{Q}]}{\sigma_{\max}[\mathbf{P}] + \sqrt{\sigma_{\max}[\mathbf{P}]\sigma_{\max}[\mathbf{P} - \mathbf{Q}]}} \,. \tag{14.213}$$

Note from Eqs.(14.213) and (14.206) that $\hat{\xi}_N[\mathbf{Q}] = 1 - \hat{\rho}_{sN}[\mathbf{Q}]$. Hence, $\hat{\xi}_N(\mathbf{Q})$ obviously is a stability margin of the perturbed system. The larger the gain $\hat{\xi}_N(\mathbf{Q})$ the better the robustness. The best choice is $\mathbf{Q} = \mathbf{I}$. In this case, norm-like and quadratic Lyapunov functions yield not only the same but also the optimal result.

14.19.6 Large-Scale Systems

Consider the interconnection function $\mathbf{h}_i$ between the subsystems of a large-scale system as the perturbation. The large-scale system with the overall state variable $\mathbf{x}(k)$ is decomposed into L subsystems with the state variable $\mathbf{x}_i(k)$

$$\mathbf{x}_i(k+1) = \mathbf{\Phi}_i\mathbf{x}_i(k) + \mathbf{h}_i[k, \mathbf{x}(k)] \qquad \mathbf{\Phi}_i = \mathcal{R}^{n_i \times n_i} \tag{14.214}$$

$$\mathbf{x}(k) = [\mathbf{x}_1^T(k)\ \ \mathbf{x}_2^T(k)\ \ \ldots\ \ \mathbf{x}_L^T(k)]^T \qquad \mathbf{x}(k) \in \mathcal{R}^{(\sum_1^L n_i)} \,. \tag{14.215}$$

Each subsystem i is dynamically weighted by a separate norm-like Lyapunov function

$$V_{Ni}[\mathbf{x}_i(k)] = \sqrt{\mathbf{x}_i^T\mathbf{P}_i\mathbf{x}_i} \,. \tag{14.216}$$

The interconnections are assumed to be bounded by

$$\|\ \mathbf{h}_i[k, \mathbf{x}(k)]\ \|_F < \sum_{j=1}^{L} \xi_{ij}\|\mathbf{x}_j(k)\|_F \qquad \xi_{ij} \ge 0 \,. \tag{14.217}$$

The number ξ_{ij} gives the influence of $\|\mathbf{x}_j\|_F$ on $\|\mathbf{h}_i\|_F$. The subsystems are assumed exponentially stable in the sense of

$$\mathbf{\Phi}_i^T\mathbf{P}_i\mathbf{\Phi}_i - \mathbf{P}_i = -\mathbf{Q}_i \quad \forall i = 1 \ldots L \,. \tag{14.218}$$

The subsystem stability margin is $\hat{\xi}_{Ni}(\mathbf{Q}_i)$. The increment $\Delta V_{Ni}[\mathbf{x}_i(k)]$ for each subsystem is structured in order to comply with norm-like Lyapunov functions

$$\Delta V_{Ni}[\mathbf{x}_i(k)] \leq -\sqrt{\sigma_{\max}[\mathbf{P}_i]}\ [\ \hat{\xi}_{Ni}(\mathbf{Q}_i)\ \|\mathbf{x}_i(k)\|_F - \|\mathbf{h}_i\|_F\ ] \,. \tag{14.219}$$

$$\leq -\sqrt{\sigma_{\max}[\mathbf{P}_i]}\ [\ \hat{\xi}_{Ni}(\mathbf{Q}_i)\ \|\mathbf{x}_i(k)\|_F - \sum_{j=1}^{L} \xi_{ij}\|\mathbf{x}_j(k)\|_F\ ] \,. \tag{14.220}$$

The overall Lyapunov function is established

$$V_N[\mathbf{x}(k)] = \sum_{i=1}^{L} d_i\ V_{Ni}[\mathbf{x}_i(k)] \,. \tag{14.221}$$

The system is stable if $\Delta V_N < 0$. Combining Eqs.(14.219) and (14.221) in abbreviated notation,

$$\Delta V_N[\mathbf{x}(k)] \leq \mathbf{d}^T \mathbf{W} \mathbf{w}(\mathbf{x}) \tag{14.222}$$

$$\mathbf{d} \triangleq (d_1\sqrt{\sigma_{\max}[\mathbf{P}_1]},\ d_2\sqrt{\sigma_{\max}[\mathbf{P}_2]}\ \ldots\ d_L\sqrt{\sigma_{\max}[\mathbf{P}_L]})^T \tag{14.223}$$

$$\mathbf{w}(\mathbf{x}) \triangleq (\|\mathbf{x}_1\|_F, \|\mathbf{x}_2\|_F\ \ldots\ \|\mathbf{x}_L\|_F)^T \tag{14.224}$$

$$\mathbf{W} \triangleq \text{matrix}[W_{ij}] \quad \text{where } W_{ij} = \begin{cases} -\hat{\xi}_{Ni}(\mathbf{Q}_i) + \xi_{ii} & \forall i = j \\ \xi_{ij} & \forall i \neq j \,. \end{cases} \tag{14.225}$$

The large-scale system is exponentially stable if $\mathbf{W}$ is a Metzler matrix, i.e., there exists a vector $\mathbf{d}$ with positive elements such that $\mathbf{d}^T\mathbf{W}$ has negative entries (*Šiljak, D.D., 1978*). Optimal stability and robustness is achieved if $\mathbf{Q}_i = \mathbf{I}_{n_i}$ with the stability margin $\hat{\xi}_N^* = \hat{\xi}_N(\mathbf{I}_{ni})$. The result in this case equals the quadratic Lyapunov function case.

Chapter 15

Robust Pole Assignment

Pole assignment is a well-known method for controller design. If the plant parameter vector is characterized by some uncertainty and if there is some flexibility in the desired closed-loop pole vector, the problem of finding a robust controller arises. The solution of this problem is not a simple one of solving linear algebraic equations as in the case of plants given in companion form. The problem is of higher complexity and requires an iterative algorithm, e.g., following a steepest descent method.

Closed-loop pole assignment by state feedback is a multilinear problem with an additional degree of freedom in the case of multivariable processes. For approximately linearized equations the concept of pseudo-inverse can be employed to successively displace selected poles and to utilize the degree of freedom (*Preuss, H.P., 1979*).

15.1 Minimizing the Distance to a Nominal Controller Parameter

15.1.1 Nominal Control System

Consider a discrete-time single-input single-output system. Using polynomials, e.g. $b(z^{-1}) = b_o + b_1 z^{-1} + b_2 z^{-2} + \ldots + b_n z^{-n}$, the transfer functions are

plant:

$$G(z^{-1}) = \frac{b(z^{-1})}{a(z^{-1})} \qquad \mathbf{b} = (b_o, \ldots b_n)^T, \quad \mathbf{a} = (a_o, \ldots a_n)^T \tag{15.1}$$

controller:

$$C(z^{-1}) = \frac{p(z^{-1})}{l(z^{-1})} \qquad \mathbf{l} = (l_o, \ldots l_{n-1})^T, \quad \mathbf{p} = (p_o, \ldots p_{n-1})^T\ . \tag{15.2}$$

The characteristic equation of the closed-loop system is

$$1 + C(z^{-1})G(z^{-1}) = 0 \quad \rightsquigarrow \quad a(z^{-1})l(z^{-1}) + p(z^{-1})b(z^{-1}) \triangleq d(z^{-1}) = 0\ . \tag{15.3}$$

The polynomial $d(z^{-1})$ is determined according to the given poles of the closed-loop system. The Diophantine equation Eq.(15.3) now is rewritten to

$$\mathbf{\Theta v} = \mathbf{d} \quad \text{where} \quad \mathbf{d} \in \mathcal{R}^{2n}, \quad \mathbf{a},\ \mathbf{b} \in \mathcal{R}^{n+1}, \quad \mathbf{l},\ \mathbf{p} \in \mathcal{R}^{n} \tag{15.4}$$

and $\mathbf{v} \triangleq (\mathbf{a}^T \vdots \mathbf{b}^T)^T \in \mathcal{R}^{2n+2}$ and $\Theta = \Theta(\mathbf{l}, \mathbf{p})$ is defined by

$$\Theta \triangleq \begin{pmatrix} l_o & 0 & \dots & \dots & 0 & p_o & 0 & \dots & \dots & 0 \\ l_1 & l_o & 0 & \dots & 0 & p_1 & p_o & 0 & \dots & 0 \\ \vdots & & \ddots & \ddots & \vdots & \vdots & & \ddots & \ddots & \vdots \\ l_{n-1} & \dots & \dots & l_o & 0 & p_{n-1} & \dots & \dots & p_o & 0 \\ 0 & l_{n-1} & & & l_o & 0 & p_{n-1} & & & p_o \\ \vdots & \vdots & \ddots & & \vdots & \vdots & & \ddots & & \vdots \\ \vdots & \vdots & & \ddots & \vdots & \vdots & & & \ddots & \vdots \\ 0 & \dots & \dots & 0 & l_{n-1} & 0 & \dots & \dots & 0 & p_{n-1} \end{pmatrix} \in \mathcal{R}^{2n \times (2n+2)} . \qquad (15.5)$$

15.1.2 Plant Uncertainty and Tolerance of Desired Poles

Designing robust control systems, the parameter vector $\mathbf{v}$ of the uncertain plant is expressed by an interval vector $[\mathbf{v}_l \;\; \mathbf{v}_h]$ symmetric to a nominal vector $\mathbf{v}_{nom}$, i.e.,

$$v_{li} \leq v_i \leq v_{hi} \quad \forall i \qquad \mathbf{v}_l = \mathbf{v}_{nom} - \mathbf{v}_e \qquad \mathbf{v}_h = \mathbf{v}_{nom} + \mathbf{v}_e \qquad (15.6)$$

or abbreviated $|\mathbf{v} - \mathbf{v}_{nom}| \leq_e \mathbf{v}_e$ where the modulus operation has to be applied componentwise, i.e., $|v_i - v_{i,nom}| \leq v_{ei} \quad \forall\, i$.

The set of all robust pole placement controllers with the composed parameter vector $\mathbf{p}_c = (\mathbf{l}^T \,\vdots\, \mathbf{p}^T)^T \in \mathcal{R}^{2n}$ is given by

$$\mathcal{S}_1 \triangleq \{\mathbf{p}_c : |\Theta(\mathbf{p}_c)\mathbf{v} - \mathbf{d}_{nom}| < \mathbf{d}_e \qquad \forall \mathbf{v} : |\mathbf{v} - \mathbf{v}_{nom}| \leq \mathbf{v}_e\} . \qquad (15.7)$$

15.1.3 Robust Pole Assignment

The robust pole assignment problem can be posed as

$$\min_{\mathbf{p}_c} \|\mathbf{p}_c - \mathbf{p}_{c,nom}\|_F^2 \qquad (15.8)$$

subject to

$$|\Theta(\mathbf{p}_c)\mathbf{v} - \mathbf{d}_{nom}| \leq_e \mathbf{d}_e \quad \text{and} \quad |\mathbf{v} - \mathbf{v}_{nom}| \leq_e \mathbf{v}_e \qquad (15.9)$$

where $\Theta_{nom}\mathbf{v}_{nom} = \mathbf{d}_{nom}$ is the desired (nominal) pole placement vector (*Soh, Y.C., et al. 1987*).

Following *Evans, R.J., et al. 1983* the set of robust controllers $\mathcal{S}_1$ can be reformulated to $\mathcal{S}_2$ and $\mathcal{S}_3$

$$\mathcal{S}_2 \triangleq \{\mathbf{p}_c : |\Theta\mathbf{v}_{nom} - \mathbf{d}_{nom}| + |\Theta|\mathbf{v}_e \leq_e \mathbf{d}_e\} \qquad (15.10)$$

$$\mathcal{S}_3 \triangleq \{\mathbf{p}_c : |\mathbf{M}\mathbf{p}_c - \mathbf{d}_{nom}| + \mathbf{N}|\mathbf{p}_c| \leq_e \mathbf{d}_e\} . \qquad (15.11)$$

The matrices $\mathbf{M}$ and $\mathbf{N}$ are partitioned matrices composed by partitions $\mathbf{W}$, see below. The numbers g_i are replaced by a_{inom}, b_{inom}, $v_{e,i+1}$, $v_{e,n+2+i}$, respectively, see Eq.(15.14).

Hence,

$$\mathbf{W}(g_i) \triangleq \begin{pmatrix} g_o & 0 & 0 & \dots & 0 \\ g_1 & g_o & 0 & \dots & 0 \\ g_2 & g_1 & g_o & \dots & 0 \\ \vdots & \ddots & \ddots & \ddots & 0 \\ g_{n-1} & g_{n-2} & \dots & \dots & g_o \\ g_n & g_{n-1} & \dots & \dots & g_1 \\ 0 & \ddots & \ddots & \ddots & \vdots \\ 0 & \vdots & 0 & 0 & g_n \end{pmatrix} = \begin{pmatrix} \mathbf{I}^{\swarrow}\mathrm{trig}(\mathbf{I}^{\swarrow}\mathbf{g}) \\ (\mathrm{trig}\,\bar{\mathbf{g}})\mathbf{I}^{\swarrow} \end{pmatrix} \quad \in \mathcal{R}^{2n \times n} \tag{15.12}$$

$$\mathbf{g} = \begin{pmatrix} g_o \\ \vdots \\ g_{n-1} \end{pmatrix}, \quad \bar{\mathbf{g}} = \begin{pmatrix} g_1 \\ \vdots \\ g_n \end{pmatrix}, \qquad \mathbf{M},\ \mathbf{N} \in \mathcal{R}^{2n \times 2n} \tag{15.13}$$

$$\mathbf{M} \triangleq \left(\mathbf{W}(a_{i,nom})\vdots\mathbf{W}(b_{i,nom})\right), \qquad \mathbf{N} \triangleq \left(\mathbf{W}(v_{e,i+1})\vdots\mathbf{W}(v_{e,n+2+i})\right) \ \forall i = 0 \dots n\,. \tag{15.14}$$

Then, the algorithm tractible for computation is

$$\min_{\mathbf{p}_c} \|\mathbf{p}_c - \mathbf{p}_{c,nom}\|_F^2 \quad \text{subject to} \quad |\mathbf{M}\mathbf{p}_c - \mathbf{d}_{nom}| + \mathbf{N}|\mathbf{p}_c| \leq_e \mathbf{d}_e\,. \tag{15.15}$$

15.1.4 Gradient Algorithm

Using the vector Lagrange multiplier $\boldsymbol{\lambda}$, find the minimum of

$$I = \|\mathbf{p}_c - \mathbf{p}_{c,nom}\|_F^2 + \boldsymbol{\lambda}^T[\ |\mathbf{M}\mathbf{p}_c - \mathbf{d}_{nom}| + \mathbf{N}|\mathbf{p}_c| \leq_e \mathbf{d}_e\ ] \tag{15.16}$$

with respect to $\mathbf{p}_c$. Then,

$$\mathbf{p}_c(k+1) = \mathbf{p}_c(k) - \alpha(k)\frac{\partial I}{\partial \mathbf{p}_c}\,|_k \tag{15.17}$$

$$\boldsymbol{\lambda}(k+1) = \mathbf{P}[\lambda(\mathrm{k}) + \beta(k)\frac{\partial I}{\partial \boldsymbol{\lambda}}\,|_k] \tag{15.18}$$

where $\alpha(k)$ and $\beta(k)$ are real scalar factors and $\mathbf{P}[\cdot]$ is the projection operator

$$(\mathbf{P}[\mathrm{x}])_i = \max[0, x_i] = \begin{cases} x_i & \text{if } x_i > 0 \\ 0 & \text{otherwise}\,. \end{cases} \tag{15.19}$$

The gradient $\frac{\partial I}{\partial \mathbf{p}_c}\,|_k$ is evaluated according to Eq.(5.23).

15.1.5 Maximally Robust Pole Placement

Introduce a positive scaling factor ρ for the unstructured plant uncertainty vector $\mathbf{v}$, i.e., $|\mathbf{v} - \mathbf{v}_{nom}| \leq_e \rho\, \mathbf{v}_e$. The condition $|\boldsymbol{\Theta}(\mathbf{p}_c)\mathbf{v} - \mathbf{d}_{nom}| \leq_e \mathbf{d}_e$ must still hold. Referring to Eq.(15.15), the maximally robust controller is given by the solution of the problem

$$\max_{\mathbf{p}_c} \rho^2 \quad \text{subject to} \quad |\mathbf{M}\mathbf{p}_c - \mathbf{d}_{nom}| + \rho\, \mathbf{N}|\mathbf{p}_c| \leq_e \mathbf{d}_e\,. \tag{15.20}$$

The Lagrangian is

$$I = \rho^2 + \boldsymbol{\lambda}_1^T(\mathbf{M}\mathbf{p}_c - \mathbf{d}_{nom} + \rho\mathbf{N}|\mathbf{p}_c| - \mathbf{d}_e) - \boldsymbol{\lambda}_2^T(\mathbf{M}\mathbf{p}_c - \mathbf{d}_{nom} - \rho\mathbf{N}|\mathbf{p}_c| + \Delta\mathbf{d}) . \quad (15.21)$$

The optimization algorithm corresponds to the aforementioned one, see Eq.(15.16), but has to follow an additional scalar update equation

$$\rho(k+1) = P[\rho(k) + \gamma(k)\frac{\partial I}{\partial \rho}\,|_k] . \quad (15.22)$$

15.2 Minimum Norm of the Output Feedback Controller

For any fixed $\mathbf{K}$ and the uncertain plant $\mathbf{A}$, $\mathbf{B}$ and $\mathbf{C}$, the closed-loop state-space representation $\mathbf{F} = \mathbf{A} + \mathbf{BKC} \in \mathcal{R}^{n\times n}$ corresponds to an interval polynomial $f(s)$. Using the four Kharitonov polynomials $f_1(s)$ through $f_4(s)$, to check stability a set of $4n$ Hurwitz determinants must be positive. The optimization problem can be defined, e.g., min $||\mathbf{K}||^2$ with respect to $\mathbf{K}$ subject to the positivity of all the $4n$ Hurwitz determinants. The performance is convex but the constraints are not (*Soh, Y.C., and Evans, R.J., 1985*).

15.3 Intersection of Parameter Regions. Multi-Model Approach

For continuous-time systems the characteristic polynomial of the closed-loop system is $a(s, \mathbf{p}, \mathbf{k})$ where $\mathbf{p}$ is the vector of the uncertain parameters of the plant and $\mathbf{k}$ is the controller. The plant parameters are assumed to be elements of Ω. The problem is to find $\mathbf{k}$ such that the zeros of $a(s, \mathbf{p}, \mathbf{k})$ are located in the left-half of the s-plane or in a certain Γ-region (nice stability) if a solution exists at all. In the case of discrete-time systems the zeros of $a(z, \mathbf{p}, \mathbf{k})$ must lie inside the unit-circle. The problem can be solved for fixed $\mathbf{k}$ or for fixed $\mathbf{p}$.

(i) For a fixed $\mathbf{k}$, find the region Ω_k such that Γ-stability is satisfied. Then, $\mathbf{k}$ is a solution if $\Omega \in \Omega_k$. Otherwise, the procedure is repeated for alternative values of $\mathbf{k}$.

(ii) For a fixed $\mathbf{p}$, find the region Σ_p in $\mathbf{k}$-space where Γ-stability is satisfied. For several $\mathbf{p}_j$, each fixed, find Σ_{pj} . The intersection $\bigcap_j \Sigma_{pj}$ of all regions, if it exists, yields the subspace of the $\mathbf{k}$-space robustly stabilizing the uncertain plant. With regard to the models $\mathbf{p}_j$ the approach is named multi-model approach (*Ackermann, J., 1980; 1984; 1985a*). As depicted in Fig. 15.1, the single-input plant $\mathbf{A}(\mathbf{p}), \mathbf{b}(\mathbf{p})$ and the controller $\mathbf{k}$ yield closed-loop poles inside Γ . In other words, for each $\mathbf{p}_j$ the state feedback controller $\mathbf{k}$ is chosen such that the characteristic polynomial $\det(s\mathbf{I} - \mathbf{A} - \mathbf{b}\mathbf{k}^T)$ is characterized by roots being located inside Γ thus yielding Σ_{pj} . For each $\mathbf{p}_j$ using $\mathbf{A}(\mathbf{p}_j)$ and $\mathbf{b}(\mathbf{p}_j)$ the state feedback controller $\mathbf{k}_j$ is given by Eq. (B.27) (*Ackermann, J., 1972; Föllinger, O., 1976*)

$$\mathbf{k}_j^T = (0\vdots 0\vdots \ldots 1)[\mathbf{b}(\mathbf{p}_j)\vdots \mathbf{A}(\mathbf{p}_j)\mathbf{b}(\mathbf{p}_j)\vdots \mathbf{A}^2(\mathbf{p}_j)\mathbf{b}(\mathbf{p}_j)\ldots\vdots \mathbf{A}^{n-1}(\mathbf{p}_j)\mathbf{b}(\mathbf{p}_j)]^{-1}\prod_{i=1}^{n}[\mathbf{A}(\mathbf{p}_j) - s_i\,\mathbf{I}] \quad (15.23)$$

where $s_i\ \forall i = 1, 2...n$ are the closed-loop eigenvalues. The product $\prod_{i=1}^n[\mathbf{A}(p_j) - s_i\mathbf{I}]$ equals the matrix given by the closed-loop polynomial $a(s)$ where s is replaced by $\mathbf{A}(p_j)$.

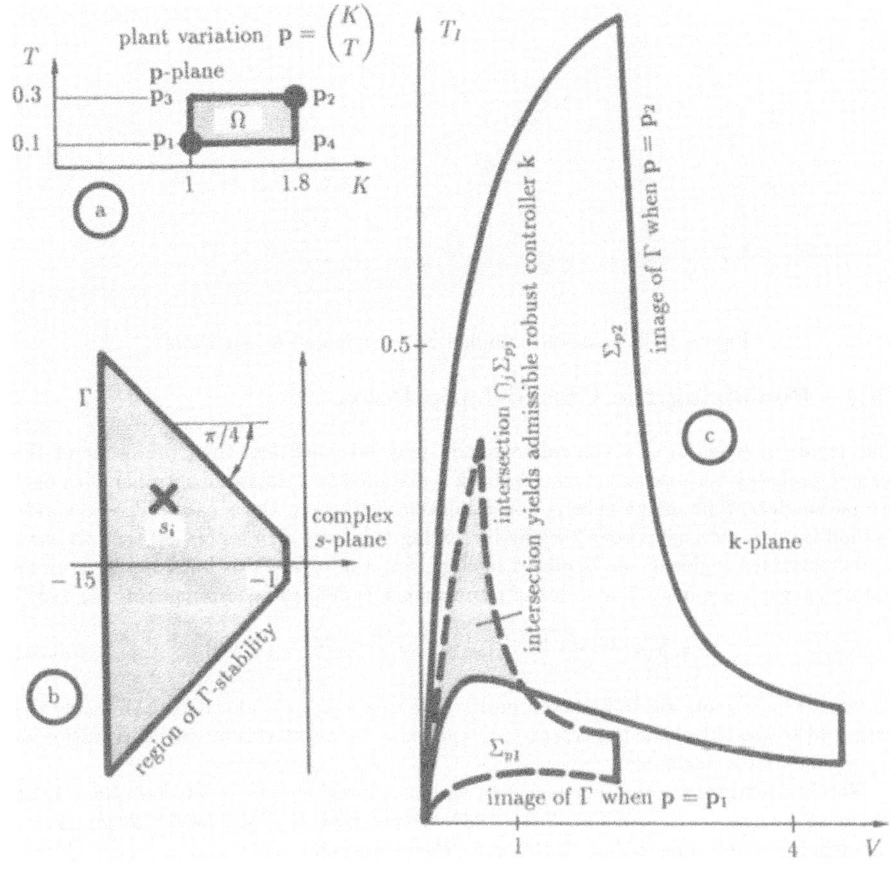

Figure 15.1: Plant parameter variation, region Γ and intersection in k-plane

Example: Given a plant $G(s) = K/(1 + sT)$ with uncertain parameters K and T. The intervals for $\mathbf{p} \triangleq (K \vdots T)^T$ are $K = [1 \ \ 1.8]$; $T = [0.1 \ \ 0.3]$ (Fig. 15.1a). Consider the PI-controller $C(s) = V[1+1/(sT_I)]$ and find the regions in the parameter plane T_I versus V such that the closed-loop poles are located in the region Γ (Fig. 15.1b). For the two $(K \ \ T)$-corner points $(1 \ \ 0.1)$ and $(1.8 \ \ 0.3)$ the admissible region for $\mathbf{k} = (V \ \ T_I)^T$ or T_I versus V is depicted in Fig. 15.1c. From $1 + G(s)C(s)$ the characteristic polynomial is $s^2 + [(KV + 1)/T]\ s + KV/(TT_I)$. Choosing $V = 0.3$ and $T_I = 0.2$, this point is located inside the intersection of Σ_{p1} and Σ_{p2} . It can easily be verified that by this choice the corner points $\mathbf{p}_3$ and $\mathbf{p}_4$ produce closed-loop poles at $-2.17 \pm j0.55$ and -2.02; -13.38, respectively, i.e., inside the region Γ.

End of Example

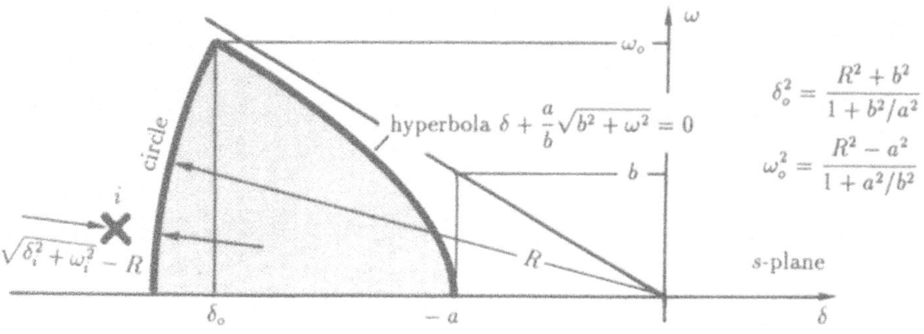

Figure 15.2: Γ-region bounded by a circle and a hyperbola

15.4 Penalizing the Closed-Loop Poles

The degree of freedom of a controller matrix may be much less than the order of the system, probably because an output feedback is installed or a classical controller with only few parameters. Pole assignment is not a suitable method in those cases. The adequate method is pole assessment operating by increasing a penalty if poles (δ_i, ω_i) tend to leave a predetermined region Γ. As depicted in Fig. 15.2, the region Γ is bounded by a circle and a hyperbola segment. The index of performance is defined as (*Konigorski, U., 1987*)

$$I = \sum_{i=1}^{n} e^{c(\delta_i + \sqrt{a^2+\omega_i^2 a^2/b^2}\ )} + e^{c(\sqrt{\delta_i^2+\omega_i^2} - R)} \qquad (c > 0) \tag{15.24}$$

where all square roots are to be taken positive. If a pole (δ_i, ω_i) is located both inside the circle and to the left of the hyperbola the exponents are negative and the contribution of the exponential is negligible.

Matrix equations for root-clustering in various subregions of the left-half plane were developed by *Lee, T.T., and Lee, S.H., 1987; Abdul-Wahab, A.A., 1990; 1990a; 1991* . Algorithms are determined that cluster the closed-loop poles inside some circular, elliptic or rectangular region. Robustness is achieved only in the sense that the poles are designed to cluster inside a desired region.

Matrix root-clustering in subregions of the complex plane, i.e., methods to find the class of eigenvalues of **A** to lie in a transformable region in the complex plane using, e.g., Kronecker products and bialternate products see *Gutman, S., and Jury, E.I., 1981*.

15.5 Robustness of Pole Assignment in a Circle Region

15.5.1 Matrix Eigenvalue Location Within a Disc

The eigenvalues $\lambda[\mathbf{A}]$ of the matrix **A** are located inside a circle $C(c, r)$ with radius r and centre c in the complex s-plane if and only if a positive definite matrix **P** exists satisfying

$$-c(\mathbf{A}^H\mathbf{P} + \mathbf{P}\mathbf{A}) + \mathbf{A}^H\mathbf{P}\mathbf{A} + (c^2 - r^2)\mathbf{P} = -\mathbf{I}\ . \tag{15.25}$$

For the proof see Eq.(15.33). The matrix **P** is given by

$$\mathbf{P} = \lim_{k\to\infty} \sum_{i=0}^{k} r^{-2(i+1)}\ [(\mathbf{A} - c\mathbf{I})^H]^i(\mathbf{A} - c\mathbf{I})^i\ . \tag{15.26}$$

The required state feedback controller **K** is derived by *Furuta, K., and Kim, S.B., 1987.*

15.5.2 Matrix Eigenvalue Location Within a Corona

Pre- and postmultiplying Eq.(15.25) by $\mathbf{a}^H$ and $\mathbf{a}$ where $\mathbf{a}$ is the eigenvector of $\mathbf{A}$ associated with $\lambda[\mathbf{A}]$ yields

$$-c(\mathbf{a}^H\mathbf{A}^H\mathbf{P}\mathbf{a}+\mathbf{a}^H\mathbf{P}\mathbf{A}\mathbf{a})+\mathbf{a}^H\mathbf{A}^H\mathbf{P}\mathbf{A}\mathbf{a}+(c^2-r^2)\mathbf{a}^H\mathbf{P}\mathbf{a}=-\mathbf{a}^H\mathbf{a} \tag{15.27}$$

$$-c(\lambda^*\mathbf{a}^H\mathbf{P}\mathbf{a}+\mathbf{a}^H\mathbf{P}\lambda\mathbf{a})+\lambda^*\mathbf{a}^H\mathbf{P}\lambda\mathbf{a}+(c^2-r^2)\mathbf{a}^H\mathbf{P}\mathbf{a}=-\mathbf{a}^H\mathbf{a} \tag{15.28}$$

$$[-c(\lambda+\lambda^*)+|\lambda|^2+c^2-r^2]\mathbf{a}^H\mathbf{P}\mathbf{a}=-\mathbf{a}^H\mathbf{a} \tag{15.29}$$

$$-2c\,\Re e\,\lambda+(\Re e\,\lambda)^2+(\Im m\,\lambda)^2+c^2-r^2=-\frac{\mathbf{a}^H\mathbf{a}}{\mathbf{a}^H\mathbf{P}\mathbf{a}} \tag{15.30}$$

$$(\Re e\,\lambda-c)^2+(\Im m\,\lambda)^2-r^2=-\frac{\mathbf{a}^H\mathbf{a}}{\mathbf{a}^H\mathbf{P}\mathbf{a}}\,. \tag{15.31}$$

Using Rayleigh's principle, see Eq.(2.5),

$$\lambda_{\min}[\mathbf{P}]\le\frac{\mathbf{a}^H\mathbf{P}\mathbf{a}}{\mathbf{a}^H\mathbf{a}}\le\lambda_{\max}[\mathbf{P}] \tag{15.32}$$

$$r_i^2=r^2-\frac{1}{\lambda_{\min}[\mathbf{P}]}\le(\Re e\,\lambda-c)^2+(\Im m\,\lambda)^2\le r^2-\frac{1}{\lambda_{\max}[\mathbf{P}]}=r_o^2\,. \tag{15.33}$$

Eq.(15.33) shows that the eigenvalues of $\mathbf{A}$ are located inside a corona with centre at c and inner and outer radius r_i and r_o, respectively (*Rachid, A., 1990*).

15.5.3 Robustness Under Structured Uncertainties

Consider a perturbed matrix $\mathbf{A}_p=\mathbf{A}+\Delta\mathbf{A}$ where $|\Delta\mathbf{A}|\le_e\mathbf{E}$. Comparison of Eq.(15.25) for $\mathbf{A}_p$ and $\mathbf{A}$ points out that the eigenvalues $\lambda[\mathbf{A}_p]$ are within $C(c,r)$ if

$$\rho_s[2(|c\mathbf{P}|\mathbf{E})_s+2(\mathbf{E}^T|\mathbf{P}\mathbf{A}|)_s+\mathbf{E}^T\mathbf{P}\mathbf{E}]<1\,. \tag{15.34}$$

If the absolute value of the coefficient matrix $\mathbf{A}$ enters affine linearly into the bound of the uncertainty, i.e., $\mathbf{E}=\varepsilon|\mathbf{A}|$ with $\varepsilon<1$, then

$$\varepsilon<\frac{1}{\rho_s[\,2(|c\mathbf{P}|\,|\mathbf{A}|)_s+2(|\mathbf{A}|^T|\mathbf{P}\mathbf{A}|)_s+|\mathbf{A}|^T|\mathbf{P}\mathbf{A}|\,]}\,. \tag{15.35}$$

The inequality Eq.(15.35) is sufficient for Eq.(15.34) since $\varepsilon|\mathbf{A}|^T|\mathbf{P}\mathbf{A}|$ has been replaced by $|\mathbf{A}|^T|\mathbf{P}\mathbf{A}|$.

15.5.4 Spectral-Norm-Bounded Uncertainty

The eigenvalues $\lambda[\mathbf{A}_p]$ remain within $C(c,r)$ if

$$[\,\|\Delta\mathbf{A}\|_s+2(|c|+\|\mathbf{A}\|_s)]\,\|\Delta\mathbf{A}\|_s<1/\lambda_{\max}[\mathbf{P}]\,. \tag{15.36}$$

Replacing $\|\Delta\mathbf{A}\|_s$ by $\|\Delta\mathbf{A}\|_s^2$ if $\|\Delta\mathbf{A}\|_s^2>\|\Delta\mathbf{A}\|_s$ (and $\|\Delta\mathbf{A}\|_s^2$ by $\|\Delta\mathbf{A}\|_s$ if $\|\Delta\mathbf{A}\|_s^2<\|\Delta\mathbf{A}\|_s$), one has

$$\max\{\|\Delta\mathbf{A}\|_s,\|\Delta\mathbf{A}\|_s^2)<\frac{1}{[1+2\,|c|+2\,\|\mathbf{A}\|_s]\,\lambda_{\max}[\mathbf{P}]}\,. \tag{15.37}$$

If $\mathbf{A}$ is Hermite

$$(\|\Delta\mathbf{A}\|_s + 2|c| + 2\rho_s[\mathbf{A}])\ \|\Delta\mathbf{A}\|_s < r^2 - \max_{\lambda} |\lambda[\mathbf{A}] - c|^2 . \tag{15.38}$$

Example: Let a discrete-time system be

$$\mathbf{x}(k+1) = (\mathbf{\Phi} + \Delta\mathbf{\Phi})\mathbf{x}(k) \quad \text{where} \quad \|\Delta\mathbf{\Phi}\|_s < 1 \tag{15.39}$$

and $\mathbf{\Phi}$ is Hermite. Then, the stability robustness condition is given by the unit circle in the z-plane, i.e., $c = 0,\ r = 1$. From Eq.(15.38) (*Rachid, A., 1990*)

$$\max\{\|\Delta\mathbf{\Phi}\|_s, \|\Delta\mathbf{\Phi}\|_s^2\} = \|\Delta\mathbf{\Phi}\|_s < \frac{1-\rho_s^2[\mathbf{\Phi}]}{1+2\rho_s[\mathbf{\Phi}]} . \tag{15.40}$$

End of Example

15.6 Robust Pole Assignment Using Condition Number

Consider the plant $(\mathbf{A}, \mathbf{B})$ and the state-feedback controller $\mathbf{u} = \mathbf{K}\mathbf{x}$

$$\dot{\mathbf{x}} = \mathbf{A}\mathbf{x} + \mathbf{B}\mathbf{u} = (\mathbf{A} + \mathbf{B}\mathbf{K})\mathbf{x} \triangleq \mathbf{F}\mathbf{x} \qquad \mathbf{B} \in \mathcal{R}^{n\times m} . \tag{15.41}$$

Assuming that the plant $(\mathbf{A}, \mathbf{B})$ is completely controllable, the controller $\mathbf{K}$ can be chosen in order to place the eigenvalues $\lambda_i[\mathbf{F}]\ \ \forall i = 1 \dots n$ at a prescribed set. If $m > 1$ multiple solutions exist. This variety can be used to provide robust properties for the closed-loop system. In addition to the pole placement, $\mathbf{K}$ has to be chosen such that $\lambda_i[\mathbf{F}]$ are as insensitive to perturbations in $\mathbf{A}$ and $\mathbf{B}$ as possible. Assume the eigenvalues $\lambda_i[\mathbf{F}]$ distinct or $\mathbf{F}$ diagonizable and let $\mathbf{T}_F = \mathbf{T}[\mathbf{F}]$ be the modal matrix associated with $\mathbf{F}$. Then, a measure of the sensitivity of the eigenvalues $\lambda[\mathbf{F}]$ is the condition number κ_F using the Frobenius norm $\|\cdot\|_F$

$$\kappa_F \triangleq \|\mathbf{T}_F\|_F \|\mathbf{T}_F^{-1}\|_F . \tag{15.42}$$

(*Bauer, F.L., and Fike, C.T., 1960; Byers, R., and Nash, S.G., 1989*). The Bauer-Fike theorem shows the importance of the condition number κ_s using the spectral norm $\|\cdot\|_s$ which can be bounded by the Frobenius norm $\|\cdot\|_F$ (*Golub, G.H., and VanLoan, C.F., 1983*): If $\mathbf{F}$ is perturbed by $\mathbf{E}$ then the eigenvalues move at a rate bounded by

$$|\ \lambda_i[\mathbf{F}] - \lambda_i[\mathbf{F} + \mathbf{E}]\ | \leq \kappa_F[\mathbf{T}_F]\ \|\mathbf{E}\|_p . \tag{15.43}$$

Per unit change in $\|\mathbf{E}\|_p$ the eigenvalues move at a rate no greater than κ_F. The change of eigenvalues as described above is true for any condition number κ_p and any Hölder norm $\|\cdot\|_p$. If multiple eigenvalues occur the system must be redesigned because multiple eigenvalues become infinitely sensitive.

In order to minimize the sensitivity of the closed-loop eigenvalues, the condition number $\kappa_F[\mathbf{T}_F]$ is minimized under the constraint of given $\mathbf{A}, \mathbf{B}, \lambda_i[\mathbf{F}]\ \ \forall i = 1 \dots n$

$$\min_{\mathbf{T}_F} \kappa_F[\mathbf{T}_F] \quad \text{where} \quad (\mathbf{A} + \mathbf{B}\mathbf{K})\mathbf{T}_F = \mathbf{T}_F \operatorname{diag}\{\lambda_i[\mathbf{F}]\} . \tag{15.44}$$

15.6.1 Parametrization of the Controller

From modal decomposition it follows

$$\mathbf{F}\mathbf{T}_F = (\mathbf{A} + \mathbf{B}\mathbf{K})\mathbf{T}_F = \mathbf{T}_F \operatorname{diag}\{\lambda_i[\mathbf{F}]\} \tag{15.45}$$

$$\mathbf{B}\mathbf{K} = \mathbf{T}_F \operatorname{diag}\{\lambda_i[\mathbf{F}]\}\mathbf{T}_F^{-1} - \mathbf{A} \,. \tag{15.46}$$

Let the control matrix $\mathbf{B}$ be factorized

$$\mathbf{B} \triangleq (\mathbf{U}_o \vdots \mathbf{U}_1)\begin{pmatrix}\mathbf{Z}\\ \mathbf{0}\end{pmatrix} \quad \text{where} \quad (\mathbf{Z}^{-1} \vdots \mathbf{0})\begin{pmatrix}\mathbf{Z}\\ \mathbf{0}\end{pmatrix} = \mathbf{I}\,. \tag{15.47}$$

The matrix $(\mathbf{U}_o \vdots \mathbf{U}_1)$ is orthogonal and $\mathbf{Z}$ nonsingular. Then,

$$\mathbf{K} = (\mathbf{Z}^{-1} \vdots \mathbf{0})(\mathbf{U}_o \vdots \mathbf{U}_1)^T(\mathbf{T}_F \operatorname{diag}\{\lambda_i[\mathbf{F}]\}\; \mathbf{T}_F^{-1} - \mathbf{A}) \tag{15.48}$$

$$\mathbf{K} = \mathbf{Z}^{-1}\mathbf{U}_o^T(\mathbf{T}_F \operatorname{diag}\{\lambda_i[\mathbf{F}]\}\; \mathbf{T}_F^{-1} - \mathbf{A})\,. \tag{15.49}$$

15.6.2 Parametrization of the Modal Matrix

Since

$$\mathbf{A}\mathbf{T}_F - \mathbf{T}_F \operatorname{diag}\{\lambda_i[\mathbf{F}]\} = -\mathbf{B}\mathbf{K}\mathbf{T}_F = -(\mathbf{U}_o \vdots \mathbf{U}_1)\begin{pmatrix}\mathbf{Z}\\ \mathbf{0}\end{pmatrix}\mathbf{K}\mathbf{T}_F \tag{15.50}$$

$$(\mathbf{U}_o \vdots \mathbf{U}_1)^T(\mathbf{A}\mathbf{T}_F - \mathbf{T}_F \operatorname{diag}\{\lambda_i[\mathbf{F}]\}) = \begin{pmatrix}\mathbf{U}_o^T\\ \mathbf{U}_1^T\end{pmatrix}\left(\mathbf{A}\mathbf{T}_F - \mathbf{T}_F \operatorname{diag}\{\lambda_i[\mathbf{F}]\}\right) = -\begin{pmatrix}\mathbf{Z}\\ \mathbf{0}\end{pmatrix}\mathbf{K}\mathbf{T}_F\,, \tag{15.51}$$

the relation

$$\mathbf{U}_1^T(\mathbf{A}\mathbf{T}_F - \mathbf{T}_F \operatorname{diag}\{\lambda_i[\mathbf{F}]\}) = \mathbf{0} \tag{15.52}$$

is a necessary and sufficient condition for solving Eq.(15.45). Referring to Eq.(B.136) and since Eq.(15.52),

$$\mathbf{U}_1^T(\mathbf{A}\mathbf{f}_i - \lambda_i\mathbf{f}_i) = \mathbf{U}_1^T(\mathbf{A} - \lambda_i\mathbf{I})\mathbf{f}_i = \mathbf{0} \tag{15.53}$$

the eigenvector $\mathbf{f}_i$ of $\mathbf{F}$ associated with $\lambda_i[\mathbf{F}]$ has to be in the nullspace (*Kautsky, J., et al. 1985*)

$$\mathbf{f}_i \in \mathcal{N}[\mathbf{U}_1^T(\mathbf{A} - \lambda_i\mathbf{I})] = \mathcal{S}\,. \tag{15.54}$$

Since

$$\mathbf{B}\mathbf{K} = \mathbf{T}_F \operatorname{diag}\{\lambda_i[\mathbf{F}]\}\mathbf{T}_F^{-1} - \mathbf{A} \tag{15.55}$$

the controller $\mathbf{K}$ exists if and only if

$$\mathcal{R}[\mathbf{T}_F \operatorname{diag}\{\lambda_i\}\mathbf{T}_F^{-1} - \mathbf{A}] \in \mathcal{R}[\mathbf{B}\mathbf{K}] = \mathcal{R}[\mathbf{B}] = \mathcal{R}[\mathbf{U}_o] \tag{15.56}$$

or $\mathcal{R}[\mathbf{T}_F \operatorname{diag}\{\lambda_i\}\mathbf{T}_F^{-1} - \mathbf{A}]$ is orthogonal to $\mathcal{N}[\mathbf{B}] = \mathcal{R}[\mathbf{U}_1]$. The m-dimensional space $\mathcal{S}_i$ is the basis of coordinates $\mathbf{s}_i$ associated with the eigenvector $\mathbf{f}_i$ according to $\mathbf{f}_i = \mathbf{S}_i\mathbf{s}_i \quad \mathbf{s}_i \in \mathcal{R}^m$ where $\mathbf{S}_i \in \mathcal{R}^{n\times m}$ is a matrix with orthogonal columns spanning the space $\mathcal{S}_i$. The modal matrix $\mathbf{T}_F$ is given by columns of the eigenvectors $\mathbf{f}_i$

$$\mathbf{T}_F = (\mathbf{f}_1\; \mathbf{f}_2 \ldots \mathbf{f}_n) = (\mathbf{S}_1 \vdots \mathbf{S}_2 \ldots \mathbf{S}_n) \text{ block diag}\{\mathbf{s}_1, \mathbf{s}_2 \ldots \mathbf{s}_n\}\,. \tag{15.57}$$

Note that $\mathbf{T}_F$ is now parametrized to mn coordinates s_i. Defining $\mathbf{s} = \text{vec } \mathbf{s}_i \in \mathcal{R}^{mn}$, the modal matrix is a function of $\mathbf{s}$ and the minimization problem is rewritten to

$$\min_{\mathbf{s}} \kappa_F[\mathbf{T}_F(\mathbf{s})] \ . \tag{15.58}$$

Byers, R., and Nash, S.G., 1989 proved that κ_F and κ_{F1} given by

$$\kappa_F(\mathbf{T}_F) = \|\mathbf{T}_F\|_F \|\mathbf{T}_F^{-1}\|_F < \kappa_{F1}(\mathbf{T}_F) = \frac{1}{2}(\|\mathbf{T}_F\|_F^2 + \|\mathbf{T}_F^{-1}\|_F^2) \tag{15.59}$$

produce equal optima. Using differentiation rules, one has

$$\frac{\partial}{\partial s_j}\|\mathbf{T}_F\|_F^2 = 2 \text{ tr } [\mathbf{T}_F^T \frac{\partial \mathbf{T}_F}{\partial s_j}] \tag{15.60}$$

$$\frac{\partial}{\partial s_j}\|\mathbf{T}_F^{-1}\|_F^2 = 2 \text{ tr } [\mathbf{T}_F^{-1,T} \frac{\partial \mathbf{T}_F^{-1}}{\partial s_j}] = -2 \text{ tr } [\mathbf{T}_F^{-1,T}\mathbf{T}_F^{-1} \frac{\partial \mathbf{T}_F}{\partial s_j}\mathbf{T}_F^{-1}] \ . \tag{15.61}$$

With the help of these functions the Newton method is applied to minimize κ_F or κ_{F1}. The constraint given by the set of prescribed $\lambda_i[\mathbf{F}]$ is implicitly taken into account via the factorization in the space $\mathcal{S}_i$. In fact, this is the key result of this section.

Chapter 16

Models for Optimal and Interconnected Systems

This chapter is devoted to several models based on least squares theory and system component interconnection. Optimal and suboptimal solutions and associated suboptimal controllers are considered. Solving the problems, multistage computational structure is employed. Large-scale interconnected systems for practical engineering purposes require decentralized controllers together with a suitable design procedure.

16.1 Optimal Control Via State Variable Matrices

16.1.1 State Variable Matrices

The state variable matrix $\mathbf{X}(t)$ is defined as the dyadic product of the classic state vector $\mathbf{x}(t) = [x_1(t)\ x_2(t)\ldots]^T$

$$\mathbf{X}(t) = \mathbf{x}(t)\mathbf{x}^T(t) = \begin{pmatrix} x_1^2(t) & x_1(t)x_2(t) & x_1(t)x_3(t) & \ldots \\ x_2(t)x_1(t) & x_2^2(t) & & \ldots \\ \vdots & \vdots & & \ddots \end{pmatrix}. \tag{16.1}$$

A linear autonomous system $\dot{\mathbf{x}}(t) = \mathbf{A}(t)\mathbf{x}(t)$ using the state variable matrix $\mathbf{X}(t)$ is represented by

$$\dot{\mathbf{X}}(t) = \mathbf{A}\mathbf{X}(t) + \mathbf{X}(t)\mathbf{A}^T. \tag{16.2}$$

The matrix $\mathbf{X}$ is symmetric. By $\mathbf{A} = \text{matrix}[A_{ij}]$, the entry $\dot{X}_{12}(t)$ in Eq.(16.2) is

$$\dot{X}_{12}(t) = A_{11}x_1x_2 + A_{12}x_2^2 + A_{13}x_3x_2 + \ldots + A_{21}x_1^2 + A_{22}x_1x_2 + A_{23}x_1x_3 + \ldots \ . \tag{16.3}$$

Separating x_2 and x_1 in the first and second part above yields

$$\begin{aligned}\dot{X}_{12}(t) &= x_2(A_{11}x_1 + A_{12}x_2 + A_{13}x_3 + \ldots) + x_1(A_{21}x_1 + A_{22}x_2 + A_{23}x_3 + \ldots) \\ &= x_2\dot{x}_1 + x_1\dot{x}_2 = \Big(\frac{d}{dt}\mathbf{x}(t)\mathbf{x}^T(t)\Big)_{12}.\end{aligned} \tag{16.4}$$

16.1.2 Linear Output Feedback and State Variable Matrices

The linear time-varying output feedback

$$\mathbf{u}(t) = \mathbf{K}(t)\mathbf{y}(t) \tag{16.5}$$

and the process

$$\dot{\mathbf{x}}(t) = \mathbf{A}(t)\mathbf{x}(t) + \mathbf{B}(t)\mathbf{u}(t) \qquad \mathbf{x}(0) = \mathbf{x}_o \qquad \mathbf{y}(t) = \mathbf{C}(t)\mathbf{x}(t) \tag{16.6}$$

yield in abbreviated notation $\mathbf{A} = \mathbf{A}(t)$ etc.

$$\dot{\mathbf{x}}(t) = (\mathbf{A} + \mathbf{BKC})\mathbf{x}(t) \qquad \mathbf{x}(0) = \mathbf{x}_o \tag{16.7}$$

$$\dot{\mathbf{X}}(t) = \dot{\mathbf{x}}\mathbf{x}^T + \mathbf{x}\dot{\mathbf{x}}^T = (\mathbf{A} + \mathbf{BKC})\mathbf{X} + \mathbf{X}(\mathbf{A} + \mathbf{BKC})^T \tag{16.8}$$

with the initial condition $\mathbf{X}(0) = \mathbf{x}_o\mathbf{x}_o^T$.

16.1.3 Optimal Linear Output Feedback Control

The objective is to minimize the performance criterion

$$I = \mathbf{x}^T(t_f)\mathbf{F}_f\mathbf{x}(t_f) + \int_{t_o}^{t_f} [\mathbf{x}^T(t)\mathbf{Q}\mathbf{x}(t) + \mathbf{u}^T(t)\mathbf{R}\mathbf{u}(t)]dt \ . \tag{16.9}$$

Referring to the relations

$$\mathbf{x}^T(t)\mathbf{x}(t) = \mathrm{tr}\ \mathbf{X}(t) \quad \text{and} \quad \mathbf{x}^T(t)\mathbf{Q}(t)\mathbf{x}(t) = \mathrm{tr}\ \mathbf{Q}(t)\mathbf{X}(t) = \mathrm{tr}\ \mathbf{X}(t)\mathbf{Q}(t)\ , \tag{16.10}$$

the performance index is rewritten

$$I = \mathrm{tr}\ \mathbf{F}_f\mathbf{X} + \int_{t_o}^{t_f} \mathrm{tr}\ (\mathbf{Q} + \mathbf{C}^T\mathbf{K}^T\mathbf{RKC})\mathbf{X}\ dt \ . \tag{16.11}$$

The Hamiltonian function is given by

$$H \triangleq \mathrm{tr}\ (\mathbf{Q} + \mathbf{C}^T\mathbf{K}^T\mathbf{RKC})\mathbf{X} + \mathrm{tr}\ \dot{\mathbf{X}}\mathbf{\Lambda}^T \ . \tag{16.12}$$

The first canonical equation is

$$\dot{\mathbf{X}}(t) = \frac{\partial H}{\partial \mathbf{\Lambda}(t)} \ . \tag{16.13}$$

The matrix $\mathbf{\Lambda}$ denotes the costate matrix and obeys to the second canonical equation

$$\dot{\mathbf{\Lambda}}(t) = -\frac{\partial H}{\partial \mathbf{X}(t)} \tag{16.14}$$

with the terminal condition (transversality condition)

$$\mathbf{\Lambda}(t_f) = \frac{\partial}{\partial \mathbf{X}(t_f)}\mathbf{F}_f[\mathbf{X}(t_f)] \ . \tag{16.15}$$

The gradient matrix (matricial gradient) of an arbitrary scalar valued and matrix structured function f with respect to $\mathbf{X}$ is denoted by $\partial f(\mathbf{X})/\partial \mathbf{X}$. Combining Eqs. (16.7) and (16.12) leads to

$$\begin{aligned} H &= \mathrm{tr}\ (\mathbf{Q} + \mathbf{C}^T\mathbf{K}^T\mathbf{RKC})\mathbf{X} + \mathrm{tr}\ (\mathbf{A} + \mathbf{BKC})\mathbf{X}\mathbf{\Lambda}^T + \mathrm{tr}\ \mathbf{X}(\mathbf{A} + \mathbf{BKC})^T\mathbf{\Lambda}^T \\ &= \mathrm{tr}\ [\mathbf{QX} + \mathbf{C}^T\mathbf{K}^T\mathbf{RKCX} + \mathbf{AX\Lambda}^T + \mathbf{BKCX\Lambda}^T + \mathbf{XA}^T\mathbf{\Lambda}^T + \mathbf{XC}^T\mathbf{K}^T\mathbf{B}^T\mathbf{\Lambda}^T] \ . \end{aligned} \tag{16.16}$$

Rearranging with respect to $\mathbf{K}$ yields

$$H = \mathrm{tr}\ [\mathbf{CXC}^T\mathbf{K}^T\mathbf{RK} + \mathbf{CX\Lambda}^T\mathbf{BK} + \mathbf{B}^T\mathbf{\Lambda}^T\mathbf{XC}^T\mathbf{K}^T] + \mathrm{tr}\ [\mathbf{QX} + \mathbf{AX\Lambda}^T + \mathbf{XA}^T\mathbf{\Lambda}^T] \ . \tag{16.17}$$

The derivative with respect to $\mathbf{K}$ is obtained by applying

$$\frac{\partial}{\partial \mathbf{M}} \text{tr } \mathbf{AM} = \mathbf{A}^T \text{and } \frac{\partial}{\partial \mathbf{M}} \text{tr } \mathbf{AM}^T = \mathbf{A}\ , \quad \frac{\partial}{\partial \mathbf{M}} \text{tr } [\mathbf{AMBM}^T] = \mathbf{A}^T\mathbf{MB}^T + \mathbf{AMB} \tag{16.18}$$

$$\text{and} \quad \frac{\partial}{\partial \mathbf{M}} \text{tr } (\mathbf{AM}^T\mathbf{BM}) = \mathbf{B}^T\mathbf{MA}^T + \mathbf{BMA}\ . \tag{16.19}$$

Hence,

$$\frac{\partial H}{\partial \mathbf{K}} = \mathbf{R}^T\mathbf{KCXC}^T + \mathbf{RKCX}^T\mathbf{C}^T + (\mathbf{CX\Lambda}^T\mathbf{B})^T + \mathbf{B}^T\mathbf{\Lambda}^T\mathbf{XC}^T \tag{16.20}$$

and since $\mathbf{X} = \mathbf{X}^T,\ \ \mathbf{\Lambda} = \mathbf{\Lambda}^T,\ \ \mathbf{Q} = \mathbf{Q}^T,\ \ \mathbf{R} = \mathbf{R}^T$

$$\frac{\partial H}{\partial \mathbf{K}} = 2\ \mathbf{RKCXC}^T + 2\ \mathbf{B}^T\mathbf{\Lambda XC}^T. \tag{16.21}$$

16.1.4 Matricial Gradients $\partial I/\partial \mathbf{K}$ and $\partial H/\partial \mathbf{K}$

Assume nonlinear relations

$$I(\mathbf{K}) \triangleq g[\mathbf{X}(t_f)] + \int_{t_o}^{t_f} f[\mathbf{X}(t), \mathbf{K}]dt \tag{16.22}$$

$$\dot{\mathbf{X}}(t) \triangleq \mathbf{F}[\mathbf{X}(t), \mathbf{K}] \qquad \mathbf{X}(0) = \mathbf{X}_o\ . \tag{16.23}$$

The Hamiltonian is

$$H(\mathbf{X}, \mathbf{\Lambda}, \mathbf{K}) \triangleq f(\mathbf{X}, \mathbf{K}) + \text{tr } \mathbf{\Lambda}^T\mathbf{F} \tag{16.24}$$

where the matrices $\mathbf{\Lambda}$ and $\mathbf{X}$ are considered as the solutions of the canonical equations. The first canonical equation in $\mathbf{X}$ is

$$\frac{\partial H}{\partial \mathbf{\Lambda}} - \dot{\mathbf{X}} = 0 \qquad \mathbf{X}(0) = \mathbf{X}_o\ . \tag{16.25}$$

The second canonical equation in $\mathbf{\Lambda}$ is

$$\frac{\partial H}{\partial \mathbf{X}} + \dot{\mathbf{\Lambda}} = 0 \qquad \mathbf{\Lambda}(t_f) = \frac{\partial g}{\partial \mathbf{X}(t_f)}\ . \tag{16.26}$$

In view of the matricial gradient $\partial I/\partial \mathbf{K}$, the following expressions are taken into consideration

$$I(\mathbf{K} + \varepsilon\ \delta\mathbf{K}) \doteq I(\mathbf{K}) + \varepsilon \text{ tr } \frac{\partial g}{\partial \mathbf{X}^T(t_f)}\ \delta\mathbf{X}(t_f) + \varepsilon \int_{t_o}^{t_f} (\text{tr } \frac{\partial f}{\partial \mathbf{X}^T}\delta\mathbf{X} + \text{tr } \frac{\partial f}{\partial \mathbf{K}^T})\delta\mathbf{K}\ \text{dt}\ . \tag{16.27}$$

From the second canonical equation one has

$$\frac{\partial H}{\partial \mathbf{X}} + \dot{\mathbf{\Lambda}} = \frac{\partial f(\mathbf{X}, \mathbf{K})}{\partial \mathbf{X}} + \frac{\partial}{\partial \mathbf{X}} \text{tr } \mathbf{\Lambda}^T\mathbf{F} + \dot{\mathbf{\Lambda}} = 0\ . \tag{16.28}$$

Transposing, postmultiplying by $\delta\mathbf{X}$ and taking the trace yields

$$\text{tr } \{\frac{\partial f(\mathbf{X}, \mathbf{K})}{\partial \mathbf{X}^T}\delta\mathbf{X}\} + \underbrace{\text{tr } \{(\frac{\partial}{\partial \mathbf{X}}\text{tr } \mathbf{\Lambda}^T\mathbf{F})^T\delta\mathbf{X}\}}_{1} + \text{tr } \{\dot{\mathbf{\Lambda}}\ \delta\mathbf{X}\} = 0\ . \tag{16.29}$$

From the process equation

$$\dot{\mathbf{X}} = \mathbf{F} \ \ \rightsquigarrow\ \ \mathbf{\Lambda}^T\dot{\mathbf{X}} = \mathbf{\Lambda}^T\mathbf{F} \ \ \rightsquigarrow\ \ \mathbf{\Lambda}^T\delta\dot{\mathbf{X}} = \mathbf{\Lambda}^T\delta\mathbf{F}\ . \tag{16.30}$$

Taking the trace of the last expression,

$$\text{tr } \Lambda^T \delta \dot{\mathbf{X}} = \text{tr } \Lambda^T \delta \mathbf{F} = \text{tr } \{\delta \text{ tr } \Lambda^T \mathbf{F}\} = \underbrace{\text{tr } \{(\frac{\partial}{\partial \mathbf{X}} \text{tr } \Lambda^T \mathbf{F})^T \delta \mathbf{X}\}}_{2} + \text{tr } \{(\frac{\partial}{\partial \mathbf{K}} \text{ tr } \Lambda^T \mathbf{F})^T \delta \mathbf{K}\} \,. \quad (16.31)$$

Separating the underbraced term 2 of Eq.(16.31) and substituting the result into Eq.(16.29) thus replacing the underbraced term 1 of Eq.(16.29) yields

$$\text{tr } \frac{\partial f(\mathbf{X},\mathbf{K})}{\partial \mathbf{X}^T} \delta \mathbf{X} + \text{tr } \Lambda^T \delta \dot{\mathbf{X}} - \text{tr } \{(\frac{\partial}{\partial \mathbf{K}} \text{tr } \Lambda^T \mathbf{F})^T \delta \mathbf{K}\} + \text{tr } \dot{\Lambda}^T \delta \mathbf{X} = 0 \,. \quad (16.32)$$

Integrating (and integrating by parts of the last expression) yields (*Geromel, J.C., and Bernussou, J., 1982; Bernussou, J., and Titli, A., 1982*)

$$\int_{t_o}^{t_f} \text{tr } \frac{\partial f}{\partial \mathbf{X}^T} \delta \mathbf{X} dt + \int_{t_o}^{t_f} \text{tr } \Lambda^T \delta \dot{\mathbf{X}} dt - \int_{t_o}^{t_f} \text{tr } \{(\frac{\partial}{\partial \mathbf{K}} \text{tr } \Lambda^T \mathbf{F})^T \delta \mathbf{K}\} + \text{tr } \Lambda^T \delta \mathbf{X} \Big|_{t_o}^{t_f} - \int_{t_o}^{t_f} \text{tr } \Lambda^T \delta \dot{\mathbf{X}} dt = 0 \,. \quad (16.33)$$

Taking into account that

$$\delta \mathbf{X}(t_o) = 0 \quad \text{and} \quad \Lambda(t_f) = \frac{\partial g}{\partial \mathbf{X}(t_f)} \quad (16.34)$$

and cancelling the second and fifth term,

$$\underbrace{\int_{t_o}^{t_f} \text{tr } (\frac{\partial f}{\partial \mathbf{X}})^T \delta \mathbf{X}}_{3} - \int_{t_o}^{t_f} \text{tr } \{(\frac{\partial}{\partial \mathbf{K}} \text{ tr } \Lambda^T \mathbf{F})^T \delta \mathbf{K}\} dt + \text{tr } \frac{\partial g}{\partial \mathbf{X}^T(t_f)} \delta \mathbf{X}(t_f) = 0 \,. \quad (16.35)$$

Substituting the underbraced expression 3 into Eq.(16.27), the term containing g^T is neutralized. Thus, from Eq.(16.27) it follows

$$I(\mathbf{K} + \varepsilon\, \delta \mathbf{K}) \doteq I(\mathbf{K}) + \varepsilon \int_{t_o}^{t_f} \text{tr } \{(\frac{\partial}{\partial \mathbf{K}} \text{tr } \Lambda^T \mathbf{F})^T \delta \mathbf{K}\} dt + \varepsilon \int_{t_o}^{t_f} \text{tr } \frac{\partial f}{\partial \mathbf{K}^T} \delta \mathbf{K} dt \quad (16.36)$$

$$I(\mathbf{K} + \varepsilon\, \delta \mathbf{K}) - I(\mathbf{K}) \doteq \varepsilon \int_{t_o}^{t_f} \text{tr } \{\frac{\partial f}{\partial \mathbf{K}} + \frac{\partial}{\partial \mathbf{K}} \text{ tr } \Lambda^T \mathbf{F}\}^T \delta \mathbf{K} \text{ dt} \,. \quad (16.37)$$

In view of this

$$\frac{\partial I(\mathbf{K})}{\partial \mathbf{K}} = \int_{t_o}^{t_f} \frac{\partial}{\partial \mathbf{K}} [f + \text{tr } \Lambda^T \mathbf{F}] dt = \int_{t_o}^{t_f} \frac{\partial}{\partial \mathbf{K}} H(\mathbf{X}, \Lambda, \mathbf{K}) dt \,. \quad (16.38)$$

The aforementioned result is very important. That is, the matricial gradient of $I(\mathbf{K})$ with respect to $\mathbf{K}$ is given by simple integration of the matricial gradient of $H(\mathbf{K})$ with respect to $\mathbf{K}$. Neither insight in $\mathbf{X}(t,\mathbf{K})$ nor knowledge of the $\mathbf{X}$-dependence with respect to $\mathbf{K}$ is required although I by the original definition depends on $\mathbf{X}$.

16.1.5 Optimal Linear Control and Lyapunov Equations

From the first canonical equation Eq.(16.25) and from Eq.(16.7)

$$\dot{\mathbf{X}} = (\mathbf{A} + \mathbf{BKC})\mathbf{X} + \mathbf{X}(\mathbf{A} + \mathbf{BKC})^T \qquad \mathbf{X}(0) = \mathbf{X}_o \,. \quad (16.39)$$

From the second canonical equation

$$\dot{\Lambda} = -\frac{\partial H}{\partial \mathbf{X}} = -[\mathbf{Q} + \mathbf{C}^T\mathbf{K}^T\mathbf{RKC} + (\Lambda^T(\mathbf{A} + \mathbf{BKC}))^T + \Lambda(\mathbf{A} + \mathbf{BKC})] \quad (16.40)$$

$$\dot{\Lambda} = -\mathbf{Q} - \mathbf{C}^T\mathbf{K}^T\mathbf{RKC} - (\mathbf{A} + \mathbf{BKC})^T\Lambda - \Lambda(\mathbf{A} + \mathbf{BKC}) \quad (16.41)$$

with the terminal condition Eq.(16.15). Repeating the result of the finite horizon case with time-variant matrices $\mathbf{A}, \mathbf{B}, \mathbf{C}, \mathbf{Q}$ and $\mathbf{R}$, the increment of the performance criterion coupled with two differential equations in $\mathbf{X}$ and $\mathbf{\Lambda}$ is optimal with respect to $\mathbf{K}$ if

$$\frac{\partial I}{\partial \mathbf{K}} = \int_{t_o}^{t_f} \frac{\partial H}{\partial \mathbf{K}} \text{ dt } = 2 \int_{t_o}^{t_f} (\mathbf{RKC} + \mathbf{B}^T \mathbf{\Lambda}) \mathbf{X} \mathbf{C}^T \text{ dt} \tag{16.42}$$

$$\dot{\mathbf{X}} = (\mathbf{A} + \mathbf{BKC})\mathbf{X} + \mathbf{X}(\mathbf{A} + \mathbf{BKC})^T, \qquad \mathbf{X}(0) = \mathbf{X}_o \tag{16.43}$$

$$\dot{\mathbf{\Lambda}} = -\mathbf{Q} - \mathbf{C}^T\mathbf{K}^T\mathbf{RKC} - (\mathbf{A} + \mathbf{BKC})^T\mathbf{\Lambda} - \mathbf{\Lambda}(\mathbf{A} + \mathbf{BKC}), \tag{16.44}$$

$$\mathbf{\Lambda}(t_f) = \frac{\partial}{\partial \mathbf{X}(t_f)} \mathbf{F}_f[\mathbf{X}(t_f)] \ . \tag{16.45}$$

16.1.6 Time-Invariant Infinite Horizon Case

In the case of time-invariant matrices $\mathbf{A}, \mathbf{B}, \mathbf{C}, \mathbf{Q}, \mathbf{R}$ and $t_o = 0, t_f = \infty$ the matrix $\mathbf{\Lambda}$ turns out constant. Defining

$$\mathbf{V} = \int_o^\infty \mathbf{X}(t) \text{ dt} \tag{16.46}$$

and integrating $\dot{\mathbf{X}}(t)$,

$$\int_o^\infty \dot{\mathbf{X}}(t) \text{dt } = \mathbf{X}(\infty) - \mathbf{X}(0) = \mathbf{0} - \mathbf{X}_o = -\mathbf{X}_o \tag{16.47}$$

which yields the result

$$\frac{\partial I(\mathbf{K})}{\partial \mathbf{K}} = 2(\mathbf{RKC} + \mathbf{B}^T\mathbf{\Lambda})\mathbf{V}\mathbf{C}^T \tag{16.48}$$

$$(\mathbf{A} + \mathbf{BKC})\mathbf{V} + \mathbf{V}(\mathbf{A} + \mathbf{BKC})^T + \mathbf{X}_o = \mathbf{0} \tag{16.49}$$

$$\dot{\mathbf{\Lambda}} = \mathbf{0} = \mathbf{Q} + \mathbf{C}^T\mathbf{K}^T\mathbf{RKC} + (\mathbf{A} + \mathbf{BKC})^T\mathbf{\Lambda} + \mathbf{\Lambda}(\mathbf{A} + \mathbf{BKC}) = \mathbf{0} \ . \tag{16.50}$$

The increment of I with respect to $\mathbf{K}$ is accompanied by two (merely algebraic) Lyapunov equations. For a special assumption $\mathbf{K}$ the quantities $\mathbf{\Lambda}, \mathbf{V}$ and $\partial I(\mathbf{K})/\partial \mathbf{K}$ can be derived.

16.1.7 Iteration

In the case of a trace function

$$I(\mathbf{K} + \beta\Delta\mathbf{K}) - I(\mathbf{K}) = \beta \text{ tr } [\mathbf{M}(\mathbf{K})\Delta\mathbf{K}], \tag{16.51}$$

according to the Kleinman lemma the increment is given by

$$\frac{\partial I(\mathbf{K})}{\partial \mathbf{K}} = \mathbf{M}^T(\mathbf{K}) \ . \tag{16.52}$$

Let the iteration be

$$\Delta\mathbf{K} = -k \ \frac{\partial I}{\partial \mathbf{K}} = \mathbf{D} \ . \tag{16.53}$$

Comparing Eqs.(16.52) and (16.53) yields $\mathbf{M}^T = -\frac{1}{k} \mathbf{D}$ and from Eq.(16.51)

$$I(\mathbf{K} + \beta\Delta\mathbf{K}) - I(\mathbf{K}) = \beta \text{ tr } [-\frac{1}{k}\mathbf{D}^T\mathbf{D}] = -\frac{\beta}{k} \text{ tr } \mathbf{D}^T\mathbf{D} < 0 \ . \tag{16.54}$$

Hence, the iteration will converge.

16.1.8 Optimal Decentralized Control

Plants often are widely distributed in space. Regarding the local distances, controllers which are installed in several separated places should remain disconnected. Separately installed controllers without interconnection are known as decentralized controllers. The decentralized control system imposes structural constraints upon the original linear system. Thus the overall control system becomes nonlinear. The matrices **B** and **K** are block diagonal: $\mathbf{B}$ = block diag $\mathbf{B}_\mu$, $\mathbf{K}$ = block diag $\mathbf{K}_\mu$. For the decentralized controller $\mathbf{K}_\mu$ only selected state variables are available. The matrix $\mathbf{B}_\mu$ tolerates selected input signals, only.

$$\mathbf{B}_\mu \in \mathcal{R}^{n_\mu \times m_\mu} \quad \mathbf{R}_\mu \in \mathcal{R}^{m_\mu \times n_\mu} \quad \mathbf{R} \in \mathcal{R}^{m \times n} \quad \mathbf{B} \in \mathcal{R}^{n \times m} \quad \mathbf{A} \in \mathcal{R}^{n \times n} \tag{16.55}$$

$$\dot{\mathbf{x}} = (\dot{\mathbf{x}}_1^T \ldots \dot{\mathbf{x}}_\mu^T \ldots \dot{\mathbf{x}}_L^T)^T = \mathbf{A}\mathbf{x} + \mathbf{B}\mathbf{u} \qquad \mathbf{u} = \mathbf{K}\mathbf{x} \,. \tag{16.56}$$

The overall system $n = \sum_1^L n_\mu$, $m = \sum_1^L m_\mu$ is a large-scale high-order system interconnecting L separated subsystems. Decentralized control reduces the design to low-order subsystem level.

If **K** has to satisfy the constraint of a block diagonal matrix, the incremental matrix is limited to the elements within the block diagonal structure. The elements beyond the block diagonal structure must be restricted to zero. For numerical examples see *Geromel, J.C., and Bernussou, J., 1979, 1982; Xinogalas, T.C., et al. 1982.*

16.1.9 Infinite Horizon and Steady-State Case

By equating Eq.(16.48) with zero

$$\frac{\partial I}{\partial \mathbf{K}} |_{\mathbf{K}^\star} = 0, \tag{16.57}$$

the optimum output controller $\mathbf{K}^\star$ is obtained as

$$\mathbf{K}^\star = -\mathbf{R}^{-1}\mathbf{B}^T \mathbf{\Lambda} \mathbf{V} \mathbf{C}^T (\mathbf{C}\mathbf{V}\mathbf{C}^T)^{-1}. \tag{16.58}$$

The matrices $\mathbf{\Lambda}$ and $\mathbf{V}$ are given by Eqs.(16.49) and (16.50) as before.

16.1.10 State Feedback, Instationarity and Finite Horizon Case

Setting $\mathbf{C} = \mathbf{I}$, all state variables are assumed available. From Eq.(16.42)

$$\mathbf{K}^\star(t) = -\mathbf{R}^{-1}(t)\mathbf{B}^T(t)\mathbf{\Lambda}(t) \,. \tag{16.59}$$

The matrix $\mathbf{X}$ no more affects the optimal solution $\mathbf{K}^\star$. At the same time $\mathbf{X}_o$ has no influence on $\mathbf{K}^\star$

$$-\dot{\mathbf{\Lambda}} = (\mathbf{A} + \mathbf{B}\mathbf{K})^T \mathbf{\Lambda} + \mathbf{\Lambda}(\mathbf{A} + \mathbf{B}\mathbf{K}) + \mathbf{Q} + \mathbf{K}^T \mathbf{R}\mathbf{K} \tag{16.60}$$

$$-\dot{\mathbf{\Lambda}} = \mathbf{A}^T\mathbf{\Lambda} - \mathbf{\Lambda}^T\mathbf{B}\mathbf{R}^{-1}\mathbf{B}^T\mathbf{\Lambda} + \mathbf{\Lambda}\mathbf{A} - \mathbf{\Lambda}\mathbf{B}\mathbf{R}^{-1}\mathbf{B}^T\mathbf{\Lambda} + \mathbf{Q} + \mathbf{\Lambda}^T\mathbf{B}\mathbf{R}^{-1}\mathbf{B}^T\mathbf{\Lambda} \tag{16.61}$$

$$-\dot{\mathbf{\Lambda}} = \mathbf{A}^T\mathbf{\Lambda} + \mathbf{\Lambda}\mathbf{A} - \mathbf{\Lambda}\mathbf{B}\mathbf{R}^{-1}\mathbf{B}^T\mathbf{\Lambda} + \mathbf{Q} \tag{16.62}$$

where $\mathbf{A} = \mathbf{A}(t)$, $\mathbf{B} = \mathbf{B}(t)$, $\mathbf{Q} = \mathbf{Q}(t)$, $\mathbf{R} = \mathbf{R}(t)$, $\mathbf{\Lambda}(t_f) = \partial \mathbf{F}_f[\mathbf{X}(t_f)]/\partial \mathbf{X}(t_f)$.

In this case, the costate matrix $\mathbf{\Lambda}(t)$ is the solution of the well-known Riccati matrix differential equation. Although the same answer is obtained just as given by the familiar vector formulation, the matrix Riccati formulation shows the dynamic behaviour of the partial derivative of the performance index with respect to the state $\mathbf{\Lambda}(t) = \partial I/\partial \mathbf{X}(t)$ in direct and physically illustrative view (*Athans, M., 1968*). In the case of the state feedback controller the derivative $\partial I/\partial \mathbf{K} = \mathbf{0}$ can be replaced by $\partial H/\partial \mathbf{K} = \partial H/\partial \mathbf{U} = \mathbf{0}$ from Eq.(16.21) and, thus, $\mathbf{K}^*$ directly can be computed from H by $\partial H/\partial \mathbf{K}|_{K^*} = \mathbf{0}$.

Remark: The optimum output feedback controller Eq.(16.58) can be "derived" from the optimum state feedback controller Eq.(16.59) postmultiplying by $\mathbf{VC}^T(\mathbf{CVC}^T)^{-1}$, although $\mathbf{\Lambda}$ in both cases is *not* the same quantity. This postmultiplicator symbolically draws the attention to the demarcation and orthogonal projection procedure and the right-pseudo-inverse.

16.2 Optimal Controller Using State Variable Vectors

Although this section does not include state variable matrices or decentralization constraints, this topic should be considered for the sake of explaning the matrix variables $\mathbf{V}$ and $\mathbf{\Lambda}$ and for the sake of comparison between state variable and state matrix derivations.

The plant

$$\dot{\mathbf{x}}(t) = \mathbf{A}\mathbf{x}(t) + \mathbf{B}\mathbf{u}(t), \quad \mathbf{x}(0) = \mathbf{x}_o \qquad \mathbf{y}(t) = \mathbf{C}\mathbf{x}(t) \tag{16.63}$$

is governed by an output feedback controller

$$\mathbf{u}(t) = \mathbf{K}\mathbf{y}(t) = \mathbf{K}\mathbf{C}\mathbf{x}(t) \tag{16.64}$$

and has to minimize the closed-loop performance

$$I = \int_o^\infty [\mathbf{x}^T(t)\mathbf{Q}\mathbf{x}(t) + \mathbf{u}^T(t)\mathbf{R}\mathbf{u}(t)]\mathrm{dt}\ . \tag{16.65}$$

Introducing the fundamental transition matrix $\mathbf{\Phi}(t)$, the dynamic behaviour of the plant and controller is given by

$$\mathbf{x}(t) = \mathbf{\Phi}(t)\mathbf{x}_o \qquad \mathbf{\Phi}(t) = e^{(\mathbf{A}+\mathbf{BKC})t}. \tag{16.66}$$

Then, the performance I is

$$I = \mathbf{x}_o^T[\int_o^\infty \mathbf{\Phi}^T(t)(\mathbf{Q} + \mathbf{C}^T\mathbf{K}^T\mathbf{RKC})\mathbf{\Phi}(t)\mathrm{dt}\ ]\mathbf{x}_o = \mathrm{tr}\ \{\mathbf{X}_o \int_o^\infty \mathbf{\Phi}^T(\mathbf{Q} + \mathbf{C}^T\mathbf{K}^T\mathbf{RKC})\mathbf{\Phi}\mathrm{dt}\ \}\ . \tag{16.67}$$

The performance I is a real scalar-valued trace function of the entries of the controller matrix $\mathbf{K}$, i.e., $I = I(\mathbf{K}) = \mathrm{tr}\ \{\mathbf{F}(\mathbf{K})\}$. The first-order expansion in β is

$$I(\mathbf{K} + \beta\ \delta\mathbf{K}) = \mathrm{tr}\ \{\mathbf{X}_o \int_o^\infty e^{(\mathbf{A}+\mathbf{BKC}+\beta\mathbf{B}\delta\mathbf{KC})^T t}[\Diamond]e^{(\mathbf{A}+\mathbf{BKC}+\beta\mathbf{B}\delta\mathbf{KC})t}\mathrm{dt}\ \} \tag{16.68}$$

$$\text{where} \quad [\Diamond] = [\mathbf{Q} + \mathbf{C}^T(\mathbf{K}^T + \beta\ \delta\mathbf{K}^T)\mathbf{R}(\mathbf{K} + \beta\ \delta\mathbf{K})\mathbf{C}]\ . \tag{16.69}$$

Simplifying the expression above (*Levine, W.S., and Athans, M., 1970*) by applying

$$\mathbf{F}(\mathbf{N}) = e^{(\mathbf{A}+\mathbf{BN})t} \tag{16.70}$$

$$\mathbf{F}(\mathbf{N} + \beta\ \delta\mathbf{N}) = e^{(\mathbf{A}+\mathbf{BN})t} + \beta \int_o^t e^{(\mathbf{A}+\mathbf{BN})(t-\sigma)}\mathbf{B}\ \delta\mathbf{N}\ e^{(\mathbf{A}+\mathbf{BN})\sigma}d\sigma \tag{16.71}$$

and choosing $\mathbf{N} := \mathbf{KC}$, $\delta\mathbf{N} = \delta\mathbf{K}\ \mathbf{C}$, it results $\mathbf{A} + \mathbf{BN} = \mathbf{A} + \mathbf{BKC}$ and

$$e^{(\mathbf{A}+\mathbf{BKC}+\beta\mathbf{B}\delta\mathbf{K}\ \mathbf{C})t} = e^{(\mathbf{A}+\mathbf{BKC})t} + \beta \int_o^t e^{(\mathbf{A}+\mathbf{BKC})(t-\sigma)}\mathbf{B}\ \delta\mathbf{K}\ \mathbf{C}\ e^{(\mathbf{A}+\mathbf{BKC})\sigma}d\sigma\ . \tag{16.72}$$

For the function structured as

$$I(\mathbf{K} + \beta\ \delta\mathbf{K}) - I(\mathbf{K}) = \beta\ \mathrm{tr}\ \{\mathbf{M}(\mathbf{K})\ \delta\mathbf{K}\} \tag{16.73}$$

with regard to the Kleinman lemma as $\beta \to 0$ the matricial gradient is given by

$$\frac{\partial I}{\partial \mathbf{K}} = \mathbf{M}^T(\mathbf{K}) . \qquad (16.74)$$

Applying the relations above and omitting some intermediate calculations,

$$\frac{\partial I}{\partial \mathbf{K}}|_{\mathbf{K}^\star} = 0 \quad \rightsquigarrow \quad \mathbf{K}^\star = -\mathbf{R}^{-1}\mathbf{B}^T\mathbf{\Lambda}\mathbf{V}\mathbf{C}^T(\mathbf{C}\mathbf{V}\mathbf{C}^T)^{-1} \qquad (16.75)$$

as given by Eq.(16.58). In this derivation use is made of the abbreviations

$$\mathbf{V} = \int_o^\infty e^{(\mathbf{A}+\mathbf{BKC})\sigma}\mathbf{x}_o\mathbf{x}_o^T e^{(\mathbf{A}+\mathbf{BKC})^T\sigma} d\sigma \qquad (16.76)$$

$$\mathbf{\Lambda} = \int_o^\infty e^{(\mathbf{A}+\mathbf{BKC})^T\tau}(\mathbf{Q}+\mathbf{C}^T\mathbf{K}^{\star T}\mathbf{R}\mathbf{K}^\star\mathbf{C})e^{(\mathbf{A}+\mathbf{BK}^\star\mathbf{C}^T)\tau} d\tau . \qquad (16.77)$$

The matrices $\mathbf{V}$ and $\mathbf{\Lambda}$ can be calculated from Eqs.(16.49) and (16.50), respectively. If $\mathbf{C} \neq \mathbf{I}$ but $\mathbf{C}^{-1}$ exists the calculations run in a similar manner including $\mathbf{C}^{-1}$ and $\mathbf{\Lambda}$ where $\mathbf{V}$ and $\mathbf{x}_o$ are of no influence in this case.

For design methods using successive pole assignment, order reduction and complete modal synthesis see *Föllinger, O., 1986*. Corresponding design problems of sampled-data systems are presented by *Yahagi, T., 1973*. Optimal *dynamical* output feedback controllers including deterministic measurement disturbances and similarities beween deterministic and stochastic controllers are discussed by *Kuhn, U., and Schmidt, G., 1987*. Design methodologies using inverse open-loop and desired closed-loop transfer matrices see *Lee, G.K.F., 1981; Hippe, P., and Wurmthaler, C., 1987*.

16.3 Optimizing in a Two-Level Computational Structure

Find an optimal decentralized controller $\mathbf{K}$ for the process in block diagonal structure

$$\dot{\mathbf{x}}(t) = \mathbf{A}\mathbf{x}(t) + \mathbf{B}\mathbf{u}(t); \quad \mathbf{u}(t) = \mathbf{K}\mathbf{x}(t); \quad \mathbf{x}(0) = \mathbf{x}_o \qquad (16.78)$$

$$I = \int_o^\infty [\mathbf{x}^T(t)\mathbf{Q}\mathbf{x}(t) + \mathbf{u}^T(t)\mathbf{R}\mathbf{u}(t)]\ \mathrm{dt} \to \min \qquad (16.79)$$

$$\dot{\mathbf{x}}(t) = (\mathbf{A}+\mathbf{BK})\mathbf{x}(t) \qquad \mathbf{x}(0) = \mathbf{x}_o \quad \rightsquigarrow \quad \mathbf{x}(t) = \mathbf{\Phi}(t)\ \mathbf{x}_o = e^{(\mathbf{A}+\mathbf{BK})t}\ \mathbf{x}_o \qquad (16.80)$$

$$I = \int_o^\infty \mathbf{x}^T(\mathbf{Q}+\mathbf{K}^T\mathbf{R}\mathbf{K})\mathbf{x}\ \mathrm{dt} = \int_o^\infty \mathbf{x}_o^T\mathbf{\Phi}^T(\mathbf{Q}+\mathbf{K}^T\mathbf{R}\mathbf{K})\mathbf{\Phi}\mathbf{x}_o\ \mathrm{dt} . \qquad (16.81)$$

If $\mathbf{x}_o$ is not known, the dependence of I on $\mathbf{x}_o$ can be eliminated assuming that $\mathbf{x}_o$ is a random vector uniformly distributed on the surface of an n-dimensional unit sphere. Thus, the mean value $\bar{\mathbf{x}} = E\{\mathbf{x}_o\}$ and the covariance cov $\mathbf{x}_o$ can lead to

$$E\{\mathbf{X}_o\} = E\{\mathbf{x}_o\mathbf{x}_o^T\} = \frac{1}{n}\ \mathbf{I}_n = \mathrm{cov}\ \mathbf{x}_o + \bar{\mathbf{x}}_o\bar{\mathbf{x}}_o^T . \qquad (16.82)$$

Applying the trace property $\mathbf{a}^T\mathbf{J}\mathbf{a} = \mathrm{tr}\ (\mathbf{J}\mathbf{a}\mathbf{a}^T)$ and $\mathbf{a} := \mathbf{\Phi}\mathbf{x}_o$, it follows

$$I = \mathrm{tr}\ \{(\mathbf{Q}+\mathbf{K}^T\mathbf{R}\mathbf{K})\mathbf{V}\} \qquad \text{and} \qquad (16.83)$$

$$\mathbf{V} = \int_o^\infty \mathbf{\Phi}\mathbf{x}_o\mathbf{x}_o^T\mathbf{\Phi}^T \mathrm{dt} . \qquad (16.84)$$

The matrix $\mathbf{V}$ can be derived in the same manner as Eq.(16.49)

$$(\mathbf{A}+\mathbf{BK})\mathbf{V} + \mathbf{V}(\mathbf{A}+\mathbf{BK})^T + \mathbf{x}\mathbf{x}_o = \mathbf{0} . \qquad (16.85)$$

The optimization procedure is now given by

$$\min_{\mathbf{K}} I = \min_{\mathbf{K}} \{\mathrm{tr}\,(\mathbf{Q} + \mathbf{K}^T\mathbf{R}\mathbf{K})\mathbf{V}\}|_{\mathbf{K}\,\in\,\text{block diag}\,\mathbf{K}_i} \tag{16.86}$$

(see *Xinogalas, T.C., et al. 1982*). The desired system matrix $\mathbf{A}_d$ is

$$\mathbf{A}_d \triangleq \text{block diag}\,\mathbf{A}_{ii}\,. \tag{16.87}$$

The complement matrix $\mathbf{A}_c$ is given by

$$\mathbf{A}_c = \mathbf{A} - \mathbf{A}_d = \text{matrix}[\mathbf{A}_{ij}] - \mathbf{A}_d\,. \tag{16.88}$$

Minimizing I with respect to the decentralized controller matrix $\mathbf{K}$ yields the conditions

$$(\mathbf{A}_d + \mathbf{B}\mathbf{K})\mathbf{V} + \mathbf{V}(\mathbf{A}_d + \mathbf{B}\mathbf{K})^T + \mathbf{x}\mathbf{x}_o^T + \mathbf{Z} = \mathbf{0} \quad \text{where} \quad \mathbf{Z} = \mathbf{A}_c\mathbf{V} + \mathbf{V}\mathbf{A}_c^T\,. \tag{16.89}$$

Employing matrix Lagrange multipliers Λ_1 and Λ_2 a new unconstrained performance index I_1 is posed

$$I_1 = \mathrm{tr}\,\{(\mathbf{Q}+\mathbf{K}^T\mathbf{R}\mathbf{K})\mathbf{V}+\Lambda_1(\mathbf{A}_c\mathbf{V}+\mathbf{V}\mathbf{A}_c^T-\mathbf{Z})+\Lambda_2[(\mathbf{A}_d-\mathbf{B}\mathbf{K})\mathbf{V}+\mathbf{V}(\mathbf{A}_d-\mathbf{B}\mathbf{K})^T+\mathbf{x}\mathbf{x}_o^T+\mathbf{Z}]\}\,. \tag{16.90}$$

The necessary conditions for achieving the minimum are given by calculating the partial derivatives. The derivatives are taken with respect to $\mathbf{K}, \Lambda_1, \Lambda_2$ and also with respect to $\mathbf{V}$ and $\mathbf{Z}$ although $\mathbf{V}$ and $\mathbf{Z}$ depend on $\mathbf{K}$. All derivatives are set to zero. Hence,

$$\frac{\partial I_1}{\partial \mathbf{Z}} = 0 \quad \rightsquigarrow \quad \Lambda_1 = \Lambda_2\,. \tag{16.91}$$

The derivative

$$\frac{\partial I_1}{\partial \mathbf{V}} = \mathbf{0} \quad (\text{using } \Lambda^T = \Lambda) \quad \text{yields} \tag{16.92}$$

$$\mathbf{Q} + \mathbf{K}^T\mathbf{R}\mathbf{K} + \mathbf{A}_c^T\Lambda_2 + \Lambda_2\mathbf{A}_c + (\mathbf{A}_d + \mathbf{B}\mathbf{K})^T\Lambda_1 + \Lambda_1(\mathbf{A}_d + \mathbf{B}\mathbf{K}) = \mathbf{0}\,. \tag{16.93}$$

From the derivation

$$\frac{\partial I_1}{\partial \mathbf{K}} = \mathbf{0} \quad \rightsquigarrow \quad \mathbf{R}^T\mathbf{K}\mathbf{V}^T + \mathbf{R}\mathbf{K}\mathbf{V} + (\mathbf{V}\Lambda_1\mathbf{B})^T + \mathbf{B}^T\Lambda_1\mathbf{V} = \mathbf{0} \quad \rightsquigarrow \quad \mathbf{K} = -\mathbf{R}^{-1}\mathbf{B}^T\Lambda_1\,. \tag{16.94}$$

This expression is replaced by matrices block diag $\Lambda_1\mathbf{V}$ and block diag $\mathbf{V}$

$$\mathbf{K} = -\mathbf{R}^{-1}\mathbf{B}^T[\text{block diag}\,(\Lambda_1\mathbf{V})]\,[\text{block diag}\,\mathbf{V}]^{-1}. \tag{16.95}$$

The matrices $\mathbf{R}$ and $\mathbf{B}$ have block diagonal structure by initial definition.

The numerical algorithm runs in a two-level structure. Assuming $\mathbf{A}_d + \mathbf{B}\mathbf{K}$ stable, an initial $\mathbf{Z}^k = \mathbf{Z}^o$ and $\Lambda_2^k = \Lambda_2^o$ is chosen. From Lyapunov Eqs.(16.89) and (16.93) $\mathbf{V}^k$ and Λ_1^k (for $k = 1$) are calculated in a first computation level. Next, it follows

$$\mathbf{Z}^{k+1} = \mathbf{A}_c\mathbf{V}^k + \mathbf{V}^k\mathbf{A}_c^T \tag{16.96}$$

$$\Lambda_2^{k+1} = \Lambda_1^k \tag{16.97}$$

$$\mathbf{K}^{k+1} = \mathbf{R}^{-1}\mathbf{B}^T(\text{block diag}\,\Lambda_1\mathbf{V})\,(\text{block diag}\,\mathbf{V})^{-1} \tag{16.98}$$

in a second level. This algorithm has to be repeated until $\mathbf{Z}^N, \Lambda_2^N$ and $\mathbf{K}^N$ satisfy a stopping condition, i.e., if the difference of the norm of the matrices between steps $N-1$ and N is smaller than a preselected tolerance. This algorithm operates significantly quicker than other approaches (see *Xinogalas, T.C., 1982*). Each iteration step must be characterized by a stable matrix $\mathbf{A}_d + \mathbf{B}\mathbf{K}$. Otherwise convergence of the iteration cannot be guaranteed.

16.4 Optimum Control Tolerance Intervals

16.4.1 Optimal Control of Time-Invariant Systems

The optimum control $\mathbf{u}(t) = \mathbf{Kx}(t)$ of a linear time-invariant system

$$\dot{\mathbf{x}}(t) = \mathbf{Ax}(t) + \mathbf{Bu}(t), \qquad \mathbf{x}(0) = \mathbf{x}_o \tag{16.99}$$

minimizing the quadratic index of performance

$$I = \int_0^{t_f} [\mathbf{x}^T(t)\mathbf{Qx}(t) + \mathbf{u}^T(t)\mathbf{Ru}(t)]dt \tag{16.100}$$

with the lower boundary $t_o = 0$ is given by the state feedback regulator $\mathbf{K}(t)$

$$\mathbf{u}(t) = \mathbf{Kx}(t) = -\mathbf{R}^{-1}\mathbf{B}^T\mathbf{Px}(t) = \mathbf{R}^{-1}\mathbf{B}^T\boldsymbol{\lambda}(t)\ . \tag{16.101}$$

The plant equation and controller equation are combined

$$\dot{\mathbf{x}}(t) = \mathbf{Ax}(t) + \mathbf{BR}^{-1}\mathbf{B}^T\boldsymbol{\lambda}(t)\ . \tag{16.102}$$

The costate variable $\boldsymbol{\lambda}(t)$ satisfies the equation

$$\dot{\boldsymbol{\lambda}}(t) = -\mathbf{A}^T\boldsymbol{\lambda}(t) + \mathbf{Qx}(t)\ , \tag{16.103}$$

with a system matrix $-\mathbf{A}^T$ skew-symmetric to $\mathbf{A}$.

16.4.2 State and Costate Equation

Combining the aforementioned equations

$$\begin{pmatrix} \dot{\mathbf{x}}(t) \\ \dot{\boldsymbol{\lambda}}(t) \end{pmatrix} = \begin{pmatrix} \mathbf{A} & \mathbf{BR}^{-1}\mathbf{B}^T \\ \mathbf{Q} & -\mathbf{A}^T \end{pmatrix} \begin{pmatrix} \mathbf{x}(t) \\ \boldsymbol{\lambda}(t) \end{pmatrix} \qquad \mathbf{x}(0) = \mathbf{x}_o;\ \boldsymbol{\lambda}(t_f) = \mathbf{0}\ . \tag{16.104}$$

Changing the independent variable and defining τ backwards in time, i.e., $\tau = t_f - t$ and $d\tau = -dt$, Eq.(16.104) is rewritten

$$\begin{pmatrix} \dot{\mathbf{x}}(\tau) \\ \dot{\boldsymbol{\lambda}}(\tau) \end{pmatrix} = -\begin{pmatrix} \mathbf{A} & \mathbf{BR}^{-1}\mathbf{B}^T \\ \mathbf{Q} & -\mathbf{A}^T \end{pmatrix} \begin{pmatrix} \mathbf{x}(\tau) \\ \boldsymbol{\lambda}(\tau) \end{pmatrix} \triangleq -\mathbf{M}\begin{pmatrix} \mathbf{x}(\tau) \\ \boldsymbol{\lambda}(\tau) \end{pmatrix}\ . \tag{16.105}$$

The transition matrix $\boldsymbol{\Phi}_N(\tau)$ associated with $\mathbf{M}$ is given in partitioned form as

$$\boldsymbol{\Phi}_N(\tau) = e^{-\mathbf{M}\,\tau} = \begin{pmatrix} \mathbf{N}_{11}(\tau) & \mathbf{N}_{12}(\tau) \\ \mathbf{N}_{21}(\tau) & \mathbf{N}_{22}(\tau) \end{pmatrix}\ . \tag{16.106}$$

Combining the system solution with the special boundary condition $\boldsymbol{\lambda}(\tau = 0) = \mathbf{0}$ yields

$$\begin{pmatrix} \mathbf{x}(\tau) \\ \boldsymbol{\lambda}(\tau) \end{pmatrix} = \boldsymbol{\Phi}_N(\tau)\begin{pmatrix} \mathbf{x}(\tau = 0) \\ \boldsymbol{\lambda}(\tau = 0) \end{pmatrix} = \begin{pmatrix} \mathbf{N}_{11}(\tau) & \mathbf{N}_{12}(\tau) \\ \mathbf{N}_{21}(\tau) & \mathbf{N}_{22}(\tau) \end{pmatrix}\begin{pmatrix} \mathbf{x}(\tau = 0) \\ \mathbf{0} \end{pmatrix} \tag{16.107}$$

$$\mathbf{x}(\tau) = \mathbf{N}_{11}(\tau)\mathbf{x}(\tau = 0) + \mathbf{0} \tag{16.108}$$

$$\boldsymbol{\lambda}(\tau) = \mathbf{N}_{21}(\tau)\mathbf{x}(\tau = 0) + \mathbf{0} \tag{16.109}$$

$$\rightsquigarrow \quad \boldsymbol{\lambda}(\tau) = \mathbf{N}_{21}(\tau)\mathbf{N}_{11}^{-1}(\tau)\mathbf{x}(\tau) \quad \rightsquigarrow \quad \mathbf{N}_{21}(\tau)\mathbf{N}_{11}^{-1}(\tau) = -\mathbf{P}(\tau)\ . \tag{16.110}$$

The optimum control then is given by

$$\mathbf{u}(t) = \mathbf{R}^{-1}\mathbf{B}^T\boldsymbol{\lambda}(t) = \mathbf{R}^{-1}\mathbf{B}^T\mathbf{N}_{21}(t_f - t)\mathbf{N}_{11}^{-1}(t_f - t)\mathbf{x}(t_f - t) \tag{16.111}$$

$$\text{where} \quad \mathbf{K}(t) = \mathbf{R}^{-1}\mathbf{B}^T\mathbf{N}_{21}(t_f - t)\mathbf{N}_{11}(t_f - t) \tag{16.112}$$

is the optimal time-varying state feedback gain matrix.

16.4.3 Riccati Equation

Combining, first, Eqs.(16.101),(16.99) and (16.110) in abbreviated notation

$$\dot{\mathbf{x}} = \mathbf{Ax} + \mathbf{Bu} = \mathbf{Ax} + \mathbf{BR}^{-1}\mathbf{B}^T\boldsymbol{\lambda} = \mathbf{Ax} - \mathbf{BR}^{-1}\mathbf{B}^T\mathbf{Px} \tag{16.113}$$

and, second, Eqs.(16.103) and (16.110) yields

$$\dot{\boldsymbol{\lambda}} = -\mathbf{A}^T\boldsymbol{\lambda} + \mathbf{Qx} \quad \text{and} \quad \dot{\boldsymbol{\lambda}} = -\mathbf{P}\dot{\mathbf{x}} - \dot{\mathbf{P}}\mathbf{x} = -\mathbf{A}^T\boldsymbol{\lambda} + \mathbf{Qx}\,. \tag{16.114}$$

Substituting $\dot{\mathbf{x}}$ of Eq.(16.113) into Eq.(16.114), the well-known Riccati differential equation is obtained

$$-\mathbf{P}(\mathbf{Ax} - \mathbf{BR}^{-1}\mathbf{B}^T\mathbf{Px}) - \dot{\mathbf{P}}\mathbf{x} = -\mathbf{A}^T(-\mathbf{Px}) + \mathbf{Qx} \tag{16.115}$$

$$-\dot{\mathbf{P}}(t) = \mathbf{A}^T\mathbf{P}(t) + \mathbf{P}(t)\mathbf{A} + \mathbf{Q} - \mathbf{P}(t)\mathbf{BR}^{-1}\mathbf{B}^T\mathbf{P}(t); \quad \mathbf{P}(t_f) = \mathbf{0}\,. \tag{16.116}$$

The matrix $\mathbf{P}(t)$ is unstable in the forward time direction.

16.4.4 Cost Matrix Associated with a State Controller

Consider a time-varying plant and the time-varying optimum controller $\mathbf{K}(t)$ according to the Riccati solution and a non-optimum controller $\mathbf{L}(t)$

$$\dot{\mathbf{x}}(t) = \mathbf{A}(t)\mathbf{x}(t) + \mathbf{B}(t)\mathbf{u}(t), \qquad \mathbf{x}(t_o) = \mathbf{x}_o \qquad \mathbf{x}(t_f) = \mathbf{x}_f \tag{16.117}$$

$$\mathbf{u}(t) = \mathbf{K}(t)\mathbf{x}(t) \ \text{ (optimal state feedback controller)} \tag{16.118}$$

$$\mathbf{u}(t) = \mathbf{L}(t)\mathbf{x}(t) \ \text{ (non-optimal state feedback controller)}\,. \tag{16.119}$$

The cost matrix $\boldsymbol{\Lambda}(t)$ is defined as a matrix associated with the feedback gain matrix $\mathbf{K}(t)$ or $\mathbf{L}(t)$ expressing the performance $I(t)$ of the control system between t and the upper optimization bound t_f:

$$I(t) = \mathbf{x}_f\mathbf{F}_f\mathbf{x}_f + \int_t^{t_f} [\mathbf{x}^T(\tau)\mathbf{Q}(\tau)\mathbf{x}(\tau) + \mathbf{u}^T(\tau)\mathbf{R}(\tau)\mathbf{u}(\tau)]d\tau \triangleq \mathbf{x}^T(t)\boldsymbol{\Lambda}(t)\mathbf{x}(t)\,. \tag{16.120}$$

The non-optimal transient is

$$\dot{\mathbf{x}}(t) = \mathbf{A}(t)\mathbf{x}(t) + \mathbf{B}(t)\mathbf{L}(t)\mathbf{x}(t) = [\mathbf{A}(t) + \mathbf{B}(t)\mathbf{L}(t)]\mathbf{x}(t)\,. \tag{16.121}$$

The state transition matrix associated with the optimal and the non-optimal controlling variable is referred to as $\boldsymbol{\Phi}_K(t,t_o)$ and $\boldsymbol{\Phi}_L(t,t_o)$. The solution in the non-optimal case is

$$\mathbf{x}(t) = \boldsymbol{\Phi}_L(t,t_o)\mathbf{x}_o \tag{16.122}$$

$$\dot{\boldsymbol{\Phi}}_L(t,t_o) = [\mathbf{A}(t) + \mathbf{B}(t)\mathbf{L}(t)]\boldsymbol{\Phi}_L(t,t_o); \qquad \boldsymbol{\Phi}_L(t_o,t_o) = \mathbf{I}\,. \tag{16.123}$$

Specializing the time-varying system to a time-invariant one yields

$$\boldsymbol{\Phi}_L(t,t_o) = e^{[\mathbf{A}+\mathbf{BL}]t}\,. \tag{16.124}$$

Note the following relations in the time-varying regulator problem

$$\mathbf{x}_f = \boldsymbol{\Phi}(t_f,t)\mathbf{x}(t) \qquad \mathbf{x}(t) = \boldsymbol{\Phi}^{-1}(t_f,t)\mathbf{x}_f = \boldsymbol{\Phi}(t,t_f)\mathbf{x}_f \tag{16.125}$$

$$\mathbf{x}(t) = \mathbf{\Phi}(t,\tau)\mathbf{x}(\tau) \qquad \mathbf{x}(\tau) = \mathbf{\Phi}(\tau,t)\mathbf{x}(t)\ . \tag{16.126}$$

Since Eq.(16.120)

$$\begin{aligned} I(t) = \quad & \mathbf{x}^T(t)\mathbf{\Phi}_L^T(t_f,t)\mathbf{F}_f\mathbf{\Phi}_L(t_f,t)\mathbf{x}(t) + \mathbf{x}^T(t)\int_t^{t_f}\mathbf{\Phi}_L^T(\tau,t)\mathbf{Q}(\tau)\mathbf{\Phi}_L(\tau,t)d\tau\ \mathbf{x}(t) \\ + \quad & \mathbf{x}^T(t)\int_t^{t_f}\mathbf{\Phi}_L^T(\tau,t)\mathbf{L}^T(\tau)\mathbf{R}(\tau)\mathbf{L}(\tau)\mathbf{\Phi}_L(\tau,t)d\tau\ \mathbf{x}(t) \end{aligned} \tag{16.127}$$

$$I(t) = \mathbf{x}^T(t)\{\mathbf{\Phi}_L^T(t_f,t)\mathbf{F}_f\mathbf{\Phi}_L(t_f,t) + \int_t^{t_f}\mathbf{\Phi}_L^T(\tau,t)[\mathbf{Q}(\tau)+\mathbf{L}^T(\tau)\mathbf{R}(\tau)\mathbf{L}(\tau)]\mathbf{\Phi}_L(\tau,t)d\tau\}\mathbf{x}(t)\ . \tag{16.128}$$

The cost matrix $\mathbf{\Lambda}$ is defined as follows

$$I(t) \triangleq \mathbf{x}^T(t)\mathbf{\Lambda}(t)\mathbf{x}(t) \tag{16.129}$$

$$\mathbf{\Lambda}(t) \triangleq \{\mathbf{\Phi}_L^T(t_f,t)\mathbf{F}_f\mathbf{\Phi}_L(t_f,t) + \int_t^{t_f}\mathbf{\Phi}_L^T(\tau,t)[\mathbf{Q}(\tau)+\mathbf{L}^T(\tau)\mathbf{R}(\tau)\mathbf{L}(\tau)]\mathbf{\Phi}_L(\tau,t)d\tau\}\ . \tag{16.130}$$

Differentiating $\mathbf{\Lambda}(t)$ with respect to t and employing

$$\dot{\mathbf{\Phi}}_L(t_f,t) = -\mathbf{\Phi}_L(t,t_f) \qquad (\mathbf{A}+\mathbf{BL})\mathbf{\Phi}_L = \mathbf{\Phi}_L(\mathbf{A}+\mathbf{BL}) \tag{16.131}$$

it results

$$\dot{\mathbf{\Lambda}} = \quad \dot{\mathbf{\Phi}}_L^T(t_f,t)\mathbf{F}_f\mathbf{\Phi}_L(t_f,t) + \mathbf{\Phi}_L^T(t_f,t)\mathbf{F}_f\dot{\mathbf{\Phi}}_L(t_f,t) \tag{16.132}$$

$$- \int_t^{t_f}\dot{\mathbf{\Phi}}_L(\tau,t)[\mathbf{Q}(\tau)+\mathbf{L}^T(\tau)\mathbf{R}(\tau)\mathbf{L}(\tau)]\mathbf{\Phi}_L(\tau,t)d\tau \tag{16.133}$$

$$- \int_t^{t_f}\mathbf{\Phi}_L(\tau,t)[\mathbf{Q}(\tau)+\mathbf{L}^T(\tau)\mathbf{R}(\tau)\mathbf{L}(\tau)]\dot{\mathbf{\Phi}}(\tau,t)d\tau \tag{16.134}$$

$$- \mathbf{\Phi}_L^T(t,t)[\mathbf{Q}(\tau)+\mathbf{L}^T(\tau)\mathbf{R}(\tau)\mathbf{L}(\tau)]\mathbf{\Phi}_L(t,t)\ . \tag{16.135}$$

Substituting the definition of $\mathbf{\Lambda}(t)$ from Eq.(16.120) yields the cost function matrix differential equation

$$-\dot{\mathbf{\Lambda}}(t) = \mathbf{\Lambda}(t)[\mathbf{A}(t)+\mathbf{B}(t)\mathbf{L}(t)]+[\mathbf{A}(t)+\mathbf{B}(t)\mathbf{L}(t)]^T\mathbf{\Lambda}(t)+\mathbf{Q}(t)+\mathbf{L}^T(t)\mathbf{R}(t)\mathbf{L}(t) \tag{16.136}$$

with the boundary condition $\mathbf{\Lambda}(t_f) = \mathbf{F}_f$. With regard to the fact that $I[\mathbf{u}_L] \geq I[\mathbf{u}_K]$ the difference $\mathbf{\Lambda}(t) - \mathbf{P}(t) \geq 0$. The cost matrix $\mathbf{\Lambda}(t)$ associated with $\mathbf{L}(t)$ corresponds to the costate matrix in Eq.(16.41).

Cost Matrix Difference

The cost matrix difference $\mathbf{\Lambda}(t) - \mathbf{P}(t)$ (*Kleinman, D.L., and Athans, M., 1968*)

$$\mathbf{\Lambda}(t) - \mathbf{P}(t) = \int_t^{t_f}\mathbf{\Phi}_L^T(\tau,t)[\mathbf{L}(\tau)-\mathbf{K}(\tau)]^T\mathbf{R}(\tau)[\mathbf{L}(\tau)-\mathbf{K}(\tau)]\mathbf{\Phi}_L(\tau,t)d\tau \tag{16.137}$$

yields a performance kernel difference (or incremental sensitivity) with respect to $t\ \forall\ (t_o,t_f)$ caused by the deviation between the real regulator matrix $\mathbf{L}(t)$ and the optimal $\mathbf{K}(t)$

$$I[t,\mathbf{u}_L] - I[t,\mathbf{u}_K] = \mathbf{x}^T(t)\mathbf{\Lambda}(t)\mathbf{x}(t) - \mathbf{x}^T(t)\mathbf{P}(t)\mathbf{x}(t) = \mathbf{x}^T(t)[\mathbf{\Lambda}(t)-\mathbf{P}(t)]\mathbf{x}(t)\ . \tag{16.138}$$

The final boundary condition is $\mathbf{\Lambda}(t_f) = \mathbf{P}(t_f) = \mathbf{F}_f(t_f)$.

Suboptimal Controller

A suboptimal controller

$$\mathbf{L}(t) = \sum_{j=1}^{M} a_j(t)\mathbf{L}_j \qquad t \in (t_o, t_f) \tag{16.139}$$

can be established using a structure of arbitrarily prechosen time functions $a_j(t)$ and matrices $\mathbf{L}_j$ unknown but prefixed constant with respect to time. For given time functions $a_j(t)$ the optimal matrices opt $\mathbf{L}_j = \mathbf{L}_j^*$ are defined by the minimum of the trace of the associated cost function

$$\frac{\partial}{\partial \mathbf{L}_j} \text{ tr } \mathbf{\Lambda}(t_o; \mathbf{L}) \mid_{\mathbf{L}=\mathbf{L}^*} = \mathbf{0} \qquad \mathbf{\Lambda}^* = \mathbf{\Lambda}(t_o, \mathbf{L}) \mid_{\mathbf{L}=\mathbf{L}^*} . \tag{16.140}$$

The solution $\mathbf{L}^*$ (proof see *Kleinman, D.L., and Athans, M., 1968*) is given by the necessary conditions

$$\frac{\partial}{\partial \mathbf{L}_j} \text{ tr } \mathbf{\Lambda}(t_o; \mathbf{L}) \mid_{\mathbf{L}_j=\mathbf{L}_j^*} = 2\int_{t_o}^{t_f} a_j(t)[\mathbf{R}(t)\mathbf{L}^*(t) - \mathbf{B}^T(t)\mathbf{\Lambda}^*(t)]\mathbf{\Phi}_{L*}(t,t_o)\mathbf{\Phi}_{L*}^T(t,t_o)dt = \mathbf{0} \tag{16.141}$$

$$\text{where} \quad \dot{\mathbf{\Phi}}_{L*}(t,t_o) = [\mathbf{A}(t) + \mathbf{B}(t)\mathbf{L}^*(t)]\mathbf{\Phi}_{L*}(t,t_o); \quad \mathbf{\Phi}_{L*}(t,t_o) = \mathbf{I} . \tag{16.142}$$

Piecewise-Constant Suboptimal Controller

For a given number N, for a given set of times t_i (of number $N-1$) between t_o and t_f and for constant matrices $\mathbf{A}, \mathbf{B}, \mathbf{Q}, \mathbf{R}$ let the time function $a_j(t)$ be

$$a_j(t) = \begin{cases} 1 & t_{i-1} < t \le t_i \\ 0 & \text{otherwise} . \end{cases} \tag{16.143}$$

From Eq.(16.141) $\forall t_i$

$$\int_{t_{i-1}}^{t_i} [\mathbf{R}\mathbf{L}_i^* - \mathbf{B}^T\mathbf{\Lambda}^*(t)]\mathbf{\Phi}_{L*}(t,t_o)\mathbf{\Phi}_{L*}(t,t_o)dt = \mathbf{0} \tag{16.144}$$

$$\mathbf{L}_i^* = \mathbf{R}^{-1}\mathbf{B}^T[\int_{t_{i-1}}^{t_i} \mathbf{\Lambda}^*(t)\mathbf{\Phi}_{L*}(t,t_o)\mathbf{\Phi}_{L*}^T(t,t_o)dt]\ [\int_{t_{i-1}}^{t_i} \mathbf{\Phi}_{L*}(t,t_o)\mathbf{\Phi}_{L*}^T(t,t_o)dt]^{-1} \tag{16.145}$$

$$\text{where } \mathbf{\Phi}_{L*}(t,t_o) = \mathbf{\Phi}_{L*}(t,t_{i-1})\mathbf{\Phi}_{L*}(t_{i-1},t_o) = e^{(\mathbf{A}+\mathbf{B}\mathbf{L}^*)(t-t_o)} \text{ for } t \in (t_{i-1}, t_i) . \tag{16.146}$$

Kleinman, D.L., Fortmann, T., and Athans, M., 1968 suggest an iterative scheme for $\mathbf{L}_i^*$ based on successive substitutions: The μth iteration of $\mathbf{\Phi}_L$ before arriving $\mathbf{\Phi}_{L*}$ is denoted $\mathbf{\Phi}_\mu$. In order to achieve a convergent algorithm, the iterate $(\mu+1)$ is obtained from

$$\mathbf{L}_i^{\mu+1} = \mathbf{L}_i^\mu + \varepsilon_\mu[\bar{\mathbf{L}}_i^\mu - \mathbf{L}_i^\mu] \tag{16.147}$$

$$\bar{\mathbf{L}}_i^\mu = \mathbf{R}^{-1}\mathbf{B}^T[\int_{t_{i-1}}^{t_i} \mathbf{\Lambda}_\mu(t)\mathbf{\Phi}_\mu(t,t_o)\mathbf{\Phi}_\mu^T(t,t_o)dt][\int_{t_{i-1}}^{t_i} \mathbf{\Phi}_\mu(t,t_o)\mathbf{\Phi}_\mu^T(t,t_o)dt]^{-1} \tag{16.148}$$

$$\mathbf{\Phi}_\mu(t,t_o) = e^{(\mathbf{A}+\mathbf{B}\mathbf{L}_i^\mu)(t-t_o)} . \tag{16.149}$$

Starting with $\mu = 1$ and an initial value $\mathbf{L}_i^1$ $\forall i = 1 \ldots N$, the matrices $\mathbf{\Phi}_1(t,t_o)$ and $\mathbf{\Lambda}_1(t)$ are calculated $\forall i = 1 \ldots N$. If $\varepsilon_1 = 1$ yields an improvement in the trace of the cost function the next iterate is computed from Eqs.(16.148) and (16.147), if not, ε_1 is reduced until improvement is obtained. Regarding a stopping condition, $\mathbf{L}_i^\mu$ is declared as $\mathbf{L}_i^*$.

16.5 Decentralized Controller Via Eigenvalue Bounds

Robust decentralized controllers for large-scale systems are designed in an attractive way when bounds and norms of system eigenvalues and Riccati matrices are taken into account. If changes in the system components and parameters are considered reliable then robust controllers can be achieved. Given the N subsystems

$$\dot{\mathbf{x}}_i = \mathbf{A}_i\mathbf{x}_i + \mathbf{B}_i\mathbf{u}_i + \sum_{j=1,j\neq i}^{N} \mathbf{H}_{ij}\mathbf{x}_j \qquad \mathbf{B}_i \in \mathcal{R}^{n_i \times m_i} \tag{16.150}$$

the overall time-invariant system can be written as

$$\dot{\mathbf{x}} = \mathbf{A}\mathbf{x} + \mathbf{B}\mathbf{u} + \mathbf{H}\mathbf{x} \qquad \mathbf{x} = \mathrm{vec}\ \mathbf{x}_i, \quad \mathbf{u} = \mathrm{vec}\ \mathbf{u}_i, \tag{16.151}$$

$$\mathbf{A} = \text{block diag}\ \mathbf{A}_i, \quad \mathbf{B} = \text{block diag}\ \mathbf{B}_i, \quad n = \sum_1^N n_i, \quad m = \sum_1^N m_i, \tag{16.152}$$

$$\mathbf{H} = \begin{pmatrix} \mathbf{0} & \mathbf{H}_{12} & \mathbf{H}_{13} & \dots & \mathbf{H}_{1N} \\ \mathbf{H}_{21} & \mathbf{0} & \mathbf{H}_{23} & \dots & \mathbf{H}_{2N} \\ \vdots & \dots & \mathbf{0} & \dots & \vdots \end{pmatrix} . \tag{16.153}$$

Let the overall index of performance be

$$I = \sum_{i=1}^{N} \int_0^\infty e^{2\alpha_i t}(\mathbf{x}_i^T\mathbf{Q}_i\mathbf{x}_i + \mathbf{u}_i^T\mathbf{R}_i\mathbf{u}_i)dt \tag{16.154}$$

where α_i and $\mathbf{Q}_i$, $\mathbf{R}_i$ are the degree of stability and adequate symmetric weighting matrices, respectively.

The original system $\mathbf{x}_i$ with the performance index above can be transformed into a new destabilized system $\mathbf{x}_i^\circ = \mathbf{x}_i e^{\alpha_i t}, \mathbf{u}_i^\circ = \mathbf{u}_i e^{\alpha_i t}$ with $\mathbf{A}_i^\circ = \mathbf{A}_i + \alpha_i \mathbf{I}_{n_i}$. The eigenvalues of $\mathbf{A}_i^\circ$ result from the eigenvalues of $\mathbf{A}_i$ being shifted to the right with α_i. The Riccati equation must be augmented by an additional term $2\alpha_i\mathbf{P}_i$. A local or decentralized controller is given by a constant state feedback $\mathbf{u}_i = \mathbf{K}_i\mathbf{x}_i$.

16.5.1 Isolated Subsystems

Assuming $\mathbf{H}_{ij} \equiv \mathbf{0}$, the overall system is completely decoupled. The subsystems are free of interaction. In order to provide stable behaviour in this isolated operation, the controllers $\mathbf{K}_i$ are designed as a Riccati type

$$\mathbf{A}_i^T\mathbf{P}_i + \mathbf{P}_i\mathbf{A}_i - \mathbf{P}_i\mathbf{B}_i\mathbf{R}_i^{-1}\mathbf{B}_i^T\mathbf{P}_i + 2\alpha_i\mathbf{P}_i + \mathbf{Q}_i = \mathbf{0} \tag{16.155}$$

$$\mathbf{K}_i = -\mathbf{R}_i^{-1}\mathbf{B}_i^T\mathbf{P}_i, \qquad \mathbf{u}_i = -\mathbf{R}_i^{-1}\mathbf{B}_i^T\mathbf{P}_i\mathbf{x}_i . \tag{16.156}$$

16.5.2 Interconnected System

The overall controlled system with $\mathbf{H}_{ij}$ as originally given obeys the equation

$$\dot{\mathbf{x}} = (\mathbf{A} - \mathbf{B}\mathbf{R}^{-1}\mathbf{B}^T\mathbf{P})\mathbf{x} + \mathbf{H}\mathbf{x} \tag{16.157}$$

where $\mathbf{u} = (\text{block diag}\ \mathbf{R}_i^{-1})(\text{block diag}\ \mathbf{B}_i)(\text{block diag}\ \mathbf{P}_i)\mathbf{x} \triangleq -\mathbf{R}^{-1}\mathbf{B}\mathbf{P}\mathbf{x}\, . \quad (16.158)$

With the definition of Eq.(16.158) the set of all Riccati equations as given in Eq.(16.155) is structured as

$$\mathbf{A}^T\mathbf{P} + \mathbf{P}\mathbf{A} - \mathbf{P}\mathbf{B}\mathbf{R}^{-1}\mathbf{B}^T\mathbf{P} + 2(\text{diag }\{\alpha_i\})\mathbf{P} + \mathbf{Q} = \mathbf{0}\,. \tag{16.159}$$

With regard to the influence of $\mathbf{H}$, the overall system behaviour differs from the isolated systems behaviour. The overall system behaviour is designed on the basis of Lyapunov theory (*Darwish, M.G., and Soliman, H.M., 1988*). The Lyapunov function $V = \mathbf{x}^T\mathbf{P}\mathbf{x}$ is chosen with a weighting matrix identical to the block diagonal Riccati matrix $\mathbf{P} > 0$. Thus, V describes the overall system of Eq.(16.157). The time derivative $\dot{V} = \mathbf{x}^T\mathbf{M}\mathbf{x}$ must be negative definite with $\mathbf{M} < 0$

$$\dot{V} = \dot{\mathbf{x}}^T\mathbf{P}\mathbf{x} + \mathbf{x}^T\mathbf{P}\dot{\mathbf{x}} = \mathbf{x}^T(\mathbf{A}^T\mathbf{P} + \mathbf{P}\mathbf{A} - 2\mathbf{P}\mathbf{B}\mathbf{R}^{-1}\mathbf{B}^T\mathbf{P} + \mathbf{H}^T\mathbf{P} + \mathbf{P}\mathbf{H})\mathbf{x}\,. \tag{16.160}$$

Eq.(16.159) is utilized to replace the terms $\mathbf{A}^T\mathbf{P} + \mathbf{P}\mathbf{A}$ in Eq.(16.160) yielding

$$\dot{V} = \mathbf{x}^T(-\mathbf{P}\mathbf{B}\mathbf{R}^{-1}\mathbf{B}^T\mathbf{P} - 2(\text{diag }\{\alpha_i\})\mathbf{P} - \mathbf{Q} + \mathbf{H}^T\mathbf{P} + \mathbf{P}\mathbf{H}\}\mathbf{x}\,. \tag{16.161}$$

In order to obtain negative $\dot{V}$ the matrix $-\mathbf{M}$ has to be positive definite, i.e.,

$$2(\text{diag }\alpha_i)\mathbf{P} + \mathbf{P}\mathbf{B}\mathbf{R}^{-1}\mathbf{B}^T\mathbf{P} + \mathbf{Q} - \mathbf{H}^T\mathbf{P} - \mathbf{P}\mathbf{H} > 0\,. \tag{16.162}$$

16.5.3 Global Calculations

Some properties are used in the sequel. Referring to Eq.(2.71), (22.25), (2.53) and since $\mathbf{P}$ is normal, for any induced matrix norm

$$|\lambda_i[\mathbf{P}\mathbf{H}]\,| \le \|\mathbf{P}\mathbf{H}\| \le \|\mathbf{P}\|\;\|\mathbf{H}\| = |\lambda_i[\mathbf{P}]\,|\;\|\mathbf{H}\|\,. \tag{16.163}$$

Invoking the singular-value property Eq.(22.68) and the property of normal matrices Eq.(22.25) (presumed that the matrices involved are normal),

$$\min_i |\lambda_i[\mathbf{A}\mathbf{B}]\,| < \min_i |\lambda_i[\mathbf{A}]\,|\;\min_i |\lambda_i[\mathbf{B}]\,|\,. \tag{16.164}$$

The eigenvalues of block diag $\mathbf{Q}_i$ result from a junction of $\lambda[\mathbf{Q}_i]$, i.e., $\lambda[\text{block diag }\mathbf{Q}_i] = \bigcup\lambda[\mathbf{Q}_i]$ since by Eq.(C.39)

$$\det\,(s\mathbf{I} - \text{block diag }\mathbf{Q}_i) = \det\prod_i(s\mathbf{I} - \mathbf{Q}_i) = \prod_i \det\,(s\mathbf{I} - \mathbf{Q}_i) = 0\,. \tag{16.165}$$

With respect to the fact that all the eigenvalues in consideration are positive the following operations prove sufficient

$$\lambda_{\min}[2(\text{diag }\{\alpha_i\})\mathbf{P}] + \lambda_{\min}[\mathbf{P}\mathbf{B}\mathbf{R}^{-1}\mathbf{B}^T\mathbf{P}] + \lambda_{\min}[\mathbf{Q}] > \lambda_{\max}[\mathbf{H}^T\mathbf{P}] + \lambda_{\max}[\mathbf{P}\mathbf{H}] \tag{16.166}$$

$$\Leftarrow \quad 2\;\alpha_{\min}\lambda_{\min}[\mathbf{P}]\lambda_{\min}[\mathbf{Q}] > 2\;\|\mathbf{H}\|\;\lambda_{\max}[\mathbf{P}] \tag{16.167}$$

$$\alpha_{\min} > \frac{\|\mathbf{H}\|\;\lambda_{\max}[\mathbf{P}] - 0.5\lambda_{\min}[\mathbf{Q}]}{\lambda_{\min}[\mathbf{P}]} = \frac{\|\mathbf{H}\|\;\max_i\lambda_{\max}[\mathbf{P}_i] - 0.5\min_i\lambda_{\min}[\mathbf{Q}_i]}{\min_i\lambda_{\min}[\mathbf{P}_i]}\,. \tag{16.168}$$

The minimum degree of stability $\alpha_{\min}$ of all the subsystems must be greater than a certain number given by the right-hand side of Eq.(16.168). If the perturbations of the overall system are carefully observed and listed up both in the norm $\|\mathbf{H}\|$ and in the extreme eigenvalues of the weighting and Riccati matrices then Eq.(16.168) robustly guarantees asymptotic stability.

16.5.4 Hierarchical Controller

There exists another approximate controller version. The control input $\mathbf{u}_i$ of Eq.(16.150) can be decomposed into $\mathbf{u}_i^l$ (local control) and into $\mathbf{u}_i^h$ (hierarchical control). The component $\mathbf{u}_i^l$ is obtained by local state feedback $\mathbf{u}_i^l = \mathbf{K}_i\mathbf{x}_i$ with uniform α_i = constant. To obtain $\mathbf{u}_i^h$ the assertion is used

$$\mathbf{B}_i\mathbf{u}_i^h + \sum_{j=1,j\neq i}^{N} \mathbf{H}_{ij}\mathbf{x}_j \rightarrow \min \ . \tag{16.169}$$

If a minimum of zero could be obtained then the controlled system would be exactly decoupled. The real minimum of the n_i-vector above is achieved via the m_i-components of $\mathbf{u}_i^h$ using linear regression, see Eq.(D.32),

$$\mathbf{u}_i^h = -\mathbf{B}_i^{\sharp L} \sum_{j=1,j\neq i}^{N} \mathbf{H}_{ij}\mathbf{x}_j \ . \tag{16.170}$$

Using this balancing and hierarchical control signal $\mathbf{u}_i^h$, a least-squares decoupled system is achieved. For the residual amount of interconnection see *Soliman, H.M., et al. 1978*.

If decentralized full-order observers for interconnected systems are applied their robustness can be investigated by employing additional noise (*Prasad, R., et al. 1988*).

16.6 Component Connection Model

The objective of this section is the eigenvalue assignment by parametrizing the interconnection between subsystems. Consider a model similar to Fig. 34.2 with a slightly simplified structure

$$\dot{\mathbf{x}} = \mathbf{A}\mathbf{x} + \mathbf{B}\mathbf{u} \qquad \mathbf{x} = \text{vec } \mathbf{x}_i, \qquad \mathbf{u} = \text{vec } \mathbf{u}_i, \tag{16.171}$$

$$\mathbf{y} = \mathbf{C}\mathbf{x} \qquad \mathbf{y} = \text{vec } \mathbf{y}_i, \tag{16.172}$$

$$\mathbf{A} = \text{block diag } \mathbf{A}_i, \quad \mathbf{B} = \text{block diag } \mathbf{B}_i, \quad \mathbf{C} = \text{block diag } \mathbf{C}_i, \quad \mathbf{K} = \text{block diag } \mathbf{K}_i. \tag{16.173}$$

Assume an interconnection and a decentralized controller model as depicted in Fig. 16.1

$$\mathbf{u} = p\mathbf{M}_y\mathbf{y} + \mathbf{K}\mathbf{x} \ . \tag{16.174}$$

The scalar parameter p is introduced in order to parametrize the interconnection. Then, p is varied from zero to one. At $p = 0$ the system is decoupled, it merely consists of isolated subsystems. At $p = 1$ the system is fully connected. Combining yields

$$\mathbf{u} = p\mathbf{M}_y\mathbf{C}\mathbf{x} + \mathbf{K}\mathbf{x} \tag{16.175}$$

$$\dot{\mathbf{x}} = \mathbf{A}\mathbf{x} + \mathbf{B}(p\mathbf{M}_y\mathbf{C}\mathbf{x} + \mathbf{K}\mathbf{x}) = (\mathbf{A} + p\mathbf{B}\mathbf{M}_y\mathbf{C} + \mathbf{B}\mathbf{K})\mathbf{x} \triangleq \mathbf{F}\mathbf{x} \ . \tag{16.176}$$

The closed-loop dynamic behaviour is given by the matrix $\mathbf{F}$. Starting with $p = 0$, interest focusses on the problem of how to vary $\mathbf{K}$ as p increases in order to satisfy a certain spectrum of the system overall eigenvalues $\Lambda_F = \{\lambda_i[\mathbf{F}]\}$. Perturbing the interconnections from $p\mathbf{M}_y$ to $(p + \Delta p)\mathbf{M}_y$, the increment $\Delta\mathbf{K}$ has to satisfy prescribed changes in

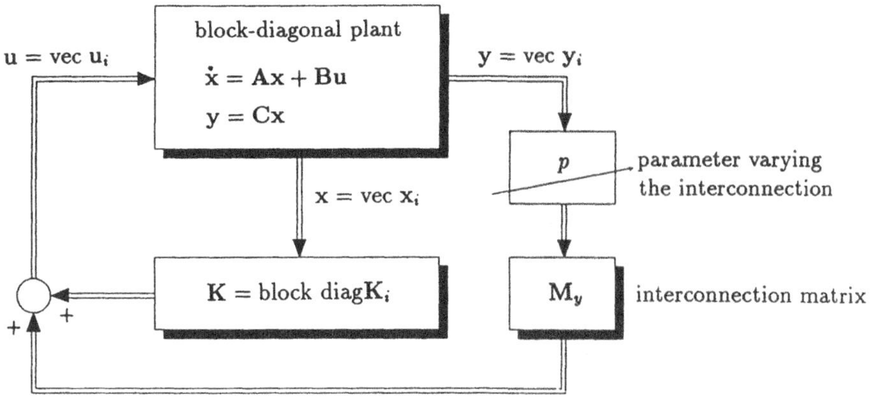

Figure 16.1: Interconnection and decentralized controller model

Λ_F (*Lefebvre, S., et al. 1982*). The right and left eigenvectors associated with $\lambda_j[\mathbf{F}]$ are denoted $\mathbf{f}_j$ and $\mathbf{f}_j^{\triangleleft}$, respectively. Thus, for each p the eigenvectors obey

$$\mathbf{F}(p)\, \mathbf{f}_j(p) = \lambda_j(p)\, \mathbf{f}_j(p)\,. \tag{16.177}$$

Differentiating with respect to p and using Eq.(16.176) yields

$$\frac{\partial \mathbf{F}}{\partial p}\, \mathbf{f}_j(p) + \mathbf{F}\frac{\partial \mathbf{f}_j}{\partial p} = \frac{\partial \lambda_j}{\partial p}\mathbf{f}_j(p) + \lambda_j(p)\frac{\partial \mathbf{f}_j}{\partial p} \tag{16.178}$$

$$(\mathbf{B}\mathbf{M}_y\mathbf{C} + \mathbf{B}\frac{\partial \mathbf{K}}{\partial p})\mathbf{f}_j(p) = \frac{\partial \lambda_j}{\partial p}\mathbf{f}_j(p) + [-\mathbf{F} + \lambda_j(p)\mathbf{I}]\frac{\partial \mathbf{f}_j}{\partial p}\,. \tag{16.179}$$

Premultiplying the above equation by $\mathbf{f}_j^{\triangleleft H}$ and referring to Eq.(B.11),

$$\mathbf{f}_j^{\triangleleft H}\mathbf{B}\mathbf{M}_y\mathbf{C}\mathbf{f}_j + \mathbf{f}_j^{\triangleleft H}\mathbf{B}\frac{\partial \mathbf{K}}{\partial p}\mathbf{f}_j = \frac{\partial \lambda_j}{\partial p}\mathbf{f}_j^{\triangleleft H}\mathbf{f}_j + [-\mathbf{f}_j^{\triangleleft H}\mathbf{F} + \mathbf{f}_j^{\triangleleft H}\lambda_j]\frac{\partial \mathbf{f}_j}{\partial p}\,. \tag{16.180}$$

Presuming that the left and right eigenvectors are normalized, its inner product is unity. The bracket vanishes with regard to the left eigenvector definition. Thus,

$$\mathbf{f}_j^{\triangleleft H}\mathbf{B}\frac{\partial \mathbf{K}}{\partial p}\mathbf{f}_j = -\mathbf{f}_j^{\triangleleft H}\mathbf{B}\mathbf{M}_y\mathbf{C}\mathbf{f}_j + \frac{\partial \lambda_j}{\partial p}\,. \tag{16.181}$$

This scalar equation holds for all $j = 1 \ldots n$. Listing up these scalar terms column-wise (*Lefebvre, S., et al. 1982*), an equivalent representation is given by

$$\begin{pmatrix} \mathbf{f}_1^T \otimes (\mathbf{f}_1^{\triangleleft H}\mathbf{B}) \\ \vdots \\ \mathbf{f}_n^T \otimes (\mathbf{f}_n^{\triangleleft H}\mathbf{B}) \end{pmatrix}^{(n \times nm)} \times \operatorname{col}\frac{\partial \mathbf{K}}{\partial p} = -\operatorname{vec}\Big(\mathbf{f}_j^{\triangleleft H}\mathbf{B}\mathbf{M}_y\mathbf{C}\mathbf{f}_j\Big) + \operatorname{vec}\frac{\partial \lambda_j}{\partial p}\,. \tag{16.182}$$

This equation can be resolved with respect to col $d\mathbf{K}/dp$ in any stage of p between $p = 0$ and $p = 1$. The admissible change in Λ_F must be predetermined at each stage. The resolving procedure requires the calculation of $\mathbf{f}_i$ and $\mathbf{f}_i^a$ at each stage. The increment in $\mathbf{K}$ is calculated by Taylor expansion

$$\text{col } \mathbf{K}(p+\Delta p) = \text{col } \mathbf{K}(p) + (\Delta p) \text{ col } \frac{d\mathbf{K}}{dp} . \tag{16.183}$$

Chapter 17

Robust State Feedback Using Ellipsoid Sets

The ellipsoidal set-theoretic approach is used to formulate the control problem arising when both input and system state matrix are perturbed. The set of perturbed parameters is bounded in a given compact convex set. The objective is to find a linear state-feedback controller **K** such that both the system state and control variables are minimally bounded by an ellipsoid and the ellipsoids satisfy a minimal performance functional. The boundary ellipsoids are minimized in a mean size given by the length sum of semi-major axes. The ellipsoidal set-theoretic approach originally was documented by *Usoro, P.B., et al. 1982*. The design method in the following was presented by *Wang, S.D., and Kuo, T.S., 1990*.

Consider the uncertain linear system

$$\dot{\mathbf{x}}(t) = (\mathbf{A} + \Delta\mathbf{A})\mathbf{x}(t) + (\mathbf{B} + \Delta\mathbf{B})\mathbf{u}(t) \tag{17.1}$$

where $\Delta\mathbf{A}$ and $\Delta\mathbf{B}$ depend on some uncertain parameters p_i . Using a state feedback $\mathbf{u}(t) = \mathbf{K}\mathbf{x}(t)$,

$$\dot{\mathbf{x}}(t) = (\mathbf{A} + \mathbf{B}\mathbf{K})\mathbf{x}(t) + (\Delta\mathbf{A} + \Delta\mathbf{B}\ \mathbf{K})\mathbf{x}(t) \triangleq (\mathbf{A} + \mathbf{B}\mathbf{K})\mathbf{x}(t) + \mathbf{g}(\mathbf{x}, t)\ . \tag{17.2}$$

The unperturbed system is characterized by the motion

$$\dot{\mathbf{x}}_C(t) = (\mathbf{A} + \mathbf{B}\mathbf{K})\mathbf{x}_C(t) \qquad \mathbf{x}_C(0) = \mathbf{x}_o\ . \tag{17.3}$$

The variable $\mathbf{x}_C(t)$ will be considered as the center of the perturbed motion $\mathbf{x}(t)$.

17.1 Bounding $\mathbf{x}(t)$ Via Matrix $\mathbf{\Gamma}(t)$

The initial state $\mathbf{x}(0)$ is assumed bounded around the center $\mathbf{x}_o$ according to

$$\mathbf{x}(0) \in \Omega_x(0) \triangleq \{\mathbf{x} :\ (\mathbf{x} - \mathbf{x}_o)^T \mathbf{\Psi}^{-1} (\mathbf{x} - \mathbf{x}_o) \le 1\} \tag{17.4}$$

where Ω_x defines an ellipsoidal set and $\mathbf{\Psi}^{-1}$ is a positive semidefinite matrix. The perturbation term $\mathbf{g}(\mathbf{x}, t)$ is considered contained within an ellipsoidal set Ω_g

$$\mathbf{g}(\mathbf{x}, t) \in \Omega_g \triangleq \{\mathbf{g} :\ \mathbf{g}^T \mathbf{Q}^{-1}(t) \mathbf{g} \le 1\} \qquad \mathbf{Q} > 0\ . \tag{17.5}$$

The perturbed motion $\mathbf{x}(t)$ is considered bounded by

$$\mathbf{x}(t) \in \Omega_x(t) \triangleq \{\mathbf{x} :\ [\mathbf{x}(t) - \mathbf{x}_C(t)]^T \mathbf{\Gamma}^{-1}(t) [\mathbf{x}(t) - \mathbf{x}_C(t)] \le 1\}\ . \tag{17.6}$$

The matrix $\mathbf{\Gamma}(t)$ results from the matrix differential equation (*Schweppe, F.C., 1973*)

$$\dot{\mathbf{\Gamma}} = (\mathbf{A}+\mathbf{BK})\mathbf{\Gamma} + \mathbf{\Gamma}(\mathbf{A}+\mathbf{BK})^T + \zeta\mathbf{\Gamma} + \frac{\mathbf{Q}}{\zeta} \quad \text{where} \quad \mathbf{\Gamma}(0) = \mathbf{\Psi} \tag{17.7}$$

and $\zeta = \zeta(t)$ is a free design parameter.

The perturbation $\mathbf{g}(\mathbf{x},t)$ is given by the vector sum of two terms depending on $\Delta\mathbf{A}$ and $\Delta\mathbf{B}\ \mathbf{K}$, respectively, bounded separately by

$$\Omega_{g1} \triangleq \{\mathbf{g}_1 : \ \mathbf{g}_1^T(\Delta\mathbf{A}\ \mathbf{\Gamma}\Delta\mathbf{A}^T)^{-1}\mathbf{g}_1 \le 1\} \tag{17.8}$$

$$\Omega_{g2} \triangleq \{\mathbf{g}_2 : \ \mathbf{g}_2^T(\Delta\mathbf{B}\ \mathbf{K}\mathbf{\Gamma}\mathbf{K}^T\Delta\mathbf{B}^T)^{-1}\mathbf{g}_2 \le 1\}\ . \tag{17.9}$$

The ellipsoid Ω_g bounds both Ω_{g1} and Ω_{g2} if

$$\mathbf{Q}(t) = \gamma_1\Delta\mathbf{A}\ \mathbf{\Gamma}\Delta\mathbf{A}^T + \gamma_2\Delta\mathbf{B}\ \mathbf{K}\mathbf{\Gamma}\mathbf{K}^T\Delta\mathbf{B}^T \quad \text{where} \quad \frac{1}{\gamma_1} + \frac{1}{\gamma_2} = 1\ . \tag{17.10}$$

Substituting $\alpha \triangleq \zeta/\gamma_1$, $\beta \triangleq \zeta/\gamma_2$ one has $\alpha + \beta = \zeta$ and

$$\dot{\mathbf{\Gamma}} = (\mathbf{A}+\mathbf{BK}+\frac{\zeta}{2}\mathbf{I})\mathbf{\Gamma} + \mathbf{\Gamma}(\mathbf{A}+\mathbf{BK}+\frac{\zeta}{2}\mathbf{I})^T + \frac{\Delta\mathbf{A}\ \mathbf{\Gamma}\Delta\mathbf{A}^T}{\alpha} + \frac{\Delta\mathbf{B}\ \mathbf{K}\mathbf{\Gamma}\mathbf{K}^T\Delta\mathbf{B}^T}{\beta}\ , \quad \mathbf{\Gamma}(0) = \mathbf{\Psi}\ . \tag{17.11}$$

The matrix $\mathbf{\Gamma}(t)$ determines both shape and size of the ellipsoid bounding the state variable $\mathbf{x}(t)$. The eigenvectors and the eigenvalues of $\mathbf{\Gamma}(t)$ provide orientation and axes, respectively. The trace of $\mathbf{\Gamma}(t)$ is a measure of the mean size given by the sum of the eigenvectors.

17.2 Bounding $\mathbf{u}(t)$ Via Matrix $\mathbf{K}\mathbf{\Gamma}\mathbf{K}^T$

The control variable $\mathbf{u}(t) = \mathbf{K}\mathbf{x}(t)$ is bounded by an ellipsoid Ω_u

$$\mathbf{u}(t) \in \Omega_u \triangleq \{\mathbf{u} : \ (\mathbf{u} - \mathbf{K}\mathbf{x}_C)^T(\mathbf{K}\mathbf{\Gamma}\mathbf{K}^T)^{-1}(\mathbf{u} - \mathbf{K}\mathbf{x}_C) \le 1\}\ . \tag{17.12}$$

17.3 Optimal Control and Ellipsoidal Constraints

The freedom given by the design parameter ζ (or α and β) and by the state-feedback matrix $\mathbf{K}$ is used to minimize the index of performance

$$I = I(\mathbf{K}, t_o) = \int_{t_o}^{t_f} \mathrm{tr}[\mathbf{\Gamma}]\ dt + \rho\ \mathrm{tr}[\mathbf{K}\mathbf{\Gamma}\mathbf{K}^T]\ dt \tag{17.13}$$

subject to the constraint Eq.(17.11).

Assume that $\Delta\mathbf{A}$ and $\Delta\mathbf{B}$ depend on uncertain parameters p_i . Then, *Wang, S.D., and Kuo, T.S., 1990* documented that $\mathrm{tr}[\mathbf{\Gamma}]$ is convex in the uncertain parameters. If the sets of the parameters are convex polyhydrons the search for maximizing the uncertainty can be reduced to the vertices of the parameter set. Hence, the maximization part of the minimax problem

$$\min_{\mathbf{K}} \max_{p_i} I(\mathbf{K}, p_i, t_o) \tag{17.14}$$

is solved by the search over the set of vertices. The minimization part is solved via Lagrange multiplier $\mathbf{\Lambda}$ and Hamiltonian function

$$H(\mathbf{\Gamma},\mathbf{K},\rho,t) = \mathrm{tr}[\mathbf{\Gamma}] + \rho\ \mathrm{tr}[\mathbf{K\Gamma K}^T] + \mathrm{tr}[\mathbf{\Lambda}^T\dot{\mathbf{\Gamma}}] \ . \tag{17.15}$$

From optimization theory it results

$$\frac{\partial H}{\partial \mathbf{\Gamma}} = -\dot{\mathbf{\Lambda}}, \quad \frac{\partial H}{\partial \mathbf{\Lambda}} = \dot{\mathbf{\Gamma}}, \quad \frac{\partial H}{\partial \mathbf{K}} = 0, \quad \frac{\partial H}{\partial \alpha} = 0, \quad \frac{\partial H}{\partial \beta} = 0 \ . \tag{17.16}$$

Then,

$$\begin{aligned}\frac{\partial H}{\partial \mathbf{\Gamma}} = -\dot{\mathbf{\Lambda}} = \mathbf{I} \ &+ \ \rho\mathbf{K}^T\mathbf{K} + \mathbf{\Lambda}(\mathbf{A}+\mathbf{BK}+\frac{\alpha+\beta}{2}\mathbf{I}) + (\mathbf{A}+\mathbf{BK}+\frac{\alpha+\beta}{2}\mathbf{I})^T\mathbf{\Lambda} \\ &+ \ \frac{1}{\alpha}\Delta\mathbf{A}^T\mathbf{\Lambda}\Delta\mathbf{A} + \frac{1}{\beta}\mathbf{K}^T\Delta\mathbf{B}^T\mathbf{\Lambda}\Delta\mathbf{B}\ \mathbf{K} \qquad \text{where} \quad \mathbf{\Lambda}(t_f) = \mathbf{0} \ . \end{aligned} \tag{17.17}$$

$$\frac{\partial H}{\partial \mathbf{\Lambda}} = \dot{\mathbf{\Gamma}} = (\mathbf{A}+\mathbf{BK}+\frac{\alpha+\beta}{2}\mathbf{I})\mathbf{\Gamma}+\mathbf{\Gamma}(\mathbf{A}+\mathbf{BK}+\frac{\alpha+\beta}{2}\mathbf{I})^T+\frac{\Delta\mathbf{A}\ \mathbf{\Gamma}\Delta\mathbf{A}^T}{\alpha}+\frac{\Delta\mathbf{B}\ \mathbf{K\Gamma K}^T\Delta\mathbf{B}^T}{\beta}, \tag{17.18}$$

$$\text{where} \quad \mathbf{\Gamma}(t_o) = \mathbf{\Psi} \ . \tag{17.19}$$

$$\frac{\partial H}{\partial \mathbf{K}} = 2\rho\ \mathbf{K\Gamma} + 2\mathbf{B}^T\mathbf{\Lambda\Gamma} + \frac{2\Delta\mathbf{B}^T\mathbf{\Lambda}\Delta\mathbf{B}\ \mathbf{K\Gamma}}{\beta} = 0 \ . \tag{17.20}$$

$$\frac{\partial H}{\partial \alpha} = \mathrm{tr}[\mathbf{\Gamma\Lambda}] - \frac{1}{\alpha^2}\mathrm{tr}[\Delta\mathbf{A}\ \mathbf{\Gamma}\Delta\mathbf{A}^T\mathbf{\Lambda}] = 0 \tag{17.21}$$

$$\frac{\partial H}{\partial \beta} = \mathrm{tr}[\mathbf{\Gamma\Lambda}] - \frac{1}{\beta^2}\mathrm{tr}[\Delta\mathbf{B}\ \mathbf{K\Gamma K}^T\Delta\mathbf{B}^T\mathbf{\Lambda}] = 0 \ . \tag{17.22}$$

From Eq.(17.20)

$$[(\rho\ \mathbf{I} + \frac{\Delta\mathbf{B}^T\mathbf{\Lambda}\Delta\mathbf{B}}{\beta})\mathbf{K} + \mathbf{B}^T\mathbf{\Lambda}]\mathbf{\Gamma} = \mathbf{0} \tag{17.23}$$

$$\mathbf{K} = -\mathbf{R}^{-1}\mathbf{B}^T\mathbf{\Lambda} \qquad \text{where} \qquad \mathbf{R} \triangleq \rho\ \mathbf{I} + \frac{\Delta\mathbf{B}^T\mathbf{\Lambda}\Delta\mathbf{B}}{\beta} \ . \tag{17.24}$$

Substituting Eq.(17.24) into Eq.(17.17) yields a matrix differential equation in $\mathbf{\Lambda}$ with terminal condition $\mathbf{\Lambda}(t_f) = \mathbf{0}$. Substituting Eq.(17.24) into Eq.(17.18) yields a matrix differential equation in $\mathbf{\Gamma}$ with initial condition $\mathbf{\Gamma}(t_o) = \mathbf{\Psi}$. From Eq.(17.21) and Eq.(17.22) the design parameters α and β are determined by

$$\alpha = \sqrt{\frac{\mathrm{tr}[\mathbf{\Gamma}\Delta\mathbf{A}^T\mathbf{\Lambda}\Delta\mathbf{A}]}{\mathrm{tr}[\mathbf{\Gamma\Lambda}]}} \qquad \text{and} \qquad \beta = \sqrt{\frac{\mathrm{tr}[\mathbf{\Gamma}(\Delta\mathbf{B}\ \mathbf{K})^T\mathbf{\Lambda}(\Delta\mathbf{B}\ \mathbf{K})]}{\mathrm{tr}[\mathbf{\Gamma\Lambda}]}} \ . \tag{17.25}$$

Numerical solution of the matrix differential equations requires considerable effort since the equations are nonlinear and coupled by the parameters α and β . In the infinite time case $t_f \to \infty$ only algebraic matrix equations in $\mathbf{\Lambda}$ must be solved but $\mathbf{\Gamma}$ remains a function of time $\mathbf{\Gamma}(t)$.

The resulting performance functional is determined by the matrix $\mathbf{\Psi}$ as required for the set Ω_x

$$I(\mathbf{K},t_o) = \mathrm{tr}[\mathbf{\Lambda}(t_o)\mathbf{\Psi}] \ . \tag{17.26}$$

Chapter 18

Robustness of Observers and Kalman-Bucy Filters

Consider a multivariable dynamic system of order n with input $\mathbf{u}(t) \in \mathcal{R}^m$ and output $\mathbf{y}(t) \in \mathcal{R}^r$. The system state variable $\mathbf{x}(t)$ is an n-vector. Observability and controllability is preassumed. Usually $\mathbf{x}(t)$ is not available. If the output $\mathbf{y}(t)$ does not provide sufficient information for high-performance control, observers are implemented to estimate either the entire vector or a part of it. In the presence of process noise and/or measurement noise observers are replaced by Kalman filters.

18.1 Reduced-Order Observer

The objective of reduced-order observers is the estimation of a variable $\mathbf{z}(t)$ of reduced dimension, i.e. $n - r$, in order to augment the output $\mathbf{y}(t)$ of a dynamic system with unavailable state variable $\mathbf{x}(t)$. With the help of $\mathbf{y}(t)$ and $\mathbf{z}(t)$ the estimate $\hat{\mathbf{x}}(t)$ of the state $\mathbf{x}(t)$ is reconstructed. The reduced-order observer assertion consists of a dynamic part and a dynamic-free part (*Leondes, C.T., and Novak, L.M., 1972*).

18.1.1 Dynamic Part of the Observer

The observer dynamics is given by

$$\dot{\mathbf{z}}(t) = \mathbf{F}_z\mathbf{z}(t) + \mathbf{L}\mathbf{y}(t) + \mathbf{H}\mathbf{u}(t) \qquad \mathbf{z}(t) \in \mathcal{R}^{n-r} \tag{18.1}$$

where $\mathbf{z}(t)$ represents an estimate. The control scheme including the observer is depicted in Fig. 18.1. The estimate $\mathbf{z}(t)$ should converge to a map of $\mathbf{x}(t)$ as $t \to \infty$

$$\lim_{t\to\infty} \mathbf{z}(t) = \mathbf{T}\mathbf{x}(t), \qquad \mathbf{T} \in \mathcal{R}^{(n-r)\times n} \tag{18.2}$$

where $\mathbf{z}(t)$ is considered as an estimate of $\mathbf{T}\mathbf{x}(t)$. The transient estimation error $\tilde{\mathbf{z}}(t)$ implicitly is defined by

$$\mathbf{T}\mathbf{x}(t) \triangleq \mathbf{z}(t) + \tilde{\mathbf{z}}(t) . \tag{18.3}$$

Combining Eqs.(18.1), (18.3) and the plant equation

$$\dot{\mathbf{x}}(t) = \mathbf{A}\mathbf{x}(t) + \mathbf{B}\mathbf{u}(t) , \qquad \mathbf{x}(0) = \mathbf{x}_o , \qquad \mathbf{y}(t) = \mathbf{C}\mathbf{x}(t) \tag{18.4}$$

yields

$$\begin{aligned}
\dot{\tilde{\mathbf{z}}}(t) &= -\dot{\mathbf{z}}(t) + \mathbf{T}\dot{\mathbf{x}}(t) = -\mathbf{F}_z\mathbf{z}(t) - \mathbf{L}\mathbf{y}(t) - \mathbf{H}\mathbf{u}(t) + \mathbf{T}[\mathbf{A}\mathbf{x}(t) + \mathbf{B}\mathbf{u}(t)] && (18.5)\\
&= -\mathbf{F}_z(\mathbf{T}\mathbf{x} - \tilde{\mathbf{z}}) - \mathbf{L}\mathbf{C}\mathbf{x} - \mathbf{H}\mathbf{u} + \mathbf{T}\mathbf{A}\mathbf{x} + \mathbf{T}\mathbf{B}\mathbf{u} && (18.6)\\
&= \mathbf{F}_z\tilde{\mathbf{z}} - (\mathbf{F}_z\mathbf{T} - \mathbf{T}\mathbf{A} + \mathbf{L}\mathbf{C})\mathbf{x} - (\mathbf{H} - \mathbf{T}\mathbf{B})\mathbf{u} . && (18.7)
\end{aligned}$$

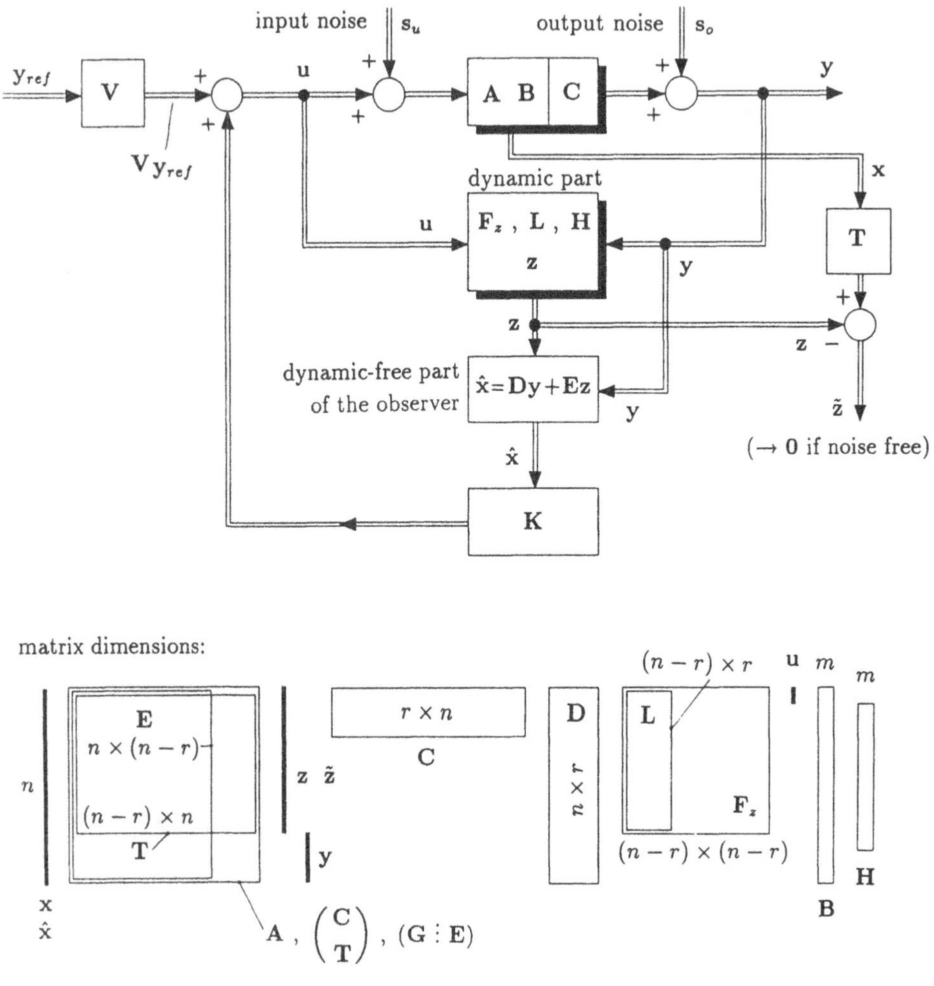

Figure 18.1: Reduced-order observer and its matrix elements

In order to obtain an estimation error $\tilde{\mathbf{z}}$ converging to zero, the matrix $\mathbf{F}_z$ must be stable. The variables $\mathbf{x}$ and $\mathbf{u}$ are of no influence on $\tilde{\mathbf{z}}$ if the following necessary and sufficient conditions are satisfied

$$\mathbf{F}_z\mathbf{T} = \mathbf{TA} - \mathbf{LC} \tag{18.8}$$

$$\mathbf{H} = \mathbf{TB}\ . \tag{18.9}$$

18.1.2 Dynamic-Free Part of the Observer

Consider an n-vector estimate $\hat{\mathbf{x}}(t)$. This variable $\hat{\mathbf{x}}(t)$ can be obliged to satisfy, first, r conditions

$$\mathbf{C}\hat{\mathbf{x}}(t) = \mathbf{y}(t) \tag{18.10}$$

of the noise-free output and, second, $n - r$ conditions of the $(n - r)$-estimation $\mathbf{z}(t)$ as derived in the previous section

$$\mathbf{T}\hat{\mathbf{x}}(t) = \mathbf{z}(t) \ . \tag{18.11}$$

Combining yields

$$\begin{pmatrix}\mathbf{C}\\ \mathbf{T}\end{pmatrix}\hat{\mathbf{x}}(t) = \begin{pmatrix}\mathbf{y}(t)\\ \mathbf{z}(t)\end{pmatrix} . \tag{18.12}$$

Assume the rows of $\mathbf{C}$ and $\mathbf{T}$ linearly independent then the inverse exists and

$$\hat{\mathbf{x}}(t) = \begin{pmatrix}\mathbf{C}\\ \mathbf{T}\end{pmatrix}^{-1}\begin{pmatrix}\mathbf{y}\\ \mathbf{z}\end{pmatrix} \triangleq \mathbf{D}\mathbf{y}(t) + \mathbf{E}\mathbf{z}(t) = \begin{pmatrix}\mathbf{D} \vdots \mathbf{E}\end{pmatrix}\begin{pmatrix}\mathbf{y}\\ \mathbf{z}\end{pmatrix} . \tag{18.13}$$

If $\mathbf{T}$ and $\mathbf{L}$ are chosen the observer matrix $\mathbf{F}_z$ is given by Eq.(18.8), $\mathbf{H}$ is determined by Eq.(18.9), $\mathbf{D}$ and $\mathbf{E}$ by

$$\begin{pmatrix}\mathbf{D} \vdots \mathbf{E}\end{pmatrix} \triangleq \begin{pmatrix}\mathbf{C}\\ \mathbf{T}\end{pmatrix}^{-1} \quad \text{or} \quad \mathbf{D}\mathbf{C} + \mathbf{E}\mathbf{T} = \mathbf{I}_n \ . \tag{18.14}$$

Combining Eqs.(18.3) and (18.11), $\mathbf{T}(\mathbf{x} - \hat{\mathbf{x}}) = \tilde{\mathbf{z}}$ is achieved.

18.1.3 Observer-Based Control System in the Presence of Noise

Assume the plant is disturbed by the input noise $\mathbf{s}_u(t)$ and measuring noise associated with the output $\mathbf{s}_o(t)$

$$\dot{\mathbf{x}}(t) = \mathbf{A}\mathbf{x}(t) + \mathbf{B}\mathbf{u}(t) + \mathbf{B}\mathbf{s}_u(t) \tag{18.15}$$

$$\mathbf{y}(t) = \mathbf{C}\mathbf{x}(t) + \mathbf{s}_o(t) \ . \tag{18.16}$$

Let the control input $\mathbf{u}(t)$ be composed of two components: first, the reference after having passed the prefilter $\mathbf{V}$ and, second, the state feedback using feedback gain $\mathbf{K}$ and the estimate $\hat{\mathbf{x}}$

$$\mathbf{u}(t) = \mathbf{K}\hat{\mathbf{x}}(t) + \mathbf{V}\mathbf{y}_{ref} \ . \tag{18.17}$$

Combination of the Eqs.(18.3), (18.1), (18.15) and (18.16) yields

$$\dot{\tilde{\mathbf{z}}}(t) = \mathbf{F}_z\tilde{\mathbf{z}}(t) - \mathbf{L}\mathbf{s}_o(t) + \mathbf{T}\mathbf{B}\mathbf{s}_u(t) \ . \tag{18.18}$$

The estimation does not converge to $\mathbf{0}$ in the disturbed case.

Combining Eqs.(18.15), (18.17), (18.13) and (18.3),

$$\dot{\mathbf{x}}(t) = (\mathbf{A} + \mathbf{B}\mathbf{K})\mathbf{x}(t) + \mathbf{B}\mathbf{K}\mathbf{D}\mathbf{s}_o(t) + \mathbf{B}\mathbf{K}\mathbf{E}\tilde{\mathbf{z}}(t) + \mathbf{B}\mathbf{s}_u(t) + \mathbf{B}\mathbf{V}\mathbf{y}_{ref} \ . \tag{18.19}$$

State variable and estimation error can be rewritten to the following combination

$$\begin{pmatrix}\dot{\mathbf{x}}\\ \dot{\tilde{\mathbf{z}}}\end{pmatrix} = \begin{pmatrix}\mathbf{A} + \mathbf{B}\mathbf{K} & \mathbf{B}\mathbf{K}\mathbf{E}\\ \mathbf{0} & \mathbf{F}_z\end{pmatrix}\begin{pmatrix}\mathbf{x}\\ \tilde{\mathbf{z}}\end{pmatrix} + \begin{pmatrix}\mathbf{B}\mathbf{K}\mathbf{D}\\ -\mathbf{L}\end{pmatrix}\mathbf{s}_o + \begin{pmatrix}\mathbf{B}\\ \mathbf{T}\mathbf{B}\end{pmatrix}\mathbf{s}_u + \begin{pmatrix}\mathbf{B}\mathbf{V}\\ \mathbf{0}\end{pmatrix}\mathbf{y}_{ref} \ . \tag{18.20}$$

In the case of imperfect observer design, i.e. if Eqs.(18.8) and (18.9) are slightly failed, the zero matrices in Eq.(18.20) are to be replaced by $(\mathbf{F}_z\mathbf{T} - \mathbf{TA} + \mathbf{LC})$ and $(\mathbf{H} - \mathbf{TB})$, respectively.

Note that the reference $\mathbf{y}_{ref}$ cannot excite the estimation error $\tilde{\mathbf{z}}(t)$ but the noise signals do. Actually, noise reaction of the control loop is not taken into consideration when the observer is designed by pole allocation of the matrix $\mathbf{F}_z$. Usually, the observer eigenvalues $\lambda_i[\mathbf{F}_z]$ are chosen sufficiently faster than $\lambda_i[\mathbf{A}]$ of the plant. Hence, noise increase as given by Eq.(18.18) may occur to a considerable extent and must be taken into account separately.

Another objective arises from the fact that the stability margin and the robustness of the overall observer-based system may be nonsatisfactory. Designing a linear controller with quadratic performance the excellent robustness properties are lost if an observer is used. The objective is to find direct and effective methods to approach robustness recovery. This matter is outlined in the following sections.

18.1.4 Full-Order Observer

Choosing $\mathbf{T} \in \mathcal{R}^{n\times n}$, from Eqs.(18.13) and (18.3) one finds $\tilde{\mathbf{z}}(t) = \mathbf{T}[\mathbf{x}(t) - \hat{\mathbf{x}}(t)]$. The matrix $\mathbf{T}$ need not be equal to $\mathbf{I}_n$. If $\mathbf{T} = \mathbf{I}_n$ then $\mathbf{z} = \hat{\mathbf{x}}$, $\mathbf{D} = \mathbf{0}$, $\mathbf{E} = \mathbf{I}$ and the well-known full-order observer is achieved.

To recover robustness and to improve the dynamic behaviour *Okada, T. et al. 1985* proposed augmenting the system by an additional feedback from the output to the reference input.

18.2 Robustness With Proportional Observers

If an observer-based controller is implemented to estimate the complete state the magnitude of tolerable plant uncertainty bounds is smaller than the magnitude when the full state is available (*Schmitendorf, W.E., 1988*). In Fig. 18.2a and b a full-state feedback system and an observer-based feedback system is depicted, respectively. The latter is based on a so-called proportional observer where the difference signal $\mathbf{y}(t) - \mathbf{C}\hat{\mathbf{x}}(t)$ is used via gain matrix $\mathbf{L}$ to reconstruct the state and obtain the estimate $\hat{\mathbf{x}}(t)$. The plant, full-order observer and estimation error equations are, respectively,

$$\dot{\mathbf{x}}(t) = \mathbf{A}\mathbf{x}(t) + \mathbf{B}\mathbf{u}(t) \qquad \mathbf{y}(t) = \mathbf{C}\mathbf{x}(t) \qquad \mathbf{\Phi} = (s\mathbf{I} - \mathbf{A})^{-1} \tag{18.21}$$

$$\dot{\hat{\mathbf{x}}}(t) = (\mathbf{A} - \mathbf{LC})\hat{\mathbf{x}}(t) + \mathbf{L}\mathbf{y}(t) + \mathbf{B}\mathbf{u}(t) \tag{18.22}$$

$$\dot{\tilde{\mathbf{x}}}(t) = (\mathbf{A} - \mathbf{LC})\tilde{\mathbf{x}}(t) \qquad \tilde{\mathbf{x}}(t) = \mathbf{x}(t) - \hat{\mathbf{x}}(t)\ . \tag{18.23}$$

The plant is assumed both observable and controllable where $\mathbf{x}, \hat{\mathbf{x}} \in \mathcal{R}^n$ and $\mathbf{y}, \mathbf{u} \in \mathcal{R}^m$ (or $r = m$). Note that Eq.(18.22) corresponds with Eqs.(18.7) and (18.8) in the previous section when $\mathbf{T} = \mathbf{I}$.

Fig. 18.2c shows an implementation of the proportional integral observer where the gain matrix $\mathbf{L}$ is replaced by a proportional plus integral element $(\mathbf{L} + \mathbf{L}_I/s)$. The proportional integral observer will be outlined later.

The implementations of Fig. 18.2a and b have identical properties as follows:

- Closed-loop property from $\mathbf{y}_{ref}$ to $\mathbf{y}$.

- Open-loop property when loop is broken at S_2 (and $S_1S_3S_4$ remain closed). This is obvious from Fig. 18.2b using $\mathbf{\Phi} = \mathbf{\Phi}(s) = (s\mathbf{I} - \mathbf{A})^{-1}$ and the following relations in Laplace domain: The error $\tilde{\mathbf{x}}(s)$ is the difference of the plant output $\mathbf{y}(s)$ and the proportional observer feedback signal $\mathbf{C\Phi h}$, that is,

$$\tilde{\mathbf{x}} = \mathbf{C\Phi Bu'} - \mathbf{C\Phi h} = \mathbf{C\Phi Bu'} - \mathbf{C\Phi}(\mathbf{Bu'} + \mathbf{L}\tilde{\mathbf{x}}) = -\mathbf{C\Phi L}\tilde{\mathbf{x}} \quad \rightsquigarrow \tilde{\mathbf{x}} = \mathbf{0}\ . \tag{18.24}$$

Hence,

$$\mathbf{\Phi Bu'} = \mathbf{\Phi h} = \hat{\mathbf{x}}\ . \tag{18.25}$$

The result equals $\mathbf{\Phi Bu'} = \mathbf{x}$ from Fig. 18.2a. The transfer function $\mathbf{H}_1$ and $\mathbf{H}_2$ play the same role in the open-loop property of Fig. 18.2a and 18.2b.

- Open-loop property when loop is broken at S_1 (where $S_2S_3S_4$ remain closed) only if the observer gain matrix satisfies (*Doyle, J.C., and Stein, G., 1979*)

$$\mathbf{L}[\mathbf{I} + \mathbf{C}(s\mathbf{I} - \mathbf{A})^{-1}\mathbf{L}]^{-1} = \mathbf{B}[\mathbf{C}(s\mathbf{I} - \mathbf{A})^{-1}\mathbf{B}]^{-1}\ . \tag{18.26}$$

Proof: The signal $\mathbf{h}$ is related to the estimate $\hat{\mathbf{x}}$ and to the inputs of the summing element in front of it

$$\mathbf{h} = \mathbf{\Phi}^{-1}\hat{\mathbf{x}} = -\mathbf{LC}\hat{\mathbf{x}} + \mathbf{Bu'} + \mathbf{LC\Phi Bu''} \tag{18.27}$$

$$\hat{\mathbf{x}} = (\mathbf{\Phi}^{-1} + \mathbf{LC})^{-1}(\mathbf{Bu'} + \mathbf{LC\Phi Bu''}) \quad \text{where} \quad \mathbf{\Phi} = (s\mathbf{I} - \mathbf{A})^{-1}\ . \tag{18.28}$$

Two identities

$$(\mathbf{\Phi}^{-1} + \mathbf{LC})^{-1} = \mathbf{\Phi} - \mathbf{\Phi L}(\mathbf{I} + \mathbf{C\Phi L})^{-1}\mathbf{C\Phi} = \mathbf{\Phi}[\mathbf{B}(\mathbf{C\Phi B})^{-1} - \mathbf{L}(\mathbf{I} + \mathbf{C\Phi L})^{-1}]\mathbf{C\Phi} \tag{18.29}$$

$$(\mathbf{\Phi}^{-1} + \mathbf{LC})^{-1}\mathbf{L} \equiv [\mathbf{\Phi} - \mathbf{\Phi L}(\mathbf{I} + \mathbf{C\Phi L})^{-1}\mathbf{C\Phi}]\mathbf{L} \equiv \mathbf{\Phi L}(\mathbf{I} + \mathbf{C\Phi L})^{-1} \tag{18.30}$$

are used to modify Eq.(18.28)

$$\hat{\mathbf{x}} = \mathbf{\Phi}[\mathbf{B}(\mathbf{C\Phi B})^{-1} - \mathbf{L}(\mathbf{I} + \mathbf{C\Phi L})^{-1}]\mathbf{C\Phi Bu'} + \mathbf{\Phi L}(\mathbf{I} + \mathbf{C\Phi L})^{-1}\mathbf{C\Phi Bu''}\ . \tag{18.31}$$

The matrix factor of $\mathbf{u'}$ in Eq.(18.31) is put to zero in order to cancel the influence of $\mathbf{u'}$ to $\hat{\mathbf{x}}$. This leads to the result of Eq.(18.26). Using it once more in the matrix factor of $\mathbf{u''}$ in Eq.(18.31), Eq.(18.31) can be rewritten to

$$\hat{\mathbf{x}} = \mathbf{\Phi L}(\mathbf{I} + \mathbf{C\Phi L})^{-1}\mathbf{C\Phi Bu''} = \mathbf{\Phi B}[\mathbf{C\Phi B}]^{-1}\mathbf{C\Phi Bu''} = \mathbf{\Phi Bu''}\ . \tag{18.32}$$

Therefore, the same result is given from Fig. 18.2b and from Fig. 18.2a and $\hat{\mathbf{x}}(t)$ is identical to $\mathbf{x}(t)$ in the open-loop system.

End of the Proof

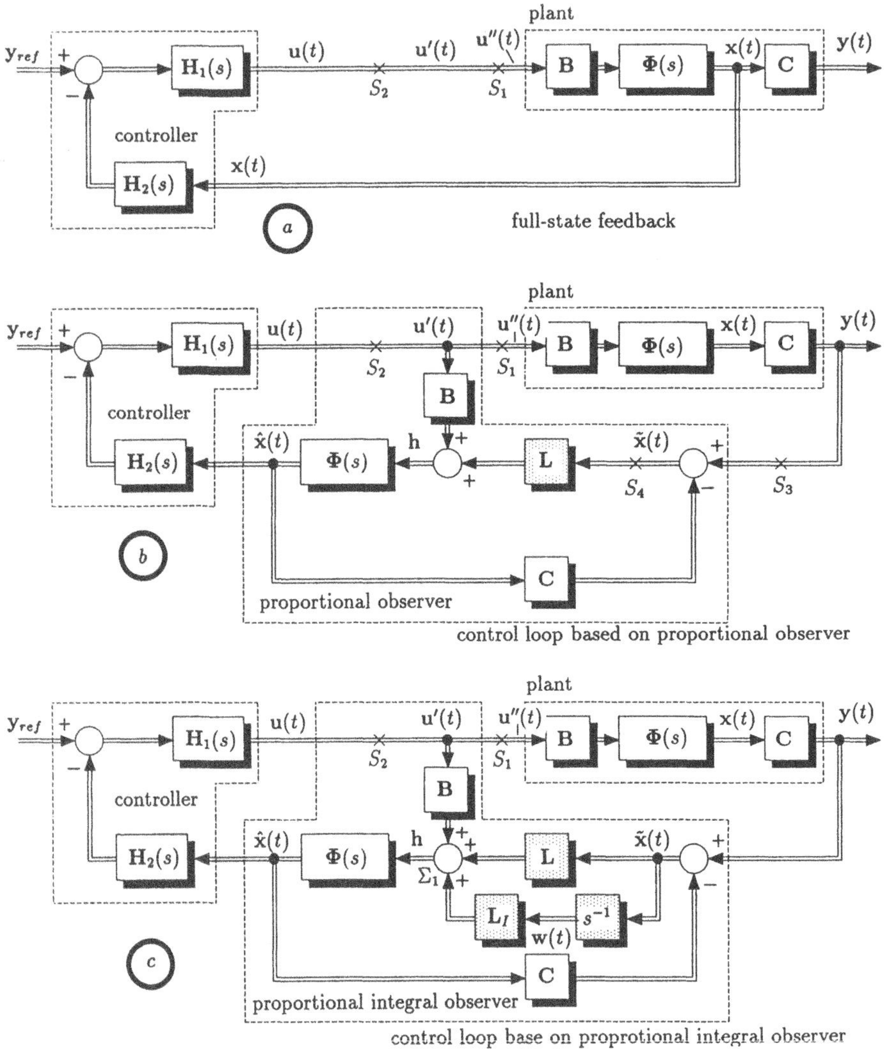

Figure 18.2: Full-state feedback (a), proportional-observer-based feedback (b) and proportional-integral-observer-based feedback (c)

18.3 Full-State Loop Transfer Recovery

The procedure of loop transfer recovery forces the return ratio at break point S_1 to approach that one at S_2 .

18.3.1 Sensitivity Recovery

From Fig. 18.2b for $\mathbf{G}(s) = \mathbf{C}(s\mathbf{I} - \mathbf{A})^{-1}\mathbf{B} = \mathbf{C\Phi B}$ and square it can be derived:

- (i) Breaking the inner loop of the observer (only) at point S_4, the loop transfer function is $-\mathbf{C}(s\mathbf{I} - \mathbf{A})^{-1}\mathbf{L}$. The matrices $\mathbf{H}_1, \mathbf{H}_2$ do not enter into this result.
- (ii) Breaking the loop at S_3 only, the loop transfer function is $\mathbf{G}(s)\mathbf{K}(s)$ where $\mathbf{G}(s)$ is the plant and $\mathbf{K}(s)$ is the observer plus controller $\mathbf{H}_2(s)$ and $\mathbf{H}_1(s)$. If $\mathbf{G}(s)\mathbf{K}(s) = \mathbf{C}(s\mathbf{I} - \mathbf{A})^{-1}\mathbf{B}\mathbf{K}(s)$ is made to approach $-\mathbf{C}(s\mathbf{I} - \mathbf{A})^{-1}\mathbf{L}$ of (i) this procedure is named sensitivity recovery.

18.3.2 Robustness Recovery

- (iii) Breaking the loop at S_2 in Fig. 18.2b, the input and output are $\mathbf{u}'$ and $\mathbf{u}$, respectively,
$$(s\mathbf{I} - \mathbf{A})^{-1}[\mathbf{L}(-\mathbf{C}\hat{\mathbf{x}} + \mathbf{G}\mathbf{u}') + \mathbf{B}\mathbf{u}'] = \hat{\mathbf{x}} \tag{18.33}$$
$$-\mathbf{H}_1\mathbf{H}_2\hat{\mathbf{x}} = \mathbf{u} \; . \tag{18.34}$$
Reduction yields the loop transfer function $-\mathbf{H}_1\mathbf{H}_2(s\mathbf{I} - \mathbf{A})^{-1}\mathbf{B}$.
- (iv) Breaking the loop at S_1 yields the loop transfer function $\mathbf{KG}$ given by the cascade of $\mathbf{G}$ and $\mathbf{K}$. The transfer function $\mathbf{KG}$ can be made equal to the expression $-\mathbf{H}_1\mathbf{H}_2(s\mathbf{I} - \mathbf{A})^{-1}\mathbf{B}$ as given in (iii). Then, this procedure corresponds to the observer design according with the "robustness recovery" procedure (*Doyle, J.C., and Stein, G., 1979*). For breaking at S_1 the open-loop transfer function is
$$\mathbf{T}_1 \triangleq -\mathbf{H}_1\mathbf{H}_2(s\mathbf{I} - \mathbf{A})^{-1}\mathbf{B} = \mathbf{K}(s)\mathbf{G}(s) \; . \tag{18.35}$$
From LQG theory, the return difference $\mathbf{T}_1$ with $\mathbf{\Phi}(s) = (s\mathbf{I}_n - \mathbf{A})^{-1}$ satisfies the so-called return difference identity (*Anderson, B.D.O., and Moore, J.B., 1971*)
$$(\mathbf{I}_m + \mathbf{T}_1)^H\mathbf{R}(\mathbf{I}_m + \mathbf{T}_1) \triangleq \mathbf{R} + (\mathbf{V\Phi B})^H(\mathbf{V\Phi B}) \tag{18.36}$$
where $\mathbf{R}$ is the control weighting matrix and $\mathbf{V}$ results from the state weighting matrix $\mathbf{Q} = \mathbf{V}^T\mathbf{V}$. Then, with $\mathbf{R} \triangleq \rho\mathbf{I}_m$ the expression for the singular values is
$$\sigma_i[\mathbf{I}_m + \mathbf{T}_1] = \sqrt{\lambda_i[\mathbf{I} + \frac{1}{\rho}(\mathbf{V\Phi B})^H(\mathbf{V\Phi B})]} = \sqrt{1 + \frac{1}{\rho}\sigma_i^2[\mathbf{V\Phi B}]} \; . \tag{18.37}$$
If $\sigma_{\min}[\mathbf{T}_1] \gg 1$ at low frequecies then
$$\sigma_i[\mathbf{I}_m + \mathbf{T}_1] \doteq \frac{1}{\sqrt{\rho}}\sigma_i[\mathbf{V\Phi}(j\omega)\mathbf{B}] \; . \tag{18.38}$$
Choosing $\sigma_{\max}[\mathbf{I}_m + \mathbf{T}_1]$ and $\sigma_{\min}[\mathbf{I} + \mathbf{T}_1]$, the performance and stability robustness properties of observer-based control sytems are given according to Eq.(24.22) (*Doyle, J.C., and Stein, G., 1981*).

Another method to recover robustness is achieved by adding noise to the system input noise model, prior to designing the observer according to Kalman filter methods.

18.3.3 Loop Transfer Recovery for a Generalized Type of Observer

Tsui, C.C., 1985, 1986, 1988, 1988a presented a method of loop transfer recovery for an observer type

$$\mathbf{K}\hat{\mathbf{x}} = \mathbf{M}\mathbf{y} + \mathbf{N}(s\mathbf{I} - \mathbf{F}_z)^{-1}(\mathbf{L}\mathbf{y} + \mathbf{H}\mathbf{u}) \qquad \text{where} \quad \mathbf{u} = \mathbf{K}\hat{\mathbf{x}} \, . \tag{18.39}$$

The loop transfer function for break points at the plant input and observer output, respectively, corresponding to S_1 and S_2, are made equal if

$$\mathbf{N}(s\mathbf{I} - \mathbf{F}_z)^{-1}\mathbf{T}\mathbf{B} \to \mathbf{0} \quad \text{or} \quad \mathbf{H} = \mathbf{T}\mathbf{B} \to \mathbf{0} \tag{18.40}$$

where $\mathbf{T}\mathbf{A} - \mathbf{F}_z\mathbf{T} = \mathbf{L}\mathbf{C}$ and $\mathbf{K} = \mathbf{N}\mathbf{T} + \mathbf{M}\mathbf{C}$.

Via observer pole allocation the design objectives of performance, observer order reduction and robustness can be considered in a unified way. Under the constraint of a minimum distance from the left of the imaginary axis, one of the following minimization procedures may be chosen

$$\|\mathbf{H}\|_F = \text{tr } \mathbf{H}^T\mathbf{H} = \text{tr } \mathbf{B}^T\mathbf{T}^T\mathbf{T}\mathbf{B} \;\to\; \min \qquad \text{or} \qquad \sup_{\omega} \|\mathbf{N}(s\mathbf{I} - \mathbf{F}_z)^{-1}\mathbf{T}\mathbf{B}\|_s \;\to\; \min \, . \tag{18.41}$$

The observer design based on decomposing a multivariable nominal system into several single-input single-output subsystems was presented by *Saberi, A., and Sannuti, P., 1990a*. Uncertainties in the plant are modeled as external blocks or in a state-space description. By this means, a theoretical basis of a separation principle for uncertain dynamical systems is pointed out. Loop transfer recovery can be obtained exactly or approximately, and for an arbitrarily specified target loop (*Saberi, A., et al., 1991*). General observer architectures can be characterized by minimizing the norm of a so-called recovery matrix (*Niemann, H.H., et al., 1991*).

18.4 Robustness of Proportional Integral Observer

An observer of the proportional integral type is characterized by an additional integration path, see Fig. 18.2c. The proportional integral observer is designed in order to recover the properties of loop transfer recovery guaranteed by a linear quadratic regulator without an observer. Moreover, robustness with respect to plant uncertainties is achieved.

18.4.1 General Design

The proportional integral observer is established as

$$\dot{\hat{\mathbf{x}}}(t) = (\mathbf{A} - \mathbf{L}\mathbf{C})\hat{\mathbf{x}}(t) + \mathbf{L}\mathbf{y}(t) + \mathbf{B}\mathbf{u}(t) + \mathbf{L}_I\mathbf{w}(t) \tag{18.42}$$

$$\dot{\mathbf{w}}(t) = \tilde{\mathbf{x}}(t) = \mathbf{y}(t) - \mathbf{C}\hat{\mathbf{x}}(t) \qquad \mathbf{y} \in \mathcal{R}^r, \quad r = m \tag{18.43}$$

where the variable $\mathbf{w}(t)$ is defined as the integral of the difference $\mathbf{y}(t) - \mathbf{C}\hat{\mathbf{x}}(t)$. Substituting $\hat{\mathbf{x}}(t) = \mathbf{x}(t) - \tilde{\mathbf{x}}(t)$ into Eq.(18.42) and combining with Eq.(18.21) yields

$$\begin{pmatrix} \dot{\tilde{\mathbf{x}}}(t) \\ \dot{\mathbf{w}}(t) \end{pmatrix} = \begin{pmatrix} \mathbf{A} - \mathbf{L}\mathbf{C} & -\mathbf{L}_I \\ \mathbf{C} & \mathbf{0} \end{pmatrix} \begin{pmatrix} \tilde{\mathbf{x}}(t) \\ \mathbf{w}(t) \end{pmatrix} \triangleq \mathbf{R}_{xw} \begin{pmatrix} \tilde{\mathbf{x}}(t) \\ \mathbf{w}(t) \end{pmatrix} \tag{18.44}$$

and

$$(s\mathbf{I}-\mathbf{A}+\mathbf{LC})\hat{\mathbf{x}}(s) = \mathbf{L}\mathbf{y}(s)+\mathbf{B}\mathbf{u}(s)+\mathbf{L}_I\mathbf{w}(s) = (\mathbf{L}+\mathbf{L}_I/s)\mathbf{y}(s)+\mathbf{B}\mathbf{u}(s)-\mathbf{L}_I\mathbf{C}\frac{1}{s}\hat{\mathbf{x}} \quad (18.45)$$

$$[s\mathbf{I} - \mathbf{A} + (\mathbf{L} + \mathbf{L}_I/s)\mathbf{C}]\hat{\mathbf{x}}(s) = (\mathbf{L} + \mathbf{L}_I/s)\mathbf{y}(s) + \mathbf{B}\mathbf{u}(s) \; . \quad (18.46)$$

For the sake of comparison, the proportional-observer equation in the s-domain, see Fig. 18.2b, is achieved by setting $\mathbf{L}_I = \mathbf{0}$. Eq.(18.26) was derived in the proportional-observer case, in order to obtain open-loop properties at S_1 identical to the full-state feedback. Repeating this derivaiton for the case of proportional integral observer, Eq.(18.26) can be rewritten substituting $\mathbf{L}$ by $\mathbf{L} + \mathbf{L}_I/s$. It results

$$(\mathbf{L} + \mathbf{L}_I/s)[\mathbf{I} + \mathbf{C}(s\mathbf{I} - \mathbf{A})^{-1}(\mathbf{L} + \mathbf{L}_I/s)]^{-1} = \mathbf{B}[\mathbf{C}(s\mathbf{I} - \mathbf{A})^{-1}\mathbf{B}]^{-1} \quad (18.47)$$

$$(s\mathbf{L} + \mathbf{L}_I)[s\mathbf{I} + \mathbf{C}(s\mathbf{I} - \mathbf{A})^{-1}(s\mathbf{L} + \mathbf{L}_I)]^{-1} = \mathbf{B}[\mathbf{C}(s\mathbf{I} - \mathbf{A})^{-1}\mathbf{B}]^{-1} \; . \quad (18.48)$$

18.4.2 Pole Placement Design

Invoking Eq.(18.44), the eigenvalues λ of the proportional integral observer can be assigned to specified numbers $\lambda_i \;\; \forall i = 1 \ldots n+m$

$$\det(\lambda\mathbf{I} - \mathbf{R}_{xw}) = \det\begin{pmatrix} \lambda\mathbf{I}_n - (\mathbf{A} - \mathbf{LC}) & \vdots & \mathbf{L}_I \\ -\mathbf{C} & & \lambda\mathbf{I}_m \end{pmatrix} = \prod_{i=1}^{n+m} (\lambda - \lambda_i) \quad (18.49)$$

$$\text{where} \quad \mathbf{L}, \mathbf{L}_I \in \mathcal{R}^{n \times m} \qquad \mathbf{C} \in \mathcal{R}^{m \times n} \qquad \mathbf{R} \in \mathcal{R}^{(n+m)\times(n+m)} \; . \quad (18.50)$$

There is a considerable degree of freedom for the designer: $2nm$ unknown elements in $\mathbf{L}$ and $\mathbf{L}_I$ and $n+m$ scalar equations in Eq.(18.49).

18.4.3 Recovery of the Full-State Feedback Design

If Eq.(18.26) is satisfied and the loop is broken at point S_1 and the open loop is excited by $\mathbf{u}''$ the observer does not get excited by $\mathbf{u}'$. At the point S_1 the full-state open-loop transfer characteristic is reconstructed. These properties (high gain margin and phase margin of 60 degrees in all channels) are also very impressive from the robustness point of view.

Regardless of the fact how Eq.(18.26) can be satisfied, Eq.(18.48) provides an easy choice of the parameters of the proportional integral observer. Eq.(18.48) is satisfied if $\mathbf{L}_I = \mathbf{B}$ is chosen and only the steady-state is considered

$$\lim_{s\to 0}(s\mathbf{L} + \mathbf{B})[s\mathbf{I} + \mathbf{C}(s\mathbf{I} - \mathbf{A})^{-1}(s\mathbf{L} + \mathbf{B})]^{-1} = \lim_{s\to 0}\mathbf{B}[\mathbf{C}(s\mathbf{I} - \mathbf{A})^{-1}\mathbf{B}]^{-1} \; . \quad (18.51)$$

Hence, proportional integral observer-based implementations are characterized by the following important property: Full-state feedback robustness is not fully recovered in the transient but is asymptotically obtained in the steady-state (*Beale, S., and Shafai, B., 1989*).

18.4.4 Robustness Against Plant Parameter Uncertainty

Consider the nominal plant matrix $\mathbf{A}$ and the perturbed plant matrix $\mathbf{A}_p$. It will be shown that the proportional integral observer-based implementation is characterized by robustness with respect to plant perturbation. The proportional integral observer is designed in order to achieve robust control. From Fig. 18.2c in the case of $\mathbf{H}_1 = \mathbf{I}$ the state variable $\mathbf{x}(s)$ of the perturbed plant is

$$\mathbf{x}(s) = \mathbf{\Phi}_p\mathbf{B}\mathbf{u}(s) = \mathbf{\Phi}_p\mathbf{B}[\mathbf{y}_{ref}(s) - \mathbf{H}_2\hat{\mathbf{x}}(s)] \quad \text{where} \quad \mathbf{\Phi}_p = (s\mathbf{I} - \mathbf{A}_p)^{-1} . \tag{18.52}$$

Let the observer be implemented by the nominal plant matrix $\mathbf{A}$, $\mathbf{B}$, $\mathbf{C}$. Then, in the s-domain the output of the summing point Σ_1 in Fig. 18.2c is

$$\mathbf{\Phi}^{-1}\hat{\mathbf{x}} = \mathbf{B}\mathbf{u} + \mathbf{L}_I\mathbf{w} + \mathbf{L}(\mathbf{y} - \mathbf{C}\hat{\mathbf{x}}) \quad \text{where} \quad \mathbf{\Phi} = (s\mathbf{I} - \mathbf{A})^{-1} . \tag{18.53}$$

From Eqs.(18.52) and (18.53) one has

$$(s\mathbf{I} - \mathbf{A}_p)\mathbf{x} + \mathbf{B}\mathbf{H}_2\hat{\mathbf{x}} = \mathbf{B}\mathbf{y}_{ref} \tag{18.54}$$

$$-\mathbf{L}\mathbf{C}\mathbf{x} + (s\mathbf{I} - \mathbf{A} + \mathbf{L}\mathbf{C})\hat{\mathbf{x}} - \mathbf{B}(\mathbf{y}_{ref} - \mathbf{H}_2\hat{\mathbf{x}}) = \mathbf{L}_I\mathbf{w} , \tag{18.55}$$

respectively. Combining yields

$$\begin{pmatrix} s\mathbf{I} - \mathbf{A}_p & \mathbf{B}\mathbf{H}_2 \\ -\mathbf{L}\mathbf{C} & \vdots\, s\mathbf{I} - \mathbf{A} + \mathbf{L}\mathbf{C} + \mathbf{B}\mathbf{H}_2 \end{pmatrix} \begin{pmatrix} \mathbf{x}(s) \\ \hat{\mathbf{x}}(s) \end{pmatrix} = \begin{pmatrix} \mathbf{B}\mathbf{y}_{ref}(s) \\ \mathbf{B}\mathbf{y}_{ref}(s) + \mathbf{L}_I\mathbf{w}(s) \end{pmatrix} . \tag{18.56}$$

Defining $\mathbf{\Gamma}$ and $\mathbf{\Omega}$ and using Eq.(C.30),

$$\mathbf{\Gamma} \triangleq \begin{pmatrix} \mathbf{A}_p & -\mathbf{B}\mathbf{H}_2 \\ \mathbf{L}\mathbf{C} & \vdots\, \mathbf{A} - \mathbf{L}\mathbf{C} - \mathbf{B}\mathbf{H}_2 \end{pmatrix} \qquad \mathbf{\Gamma}^{-1} \triangleq \begin{pmatrix} \mathbf{E}_1 & \mathbf{E}_2 \\ \mathbf{E}_3 & \mathbf{E}_4 \end{pmatrix} \tag{18.57}$$

$$\mathbf{\Gamma}^{-1} = \begin{pmatrix} \mathbf{A}_p^{-1} - \mathbf{A}_p^{-1}\mathbf{B}\mathbf{H}_2\mathbf{\Omega}^{-1}\mathbf{L}\mathbf{C}\mathbf{A}_p^{-1} & \vdots & \mathbf{A}_p^{-1}\mathbf{B}\mathbf{H}_2\mathbf{\Omega}^{-1} \\ -\mathbf{\Omega}^{-1}\mathbf{L}\mathbf{C}\mathbf{A}_p^{-1} & \vdots & \mathbf{\Omega}^{-1} \end{pmatrix} \tag{18.58}$$

where $\mathbf{\Omega} \triangleq \mathbf{A} - \mathbf{L}\mathbf{C} - \mathbf{B}\mathbf{H}_2 + \mathbf{L}\mathbf{C}\mathbf{A}_p^{-1}\mathbf{B}\mathbf{H}_2$ it results

$$(s\mathbf{I} - \mathbf{\Gamma})\begin{pmatrix} \mathbf{x}(s) \\ \hat{\mathbf{x}}(s) \end{pmatrix} = \begin{pmatrix} \mathbf{B} \\ \mathbf{B} \end{pmatrix}\mathbf{y}_{ref}(s) + \begin{pmatrix} \mathbf{0} \\ \mathbf{L}_I \end{pmatrix}\mathbf{w}(s) \tag{18.59}$$

$$\begin{pmatrix} \mathbf{x}(s) \\ \hat{\mathbf{x}}(s) \end{pmatrix} = (s\mathbf{I} - \mathbf{\Gamma})^{-1}\begin{pmatrix} \mathbf{B} \\ \mathbf{B} \end{pmatrix}\mathbf{y}_{ref}(s) + (s\mathbf{I} - \mathbf{\Gamma})^{-1}\begin{pmatrix} \mathbf{0} \\ \mathbf{L}_I \end{pmatrix}\mathbf{w}(s) . \tag{18.60}$$

The steady-state is given by

$$\begin{pmatrix} \mathbf{x}(\infty) \\ \hat{\mathbf{x}}(\infty) \end{pmatrix} = -\mathbf{\Gamma}^{-1}\begin{pmatrix} \mathbf{B} \\ \mathbf{B} \end{pmatrix}\mathbf{y}_{ref}(\infty) - \mathbf{\Gamma}^{-1}\begin{pmatrix} \mathbf{0} \\ \mathbf{L}_I \end{pmatrix}\mathbf{w}(\infty) \tag{18.61}$$

$$= -\begin{pmatrix} \mathbf{E}_1 & \mathbf{E}_2 \\ \mathbf{E}_3 & \mathbf{E}_4 \end{pmatrix}\begin{pmatrix} \mathbf{B} \\ \mathbf{B} \end{pmatrix}\mathbf{y}_{ref}(\infty) - \begin{pmatrix} \mathbf{E}_1 & \mathbf{E}_2 \\ \mathbf{E}_3 & \mathbf{E}_4 \end{pmatrix}\begin{pmatrix} \mathbf{0} \\ \mathbf{L}_I \end{pmatrix}\mathbf{w}(\infty) . \tag{18.62}$$

Referring to Fig. 18.2a, the full-state feedback system obeys the equations

$$\mathbf{x}(s) = (s\mathbf{I} - \mathbf{A}_p + \mathbf{B}\mathbf{H}_2)^{-1}\mathbf{B}\mathbf{y}_{ref}(s) \tag{18.63}$$

$$\mathbf{x}(\infty) = (-\mathbf{A}_p + \mathbf{BH}_2)^{-1}\mathbf{By}_{ref}(\infty) . \tag{18.64}$$

In order to recover the behaviour of the full-state feedback system by the system based on the proportional integral observer, $\hat{\mathbf{x}}(\infty)$ in Eq.(18.62) has to be identical to $\mathbf{x}(\infty)$ in Eq.(18.64). Thus,

$$-(\mathbf{E}_3 + \mathbf{E}_4)\mathbf{By}_{ref}(\infty) - \mathbf{E}_4\mathbf{L}_I\mathbf{w}(\infty) = (-\mathbf{A}_p + \mathbf{BH}_2)^{-1}\mathbf{By}_{ref}(\infty) \tag{18.65}$$

$$[-\boldsymbol{\Omega}^{-1}\mathbf{LCA}_p^{-1} + \boldsymbol{\Omega}^{-1} + (-\mathbf{A}_p + \mathbf{BH}_2)^{-1}]\,\mathbf{By}_{ref}(\infty) = -\boldsymbol{\Omega}^{-1}\mathbf{L}_I\mathbf{w}(\infty) . \tag{18.66}$$

Applying simple matrix algebra operations yields

$$(\mathbf{A}_p - \mathbf{A})(-\mathbf{A}_p + \mathbf{BH}_2)^{-1}\mathbf{By}_{ref}(\infty) = \mathbf{L}_I\mathbf{w}(\infty) . \tag{18.67}$$

Using the identity $(\mathbf{A}_p - \mathbf{A})(-\mathbf{A}_p + \mathbf{BH}_2)^{-1} \equiv (\mathbf{A} - \mathbf{BH}_2)(\mathbf{A}_p - \mathbf{BH}_2)^{-1} - \mathbf{I}$, one has

$$[(\mathbf{A} - \mathbf{BH}_2)(\mathbf{A}_p - \mathbf{BH}_2)^{-1} - \mathbf{I}]\,\mathbf{By}_{ref}(\infty) \triangleq \mathbf{Qy}_{ref}(\infty) = \mathbf{L}_I\mathbf{w}(\infty) . \tag{18.68}$$

A solution is achieved for special cases, e.g., multiple-input single-output system, applying minimum norm relations (*Beale, S., and Shafai, B., 1989*). The design is completed by calculating $\mathbf{L}$ from Eq.(18.49).

An equivalent design approach is given from Eq.(18.62)

$$\mathbf{x}(\infty) = (-\mathbf{E}_1 - \mathbf{E}_2)\mathbf{By}_{ref}(\infty) - \mathbf{E}_2\mathbf{L}_I\mathbf{w}(\infty) \tag{18.69}$$

$$\hat{\mathbf{x}}(\infty) = (-\mathbf{E}_3 - \mathbf{E}_4)\mathbf{By}_{ref}(\infty) - \mathbf{E}_4\mathbf{L}_I\mathbf{w}(\infty) . \tag{18.70}$$

Subtracting both equations and requiring $\mathbf{x}(\infty) - \hat{\mathbf{x}}(\infty) = \tilde{\mathbf{x}}(\infty) \rightarrow \mathbf{0}$ yields

$$(\mathbf{E}_2 - \mathbf{E}_4)^{-1}(-\mathbf{E}_1 - \mathbf{E}_2 + \mathbf{E}_3 + \mathbf{E}_4)\mathbf{By}_{ref}(\infty) = \mathbf{L}_I\mathbf{w}(\infty) . \tag{18.71}$$

Omitting tedious matrix calculations a result is achieved which is equivalent to Eq.(18.67).

A nominally linear system with uncertainty modeled as cone-bounded nonlinearities is outlined by *Saberi, A., and Sannuti, P., 1990*.

18.5 Robust Kalman-Bucy Filters

18.5.1 Unperturbed Kalman-Bucy Filters

Consider the plant in the nominal case with additive noise

$$\dot{\mathbf{x}}(t) = \mathbf{Ax}(t) + \mathbf{w}(t) \qquad \mathbf{x}(0) = \mathbf{x}_o \qquad \mathbf{y}(t) = \mathbf{Cx}(t) + \mathbf{v}(t) . \tag{18.72}$$

The vectors $\mathbf{w}(t)$ and $\mathbf{v}(t)$ denote process and measurement noise, respectively. They are considered uncorrelated, stationary, zero mean with Gaussian distribution. The Kalman-Bucy filter is given by the equation for the estimate (*Brammer, K., and Siffling, G., 1975; Eitelberg, E., 1986*)

$$\dot{\hat{\mathbf{x}}}(t) = \mathbf{A}\hat{\mathbf{x}}(t) + \mathbf{K}_K[\mathbf{y}(t) - \mathbf{C}\hat{\mathbf{x}}(t)] . \tag{18.73}$$

Excitation by $\mathbf{u}(t)$ is omitted for abbreviation. See also Eq.(18.100). The main problem is to design a robust Kalman-Bucy filter such that it tracks the state variables even in the presence of parametrical and noise uncertainties. In order to minimize $E\{(\mathbf{x}-\hat{\mathbf{x}})^T(\mathbf{x}-\hat{\mathbf{x}})\}$ the steady-state Kalman gain matrix $\mathbf{K}_K$ has to be chosen according to

$$\mathbf{K}_K = \mathbf{P}_K\mathbf{C}^T\mathbf{V}^{-1} \tag{18.74}$$

$$\mathbf{A}\mathbf{P}_K + \mathbf{P}_K\mathbf{A}^T + \mathbf{W} - \mathbf{P}_K\mathbf{C}^T\mathbf{V}^{-1}\mathbf{C}\mathbf{P}_K = \mathbf{0} \qquad (18.75)$$

where $\mathbf{W}$ and $\mathbf{V}$ are symmetric positive definite matrices representing the covariances of the system noise $\mathbf{w}$ and measurement noise $\mathbf{v}$, respectively,

$$\mathbf{W}\delta(t-\tau) = \mathrm{cov}[\mathbf{w}(t), \mathbf{w}(\tau)]\,, \quad \mathbf{V}\delta(t-\tau) = \mathrm{cov}[\mathbf{v}(t), \mathbf{v}(\tau)]\,, \quad \mathrm{cov}[\mathbf{w}(t), \mathbf{v}(\tau)] = \mathbf{0}\,. \qquad (18.76)$$

For abbreviation, the notation

$$\|\mathbf{x} - \hat{\mathbf{x}}\|_{eF}^2 = E\{(\mathbf{x} - \hat{\mathbf{x}})^T(\mathbf{x} - \hat{\mathbf{x}})\} \qquad (18.77)$$

is introduced. The subscript $_{eF}$ denotes the expected function of the Frobenius vector norm.

18.5.2 Kalman-Bucy-Filter for Perturbed Plants

In practical systems, it cannot be assumed that $\mathbf{A}, \mathbf{V}, \mathbf{W}$ are known exactly. The actual matrices are unavailable. Only nominal ones are given. It is assumed that norm bounds for the uncertainty are known. Thus, the implementation of the Kalman-Bucy filter will deviate to a certain amount from its optimal behaviour.

The plant and the Kalman-Bucy filter with parametrical uncertainties and least favourable noise uncertainties can be written as follows:

$$\dot{\mathbf{x}}(t) = (\mathbf{A} + \Delta\mathbf{A})\mathbf{x}(t) + \mathbf{w}(t) \qquad (18.78)$$

$$\mathbf{y}(t) = (\mathbf{C} + \Delta\mathbf{C})\mathbf{x}(t) + \mathbf{v}(t) \qquad (18.79)$$

$$\dot{\hat{\mathbf{x}}}(t) = \mathbf{A}\hat{\mathbf{x}}(t) + \hat{\mathbf{K}}_K[\mathbf{y}(t) - \mathbf{C}\hat{\mathbf{x}}(t)] \qquad (18.80)$$

where the robust steady-state Kalman gain matrix is

$$\hat{\mathbf{K}}_K = \mathbf{P}_K\mathbf{C}^T(\mathbf{V}_o + \eta_v\mathbf{I})^{-1} \qquad (18.81)$$

$$\mathbf{A}\mathbf{P}_K + \mathbf{P}_K\mathbf{A}^T + \mathbf{W}_o + \eta_w\mathbf{I} - \mathbf{P}_K\mathbf{C}^T(\mathbf{V}_o + \eta_v\mathbf{I})^{-1}\mathbf{C}\mathbf{P}_K = \mathbf{0}\,. \qquad (18.82)$$

The expressions $\mathbf{V}_o + \eta_v\mathbf{I}$ and $\mathbf{W}_o + \eta_w\mathbf{I}$ denote the least favourable covariances. The estimation error is given by $\tilde{\mathbf{x}}(t) = \mathbf{x}(t) - \hat{\mathbf{x}}(t)$. Now, subtracting Eq.(18.80) from Eq.(18.78) and combining with Eq.(18.78) yields

$$\underbrace{\begin{pmatrix} \dot{\mathbf{x}} \\ \dot{\tilde{\mathbf{x}}} \end{pmatrix}}_{\dot{\mathbf{x}}_{pe}(t)} = \underbrace{\begin{pmatrix} \mathbf{A} & \mathbf{0} \\ \mathbf{0} & \mathbf{A} - \hat{\mathbf{K}}_K\mathbf{C} \end{pmatrix}}_{\mathbf{A}_{pe}} \begin{pmatrix} \mathbf{x} \\ \tilde{\mathbf{x}} \end{pmatrix} + \underbrace{\begin{pmatrix} \Delta\mathbf{A} & \mathbf{0} \\ \Delta\mathbf{A} - \hat{\mathbf{K}}_K\Delta\mathbf{C} & \mathbf{0} \end{pmatrix}}_{\Delta\mathbf{A}_{pe}} \begin{pmatrix} \mathbf{x} \\ \tilde{\mathbf{x}} \end{pmatrix} + \underbrace{\begin{pmatrix} \mathbf{I} & \mathbf{0} \\ \mathbf{I} & -\hat{\mathbf{K}}_K \end{pmatrix}}_{\mathbf{G}} \underbrace{\begin{pmatrix} \mathbf{w} \\ \mathbf{v} \end{pmatrix}}_{\mathbf{n}}\,. \qquad (18.83)$$

With the abbreviations cited above this type of perturbed plant including Kalman-Bucy estimation error is investigated and can readily be solved (*Chen, B.S., and Dong, T.Y., 1988; Kay, S.M., 1988; Staudte, R.G., and Sheather, S.J., 1990*). Using the matrix $\mathbf{\Phi}_{pe}(t) = \exp \mathbf{A}_{pe}t$ and the expectation of the Frobenius vector, the solution of Eq.(18.83) in norm-bounded form is

$$\|\mathbf{x}_{pe}(t)\|_{eF} \le \|\mathbf{\Phi}_{pe}(t)\|_s\|\mathbf{x}_{peo}\|_{eF} + \int_0^t \|\mathbf{\Phi}_{pe}(t-\tau)\|_s\Big(\|\Delta\mathbf{A}_{pe}(\tau)\|_s\|\mathbf{x}_{pe}(\tau)\|_{eF} + \|\mathbf{G}\mathbf{n}(\tau)\|_{eF}\Big)d\tau\,. \qquad (18.84)$$

From the abbreviations in Eq.(18.83) it follows

- (i)

$$\|\Delta \mathbf{A}_{pe}\|_s \leq 2\|\Delta \mathbf{A}(t)\|_s + \|\hat{\mathbf{K}}_K\|_s\|\Delta \mathbf{C}(t)\|_s \leq 2\beta_A + \beta_C\|\hat{\mathbf{K}}_K\|_s \tag{18.85}$$

where β_A and β_C are the bounds of $\Delta\mathbf{A}$ and $\Delta\mathbf{C}$, respectively,

$$\Delta \mathbf{A}(t) \in \{\|\Delta \mathbf{A}(t)\|_s < \beta_A\} \qquad \Delta \mathbf{C}(t) \in \{\|\Delta \mathbf{C}(t)\|_s < \beta_C\} \quad \text{and} \tag{18.86}$$

- (ii)

$$\|\mathbf{G}\mathbf{n}(t)\|_{eF} \leq \|\mathbf{G}\|_s\|\mathbf{n}(t)\|_{eF} \leq (2 + \|\mathbf{K}_K\|_s)\big(\|\mathbf{w}(t)\|_{eF} + \|\mathbf{v}(t)\|_{eF}\big) \tag{18.87}$$

$$\leq (2 + \|\mathbf{K}_K\|_s)\{\sqrt{\mathrm{tr}[\mathbf{W}_o + \eta_w \mathbf{I}]} + \sqrt{\mathrm{tr}[\mathbf{V}_o + \eta_v \mathbf{I}]}\} \triangleq \beta_1 \, . \tag{18.88}$$

Now, the transition matrix $\mathbf{\Phi}_{pe}(t)$ is considered bounded by $\|\mathbf{\Phi}_{pe}(t)\|_s \leq m_o e^{-\alpha t}$. From Eq.(18.84) combining both integrals it follows

$$\|\mathbf{x}_{pe}(t)\|_{eF} e^{\alpha t} \leq m_o\|\mathbf{x}_{peo}\|_{eF} + m_o(2\beta_A + \beta_C\|\hat{\mathbf{K}}_K\|_s)\int_0^t \|\mathbf{x}_{pe}(\tau)\|_{eF} e^{\alpha\tau} d\tau + m_o\beta_1 \int_0^t e^{\alpha\tau} d\tau \, . \tag{18.89}$$

Applying Bellman-Gronwall lemma Eq.(11.20),

$$f(t) := m_o\|\mathbf{x}_{peo}\|_F + m_o\beta_1 \frac{e^{\alpha t} - 1}{\alpha} \qquad g(t) := m_o(2\beta_A + \beta_C\|\hat{\mathbf{K}}_K\|_s) = \beta_2 \tag{18.90}$$

$$k(\tau) := 1 \qquad h(t) := \|\bar{\mathbf{x}}_{pe}\|_{eF}\, e^{\alpha t} \tag{18.91}$$

it can be observed that the particular part of the solution relevant for stability is given by

$$e^{-\alpha t}\int_0^t f(\tau) e^{\int_\tau^t g(\eta)d\eta}\, d\tau = e^{-\alpha t}\int_0^t m_o\beta_1 \frac{e^{\alpha\tau} - 1}{\alpha} e^{\beta_2(t-\tau)}\, d\tau + \dots \, . \tag{18.92}$$

Stability is guaranteed if $\alpha - \beta_2 > 0$ or

$$\alpha > m_o(2\beta_A + \beta_C\|\hat{\mathbf{K}}_K\|_s) \, . \tag{18.93}$$

18.5.3 Improving Stability

If the inequality Eq.(18.93) cannot be satisfied the degree of stability must be increased. This can be achieved by replacing $\mathbf{A}$ in Eq.(18.82) by $\mathbf{A} + \alpha_a\mathbf{I}$. It is well known that this operation improves the stability to the same extent as if the matrix $\mathbf{P}$ is artificially multiplied by $e^{-2\alpha_a t}$ (*Anderson, B.D.O., 1973*).

18.6 Plant Perturbation and Noise Uncertainty

The objective is to design a robust controller with respect to both plant perturbation and noise uncertainty.

Apart from deterministic models uncertainty is described by means of stochastic models. Sensor noise as cited above in most cases is treated as frequency-rich or as a stochastic process. Stochastic models yield an adequate possibility to overcome uncertainty in real processes, see e.g. *Peterka, V., 1986*. State-dependent noise

$$\dot{\mathbf{x}}(t) = \mathbf{A}\mathbf{x} + \sum_i x_i(t)\mathbf{A}_{wi}\dot{\mathbf{w}}(t) \tag{18.94}$$

where $\dot{\mathbf{w}}$ is white is considered as a special case of uncertainty (*Wonham, M.W., 1967*). Noise parameters are considered fixed or variable within predetermined bounds.

Another version of stochastic uncertainty models considers sudden changes of the system matrices $\mathbf{A}$ or $\mathbf{B}$. In these cases, the consecutive values $\mathbf{A}_i$ are treated both with Markov chain parameters and additionally with white noise parameters (*Yaz, E., 1990*).

Even though nonstatistic, an additional quality of robust control algorithm may be obtained by identifying the uncertainty by means of fuzzy logic (*Zadeh, L.A., 1965; Dubois, D., and Prade, H., 1980; Yager, R.R., et al. 1987; Rommelfanger, H., 1988; Pal, S.K., 1991*). Analyzing the knowledge and skill of planning engineers and operating personnel, as far as uncertainties are concerned, the parameters of robust controllers can be improved considerably.

18.6.1 Unperturbed LQG System

Consider a linear time-invariant continuous process under excitation of stationary white Gaussian noise $\mathbf{w}(t)$ and $\mathbf{v}(t)$

$$\dot{\mathbf{x}}(t) = \mathbf{A}\mathbf{x}(t) + \mathbf{B}\mathbf{u}(t) + \mathbf{w}(t) \qquad \mathbf{y}(t) = \mathbf{C}\mathbf{x}(t) + \mathbf{v}(t) \tag{18.95}$$

$$E\{\mathbf{w}(t)\mathbf{w}^T(\tau)\} = \mathrm{cov}\,[\mathbf{w}(t), \mathbf{w}(\tau)] = \mathbf{W}\delta(t-\tau), \quad E\{\mathbf{v}(t)\mathbf{v}^T(\tau)\} = \mathrm{cov}\,[\mathbf{v}(t), \mathbf{v}(\tau)] = \mathbf{V}\delta(t-\tau) \tag{18.96}$$

$$E\{\mathbf{w}(t)\} = 0 \qquad E\{\mathbf{v}(t)\} = 0 \qquad \mathbf{W} > 0 \qquad \mathbf{V} > 0\,. \tag{18.97}$$

Process noise $\mathbf{w}(t)$ and measurement noise $\mathbf{v}(t)$ are assumed independent of each other.

The optimal feedback law under square structured index of performance I is well-known and given by $\mathbf{u}(t) = \mathbf{K}\hat{\mathbf{x}}(t)$

$$I = \lim_{T\to\infty} \frac{1}{T} E\{\int_0^T (\mathbf{x}^T\mathbf{Q}\mathbf{x} + \rho\ \mathbf{u}^T\mathbf{R}\mathbf{u})dt\} \to \min \tag{18.98}$$

where $\mathbf{Q} = \mathbf{Q}^T \geq 0$, $\mathbf{R} = \mathbf{R}^T > 0$. The optimal feedback controller $\mathbf{K}$ is given by

$$\mathbf{K} = -\frac{1}{\rho}\mathbf{R}^{-1}\mathbf{B}^T\mathbf{P} \quad \text{and} \quad \mathbf{P}\mathbf{A} + \mathbf{A}^T\mathbf{P} + \mathbf{Q} - \frac{1}{\rho}\mathbf{P}\mathbf{B}\mathbf{R}^{-1}\mathbf{B}^T\mathbf{P} = \mathbf{0}\,. \tag{18.99}$$

The estimate $\hat{\mathbf{x}}(t)$ is the output of the Kalman-Bucy filter given by

$$\dot{\hat{\mathbf{x}}}(t) = \mathbf{A}\hat{\mathbf{x}}(t) + \mathbf{B}\mathbf{u}(t) + \mathbf{K}_K[\mathbf{y}(t) - \mathbf{C}\hat{\mathbf{x}}(t)] \tag{18.100}$$

$$\mathbf{K}_K = \mathbf{P}_K\mathbf{C}^T\mathbf{V}^{-1} \tag{18.101}$$

$$\mathbf{A}\mathbf{P}_K + \mathbf{P}_K\mathbf{A}^T + \mathbf{W} - \mathbf{P}_K\mathbf{C}^T\mathbf{V}^{-1}\mathbf{C}\mathbf{P}_K = \mathbf{0}\,. \tag{18.102}$$

18.6.2 Perturbed System

The problem of perturbed plant parameters and noise uncertainty is stated if $\mathbf{A}$ is substituted by $\mathbf{A} + \Delta\mathbf{A}$ where $\|\Delta\mathbf{A}\|_s \leq \beta_A$ etc. The perturbed covariance $\mathbf{W}_p$ is assumed bounded

$$\mathbf{W}_p = \mathbf{W} + \Delta\mathbf{W} \qquad \|\Delta\mathbf{W}\|_s \leq \eta_w\,. \tag{18.103}$$

Then, the covariance $\mathbf{W}_p$ is considered to be a unit of the set $\mathcal{S}_w$

$$\mathbf{W}_p \in \mathcal{S}_w = \{\mathbf{W}_p\ :\ \ \|\Delta\mathbf{W}\|_s \leq \eta_w,\ \ \mathbf{W}_p > 0\}\,. \tag{18.104}$$

For $\Delta\mathbf{B}$, $\Delta\mathbf{C}$ and $\mathbf{V}$ there exist corresponding bound assumptions.

18.6.3 Min-Max Problem and Saddle-Point Problem

Assume the optimal control $\mathbf{u}^\star$ as a member of $\mathcal{U}$. Then, the minimax problem is stated

$$\min_{\mathbf{u}\in\mathcal{U}} \max_{\mathbf{W}_p\in\mathcal{S}_w,\ \mathbf{V}_p\in\mathcal{S}_v} I(\mathbf{u},\mathbf{W}_p,\mathbf{V}_p\} \ . \tag{18.105}$$

If the order of the min-max operations in Eq.(18.105) may be permuted then a triplet $(\mathbf{u}^\star,\mathbf{W}_p^\star,\mathbf{V}_p^\star)$ exists satisfying the saddle-point condition (*Looze, D.P., et al. 1983*)

$$I(\mathbf{u}^\star,\mathbf{W}_p,\mathbf{V}_p) \le I(\mathbf{u}^\star,\mathbf{W}_p^\star,\mathbf{V}_p^\star) \le I(\mathbf{u},\mathbf{W}_p^\star,\mathbf{V}_p^\star) \tag{18.106}$$

$$\text{where} \qquad \mathbf{u}\in\mathcal{U} \qquad \mathbf{W}_p\in\mathcal{S}_w \qquad \mathbf{V}_p\in\mathcal{S}_v \ . \tag{18.107}$$

18.6.4 Unperturbed Plant. Uncertain Noise

The robust controller for certain plant but uncertain noise is given by solving the Eqs.(18.99) through (18.102) with the least-favourable noise covariances $\mathbf{W}+\eta_w\mathbf{I}$ and $\mathbf{V}+\eta_v\mathbf{I}$ where from Eq.(18.104)

$$\mathcal{S}_w = \{\mathbf{W}_p \ : \quad \mathbf{W}_p = \mathbf{W}_p^T > 0,\ (\mathbf{W}-\eta_w\mathbf{I}) \le \mathbf{W}_p \le (\mathbf{W}+\eta_w\mathbf{I})\} \ . \tag{18.108}$$

Hence, using the (additional) index $_L$ for least-favourable,

$$\dot{\hat{\mathbf{x}}}(t) = \mathbf{A}\hat{\mathbf{x}}(t) + \mathbf{B}\mathbf{u}(t) + \mathbf{K}_{KL}[\mathbf{y}(t) - \mathbf{C}\hat{\mathbf{x}}(t)] \tag{18.109}$$

$$\mathbf{K}_{KL} = \mathbf{P}_{KL}\mathbf{C}^T(\mathbf{V}+\eta_v\mathbf{I})^{-1} \tag{18.110}$$

$$\mathbf{A}\mathbf{P}_{KL} + \mathbf{P}_{KL}\mathbf{A}^T + (\mathbf{W}+\eta_w\mathbf{I}) - \mathbf{P}_{KL}\mathbf{C}^T(\mathbf{V}+\eta_v\mathbf{I})^{-1}\mathbf{C}\mathbf{P}_{KL} = \mathbf{0} \ . \tag{18.111}$$

18.6.5 Algebraic Formulation of I

Following *Looze, D.P., et al. 1983* the performance I can be reformulated

$$I = \mathrm{tr}\ [\mathbf{W}\mathbf{P}] + \mathrm{tr}[(\mathbf{W} + \mathbf{K}_K\mathbf{V}\mathbf{K}_K^T)\mathbf{X}_a] \tag{18.112}$$

where $\mathbf{X}_a \ge 0$ is the solution of

$$(\mathbf{A}-\mathbf{K}_K\mathbf{C})^T\mathbf{X}_a + \mathbf{X}_a(\mathbf{A}-\mathbf{K}_K\mathbf{C}) + \rho\mathbf{K}^T\mathbf{R}\mathbf{K} = \mathbf{0} \ . \tag{18.113}$$

18.6.6 Both Perturbed Plant and Uncertain Noise

Now, the difference $\mathbf{x}_d$ of two cases is studied. The difference $\mathbf{x}_d$ between, first, the state variable $\mathbf{x}(t)$ in the worst case with plant uncertainty, only, i.e. Eq.(18.114), and, second, the worst cast $\hat{\mathbf{x}}(t)$ where noise uncertainty occurs, only, as given in Eq.(18.109). Thus, using $\mathbf{y} = (\mathbf{C}+\Delta\mathbf{C})\mathbf{x}+\mathbf{v}$ and $\mathbf{u} = \mathbf{K}\hat{\mathbf{x}}$,

$$\dot{\mathbf{x}} = (\mathbf{A}+\Delta\mathbf{A})\mathbf{x} + (\mathbf{B}+\Delta\mathbf{B})\mathbf{u} + \mathbf{w} \tag{18.114}$$

$$\dot{\mathbf{x}}_d = \dot{\mathbf{x}} - \dot{\hat{\mathbf{x}}} = (\mathbf{A}-\mathbf{K}_{KL}\mathbf{C})\mathbf{x}_d + \Delta\mathbf{A}\ \mathbf{x} + \Delta\mathbf{B}\ \mathbf{u} - \mathbf{K}_{KL}\Delta\mathbf{C}\ \mathbf{x} + \mathbf{w} - \mathbf{K}_{KL}\mathbf{v} \tag{18.115}$$

$$\bar{\mathbf{x}}(t) \triangleq \begin{pmatrix}\mathbf{x}(t)\\ \mathbf{x}_d(t)\end{pmatrix} = \begin{pmatrix}\mathbf{x}(t)\\ \mathbf{x}(t)-\hat{\mathbf{x}}(t)\end{pmatrix} \qquad \mathbf{n}(t) \triangleq \begin{pmatrix}\mathbf{w}(t)\\ \mathbf{v}(t)\end{pmatrix} \qquad \bar{\mathbf{G}} \triangleq \begin{pmatrix}\mathbf{I} & \mathbf{0}\\ \mathbf{I} & -\mathbf{K}_{KL}\end{pmatrix} \tag{18.116}$$

$$\bar{\mathbf{A}} = \begin{pmatrix} \mathbf{A}+\mathbf{B}\mathbf{K} & -\mathbf{B}\mathbf{K} \\ \mathbf{0} & \vdots\ \mathbf{A}-\mathbf{K}_{KL}\mathbf{C} \end{pmatrix} \qquad \Delta\bar{\mathbf{A}} = \begin{pmatrix} \Delta\mathbf{A}+\Delta\mathbf{B}\ \mathbf{K} & -\Delta\mathbf{B}\ \mathbf{K} \\ \Delta\mathbf{A}+\Delta\mathbf{B}\ \mathbf{K}-\mathbf{K}_{KL}\Delta\mathbf{C}\ \vdots & -\Delta\mathbf{B}\ \mathbf{K} \end{pmatrix} \tag{18.117}$$

$$\dot{\bar{\mathbf{x}}}(t) = \bar{\mathbf{A}}\bar{\mathbf{x}}(t) + \Delta\bar{\mathbf{A}}\bar{\mathbf{x}}(t) + \bar{\mathbf{G}}\mathbf{n}(t)\ . \tag{18.118}$$

In order to preserve stability, $\max_i \Re e\ \lambda_i[\bar{\mathbf{A}}] < 0$ must be satisfied. Then, bounding the transition matrix by

$$\|e^{\bar{\mathbf{A}}t}\| = \|\bar{\mathbf{\Phi}}(t)\| \leq \phi_o e^{\max_i \Re e\ \lambda_i[\bar{\mathbf{A}}]} \tag{18.119}$$

and applying the Bellman-Gronwall lemma, see e.g. Eq.(11.111) (*Chen, B.S., and Dong, T.Y., 1989*) stability robustness is given if

$$-\max_i \Re e\ \lambda_i[\bar{\mathbf{A}}] > \phi_o(2\beta_A + 2\beta_B\|\bar{\mathbf{G}}\|_s + \beta_C\|\mathbf{K}_{KL}\|_s)\ . \tag{18.120}$$

Chapter 19

Initial Condition Perturbation, Overshoot and Robustness

Overshoot and speed requirements of a control system are important performance objectives. There is a strong need for balancing stability robustness versus the performance. This short chapter is devoted to optimizing stability robustness and performance as cited above. The overshoot behaviour of a closed-loop system is defined as the maximum value of the spectral norm of its transition matrix $\mathbf{\Phi}_{cl}(t)$ weighted by some weighting matrices $\mathbf{W}_1, \mathbf{W}_2$

$$M_{\max} = \max_t \|\mathbf{W}_1\mathbf{\Phi}_{cl}(t)\mathbf{W}_2\|_s = \|\mathbf{W}_1\mathbf{\Phi}_{cl}(t_{\max})\mathbf{W}_2\|_s \quad \text{where} \quad \mathbf{\Phi}_{cl}(t) = e^{\mathbf{F}t} \ . \tag{19.1}$$

The matrix $\mathbf{F}$ is the coefficient matrix of the closed-loop system containing plant and (observer-based dynamic) controller. The unified state representation as given by $\mathbf{F}_{pr}$ in Eq.(11.96) is renamed to $\mathbf{F}$.

The pair of weighting matrices $(\mathbf{W}_1, \mathbf{W}_2)$ is chosen as $(\mathbf{C}\mathbf{F}^{-1}, \mathbf{b})$ and $(\mathbf{C}, \mathbf{b})$ in the case of step input and Dirac impulse input, respectively. This will be proved in what follows. Setting the weighting matrices $\mathbf{W}_i = \mathbf{I}$, the overshoot behaviour of all the states is measured when the initial conditions are perturbed.

19.1 Dirac Input

Consider a system with an external scalar input $r(t)$ where $\mathbf{b}$ corresponds to the input matrix depending on the entry point of $r(t)$. Then,

$$\mathbf{x}(t) = \mathbf{\Phi}_{cl}(t)\mathbf{x}(0) + \int_0^\infty \mathbf{\Phi}_{cl}(t-\tau)\ \mathbf{b}\ r(\tau)d\tau \ . \tag{19.2}$$

If the input is a Dirac delta function $r(t) = \delta(t)$ then

$$\mathbf{y}(t) = \mathbf{C}\mathbf{x}(t) = \mathbf{C}\mathbf{\Phi}_{cl}(t)\mathbf{x}(0) + \mathbf{C}\mathbf{\Phi}_{cl}(t)\mathbf{b} = \mathbf{C}\mathbf{\Phi}_{cl}(t)\mathbf{x}'(0) \quad \text{where} \quad \mathbf{x}'(0) \triangleq \mathbf{x}(0) + \mathbf{b} \ . \tag{19.3}$$

The effect of the selected input function is incorporated into an initial condition perturbation represented by the vector $\mathbf{x}'(0)$ which is an important relation.

19.2 Step Input

The impulse response of the closed-loop system is given by

$$\mathbf{y}(s) = \mathbf{C}(s\mathbf{I} - \mathbf{F})^{-1}\mathbf{b} \qquad \mathbf{y}(t) = \mathbf{C}\mathbf{\Phi}_{cl}(t)\mathbf{b} = \mathbf{C}e^{\mathbf{F}t}\mathbf{b} \ . \tag{19.4}$$

The step response is

$$\int_0^t \mathbf{y}(t)dt = \mathbf{CF}^{-1}e^{\mathbf{F}t}\mathbf{b} \quad |_o^t = \mathbf{CF}^{-1}[\mathbf{\Phi}_{cl}(t) - \mathbf{I}]\mathbf{b} \; . \tag{19.5}$$

Subtraction of the steady-state $-\mathbf{CF}^{-1}\mathbf{b}$ yields $\mathbf{CF}^{-1}\mathbf{\Phi}_{cl}(t)\mathbf{b}$.

19.3 Stability Robustness Measure

Following *Doraiswami, R., 1990* for stable systems, the transition matrix always satisfies

$$\|\mathbf{\Phi}_{cl}(t)\|_s = \sqrt{\lambda_{\max}[\mathbf{\Phi}_{cl}^H(t)\mathbf{\Phi}_{cl}(t)]} \leq m_o e^{-\lambda_o t} \tag{19.6}$$

where m_o and λ_o are positive scalars and $-\lambda_o$ is larger than the spectral abscissa of $\mathbf{A}$. The factor m_o is obtained from

$$m_o = \max_t\{\|\mathbf{\Phi}_{cl}(t)\|_s e^{\lambda_o t}\} \; . \tag{19.7}$$

Solving the closed-loop differential equation

$$\dot{\mathbf{x}} = \mathbf{Fx} + \mathbf{g}(t,\mathbf{x}) \qquad \text{where} \qquad \|\mathbf{g}(t,\mathbf{x})\|_F \leq \beta \;\|\mathbf{x}\|_F \tag{19.8}$$

using Bellman-Gronwall lemma, Eq.(11.19), yields

$$\|\mathbf{x}(t)\| \leq m_o\|\mathbf{x}(0)\|e^{(-\lambda_o + m_o\beta)t} \; . \tag{19.9}$$

Asymptotic stability is given if

$$\lambda_o > m_o\beta \qquad \text{or} \qquad \beta_o \triangleq \lambda_o/m_o > \beta \; . \tag{19.10}$$

The number β_o is considered as a measure of robustness. A similar problem was solved on operator techniques by *Jayasuriya, S., 1984*.

19.4 Relation m_o/λ_o Versus $\mathbf{\Phi}_{cl}(j\omega)$

Parseval's theorem, see Eq.(3.50), for matrix-valued functions is given by

$$\int_0^\infty \mathbf{\Phi}_{cl}^H(t)\mathbf{\Phi}_{cl}(t)dt = \frac{1}{2\pi}\int_0^\infty \mathbf{\Phi}_{cl}^H(j\omega)\mathbf{\Phi}_{cl}(j\omega)d\omega \tag{19.11}$$

where $\mathbf{\Phi}_{cl}(j\omega) = (j\omega\mathbf{I} - \mathbf{F})^{-1} = \mathcal{F}\{\mathbf{\Phi}_{cl}(t)\}$ preassuming that all the entries of $\mathbf{\Phi}_{cl}(t)$ only exist for $t \geq 0$. Taking the square root of the maximum eigenvalue of both sides Eq.(19.11),

$$\sqrt{\lambda_{\max}[\int_0^\infty \mathbf{\Phi}_{cl}^H(t)\mathbf{\Phi}_{cl}(t)dt]} = \sqrt{\lambda_{\max}[\frac{1}{2\pi}\int_0^\infty \mathbf{\Phi}_{cl}^H(j\omega)\mathbf{\Phi}_{cl}(j\omega)d\omega]} \; . \tag{19.12}$$

Since the integrands are positive-semidefinite matrices the relation

$$\sqrt{\lambda_{\max}[\sum_i \mathbf{R}_i]} \leq \sum_i \sqrt{\lambda_{\max}[\mathbf{R}_i]} \qquad\qquad \mathbf{R}_i \geq 0 \tag{19.13}$$

holds and is applied to the left-hand side of Eq.(19.12). Hence,

$$\int_0^\infty \sqrt{\lambda_{\max}[\mathbf{\Phi}_{cl}^H(t)\mathbf{\Phi}_{cl}(t)]}dt \geq \sqrt{\lambda_{\max}[\frac{1}{2\pi}\int_0^\infty \mathbf{\Phi}_{cl}^H(j\omega)\mathbf{\Phi}_{cl}(j\omega)d\omega]} \; . \tag{19.14}$$

Integrating both sides of Eq.(19.6) with respect to t and combining with Eq.(19.14) yields

$$\eta \triangleq \sqrt{\lambda_{\max}[\frac{1}{2\pi}\int_0^\infty \mathbf{\Phi}_{cl}^H(j\omega)\mathbf{\Phi}_{cl}(j\omega)d\omega]} \leq \int_0^\infty \sqrt{\lambda_{\max}[\mathbf{\Phi}_{cl}^H(t)\mathbf{\Phi}_{cl}(t)]}dt \leq m_o \int_0^\infty e^{-\lambda_o t}dt = \frac{m_o}{\lambda_o}. \quad \square \tag{19.15}$$

19.5 Relation of Robustness Measure β_o Versus $\|\mathbf{\Phi}_{cl}(\omega)\|_\infty$

From Eq.(19.13) applied to the matrix $\mathbf{R}_i = \mathbf{\Phi}_{cl}^H(j\omega)\mathbf{\Phi}_{cl}(j\omega)$

$$\sqrt{\lambda_{\max}[\int_{-\infty}^\infty \mathbf{\Phi}_{cl}^H(j\omega)\mathbf{\Phi}_{cl}(j\omega)d\omega]} \leq \int_{-\infty}^\infty \sqrt{\lambda_{\max}[\mathbf{\Phi}_{cl}^H(j\omega)\mathbf{\Phi}_{cl}(j\omega)]}d\omega = \int_{-\infty}^\infty \|\mathbf{\Phi}_{cl}(j\omega)\|_s d\omega \; . \; \square \tag{19.16}$$

Rewrite Eq.(19.15) with the definition of β_o as given in Eq.(19.10). Then, comparison with Eq.(19.16) yields

$$\frac{1}{\sqrt{2\pi}}\int_{-\infty}^\infty \|\mathbf{\Phi}_{cl}(j\omega)\|_s d\omega > \eta < \frac{m_o}{\lambda_o} \triangleq \frac{1}{\beta_o} \; . \tag{19.17}$$

The relation cited above shows the *loose relation* between $1/\beta_o$ and $\|\mathbf{\Phi}_{cl}(j\omega)\|_s$ and, furthermore, with the H_∞-norm $\|\mathbf{\Phi}_{cl}(j\omega)\|_\infty$.

19.6 Lyapunov Approach

Consider the algebraic Lyapunov equation $\mathbf{F}^T\mathbf{P} + \mathbf{P}\mathbf{F} = -\mathbf{Q} = -2\mathbf{I}$ for some $\mathbf{P} > 0$ and $\mathbf{Q} > 0$ and its solution, see Eq.(13.16),

$$\mathbf{P} = 2\int_0^\infty e^{\mathbf{F}^T\tau}e^{\mathbf{F}\tau}d\tau \tag{19.18}$$

$$\begin{aligned} \lambda_{\max}[\mathbf{P}] &\leq 2\int_0^\infty \lambda_{\max}[e^{\mathbf{F}^T\tau}e^{\mathbf{F}\tau}]d\tau = 2\int_0^\infty \lambda_{\max}[\mathbf{\Phi}_{cl}^H(\tau)\mathbf{\Phi}_{cl}(\tau)]d\tau & (19.19) \\ &= 2\int_0^\infty \|\mathbf{\Phi}_{cl}(\tau)\|_s^2 d\tau \leq 2m_o^2 \int_0^\infty e^{-2\lambda_o\tau}d\tau = \frac{m_o^2}{\lambda_o} & (19.20) \end{aligned}$$

where Eq.(19.6) has been employed. Comparison of $\beta_o > \beta$ from Eq.(19.10) with Eq.(19.20) yields

$$\beta_o \triangleq \frac{\lambda_o}{m_o} < \frac{m_o}{\lambda_{\max}[\mathbf{P}]} \; . \tag{19.21}$$

Since β_o is overestimated by the inequality above, the condition $m_o/\lambda_{\max}[\mathbf{P}] > \beta$ is only a necessary one.

19.7 Relation Cost Function and Robustness Measure

Referring to Eq.(19.9), the quadratic cost function is related to the robustness measure β_o as follows

$$\int_0^\infty \mathbf{x}^T(\tau)\mathbf{x}(\tau)d\tau = \int_0^\infty \|\mathbf{x}(\tau)\|_F^2 d\tau \leq m_o^2\|\mathbf{x}(0)\|_F^2 \int_0^\infty e^{2(-\lambda_o+m_o\beta)\tau}\, d\tau \quad (19.22)$$

$$= \frac{m_o^2\|\mathbf{x}(0)\|_F^2}{2(\lambda_o - m_o\beta)} = \frac{m_o\|\mathbf{x}(0)\|_F^2}{2(\beta_o - \beta)} \ . \quad (19.23)$$

19.8 Ideal Controller

Consider the measure for the overshoot

$$M_{\max} = \max_t \|\mathbf{\Phi}_{cl}(t)\|_s = \max_t \sqrt{\lambda_{\max}[e^{\mathbf{F}^T t} e^{\mathbf{F}t}]} \ . \quad (19.24)$$

If $\mathbf{F}$ is normal and obeys the Lyapunov equation

$$\mathbf{F}^T + \mathbf{F} = -\mathbf{Q} \quad (19.25)$$

for $\mathbf{Q} > 0$, then

$$M_{\max} = \max_t \sqrt{\lambda_{\max}[e^{(\mathbf{F}^T+\mathbf{F})t}]} = \max_t \sqrt{\lambda_{\max}[e^{-\mathbf{Q}t}]} = \max_t \sqrt{e^{\lambda_{\max}[-\mathbf{Q}]t}} = \max_t e^{\lambda_{\max}[-\mathbf{Q}]t/2} \ . \quad (19.26)$$

If $\mathbf{F}$ is normal and $t_{\max}$ is replaced by 0, for simplification, then $m_o = M_{\max} = 1$.

The speed of the control system is considered given by λ_o. For a given λ_o the robustness measure $\beta_o = \lambda_o/m_o$ is maximum when m_o is minimum. The minimum value is $m_o = 1$. Hence, a controller is an ideal one if it guarantees the closed-loop system matrix being normal.

19.9 Algorithm for Optimum Envelope

The optimum envelope of the transient behaviour is given by choosing m_o, λ_o such that minimum area $\int_0^\infty m_o e^{-\lambda_o\tau} d\tau$ is obtained and both $\|\mathbf{\Phi}_{cl}(t)\|_s \leq m_o e^{-\lambda_o t}$ and $m_o = \max_t\{\|\mathbf{\Phi}_{cl}(t)\| e^{\lambda_o t}\}$ are satisfied. Minimum area requires $m_o/\lambda_o \rightarrow$ min. Using the constraints above, the minimum area versus λ is achieved by

$$\lambda_o = \arg\min_\lambda \frac{m_o}{\lambda} = \min_\lambda\{\max_t[\ \|\mathbf{\Phi}_{cl}(t)\|_s e^{\lambda t}]/\lambda\} \ . \quad (19.27)$$

Chapter 20

L_p^n-Stability and Robust Nonlinear Control

Stability and stability robustness are important properties of control systems. Hence, particular attention is paid to this topic. However, control system performance, disturbance rejection and tracking facilities (*Jayasuriya, S., 1987*) may not be neglected. Function spaces and L_p^n-norms yield a very useful tool both for investigating stability and disturbance rejection and tracking problems in the face of uncertainties.

Operators are utilized throughout this chapter. An operator, e.g. $\mathbf{G}^{op}$, indicates a general operation to be executed with a variable following the operator. Examples for an operator are: nonlinear characteristics, fields of characteristics, general mapping, truncation, matrix multiplication, data transmission via continuous-time and discrete-time systems in parallel.

20.1 Robust L_2^n-Stability

20.1.1 Time-Varying Nonlinear Uncertainty

The system under investigation is depicted in Fig. 20.1. The plant including uncertainty is structured as follows

$$\begin{aligned} \mathbf{G}_p^{op} &= \mathbf{G}^{op} + \Delta\mathbf{G}^{op} && \text{additive} && (20.1)\\ \text{or } \mathbf{G}_p^{op} &= \mathbf{G}^{op}(\mathbf{I} + \Delta\mathbf{L}_i) && \text{input associated multiplicative} && (20.2)\\ \text{or } \mathbf{G}_p^{op} &= (\mathbf{I} + \Delta\mathbf{L}_o)\mathbf{G}^{op} && \text{output associated multiplicative .} && (20.3) \end{aligned}$$

Let the uncertainty $\Delta\mathbf{G}^{op}$ be time-varying $\Delta\mathbf{G}^{op} = \Delta\mathbf{G}^{op}(t)$. Consider the uncertainty being bounded by Eq.(3.3) for $p = 2$

$$\begin{aligned} \|\Delta\mathbf{G}^{op}\mathbf{u}(t)\|_2 &\le \rho_1\|\mathbf{u}(t)\|_2 \text{ absolutely} && (20.4)\\ \text{or } \|\Delta\mathbf{G}^{op}\mathbf{u}_2(t) - \Delta\mathbf{G}^{op}\mathbf{u}_1(t)\|_2 &\le \rho_2\|\mathbf{u}_2(t) - \mathbf{u}_1(t)\|_2 \text{ incrementally .} && (20.5) \end{aligned}$$

20.1.2 Robust Stability Conditions

A system is closed-loop L_2^n-stable if the norm of input and output (see Fig. 20.2) are related as follows

$$\begin{aligned} \|\mathbf{y}\|_2 &\le \alpha_1\|\mathbf{y}_{ref}\|_2 + c_1 && \forall\, \mathbf{y}_{ref} \in L_2^n && (L_2^n\text{-stable}) && (20.6)\\ \|\mathbf{y}_2 - \mathbf{y}_1\|_2 &\le \alpha\|\mathbf{y}_{ref2} - \mathbf{y}_{ref1}\|_2 && \forall \mathbf{y}_{ref1}, \mathbf{y}_{ref2} \in L_2^n && (\text{incrementally } L_2^n\text{-stable}) && (20.7)\\ \|\mathbf{u}\|_2 &\le \beta_1\|\mathbf{y}_{ref}\|_2 + d_1 && \forall\, \mathbf{y}_{ref} \in L_2^n && (L_2^n\text{-stable}). && (20.8) \end{aligned}$$

Eq.(20.8) presents an example for a bounded control variable.

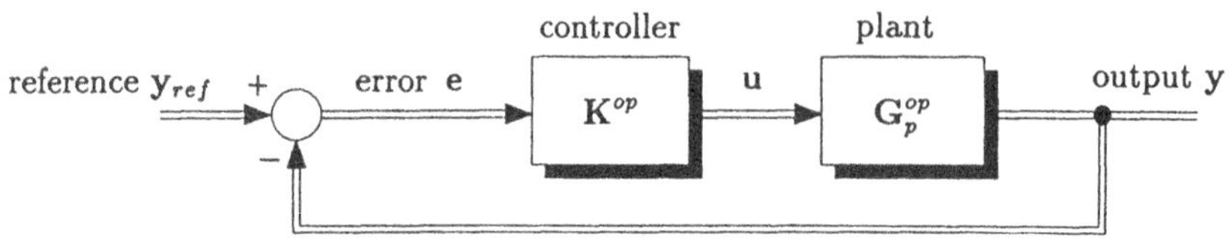

Figure 20.1: Multivariable control system

20.1.3 Robust Stability Assessment

Suppose the nominal closed-loop transfer matrix **KS** is asymptotically stable where $\mathbf{S}(s)$ is the sensitivity matrix in the nominal feedback system, i.e. $\mathbf{S}(s) = [\mathbf{I} + \mathbf{G}(s)\mathbf{K}(s)]^{-1}$. Considering $\mathbf{u} = \mathbf{K}\mathbf{e}$ and $\Delta\mathbf{G} = \mathbf{G}_p - \mathbf{G}$ and extending the case of additive uncertainty yields

$$\begin{aligned}
\mathbf{u} &= \mathbf{K}\mathbf{e} = \mathbf{K}(\mathbf{I}+\mathbf{G}_p\mathbf{K})^{-1}\mathbf{y}_{ref} && (20.9)\\
&= \mathbf{K}\{(\mathbf{I}+\mathbf{G}\mathbf{K})^{-1}[(\mathbf{I}+\mathbf{G}\mathbf{K})-(\mathbf{I}+\mathbf{G}_p\mathbf{K})](\mathbf{I}+\mathbf{G}_p\mathbf{K})^{-1}+(\mathbf{I}+\mathbf{G}\mathbf{K})^{-1}\}\mathbf{y}_{ref} \\
&= \mathbf{K}\{(\mathbf{I}+\mathbf{G}\mathbf{K})^{-1}[-\Delta\mathbf{G}\ \mathbf{K}](\mathbf{I}+\mathbf{G}_p\mathbf{K})^{-1}+(\mathbf{I}+\mathbf{G}\mathbf{K})^{-1}\}\ \mathbf{y}_{ref} && (20.10)\\
\mathbf{u} &= -\mathbf{K}\mathbf{S}\ \Delta\mathbf{G}\ \mathbf{K}(\mathbf{I}+\mathbf{G}_p\mathbf{K})^{-1}\mathbf{y}_{ref}+\mathbf{K}\mathbf{S}\mathbf{y}_{ref} = -\mathbf{K}\mathbf{S}\ \Delta\mathbf{G}\ \mathbf{u}+\mathbf{K}\mathbf{S}\mathbf{y}_{ref}\ . && (20.11)
\end{aligned}$$

Recall the fact that the transfer matrix and its spectral norm are required to obtain the relation between function norms on L_2^n, see Eq.(20.90). Using Eq.(20.101) or (20.90), the function norm $\|\mathbf{u}\|_2$ is bounded by (*Chen, B.S., and Chou, H.H., 1988; Chiang, C.C., and Chen, B.S., 1988*)

$$\|\mathbf{u}\|_2 \le \left(\sup_\omega \sigma_{\max}[\mathbf{K}(j\omega)\mathbf{S}(j\omega)]\right)\|\Delta\mathbf{G}\ \mathbf{u}\|_2 + \left(\sup_\omega \sigma_{\max}[\mathbf{K}(j\omega)\mathbf{S}(j\omega)]\right)\|\mathbf{y}_{ref}\|_2\ . \quad (20.12)$$

Referring to a given bound $\|\Delta\mathbf{G}(t)\mathbf{u}(t)\|_2 \le \rho_1\|\mathbf{u}(t)\|_2$ and supposing

$$\sup_\omega \sigma_{\max}[\mathbf{K}(j\omega)\mathbf{S}(j\omega)] \le \alpha_3\ , \quad (20.13)$$

it follows

$$\|\mathbf{u}\|_2 \le \alpha_3\rho_1\|\mathbf{u}\|_2+\alpha_3\|\mathbf{y}_{ref}\|_2 \quad \rightsquigarrow \quad \|\mathbf{u}\|_2 \le (1-\alpha_3\rho_1)^{-1}\alpha_3\|\mathbf{y}_{ref}\|_2 \ \text{ if } \ \alpha_3\rho_1<1\ . \quad (20.14)$$

Thus, the L_2^n-stability condition Eq.(20.8) is complied with if

$$\beta_1 = (1-\alpha_3\rho_1)^{-1}\alpha_3, \qquad d_1 = 0 \qquad \text{and} \qquad \alpha_3\rho_1 < 1\ . \quad (20.15)$$

Assume β_1 and ρ_1 given then

$$\sup_\omega \sigma_{\max}[\mathbf{K}(j\omega)\mathbf{S}(j\omega)] \le \alpha_3 = \frac{\beta_1}{1+\beta_1\rho_1} < 1/\rho_1\ . \quad (20.16)$$

In a similar way, bounds can be given on $\|\mathbf{y}\|_2$ versus $\|\mathbf{y}_{ref}\|_2$ or incrementally L_2^n-stable feedback systems are treated. If the multiplicative type of nonlinear uncertainty is taken into consideration results are obtained analogously to the above derivations applying the closed-loop sensitivity matrix $\mathbf{S}_o(s) = [\mathbf{I}+\mathbf{K}(s)\mathbf{G}(s)]^{-1}$. Incremental L_2^n-stability is given if

$$\sup_\omega \sigma_{\max}[\mathbf{I}-\mathbf{S}_o(j\omega)] < 1/\rho_2\ . \quad (20.17)$$

As shown by *Chou, H.H., et al. 1990* the criterion is sufficient *and* necessary.

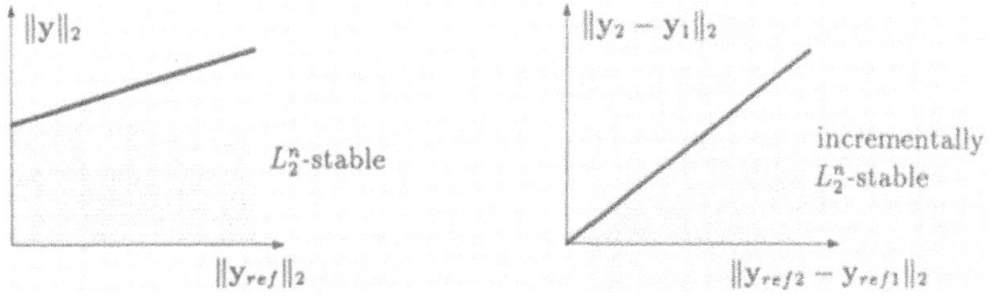

Figure 20.2: Relation between norms L_2^n

20.1.4 Robust L_2^n-Stabilizer Design

Continuing Eq.(20.16) and introducing $\mathbf{KS} = (\mathbf{Y} + \mathbf{A}_1\mathbf{Z})\mathbf{A}$ in parametrized form as given by Eq.(30.65), one has

$$\sup_\omega \sigma_{\max}[\mathbf{K}(j\omega)\mathbf{S}(j\omega)] = \sup_\omega \sigma_{\max}[\{\mathbf{Y}(j\omega) + \mathbf{A}_1(j\omega)\mathbf{Z}(j\omega)\}\mathbf{A}(j\omega)] < 1/\rho_1 \qquad (20.18)$$

where $\mathbf{A}, \mathbf{A}_1, \mathbf{B}, \mathbf{B}_1, \mathbf{X}, \mathbf{Y}$ and $\mathbf{Z}$ are rational matrices satisfying

$$\mathbf{G} = \mathbf{A}^{-1}\mathbf{B} = \mathbf{B}_1\mathbf{A}_1^{-1} \quad \text{and} \quad \mathbf{AX} + \mathbf{BY} = \mathbf{I}\,. \qquad (20.19)$$

With the definition

$$\mathbf{\Omega}(j\omega) \triangleq \rho_1\{\mathbf{Y}(j\omega) + \mathbf{A}_1(j\omega)\mathbf{Z}(j\omega)\}\mathbf{A}(j\omega) \qquad (20.20)$$

the stability condition is rewritten to

$$\sigma_{\max}[\mathbf{\Omega}(j\omega)] < 1 \quad \forall\, \omega\,. \qquad (20.21)$$

An equivalent condition (*Chiang, C.C., and Chen, B.S., 1988*), see Eq.(2.26), is

$$\mathbf{I} - \mathbf{\Omega}^H(j\omega)\mathbf{\Omega}(j\omega) > 0 \quad \text{(i.e. positive definite)}\,. \qquad (20.22)$$

The design algorithm has to be executed in this order: Solve $\mathbf{A}, \mathbf{B}, \mathbf{A}_1, \mathbf{B}_1, \mathbf{X}, \mathbf{Y}$ from Eq.(20.19). The denominator of $\mathbf{Z}$ can be determined by assigning of the closed-loop poles. The numerator of $\mathbf{Z}$ is chosen to comply with the robustness standards.

20.2 Weak L_p^n-Stability

Consider a system with an operator $\mathbf{G}^{op}$ and the input $\mathbf{u}$ given. If the input $\mathbf{u}$ and output $\mathbf{y}$ belong to L_p^n and a relation

$$\|\mathbf{y}\|_p = g[\ \|\mathbf{u}\|_p\ ] \qquad (20.23)$$

can be found the system is said to be weakly L_p^n-stable.

Weak L_∞^n-Stability is given if functions $\mathbf{f}$ on L_∞^n are considered, only. Each component $f_i(t)$ of $\mathbf{f}(t)$ is bounded in the sense of $\sup|f_i(t)|$. If a relation $g[\cdot]$ in the case $p = \infty$ can be found in Eq.(20.23) then this is one of the cases where the system is denoted as bounded input-bounded output stable.

20.3 L_p^n-Stability

L_p^n-stability is given if the function $g[\cdot]$ can be replaced by a linear relation between the input and output norm using the constant c

$$\|\mathbf{y}\|_p \le c\ \|\mathbf{u}\|_p \qquad \mathbf{y},\mathbf{u} \in L_p^n\ . \tag{20.24}$$

If a system $\mathbf{y} = \mathbf{G}^{op}\mathbf{u}$ with the mapping function (operator) $\mathbf{G}^{op}$ is L_p^n-stable the smallest number c can be considered the operator norm or the operator bound (*Sandberg, I.W., 1964; Ludyk, G., 1985*).

20.4 Small Gain Theorem

The multivariable closed-loop system as depicted in Fig. 20.4 can be described in operator form

$$\mathbf{e} = \mathbf{y}_{ref} - \mathbf{y} \tag{20.25}$$

$$\mathbf{u} = \mathbf{w} + \mathbf{N}^{op}\mathbf{e} \tag{20.26}$$

$$\mathbf{y} = \mathbf{G}^{op}\mathbf{u}\ . \tag{20.27}$$

Under the assumption of causal operators $\mathbf{G}^{op}$, $\mathbf{N}^{op} \in L_{pe}^n$ and under the limitations

$$\|(\mathbf{N}^{op}\mathbf{e})_T\|_p \le \gamma_N\|\mathbf{e}_T\|_p + \beta_N \qquad \|(\mathbf{G}^{op}\mathbf{u})_T\|_p \le \gamma_G\|\mathbf{u}_T\| + \beta_G \tag{20.28}$$

including gain terms γ_G and γ_N (Fig. 20.3) the system equations Eqs.(20.25) to (20.27) are solved. The output $\mathbf{y}$ is

$$\mathbf{y} = \mathbf{G}^{op}(\mathbf{w} + \mathbf{N}^{op}\mathbf{e}) = \mathbf{G}^{op}\mathbf{w} + \mathbf{G}^{op}\mathbf{N}^{op}(\mathbf{y}_{ref} - \mathbf{y})\ . \tag{20.29}$$

The norm of its truncated modification $\mathbf{y}_T$ in L_{pe}^n is given by

$$\|\mathbf{y}_T\|_p = \gamma_G\|\mathbf{w}_T\|_p + \beta_G + \gamma_G\|\mathbf{N}^{op}(\mathbf{y}_{ref} - \mathbf{y})_T\|_p \tag{20.30}$$

$$= \gamma_G\|\mathbf{w}_T\|_p + \beta_G + \gamma_G[\gamma_N\|(\mathbf{y}_{ref} - \mathbf{y})_T\|_p + \beta_N]\ . \tag{20.31}$$

Assuming $\gamma_G\gamma_N < 1$, weak L_p^n-stability is given and

$$\|\mathbf{y}_T\|_p \le \gamma_G\|\mathbf{w}_T\|_p + \beta_G + \gamma_G\gamma_N\|\mathbf{y}_{ref\ T}\|_p + \gamma_G\gamma_N\|\mathbf{y}_T\|_p + \gamma_G\beta_N \tag{20.32}$$

$$\|\mathbf{y}_T\|_p \le (1 - \gamma_G\gamma_N)^{-1}(\gamma_G\gamma_N\|\mathbf{y}_{ref\ T}\|_p + \gamma_G\|\mathbf{w}_T\|_p + \beta_G + \beta_N\gamma_G)\ . \tag{20.33}$$

If $\|\mathbf{y}_{ref}\| < \infty$ and $\|\mathbf{w}\| < \infty$ the truncation of $\mathbf{y}_T$ within the norm is no more essential and can be omitted. The solution $\mathbf{y}$ is $\in L_p^n$ and a norm $\|\mathbf{y}\|_p$ exists (*Zames, G., 1966; Wu, M.Y., and Desoer, C.A., 1969*).

In a similar way, dissolving with respect to $\mathbf{e}$ yields

$$\mathbf{e} = \mathbf{y}_{ref} - \mathbf{G}^{op}\mathbf{u} = \mathbf{y}_{ref} - \mathbf{G}^{op}(\mathbf{w} + \mathbf{N}^{op}\mathbf{e}) \tag{20.34}$$

$$\|\mathbf{e}\|_p = (1 - \gamma_G\gamma_N)^{-1}(\|\mathbf{y}_{ref}\|_p + \gamma_G\|\mathbf{w}\|_p + \beta_G + \beta_N\gamma_G)\ . \tag{20.35}$$

Separating $\mathbf{u}$, it results

$$\mathbf{u} = \mathbf{w} + \mathbf{N}^{op}\mathbf{y}_{ref} - \mathbf{N}^{op}\mathbf{G}^{op}\mathbf{u} \tag{20.36}$$

$$\|\mathbf{u}\|_p = (1 - \gamma_G\gamma_N)^{-1}(\gamma_N\|\mathbf{y}_{ref}\|_p + \|\mathbf{w}\|_p + \beta_N + \beta_G\gamma_N)\ . \tag{20.37}$$

The aforementioned equations are known as the solution of the *Small gain theorem*. This theorem holds in the case of continuous-time and discrete-time systems and in the case of single-input single-output systems as well as in the case of multiple-input multiple-output systems and in robust stabilization of additively perturbed plants (*MacCluer, C.R., and Chait, Y., 1990*).

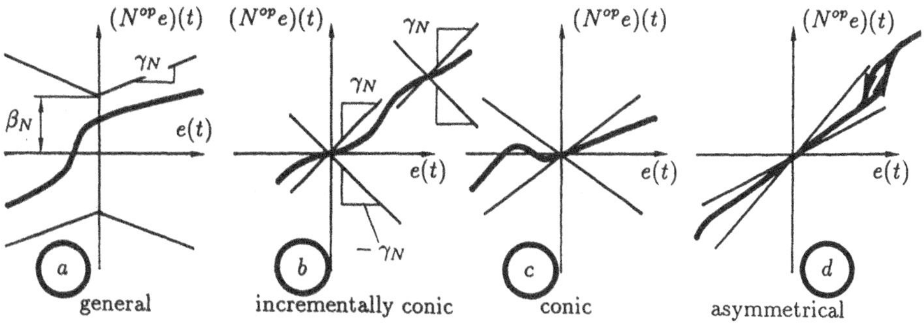

Figure 20.3: Nonlinear characteristics and associated bounds

20.5 Small Gain Theorem With Incrementally Conic Nonlinearities

With the assumptions of a linear operator $\mathbf{G}^{op}$

$$\|\mathbf{G}^{op}\mathbf{u}\|_p \leq \gamma_G \|\mathbf{u}\|_p \quad \text{or} \quad \|\mathbf{G}^{op}\|_p \leq \gamma_G \qquad \mathbf{u} \in L_p^n \tag{20.38}$$

and an incrementally conic nonlinearity , i.e.,

$$\mathbf{N}^{op}\mathbf{0} = \mathbf{0} \tag{20.39}$$

$$\|\mathbf{N}^{op}\mathbf{e}_1 - \mathbf{N}^{op}\mathbf{e}_2\|_p \leq \gamma_N \|\mathbf{e}_1 - \mathbf{e}_2\|_p \qquad \mathbf{e}_1,\ \mathbf{e}_2 \in L_p^n\ , \tag{20.40}$$

the results for the Small gain theorem in incremental form can be deduced as (*Böcker, J., et al. 1986*)

$$\|\Delta\mathbf{y}\|_p \;\leq\; (1-\gamma_G\gamma_N)^{-1}(\gamma_G\gamma_N\|\Delta\mathbf{y}_{ref}\|_p + \gamma_G\|\Delta\mathbf{w}\|_p) \tag{20.41}$$

$$\|\Delta\mathbf{e}\|_p \;\leq\; (1-\gamma_G\gamma_N)^{-1}(\|\Delta\mathbf{y}_{ref}\|_p + \gamma_G\|\Delta\mathbf{w}\|_p) \tag{20.42}$$

$$\|\Delta\mathbf{u}\|_p \;\leq\; (1-\gamma_G\gamma_N)^{-1}(\gamma_N\|\Delta\mathbf{y}_{ref}\|_p + \|\Delta\mathbf{w}\|_p)\ . \tag{20.43}$$

In the incremental form either existence, uniqueness, boundedness (stability) and continuous dependence are guaranteed.

In an abbreviated manner of representation: Specializing the results of Eq.(20.33) to $\beta_G = \beta_N = 0$ yields stability results in L_p^n given by Eq.(20.41) etc.

20.6 Exponential L_p^n-Stability

Multiplying the output function $\mathbf{y}$ by an increasing exponential function $e^{\alpha t}(\alpha > 0)$, a new (transformed) function denoted as $\mathbf{y}_\alpha$ is obtained. If the norm of the "additionally excited" signal $\|\mathbf{y}_\alpha\|_p$ exists this indicates that the original signal $\mathbf{y}$ converges better than $e^{-\alpha t}$. Next, multiplication by the scalar $e^{\alpha t}$ is substituted by an additional operator $\mathbf{P}_\alpha^{op}$, i.e.,

$$\mathbf{y}_\alpha \triangleq \mathbf{P}_\alpha^{op}\mathbf{y} \triangleq \mathbf{y}e^{\alpha t} \tag{20.44}$$

$$\|\mathbf{y}_\alpha\|_p = [\sum_{i=1}^{n} \int_R |y_i(t)e^{\alpha t}|^p \, dt]^{1/p}\ . \tag{20.45}$$

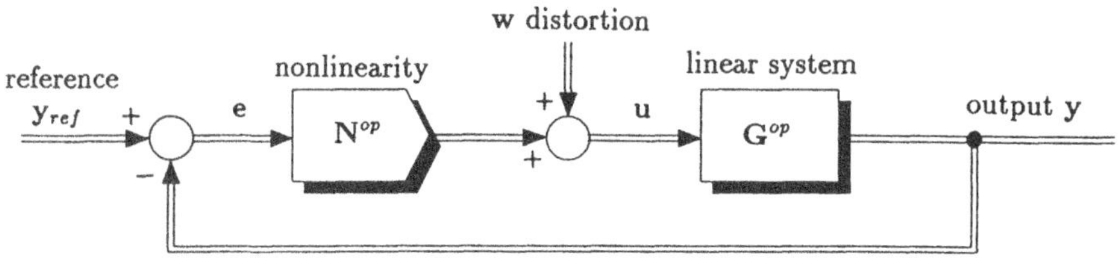

Figure 20.4: Nonlinear multivariable control system

The operator $\mathbf{P}_\alpha^{op}$ can be applied to each signal with the same notation. Note that the same system and the same input signal is maintained. The system equations are multiplied by $e^{\alpha t}$, thus yielding a transformed system the signals of which (labelled by a subscript α) correspond to the original signals subject to a multiplication operation. Then, the multiplication with $e^{\alpha t}$ or application of the operator $\mathbf{P}_\alpha^{op}$ to each equation of the control system yields

$$\left.\begin{aligned} \mathbf{e} &= \mathbf{y}_{ref} - \mathbf{y} \\ \mathbf{u} &= \mathbf{N}^{op}\mathbf{e} + \mathbf{w} \\ \mathbf{y} &= \mathbf{G}^{op}\mathbf{u} \end{aligned}\right\} \quad \leadsto \quad \left\{\begin{aligned} \mathbf{e}_\alpha &= \mathbf{y}_{ref\alpha} - \mathbf{y}_\alpha \\ \mathbf{u}_\alpha &= \mathbf{P}_\alpha^{op}\mathbf{N}^{op}\mathbf{e} + \mathbf{w}_\alpha \\ \mathbf{y}_\alpha &= \mathbf{P}_\alpha^{op}\mathbf{G}^{op}\mathbf{u} \ . \end{aligned}\right. \tag{20.46}$$

Referring to the simple relation

$$\mathbf{y}_{ref\alpha} = \mathbf{y}_{ref} e^{\alpha t} \quad \leadsto \quad \mathbf{y}_{ref} = \mathbf{y}_{ref\alpha}\, e^{-\alpha t} = \mathbf{P}_{-\alpha}^{op}\, \mathbf{y}_{ref\alpha} \quad \leadsto \quad \mathbf{P}_\alpha^{op}\mathbf{P}_{-\alpha}^{op} = 1 \tag{20.47}$$

and using throughout transformed variables, $\mathbf{e}_\alpha$, $\mathbf{u}_\alpha$, $\mathbf{y}_\alpha$, $\mathbf{y}_{ref\alpha}$, the equations can be rewritten to a transformed system

$$\mathbf{e}_\alpha = \mathbf{y}_{ref\alpha} - \mathbf{y}_\alpha \tag{20.48}$$

$$\mathbf{u}_\alpha = \mathbf{P}_\alpha^{op}\mathbf{N}^{op}\mathbf{P}_{-\alpha}^{op}\, \mathbf{e}_\alpha + \mathbf{w}_\alpha \triangleq \mathbf{N}_\alpha^{op}\mathbf{e}_\alpha + \mathbf{w}_\alpha \tag{20.49}$$

$$\mathbf{y}_\alpha = \mathbf{P}_\alpha^{op}\mathbf{G}^{op}\mathbf{P}_{-\alpha}^{op}\, \mathbf{u}_\alpha \triangleq \mathbf{G}_\alpha^{op}\mathbf{u}_\alpha \ . \tag{20.50}$$

New operators $\mathbf{N}_\alpha^{op}$ and $\mathbf{G}_\alpha^{op}$ have been defined. Hence, the same structure was achieved as the original system Eqs.(20.25) through (20.27) (*Ljusternik, L.A., and Sobolew, W.I., 1979*). A general concept of exponential rate of convergence for uncertain nonlinear systems is presented by *Corless, M., 1989*.

20.7 Simple L_2-Stability and Circle Criterion

Assume N^{op} and G^{op} complies with the conditions

$$|(N^{op}e)| < \gamma_N |e(t)| \qquad (\gamma_N > 0) \tag{20.51}$$

$$\|y\|_2 = \|G^{op}u\|_2 \le \gamma_G \|u\|_2 \quad \text{where} \quad \|G^{op}\|_2 \le \gamma_G > 0 \ . \tag{20.52}$$

The L_2-stability of the system as depicted in Fig. 20.4 is presented by a method applying the well-known transfer function. Suppose the region R both in time domain and in frequency domain is the interval $(-\infty, +\infty)$. Within R the functions are defined. Using Parseval theorem yields $(\|y\|_2)^2 = \frac{1}{2\pi}(\|\hat{y}\|_2)^2$. Applying the Fourier transformation to the linear part $\hat{y} = \mathcal{F}y = G(j\omega)\hat{u}(j\omega)$ and calculating the norm of the second power Eq.(3.18) yields

$$(\|y\|_2)^2 = \frac{1}{2\pi}(\|G(j\omega)\hat{u}(j\omega)\|_2)^2 = \frac{1}{2\pi}\|G^2(j\omega)\hat{u}^2(j\omega)\|_1 \, . \tag{20.53}$$

Applying Hölder equation Eq. (3.20) and Eq.(3.18), $(\|y\|_2)^2$ can be rewritten as

$$(\|y\|_2)^2 \leq \frac{1}{2\pi}\|G^2(j\omega)\|_\infty \; \|\hat{u}^2(j\omega)\|_1 = \frac{1}{2\pi}\|G^2(j\omega)\|_\infty(\|\hat{u}(j\omega)\|_2)^2 = \|G^2(j\omega)\|_\infty(\|u\|_2)^2 \tag{20.54}$$

$$\text{or} \qquad \|y\|_2 \leq \|G(j\omega)\|_\infty \|u\|_2 \, . \tag{20.55}$$

It is well known that the Hölder inequality is the smallest estimate. Referring to Eq.(3.39), the operator norm $\|G^{op}\|_p$ is also known as the smallest scalar. Hence,

$$\|G^{op}\|_2 = \|G(j\omega)\|_\infty = \sup_\omega |G(j\omega)| \, . \tag{20.56}$$

As outlined in Eq.(20.24), a system is stable if a constant number c can be found which gives a fixed relation between output and input norm. From Small gain theorem it is known that this is true if $\gamma_G\gamma_N < 1$. With the help of Eq.(20.52), γ_G is given by $\|G^{op}\|$

$$\gamma_G\gamma_N < 1 \qquad \leadsto \qquad \|G^{op}\|_2\gamma_N < 1 \, . \tag{20.57}$$

Using Eq. (20.56) yields

$$\sup_\omega |G(j\omega)| < \frac{1}{\gamma_N} \, . \tag{20.58}$$

The graphic interpretation of Eq. (20.58) is referred to as Circle criterion (*Zames, G., 1966; Vidyasagar, M., 1978*).

Eq.(20.58) can be considered as a robustness condition for variations occurring in the plant or in the nonlinearity. Note that only the supremum on ω and the sector gain of the conic nonlinearity determine the L_2-stability condition. It must be emphasized that the L_2-stability statement is based on the integral of squared signals but only the supremum versus ω of the modulus of $G(j\omega)$ is taken into consideration by the stability criterion.

In the case of discrete-time systems the transfer function $G(j\omega)$ has to be replaced by the corresponding z-transfer function $G_z(z)$. This function is written abbreviated as $G(z)$ where $z = \exp(j\omega T)$ with the sampling period T.

20.8 Exponential L_2-Stability and Circle Criterion

In the case of $n = 1$ and $p = 2$ the results of exponential L_p^n- stability are specialized. If $\alpha = 0$ there results $\|G^{op}\|_2 = \|G(j\omega)\|_\infty$ where $G(j\omega)$ relates the Fourier transforms of input and output. Now, an interrelation between the $e^{\alpha t}$-multiplied input $e^{\alpha t}u$ to the multiplied output $e^{\alpha t}y$ is required. From Laplace and Fourier transformation it is known that multiplication with $e^{\alpha t}$ in t-domain ($\alpha > 0$) corresponds to a shifting operation in the s-domain, shifting to the right in the direction of real axis with an amount α. Hence, the multiplied signals in s-domain are given by $\hat{u}(j\omega - \alpha)$ and $\hat{y}(j\omega - \alpha)$, respectively. The quotient is $G(j\omega - \alpha)$ and

$$\|G_\alpha^{op}\|_\infty = \|P_\alpha^{op}G^{op}P_{-\alpha}^{op}\| = \sup_\omega |G(j\omega - \alpha)| \, . \tag{20.59}$$

Studying the sector condition of the nonlinear element, the result turns out quite different although the notation N_α^{op} looks very similar to G_α^{op}. This fact will be shown by the following calculations. The sector condition in the case of multiplied signals u_α and y_α can be derived from the original sector condition

$$(N^{op}e)(t) \le \gamma_N e(t)\ . \tag{20.60}$$

Multiplying by $e^{\alpha t}$, there results

$$P_\alpha^{op} N^{op} e = P_\alpha^{op} N^{op} P_{-\alpha}^{op}\ e_\alpha = N_\alpha^{op} e_\alpha \le e^{\alpha t}\gamma_N e = e^{\alpha t}\gamma_N e^{-\alpha t} e_\alpha = \gamma_N e_\alpha \tag{20.61}$$

$$\text{or} \qquad N_\alpha^{op} e_\alpha \le \gamma_N e_\alpha \tag{20.62}$$

which, in fact, is the same relation as in the simple L_2-stability case. Finally, the system of Fig. 20.4 is exponentially L_2-stable if

$$\sup_\omega |G(j\omega - \alpha)| < 1/\gamma_N \tag{20.63}$$

where α may be considered a stability degree. Simple L_2-stability is obtained with $\alpha = 0$.

From linear theory it follows that the Nyquist plot $G(j\omega - \alpha)$ is "enlarged" in comparison with $G(j\omega)$ which reveals that the exponential L_2-stability is more difficult to satisfy than simple L_2-stability. A graphic interpretation can easily be given by the Circle criterion.

For incrementally conic nonlinearities the sufficient condition of the Circle criterion can be modified. The center of the circle in such cases need not lie on the real axis (*Cho, Y.S., and Narendra, K.S., 1968*).

20.9 Asymmetric Nonlinearity Bounds and L_2-Stability

In the field of industrial applications, nonlinear characteristics with the wide range of a conic nonlinearity are not very frequent. Symmetric characteristics are comfortable to calculate but rather unrealistic. In the case of Fig. 20.3a and c, even the control action with opposite direction is admissible. If the characteristic bounds are spread very far, only weak stability statements can be expected. These conditions are sufficient conditions but in most cases are too conservative for industrial purposes. Asymmetric nonlinearity bounds are suitable for industrial components including uncertainties thus yielding applicable design methods (*Sandberg, I.W., 1964*).

Asymmetric nonlinearities are bounded by the following inequalities

$$\gamma_{N\min}\ e(t) < (Ne)(t) < \gamma_{N\max}\ e(t) \tag{20.64}$$

$$\text{where} \quad \gamma_{Nm} \triangleq 0.5(\gamma_{N\max} + \gamma_{N\min}) \quad \text{and} \quad \gamma_{Nr} \triangleq 0.5(\gamma_{N\max} - \gamma_{N\min})\ . \tag{20.65}$$

Defining an average gain γ_{Nm} and a gain range $2\gamma_{Nr}$, the original system of Fig.20.4 can be transposed into a system with symmetric nonlinearity N_{sym}. This result is obtained by several stages as depicted in Fig. 20.5. By transposing one of the gain blocks γ_{Nm} as a feedback close to G, a transformed linear system G_{sym} is introduced (*Unbehauen, H., 1983*). Using γ_{Nm} and γ_{Nr} , three simple relations can be stated

$$\gamma_{Nm}^2 - \gamma_{Nr}^2 = \gamma_{N\max}\,\gamma_{N\min} \tag{20.66}$$

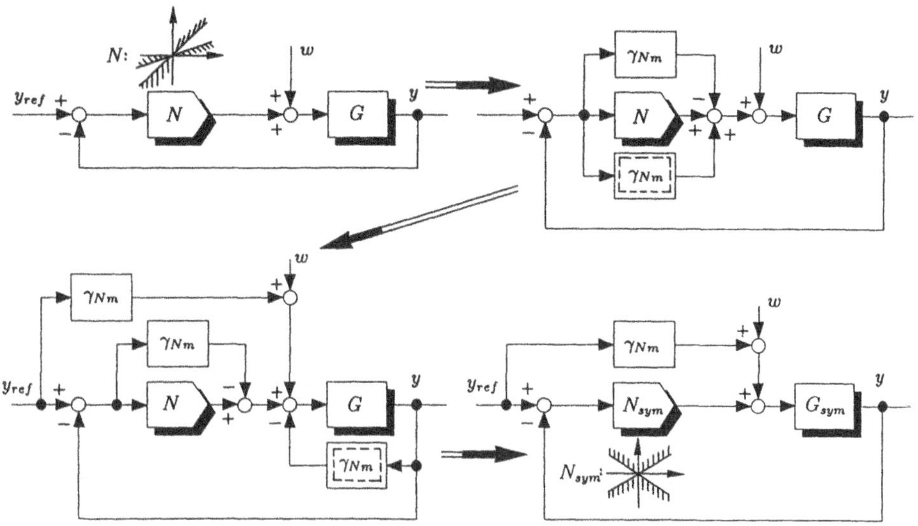

Figure 20.5: Symmetrizing the nonlinear characteristic

$$\Big(\frac{\gamma_{Nm}}{\gamma_{N\max}\,\gamma_{N\min}}\Big)^2 - \frac{1}{\gamma_{N\max}\,\gamma_{N\min}} = 0.25\Big(\frac{1}{\gamma_{N\max}} - \frac{1}{\gamma_{N\min}}\Big)^2 \triangleq r_c^2 \tag{20.67}$$

$$\frac{\gamma_{Nm}}{\gamma_{N\min}\gamma_{N\max}} = 0.5\Big(\frac{1}{\gamma_{N\min}} + \frac{1}{\gamma_{N\max}}\Big) \triangleq g_c\,. \tag{20.68}$$

Assuming both $\gamma_{N\min}$ and $\gamma_{N\max}$ positive, there results $g_c > 0$. Now, applying the simple exponential L_2-stability condition Eq.(20.63) to the transformed system,

$$\sup_{\omega} |G_{sym}(j\omega - \alpha)| < 1/\gamma_{Nr} \tag{20.69}$$

$$\sup_{\omega} \Big|\frac{G(j\omega-\alpha)}{1+\gamma_{Nm}G(j\omega-\alpha)}\Big| < 1/\gamma_{Nr}\,. \tag{20.70}$$

Extension to any ω and omitting $(j\omega - \alpha)$, for convenience, yields

$$|1 + \gamma_{Nm}G(j\omega - \alpha)| > \gamma_{Nr}|G(j\omega - \alpha)| \tag{20.71}$$

$$(1+\gamma_{Nm}G)(1+\gamma_{Nm}G^*) > \gamma_{Nr}^2 GG^* \tag{20.72}$$

$$1+\gamma_{Nm}(G+G^*) + \gamma_{Nm}^2 GG^* > \gamma_{Nr}^2 GG^* \tag{20.73}$$

$$1+\gamma_{Nm}(G+G^*) + (\gamma_{Nm}^2 - \gamma_{Nr}^2)GG^* > 0\,. \tag{20.74}$$

Substituting the results of Eq.(20.66) to (20.68), it results

$$1+\gamma_{Nm}(G+G^*) + \gamma_{N\max}\,\gamma_{N\min}\,GG^* > 0 \tag{20.75}$$

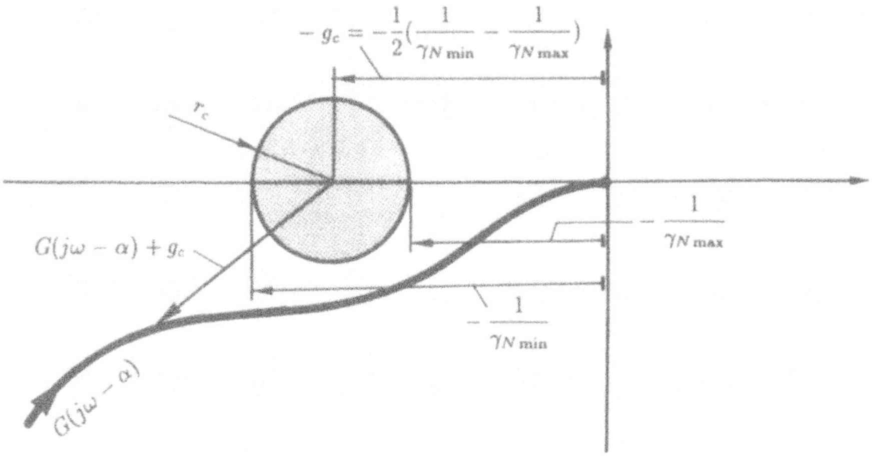

Figure 20.6: Circle criterion using Nyquist plot $G(j\omega-\alpha)$ in the case of asymmetrical nonlinearity

$$\Big(G+\frac{\gamma_{Nm}}{\gamma_{N\,min}\gamma_{N\,max}}\Big)\Big(G^*+\frac{\gamma_{Nm}}{\gamma_{N\,min}\,\gamma_{N\,max}}\Big)>\Big(\frac{\gamma_{Nm}}{\gamma_{N\,max}\gamma_{N\,min}}\Big)^2-\frac{1}{\gamma_{N\,max}\gamma_{N\,min}}=r_c^2 \quad (20.76)$$

$$(G+g_c)(G^*+g_c)>r_c^2 \quad (20.77)$$

$$|G(j\omega-\alpha)+g_c|>r_c\,. \quad (20.78)$$

The graphic interpretation of Eq. (20.78) is referred to as Circle criterion again and is depicted in Fig. 20.6. In order to guarantee exponential L_2-stability, the Nyqyist plot $G(j\omega-\alpha)$ must not touch or enter the dotted disc.

If the transformed transfer system $G_{sym}(s)$ (see Fig.20.5) turns out with n_p^+ poles in the right half of the s-plane then the Nyquist plot of $G_{sym}(j\omega)$ has to perform n_p^+ encirclements on the critical disc (centered at g_c with radius r_c)

$$\Delta\ \arg_{-\infty<\omega<\infty}\ [G(\alpha+j\omega)+g_c]=2\pi n_p^+ \quad (20.79)$$

in addition to $|G(\alpha+j\omega)+g_c|>r_c\ \forall\omega$. Then, the system is exponentially L_2-stable.

To a certain extent there is a relation of this matter to Popov criterion for absolute stability. *Tesi, A., and Vicino, A., 1991a* applied the Popov criterion to determine the maximal interval variation around a fixed nominal value of the transfer function of the plant.

20.10 Exponential L_∞-Stability

With the assumptions $\|N_\alpha^{op}\|_\infty \le \gamma_N$ and $\|\mathrm{G}_\alpha^{op}\| \le \gamma_G$ and following Eq.(20.33), the system of Fig. 20.4 is exponentially L_∞-stable if the product $\gamma_G\gamma_N<1$. Using Eq.(3.41), from $\gamma_G\gamma_N<1$ the condition follows

$$\gamma_G=\|G_\alpha^{op}\|_\infty=\int_{-\infty}^{\infty}|g_\alpha|dt=\int_{-\infty}^{\infty}|g(t)|e^{\alpha t}dt<1/\gamma_N\,. \quad (20.80)$$

This condition guarantees the system to be exponentially L_∞-stable with degree of stability α. Referring to Eq.(20.33), the norm $\|y(t)e^{\alpha t}\|_\infty$ can be estimated

$$\sup_t |y(t)e^{\alpha t}| \leq (1-\gamma_G\gamma_N)^{-1}(\gamma_G\gamma_N \sup_t |y_{ref}(t)e^{\alpha t}|+\gamma_G \sup_t |w(t)e^{\alpha t}|+\beta_G+\beta_G\beta_N) \quad (20.81)$$

$$|y(t)| < e^{-\alpha t}(1-\gamma_G\gamma_N)^{-1}(\gamma_G\gamma_N \sup_t |y_{ref}(t)e^{\alpha t}|+\gamma_G \sup_t |w(t)e^{\alpha t}|+\beta_G+\beta_G\beta_N)\ . \quad (20.82)$$

Similarly, $e(t)$ or $u(t)$ can be estimated.

L_∞-stability is used if, irrespective of the sign, the maximum values of control signals must be limited (e.g., maximum torque, current, acceleration) as given by the L_∞-norm.

Zhu, G., and Skelton, R., 1991 present minimization approaches for quadratic functions (for the linear quadratic impulse) subject to multiple inequality constraints.

20.11 L_2^n-Norm of $\mathbf{y}(t)$ and $\hat{\mathbf{y}}(j\omega)$

Suppose a multivariable system as depicted in Fig. 20.4. Applying the norm on L_p^n-space, i.e. Eq.(3.3), to vector-valued functions on L_2^n and substituting the Fourier transform $\hat{\mathbf{y}} = \hat{\mathbf{y}}(j\omega) = \mathcal{F}\{\mathbf{y}(t)\} = \mathcal{F}\mathbf{y}$ yields

$$\begin{aligned}
\|\mathbf{y}\|_2^2 &= \sum_{i=1}^n \|y_i\|_2^2 = \frac{1}{2\pi}\sum_{i=1}^n \|\hat{y}_i\|_2^2 = \frac{1}{2\pi}\sum_{i=1}^n \int_{-\infty}^{\infty} |\hat{y}_i(j\omega)|^2 d\omega = \frac{1}{2\pi}\int_{-\infty}^{\infty}\sum_{i=1}^n |\hat{y}_i(j\omega)|^2 d\omega \\
&= \frac{1}{2\pi}\int_{-\infty}^{\infty} \hat{\mathbf{y}}^H(j\omega)\hat{\mathbf{y}}(j\omega)d\omega = \frac{1}{2\pi}\|\hat{\mathbf{y}}^H(j\omega)\hat{\mathbf{y}}(j\omega)\|_1 \qquad (20.83) \\
\|\mathbf{y}\|_2^2 &= \frac{1}{2\pi}\|\ \|\hat{\mathbf{y}}\|_F^2(j\omega)\ \|_1\ . \qquad (20.84)
\end{aligned}$$

The subscript $_F$ denotes the Frobenius *vector* norm. As mentioned in a former section, suppose the time and frequency region $(-\infty, +\infty)$.

20.12 L_2^n-Norms and Spectral Norm of the Transfer Matrix

Continuing the derivation of Eq.(20.84), an estimate of the system output based on the spectral norm of the transfer matrix is used, i.e., Eq.(2.50) is substituted

$$(\|\mathbf{y}\|_2)^2 \leq \frac{1}{2\pi}\|\ [\ \|\mathbf{G}(j\omega)\|_s^2\ \|\hat{\mathbf{u}}(j\omega)\|_F^2\ ](j\omega)\ \|_1\ . \qquad (20.85)$$

Note that the expression within the brackets [ ] is the squared spectral *matrix* norm $\|\mathbf{G}(j\omega)\|_s$ and Frobenius *vector* norm $\|\hat{\mathbf{u}}(j\omega)\|_F$, both scalar-valued functions of ω. Finally, the product of these functions is subject to the *function* norm L_1. Using Hölder inequality Eq.(3.20) (with $p = \infty, q = 1$), it results

$$\begin{aligned}
\|\mathbf{y}\|_2^2 &\leq \frac{1}{2\pi}\|\ (\|\mathbf{G}(j\omega)\|_s)^2\ \|_\infty\ \|\ (\|\hat{\mathbf{u}}(j\omega)\|_2)^2\ \|_1 \qquad (20.86) \\
\|\mathbf{y}\|_2^2 &\leq \frac{1}{2\pi}\|\ (\|\mathbf{G}(j\omega)\|_s)^2\ \|_\infty\ (\|\hat{\mathbf{u}}(j\omega)\|_2)^2 = \|\ (\|\mathbf{G}\|_s)^2\ \|_\infty\ (\|\mathbf{u}\|_2)^2 \qquad (20.87) \\
\|\mathbf{y}\|_2 &\leq \|\ \|\mathbf{G}(j\omega)\|_s\ \|_\infty\ \|\mathbf{u}\|_2 \qquad (20.88)
\end{aligned}$$

$$\frac{\|\mathbf{y}\|_2}{\|\mathbf{u}\|_2} = \frac{[\int_0^\infty \mathbf{y}^H(t)\mathbf{y}(t)dt]^{0.5}}{[\int_0^\infty \mathbf{u}^H(t)\mathbf{u}(t)dt]^{0.5}} \le \| \, \|\mathbf{G}(j\omega)\|_s \, \|_\infty = \sup_\omega \|\mathbf{G}(j\omega)\|_s \tag{20.89}$$

$$\frac{\|\mathbf{y}\|_2}{\|\mathbf{u}\|_2} \le \sup_\omega \sigma_{\max}[\mathbf{G}(j\omega)] \quad \rightsquigarrow \quad \frac{\|\mathbf{y}\|_2}{\|\mathbf{u}\|_2} \le \sup_\omega \|\mathbf{G}(j\omega)\|_F \, . \tag{20.90}$$

(*Sevaston, G.E., and Longman, R.W., 1988*). In view of Eq.(2.63), the statement given above can be replaced by a more conservative inequation using Frobenius norm which requires significantly less computational effort.

Example: Single-input single-output system and Eq.(20.90):

$$\text{Input } u(t) = \mathcal{L}^{-1}\frac{1}{s+b}, \quad \text{transfer function } G(s) = \frac{1}{s+a} \, , \tag{20.91}$$

$$\text{output } y(t) = \mathcal{L}^{-1}\frac{1}{s+a}\frac{1}{s+b} = (e^{-at} - e^{-bt})/(b-a); \; b > 0, a > 0 \, . \tag{20.92}$$

$$G(j\omega) = \frac{1}{j\omega + a}, \quad \sigma[G(j\omega)] = \sqrt{\frac{1}{a^2+\omega^2}}, \quad \max_\omega \sigma[G(j\omega)] = \frac{1}{a} \, . \tag{20.93}$$

$$\int_0^\infty u^2 dt = \frac{1}{2b} \qquad \int_0^\infty y^2 dt = \frac{1}{2ab(a+b)} \tag{20.94}$$

$$\frac{(\int_0^\infty y^2 dt)^{0.5}}{(\int_0^\infty u^2 dt)^{0.5}} = \sqrt{\frac{2b}{2ab(a+b)}} = \frac{1}{\sqrt{a^2+ab}} \le \max_\omega \sigma = \frac{1}{a} \, . \tag{20.95}$$

End of Example

20.13 L_2^n-Norms, Spectral Norm and Parseval Theorem

Consider a system with impulse response matrix $\mathbf{G}(t)$, with input vector $\mathbf{u}(t)$ and output $\mathbf{y}(t)$. Convolution yields

$$\mathbf{y}(t) = \int_0^t \mathbf{G}(t-\tau)\mathbf{u}(\tau)d\tau \, . \tag{20.96}$$

Define $\mathbf{y}(t) \triangleq \mathbf{G}^{op}\mathbf{u}(t)$ where $\mathbf{G}^{op}$ is a linear operator mapping the input $\mathbf{u}(t)$ on to the output $\mathbf{y}(t)$. Next, applying the Fourier transform $\hat{\mathbf{y}}(j\omega) = \mathbf{G}(j\omega)\hat{\mathbf{u}}(j\omega)$ and ^{H} for the conjugate transpose the real quadratic form is

$$\int_{-\infty}^{\infty} [\mathcal{F}\{\mathbf{y}(t)\}]^H \, [\mathcal{F}\{\mathbf{y}(t)\}] \, d\omega = \int_{-\infty}^{\infty} \hat{\mathbf{u}}^H(j\omega)\mathbf{G}^H(j\omega)\mathbf{G}(j\omega)\hat{\mathbf{u}}(j\omega) \le \tag{20.97}$$

$$\le \int_{-\infty}^{\infty} \sigma_{\max}^2[\mathbf{G}(j\omega)]\hat{\mathbf{u}}^H(j\omega)\hat{\mathbf{u}}(j\omega)d\omega \le \left(\sup_\omega \sigma_{\max}^2[\mathbf{G}(j\omega)]\right) \int_{-\infty}^{\infty} \hat{\mathbf{u}}^H(j\omega)\hat{\mathbf{u}}(j\omega)d\omega \, . \tag{20.98}$$

Utilizing Parseval theorem and the normed function space L_2^n as given by Eq.(3.3) where $\|\mathbf{x}\|_2 = (\int_0^\infty \mathbf{x}^T(t)\mathbf{x}(t)dt)^{1/2}$, it results from Eq.(20.98)

$$\int_{-\infty}^{\infty} [\mathcal{F}\{\mathbf{u}(t)\}]^H \, [\mathcal{F}\{\mathbf{u}(t)\}]d\omega = \int_{-\infty}^{\infty} \hat{\mathbf{u}}^H(j\omega)\hat{\mathbf{u}}(j\omega)d\omega = 2\pi \int_0^\infty \mathbf{u}^T(t)\mathbf{u}(t)dt = 2\pi\|\mathbf{u}\|_2^2 \, . \tag{20.99}$$

Combining with Eq.(20.98),

$$2\pi\|\mathbf{y}\|_2^2 \le \left(\sup_\omega \sigma_{\max}^2[\mathbf{G}(j\omega)]\right)2\pi\|\mathbf{u}\|_2^2 \quad \rightsquigarrow \quad \|\mathbf{y}\|_2 \le \|\mathbf{u}\|_2 \sup_\omega \sigma_{\max}[\mathbf{G}(j\omega)] \tag{20.100}$$

$$\frac{\|\mathbf{y}(t)\|_2}{\|\mathbf{u}(t)\|_2} = \frac{\|\mathbf{G}^{op}\mathbf{u}(t)\|_2}{\|\mathbf{u}(t)\|_2} \le \sup_\omega \sigma_{\max}[\mathbf{G}(j\omega)] \, . \tag{20.101}$$

By Eq.(20.101), an input-output bound on L_2^n is achieved where the maximum singular value and its supremum versus ω plays a key role.

20.14 Norm of the Operator $\mathbf{G}^{op}$

Since both the Hölder inequality Eq.(3.20) and Eq.(2.50) are minimal estimates, Eq.(20.101) can be used to determine the norm of the operator $\mathbf{G}^{op}$

$$\|\mathbf{G}^{op}\|_2 = \| \ \|\mathbf{G}(j\omega)\|_s \ \|_\infty = \sup_\omega \|\mathbf{G}(j\omega)\|_s \ . \tag{20.102}$$

20.15 L_2^n-Stability and Exponential L_2^n-Stability

With the assumption $\|(\mathbf{N}^{op}\mathbf{e})(t)\|_F < \gamma_N \|\mathbf{e}(t)\|_F$ the condition $\gamma_G \gamma_N < 1$ is satisfied if

$$\sup_\omega \|\mathbf{G}(j\omega)\|_s < 1/\gamma_N \tag{20.103}$$

and simple L_2^n-stability is given. Exponential L_2^n-stability with degree of stability α is achieved if

$$\sup_\omega \|\mathbf{G}(j\omega - \alpha)\|_s < 1/\gamma_N \ . \tag{20.104}$$

For numerical or graphical control system design, the approach as follows is suitable. Suppose $\mathbf{N}^{op}$ is specialized to a decoupled (diagonalized) operator. Premultiplying and postmultiplying $\mathbf{N}^{op}$ by matrices diag $\chi_i^{-1}\gamma_{Ni}^{-0.5}$ and diag $\chi_i\gamma_{Ni}^{-0.5}$, respectively, a transformed nonlinear system with standardized sector gains can be achieved. The new sector gains are determined by the arbitrary and positive constants χ_i. Asymmetric characteristics can be transformed to symmetrical ones as shown in the L_2-stability case. Exponential L_2^n-stability is guaranteed if

$$|G_{ii}(j\omega - \alpha) + g_{ci}| - r_i(j\omega - \alpha) > r_{ci} \qquad \forall i \tag{20.105}$$

$$\text{where} \quad r_i(j\omega - \alpha) = 0.5 \sum_{j \neq i} \frac{\chi_i}{\chi_j} |G_{ij}(j\omega - \alpha)| + \frac{\chi_j}{\chi_i} |G_{ji}(j\omega - \alpha)| \tag{20.106}$$

$$g_{ci} = 0.5\Big(\frac{1}{\gamma_{Nmax\ i}} + \frac{1}{\gamma_{Nmin\ i}}\Big) \qquad r_{ci} = 0.5\Big(\frac{1}{\gamma_{Nmin\ i}} - \frac{1}{\gamma_{Nmax\ i}}\Big) \ . \tag{20.107}$$

Eq.(20.105) can be discussed graphically for all i. This procedure is known as *multivariable circle criterion* (Fig. 20.7). None of the modified Gershgorin bands may enter the disc centred at $-g_{ci}$ with radius r_{ci} (*Freeman, E.A., 1972; Rosenbrock, H.H., 1972; Safonov, M.G., and Athans, M., 1981*).

Approximating the spectral norm by the Frobenius norm, see Eq.(20.104), computational effort becomes less but a stronger condition is achieved.

20.16 Time-Varying Open-Loop Systems and L_p-Stability

The input output relation of linear time-invariant systems is given by the convolution using the weighting function $g(t)$ where $g = 0 \quad \forall t < 0$

$$y(t) = \int_{-\infty}^{\infty} g(t - \tau) u(\tau) d\tau \ . \tag{20.108}$$

Linear time-varying systems are represented by

$$y(t) = \int_{-\infty}^{\infty} g(t, \tau) u(\tau) d\tau \qquad \tau > 0 \tag{20.109}$$

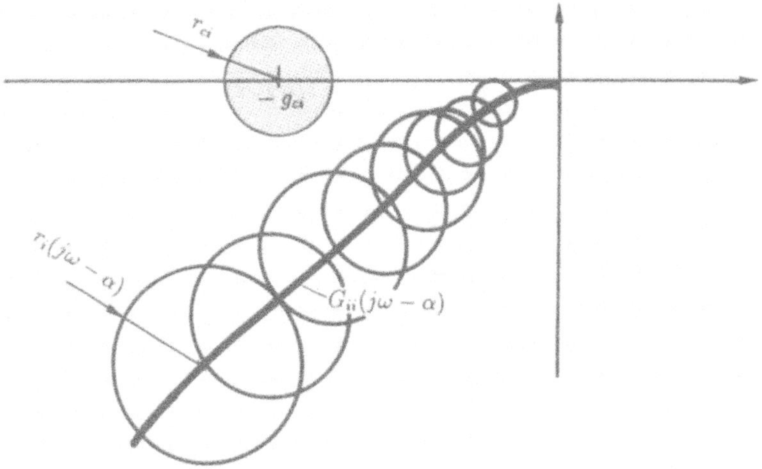

Figure 20.7: Multivariable circle criterion using modified Gershgorin band

where $u \in L_1$ and bounded $g(t,\cdot)$ $\forall t$ is supposed. Then, there exists a fixed relation (*Desoer, C.A., and Vidyasagar, M., 1975, p. 112*)

$$\|y\|_1 \leq c\|u\|_1 \quad \text{if and only if} \quad \sup_{\tau\geq 0} \int_{-\infty}^{\infty} |g(t,\tau)|dt = c_1 < \infty\ . \tag{20.110}$$

The constant c_1 is referred to as the induced norm of the operator G^{op}. Considering $g(t,\tau)$ and $u(\tau)$ locally integrable with respect to τ then y is L_p-stable $\forall p \in [1,\infty]$, i.e.,

$$\|y\|_p \;\leq\; (c_1)^{1/p}\,(c_\infty)^{1/q}\,\|u\|_p \quad \text{where} \quad \frac{1}{p}+\frac{1}{q}=1 \tag{20.111}$$

$$\text{if both} \quad c_1 \triangleq \int_{-\infty}^{\infty} |g(t,\tau)|dt = \|g(\cdot,\tau)\|_1 < \infty \ \ \forall\tau\geq 0 \tag{20.112}$$

$$\text{and} \quad c_\infty \triangleq \int_{-\infty}^{\infty} |g(t,\tau)|d\tau = \|g(t,\cdot)\|_1 < \infty \ \ \forall t\geq 0\ . \tag{20.113}$$

Considering $p=1$ it results $q=\infty$. Hence, c_∞ is of no relevance and the existence of c_1 is necessary and sufficient for L_1-stability. On the other hand, $p=\infty$ and $q=1$ shows that c_∞ is necessary and sufficient for L_∞-stability.

20.17 Time-Varying Closed-Loop Systems and L_p-Stability

Consider the system as depicted in Fig. 20.8. In operator notation

$$u = h * e \qquad\qquad y = k(t)\,u$$

$$e = y_{ref} - y = y_{ref} - k(t)(h*e)(t)\ . \tag{20.114}$$

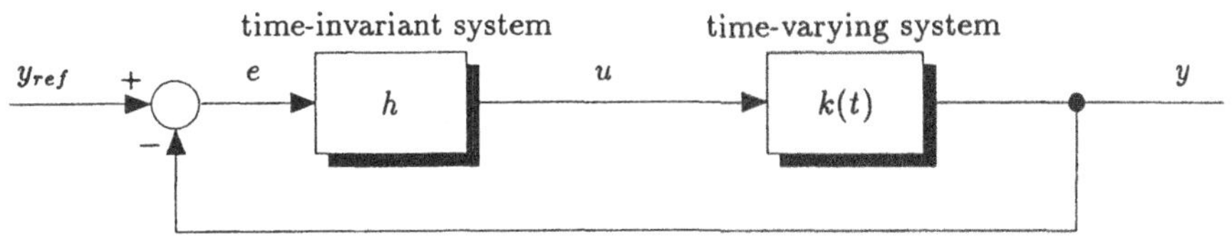

Figure 20.8: Control loop with a time-varying and a time-invariant system

A stability test can be derived from the product of the convolution operator h and the time-varying operator k. The function $g(t,\tau)$ in Eqs.(20.112) and (20.113) has to be replaced by $h(t-\tau)k(t)$. Applying Eq.(20.109) to Eq.(20.114), it must be taken into consideration that Eq.(20.114) corresponds to a closed loop whereas Eq.(20.109) belongs to an open-loop system.

Hence, the stability conditions have to be applied as given by the Small gain theorem, with the main result that the loop return gain may not exceed unity, see Eq.(20.33). In the case of a time-varying closed-loop process the supremum gain for all t must not exceed unity. Thus, applying Eq.(20.113), the L_∞-stability condition is

$$\sup_{t\geq 0}\int_{-\infty}^{\infty}|h(t-\tau)k(t)|\,d\tau=\sup_{t\geq 0}|k(t)|\int_0^t|h(t-\tau)|\,d\tau=\sup_{t\geq 0}\;|k(t)|\int_0^t|h(\tau|\,d\tau\;<\;1\,. \tag{20.115}$$

Applying Eq.(20.112),

$$\sup_{\tau\geq 0}\int_{-\infty}^{\infty}|h(t-\tau)k(t)|\,dt=\sup_{\tau\geq 0}\int_{\tau}^{\infty}|h(t-\tau)k(t)|\,dt\;<\;1\,. \tag{20.116}$$

If both Eqs.(20.115) and (20.116) are satisfied then L_p-stability $\forall p=[1,\infty]$ is guaranteed.

20.18 Control of Norm Uncertain Plants

Consider signals belonging to normed spaces defined in the time interval $(-\infty,\infty)$. The signals are inputs and outputs of the plant and the controller, the dynamic behaviour of which is not taken into consideration (Fig. 20.9). The plant is considered partially affected by uncertainties and approximated by a model $\mathbf{G}$ with additive and unknown noise limited in norm. The output of the real plant and of the model are denoted $\mathbf{y}$ and $\mathbf{G}^{op}\mathbf{u}$, respectively.

20.18.1 Plant Uncertainty in Norm

A plant is defined uncertain in norm (*Carlucci, D., and Donati, F., 1975*) if the difference $\mathbf{y}-\mathbf{G}^{op}\mathbf{u}$ obeys the inequality in the function norm $\|\cdot\|$

$$\|\mathbf{v}\|=\|\mathbf{y}-\mathbf{G}^{op}\mathbf{u}\|\leq\gamma_G\|\mathbf{u}\|+\gamma_{Go} \tag{20.117}$$

where γ_G and γ_{Go} are constants. The control signal $\mathbf{u}$ is assumed to belong to a set of admissible inputs $\mathcal{B}_u$, the output $\mathbf{y}\in\mathcal{B}_y$ is known only with respect to the statement

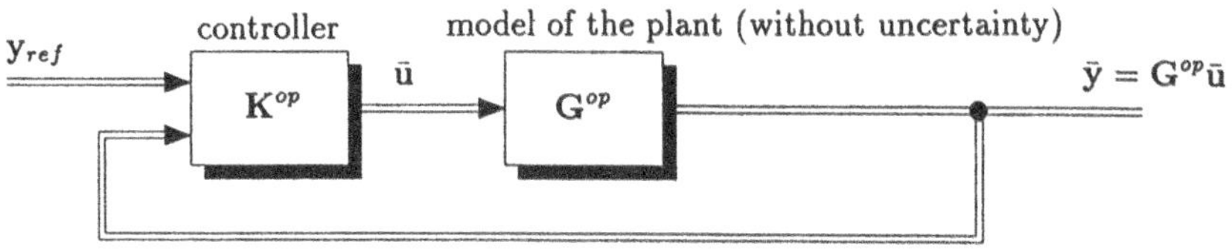

Figure 20.9: Controller and model without uncertainty

of inequality above. Suppose the L_2-norm is considered. Then, the output $\mathbf{y}$ belongs to a sphere centered at the model output $\mathbf{G}^{op}\mathbf{u}$. The radius of the sphere is determined by $\gamma_G\|\mathbf{u}\| + \gamma_{G_o}$ and is linearly dependent on the input norm.

20.18.2 Closed-Loop Uncertainty Definitions

In the case of the controlled unperturbed model, the control signal and system output are termed with an overlined bar $\bar{\mathbf{u}}$ and $\bar{\mathbf{y}}$ (Fig.20.9). Controlling the actual perturbed plant the bar is omitted, see Fig. 20.10. Now, for a given set of input signals $\mathbf{y}_{ref}$ the output signal $\mathbf{y}$ is treated incorporating an additional noise. The norm of the difference signals $\|\mathbf{y}-\bar{\mathbf{y}}\|$ and $\|\mathbf{u}-\bar{\mathbf{u}}\|$ in relation with an uncertainty signal $\|\mathbf{v}\|$ is considered

$$\sigma_{yv} \triangleq \sup_{\mathbf{v}} \frac{\|\mathbf{y}-\bar{\mathbf{y}}\|}{\|\mathbf{v}\|} \qquad \|\mathbf{v}\| > 0 \tag{20.118}$$

$$\sigma_{uv} \triangleq \sup_{\mathbf{v}} \frac{\|\mathbf{u}-\bar{\mathbf{u}}\|}{\|\mathbf{v}\|} \qquad \|\mathbf{v}\| > 0\,. \tag{20.119}$$

The quantities σ_{yv} and σ_{uv} are designated as the maximum plant sensitivity and the maximum controller sensitivity with respect to the uncertainty, respectively. On the basis of the output and control signal differences an open-loop uncertainty gain γ_o is defined as

$$\gamma_o \triangleq \sup_{y,\bar{y}} \frac{\|\mathbf{G}^{op}\mathbf{K}^{op}\mathbf{y} - \mathbf{G}^{op}\mathbf{K}^{op}\bar{\mathbf{y}}\|}{\|\mathbf{y}-\bar{\mathbf{y}}\|} \qquad \text{where } \|\mathbf{y}-\bar{\mathbf{y}}\| > 0 \text{ and } \mathbf{y},\bar{\mathbf{y}} \in \mathcal{B}_y\,. \tag{20.120}$$

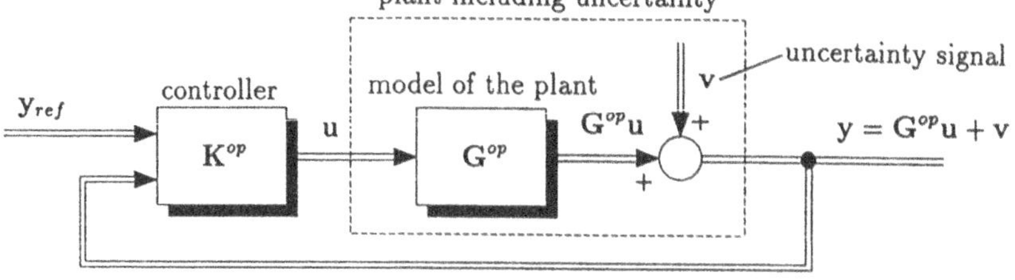

Figure 20.10: Controller and norm uncertain plant

20.18.3 Closed-Loop Uncertainty Gain Properties

Maximum Model Uncertainty Gain

With the definition of a maximum model uncertainty gain

$$\gamma_{u\,\sup} \triangleq \sup_{\mathbf{u},\bar{\mathbf{u}}} \frac{\|\mathbf{G}^{op}\mathbf{u} - \mathbf{G}^{op}\bar{\mathbf{u}}\|}{\|\mathbf{u} - \bar{\mathbf{u}}\|} \quad \text{where } \|\mathbf{u} - \bar{\mathbf{u}}\| > 0 \text{ and } \mathbf{u}, \bar{\mathbf{u}} \in \mathcal{B}_u \tag{20.121}$$

by subtracting $\mathbf{y} = \mathbf{G}^{op}\mathbf{u} + \mathbf{v}$ and $\bar{\mathbf{y}} = \mathbf{G}^{op}\bar{\mathbf{u}}$, it results

$$\mathbf{y} - \bar{\mathbf{y}} - \mathbf{v} = \mathbf{G}^{op}\mathbf{u} - \mathbf{G}^{op}\bar{\mathbf{u}} \,. \tag{20.122}$$

Using Eq.(20.121) and, next, dividing by $\|\mathbf{v}\|$ yields

$$|\,\|\mathbf{y}-\bar{\mathbf{y}}\|-\|\mathbf{v}\|\,| \le \|\mathbf{G}^{op}\mathbf{u}-\mathbf{G}^{op}\bar{\mathbf{u}}\| \le \gamma_{u\,\sup}\,\|\mathbf{u}-\bar{\mathbf{u}}\| \quad \rightsquigarrow \quad |1-\frac{\|\mathbf{y}-\bar{\mathbf{y}}\|}{\|\mathbf{v}\|}| \le \gamma_{u\,\sup}\,\frac{\|\mathbf{u}-\bar{\mathbf{u}}\|}{\|\mathbf{v}\|} \,. \tag{20.123}$$

Applying Eqs.(20.118) and (20.119), it follows

$$|1 - \sigma_{yv}| \le \gamma_{u\,\sup}\;\; \sigma_{uv} \,. \tag{20.124}$$

Minimum Model Uncertainty Gain

Defining a minimum model uncertainty gain

$$\gamma_{u\,\inf} \triangleq \inf_{\mathbf{u},\bar{\mathbf{u}}} \frac{\|\mathbf{G}^{op}\mathbf{u} - \mathbf{G}^{op}\bar{\mathbf{u}}\|}{\|\mathbf{u} - \bar{\mathbf{u}}\|} \qquad \text{where} \qquad \|\mathbf{u} - \bar{\mathbf{u}}\| > 0 \,, \tag{20.125}$$

from Eq.(20.122) there results

$$\|\mathbf{y} - \bar{\mathbf{y}}\| + \|\mathbf{v}\| \ge \gamma_{u\,\inf}\,\|\mathbf{u} - \bar{\mathbf{u}}\| \quad \rightsquigarrow \quad \frac{\|\mathbf{y} - \bar{\mathbf{y}}\|}{\|\mathbf{v}\|} + 1 \ge \gamma_{u\,\inf}\,\frac{\|\mathbf{u} - \bar{\mathbf{u}}\|}{\|\mathbf{v}\|} \tag{20.126}$$

$$\sigma_{yv} + 1 \ge \gamma_{u\,\inf}\;\; \sigma_{uv} \,. \tag{20.127}$$

Open-Loop Uncertainty Gain

Substracting $\mathbf{y} = \mathbf{v} + \mathbf{G}^{op}\mathbf{K}^{op}\mathbf{y}$ and $\bar{\mathbf{y}} = \mathbf{G}^{op}\mathbf{K}^{op}\bar{\mathbf{y}}$ yields

$$\mathbf{y} - \bar{\mathbf{y}} - \mathbf{G}^{op}\mathbf{K}^{op}\mathbf{y} + \mathbf{G}^{op}\mathbf{K}^{op}\bar{\mathbf{y}} = \mathbf{v} \tag{20.128}$$

$$\|\mathbf{y} - \bar{\mathbf{y}}\| + \|\mathbf{G}^{op}\mathbf{K}^{op}\mathbf{y} - \mathbf{G}^{op}\mathbf{K}^{op}\bar{\mathbf{y}}\| \ge \|\mathbf{v}\| \tag{20.129}$$

$$\frac{\|\mathbf{y} - \bar{\mathbf{y}}\|}{\|\mathbf{v}\|} + \frac{\|\mathbf{G}^{op}\mathbf{K}^{op}\mathbf{y} - \mathbf{G}^{op}\mathbf{K}^{op}\bar{\mathbf{y}}\|}{\|\mathbf{y} - \bar{\mathbf{y}}\|}\;\frac{\|\mathbf{y} - \bar{\mathbf{y}}\|}{\|\mathbf{v}\|} \ge 1 \,. \tag{20.130}$$

By Eq.(20.120), it results

$$\sigma_{yv} + \gamma_o\sigma_{yv} = \sigma_{yv}(1 + \gamma_o) \ge 1 \,. \tag{20.131}$$

20.18.4 Control Uncertainty Versus Plant Uncertainty

Consider the same controller being applied to a model **G** and to the plant with uncertainty, respectively. The designer is interested to which extent the plant uncertainty in norm is reduced (or probably enlarged) when the controller is operating. Combining

$$\|\mathbf{u}\| \leq \|\mathbf{u} - \bar{\mathbf{u}}\| + \|\bar{\mathbf{u}}\| \quad \text{and} \quad \|\mathbf{u} - \bar{\mathbf{u}}\| \leq \sigma_{uv}\|\mathbf{v}\| \tag{20.132}$$

as derived from Eq.(20.119) yields

$$\|\mathbf{u}\| \leq \|\bar{\mathbf{u}}\| + \sigma_{uv}\|\mathbf{v}\| \, . \tag{20.133}$$

When $\|\mathbf{u}\|$ from this inequality is substituted into the plant uncertainty definition Eq.(20.117)

$$\|\mathbf{v}\| \leq \gamma_G(\|\bar{\mathbf{u}}\| + \sigma_{uv}\|\mathbf{v}\|) + \gamma_{Go} \tag{20.134}$$

$$\|\mathbf{v}\|(1 - \gamma_G\sigma_{uv}) \leq \gamma_G\|\bar{\mathbf{u}}\| + \gamma_{Go} \tag{20.135}$$

is achieved. If $\gamma_G\sigma_{uv} < 1$ is true then by substitution of $\|\mathbf{v}\|$ from the inequality above into the statement $\sigma_{yv}\|\mathbf{v}\| \geq \|\mathbf{y} - \bar{\mathbf{y}}\|$ as derived from Eq.(20.118), finally, yields the sufficient condition

$$\|\mathbf{y} - \bar{\mathbf{y}}\| \leq \sigma_{yv}(1 - \gamma_G\sigma_{uv})^{-1}(\gamma_G\|\bar{\mathbf{u}}\| + \gamma_{Go}) \, . \tag{20.136}$$

Thus,

$$k_u = \sigma_{yv}(1 - \gamma_G\sigma_{uv})^{-1} \tag{20.137}$$

is an attenuation factor which relates the open-loop uncertainty $\gamma_G\|\bar{\mathbf{u}}\| + \gamma_{Go}$ with the output uncertainty $\|\mathbf{y} - \bar{\mathbf{y}}\|$ in the worst case.

20.18.5 Output Uncertainty Versus Reference Norm

Defining a maximum gain of the control signal $\bar{\mathbf{u}}$ versus reference $\mathbf{y}_{ref}$ for a given set of reference signals

$$\gamma_{uy_{ref}} \triangleq \sup_{y_{ref}} \frac{\|\bar{\mathbf{u}}\|}{\|\mathbf{y}_{ref}\|} \quad \text{where} \quad \|\mathbf{y}_{ref}\| > 0 \, , \tag{20.138}$$

an upper limit for the output uncertainty norm $\|\mathbf{y} - \bar{\mathbf{y}}\|$ versus reference norm is obtained

$$\|\mathbf{y} - \bar{\mathbf{y}}\| \leq \sigma_{yv}(1 - \gamma_G\sigma_{uv})^{-1}(\gamma_G\gamma_{uy_{ref}}\|\mathbf{y}_{ref}\| + \gamma_{Go}) = k_u(\gamma_G\gamma_{uy_{ref}}\|\mathbf{y}_{ref}\| + \gamma_{Go}) \, . \tag{20.139}$$

If the attenuation factor k_u is larger than unity the feedback control is worse than the open loop. A good uncertainty attenuation is obtained if σ_{yv} is small. This condition is given if the feedback system is already characterized by a good noise reduction. On the other hand, a large plant uncertainty γ_G or a high feedback gain σ_{uv} (caused by the controller) may deteriorate or cancel the attenuation, see *Carlucci, D., and Donati, F., 1975.*

20.18.6 Large Uncertainty Parameter γ_G

Assume that the control system guarantees good additive noise attenuation, i.e. $\sigma_{yv} < 1$. Applying Eq.(20.131) yields

$$1 > \sigma_{yv} \geq \frac{1}{1 + \gamma_o} \quad \rightsquigarrow \quad \sigma_{yv} = \frac{1}{1 + \xi\gamma_o} \quad \text{where} \quad 0 < \xi \leq 1 \, . \tag{20.140}$$

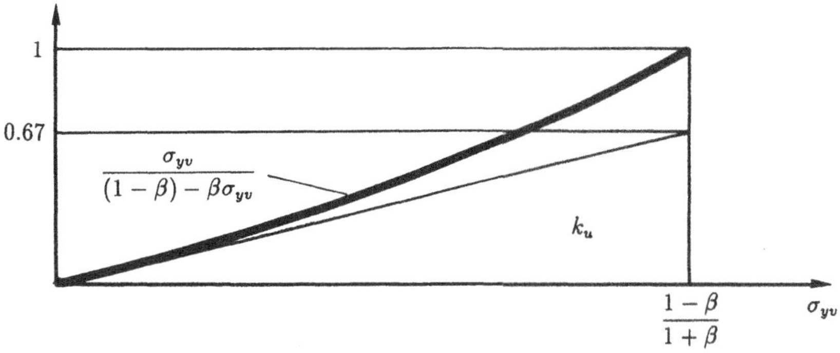

Figure 20.11: Upper limit for the uncertainty attenuation factor k_u (when $\beta = 0.5$)

Substituting into Eq.(20.124),

$$1 - \frac{1}{1+\xi\gamma_o} < \gamma_{u\,\sup}\ \sigma_{uv} \quad \leadsto \quad \sigma_{uv} > \frac{\xi\gamma_o}{(1+\xi\gamma_o)\gamma_{u\,\sup}} \tag{20.141}$$

is achieved. Substituting Eqs.(20.140) and (20.141) into Eq.(20.137),

$$k_u \geq \frac{\frac{1}{1+\xi\gamma_o}}{1-\gamma_G\frac{\xi\gamma_o}{(1+\xi\gamma_o)\gamma_{u\,\sup}}} = \frac{1}{1-\xi\gamma_o(\gamma_G\gamma_{u\,\sup}^{-1}-1)}\,. \tag{20.142}$$

Assuming $\gamma_G > \gamma_{u\,\sup}$, the attenuation factor k_u exceeds unity and, moreover, increases with the factor γ_o. The feedback control is worse than the open loop.

20.18.7 Small Uncertainty Parameter γ_G

Rewriting Eq.(20.127) yields

$$\sigma_{uv} \leq (1+\sigma_{yv})\gamma_{u\,\inf}^{-1}\,. \tag{20.143}$$

If $\gamma_G\sigma_{uv} < 1$ then the substitution of Eq.(20.143) into Eq.(20.137) results in the inequality

$$k_u \leq \sigma_{yv}[1-\beta(1+\sigma_{yv})]^{-1} \qquad \text{where} \qquad \beta \triangleq \frac{\gamma_G}{\gamma_{u\,\inf}}\,. \tag{20.144}$$

If the uncertainty parameter γ_G is smaller than $\gamma_{u\,\inf}$ then $\beta < 1$ and

$$k_u \leq \frac{\sigma_{yv}}{1-\beta(1+\sigma_{yv})} = \frac{\sigma_{yv}}{(1-\beta)-\beta\sigma_{yv}} \quad \left(\text{e.g. } < 1 \ \text{ if } \ \sigma_{yv} < \frac{1-\beta}{1+\beta}\right). \tag{20.145}$$

Part IV

Robustness in the Frequency Domain

Chapter 21

Uncertain Polynomials. Interval Polynomials

21.1 Preliminary Considerations

Let the set $\mathcal{U}$ be the closed set of values s which are not located in a desired stability region Γ, e.g., $\mathcal{U}$ is the closed right-half plane. The polynomial $a(s, \mathcal{Q})$ is denoted $\mathcal{U}$-stable if and only if no root of $a(s, \mathcal{Q})$ is located in the undesired region, i.e.,

$$a(s, \mathcal{Q}) = \{a(s, q) \ : \ q \in \mathcal{Q}\} \neq 0 \quad \forall s \in \mathcal{U} \ . \tag{21.1}$$

The following assumptions must be satisfied by the parametrization of the polynomial coefficients

(i) $q \in \mathcal{Q} \subset \mathcal{R}^m$ where $\mathcal{Q}$ is a non-empty, connected and compact set.

(ii) The coefficient parametrizations are continuous.

(iii) $a_n(q) > 0 \ \ \forall q \in \mathcal{Q}$ where a_n is the coefficient of the highest power s^n .

Then, $\mathcal{U}$-stability of $a(s, \mathcal{Q})$ is given if and only if for some $q_o \in \mathcal{Q}$ the polynomial $a(s, q_o)$ is $\mathcal{U}$-stable and $0 \notin \ a(s, \mathcal{Q}) \ \ \forall s \in \partial\mathcal{U}$ where $\partial\mathcal{U}$ is the boundary of $\mathcal{U}$.

In the case of a lumped system the polynomial is $a(s, q) = \sum_{i=0}^{n} a_i(q) s^i \quad q \in \mathcal{Q} \subset \mathcal{R}^m$. Replacing $a(s, \mathcal{Q})$ by $\chi(s, \mathcal{Q})$, the stability condition can be extended to distributed control systems. The characteristic function $\chi(s, \mathcal{Q})$ is the sum of products of the elements of matrix factors given by the coprime matrix factorization of the matrix transfer function (*Anagnost, J.J., et al. 1989; Desoer, C.A., et al. 1980*).

Considering the stability or performance robustness of linear continuous-time or discrete-time systems, the key problem is to check the system stability via the zeros of the characteristic polynomial in the face of uncertain coefficients of the polynomial or as a function of another kind of uncertain parameters. It is also of interest to localize system errors such that stability remains satisfied. A key result concerning the stability of the roots of a polynomial was presented by *Kharitonov, V.L., 1979*.

The characteristic polynomials of continuous-time systems are termed $a(s)$, $b(s)$, $c(s)$ or $p(s)$. Discrete-time system's polynomials are normally termed $f(z)$ or $g(z)$. Error polynomials are given by the notation $e(s)$ or $e(z)$. The index $_p$ denotes perturbed polynomials.

21.2 Kharitonov's Theorem for Continuous-Time Systems

21.2.1 Polytopes of the Perturbation

A polytope is considered a convex hull of a finite number of points in an m-dimensional space. Consider the system parameters a_i located inside the region

$$l_i = a_{i,nom} - \Delta a_{li} \leq a_i \leq a_{i,nom} + \Delta a_{hi} = h_i, \quad l_i > 0 \quad \forall i = [1, n] \tag{21.2}$$

where $a_{i,nom}$ is the nominal parameter and l, h denotes low, high, respectively. The equations above state a rectangular polytope in the space of a_i with edges parallel to the coordinate axes.

21.2.2 Extreme Polynomials and Corner or Vertex Polynomials

Consider the nominal polynomial

$$a_{nom}(s) = a_{o,nom} + a_{1,nom}s + \ldots + a_{n,nom}s^n \tag{21.3}$$

and an infinite family $S[l_i, h_i]$ of perturbed polynomials

$$a_p(s) = a_o + a_1 s + \ldots + a_n s^n \tag{21.4}$$

with each a_i varying inside the given bounds $0 < l_i \leq a_i \leq h_i \quad \forall i = [1, n]$. Then, the entire family of polynomials is Hurwitz stable if and only if four extreme polynomials $a_{e1}(s)$ through $a_{e4}(s)$ of degree n are Hurwitz, i.e., all the roots have negative real part,

$$\begin{aligned}
a_{e1}(s) &= h_o + l_1 s + l_2 s^2 + h_3 s^3 + h_4 s^4 + l_5 s^5 + l_6 s^6 + \ldots & (21.5)\\
a_{e2}(s) &= h_o + h_1 s + l_2 s^2 + l_3 s^3 + h_4 s^4 + h_5 s^5 + l_6 s^6 + \ldots & (21.6)\\
a_{e3}(s) &= l_o + h_1 s + h_2 s^2 + l_3 s^3 + l_4 s^4 + h_5 s^5 + h_6 s^6 + \ldots & (21.7)\\
a_{e4}(s) &= l_o + l_1 s + h_2 s^2 + h_3 s^3 + l_4 s^4 + l_5 s^5 + h_6 s^6 + \ldots \quad . & (21.8)
\end{aligned}$$

This set is termed $T[l_i, h_i]$ or $T[\mathrm{l}, \mathrm{h}]$. Note that *only four* polynomials as given by the extreme polynomials above are required for any n. If the degree n of the polynomial is less, fewer corner polynomials are required: For $n = 2$ the coefficients are required to be of like sign, only. If $n = 3$ only $a_{e1}(s)$ is needed, the other polynomials can be omitted; if $n = 4$ only $a_{e1}(s)$ and $a_{e2}(s)$; if $n = 5$ only $a_{e1}(s)$, $a_{e2}(s)$ and $a_{e3}(s)$ (*Anderson, B.D.O., et al. 1987*).

The coefficients are considered independent. Hurwitz stability can be investigated, only, i.e., one can only check the location of all roots in the open left half-plane. Note that the interval *polynomial* result *must not* be generalized to the interval *matrix* case. This is only true in the case of $n \leq 2$ (*Bialas, S., 1983; Karl, W.C., et al. 1984; Barmish, B.R., and Hollot, C.V., 1984*).

In the case of complex coefficients the Kharitonov theorem requires eight corner polynomials (*Bose, N.K., and Shi, Y.Q., 1987*).

The weak version of Kharitonov's theorem is given by the statement that it is necessary and sufficient that all corner polynomials are stable. Corner polynomials are obtained from polynomials corresponding to the corners of the parameter polytope, i.e. when $a_i \in \{l_i, h_i\} \ \forall\, i$. There exist 2^{n+1} corner polynomials. The strong version documents that only four (or less) out of the 2^{n+1} corners are both necessary and sufficient for robust stability.

Example: Robust Controller Design Using Interval Polynomials.

Consider the characteristic polynomial $\det(s\mathbf{I} - \mathbf{A} - \mathbf{BKC})$ of a multivariable output feedback system given by the diagram of Fig. 8.1. Let $\mathbf{A}$ be uncertain, $\mathbf{B}$ and $\mathbf{C}$ fixed. For a most likely $\mathbf{A} = \mathbf{A}_o$ a controller $\mathbf{K}_o$ can be found according to any conventional design method. On calculating the characteristic polynomial and comparing with Eq.(21.3), the characteristic polynomial coefficient vector is obtained.

Now, assume, first, the matrix $\mathbf{A}$ varying in the vicinity of $\mathbf{A}_o$ within the narrow bounds $\mathbf{A}_1$ and $\mathbf{A}_2$ and, second, assume a controller $\mathbf{K}_R$ near $\mathbf{K}_o$. From these data the vectors $\mathbf{l}$ and $\mathbf{h}$ are deduced. Using $T[\mathbf{l}, \mathbf{h}]$ the stability of the system is investigated. If stability is satisfied, then $\mathbf{K}_R$ is stable *and* robust. Next, using a computer-aided method, the matrix difference $\mathbf{A}_2 - \mathbf{A}_1$ and/or the controller matrix are changed appropriately. In order to keep the performance as close as possible to the most likely $\mathbf{K}_o$, the minimization of $\|\mathbf{K}_R - \mathbf{K}_o\|$ seems adequate (*Evans, R., and Xianya, X., 1985*). If $\mathbf{K}_R$ is unstable then $\mathbf{A}_2 - \mathbf{A}_1$ must be reduced or an alternative $\mathbf{K}_R$ has to be selected.

End of the Example

Testing Hurwitz stability the case arises in which the coefficients of the polynomials are polynomials in some free parameters, as a mild form of nonlinear dependence. These parameters are located in a certain region in the parameter space. If the number of free parameters is not large *Mansour, M., and Kraus, F.J., 1989; Kraus, F.J., et al. 1989* presented useful algorithms, see also Eqs.(21.91) and (21.176).

Kharitonov's theorem is based on the interlacing property of the roots of the odd and even parts of Hurwitz polynomials (*Kharitonov, V.L., 1979; Guiver, J.P., and Bose, N.K., 1983; Yeung, K.S., 1983; Barmish, B.R., 1984*). The polynomial $a(s) = a_{even}(s^2) + s\ a_{odd}(s^2)$, see Eqs. (21.15) and (21.16), is Hurwitz if and only if its odd and even parts $a_{odd}(s^2)$ and $a_{even}(s^2)$, respectively, are characterized by the following properties:

The leading coefficients of $a_{odd}(s^2)$ and $a_{even}(s^2)$ have the same sign. Furthermore, replacing s^2 by λ, the roots λ_{even} and λ_{odd} of $a_{even}(\lambda)$ and $a_{odd}(\lambda)$, respectively, are nonrepeated, real, negative and interlace, i.e., one has $a_{even}(\lambda_{even}) = 0$, $a_{odd}(\lambda_{odd}) = 0$ and

$$n = 2\,\nu \quad \lambda_{even,1} < \lambda_{odd,1} < \lambda_{even,2} < \lambda_{odd,2} < \;....\; < \lambda_{even,\nu} < 0 \tag{21.9}$$

$$n = 2\,\nu + 1 \quad \lambda_{odd,1} < \lambda_{even,1} < \lambda_{odd,2} < \lambda_{even,2} < \;....\; < \lambda_{odd,\nu} < \lambda_{even,\nu} < 0\,. \tag{21.10}$$

E.g.,

$$a(s) = (s+1)(s+2)(s+3)(s+4)(s+5) = 15(8 + 15s^2 + s^4) + s(274 + 85s^2 + s^4) \tag{21.11}$$

$\nu = 2$, $n = 5$

$$a_{even}(\lambda) = 8 + 15\lambda + \lambda^2 \quad \leadsto \quad \lambda_{even,1} = -14.446\ ;\ \lambda_{even,2} = -0.554 \tag{21.12}$$

$$a_{odd}(\lambda) = 274 + 85\lambda + \lambda^2 \quad \leadsto \quad \lambda_{odd,1} = -81.644\ ;\ \lambda_{odd,2} = -3.356\,. \tag{21.13}$$

21.2.3 Extensions

Kharitonov's theorem was extended to stability conditions of numerous families of polynomials, e.g., to the family of polynomials with coefficients multilinearly depending on the perturbations (*Barmish, B.R., and Shi, Z., 1990; Pujara, L.R., 1990*), to coefficients varying in a given diamond (*Tempo, R., 1990*), in order to guarantee stability margin (*Petersen, I.R., 1990*), furthermore, to the zeros of a polynomial located inside a left sector in the left half-plane (*Foo, Y.K., and Soh, Y.C., 1990; Soh, Y.C., and Foo, Y.K., 1990*) and to linear correlated parametric uncertainties where the admissible uncertainty set is a polytope with predetermined radius (*Tesi, A., and Vicino, A., 1990*). Conditions under which an interval polynomial possesses a certain number of roots in the open left-half plane are presented by *Kokame, H., and Mori, T., 1991*.

21.3 Frequency Domain Representation

Specializing $s = j\omega$ in $a(s) = \sum_{i=0}^{n} a_i s^i$ where $0 < l_i \leq a_i \leq h_i$ and setting

$$a(s) = a_{even}(s^2) + s\ a_{odd}(s^2) \tag{21.14}$$

an even and an odd part is defined, respectively,

$$a_{even}(s^2) \triangleq a_o + a_2 s^2 + a_4 s^4 + \dots\ , \tag{21.15}$$

$$a_{odd}(s^2) \triangleq a_1 + a_3 s^2 + a_5 s^4 + \dots\ . \tag{21.16}$$

One realizes that

$$a_{even}(-\omega^2) = \Re e\ a(j\omega) \tag{21.17}$$

$$a_{odd}(-\omega^2) = \frac{1}{\omega} \Im m\ a(j\omega)\ . \tag{21.18}$$

Documented by *Dasgupta, S., 1988*, for any fixed ω the box of coefficient space a_i is mapped to the rectangle in the complex $a(j\omega)$-plane. The corners of this rectangle are given by

$$a_{e1}(j\omega) = a_{even}(-\omega^2)_{\max} + j\omega\ a_{odd}(-\omega^2)_{\min} \tag{21.19}$$

$$a_{e2}(j\omega) = a_{even}(-\omega^2)_{\max} + j\omega\ a_{odd}(-\omega^2)_{\max} \tag{21.20}$$

$$a_{e3}(j\omega) = a_{even}(-\omega^2)_{\min} + j\omega\ a_{odd}(-\omega^2)_{\max} \tag{21.21}$$

$$a_{e4}(j\omega) = a_{even}(-\omega^2)_{\min} + j\omega\ a_{odd}(-\omega^2)_{\min} \tag{21.22}$$

where $a_{even}(-\omega^2)_{\max}$ and $a_{odd}(-\omega^2)_{\min}$ denote polynomials which are obtained by maximizing and minimizing its value, respectively, i.e.,

$$a_{even}(-\omega^2)_{\max} = h_o + l_2(-\omega^2) + h_4\omega^4 + \dots \tag{21.23}$$

$$a_{odd}(-\omega^2)_{\min} = l_1 + h_3(-\omega^2) + l_5\omega^4 + \dots \quad \text{etc.} \tag{21.24}$$

The corners cited above are identical with the Kharitonov extreme polynomials Eqs.(21.5) through (21.8).

21.3.1 Combination of Two Polynomials

Consider two polynomials $a(s)$ and $b(s)$ where

$$a(s) = \sum_0^n a_i s^i \quad a_n \neq 0 \qquad b(s) = \sum_0^n b_i s^i \quad b_n \neq 0\ . \tag{21.25}$$

The question arises whether the linear combination

$$c(s) = (1-\rho)\ a(s) + \rho\ b(s) \qquad \rho \in [0,1] \tag{21.26}$$

is Hurwitz stable. Select the even and odd part of the polynomials $a(s)$ and $b(s)$, respectively, i.e.,

$$a(s) = a_{even}(s^2) + s\ a_{odd}(s^2) \qquad b(s) = b_{even}(s^2) + s\ b_{odd}(s^2), \tag{21.27}$$

and assume that $a(s)$ and $b(s)$ are Hurwitz. Then, the polynomial $c(s)$ is stable if (*Bialas, S., and Garloff, J., 1985*)

$$\Re e\ a(j\omega) = a_{even}(-\omega^2) = \Re e\ b(j\omega) = b_{even}(-\omega^2) \quad \text{or} \tag{21.28}$$

$$\Im m\ a(j\omega) = \omega\ a_{odd}(-\omega^2) = \Im m\ b(j\omega) = \omega\ b_{odd}(-\omega^2) \tag{21.29}$$

or in a simplified version

$$a_{even}(s^2) = b_{even}(s^2) \quad \text{or} \quad a_{odd}(s^2) = b_{odd}(s^2)\ . \tag{21.30}$$

21.4 Two-Polynomial Combination and Hurwitz Testing Matrix

The objective is to find maximal bounds on r such that $p_o(s)+rp_1(s)$ remains stable. The coefficients of the polynomial $p_k(s)$ are termed p_{ki} . For the nth order polynomial

$$p(s) = \sum_{i=0}^{n} a_i s^i \qquad a_n > 0 \tag{21.31}$$

the Hurwitz testing matrix is

$$\mathbf{H}(p) = \begin{pmatrix} a_{n-1} & a_{n-3} & a_{n-5} & \dots & 0 \\ a_n & a_{n-2} & a_{n-4} & \dots & 0 \\ 0 & a_{n-1} & a_{n-3} & \dots & 0 \\ 0 & a_n & a_{n-2} & \dots & 0 \\ 0 & 0 & a_{n-1} & \dots & 0 \\ \vdots & \vdots & \vdots & \ddots & 0 \\ 0 & \dots & \dots & a_2 & a_o \end{pmatrix}, \tag{21.32}$$

e.g., for $n = 6$ and $n = 7$, respectively,

$$\begin{pmatrix} a_5 & a_3 & a_1 & 0 & 0 & 0 \\ a_6 & a_4 & a_2 & a_o & 0 & 0 \\ 0 & a_5 & a_3 & a_1 & 0 & 0 \\ 0 & a_6 & a_4 & a_2 & a_o & 0 \\ 0 & 0 & a_5 & a_3 & a_1 & 0 \\ 0 & 0 & a_6 & a_4 & a_2 & a_o \end{pmatrix} \qquad \begin{pmatrix} a_6 & a_4 & a_2 & a_o & 0 & 0 & 0 \\ a_7 & a_5 & a_3 & a_1 & 0 & 0 & 0 \\ 0 & a_6 & a_4 & a_2 & a_o & 0 & 0 \\ 0 & a_7 & a_5 & a_3 & a_1 & 0 & 0 \\ 0 & 0 & a_6 & a_4 & a_2 & a_o & 0 \\ 0 & 0 & a_7 & a_5 & a_3 & a_1 & 0 \\ 0 & 0 & 0 & a_6 & a_4 & a_2 & a_o \end{pmatrix}. \tag{21.33}$$

Let $\mathbf{H}_k$ be the testing matrix corresponding to the polynomial $p_k(s)$, $k = \{1,2\}$.

Suppose that a matrix $\mathbf{M}$ possesses only real eigenvalues. Then,

$$\lambda_{\max+}[\mathbf{M}] \triangleq \begin{cases} \lambda_{\max} & \text{if} \quad \lambda_{\max}[\mathbf{M}] > 0 \\ 0^+ & \text{if} \quad \lambda_{\max}[\mathbf{M}] < 0 \, . \end{cases} \tag{21.34}$$

Similarly, $\lambda_{\max-}[\mathbf{M}]$ is defined.

21.4.1 Stability Conditions of a Convex Combination of Polynomials

Given $p_o(s)$ and $p_1(s)$ of nth order. The convex combination

$$p_r(s) = p_o(s) + rp_1(s) \qquad \forall r \in [0,1] \tag{21.35}$$

is strictly Hurwitz stable if and only if p_o is strictly Hurwitz and

$$\lambda_i[\mathbf{H}_o^{-1}\mathbf{H}_1] \notin (-\infty, 0] \tag{21.36}$$

(Fu, M., and Barmish, B.R., 1987).

21.4.2 Range of Stability of a Convex Combination of Polynomials

Given an nth order polynomial $p_o(s)$ which is strictly Hurwitz and an mth order polynomial $p_1(s)$, $m \leq n$. Then, the family

$$p_r(s) = p_o(s) + rp_1(s) \qquad \text{is stable} \qquad \forall\, r = (r_{\min}, r_{\max}) \tag{21.37}$$

$$\text{where} \quad r_{\min} = \max\{r_1, \lambda^{-1}_{\min -}[-\mathbf{H}_o^{-1}\mathbf{H}_1]\}, \quad r_1 = \begin{cases} -\frac{a_{on}}{a_{1n}} & m = n \\ -\infty & m < n \end{cases} \tag{21.38}$$

$$r_{\max} = \lambda^{-1}_{\max +}[-\mathbf{H}_o^{-1}\mathbf{H}_1] \tag{21.39}$$

and $\mathbf{H}_1$ is given by treating $p_1(s)$ as an nth order polynomial.

21.5 Kharitonov's Theorem for Discrete-Time Systems

The continuous-time results do not apply directly to discrete-time problems (*Hollot, C.V., and Bartlett, A.C., 1986; Cieslik, J., 1987; Kraus, F.J., et al. 1988a*). But there are several analogs and counterparts to these results. Most of these methods are presented in what follows.

For obvious reasons, stability of discrete-time systems may be, e.g., deduced from Routh's criterion in the s-domain using bilinear transformation. This leads to a recursion of symmetric polynomials in the z-domain (*Benidir, M., and Picinbono, B., 1988*).

21.5.1 Bound Identity in Parts

Let a discrete-time system be described by the polynomial

$$f(z) = \sum_{i=0}^{n} a_i z^{n-i} \qquad a_i \in [l_i,\ h_i] \tag{21.40}$$

where the special case of $l_i = h_i \quad \forall i \in [0, 1, 2 ... (n-1)/2 \downarrow]$ is considered, i.e., the lower part $0, 1, 2 ... n/2$ shrinks to a point. The downarrow denotes the next lower integer. Then, the system is stable if and only if the *corner* points are stable (*Hollot, C.V., and Bartlett, A.C., 1986*). Note that the result using corner points and rectangular boxes parallel to the axes is only valid for the partial bound identity as cited above.

21.5.2 Corners and Edges

Consider a polynomial $f(z) = \sum_{i=0}^{n} a_i z^i$ of any degree n. The coefficients a_i only vary inside a polygon in planes a_n versus a_o, a_{n-1} versus a_1, a_{n-i} versus a_i etc. All the coefficients are located in polygons bounded by edges of slope 45, 90 and 135 degrees as depicted in Fig. 21.1. Then, the polynomial $f(z)$ is robustly Schur stable if and only if every possible combination of corner polynomials is Schur stable (see Fig. 21.1). If n is even, $a_{n/2}$ varies in an interval, only.

The above-mentioned theorem (*Kraus, F.J., et al. 1988*) is a close parallel to the weak Kharitonov theorem. The only difference is that the rectangular boxes are not located parallel to the axes but are tilted by 45 degrees. However, the coefficients of $g_e(z)$ and $g_o(z)$ in Eqs.(21.41) and (21.42) are rectangles with edges parallel to the axes. For the corresponding theorem to the strong Kharitonov theorem see Fig. 21.5.

21.5.3 Sufficient Conditions

Consider the case that the parameters a_i and a_{n-i} are uncertain with respect to the axis-parallel rectangular box B_i. Then, a square box A_{is} can be found such that each point of B_i is contained by A_{is}, see Fig. 21.2. Hence, A_{is} provides a sufficient condition for the desired uncertainty box B_i. There are boxes A_{is} $\forall i = 0, 1...n/2 \downarrow$ with four corner points. For even n there are additional two interval end-points. In Fig. 21.2 different methods are depicted. There exist various possibilities to choose corner point combinations covering the interesting parameter variation box B_i. The sufficient region A_{is} is the less conservative the more rectangular boxes are selected, at the expense of introducing more and more corner points. The quality of approximation is limited by the computational effort given by the exponentially increasing number of corner point combinations that have to be tested computationally for Schur stability.

21.5.4 Necessary Condition

Necessary stability conditions are achieved by inscribing the boxes B_i with squares or rectangles A_{in} as depicted in Fig. 21.3. The corner points of the regions A_{in} yield necessary conditions for the Schur stability of $f(z)$ since the shaded region is sufficient and necessary for stability.

21.5.5 Lossless Positive Real Function

With regard to $f(z) = \sum_{i=0}^{n} a_i z^i$ two polynomials $g_e(z)$ and $g_o(z)$ are defined by the symmetric and antisymmetric part of $f(z)$

$$\begin{aligned} g_e(z) \triangleq \frac{1}{2}[f(z) + z^n f(z^{-1})] &= \frac{a_o + a_n}{2} z^n + \frac{a_1 + a_{n-1}}{2} z^{n-1} + \ldots + \frac{a_{n-1} + a_1}{2} z + \frac{a_n + a_o}{2} \\ &= \alpha_o z^n + \alpha_1 z^{n-1} + \ldots + \alpha_1 z + \alpha_o \end{aligned} \quad (21.41)$$

$$\begin{aligned} g_o(z) \triangleq \frac{1}{2}[f(z) - z^n f(z^{-1})] &= \frac{a_o - a_n}{2} z^n + \frac{a_1 - a_{n-1}}{2} z^{n-1} + \ldots + \frac{a_{n-1} - a_1}{2} z + \frac{a_n - a_o}{2} \\ &= \beta_o z^n + \beta_1 z^{n-1} + \ldots + \beta_1 z + \beta_o \, . \end{aligned} \quad (21.42)$$

The interrelation between a_i, a_{n-i}, α_i, β_i is briefly sketched in Fig. 21.4. The g_o, g_e-plane is depicted in Fig. 21.5.

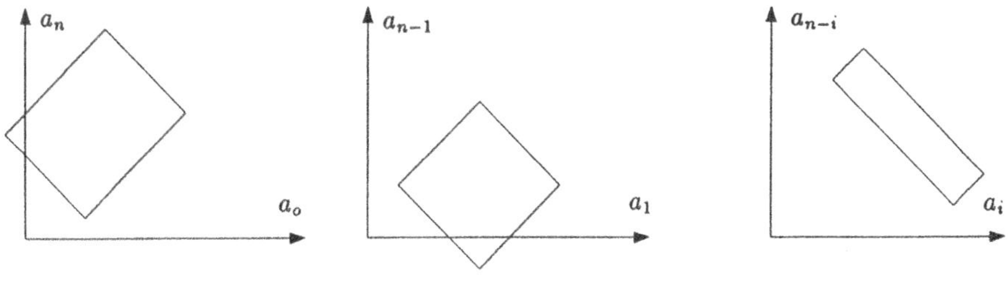

Figure 21.1: Rectangular boxes for stable variations of a discrete-time system

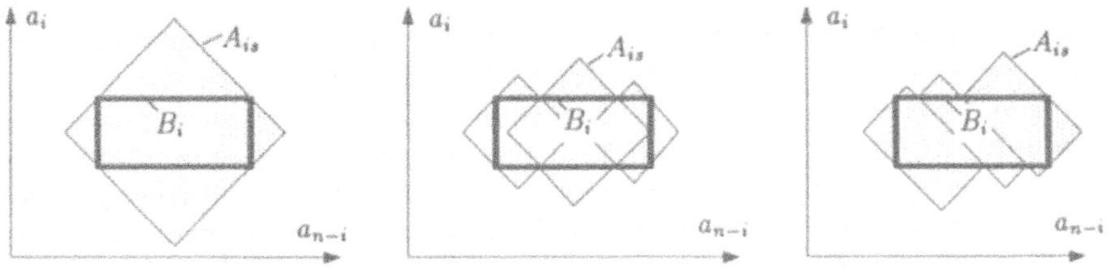

Figure 21.2: Sufficiency tests for axis-parallel box B_i with different degree of conservativeness

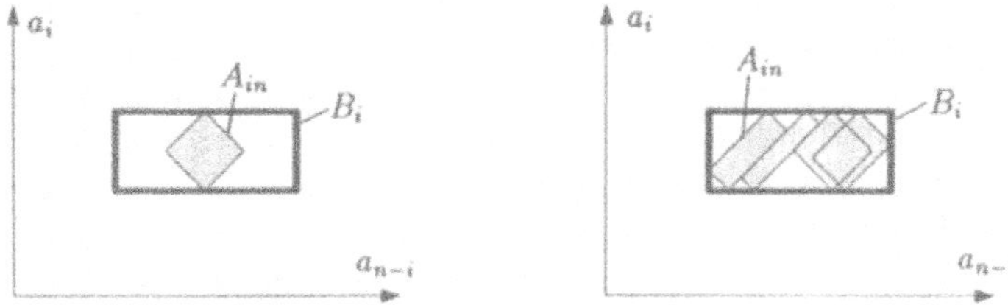

Figure 21.3: Necessity tests

Given a polynomial $s_n(z) = \sum_{i=0}^{n} d_i z^i$ the reciprocated polynomial $s_n^R(z)$ is defined as

$$s_n^R(z) = z^n s_n(z^{-1}) = d_n + d_{n-1}z + \ldots + d_o z^n \ . \tag{21.43}$$

A polynomial of degree n is called symmetric or mirror polynomial if $s_n^R(z) = s_n(z)$ or $s_i = s_{n-i} \quad \forall\, i = 0, 1 \ldots n$ and antisymmetric or antimirror if $s_n^R(z) = -s_n(z)$ or $s_i = -s_{n-i} \quad \forall\, i = 0, 1 \ldots n$. The polynomial $s_n(z)$ can be decomposed into $s_n(z) = u_1(z) + u_2(z)$ where $u_1(z)$ and $u_2(z)$ are the mirror-image and antimirror-image polynomial, respectively, i.e.

$$u_1(z) = \sum_{i=0}^{n} 0.5(a_i + a_{n-i})z^i \qquad u_2(z) = \sum_{i=0}^{n} 0.5(a_i - a_{n-i})z^i \ . \tag{21.44}$$

Then, the given $f(z)$ has zeros only in $|z| < 1$ if and only if $g_e(z)/g_o(z)$ is discrete lossless positive real (*Anderson, B.D.O., et al. 1989*) or, equivalently, the zeros of $g_e(z)$ and

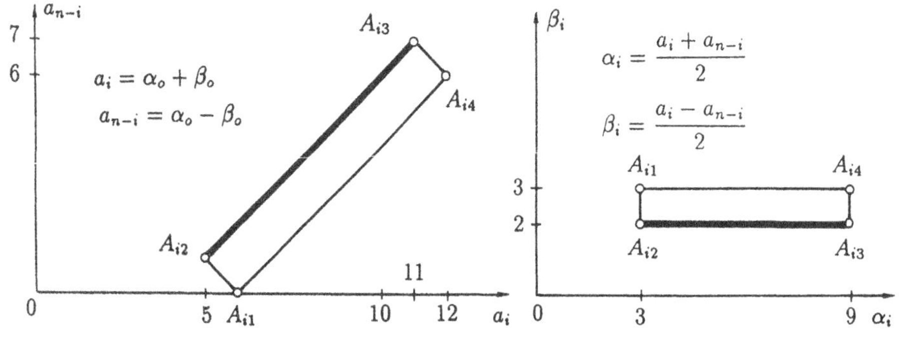

Figure 21.4: Interrelation between polynomial coefficients

$g_o(z)$ are simple, all lie on the unit circle $|z| = 1$ and alternate on $|z| = 1$. Moreover, $|a_n/a_o| > 1$. Discrete lossless positive real functions form a convex set.

As far as this matter is concerned, *Nour Eldin, H.A., 1971* presented basic considerations using Cauchy's index. *Schüssler, H.W. 1976* additionally pointed out conditions on the slope of $\Im m\, g_e(z)/g_o(z)$ on the unit circle. *Gnanasekaran, R., 1981* introduced sufficient conditions for stability of two-dimensional discrete-time systems with polynomials of the type $\sum_{i=0}^{n} d_i(x)z^i$ where $d_i(x)$ is a real polynomial in x derived from the coefficients of $a(z)$, thus completing *Schüssler, H.W., 1976 and Bose, N.K., 1977*. Moreover, the location of the zeros with respect to the z-plane unit circle can be carried out in a tabular formulation where the table has reduced size and needs less computation than other tables (*Bistritz, Y., 1984*). *Kraus, F.J., et al. 1988a* presented robust stability conditions for low-order polynomials.

21.5.6 Generalized Polynomial

Given two polynomials $f_1(z)$ and $f_2(z)$, by Eqs.(21.41) and (21.42), $g_{e1}(z)$, $g_{e2}(z)$, $g_{o1}(z)$, $g_{o2}(z)$ are defined. Then, if and only if $g_{ei}(z)$ and $g_{oi}(z)$ are stable $\forall\, i = 1,2$, the following four polynomials are stable $\forall\ \lambda \in [0,1]$ (*Bose, N.K., 1985*)

$$\lambda g_{e1} + (1-\lambda)g_{e2} + g_{o1} \qquad G_1G_3 \tag{21.45}$$
$$\lambda g_{e1} + (1-\lambda)g_{e2} + g_{o2} \qquad G_2G_4 \tag{21.46}$$
$$\lambda g_{o1} + (1-\lambda)g_{o2} + g_{e1} \qquad G_1G_4 \tag{21.47}$$
$$\lambda g_{o1} + (1-\lambda)g_{o2} + g_{e2} \qquad G_2G_3 \ . \tag{21.48}$$

The polynomials mentioned above can be transposed into the diagram of Fig. 21.5a, i.e., the polynomial

$$\lambda g_{e1} + (1-\lambda)g_{e2} + g_{o1} \quad \text{for} \quad \lambda = 1 \quad \text{corresponds to} \quad G_1 \quad \text{or to} \quad f_1(z)\ . \tag{21.49}$$

For $\lambda \in [0,1]$ it corresponds to the line G_1G_3. Finally, the four polynomials can be replaced by a single generalized polynomial of a stable family

$$\lambda g_{e1} + (1-\lambda)g_{e2} + \mu g_{o1} + (1-\mu)g_{o2} \qquad \forall\ \lambda \in [0,1],\ \mu \in [0,1] \tag{21.50}$$

which corresponds to the entire box G_1, G_2, G_3, G_4 in Fig. 21.5a. Hence, the result is a sufficient condition for stability.

At the expense of introducing more points between the end-points G_1 and G_2, tighter sufficiency conditions can be achieved, see Fig. 21.5b (*Kraus, F.J., et al. 1988*).

21.5.7 Positive Realness and Schur Stability

Positive realness and Schur stability are related by the following: The real coefficient transfer function $G(z) = p_1(z)/p_2(z)$ is strictly positive real if and only if both polynomials $p_1(z)$ and $p_2(z)$ are strictly Schur stable and $\Re e\ G(e^{j\omega}) > 0$ for all real ω .

In the presence of interval variations of the coefficients of $p_1(z)$ and $p_2(z)$ according with a Kharitonov-type box domain the conditions for positive realness of $G(z)$ are presented by *Katbab, A., and Jury, E.I., 1990a; Chapellat, H., et al. 1991*.

Table 21.1: Number N of corners versus degree n of $f(z)$

n	2	3	4	5	6	7	8	9	10	30
N	2	4	10	22	32	52	48	76	80	576

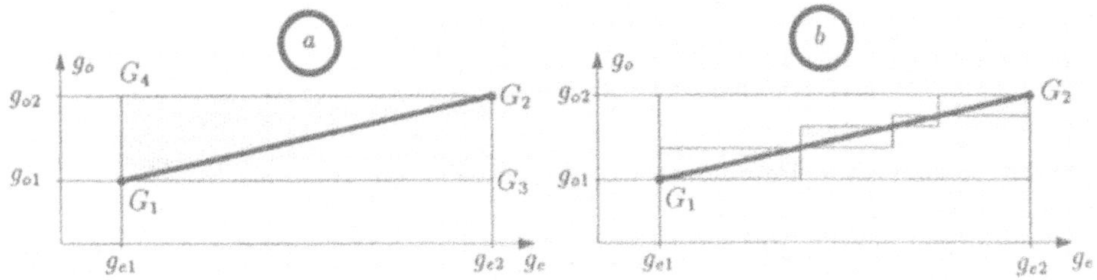

Figure 21.5: Generalized polynomial in g_o, g_e-plane

21.5.8 Discrete Analog to Kharitonov's Strong Theorem

A theorem analogous to the strong version of Kharitonov's theorem was presented by *Mansour, M., Kraus, F.J., and Anderson, B.D.O., 1989; Mansour, M., 1989*. For the reduction of the number of corner points the interlacing property on the unit circle is used. The zeros of $g_e(z)$ and $g_o(z)$ are projected onto the horizontal line $[-1,+1]$. The interlacing property of the zeros is not affected. The projections $g_e'(\lambda)$ and $g_o'(\lambda)$ turn out Chebyshev polynomials. The interlacing property of the zeros of $g_e'(\lambda)$ and $g_o'(\lambda)$ on the line $[-1,+1]$ guarantees stability. The interlacing property is satisfied if in every interval $[-1,+1]$ the four polynomial pairs $(g'_{e\max}, g'_{o\max})$, $(g'_{e\max}, g'_{o\min})$, $(g'_{e\min}, g'_{o\max})$ and $(g'_{e\min}, g'_{o\min})$ are Schur stable where $(g'_{e\max}, g'_{e\min})$ etc. are defined by choosing α_i to equal its largest, smallest etc. value, respectively.

The number of corners N versus degree n of the polynomial $f(z)$ is given in Table 21.1. For large N the number of corners to be checked increases less than quadratic. For comparison, the number of corners as given by the weak version is 2^{n+1}. Note that N does not increase smoothly with n.

Fig. 21.6 presents the list of corner points to be checked for $n = 2$ through $n = 8$ to obtain necessary and sufficient conditions for Schur stability. The first column of $n = 5$ means: $\alpha_{o\min}, \alpha_{1\min}, \alpha_{2\min}, \beta_{o\min}, \beta_{1\max}, \beta_{2\min}$. Note that for $n = 6$, 7, 8 all the four polynomial combinations $(g'_{e\max}, g'_{o\max})$, $(g'_{e\max}, g'_{o\min})$, $(g'_{e\min}, g'_{o\max})$ and $(g'_{e\min}, g'_{o\min})$ with block length 4×8, 4×13, 4×12 must be considered. In Fig. 21.6 only one block is depicted. The polynomial coefficients of $g'_{o\max}$ and $g'_{o\min}$ are inverse, i.e., ■ and □ are interchanged.

21.5.9 Polynomials With Coefficients Symmetric to the Origin

Consider a Schur stable polynomial with real coefficients $f(z) = \sum_0^n a_i z^i = 0, \quad a_n = 1$. For Schur stability it is sufficient that $\sum_0^{n-1} |a_i| < 1$. Allowing parameter variations in a perturbed polynomial $f_p(z) = z^n + \sum_0^{n-1} a_{ip} z^i$ where $|a_{ip}| \le a_i \quad \forall\, i \in [1, 2 \ldots n-1]$, this perturbed polynomial $f_p(z)$ is Schur stable if and only if the unperturbed parameters

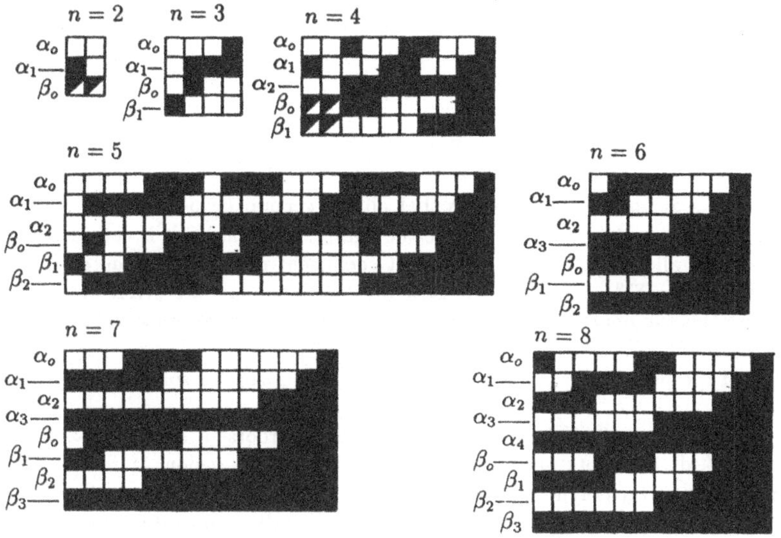

Figure 21.6: Corners to be checked for $n = 2$ through $n = 5$ where ■ denotes maximum value, □ minimum and ◩ maximum or minimum

obey the condition

$$\sum_{0}^{n-1} a_i < 1 . \tag{21.51}$$

For the proof see *Mori, T., and Kokame, H., 1986*.

21.6 Edge Theorem and Minimum Number of Edges

Consider some general polytope of coefficients. The corresponding polynomial is

$$a_p(s) = a_o(s) + \sum_{l=1}^{N} \mu_l a_l(s) \qquad 0 \le \mu_l \le 1 \qquad l = 1, 2 \ldots N \tag{21.52}$$

where $a_l(s)$ are the corresponding corner polynomials. The coefficient perturbations are considered polytopic, which means that the family of polynomials given by these coefficients is expressed as a convex hull of the generating polynomials $a_1(s)$ through $a_N(s)$. Varying μ_l the edge in direction $a_l(s)$ is obtained. The edge polynomials are given by

$$a_{ik}(s) = a_i(s) + \lambda[a_k(s) - a_i(s)] = (1-\lambda)a_i(s) + \lambda a_k(s) \quad \forall i, \quad k > i, \quad \lambda \in [0,1] . \tag{21.53}$$

Another representation is: The polytope P of nth order polynomial is a set of polynomials $a(s)$ generated by polynomials $a_l(s)$

$$\mathrm{P} = \{ \; a(s) = \sum_{l=1}^{N} \lambda_l a_l(s) \; : \; \sum_{1}^{N} \lambda_l = 1, \quad \lambda_l \ge 0 \;\; \forall \, l \; \} . \tag{21.54}$$

When $a(s) = \sum_0^{n-1} a_i s^i + s^n$ then the polynomial is monic.

A general nth order polynomial $p(s)$ is described by $p(s) = \sum_{i=0}^{n} a_i s^i \quad a_n \neq 0$. A polytope of polynomials $p(s)$, based on N given polynomials $p_l(s)$ with coefficients a_{li}, is characterized by the relation

$$a_i = \sum_{l=1}^{N} \lambda_l a_{li} \qquad \forall\, \lambda_l \,:\, \lambda_l \geq 0\,,\; \sum_{l=1}^{N} \lambda_l = 1\,. \tag{21.55}$$

A polytope of real polynomials requires that all the generating polynomials $p_l(s)$ are of nth order with the same sign of a_{ln} . Since the family of polynomials is considered as the convex hull this allows for linearly dependent coefficient perturbations $a_{li}(\mathbf{q})$ in the lth polynomial $a_l(s) = \sum_{i=0}^{n} a_{li}(\mathbf{q}) s^i$. The generating polynomials are given by coefficients $a_{li}(\mathbf{q})$ where the components of the vector $\mathbf{q}$ coincide with its extreme values $q_{i\,\max}$ and $q_{i\,\min}$.

A zero of the polytope is given if (*not only if*) it is a zero of some of the generating polynomials.

The objective is to find the region in which all the roots of $a_p(s) = 0$ are located, or to check if all the roots are located in a predetermined region. It is important to focus one's interest on polynomial families which are polytopic in the coefficient space, e.g., when the coefficients of the polynomial depend linearly on uncertain parameters and these parameters are confined to a polytopic set, see Eq.(21.109).

Edge Theorem: Shown by *Bartlett, A.C., Hollot, C.V., and Lin, H., 1987; Kraus, F.J., Mansour, M., and Jury, E.I., 1990; Kraus, F.J., and Mansour, M., 1990* a polytope of nth order real polynomials is Γ-stable, i.e., the roots are located inside Γ in the complex plane, for a connected set Γ, if all the exposed edges are Γ-stable. *Exposed sets* are those sets of the polytope of coefficients which are elements of a nontrivial supporting hyperplane. Exposed *edges* and *faces* are one-dimensional and two-dimensional exposed sets, respectively. The exposed edges of the polytope of polynomials completely guarantee the roots of every member of the polynomial set. The exposed edges are parametrized in only one single variable, see λ above. As a consequence of the one-dimensional dependence the computational effort is low. Linear dependence of polynomial coefficients and, in addition, arbitrary connected regions Γ in the complex plane are admissible, both in contrast to Kharitonov's theorem. Dependent coefficient variations are arising in practical control problems, e.g., control systems with a fixed controller and a plant with uncertain parameters, see e.g. Eq.(21.71). Since the result of the Edge theorem is given in terms of root locations the Edge theorem is applicable to *both* continuous-time and discrete-time systems. Moreover, it can be used to determine system stability and system performance.

The result of the Edge theorem can be executed, e.g., by using the root locus method (*Kraus, F.J., und Mansour, M., 1987*). For practicality, the root locus is generated by $a_i(s) + \lambda/(1-\lambda)\; a_k(s) = a_i(s) + v a_k(s)$, varying v from 0 to ∞ . When Γ is the open left half-plane then the v-sweep stability test can be replaced by a one-shot-test (*Fu, M., and Barmish, B.R., 1988*).

Since the number of edges increases exponentially with n, possibilities to reduce the number of edges to be checked are very important (*Kraus, F.J., and Mansour, M., 1990*).

For a generalized Edge theorem which extends the well-known Edge theorem to polytopes of functions that are not necessarily polynomials see *Dasgupta, S., et al. 1988*. As a modification, domains which approximate the unit circle from inside or outside can be used to check the Schur stability of a polynomial. It turns out that these conditions can be made either sufficient or necessary and even both. The conditions are quadratic in the coefficients of the polynomial under consideration (*Fam, A.T., 1989*).

The Edge theorem states that for the case of affine coefficient functions $a_i(\mathbf{q})$ it suffices to check the exposed edges. For coefficient functions more complicated than affine (e.g., multilinear, polynomial), an unstable enclave inside a box of uncertain parameters might exist and an edge test does not detect this enclave (*Ackermann, J., et al. 1990*).

If and only if the zeros of the exposed edges are contained in Γ then the zeros of the entire polytope are located inside Γ . This region Γ need not satisfy the restriction of simple connectedness. Any region Γ is admissible if its complement Γ^c has the property that every point $d \in \Gamma^c$ lies on some continuous path which remains within Γ^c and is unbounded (*Fu, M., and Barmish, B.R., 1988a*), i.e., Γ^c is pathwise connected on the Riemann sphere. This requirement is less restrictive than simple connectedness of Γ and allows for a large class of disconnected regions, e.g., for the dominant pole location problem. The so-called dominant pole location problem is a special case of Γ-stability. Two closed-loop poles are assigned within some prescribed region of a given pair of conjugate complex poles and the remaining poles are assigned far in left of the complex s-plane. The region Γ of the assigned poles, thus, is not a simple connected region.

21.7 Stability of Disc Polynomials

Consider the nominal and the perturbed polynomial of a continuous-time dynamic system, respectively,

$$\begin{aligned} a(s) &= a_o + a_1 s + \ldots + a_n s^n && (21.56) \\ a_p(s) &= a_o + \delta_o + (a_1 + \delta_1)s + \ldots + (a_n + \delta_n)s^n \triangleq a(s) + \delta(s) \,. && (21.57) \end{aligned}$$

The polynomial $a_p(s)$ is called disc polynomial if the coefficients $a_{pi} = a_i + \delta_i$ are located inside a disc D_i with center a_i and radius r_i, i.e.,

$$|(a_i + \delta_i) - a_i| = |\delta_i| \leq r_i \,. \tag{21.58}$$

The disc D_n may not contain the origin of the complex plane. For comparison, Kharitonov's theorem yields necessary and sufficient conditions that the roots of $a_p(s)$ are located in the left-half s-plane when the coefficients a_i are confined to a rectangular.

The perturbed system is stable if and only if $a_p(j\omega) \neq 0$. Hence, from Eq.(21.57)

$$a_p(s) \mid_{s=j\omega} = a_p(j\omega) = a(j\omega) + \delta(j\omega) \neq 0 \quad \rightsquigarrow \quad 1 + \frac{\delta(j\omega)}{a(j\omega)} \neq 0 \,. \tag{21.59}$$

Dividing the above equation by $a(j\omega)$ is admissible since the unperturbed system is preassumed stable. Referring to Nyquist's theorem the plot $\delta(j\omega)/a(j\omega)$ may not touch $(-1, j0)$ or

$$\sup_\omega \left|\frac{\delta(j\omega)}{a(j\omega)}\right| = \left\|\frac{\delta(s)}{a(s)}\right\|_\infty < 1 \,. \tag{21.60}$$

Let $\delta_i = \varepsilon_i r_i e^{j\theta_i}$, $\varepsilon_i \in [0,1]$, $\theta_i \in [0, 2\pi)$ then

$$\delta(j\omega) = \sum_{i=0}^{n} \delta_i (j\omega)^i = \sum_{i=0}^{n} \varepsilon_i r_i e^{j\theta_i} (j\omega)^i \,. \tag{21.61}$$

The worst case or largest perturbation results from selecting $\varepsilon_i = 1$ and $\theta_i = -i\pi/2$ which leads to

$$\gamma(j\omega) \triangleq r_o + r_1\omega + r_2\omega^2 + r_3\omega^3 + r_4\omega^4 + \ \ldots\ + r_n\omega^n \tag{21.62}$$

or

$$\gamma(s) = r_o - jr_1 s - r_2 s^2 + jr_3 s^3 + r_4 s^4 - .... s^n . \tag{21.63}$$

Hence, $|\delta(j\omega)| \leq \gamma(j\omega)$ and necessary and sufficient for the Hurwitz stability of $a_p(s)$ is (*Chapellat, H., et al. 1990*)

$$\|\frac{\gamma(s)}{a(s)}\|_\infty = \sup_\omega |\frac{\gamma(j\omega)}{a(j\omega)}| < 1 . \tag{21.64}$$

In the derivation above it was preassumed that the centers a_i are located on the real axis which corresponds to physically realizable systems.

21.7.1 Unidirectional Uncertainty Margin η

If a disc polynomial is stable the largest admissible positive number η enlarging the discs from r_i to ηr_i is given by $\eta = \|\gamma(s)/a(s)\|_\infty^{-1}$.

21.7.2 Schur Stability

Recall the Nyquist criterion in the discrete-time case. Hence, analogously to Eqs.(21.57) and (21.64), the perturbed polynomial $f_p(z)$ is Schur stable if the unperturbed polynomial $f(z)$ is Schur and

$$\gamma(e^{j\theta}) = \sum_{i=0}^{n} \varepsilon_i r_i (e^{j\theta})^i \qquad \theta = [0,\ 2\pi) \tag{21.65}$$

$$\sup_\theta |\frac{\gamma(e^{j\theta})}{f(e^{j\theta})}| = \sup_\theta |\frac{\sum_{i=0}^n r_i}{f(e^{j\theta})}| < 1 \qquad \rightsquigarrow \qquad \sum_{i=0}^{n} r_i < \inf_\theta f(e^{j\theta}) . \tag{21.66}$$

21.7.3 Interdependent Perturbation

Consider a set of perturbed polynomials

$$a_p(s) = p_o(s) + \sum_{l=1}^{N} \delta_l p_l(s) \tag{21.67}$$

where $\delta_l \in D_l \ \ \forall l = 1, 2 \ldots N$ and D_l are discs with center d_l and radius r_l . The entire family $a_p(s)$ is Hurwitz stable if the unperturbed polynomial $a(s) = p_o(s) + \sum_{l=1}^N d_l p_l(s)$ is Hurwitz stable and

$$\sup_\omega \frac{\sum_{l=1}^n r_l |p_l(j\omega)|}{|p_o(j\omega) + \sum_{l=1}^N d_l p_l(s)|} < 1 . \tag{21.68}$$

21.8 Robustly Stabilizing Interval Plants

This section is devoted to a generalization of Kharitonov's theorem. The application of Kharitonov's theorem to control systems with interval plants in some cases provides sufficient results, only. An example is given by the following case. Suppose that in the coefficient space of the polynomial of an interval plant a segment is needed which is a line, only. Kharitonov's theorem provides necessary and sufficient conditions for an entire rectangular. One has to overbound the line with a rectangle and to apply the Kharitonov bound to its vertices. Hence, points in coefficient space are enclosed by Kharitonov's theorem that are not necessary to be included from the interval point of view. Thus, the

Kharitonov conditions are necessary and sufficient for the rectangular but only sufficient for the line segment.

Let $\mathcal{F}$ denote the family of all possible m-tuples $\mathcal{P}$ where the coefficients vary in the intervals cited in Eq.(21.2). For each individual polynomial $p_i(s)$ according to the given interval of its coefficients the four Kharitonov polynomials k_i^r are stated, see Eqs. (21.5) through (21.8) $\forall i = 1, 2 ... m$, $\forall r = 1, 2, 3, 4$. The segment $[k_l^1(s), k_l^2(s)]$ is defined by all convex combinations

$$(1-\lambda)k_l^1(s) + \lambda k_l^2(s) \qquad \lambda \in [0,1] \; . \tag{21.69}$$

Referring to the four Kharitonov polynomials the Kharitonov segments are defined by

$$[k_l^1(s), k_l^2(s)] \quad [k_l^1(s), k_l^3(s)] \quad [k_l^2(s), k_l^4(s)] \quad [k_l^3(s), k_l^4(s)] \; . \tag{21.70}$$

Consider the single-input multiple-output plant $\mathbf{G}(s)$ and the appropriate multiple-input single-output controller $\mathbf{C}(s)$, characterized by the polynomials in the following scheme

$$\mathbf{G}(s) = \frac{1}{p_1(s)} \begin{pmatrix} p_2(s) \\ p_3(s) \\ \vdots \\ p_m(s) \end{pmatrix} \qquad \mathbf{C}(s) = \frac{1}{c_1(s)} \Big(c_2(s) \; \vdots \; c_3(s) \; \vdots \; ..... \vdots \; c_m(s) \Big) \; . \tag{21.71}$$

Then, the characteristic polynomial of the closed-loop system is

$$\det[\mathbf{I} + \mathbf{G}(s)\mathbf{C}(s)] = \sum_1^m c_i(s) p_i(s) \; . \tag{21.72}$$

Let $\mathcal{P}$ denote an m-tuple of real polynomials

$$\mathcal{P} \triangleq \{p_1(s), \; p_2(s), ... \; p_m(s)\} \tag{21.73}$$

where the coefficient $p_{i,j}$ of s^j in the polynomial $p_i(s)$ varies in a predetermined interval $p_{i,j} \in [l_{i,j} \;\; h_{i,j}]$ $i = 1, 2 ... m$, $j = 1, 2 ... \partial p_i(s)$. Collecting all possible segments the entire family is denoted $\mathcal{S}_m$. Any one of these segments can be selected to state the set of all convex combinations of two m-tuples of polynomials. In general,

$$\mathcal{P}_\lambda = \Big(k_1^{r_1}(s), \; k_2^{r_2}(s), \; .... \; k_{l-1}^{r_{l-1}}(s), \; (1-\lambda)k_l^1(s) + \lambda k_l^2(s), \; k_{l+1}^{r_{l+1}}(s), ... \; k_m^{r_m}(s)\Big) \tag{21.74}$$

$$\forall l = 1, 2 ... m, \quad r_\nu = 1, 2, 3, 4 \qquad \lambda \in [0,1] \; . \tag{21.75}$$

The total number of segments is given by the number of combinations. Four Kharitonov polynomials per position, m positions, $l \in [1, m]$ yields $m4^m$ segments.

Let $\mathcal{K}_m$ be the finite set of all possible m-tuples where each polynomial equals one of the four Kharitonov polynomials. Then, the set $\mathcal{K}_m$ possesses 4^m m-tuples. Now, let $C = \{c_1(s), \; c_2(s), ... \; c_m(s)\}$ be any m-tuple of polynomials corresponding to the controller $\mathbf{C}(s)$. The m-tuple C stabilizes $\mathcal{P}$ if the polynomial Eq.(21.72) is stable. The m-tuple C stabilizes the entire segment if C stabilizes $\mathcal{P}_\lambda$ $\forall \, \lambda = [0 \;\; 1]$.

21.8.1 Box Theorem

Following *Chapellat, H., and Bhattacharyya, S.P., 1989*, C stabilizes the entire family $\mathcal{F}$ if and only if C stabilizes every segment $\mathcal{S}_m$. Under the restriction that every $c_i(s)$ is

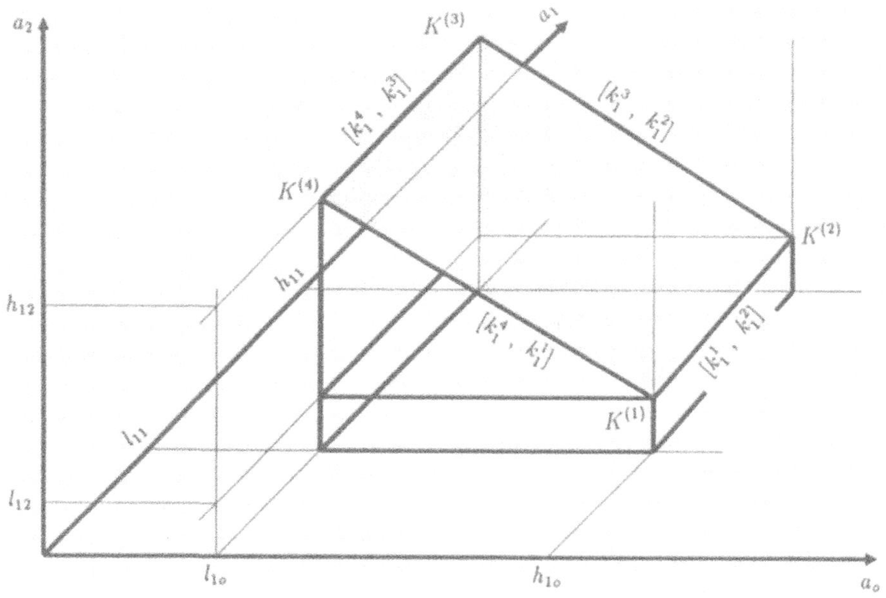

Figure 21.7: Kharitonov segments for the second-order polynomial $p_1(s)$

either an even or an odd polynomial, C stabilizes $\mathcal{F}$ if every m-tuple $\mathcal{K}_m$ is stabilized.

Example: The four Kharitonov segments for second-order $p_1(s) = p_{1o} + p_{11}s + p_{12}s^2$ are

$$a_{e1} \quad k_1^1 = h_{10} + l_{11}s + l_{12}s^2 \quad K^{(1)} \tag{21.76}$$
$$a_{e2} \quad k_1^2 = h_{10} + h_{11}s + l_{12}s^2 \quad K^{(2)} \tag{21.77}$$
$$a_{e3} \quad k_1^3 = l_{10} + h_{11}s + h_{12}s^2 \quad K^{(3)} \tag{21.78}$$
$$a_{e4} \quad k_1^4 = l_{10} + l_{11}s + h_{12}s^2 \quad K^{(4)} \tag{21.79}$$

The corresponding Kharitonov segments are depicted in Fig. 21.7. □

The average degree of $\partial p_i(s)$ of the polynomial $p_i(s)$ be denoted q . Then, the edge theorem requires $m(q+1)2^{m(q+1)-1}$ edges. The Box theorem only needs $m4^m$ line segments independent of q .

The stability margin on the basis of the above derivation is an important design parameter since it estimates the distance from instability (*Chapellat, H., and Bhattacharyya, S.P., 1989a*).

A generalization of the Box theorem for determining robust Schur stability of linear time-invariant discrete-time systems for interval plants is presented by *Katbab, A., and Jury, E.I., 1990*.

21.8.2 Unstructured Stability Margin

Consider the family of perturbed plants with scalar transfer function $G(s) = n(s)/d(s)$ where $n(s)$ and $d(s)$ belong to a family of real interval polynomials with coefficients in a predetermined box. The associated Kharitonov polynomials are denoted $k_n^i(s)$ and $k_d^j(s)$, respectively, $\forall\ i,j = 1,2,3,4$. The transfer functions associated with $k_n^i(s)$ and $k_d^j(s)$ are referred to as Kharitonov systems $G_K(s)$ (*Chapellat, H., et al. 1990a*). There are 16 Kharitonov systems $G_K(s)$. Then, referring to the Small gain theorem, a single-loop control system which consists of the perturbed plant cited above and a controller with transfer function $C(s)$ is stable if and only if

$$\|C(s)\|_\infty < \alpha \quad \text{and} \quad \alpha \leq \frac{1}{\max_{G\in G_K} \|G(s)\|_\infty} . \tag{21.80}$$

So it is sufficient to compute the maximum H_∞-norm of the Kharitonov systems. The computational effort is reduced considerably.

21.9 Distance from Hurwitz Stability

Consider the polynomial $p(s,\mathbf{a}) = \sum_{i=0}^{n} a_i s^i$ where $\mathbf{a} = \mathbf{a}(\mathbf{q})$ is the polynomial coefficient. The vector $\mathbf{q}$ represents the uncertainty parameter vector. Interest arises in the minimal distance of a given nominal polynomial from the stability boundary, particularly the Hurwitz stability boundary. There are three root boundaries: (i) a real root boundary $s = 0 \quad p(0,\mathbf{a}_R) = 0$, (ii) a degree-dropping boundary for roots $s = \infty$ characterized by $a_n = 0$ and (iii) a complex root boundary for $s = j\omega \quad p(j\omega,\mathbf{a}_C) = 0$.

The perturbed and unperturbed parameter is termed $\mathbf{a}_p = \mathbf{a} + \mathbf{F}\mathbf{q}$ and $\mathbf{a}$, respectively. For $\mathbf{q} = \mathbf{0}$ the system is considered unperturbed. Find $\mathbf{q}_{\min} = \arg\min_q \|\mathbf{q}\|_F$ such that $p(s,\mathbf{a}_p)$ is Hurwitz. The minimum distance is termed $r = \|\mathbf{q}_{\min}\|_F$. Then, the largest hypersphere in q-space has the radius r. The polynomial is Hurwitz for all $\|\mathbf{q}\|_F < r$ (*Kaesbauer, D., and Ackermann, J., 1990*).

Preliminary 1: The largest hypersphere with the center in the origin and touching the hyperplane $a_1q_1 + a_2q_2 + \ \ldots\ + a_{l_q}q_{l_q} + e_o = 0$ in the q-space is given by solving

$$\|\mathbf{q}\|_F^2 + \lambda(\sum_{i=0}^{l_q} a_i q_i + e_o) \triangleq \mathbf{q}^T\mathbf{q} + \lambda(\mathbf{a}^T\mathbf{q} + e_o) \to \min \tag{21.81}$$

$$\text{for} \quad \lambda = \frac{2e_o}{\mathbf{a}^T\mathbf{a}} , \quad \mathbf{q}_{\min} = -\frac{\lambda}{2}\mathbf{a} , \quad \mathbf{q}_{\min}^T\mathbf{q}_{\min} = \|\mathbf{q}_{\min}\|_F^2 = \frac{e_o^2}{\mathbf{a}^T\mathbf{a}} . \ \square \tag{21.82}$$

Preliminary 2: Find $\min\|\mathbf{q}\|_F$ subject to $\mathbf{Da} + \mathbf{DFq} = \mathbf{0}$. Referring to the definitions given in Eq.(21.87) and $\mathbf{DF} \triangleq \binom{\mathbf{g}_1^T}{\mathbf{g}_2^T}$, the vectors $\mathbf{g}_1$ and $\mathbf{g}_2$ are orthogonal. The problem leads to

$$\mathbf{q}^T\mathbf{q} + \lambda_1(\mathbf{d}_1^T\mathbf{a} + \mathbf{g}_1^T\mathbf{q}) + \lambda_2(\mathbf{d}_2^T\mathbf{a} + \mathbf{g}_2^T\mathbf{q}) \to \min \tag{21.83}$$

where λ_1 and λ_2 are Lagrange multipliers. Differentiating with respect to $\mathbf{q}$, λ_1, λ_2 yields

$$\mathbf{q}_{\min} = -\frac{\mathbf{d}_1^T\mathbf{a}}{\mathbf{g}_1^T\mathbf{g}_1}\mathbf{g}_1 - \frac{\mathbf{d}_2^T\mathbf{a}}{\mathbf{g}_2^T\mathbf{g}_2}\mathbf{g}_2 \quad \text{and} \quad \|\mathbf{q}_{\min}\|_F = \sqrt{\frac{(\mathbf{d}_1^T\mathbf{a})^2}{\mathbf{g}_1^T\mathbf{g}_1} + \frac{(\mathbf{d}_2^T\mathbf{a})^2}{\mathbf{g}_2^T\mathbf{g}_2}} . \ \square \tag{21.84}$$

21.9.1 Simple Perturbation

In this subsection root boundaries $s = 0$ and $s = j\omega$ are considered, only, $\mathbf{F}$ is selected $\mathbf{F} \triangleq \binom{\mathbf{I}_{n\times n}}{\mathbf{0}_{n\times 1}}$, and $\mathbf{q} \in \mathcal{R}^n$, $\mathbf{a}, \mathbf{a}_p \in \mathcal{R}^{n+1}$, $\mathbf{F} \in \mathcal{R}^{(n+1)\times n}$. The perturbed parameter is $\mathbf{a}_p = \mathbf{a} + \mathbf{F}\mathbf{q}$.

For the real root boundary one has $s = 0$ and

$$p(0, \mathbf{a}_p) = p(0, \mathbf{a} + \mathbf{F}\mathbf{q}) = 0 \quad \rightsquigarrow \quad (\mathbf{a} + \mathbf{F}\mathbf{q})_o = 0 \quad \rightsquigarrow \quad a_o + (\mathbf{F}\mathbf{q})_o = 0\,. \tag{21.85}$$

When $\mathbf{F} = \binom{\mathbf{I}_{n\times n}}{\mathbf{0}_{n\times 1}}$ it results $a_o + q_1 = 0$ and $q_{R\min} = q_1 = -a_o$ and $r_R = a_o$.

For the complex root boundary

$$p(j\omega, \mathbf{a}_p) = p(j\omega, \mathbf{a} + \mathbf{F}\mathbf{q}) = \sum_{i=1}^{n} (\mathbf{a} + \mathbf{F}\mathbf{q})_i (j\omega)^i = 0\,. \tag{21.86}$$

The real and imaginary part of $p(j\omega, \mathbf{a} + \mathbf{F}\mathbf{q}) = 0$ is rephrased

$$\mathbf{D}(\mathbf{a} + \mathbf{F}\mathbf{q}) = 0 \quad \text{where} \quad \mathbf{D} = \begin{pmatrix} 1 & 0 & -\omega^2 & 0 & \omega^4 & \cdots \\ 0 & 1 & 0 & -\omega^2 & 0 & \cdots \end{pmatrix} \triangleq \begin{pmatrix} \mathbf{d}_1^T \\ \mathbf{d}_2^T \end{pmatrix} \in \mathcal{R}^{2\times(n+1)}\,. \tag{21.87}$$

Defining $\mathbf{DF} \triangleq \binom{\mathbf{g}_1^T}{\mathbf{g}_2^T}$ one has the problem of finding $\min \|\mathbf{q}\|_F$ such that

$$\mathbf{Da} + \mathbf{DFq} = \mathbf{0}, \quad \text{i.e.,} \quad \mathbf{d}_1^T\mathbf{a} + \mathbf{g}_1^T\mathbf{q} = 0 \quad \text{and} \quad \mathbf{d}_2^T\mathbf{a} + \mathbf{g}_2^T\mathbf{q} = 0\,. \tag{21.88}$$

For the specific case of $\mathbf{F} = \binom{\mathbf{I}_{n\times n}}{\mathbf{0}_{n\times 1}}$

$$\mathbf{DF} = \begin{pmatrix} 1 & 0 & -\omega^2 & 0 & \cdots \\ 0 & 1 & 0 & -\omega^2 & \cdots \end{pmatrix} \in \mathcal{R}^{2\times n} \tag{21.89}$$

is achieved. In view of this, $\mathbf{g}_1$ and $\mathbf{g}_2$ are orthogonal. Since $\mathbf{D}$ depends on ω, the result $\mathbf{q}_{C\min}(\omega)$ also is a function of ω. Finally, selecting the norm r_C of $\mathbf{q}_{C\min}$

$$\rho_H = \min\{r_R\,,\; r_C(\omega)\}\,. \tag{21.90}$$

21.9.2 Affine Linear Dependence

Assume that the polynomial $p(s, \mathbf{q})$ of a closed-loop system is of degree n and the coefficients a_i depend affine linearly on the parameter vector $\mathbf{q} \in \mathcal{R}^{l_q}$. The parameter vector contains parameters $\mathbf{a}$ of the plant and/or controller. The unperturbed system is considered given by $\mathbf{a}$ where $\mathbf{q} = \mathbf{0}$. The coefficient vector of the perturbed system is

$$\mathbf{a}_p = \mathbf{a} + \mathbf{F}\mathbf{q} \qquad \text{where} \qquad \mathbf{F} \in \mathcal{R}^{(n+1)\times l_q}\,. \tag{21.91}$$

The unperturbed system for $\mathbf{q} = \mathbf{0}$ is preassumed stable. Increasing $\mathbf{q}$, there are three possibilities for the solutions of the polynomial to get unstable:

(i) a real zero passes the imaginary axis in the origin at $s = 0$, $\omega = 0$,

(ii) a real zero enters the right half-plane via infinity or

(iii) a pair of conjugate complex zeros passes the imaginary axis at $s = \pm j\omega$.

Each case determines a hypersurface and yields a *largest hypersphere* with radius r_R, r_∞ and r_C, respectively.

Zeros in the Origin

The real zero in the origin is given by $p(s,\mathbf{q}) = p(0,\mathbf{q}) = 0 \quad \leadsto \quad a_{po} = 0$

$$a_{po} = (\mathbf{a} + \mathbf{F}\mathbf{q})_o = a_o + f_{11}q_1 + f_{12}q_2 + \ \dots \ + f_{1l_q}q_{l_q} = 0 \ . \tag{21.92}$$

Invoking Eq.(21.82)

$$\|\mathbf{q}_{R\min}\|_F^2 = r_R^2 = \frac{a_o^2}{\mathbf{f}_1^T\mathbf{f}_1} \quad \text{where} \quad \mathbf{f}_1 \triangleq (f_{11} \quad f_{12} \quad \dots\dots \quad f_{1l_q})^T \ . \tag{21.93}$$

If a_o is constant the hyperplane does not exist. It is impossible that a zero enters the right half-plane via origin. Hence, r_R is considered infinity.

Zeros at Infinity

A zero at infinity only can arise if the coefficient a_n of s^n vanishes. Rewriting the polynomial in powers of s^{-1} the coefficient of $(s^{-1})^0$ equals a_n , e.g., for a second-order polynomial $p(s) = \frac{1}{a}(s+a)(Ts+1) = \frac{T}{a}s^2 + (T+\frac{1}{a})s + 1 = 0$ a zero near infinity $(a \to \infty)$ and another one at $1/T$ exists. The zero at infinity is represented by the expression $p(\infty,\mathbf{q}) = 0 \quad \leadsto \quad a_{pn} = 0$

$$a_{pn} = (\mathbf{a}+\mathbf{F}\mathbf{q})_{n+1} = a_n + f_{n+1,1}\, q_1 + \dots + f_{n+1,l_q}\, q_{l_q} = 0 \quad \leadsto \quad \|\mathbf{q}_{\infty\min}\|_F^2 = r_\infty^2 = \frac{a_n^2}{\mathbf{f}_{n+1}^T\mathbf{f}_{n+1}} \ . \tag{21.94}$$

Imaginary Zeros

The imaginary zeros are determined by $p(j\omega,\mathbf{a}_p) = p(j\omega,\mathbf{a}+\mathbf{F}\mathbf{q}) = 0$. Separating real and imaginary part yields

$$\mathbf{D}(\mathbf{a}+\mathbf{F}\mathbf{q}) = \begin{pmatrix} \mathbf{d}_1^T \\ \mathbf{d}_2^T \end{pmatrix} (\mathbf{a}+\mathbf{F}\mathbf{q}) = \mathbf{0} \ . \tag{21.95}$$

In the simple case of Eq.(21.89) numerous elements f_{ij} of $\mathbf{F}$ are zero. Now, $\mathbf{F} = \text{matrix}[f_{ij}]$ has nonzero elements. Hence,

$$\begin{aligned} \mathbf{DF} &= \begin{pmatrix} 1 & 0 & -\omega^2 & 0 & \dots \\ 0 & 1 & 0 & -\omega^2 & \dots \end{pmatrix} \begin{pmatrix} f_{11} & f_{12} & f_{13} & \dots \\ \vdots & \vdots & \vdots & \dots \end{pmatrix} \\ &= \begin{pmatrix} f_{11} - \omega^2 f_{31} + \dots \vdots & f_{12} - \omega^2 f_{32} + \dots \vdots & f_{13} - \omega^2 f_{33} + \dots \vdots & \dots \\ f_{21} - \omega^2 f_{41} + \dots \vdots & f_{22} - \omega^2 f_{42} + \dots \vdots & f_{23} - \omega^2 f_{43} + \dots \vdots & \dots \end{pmatrix} \in \mathcal{R}^{2\times l_q} \end{aligned} \tag{21.96}$$

consists of two rows representing two hyperplanes which are not orthogonal. Denoting these rows as $\mathbf{e}_1^T$ and $\mathbf{e}_2^T$, consider

$$\mathbf{e}_3 \triangleq (1-\lambda)\mathbf{e}_1 + \lambda\mathbf{e}_2 \ . \tag{21.97}$$

For λ such that $\mathbf{e}_3 \perp \mathbf{e}_1$ or $\mathbf{e}_3^T\mathbf{e}_1 = 0$ this leads to $\lambda_o = (\mathbf{e}_1^T\mathbf{e}_1)/(\mathbf{e}_1^T\mathbf{e}_1 - \mathbf{e}_2^T\mathbf{e}_1)$. From

$$\mathbf{Da} + \mathbf{DFq} \triangleq \begin{pmatrix} b_1 \\ b_2 \end{pmatrix} + \begin{pmatrix} \mathbf{e}_1^T \\ \mathbf{e}_2^T \end{pmatrix} \mathbf{q} \triangleq \begin{pmatrix} b_1 + \mathbf{e}_1^T\mathbf{q} \\ b_2 + \mathbf{e}_2^T\mathbf{q} \end{pmatrix} = \mathbf{0} \tag{21.98}$$

and by defining

$$b_3 + \mathbf{e}_3^T\mathbf{q} \triangleq (1-\lambda)(b_1 + \mathbf{e}_1^T\mathbf{q}) + \lambda(b_2 + \mathbf{e}_2^T\mathbf{q}) \tag{21.99}$$

the expression $b_3 + \mathbf{e}_3^T\mathbf{q}$ is 0, for any λ if both $b_1 + \mathbf{e}_1^T\mathbf{q} = 0$ and $b_2 + \mathbf{e}_2^T\mathbf{q} = 0$. In order to utilize the result given in Eq.(21.84), $b_2 + \mathbf{e}_2^T\mathbf{q}$ is replaced by $b_3 + \mathbf{e}_3^T\mathbf{q}$ for $\lambda = \lambda_o$ (*Kaesbauer, D., and Ackermann, J., 1990*). Eq.(21.84) leads to $\|\mathbf{q}_{C\,\min}\|_F = r_C$. Note that $\lambda = \lambda(\omega)$. Finally, the distance from Hurwitz stability is given by

$$\rho_H = \min\{r_R,\ r_\infty,\ r_C(\omega)\ \}\ . \tag{21.100}$$

For general boundaries $\partial\Gamma$, other than the imaginary axis, see the *boundary representation theorem* or *Cavallo, A., et al. 1991*.

21.10 Boundary Representation Theorem

Consider a linear time-invariant system with the characteristic polynomial

$$a(s,\mathbf{q}) = a_o(\mathbf{q}) + a_1(\mathbf{q})s + \ \dots\ + a_n(\mathbf{q})s^n \tag{21.101}$$

where $\mathbf{q}$ is an l_q-vector of real uncertain parameters in a set Q of admissible perturbations, and $\mathbf{a}(\mathbf{q}) = [a_o(\mathbf{q})\ a_1(\mathbf{q}) \dots a_n(\mathbf{q})]^T$ is the coefficient vector of the polynomial. The question is if $a(s,\mathbf{q})$ is Γ-stable. Parametrizing all the complex branches of the boundary $\partial\Gamma$ by a parameter α, i.e. $s(\alpha) = \sigma(\alpha) + j\omega(\alpha)$, then $Q_{Im}(\alpha)$ is defined as the set of parameters $\mathbf{q}$ such that $a(s,\mathbf{q}) = 0$.

Then, the Boundary representation theorem for the complex root boundary states that $\mathbf{q} \in Q_{Im}(\alpha)$ if and only if

$$\begin{pmatrix} d_o(\alpha) & d_1(\alpha) & \dots & d_n(\alpha) \\ 0 & d_o(\alpha) & \dots & d_{n-1}(\alpha) \end{pmatrix} \mathbf{a}(\mathbf{q}) = \begin{pmatrix} 0 \\ 0 \end{pmatrix} \tag{21.102}$$

$$\begin{aligned} \text{where}\ d_o(\alpha) &= 1 & (21.103)\\ d_1(\alpha) &= 2\,\sigma(\alpha) & (21.104)\\ d_{i+1}(\alpha) &= 2\,\sigma(\alpha)d_i(\alpha) - [\sigma^2(\alpha) + \omega^2(\alpha)]d_{i-1}(\alpha) \quad \forall\, i = 1,2,\dots n-1\ . & (21.105) \end{aligned}$$

See *Ackermann, J., et al. 1991* for the proof. The method is suitable for a small number l_q of uncertain parameters and polynomial coefficient functions $a_i(\mathbf{q})$.

21.11 Maximum Γ-Stability Bounds for Polytopes of Polynomials

Consider the family of perturbed polynomials in the complex variable s

$$\mathcal{P}_p^\varepsilon \triangleq \{p_p(s,\mathbf{q}) = p(s) + \sum_{i=0}^{n} a_i(\mathbf{q})s^i\ \} \qquad \mathbf{q} \in \varepsilon Q \subset \mathcal{R}^m \tag{21.106}$$

where m corresponds to l_q in Eq.(21.91),

$p(s)$ is the nominal polynomial of nth order,

$\mathbf{q} \in \mathcal{R}^m$ is the vector of perturbation parameters located in the polytope Q,

$a_i(\mathbf{q})$ are the coefficients of the ith power of s as a linear or nonlinear function of the perturbation parameter,

$\varepsilon \geq 0$ is a parameter determining the size of the polytope.

The objective is to find ε such that $p_p(s, \mathbf{q})$ is Γ-stable for all $\varepsilon < \varepsilon_{\max}$. This bound is called maximal Γ-stability bound (*Fu, M., 1989*).

In many cases $a_i(\mathbf{q})$ is linear and q_j belongs to a hyper-rectangle $q_j \in [l_j,\ h_j]$ where the bounds low and high are $l_j < 0 \quad h_j > 0$. Observe that $l_j,\ h_j$ denote the q_j-box not the a_i-box. Rewriting

$$p_o + \sum_1^n a_i(\mathbf{q})s^i = p_o + a_1(\mathbf{q})s + a_2(\mathbf{q})s^2 + ... + a_n(\mathbf{q})s^n \tag{21.107}$$

$$= p_o + a_1(q_1)s + a_2(q_1)s^2 + ... + a_n(q_1)s^n \tag{21.108}$$
$$+a_1(q_2)s + a_2(q_2)s^2 + ... + a_n(q_2)s^n + ... + ... + a_n(q_m)s^n$$

$$\triangleq p_o + \sum_{j=1}^m q_j\ \delta p_j(s) \quad \text{where} \quad q_j \in [l_j,\ h_j] \quad j = 1,2...m\ . \tag{21.109}$$

The polynomials $\delta p_j(s)$ are named perturbation polynomials. When $\delta p_j(s)$ are linear independent this polytope $p_p(s, \mathbf{q})$ is denoted parallelotope.

The family of perturbation polynomials is defined by

$$\delta\mathcal{P}_p \triangleq \{p_p(s) - p(s)\ :\ \ p_p(s) \in \mathcal{P}_p\}\ . \tag{21.110}$$

Introducing a variable size using the parameter ε the family of these polynomials is

$$\mathcal{P}_p^\varepsilon \triangleq \{\ p(s) + \sum_{j=1}^m q_j\ \delta p_j(s)\ :\quad q_j \in [\varepsilon l_j, \varepsilon h_j]\ ,\ \ j = 1,2...m\}\ . \tag{21.111}$$

21.11.1 Boundary Theorem

For any $d \in \partial\Gamma$, i.e., for any path d inside the boundary of Γ in the complex plane, and $p(d) \neq 0$ the normalized value set is defined as

$$V(\delta\mathcal{P}_p, d) \triangleq \{\frac{\delta p_p(d)}{p(d)}\ :\quad \delta p_p \in \delta\mathcal{P}_p\ \}\ . \tag{21.112}$$

If $\mathcal{P}_p$ is a polytope of polynomials then $V(\delta\mathcal{P}_p, d)$ is a convex polygon. The system is Γ-stable if $p(d) \neq 0$ for all $d \in \Gamma^c$ where Γ^c is the complement of Γ . The set Γ, e.g., is the open left half-plane or the open unit disc. The family $\mathcal{P}_p$ of polynomials is Γ-stable if and only if the convex polygon V satisfies

$$-1\ \notin V(\delta\mathcal{P}_p, d) \qquad \forall d \in \partial\Gamma^c\ . \tag{21.113}$$

Necessity is obvious: For $d \in \partial\Gamma \subset \Gamma^c$

$$p_p(d) \neq 0 \quad \rightsquigarrow \quad [p_p(d) - p(d)]/p(d) \neq -1\ . \tag{21.114}$$

For sufficiency see *Fu, M., 1989*.

21.11.2 Minimum Intersection Theorem

Applying Eq.(21.113) to the set $\mathcal{P}_p^\varepsilon$ the minimum intersection is achieved. Note that $\delta\mathcal{P}_p \equiv \delta\mathcal{P}_p^\varepsilon \mid_{\varepsilon=1}$. The relation $-1 \notin V(\delta\mathcal{P}_p^\varepsilon, d)$ states that the intersection of the real axis with the convex polygon $V(\delta\mathcal{P}_p^\varepsilon, d)$ must not reach -1 (*Fu, M., 1989*) .

For a parallelotope of perturbed polynomials the convex polygon is given by

$$V(\delta\mathcal{P}_L, d) = \{\sum_{j=1}^{m} q_j \frac{\delta p_j(d)}{p(d)} \quad : \quad q_j \in [l_j,\ h_j]\ ,\ l_j < 0,\ h_j > 0,\ \ j = 1,2...m\ \}\ . \tag{21.115}$$

The normalized value set has at most $2m$ vertices v_i . Abbreviating $\rho_j(s) \triangleq \delta p_j(s)/p(s)$ for any fixed s, the vertices v_i are given by a simple shifting operation as follows

$$v_1 = l_1\rho_1 + l_2\rho_2 + ... + l_{m-1}\rho_{m-1} + l_m\rho_m \quad \text{where} \quad \rho_j(s) \triangleq \frac{\delta p_j(s)}{p(s)} \tag{21.116}$$

$$v_2 = l_1\rho_1 + l_2\rho_2 + ... + l_{m-1}\rho_{m-1} + h_m\rho_m \tag{21.117}$$

$$\vdots$$

$$v_m = l_1\rho_1 + h_2\rho_2 + ... + h_{m-1}\rho_{m-1} + h_m\rho_m \tag{21.118}$$

$$v_{m+1} = h_1\rho_1 + h_2\rho_2 + ... + h_{m-1}\rho_{m-1} + h_m\rho_m \tag{21.119}$$

$$\vdots$$

$$v_{2m} = h_1\rho_1 + l_2\rho_2 + ... + l_{m-1}\rho_{m-1} + l_m\rho_m\ . \tag{21.120}$$

The edges of $V(\delta\mathcal{P}_L, d)$ are achieved by

$$\text{conv}\{v_i, v_{i+1}\} \qquad i = 1,2....(2m-1) \qquad \text{and} \qquad \text{conv}\{v_{2m}, v_1\}\ . \tag{21.121}$$

The complex numbers ρ_i for any fixed s are arranged such that they obey

$$0 \le \angle(\rho_1) \le \angle(\rho_2) \le .... \le \angle(\rho_m) < 180^o\ . \tag{21.122}$$

Since $l_j < 0$ and $h_j > 0$, and the vertices, per definition, satisfy $\angle(v_1) \le 0$, $\angle(v_{m+1}) \ge 0$, the vertices $v_1, v_2, ...v_{2m}, v_1$ traverse in a clockwise direction.

Summarizing and generalizing: The polytope of polynomials

$$\mathcal{P}_p^\varepsilon \triangleq \{p(s) + \sum_{i=1}^{n} a_i(\mathbf{q})s^i \quad \mathbf{q} \in \varepsilon Q\ \} \tag{21.123}$$

is Γ-stable for $0 < \varepsilon < \varepsilon_{\max}$, $p(s)$ being Γ-stable and $\partial\Gamma$ being the union of a finite number of continuous paths in the complex s-plane, if, selecting $\partial\Gamma \subset \Gamma^c$,

$$-1 \notin V(\delta\mathcal{P}_p^\varepsilon, d) \qquad \text{or} \qquad -\frac{1}{\varepsilon} \notin V(\delta\mathcal{P}_p, d) \qquad \forall d \in \partial\Gamma \subset \Gamma^c \tag{21.124}$$

$$\text{where} \quad \delta\mathcal{P}_p \triangleq \{\delta p_p(s) = p_p(s) - p(s) \qquad p_p(s) \in \mathcal{P}_p\}\ . \tag{21.125}$$

Hence, the intersection of $V(\delta\mathcal{P}_p, d)$ with the real axis must be greater than $-1/\varepsilon$. The intersection depends on d and is termed $I(d)$. Then, the maximum value $\varepsilon_{\max}$ is $\varepsilon_{\max} = 1/|\inf\{I(d)\}|$. The minimum intersection is given either by a vertex or by an edge, and is found by an appropriate computing algorithm. For parallelotopes of polynomials the number of vertices or edges is at most $2m$. Thus, the computational effort is low. A robustness measure can be defined by $(\varepsilon_{\max} - \varepsilon)/\varepsilon_{\max}$, indicating the relative robustness of a system operating with the parameter ε .

21.12 Varying Uncertainty Bounds While Maintaining Stability

The objective is to find the set of allowable perturbations in the vicinity of a nominal polynomial for which stability is not yet affected, i.e., the polynomial will have at least one purely imaginary root (*Argoun, M.B., 1986a*). Single parameter variations can be considered as well as combinations of parameter variations.

The polynomial $a(s) = \sum_0^n a_i s^i$ is perturbed in its coefficients $a_i + \Delta a_i$. They may vary from $a_i - \bar{a}_i$ to $a_i + \bar{a}_i$ or $|\Delta a_i| \leq \bar{a}_i$. The polynomial $a(s)$ is decomposed into even-power terms $r(s)$ and odd-power ones, i.e., $a(s) = r(s) + q(s)$. For odd coefficients only to be perturbed, the perturbed polynomial $a_p(s) = \sum_0^n (a_i + \Delta a_i)s^i$ remains stable if and only if

$$\max_{\bar{a}_i,\ i\ odd} |\Delta q(\omega_k)| \leq |q(j\omega_k)| \qquad \forall\ \omega_k \tag{21.126}$$

where ω_k results from $r(j\omega_k) = 0$ and $\Delta q(\omega_k)$ is a polynomial

$$\Delta q(\omega_k) \triangleq \bar{a}_1 \omega_k + \bar{a}_3 \omega_k^3 + \dots . \tag{21.127}$$

Sufficiency results from the principle of argument (*Yeung, K.S., 1983*). For proof of necessity see *Argoun, M.B., 1986c and 1987*.

Additional even-coefficient perturbations can be studied separately. In view of the results presented above, this theorem enables the designer to indicate how close $a_p(s)$ is to instability.

21.13 Stability Hypersphere

21.13.1 Largest Hypersphere and Hurwitz Region

Given the parameter vector $\mathbf{a} \triangleq (a_n,\ a_{n-1}, \dots\ a_o)^T \in \mathcal{R}^{n+1}$ with descending indices, the space of coefficients a_i is used to state another important result. Defining three vector sets of polynomial coefficients

$$\Delta_o \triangleq \{\mathbf{a} \mid \mathbf{a} \in \mathcal{R}^{n+1}, a_o = 0\} \tag{21.128}$$

$$\Delta_n \triangleq \{\mathbf{a} \mid \mathbf{a} \in \mathcal{R}^{n+1}, a_n = 0\} \tag{21.129}$$

$$\Delta_\omega(\omega) \triangleq \{\mathbf{a} \mid \mathbf{a} \in \mathcal{R}^{n+1}, \mathbf{a} = \mathbf{a}(s) = (s^2 + \omega^2)\mathbf{l}(s)\} \tag{21.130}$$

where $\mathbf{l}(s)$ arbitrary and ω real. The coefficient vectors $\Delta_o, \Delta_n, \Delta_\omega(\omega)$ correspond to polynomials with roots at $s = 0$, $s = \infty$ and $s = \pm j\omega$, respectively. For comparison, see Eq.(21.100). Next, three scalars are defined as the Euclidian distances between the nominal vector $\mathbf{a}_{nom}$ and the subspaces given by Δ_o, Δ_n and Δ_ω

$$d_o \triangleq \|\mathbf{a}_{nom} - \Delta_o\|_F, \quad d_n \triangleq \|\mathbf{a}_{nom} - \Delta_n\|_F, \tag{21.131}$$

$$d_\omega(\omega) \triangleq \|\mathbf{a}_{nom} - \Delta_\omega(\omega)\|_F \quad \rightsquigarrow \quad d \triangleq \inf_\omega d_\omega(\omega) . \tag{21.132}$$

Following *Soh, C.B., et al. 1985; Bhattacharyya, S.P., 1987*, the size of the Hurwitz region can be estimated by the largest $\mathbf{a}$-space hypersphere. This hypersphere is centered at the nominal coefficient vector $\mathbf{a}_{nom}$ and has the radius

$$\rho_H(\mathbf{a}_{nom}) = \min\{d_o, d_n, d\} . \tag{21.133}$$

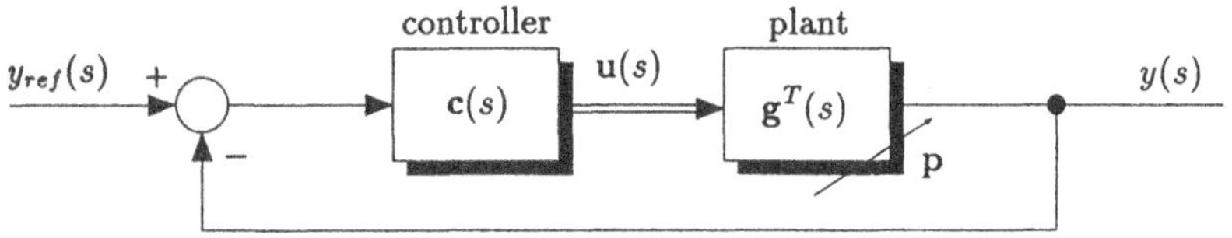

Figure 21.8: Control loop containing a multiple-input single-output plant with perturbed parameter **p**

If changes $\Delta\mathbf{a}$ in the coefficient vector do not exceed a limit, i.e., $\|\Delta\mathbf{a}\|_F < \rho_H(\mathbf{a}_{nom})$, then the closed-loop stability is guaranteed. Hence, $\rho_H(\mathbf{a}_{nom})$ is a measure of stability margin.

For polynomials which have zeros only within or on the unit circle see *Soh, C.B., et al. 1985*.

21.13.2 Stability Hyperellipsoid

In practical control systems the perturbations Δa_i strongly differ in magnitude. Let the perturbations lie within a rectangle. In order to comply with wide range perturbations, hyperellipsoids are employed. The largest stability hyperellipsoid in the parameter space is considered centered at the nominal parameter point $\mathbf{a}_{nom}$. The ratios α_i of the principal axes of the hyperellipsoid are considered fixed, for the present. Keeping the shape fixed the largest possible hyperellipsoid is obtained by enlarging the magnitude by a factor η without affecting the stability. After having enlarged all the ratios from α_i to $\eta\alpha_i$ the stability hypersphere is achieved as follows. Defining the matrix (diag $\alpha_i)^{-1}$ the hyperellipsoid in the **a**-space is mapped onto a hypersphere in $\bar{\mathbf{a}}$-space. The linear transformation $\mathbf{a} = (\text{diag } \alpha_i)\bar{\mathbf{a}}$ maps hyperspheres of the $\bar{\mathbf{a}}$-space into hyperellipsoids in the **a**-space. Then, referring to Eq.(21.133), the optimum η is obtained as

$$\eta^* = \bar{\rho}_H(\bar{\mathbf{a}}_{nom}) = \bar{\rho}_H(\bar{d}_o, \bar{d}_n, \bar{d}) \ . \tag{21.134}$$

By consideration of expediency, the choice of α_i will be assisted by some appropriate optimization calculus. Application to control systems yields the following observation. For a certain fixed assumption of the controller $\mathbf{C}(s)$ the largest stability hypersphere S_h, centered at the nominal plant coefficient vector $\mathbf{p}_{nom}$ with radius $\rho_H(\mathbf{p}_{nom})$, has to be determined such that the closed-loop system remains stable. Then, the stability preserving perturbation $\Delta\mathbf{p}$ of the plant is bounded by $\|\Delta\mathbf{p}\|_F < \rho_H(\mathbf{p}_{nom})$.

21.13.3 Stability Hypersphere and Multiple-Input Plant

Consider a system with multiple-input single-output plant, see Fig. 21.8. The general case of multiple-input multiple-output plant, the general case of functional dependence plant transfer coefficients over perturbation, was investigated by *Bhattacharyya, S.P., 1987* .

Multiple-Input Plant

Let the plant be of order q and the control variable $\mathbf{u} \in \mathcal{C}^m$. The least common denominator of all entries of $\mathbf{g}(s)$ is termed $d(s)$. Then,

$$\mathbf{G}(s) = \mathbf{g}^T(s) \triangleq d^{-1}(s)\mathbf{n}^T(s) \tag{21.135}$$

$$\mathbf{n}^T(s) \triangleq \sum_{i=0}^{q} \mathbf{n}_i^T s^i \qquad d(s) \triangleq d_q s^q + \ldots + d_o \qquad \mathbf{n}_i \in \mathcal{R}^m \ . \tag{21.136}$$

Assuming $\mathbf{n}(s)$ and $d(s)$ coprime, the plant parameter vector is

$$\mathbf{p} \triangleq (\mathbf{n}_o^T, d_o, \mathbf{n}_1^T, d_1, \ldots \mathbf{n}_q^T, d_q) \in \mathcal{R}^{(m+1)(q+1)} \ . \tag{21.137}$$

With $\mathbf{p}_{nom}$ denoting the nominal parameter, the perturbation vector is $\Delta\mathbf{p} = \mathbf{p} - \mathbf{p}_{nom}$.

Controller

Assume that the controller $\mathbf{C}(s)$ stabilizes the closed-loop system at the nominal parameter $\mathbf{p}_{nom}$. With the definitions

$$\mathbf{C}(s) = \mathbf{c}(s) \triangleq \mathbf{n}_c(s) d_c^{-1}(s) \tag{21.138}$$

$$\mathbf{n}_c(s) \triangleq \sum_{i=0}^{p} \mathbf{n}_{ci} s^i \qquad d_c(s) = d_{cp} s^p + \ldots + d_{co} \qquad \mathbf{n}_{ci} \in \mathcal{R}^m \tag{21.139}$$

and $d_c, \mathbf{n}_c$ coprime the closed-loop parameters can be evaluated.

21.13.4 Closed-Loop System

The characteristic polynomial of the closed-loop system is given by the polynomial of degree $n = p + q$

$$\begin{aligned} a(s) &\triangleq d_c(s)d(s) + \mathbf{n}^T(s)\mathbf{n}_c(s) = d_c(s)d(s) + \mathbf{n}_c^T(s)\mathbf{n}(s) && (21.140)\\ &= a_o + a_1 s + a_2 s^2 + \ldots + a_n s^n && (21.141) \end{aligned}$$

with the parameter vector $\mathbf{a} \triangleq (a_n \ \ a_{n-1} \ \ldots \ a_o)^T$.

21.13.5 Mapping p into a

Now, the controller $\mathbf{c}(s)$ is considered as an algebraic operator that linearly maps the plant parameter vector $\mathbf{p}$ into the characteristic vector $\mathbf{a}$. In order to satisfy Eq.(21.141) this map is achieved by premultiplying $\mathbf{p}$ by a matrix $\mathbf{M}_c$, i.e.

$$\mathbf{M}_c\mathbf{p} \triangleq \begin{pmatrix} 0 & 0 & 0 & 0 & \ldots & \mathbf{n}_{cp}^T & d_{cp} \\ \vdots & \vdots & 0 & 0 & \ldots & \vdots & \vdots \\ 0 & 0 & \mathbf{n}_{cp}^T & d_{cp} & \ldots & \mathbf{n}_{co}^T & d_{co} \\ \mathbf{n}_{cp}^T & d_{cp} & \vdots & \vdots & \ldots & 0 & 0 \\ \vdots & \vdots & \mathbf{n}_{co}^T & d_{co} & \ldots & \vdots & \vdots \\ \mathbf{n}_{co}^T & d_{co} & 0 & 0 & 0 & 0 & 0 \end{pmatrix} \begin{pmatrix} \mathbf{n}_o \\ d_o \\ \mathbf{n}_1 \\ \vdots \\ \mathbf{n}_q \\ d_q \end{pmatrix} = \mathbf{a} \triangleq \begin{pmatrix} a_n \\ a_{n-1} \\ \vdots \\ \vdots \\ \vdots \\ a_o \end{pmatrix} . \tag{21.142}$$

Recalling and applying the definition in the a-space given in Eqs.(21.128) through (21.130), the inverse map in the p-space is defined as

$$\mathbf{p}_o \in \mathbf{\Pi}_o \triangleq \mathbf{M}_c^{-1}(\Delta_o) \triangleq \{\mathbf{p}:\ \mathbf{p} \in \mathcal{R}^{(1+r)(1+q)}\ ,\ \mathbf{M}_c\mathbf{p} \in \Delta_o\} \tag{21.143}$$

$$\mathbf{p}_n \in \mathbf{\Pi}_n \triangleq \mathbf{M}_c^{-1}(\Delta_n) \triangleq \{\mathbf{p}:\ \mathbf{p} \in \mathcal{R}^{(1+r)(1+q)}\ ,\ \mathbf{M}_c\mathbf{p} \in \Delta_n\} \tag{21.144}$$

$$\mathbf{p}_\omega \in \mathbf{\Pi}_\omega(\omega) \triangleq \mathbf{M}_c^{-1}[\Delta_\omega(\omega)] \triangleq \{\mathbf{p}:\ \mathbf{p} \in \mathcal{R}^{(1+r)(1+q)}\ ,\ \mathbf{M}_c\mathbf{p} \in \Delta_\omega(\omega)\}\ . \tag{21.145}$$

Applying Eq.(21.133), the radius of the largest hypersphere centered at $\mathbf{p}_{nom}$ is

$$\rho_H(\mathbf{p}_{nom}) = \min\{\|\mathbf{p}_{nom} - \mathbf{p}_o\|_F,\ \|\mathbf{p}_{nom} - \mathbf{p}_n\|_F,\ \| \inf_\omega \|\mathbf{p}_{nom} - \mathbf{p}_\omega(\omega)\|_F\}\ . \tag{21.146}$$

If perturbations $\Delta\mathbf{p}$ do not intersect $\mathbf{\Pi}_o, \mathbf{\Pi}_n$ and $\mathbf{\Pi}_\omega(\omega)$ for any ω , all its maps into a-space will remain within Δ_o, Δ_n and $\Delta_\omega(\omega)$ and the closed-loop characteristic polynomial is Hurwitz.

21.13.6 Robust Root Locus

The locus of points in the complex plane where the characteristic equation of the control system is satisfied for some open-loop gain *and* some bounded plant uncertainty $\Delta\mathbf{p}$ is called robust root locus. The root loci including uncertainty are described by *Barmish, B.R., and Tempo, R., 1990* in the case where the perturbations enter affine linearly into the coefficients of the plant. The method is based on Kharitonov's theorem.

21.13.7 Design Procedure. Robustification

The problem is to find a controller with nominal parameter $\mathbf{p}_c$ such that the plant parameter perturbations can be maximized (*Bhattacharyya, S.P., 1987*). This procedure is known as robustification of a controller. Hence, the design algorithm can be stated as to maximize the hypersphere radius $\rho_H(\mathbf{p}_c)$ over all stabilizing parameters $\mathbf{p}_c$ of the controller

$$\max_{\mathbf{p}_c} \rho_H(\mathbf{p}_c) = \max_{\mathbf{p}_c} \min\{d_o(\mathbf{p}_c), d_n(\mathbf{p}_c), d(\mathbf{p}_c)\}\ . \tag{21.147}$$

21.14 Maximum Perturbation of Stable Polynomials

Suppose the polynomials $a(s) = s^n + \mathbf{a}^T\mathbf{s}$, written in reciprocated form, is stable where

$$\mathbf{a} \triangleq \begin{pmatrix} a_1 \\ \vdots \\ a_n \end{pmatrix} \quad \mathbf{s} \triangleq \begin{pmatrix} s^{n-1} \\ \vdots \\ s^0 \end{pmatrix} . \quad \text{Let} \quad \mathbf{b} \triangleq \begin{pmatrix} b_1 \\ \vdots \\ b_n \end{pmatrix} \tag{21.148}$$

then the polynomial $b(s) = s^n + \mathbf{b}^T\mathbf{s}$ is also stable if the vector set of coefficients $\mathbf{b}$ satisfies

$$\mathbf{b} \in \{\mathbf{b} \in \mathcal{R}^n : \|\mathbf{b} - \mathbf{a}\|_F \le R^2\}\ . \tag{21.149}$$

The maximum of the perturbation radius R is given for $n \ge 3$ by

$$R^2 = \min\{a_n^2, d_2^2\} \tag{21.150}$$

$$d_2^2 = \min_{v \in (0,\infty)} (\mathbf{y} - \mathbf{a})^T[\mathbf{I}_n - \mathbf{\Theta}(\mathbf{\Theta}^T\mathbf{\Theta})^{-1}\mathbf{\Theta}^T](\mathbf{y} - \mathbf{a}) \tag{21.151}$$

where (*Soh, Y.C., and Evans, R.J., 1988*)

$$\mathbf{y} = (0 \quad v^2 \quad 0 \quad 0 \ldots 0)^T \in \mathcal{R}^n \tag{21.152}$$

$$\boldsymbol{\Theta} = \begin{pmatrix} 1 & 0 & \ldots & 0 \\ 0 & 1 & & \vdots \\ v^2 & 0 & \ddots & 0 \\ 0 & v^2 & & 1 \\ \vdots & 0 & \ddots & 0 \\ 0 & \ldots & & v^2 \end{pmatrix} \in \mathcal{R}^{n\times(n-2)} . \tag{21.153}$$

21.15 Maximum Perturbation Parameter e_{max}

Given a polynomial $a_p(s) = \sum_{i=0}^{n-1} a_i s^i + s^n$ where $a_{i,nom} - el_i < a_i < a_{i,nom} + eh_i$. The weights l_i and h_i must be non-negative. Find e such that $a_p(s)$ is stable. Denote this value e_{max} . Then, four Kharitonov polynomials $a_{ej}(s;e)$ $j = 1,2,3,4$ can be determined from Eq.(21.5) through Eq.(21.8) where the coefficients depend continuously on e . The stability of each Kharitonov polynomial $a_{ej}(s;e)$ is proved by using Hurwitz criterion. Assume $a_p(s)$ stable for $e = 0$. If one of the four Hurwitz testing matrices ceases to have all its leading principal minors positive determine that value e_j . The largest value of e for strict Hurwitz stability of the set of perturbed polynomials then is given by $e_{max} = \min_j\{e_j\}$ (*Barmish, B.R., 1984*).

Remember the Hurwitz testing matrix of dimension $n \times n$ for the nth order polynomial $a(s) = \sum_{i=0}^{n} a_i s^i$ as given in Eq.(21.32). The conditions using the leading principal minors are $a_{n-1} > 0$, $a_{n-1}a_{n-2} - a_n a_{n-3} > 0$ etc.

21.16 Graphical Mikhailov Robustness Criterion

The Mikhailov criterion states that the unperturbed (nominal) system with the characteristic polynomial $p(s) = \sum_{i=0}^{n} a_{i,nom} s^i$ is strictly Hurwitz stable if the hodograph of $p(s)$ for $s = j\omega$ and $\omega = [0,\infty]$ consecutively crosses n quadrants of the complex plane in strictly counterclockwise direction and does not touch the origin. Furthermore, define $\nu = \omega^2$

$$\begin{aligned} p(s) &\triangleq p_{even,nom}(\nu) + j\omega p_{odd,nom}(\nu) \\ &= (a_{o,nom} - a_{2,nom}\nu + a_{4,nom}\nu^2 - \ldots) + j\omega(a_{1,nom} - a_{3,nom}\nu + a_{5,nom}\nu^2 - \ldots) . \end{aligned} \tag{21.154}$$

Let the nth order perturbed polynomial be $p_p(s) = \sum_{i=0}^{n} a_i s^i$ where the coefficients lie in the box $|a_i - a_{i,nom}| < \gamma\alpha_i \quad \forall i = 0,1,2 \ldots n$.

For the perturbed system a modified polynomial $p_M(j\omega)$ is defined

$$p_M(j\omega) \triangleq \frac{p_{even,nom}(\nu)}{p_{even,\alpha}(\nu)} + j\frac{p_{odd,nom}(\nu)}{p_{odd,\alpha}(\nu)} \tag{21.155}$$

where the unperturbed (nominal) system is preassumed stable and where

$$p_{even,\alpha}(\nu) = \alpha_o + \alpha_2\,\nu + \alpha_4\,\nu^2 + \ldots \quad \text{and} \quad p_{odd,\alpha}(\nu) = \alpha_1 + \alpha_3\,\nu + \alpha_5\,\nu^2 + \ldots \,. \tag{21.156}$$

By Mikhailov criterion, the perturbed system as cited above is strictly Hurwitz stable if and only if, for a certain scalar factor γ, see Fig. 21.9,

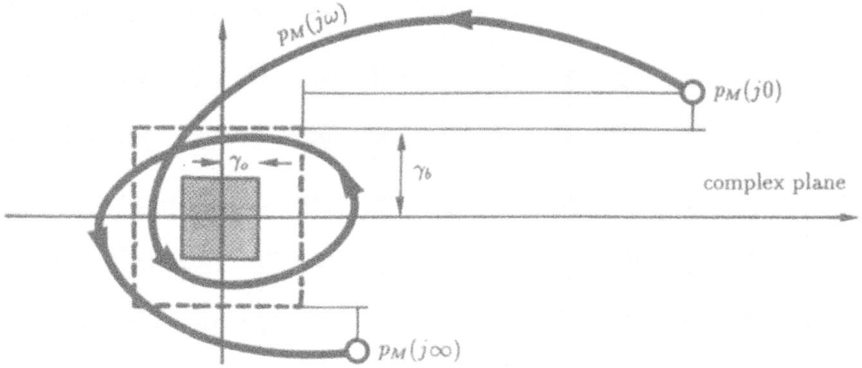

Figure 21.9: Modified Mikhailov hodograph for perturbed polynomials

(i) the hodograph of the modified polynomial $p_M(j\omega)$ crosses exactly n consecutive quadrants of the complex plane in strictly counterclockwise direction and does not intersect a square centered at the origin with sides equal to $2\,\gamma_o$. For the present, assume γ_o arbitrary but sufficiently small.

(ii) The boundary points $p_M(j0)$ and $p_M(j\infty)$ are such that their real and imaginary parts lie outside a square with sides $2\,\gamma_b$.

(iii) The condition $\gamma < \min\{\gamma_o\,,\,\gamma_b\}$ is satisfied.

In order to maximize the scalar γ, the numbers γ_o and γ_b are selected such that the squares touch the hodograph (*Tsypkin, Y.Z., and Polyak, B.T., 1991; Katbab, A., and Jury, E.I., 1991*).

21.17 Stability Robustness of Characteristic Polynomials Via Nyquist Criterion

Consider a dynamic linear time-invariant system and its characteristic polynomial $a(s)$ or $f(z)$. Let the coefficients of the polynomial be perturbed. There is a unified approach for Hurwitz and Schur polynomials and even for Γ-stability properties. The Nyquist criterion is used to check the stability of the perturbed polynomial and to find upper bounds for the perturbed coefficients (*Lin, S.H., et al. 1989; Zhao, K.Y., and Barmish, B.R., 1990; Djaferis, T.E.,. 1991; Kraus, F.J., and Truöl, W., 1991*).

21.17.1 Continuous-Time Systems

Let the nominal characteristic polynomial of a continuous-time system be $a(s)$ and strictly Hurwitz. The perturbed polynomial $a_p(s)$ is obtained by the sum of the nominal polynomial $a(s)$ and the error polynomial $e(s) = \sum_0^n e_i s^i$

$$a_p(s) = a(s) + e(s) = \sum_0^n a_i s^i + \sum_0^n e_i s^i \ . \tag{21.157}$$

Rewriting

$$\frac{a_p(s)}{a(s)} = 1 + \frac{e(s)}{a(s)} = 1 + \frac{\sum_0^n e_i s^i}{\sum_0^n a_i s^i} \triangleq 1 + G(s) \ . \tag{21.158}$$

If and only if the Nyquist plot of the faked transfer function $G(j\omega)$ does not enclose the critical point $(-1, j0)$ $\forall\ \omega = 0 \ldots \infty$ the perturbed system $a_p(s)$ remains Hurwitz.

If $a(s)$ is unstable, i.e., $a(s)$ has n_z zeros in the open right-half s-plane, but the Nyquist plot $G(j\omega)$ does not enclose the critical point then $a_p(s)$ also has n_z unstable zeros.

When $a(s)$ is preassumed stable, a sufficient condition to guarantee stable $a_p(s)$ is given by the infinity function norm $\|G(j\omega)\|_\infty < 1$.

A structured perturbation bound is obtained if a structural information of the error coefficients is given. Assume that positive (and real) numbers e_{oi} exist and that the error coefficient $|e_i| \le \varepsilon e_{oi}$. Then,

$$\|G(j\omega)\|_\infty = \max_{\omega \ge 0} |\frac{\sum_0^n e_i s^i}{a(s)}|_{s=j\omega} = \max_{\omega \ge 0} \frac{\varepsilon |\sum_0^n e_{oi}(j\omega)^i|}{|a(j\omega)|} < 1 \tag{21.159}$$

$$\varepsilon < \left(\max_{\omega \ge 0} \frac{|\sum_0^n e_{oi}(j\omega)^i|}{|a(j\omega)|}\right)^{-1} \le \min_{\omega \ge 0} \frac{|a(j\omega)|}{\sum_0^n e_{oi}\omega^i} \ . \tag{21.160}$$

21.17.2 Discrete-Time Systems

The results outlined in the previous subsection can be reformulated in the case of discrete-time systems. Given a Schur polynomial $f(z)$ in the nominal case, the perturbed polynomial $f_p(z) = f(z) + e(z)$ remains Schur if the Nyquist plot $G(z) = e(z)/f(z)$ does not encircle the critical point $(-1, j0)$ where $z = \exp(j\omega T)$, T is the sampling period and ωT varies from 0 to π. A sufficient condition is

$$\max_{\omega T \in [0,\pi]} |\frac{e(z)}{f(z)}|_{z=e^{j\omega T}} < 1 \ . \tag{21.161}$$

Replacing z by unity and e_i by $|e_i|$ in $e(z)$, an upper bound statement is

$$\max_{\omega T \in [0,\pi]} |e(z)|_{z=e^{j\omega T}} \le \sum_0^n |e_i| \ . \tag{21.162}$$

From Eq.(21.161)

$$\max_{\omega T \in [0,\pi]} |e(z)|_{z=e^{j\omega T}} < \min |f(z)|_{z=e^{j\omega T}} \ . \tag{21.163}$$

Combining Eqs.(21.162) and (21.163) yields another sufficient condition for the error polynomial

$$\sum_0^n |e_i| < \Omega \triangleq \min_{\omega T \in [0,\pi]} |f(z)|_{z=e^{j\omega T}} \ . \tag{21.164}$$

If a structural information $|e_i| < \varepsilon e_{oi}$ $\forall\ i = [1, 2 \ldots n]$ is available where e_{oi} is positive, an upper bound results

$$\varepsilon \sum_{i=0}^n e_{oi} < \min_{\omega T \in [0,\pi]} |f(z)|_{z=e^{j\omega T}} \quad \leadsto \quad \varepsilon < \varepsilon_{\max} \triangleq \frac{\Omega}{\sum_{i=0}^n e_{oi}} \ . \tag{21.165}$$

The larger $\varepsilon_{\max}$ or Ω the better the stability robustness. The quantity Ω can be interpreted by the product of the distances from the roots λ_i of $f(z)$ to the unit circle

$$\Omega = \min_{\omega T \in [0,\pi]} |f(z)|_{z=e^{j\omega T}} = \min_{\omega T \in [0,\pi]} \prod_i |z - \lambda_i|_{z=e^{j\omega T}} \ . \tag{21.166}$$

21.17.3 Discrete-Time Interval System

Discrete-time interval systems can be reduced to perturbed systems as follows. Define an interval system

$$f_I(z) \qquad \text{where} \qquad f_{Ii} \in [l_i, h_i] \ \forall\, i \ . \tag{21.167}$$

The polynomial set

$$f(z) + e(z) \qquad \text{where} \qquad |e_i| \le \frac{1}{2}(h_i - l_i) \qquad f_i = \frac{1}{2}(h_i + l_i) \tag{21.168}$$

is identical to the interval polynomial owing to the bounds chosen for $|e_i|$. Hence, referring to Eq.(21.164) the interval system is Schur if

$$0.5 \sum_{i=0}^{n} (h_i - l_i) < \min_{\omega T \in [0,\pi]} |f(z)|_{z=e^{j\omega T}} \ . \tag{21.169}$$

Example: Given an arbitrary polynomial $a(z) = \sum_{i=0}^{n} a_i z^i$, $f(z) + e(z) := a(z)$ and $f(z) := a_n z^n$. Then,

$$e(z) = \sum_{i=0}^{n-1} a_i z^i + 0 z^n \tag{21.170}$$

and from Eq.(21.164)

$$\sum_{i=0}^{n} |e_i| < \Omega = \min_{\omega T \in [0,\pi]} |a_n z^n|_{z=e^{j\omega T}} = |a_n| \ . \tag{21.171}$$

Referring to Eq.(21.170) leads to

$$\sum_{i=0}^{n} |e_i| = \sum_{i=0}^{n-1} |a_i| < |a_n| \ . \tag{21.172}$$

The polynomial $f(z)$ is Schur if the condition above is satisfied.

End of Example

21.17.4 Γ-Stability Robustness

A system is defined Γ-stable if and only if the zeros of its characteristic polynomial lie inside a region Γ. If Γ coincides with the open left half-plane then Γ-stability becomes simple Hurwitz stability. By the choice of an appropriate region Γ, the control system performance is specified (*Ackermann, J., 1980; Vicino, A.,1989*). Owing to the Cauchy theorem (argument principle) a perturbed system remains Γ-stable if $a(s)$ [or $f(z)$] is Γ-stable and the Nyquist path Γ_N defined by the boundary curve of Γ does not encircle the critical point (*Sondergeld, K.P., 1983; Lin, S.H., et al. 1989*).

Referring to Eq.(21.158), a sufficient condition for Γ-stability is

$$\max_{s \in \Gamma} \left|\frac{e(s)}{a(s)}\right| < 1 \ . \tag{21.173}$$

If structural information is available $|e_i| \leq \delta_i \quad \forall\, i \in \; [0, 1 ... n]$ then $\max | \sum_{i=0}^{n} e_i s^i |$ can be replaced by the sum of the maximum moduli and it results

$$\sum_{i=0}^{n} \delta_i (\max_{s \in \Gamma} |s|)^i < \min_{s \in \Gamma} |a(s)| \,. \tag{21.174}$$

It must be emphasized that the first step in applying this method is the appropriate choice of $a(s)$. If the above sufficient condition for Γ-stability is failed other Γ-stable polynomials $a(s)$ may satisfy the robust Γ-stability with respect to the perturbation polynomial $e(s)$.

21.18 Robust Schur Stability of a Polytope of Polynomials

Let $\mathcal{P}$ denote a polytope of polynomials and $f_l(z) \quad i = 1, 2 \; ... \; N$ its vertices. Then, the polytope $\mathcal{P}$ is the set of all convex combinations of $f_l(z)$

$$\mathcal{P} = \mathrm{conv}\{f_1(z),\; f_2(z), ...\; f_N(z)\} \,. \tag{21.175}$$

Consider the polynomial

$$f_p(z, \mathbf{q}) = z^n + \sum_{i=0}^{n-1} a_i(\mathbf{q}) z^i \qquad \mathbf{q} \in \mathcal{Q}. \tag{21.176}$$

Assume that $\mathbf{a}(\mathbf{q}) = \{a_o(\mathbf{q}),\; a_1(\mathbf{q}), ...\; a_{n-1}(\mathbf{q})\}$ is a polytope which is true if the $a_i(\mathbf{q})$ are linearly dependent on $\mathbf{q}$ and there exists an upper and lower bound for each q_i of $\mathbf{q}$. Hence, the set of polynomials $f_p(z, \mathbf{q})$ is also a polytope generated by the extreme vertices of $\mathcal{Q}$. The set $f_p(z, \mathbf{q})$ is a convex hull of the *finite* set of polynomials

$$f_j(z) \triangleq z^n + \sum_{i=0}^{n-1} a_i(q^j) z^i \tag{21.177}$$

where q^j denotes the jth extreme point. For a polynomial

$$f(z) = \sum_{i=0}^{n} a_i z^i = a_n \prod_{i=0}^{n} (z - z_i) \tag{21.178}$$

$$\mathbf{S}(f) \triangleq \begin{pmatrix} a_n & a_{n-1} & a_{n-2} & \cdots & a_3 & a_2 - a_o \\ 0 & a_n & a_{n-1} & \cdots & a_4 - a_o & a_3 - a_1 \\ \vdots & \vdots & \vdots & \vdots & \vdots & \vdots \\ 0 & -a_o & -a_1 & \cdots & a_n - a_{n-4} & a_{n-1} - a_{n-3} \\ -a_o & -a_1 & -a_2 & \cdots & -a_{n-3} & a_n - a_{n-2} \end{pmatrix} \in \mathcal{R}^{(n-1)\times(n-1)} \tag{21.179}$$

then $\det \mathbf{S}(f) = a_n^{n-1} \prod_{k=1, i<k}^{n} \; (1 - z_i z_k) = 0$ if a complex pair of roots crosses the unit circle (*Jury, E.I., and Pavlidis, T., 1963*). Based on this fact the set of monic polynomials given by $f_p(z, \mathbf{q})$ is Schur stable if the generating points $f_j(z) \;\; \forall j = 1, 2....N$ have zeros inside the unit circle and if the matrix $\mathbf{S}(f_j)\mathbf{S}^{-1}(f_l) \; \forall j, l = 1, 2...\; N$ has no real eigenvalue in $(-\infty, 0)$. See *Ackermann, J.E., and Barmish, B.R., 1988* for the proof. An algorithm to achieve maximum intervals is presented by *Soh, C.B., 1991*.

21.19 Aperiodicity Condition for Discrete-Time Systems

In the preceding sections *stability* conditions have been considered. Now, the condition for *aperiodicity* is regarded. The condition for aperiodicity is that all the roots of a real polynomial are real and positive in the interval $[0,1)$. *Strict* aperiodicity is given if the roots are *distinct*, additionally.

Robustness conditions for aperiodicity were investigated by *Soh, C.B.*, in the case of polynomial coefficients with alternating sign. The polynomial

$$f(z) = \sum_{i=0}^{n} a_i(-1)^i z^{n-i} \qquad a_i \in [l_i,\ h_i] \tag{21.180}$$

is strictly aperiodic if and only if two polynomials $f_1(z)$ and $f_2(z)$ are strictly aperiodic

$$\left.\begin{array}{l} f_1(z) \triangleq \sum_{i=1}^{n} a_{1i}(-1)^i z^{n-i} \\ f_2(z) \triangleq \sum_{i=1}^{n} a_{2i}(-1)^i z^{n-i} \end{array}\right\} \text{ where } \left\{\begin{array}{llll} a_{1i} = l_i & a_{2i} = h_i & \text{for } i \text{ even} & \forall\, i = [0,n] \\ a_{1i} = h_i & a_{2i} = l_i & \text{for } i \text{ odd} & \forall\, i = [0,n]\ . \end{array}\right. \tag{21.181}$$

Similar Kharitonov-like conditions hold for simple aperiodicity and for continuous-time systems (*Jury, E.I., 1990*).

21.20 Hurwitz Antisymmetric Property

A rectangular matrix is said antisymmetric if it can be generated from some uncertain core vector $\mathbf{c}(\mathbf{q}) = (c_o \quad c_1 .... c_\nu)^T$ via a sequence of augmentations. A core is Hurwitz if the corresponding polynomial $\sum_{i=0}^{\nu} c_i s^{n-i}$ is Hurwitz invariant. Suppose that the leading coefficient $c_o(\mathbf{q})$ is sign invariant. Augmentations are obtained by attaching a row or a column where each diagonal term is a sign invariant uncertain function of $\mathbf{q}$. A matrix function $\mathbf{\Phi}(\mathbf{q})$ is called Hurwitz antisymmetric if it is antisymmetric and generated from a Hurwitz core $\mathbf{c}(\mathbf{q})$. An uncertain system with state-space representation $\mathbf{A}(\mathbf{q}), \mathbf{B}(\mathbf{q})$ is stabilizable by a linear state feedback controller if in the first $n-m$ rows of the matrix $s\mathbf{I} - \mathbf{A}(\mathbf{q})$ there is a submatrix of dimension $(n-m) \times (n-m-1)$ having Hurwitz antisymmetric property (*Wei, K., 1989*).

21.21 Sufficient Stability Condition

The polynomial

$$a(s) = \sum_{i=0}^{n} a_i s^{n-i} \qquad a_i > 0 \qquad a_i \in [l_i,\ h_i] \tag{21.182}$$

is Hurwitz stable if the conditions

$$h_{i-1} h_{i+2} \leq 0.4655\ l_i l_{i+1} \qquad \forall i = 1,2\ ...\ n-2 \tag{21.183}$$

are satisfied (*Bose, N.K., et al. 1988*).

21.22 Markov's Theorem on Determinants

Decompose the polynomial $a(s)$

$$a(s) = \sum_{i=0}^{n} a_i s^{n-i} \triangleq a_{even}(s^2) + s\ a_{odd}(s^2)\ , \tag{21.184}$$

assuming $a_{even}(s)$ and $a_{odd}(s)$ coprime. Then, a rational function is given by

$$\frac{a_{odd}(s^2)}{a_{even}(s^2)} = \alpha_{-1} + \frac{\alpha_o}{s} - \frac{\alpha_1}{s^2} + \frac{\alpha_2}{s^3} - \ldots\ldots \ . \tag{21.185}$$

The n leading coefficients in this expansion are named Markov parameters. If $n = 2m$ the Markov parameters are $\alpha_o,\ \alpha_1, \ldots\ldots\ \alpha_{2m-1}$, if $n = 2m+1$ the Markov parameters are $\alpha_{-1},\ \alpha_o, \ldots\ldots\ \alpha_{2m-1}$. Define a box in the space of Markov parameters, e.g., for n even

$$\mathcal{M}_\varepsilon \triangleq \{\ \sigma\ :\ \ |\alpha_i - \sigma_i| < \varepsilon\ ,\quad i = 0,1,2....(2m-1)\ \}\ . \tag{21.186}$$

Markov's Theorem on determinants requires only *two* members of the family of Markov parameters to guarantee Hurwitz stability. Let ε_1 and ε_2 be the largest numbers ε ($\varepsilon > 0$) for which the two sets of Markov parameters are Hurwitz. The two sets of Markov parameters are

$$a_1 \ \triangleq\ (\alpha_o - \varepsilon,\ \alpha_1 + \varepsilon,\ \alpha_2 - \varepsilon,\ \ldots\ldots\ \alpha_{2m-1} + \varepsilon) \tag{21.187}$$

$$a_2 \ \triangleq\ (\alpha_o + \varepsilon,\ \alpha_1 - \varepsilon,\ \alpha_2 + \varepsilon,\ \ldots\ldots\ \alpha_{2m-1} - \varepsilon)\ . \tag{21.188}$$

Then, the largest stability box in the space of Markov parameters for which $\mathcal{M}_\varepsilon \mid_{\varepsilon=\varepsilon_{max}}$ contains only Hurwitz Markov parameters is given by $\varepsilon_{max} = \min\{a_1,\ a_2\}$. See *Hollot, C.V., 1989* for the proof.

21.23 Suggestions for Further Reading

Boundary implications on interval rational functions for strict positive realness, positive realness and positiveness can be derived by Kharitonov polynomials associated with the numerator and denominator polynomial (*Dasgupta, S., 1987; Bose, N.K., and Delansky, J.F., 1989*).

Chapter 22

Eigenvalues and Singular Values of Complex Matrices

Real matrices are matrices with elements of real numbers or functions; they are often applied in the state-space approach of control systems, various regression techniques, convolution sums but also in some other topics related to the time domain, e.g. stability radius techniques. If design problems of multivariable systems are treated in the s-plane or in the frequency domain use is made of transfer matrices with entries of complex numbers and functions. To avoid confusion, symbols of a different kind are chosen in most cases, e.g., $\mathbf{G}$ for a complex matrix and $\mathbf{A}$ for a real one.

As far as differences between complex and real matrices do not arise, properties of real matrices are not repeated in this chapter.

22.1 Some Simple Properties on Complex Matrices

For scalars $a \in C$ and $b \in C$ note that $ab^* \neq a^*b$ but $ab^* = (a^*b)^*$.

Note the properties on complex vectors $\mathbf{a} \in C^n, \mathbf{b} \in C^n$: Changing the order of $\mathbf{a}^*$ and $\mathbf{b}$ yields $\mathbf{a}^{*T}\mathbf{b} = \mathbf{b}^T\mathbf{a}^*$. The asterix * must be kept linked to $\mathbf{a}$. Calculating the complex-conjugate of the product $\mathbf{a}^{*T}\mathbf{b}$ yields $\mathbf{a}^T\mathbf{b}^* = (\mathbf{a}^{T*}\mathbf{b})^*$. Or $(\mathbf{a}^T\mathbf{b})^* = \mathbf{a}^{T*}\mathbf{b}^*$.

$$\mathbf{a}^T\mathbf{b} = \mathbf{b}^T\mathbf{a} \qquad \mathbf{a}^H\mathbf{b} \neq \mathbf{b}^H\mathbf{a} \qquad \mathbf{a}^H\mathbf{b} = (\mathbf{b}^H\mathbf{a})^* \qquad \mathbf{a}^T\mathbf{b}^* = (\mathbf{a}^H\mathbf{b})^* \tag{22.1}$$

$$\mathbf{a}^T\mathbf{b}^* = \mathbf{b}^{T*}\mathbf{a} = \mathbf{b}^H\mathbf{a} \qquad \mathbf{a}^H\mathbf{b} = \mathbf{a}^{T*}\mathbf{b} = (\mathbf{a}^T\mathbf{b}^*)^* = (\mathbf{b}^{T*}\mathbf{a})^* = (\mathbf{b}^H\mathbf{a})^* \ . \tag{22.2}$$

If $\mathbf{b} = \mathbf{a}$, since the result is real,

$$\mathbf{a}^{*T}\mathbf{a} = \mathbf{a}^T\mathbf{a}^* = \mathbf{a}^H\mathbf{a} \quad \text{and} \quad (\mathbf{a}^H\mathbf{a})^* = (\mathbf{a}^H\mathbf{a}) \ . \tag{22.3}$$

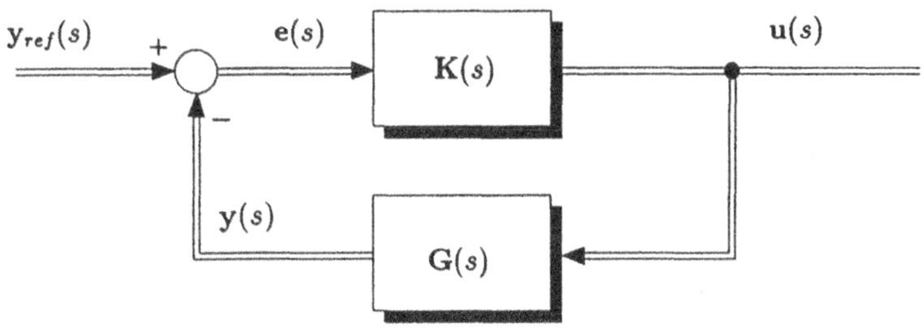

Figure 22.1: Multivariable control loop containing controller $\mathbf{K}(s)$ and plant $\mathbf{G}(s)$

Frequently the inner product notation $\langle \mathbf{b}, \mathbf{a} \rangle = (\mathbf{b}, \mathbf{a}) = \sum b_i a_i^* = \mathbf{a}^H \mathbf{b}$ is used. Note that $\langle \mathbf{Aa}, \mathbf{b} \rangle = \langle \mathbf{a}, \mathbf{A}^H \mathbf{b} \rangle$.

The conjugate of a complex matrix $\mathbf{G}$ is referred to as $\mathbf{G}^*$

$$\mathbf{G} = \mathbf{B} + j\mathbf{C} \quad \rightsquigarrow \quad \mathbf{G}^* = \mathbf{B} - j\mathbf{C}. \tag{22.4}$$

The matrices $\mathbf{B} = \Re e\ \mathbf{G}$ and $\mathbf{C} = \Im m\ \mathbf{G}$ are the rectangular representation of the matrix $\mathbf{G}$. An alternative rectangular decomposition (*Householder, A.S., 1964*) of $\mathbf{G}$ is

$$\mathbf{G} = \frac{\mathbf{G} + \mathbf{G}^H}{2} + j\left(\frac{\mathbf{G} - \mathbf{G}^H}{2j}\right). \tag{22.5}$$

The conjugate transpose $\mathbf{G}^H$ of a complex matrix $\mathbf{G}$ is designated with the superscript ^{H} and is defined as

$$\mathbf{G} = \mathbf{B} + j\mathbf{C} \quad \rightsquigarrow \quad \mathbf{G}^H = \mathbf{B}^T - j\mathbf{C}^T = (\mathbf{G}^*)^T. \tag{22.6}$$

The complex conjugate of a sum (product) matrix can be taken of the matrix summands (multiplicands) before operation. Since $\det[(\lambda\mathbf{I} - \mathbf{G})^*] = [\det(\lambda\mathbf{I} - \mathbf{G})]^*$, the eigenvalues $\mu = \lambda[\mathbf{G}^*]$ of the complex conjugate $\mathbf{G}^*$ and the conjugate transpose $\mathbf{G}^H$ are related as follows

$$\mu = \lambda[\mathbf{G}^*] = \lambda[\mathbf{G}^H] = \left(\lambda[\mathbf{G}]\right)^* = \lambda^*[\mathbf{G}]. \tag{22.7}$$

The *field of values* of $\mathbf{G}$ is defined as a set of complex numbers $\mathbf{x}^H\mathbf{G}\mathbf{x}$ where $\mathbf{x}^H\mathbf{x} = 1$ (*Bauer, F.L., 1962*). In other words,

$$\mathbf{x}^H\mathbf{G}\mathbf{x} = \sum_{i,j=1}^{n} G_{ij} x_i x_j^* \qquad \mathbf{x} = (x_1\ x_2 \ldots x_n)^T \qquad \mathbf{x}^H\mathbf{x} = \sum_{i=1}^{n} |x_i|^2 = 1 \tag{22.8}$$

is a function on the unit sphere.

22.2 Hermite Matrices

A Hermite matrix $\mathbf{G}$ has the property $\mathbf{G}^H = \mathbf{G}$. Thus, a Hermite matrix has symmetric real value and skew-symmetric imaginary value. A Hermite matrix corresponds to a symmetric real matrix (*Horn, R.A., and Johnson, C.R., 1990*).

A matrix is said to be skew-Hermite if $\mathbf{G}^H = -\mathbf{G}$. A skew-Hermite matrix has skew-symmetric real part and a symmetric imaginary part. Hence, G_{ii} is imaginary. Skew Hermite matrices correspond to skew-symmetric real matrices ($\mathbf{G}^T = -\mathbf{G}$, $G_{ii} = 0$) (*Thrall, R.M., and Tornheim, L., 1957*).

22.2.1 Eigenvalues of Hermitian Matrices

For complex matrices in general $\lambda[\mathbf{A}^*] = \lambda[\mathbf{A}^H] = \lambda^*[\mathbf{A}]$. The spectrum of a Hermitian matrix is real. If $\mathbf{G}$ is Hermitian the equality $\lambda[\mathbf{G}] = \lambda^*[\mathbf{G}]$ follows from $\det(\lambda\mathbf{I} - \mathbf{G}) = 0$ and

$$\det(\lambda\mathbf{I} - \mathbf{G}^H) = 0 \quad \rightsquigarrow \quad \det(\lambda^*\mathbf{I} - \mathbf{G}^T) = 0 \quad \rightsquigarrow \quad \det(\lambda^*\mathbf{I} - \mathbf{G}) = 0\ . \tag{22.9}$$

The Hermite form is $c = \mathbf{x}^H\mathbf{G}\mathbf{x}$. Taking the conjugate transpose

$$c^* = (\mathbf{x}^H\mathbf{G}\mathbf{x})^* = \mathbf{x}^T\mathbf{G}^*\mathbf{x}^* = (\mathbf{x}^T\mathbf{G}^*\mathbf{x}^*)^T = \mathbf{x}^H\mathbf{G}^H\mathbf{x} = \mathbf{x}^H\mathbf{G}\mathbf{x} = c \tag{22.10}$$

shows that c must be real. Replacing $\mathbf{x}$ by the eigenvector $\mathbf{g}_i$ and using $\mathbf{G}\mathbf{g}_i = \lambda_i[\mathbf{G}]\mathbf{g}_i$ again proves that $\lambda_i[\mathbf{G}] \in \mathcal{R}$ if $\mathbf{G}$ is Hermite. □

A normal matrix $\mathbf{G}$ (i.e. $\mathbf{G}^H\mathbf{G} = \mathbf{G}\mathbf{G}^H$) is Hermitian if and only if its spectrum is real. □

Note the following properties for Hermite matrices $\mathbf{G}$

$$\lambda_{\min}[\mathbf{G}] = \min_{\mathbf{x}^H\mathbf{x}=1} \mathbf{x}^H\mathbf{G}\mathbf{x} = \min_{\mathbf{x}\neq 0} \frac{\mathbf{x}^H\mathbf{G}\mathbf{x}}{\mathbf{x}^H\mathbf{x}} \quad \text{and} \quad \lambda_{\max}[\mathbf{G}] = \max_{\mathbf{x}^H\mathbf{x}=1} \mathbf{x}^H\mathbf{G}\mathbf{x}\,. \tag{22.11}$$

$$\sigma_{\min}[\mathbf{F}] = \min_{\mathbf{x}^H\mathbf{x}=1} \mathbf{x}^H\mathbf{F}^H\mathbf{F}\mathbf{x} \quad \text{and} \quad \sigma_{\max}[\mathbf{F}] = \max_{\mathbf{x}^H\mathbf{x}=1} \mathbf{x}^H\mathbf{F}^H\mathbf{F}\mathbf{x} \qquad \text{where} \quad \mathbf{F} \in \mathcal{C}^{n\times n}\,. \;\square \tag{22.12}$$

A negative [positive] definite Hermitian matrix $\mathbf{G}$ has the property $\lambda_i[\mathbf{G}] < 0 \quad [> 0]$.

For $\mathbf{A} = \mathbf{A}^H \in \mathcal{C}^{n\times n}$ if $\mathbf{A} > 0$ and rank $\mathbf{A} = n$ then $\lambda[\mathbf{A}] > 0$. If $\mathbf{A} \geq 0$ and rank $\mathbf{A} = r$ then $\lambda[\mathbf{A}] = 0$ with multiplicity $n - r$. □

If $\mathbf{A} = \mathbf{A}^H \in \mathcal{C}^{n\times n}$ then $\lambda_i[\mathbf{A}] \in \mathcal{R}$ but $\mathbf{a}_i \in \mathcal{C}^n$, if $\mathbf{A} = \mathbf{A}^T \in \mathcal{R}^{n\times n}$ then $\lambda_i[\mathbf{A}] \in \mathcal{R}$ and $\mathbf{a}_i \in \mathcal{R}^n$ (see Table 2.1).

If $c = \mathbf{x}^H\mathbf{A}\mathbf{x}$ is $> 0[\geq 0]$ then $\mathbf{A}$ is Hermitian and positive definite [positive semidefinite] and all the eigenvalues are positive [non-negative]. □

The symmetric part $\mathbf{A}_s$ of every real matrix $\mathbf{A}$ is also Hermitian and real symmetric. Hence, the symmetric part $\mathbf{A}_s < 0$ (negative definite) $\leadsto \lambda_i[\mathbf{A}_s] < 0$ and since Eq.(13.100), i.e. $\Re e\ \lambda_i[\mathbf{A}] < \lambda_i[\mathbf{A}_s]$, it follows $\Re e\ \lambda_i[\mathbf{A}] < 0$ or $\mathbf{A}$ stable.

For Hermite matrices $\mathbf{A} = \mathbf{A}^H$ the inner product has the property $\langle \mathbf{A}\mathbf{a}, \mathbf{b}\rangle = \langle \mathbf{a}, \mathbf{A}\mathbf{b}\rangle$ since $\langle \mathbf{A}\mathbf{a}, \mathbf{b}\rangle = \mathbf{a}^T\mathbf{A}^T\mathbf{b}^* = \langle \mathbf{a}, \mathbf{A}^{T*}\mathbf{b}\rangle = \langle \mathbf{a}, \mathbf{A}^H\mathbf{b}\rangle = \langle \mathbf{a}, \mathbf{A}\mathbf{b}\rangle$.

22.2.2 Eigenvectors of Hermitian Matrices

The eigenvectors of Hermite matrices associated with different eigenvalues are orthogonal and can be modified to orthogonal unit vectors which can easily be verified. Taking the conjugate transpose of $\mathbf{A}\mathbf{a} = \lambda\mathbf{a}$ yields $\mathbf{a}^H\mathbf{A}^H = \mathbf{a}^H\mathbf{A} = \lambda\mathbf{a}^H$. Postmultiplying $\mathbf{a}_1^H\mathbf{A} = \lambda_1\mathbf{a}_1^H$ by $\mathbf{a}_2$ and premultiplying $\mathbf{A}\mathbf{a}_2 = \lambda_2\mathbf{a}_2$ by $\mathbf{a}_1^H$, two identical terms are achieved

$$\mathbf{a}_1^H\mathbf{A}\mathbf{a}_2 = \lambda_1\mathbf{a}_1^H\mathbf{a}_2 \qquad \mathbf{a}_1^H\mathbf{A}\mathbf{a}_2 = \lambda_2\mathbf{a}_1^H\mathbf{a}_2 \tag{22.13}$$

which only can be satisfied if $\mathbf{a}_1^H\mathbf{a}_2 = 0$.

If a Hermitian matrix $\mathbf{A}$ has distinct eigenvalues then $\mathbf{A}$ can be diagonalized by a similarity transformation with a unitary matrix. If a Hermitian matrix has eigenvalues which are *not* distinct, then, with regard to the orthogonality of the eigenvectors, the diagonalization is *also* possible. Invoking the property that eigenvalues and eigenvectors change continuously if $\mathbf{A}$ is changed, slight modifications in $\mathbf{A}$ may cause eigenvalues to coincide but the eigenvectors with regard to the orthogonality cannot (*Franklin, J.N., 1968*). Remember the identity matrix $\mathbf{I}_n$ characterized by n identical eigenvalues unity and n distinct orthogonal unit eigenvectors.

22.2.3 Inclusion principle

Given a Hermitian matrix $\mathbf{A} \in \mathcal{C}^{n\times n}$ and a matrix $\mathbf{B} \in \mathcal{C}^{(n-1)\times(n-1)}$ evolved from $\mathbf{A}$ by deleting the last row and column. Then (*Franklin, J.N., 1968*),

$$\lambda_{\max}[\mathbf{A}] \triangleq \lambda_1[\mathbf{A}] \geq \lambda_{\max}[\mathbf{B}] \triangleq \lambda_1[\mathbf{B}] \geq \lambda_2[\mathbf{A}] \geq \lambda_2[\mathbf{B}] \ldots \geq \lambda_n[\mathbf{A}]\,. \tag{22.14}$$

22.3 Unitary Matrices and Vectors

An unitary matrix has the property

$$\mathbf{G}^H = \mathbf{G}^{-1}. \tag{22.15}$$

Moreover, $|\det \mathbf{G}| = 1$. If a special case of a unitary matrix with real-valued elements is considered, this matrix is an orthogonal one. Normal matrices $\mathbf{G}$ are defined by the property $\mathbf{G}^H\mathbf{G} = \mathbf{G}\mathbf{G}^H$. Unitary matrices always are normal. A unitary matrix $\mathbf{U}$ (i.e. $\mathbf{U}^H\mathbf{U} = \mathbf{U}\mathbf{U}^H = \mathbf{I}$) satisfies the property

$$|\lambda_i[\mathbf{U}]|^2 = 1 \qquad \forall i, \tag{22.16}$$

i.e., the spectrum of an unitary matrix lies on the unit circle.

Proof: Consider the unitary matrix $\mathbf{U}$, its eigenvalues λ and normalized eigenvector $\mathbf{u}$

$$\mathbf{U}\mathbf{u} = \lambda\mathbf{u} \qquad\qquad \mathbf{u}^H\mathbf{u} = 1\ . \tag{22.17}$$

$$\text{Then,} \qquad (\mathbf{U}\mathbf{u})^H(\mathbf{U}\mathbf{u}) = \mathbf{u}^H\mathbf{U}^H\mathbf{U}\mathbf{u} = \mathbf{u}^H\mathbf{u} = 1 \tag{22.18}$$

$$(\mathbf{U}\mathbf{u})^H(\mathbf{U}\mathbf{u}) = (\lambda\mathbf{u})^H(\lambda\mathbf{u}) = \mathbf{u}^H\lambda^*\lambda\mathbf{u} = \lambda^*\lambda\mathbf{u}^H\mathbf{u} = |\lambda|^2\ . \ \square \tag{22.19}$$

Unitary vectors have the property

$$\mathbf{q}_i^H\mathbf{q}_k = \delta_{ik}\ . \tag{22.20}$$

They are normalized to unity as presumed when applying the Kronecker symbol. □

If $\mathbf{G}$ is normal, i.e. $\mathbf{G}\mathbf{G}^H = \mathbf{G}^H\mathbf{G}$, then its modal matrix is unitary

$$\mathbf{G} = \mathbf{T}\mathbf{\Lambda}\mathbf{T}^{-1} \quad \text{where} \quad \mathbf{T}^{-1} = \mathbf{T}^H \quad \text{and} \quad \mathbf{\Lambda} = \text{diag}\{\lambda_i[\mathbf{G}]\}\ . \tag{22.21}$$

This can easily be verified by

$$\mathbf{G}\mathbf{G}^H = \mathbf{T}\mathbf{\Lambda}\mathbf{T}^{-1}\mathbf{T}\mathbf{\Lambda}^H\mathbf{T}^H = \mathbf{T}\mathbf{\Lambda}\mathbf{\Lambda}^H\mathbf{T}^H \tag{22.22}$$

$$\mathbf{G}^H\mathbf{G} = \mathbf{T}\mathbf{\Lambda}^H\mathbf{T}^H\mathbf{T}\mathbf{\Lambda}\mathbf{T}^{-1} = \mathbf{T}\mathbf{\Lambda}^H\mathbf{\Lambda}\mathbf{T}^H \tag{22.23}$$

since for $\mathbf{\Lambda} = \text{diag}\{\lambda_i[\mathbf{G}]\}$ the equality $\mathbf{\Lambda}\mathbf{\Lambda}^H = \mathbf{\Lambda}^H\mathbf{\Lambda}$ always is true. □

The Kronecker product $\mathbf{A} \otimes \mathbf{B}$ is unitary if $\mathbf{A},\ \mathbf{B} \in \mathcal{C}^{n\times n}$ are unitary. □

22.4 Normal Matrices

What is the condition for a matrix $\mathbf{A}$ that $\mathbf{a}_i$ equals $\mathbf{a}_i^q$ as given by $\mathbf{A}\mathbf{a}_i = \lambda_i\mathbf{a}_i$ and $\mathbf{A}^H\mathbf{a}_i^q = \lambda_i^*\mathbf{a}_i^q$? By premultiplying the last but one equation by $\mathbf{A}^H$ (and the last equation by $\mathbf{A}$) and by equating $\mathbf{a}_i^q = \mathbf{a}_i$ it follows

$$\mathbf{A}^H\mathbf{A}\mathbf{a}_i = \lambda_i\mathbf{A}^H\mathbf{a}_i = \lambda_i\lambda_i^*\mathbf{a}_i = |\lambda_i|^2\mathbf{a}_i \quad \text{and} \quad \mathbf{A}\mathbf{A}^H\mathbf{a}_i = |\lambda_i|^2\mathbf{a}_i\ . \tag{22.24}$$

In order to satisfy both equations above, i.e., to obtain the result that $\mathbf{a}_i$ is an eigenvector associated with the eigenvalue $|\lambda_i|^2$ it is required that $\mathbf{A}^H\mathbf{A} = \mathbf{A}\mathbf{A}^H$ or $\mathbf{A}$ is normal. Invoking the definition of the singular value,

$$\sigma_i[\mathbf{A}] = +\sqrt{\lambda_i[\mathbf{A}^H\mathbf{A}]} = +\sqrt{\lambda_i[\mathbf{A}\mathbf{A}^H]} = |\lambda_i| \quad \text{and} \quad \|\mathbf{A}\|_s = \max_i |\lambda_i|\ . \tag{22.25}$$

Important representatives of normal matrices are symmetric, skew-symmetric, Hermite, skew-Hermite and unitary matrices.

Example: Consider a Hermitian matrix $\mathbf{A} = \mathbf{B}\mathbf{B}^H$ where

$$\mathbf{B} = \begin{pmatrix} 0 & 1 \\ j & 2j \end{pmatrix} . \quad \text{Then,} \quad \mathbf{A} = \begin{pmatrix} 1 & -2j \\ 2j & 5 \end{pmatrix} \tag{22.26}$$

$$\sigma_i[\mathbf{A}] = \sqrt{\lambda_i[\mathbf{A}\mathbf{A}^H]} = \sqrt{\lambda_i[\begin{pmatrix} 5 & -12j \\ 12j & 29 \end{pmatrix}]} = \sqrt{17 \pm 12\sqrt{2}} \qquad \lambda_i[\mathbf{A}] = 3 \pm 2\sqrt{2} . \tag{22.27}$$

Note that $3 \pm 2\sqrt{2} \equiv \sqrt{17 \pm 12\sqrt{2}}$. Since the matrix $\mathbf{A}$ of the example above is Hermite and since every Hermite matrix is normal, Eq.(22.25) is satisfied.

End of Example

If $\mathbf{A} \in \mathcal{C}^{n \times n}$ denotes a normal matrix then there exists a unitary matrix $\mathbf{U}_\lambda$ and a diagonal matrix $\mathbf{\Lambda}_n = \text{diag}_n \lambda_i[\mathbf{A}]$ such that

$$\mathbf{A} = \mathbf{U}_\lambda(\text{diag}_n \lambda_i[\mathbf{A}])\mathbf{U}_\lambda^H = \mathbf{U}_\lambda \mathbf{\Lambda}_n \mathbf{U}_\lambda^H \tag{22.28}$$

which can easily be verified by

$$\begin{aligned} \mathbf{A}\mathbf{A}^H &= \mathbf{U}_\lambda \mathbf{\Lambda}_n \mathbf{U}_\lambda^H (\mathbf{U}_\lambda \mathbf{\Lambda}_n \mathbf{U}_\lambda^H)^H = \mathbf{U}_\lambda \mathbf{\Lambda}_n \mathbf{U}_\lambda^H \mathbf{U}_\lambda \mathbf{\Lambda}_n^H \mathbf{U}_\lambda^H = \mathbf{U}_\lambda \mathbf{\Lambda}_n \mathbf{\Lambda}_n^H \mathbf{U}_\lambda^H & (22.29) \\ &= \mathbf{U}_\lambda \mathbf{\Lambda}_n \mathbf{\Lambda}_n^* \mathbf{U}_\lambda^H = \mathbf{U}_\lambda \mathbf{\Lambda}_n^* \mathbf{\Lambda}_n \mathbf{U}_\lambda^H = (\mathbf{U}_\lambda \mathbf{\Lambda}_n^* \mathbf{U}_\lambda^H)\mathbf{U}_\lambda \mathbf{\Lambda}_n \mathbf{U}_\lambda^H = \mathbf{A}^H\mathbf{A} . \ \square & (22.30) \end{aligned}$$

For contrast to Eq.(22.28), note the singular-value decomposition of $\mathbf{A}\mathbf{A}^H$ in the case of normal $\mathbf{A}$ as given in Eq.(22.96)

$$\mathbf{A}\mathbf{A}^H = \mathbf{U}(\text{diag}\{\sigma_i^2[\mathbf{A}]\})\mathbf{U}^H = \mathbf{A}^H\mathbf{A} . \tag{22.31}$$

For determining the matrix $\mathbf{U}_\lambda$ in Eq.(22.28), employ

$$\mathbf{A}\mathbf{U}_\lambda = \mathbf{U}_\lambda \mathbf{\Lambda}_n \quad \rightsquigarrow \quad (\mathbf{I} \otimes \mathbf{A} - \mathbf{\Lambda}_n \otimes \mathbf{I})\text{col}\mathbf{U}_\lambda = \mathbf{0} . \tag{22.32}$$

Remark: Note the fact that the singular-value decomposition , see Eq.(22.88), of a normal matrix does not coincide with its eigenvalue decomposition as given by Eq.(22.28) although comparison of Eqs.(22.96) and (22.97) may cause a misleading consideration. In the case of normal matrices the matrix $\mathbf{U}$ does not in general equal matrix $\mathbf{V}$ in Eqs.(22.96) and (22.97), see Examples 1 and 3 below.

Example 1: $\mathbf{A}$ skew-Hermite. Then, the eigenvalues are always purely imaginary.

$$\mathbf{A} = \begin{pmatrix} j & 1 \\ -1 & j \end{pmatrix} \quad \lambda_i[\mathbf{A}] = 2j;\ 0 \quad \lambda_i[\mathbf{A}^H] = -2j;\ 0 \quad \sigma_i[\mathbf{A}] = 2;\ 0 . \tag{22.33}$$

Eigenvalue decomposition:

$$\mathbf{U}_\lambda = \frac{\sqrt{2}}{2}\begin{pmatrix} 1 & j \\ j & 1 \end{pmatrix} \quad \mathbf{A} = \mathbf{U}_\lambda \mathbf{\Lambda}_n \mathbf{U}_\lambda^H \rightsquigarrow \mathbf{A} = \frac{\sqrt{2}}{2}\begin{pmatrix} 1 & j \\ j & 1 \end{pmatrix}\begin{pmatrix} 2j & 0 \\ 0 & 0 \end{pmatrix}\frac{\sqrt{2}}{2}\begin{pmatrix} 1 & -j \\ -j & 1 \end{pmatrix} . \tag{22.34}$$

Singular-value decomposition, see Eq.(22.105):

$$\mathbf{A} = \mathbf{U}\mathbf{\Sigma}\mathbf{V}^H = \mathbf{u}_1\ \sigma_1\ \mathbf{v}_1^H = \frac{1}{\sqrt{2}}\begin{pmatrix} j \\ -1 \end{pmatrix} 2 \left(\frac{1}{\sqrt{2}} \ \vdots\ \frac{-j}{\sqrt{2}}\right) \tag{22.35}$$

$$\mathbf{v}_1 = \frac{1}{\sqrt{2}}\begin{pmatrix} 1 \\ j \end{pmatrix} \quad \text{(eigenvector of } \mathbf{A}\mathbf{A}^H \text{ associated with } \sigma_1^2 = 4) \qquad \mathbf{u}_1 = \frac{1}{\sqrt{2}}\begin{pmatrix} j \\ -1 \end{pmatrix} \tag{22.36}$$

There is a remarkable difference between eigenvalue decomposition and singular-value decomposition. □

Example 2: A skew-symmetric:

$$\mathbf{A} = \begin{pmatrix} 0 & 1 \\ -1 & 0 \end{pmatrix} \qquad \lambda_i[\mathbf{A}] = \pm j \qquad \sigma_i[\mathbf{A}] = 1;\ 1\ . \ \square \tag{22.37}$$

Example 3: A unitary:

$$\mathbf{A} = \begin{pmatrix} 0 & e^{j\varphi} \\ e^{j\psi} & 0 \end{pmatrix} \qquad \lambda_{1,2}[\mathbf{A}] = \pm e^{j\frac{\varphi+\psi}{2}} \qquad \mathbf{A}\mathbf{A}^H = \mathbf{I} \qquad \sigma_{1,2}[\mathbf{A}] = 1;\ 1\ . \tag{22.38}$$

Eq.(22.28) $\mathbf{U}_\lambda \mathrm{diag}\{\lambda_i\}\mathbf{U}_\lambda^H = \mathbf{A}$ yields

$$\begin{pmatrix} \frac{\sqrt{2}}{2}e^{j\frac{\varphi-\psi}{2}} & \frac{\sqrt{2}}{2} \\ \frac{\sqrt{2}}{2} & -\frac{\sqrt{2}}{2}e^{-j\frac{\varphi-\psi}{2}} \end{pmatrix} \begin{pmatrix} e^{j\frac{\varphi+\psi}{2}} & 0 \\ 0 & -e^{j\frac{\varphi+\psi}{2}} \end{pmatrix} \begin{pmatrix} \frac{\sqrt{2}}{2}e^{-j\frac{\varphi-\psi}{2}} & \frac{\sqrt{2}}{2} \\ \frac{\sqrt{2}}{2} & -\frac{\sqrt{2}}{2}e^{j\frac{\varphi-\psi}{2}} \end{pmatrix} = \begin{pmatrix} 0 & e^{j\varphi} \\ e^{j\psi} & 0 \end{pmatrix} . \tag{22.39}$$

Anticipating Eq.(22.88), the singular-value decomposition of the normal matrix **A** is given by the following steps

$$\mathbf{A}^H\mathbf{A} = \begin{pmatrix} 0 & e^{-j\psi} \\ e^{-j\varphi} & 0 \end{pmatrix} \begin{pmatrix} 0 & e^{j\varphi} \\ e^{j\psi} & 0 \end{pmatrix} = \mathbf{I}, \quad \mathbf{V} = \mathbf{I}, \quad \mathbf{U} = \mathbf{A}\mathbf{V}\mathbf{\Sigma}^{-1} = \begin{pmatrix} 0 & e^{j\varphi} \\ e^{j\psi} & 0 \end{pmatrix}, \tag{22.40}$$

$$\mathbf{A} = \mathbf{U}\mathbf{\Sigma}\mathbf{V}^H = \begin{pmatrix} 0 & e^{j\varphi} \\ e^{j\psi} & 0 \end{pmatrix} \begin{pmatrix} 1 & 0 \\ 0 & 1 \end{pmatrix} \mathbf{I} = \begin{pmatrix} 0 & e^{j\varphi} \\ e^{j\psi} & 0 \end{pmatrix} . \ \square \tag{22.41}$$

22.5 Involutory Matrices

An involutory matrix is both Hermitian and unitary. In view of $\mathbf{G}^H = \mathbf{G}$ and $\mathbf{G}^H = \mathbf{G}^{-1}$,

$$\mathbf{G} = \mathbf{G}^{-1} \quad \text{and} \quad \mathbf{G}^2 = \mathbf{I}\ . \tag{22.42}$$

A matrix is said to be half involutory if it is both skew-Hermitian and unitary. Thus,

$$\mathbf{G}^2 = -\mathbf{I}\ . \tag{22.43}$$

22.6 Unitarity of a Modal Matrix Associated with a Hermite Matrix

Assume a given Hermite matrix $\mathbf{Q}$ has distinct eigenvalues $\lambda_i[\mathbf{Q}]$ and (right) eigenvectors $\mathbf{q}_i$. The eigenvectors $\mathbf{q}_i$ and $\mathbf{q}_k$ associated with different eigenvalues λ_i and λ_k are defined by

$$\mathbf{Q}\mathbf{q}_i = \lambda_i\mathbf{q}_i \tag{22.44}$$

$$\mathbf{Q}\mathbf{q}_k = \lambda_k\mathbf{q}_k. \tag{22.45}$$

Taking the conjugate transpose of Eq.(22.44),

$$\mathbf{q}_i^H\mathbf{Q}^H = \lambda_i\mathbf{q}_i^H. \tag{22.46}$$

Premultiplying Eq.(22.45) by q_i^H, postmultiplying Eq.(22.46) by q_k and taking the difference yields

$$q_i^H Q q_k - q_i^H Q^H q_k = \lambda_k q_i^H q_k - \lambda_i q_i^H q_k \tag{22.47}$$

$$q_i^H (Q - Q^H) q_k = (\lambda_k - \lambda_i) q_i^H q_k \quad \leadsto \quad 0 = (\lambda_k - \lambda_i) q_i^H q_k \; . \tag{22.48}$$

Eigenvectors of a Hermite matrix Q associated with various eigenvalues are unitary as noticed in Eq.(22.20).

The modal matrix $T[Q]$ is defined as a matrix built of eigenvectors q_i . Premultiplying T by its conjugate transpose T^H gives

$$T^H T = \begin{pmatrix} q_1^H \\ q_2^H \\ \vdots \end{pmatrix} (q_1 \; q_2 \ldots) = I \quad \leadsto \quad T^H = T^{-1}. \tag{22.49}$$

The modal matrix $T[Q]$ associated with a Hermite matrix Q is unitary.

22.7 Derivatives of trG^HG and $||G||_F$

Note the following properties of the derivatives of trG^HG and $||G||_F$ with respect to the real and imaginary part

$$\frac{\partial \text{ tr } G^H G}{\partial(\Re e \; G)} = 2 \; \Re e \; G \qquad \frac{\partial \text{ tr } G^H G}{\partial(\Im m \; G)} = 2 \; \Im m \; G \tag{22.50}$$

$$\frac{\partial ||G||_F}{\partial(\Re e \; G)} = \frac{\Re e \; G}{||G||_F} \qquad \frac{\partial ||G||_F}{\partial(\Im m \; G)} = \frac{\Im m \; G}{||G||_F} \; . \tag{22.51}$$

22.8 Hermite Form

Assume that a complex (n,n)-matrix G has full rank, i.e. rank $G = n$. A Hermite matrix Q is defined

$$Q = G^H G, \qquad Q^H = Q \; . \tag{22.52}$$

The Hermite form is written as

$$c = v^H Q v = v^H G^H G v = (Gv)^H (Gv) \tag{22.53}$$

where Gv is a vector with complex elements. The expression on the right hand of Eq.(22.53) defines the inner product of the conjugate $(Gv)^*$ and (Gv). This term is always real and positive. If rank $G = n$ the Hermite form c is positive definite, if rank $G = r < n$ then c is positive semidefinite.

Proof: The transformation $v = Tw$ is introduced where T is the modal matrix $T[Q]$ associated with Q. The matrix T is composed of the eigenvectors of Q. Then, including the fact $T^H = T^{-1}$

$$c = w^H T^H Q T w = w^H (\text{diag } \lambda_i[Q]) w = \sum_{i=1}^{n} \lambda_i[Q] \; w_i^2 \; . \tag{22.54}$$

The eigenvalues λ_i are positive because c is positive definite due to Eq.(22.53). If rank $G = r < n$ then $(n-r)$ eigenvalues λ_i of G are zero and c is positive semidefinite. Since $\lambda_i \geq 0$

$$\lambda_i \triangleq \sigma_i^2 \qquad (\sigma_i \ldots \text{real}), \qquad T^H Q T = \text{diag } \sigma_i^2 = (\text{diag } \sigma_i)^2 \tag{22.55}$$

$$\mathbf{Q} = \mathbf{T}(\text{diag } \sigma_i)^2\mathbf{T}^H = (\mathbf{T} \text{ diag } \sigma_i)(\mathbf{T} \text{ diag } \sigma_i)^H \tag{22.56}$$

$$\mathbf{G} = (\text{diag } \sigma_i)\mathbf{T}^H \quad \text{and} \quad \mathbf{G}^H = \mathbf{T}(\text{diag } \sigma_i). \tag{22.57}$$

Per definition,

$$\sigma_i[\mathbf{G}] \triangleq +\sqrt{\lambda_i[\mathbf{G}^H\mathbf{G}]} \tag{22.58}$$

are the singular values (principal gains) of $\mathbf{G}$. The matrix $\mathbf{T}[\mathbf{G}^H\mathbf{G}]$ is the modal matrix associated with $\mathbf{Q} = \mathbf{G}^H\mathbf{G}$. Since $\mathbf{Q}$ is Hermite the matrices $\mathbf{Q}$ and $\mathbf{Q}^H$ share the eigenvectors λ_i .

22.9 Singular Values

According with Eq.(22.58), the singular values σ_i of a (complex) matrix $\mathbf{G}$ are defined as the positive square root of the eigenvalues $\lambda_i[\mathbf{G}^H\mathbf{G}]$. An (n,n)-matrix $\mathbf{G}$ possesses n singular values. If $\mathbf{G}$ is specialized to a complex scalar the singular value turns out as the absolute value. The singular value is the generalization of the absolute value in matrix case. The brackets are used in $\sigma_i[\cdot]$ for denoting the operator-like function of σ_i. The largest singular value $\sigma_{i\,\max}[\mathbf{G}]$ is identical to the spectral norm $||\mathbf{G}||_s$ for the complex matrix $\mathbf{G}$.

The singular values $\sigma_{\max}[\mathbf{G}]$ and $\sigma_{\min}[\mathbf{G}]$ authentically represent the maximum and minimum gain given by the transfer matrix $\mathbf{G}$. This property cannot be provided by the eigenvalues $\lambda_i[\mathbf{G}]$.

22.9.1 Properties of Singular Values

Since for $\mathbf{G} \in \mathcal{C}^{n \times m}$

$$\det(\mu_i^2\mathbf{I} - \mathbf{G}^H\mathbf{G}) = \det[(\mu_i\mathbf{I} - \sqrt{\mathbf{G}^H\mathbf{G}})(\mu_i\mathbf{I} + \sqrt{\mathbf{G}^H\mathbf{G}})] \tag{22.59}$$

$$= \det(\mu_i\mathbf{I} - \sqrt{\mathbf{G}^H\mathbf{G}}) \det(\mu_i\mathbf{I} + \sqrt{\mathbf{G}^H\mathbf{G}}) = 0 \tag{22.60}$$

$$\mu_i^2 = \lambda_i[\mathbf{G}^H\mathbf{G}], \quad \mu_i = \lambda_i[\sqrt{\mathbf{G}^H\mathbf{G}}] \quad \text{and} \quad +\sqrt{\lambda_i[\mathbf{G}^H\mathbf{G}]} = \lambda_i[+\sqrt{\mathbf{G}^H\mathbf{G}}] \quad \forall\, i = 1 \dots m\,. \tag{22.61}$$

Calculating the singular values via eigenvalues the matrices $\mathbf{G}$ and $\mathbf{G}^H$ commute, i.e.

$$\sigma_i[\mathbf{G}] = \lambda_i[\sqrt{\mathbf{G}^H\mathbf{G}}] = \sqrt{\lambda_i[\mathbf{G}^H\mathbf{G}]} = \lambda_i[\sqrt{\mathbf{G}\mathbf{G}^H}] = \sqrt{\lambda_i[\mathbf{G}\mathbf{G}^H]} \quad \forall \lambda_i \neq 0\,, \tag{22.62}$$

only the number (multiplicity) of zero eigenvalues of $\sqrt{\mathbf{G}^H\mathbf{G}}$ and $\sqrt{\mathbf{G}\mathbf{G}^H}$ is different and depends on the rank $(\mathbf{G}^H\mathbf{G}) = \text{rank}\,(\mathbf{G}\mathbf{G}^H) = r$. The multiplicity of zero eigenvalues for $\mathbf{G}^H\mathbf{G}$ and $\mathbf{G}\mathbf{G}^H$ is $m-r$ and $n-r$, respectively. □

If $\mathbf{G} \in \mathcal{C}^{n \times n}$ is square then the singular values are invariant under unitary transformation

$$\sigma_i[\mathbf{G}] = \sigma_i[\mathbf{U}\mathbf{G}] = \sigma_i[\mathbf{G}\mathbf{U}] \quad \forall i = 1 \dots n \tag{22.63}$$

where $\mathbf{U}$ is any unitary matrix $\mathbf{U} \in \mathcal{C}^{n \times n}$. □

If $\mathbf{G} \in \mathcal{C}^{n \times n}$ is normal then referring to Eq.(22.25)

$$\sigma_i[\mathbf{G}] = |\ \lambda_i[\mathbf{G}]\ | \quad \forall i = 1 \dots n\ . \ \square \tag{22.64}$$

Using Rayleigh's theorem, one has

$$\sigma_{\max}^2[\mathbf{G}] = \max_{||\mathbf{x}||_F = 1} \mathbf{x}^H\mathbf{G}^H\mathbf{G}\mathbf{x}\ . \ \square \tag{22.65}$$

Among the n singular values of an (n,n)-matrix the maximum and minimum singular value $\sigma_{\max}$ and $\sigma_{\min}$ play an outstanding role. Important properties are as follows.

Sum of matrices:

$$\sigma_{\min}[\mathbf{E}+\mathbf{F}] \geq \sigma_{\min}[\mathbf{E}] - \sigma_{\max}[\mathbf{F}] \tag{22.66}$$

$$\sigma_{\max}[\mathbf{E}+\mathbf{F}] \leq \sigma_{\max}[\mathbf{E}] + \sigma_{\max}[\mathbf{F}] \ . \tag{22.67}$$

Product of matrices:

$$\sigma_{\min}[\mathbf{EF}] \geq \sigma_{\min}[\mathbf{E}]\sigma_{\min}[\mathbf{F}] \tag{22.68}$$

$$\sigma_{\max}[\mathbf{EF}] \leq \sigma_{\max}[\mathbf{E}]\sigma_{\max}[\mathbf{F}] \ . \tag{22.69}$$

Inverse of a matrix:

$$\sigma_{\max}[\mathbf{E}^{-1}] = \frac{1}{\sigma_{\min}[\mathbf{E}]} \quad \text{if } \mathbf{E}^{-1} \text{ exists} \tag{22.70}$$

or

$$\sigma_{\min}[\mathbf{A}] = \min_{\mathbf{x}\neq 0} \frac{\|\mathbf{Ax}\|_F}{\|\mathbf{x}\|_F} = \begin{cases} \|\mathbf{A}^{-1}\|_s^{-1} = \sigma_{\max}^{-1}[\mathbf{A}^{-1}] & \text{if } \det\mathbf{A} \neq 0 \\ 0 & \text{if } \det\mathbf{A} = 0 \ . \end{cases} \tag{22.71}$$

If $\mathbf{X}$ and $\mathbf{Y}$ are all-pass or inner, i.e. $\mathbf{XX}^R = \mathbf{I}$, $\mathbf{YY}^R = \mathbf{I}$, where $\mathbf{X}^R(s) = \mathbf{X}^T(-s)$ then $\sigma_i[\mathbf{XAY}] = \sigma_i[\mathbf{A}]$. □

For any $\mathbf{A}$ one has $\sigma_i[\mathbf{A}^R] = \sigma_i[\mathbf{A}]$. □

The maximum singular value is positive definite

$$\sigma_{\max}[\mathbf{E}] > 0 \ , \qquad \sigma_{\max}[\mathbf{E}] = 0 \quad \text{if} \quad \mathbf{E} = \mathbf{0} \ . \tag{22.72}$$

Bounding relations for singular values and eigenvalues admissible for any eigenvalue $\lambda_i[\mathbf{E}]$, i.e., the eigenvalues are bounded in modulus by the singular values

$$\sigma_{\min}[\mathbf{E}] \leq | \ \lambda_i[\mathbf{E}] \ | \leq \sigma_{\max}[\mathbf{E}] \quad \forall i = 1 \ldots n \tag{22.73}$$

$$\sigma_{\min}[\mathrm{G}] \ \|\mathbf{u}\| \leq \|\mathbf{Gu}\| \leq \sigma_{\max}[\mathrm{G}] \ \|\mathbf{u}\| \qquad \textit{(Morari, M., 1983)}. \tag{22.74}$$

The sum of squared singular values equals the Frobenius norm of $\mathbf{G} \in \mathcal{C}^{n\times m}$

$$\|\mathrm{G}\|_F^2 = \mathrm{tr}\ \mathrm{G}^H\mathrm{G} = \sum_{i=1}^{\min\{n,m\}} \lambda_i[\mathrm{G}^H\mathrm{G}] = \sum_{i=1}^{\min\{n,m\}} \sigma_i^2[\mathrm{G}] \ . \ \square \tag{22.75}$$

The product of singular values of a square matrix equals the modulus of $\det \mathbf{G}$

$$\prod_i \sigma_i[\mathrm{G}] = \sqrt{\prod_i \lambda_i[\mathrm{G}^H\mathrm{G}]} = \sqrt{\det(\mathrm{G}^H\mathrm{G})} = \sqrt{(\det \mathrm{G})^* \det \mathbf{G}} = \sqrt{|\det \mathrm{G}|^2} = |\det \mathrm{G}| \ . \ \square \tag{22.76}$$

Example 1: Consider a perturbed system $\mathbf{G}_p = \mathbf{G} + \Delta\mathbf{G}$. Then, invoking Eq.(22.66) and with regard to

$$\sigma_{\min}[\mathrm{G}_p] = \sigma_{\min}[\mathrm{G}+\Delta\mathrm{G}] \geq \sigma_{\min}[\mathrm{G}] - \sigma_{\max}[\Delta\mathrm{G}] \ , \tag{22.77}$$

the perturbation $\Delta\mathbf{G}$ in the worst case affects the minimum singular value to an extent given by the maximum singular value of the perturbation. Noting the fact that a matrix becomes singular when its minimum singular value vanishes, Eq.(22.77) is an important rule to assess the perturbation in an open-loop system.

Example 2: If $\sigma_{\min}[\mathbf{G}] \neq 0$ then $\sigma_{\min}[\mathbf{G}]$ is a measure how close $\mathbf{G}$ is to the singular case, i.e. where $\mathbf{G} + \Delta\mathbf{G} = \mathbf{G}_p$ is singular. Proof: By Eq.(22.66),

$$\det(\mathbf{G} + \Delta\mathbf{G}) = 0 \iff \sigma_{\min}[\mathbf{G} + \Delta\mathbf{G}] = 0 \text{ and } \Leftarrow \|\Delta\mathbf{G}\|_s = \sigma_{\max}[\Delta\mathbf{G}] \geq \sigma_{\min}[\mathbf{G}] \,. \tag{22.78}$$

Example 3: In the case of multivariable control, the singular value $\sigma_{\min}[\mathbf{I} + \mathbf{GK}]$ corresponds with the stability radius $\min|1 + GK|$ of a single input-single output control system. If the plant $\mathbf{G}(s)$ (see Fig. 22.1) is augmented by an unstructured uncertainty $\mathbf{L}$ in the so-called multiplicative form $(\mathbf{I} + \mathbf{L})\mathbf{G}$ using Eq.(22.66)

$$\sigma_{\min}[\mathbf{I} + \mathbf{GK} + \mathbf{LGK}] \geq \sigma_{\min}[\mathbf{I} + \mathbf{GK}] - \sigma_{\max}[\mathbf{LGK}] \,. \tag{22.79}$$

The uncertainty $\mathbf{L}$ reduces the singular value $\sigma_{\min}[\mathbf{I} + \mathbf{GK}]$ given by the plant without uncertainty. Using the so-called characteristic locus framework an optimization or maximin problem arises: Maximizing $\sigma_{\min}[\mathbf{I} + \mathbf{GK} + \mathbf{LGK}]$ with respect to $\mathbf{K}$ while mastering the fact that it is minimized with respect to the uncertainty $\mathbf{L}$ (*Kouvaritakis, B., and Trimboli, M.S., 1989*).

End of Examples

22.9.2 Derivatives of the Singular Value

Referring to Eq.(5.27),

$$\frac{\partial \sigma_i}{\partial \mathbf{G}} = \frac{\partial \mathbf{u}_i^H \mathbf{G} \mathbf{v}_i}{\partial \mathbf{G}} = \mathbf{u}_i^* \mathbf{v}_i^T \tag{22.80}$$

$$\frac{\partial \sigma_i}{\partial (\Re e\, \mathbf{G})} = \Re e\; \mathbf{u}_i^* \mathbf{v}_i^T = \Re e\; \mathbf{u}_i \mathbf{v}_i^H \qquad \frac{\partial \sigma_i}{\partial (\Im m\, \mathbf{G})} = -\Im m\; \mathbf{u}_i^* \mathbf{v}_i^T = \Im m\; \mathbf{u}_i \mathbf{v}_i^H \tag{22.81}$$

$$\frac{\partial (\sigma_i^2)}{\partial (\Re e\, \mathbf{G})} = -(\Re e\, \mathbf{G})(\mathbf{\Lambda}_1^T + \mathbf{\Lambda}_1) - (\Im m\, \mathbf{G})(\mathbf{\Lambda}_2^T - \mathbf{\Lambda}_2) \tag{22.82}$$

$$\frac{\partial (\sigma_i^2)}{\partial (\Im m\, \mathbf{G})} = -(\Im m\, \mathbf{G})(\mathbf{\Lambda}_1^T + \mathbf{\Lambda}_1) + (\Re e\, \mathbf{G})(\mathbf{\Lambda}_2^T - \mathbf{\Lambda}_2) \tag{22.83}$$

$$\text{where } \mathbf{\Lambda}_1 = -\frac{\partial \lambda[\mathbf{G}^H \mathbf{G}]}{\partial (\Re e\, \mathbf{G}^H \mathbf{G})} = -\Re e\; \mathbf{v}_i^* \mathbf{v}_i^T, \quad \mathbf{\Lambda}_2 = -\frac{\partial \lambda[\mathbf{G}^H \mathbf{G}]}{\partial (\Im m\, \mathbf{G}^H \mathbf{G})} = \Im m\; \mathbf{v}_i^* \mathbf{v}_i^T \tag{22.84}$$

(*Sevaston, G.E., and Longman, R.W., 1988; Freudenberg, J.S., et al. 1982*).

22.10 Singular Vectors

Assume $\mathbf{G} \in \mathcal{C}^{n \times m}$ ($\rightsquigarrow \mathbf{G}^H \mathbf{G} \in \mathcal{C}^{m \times m}$ and $\mathbf{G}\mathbf{G}^H = \mathcal{C}^{n \times n}$) with distinct singular values. Then, $\sigma_i[\mathbf{G}] = +(\lambda_i[\mathbf{G}^H \mathbf{G}])^{0.5}$ are real non-negative numbers $\forall i = 1 \ldots r$ since

$$\lambda[\mathbf{G}^H \mathbf{G}] = \lambda^*[(\mathbf{G}^H \mathbf{G})^H] = \lambda^*[\mathbf{G}^H \mathbf{G}] \tag{22.85}$$

which only can be realized if λ are real. The right eigenvector $\mathbf{v}_i$ of $\mathbf{G}^H \mathbf{G}$ associated with σ_i^2 is given by

$$(\mathbf{G}^H \mathbf{G})\mathbf{v}_i = \sigma_i^2 \mathbf{v}_i \,. \tag{22.86}$$

Taking the conjugate transpose of the equation above,

$$\mathbf{v}_i^H (\mathbf{G}^H \mathbf{G})^H = \mathbf{v}_i^H (\mathbf{G}^H \mathbf{G}) = \sigma_i^2 \mathbf{v}_i^H \tag{22.87}$$

this equation already defines the left eigenvector. Thus, the right and left eigenvector are the same.

The matrix $\mathbf{G}\mathbf{G}^H$ has the same eigenvalues except some that are zero. The right (and left) eigenvectors of $\mathbf{G}\mathbf{G}^H$ are denoted $\mathbf{u}_i$.

22.11 Singular-Value Decomposition

Consider $\mathbf{G} \in \mathcal{C}^{n\times m}$ and its non-zero singular values $\sigma_i = \sigma_i[\mathbf{G}]\ \forall i = 1\ldots r$ then the singular-value decomposition of $\mathbf{G}$ is (*Golub, G.H., and Van Loan, C.F., 1990*)

$$\mathbf{G} = \mathbf{U}\mathbf{\Sigma}\mathbf{V}^H \tag{22.88}$$

where $\mathbf{\Sigma} = \text{matrix}[\Sigma_{ij}]$, $\quad \mathbf{\Sigma} \in \mathcal{R}^{n\times m}$,

$$\Sigma_{ii} = \sigma_i \quad \forall i = 1\ldots r\ ;\ r = \text{rank}\ (\mathbf{G}^H\mathbf{G}) = \text{rank}\ (\mathbf{G}\mathbf{G}^H) \tag{22.89}$$

$$\Sigma_{ij} = 0 \quad \forall i \neq j,\ i = 1\ldots n,\ j = 1\ldots m \tag{22.90}$$

and the matrices $\mathbf{U}$ and $\mathbf{V}$ are unitary, $\mathbf{U} \in \mathcal{C}^{n\times n}, \mathbf{V} \in \mathcal{C}^{m\times m}$.

The left and right singular vectors or eigenvectors for $\mathbf{G}\mathbf{G}^H$ and $\mathbf{G}^H\mathbf{G}$ are termed $\mathbf{u}_i \in \mathcal{C}^n\ \forall i = 1\ldots n$ and $\mathbf{v}_i \in \mathcal{C}^m\ \forall i = 1\ldots m$, respectively. The unitary matrices $\mathbf{U}$ and $\mathbf{V}$ are given by columns equal to the left and right singular vectors

$$\mathbf{U} = (\mathbf{u}_1\ \mathbf{u}_2 \ldots\ldots \mathbf{u}_n) \qquad \mathbf{V} = (\mathbf{v}_1\ \mathbf{v}_2 \ldots\ldots \mathbf{v}_m) \tag{22.91}$$

(*Klema, V.C., and Laub, A.J., 1980*). According to Eq.(22.88),

$$\mathbf{G}\mathbf{V} = \mathbf{U}\mathbf{\Sigma} = (\sigma_1\mathbf{u}_1 \vdots \sigma_2\mathbf{u}_2 \ldots \sigma_r\mathbf{u}_r \vdots \mathbf{0}\ldots\mathbf{0}) \tag{22.92}$$

$$\text{or} \qquad \mathbf{G}\mathbf{v}_i = \sigma_i\mathbf{u}_i \qquad \forall i = 1\ldots r \qquad \text{or} \qquad \mathbf{G} = \sum_{i=1}^{\min(n,m)} \sigma_i\mathbf{u}_i\mathbf{v}_i^H\ . \tag{22.93}$$

The eigenvectors $\mathbf{v}_i$ and $\mathbf{u}_i$ are orthogonal eigenbases for $\mathbf{G}^H\mathbf{G}$ and $\mathbf{G}\mathbf{G}^H$, respectively, i.e. $\mathbf{u}_k^H\mathbf{u}_i = \delta_{ki}$ and $\mathbf{v}_j^H\mathbf{v}_l = \delta_{jl}$. Hence,

$$\sigma_i = \mathbf{u}_i^H\mathbf{G}\mathbf{v}_i, \quad \sigma_i^2 = \mathbf{v}_i^H(\mathbf{G}^H\mathbf{G})\mathbf{v}_i, \quad \sigma_i^2 = \mathbf{u}_i^H(\mathbf{G}\mathbf{G}^H)\mathbf{u}_i \tag{22.94}$$

and

$$\mathbf{G}^H\mathbf{G} = \sum_{i=1}^{r}\sigma_i^2\mathbf{v}_i\mathbf{v}_i^H, \qquad \mathbf{G}\mathbf{G}^H = \sum_{i=1}^{r}\sigma_i^2\mathbf{u}_i\mathbf{u}_i^H \tag{22.95}$$

or

$$\mathbf{G}\mathbf{G}^H = \mathbf{U}\mathbf{\Sigma}\mathbf{V}^H\ \mathbf{V}\mathbf{\Sigma}^H\mathbf{U}^H = \mathbf{U}\mathbf{\Sigma}\mathbf{\Sigma}^H\mathbf{U}^H = \mathbf{U}\mathbf{\Sigma}^2\mathbf{U}^H \tag{22.96}$$

$$\mathbf{G}^H\mathbf{G} = \mathbf{V}\mathbf{\Sigma}^H\mathbf{U}^H\ \mathbf{U}\mathbf{\Sigma}\mathbf{V}^H = \mathbf{V}\mathbf{\Sigma}^H\mathbf{\Sigma}\mathbf{V}^H = \mathbf{V}\mathbf{\Sigma}^2\mathbf{V}^H\ . \tag{22.97}$$

The singular vectors $\mathbf{v}_i$ and $\mathbf{u}_i$ are also termed input and output principal directions. The singular vectors associated with the largest or smallest singular values are termed major or minor principal direction, respectively.

Note the property rank$\mathbf{G} = r$. The range space $\mathcal{R}[\mathbf{G}]$ is the subspace spanned by $\mathbf{u}_1, \mathbf{u}_2 \ldots \mathbf{u}_r$. The null space $\mathcal{N}[\mathbf{G}]$ is spanned by $\mathbf{v}_{r+1}, \mathbf{v}_{r+2} \ldots \mathbf{v}_m$ (*Reid, J.G., 1983; Stewart, G.W., 1973*). The subspace span$[\mathbf{a}_1, \mathbf{a}_2]$ consists of all linear combinations of the vectors $\mathbf{a}_1$ and $\mathbf{a}_2$.

Example: Singular-value decomposition

$$\mathbf{G} = \begin{pmatrix} 1 & 0 & 2 \\ 0 & 3 & 0 \end{pmatrix} \qquad n = 2,\ m = 3 \tag{22.98}$$

$$\mathbf{G}\mathbf{G}^H = \begin{pmatrix} 5 & 0 \\ 0 & 9 \end{pmatrix} \qquad \sqrt{\lambda_i[\mathbf{G}\mathbf{G}^H]} = \sqrt{5}\ ;\ 3 \qquad \mathbf{U} = \begin{pmatrix} 1 & 0 \\ 0 & 1 \end{pmatrix} \tag{22.99}$$

$$\mathbf{G}^H\mathbf{G} = \begin{pmatrix} 1 & 0 & 2 \\ 0 & 9 & 0 \\ 2 & 0 & 4 \end{pmatrix} \qquad \sqrt{\lambda_i[\mathbf{G}^H\mathbf{G}]} = \sqrt{5}\ ;\ 3\ ;\ 0 \qquad \mathbf{V} = \begin{pmatrix} 1/\sqrt{5} & 0 & 2/\sqrt{5} \\ 0 & 1 & 0 \\ 2/\sqrt{5} & 0 & -1/\sqrt{5} \end{pmatrix} \tag{22.100}$$

$$\sigma_1[\mathbf{G}] = \sqrt{5}\ ; \quad \sigma_2[\mathbf{G}] = 3 \tag{22.101}$$

$$\mathbf{G} = \mathbf{U}\mathbf{\Sigma}\mathbf{V}^H = \begin{pmatrix} 1 & 0 \\ 0 & 1 \end{pmatrix}\begin{pmatrix} \sqrt{5} & 0 & 0 \\ 0 & 3 & 0 \end{pmatrix}\begin{pmatrix} 1/\sqrt{5} & 0 & 2/\sqrt{5} \\ 0 & 1 & 0 \\ 2/\sqrt{5} & 0 & -1/\sqrt{5} \end{pmatrix} = \begin{pmatrix} 1 & 0 & 2 \\ 0 & 3 & 0 \end{pmatrix} . \tag{22.102}$$

End of Example

To emphasize the relations when $r \neq n$, decompose $\mathbf{V}, \mathbf{U}$ and $\mathbf{\Sigma}$

$$\mathbf{V} = (\mathbf{V}_1^{(m\times r)} \ \vdots\ \mathbf{V}_2^{m\times(m-r)}), \quad \mathbf{U} = (\mathbf{U}_1^{(n\times r)} \ \vdots\ \mathbf{U}_2^{n\times(n-r)}), \quad \mathbf{\Sigma} = \begin{pmatrix} \mathbf{\Sigma}_r & \mathbf{0} \\ \mathbf{0} & \mathbf{0}^{(n-r)\times(m-r)} \end{pmatrix} . \tag{22.103}$$

Then,

$$\mathbf{U}^H\mathbf{G}\mathbf{V} = \begin{pmatrix} \mathbf{\Sigma}_r & \mathbf{0} \\ \mathbf{0} & \mathbf{0} \end{pmatrix} \tag{22.104}$$

$$\mathbf{G} = \mathbf{U}\begin{pmatrix} \mathbf{\Sigma}_r & \mathbf{0} \\ \mathbf{0} & \mathbf{0} \end{pmatrix}\mathbf{V}^H = (\mathbf{U}_1 \ \ \mathbf{U}_2)\begin{pmatrix} \mathbf{\Sigma}_r & \mathbf{0} \\ \mathbf{0} & \mathbf{0} \end{pmatrix}(\mathbf{V}_1 \ \ \mathbf{V}_2)^H = \mathbf{U}_1\mathbf{\Sigma}_r\mathbf{V}_1^H . \tag{22.105}$$

Calculating $\mathbf{U}_1$ and $\mathbf{V}_1$: First, the matrix $\mathbf{V}_1$ is chosen as an orthonormal basis given by the eigenvectors of $\mathbf{G}^H\mathbf{G}$ associated with $\sigma_i^2\ \forall i = 1, 2 ... r$ and, second, $\mathbf{U}_1$ results from $\mathbf{U}_1 = \mathbf{G}\mathbf{V}_1\mathbf{\Sigma}_r^{-1}$.

The singular-value decompositions of $\mathbf{A}$, $\mathbf{B}$ and $\mathbf{A} \otimes \mathbf{B}$ are $\mathbf{U}_A\mathbf{\Sigma}_A\mathbf{V}_A^H$, $\mathbf{U}_B\mathbf{\Sigma}_B\mathbf{V}_B^H$ and $(\mathbf{U}_A \otimes \mathbf{U}_B)(\mathbf{\Sigma}_A \otimes \mathbf{\Sigma}_B)(\mathbf{V}_A \otimes \mathbf{V}_B)^H$, respectively. □

22.12 Row and Column Compression

Pre- and postmultiplication of the singular-value decomposition by $\mathbf{U}^H$ and $\mathbf{V}$, respectively, yields a

$$\text{row compression} \quad \mathbf{U}^H\mathbf{G} = \mathbf{\Sigma}\mathbf{V}^H = \begin{pmatrix} \mathbf{\Sigma}_r & \mathbf{0} \\ \mathbf{0} & \mathbf{0} \end{pmatrix}\mathbf{V}^H = \begin{pmatrix} \mathbf{G}_R \\ \mathbf{0} \end{pmatrix} \tag{22.106}$$

$$\text{and a column compression} \quad \mathbf{G}\mathbf{V} = \mathbf{U}\mathbf{\Sigma} = \mathbf{U}\begin{pmatrix} \mathbf{\Sigma}_r & \mathbf{0} \\ \mathbf{0} & \mathbf{0} \end{pmatrix} = (\mathbf{G}_C \ \vdots\ \mathbf{0}) \tag{22.107}$$

where $\mathbf{G}_R$ is full row rank, i.e., $\mathbf{G}_R$ consists of r linearly independent rows and $\mathbf{G}_C$ is full column rank, i.e., $\mathbf{G}_C$ has r linearly independent columns.

22.13 Minimum Destabilizing Perturbation Matrix

Using the inner product for matrices

$$\langle \mathbf{A}, \mathbf{B} \rangle = \text{tr}\,[\mathbf{A}^H \mathbf{B}]\,, \tag{22.108}$$

the projection of a matrix $\mathbf{E}$ onto the subspace spanned by $\mathbf{u}_i \mathbf{v}_j^H$ is given by the matrix $(\mathbf{u}_i \mathbf{v}_j^H)$ premultiplied by the inner product $\langle \mathbf{u}_i \mathbf{v}_j^H, \mathbf{E} \rangle$

$$\text{tr}\,[(\mathbf{u}_i \mathbf{v}_j^H)^H \mathbf{E}]\mathbf{u}_i \mathbf{v}_j^H = \text{tr}\,[\mathbf{v}_j \mathbf{u}_i^H \mathbf{E}]\mathbf{u}_i \mathbf{v}_j^H \tag{22.109}$$

where $\mathbf{u}_i, \mathbf{v}_i$ are the left and right singular vectors. The matrices $\mathbf{u}_i \mathbf{v}_j^H$ form an orthogonal basis in which any nonsquare matrix $\mathbf{E}$ can be expressed as

$$\mathbf{E} = \sum_{i=1}^{n} \sum_{j=1}^{n} \langle \mathbf{u}_i \mathbf{v}_j^H, \mathbf{E} \rangle \mathbf{u}_i \mathbf{v}_j^H\,. \tag{22.110}$$

If the perturbed matrix $\mathbf{A}_p = \mathbf{A} + \Delta\mathbf{A}$ is singular *and* $\|\Delta\mathbf{A}\|_s = \sigma_{\min}[\mathbf{A}]$ then $\langle \mathbf{u}_n \mathbf{v}_n^H, \Delta\mathbf{A} \rangle = -\sigma_n = -\sigma_{\min}[\mathbf{A}]$ where $\mathbf{u}_i, \mathbf{v}_i$ are the left and right singular vector of $\mathbf{A}$ corresponding to σ_i and the singular values are arranged in descending order (i.e. $\sigma_i > \sigma_{i+1}$). Hence, if $\|\Delta\mathbf{A}\|_s < \sigma_{\min}[\mathbf{A}]$ then $(\mathbf{A} + \Delta\mathbf{A})$ must be nonsingular. The direction $\mathbf{u}_n \mathbf{v}_n^H$ is called the most sensitive direction. If a constraint on $\Delta\mathbf{A}$ is considered such that $\Delta\mathbf{A}$ cannot have a projection of magnitude $\sigma_n = \sigma_{\min}[\mathbf{A}]$ in the direction $\mathbf{u}_n \mathbf{v}_n^H$ then $\Delta\mathbf{A}$ must be increased along another direction in order to get $\mathbf{A} + \Delta\mathbf{A}$ singular. This problem of finding $\min_{\Delta\mathbf{A}} \|\Delta\mathbf{A}\|_s$ subject to $\det(\mathbf{A} + \Delta\mathbf{A}) = 0$ and $|\langle \mathbf{u}_n \mathbf{v}_n^H, \Delta\mathbf{A} \rangle| \le \phi < \sigma_{\min}[\mathbf{A}]$ has been solved by *Lehtomaki, N.A., et al. 1984*. The simplest case is given in what follows. The matrix $(\mathbf{A} + \Delta\mathbf{A})$ is nonsingular if

$$\sigma_{\max}[\Delta\mathbf{A}] < \sqrt{\sigma_n \sigma_{n-1} + \phi(\sigma_n - \sigma_{n-1})} \quad \text{and} \quad |\langle \mathbf{u}_n \mathbf{v}_n^H, \Delta\mathbf{A} \rangle| \le \phi < \sigma_n \tag{22.111}$$

where σ_n and σ_{n+1} are the two smallest singular values of $\mathbf{A}$ and $\mathbf{u}_n, \mathbf{v}_n$ are the left and right singular vectors, respectively, corresponding to σ_n.

Arbitrary constraints on $\Delta\mathbf{A}$ lead to a numerically complex searching procedure.

22.14 Smallest Singular Value and Rank Reduction of G

The matrix $\mathbf{G}$ is considered of rank $\mathbf{G} = r$. Perturbation $\Delta\mathbf{G}$ may cause reduction of $\mathbf{G}$ to rank $(\mathbf{G} + \Delta\mathbf{G}) = r - 1$. Then $\|\Delta\mathbf{G}\|_F \ge \sigma_{\min}[\mathbf{G}] = \sigma_r[\mathbf{G}]$ where the matrix $\mathbf{\Sigma}_r = \text{diag}\{\sigma_{\max}, \sigma_2, \sigma_3 \ldots \sigma_r\}$ and $\mathbf{\Sigma} = \text{block diag}\{\mathbf{\Sigma}_r, \mathbf{0}\}$ and $\sigma_1 \ge \sigma_2 \ldots \ge \sigma_r$. Thus, $\sigma_r[\mathbf{G}]$ is a lower bound for the perturbation norm which causes rank reduction.

22.15 Polar Decomposition

By analogy to $\lambda = |\lambda| \exp\,(j\gamma)$, complex matrices can be decomposed in what follows.

22.15.1 Using Unitary Matrices

Since the spectrum of an unitary matrix lies on the unit circle the matrix $\mathbf{G} \in \mathcal{C}^{n \times n}$ can be represented in the form

$$\mathbf{G} = \mathbf{H}_L \mathbf{U}_L \qquad \mathbf{H}_L = (\mathbf{G}\mathbf{G}^H)^{1/2} \tag{22.112}$$

i.e. the polar decomposition, where $\mathbf{U}_L$ is unitary and $\mathbf{H}_L \geq 0$ is positive semi-definite and given by $\mathbf{H}_L = (\mathbf{G}\mathbf{G}^H)^{1/2}$. If G is nonsingular, $\mathbf{H}_L > 0$.

An alternative polar decomposition is

$$\mathbf{G} = \mathbf{U}_R\mathbf{H}_R \qquad \mathbf{H}_R = (\mathbf{G}^H\mathbf{G})^{1/2} \, . \tag{22.113}$$

If $\mathbf{U}$ and $\mathbf{H}$ commute then $\mathbf{H}_L\mathbf{U}_L = \mathbf{U}_R\mathbf{H}_R$, $\mathbf{U}_L = \mathbf{U}_R$, $\mathbf{H}_L = \mathbf{H}_R$ and the matrix $\mathbf{G}$ is normal.

22.15.2 Direct Analogy to Scalars

Another polar representation is

$$\mathbf{G} = \mathbf{G}_{od} \exp j\mathbf{\Phi}_d \quad \in \mathcal{C}^{n\times n} \tag{22.114}$$

$$\text{where} \quad \mathbf{G}_{od} = [(\Re e\ \mathbf{G})^2 + (\Im m\ \mathbf{G})^2]^{1/2} \ \text{ and } \ \mathbf{\Phi}_d = \tan^{-1}[(\Re e\ \mathbf{G})^{-1}\Im m\ \mathbf{G}] \, . \tag{22.115}$$

Note that the definition $\mathbf{G} = (\exp j\mathbf{\Phi}'_d)\mathbf{G}'_{od}$ yields different results.

The matrix $\mathbf{M}^{1/m}$ is the principal mth root of $\mathbf{M}$ where m is a positive integer. The spectrum of $\mathbf{M}^{1/m}$ is given by $\lambda_i[\mathbf{M}^{1/m}] = \lambda_i^{1/m}[\mathbf{M}]\ \forall i = 1\ldots n$ where the arguments are $\arg\lambda_i \in (-\pi,\pi)$ and $\arg\lambda_i[\mathbf{M}^{1/m}] \in \left(-\frac{\pi}{m}, \frac{\pi}{m}\right)$.

The matrix $\tan^{-1}\mathbf{M}$ has the property

$$\lambda_i[\tan^{-1}\mathbf{M}] = \tan^{-1}\frac{\Im m\ \lambda_i[\mathbf{M}]}{\Re e\ \lambda_i[\mathbf{M}]} \, . \tag{22.116}$$

22.15.3 Using Modal Matrix

Consider $\mathbf{G} \in \mathcal{C}^{n\times n}$ and its modal representation

$$\mathbf{G} = \mathbf{T}\mathbf{J}\mathbf{T}^{-1} = \mathbf{G}_R + j\mathbf{G}_I = \mathbf{T}(\Re e\ \mathbf{J})\mathbf{T}^{-1} + j\mathbf{T}(\Im m\ \mathbf{J})\mathbf{T}^{-1} \tag{22.117}$$

where $\mathbf{J}$ is the Jordan form and $\mathbf{T}$ the modal matrix for *both* $\Re e\ \mathbf{J}$ and $\Im m\ \mathbf{J}$. Then,

$$\mathbf{G} = (\mathbf{G}_R^2 + \mathbf{G}_I^2)^{1/2}(\exp j\mathbf{\Phi}) = (\exp j\mathbf{\Phi})(\mathbf{G}_R^2 + \mathbf{G}_I^2)^{1/2} \tag{22.118}$$

$$\text{where} \quad \mathbf{G}_o = (\mathbf{G}_R^2 + \mathbf{G}_I^2)^{1/2} \qquad \mathbf{\Phi} = \tan^{-1}(\mathbf{G}_I\mathbf{G}_R^{-1}) = \tan^{-1}(\mathbf{G}_R^{-1}\mathbf{G}_I) \tag{22.119}$$

are the amplitude and the phase of $\mathbf{G}$, respectively. The real and imaginary part is

$$\Re e\ \mathbf{G} = \mathbf{G}_o \cos\mathbf{\Phi} = (\cos\mathbf{\Phi})\mathbf{G}_o \qquad \Im m\ \mathbf{G} = \mathbf{G}_o \sin\mathbf{\Phi} = (\sin\mathbf{\Phi})\mathbf{G}_o \, , \tag{22.120}$$

respectively. The matrices commute since both the real and imaginary part of $\mathbf{G}$ contain the same modal matrix $\mathbf{T}[\mathbf{G}]$ (*Shieh, L.S., et al. 1987*).

22.16 Minimax Optimization of the Spectral Radius

Consider the nominal plant $\mathbf{G}$, the perturbed plant $\mathbf{G} + \Delta\mathbf{G}$ and the controller $\mathbf{K}$. The problem might arise of the form

$$\max_{\Delta\mathbf{G}}\min_{\mathbf{K}} \rho_s[(\mathbf{G} + \Delta\mathbf{G})\mathbf{K}] = \frac{1}{k_{\max}} \tag{22.121}$$

where $k_{\max}$ corresponds to the amplitude margin. Then, for a certain perturbation function $\mathbf{M} := \mathbf{G} + \Delta\mathbf{G}$ at a particular frequency, the minimum spectral radius ρ_s of the open-loop transfer matrix $\mathbf{MK}$ with respect to $\mathbf{K}$, such that $\sigma_{\min}[\mathbf{K}] \geq \kappa$, is given by

$$\min_{\mathbf{K}} \rho_s[\mathbf{MK}] = \kappa \Big(\prod_{i=1}^{m} \sigma_i[\mathbf{M}] \Big)^{1/m} \qquad \mathbf{M}, \mathbf{K} \in \mathcal{C}^{m \times m} . \tag{22.122}$$

(Proof see *Kouvaritakis, B., and Trimboli, M.S., 1989*).

22.17 Gershgorin's Theorem

If $\mathbf{G}$ is a matrix with complex elements G_{ij} and eigenvalues $\lambda_k[\mathbf{G}]$. Then, each eigenvalue is located in at least one of the discs centered at the main diagonal element G_{ii} with radius $\sum_{j=1, j \neq i}^{n} \; |G_{ij}|$ (row sum)

$$|s - G_{ii}| \leq \sum_{j=1, j \neq i}^{n} \; |G_{ij}| \tag{22.123}$$

or in the junction set of the discs. The radius can be replaced by $\sum_{i=1, i \neq j}^{n} \; |G_{ij}|$ (column sum) which gives an additional bound. The eigenvalues are located in the intersection of the aforementioned junction sets.

If a set of d discs has no intersection with the other $n - d$ discs, then d eigenvalues are located within this set of d discs.

A matrix $\mathbf{G}$ is called diagonally dominant if $\forall i = 1 \ldots n$

$$|G_{ii}| > \sum_{j=1, j \neq i}^{n} \; |G_{ij}| \; . \tag{22.124}$$

Chapter 23

Resolvent Matrix and Stability Radius

Consider the homogeneous equation of a dynamic system $\dot{\mathbf{x}}(t) = \mathbf{A}\mathbf{x}(t) \quad \in \mathcal{R}^n$ where $\mathbf{A}$ represents the nominal system which is assumed stable. Let the system be linearly perturbed by an additive error matrix $\Delta\mathbf{A}$

$$\dot{\mathbf{x}}(t) = (\mathbf{A} + \Delta\mathbf{A})\mathbf{x}(t) \tag{23.1}$$

where $\Delta\mathbf{A}$ is spectral norm bounded, i.e. $\|\Delta\mathbf{A}\|_s < c$. *Martin, J.M., 1987* proposed a measure m_M for the stability robustness of the linear state-space system model

$$m_M \triangleq \min_{\omega \geq 0} \frac{1}{\|\mathbf{L}(\omega)\|_s} = \frac{1}{\max_{\omega \geq 0} \|\mathbf{L}(\omega)\|_s} \tag{23.2}$$

$$\text{where } \mathbf{L}(\omega) \triangleq (j\omega\mathbf{I} - \mathbf{A})^{-1} . \tag{23.3}$$

The matrix $(j\omega\mathbf{I} - \mathbf{A})^{-1}$ is denoted resolvent matrix. Since $\mathbf{A}$ is stable all the poles are located in the open left half of the s-plane. Poles on the imaginary axis are excluded. Hence, no zero $\det(j\omega\mathbf{I} - \mathbf{A}) = 0$ occurs and $\mathbf{L}(\omega) = (j\omega\mathbf{I} - \mathbf{A})^{-1}$ always exists.

Remember the following properties:

$$\lambda\mathbf{I} - \mathbf{L}^H\mathbf{L} = \mathbf{L}^H\mathbf{L}^{H,-1}(\lambda\mathbf{I} - \mathbf{L}^H\mathbf{L})\mathbf{L}^{-1}\mathbf{L} = \mathbf{L}^H(\lambda\mathbf{L}^{H,-1}\mathbf{L}^{-1} - \mathbf{I})\mathbf{L} \tag{23.4}$$

$$\det(\lambda\mathbf{I} - \mathbf{L}^H\mathbf{L}) = (\det\mathbf{L}^H)(\det\mathbf{L})(-\lambda)\det(\frac{1}{\lambda}\mathbf{I} - \mathbf{L}^{H,-1}\mathbf{L}^{-1}) \quad \rightsquigarrow \quad \frac{1}{\lambda} = \frac{1}{\lambda[\mathbf{L}^H\mathbf{L}]} = \lambda[\mathbf{L}^{H,-1}\mathbf{L}^{-1}] \tag{23.5}$$

$$\frac{1}{\lambda_{\max}[\mathbf{L}^H\mathbf{L}]} = \lambda_{\min}[\mathbf{L}^{H,-1}\mathbf{L}^{-1}] \quad \rightsquigarrow \quad \frac{1}{\sigma_{\max}[\mathbf{L}]} = \sigma_{\min}[\mathbf{L}^{-1}] . \ \square \tag{23.6}$$

Combining Eqs.(23.3) and (23.6) yields

$$m_M = \min_{\omega \geq 0} \frac{1}{\sigma_{\max}[\mathbf{L}(\omega)]} = \min_{\omega \geq 0} \sigma_{\min}[\mathbf{L}^{-1}(\omega)] = \min_{\omega \geq 0} \sigma_{\min}[j\omega\mathbf{I} - \mathbf{A}] . \tag{23.7}$$

Using the measure m_M and the norm-bounded perturbation, a necessary and sufficient condition for the stability of the system can be given. Thus, the stability robustness is estimated.

23.1 Preliminary Properties

23.1.1 Gastinel-Kahan Theorem

If $\mathbf{P}$ is a nonsingular matrix and $\|\cdot\|$ is any induced norm then

$$\frac{1}{\|\mathbf{P}^{-1}\|} \leq \|\mathbf{R}\| \quad \text{where } (\mathbf{P} + \mathbf{R}) \text{ is singular ,} \tag{23.8}$$

i.e., a nonsingular matrix $\mathbf{P}$ differs in norm from a singular matrix by no more than $\|\mathbf{P}^{-1}\|^{-1}$ (*Soh, C.B., and Chan, C.K., 1989*).

Proof: If $\mathbf{P}+\mathbf{R}$ is singular there exists $\mathbf{x} \neq \mathbf{0}$ such that $(\mathbf{P}+\mathbf{R})\mathbf{x} = \mathbf{0}$ or $\mathbf{P}\mathbf{x} = -\mathbf{R}\mathbf{x}$. From $\|\mathbf{P}\|\,\|\mathbf{x}\| \geq \|\mathbf{P}\mathbf{x}\|$ results

$$\|\mathbf{R}\| \geq \frac{\|\mathbf{R}\mathbf{x}\|}{\|\mathbf{x}\|} = \frac{\|-\mathbf{P}\mathbf{x}\|}{\|\mathbf{x}\|} = \frac{\|\mathbf{P}\mathbf{x}\|}{\|\mathbf{P}^{-1}\mathbf{P}\mathbf{x}\|} \geq \frac{\|\mathbf{P}\mathbf{x}\|}{\|\mathbf{P}^{-1}\|\,\|\mathbf{P}\mathbf{x}\|} = \frac{1}{\|\mathbf{P}^{-1}\|} \ . \ \square \tag{23.9}$$

Gastinel-Kahan theorem modification: For any induced norm $\|\cdot\|$ and nonsingular $\mathbf{A}$

$$\frac{1}{\|\mathbf{A}^{-1}\|} = \min\{\|\Delta\mathbf{A}\| : \quad \mathbf{A}+\Delta\mathbf{A} \quad \text{is singular } \} \tag{23.10}$$

or

$$\|\mathbf{A}^{-1}\|^{-1} \leq \|\Delta\mathbf{A}\| \quad \text{where} \quad \det(\Delta\mathbf{A}+\mathbf{A}) = 0 \ . \tag{23.11}$$

Example 1: $\mathbf{A} = \mathbf{A}(\omega) = j\omega\mathbf{I}$, $\Delta\mathbf{A} = ?$ such that $\mathbf{A}+\Delta\mathbf{A}$ nonsingular:

$$\|(j\omega\mathbf{I})^{-1}\|^{-1} \leq \|\Delta\mathbf{A}\| \tag{23.12}$$

$$\left(\max_{\|\mathbf{x}\|=1} \|(j\omega\mathbf{I})^{-1}\mathbf{x}\|\right)^{-1} = \left(\max_{\|\mathbf{x}\|=1} \|-j\omega^{-1}\mathbf{x}\|\right)^{-1} = |\omega| \leq \|\Delta\mathbf{A}\| \ . \ \square \tag{23.13}$$

Example 2: If $(j\omega\mathbf{I}-\mathbf{G})$ is singular then $|\omega| \leq \|\mathbf{G}\|$. $\square$

23.1.2 Martin's Theorem

For continuous-time systems: Let $\mathbf{M}$ and $\mathbf{E}$ be constant (n,n)-matrices. If the spectral abscissa $\alpha_s[\mathbf{M}] < 0$ and $\alpha_s[\mathbf{M}+\mathbf{E}] \geq 0$ then there exists $\delta_o \in (0,1]$ and $\omega \in \mathcal{R}$ such that $j\omega\mathbf{I}-\mathbf{M}-\delta_o\mathbf{E}$ is singular.

In the case of time-discrete systems: If the spectral radius $\rho_s[\mathbf{\Phi}] < 1$ and $\rho_s[\mathbf{\Phi}+\mathbf{E}_d] \geq 1$ then there exists $\delta_o \in (0,1]$ and $\omega \in [0,2\pi]$ such that $e^{j\omega}\mathbf{I}-\mathbf{\Phi}-\delta_o\mathbf{E}_d$ is singular. (Proof see *Martin, J.M., 1987; Martin, J.M., and Hewer, G.A., 1987.*)

23.2 Stability Robustness. Continuous-Time Systems

Consider a nominal system $\mathbf{A}$ with additive perturbation matrix $\Delta\mathbf{A}$ and the perturbed system $\mathbf{A}_p = \mathbf{A}+\Delta\mathbf{A}$. If the perturbation is bounded by any norm $\|\Delta\mathbf{A}\| \leq c$ and $\|\Delta\mathbf{A}\|$ satisfies $\|\Delta\mathbf{A}\| \leq c < m_M$, the perturbed system Eq.(23.1) is stable (*Martin, J.M., 1987*). If $\mathbf{A}$ is stable, i.e. $\alpha_s[\mathbf{A}] < 0$, then the perturbed system is also stable, i.e. $\alpha_s[\mathbf{A}_p] < 0$, if

$$\|\Delta\mathbf{A}\| \leq c \leq m_M = \frac{1}{\|(s\mathbf{I}-\mathbf{A})^{-1}\|} \qquad s = j\omega \quad \forall\omega \in \mathcal{I}_\omega \tag{23.14}$$

where $\mathcal{I}_\omega \triangleq \{\omega \in \mathcal{R} : |\omega| \leq \|\mathbf{A}\| + \|\Delta\mathbf{A}\|\}$. The set of all perturbation matrices $\Delta\mathbf{A}$ of minimal norm, destabilizing the stable nominal system $\mathbf{A}$, is characterized in terms of the norm of $(j\omega\mathbf{I}-\mathbf{A})^{-1}$ evaluated for some point on the imaginary axis.

Proof by contradiction: Consider the opposite relation $\alpha_s[\mathbf{A}_p] \geq 0$ and apply Martin's theorem where $\mathbf{M} := \mathbf{A}$ and Gastinel-Kahan theorem with $\mathbf{P} = j\omega\mathbf{I} - \mathbf{A}$ and $\mathbf{R} = \varepsilon\Delta\mathbf{A}$. Then there exists $\varepsilon \in (0,1]$ and $\omega \in \mathcal{R}$ such that $j\omega\mathbf{I} - \mathbf{M} - \Delta\mathbf{A} = \mathbf{P} + \mathbf{R}$ is singular. Since $\alpha_s[\mathbf{A}] < 0$ the matrix $\mathbf{P} = j\omega\mathbf{I} - \mathbf{A}$ is nonsingular. Referring to Gastinel-Kahan theorem,

$$\|\mathbf{R}\| = \| - \varepsilon\Delta\mathbf{A}\| \geq \frac{1}{\|\mathbf{P}^{-1}\|} = \frac{1}{\|(j\omega\mathbf{I} - \mathbf{A})^{-1}\|} \qquad \forall\omega \tag{23.15}$$

$$\text{or} \qquad \|\Delta\mathbf{A}\| \geq \frac{1}{\|(j\omega\mathbf{I} - \mathbf{A})^{-1}\|} \qquad \forall\omega \tag{23.16}$$

which contradicts $\omega \in \mathcal{I}_\omega$ in the range $|\omega| > \|\mathbf{A}\| + \|\Delta\mathbf{A}\|$. Since $j\omega\mathbf{I} - \mathbf{M} - \varepsilon\Delta\mathbf{A}$ is singular it results $|\omega| \leq \|\mathbf{A} + \varepsilon\Delta\mathbf{A}\| \leq \|\mathbf{A}\| + \|\Delta\mathbf{A}\|$ which is the definition of $\mathcal{I}_\omega$ and is also in contradiction. Hence, $\alpha_s[\mathbf{A}_p] \geq 0$ must be false and the theorem is proved. If ω is limited by $|\omega| \leq \|\mathbf{A}\| + c$ stability is given if and only if $c < m_M$. If $0 \leq \omega < \infty$ then $c < m_M$ is a sufficient stability condition, only. □

Combining Eq.(23.14) with the range of $\mathcal{I}_\omega$ in the general case,

$$\|(j\omega\mathbf{I} - \mathbf{A})^{-1}\|^{-1} \geq |\omega| - \|\mathbf{A}\| \qquad \forall\omega \in \mathcal{R} \tag{23.17}$$

for all kind of induced matrices (*Martin, J.M., and Hewer, G.A., 1987*).

Rewriting Eq.(23.14): If $\alpha_s[\mathbf{A}] < 0$ then

$$\alpha_s[\mathbf{A} + \mathbf{E}] < 0 \;\text{ if }\; \|\mathbf{E}\| < \|(j\omega\mathbf{I} - \mathbf{A})^{-1}\|^{-1} \quad \forall\, |\omega| \leq \|\mathbf{A}\| + \|\mathbf{E}\| \,. \tag{23.18}$$

Referring to Eq.(23.18), the result $\|\mathbf{E}\|$ depends on ω. The minimum $\|\mathbf{E}\|$ versus ω is obtained only in the case where $\inf_\omega[\, \|(j\omega\mathbf{I} - \mathbf{A})^{-1}\|^{-1}]$ occurs at $\omega = 0$ (*Martin, J.M, and Hewer, G.A., 1987*) but it may happen that the perturbing matrices of minimal norm are not real. In the case of spectral norm the results above can be rewritten, e.g. from Eq.(23.18)

$$\|\mathbf{E}\|_s < \sigma_{\min}[j\omega\mathbf{I} - \mathbf{A}] \,. \tag{23.19}$$

Eq.(23.18) can be rewritten with modified range of $|\omega|$. Using the fact that

$$\min \|\mathbf{E}\| = \|(j\omega\mathbf{I} - \mathbf{A})^{-1}\|^{-1} \,|_{\omega=0} = \|\mathbf{A}^{-1}\|^{-1} \,, \tag{23.20}$$

the range of $|\omega|$ as given by Eq.(23.18) is implied by

$$|\omega| < \|\mathbf{A}\| + \|\mathbf{A}^{-1}\|^{-1} \,. \;\square \tag{23.21}$$

Another property is (*Martin, J.M., and Hewer, G.A., 1987*)

$$\min_\omega \, \sigma_{\min}[j\omega\mathbf{I} - \mathbf{A}] \neq \Im m\ \lambda_i[\mathbf{A}] \,. \;\square \tag{23.22}$$

23.2.1 Using Symmetric Part of Matrices

If $\lambda_i[\mathbf{A}_s] < 0$ and $\mathbf{A}_{ps}$ is nonsingular where $\mathbf{A}_p = \mathbf{A} + \Delta\mathbf{A}$ and $_s$ denotes the symmetric part and $\|\cdot\|_s$ is the spectral matrix norm then $\Re e\ \lambda_i[\mathbf{A}_p] < 0$. □

If $\mathbf{A}$ and $\mathbf{A}_s$ are stable coefficient matrices of a continuous system then $\mathbf{A}_p$ is stable if

$$\|(\Delta\mathbf{A})_s\|_s < \|\mathbf{A}_s^{-1}\|_s^{-1} \qquad \Leftarrow \qquad \|\Delta\mathbf{A}\|_s < \|\mathbf{A}_s^{-1}\|_s^{-1} = \sigma_{\min}[\mathbf{A}_s] \,. \;\square \tag{23.23}$$

Proof see *Soh, C.B., and Chan, C.K., 1989.*

23.2.2 Direct Derivation

Consider the linear perturbed system with the following perturbation bounds

$$\dot{\mathbf{x}}(t) = (\mathbf{A} + \Delta\mathbf{A})\mathbf{x}(t) \qquad |\Delta\mathbf{A}| \leq_e \mathbf{E}, \qquad ||\mathbf{E}||_s = \alpha \,. \tag{23.24}$$

In the Laplace domain the well-known solution is

$$\mathbf{X}(s) = (s\mathbf{I} - \mathbf{A} - \Delta\mathbf{A})^{-1}\mathbf{x}(0) = [\mathbf{I} - (s\mathbf{I} - \mathbf{A})^{-1}\Delta\mathbf{A}]^{-1}(s\mathbf{I} - \mathbf{A})^{-1}\mathbf{x}(0) \,. \tag{23.25}$$

The assumption below is a necessary and sufficient stability condition

$$\sigma_{\max}[(j\omega\mathbf{I} - \mathbf{A})^{-1}] < \frac{1}{\alpha} \tag{23.26}$$

and corresponds to Eq.(23.14). Relations beween Eq.(23.26) and time domain results on quadratic stabilization, Eq.(13.153), are presented by *Khargonekar, P.P., et al. 1990.*

Proof Sufficiency: From Eqs.(22.69), (23.24) and (23.26)

$$\begin{aligned}\sigma_{\max}[(j\omega\mathbf{I} - \mathbf{A})^{-1}\Delta\mathbf{A}] &\leq \sigma_{\max}[(j\omega\mathbf{I} - \mathbf{A})^{-1}]\sigma_{\max}[\Delta\mathbf{A}] \leq \sigma_{\max}[(j\omega\mathbf{I} - \mathbf{A})^{-1}]\sigma_{\max}[\mathbf{E}] &(23.27)\\ &\leq \sigma_{\max}[(j\omega\mathbf{I} - \mathbf{A})^{-1}] \times \alpha < 1 &(23.28)\end{aligned}$$

Now, consider a rational proper stable $\mathbf{R}(s)$ and assume $\sigma_{\max}[\mathbf{R}(j\omega)] < 1, \quad \omega \in [0, \infty)$. Then, by $|\ \lambda_{\max}[\mathbf{A}]\ | \leq \sigma_{\max}[\mathbf{A}]$

$$|\ \lambda_i[\mathbf{R}(s)]\ | \leq |\ \lambda_{\max}[\mathbf{R}(s)]\ | \leq \sigma_{\max}[\mathbf{R}(s)] < 1 \quad \forall s \in \mathcal{C}^+ \quad \forall i \tag{23.29}$$

where $\mathcal{C}^+$ is the closed right-half s-plane ($\Re e\ s \geq 0$) it results $\det[\mathbf{I} - \mathbf{R}(s)] \neq 0$ and $[\mathbf{I} - \mathbf{R}(s)]^{-1}$ stable. Substituting $\mathbf{R}(s) = (s\mathbf{I} - \mathbf{A})^{-1}\Delta\mathbf{A}$ in Eq.(23.25) with the assumption $\sigma_{\max}[\mathbf{R}(j\omega)] < 1$ agrees with the result in Eq.(23.28) and the perturbed system is stable.

Proof Necessity: If the system is stable then $\alpha\sigma_{\max}[(j\omega\mathbf{I} - \mathbf{A})^{-1}] < 1\ \forall\omega \in [0, \infty)$ or, proving by contradiction, the system is stable for $\alpha\sigma_{\max}[(j\omega\mathbf{I}-\mathbf{A})^{-1}] > 1$ for some $\omega = \omega_o$. Performing singular-value decomposition,

$$(j\omega_o\mathbf{I} - \mathbf{A}) = \mathbf{U} \text{ diag } \sigma_i[j\omega_o\mathbf{I} - \mathbf{A}]\ \mathbf{V} \qquad \sigma_1 \geq \sigma_2 \geq \ldots\ldots \geq \sigma_n \ , \ \mathbf{U},\ \mathbf{V} \text{ unitary} \,. \tag{23.30}$$

Choosing a special case of perturbation with identical singular values $\Delta\mathbf{A}_1 = \mathbf{U}$ diag σ_n $\mathbf{V}$, one has

$$||\Delta\mathbf{A}_1||_s = \sigma_{\max}[\Delta\mathbf{A}_1] = \sigma_n = \sigma_{\min}[j\omega_o\mathbf{I} - \mathbf{A}] = \sigma_{\max}^{-1}[(j\omega_o\mathbf{I} - \mathbf{A})^{-1}] < \alpha \,. \tag{23.31}$$

The inequality above results from the contradicting assumption $\alpha\sigma_{\max}[(j\omega\mathbf{I} - \mathbf{A})^{-1}] > 1$. Regarding $|\Delta\mathbf{A}_1| \leq_e \mathbf{E}$ and $||\mathbf{E}||_s < \alpha$, the matrix $\Delta\mathbf{A}_1$ is a permissible perturbation. From Eq.(23.25) at $s = j\omega_o$, using Eq.(23.30), it follows

$$\mathbf{U} \text{ diag}\{\sigma_1 - \sigma_n,\ \sigma_2 - \sigma_n,\ \ldots.\ \sigma_n - \sigma_n\}\mathbf{V}\mathbf{X}(s) = \mathbf{U} \text{ diag}\{\sigma_1 - \sigma_n,\ \sigma_2 - \sigma_n,\ \ldots.\ 0\}\mathbf{V}\mathbf{X}(s) = \mathbf{x}(0) \tag{23.32}$$

and the fact that $\mathbf{X}(s)$ has a pole at $s = j\omega_o$. This contradicts the asymptotic-stability assumption of the perturbed system and necessity is proven (*Chen, B.S., and Kung, J.Y., 1988*).

End of the Proofs

23.3 Complex Stability Radius for Continuous-Time Systems

The complex stability radius is defined as

$$r_c[\mathbf{A}] \triangleq \min_{\omega} \sigma_{\min}[j\omega\mathbf{I} - \mathbf{A}] \qquad \mathbf{A} \in \mathcal{C}^{n \times n} \text{ and stable, } \omega \text{ real} \,. \tag{23.33}$$

Comparison with Eq.(23.14) yields

$$\|\mathbf{E}\|_s \le \sigma_{\min}[j\omega\mathbf{I} - \mathbf{A}] \quad (\forall\omega \in \mathcal{I}_\omega) \quad \ge r_c[\mathbf{A}] \tag{23.34}$$

$$\rightsquigarrow \quad r_c[\mathbf{A}] = \inf_{\mathbf{E}} \{ \|\mathbf{E}\|_s \ : \ \lambda[\mathbf{A}+\mathbf{E}] \cap \mathcal{C}^+ \neq \emptyset \ , \ \mathbf{E} \in \mathcal{C}^{n\times n} \} \tag{23.35}$$

where $\mathbf{E}$ is an admissible additive perturbation for which $\alpha_s[\mathbf{A}+\mathbf{E}] \le 0$.

Since $r_c[\mathbf{A}]$ is the minimum of $\sigma_{\min}[j\omega\mathbf{I} - \mathbf{A}]$ versus ω it follows: Setting arbitrarily $\omega = 0$ in Eq.(23.33), the minimum versus ω will be failed. Hence,

$$r_c[\mathbf{A}] \le \sigma_{\min}[0 - \mathbf{A}] = \sigma_{\min}[\mathbf{A}] \ . \tag{23.36}$$

Using generalized stability, some region Γ_U (including right-half s-plane) is defined. Then, the distance from instability is

$$\min \|\mathbf{E}\|_s = \min_{\gamma\in\Gamma_U} \ \sigma_{\min}[\gamma\mathbf{I} - \mathbf{A}] = \min_{\gamma\in\partial\Gamma_U} \ \sigma_{\min}[\gamma\mathbf{I} - \mathbf{A}] \tag{23.37}$$

where $\mathbf{A} \in \mathcal{C}^{n\times n}$ is a matrix with no eigenvalue in Γ_U, $\partial\Gamma_U$ is the boundary of Γ_U. The perturbation matrix is termed $\mathbf{E} \in \mathcal{C}^{n\times n}$, $\mathbf{A}_p = \mathbf{A} + \mathbf{E}$ has at least a single eigenvalue inside Γ_U (*Hinrichsen, D., and Pritchard, A.J., 1986*). A combination of Eqs.(23.33) and (23.37) is written in Eq.(23.46).

Note that by consideration of the complex stability radius the perturbation $\mathbf{E}$ may turn out complex even when $\mathbf{A}$ is real. The real stability radius defined below is more important from the practitioner's point of view but the calculation of the real stability radius is more difficult.

Maximizing the complex stability radius utilizing a linear state feedback controller is a design objective investigated by *Hinrichsen, D., et al. 1990* where the optimal stability radius is characterized by parametrized Riccati equations. A computing algorithm is outlined in *Motscha, M., 1988*.

23.4 Real Stability Radius for Continuous-Time Systems

The real stability radius is defined as the minimum spectral norm of the *real* perturbation matrix $\mathbf{E}$

$$r_R[\mathbf{A}] \triangleq \min_{\mathbf{E}} \|\mathbf{E}\|_s \qquad \mathbf{E} \in \mathcal{R}^{n\times n} \qquad \lambda[\mathbf{A}+\mathbf{E}] = j\alpha, \tag{23.38}$$

i.e., at least one eigenvalue of the perturbed matrix $\mathbf{A}+\mathbf{E}$ is imaginary. The unperturbed matrix $\mathbf{A} \in \mathcal{R}^{n\times n}$ is assumed Hurwitz stable.

Properties:

In general, $r_c[\mathbf{A}] \le r_R[\mathbf{A}] \le \sigma_{\min}[\mathbf{A}]$. □

If $\mathbf{A}$ is real, stable and normal $r_c[\mathbf{A}] = r_R[\mathbf{A}] = -\alpha_s[\mathbf{A}]$ and, hence, the spectral abscissa $\alpha_s[\mathbf{A}]$ or the minimum distance from the imaginary axis is a good indicate for stability robustness. □

If $\mathbf{A}$ is symmetric then $r_c[\mathbf{A}] = r_R[\mathbf{A}] = -\alpha_s[\mathbf{A}] = \sigma_{\min}[\mathbf{A}]$. □

If $\mathbf{A}$ is not normal then the value $\alpha_s[\mathbf{A}]$ is a deceiving indicator for stability robustness since $r_R[\mathbf{A}] \le \sigma_{\min}[\mathbf{A}] \le |\ \alpha_s[\mathbf{A}]\ | \le |\ \lambda_i[\mathbf{A}]\ |$. □

If $\mathbf{A} \in \mathcal{C}^{n\times n}$ and stable and if $\mathbf{E} = r_K[\mathbf{A}]\ \mathbf{I}$ (where $_K$ stands for $_c$ or $_R$) and if the sum $\mathbf{A}+\mathbf{E}$ is unstable it results $r_K[\mathbf{A}] \le -\alpha_s[\mathbf{A}]$. □

For $\mathbf{A} \in \mathcal{R}^{n\times n}$ and $\mathbf{A}_a \triangleq 0.5\,(\mathbf{A} - \mathbf{A}^T)$, denoting the antisymmetric part, one has

$$\sigma^2_{\min}[\mathbf{A}] - \|\mathbf{A}_a\|^2_s \leq r_c^2[\mathbf{A}] \,. \tag{23.39}$$

This relation is useful for nearly symmetric matrices yielding a lower bound for $r_c[\mathbf{A}]$. An upper bound is always given by $\sigma_{\min}[\mathbf{A}]$.

23.5 Lower Bound for the Real Stability Radius Using Kronecker Sum

23.5.1 Continuous-Time Systems

Assume that there exists an intersection of the spectrum $\Lambda[\mathbf{A}]$ of a matrix $\mathbf{A} \in \mathcal{R}^{n\times n}$ and the imaginary axis, the origin excluded, then $\text{rank}(\mathbf{A} \oplus \mathbf{A}) \leq n^2 - 2$. Moreover, for $\Delta\mathbf{A} \in \mathcal{R}^{n\times n}$

$$\min\{\|\Delta\mathbf{A}\|_s \;:\; \text{rank}(\mathbf{A} + \Delta\mathbf{A}) \leq r\;\} = \sigma_{r+1}[\mathbf{A}] \tag{23.40}$$

and for $\Gamma_c = \{j\omega \;:\; \omega \in \mathcal{R}\;\}$

$$\inf\{\|\Delta\mathbf{A}\|_s \;:\; \Lambda[\mathbf{A} + \Delta\mathbf{A}] \cap \Gamma_c \neq \emptyset\} \geq \min\{\sigma_{\min}[\mathbf{A}],\; \frac{1}{2}\sigma_{n^2-1}[\mathbf{A} \oplus \mathbf{A}]\} \tag{23.41}$$

(*Qiu,L., and Davison, E.J., 1988*). Note that $\sigma_1[\cdot]$ corresponds to $\sigma_{\max}[\cdot]$. This is a lower bound for the distance of a stable matrix $\mathbf{A}$ from the set of unstable matrices. This bound is tighter than other bounds as known e.g. from Eq.(13.73), i.e., $1/\sigma_{\max}[\mathbf{P}]$ where $\mathbf{P}$ results from $\mathbf{A}^T\mathbf{P} + \mathbf{P}\mathbf{A} = -2\mathbf{I}$ or from $r_c[\mathbf{A}] = \inf_\omega \sigma_{\min}[j\omega\mathbf{I} - \mathbf{A}]$ outlined in Eq.(23.33). Moreover, there is less computational effort. Interesting special cases are obtained for $\mathbf{A}$ normal or $\in \mathcal{R}^{2\times 2}$. If $\mathbf{A}$ is normal and stable

$$\inf\{\|\Delta\mathbf{A}\|_s \;:\; \Lambda[\mathbf{A} + \Delta\mathbf{A}] \cap \Gamma_c \neq \emptyset\} = \min\{-\Re e\;\lambda_i[\mathbf{A}]\} = -\alpha_s[\mathbf{A}] \,. \tag{23.42}$$

For $\mathbf{A} \in \mathcal{R}^{2\times 2}$ since $\sigma_{n^2-1} = \sigma_3[\mathbf{A} \oplus \mathbf{A}] = -\text{tr}\mathbf{A}$ the result for the lower bound is $\min\{\sigma_{\min}[\mathbf{A}], -\frac{1}{2}\text{tr}\mathbf{A}\}$. For other kind of composite matrices, formulae are presented by *Qiu, L., and Davison, E.J., 1988*.

23.5.2 Discrete-Time Systems

When the discrete-time system matrix $\mathbf{\Phi}$ has eigenvalues on the unit circle except at $(-1,1)$ then $\text{rank}(\mathbf{\Phi} \otimes \mathbf{\Phi} - \mathbf{I}_{n^2}) < n^2 - 2$. For $\Delta\mathbf{\Phi} \in \mathcal{R}^{n\times n}$, $\Gamma_d = \{e^{j\omega} \;:\; \omega \in \mathcal{R}\;\}$

$$\begin{aligned}\inf\{\|\Delta\mathbf{\Phi}\|_s \;&:\; \Lambda[\mathbf{\Phi} + \Delta\mathbf{\Phi}] \cap \Gamma_d \neq \emptyset\} \\ &\geq \min\{\sigma_{\min}[\mathbf{\Phi} - \mathbf{I}],\; \sigma_{\min}[\mathbf{\Phi} + \mathbf{I}],\; \sqrt{\sigma_{n^2-1}[\mathbf{\Phi} \otimes \mathbf{\Phi} - \mathbf{I}_{n^2}] + \sigma^2_{\max}[\mathbf{\Phi}]} - \sigma_{\max}[\mathbf{\Phi}]\} \,.\end{aligned} \tag{23.43}$$

If $\mathbf{\Phi}$ is normal the right-hand side of Expr.(23.43) is an equality, i.e., it results the equality $= \min_i\{1 - |\lambda_i[\mathbf{\Phi}]\,|\;,\; i = 1,2...n\}$. For $\mathbf{\Phi} \in \mathcal{R}^{2\times 2}$ one has (*Qiu, L., and Davison, E.J., 1988*)

$$\geq \min\{\sigma_{\min}[\mathbf{\Phi} - \mathbf{I}],\; \sigma_{\min}[\mathbf{\Phi} + \mathbf{I}],\; \sqrt{1 - \det\mathbf{\Phi} + \sigma^2_{\max}[\mathbf{\Phi}]} - \sigma_{\max}[\mathbf{\Phi}]\} \,. \tag{23.44}$$

23.6 Structured Stability Radius

Consider the autonomous perturbed system

$$\dot{\mathbf{x}}(t) = \mathbf{A}_p \mathbf{x}(t), \qquad \mathbf{A}_p = \mathbf{A} + \mathbf{LER} \tag{23.45}$$

where $\mathbf{A}$ is the nominal system, $\mathbf{L}$ and $\mathbf{R}$ are scaling matrices and $\mathbf{E}$ is an unknown perturbation matrix. Let all $\lambda[\mathbf{A}] \in \mathcal{C}^-$, i.e., in the open left half-plane.

The objective of the complex structured stability radius is to find that complex perturbation matrix $\mathbf{E}$ with smallest value versus ω of the spectral norm (maximum singular value) which makes $\mathbf{A}_p$ unstable. The scaling matrices $\mathbf{L}$ and $\mathbf{R}$ are utilized to structure the influence of $\mathbf{E}$ on the nominal system matrix $\mathbf{A}$. The unstructured case is given by $\mathbf{L} = \mathbf{R} = \mathbf{I}$. The structured complex stability radius is defined

$$r_c[\mathbf{A}; \mathbf{L}, \mathbf{R}] \triangleq \inf_{\mathbf{E}} \{\|\mathbf{E}\|_s \; : \; \lambda[\mathbf{A} + \mathbf{LER}] \cap \mathcal{C}^+ \neq \emptyset\} \tag{23.46}$$

$$\mathbf{A} \in \mathcal{C}^{n \times n}, \qquad \mathbf{L} \in \mathcal{C}^{n \times m}, \qquad \mathbf{R} \in \mathcal{C}^{r \times n}, \qquad \mathbf{E} = \mathcal{C}^{m \times r} \tag{23.47}$$

where $\mathcal{C}^+$ denotes the closed right-half s-plane.

Structured real perturbations of the system $\mathbf{A}$ to $\mathbf{A} + \mathbf{LER}$ are investigated by *Hinrichsen, D., and Pritchard, A.J., 1988*. Evaluation is difficult. When $\mathbf{L}$ and $\mathbf{R}$ are column and row matrices, respectively, the computation is facilitated.

23.7 Complex Stability Radius and Transfer Matrix

Let the matrices $\mathbf{A}, \mathbf{L}, \mathbf{R}$ be used as the system, input and output matrix, respectively, of a multivariable system in state-space representation, i.e.

$$\mathbf{G}(s) = \mathbf{R}(s\mathbf{I}_n - \mathbf{A})^{-1}\mathbf{L} \; . \tag{23.48}$$

Then, an important property relating Eq.(23.46) and $\mathbf{G}(s)$ for $s = j\omega, \omega \in \mathcal{R}$ is

$$r_c[\mathbf{A}; \mathbf{L}, \mathbf{R}] = \min_{\omega} \|\mathbf{G}(j\omega)\|_s^{-1} = [\sup_{\omega} \|\mathbf{G}(j\omega)\|_s]^{-1} = \frac{1}{\|\mathbf{G}(j\omega)\|_\infty} \tag{23.49}$$

where $\|\mathbf{G}(j\omega)\|_\infty$ is the H_∞-norm (*Hinrichsen, D., and Pritchard, A.J., 1986a and 1988*).

23.8 Complex Stability Radius and Hamiltonian Matrices

An algorithm for computation of the stability radius $r_c(\mathbf{A}; \mathbf{L}, \mathbf{R})$ is based on a Hamiltonian matrix $\mathbf{H}_\rho$. Define the Hamiltonian matrix

$$\mathbf{H}_\rho = \mathbf{H}_\rho(\mathbf{A}; \mathbf{L}, \mathbf{R}) \triangleq \begin{pmatrix} \mathbf{A} & \mathbf{LL}^H \\ -\rho \mathbf{R}^H \mathbf{R} & -\mathbf{A}^H \end{pmatrix} \tag{23.50}$$

where ρ is a non-negative number. Then, the point $j\omega$ on the imaginary axis of the complex s-plane is an element of the spectrum $\Lambda[\mathbf{H}_\rho]$ if and only if

$$\rho = \frac{1}{\sigma^2[\mathbf{G}(j\omega)]} \; . \tag{23.51}$$

Proof: Using Eq.(C.39), the eigenvalues $\lambda_i[\mathbf{H}_\rho]$ are given by

$$\begin{aligned}
\det(s\mathbf{I}_n - \mathbf{H}_\rho) &= \det\begin{pmatrix} s\mathbf{I}_n - \mathbf{A} & -\mathbf{L}\mathbf{L}^H \\ \rho\mathbf{R}^H\mathbf{R} & s\mathbf{I}_n + \mathbf{A}^H \end{pmatrix} && (23.52)\\
&= \det(s\mathbf{I}_n - \mathbf{A})\det[s\mathbf{I}_n + \mathbf{A}^H + \rho\mathbf{R}^H\mathbf{R}(s\mathbf{I}_n - \mathbf{A})^{-1}\mathbf{L}\mathbf{L}^H] && (23.53)\\
&= \det(s\mathbf{I}_n - \mathbf{A})\det(s\mathbf{I}_n + \mathbf{A}^H)\det[\mathbf{I}_n + (s\mathbf{I}_n + \mathbf{A}^H)^{-1}\rho\mathbf{R}^H\mathbf{G}(s)\mathbf{L}^H]\,, && (23.54)
\end{aligned}$$

by Eq.(A.18),

$$\begin{aligned}
\det(s\mathbf{I}_n - \mathbf{H}_\rho) &= (-1)^n\det(s\mathbf{I}_n - \mathbf{A})\det(-s\mathbf{I}_n - \mathbf{A}^H)\det[\mathbf{I}_m - \mathbf{L}^H(-s\mathbf{I}_n - \mathbf{A}^H)^{-1}\rho\mathbf{R}^H\mathbf{G}(s)] && (23.55)\\
&= (-1)^n(-\rho)^m\det(s\mathbf{I}_n - \mathbf{A})\det(-s\mathbf{I}_n - \mathbf{A}^H)\det[\mathbf{I}_m \times (-\rho^{-1}) + \mathbf{G}^H(-s^*)\mathbf{G}(s)] = 0 &&\\
\text{by } s = j\omega_o\,, \quad &= (-1)^n(-\rho)^m\det(j\omega_o\mathbf{I}_n - \mathbf{A})\det(-j\omega_o\mathbf{I}_n - \mathbf{A}^H)\det[-\rho^{-1}\mathbf{I}_m + \mathbf{G}^H(j\omega_o)\mathbf{G}(j\omega_o)] = 0 &&
\end{aligned}$$

$$\rho^{-1} = \lambda[\mathbf{G}^H(j\omega_o)\ \mathbf{G}(j\omega_o)] = \sigma^2[\mathbf{G}(j\omega_o)]\,. \tag{23.56}$$

The reciprocal ρ equals the singular values of $\mathbf{G}(j\omega_o)$. □

23.9 Complex Stability Radius r_c and Parameter ρ

From Eq.(23.51)

$$\rho^{-1} = \sigma^2[\mathbf{G}(j\omega)] \le \sigma^2_{\max}[\mathbf{G}(j\omega)] = \|\mathbf{G}(j\omega)\|_s^2\,. \tag{23.57}$$

From Eq.(23.49) it follows

$$r_c^{-1}(\mathbf{A};\mathbf{R},\mathbf{L}) = \sup_\omega \|\mathbf{G}(j\omega)\|_s \ge \|\mathbf{G}(j\omega)\|_s\,. \tag{23.58}$$

Hence, the spectral norm $\|\mathbf{G}(j\omega)\|_s^2$ is bounded by

$$\|\mathbf{G}(j\omega)\|_\infty^2 = r_c^{-2}(\mathbf{A};\mathbf{L},\mathbf{R}) \ge \|\mathbf{G}(j\omega)\|_s^2 \ge \rho^{-1}\,. \tag{23.59}$$

From the inequation above it results that the spectrum $\Lambda[\mathbf{H}_\rho]$ contains eigenvalues in the right-half s-plane if any ρ is chosen greater than the squared stability radius r_c^2, i.e. if $\rho > r_c^2$. The derivation above can be illustrated graphically. The plot of eigenloci $\lambda_i[\mathbf{H}_\rho]$ for ρ increasing from 0 to ∞ hits the imaginary axis for the first time for $\rho = r_c^2$ and at a certain frequency ω_o. Hence,

$$j\omega_o = \lambda[\mathbf{H}_\rho] \ \mid\ \rho = r_c^2 \quad \Longleftrightarrow \quad \|\mathbf{G}(j\omega_o)\|_s = \|\mathbf{G}(j\omega)\|_\infty\,. \tag{23.60}$$

The H_∞-norm of $\mathbf{G}(s)$ can be evaluated, first, by computing $\|\mathbf{G}(j\omega\|_s$ for many values of ω and by numerically approaching the local maxima. Only those maxima are approached where the starting value is in a sufficiently small neighbourhood. Second, the global maximum is selected from all evaluated local ones. However, the method presented above (*Hinrichsen, D., et al. 1989*) immediately yields the global maximum which is an important fact.

23.10 Stability Radius of Parametrized Systems

Consider a system with parameterized matrix $\mathbf{A}(p_1,p_2)$. Let p_1 and p_2 denote the parameters within a given range. The stability radius can be calculated for a number of suitably chosen parameter pairs (p_1,p_2). With regard to $\mathbf{A}(p_1,p_2) + \mathbf{LER}$ the matrix $\mathbf{E} \in \mathcal{R}^{1\times 2} \triangleq (\Delta p_1, \Delta p_2)$ and $\mathbf{L} \in \mathcal{R}^{n\times 1}$, $\mathbf{R} \in \mathcal{R}^{2\times n}$. Then, the real structured stability radius is replaced by an ellipsoid centred at (p_1,p_2). This ellipsoid covers that region of parameters $(\Delta p_1, \Delta p_2)$ where stability is guaranteed.

23.11 Stability Radius of Polynomials

Consider the monic Hurwitz polynomial and its coefficient vector $\mathbf{a} = (a_o, \dots a_{n-1})^T$

$$a(s, \mathbf{a}) = s^n + a_{n-1}s^{n-1} + \dots + a_1 s + a_o \, . \tag{23.61}$$

Let the unperturbed coefficients a_{j-1} be perturbed to $a_{p,j-1}$

$$a_{p,j-1}(\mathbf{d}) = a_{j-1} - \sum_{i=1}^{r} d_i R_{ij} \qquad j \in [1, 2 \dots n] \tag{23.62}$$

where $\mathbf{d} = (d_1, d_2, \dots d_r)$ is a disturbance row matrix and $\mathbf{R} = \text{matrix}[R_{ij}]$ denotes the perturbation structure matrix. The elements of the perturbation matrix are termed d_i. Then, the stability radius of $a(s, \mathbf{a})$ can be defined as

$$r_c[a; \mathbf{R}] \triangleq \inf_{\mathbf{d}}\{\|\mathbf{d}\|; \;\; \mathbf{d} \in \mathcal{R}^{1\times r}, \;\; \exists \lambda \in \mathcal{C}_+ : a(\lambda, \mathbf{a}[\mathbf{d}]) = 0\} \, . \tag{23.63}$$

If the matrix $\mathbf{L} = (0, 0 \dots, 1)^T$ is used and $\mathbf{A}$ is selected a companion matrix associated with the polynomial $a(s, \mathbf{a})$, i.e., the last row of $\mathbf{A}$ consists of entries identical to the coefficients a_{j-1} , then

$$r_c[a; \mathbf{R}] = r_c[\mathbf{A}; \mathbf{L}, \mathbf{R}] \, . \tag{23.64}$$

Stability radii of Hurwitz polynomials under structured perturbations of the coefficient vector are presented by *Hinrichsen, D., and Pritchard, A.J., 1989*.

23.12 Stability Robustness. Discrete Systems

By analogy to Eq.(23.14): If the spectral radius $\rho_s[\mathbf{\Phi}] < 1$ then $\rho_s[\mathbf{\Phi}_p] = \rho_s[\mathbf{\Phi} + \Delta\mathbf{\Phi}] < 1$ if

$$\|\Delta\mathbf{\Phi}\| < \|(e^{j\omega}\mathbf{I} - \mathbf{\Phi})^{-1}\|^{-1} \qquad \forall\omega \in [0, 2\pi] \, . \tag{23.65}$$

Proof by contradiction: If $\rho_s[\mathbf{\Phi}_p] \geq 1$ then there exists $\omega \in [0, 2\pi]$ such that the matrix $e^{j\omega}\mathbf{I} - \mathbf{\Phi} - \varepsilon\Delta\mathbf{\Phi}$ is singular. Since $e^{j\omega}\mathbf{I} - \mathbf{\Phi}$ is nonsingular $\|\varepsilon\Delta\mathbf{\Phi}\| \geq \|(e^{j\omega}\mathbf{I} - \mathbf{\Phi})^{-1}\|^{-1}$ which is in contradiction. Hence, $\rho_s[\mathbf{\Phi}_p] < 1$ must be true.

23.12.1 Using Modulus Matrix

If $|\mathbf{\Phi}| + |\Delta\mathbf{\Phi}| - \mathbf{I}$ is nonsingular and $\rho_s[\,|\mathbf{\Phi}|\,] < 1$ then $\rho_s[\mathbf{\Phi}_p] < 1$. □

If the discrete-time system with coefficient matrix $\mathbf{\Phi}$ and $|\mathbf{\Phi}|$ is stable then the perturbed system $\mathbf{\Phi} + \Delta\mathbf{\Phi}$ is stable if

$$\|\,|\Delta\mathbf{\Phi}|\,\|_s < \|\,(\,|\mathbf{\Phi}| - \mathbf{I}\,)^{-1}\|_s^{-1} \;\square \tag{23.66}$$

(Proof see *Soh, C.B., and Chan, C.K., 1989*).

23.13 Interrelation to the Lyapunov Matrix. Continuous Systems

If $\mathbf{A}$ is assumed Hurwitz stable there exists a positive definite solution $\mathbf{P}$ of the Lyapunov matrix equation

$$\mathbf{A}^T\mathbf{P} + \mathbf{P}\mathbf{A} = -\mathbf{Q} = -2\mathbf{I} \tag{23.67}$$

where $\mathbf{Q}$ is a positive definite matrix, e.g. $\mathbf{Q} = 2\mathbf{I}$. This Lyapunov matrix $\mathbf{P}$ can be used to state upper and lower bounds for m_M (*Mori, T., 1989*), see below.

23.13.1 Lower Bound

From Eq.(23.67), subtracting and adding $j\omega\mathbf{P}$ and using Eq.(23.3),

$$-j\omega\mathbf{P} - \mathbf{A}^T\mathbf{P} - \mathbf{P}\mathbf{A} + j\omega\mathbf{P} = 2\mathbf{I} \tag{23.68}$$

$$(-j\omega\mathbf{I} - \mathbf{A}^T)^{-1} \times | \quad (-j\omega\mathbf{I} - \mathbf{A}^T)\mathbf{P} + \mathbf{P}(-\mathbf{A} + j\omega\mathbf{I}) = 2\mathbf{I} \quad | \times (j\omega\mathbf{I} - \mathbf{A})^{-1} \tag{23.69}$$

$$\mathbf{P}(j\omega\mathbf{I} - \mathbf{A})^{-1} + (-j\omega\mathbf{I} - \mathbf{A}^T)^{-1}\mathbf{P} = 2\mathbf{I}(-j\omega\mathbf{I} - \mathbf{A}^T)^{-1}(j\omega\mathbf{I} - \mathbf{A})^{-1} \tag{23.70}$$

$$\mathbf{P}\mathbf{L} + \mathbf{L}^H\mathbf{P} = \mathbf{L}^H\mathbf{P} + \mathbf{P}\mathbf{L} = 2\mathbf{L}^H\mathbf{L}\ . \tag{23.71}$$

Applying the spectral norm and the triangle inequality,

$$\|\mathbf{L}^H\mathbf{P} + \mathbf{P}\mathbf{L}\|_s = 2\|\mathbf{L}^H\mathbf{L}\|_s \tag{23.72}$$

$$\|\mathbf{L}^H\mathbf{P}\|_s + \|\mathbf{P}\mathbf{L}\|_s \geq \|\mathbf{L}^H\mathbf{P} + \mathbf{P}\mathbf{L}\|_s = 2\|\mathbf{L}^H\mathbf{L}\|_s \tag{23.73}$$

$$\|\mathbf{L}^H\|_s\|\mathbf{P}\|_s + \|\mathbf{P}\|_s\|\mathbf{L}\|_s = 2\|\mathbf{L}\|_s\|\mathbf{P}\|_s \geq 2\|\mathbf{L}\|_s^2 \tag{23.74}$$

$$\|\mathbf{P}\|_s \geq \|\mathbf{L}(\omega)\|_s \qquad \forall\omega\ . \tag{23.75}$$

Since the inequality above holds for all ω it also holds for the maximum with respect to ω

$$\|\mathbf{P}\|_s \geq \max_{\omega} \|\mathbf{L}(\omega)\|_s \tag{23.76}$$

$$\|\mathbf{P}\|_s^{-1} \leq \frac{1}{\max_\omega \|\mathbf{L}(\omega)\|_s} = m_M\ . \tag{23.77}$$

Comparison with Eq.(23.2) shows that $\|\mathbf{P}\|_s^{-1}$ is a lower bound of m_M.

23.13.2 Upper Bound

Pre- and postmultiplying Eq.(23.71) by $\mathbf{P}^{-1/2}$ yields

$$\mathbf{P}^{-1/2}\mathbf{L}^H\mathbf{P}^{1/2} + \mathbf{P}^{1/2}\mathbf{L}\mathbf{P}^{-1/2} = 2\mathbf{P}^{-1/2}\mathbf{L}^H\mathbf{L}\mathbf{P}^{-1/2}\ . \tag{23.78}$$

Since $\mathbf{P}$ is assumed symmetric

$$(\mathbf{P}^{-1/2}\mathbf{L}^H\mathbf{P}^{1/2})^H = \mathbf{P}^{1/2}\mathbf{L}\mathbf{P}^{-1/2} \triangleq \mathbf{G}\ . \tag{23.79}$$

Then, from Eq.(23.78)

$$\mathbf{G}^H + \mathbf{G} = 2\mathbf{P}^{-1/2}\mathbf{L}^H\mathbf{L}\mathbf{P}^{-1/2}\ . \tag{23.80}$$

Taking the maximum eigenvalue of both sides and applying Eq.(B.74),

$$\max\ \lambda_i[\frac{\mathbf{G} + \mathbf{G}^H}{2}] = \max\ \lambda_i[\mathbf{P}^{-1/2}\mathbf{L}^H\mathbf{L}\mathbf{P}^{-1/2}] \tag{23.81}$$

$$\max\ \Re e\ \lambda_i[\mathbf{G}] = \max\ \lambda_i[\mathbf{P}^{-1/2}\mathbf{L}^H\mathbf{L}\mathbf{P}^{-1/2}] \tag{23.82}$$

$$\Re e\ \lambda_i[\mathbf{P}^{1/2}\mathbf{L}\mathbf{P}^{-1/2}] \leq \|\mathbf{L}\mathbf{P}^{-1/2}\|_s^2 \leq \|\mathbf{L}\|_s^2\|\mathbf{P}^{-1/2}\|_s^2\ . \tag{23.83}$$

The eigenvalues $\lambda_i[\mathbf{P}^{1/2}\mathbf{L}\mathbf{P}^{-1/2}] \triangleq \lambda_{iL}$ are given by

$$\det[\lambda_{iL}\mathbf{I} - \mathbf{P}^{1/2}(j\omega\mathbf{I} - \mathbf{A})^{-1}\mathbf{P}^{-1/2}] = \det[\mathbf{P}^{-1/2}\lambda_{iL}\mathbf{P}^{1/2}(j\omega\mathbf{I} - \mathbf{A}) - \mathbf{I}] = 0 \tag{23.84}$$

$$\det[j\omega\mathbf{I} - \mathbf{A} - \mathbf{I}/\lambda_{iL}] = \det[(j\omega - 1/\lambda_{iL})\mathbf{I} - \mathbf{A}] = 0 \tag{23.85}$$

$$j\omega - 1/\lambda_{iL} = \lambda_i[\mathbf{A}] \triangleq a_i + jb_i \tag{23.86}$$

$$\lambda_{iL} = \frac{-1}{a_i + j(b_i - \omega)} \quad \rightsquigarrow \quad \Re e\ \lambda_{iL} = \frac{-a_i}{a_i^2 + (b_i - \omega)^2}\ . \tag{23.87}$$

The maximum with respect to ω is obtained as

$$\max_{\omega} \Re e\ \lambda_{iL} = -\frac{1}{a_i}\ . \tag{23.88}$$

The maximum with respect to i is given by the minimum of the term $(-a_i)$. This is met by the eigenvalue $\lambda_i[\mathbf{A}]$ with maximum $\Re e\ \lambda_i[\mathbf{A}]$ in the left-half s-plane

$$\max_i \max_{\omega} \Re e\ \lambda_{iL} = \max_i \frac{1}{(-a_i)} = \frac{1}{(-\max_i \Re e\ \lambda_i[\mathbf{A}])}\ . \tag{23.89}$$

Combining Eqs.(23.83) and (23.89) yields

$$\frac{1}{-\max_i \Re e\ \lambda_i[\mathbf{A}]} \leq \|\mathbf{L}\|_s^2 \|\mathbf{P}^{-1/2}\|_s^2\ . \tag{23.90}$$

Replacing $\|\mathbf{L}(\omega)\|_s$ by $\max_{\omega \geq 0} \|\mathbf{L}(\omega)\|_s$ and inverting the last equation yields

$$-\max_i \Re e\ \lambda_i[\mathbf{A}] \geq \frac{1}{\max_{\omega \geq 0} \|\mathbf{L}(\omega)\|_s^2} \frac{1}{\|\mathbf{P}^{-1/2}\|_s^2} \tag{23.91}$$

$$\|\mathbf{P}^{-1/2}\|_s \big(-\max_i \Re e\ \lambda_i[\mathbf{A}]\big)^{1/2} \geq \frac{1}{\max_{\omega \geq 0} \|\mathbf{L}(\omega)\|_s} = m_M\ . \tag{23.92}$$

Referring to Eq.(23.2), the equation above is an upper bound for m_M.

23.14 Eigenvalue Exclusion Lemma and Stability Robustness

23.14.1 Induced Norm of the Uncertainty. Continuous-Time Systems

Consider the perturbed system matrix $\mathbf{A}_p = \mathbf{A} + \Delta\mathbf{A} \in \mathcal{R}^{n \times n}$. Assuming $\mathbf{A}$ stable the condition is required that $\mathbf{A}_p$ does not possess an eigenvalue s_o, i.e. $\lambda[\mathbf{A}_p] \neq s_o$. From $\det(s_o\mathbf{I} - \mathbf{A} - \Delta\mathbf{A}) \neq 0$

$$\iff \quad \det\{(s_o\mathbf{I} - \mathbf{A})[\mathbf{I} - (s_o\mathbf{I} - \mathbf{A})^{-1}\Delta\mathbf{A}]\} \neq 0 \tag{23.93}$$

$$\iff \quad \det[\mathbf{R}^{-1}\mathbf{R} - \mathbf{R}^{-1}(s_o\mathbf{I} - \mathbf{A})^{-1}\Delta\mathbf{A}\ \mathbf{R}] \neq 0 \quad \text{for any nonsingular } \mathbf{R} \tag{23.94}$$

$$\Leftarrow \quad \|\mathbf{R}^{-1}(s_o\mathbf{I} - \mathbf{A})^{-1}\Delta\mathbf{A}\ \mathbf{R}\|_p < 1\ . \tag{23.95}$$

For $s_o = j\omega$ Hurwitz stability can be checked (*Qiu, L., and Davison, E.J., 1986*). For $\mathbf{R} = \mathbf{I}$ and $s_o = j\omega_o$

$$\|(j\omega_o\mathbf{I} - \mathbf{A})^{-1}\Delta\mathbf{A}\|_p < 1 \tag{23.96}$$

$$\Leftarrow \quad \|(j\omega_o\mathbf{I} - \mathbf{A})^{-1}\|_p \|\Delta\mathbf{A}\|_p < 1\ . \tag{23.97}$$

Since $|\Delta\mathbf{A}| \leq_e \mathbf{E}$

$$\Leftarrow \quad \|(j\omega_o\mathbf{I} - \mathbf{A})^{-1}\|_p \|\mathbf{E}\|_p < 1\ . \tag{23.98}$$

In the case $p = 2$ spectral norms are used and

$$\sigma_{\max}[\Delta\mathbf{A}] \leq \sigma_{\max}[\mathbf{E}] \leq \inf_{\omega}\ \sigma_{\min}[j\omega_o\mathbf{I} - \mathbf{A}]\ . \tag{23.99}$$

This bound is tighter than $\sigma_{\max}[\Delta\mathbf{A}] \leq 1/\sigma_{\max}[\mathbf{P}]$ as given by Eq.(13.73) since it gives a higher bound $\sigma_{\max}[\Delta\mathbf{A}]$ (*Qiu, L., and Davison, E.J., 1986*).

Proof: From $\mathbf{A}^T\mathbf{P} + \mathbf{P}\mathbf{A} = -2\,\mathbf{I}$

$$(-j\omega\mathbf{I} - \mathbf{A}^T)\mathbf{P} + \mathbf{P}(j\omega\mathbf{I} - \mathbf{A}) = 2\,\mathbf{I}\,. \tag{23.100}$$

Premultiplying and postmultiplying by $(-j\omega\mathbf{I} - \mathbf{A}^T)^{-1}$ and $(j\omega\mathbf{I} - \mathbf{A})^{-1}$, respectively, yields

$$\mathbf{P}(j\omega\mathbf{I} - \mathbf{A})^{-1} + (-j\omega\mathbf{I} - \mathbf{A}^T)\mathbf{P} = 2(-j\omega\mathbf{I} - \mathbf{A}^T)^{-1}(j\omega\mathbf{I} - \mathbf{A})^{-1}\,. \tag{23.101}$$

Taking the maximum eigenvalue of both sides, employing $\lambda_{\max}[\mathbf{PG}] \leq \sigma_{\max}[\mathbf{PG}]$ and Eq.(22.69)

$$2\sigma_{\max}[\mathbf{P}]\sigma_{\max}[(j\omega\mathbf{I} - \mathbf{A})^{-1}] \geq 2\sigma^2_{\max}[(j\omega\mathbf{I} - \mathbf{A})^{-1}] \tag{23.102}$$

$$\sigma_{\max}[\mathbf{P}] \geq \sigma_{\max}[(j\omega\mathbf{I} - \mathbf{A})^{-1}]\,. \tag{23.103}$$

Selecting the supremum of $\sigma_{\max}[(j\omega\mathbf{I} - \mathbf{A})^{-1}]$ versus ω,

$$\sup_{\omega}\ \sigma_{\max}[\mathbf{P}] = \sigma_{\max}[\mathbf{P}] \geq \sup_{\omega}\ \sigma_{\max}[(j\omega\mathbf{I} - \mathbf{A})^{-1}] = \sup_{\omega}\ \sigma^{-1}_{\min}[j\omega\mathbf{I} - \mathbf{A}] \tag{23.104}$$

$$\frac{1}{\sigma_{\max}[\mathbf{P}]} \leq \inf_{\omega}\ \sigma_{\min}[j\omega\mathbf{I} - \mathbf{A}]\ .\ \Box \tag{23.105}$$

23.14.2 Normal Matrix A. Sufficient and Necessary Conditions

Assume $\mathbf{A}$ is normal then using Eq.(22.28) $\mathbf{A} = \mathbf{U}^{-1}\mathrm{diag}_n\{\lambda_i[\mathbf{A}]\}\mathbf{U}$ where $\mathbf{U}$ is unitary. Then, from Eq.(23.99)

$$\inf_{\omega}\ \sigma_{\min}[j\omega_o\mathbf{I} - \mathbf{A}] = \inf_{\omega}\ \sigma_{\min}[\mathbf{U}^{-1}\mathrm{diag}_n\{j\omega_o - \lambda_1, \ldots, j\omega_o - \lambda_n\}\mathbf{U}] \tag{23.106}$$

$$\inf_{\omega}\ \sigma_{\min}[\mathrm{diag}_n\{j\omega_o - \lambda_i\}] = \min_i |\Re e\ \lambda_i| = \min_i\{-\Re e\ \lambda_i\}\ . \tag{23.107}$$

Adjoining a unitary matrix $\mathbf{U}$ in the argument of $\sigma_{\min}$ is admissible. Hence, the sufficient condition is

$$\sigma_{\max}[\Delta\mathbf{A}] \leq \min_i\{-\Re e\ \lambda_i\}\ . \tag{23.108}$$

Let $\Delta\mathbf{A} = \min_i\{-\Re e\ \lambda_i\}\ \mathbf{I}$ then

$$\mathbf{A} + \Delta\mathbf{A} = \mathbf{U}^{-1}[\mathrm{diag}_n\{\lambda_i\} + \min_i\{-\Re e\ \lambda_i\}\ \mathbf{I}\ ]\ \mathbf{U} \tag{23.109}$$

(*Qiu, L., and Davison, E.J., 1986*) is an unstable matrix. Thus, Eq.(23.108) also is necessary.

23.14.3 Induced Norm of the Uncertainty. Discrete-Time Systems

For discrete-time systems the derivation runs similarly to the derivation of Eq.(23.98). Hence, Schur stability of a matrix $\mathbf{\Phi}$ perturbed to $\mathbf{\Phi} + \mathbf{E}_d$ is given if

$$\|\mathbf{E}_d\|_p\|(e^{j\theta}\ \mathbf{I} - \mathbf{\Phi})^{-1}\|_p < 1\ . \tag{23.110}$$

23.14.4 Perron-Frobenius Eigenvalue

Alternatively, from Eq.(23.93) $\det[\mathbf{I}-(s_o\mathbf{I}-\mathbf{A})^{-1}\Delta\mathbf{A}] \neq 0$

$$\Leftarrow \quad \rho_s[(s_o\mathbf{I}-\mathbf{A})^{-1}\Delta\mathbf{A}] < 1 \quad \Leftarrow \quad \rho_s[\ |(s_o\mathbf{I}-\mathbf{A})^{-1}|\ |\Delta\mathbf{A}|\ ] < 1\ . \tag{23.111}$$

For Hurwitz stability $s_o = j\omega_o$

$$\sup_{\omega} \pi[\ |(j\omega_o\mathbf{I}-\mathbf{A})^{-1}|\ |\Delta\mathbf{A}|\ ] < 1 \tag{23.112}$$

and if $|\Delta a_{ij}| < e_{ij}$, defining $\mathbf{E} = \text{matrix}[e_{ij}]$,

$$\sup_{\omega} \pi[\ |(j\omega_o\mathbf{I}-\mathbf{A})^{-1}|\ \mathbf{E}] < 1\ . \tag{23.113}$$

For Schur stability

$$\sup_{\theta} \pi[\ |(e^{j\theta}\mathbf{I}-\mathbf{\Phi})^{-1}|\ \mathbf{E}_d] < 1\ . \tag{23.114}$$

23.15 Generalized Resolvent Matrix $q\mathbf{I}-\mathbf{A}_\varepsilon$

Let $\mathbf{A}$ and $\mathbf{A}_p$ denote the nominal and the perturbed system matrix $\in \mathcal{R}^{n\times n}$, respectively. The matrix $\mathbf{A}_\varepsilon$ is introduced varying continuously from $\mathbf{A}$ to $\mathbf{A}_p$ when ε changes from 0 to 1. Hence, $\mathbf{A}_\varepsilon(\varepsilon=0) = \mathbf{A}$ and $\mathbf{A}_\varepsilon(\varepsilon=1) = \mathbf{A}_p$. Assume $\mathbf{A}$ possesses m eigenvalues in the domain D enclosed by the continuous curve Q where $0 \le m \le n$. Let the parameter q vary along the curve Q . If

$$\det(q\mathbf{I}-\mathbf{A}_\varepsilon) \neq 0 \qquad \forall q \in \mathrm{Q},\ \forall \varepsilon \in [0,1] \tag{23.115}$$

is satisfied the complex number q cannot be an eigenvalue of $\mathbf{A}_\varepsilon$. The eigenvalue loci or root loci will never pass or touch Q . Hence, the perturbed matrix $\mathbf{A}_p$ has exactly m eigenvalues in the domain D (*Juang, Y.T., 1989*).

If Q is chosen as the imaginary axis (D the open left-half complex plane) and $m = n$ then Hurwitz stability robustness of $\mathbf{A}$ is checked. If Q is considered the unit circle (D the unit circle disc) and $m = n$, Schur stability is proved. If the boundary of specified regions consists of several complex curves then Q is the union of all boundaries.

23.15.1 Additive Perturbation

From Eq.(23.115) and additive perturbations $\mathbf{A}_p = \mathbf{A}+\Delta\mathbf{A}$, $\mathbf{A}_\varepsilon = \mathbf{A}+\varepsilon\mathbf{E}$ is chosen where $\varepsilon \in [0,1]$

$$\det\{(q\mathbf{I}-\mathbf{A})[\mathbf{I}-\varepsilon(q\mathbf{I}-\mathbf{A})^{-1}\mathbf{E}]\} \neq 0 \tag{23.116}$$

$$\det[\mathbf{I}-\varepsilon(q\mathbf{I}-\mathbf{A})^{-1}\mathbf{E}] \neq 0 \tag{23.117}$$

$$\Leftarrow \qquad 1 > |\lambda_i[\varepsilon(q\mathbf{I}-\mathbf{A})^{-1}\mathbf{E}]\ | = \varepsilon|\lambda_i[(q\mathbf{I}-\mathbf{A})^{-1}\mathbf{E}]\ | \tag{23.118}$$

$$|\lambda_i[(q\mathbf{I}-\mathbf{A})^{-1}\mathbf{E}]\ | < 1 \qquad \forall q \in \mathrm{Q},\ i = 1,2...n\ . \tag{23.119}$$

Hence, the perturbed system $\mathbf{A}_p$ has m eigenvalues inside D if Eq.(23.119) is satisfied.

Remark: Since Eq.(A.18) and the involved matrices are square, $(q\mathbf{I}-\mathbf{A})^{-1}$ and $\mathbf{E}$ may be permuted as argument of λ_i .

23.15.2 Multiplicative Perturbation

Selecting the case of input-associated multiplicative perturbation $\mathbf{A}_p = \mathbf{A}(\mathbf{I} + \mathbf{L}_i)$ and $\mathbf{A}_\varepsilon = \mathbf{A}(\mathbf{I} + \varepsilon\mathbf{L}_i)$, from Eq.(23.115)

$$\det[\mathbf{I} - \varepsilon(q\mathbf{I} - \mathbf{A})^{-1}\mathbf{A}\mathbf{L}_i] \neq 0 \quad \Leftarrow \quad |\lambda_i[(q\mathbf{I} - \mathbf{A})^{-1}\mathbf{A}\mathbf{L}_i]\,| < 1\,. \tag{23.120}$$

Selecting output-associated multiplicative perturbations, then $\mathbf{A}_p = (\mathbf{I} + \mathbf{L}_o)\mathbf{A}$ and $\mathbf{A}_\varepsilon = (\mathbf{I} + \varepsilon\mathbf{L}_o)\mathbf{A}$ a sufficient condition is

$$|\lambda_i[\mathbf{L}_o\mathbf{A}(q\mathbf{I} - \mathbf{A})^{-1}]\,| = |\lambda_i[\mathbf{A}(q\mathbf{I} - \mathbf{A})^{-1}\mathbf{L}_o]\,| < 1\,. \tag{23.121}$$

23.15.3 Replacing the Eigenvalue by any Norm

For any p and any invertible scaling matrix $\mathbf{L}$ and $\mathbf{R}$, from Eq.(23.119)

$$|\lambda_i[\mathbf{L}^{-1}(q\mathbf{I} - \mathbf{A})^{-1}\mathbf{R}^{-1}\mathbf{R}\mathbf{E}\mathbf{L}]\,| < 1 \qquad \forall q \in Q \tag{23.122}$$

$$\Leftarrow \quad \|\mathbf{L}^{-1}(q\mathbf{I} - \mathbf{A})^{-1}\mathbf{R}^{-1}\mathbf{R}\mathbf{E}\mathbf{L}\|_p < 1 \qquad \forall q \in Q \tag{23.123}$$

$$\Leftarrow \quad \|\mathbf{L}^{-1}(q\mathbf{I} - \mathbf{A})^{-1}\mathbf{R}^{-1}\|_p\|\mathbf{R}\mathbf{E}\mathbf{L}\|_p < 1 \qquad \forall q \in Q\,. \tag{23.124}$$

For alternative kind of perturbations similar results can be derived.

23.15.4 Upper Bound of Structured Perturbations and Common Factor μ

Let $\mathbf{A}_p = \mathbf{A} + \Delta\mathbf{A}$ and $|\Delta\mathbf{A}| \leq_e \mu\mathbf{E}$. The matrices $\mathbf{L}$ and $\mathbf{R}$ represent non-negative invertible scaling matrices. Then,

$$|\mathbf{L}\Delta\mathbf{A}\ \mathbf{R}| \leq_e \mu|\mathbf{L}\mathbf{E}\mathbf{R}|\,. \tag{23.125}$$

Using p-Norm

For norms with $p = 1,\ \infty,\ s,\ \mathrm{F}$

$$|\mathbf{M}_1| \leq_e |\mathbf{M}_2| \quad \rightsquigarrow \quad \|\mathbf{M}_1\|_p \leq \|\ |\mathbf{M}_2|\ \|_p \tag{23.126}$$

is satisfied. From Eqs.(23.124), (23.125) and (23.126) it follows

$$\|\mathbf{L}^{-1}(q\mathbf{I} - \mathbf{A})^{-1}\mathbf{R}^{-1}\|_p\|\mathbf{R}\Delta\mathbf{A}\ \mathbf{L}\|_p \leq \mu\ \|\mathbf{L}^{-1}(q\mathbf{I} - \mathbf{A})^{-1}\mathbf{R}^{-1}\|_p\|\mathbf{R}\mathbf{E}\mathbf{L}\|_p < 1 \tag{23.127}$$

or

$$\mu < \frac{1}{\|\mathbf{L}^{-1}(q\mathbf{I} - \mathbf{A})^{-1}\mathbf{R}^{-1}\|_p\|\mathbf{R}\mathbf{E}\mathbf{L}\|_p} \qquad \forall q \in Q \tag{23.128}$$

(*Juang, Y.T., 1989*).

Using Spectral Radius

$$\mathbf{A}_p = \mathbf{A} + \Delta\mathbf{A} \qquad |\Delta\mathbf{A}| \leq_e \mu\mathbf{E} \tag{23.129}$$

Two obvious facts are used:

$$(i) \qquad |\mathbf{M}\ \Delta\mathbf{A}| \leq_e |\mathbf{M}|\ |\Delta\mathbf{A}| \leq_e |\mathbf{M}|\mu|\mathbf{E}| \tag{23.130}$$

$$(ii) \qquad |\mathbf{M}_1| \leq_e |\mathbf{M}_2| \quad \rightsquigarrow \quad \rho_s[\mathbf{M}_1] \leq \rho_s[\ |\mathbf{M}_2|\ ] \tag{23.131}$$

(*Ortega, J.M., 1972*). From Eq.(23.119) replacing $\mathbf{E}$ by $\Delta\mathbf{A}$ and choosing $\varepsilon = 1$, the perturbed system has m eigenvalues inside D if

$$|\lambda_i| \leq \rho_s[(q\mathbf{I} - \mathbf{A})^{-1}\Delta\mathbf{A}] < 1 \qquad \forall q \in \mathrm{Q}\ . \tag{23.132}$$

Choosing $\mathbf{M} = (q\mathbf{I} - \mathbf{A})^{-1}$ then from Eq.(23.130)

$$|\underbrace{(q\mathbf{I} - \mathbf{A})^{-1}\Delta\mathbf{A}}_{\mathbf{M}_1}| \leq_e |(q\mathbf{I} - \mathbf{A})^{-1}|\ |\Delta\mathbf{A}| \leq_e \underbrace{|(q\mathbf{I} - \mathbf{A})^{-1}|}_{|\mathbf{M}_2|}\mu\mathbf{E}\ . \tag{23.133}$$

Using the underbrace abbreviation above and Eq.(23.131),

$$\rho_s[(q\mathbf{I} - \mathbf{A})^{-1}\Delta\mathbf{A}] \leq \rho_s[\ |(q\mathbf{I} - \mathbf{A})^{-1}|\mu\mathbf{E}\ ] = \mu\rho_s[\ |(q\mathbf{I} - \mathbf{A})^{-1}|\mathbf{E}\ ]\ . \tag{23.134}$$

If $\mu\rho_s[\ |(q\mathbf{I} - \mathbf{A})^{-1}|\mathbf{E}\ ] < 1$ then $\rho_s[\ |(q\mathbf{I} - \mathbf{A})^{-1}|\Delta\mathbf{A}\ ] < 1$ as required by Eq.(23.119). Hence, for robust stability (*Juang, Y.T., 1989*)

$$\mu < \frac{1}{\rho_s[\ |(q\mathbf{I} - \mathbf{A})^{-1}|\mathbf{E}\ ]} \qquad \forall q \in \mathrm{Q}\ . \tag{23.135}$$

23.16 Stability Margin for Diagonally Perturbed Systems

Consider a system as depicted in Fig. 25.1 where $\mathbf{G}$ is replaced by $\mathbf{G}_p = \mathbf{G}(\mathbf{I} + \Delta\mathbf{L}_o)$. The transfer function of the perturbed system contains a multiplicative diagonal uncertainty associated with the output $\Delta\mathbf{L}_o = \mathrm{diag}\{\Delta_1,\ \Delta_2, ...\Delta_m\}$. Then, stability is guaranteed if $\det\{\mathbf{I} + [\mathbf{I} + \Delta\mathbf{L}_o(s)]\mathbf{G}(s)\}\ |_{\ s=j\omega} \neq 0$

$$\det\{[\mathbf{I} + \Delta\mathbf{L}_o\mathbf{G}(\mathbf{I} + \mathbf{G})^{-1}](\mathbf{I} + \mathbf{G})\}\ |_{\ s=j\omega} \neq 0 \tag{23.136}$$

$$\det[\mathbf{I} + \Delta\mathbf{L}_o(\mathbf{I} + \mathbf{G}^{-1})^{-1}] \neq 0 \quad \rightsquigarrow \quad \|\Delta\mathbf{L}_o(\mathbf{I} + \mathbf{G}^{-1})^{-1}\| < 1 \tag{23.137}$$

$$\Leftarrow \quad \|\Delta\mathbf{L}_o\|\ \|(\mathbf{I} + \mathbf{G}^{-1})^{-1}\| < 1 \quad \rightsquigarrow \quad \|\Delta\mathbf{L}_o(j\omega)\| < \|\ [\mathbf{I} + \mathbf{G}^{-1}(j\omega)]\ \|^{-1} \tag{23.138}$$

where $\|\cdot\|$ is any induced norm. Assume that each element of the diagonal perturbation is individually bounded by $|\Delta_i(j\omega)| < |r_i(j\omega)| \quad \forall i,\ \omega$. The stability margin is the largest positive number $k_m(\omega)$ such that the system is stable for every stable and diagonal $\Delta\mathbf{L}_o$, i.e., $|\Delta_i(j\omega)| < k_m(\omega)\ |r_i(j\omega)| \quad \forall i,\ \omega$. Defining $\mathbf{R}(s) \triangleq \mathrm{diag}\{r_i(s)\}$ the stability margin is bounded *below* by

$$k_m(\omega) \geq \|\ [\mathbf{I} + \mathbf{G}^{-1}(j\omega)]^{-1}\mathbf{R}(j\omega)\|^{-1}_{pD_p^\star(j\omega)} \tag{23.139}$$

$$\text{where} \qquad \|\mathbf{A}\|_{pD_p^\star(j\omega)} = \sup_{\mathbf{x}\neq\mathbf{0}} \frac{\|\mathbf{A}\mathbf{x}\|_{pD_p^\star(j\omega)}}{\|\mathbf{x}\|_{pD_p^\star(j\omega)}} \begin{cases} = \pi[\ |\mathbf{A}|\ ] & p = 1,\ \infty \\ < \pi[\ |\mathbf{A}|\ ] & 1 < p < \infty \end{cases} \tag{23.140}$$

$$\text{and} \qquad \mathbf{D}_p^\star = \mathrm{diag}\{y_1^{1/p}x_1^{-1/q},\ .....\ ,\ y_m^{1/p}x_m^{-1/q}\} \qquad \frac{1}{q} + \frac{1}{p} = 1 \tag{23.141}$$

and x_i and y_i are the components of the right and left Perron eigenvectors of $|\mathbf{A}|$. See Eq.(25.60) and *Safonov, M.G., 1982* for the proof.

Chapter 24

Robustness Via Singular-Value Analysis

Plant uncertainties and plant model variations must be taken into consideration in almost every application. With regard to plant uncertainties the global control system performance decreases. The subject of this chapter is the application of singular values to the overview problem of balancing system stability robustness, system performance and plant uncertainty.

24.1 Control System Performance and Minimum Singular Value

24.1.1 Tracking Requirements

The configuration of a multivariable control system is depicted in Fig. 24.1. The matrices $\mathbf{K}$ and $\mathbf{G}$ are the transfer matrices of the controller and plant, respectively. If the error is defined as the difference between the reference and the noise free output, see Fig. 24.1 and Eq.(30.50),

$$\mathbf{e} = \mathbf{y}_{ref} - \mathbf{y} = (\mathbf{I}+\mathbf{GK})^{-1}(\mathbf{y}_{ref} - \mathbf{w}) + (\mathbf{I}+\mathbf{GK})^{-1}\mathbf{GKn} \tag{24.1}$$

Referring to Eq.(22.66) and applying appropriate substitutions,

$$\sigma_{\min}[\mathbf{GK}] - 1 \leq \sigma_{\min}[\mathbf{I}+\mathbf{GK}] \leq \sigma_{\min}[\mathbf{GK}] + 1\ . \tag{24.2}$$

Hence, large values $\sigma_{\min}[\mathbf{GK}]$ imply large values of $\sigma_{\min}[\mathbf{I}+\mathbf{GK}]$ or small values of $\sigma_{\max}[(\mathbf{I}+\mathbf{GK})^{-1}]$, i.e., from Eq.(24.1) small error values due to $\mathbf{y}_{ref}$ and $\mathbf{w}$. But, on the other hand, $\mathbf{e}$ due to $\mathbf{n}$ may be large. Summarizing, the performance objectives in the active frequency range are $\sigma_{\max}[(\mathbf{I}+\mathbf{GK})^{-1}]$ small or $\sigma_{\min}[\mathbf{I}+\mathbf{GK}]$ large. From single-input single-output systems the necessary condition $|1 + G(j\omega)K(j\omega)| \geq p_o(\omega_o) \gg 1$ is well known where $p_o(\omega)$ is a performance objective measure. Even if the poles of the closed-loop control systems have been carefully chosen and are located obviously favourable, it may happen that $\sigma_{\min}[\mathbf{I}+\mathbf{G}(j\omega)\mathbf{K}(j\omega)]$ turns out very small and the system fails to be robust. Cases where small values of stability margin may occur are characterized, first, by the choice of closed-loop poles merely a little distance to the left of the dominant pole of the plant and, second, by the existence of additional plant poles far to the left of the s-plane or, third, by the existence of additional plant zeros with small magnitude (*Noisser, R., 1988*).

24.1.2 Control Variable Requirements

Appropriate dimensions of the signals are preassumed. Since

$$\mathbf{u} = \mathbf{K}(\mathbf{I}+\mathbf{GK})^{-1}(\mathbf{y}_{ref} - \mathbf{n} - \mathbf{w}) \doteq \mathbf{G}^{-1}(j\omega)(\mathbf{y}_{ref} - \mathbf{n} - \mathbf{w}) \tag{24.3}$$

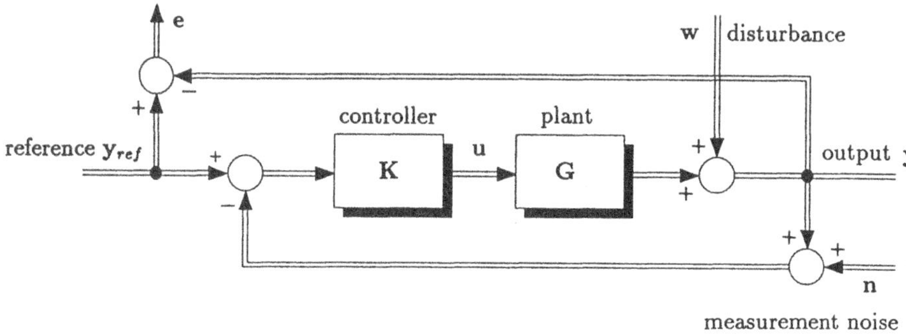

Figure 24.1: Multivariable system with reference, disturbance and noise input

the reference signal, noise and disturbance are strongly amplified and the control signal $\mathbf{u}$ is large if in the active frequency range there is $\sigma_{\max}[\mathbf{G}] \ll 1$ or $\sigma_{\min}[\mathbf{G}^{-1}] \gg 1$.

A condition for the power of the control signal $\mathbf{u}$ in the case $\mathbf{y}_{ref} = \mathbf{0}$ and $\mathbf{w} = \mathbf{0}$ can be formulated as follows. Using

$$\mathbf{S} = (\mathbf{I} + \mathbf{GK})^{-1} \qquad \mathbf{u}(s) = -\mathbf{K}(s)\mathbf{S}(s)\mathbf{n}(s) \ , \tag{24.4}$$

in order to satisfy

$$\max_{\|\mathbf{n}(j\omega)\|_F=1} \|\mathbf{u}(j\omega)\|_F \leq b_u(\omega) \qquad \forall\omega \geq 0 \tag{24.5}$$

the condition is

$$\sigma_{\max}[\mathbf{KS}] = \sigma_{\max}[\mathbf{K}(\mathbf{I} + \mathbf{GK})^{-1}] \leq b_u(\omega) \qquad \forall\omega \geq 0 \ . \tag{24.6}$$

The minimal sensitivity is given by $\inf_{\mathbf{K}} \|\mathbf{S}(s)\|_\infty$. For scalar systems the gain margin problem is to find $\sup_{K(s)} b/a$ such that $K(s)$ stabilizes $kG(s)$ for each $k \in [a, b]$ and $0 < a < 1 < b$. Sensitivity minimization problem subject to a gain margin constraint is always solvable. However, for nonminimum-phase plant the gain margin maximization is not compatible with sensitivity minimization (*Yan, W., and Anderson, B.D.O., 1990*).

24.1.3 Disturbance Rejection Requirements

In the case $\mathbf{y}_{ref} = \mathbf{0}$ and $\mathbf{n} = \mathbf{0}$, in order to satisfy disturbance rejection according to

$$\max_{\|\mathbf{w}(j\omega)\|_F=1} \|\mathbf{y}(j\omega)\|_F \leq b_y(\omega) \qquad \forall\omega \geq 0 \ , \tag{24.7}$$

using $\mathbf{y}(s) = \mathbf{S}(s)\mathbf{w}(s)$, one has to guarantee

$$\sigma_{\max}[\mathbf{S}(j\omega)] = \sigma_{\min}^{-1}[\mathbf{I} + \mathbf{GK}] \leq b_y(\omega) \qquad \forall\omega \geq 0 \ . \tag{24.8}$$

24.2 System Stability and Multiplicative Uncertainty

If high loop gains are implemented then the requirement for tolerance to uncertainty becomes very important. Standard requirements for the robust controller $\mathbf{K}(s)$ are the stability of the perturbed system as well as the nominal system. The stability conditions for the system are known from the multivariable generalization of the Nyquist criterion. The number of (clockwise) encirclements of the plot det[$\mathbf{I}+\mathbf{GK}$] evaluated on the standard Nyquist D-contour (imaginary axis and semicircle with infinite diameter round the right-half s-plane in clockwise direction) must be equal to the negative number of unstable open-loop poles of $\mathbf{GK}$ (*Rosenbrock, H.H., 1974*).

To guarantee robust stability this requirement must be satisfied for every $\mathbf{G}_p$ which obeys the perturbation bounds. E.g., for output associated multiplicative uncertainty $\mathbf{G}_p = (\mathbf{I} + \Delta\mathbf{L}_o)\mathbf{G}$ the tolerance $\Delta\mathbf{L}_o$ is considered bounded by $\sigma_{\max}[\Delta\mathbf{L}_o] \leq l_o(\omega)$. Only little structural information about a matrix is reflected in its singular values. Hence, conservatism of such robustness test may arise. However, in some cases this fact turns out desirable in order to obtain wide range robustness for unstructured uncertainty.

The stability of the system is not affected by the perturbation if the number of encirclements (in clockwise direction) of det $(\mathbf{I} + \mathbf{G}_p\mathbf{K})$ equals the negative number of unstable open-loop modes of $\mathbf{G}_p\mathbf{K}$ and $\mathbf{GK}$. In other words, the encirclements of the scalar function det$(\mathbf{I}+\mathbf{G}_p\mathbf{K})$ must remain unchanged for all perturbations $\Delta\mathbf{L}_o$ for which $\sigma_{\max}[\Delta\mathbf{L}_o] \leq l_o(\omega)$. Replacing $\Delta\mathbf{L}_o$ by $\varepsilon\mathbf{L}_o$ this is guaranteed if and only if

$$\det\,[\mathbf{I} + (\mathbf{I} + \varepsilon\mathbf{L}_o)\mathbf{GK}(s)] \neq 0 \qquad \forall\varepsilon = [0,1], \quad s \in \text{D-contour} \tag{24.9}$$

where $\mathbf{L}_o$ is given by $\sigma_{\max}[\mathbf{L}_o] \triangleq l_o(\omega)$, or, equivalently, if and only if

$$\sigma_{\min}[\mathbf{I} + (\mathbf{I} + \varepsilon\mathbf{L}_o)\mathbf{GK}(s)] > 0 \qquad \forall\varepsilon = [0,1], \quad s \in \text{D-contour}\,. \tag{24.10}$$

The amount of the minimum singular value cited above may be considered as the "minimum distance" of the determinant to the origin.

With regard to $\sigma_{\min}[\mathbf{A} + \mathbf{\Delta}] \geq \sigma_{\min}[\mathbf{A}] - \sigma_{\max}[\mathbf{\Delta}]$ the uncertainty $\mathbf{\Delta}$ reduces the minimum singular value of $\mathbf{A}$ if changed to $\mathbf{A} + \mathbf{\Delta}$, at a rate no more than the amount given by the maximum singluar value of the uncertainty $\mathbf{\Delta}$. Since

$$\det[\mathbf{I} + \mathbf{GK} + \varepsilon\mathbf{L}_o\mathbf{GK}] \equiv \det(\mathbf{I} + \mathbf{GK})\ \det[\mathbf{I} + \varepsilon\mathbf{L}_o\mathbf{GK}(\mathbf{I} + \mathbf{GK})^{-1}] \neq 0 \tag{24.11}$$

and since $\det(\mathbf{I} + \mathbf{GK}) \neq 0 \quad \forall s = j\omega$

$$\sigma_{\min}[\mathbf{I} + \varepsilon\mathbf{L}_o\mathbf{GK}(\mathbf{I} + \mathbf{GK})^{-1}] > 0 \tag{24.12}$$

$$\text{"}\Leftarrow\text{"} \qquad \sigma_{\min}[\mathbf{I}] - \varepsilon\sigma_{\max}[\mathbf{L}_o\mathbf{GK}(\mathbf{I} + \mathbf{GK})^{-1}] > 0\,. \tag{24.13}$$

The expression[1] above is implied by the following one if the worst case uncertainty case $\varepsilon = 1$ is selected (*Doyle, J.C., and Stein, G., 1981*)

$$\text{"}\Leftarrow\text{"} \qquad 1 - \sigma_{\max}[\mathbf{L}_o\mathbf{GK}(\mathbf{I} + \mathbf{GK})^{-1}] > 0 \tag{24.14}$$

$$\text{"}\Leftarrow\text{"} \qquad 1 - \sigma_{\max}[\mathbf{L}_o]\sigma_{\max}[\mathbf{GK}(\mathbf{I} + \mathbf{GK})^{-1}] > 0 \tag{24.15}$$

$$\frac{1}{\sigma_{\max}[\mathbf{L}_o]} > \sigma_{\max}[\mathbf{GK}(\mathbf{I} + \mathbf{GK})^{-1}] \tag{24.16}$$

$$\sigma_{\max}[\mathbf{L}_o] < \sigma_{\max}^{-1}[\mathbf{GK}(\mathbf{I} + \mathbf{GK})^{-1}] = \sigma_{\min}[\mathbf{I} + (\mathbf{GK})^{-1}]\,. \tag{24.17}$$

[1]Note that " $\Leftarrow$ " means that from the singular-value-property point of view the meaning equals $\Leftarrow$. However, it was documented by *Chen, M.J., and Desoer, C.A., 1982* that the conditions are sufficient *and* necessary.

In the frequency range beyond the crossover frequency ω_D the entries of **GK** are small, hence $(\mathbf{I}+\mathbf{GK}) \to \mathbf{I}$ and the result is

$$\sigma_{\max}[\mathbf{GK}] < \frac{1}{\sigma_{\max}[\mathbf{L}_o]} = \frac{1}{l_o(\omega)} \qquad \omega \gg \omega_D \; . \tag{24.18}$$

The maximum singular value of the loop transfer function matrix beyond crossover frequency is bounded by a number which is inversely proportional to the uncertainty.

The necessary and sufficient conditions for robust stability hold (i) for stable uncertainty or (ii) for unstable plant uncertainty with the same number of unstable open-loop poles in the perturbed and unperturbed case (*Lehtomaki, N.A., et al. 1981;. Chen, M.J., and Desoer, C.A., 1982; Vidyasagar, M., and Kimura, H., 1986*). If the number of unstable poles of the plant to be robustly stabilized is not precisely known the method of graph metric must be used, see Eq.(31.134).

24.3 System Performance and Multiplicative Uncertainty

Postulating that the origin distance associated with Eq.(24.10) is larger than a low-frequency performance objective $p_o(\omega)$,

$$\sigma_{\min}[\mathbf{I}+(\mathbf{I}+\varepsilon\mathbf{L}_o)\mathbf{GK}(s)]_{s=j\omega} \geq p_o(\omega) \tag{24.19}$$

$$\Leftarrow \; \sigma_{\min}[\mathbf{I}+\mathbf{GK}+\varepsilon\mathbf{L}_o\mathbf{GK}] \geq \sigma_{\min}[\mathbf{I}+\varepsilon\mathbf{L}_o\mathbf{GK}(\mathbf{I}+\mathbf{GK})^{-1}]\sigma_{\min}[\mathbf{I}+\mathbf{GK}] \geq p_o(\omega) \tag{24.20}$$

$$\Leftarrow \; \left(1-\varepsilon l_o\sigma_{\max}[\mathbf{GK}(\mathbf{I}+\mathbf{GK})^{-1}]\right)\left(\sigma_{\min}[\mathbf{GK}]-1\right) \geq p_o \; . \tag{24.21}$$

In the low-frequency range $\omega < \omega_o$ one has $\sigma_{\min}[\mathbf{GK}] \gg 1$ and with $\varepsilon = 1$ (worst case)

$$(1-l_o)\sigma_{\min}[\mathbf{GK}] \geq p_o(\omega) \quad \rightsquigarrow \quad \sigma_{\min}[\mathbf{GK}] \geq \frac{p_o(\omega)}{1-l_o(\omega)} \qquad \omega < \omega_o \; . \tag{24.22}$$

24.4 Generalized Stability Margin

24.4.1 Uncertainty Associated with the Output

Referring to $\sigma_{\max}[\mathbf{A}^{-1}] = 1/\sigma_{\min}[\mathbf{A}]$, Eq.(24.16) is inverted

$$\sigma_{\max}[\mathbf{L}_o] = l_o(\omega) < \sigma_{\min}[(\mathbf{I}+\mathbf{GK})(\mathbf{GK})^{-1}] = \sigma_{\min}[\mathbf{I}+(\mathbf{GK})^{-1}] \; . \tag{24.23}$$

The expression $\sigma_{\min}[\mathbf{I}+(\mathbf{GK})^{-1}]$ is a generalization of stability margins as known from single-input single-output systems, e.g. gain and phase margin. Note that the right-hand side is a function of the *nominal* plant parameters.

24.4.2 Uncertainty Associated with the Input

The derivation of the robust stability condition is *repeated* in the case of the uncertainty associated to the input of the plant. Similarly to Eq.(24.11) and for $\Delta\mathbf{L}_i = \varepsilon\mathbf{L}_i$

$$\det\,[\mathbf{I}+\mathbf{G}(\mathbf{I}+\Delta\mathbf{L}_i)\mathbf{K}] \equiv \det\,[\mathbf{I}+\mathbf{KG}(1+\Delta\mathbf{L}_i)] \neq 0 \tag{24.24}$$

$$\det\,(\mathbf{I}+\mathbf{KG})\,\det\,[\mathbf{I}+(\mathbf{I}+\mathbf{KG})^{-1}\mathbf{KG}\Delta\mathbf{L}_i] \neq 0 \tag{24.25}$$

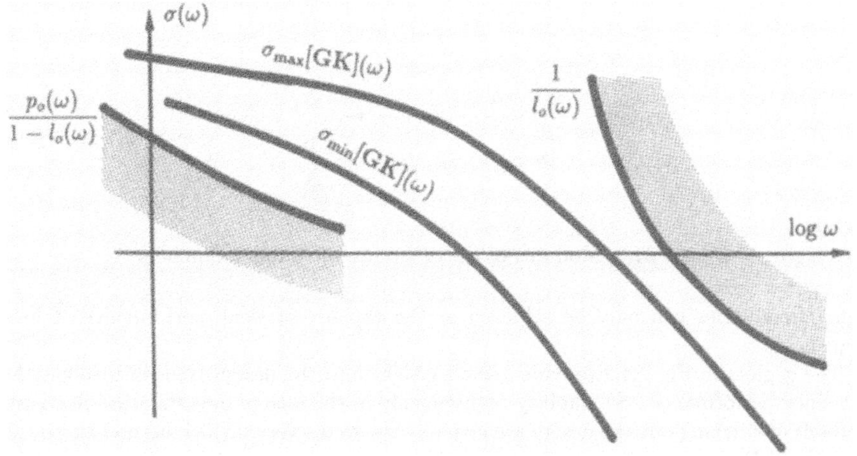

Figure 24.2: Constraints for singular values given by low-frequency performance and plant uncertainty at high frequencies

$$\Leftrightarrow \quad \sigma_{min}[\mathbf{I} + (\mathbf{I} + \mathbf{KG})^{-1}\mathbf{KG}\Delta\mathbf{L}_i] > 0 \tag{24.26}$$

$$" \Leftarrow " \quad 1 - \sigma_{max}[\mathbf{L}_i]\ \sigma_{max}[(\mathbf{I} + \mathbf{KG})^{-1}\mathbf{KG}] > 0 \tag{24.27}$$

$$" \Leftarrow " \quad \frac{1}{\sigma_{max}[\mathbf{L}_i]} \triangleq \frac{1}{l_i(\omega)} > \sigma_{max}[(\mathbf{I} + \mathbf{KG})^{-1}\mathbf{KG}] \tag{24.28}$$

$$\sigma_{max}[\mathbf{L}_i] = l_i(\omega) < \sigma_{min}[\mathbf{I} + (\mathbf{KG})^{-1}] \tag{24.29}$$

Comparing Eqs.(24.23) and (24.29) shows that the stability robustness measure uses the loop transfer matrix from the same loop breaking point the uncertainty is associated with. For several examples see Table 26.1.

If multiplicative uncertainties occur associated both with the input and the output, the method above is applicable; spectral norm of matrices must be applied repeatedly.

In Fig. 24.2 the boundaries are depicted as given by Eqs.(24.18) and (24.22). The maximum and minimum singular value of the open-loop transfer matrix has to satisfy the appropriate boundary condition given by the uncertainty stability condition and the low-frequency performance condition, respectively. The design of closed-loop control systems requires that several performance and stability requirements are satisfied. The maximum singular values are an appropriate tool to define these requirements. Rewriting the *closed-loop* singular-value requirements to the *open-loop* singular values is one of the characteristics of loop shaping.

The chosen open-loop transfer matrix must also satisfy the nominal stability condition in the vicinity of the crossover frequency. This makes clear that a strong dependence exists between three performance statements: (i) frequency range of good low-frequency performance, (ii) nominal stability quality (steepness) near crossover and (iii) high-frequency constraint determined by the plant uncertainty.

Hence, the low-frequency range over which the performance objectives must be satisfied are limited by the frequency range given by the uncertainties. In most cases of applications (finite dimensional linear time-invariant systems) the stability robustness at

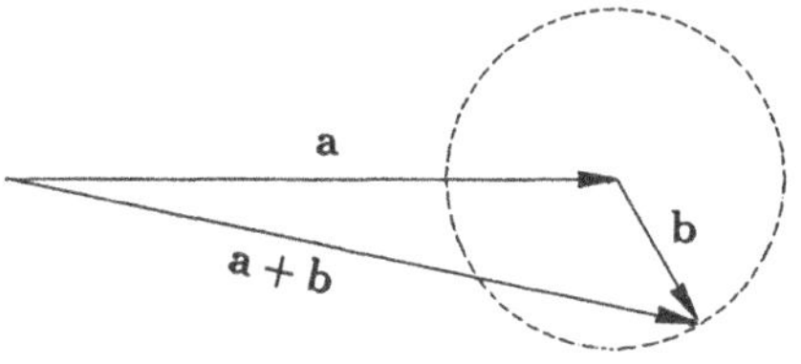

Figure 24.3: Modulus of the difference of complex numbers

higher frequencies can only be achieved at the expense of weak performance at lower frequencies.

The σ-plots in Fig. 24.2 provide a good insight into the design tradeoff involving low frequency performance and stability requirements in the face of uncertainties. Successful methods of carrying out the design are given by the characteristic loci method and by the generalized Nyqyist method.

The characteristic loci method is based on the frequency dependent eigenvalues of the loop transfer matrix, see e.g. Eq.(25.92) and *MacFarlane, A.G.J., and Kouvaritakis, B., 1977*. By the inverse-Nyquist-array method the multivariable design problem is solved by appropriatedly pre- and post-compensating the loop transfer function matrix in order to obtain a diagonally dominant system (*Rosenbrock, H.H., 1974; Maciejowski, J.M., 1989*). Using, e.g., the Gershgorin method the design problem is reduced to a sequence of scalar problems.

24.5 Performance and Plant Uncertainty. Single-Input Single-Output System

The standard single-variable feedback system is illustrated in Fig. 24.4. Consider the plant uncertainty be given by the transfer function additive term $\Delta G(s)$ or by the multiplicative term $1 + \Delta L(s)$, i.e. $\Delta G(s) = \Delta L(s)\, G(s)$. The system error $e(t)$ is described by the Laplace transform $e(s) = \mathcal{L}\{e(t)\}$

$$e(s) = \frac{1}{1 + C(s)G(s)}\, y_{ref}(s) \quad \text{and} \quad e(s) = \frac{1}{1 + C(s)[1 + \Delta L(s)]G(s)}\, y_{ref}(s) \tag{24.30}$$

in the nominal case and in the case including uncertainty, respectively.

The amount of control performance deterioration (or sensitivity of control system quality) can be expressed by the quantity $D(\omega)$

$$D(\omega) \triangleq |1 + C(j\omega)G(j\omega)| - |1 + C(j\omega)[1 + \Delta L(j\omega)]G(j\omega)| \,. \tag{24.31}$$

Since $|a| - |a + b| \leq |b|$, see Fig. 24.3, the difference $D(\omega)$ in the equation above can be bounded as

$$D(\omega) \;<\; |C(j\omega)\Delta L(j\omega)G(j\omega)| \,. \tag{24.32}$$

In order to guarantee a certain value of robustness the magnitude of $D(\omega)$ must not exceed a boundary value $D_o(\omega)$

$$D(\omega) \;<\; D_o(\omega) \,. \tag{24.33}$$

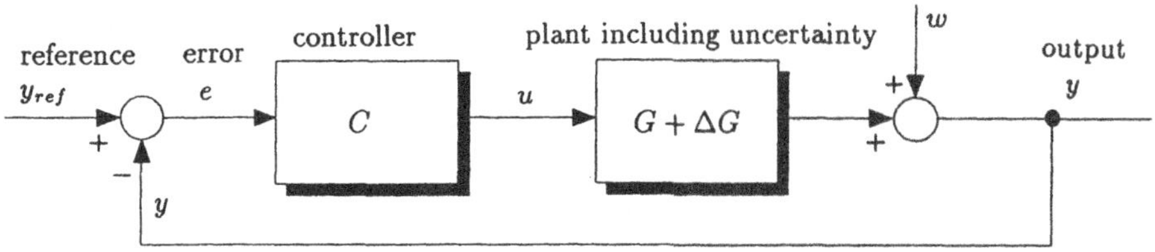

Figure 24.4: Standard single variable control system

This relationship can be substituted by

$$|C(j\omega)\ \Delta L(j\omega)\ G(j\omega)\ |\ <\ D_o \tag{24.34}$$

which is a sufficient condition to satisfy Eq.(24.32) and Eq.(24.33). Within Eq.(24.34) the relation Eq.(24.33) is satisfied, beyond Eq.(24.34) the relation Eq.(24.33) *may be* true but there is no guarantee.

The quantity $|C(j\omega)G(j\omega)|$ gives an absolute performance specification as a function of the frequency ω, well known by the commonly used Bode plots. The quantity $|\Delta L(j\omega)|$ can be bounded by an uncertainty radius $r_L(\omega)$. These definitions yield (as a sufficient condition)

$$r_L(\omega)\ |C(j\omega)G(j\omega)|\ <\ D_o(\omega)\ . \tag{24.35}$$

A larger uncertainty radius $r_L(\omega)$ requires less performance $|C(j\omega)G(j\omega)|$ of the unperturbed control system if the performance deterioration is bounded within a given amount $D_o(\omega)$. The control system performance is less sensitive to uncertainties if the process itself has the property of $|C(j\omega)G(j\omega)|$ being small.

Furthermore, single-input single-output systems are characterized by limitations of the sensitivity function imposed by right-half-plane poles p_i of the open-loop system. The sensitivity function $S(s)$ must satisfy the integral relation (*Freudenberg, J.S., and Looze, D.P., 1985*)

$$\pi \sum_{i=1}^{N} \Re e\ p_i = \int_0^\infty \log|S(j\omega)|d\omega \qquad \text{where} \qquad S(s) = \frac{1}{1 + C(s)G(s)}\ . \tag{24.36}$$

24.6 Sensitivity Relations

The differential sensitivity of the closed-loop system transfer function $F(s)$ on the process uncertainty is derived in Eq.(8.37)

$$S_G^F = \frac{\Delta F/F}{\Delta G/G} = 1 - F(s) = 1 - \frac{CG}{1 + CG}\ . \tag{24.37}$$

With the uncertainty $\Delta G = \Delta L\ G$

$$\frac{\Delta F}{F}(j\omega) = \Delta \ln\ F(j\omega) = [1 - F(j\omega)]\Delta G/G = [1 - F(j\omega)]\Delta L(j\omega) \tag{24.38}$$

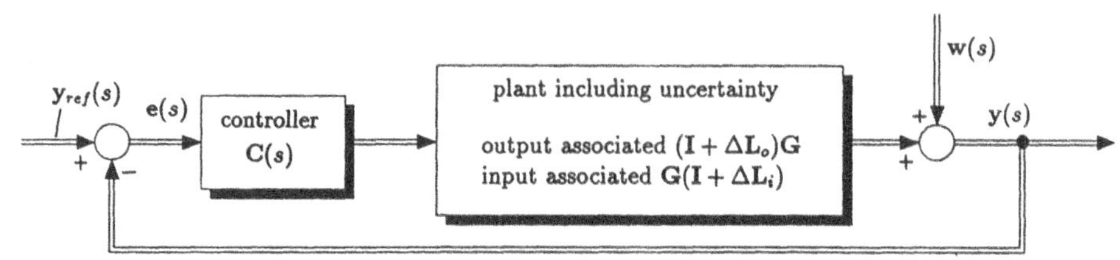

Figure 24.5: Standard multivariable control system

$$|\Delta \ln F(j\omega)| = |1 - F(j\omega)|\, r_L(\omega) = \frac{1}{|1 + C(j\omega)G(j\omega)|}\, r_L(\omega) \tag{24.39}$$

is obtained with the following result. The less the uncertainty radius $r_L(\omega)$ or the closer $F(j\omega)$ to unity the less is the local influence on ΔF.

24.7 Performance and Plant Uncertainty. Multivariable System

In the case of multivariable plants there exist two multiplicative forms for uncertainty representation. Uncertainty associated with the process outputs yields an overall transfer function matrix

$$[\mathbf{I} + \Delta\mathbf{L}_o(j\omega)]\mathbf{G}(j\omega)\ . \tag{24.40}$$

The output of $\mathbf{G}(j\omega)$ has to pass the multiplicative transfer function matrix $\mathbf{I}+\Delta\mathbf{L}_o(j\omega)$. When the network $\Delta\mathbf{L}_i(j\omega)$ modelling the uncertainty is a part of the postmultiplication matrix of $\mathbf{G}(j\omega)$ the uncertainty is associated with the process inputs (or controlling variables) and the transfer matrix is

$$\mathbf{G}(j\omega)[\mathbf{I} + \Delta\mathbf{L}_i(j\omega)]\ . \tag{24.41}$$

The dynamic representation of the standard multivariable control loop is given by

$$\mathbf{e}(s) = \{\mathbf{I} + [\mathbf{I} + \Delta\mathbf{L}_o(s)]\mathbf{G}(s)\mathbf{C}(s)\}^{-1}[\mathbf{y}_{ref}(s) - \mathbf{w}(s)] \tag{24.42}$$

when the uncertainty is associated with the output of the plant. The performance divergence can be expressed by employing the singular values $\sigma[\mathbf{G}]$. The singular values of a matrix $\mathbf{G}$ are defined as the positive square roots of the eigenvalues of the Hermitian matrix $\mathbf{G}^H\mathbf{G}$ where $\mathbf{G}^H$ is the complex conjugate transpose of $\mathbf{G}$, i.e. $\sigma[\mathbf{G}] = +\sqrt{\lambda[\mathbf{G}^H\mathbf{G}]}$.

The performance deterioration in the multivariable case is defined as the difference of the minimum singular values (*Johnston, R.D., and Barton, G.W., 1987*)

$$D(\omega) \triangleq \sigma_{\min}[\mathbf{I} + \mathbf{G}(j\omega)\mathbf{C}(j\omega)] - \sigma_{\min}[\mathbf{I} + \{\mathbf{I} + \Delta\mathbf{L}_o(j\omega)\}\mathbf{G}(j\omega)\mathbf{C}(j\omega)]\ . \tag{24.43}$$

Employing

$$\sigma_{\min}[\mathbf{A}] - \sigma_{\max}[\mathbf{B}] \ \le \sigma_{\min}[\mathbf{A} + \mathbf{B}] \tag{24.44}$$

$$\sigma_{\max}[\mathbf{AB}] \ \le \sigma_{\max}[\mathbf{A}]\sigma_{\max}[\mathbf{B}]\ , \tag{24.45}$$

Eq.(24.43) is implied by

$$\sigma_{\max}[\Delta\mathbf{L}_o(j\omega)\mathbf{G}(j\omega)\mathbf{C}(j\omega)] < D_o(\omega) \tag{24.46}$$

$$\Leftarrow \quad \sigma_{\max}[\Delta\mathbf{L}_o(j\omega)]\ \sigma_{\max}[\mathbf{G}(j\omega)\mathbf{C}(j\omega)] < D_o(\omega) . \tag{24.47}$$

In the case of input associated uncertainty a similar result is obtained. By applying Eq.(24.41) and rearranging,

$$\sigma_{\max}[\Delta\mathbf{L}_i(j\omega)]\ \sigma_{\max}[\mathbf{G}(j\omega)]\ \sigma_{\max}[\mathbf{C}(j\omega)] < D_o(\omega) . \tag{24.48}$$

If both uncertainty models $\Delta\mathbf{L}_o$ and $\Delta\mathbf{L}_i$ are of the same matrix-valued quantity $\Delta\mathbf{L}(j\omega)$ Eq.(24.47) is implied by Eq.(24.48). Thus, applying the uncertainty radius $r_L(\omega)$ for every

$$\Delta\mathbf{L} \in \{\Delta\mathbf{L} \ :\ \sigma_{\max}[\Delta\mathbf{L}] \le r_L(\omega) \triangleq \sigma_{\max}[\mathbf{L}(j\omega)]\} \tag{24.49}$$

the following general and sufficient relation is obtained

$$r_L(\omega)\ \sigma_{\max}[\mathbf{G}(j\omega)]\ \sigma_{\max}[\mathbf{C}(j\omega)] < D_o(\omega) . \tag{24.50}$$

Normalized performance deterioration quantities can be defined if $D(\omega)$ in Eq.(24.43) is put in relation to the singular value of the actual or nominal loop return difference.

24.8 Assessment of Measurement Noise Rejection

In Fig. 24.6 a multivariable system is depicted with sensitivity $\mathbf{S}(s)$ and closed-loop transfer matrix $\mathbf{F}(s)$

$$\mathbf{S}(s) \triangleq [\mathbf{I} + \mathbf{G}(s)\mathbf{C}(s)]^{-1} \qquad \mathbf{T}(s) \triangleq \mathbf{I} - \mathbf{S}(s) = \mathbf{S}(s)\mathbf{G}(s)\mathbf{C}(s) . \tag{24.51}$$

Omitting the operator s the output $\mathbf{y}$ as a function of disturbance, reference and measurement noise is

$$\mathbf{y} = \mathbf{S}\mathbf{w} + (\mathbf{I} - \mathbf{S})\mathbf{V}\mathbf{y}_{ref} - (\mathbf{I} - \mathbf{S})\mathbf{n} = (\mathbf{I} - \mathbf{T})\mathbf{w} + \mathbf{T}\mathbf{V}\mathbf{y}_{ref} - \mathbf{T}\mathbf{n} . \tag{24.52}$$

The aim of designing control loops is to minimize $\mathbf{T}(s)$ (in some scalar sense) in order to minimize measurement noise propagation assuming that there is no need for tracking accuracy in the frequency region of measurement noise. An assessment of the measurement noise reaction in the contol loop can be obtained by using the maximum singular value (principal gain) as an upper boundary of the matrix-valued gain. From $\mathbf{T} = \mathbf{I} - \mathbf{S}$ it follows

$$\sigma_{\max}[\mathbf{T}] = \sigma_{\max}[\mathbf{I} - \mathbf{S}] \le \sigma_{\max}[\mathbf{I}] + \sigma_{\max}[-\mathbf{S}] = 1 + \sigma_{\max}[\mathbf{S}] . \tag{24.53}$$

Using Eq.(2.26) let $\mathbf{x}$ be that vector for which the supremum is obtained, i.e.,

$$\sigma_{\max}[\mathbf{S}] = \frac{\|\mathbf{S}\mathbf{x}\|_F}{\|\mathbf{x}\|_F} . \tag{24.54}$$

From $\mathbf{T} = \mathbf{I} - \mathbf{S}$ and from the triangle inequality it results

$$\|\mathbf{x} - \mathbf{S}\mathbf{x}\|_F \ge \|\mathbf{x}\|_F - \|\mathbf{S}\mathbf{x}\|_F \tag{24.55}$$

$$\|\mathbf{x} - \mathbf{S}\mathbf{x}\|_F = \| - (\mathbf{x} - \mathbf{S}\mathbf{x})\|_F = \|\mathbf{S}\mathbf{x} - \mathbf{x}\|_F \ge \|\mathbf{S}\mathbf{x}\|_F - \|\mathbf{x}\|_F . \tag{24.56}$$

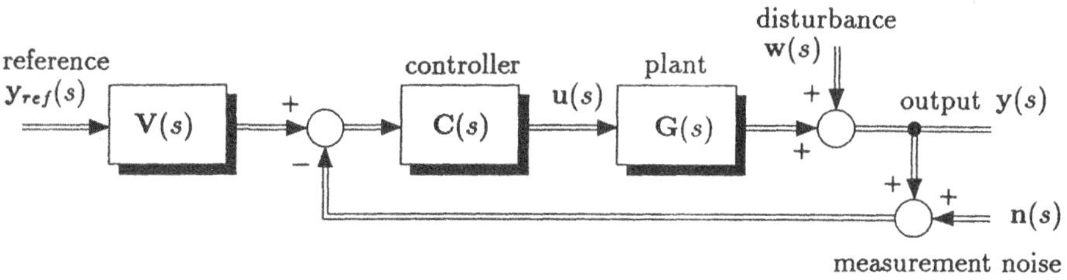

Figure 24.6: Reference, disturbance and measurement noise inputs to a multivariable control system

From both equations above

$$\|\mathbf{Tx}\|_F \geq \mid \|\mathbf{x}\|_F - \|\mathbf{Sx}\|_F \mid \qquad \mid \times 1/\|\mathbf{x}\|_F \tag{24.57}$$

$$\frac{\|\mathbf{Tx}\|_F}{\|\mathbf{x}\|_F} \geq |1 - \frac{\|\mathbf{Sx}\|_F}{\|\mathbf{x}\|_F} | \tag{24.58}$$

$$\sigma_{\max}[\mathbf{T}] \geq \mid 1 - \sigma_{\max}[\mathbf{S}] \mid . \tag{24.59}$$

Combining Eqs.(24.53) and (24.59) yields (*Maciejowski, J.M., 1989*) an upper and a lower bound for the maximum singular value of the overall closed-loop transfer matrix **T** in terms of the singular values of the plant

$$1 + \sigma_{\max}[\mathbf{S}(s)] \geq \sigma_{\max}[\mathbf{T}(s)] \geq |1 - \sigma_{\max}[\mathbf{S}(s)]| \qquad \forall s . \tag{24.60}$$

24.9 Assessment of Control Signals Magnitude

From Fig. 24.6 the control signal is

$$\mathbf{u} = \mathbf{C}(\mathbf{I} - \mathbf{T})(\mathbf{V}\mathbf{y}_{ref} - \mathbf{w} - \mathbf{n}) = (\mathbf{I} + \mathbf{CG})^{-1}\mathbf{C}(\mathbf{V}\mathbf{y}_{ref} - \mathbf{w} - \mathbf{n}) . \tag{24.61}$$

In order to keep **u** small, the maximum singular value of the premultiplication matrix above should be as small as possible

$$\sigma_{\max}[(\mathbf{I} + \mathbf{CG})^{-1}\mathbf{C}] = \frac{\sigma_{\max}[\mathbf{C}]}{\sigma_{\min}[\mathbf{I} + \mathbf{CG}]} \ll 1 \tag{24.62}$$

$$\sigma_{\max}[\mathbf{C}] \ll \sigma_{\min}[\mathbf{I} + \mathbf{CG}] \leq \sigma_{\max}[\mathbf{I} + \mathbf{CG}] \leq 1 + \sigma_{\max}[\mathbf{C}]\sigma_{\max}[\mathbf{G}] \tag{24.63}$$

$$1 \ll \frac{1}{\sigma_{\max}[\mathbf{C}]} + \sigma_{\max}[\mathbf{G}] . \tag{24.64}$$

Thus, a sufficient condition for keeping the control signals small is to choose $\sigma_{\max}[\mathbf{C}]$ small in that range of frequency where $\sigma_{\max}[\mathbf{G}]$ is not large. This depends upon the spectral richness of the noise signal. The control signals being excited by the reference signal must be assessed including the prefilter transfer matrix **V**.

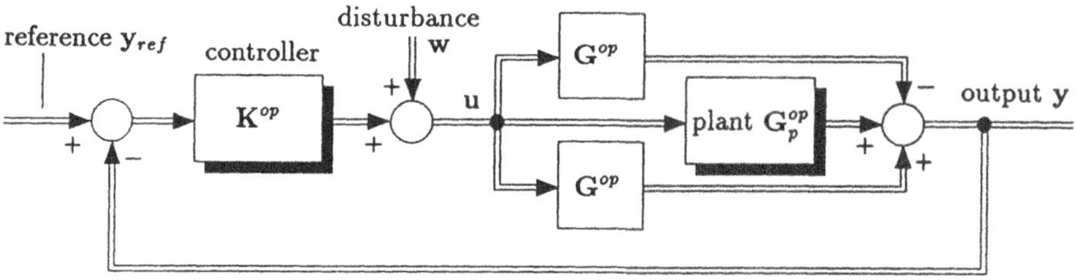

Figure 24.7: Multivariable control systems with additional $\mathbf{G}^{op}$

24.10 Robust Dynamic Controller Based on Induced Gain and Singular Values

24.10.1 Plant Perturbation and Stability Condition

Using functional analysis methods and the Small gain theorem, the block diagram of Fig. 24.7 is described by

$$\mathbf{u} = \mathbf{w} + \mathbf{K}^{op}(\mathbf{y}_{ref} - \mathbf{G}_p^{op}\mathbf{u}) = \mathbf{w} + \mathbf{K}^{op}\mathbf{y}_{ref} - \mathbf{K}^{op}\mathbf{G}_p^{op}\mathbf{u} \,. \tag{24.65}$$

L_p^n-stability is guaranteed if

$$\gamma[\mathbf{G}_p^{op}]\ \gamma[\mathbf{K}^{op}] < 1 \tag{24.66}$$

where $\gamma[\cdot]$ is the induced gain associated with an operator within the brackets. By adding and subtracting an additional element $\mathbf{G}^{op}$ to the perturbed plant operator $\mathbf{G}_p^{op}$, namely in parallel to the plant, the control system is transformed (*Zames, G., 1966*) into the block diagram of Fig. 24.8 (*Zafiriou, E., and Morari, M., 1985*). The operator $\mathbf{G}^{op}$ is the nominal operator of the plant and is assumed asymptotically stable, proper and rational. Then, $\mathbf{G}_p^{op} - \mathbf{G}^{op} = \Delta\mathbf{G}^{op}$ corresponds to the plant perturbation or tolerance. From Eq.(24.66) the equivalent stability condition is

$$\gamma[\Delta\mathbf{G}^{op}]\ \gamma[(\mathbf{I}+\mathbf{K}\mathbf{G})^{-1}\mathbf{K}] < 1 \,. \tag{24.67}$$

With regard to $r = m$ the matrices $\mathbf{K}, \mathbf{G}_p, \mathbf{G} \in \mathcal{C}^{m \times m}$ are associated with the operators $\mathbf{K}^{op}, \mathbf{G}_p^{op}$ and $\mathbf{G}^{op}$, respectively. Referring to Eq.(20.90), the condition

$$\sigma_{\max}[\Delta\mathbf{G}]\ \sigma_{\max}[(\mathbf{I}_m + \mathbf{K}\mathbf{G})^{-1}\mathbf{K}] < 1 \quad \forall\omega = [0, \infty) \tag{24.68}$$

is a sufficient stability condition for Eq.(24.67) preassuming that the nominal system $(\mathbf{I}_m + \mathbf{K}\mathbf{G})^{-1}$ is stabilized by $\mathbf{K}$. More conservative than the above inequality is

$$\sup_{\omega}\ \sigma_{\max}[\Delta\mathbf{G}]\ \sup_{\omega}\sigma_{\max}[(\mathbf{I}_m + \mathbf{K}\mathbf{G})^{-1}\mathbf{K}] < 1 \tag{24.69}$$

because $\sup_\omega$ is selected separately.

If an additive plant variation $\Delta\mathbf{G}$ is not adequate a multiplicative one may be used. The multiplicative perturbation can be associated with the output or with the input and is termed $(\mathbf{I} + \Delta\mathbf{L}_o)\mathbf{G}$ and $\mathbf{G}(\mathbf{I} + \Delta\mathbf{L}_i)$, respectively. Note the simple relation

$$\Delta\mathbf{G} = \Delta\mathbf{L}_o\ \mathbf{G} = \mathbf{G}\ \Delta\mathbf{L}_i \,. \tag{24.70}$$

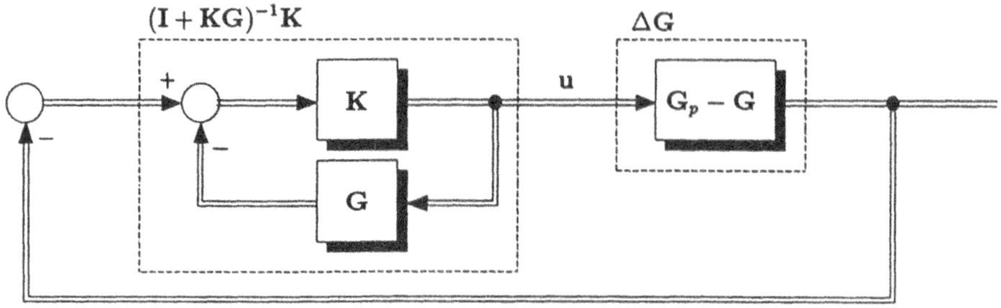

Figure 24.8: Transformed multivariable control system

Substitution into Eq.(24.68) yields for output perturbation

$$\sigma_{\max}[\Delta \mathbf{L}_o]\ \sigma_{\max}[\mathbf{G}(\mathbf{I}_m + \mathbf{KG})^{-1}\mathbf{K}] < 1 \qquad \forall\omega \tag{24.71}$$

$$\Leftarrow \quad \sup_\omega\ \sigma_{\max}[\Delta \mathbf{L}_o]\ \sup_\omega \sigma_{\max}[\mathbf{G}(\mathbf{I}_m + \mathbf{KG})^{-1}\mathbf{K}] < 1 \tag{24.72}$$

and for input perturbation

$$\sigma_{\max}[\Delta \mathbf{L}_i]\ \sigma_{\max}[(\mathbf{I}_m + \mathbf{KG})^{-1}\mathbf{KG}] < 1 \qquad \forall\omega \tag{24.73}$$

$$\Leftarrow \quad \sup_\omega\ \sigma_{\max}[\Delta \mathbf{L}_i]\ \sup_\omega \sigma_{\max}[(\mathbf{I}_m + \mathbf{KG})^{-1}\mathbf{KG}] < 1\ . \tag{24.74}$$

The terms

$$\sup_\omega\ \sigma_{\max}[\Delta \mathbf{L}_i] \qquad \text{and} \qquad \sup_\omega\ \sigma_{\max}[\Delta \mathbf{L}_o] \tag{24.75}$$

are denoted *associated degree of robustness (Sevaston, G.E., and Longman, R.W., 1988)*.

24.10.2 Optimal Robust Dynamic Feedback Controller

Consider a plant and a dynamic controller as depicted in Fig. 24.9. Combining plant and controller equation yields

$$\begin{pmatrix}\dot{\mathbf{x}}\\ \dot{\mathbf{z}}\end{pmatrix} = \begin{pmatrix}\mathbf{A} + \mathbf{BK}_r\mathbf{C} & \vdots & \mathbf{BC}_r\\ \mathbf{B}_r\mathbf{C} & & \mathbf{A}_r\end{pmatrix}\begin{pmatrix}\mathbf{x}\\ \mathbf{z}\end{pmatrix} + \begin{pmatrix}\mathbf{B}\\ \mathbf{0}\end{pmatrix}\mathbf{v}_{ref} \triangleq \bar{\mathbf{A}}\begin{pmatrix}\mathbf{x}\\ \mathbf{z}\end{pmatrix} + \begin{pmatrix}\mathbf{B}\\ \mathbf{0}\end{pmatrix}\mathbf{v}_{ref} \tag{24.76}$$

$$\text{where} \quad \mathbf{x} \in \mathcal{R}^n; \quad \mathbf{v}_{ref}, \mathbf{v}, \mathbf{u} \in \mathcal{R}^m; \quad \mathbf{y} \in \mathcal{R}^r; \quad \mathbf{z} \in \mathcal{R}^{n_z}\ . \tag{24.77}$$

The objective is to minimize a quadratic index of performance I under the functional inequality constraint given by the tolerance matrix $\Delta\mathbf{G}(s)$. Using the weighting matrices $\mathbf{R} > 0$; $\mathbf{Q}, \mathbf{Q}_z \geq 0$ the index I is considered the expectation over all possible initial vectors $\mathbf{x}_o$

$$I = E\{\int_o^\infty [\mathbf{x}^T(t)\mathbf{Q}\mathbf{x}(t) + \mathbf{u}^T(t)\mathbf{R}\mathbf{u}(t) + \mathbf{z}^T(t)\mathbf{Q}_z\mathbf{z}(t)]dt\}\ . \tag{24.78}$$

Assume that the tolerance $\Delta\mathbf{G}$ of the plant matrix is bounded by the scalar-valued function $\alpha(\omega)$

$$\sigma_{\max}[\Delta\mathbf{G}(j\omega)] = \sigma_{\max}[\mathbf{G}_p(s) - \mathbf{G}(s)] < \alpha(\omega) \quad \forall\ \omega = [0, \infty)\ . \tag{24.79}$$

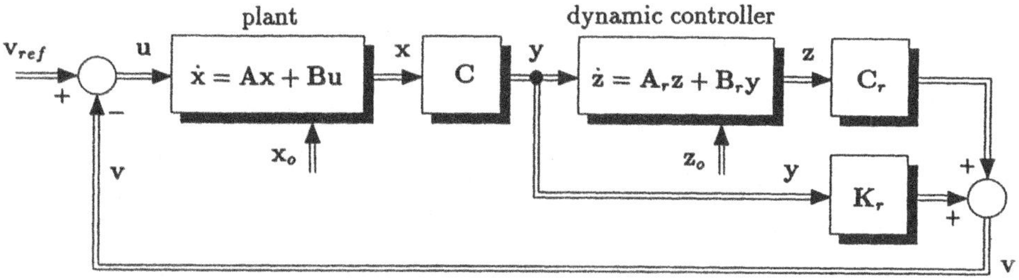

Figure 24.9: Plant and dynamic output feedback controller

24.10.3 Unconstrained Optimization Algorithm

The results given by state matrices, see Eqs.(16.12), (16.83),(16.90), are applied to the subject of this section. Define block diagonal matrices

$$\bar{\mathbf{C}} \triangleq \text{block diag } \{\mathbf{C}, \mathbf{I}\}, \quad \bar{\mathbf{B}} \triangleq \text{block diag } \{\mathbf{B}, \mathbf{I}\}, \quad \bar{\mathbf{R}} \triangleq \text{block diag } \{\mathbf{R}, \mathbf{0}\}, \tag{24.80}$$

$$\bar{\mathbf{Q}} \triangleq \text{block diag } \{\mathbf{Q}, \mathbf{Q}_z\} + \bar{\mathbf{C}}^T \bar{\mathbf{K}}_r^T \bar{\mathbf{R}} \bar{\mathbf{K}}_r \bar{\mathbf{C}} \qquad \bar{\mathbf{K}}_r \triangleq \begin{pmatrix} \mathbf{K}_r & \mathbf{C}_r \\ -\mathbf{B}_r & -\mathbf{A}_r \end{pmatrix} . \tag{24.81}$$

Using unknown matrices $\mathbf{L}$ and $\mathbf{\Lambda}$, the unconstrained problem of $I \to$ min can be rewritten to the problem of minimizing a modified index I_a

$$I_a = \text{tr } [E\{\begin{pmatrix} \mathbf{x}_o \\ \mathbf{z}_o \end{pmatrix}(\mathbf{x}_o^T \vdots \mathbf{z}_o^T)\}\mathbf{L} + \mathbf{\Lambda}(\bar{\mathbf{A}}\mathbf{L} + \mathbf{L}\bar{\mathbf{A}}^T + \bar{\mathbf{Q}})] \tag{24.82}$$

$$\mathbf{A}_r \in \mathcal{R}^{n_z \times n_z}, \quad \mathbf{B}_r \in \mathcal{R}^{n_z \times r}, \quad \mathbf{C}_r \in \mathcal{R}^{m \times n_z}, \quad \mathbf{z}_o \in \mathcal{R}^{n_z}, \tag{24.83}$$

$$\mathbf{K}_r \in \mathcal{R}^{m \times r}, \quad \bar{\mathbf{K}}_r \in \mathcal{R}^{(m+n_z) \times (r+n_z)}, \quad \bar{\mathbf{A}}, \mathbf{\Lambda}, \mathbf{L} \in \mathcal{R}^{(n+n_z) \times (n+n_z)} . \tag{24.84}$$

24.10.4 Robust Optimization

The solution satisfying both min I_a and $\sigma[\Delta \mathbf{G}] < \alpha(\omega)$ is obtained by a two-stage feasible direction search as presented by *Sevaston, G.E., and Longman, R.W., 1988.*

First stage: Taking the $\Delta \mathbf{G}_i$ case into consideration, only, steer $\bar{\mathbf{K}}_r$ so that Eq.(24.74) is satisfied. This can be obtained by modifying $\bar{\mathbf{K}}_r$ at the ith step with the step size Δ_i according to

$$\bar{\mathbf{K}}_{r,i+1} = \bar{\mathbf{K}}_{r,i} - \frac{\partial \sigma[\mathbf{F}]}{\partial \bar{\mathbf{K}}_r}\Delta_i \qquad \mathbf{F} = (\mathbf{I}_m + \mathbf{K}\mathbf{G})^{-1}\mathbf{K}\mathbf{G} . \tag{24.85}$$

The matrices $\mathbf{K}(j\omega)$ and $\mathbf{G}(j\omega)$ in $j\omega$-domain are the dynamic controller and nominal plant transfer function, respectively. The derivative of the singular value $\sigma[\mathbf{F}]$ with respect to $\bar{\mathbf{K}}_r$ can be expressed with the help of singular vectors. The singular value may be replaced by the Frobenius norm for computational ease.

Second stage: Calculating the gradients of I_a with respect to $\mathbf{L}$ and $\mathbf{\Lambda}$ and equating with zero yields a first-order approximation of the necessary optimality condition. Then,

$\bar{\mathbf{K}}_r$ and $\mathbf{z}_o$ are modified according to

$$\bar{\mathbf{K}}_{r,i+1} = \bar{\mathbf{K}}_{r,i} - \Big(\frac{\partial I_a}{\partial \bar{\mathbf{K}}_r}\Big)_i \Delta_i \qquad \mathbf{z}_{o,i+1} = \mathbf{z}_{o,i} - \Big(\frac{\partial I_a}{\partial \mathbf{z}_o}\Big)_i \Delta_i \, . \tag{24.86}$$

By the preceding matrix/vector equations, as many scalar equations are provided as unknowns. Returning to the first stage, the preceding algorithm is repeated in a computer-aided version until some stopping condition is achieved. Interactive supervision of the search direction and step size enables the engineer to improve robustness or system performance according to his experience.

24.11 Robust Minimum Norm Controller

Design a minimum norm controller $\mathbf{K}$ for an additively perturbed system $\mathbf{A}$ such that $\|\mathbf{WK}\| \to \min$ subject to $\Theta \neq 0$ where $\|\cdot\|$ is any norm, $\mathbf{W}$ is a weighting matrix and $\delta = -\alpha_s[\mathbf{A}_p + \mathbf{BK}] > 0$ equals the maximum of the negative value of the spectral abscissa of the perturbed closed-loop control system. Finally, for a predetermined δ and $\mathbf{E}$ ($\geq_e \Delta\mathbf{A}$) (*Weinmann, A., 1991*)

$$\Theta = \det[s\mathbf{I} - \mathbf{A} - \mathbf{BK} - \Delta\mathbf{A}] \mid_{s=-\delta+j\omega} \neq 0 \tag{24.87}$$

$$\Leftrightarrow \qquad \sigma_{\min}[\mathbf{I} - (s\mathbf{I} - \mathbf{A} - \mathbf{BK})^{-1}\Delta\mathbf{A}] \mid_{s=-\delta+j\omega} \geq 0 \tag{24.88}$$

$$\Leftarrow \qquad 1 - \sigma_{\max}[(s\mathbf{I} - \mathbf{A} - \mathbf{BK})^{-1}] \mid_{s=-\delta+j\omega} \sigma_{\max}[\Delta\mathbf{A}] \geq 0 \tag{24.89}$$

$$\sigma_{\max}[\mathbf{E}] \leq \sigma_{\min}[s\mathbf{I} - \mathbf{A} - \mathbf{BK}] \mid_{s=-\delta+j\omega} . \tag{24.90}$$

It was preassumed that open-loop poles do not coincide with closed-loop ones at $-\delta + j\omega$ and that the number of unstable open-loop poles of $\mathbf{A}$ may not be changed by $\Delta\mathbf{A}$.

24.12 Uncertainty Assessment Via Internal Parallel Model

24.12.1 Equivalent Internal Parallel Model

Consider a multivariable system as given in Fig. 24.10a where $\mathbf{C}(s)$ is the controller and $\mathbf{G}_p(s)$ the perturbed plant. Omitting the argument s,

$$\mathbf{y} = (\mathbf{I} + \mathbf{G}_p\mathbf{C})^{-1}(\mathbf{G}_p\mathbf{C}\mathbf{y}_{ref} + \mathbf{w}) \tag{24.91}$$

it can easily be shown that the control system as given in Fig. 24.10b is equivalent to the system in Fig. 24.10a. The matrix $\mathbf{G}$ is the unperturbed plant (or model of the nominal plant) and $\mathbf{K}(s)$ is another kind of controller. Both representations are equivalent if

$$\mathbf{C} = \mathbf{K}(\mathbf{I} - \mathbf{GK})^{-1} \equiv (\mathbf{I} - \mathbf{KG})^{-1}\mathbf{K} \quad \text{or} \quad \mathbf{K} = (\mathbf{I} + \mathbf{CG})^{-1}\mathbf{C} \equiv \mathbf{C}(\mathbf{I} + \mathbf{GC})^{-1} \, . \tag{24.92}$$

Eq.(24.92) and Fig. 24.10b can be obtained from Fig 24.10a by adding *and* subtracting the effect of the controlling variable $\mathbf{u}$ on the measurement signal $\mathbf{y}$. Adding is done at an internal summing point within the block element $\mathbf{K}(s)$, subtracting at the summing point S_1. In the system structure in Fig. 24.10b as originally given by *Morari, M., 1983* the perturbed plant $\mathbf{G}_p$ is considered as an unperturbed plant and $\mathbf{G}$ as the transfer matrix of an uncertain model and the assessment of the uncertainty to the closed-loop behaviour is examined.

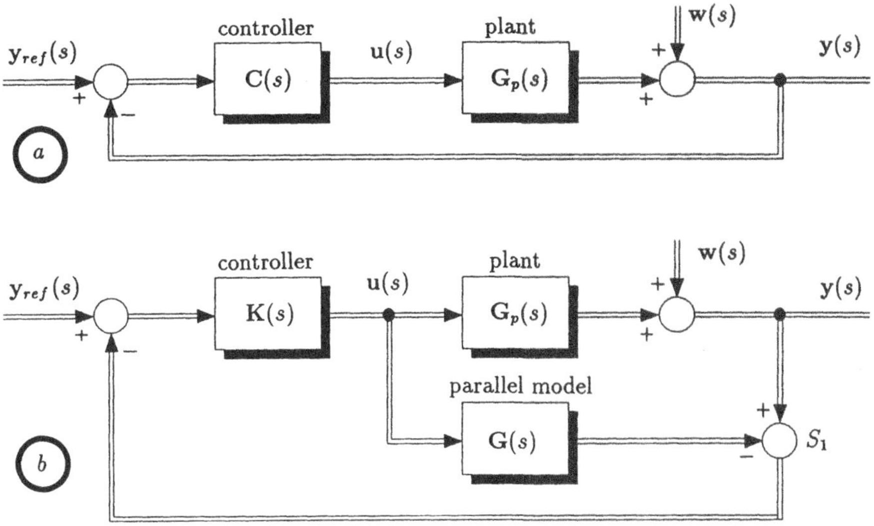

Figure 24.10: Multivariable control (a) and equivalent system using a parallel internal model (b)

24.12.2 Closed-Loop Controller Transfer Matrix

From Fig.24.10a it follows $\mathbf{C}(\mathbf{y}_{ref} - \mathbf{y}) = \mathbf{u}$ and $\mathbf{G}_p\mathbf{u} + \mathbf{w} = \mathbf{y}$ or

$$\mathbf{u} = (\mathbf{I} + \mathbf{C}\mathbf{G}_p)^{-1}\mathbf{C}(\mathbf{y}_{ref} - \mathbf{w}) \equiv \mathbf{C}(\mathbf{I} + \mathbf{G}_p\mathbf{C})^{-1}(\mathbf{y}_{ref} - \mathbf{w}) \ . \tag{24.93}$$

The matrix $\mathbf{C}(\mathbf{I} + \mathbf{G}_p\mathbf{C})^{-1}$ is denoted closed-loop controller transfer matrix. It is a well-known fact that a large controller matrix $\mathbf{C}$ provides perfect control. Neglecting realizability constraints and assuming suitable dimensions and $\mathbf{C}$ large then the expression $\mathbf{C}(\mathbf{I} + \mathbf{G}_p\mathbf{C})^{-1}$ yields $\mathbf{G}_p^{-1}$. From Eq.(24.93) it results

$$\mathbf{u} = \mathbf{G}_p^{-1}(\mathbf{y}_{ref} - \mathbf{w}) \qquad \text{or} \qquad \mathbf{G}_p\mathbf{u} = \mathbf{y}_{ref} - \mathbf{w} \ . \tag{24.94}$$

From Fig.24.10b $\mathbf{G}_p\mathbf{u} = \mathbf{y} - \mathbf{w}$ is obtained. Comparison shows that $\mathbf{y} = \mathbf{y}_{ref}$, irrespective of the disturbance $\mathbf{w}$. Obviously, this is a perfect control.

24.12.3 Stability

If the plant equals the model exactly, i.e. $\mathbf{G}_p = \mathbf{G}$, then $\mathbf{K} = (\mathbf{I} + \mathbf{C}\mathbf{G}_p)^{-1}\mathbf{C}$. Invoking Eq.(25.5), the closed-loop multivariable system turns out open-loop, see Fig. 24.10b when $\mathbf{G}_p = \mathbf{G}$, and the system is stable if both $\mathbf{G}_p$ and $\mathbf{K}$ are stable.

24.12.4 Realizability

Those system characteristics which prohibit to make the closed-loop controller transfer matrix equal to $\mathbf{G}_p^{-1}$ can serve as so-called resilience measures. Resilience describes the ability of the plant to move fast but smoothly and to provide disturbance rejection (*Morari, M., 1983*). Since $\mathbf{G}_p$ is unavailable the optimal controller $\mathbf{K}$ can only be made close to $\mathbf{G}^{-1}$. With regard to the realizability of the inverse, only the invertible part of $\mathbf{G}$ may be considered. There are three factors responsible for limited resilience: Non-minimum phase elements (right-half-plane zeros), physical constraints on the controlling variable and perturbation of the plant (or model/plant mismatch). Consider the case that the model or unperturbed plant contains dead time delay and/or right-half-plane zeros. Then, $\mathbf{G}$ is factorized into a noninvertible and an invertible part

$$\mathbf{G} = \mathbf{G}_n \mathbf{G}_i \ , \tag{24.95}$$

in this order. The choice $\|\mathbf{G}_n\|_s = 1$ proves optimal, see Eq.(24.96). Choosing $\mathbf{K} = \mathbf{G}_i^{-1}$, from Fig.24.10b and Eqs.(24.93) and (24.92) it follows with $\mathbf{G}_p = \mathbf{G}$

$$\mathbf{y} - \mathbf{w} = \mathbf{G}\mathbf{K}(\mathbf{y}_{ref} - \mathbf{w}) = \mathbf{G}_n \mathbf{G}_i \mathbf{G}_i^{-1}(\mathbf{y}_{ref} - \mathbf{w}) = \mathbf{G}_n(\mathbf{y}_{ref} - \mathbf{w}) \ . \tag{24.96}$$

Hence, the closed-loop transfer function is given by $\mathbf{G}_n$. To achieve $\mathbf{y}$ near $\mathbf{y}_{ref}$, $\|\mathbf{G}_n\|_s$ close to unity is appropriate.

24.12.5 Control Variable Constraint

From Fig.24.10b if $\mathbf{G} = \mathbf{G}_p$

$$\mathbf{u} = \mathbf{K}(\mathbf{y}_{ref} - \mathbf{w}) = \mathbf{G}^{-1}(\mathbf{y}_{ref} - \mathbf{w}) \quad \leadsto \quad (\mathbf{y}_{ref} - \mathbf{w}) = \mathbf{G}\mathbf{u} \ . \tag{24.97}$$

In order to obtain $\|\mathbf{u}\|_F \leq \|\mathbf{u}\|_{F\max}$ there results the sufficient condition

$$\|\mathbf{y}_{ref} - \mathbf{w}\|_F \leq \sigma_{\min}[\mathbf{G}] \ \|\mathbf{u}\|_{F\max} \ . \tag{24.98}$$

The controlling variable $\mathbf{u}$ can be kept within the bound $\|\mathbf{u}\|_{F\ \max}$ if the reference and disturbance are bounded according to the inequality above.

The minimum singular value of $\mathbf{G}$ plays the role of a gain factor for the multivariable system from the controlling variable bound to the admissible input bounds. Following given constraints on the controlling variable $\mathbf{u}$, good control performance can be obtained if the minimum singular value of the nominal plant $\mathbf{G}$ is large.

24.12.6 Model/Plant Mismatch and Control Quality

Assume that the uncertainty of the plant is given by an output uncertainty. Then,

$$\mathbf{G}_p(s) = [\mathbf{I} + \Delta\mathbf{L}_o(s)]\mathbf{G}(s) \tag{24.99}$$

and Fig. 24.10b can be redrawn to Fig. 24.11. From Eq.(24.99) for bounded uncertainty

$$\frac{\|\mathbf{G}_p(j\omega) - \mathbf{G}(j\omega)\|_s}{\|\mathbf{G}(j\omega)\|_s} \leq \|\Delta\mathbf{L}_o(j\omega)\|_s \leq r_o(\omega) \ . \tag{24.100}$$

Stability of the system requires $\forall\omega$

$$\|\Delta\mathbf{L}_o(j\omega)\mathbf{G}(j\omega)\mathbf{K}(j\omega)\|_s = \|\Delta\mathbf{L}_o(j\omega)\mathbf{G}_n(j\omega)\mathbf{G}_i(j\omega)\mathbf{K}(j\omega)\|_s < 1 \ . \tag{24.101}$$

Let $\mathbf{K} = \mathbf{G}_i^{-1}$. Referring to $\|\mathbf{G}_n(j\omega)\|_s = 1$, Eq.(24.101) is implied by

$$\|\Delta\mathbf{L}_o(j\omega)\|_s \leq r_o(\omega) < 1 \ . \tag{24.102}$$

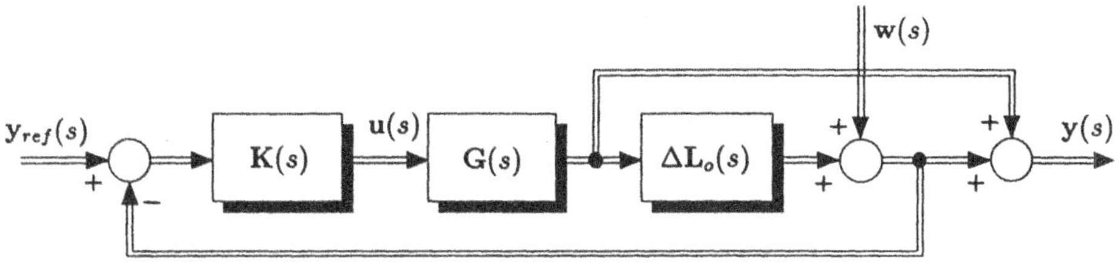

Figure 24.11: Plant with output uncertainty

Hence, the uncertainty radius $r_o(\omega)$ may not exceed unity. However, for practical plants $r_o(\omega)$ may be expected beyond unity for high frequencies. Then, the controller $\mathbf{K}(s)$ must be detuned for high frequencies. This can be obtained by establishing

$$\mathbf{K}(s) = \mathbf{G}_i^{-1}(s)\mathbf{K}_f(s) \tag{24.103}$$

where $\mathbf{K}_f(s)$ is a low-pass filter. From left-hand side of Eq.(24.101) and Eq.(24.103)

$$\|\Delta\mathbf{L}_o(j\omega)\|_s\|\mathbf{G}(j\omega)\|_s\|\mathbf{G}_i^{-1}(j\omega)\|_s\|\mathbf{K}_f(j\omega)\|_s < 1 \ . \tag{24.104}$$

Since

$$\|\mathbf{G}\|_s \le \|\mathbf{G}_n\|_s\|\mathbf{G}_i\|_s \qquad \leadsto \qquad \|\mathbf{G}_i^{-1}\|_s \le \|\mathbf{G}^{-1}\|_s = 1/\sigma_{\min}[\mathbf{G}] \tag{24.105}$$

the inequality above is implied by (*Morari, M., 1983a*)

$$r_o(\omega)\ \|\mathbf{G}(j\omega)\|_s\ \sigma_{\min}^{-1}[\mathbf{G}]\ \|\mathbf{K}_f(j\omega)\|_s < 1 \tag{24.106}$$

$$\kappa_s(\omega) \triangleq \frac{\sigma_{\max}[\mathbf{G}(j\omega)]}{\sigma_{\min}[\mathbf{G}(j\omega)]} < \frac{1}{\|\mathbf{K}_f(j\omega)\|_s\ r_o(\omega)} \ . \tag{24.107}$$

If $\mathbf{G}(j\omega)$ is ill-conditioned and if $r_o(\omega)$ is large, especially at high frequencies, the inequality above forces $\|\mathbf{K}_f(j\omega)\|_s$ to be small. This restriction reduces the overall transfer matrix $\mathbf{G}_n(s)\mathbf{K}_f(s)$ to be obtained from Eq.(24.96). Reduction of the norm $\|\mathbf{G}_n(s)\mathbf{K}_f(s)\|_s$ below unity affects good closed-loop performance at high frequencies.

From Eq.(24.107), examining frequency-dependent and high condition numbers it results: At high frequencies the spectral norm of the low-pass filter, $\sigma_{\max}[\mathbf{K}_f]$, must be reduced to ensure stability, especially when $r_o(\omega)$ is not a small number. In other words, control quality must be sacrificed in order to guarantee stability.

Emphasizing, a plant is amenable to good control if (i) the condition number $\kappa_s(\omega)$ is low over a wide frequency range, preferably lower than 10, see Eq.(24.107), and (ii) if the minimum singular value of the plant is large, see Eq.(24.98).

24.12.7 Improved Process Conditioning

If the number of manipulatable variables of the plant is higher than the number of specified measured variables the conditioning may be improved considerably. Consider e.g. three controlling variables u_1, u_2, u_3 and two output variables y_1, y_2. Then, there exist three feasible sets of controlling variables $(u_1, u_2), (u_1, u_3)$ and (u_2, u_3). Preferable for the best control performance is that pair which guarantees both a low condition number and a large minimum singular value over a wide frequency range.

Further improvement can be obtained by augmenting the control structure. If there are excess manipulatable variables available then some state variables can be fed back to that input not being used in the control structure (*Johnston, R.D., and Barton, G.W., 1984*) thus changing the plant dynamics. Appropriate feedback may cause considerable system improvement.

Chapter 25

Generalized Nyquist Stability of Perturbed Systems

Analysis and design of control systems in the frequency domain is one of the best known methods in control engineering. The stability of single-input single-output systems is investigated via transfer function calculus and Nyquist criterion (*Leithead, W.E., and O'Reilly, J., 1991*). Multi-input multi-output systems require the transfer matrix operations and characteristic loci. When plants are perturbed characteristic loci can be still used if they are augmented by E-contours or structured E-contours.

25.1 Stability of Multivariable Feedback Systems

The multivariable unit feedback system of Fig. 25.1 is stable if the overall transfer matrix

$$\mathbf{F}(s) = [\mathbf{I} + \mathbf{G}(s)]^{-1}\mathbf{G}(s) = \mathbf{G}(s)[\mathbf{I} + \mathbf{G}(s)]^{-1} = \frac{\mathrm{adj}[\mathbf{I} + \mathbf{G}(s)]}{\det[\mathbf{I} + \mathbf{G}(s)]}\,\mathbf{G}(s) \qquad (25.1)$$

does not contain any pole in the closed right-half s-plane. This is true if neither $n(s)$ nor $m(s)$ has a right-half s-plane zero. The polynomial $n(s)$ results from

$$\det[\mathbf{I} + \mathbf{G}(s)] \triangleq \frac{n(s)}{d(s)} \qquad (25.2)$$

where $n(s)$ and $d(s)$ are coprime. The polynomial $m(s)$ is the least common denominator of all components of $[\mathrm{adj}(\mathbf{I} + \mathbf{G})]\mathbf{G}$

$$\frac{\mathbf{F}_1(s)}{m(s)} \triangleq [\mathrm{adj}(\mathbf{I} + \mathbf{G})]\mathbf{G}(s) \qquad (25.3)$$

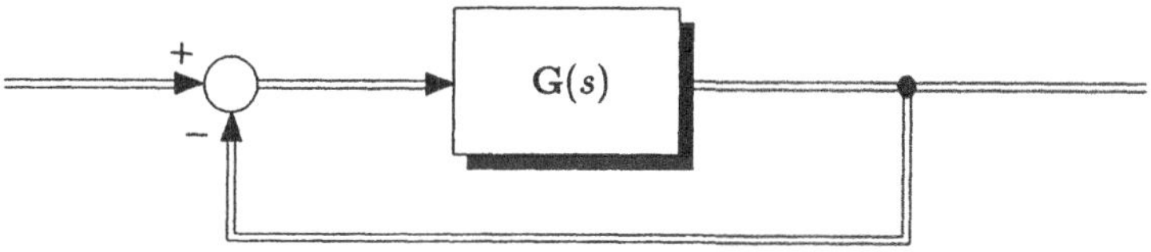

Figure 25.1: Multivariable unit feedback system

where $\mathbf{F}_1(s)$ is a polynomial matrix. Hence,

$$\mathbf{F}(s) = \frac{d(s)}{n(s)m(s)}\,\mathbf{F}_1(s)\ . \tag{25.4}$$

Let $\mathbf{G}(s)$ be a proper matrix. Then, the characteristic polynomial $g(s)$ of $\mathbf{G}(s)$ is defined to be the least common denominator of all the minors of $\mathbf{G}(s)$. The characteristic polynomial $g(s)$ is divisible by $d(s)$ without remainder (*Chen, C.T., 1968*)

$$n_1(s) \triangleq g(s)/d(s)\ . \qquad \text{Hence,} \quad \mathbf{F}(s) = \frac{g(s)}{n_1(s)\,n(s)\,m(s)}\mathbf{F}_1(s)\ . \tag{25.5}$$

The closed-loop system of Fig. 25.1 is stable if all the zeros of $n(s)$ and $n_1(s)$ are in the open left-half s-plane. If the open-loop system matrix $\mathbf{G}(s)$ is a stable matrix then no zeros of $n_1(s)$ exist. The zeros of $n_1(s)$ determine modes both of the open-loop and the closed-loop system. The zeros of $n(s)$ can be checked by the Nyquist plot, i.e., the encirclements of the Nyquist plot $\det[\mathbf{I} + \mathbf{G}(s)] \quad s = j\omega$ round the origin. The stability boundary of the system is characterized by $\det[\mathbf{I}+\mathbf{G}(j\omega)] = 0$. To avoid system behaviour like this, the condition

$$\det[\mathbf{I} + \mathbf{G}(s)] \neq 0 \qquad s = j\omega \tag{25.6}$$

must be satisfied.

The stability of a dynamic system as depicted in Fig. 25.1, comprising a nominal plant transfer matrix $\mathbf{G}(s) \in \mathcal{C}^{m\times m}$ and unity feedback $\mathbf{I}_m$, is based on the condition that $\mathbf{G}(s)$ does not contain $-1 + j0$ as an eigenvalue, i.e., the matrix $\mathbf{I} + \mathbf{G}(s)$ is nonsingular or $\det[\mathbf{I} + \mathbf{G}(s)] \neq 0\ |_{s=j\omega} \quad \forall\, \omega$.

If one of the eigenvalues of $\mathbf{G}(s)$ intersects the critical point $(-1+j0)$, in any case the stability of the closed-loop multivariable system is affected, see Eq.(25.103). Hence, the inequality Expr.(25.6) is posed.

25.2 Perturbations

If $\mathbf{G}(s)$ is perturbed to $\mathbf{G}_p(s)$, i.e. either

$$\mathbf{G}_p(s) = \mathbf{G}(s) + \Delta\mathbf{G}(s) \qquad \text{or} \qquad \mathbf{G}_p(s) = \mathbf{G}(s)[\mathbf{I} + \Delta\mathbf{L}_i(s)]\ , \tag{25.7}$$

then Eq.(25.6) is satisfied if both $\mathbf{G}(s)$ is stable and the perturbation $\Delta\mathbf{G}(s)$ or $\Delta\mathbf{L}_i$ are sufficiently small.

25.2.1 Additive Perturbation

In the case of additive perturbation $\Delta\mathbf{G}$

$$\det[\mathbf{I} + \mathbf{G} + \Delta\mathbf{G}] = \det(\mathbf{I} + \mathbf{G})[\mathbf{I} + (\mathbf{I} + \mathbf{G})^{-1}\Delta\mathbf{G}] \tag{25.8}$$

$$= \det(\mathbf{I} + \mathbf{G})\det[\mathbf{I} + (\mathbf{I} + \mathbf{G})^{-1}\Delta\mathbf{G}] \neq 0 \tag{25.9}$$

$$(\mathbf{I} + \mathbf{G})^{-1}\Delta\mathbf{G} \neq -\mathbf{I}\ . \tag{25.10}$$

Note that $\mathbf{A} \neq \mathbf{I}$ is satisfied if $\sigma_{\max}[\mathbf{A}] \neq \sigma_{\max}[\mathbf{I}]$ but not only if $\sigma_{\max}[\mathbf{A}] \neq \sigma_{\max}[\mathbf{I}]$. To satisfy $\mathbf{A} = \mathbf{I}$ a necessary condition is $\sigma_{\max}[\mathbf{A}] = \sigma_{\max}[\mathbf{I}] = 1$. □

The inequality Eq.(25.10) is implied by

$$\sigma_{\max}[(\mathbf{I}+\mathbf{G})^{-1}\,\Delta\mathbf{G}] < 1\ . \tag{25.11}$$

Expr.(25.11) is a sufficient condition for guaranteeing the inequality of Eq.(25.10): If the maximum singular value does not match unity then every singular value is smaller than unity. Referring to Eq.(24.45), the Eq.(25.11) is implied by

$$\sigma_{\max}[(\mathbf{I}+\mathbf{G})^{-1}]\,\sigma_{\max}[\Delta\mathbf{G}] < 1 \tag{25.12}$$

$$\sigma_{\max}[\Delta\mathbf{G}] < \frac{1}{\sigma_{\max}[(\mathbf{I}+\mathbf{G})^{-1}]} = \sigma_{\min}[\mathbf{I}+\mathbf{G}]\ . \tag{25.13}$$

When the only information available about $\Delta\mathbf{G}$ is $\sigma_{\max}[\Delta\mathbf{G}]$, the condition Eq.(25.13) provides the least conservative possible characterization of the stability margin (*Safonov, M.G., 1982; Postlethwaite, I., and Foo, Y.K., 1985*).

Let $\mathbf{G}(j\omega)$ be decomposed into $\mathbf{G}_1(j\omega)\mathbf{G}_2(j\omega)$ and consider only $\mathbf{G}_2$ additively perturbed. Then, the robust stability condition is

$$\sigma_{\max}[\Delta\mathbf{G}_2] < \sigma_{\min}[\mathbf{G}_1^{-1}+\mathbf{G}_2]\ . \tag{25.14}$$

Suppose $\mathbf{G}$ is substituted by its state-space representation $-(s\mathbf{I}-\mathbf{A})^{-1}\mathbf{BK}$ and only $\mathbf{A}$ is considered additively perturbed then the result is (*Lee, W.H., et al. 1982*)

$$\sigma_{\max}[\Delta\mathbf{A}] < \sigma_{\min}[j\omega\mathbf{I}-\mathbf{A}-\mathbf{BK}]\ . \tag{25.15}$$

25.2.2 Multiplicative Perturbation

In the case of multiplicative and input associated perturbation and for stable $\mathbf{G}$

$$\det[\mathbf{I}+\mathbf{G}(\mathbf{I}+\Delta\mathbf{L}_i)] \neq 0 \tag{25.16}$$

$$\det(\mathbf{I}+\mathbf{G})[\mathbf{I}+(\mathbf{I}+\mathbf{G})^{-1}\mathbf{G}\Delta\mathbf{L}_i] = \det(\mathbf{I}+\mathbf{G})\ \det[\mathbf{I}+(\mathbf{I}+\mathbf{G})^{-1}\mathbf{G}\Delta\mathbf{L}_i] \neq 0 \tag{25.17}$$

$$\Leftarrow \quad \det[\mathbf{I}+(\mathbf{G}^{-1}+\mathbf{I})^{-1}\Delta\mathbf{L}_i] \neq 0 \tag{25.18}$$

$$\Leftarrow \quad \sigma_{\max}[(\mathbf{I}+\mathbf{G}^{-1})^{-1}\Delta\mathbf{L}_i] \leq \sigma_{\max}[(\mathbf{I}+\mathbf{G}^{-1})^{-1}]\,\sigma_{\max}[\Delta\mathbf{L}_i] < 1 \tag{25.19}$$

$$\sigma_{\max}[\Delta\mathbf{L}_i] < \frac{1}{\sigma_{\max}[(\mathbf{I}+\mathbf{G}^{-1})^{-1}]} = \sigma_{\min}[\mathbf{I}+\mathbf{G}^{-1}]\ . \tag{25.20}$$

For comparison see Eq.(24.29).

25.2.3 Perturbation Matrix Yielding Unitary Eigenvectors

Assume a control loop with the return difference matrix $\mathbf{I}+\mathbf{G}(j\omega)$ and an additive perturbation $\Delta\mathbf{G}(j\omega)$ to $\mathbf{G}(j\omega)$. Then, the system is most sensitive to changes in the minimum singular value if $\mathbf{G}$ changes in the direction of $\mathbf{p}(j\omega)\mathbf{q}^H(j\omega)$, i.e., if perturbation $\Delta\mathbf{G}$ yields a considerable projection onto $\mathbf{pq}^H$. The vectors $\mathbf{p}$ and $\mathbf{q}$ denote the right and left singular vector of $\mathbf{I}+\mathbf{G}(j\omega)$, respectively, corresponding to the smallest singular value, see Eqs.(5.132), (5.27), (22.88) and (22.80). *Lee, W.H., et al. 1982* pointed out that the system of Fig. 25.1 is stable if

$$\sigma_{\max}[\Delta\mathbf{G}(j\omega)] < \sqrt{\sigma_{\min}[\mathbf{I}+\mathbf{G}(j\omega)]\sigma_{\min 2}[\mathbf{I}+\mathbf{G}(j\omega)]} \tag{25.21}$$

is satisfied, presupposed that the perturbation $\Delta\mathbf{G}$ is such that the eigenvectors $\mathbf{q}$ and $\mathbf{p}$ are unitary with respect to the perturbation matrix

$$\mathbf{q}^H(j\omega)\Delta\mathbf{G}(j\omega)\mathbf{p}(j\omega) = 0 \qquad \forall\omega\ . \tag{25.22}$$

If the smallest singular value is very small, indeed, and the smallest but one is significantly larger this method is very important. Selecting the smallest singular value for designing the control system, as given by Eq.(25.13), less performance would be the result.

25.2.4 Uncertainty Plot for Multiplicative Uncertainty

Another approach similar to the result of Eq.(25.20) is given by the following phase properties in the single-input single-output case. Consider $\Delta\arg[1+G_p(j\omega)]$ when $-\infty < \omega < \infty$ and $G_p = G(1+\Delta L_i)$. Evaluating yields

$$\Delta\arg[1+1/G(j\omega)+\Delta L_i(j\omega)] + \Delta\arg G\ . \tag{25.23}$$

The first term above only causes a change in the argument if the bracket term vanishes for any ω. This is given if

$$|1+\Delta L_i(j\omega)| \geq 1/|G(j\omega)|\ , \tag{25.24}$$

i.e., if the plot of the uncertainty in the vicinity of the Nyquist point intersects the inverse open-loop plot.

25.3 Scaling Techniques

Using singular-value techniques associated with conventional frequency domain criteria, sufficient conditions are obtained. But in many cases they are "more than sufficient" or too conservative. In order to avoid low performance, several methods reducing the conservatism are applied. They are important for every kind of application.

25.3.1 Unstructured and Structured Perturbation of Transfer Matrices

Let the class of *unstructured* perturbation $\Delta\mathbf{G}$ of a multivariable plant be given by

$$\mathcal{D}_u = \{\Delta\mathbf{G}:\ \sigma_{\max}[\Delta\mathbf{G}] \leq \sigma_{\max}[\mathbf{E}_P]\} \qquad \Delta\mathbf{G}, \mathbf{E}_P \in \mathcal{C}^{m\times m}\ . \tag{25.25}$$

where $\Delta\mathbf{G}$ is the additive perturbation of the plant transfer matrix $\mathbf{G}$ in frequency domain and $\mathbf{E}_P$ is the positive matrix of the perturbation bound. The class of *structured* perturbation is termed

$$\mathcal{D}_s = \{\Delta\mathbf{G}\ :\quad |\Delta\mathbf{G}| \leq_e \mathbf{E}_P\}\ . \tag{25.26}$$

The class $\mathcal{D}_s$ is a subset of $\mathcal{D}_u$, see Fig. 25.2 (*Kouvaritakis, B., and Latchman, H., 1985*).

To prove $\mathcal{D}_s \subset \mathcal{D}_u$ it must be guaranteed, first, that there exist matrices $\Delta\mathbf{G}_1$ with the property $\Delta\mathbf{G}_1 \in \mathcal{D}_u$ but $\Delta\mathbf{G}_1 \notin \mathcal{D}_s$ and, second, that all matrices $\Delta\mathbf{G}_2 \in \mathcal{D}_s$ are within the set $\mathcal{D}_u$, i.e., that $|\Delta\mathbf{G}_2| \leq_e \mathbf{E}_P$ implies $\sigma_{\max}[\Delta\mathbf{G}_2] \leq \sigma_{\max}[\mathbf{E}_P]$.

The first part of the proof is given when considering the very simple case

$$\mathbf{E}_P = \text{matrix}[p_{ij}] \quad \text{where} \quad p_{ij} = p > 0 \quad \forall i,j \qquad \text{and} \qquad |\Delta\mathbf{G}_1| = mp\mathbf{E}_{ij}\ . \tag{25.27}$$

The matrix $\mathbf{E}_{ij}$ is a matrix with 1 at the i,j-position and zero elsewhere. Then, since rank $\mathbf{E}_P^T\mathbf{E}_P = 1$ there are $m-1$ eigenvalues zero. Only a single eigenvalue $\lambda[\mathbf{E}_P^T\mathbf{E}_P] \neq 0$ is equal to $\sigma[\mathbf{E}_P]$. In this case, with regard to $\mathrm{tr}\mathbf{E}_P^T\mathbf{E}_P = \lambda[\mathbf{E}_P^T\mathbf{E}_P] = \sigma^2[\mathbf{E}_P]$ and $\mathrm{tr}\mathbf{E}_P^T\mathbf{E}_P = m(mp^2)$ it follows $\sigma[\mathbf{E}_P] = mp$. The singular values of $\Delta\mathbf{G}_1$ are

$$\lambda[\Delta\mathbf{G}_1^T\Delta\mathbf{G}_1] = \sigma^2[\Delta\mathbf{G}_1] = (mp)^2 \ . \qquad (25.28)$$

Hence, $\Delta\mathbf{G}_1 \in \mathcal{D}_u$. But note that

$$|\Delta\mathbf{G}_1|_{ij} = mp \not\leq (\mathbf{E}_P)_{ij} = p \quad \text{and} \quad \Delta\mathbf{G}_1 \notin \mathcal{D}_s \ . \qquad (25.29)$$

Regarding the second part of the proof, the matrix $\mathbf{E}_P$ is assumed non-negative, i.e. $\mathbf{E}_P \geq_e \mathbf{0}$. Then, $\mathbf{E}_P^T\mathbf{E}_P \geq_e \mathbf{0}$. Referring to Eq.(6.73), with the scalar e

$$\frac{\partial\sigma^2[\mathbf{E}_P]}{\partial e} = \frac{\partial\lambda[\mathbf{E}_P^T\mathbf{E}_P]}{\partial e} = \mathbf{w}^T\frac{\partial(\mathbf{E}_P^T\mathbf{E}_P)}{\partial e}\mathbf{w} = \mathbf{w}^T\left(\frac{\partial\mathbf{E}_P^T}{\partial e}\mathbf{E}_P + \mathbf{E}_P^T\frac{\partial\mathbf{E}_P}{\partial e}\right)\mathbf{w} \ . \qquad (25.30)$$

Replacing e by $p_{ij} = (\mathbf{E}_P)_{ij}$,

$$\frac{\partial\sigma^2_{\max}[\mathbf{E}_P]}{\partial p_{ij}} = \mathbf{w}^T(\mathbf{E}_{ij}^T\mathbf{E}_P + \mathbf{E}_P\mathbf{E}_{ij})\mathbf{w} \geq 0 \ . \qquad (25.31)$$

Hence, any increase in p_{ij} is followed by an increase of σ and

$$\mathbf{Q} \leq_e \mathbf{E}_P \quad \rightsquigarrow \quad \sigma_{\max}[\mathbf{Q}] \leq \sigma_{\max}[\mathbf{E}_P] \ . \qquad (25.32)$$

If $|\Delta\mathbf{G}_2| = \mathbf{Q}$ it follows $\sigma[\,|\Delta\mathbf{G}_2|\,] < \sigma[\mathbf{E}_P]$. Using the definition of the spectral norm Eq.(2.26),

$$\sigma_{\max}[\Delta\mathbf{G}_2] \leq \sigma_{\max}[\,|\Delta\mathbf{G}_2|\,] \leq \sigma_{\max}[\mathbf{E}_P] \qquad (25.33)$$

is always satisfied if $\Delta\mathbf{G}_2 \leq_e |\Delta\mathbf{G}_2| \leq_e \mathbf{E}_P$. The boundaries in Fig. 25.2 coincide for $\Delta\mathbf{G} = \mathbf{E}_P$. If $\Delta\mathbf{G} = \mathbf{E}_P$ then $\Delta\mathbf{G}$ lies on the boundary of both $\mathcal{D}_u$ and $\mathcal{D}_s$.

The perturbation class $\mathcal{D}_u$ disregards any structural information. Analyzing the robustness of closed-loop systems with structured perturbation by means of singular values, the result may be very conservative, see Fig. 25.2. In order to reduce the degree of conservativeness, scaling techniques are employed.

25.3.2 Similarity Scaling. Diagonal Scaling

Applying the matrix $\mathbf{S}$ to the perturbed plant or open-loop system such as $\mathbf{S}\mathbf{G}_p\mathbf{S}^{-1}$, similarity scaling is achieved. Note that similarity scaling preserves the eigenvalues. Employing only diagonal scaling matrices $\mathbf{D} \in \mathcal{R}^{m\times m}$, there is practically no loss in generality but high return in tractability (*Kouvaritakis, B., and Latchman, H., 1985*).

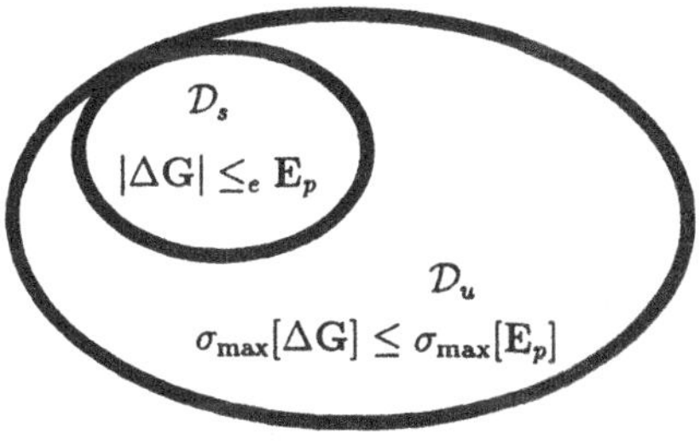

Figure 25.2: Regions of structured and unstructured uncertainty

25.3.3 Non-Diagonal Similarity Scaling

For the general induced norm $\|\mathbf{XMX}^{-1}\|_p$ the infimum over the set of complex matrices $\mathbf{X} \in \mathcal{C}^{m\times m}$ is given by the spectral radius (*Safonov, M.G., and Doyle, J.C., 1984*)

$$\inf_{\mathbf{X}} \|\mathbf{XMX}^{-1}\|_p = \rho_s[\mathbf{M}] \tag{25.34}$$

$$\text{where} \quad \|\mathbf{M}\|_p = \sup_{\|\mathbf{x}\|_p=1} \|\mathbf{Mx}\|_p\,, \qquad \|\mathbf{x}\|_p = \Big(\sum_{i=1}^{m} |x_i|^p\Big)^{1/p}\,, \qquad p \in [1,\infty]\,. \tag{25.35}$$

25.3.4 Minimizing Frobenius Norm

Reducing the norm of a matrix by scaling is an important matter in optimal control system design. For any $\mathbf{G} \in \mathcal{C}^{m\times m}$ find any real diagonal scaling matrix $\mathbf{D}$ in order to minimize $\|\mathbf{DGD}^{-1}\|_F^2$. Using $\|\mathbf{G}\|_F^2 = \sum_{i,j} |G_{ij}|^2$ and $\mathbf{D} = \operatorname{diag} d_i$

$$\|\mathbf{DGD}^{-1}\|_F^2 = \sum_{i,j=1}^{m} (d_{ii}|G_{ij}|d_{jj}^{-1})^2 = \sum_{i,j=1}^{m} d_{ii}^2|G_{ij}|^2 d_{jj}^{-2}\,. \tag{25.36}$$

Note that postmultiplying a matrix $\mathbf{A} \in \mathcal{C}^{m\times m}$ by the sum vector $\mathbf{1}$ yields a vector with components equal to the row sums of the matrix $\mathbf{A}$

$$\mathbf{A1} = (\sum_i A_{1i}\,,\ \sum_i A_{2i}\,,\ \dots\ \sum_i A_{mi})^T\,, \tag{25.37}$$

the product $\mathbf{A}^T\mathbf{1}$ equals a vector of column sums of $\mathbf{A}$, the product $\mathbf{1}^T\mathbf{A}$ yields a row matrix with entries equal to the column sums of $\mathbf{A}$ and, finally, $\mathbf{1}^T\mathbf{A1} = \sum_{i,j} A_{ij}$. □

Using the Hadamard product $\mathbf{A}^D$ = matrix[$|A_{ij}|^2$], the Frobenius norm can be rewritten to

$$\|\mathbf{DGD}^{-1}\|_F^2 = \mathbf{1}^T\mathbf{D}^2\mathbf{G}^D\mathbf{D}^{-2}\mathbf{1}\,. \tag{25.38}$$

Minimization with respect to an element of $\mathbf{D}$ yields

$$\frac{\partial}{\partial d_{ii}}\mathbf{1}^T\mathbf{D}^2\mathbf{G}^D\mathbf{D}^{-2}\mathbf{1} = 2(d_{ii}\mathbf{e}_i^T\mathbf{G}^D\mathbf{D}^{-2}\mathbf{1} - \mathbf{1}^T\mathbf{D}^2\mathbf{G}^D\mathbf{e}_i d_{ii}^{-3}) = 0 \tag{25.39}$$

$$d_{ii}^4\mathbf{e}_i^T\mathbf{G}^D\mathbf{D}^{-2}\mathbf{1} = \mathbf{1}^T\mathbf{D}^2\mathbf{G}^D\mathbf{e}_i = \mathbf{e}_i^T\mathbf{G}^{D,T}\mathbf{D}^2\mathbf{1}\,. \tag{25.40}$$

In general, $\forall\ i$

$$\mathbf{D}^4\mathbf{G}^D\mathbf{D}^{-2}\mathbf{1} = \mathbf{G}^{D,T}\mathbf{D}^2\mathbf{1} \tag{25.41}$$

$$\mathbf{D}^2\mathbf{G}^D\mathbf{D}^{-2}\mathbf{1} = \mathbf{D}^{-2}\mathbf{G}^{D,T}\mathbf{D}^2\mathbf{1} = (\mathbf{D}^2\mathbf{G}^D\mathbf{D}^{-2})^T\mathbf{1}\,. \tag{25.42}$$

Thus, the minimum is achieved if the row and column sums are equal. Note that an upper bound on the minimal Frobenius norm is given by

$$\min_{\mathbf{D}} \|\mathbf{DGD}^{-1}\|_F \le \sqrt{n\ \pi[\mathbf{G}^D]} \tag{25.43}$$

where $\pi[\mathbf{G}^D]$ is the Perron eigenvalue of the Hadamard product of $\mathbf{G}$. The above estimate is obtained from

$$\|\mathbf{DGD}^{-1}\|_F^2 = \sum_{i,j}(\mathbf{D}^2\mathbf{G}^D\mathbf{D}^{-2})_{ij} = \sum_{i,j}(\mathbf{D}\ |\mathbf{G}|\ \mathbf{D}^{-1})_{ij}^2 \le n\|\mathbf{D}^2\mathbf{G}^D\mathbf{D}^{-2}\|_1 \tag{25.44}$$

$$\min_{\mathbf{D}} \|\mathbf{DGD}^{-1}\|_F \leq \sqrt{n} \; \min_{\mathbf{D}} \|\mathbf{D}^2\mathbf{G}^D\mathbf{D}^{-2}\|_1 \tag{25.45}$$

where the following result has been used (*Limebeer, D.J.N., 1982*)

$$\inf_{\mathbf{T}} \|\mathbf{TAT}^{-1}\|_1 = \pi[\mathbf{A}] \qquad \mathbf{T} = \text{diag}\, T_{ii} \,. \tag{25.46}$$

25.3.5 Minimum Variance Scaling

Minimizing the expression

$$\sum_{i,j} [\, (\mathbf{DE}_P\mathbf{D}^{-1})_{ij} - \frac{1}{m^2}\sum_{i,j}(\mathbf{DE}_P\mathbf{D}^{-1})_{ij}\,]^2 \tag{25.47}$$

with respect to the diagonal real matrix $\mathbf{D}$ yields the result in $\mathbf{D}$ for a given real matrix $\mathbf{E}_P$ (*Kouvaritakis, B., and Latchman, H., 1985*)

$$(\mathbf{D}^2\mathbf{E}_P^D\mathbf{D}^{-2} - \mathbf{D}^{-2}\mathbf{E}_P^D\mathbf{D}^2)\mathbf{1} = \frac{\mathbf{1}^T\mathbf{DE}_P\mathbf{D}^{-1}\mathbf{1}}{m^4}(\mathbf{DE}_P\mathbf{D}^{-1} - \mathbf{D}^{-1}\mathbf{E}_P^T\mathbf{D})\mathbf{1} \,. \tag{25.48}$$

25.3.6 Minimizing $\sigma_{\max}[\mathbf{DGD}^{-1}]$. Maximizing $\sigma_{\min}[\mathbf{DGD}^{-1}]$

Consider $\mathbf{DGD}^{-1}$ with diagonal $\mathbf{D}$ and its singular value $\sigma = \sigma[\mathbf{DGD}^{-1}]$ given by

$$\mathbf{Hv} \triangleq (\mathbf{DGD}^{-1})^H(\mathbf{DGD}^{-1})\mathbf{v} = \sigma^2\mathbf{v} \,. \tag{25.49}$$

Differentiating and premultiplying by $\mathbf{v}^H$ yields

$$\frac{\partial \mathbf{H}}{\partial d_{ii}}\mathbf{v} + \mathbf{H}\frac{\partial \mathbf{v}}{\partial d_{ii}} = \sigma^2\frac{\partial \mathbf{v}}{\partial d_{ii}} + \mathbf{v}\frac{\partial(\sigma^2)}{\partial d_{ii}} \tag{25.50}$$

$$\mathbf{v}^H \times | \quad \frac{\partial \mathbf{H}}{\partial d_{ii}}\mathbf{v} + (\mathbf{H} - \sigma^2\mathbf{I})\frac{\partial \mathbf{v}}{\partial d_{ii}} = \mathbf{v}\frac{\partial(\sigma^2)}{\partial d_{ii}} \tag{25.51}$$

$$\mathbf{v}^H\frac{\partial \mathbf{H}}{\partial d_{ii}}\mathbf{v} + (\mathbf{v}^H\mathbf{H} - \mathbf{v}^H\sigma^2)\frac{\partial \mathbf{v}}{\partial d_{ii}} = \mathbf{v}^H\mathbf{v}\frac{\partial(\sigma^2)}{\partial d_{ii}} \tag{25.52}$$

$$\mathbf{v}^H\frac{\partial \mathbf{H}}{\partial d_{ii}}\mathbf{v} + 0 = 1 \times \frac{\partial(\sigma^2)}{\partial d_{ii}} = \frac{\partial(\sigma^2)}{\partial d_{ii}} \,. \tag{25.53}$$

By setting $\partial(\sigma^2)/\partial d_{ii} = 0$, it results $\mathbf{v}^H(\partial\mathbf{H}/\partial d_{ii})\mathbf{v} = 0$. Separating the d_{ii} element within the matrices $\mathbf{D}$ and $\mathbf{D}^{-1}$ yields

$$\mathbf{D} = \mathbf{D}_o + d_{ii}\mathbf{E}_{ii} \qquad \mathbf{D}^{-1} = \bar{\mathbf{D}}_o + d_{ij}^{-1}\mathbf{E}_{ii} \tag{25.54}$$

where $\mathbf{D}_o$ and $\bar{\mathbf{D}}_o$ are achieved from the matrices $\mathbf{D}$ and $\mathbf{D}^{-1}$ with the element d_{ii} and d_{ii}^{-1} replaced by zero. Then,

$$\begin{aligned}\frac{\partial \mathbf{H}}{\partial d_{ii}} = \frac{\partial \mathbf{D}^{-1}\mathbf{G}^H\mathbf{D}^2\mathbf{GD}^{-1}}{\partial d_{ii}} &= - \; d_{ii}^2\mathbf{E}_{ii}\mathbf{G}^H\mathbf{D}^2\mathbf{GD}^{-1} \\ &+ \; 2\mathbf{D}^{-1}\mathbf{G}^H d_{ii}\mathbf{E}_{ii}\mathbf{GD}^{-1} - \mathbf{D}^{-1}\mathbf{G}^H\mathbf{D}^2 d_{ii}^2\mathbf{E}_{ii} \,. \end{aligned} \tag{25.55}$$

Omitting calculations, the result finally is given by that matrix $\mathbf{D}$ which satisfies

$$\mathbf{v}^H[(\mathbf{DAD}^{-1})^H\mathbf{E}_{ii}(\mathbf{DAD}^{-1}) - \sigma^2\mathbf{E}_{ii}]\mathbf{v} = 0 \,. \tag{25.56}$$

25.3.7 Optimal and Suboptimal D-Weighted Norms

For the real constant parameters p and q obeying $1/q + 1/p = 1$ the matrix

$$\mathbf{D}_p^\star \triangleq \text{diag}\ \{y_1^{1/p} x_1^{-1/q}, \ldots, y_m^{1/p} x_m^{-1/q}\} \tag{25.57}$$

denotes the optimal p-norm weight. Associated with the matrix $\mathbf{G}(j\omega)$ the Perron eigenvalue of $|\mathbf{G}(j\omega)|$ is termed $\pi[\ |\mathbf{G}(j\omega)|\ ]$ and hence the right and left Perron eigenvectors are given

$$|\mathbf{G}|\mathbf{x} = \pi[\ |\mathbf{G}(j\omega)|\ ]\mathbf{x} \qquad \mathbf{y}^T|\mathbf{G}| = \pi[\ |\mathbf{G}(j\omega)|\ ]\mathbf{y}^T\ . \tag{25.58}$$

For $p = 1$ and $p = \infty$ the p-norm weight $\mathbf{D}_p^\star$ is optimal, i.e.,

$$\|\mathbf{G}(j\omega)\|_{pD_p^\star} = \|\ |\mathbf{G}(j\omega)|\ \|_{pD_p^\star} = \inf_D \|\ |\mathbf{G}(j\omega)|\ \|_{pD} = \pi[\ |\mathbf{G}(j\omega)|\ ]\ . \tag{25.59}$$

For $1 < p < \infty$ the p-norm weight $\mathbf{D}_p^\star$ is only suboptimal in the sense

$$\|\mathbf{G}(j\omega)\|_{pD_p^\star} \leq \|\ |\mathbf{G}(j\omega)|\ \|_{pD_p^\star} = \inf_D \|\ |\mathbf{G}(j\omega)|\ \|_{pD} = \pi[\ |\mathbf{G}(j\omega)|\ ]\ . \tag{25.60}$$

Usually, it suffices to compute $\pi[\ |\mathbf{G}(j\omega)|\ ]$ (an easy numerical task)

$$\|\mathbf{G}(j\omega)\|_{pD_p^\star}(j\omega) \leq \pi[\ |\mathbf{G}(j\omega)|\ ]\ , \tag{25.61}$$

the equality holding for $p = 1, \infty$, the inequality elsewhere.

Moreover, $\forall\ p = [2, \infty)$

$$\inf_{\mathbf{D}} \|\mathbf{G}\|_{1D} = \inf_{\mathbf{D}} \|\mathbf{G}\|_{\infty D} \geq \inf_{\mathbf{D}} \|\mathbf{G}\|_{pD} \quad \square \tag{25.62}$$

(*Stoer, J., and Witzgall, C., 1962; Bauer, F.L., 1963; Seneta, E., 1973; Safonov, M.G., 1982*).

Summarizing: Optimal solutions of the infimum are given for diagonal matrices in terms of the Perron eigenvalue

$$\inf_{\mathbf{D}} \|\mathbf{D}\mathbf{G}\mathbf{D}^{-1}\|_p = \pi[\mathbf{G}] \tag{25.63}$$

when one of the following three cases is given, i.e., if $\mathbf{G}$ is a positive matrix or when $p = 1$ or $p = \infty$ (*Safonov, M.G., 1982; Safonov, M.G., and Doyle, J.C., 1984*) :

$$\inf_{\mathbf{D}} \|\ |\mathbf{G}(j\omega)|\ \|_{pD} = \pi[\ |\mathbf{G}(j\omega)|\ ] \quad p = [1, \infty] \quad \square \tag{25.64}$$

$$\|\mathbf{G}\|_{1D} = \|\ |\mathbf{G}|\ \|_{1D}\ . \quad \square \tag{25.65}$$

Finally,

$$\|\mathbf{G}\|_{\infty D} = \|\ |\mathbf{G}|\ \|_{\infty D} = \max_i \sum_j |d_i g_i d_i^{-1}| \tag{25.66}$$

since

$$\|\mathbf{G}\|_{\infty D} = \sup_{\|\mathbf{x}\|_\infty} \frac{\|\mathbf{G}\mathbf{x}\|_{\infty D}}{\|\mathbf{x}\|_{\infty D}} = \sup_{\|\mathbf{x}\|_\infty} \|\mathbf{D}\mathbf{G}\mathbf{D}^{-1}\mathbf{x}\|_\infty \tag{25.67}$$

$$= \max_i \sum_j |d_i g_{ij} d_j^{-1}| = \max_i \sum_j |d_i| g_{ij} |d_j^{-1}| = \|\ |\mathbf{G}|\ \|_{\infty D}\ . \quad \square \tag{25.68}$$

Example: Perron scaling in the case $m = 2$:

$$\text{Let } m = 2, \qquad \mathbf{G} = \begin{pmatrix} a & b \\ c & d \end{pmatrix}, \qquad \mathbf{D}_p = \text{ diag } \{1, \alpha\} \ . \tag{25.69}$$

Find $\inf \sigma_{\max}[\mathbf{D}_p \ |\mathbf{G}| \ \mathbf{D}_p^{-1}]$ with respect to $\mathbf{D}_p$.

$$(\mathbf{D}_p \ |\mathbf{G}| \ \mathbf{D}_p^{-1})^T (\mathbf{D}_p \ |\mathbf{G}| \ \mathbf{D}_p^{-1}) = \begin{pmatrix} 1 & 0 \\ 0 & \alpha^{-1} \end{pmatrix} \begin{pmatrix} |a| & |c| \\ |b| & |d| \end{pmatrix} \begin{pmatrix} 1 & 0 \\ 0 & \alpha \end{pmatrix} \begin{pmatrix} 1 & 0 \\ 0 & \alpha \end{pmatrix} \begin{pmatrix} |a| & |b| \\ |c| & |d| \end{pmatrix} \begin{pmatrix} 1 & 0 \\ 0 & \alpha^{-1} \end{pmatrix} \tag{25.70}$$

$$= \begin{pmatrix} |a|^2 + |c|^2\alpha^2 & |a| \ |b|\alpha^{-1} + |c| \ |d|\alpha \\ |a| \ |b|\alpha^{-1} + |c| \ |d|\alpha & |b|^2\alpha^{-2} + |d|^2 \end{pmatrix} \tag{25.71}$$

$$h \triangleq (\sigma^2 - |a|^2 - |c|^2\alpha^2)(\sigma^2 - |b|^2\alpha^{-2} - |d|^2) - (|a| \ |b|\alpha^{-1} + |c| \ |d|\alpha)^2 = 0 \tag{25.72}$$

$$\frac{\partial h}{\partial(\sigma^2)} \frac{\partial(\sigma^2)}{\partial \alpha} = 0 \quad \leadsto \quad (\sigma^2 - |b|^2\alpha^{-2} - |d|^2 + \sigma^2 - |a|^2 - |c|^2\alpha^2)\frac{\partial(\sigma^2)}{\partial \alpha} = 0 \tag{25.73}$$

$$\sigma^2 = 0.5(|b|^2\alpha^{-2} + |c|^2\alpha^2 + |a|^2 + |d|^2) \tag{25.74}$$

$$\frac{\partial(\sigma^2)}{\partial \alpha} = 0 \quad \leadsto \quad -2|b|^2\alpha^{-3} + 2|c|^2\alpha = 0 \quad \leadsto \quad \alpha = \sqrt{\frac{|b|}{|c|}} \ . \tag{25.75}$$

Specializing $a = d = 0$, one has

$$\sigma_{\max}[\mathbf{D}_P \ |\mathbf{G}| \ \mathbf{D}_P^{-1}] = |b| \ \text{(scaled)} \qquad \text{and} \qquad \sigma_{\max}[\ |\mathbf{G}|\ ] = \sqrt{|b|^2 + |c|^2} \ \text{(unscaled)} \ . \tag{25.76}$$

Note the important fact that, following the steps in arriving at Eq.(25.75), maximizing the minimum singular value will lead to the same result. Moreover, minimizing the condition number $\kappa_s[\mathbf{DGD}^{-1}]$ with respect to $\mathbf{D}$ is characterized by the identical result (*Kouvaritakis, B., and Latchman, H., 1985*).

End of Example

25.3.8 Non-Similarity Perron Scaling

The following lemma is very useful for getting insight into the structural information of the perturbation. For $\mathbf{G}, \mathbf{H} \in \mathcal{C}^{m \times m}$ and real diagonal scaling matrices $\mathbf{L}$ and $\mathbf{R}$

$$\inf_{\mathbf{L},\mathbf{R}} \{\sigma_{\max}[\mathbf{LGR}] \ \sigma_{\max}[\mathbf{R}^{-1}\mathbf{HL}^{-1}]\} < \pi[\ |\mathbf{G}| \ |\mathbf{H}|\ ] \ . \tag{25.77}$$

The optimal $\mathbf{L}$ and $\mathbf{R}$ are given by

$$\mathbf{L} = \sqrt{\mathbf{Y}_{GH}\mathbf{X}_{GH}^{-1}} \qquad \qquad \mathbf{R} = \sqrt{\mathbf{X}_{HG}\mathbf{Y}_{HG}^{-1}} \tag{25.78}$$

where $\mathbf{X}_{GH}\mathbf{1}$, $\mathbf{1}^T Y_{GH}$, $\mathbf{X}_{HG}\mathbf{1}$ and $\mathbf{1}^T Y_{HG}$ are the right and left Perron eigenvectors of $\mathbf{GH}$ and $\mathbf{HG}$, respectively (*Bauer, F.L., 1963*). When $\mathbf{G}, \mathbf{H} \in \mathcal{R}_+^{m \times m}$ the equality holds in Eq.(25.77). Note that

$$\inf_{\mathbf{L},\mathbf{R}} \sigma_{\max}[\mathbf{LGR}] \quad \text{with constraint} \quad \mathbf{LR} = \mathbf{I} \tag{25.79}$$

yields the optimal similarity scaling approach with diagonal scaling matrices. For optimal scalings by means of nonlinear programming see *Wismer, D.A., and Chattergy, R., 1978*.

25.3.9 Numerical Aspects

The scaling methods outlined in the preceding sections in most cases can only be treated numerically. Steepest descent or gradient search algorithms are applied. The problem of the existence of local and global minimum is very important.

The function $\sigma_{\max}[\mathbf{D}\mathbf{G}\mathbf{D}^{-1}]$ is convex in $\mathbf{D}$ for any diagonal scaling matrix $\mathbf{D}$. Note that $\sigma^2_{\max}[e^{\mathbf{D}a}\mathbf{G}e^{-\mathbf{D}a}]$ also is a convex function for any $\mathbf{G}$, any diagonal $\mathbf{D}$ and a real parameter a. Hence, a minimizing algorithm for $\sigma^2_{\max}[\mathbf{D}\mathbf{G}\mathbf{D}^{-1}]$ will converge to the global minimum (*Safonov, M.G., and Doyle, J.C., 1984*). Convergence is also proved for some classes of block diagonal matrices.

25.4 Weighted Matrix Norms in Singular-Value-Based Formulas

Reducing conservatism is an important objective in stability robustness theory. Weighted matrix norms enable the designer to incorporate structural information both of the return difference and the perturbation matrices. The key results using weighting matrix norms are presented in what follows (*Yeh, H.H., et al. 1985*). Referring to Eq.(25.13) the stability robustness criterion based on maximum singular values is generalized

$$\|\mathbf{\Delta}\|_p < \frac{1}{\|\,[\mathbf{I}+\mathbf{G}(s)]^{-1}\|_p}\,. \tag{25.80}$$

The definitions of weighted norms $\|\mathbf{x}\|_{\infty q} = \|\mathbf{Q}^{-1}\mathbf{x}\|_\infty = \max_i(|x_i|/q_i)$ where

$$\mathbf{Q} = \operatorname{diag} q_i \quad \text{and} \quad \|\mathbf{A}\|_{\infty Q} = \|\,|\mathbf{A}|\mathbf{q}\,\|_{\infty q} = \|\mathbf{Q}^{-1}|\mathbf{A}|\mathbf{q}\|_\infty \tag{25.81}$$

are applied, see Eq.(2.119) when $\mathbf{D} = \mathbf{Q}^{-1}$. Then, the system is considered asymptotically stable if there exists a vector $\mathbf{q}(s)$ such that its *vector norm* $\|\cdot\|_\infty$ obeys

$$\|\mathbf{Q}^{-1}(s)|\mathbf{\Delta}(s)|\mathbf{q}(s)\|_\infty < \frac{1}{\|\mathbf{Q}^{-1}(s)\mid[\mathbf{I}+\mathbf{G}(s)]^{-1}|\,\mathbf{q}(s)\|_\infty} \tag{25.82}$$

$$\text{where} \quad \mathbf{Q} = \operatorname{diag}_n q_i(s) \quad \mathbf{q} = (q_1(s)\ q_2(s)\dots q_n(s))^T\,. \tag{25.83}$$

Assume that the matrix $[\mathbf{I}+\mathbf{G}(s)]^{-1}$ is irreducible, i.e., a permutation matrix $\mathbf{P}$ (with the property $\mathbf{P}^{-1} = \mathbf{P}^T$) does not exist such that

$$\mathbf{A} \triangleq [\mathbf{I}+\mathbf{G}(s)]^{-1}, \qquad \mathbf{P}^T[\mathbf{I}+\mathbf{G}(s)]^{-1}\mathbf{P} \triangleq \mathbf{P}^T\mathbf{A}\mathbf{P} = \begin{pmatrix} \mathbf{A}_{11} & \mathbf{A}_{12} \\ 0 & \mathbf{A}_{22} \end{pmatrix}\,. \tag{25.84}$$

For an irreducible and non-negative matrix $|\mathbf{A}|$ the Perron-Frobenius eigenvalue $\pi[\,|\mathbf{A}|\,]$ exists. Choosing the Perron eigenvector $\mathbf{q}_P$ and establishing the associated matrix $\mathbf{Q}_P$, the right-hand side of Eq.(25.82) can be evaluated. Since $\mathbf{Q}_P^{-1}\mathbf{q}_P = 1_n$ one finds

$$\begin{aligned}
\|\mathbf{Q}_P^{-1}(s)\ \ |\mathbf{\Delta}(s)|\ \ \mathbf{q}_P(s)\|_\infty &< \frac{1}{\|\mathbf{Q}_P^{-1}(s)\,|\mathbf{A}(s)|\,\mathbf{q}_P(s)\|_\infty} = \frac{1}{\|\mathbf{Q}_P^{-1}(s)\,\pi[\,|\mathbf{A}(s)|\,]\,\mathbf{q}_P(s)\|_\infty} \\
&= \frac{1}{\|\pi[\,|\mathbf{A}(s)|\,]\,\mathbf{Q}_P^{-1}(s)\mathbf{q}_P(s)\|_\infty} = \frac{1}{\pi[\,|\mathbf{A}(s)|\,]} = \frac{1}{\pi[\,|\,\{\mathbf{I}+\mathbf{G}(s)\}^{-1}|\,]}
\end{aligned} \tag{25.85}$$

where $\|\cdot\|_\infty$ denotes the *infinity vector norm*. Evaluating the left-hand side of Eq.(25.82), similarly, one has

$$\pi[\,|\mathbf{\Delta}(s)|\,] < \frac{1}{\pi[\,|\,\{\mathbf{I}+\mathbf{G}(s)\}^{-1}|\,]}\,. \tag{25.86}$$

25.5 E-Contours

The E-contours are defined as eigenvalue inclusion regions (characteristic loci inclusion regions) and are given by the behaviour of the eigenvalues λ_p of the perturbed plant transfer matrix $\mathbf{G}_p$. The condition for λ_p to be an eigenvalue is determined by

$$\det(\lambda_p \mathbf{I}_m - \mathbf{G}_p) = \det(\lambda_p \mathbf{I}_m - \mathbf{G} - \Delta\mathbf{G}) = 0 \qquad \mathbf{G}_p \in \mathcal{C}^{m\times m} \tag{25.87}$$

$$\det(\lambda_p \mathbf{I}_m - \mathbf{G})\ \det[\mathbf{I}_m - (\lambda_p \mathbf{I}_m - \mathbf{G})^{-1}\Delta\mathbf{G}] = 0\ . \tag{25.88}$$

The first determinant above cannot be zero, hence, the second must be. Now, the expression

$$(\lambda_p \mathbf{I}_m - \mathbf{G})^{-1}\Delta\mathbf{G} = \mathbf{I}_m \tag{25.89}$$

is *replaced* by another one substituting λ_p by z and employing the maximum singular value, i.e.,

$$\sigma_{\max}[(z\mathbf{I}_m - \mathbf{G})^{-1}\Delta\mathbf{G}] = 1\ . \tag{25.90}$$

Once more replacing z by a different quantity, one has

$$\sigma_{\max}[(z\mathbf{I}_m - \mathbf{G})^{-1}]\sigma_{\max}[\Delta\mathbf{G}] = 1 \quad \text{or} \quad \sigma_{\min}[(z\mathbf{I}_m - \mathbf{G})] = \sigma_{\max}[\Delta\mathbf{G}]\ . \tag{25.91}$$

When $\Delta\mathbf{G}$ is specialized $\Delta\mathbf{G} = \mathbf{0}$ then $z = \lambda$ (unperturbed) must be a solution. This is easily verified since $\sigma_{\max}[\Delta\mathbf{G} = \mathbf{0}] = 0$ and $\sigma_{\min}[z\mathbf{I}_m - \mathbf{G}] = 0$ implies $\det(z\mathbf{I}_m - \mathbf{G}) = 0$. See also Figs. 25.3 and 25.5. For a given uncertainty $\Delta\mathbf{G}$, Eq.(25.91) yields the boundary of the E-contour. Suppose that $\Delta\mathbf{G}$ is changed such that $\sigma_{\max}[\Delta\mathbf{G}]$ decreases then z should remain within the E-contour, see eigenvalue inclusion theorem Eq.(25.97). Hence, the complete region within and on the E-contour is given by

$$\sigma_{\min}[\mathbf{G} - z\mathbf{I}_m] \le \sigma_{\max}[\Delta\mathbf{G}]\ . \tag{25.92}$$

In the case of multiplicative uncertainty (calculations omitted) the result is

$$\sigma_{\min}[\mathbf{I}_m - z\mathbf{G}^{-1}] \le \sigma_{\max}[\Delta\mathbf{L}_i]\ . \tag{25.93}$$

Note that the sufficient stability conditions of Eq.(25.13) and (25.20) can be obtained from the E-contours presented above by setting $z = -1$ and by adapting the inequalities.

25.5.1 Eigenvalue Exclusion Circle

Consider $\sigma_{\min}[\mathbf{G} - z\mathbf{I}_m] \ge \delta$ for any $z \in \mathcal{C}$ and a perturbation matrix $\Delta\mathbf{G}$ such that $\sigma_{\max}[\Delta\mathbf{G}] \le \delta$. Referring to the definition that λ_p is an eigenvalue of $\mathbf{G} + \Delta\mathbf{G}$, from

$$\det(\lambda_p \mathbf{I}_m - \mathbf{G} - \Delta\mathbf{G}) = \det[(\lambda_p - z)\mathbf{I}_m - (\mathbf{G} - z\mathbf{I}_m) - \Delta\mathbf{G}] = 0 \tag{25.94}$$

the quantity $\lambda_p - z$ is an eigenvalue of $\mathbf{G} - z\mathbf{I} + \Delta\mathbf{G}$. With regard to Eq.(22.66) and (2.55),

$$|\lambda_p - z| \ge \sigma_{\min}[\mathbf{G} - z\mathbf{I}_m + \Delta\mathbf{G}] \tag{25.95}$$

$$\Leftarrow \quad |\lambda_p - z| \ge \sigma_{\min}[\mathbf{G} - z\mathbf{I}_m] - \delta\ . \tag{25.96}$$

Hence, the eigenvalue $\lambda_p = \lambda[\mathbf{G} + \Delta\mathbf{G}]$ cannot lie inside a circle centered at z with radius $\sigma_{\min}[\mathbf{G} - z\mathbf{I}] - \delta$. □

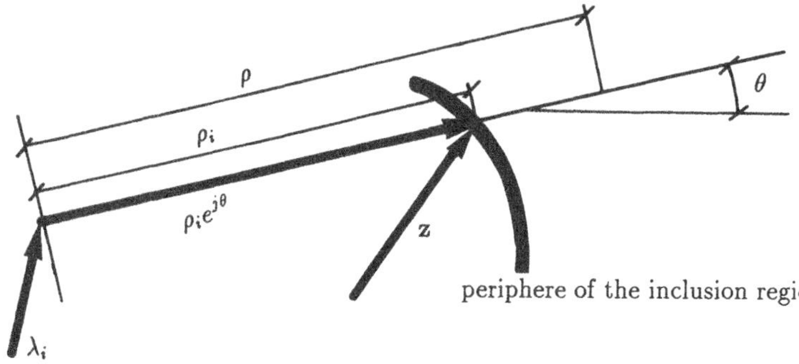

Figure 25.3: Eigenvalue inclusion. Periphere of the inclusion region

25.5.2 Predetermined Eigenvalue of the Perturbed System

Consider $\mathbf{G}$ and a perturbation $\Delta\mathbf{G}$ where $\sigma_{\max}[\Delta\mathbf{G}] \leq \delta \geq 0$. Then, solving the equation $\sigma_{\min}[\mathbf{G} - z\mathbf{I}_m] = \delta$ with respect to z this complex number z can be made an eigenvalue of $\mathbf{G} + \Delta\mathbf{G}$, i.e. $z = \lambda[\mathbf{G} + \Delta\mathbf{G}]$ (*Daniel, R.W., and Kouvaritakis, B., 1985*).

25.5.3 Eigenvalue Inclusion Theorem

Let $\lambda_i[\mathbf{G}]\ \forall i = 1 \ldots m$ be distinct eigenvalues of $\mathbf{G}$ and let the uncertainty $\Delta\mathbf{G}$ be bounded $\sigma_{\max}[\Delta\mathbf{G}] \leq \delta$. Then, exploring the eigenvalue behaviour of the perturbed plant,

$$\sigma_{\min}[\mathbf{G} - \lambda_i\mathbf{I}_m - \rho\ e^{j\theta}\mathbf{I}_m] = \delta < \delta_o > 0\ . \tag{25.97}$$

When the smallest value of ρ is termed ρ_i ($\rho_i < \rho \geq 0$) then $z_i = \lambda_i + \rho_i e^{j\theta}$ form closed simple curves in the vicinity of the nominal eigenvalues, see Fig. 25.3 (*Daniel, R.W., and Kouvaritakis, B., 1985*).

Repeating this procedure for each eigenvalue, m simply connected regions are obtained. Each region defines a tight inclusion region for precisely one eigenvalue of $\mathbf{G} + \Delta\mathbf{G}$ where the perturbation has to be bounded by $\sigma_{\max}[\Delta\mathbf{G}] \leq \delta$.

25.6 Nearest Normal Approximation

The nearest normal approximation $\mathbf{G}_N$ of the given matrix $\mathbf{G} \in \mathcal{C}^{m\times m}$ is defined by the operation $\mathbf{G} = \mathbf{G}_N + \mathbf{E}$ where $\sigma_{\max}[\mathbf{E}] \triangleq \delta_E \rightarrow \min$, i.e.,

$$\mathbf{G}_N = \arg\min_{\mathbf{G}_N} \sigma_{\max}[\mathbf{G} - \mathbf{G}_N]\ |_{\mathbf{G}_N\ normal}\ . \tag{25.98}$$

Then, the eigenvalues $\lambda[\mathbf{G}_p]$ are located inside a disc of radius $\delta_E + \delta$ centered at $\lambda[\mathbf{G}_N]$ where $\sigma_{\max}[\Delta\mathbf{G}] \leq \delta$. Moreover, all $\lambda[\mathbf{G}_p]$ are contained within the envelope of circles of radius δ centered in the boundary points of the numerical range of $\mathbf{G}$. The numerical range of $\mathbf{G}$, see Eq.(2.3), is given by the set of all complex numbers resulting from $\mathbf{x}^H\mathbf{G}\mathbf{x}/\mathbf{x}^H\mathbf{x}\quad \forall\mathbf{x} \in \mathcal{C}^m,\ \mathbf{x} \neq \mathbf{0}$ (*Daniel, R.W., and Kouvaritakis, B., 1985*).

25.7 Nyquist Criterion for Multivariable Systems

25.7.1 Derivation Using State Space

Employing the property $\det(\mathbf{I}+\mathbf{XY}) = \det(\mathbf{I}+\mathbf{YX})$, one has

$$\det(\mathbf{I}+\mathbf{G}) = \det[\mathbf{I}+\mathbf{C}(s\mathbf{I}-\mathbf{A})^{-1}\mathbf{B}] = \det[\mathbf{I}+\mathbf{BC}(s\mathbf{I}-\mathbf{A})^{-1}] = \frac{\det(s\mathbf{I}-\mathbf{A}-\mathbf{BC})}{\det(s\mathbf{I}-\mathbf{A})} . \tag{25.99}$$

To obtain a stable closed-loop system the zeros of $\det(s\mathbf{I}-\mathbf{A}-\mathbf{BC})$ must not be located in the closed right-half s-plane. Assume the number of unstable poles of the open-loop system equal to P. Then, following the principle of argument of the complex variable theory, the encirclements of the origin by the locus $\det(\mathbf{I}+\mathbf{G})$ equals the number of zeros minus the number of poles of $\det(\mathbf{I}+\mathbf{G})$ as s traverses the Nyquist D-contour. Hence, the encirclements

$$\begin{aligned} \text{circ } \det(\mathbf{I}+\mathbf{G}) &= \text{number of zeros of } \det(s\mathbf{I}-\mathbf{A}-\mathbf{BC})/\det(s\mathbf{I}-\mathbf{A}) \\ &\quad \text{minus number of poles of the same} \\ &\quad \text{function in the closed right-half } s\text{-plane} \\ &= \text{number of zeros of } \det(s\mathbf{I}-\mathbf{A}+\mathbf{BC}) \text{ minus} \\ &\quad \text{number of zeros of } \det(s\mathbf{I}-\mathbf{A}) \\ &= 0-P=-P . \end{aligned} \tag{25.100}$$

Use has been made of the fact that neither $\det(s\mathbf{I}-\mathbf{A})$ nor $\det(s\mathbf{I}-\mathbf{A}-\mathbf{BC})$ contains poles. Note that circ $\det(\mathbf{I}+\mathbf{G})$ provides the number of clockwise encirclements of the origin by the locus $\det(\mathbf{I}+\mathbf{G})$ as s traverses the closed D-contour in the complex s-plane in the clockwise sense. Alternatively, circ $\{-1+\det[\mathbf{I}+\mathbf{G}(s)]\}$ of the point $(-1, j0)$ can be used. An improved version is given when D is replaced by D_R where the contour D_R is the Nyquist contour enclosing the right half of the s-plane with radius R toward infinity and indentations with radius $1/R$ along the imaginary axis such that open-loop poles are enclosed in the closed right half-plane.

25.7.2 Derivation Using Characteristic Loci

Consider the case that $F(s) = \det[\mathbf{I}+\mathbf{G}(s)]$ has P poles and N zeros in the closed right half s-plane. The argument of $F(s)$ changes when s is changed. If s traverses a closed path in the s-plane then with regard to the principle of the argument the change of argument is

$$\Delta \arg F(s) = N - P . \tag{25.101}$$

The change of argument is counted in clockwise direction if the path in the s-plane is traversed clockwise.

Clockwise direction is used since the usual Nyquist contour is a path consisting of the imaginary axis and a semicircle round the right half s-plane with infinite diameter. Then, the stability condition is $N = 0$ or

$$\Delta \arg \det\, [\mathbf{I}+\mathbf{G}(s)] = -P \tag{25.102}$$

where the change of argument is counted clockwise. Invoking the fact that the determinant is the product of eigenvalues, it results

$$\Delta \arg \prod_{i=1}^{m} \lambda_i[\mathbf{I}+\mathbf{G}(s)] = \Delta \arg \prod_{i=1}^{m}\{1+\lambda_i[\mathbf{G}(s)]\} = \sum_{i=1}^{m} \Delta \arg \{1+\lambda_i[\mathbf{G}(s)]\} = -P . \tag{25.103}$$

The plots of $\lambda_i[\mathbf{G}(s)]$ are the well-known characteristic loci when s traverses the Nyquist contour once. If the system is augmented by a diagonal feedback $k\mathbf{I}$, the criterion can be rewritten to

$$\sum_{i=1}^{m} \Delta \arg \{1 + \lambda_i[k\mathbf{G}(s)]\} = \sum_{i=1}^{m} \Delta \arg \{1 + k\lambda_i[\mathbf{G}(s)]\} \quad (25.104)$$

$$= \sum_{i=1}^{m} \Delta \arg \{1/k + \lambda_i[\mathbf{G}(s)]\} = -P \quad (25.105)$$

and the total amount of encirclements of the critical point $(-1/k, j0)$ is counted.

25.8 Stability of the Sum of Matrix Polynomials

25.8.1 Matrix A Stable, Matrix A + B Stable

Let $\mathbf{A}(s)$ and $\mathbf{B}(s)$ be (n,n)-matrix polynomials. If both det $\mathbf{A}(s)$ and det $[\mathbf{A}(s)+\mathbf{B}(s)]$ are stable, then the polynomials det $[\mathbf{A}(s) + (1-\rho)\mathbf{B}(s)]$ are stable for any $0 \le \rho \le 1$.

Proof: Since

$$\det\,[\mathbf{A}(s) + \mathbf{B}(s)] = \det\,[\mathbf{I} + \mathbf{B}(s)\mathbf{A}^{-1}(s)]\det\,\mathbf{A}(s) \quad (25.106)$$

is stable, the vector loci of λ_i as given by

$$\det\,[\mathbf{I} + \mathbf{B}(s)\mathbf{A}^{-1}(s)] = \prod_{i=1}^{n} \lambda_i[\mathbf{I} + \mathbf{B}(s)\mathbf{A}^{-1}(s)] = \prod_{i=1}^{n}\{1 + \lambda_i[\mathbf{B}(s)\mathbf{A}^{-1}(s)]\} \quad (25.107)$$

for all $i = 1, 2 \ldots n$ do not encircle the Nyquist point $(-1, j0)$. Since the vector loci λ_i do not encircle $(-1, j0)$, the same is true for the vector loci $(1-\rho)\lambda_i$ for $0 \le \rho \le 1$. Hence, following

$$\det\,[\mathbf{A}(s)+(1-\rho)\mathbf{B}(s)] = \det\,[\mathbf{I}+(1-\rho)\mathbf{B}(s)\mathbf{A}^{-1}(s)]\det\,\mathbf{A}(s) = \prod_{i=1}^{n}\{1+(1-\rho)\lambda_i[\mathbf{B}(s)\mathbf{A}^{-1}(s)]\}\det\,\mathbf{A}(s) \quad (25.108)$$

the polynomial det $[\mathbf{A}(s) + (1-\rho)\mathbf{B}(s)]$ is stable $\forall \rho = [0, 1]$. □

25.8.2 Matrix A Stable, Matrix A + B Unstable

If det $\mathbf{A}(s)$ is stable but det $[\mathbf{A}(s) + \mathbf{B}(s)]$ is unstable then at least one of the vector loci of $(1-\rho)\lambda_i[\mathbf{B}(s)\mathbf{A}^{-1}(s)]$ crosses $(-1, j0)$ at some critical value $\rho = \rho_c$. Then, for this i and for some ω_c of $s = j\omega$

$$(1-\rho_c)\lambda_i[\mathbf{B}(j\omega_c)\mathbf{A}^{-1}(j\omega_c)] = -1 \quad \rightsquigarrow \quad \det\,[\mathbf{A}(j\omega_c) + (1-\rho_c)\mathbf{B}(j\omega_c)] = 0 \quad (25.109)$$

and $\mathbf{A} + (1-\rho)\mathbf{B}$ is stable for $\rho > \rho_c$ *(Okada, T., et al. 1988)* □.

25.9 Perturbed System Nyquist Stability

Consider the perturbed system $\mathbf{G}_p = \mathbf{G} + \Delta\mathbf{G}$ with P unstable open-loop poles and with $\sigma_{\max}[\Delta\mathbf{G}] \le \delta$. If in each of the m control loops a scalar gain k is implemented then the closed-loop transfer function is stable if the solutions in s of

$$\det[\mathbf{I} + k\mathbf{G}_p(s)] = \det[\frac{1}{k}\mathbf{I} + \mathbf{G}(s) + \Delta\mathbf{G}(s)] = 0 \quad (25.110)$$

are located in the open left half-plane. Then, from generalized Nyquist criterion the characteristic loci of $\mathbf{G}(s) + \Delta\mathbf{G}(s)$ must give a sum of counterclockwise encirclements of the critical point $-1/k + j0$ equal to P when $s = j\omega$ runs from $\omega = -\infty$ to $\omega = \infty$ (*MacFarlane, A.G.J., and Postlethwaite, I., 1977*). Since the characteristic loci must lie inside the E-contours and cannot lie outside it follows that the E-contours for $\sigma_{\max}[\Delta\mathbf{G}] \leq \delta$ for the same frequency region must give a sum of counterclockwise encirclements of the critical point equal to P.

If the feedback gain k is given then the system tolerance to uncertainties is determined by the maximum value of δ for which the corresponding E-contour just avoids the point $-1/k + j0$.

The eigenvalues of $\mathbf{G}_p$ are located inside the disc with radius $\delta_E + \delta$ and centered at $\lambda_{\min}[\mathbf{G}_N]$ where δ_E and δ are given by Eq.(25.98). For some frequency $s = j\omega$ let $\gamma_{\min}$ be the minimum distance from the critical point $-1/k + j0$ to the inclusion region then

$$\sigma_{\min}[\mathbf{I} + k\mathbf{G}_p] \geq k\gamma_{\min} , \tag{25.111}$$

see Fig. 25.4 (*Daniel, R.W., and Kouvaritakis, B., 1985*).

Proof: Let $\lambda_{\min}[\mathbf{G}_N]$ denote the eigenvalue of $\mathbf{G}_N$ nearest to $-1/k + j0$. Then, from Eq.(2.55)

$$|1/k + \lambda_{\min}[\mathbf{G}_N]\,| = |\lambda_{\min}[\mathbf{I}/k + \mathbf{G}_N]\,| \geq \sigma_{\min}[\mathbf{I}/k + \mathbf{G}_N] . \tag{25.112}$$

Since $\mathbf{G}_p = \mathbf{G} + \Delta\mathbf{G} = \mathbf{G}_N + \mathbf{E} + \Delta\mathbf{G}$, $\sigma_{\max}[\Delta\mathbf{G}] \leq \delta$ and $\sigma_{\max}[\mathbf{E}] = \delta_E$

$$\sigma_{\min}[\mathbf{I}/k + \mathbf{G}_p] \geq \sigma_{\min}[\mathbf{I}/k + \mathbf{G}_N] - \sigma_{\max}[\mathbf{E} + \Delta\mathbf{G}] \geq \sigma_{\min}[\mathbf{I}/k + \mathbf{G}_N] - \delta_E - \delta . \tag{25.113}$$

Combining Eqs.(25.113) and (25.112) yields a sufficient condition

$$\sigma_{\min}[\mathbf{I}/k + \mathbf{G}_p] \geq |\lambda_{\min}[\mathbf{I}/k + \mathbf{G}_N]\,| - \delta_E - \delta \tag{25.114}$$

$$\sigma_{\min}[\mathbf{I}/k + \mathbf{G}_p] \geq \;|\lambda_{\min}[\mathbf{G}_N] - (-1/k)| \; - (\delta_E + \delta) \triangleq \gamma_{\min} \tag{25.115}$$

$$\sigma_{\min}[\mathbf{I}/k + \mathbf{G}_p] \geq \gamma_{\min} \quad \text{or} \quad \sigma_{\min}[\mathbf{I} + k\mathbf{G}_p] \geq k\gamma_{\min} . \tag{25.116}$$

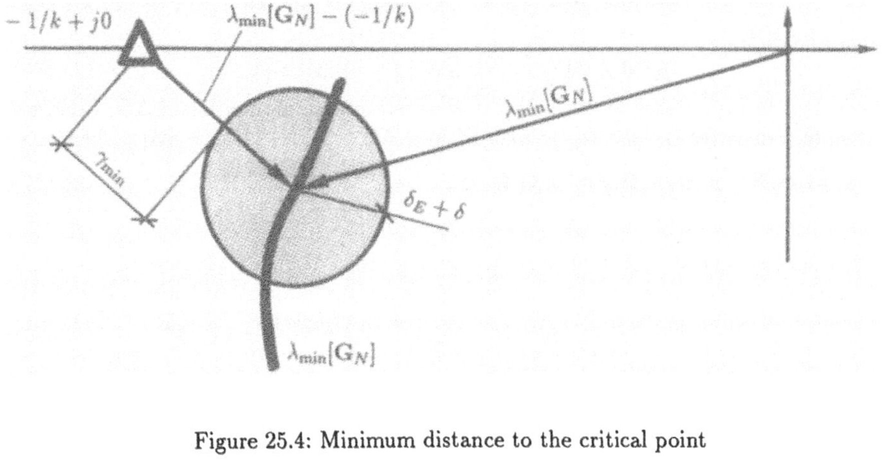

Figure 25.4: Minimum distance to the critical point

25.10 Similarity Scaling of Perturbed Systems

The eigenvalues of the perturbed system $\lambda_i[\mathbf{G}_p(s)]$ are given by

$$\det\{\ \lambda_i[\mathbf{G}_p(s)]\mathbf{I}_m - \mathbf{G}_p(s)\ \} = \det[\lambda_i\mathbf{I}_m - \mathbf{G} - \Delta\mathbf{G}] = 0 \tag{25.117}$$

and after similarity transformation

$$\det[\lambda_i\mathbf{I}_m - \mathbf{S}\mathbf{G}\mathbf{S}^{-1} - \mathbf{S}\ \Delta\mathbf{G}\ \mathbf{S}^{-1}] = 0\ . \tag{25.118}$$

The point z is contained within or on the E-contour if from Eq.(25.92)

$$\det[\mathbf{S}(\mathbf{G} - z\mathbf{I}_m)\mathbf{S}^{-1} + \mathbf{S}\ \Delta\mathbf{G}\ \mathbf{S}^{-1}] = 0 \tag{25.119}$$

$$\Leftarrow \quad \sigma_{\min}[\mathbf{S}(\mathbf{G} - z\mathbf{I}_m)\mathbf{S}^{-1}] \le \sigma_{\max}[\mathbf{S}\ \Delta\mathbf{G}\ \mathbf{S}^{-1}] \le \sigma_{\max}[\mathbf{S}\ \mathbf{E}_P\ \mathbf{S}^{-1}]\ . \tag{25.120}$$

There are several possibilities to choose the scaling matrix $\mathbf{S}$ in order to reduce conservatism for $\Delta\mathbf{G} \in \mathcal{D}_s$:

(i) Minimizing $\sigma_{\max}[\mathbf{S}\ \mathbf{E}_P\ \mathbf{S}^{-1}]$ with minimum at $\mathbf{S} = \mathbf{S}^\star$. But it may occur that the left-hand side of Eq.(25.120) gets smaller

$$\sigma_{\min}[\mathbf{S}^\star(\mathbf{G} - z\mathbf{I}_m)\mathbf{S}^{\star -1}] < \sigma_{\min}[\mathbf{G} - z\mathbf{I}_m]\ . \tag{25.121}$$

(ii) Maximizing for each z

$$\sigma_{\min}[\mathbf{S}(\mathbf{G} - z\mathbf{I}_m)\mathbf{S}^{-1}] \tag{25.122}$$

at $\mathbf{S} = \mathbf{S}^\star$ but it may occur that the right-hand side of Eq.(25.120) increases in comparison to the unscaled case

$$\sigma_{\max}[\mathbf{S}\mathbf{E}_P\mathbf{S}^{-1}] > \sigma_{\max}[\mathbf{E}_P]\ . \tag{25.123}$$

25.11 Bound for the Spectral Radius of the Uncertainty

Let $\mathbf{M}$ be any matrix $\mathbf{M} \in \mathcal{C}^{m\times m}$ and $\Delta\mathbf{G}$ be a structured perturbation bounded by $|\Delta\mathbf{G}| \le_e \mathbf{E}_P$. Then, the spectral radius of the matrix weighted[1] uncertainty $\mathbf{M}\ \Delta\mathbf{G}$ is

$$\begin{aligned} \rho_s[\mathbf{M}\ \Delta\mathbf{G}] &= \rho_s[\mathbf{L}\mathbf{M}\ \Delta\mathbf{G}\ \mathbf{L}^{-1}] = \rho_s[\mathbf{L}\mathbf{M}\mathbf{R}\mathbf{R}^{-1}\Delta\mathbf{G}\ \mathbf{L}^{-1}] && (25.124)\\ &\le \sigma_{\max}[\mathbf{L}\mathbf{M}\mathbf{R}\mathbf{R}^{-1}\Delta\mathbf{G}\ \mathbf{L}^{-1}] \le \sigma_{\max}[\mathbf{L}\mathbf{M}\mathbf{R}]\sigma_{\max}[\mathbf{R}^{-1}\Delta\mathbf{G}\ \mathbf{L}^{-1}]\ . && (25.125)\end{aligned}$$

Using Eq.(25.77),

$$\rho_s[\mathbf{M}\ \Delta\mathbf{G}] \le \pi[\ |\mathbf{M}|\ |\Delta\mathbf{G}|\ ] \le \pi[\ |\mathbf{M}|\ \mathbf{E}_P]\ . \tag{25.126}$$

The matrix $\mathbf{M}$ can be used to get insight into the structural information of $\Delta\mathbf{G}$. Consider two unitary matrices $\mathbf{\Omega}_L$ and $\mathbf{\Omega}_R$ to map $\mathbf{M}$ into $|\mathbf{M}|$, i.e. $\mathbf{\Omega}_L\mathbf{M}\mathbf{\Omega}_R = |\mathbf{M}|$. Then,

$$\begin{aligned} \rho_s[\mathbf{M}\ \Delta\mathbf{G}] = \rho_s[\mathbf{\Omega}_L\mathbf{M}\mathbf{\Omega}_R\mathbf{\Omega}_R^{-1}\ \Delta\mathbf{G}\ \mathbf{\Omega}_L^{-1}] &= \rho_s[\ |\mathbf{M}|\ \mathbf{\Omega}_R^{-1}\Delta\mathbf{G}\ \mathbf{\Omega}_L^{-1}] && (25.127)\\ &< \rho_s[\ |\mathbf{M}|\ \mathbf{E}_P] = \pi[\ |\mathbf{M}|\ \mathbf{E}_P]\ . && (25.128)\end{aligned}$$

The inequality above is obtained from the fact that $\mathbf{\Omega}_L$ and $\mathbf{\Omega}_R$ are unitary and $\mathbf{\Omega}_R^{-1}\ \Delta\mathbf{G}\ \mathbf{\Omega}_L^{-1}$ belongs to the class of structured perturbation if $\Delta\mathbf{G}$ does. The maximum spectral radius of $\mathbf{M}\ \Delta\mathbf{G}$ is given by the Perron eigenvalue of $|\mathbf{M}|\ \mathbf{E}_P$. The Perron eigenvalue of $|\mathbf{M}|\ \mathbf{E}_P$ can be used as an upper bound which proves advantageous since the largest eigenvalue is easy to compute.

[1]Note that $\Delta\mathbf{G}$ and $\Delta\mathbf{L}_o$ or $\Delta\mathbf{L}_i$ are the additive and multiplicative uncertainty (associated with the output or input), respectively, and $\mathbf{L}$ and $\mathbf{R}$ are special cases for the scaling matrix $\mathbf{S}$.

25.12 Minimizing the Cross-Condition Number

Rewriting Eq.(25.125) yields

$$\rho_s[\Delta G\ M] \le \inf_{L,R} \sigma_{max}[LE_PR]\ \sigma_{max}[R^{-1}ML^{-1}] = \inf_{L,R} \frac{\sigma_{max}[LE_PR]}{\sigma_{min}[LM^{-1}R]} \,. \tag{25.129}$$

Comparing with the cross-condition number

$$\kappa_s(A,B) \triangleq \frac{\sigma_{max}[A]}{\sigma_{min}[B]} \quad \rightsquigarrow \quad \rho_s[\Delta G\ M] \le \inf_{L,R} \kappa_s[LE_PR, LM^{-1}R] \,. \tag{25.130}$$

For given $M^{-1}, E_P \in \mathcal{C}^{m\times m}$ the optimal L and R minimizing the cross-condition number above must be such that

$$u_p^H E_{ii} u_p = u_m^H E_{ii} u_m \qquad w_p E_{ii} w_p = w_m^H E_{ii} w_m \qquad \forall i = 1 \dots n \tag{25.131}$$

where u_p and u_m are the singular vectors corresponding to $\sigma_{max}[LE_PR]$ and $\sigma_{min}[LM^{-1}R]$. The vectors w_p and w_m are given by

$$w_p = \frac{1}{\sigma_{max}[LE_PR]} LE_PRu_p \quad \text{and} \quad w_m = \frac{1}{\sigma_{min}[LM^{-1}R]} LM^{-1}Ru_m \,. \tag{25.132}$$

All vectors mentioned above are of Frobenius norm unity. Proof see *Kouvaritakis, B., and Latchman, H., 1985.*

25.13 Stability Margin

Recalling the stability condition from Eq.(25.9) and focussing on $z = 1$

$$\det(I + G_p) = \det(I + G + \Delta G) \ne 0 \quad \rightsquigarrow \quad \det(G - zI + \Delta G) \ne 0 \,. \tag{25.133}$$

Using the definition M and similarity scaling yields

$$M \triangleq (G - zI)^{-1} \qquad\qquad \det(M^{-1} + \Delta G) \ne 0 \tag{25.134}$$

and for $\det S \ne 0$ it results

$$\det(SM^{-1}S^{-1} + S\ \Delta G\ S^{-1}) \ne 0 \tag{25.135}$$

$$\det(SM^{-1}S^{-1}) \det[I + (SM^{-1}S^{-1})^{-1} S\ \Delta G\ S^{-1}] \ne 0 \tag{25.136}$$

$$\det[I + (SM^{-1}S^{-1})^{-1} S\ \Delta G\ S^{-1}] \ne 0 \tag{25.137}$$

$$\Leftarrow \qquad \sigma_{max}[(SM^{-1}S^{-1})^{-1}]\ \sigma_{max}[S\ \Delta G\ S^{-1}] < 1 \tag{25.138}$$

$$\frac{\sigma_{max}[S\ \Delta G\ S^{-1}]}{\sigma_{min}[SM^{-1}S^{-1}]} = \frac{\sigma_{max}[S\ \Delta G\ S^{-1}]}{\sigma_{min}[S(G - zI)S^{-1}]} < 1 \,. \tag{25.139}$$

Employing non-similarity scaling,

$$\frac{\sigma_{max}[L\ \Delta G\ R]}{\sigma_{min}[L(G - zI)R]} < 1 \qquad \det L \ne 0,\ \det R \ne 0 \,. \tag{25.140}$$

The smaller the fraction on the left-hand side of Eq.(25.139) and (25.140) the better is the stability margin. For the case of diagonal perturbation and diagonal scaling matrix one has $\sigma[\mathbf{D}\ \Delta\mathbf{G}\ \mathbf{D}^{-1}] = \sigma[\Delta\mathbf{G}]$. Then, the maximum singular value $\sigma_{\max}[\Delta\mathbf{G}]$ with $z = -1$ is given by

$$\sigma_{\max}[\Delta\mathbf{G}] \le \{\inf_{\mathbf{D}} \sigma_{\max}[\mathbf{D}(\mathbf{G}+\mathbf{I})^{-1}\mathbf{D}^{-1}]\}^{-1} \ . \tag{25.141}$$

This quantity provides an estimate of the perturbation bound or stability margin.

In order to avoid the point $(-1+j0)$ in Eq.(25.9)

$$\mathbf{M} = (\mathbf{I}+\mathbf{G})^{-1} \qquad \det(\mathbf{I}+\mathbf{M}\ \Delta\mathbf{G}) \neq 0 \ , \tag{25.142}$$

one has $\lambda[\mathbf{M}\ \Delta\mathbf{G}] \neq -1$ and the bounds

$$\sigma_{\max}[\mathbf{M}]\sigma_{\max}[\Delta\mathbf{G}] \le 1 \quad \text{or} \quad |\lambda[\mathbf{M}\ \Delta\mathbf{G}]\ |_{\max} = \rho_s[\mathbf{M}\ \Delta\mathbf{G}] < 1 \ . \tag{25.143}$$

Applying Eq.(25.77), i.e. *LR*-Perron scaling, and using Eq.(25.126),

$$\rho_s[\mathbf{M}\ \Delta\mathbf{G}] \le \pi[\ |\mathbf{M}|\ \mathbf{E}_P] < 1 \ . \tag{25.144}$$

Note that from Eq.(25.9) $\Delta\mathbf{G}$ and $\mathbf{M}$ can be permuted resulting in

$$\rho_s[\Delta\mathbf{G}\ \mathbf{M}] \le \pi[\mathbf{E}_P|\mathbf{M}|\ ] < 1 \ . \tag{25.145}$$

Use is made of the better bound.

25.14 Stability Margin Using the Cross-Condition Number

Suppose that the open-loop transfer matrix $\mathbf{G}(s)$ or plant transfer matrix with unity feedback is nominally closed-loop stable. Under additive perturbation $\Delta\mathbf{G}(s) \in \mathcal{D}_s$ the perturbed closed-loop system is stable if the optimal cross-condition number remains smaller than unity

$$\inf_{\mathbf{L},\mathbf{R}} \kappa_s[\mathbf{L}\mathbf{E}_P\mathbf{R}, \mathbf{L}\mathbf{M}^{-1}\mathbf{R}] = \inf_{\mathbf{L},\mathbf{R}} \frac{\sigma_{\max}[\mathbf{L}\mathbf{E}_P\mathbf{R}]}{\sigma_{\min}[\mathbf{L}\mathbf{M}^{-1}\mathbf{R}]} < 1 \quad \forall\ s \quad \text{where} \quad \mathbf{M} = [\mathbf{I}+\mathbf{G}(s)]^{-1} \ . \tag{25.146}$$

Proof:

$$\det[\mathbf{I}+\mathbf{G}(s)+\Delta\mathbf{G}(s)] = \det[\mathbf{M}^{-1}+\Delta\mathbf{G}(s)] \neq 0 \tag{25.147}$$

$$\det[\mathbf{L}\mathbf{M}^{-1}\mathbf{R}+\mathbf{L}\ \Delta\mathbf{G}\ \mathbf{R}] \neq 0 \quad \det\mathbf{L} \neq 0,\ \det\mathbf{R} \neq 0 \tag{25.148}$$

$$\det[\mathbf{L}\mathbf{M}^{-1}\mathbf{R}]\det[\mathbf{I}+(\mathbf{L}\mathbf{M}^{-1}\mathbf{R})^{-1}\mathbf{L}\ \Delta\mathbf{G}\ \mathbf{R}] \neq 0 \tag{25.149}$$

$$\Leftarrow \quad \sigma_{\max}[(\mathbf{L}\mathbf{M}^{-1}\mathbf{R})^{-1}]\sigma_{\max}[\mathbf{L}\ \Delta\mathbf{G}\ \mathbf{R}] = \frac{1}{\sigma_{\min}[\mathbf{L}\mathbf{M}^{-1}\mathbf{R}]}\sigma_{\max}[\mathbf{L}\ \Delta\mathbf{G}\ \mathbf{R}] < 1 \tag{25.150}$$

$$\Leftarrow \quad \frac{\sigma_{\max}[\mathbf{L}\mathbf{E}_P\mathbf{R}]}{\sigma_{\min}[\mathbf{L}\mathbf{M}^{-1}\mathbf{R}]} < 1 \quad \text{and} \quad \inf_{\mathbf{L},\mathbf{R}} \frac{\sigma_{\max}[\mathbf{L}\mathbf{E}_P\mathbf{R}]}{\sigma_{\min}[\mathbf{L}\mathbf{M}^{-1}\mathbf{R}]} < 1 \ . \ \square \tag{25.151}$$

A suboptimal stability criterion is obtained by applying the suboptimal scaling techniques of Eq.(25.77). Hence, suboptimally

$$\pi[\ |\mathbf{M}|\ \mathbf{E}_P] = \pi[\ |\{\mathbf{I}+\mathbf{G}(s)\}^{-1}|\ \mathbf{E}_P] < 1 \qquad \forall s \ . \tag{25.152}$$

Using optimal scaling, **L** and **R** are preferably chosen diagonal. But note that the suboptimal criterion does not require either **L** and **R**. If the criterion is satisfied and the plant is subject to any perturbation $|\Delta\mathbf{G}| < \mathbf{E}_P$ then the closed-loop system remains stable.

25.15 Structured E-Contours

Inclusion bands for the characteristic loci of the perturbed multivariable system can be obtained by the E-contour technique. For the class of unstructured perturbations the results are exact. For the class of structured perturbations the optimum information is reconstructed by application of scaling methods.

For z any complex number and $\lambda_p = \lambda[\mathbf{G} + \Delta\mathbf{G}]$ any eigenvalue of the perturbed open-loop system,

$$\det(\lambda_p \mathbf{I} - \mathbf{G} - \Delta\mathbf{G}) = 0 \tag{25.153}$$

$$\iff \quad \det[\mathbf{L}(\mathbf{G} - \lambda_p \mathbf{I})\mathbf{R} + \mathbf{L}\ \Delta\mathbf{G}\ \mathbf{R}] = 0\ . \tag{25.154}$$

If the condition

$$\sigma_{\min}[\mathbf{L}(\mathbf{G} - z\mathbf{I})\mathbf{R} + \mathbf{L}\ \Delta\mathbf{G}\ \mathbf{R}] > 0 \tag{25.155}$$

holds then any $\sigma > 0$. Then, Eq.(25.154) cannot be satisfied by $z = \lambda_p$. Since Eq.(22.66) the condition

$$\sigma_{\min}[\mathbf{L}(\mathbf{G} - z\mathbf{I})\mathbf{R}] - \sigma_{\max}[\mathbf{L}\ \Delta\mathbf{G}\ \mathbf{R}] > 0 \tag{25.156}$$

is sufficient for Eq.(25.155). Rewriting yields a condition that z *cannot* be an eigenvalue λ_p of the perturbed system

$$\frac{\sigma_{\max}[\mathbf{L}\ \Delta\mathbf{G}\ \mathbf{R}]}{\sigma_{\min}[\mathbf{L}(\mathbf{G} - z\mathbf{I})\mathbf{R}]} < 1\ . \tag{25.157}$$

Since $|\Delta\mathbf{G}| \leq_e \mathbf{E}_P$

$$\frac{\sigma_{\max}[\mathbf{L}\mathbf{E}_P\mathbf{R}]}{\sigma_{\min}[\mathbf{L}(\mathbf{G} - z\mathbf{I})\mathbf{R}]} < 1\ . \tag{25.158}$$

Hence, for any $z \in C$ such that Eq.(25.158) is satisfied the eigenvalue λ_p cannot attain z. This is the exclusion argument presented by *Kouvaritakis, B., and Latchman, H., 1985*.

Now, the structured E-contours are given by combination with the conventional E-contours (*Daniel, R.W., and Kouvaritakis, B., 1985*). For each eigenvalue λ_p the following algorithm can be used to obtain the structured E-contour. First, for a fixed θ from λ_p radiate outwards until the E-contour $\sigma_{\min}[\mathbf{G} - z\mathbf{I}] = \sigma_{\max}[\mathbf{E}_P]$ is intersected, see Fig. 25.5. The complex number z determines the distance ρ_E. Second, if the infimum condition number as given in Eq.(25.146) is less than unity then incrementally step inwards until at z_s

$$\inf_{\mathbf{L},\mathbf{R}} \frac{\sigma_{\max}[\mathbf{L}\mathbf{E}_P\mathbf{R}]}{\sigma_{\min}[\mathbf{L}(\mathbf{G} - z_s\mathbf{I})^{-1}\mathbf{R}]} = 1 \tag{25.159}$$

is satisfied.

25.16 Kalman Inequality. Approximation of $\sigma_{\min}[\mathbf{I} + \mathbf{G}]$

Single-input single-output systems with an optimal linear-quadratic state feedback regulator have the property that the return difference transfer function $1 + G(s)$ satisfies the inequality $|1 + G(j\omega)| \geq 1 \ \forall\omega$. A unit disc with centre $(-1, j0)$ and radius 1 in the complex s-plane is avoided by the locus of $G(j\omega)$. The guaranteed minimum phase margin of the system is ± 60 degrees (*Kalman, R.E., 1964*).

The multivariable version of the condition above is developed from algebraic Riccati equation by straightforward manipulation (*Anderson, B.D.O., 1969; Lehtomaki, N.A., et al. 1981*).

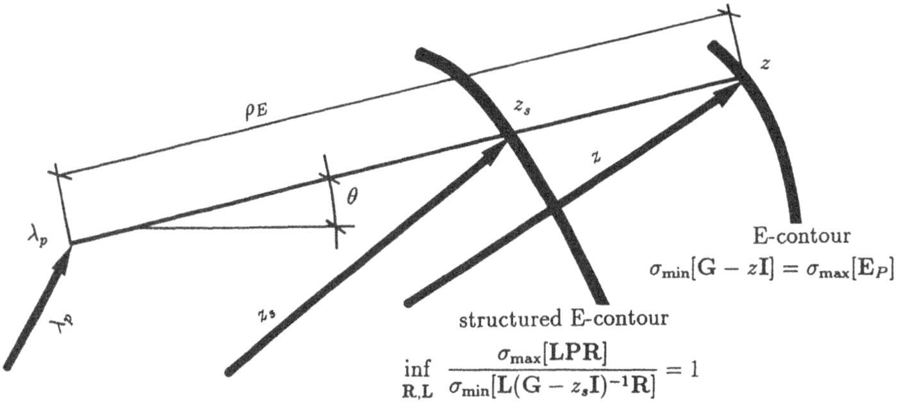

Figure 25.5: E-contour and structured E-contour

The Riccati equation arises from the optimization problem

$$I = \int_0^\infty [\mathbf{x}^T(t)\mathbf{Q}\mathbf{x}(t) + \mathbf{u}^T(t)\mathbf{R}\mathbf{u}(t)]dt \;\rightarrow\; \min, \qquad \mathbf{Q} \geq 0, \;\; \mathbf{R} > 0 \tag{25.160}$$

applied to the plant

$$\dot{\mathbf{x}}(t) = \mathbf{A}\mathbf{x}(t) + \mathbf{B}\mathbf{u}(t), \qquad \mathbf{x}(0) = \mathbf{x}_o \,. \tag{25.161}$$

The optimal state feedback controller (LQ-regulator) is given by

$$\mathbf{u}(t) = -\mathbf{R}^{-1}\mathbf{B}^T\mathbf{P}\mathbf{x}(t) \quad \text{and} \quad \mathbf{A}^T\mathbf{P} + \mathbf{P}\mathbf{A} + \mathbf{Q} - \mathbf{P}\mathbf{B}\mathbf{R}^{-1}\mathbf{B}^T\mathbf{P} = \mathbf{0} \tag{25.162}$$

where $\mathbf{P}$ is positive definite. Expanding with $(s^* + s)\mathbf{P}$,

$$(s^* - \mathbf{A}^T)\mathbf{P} + \mathbf{P}(s\mathbf{I} - \mathbf{A}) + \mathbf{P}\mathbf{B}\mathbf{R}^{-1}\mathbf{B}^T\mathbf{P} = \mathbf{Q} + 2\mathbf{P}\,\Re e\; s\,. \tag{25.163}$$

Taking into consideration that the open-loop transfer matrix $\mathbf{G}(s)$ is given by

$$\mathbf{G}(s) = \mathbf{R}^{-1}\mathbf{B}^T\mathbf{P}(s\mathbf{I} - \mathbf{A})^{-1}\mathbf{B}\,, \tag{25.164}$$

using the definition

$$\mathbf{H}(s) \triangleq [(s\mathbf{I} - \mathbf{A})^{-1}\mathbf{B}]^H(\mathbf{Q} + 2\mathbf{P}\,\Re e\; s)[(s\mathbf{I} - \mathbf{A})^{-1}\mathbf{B}]\,, \tag{25.165}$$

premultiplying and postmultiplying Eq.(25.163) by $[(s\mathbf{I} - \mathbf{A})^{-1}\mathbf{B}]^H$ and $[(s\mathbf{I} - \mathbf{A})^{-1}\mathbf{B}]$, respectively, yields

$$\underbrace{[(s\mathbf{I} - \mathbf{A})^{-1}\mathbf{B}]^H(s^*\mathbf{I} - \mathbf{A}^T)}_{\mathbf{B}^H}\mathbf{P}[(s\mathbf{I} - \mathbf{A})^{-1}\mathbf{B}] + [(s\mathbf{I} - \mathbf{A})^{-1}\mathbf{B}]^H\mathbf{P}\underbrace{(s\mathbf{I} - \mathbf{A})[(s\mathbf{I} - \mathbf{A})^{-1}\mathbf{B}]}_{\mathbf{B}}$$

$$+ \underbrace{[(s\mathbf{I} - \mathbf{A})^{-1}\mathbf{B}]^H\mathbf{P}\mathbf{B}}_{(\mathbf{R}\mathbf{G})^H}\underbrace{\mathbf{R}^{-1}\mathbf{B}^T\mathbf{P}[(s\mathbf{I} - \mathbf{A})^{-1}\mathbf{B}]}_{\mathbf{G}} = \mathbf{H}(s) \tag{25.166}$$

$$\mathbf{R}\mathbf{G}(s) + \mathbf{G}^H(s)\mathbf{R} + \mathbf{G}^H(s)\mathbf{R}\mathbf{G}(s) = \mathbf{H}(s) \qquad | + \mathbf{R} \tag{25.167}$$

or

$$[\mathbf{I}+\mathbf{G}(s)]^H\mathbf{R}[\mathbf{I}+\mathbf{G}(s)] = \mathbf{R}+\mathbf{H}(s)\ . \tag{25.168}$$

If $\mathbf{Q} > 0$, $\mathbf{B}$ full rank, $\mathbf{P} \geq 0$ and the Nyquist contour D_R is indented along $s = j\omega$ and the indentations are sufficiently small then $2\mathbf{P}\ \Re e\ s \geq 0$ and Eq.(25.168) implies $>$ in Eq.(25.169). If $\det(j\omega\mathbf{I}-\mathbf{A}) \neq 0\ \ \forall\omega$ or $\Re e\ s \geq 0,\ \ s \in D_R$, Eq.(25.168) implies $\geq$ in Eq.(25.169). Hence,

$$[\mathbf{I}+\mathbf{G}(j\omega)]^H\mathbf{R}[\mathbf{I}+\mathbf{G}(j\omega)] \geq \mathbf{R} \qquad \forall\,\omega \tag{25.169}$$

which is the multivariable version of Kalman inequality. Defining $\bar{\mathbf{G}}(s) \triangleq \mathbf{R}^{1/2}\mathbf{G}(s)\mathbf{R}^{-1/2}$, Eq.(25.169) can be rewritten to

$$[\mathbf{I}+\bar{\mathbf{G}}(s)]^H[\mathbf{I}+\bar{\mathbf{G}}(s)] \geq \mathbf{I} \quad \leadsto \quad \sigma_{\min}[\mathbf{I}+\bar{\mathbf{G}}(s)] \geq 1 \quad s \in D_R\ . \tag{25.170}$$

A similar result is obtained in the case $\mathbf{R} = \rho\mathbf{I}\ \ (\rho > 0)$ and $\mathbf{Q} = \mathbf{Q}_1^T\mathbf{Q}_1$

$$\sigma_i[\mathbf{I}+\mathbf{G}(j\omega)] = \{1+\frac{1}{\rho}\sigma_i^2[\mathbf{Q}_1(j\omega\mathbf{I}-\mathbf{A})^{-1}\mathbf{B}]\}^{1/2} \tag{25.171}$$

which is implied by Eqs.(25.165) and (25.168). If $\mathbf{G}(j\omega)$ is large then $\sigma_{\min}[\mathbf{G}(j\omega)] \gg 1$ and

$$\sigma_i[\mathbf{I}+\mathbf{G}(j\omega)] \doteq \frac{1}{\sqrt{\rho}}\ \sigma_i[\mathbf{Q}_1(j\omega\mathbf{I}-\mathbf{A})^{-1}\mathbf{B}]\ . \tag{25.172}$$

Eqs.(25.170) and (25.172) can be employed in Eq.(25.13) approximating $\sigma_{\min}[\mathbf{I}+\mathbf{G}]$ in a variety of selected cases yielding the basis for numerical algorithms (*Birdwell, J.D., and Laub, A.J., 1987*).

25.17 Reducing Conservatism

25.17.1 Reducing Conservatism Via ρ-System

Consider an unperturbed multivariable closed-loop system as depicted in the left-hand side of Fig. 25.6. It can easily be verified that the overall transfer system in the right-hand side of the same figure which consists of three blocks is equivalent if the transfer matrix $\mathbf{H}(s)$ is expressed by

$$\mathbf{H}(s) = [\mathbf{I}+(1-\rho)\mathbf{G}(s)]^{-1}\ . \tag{25.173}$$

This equivalent system is denoted ρ-system. The transfer matrix with input $\mathbf{y}_{ref}$ and output $\mathbf{y}$ turns out identical for any ρ. The sensitivity matrix of the system in Fig. 25.6, left-hand side, is given by $[\mathbf{I}+\mathbf{G}(s)]^{-1}$.

Now, the system is considered with perturbed matrix $\mathbf{G}_p(s)$ but $\mathbf{H}(s)$ and ρ are kept unchanged, see Fig. 25.7. Then, the transfer matrix of the right-hand side of Fig. 25.7 must be modified to

$$\mathbf{G}_p'(s) = \mathbf{G}_p(s)[\mathbf{I}+(1-\rho)\mathbf{G}_p(s)]^{-1}[\mathbf{I}+(1-\rho)\mathbf{G}(s)] \tag{25.174}$$

in order to guarantee the same input output properties. This can easily be verified by the simple closed-loop properties (*Okada, T., et al. 1988*).

The key idea of defining the perturbation is shown by the following equations. The open-loop transfer matrix, see right-hand side of Fig. 25.7, i.e. $\rho\mathbf{G}'(s)\mathbf{H}(s)$, is related to

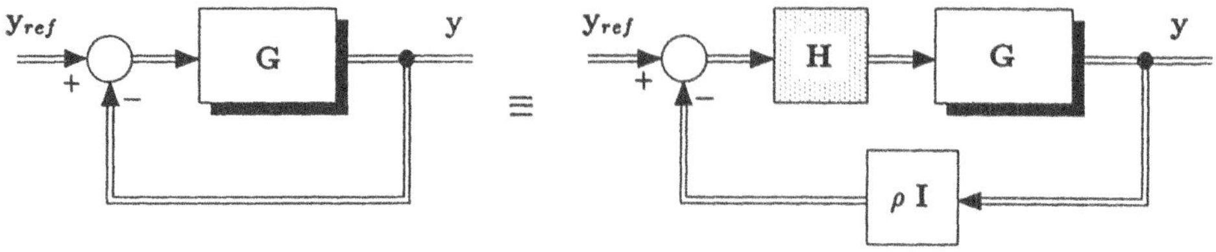

Figure 25.6: Unperturbed system and equivalent ρ-system

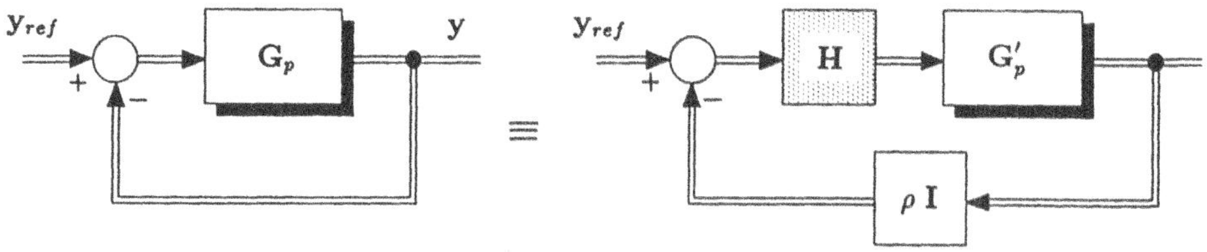

Figure 25.7: Perturbed system and equivalent perturbed ρ-system

the open-loop transfer matrix $\rho\mathbf{G}(s)\mathbf{H}(s)$ of the *original* system, see left-hand side of Fig. 25.7. Thus, additive and input-associated multiplicative uncertainty are given by

$$\rho\mathbf{G}_p'(s)\mathbf{H}(s) = \rho\mathbf{G}(s)\mathbf{H}(s) + \mathbf{\Delta}_a(s) \quad \rightsquigarrow \quad \mathbf{\Delta}_a(s) \tag{25.175}$$

$$\text{or} = [\mathbf{I} + \mathbf{\Delta}_{mi}(s)]\rho\mathbf{G}(s)\mathbf{H}(s) \quad \rightsquigarrow \quad \mathbf{\Delta}_{mi}(s)\ , \tag{25.176}$$

respectively. Comparison with the corresponding result of Eq.(25.13) yields

$$\sigma_{\max}[\mathbf{\Delta}_a(s)] \leq \sigma_{\min}[\mathbf{I} + \rho\mathbf{G}(s)\mathbf{H}(s)]\ . \tag{25.177}$$

Thus, $\sigma_{\max}[\mathbf{\Delta}_a(s)]$ as given by Eq.(25.175) is bounded by the minimum singular value determined by ρ, $\mathbf{G}(s)$ and $\mathbf{H}(s)$. Additionally, it is required that the open-loop system is stable which equals the condition that the peak value of $\sigma_{\max}[\mathbf{\Delta}_a(j\omega)]$ or $\|\mathbf{\Delta}_a(j\omega)\|_\infty$ is finite for all $0 \leq \rho \leq 1$. For an exact proof see *Okada, T., et al. 1988.*

Comparing Eq.(25.173) with the sensitivity matrix $[\mathbf{I}+\mathbf{G}(s)]^{-1}$ shows that the closed-loop behaviour is given by $\rho = 0$. (The open-loop properties correspond to $\rho = 1$.) With regard to the special arrangement of Eqs.(25.175) and (25.176) as the loop is closed both ρ and $\mathbf{\Delta}_a(s)$ become small. (The same is for $\mathbf{\Delta}_{mi}$.) Conservatism is reduced because *both* parameters ρ and $\sigma_{\max}[\mathbf{\Delta}_a(s)]$ become small with decreasing ρ. Typical values for computational purpose are $0 \leq \rho \leq 0.1$.

25.17.2 Singular-Value Conditions For Reduced Sensitivity

Consider the comparison of the open-loop system and open-loop perturbed system as depicted in Fig. 25.8. In the right-hand side of the figure the corresponding closed-loop systems are compared.

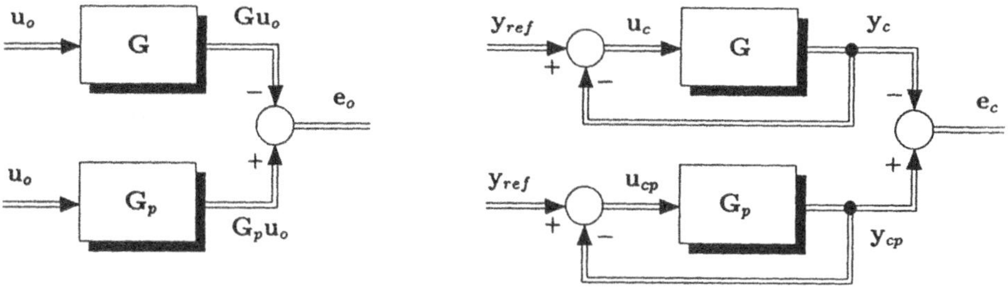

Figure 25.8: Comparison of open-loop and closed-loop systems with unperturbed and perturbed plant

For a perturbed system the open-loop deviation is $\mathbf{e}_o \triangleq (\mathbf{G}_p - \mathbf{G})\mathbf{u}_o$, the closed-loop deviation is $\mathbf{e}_c \triangleq \mathbf{G}_p\mathbf{u}_{cp} - \mathbf{G}\mathbf{u}_c$. Using $\mathbf{S}(s) \triangleq [\mathbf{I} + \mathbf{G}_p(s)]^{-1}$ and setting $\mathbf{u}_c = \mathbf{u}_o$, then simplification yields $\mathbf{e}_c = (\mathbf{I} + \mathbf{G}_p)^{-1}\mathbf{e}_o = \mathbf{S}\mathbf{e}_o$. Reduced sensitivity is given if

$$\int_0^\infty \mathbf{e}_c^T(t)\mathbf{e}_c(t)\, dt \le \int_0^\infty \mathbf{e}_o^T(t)\mathbf{e}_o(t)\, dt \tag{25.178}$$

$$\mathbf{S}^T(-j\omega)\mathbf{S}(j\omega) \le \mathbf{I} \quad \leadsto \quad \sigma_{\max}[\mathbf{S}(j\omega)] \le 1 \tag{25.179}$$

where Parseval's theorem has been employed. For the equivalent ρ-systems as given in the right-hand sides of Figs. 25.6 and 25.7 the sensitivity function is

$$\mathbf{S}'(s) = [\mathbf{I} + \rho\mathbf{G}_p'(s)\mathbf{H}(s)]^{-1} . \tag{25.180}$$

Reduced sensitivity is given if $\sigma_{\max}[\mathbf{S}'(j\omega)] \le 1$ is satisfied. Hence,

$$\sigma_{\max}[\mathbf{S}'(j\omega)] = \sigma_{\max}[\{\mathbf{I} + \rho\mathbf{G}_p'(s)\mathbf{H}(s)\}^{-1}] < 1 \tag{25.181}$$

$$\sigma_{\min}[\mathbf{I} + \rho\mathbf{G}_p'(s)\mathbf{H}(s)] = \sigma_{\min}[\mathbf{I} + \rho\mathbf{G}(s)\mathbf{H}(s) + \mathbf{\Delta}_a(s)] > 1 . \tag{25.182}$$

By Eq.(22.66),

$$\sigma_{\min}[\mathbf{I} + \rho\mathbf{G}(s)\mathbf{H}(s) + \mathbf{\Delta}] \ge \sigma_{\min}[\mathbf{I} + \rho\mathbf{G}(s)\mathbf{H}(s)] - \sigma_{\max}[\mathbf{\Delta}_a(s)] > 1 \tag{25.183}$$

$$\sigma_{\max}[\mathbf{\Delta}_a(s)] < \sigma_{\min}[\mathbf{I} + \rho\mathbf{G}(s)\mathbf{H}(s)] - 1 . \tag{25.184}$$

If $\mathbf{\Delta}_a(s)$ satisfies the condition above in the effective frequency range, then the closed-loop system satisfies the condition for sensitivity reduction (*Okada, T., et al. 1988*).

25.18 Robustness on the Riemann Sphere

Transfer function and frequency response plots, projected onto the Riemann sphere, provide a simple method of giving insight into the stability and robustness of control systems. Consider single-input single-output systems, only. The unperturbed and perturbed closed-loop transfer function is termed $f(s)$ and $h(s)$, respectively. When changing from f to h, the robustness of the closed-loop system is investigated (*El-Sakkary, A.K., 1989*).

25.18.1 Stereographic Projection and Chordal Metric

Consider the Gaussian plane and the Riemann unit sphere being placed onto the Gaussian plane such that the south pole S touches the plane at the origin 0 as depicted in Fig. 25.9. Any point s of the complex plane is stereographically projected onto the Riemann sphere by intersecting the straight line between s and the north pole N with the Riemann sphere. The intersection point is denoted q. The chordal metric $d[s_1,\ s_2]$ between two points s_1 and s_2 in the complex plane is defined as the euclidean distance between their stereographic projections q_1 and q_2 and is given by

$$d[s_1,\ s_2] = \frac{|s_2 - s_1|}{\sqrt{1+|s_1|^2}\,\sqrt{1+|s_2|^2}} \,. \tag{25.185}$$

25.18.2 Properties of the Chordal Metric

The chordal metric $d[s_1,\ s_2]$ between two points of the complex plane is characterized by the properties

$$d[s,\ \infty] = \frac{1}{\sqrt{1+|s|^2}}; \qquad d[s,\ 0] = \frac{|s|}{\sqrt{1+|s|^2}}; \qquad d[s_1,\ s_2] = d[\frac{1}{s_1},\frac{1}{s_2}] \,. \tag{25.186}$$

Any transfer function $f(s)$ $\forall s$ where $\Re e\ s \geq 0$ is considered as a subset of the complex plane. The transfer function, e.g., $1/(1+s)$ where $s = \sigma + j\omega$ yields a disc c for some $\sigma > 0$ enclosed by the circle $1/[(1+\sigma)+j\omega]$ $\forall\omega = (-\infty,\ \infty)$. The maximum chordal metric between two transfer functions g and h $\forall\ \Re e\ s \geq 0$ is denoted chordal distance between the systems g and h and is given by

$$\sup_{\Re e\ s \geq 0} \frac{|f-h|}{\sqrt{1+|f|^2}\,\sqrt{1+|h|^2}} \quad \text{where} \quad f = f(s),\ h = h(s) \,. \tag{25.187}$$

25.18.3 Stability

A system with transfer function $f(s)$ is stable if and only if the chordal metric $d[f(s),\ 0] < 1\ \ \forall\ \Re e\ s \geq 0$. In other words, a system is stable if the stereographic projection of the mapping $f(s)$ onto the Riemann sphere does not contain the north pole N.

Proof: For systems with strictly proper transfer functions $\lim_{\sigma,\omega\to\infty}\ f(s) = 0$ where $s = \sigma + j\omega$ the south pole S always belongs to the stereographic projection of $f(s)$ onto the Riemann sphere. If $f(s)$ has no right half-plane poles the image does not contain the north pole. Hence, the chordal metric is smaller than unity. □

Example: Consider the system $f = 1/(s+a)$. If $a > 0$ the system is stable, if $a < 0$ unstable. If $a > 0$ the frequency response plots $\forall\sigma$ are located in the interior of the disc c in Fig. 25.9, the stereographic projection q_c does not meet the north pole N. If $a < 0$ there is a $\Re e\ s = \sigma = -a > 0$ where the frequency plot l in Fig. 25.9 reaches infinity at $\omega = 0$ and the projection passes through the north pole. □

25.18.4 Properties of the Chordal Metric Between Two Transfer Functions

There are several inequalities which are useful for estimating the robustness properties of a control system with transfer functions $f(s)$ and $h(s)$. In particular,

$$d[1+f,\ 1+h] \leq 3\ d[f,h] \qquad \forall\ \Re e\ s \geq 0 \,. \tag{25.188}$$

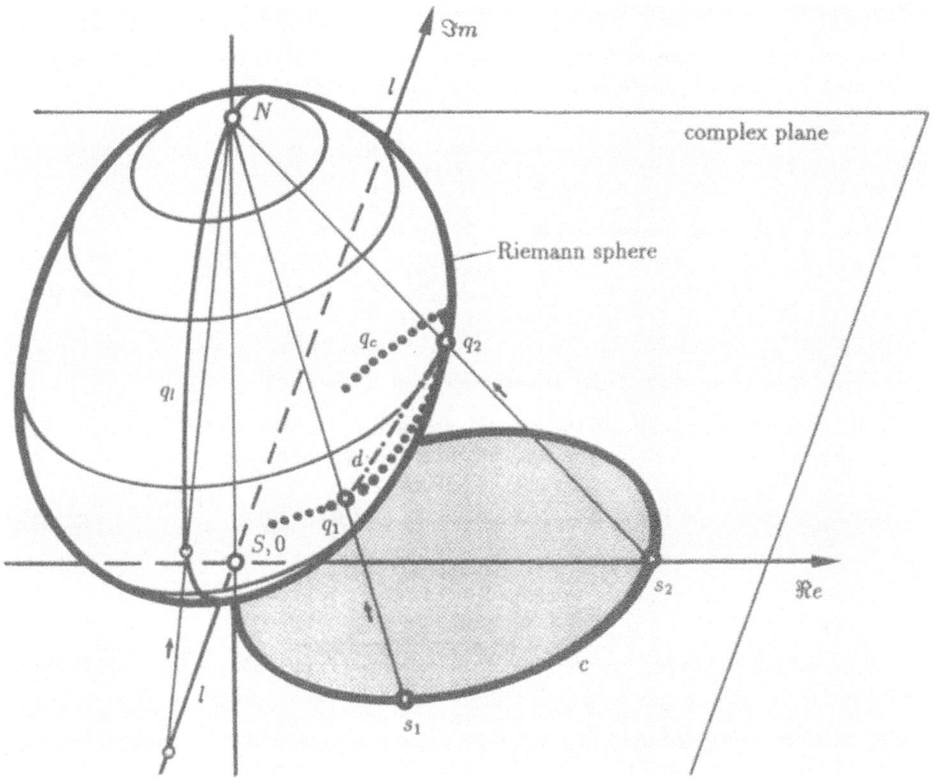

Figure 25.9: Complex plane and Riemann sphere

Transfer Functions

Consider two systems in their mappings f and h, i.e. their transfer functions $f(s)$ and $h(s)$. If f is stable and h is such that the chordal metric

$$d[f,h] < \frac{1}{\sqrt{1+|f|^2}} \quad \forall\, \Re e\ s \geq 0 \tag{25.189}$$

then h is stable and the modulus $|f-h|$ is bounded by

$$|f-h| \leq \frac{(1+|f|^2)\, d[f,h]}{1-d[f,h]\sqrt{1+|f|^2}} \quad \forall\, \Re e\ s \geq 0\,. \tag{25.190}$$

For the proof see *El-Sakkary, A.K., 1989*.

Transfer Functions and Their Inverse

Again consider two systems in their mappings f and h, i.e. their transfer functions $f(s)$ and $h(s)$. If f is stable and h is such that the chordal metric

$$d[f,h] < \frac{1}{\sqrt{1+|f^{-1}|^2}} \quad \forall\, \Re e\ s \geq 0\ , \tag{25.191}$$

then h^{-1} is stable and the modulus $|f^{-1} - h^{-1}|$ is bounded by

$$|f^{-1} - h^{-1}| \leq \frac{(1+|f^{-1}|^2)\ d[f,h]}{1 - d[f,h]\sqrt{1+|f^{-1}|^2}} \quad \forall\, \Re e\ s \geq 0\ . \tag{25.192}$$

Transfer Functions and Their Inverse Return Difference

Finally, consider two systems in their mappings g and h, i.e. their transfer functions $g(s)$ and $h(s)$. If $(1+g)^{-1}$ is stable and h is such that the chordal metric

$$d[g,h] < \frac{1}{3\sqrt{1+|(1+g)^{-1}|^2}} \quad \forall\, \Re e\ s \geq 0\ , \tag{25.193}$$

then $(1+h)^{-1}$ is stable and the modulus $|(1+g)^{-1} - (1+h)^{-1}|$ is bounded by

$$|(1+g)^{-1} - (1+h)^{-1}| \leq \frac{3(1+|(1+g)^{-1}|^2)\ d[g,h]}{1 - 3d[g,h]\sqrt{1+|(1+g)^{-1}|^2}} \quad \forall\, \Re e\ s \geq 0\ . \tag{25.194}$$

If g and h are considered open-loop transfer functions then closed-loop transfer functions are given by $(1+g)^{-1}$ and $(1+h)^{-1}$. If stability of $(1+g)^{-1}$ is known Eqs.(25.193) and (25.194) provide stability and robustness estimation of the perturbed system with transfer functions in the range between g and h.

25.19 Dead Time System

Consider the scalar open-loop transfer function $e^{-sT}L(s)$. Then, the closed-loop system is robustly stable with respect to the dead time T if

$$T < \frac{2}{\omega}\sin^{-1}[\frac{1}{2\ M(\omega)}] \quad \text{where} \quad M(\omega) = |\frac{L(j\omega)}{1+L(j\omega)}| \tag{25.195}$$

the frequency ω being chosen positive and such that $M(\omega) > 0.5$. The function $M(\omega)$ only contains functions of the system with dead time zero. Eq.(25.195) is an estimate, only, but separates the dead time T from $M(\omega)$ (*El-Sakkary, A.K., 1988*). The exact solution (*Walton, K., and Marshall, J.E., 1987; Zhou, L., and Jong, M.T., 1989*) is sacrificed to the separation.

Chapter 26

Block-Structured Uncertainty and Structured Singular Value

Control theory is characterized by a tradeoff between the level of control performance and the tolerance with respect to plant uncertainties. The structured singular value (*Doyle, J.C., 1982; Doyle, J.C., et al. 1982*) provides a reliable measure to satisfy both performance and robustness requirements. The structured singular value can be considered as a measure generalizing those methods which are based on the ordinary-singular-value decomposition.

26.1 Single Additive Uncertainty

Consider the multivariable control loop as depicted in Fig. 26.1a. The plant $\mathbf{G}(s)$ is characterized by an additive perturbance $\Delta\mathbf{G}(s)$ which is expressed by a norm-bound $\mathbf{\Theta}(s)$ and by the transfer matrices $\mathbf{L}^{-1}(s)$ and $\mathbf{R}(s)$, i.e.,

$$\Delta\mathbf{G}(s) = \mathbf{L}^{-1}(s)\mathbf{\Theta}(s)\mathbf{R}(s) \quad \text{where} \quad \|\mathbf{\Theta}(j\omega)\|_s \le 1 \ \ \forall\omega \ge 0 \, . \tag{26.1}$$

In Fig. 26.1b $\Delta\mathbf{G}$ is partitioned according to Eq.(26.1). Fig. 26.1c is redrawn from Fig. 26.1b. The matrices $\mathbf{L}(s)$ and $\mathbf{R}(s)$ are considered free of poles and zeros in the right half-plane. Since $\mathbf{\Theta}(s)$ is assumed stable $\Delta\mathbf{G}(s)$ has no right-half-plane poles and belongs to cone-bounded transfer functions. From $\mathbf{u} = \mathbf{K}(\mathbf{y}_{ref} - \mathbf{y})$ and $\mathbf{y} = \mathbf{G}\mathbf{u} + \mathbf{L}^{-1}\tilde{\mathbf{v}}$ it results

$$\mathbf{y} = (\mathbf{I}+\mathbf{GK})^{-1}\mathbf{GK}\mathbf{y}_{ref} + (\mathbf{I}+\mathbf{GK})^{-1}\mathbf{L}^{-1}\tilde{\mathbf{v}} \qquad \text{and} \tag{26.2}$$

$$\tilde{\mathbf{u}} = \mathbf{RK}(\mathbf{I}+\mathbf{GK})^{-1}\mathbf{y}_{ref} - \mathbf{RK}(\mathbf{I}+\mathbf{GK})^{-1}\mathbf{L}^{-1}\tilde{\mathbf{v}} \, . \tag{26.3}$$

Using a partitioned overall transfer matrix

$$\begin{pmatrix}\mathbf{y}\\ \tilde{\mathbf{u}}\end{pmatrix} = \begin{pmatrix}(\mathbf{I}+\mathbf{GK})^{-1}\mathbf{GK} & (\mathbf{I}+\mathbf{GK})^{-1}\mathbf{L}^{-1}\\ \mathbf{RK}(\mathbf{I}+\mathbf{GK})^{-1} & -\mathbf{RK}(\mathbf{I}+\mathbf{GK})^{-1}\mathbf{L}^{-1}\end{pmatrix}\begin{pmatrix}\mathbf{y}_{ref}\\ \tilde{\mathbf{v}}\end{pmatrix} \triangleq \begin{pmatrix}\mathbf{J} & \mathbf{H}\\ \mathbf{N} & \mathbf{M}\end{pmatrix}\begin{pmatrix}\mathbf{y}_{ref}\\ \tilde{\mathbf{v}}\end{pmatrix} . \tag{26.4}$$

Note the identity $\mathbf{K}(\mathbf{I}+\mathbf{GK})^{-1} \equiv (\mathbf{I}+\mathbf{KG})^{-1}\mathbf{K}$. By applying $\tilde{\mathbf{v}}(s) = \mathbf{\Theta}(s)\tilde{\mathbf{u}}(s)$ to Eq.(26.4) as given by the feedforward path of the uncertainty $\mathbf{\Theta}(s)$,

$$\mathbf{y} = \mathbf{J}\mathbf{y}_{ref} + \mathbf{H}\mathbf{\Theta}\tilde{\mathbf{u}} \quad \text{and} \quad \tilde{\mathbf{u}} = \mathbf{N}\mathbf{y}_{ref} + \mathbf{M}\mathbf{\Theta}\tilde{\mathbf{u}} \quad \leadsto \quad \tilde{\mathbf{u}} = (\mathbf{I}-\mathbf{M}\mathbf{\Theta})^{-1}\mathbf{N}\mathbf{y}_{ref} \tag{26.5}$$

is obtained. The right-hand side equation above is implemented in Fig. 26.1d.

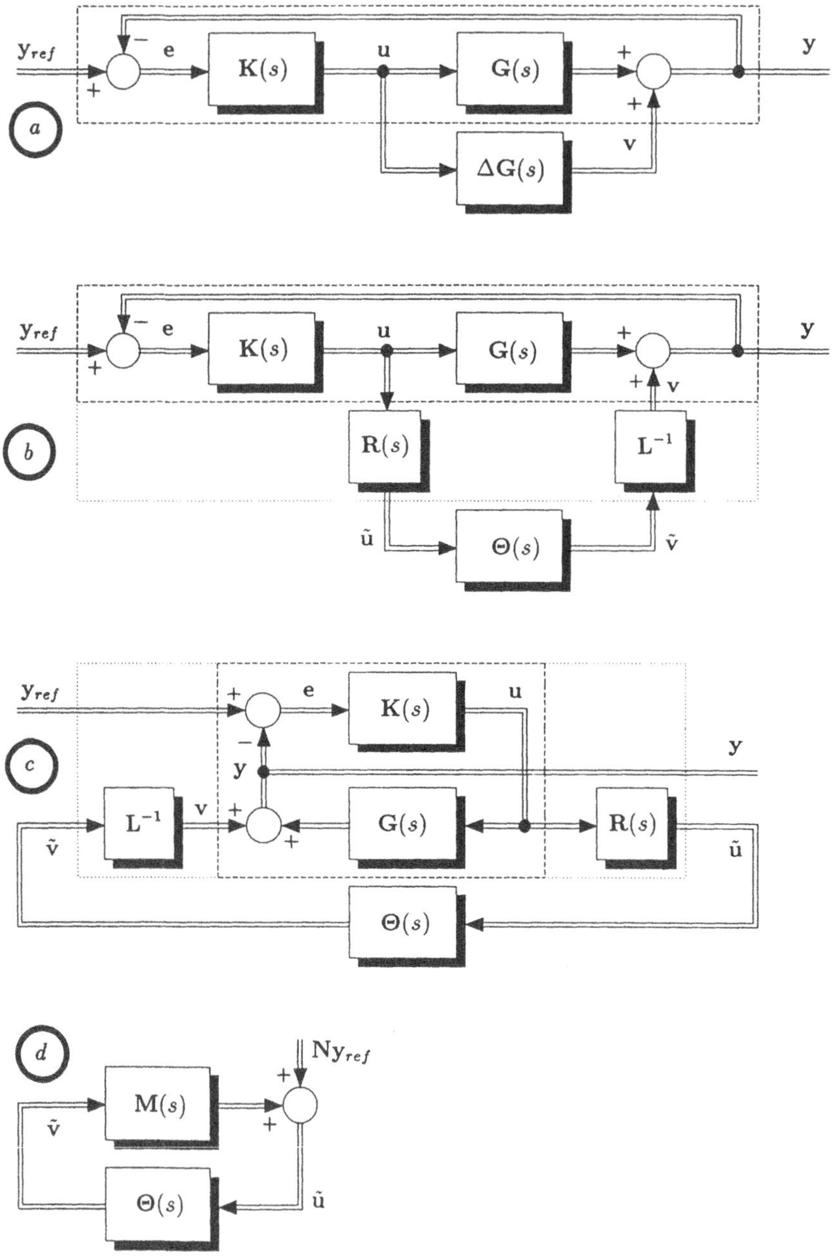

Figure 26.1: Multivariable control and isolating the plant uncertainty by rearranging the block diagram

From Fig. 26.1b or c a simplified Fig. 26.2 is redrawn. From Fig. 26.2

$$-\mathbf{K}(\mathbf{L}^{-1}\mathbf{x}_{in} + \mathbf{G}\mathbf{u}) = \mathbf{u} \qquad \text{and} \qquad \mathbf{R}\mathbf{u} = \mathbf{x}_{out} \tag{26.6}$$

can be derived directly with the result $\mathbf{M} = -\mathbf{R}(\mathbf{I}+\mathbf{K}\mathbf{G})^{-1}\mathbf{K}\mathbf{L}^{-1}$ as given in Eq.(26.4).

The uncertainty matrix $\Theta(s)$ is pulled out from the system and is considered as an external part or as a part of the environment. Thus, the original system is free of uncertainty but an additional input and output is required to receive information from the uncertainty and to submit signals to the uncertainty. From the point of view of the uncertainty, the unperturbed control system operates with the transfer matrix $\mathbf{M}(s)$.

Referring to the stability problem posed in Fig. 25.1 the encirclements of all the characteristic loci of $-\mathbf{M}\Theta$ round the Nyquist point $-1+j0$ must be zero. This is true since Θ is assumed stable and since $\mathbf{M}$ is stable because it can be assumed that $\mathbf{K}$ stabilizes the plant without uncertainty. The characteristic loci of $-\mathbf{M}\Theta$ are given by $\lambda[-\mathbf{M}\Theta]$. Referring to Eq.(2.53),

$$|\ \lambda[\mathbf{M}\Theta]\ | \le \rho_s[\mathbf{M}\Theta] \le \sigma_{\max}[\mathbf{M}\Theta] \tag{26.7}$$

and no encirclement exists if

$$\sigma_{\max}[\mathbf{M}(s)\Theta(s)] < 1 \qquad s = j\omega \quad \forall\omega = 0 \dots \infty \tag{26.8}$$

$$\sup_{\omega}\ \sigma_{\max}[\mathbf{M}(s)\Theta(s)] = \|\mathbf{M}\Theta\|_\infty < 1 \tag{26.9}$$

$$\Leftarrow \qquad \|\mathbf{M}\|_\infty \|\Theta\|_\infty < 1\ . \tag{26.10}$$

Since $\|\Theta\|_\infty \le 1$, a sufficient stability condition is

$$\|\mathbf{M}\|_\infty < 1\ . \tag{26.11}$$

The principal gain or singular value of $\mathbf{M}(s)$ is given by $\sigma_i[\mathbf{M}(j\omega)]$. Referring to this definition, none of the principal gains of $\mathbf{M}(s)$ may exceed unity. In the case of stability $\|\mathbf{M}\|_\infty^{-1}$ is denoted gain margin excess.

In Table 26.1 six rows of control systems with various plant structures are presented. Output, input and additive uncertainties in a parallel feedforward and in a divisive feedback structure and the admissible maximum singular value of control system transfer functions are outlined in separate rows. Representative types of uncertainty sources and performance specifications are listed in columns, correspondingly.

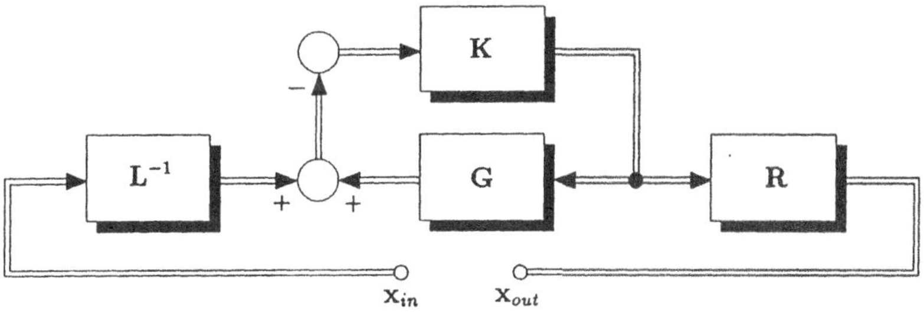

Figure 26.2: Control system with input signal from the uncertainty and output signal supplying the uncertainty

Table 26.1: Stability robustness and performance conditions if only a single uncertainty occurs at a certain location

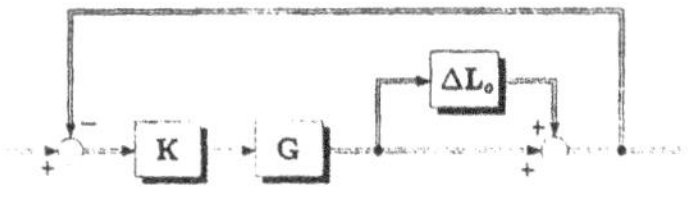 $\sigma_{\max}[\mathbf{RGK(I+GK)^{-1}L^{-1}}] < 1$	sensor errors changing number of rhp zeros neglected high frequency dynamics	sensor noise attenuation output response to output commands
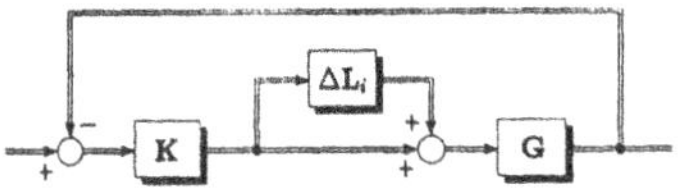 $\sigma_{\max}[\mathbf{RKG(I+KG)^{-1}L^{-1}}] < 1$	actuator errors neglected high frequency dynamics changing number of rhp poles	input response to input commands
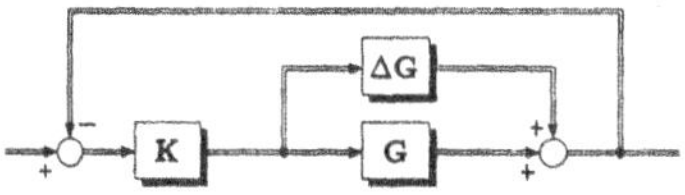 $\sigma_{\max}[\mathbf{RK(I+GK)^{-1}L^{-1}}] < 1$	uncertain rhp poles	input errors to output commands
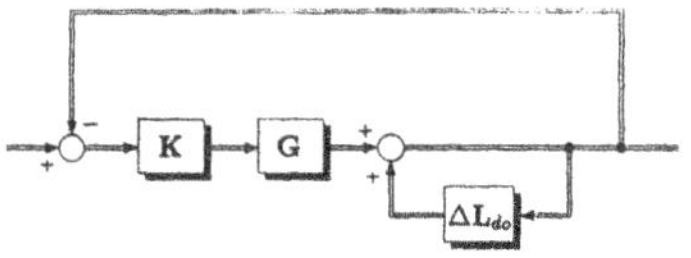 $\sigma_{\max}[\mathbf{R(I+GK)^{-1}L^{-1}}] < 1$	divisive output uncertainty changing number of rhp poles low-frequency plant parameter errors	output errors to output commands
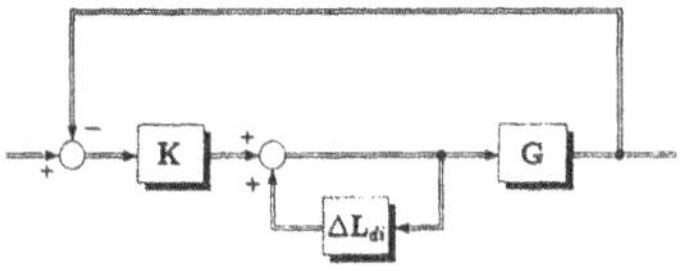 $\sigma_{\max}[\mathbf{R(I+KG)^{-1}L^{-1}}] < 1$	divisive output uncertainty changing number of rhp poles low-frequency plant parameter errors	input errors to input commands
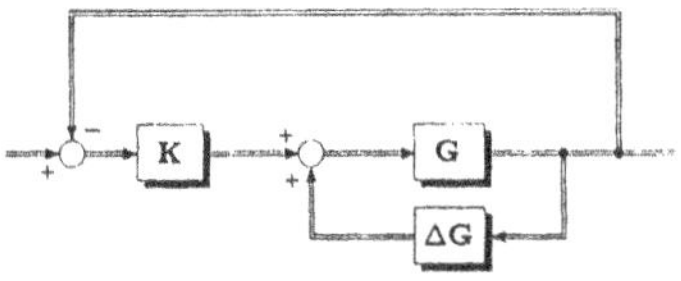 $\sigma_{\max}[\mathbf{R(I+GK)^{-1}GL^{-1}}] < 1$	divisive output uncertainty uncertain rhp poles low-frequency plant parameter errors	output errors to input commands

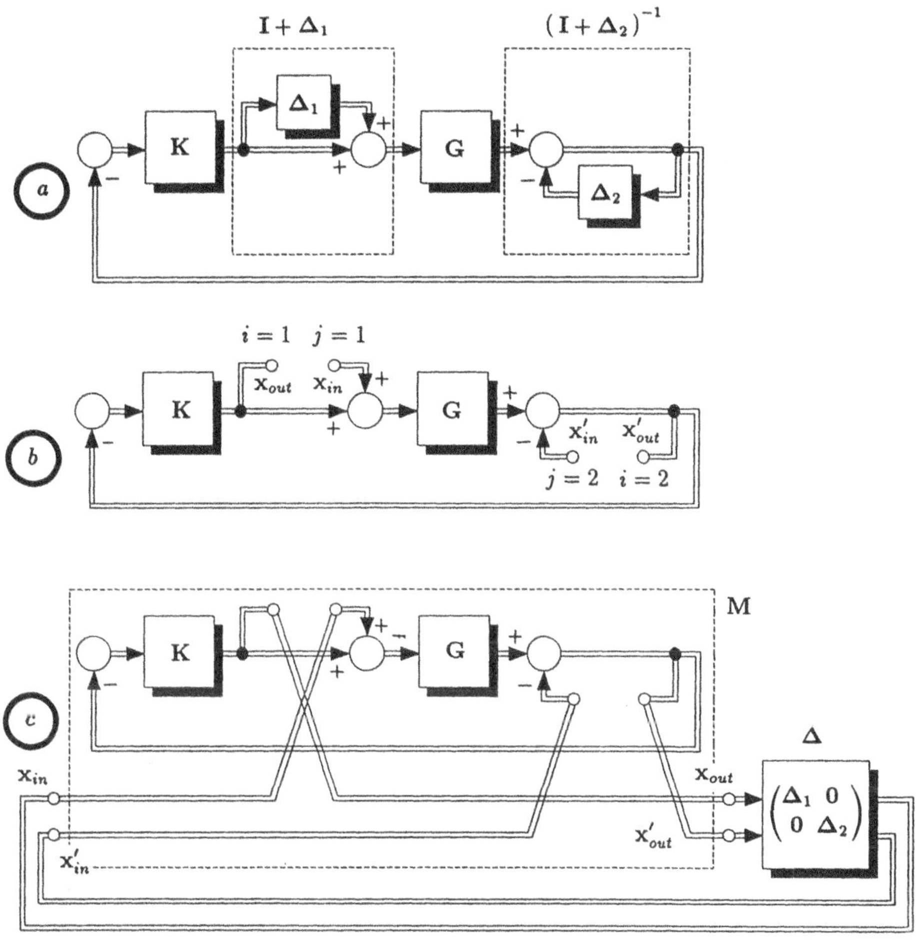

Figure 26.3: Multivariable control system with input and output uncertainty (a), when uncertainties are removed (b) and pulled out (c)

26.2 Block-Structured Uncertainty

If uncertainties occur at several different locations then the uncertainties are denoted $\Delta_i \; \forall \, i = 1 \ldots m$. There is a vector input to the control system at location j where signals of the uncertainty output are received. There is a vector output i of the system where signals are supplied to $\mathbf{\Delta}_i$.

In Fig. 26.3a a system is sketched with both an input and a divisive output uncertainty. In Fig. 26.3b the uncertainty is removed and the system is augmented by appropriate input and output signals.

Calculating the input output relations one has $-\mathbf{K}\mathbf{x}'_{out} = \mathbf{x}_{out}$ from the left side of

Fig. 26.3b and $\mathbf{G}(\mathbf{x}_{out}+\mathbf{x}_{in})-\mathbf{x}'_{in} = \mathbf{x}'_{out}$ which leads to the resulting two-input two-output system

$$\mathbf{M} = \begin{pmatrix} \mathbf{M}_{11} & \mathbf{M}_{12} \\ \mathbf{M}_{21} & \mathbf{M}_{22} \end{pmatrix} = \begin{pmatrix} -(\mathbf{I}+\mathbf{KG})^{-1}\mathbf{KG} & (\mathbf{I}+\mathbf{KG})^{-1}\mathbf{K} \\ (\mathbf{I}+\mathbf{GK})^{-1}\mathbf{G} & -(\mathbf{I}+\mathbf{GK})^{-1} \end{pmatrix} . \tag{26.12}$$

The block-diagonal uncertainty $\mathbf{\Delta} = \text{diag}\mathbf{\Delta}_i$ yields

$$\mathbf{\Delta} \times \begin{pmatrix} \mathbf{x}_{out} \\ \mathbf{x}'_{out} \end{pmatrix} = \begin{pmatrix} \mathbf{\Delta}_1 & 0 \\ 0 & \mathbf{\Delta}_2 \end{pmatrix} \begin{pmatrix} \mathbf{x}_{out} \\ \mathbf{x}'_{out} \end{pmatrix} = \begin{pmatrix} \mathbf{x}_{in} \\ \mathbf{x}'_{in} \end{pmatrix} , \tag{26.13}$$

see Fig. 26.3c .

The partitioned matrix $\mathbf{M} = \text{matrix}[\mathbf{M}_{ij}]$ represents the control system with inputs and outputs connected to the block-diagonal matrix $\mathbf{\Delta}$. Considering m different uncertainties, then $\mathbf{\Delta} = \text{block diag } \{\mathbf{\Delta}_1, \mathbf{\Delta}_2 \ldots \mathbf{\Delta}_m\} = \text{diag}\mathbf{\Delta}_i$ which is connected to the plant via scaling matrices $\mathbf{R} = \text{block diag } \mathbf{R}_i$ and $\mathbf{L} = \text{block diag } \mathbf{L}_i \quad \forall i = 1 \ldots m$. The general robustness condition

$$\sigma_{\max} \left[\mathbf{R}(j\omega)\mathbf{M}(j\omega)\mathbf{L}^{-1}(j\omega)\right] < 1 \qquad \forall\omega = 0 \ldots \infty \tag{26.14}$$

ignores the block-diagonal structure of uncertainties. Hence, the result may turn out very conservative.

In Fig. 26.3 no scaling matrices are included. Hence, comparing Eqs.(26.11) and (26.14) shows that $\mathbf{R}(s)$ and $\mathbf{L}^{-1}(s)$ are included in $\mathbf{M}(s)$ in the former and kept separated in the latter. It depends on the particular application whether it is suitable to take scaling matrices into account.

If each perturbation is considered separately, a lower bound of the robustness measure is obtained. If all the uncertainties are taken into consideration but the block-diagonal structure is neglected, an upper limit is achieved. These limits are given versus frequency. The lower limit is too optimistic, the upper limit too pessimistic. To reduce the gap between the limits the structured singular value was introduced.

26.3 Structured Singular Value

The structured singular value yields a method to analyze simultaneous occurrence of the uncertainties anywhere in the control system. Useless conservativeness is avoided.

Applying the standard singular-value method to block-diagonal bounded perturbation may be excessively conservative and, thus, yields weak performance.

Doyle, J.C., 1982 introduced the following sets relevant to control system's theory:

(i) Set of δ-norm-bounded block-diagonal matrices $\in \mathcal{R}^{k\times k}$

$$\mathcal{X}_\delta = \{\mathbf{\Delta}_{bd}\} = \{\text{block diag}[\underbrace{\mathbf{\Delta}_1, \mathbf{\Delta}_1, \ldots, \mathbf{\Delta}_1}_{m_1}, \underbrace{\mathbf{\Delta}_2, \mathbf{\Delta}_2, \ldots, \mathbf{\Delta}_2}_{m_2}, \quad \ldots\ldots \quad , \underbrace{\mathbf{\Delta}_n, \mathbf{\Delta}_n, \ldots, \mathbf{\Delta}_n}_{m_n}]\} \tag{26.15}$$

$$\text{where} \quad \mathbf{\Delta}_i \in \mathcal{R}^{k_i \times k_i}, \quad \|\mathbf{\Delta}_i\|_\infty \leq \delta \ \forall\, i = 1, 2 \ldots n; \quad k \triangleq \sum_{j=1}^{n} m_j k_j \,; \quad m = \sum_{j=1}^{n} m_j \,. \tag{26.16}$$

(ii) Set $\mathcal{X}_\infty$ of block-diagonal matrices with no restriction on the norm.

(iii) Set $\mathcal{U}$ of block-diagonal unitary matrices of the same structure as $\mathbf{\Delta}_{bd}$.

(iv) Set $\mathcal{D}$

$$\mathcal{D} = \{\text{block diag}\Big(d_1\mathbf{I}_{k_1},\ d_2\mathbf{I}_{k_1}, ...d_{m_1}\mathbf{I}_{k_1},\ d_{m_1+1}\mathbf{I}_{k_2},\ \ldots\ ,d_m\mathbf{I}_{k_m}\Big)\} \tag{26.17}$$

where d_j is real and positive.

The main definition is the structured singular value μ_D as a scalar-valued frequency dependent function

$$\mu_D[\mathbf{M}(j\omega)] = \begin{cases} 0 & \text{if } \det(\mathbf{I}-\mathbf{M\Delta}_{bd}) \neq 0,\ \ \mathbf{\Delta}_{bd} \in \mathcal{X}_\infty, \\ & \text{i.e., no } \mathbf{\Delta}_{bd} \text{ solves } \det(\mathbf{I}-\mathbf{M\Delta}_{bd}) = 0 \\ \left(\min\ \{\sigma_{\max}[\mathbf{\Delta}_{bd}]\}\right)^{-1} & \\ \text{minimum with respect} & \\ \text{to } \mathbf{\Delta}_{bd} \text{ where } \mathbf{\Delta}_{bd} \in \mathcal{X}_\infty & \text{otherwise.} \\ \text{and } \det(\mathbf{I}-\mathbf{M\Delta}_{bd}) = 0 & \end{cases} \tag{26.18}$$

For a given matrix $\mathbf{M}$ one has, first, to find a set of matrices $\mathbf{\Delta}_{bd}$ from the set $\mathcal{X}_\infty$ which satisfies $\det(\mathbf{I}-\mathbf{M\Delta}_{bd}) = 0$ and, second, to select $\mathbf{\Delta}_{bd}$ with minimum value of $\sigma_{\max}[\mathbf{\Delta}_{bd}]$. The structured singular value, finally, is inversely proportional to the selected minimum $\sigma_{\max}[\mathbf{\Delta}_{bd}]$.

By following steps similar to those in arriving at Eq.(26.11), the robust stability condition is given by

$$\sup_{\omega}\ \mu_D[\mathbf{M}(j\omega)] < 1\ . \tag{26.19}$$

The minus sign in the expression $\det(\mathbf{I}-\mathbf{M\Delta}_{bd}) = 0$ results from the fact that there is no sign inversion in the loop given by $\mathbf{M}$ and $\mathbf{\Delta}$, see Fig. 26.3c . With the slight difference given by the scaling operation, $\det(\mathbf{I}-\mathbf{M\Delta}_{bd}) = 0$ corresponds to the condition that $(\mathbf{I}-\mathbf{M\Theta})$ is singular in Eq.(26.5). Referring to the derivation of Eq.(26.11) the *maximum* singular value is required to check instability condition, i.e., maximum with respect to the set of singular values of $\mathbf{\Theta}$ or $\mathbf{\Delta}_{bd}$, respectively. Selection of the *minimum* versus $\mathbf{\Delta}_{bd}$ in Eq.(26.18) is required to attain maximum inverse which leads to maximum and, hence, sufficient $\mu_D[\mathbf{M}(j\omega)]$.

Roughly speaking, $\mu_D[\mathbf{M}(j\omega)]$ is inversely proportional to a frequency dependent stability margin against uncertainty. This definition is analogous to the following stability margin definition given by *deGaston, R.R.E., and Safonov, M.G., 1986*

$$k_s \triangleq \min_{\mathbf{\Delta}}\{k \in [0,\infty)\ :\ \det\left(\mathbf{I}-k\mathbf{\Delta}\ \mathbf{H}\right) = 0\} \tag{26.20}$$

where $\mathbf{\Delta}$ = block diag $\{\mathbf{\Delta}_1, \mathbf{\Delta}_2, \ldots, \mathbf{\Delta}_n\}$ is a matrix of frequency-dependent uncertainties $\mathbf{\Delta}_i$ and $\mathbf{H}(s) \in \mathcal{C}^{n\times n}$ is the stable nominal unperturbed loop transfer matrix.

Considering a system with *any* number of *real* uncertain parameters which may be repeated and related to each other, *Sánchez Peña, R.S., and Sideris, A., 1990* introduced the definition of robustness margin.

26.4 Small μ-Theorem

First of all, if the maximum spectral radius

$$\sup_{\Re e\ s\geq 0}\ \rho_s[\mathbf{M\Delta}] = \sup_{\Re e\ s\geq 0}\ \rho_s[\mathbf{ML}^{-1}\mathbf{\Theta R}] < 1\ , \tag{26.21}$$

then $\mathbf{I}+\mathbf{M\Delta}=\mathbf{I}+\mathbf{ML}^{-1}\mathbf{\Theta R}$ is nonsingular for all $\Re e\ s \geq 0$, i.e.,

$$\det(\mathbf{I}+\mathbf{M\Delta}) = \det(\mathbf{I}+\mathbf{ML}^{-1}\mathbf{\Theta R}) = \det(\mathbf{I}+\mathbf{RML}^{-1}\mathbf{\Theta}) \neq 0 \ . \tag{26.22}$$

Assuming $\mathbf{M}$ and $\mathbf{\Delta}$ stable,

$$\sup_{\omega}\ \rho_s[\mathbf{M\Delta}] = \sup_{\omega}\ \rho_s[\mathbf{RML}^{-1}\mathbf{\Theta}] < 1 \tag{26.23}$$

is sufficient. Since $\rho_s[\mathbf{M\Delta}] \leq \mu_D[\mathbf{M\Delta}]$, see Eq.(26.37), Eq.(26.23) is implied by

$$\sup_{\omega}\ \mu_D[\mathbf{RML}^{-1}\mathbf{\Theta}] < 1 \ . \tag{26.24}$$

Using the property Eq.(26.34), i.e. $\mu_D[\mathbf{RML}^{-1}\mathbf{\Theta}] < \sigma_{\max}[\mathbf{\Theta}]\mu_D[\mathbf{RML}^{-1}]$, and the upper bound for the uncertainty $\sigma_{\max}[\mathbf{\Theta}] < 1$ the inequality Eq.(26.24) is implied by

$$\sup_{\omega}\ \mu_D[\mathbf{RML}^{-1}] < 1 \tag{26.25}$$

$$\mathbf{R} = \text{diag}\ \mathbf{R}_i,\quad \mathbf{L} = \text{diag}\ \mathbf{L}_i,\quad \mathbf{\Delta} = \text{diag}\ \mathbf{\Delta}_i(s),\quad \mathbf{\Delta}_i(s) = \mathbf{L}_i^{-1}(s)\mathbf{\Theta}(s)\mathbf{R}_i(s) \ . \tag{26.26}$$

For comparison with the singular-value uncertainty assessment see Eq.(26.14).

When $\mathbf{G}_p$ is selected from a set of arcwise connected transfer functions and several nominal plant models are used, a generalized small-μ-test including possibly unstable perturbations has been developed by *Foo, Y.K., and Postlethwaite, I., 1988*.

26.5 Properties of the Structured Singular Value

The structured singular value is a generalization of the maximum singular value which can be emphasized by the following properties.

26.5.1 Block-Diagonal Matrix With One Block, Only

Consider $\mathcal{X}_\infty$ in the case of $n=1$, $m_1=1$ and any k_1. Then $\mathcal{X}_\infty$ is given by the special case that the block diagonal matrix consists of one block, only: $\mathbf{\Delta}_{bd} = \mathbf{\Delta}_1 = \mathbf{\Delta}$. Since the singular value σ of the matrix $\mathbf{I}-\mathbf{M\Delta}$ is given by

$$\det[\sigma^2\mathbf{I}-(\mathbf{I}-\mathbf{M\Delta})^H(\mathbf{I}-\mathbf{M\Delta})] = \det[\sigma\mathbf{I}-(\mathbf{I}-\mathbf{M\Delta})]\ \det[\sigma\mathbf{I}+(\mathbf{I}-\mathbf{M\Delta})] = 0\ , \tag{26.27}$$

with respect to Eq.(26.18), i.e., $\det(\mathbf{I}-\mathbf{M\Delta})=0$, Eq.(26.27) has only the solution given by $\sigma=\sigma[\mathbf{I}-\mathbf{M\Delta}]=0$. Using the singular value properties for the sum and product of matrices, and since the minimum $\sigma_{\max}[\cdot]$ has to be selected it follows

$$\sigma_{\max}[\mathbf{M}]\,\sigma_{\max}[\mathbf{\Delta}] = 1 \quad \leadsto \quad \mu_D[\mathbf{M}] = (\sigma_{\max}[\mathbf{\Delta}])^{-1} = \sigma_{\max}[\mathbf{M}] \quad \leadsto \quad \mu_D[\mathbf{M}] = \sigma_{\max}[\mathbf{M}] \ . \tag{26.28}$$

The structured singular value of a block diagonal matrix with only one block equals the maximum singular value of this block.

Since

$$1/\mu_D[\mathbf{M}(j\omega)] \leq \frac{1}{\max_i\{\sigma_{\max}[\mathbf{M}_{ii}]\}}\ , \tag{26.29}$$

the stability margin against simultaneous uncertainty is never greater than the stability margin against each uncertainty acting alone. If only one uncertainty is present, the appropriate singular value yields a reliable measure of the robustness properties. When more than one uncertainty occurs the singular value approach may provide results that are arbitrarily more conservative.

26.5.2 Submatrices are Identity Matrices

Consider the case when the block-diagonal matrix Δ_{bd} contains only m_1 submatrices Δ_1 equal to α times unity, i.e.,

$$n = 1, \quad k_1 = 1, \quad k = m_1, \quad \mathbf{\Delta}_{bd} = \mathbf{\Delta}_1 = \begin{pmatrix} \alpha & 0 & \dots & 0 \\ 0 & \alpha & \dots & 0 \\ \vdots & \vdots & \ddots & \vdots \\ 0 & 0 & \dots & \alpha \end{pmatrix} = \alpha \mathbf{I}_{m_1} \ , \quad \alpha \leq \delta \ . \qquad (26.30)$$

Then,

$$\mu_D[\mathbf{M}] = \frac{1}{|\alpha|_{\min}} \quad | \quad \det(\mathbf{I} - \mathbf{M}\alpha\mathbf{I}_{m_1}) = 0 \ . \qquad (26.31)$$

Hence,

$$\frac{1}{\alpha} = \lambda[\mathbf{M}] \quad \rightsquigarrow \quad \frac{1}{|\alpha|} = |\ \lambda[\mathbf{M}]\ | \quad \rightsquigarrow \quad \frac{1}{|\alpha|_{\min}} = |\ \lambda[\mathbf{M}]\ |_{\max} = \rho_s[\mathbf{M}] = \mu_D[\mathbf{M}] \ . \quad (26.32)$$

The structured singular value of a square matrix $\mathbf{M}$ where $\mathbf{\Delta}_{bd} = \alpha\mathbf{I}_{m_1}$ equals the spectral radius of $\mathbf{M}$.

26.5.3 Block-Diagonal Unitary and Identity Matrices

The following properties of μ_D are proven (*Doyle, J.C., 1982*)

$$\mu_D[\alpha\mathbf{G}] = |\alpha|\ \mu_D[\mathbf{G}] \qquad\qquad \mu_D[\mathbf{I}] = 1 \qquad (26.33)$$

$$\mu_D[\mathbf{AB}] \leq \sigma_{\max}[\mathbf{A}]\mu_D[\mathbf{B}] \qquad\qquad \mu_D[\mathbf{X}] - \sigma_{\max}[\mathbf{X}] \quad \forall \mathbf{X} \subset \mathcal{X}_\infty \qquad (26.34)$$

$$\mu_D[\mathbf{GU}] = \mu_D[\mathbf{UG}] = \mu[\mathbf{G}] \qquad \mathbf{U}\dots \text{ block-diagonal unitary matrix} \ . \qquad (26.35)$$

If $\mathbf{\Delta}_{bd} \in \mathcal{X}_\delta$ and $\mathbf{U} \in \mathcal{U}$ then $\mathbf{U\Delta} \in \mathcal{X}_\delta$ and $\mathbf{\Delta\ U} \in \mathcal{X}_\delta$. □

If $\mathbf{\Delta}_{bd} \in \mathcal{X}_\infty$ and $\mathbf{D} \in \mathcal{D}$ then $\mathbf{D\Delta}_{bd}\mathbf{D}^{-1} = \mathbf{\Delta}_{bd}$ and

$$\mu_D[\mathbf{DMD}^{-1}] = \mu_D[\mathbf{M}] \ . \ \square \qquad (26.36)$$

26.5.4 Lower and Upper Bound

The property

$$\max_{\mathbf{U}\in\mathcal{U}} \rho_s[\mathbf{UM}] = \mu_D[\mathbf{M}] \leq \inf_{\mathbf{D}\in\mathcal{D}} \sigma_{\max}[\mathbf{DMD}^{-1}] \qquad (26.37)$$

can be used to estimate $\mu_D[\mathbf{M}]$. Since the stability condition is $\sup_\omega \mu_D[\mathbf{M}(j\omega)] < 1$ it must be avoided that an estimate relation $\hat{\mu}_D$ smaller than μ_D is obtained. If $\hat{\mu}_D > \mu_D$ is obtained, then $\hat{\mu}_D[\mathbf{M}(j\omega)] < 1$ is sufficient . Hence, the right-hand inequality in Eq.(26.37) is a safer estimate than the left-hand equality because the left-hand side maximization with respect to the unitary matrix $\mathbf{U}$ is likely to achieve local maxima and to fail the global maximum.

The right-hand inequality in Eq.(26.37) becomes an equality if there are only three or fewer blocks ($n \leq 3$) and the blocks are not repeated ($m_j = 1\ \forall j = [1, n]$) , see Eq.(26.15). It can be shown that the minimization problem involved is always convex. The minimization has to be taken with respect to $n - 1$ parameters d_j, only, see Eq.(26.17), irrespective of the block size k_j.

The transformation $\mathbf{DMD}^{-1}$ is a scaling operation and Eq.(26.37) makes clear that the minimum singular value of the scaled matrix $\mathbf{M}$ equals the structured singular value. The structured singular value is independent of rescaling operations, see Eq.(26.36).

Improved and computable upper bounds to the structured singular value can also be obtained as the solution of a constrained optimization problem, providing a sufficient condition (*Fan, M.K.H., et al. 1991*).

The structured singular value yields improved results, compared with the performance robustness trade-off via loop-shaping techniques using simple singular values. Furthermore, the structured singular value allows for detailed integration of physical uncertainty descriptions (*Skogestad, S., et al. 1988*).

Note the encouraging fact (*Doyle, J.C., et al. 1982*) that even for cases of $n > 3$ the result of the right-inequality in Expr.(26.37) is quite reliable since the inequality deviates from the equality no more than roughly 15 per cent.

26.5.5 Bounds Given by Singular Values of the Elements of M

An upper bound of the structured singular value is given by ignoring the block diagonal structure and computing $\sigma_{\max}[\mathbf{M}]$. A lower bound is the largest maximum singular value when each perturbation is considered isolated. There is no guarantee that the bounds mentioned above are tight.

In the case of two uncertainty sources there exist additional bounds given by *Freudenberg, J.S., 1989*

$$\alpha \triangleq \sqrt{\|\mathbf{M}_{12}\|_s \|\mathbf{M}_{21}\|_s} \qquad \beta \triangleq \max\{\|\mathbf{M}_{11}\|_s, \|\mathbf{M}_{22}\|_s\} \tag{26.38}$$

$$\alpha - \beta \le \mu_D[\mathbf{M}] \le \alpha + \beta \; . \tag{26.39}$$

Numerical algorithms for the computation of $\mu_D[\mathbf{M}]$ are presented by *Fan, M.K.H., and Tits, A.L., 1986 and 1988*.

The right-hand infimum in Eq.(26.37) can be overestimated by the Perron-Frobenius eigenvalue for scalar perturbations, even repeated, i.e. $k_i = 1$

$$\inf_{\mathbf{D}} \sigma_{\max}[\mathbf{DMD}^{-1}] \le \inf_{\mathbf{D}} \sigma_{\max}[\,|\mathbf{DMD}^{-1}|\,] = \pi[\mathbf{M}] \; . \tag{26.40}$$

For detailed evaluation see Eq.(25.64).

An extension to $k_i > 1$ is given by partitioning according to the block-diagonal structure of $\mathbf{\Delta}$ (*Maciejowski, J.M., 1989*)

$$\mathbf{M}_N \triangleq \begin{pmatrix} \sigma_{\max}[\mathbf{M}_{11}] & \dots & \sigma_{\max}[\mathbf{M}_{1m}] \\ \sigma_{\max}[\mathbf{M}_{12}] & \dots & \vdots \\ \vdots & \ddots & \end{pmatrix} \tag{26.41}$$

$$\inf_{\mathbf{D}} \sigma_{\max}[\mathbf{DMD}^{-1}] \le \pi[\mathbf{M}_N] \; . \; \square \tag{26.42}$$

Osborne, E.E., 1960 presented an estimate applicable if $k_i = 1$ which leads to

$$\inf_{\mathbf{D}} \sigma_{\max}[\mathbf{DMD}^{-1}] \le \sigma_{\max}[\mathbf{D}^{\star}\mathbf{MD}^{\star,-1}] \tag{26.43}$$

$$\text{where} \qquad \mathbf{D}^{\star} = \arg \min_{\mathbf{D} \in \mathcal{D}} \sum_i \sigma_i^2[\mathbf{DMD}^{-1}] \; . \; \square \tag{26.44}$$

A computational technique is presented by *deGaston, R.R.E., and Safonov, M.G., 1988* with the objective of calculating the stability margin of feedback control systems with diagonally structured perturbation. The method holds for real and complex parameter variations. The calculation yields the stability margin k_m without conservativeness

$$k_m \triangleq \min_{\Delta} \{k \in [0, \infty) \; : \; \det(\mathbf{I} - k \, \mathbf{\Delta} \, \mathbf{G}) = 0 \} \tag{26.45}$$

where $\mathbf{G}(s) \in C^{n \times n}$ is the stable nominal unperturbed loop transfer matrix and $\mathbf{\Delta} \triangleq \text{diag}\mathbf{\Delta}_i$ is a block-diagonal perturbation matrix. The problem of largest stability margin is discussed by *Dahleh, M.A., 1990*.

26.6 Diagonally Scaled H_∞-Optimal Control Problem

Based on the fact that an upper bound on the μ_D-function is given by the H_∞-norm, the condition $\mu_D[\mathbf{\Omega}_{ij}(s)] < 1$ can be replaced by the sufficient condition $\|\mathbf{\Omega}_{ij}(s)\|_\infty < 1$ which is a link to H_∞-theory. Since this bound is not a strict one, in order to reduce conservativeness diagonal scaling is introduced. Invoking Eq.(30.50) and the transfer matrix $\mathbf{\Omega}_{ij}(s)$ from any plant input $\mathbf{z}_j$ to output $\mathbf{v}_i$, the diagonally scaled H_∞-optimal control problem is stated by

$$\mu_D[\mathbf{\Omega}_{ij}(s)] < \inf_{\mathbf{D}} \; \min_{\mathbf{K}} \|\mathbf{D}(s)\mathbf{\Omega}_{ij}(s)\mathbf{D}^{-1}(s)\|_\infty \qquad \mathbf{D} \in H_\infty \, . \tag{26.46}$$

Iteratively fixing $\mathbf{D}(s)$ and minimizing $\|\mathbf{D}(s)\mathbf{\Omega}_{ij}(s)\mathbf{D}^{-1}(s)\|_\infty$ over the set of controllers $\mathbf{K}(s)$ and, then, fixing $\mathbf{K}(s)$ and minimizing $\|\mathbf{D}(s)\mathbf{\Omega}_{ij}(s)\mathbf{D}^{-1}(s)\|_\infty$ over $\mathbf{D}(s)$ provides an effective procedure for synthesis (*Safonov, M.G., 1990*).

26.7 Mapping Theorem

Highly structured uncertainties and their influence on the stability of a system can be investigated by accurate methods different from singular values or structured singular values. A mapping operation using highly structured uncertainty data, outlined below, is a basis for nonconservative robustness. In Eq.(26.4), omitting $\mathbf{R}$ and $\mathbf{L}$, the transfer matrix $\mathbf{M}(s) = [\mathbf{I} + \mathbf{K}(s)\mathbf{G}(s)]^{-1}\mathbf{K}(s)$ represents all interactions of the feedback loop with its uncertainties. In Eq.(26.5), $[\mathbf{I} - \mathbf{M}(s)\mathbf{\Theta}(s)]^{-1}$ is responsible for the overall transition. Hence, the map $\det[\mathbf{I} - \mathbf{M}(s)\mathbf{\Theta}(s)]$ for $s = j\omega_o$ determines the stability robustness where $\mathbf{\Theta}(s)$ is the uncertainty matrix the elements of which lie inside a polygon p_y at each frequency ω_o . For ease of presentation, the image of this map is termed by an operator T^{op} . With respect to the continuity of the function represented by the determinant, the image is a connected region.

Foo, Y.K., 1990 pointed out an extension to basic stability considerations which slightly reminds one of the Edge theorem: The exterior boundary of the image of the Cartesian product (P_c^{op}) of the polygons $p_y(\omega_o)$ under the map T^{op} is contained in the boundary of the image of the Cartesian product of the *boundaries* of the polygons, i.e.,

$$\partial T^{op} P_c^{op} p_y(\omega_o) \subseteq \partial T^{op} P_c^{op} \partial p_y(\omega_o) \tag{26.47}$$

where ∂D denotes the boundary of D . Thus, the image can be constructed from the image of the *edges* and the perturbed system is stable if and only if all the edge perturbations yield a stable system. Numerically, the technique presented by *deGaston, R.R.E., and Safonov, M.G., 1988* is applicable.

Chapter 27

Performance Robustness

Robustness is usually considered as stability robustness in the face of perturbed plant dynamics. If, additionally, a certain level of performance has to be guaranteed, robustness is modified to performance robustness. In the s-domain the method of maximum singular values and structured singular values can easily be reformulated to obtain acceptable performance in the presence of plant uncertainties.

27.1 Singular-Value-Based Method

27.1.1 Single Uncertainty

From Fig. 27.1

$$\mathbf{y} = [\mathbf{I} + (\mathbf{G} + \boldsymbol{\Delta})\mathbf{K}]^{-1}(\mathbf{G} + \boldsymbol{\Delta})\mathbf{K}\mathbf{y}_{ref} \triangleq \mathbf{T}\mathbf{y}_{ref} \ . \tag{27.1}$$

The performance is considered acceptable if $\|\mathbf{T}\|_\infty < 1$ or

$$\|\mathbf{T}\|_\infty = \| \, [\mathbf{I} + (\mathbf{G} + \boldsymbol{\Delta})\mathbf{K}]^{-1}(\mathbf{G} + \boldsymbol{\Delta})\mathbf{K}\mathbf{y}_{ref}\|_\infty < 1 \ . \tag{27.2}$$

Stability robustness is given if, with uncertainty removed, the transfer function matrix from the system input $\mathbf{v}$ to the output $\mathbf{u}$, i.e., $-(\mathbf{I} + \mathbf{KG})^{-1}\mathbf{K} = \mathbf{M}(s)$, applied to Eq.(26.11) or (26.25), yields

$$\|\mathbf{M}(s)\|_\infty < 1 \qquad \text{or} \qquad \sup_\omega \mu_D[\mathbf{M}(s)] < 1 \qquad s = j\omega \qquad \forall\omega = 0 \ldots \infty \ . \tag{27.3}$$

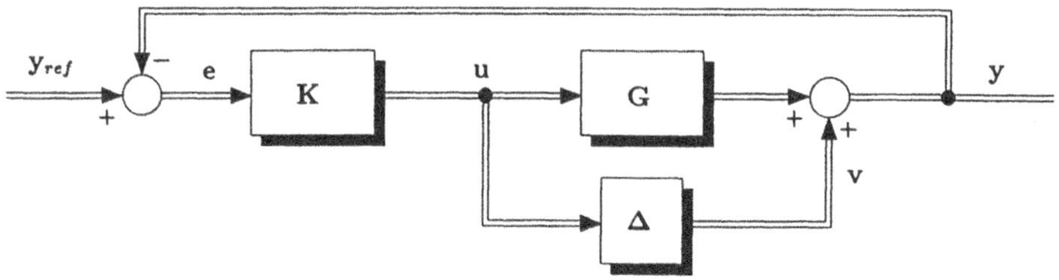

Figure 27.1: Multivariable control system with a single uncertainty

27.1.2 General Uncertainty

From Fig. 27.1 one has $\mathbf{K}(\mathbf{y}_{ref} - \mathbf{y}) = \mathbf{u}$ and $\mathbf{v} + \mathbf{G}\mathbf{u} = \mathbf{y}$, hence, two equations and four variables $\mathbf{y}_{ref}, \mathbf{y}, \mathbf{u}, \mathbf{v}$. The outputs $\mathbf{y}$ and $\mathbf{u}$ can be considered as a function of the inputs $\mathbf{y}_{ref}$ and $\mathbf{v}$ to the system given by the reference source and the uncertainty, respectively. Rewriting yields

$$\begin{pmatrix} \mathbf{y} \\ \mathbf{u} \end{pmatrix} = \begin{pmatrix} (\mathbf{I}+\mathbf{G}\mathbf{K})^{-1}\mathbf{G}\mathbf{K} & (\mathbf{I}+\mathbf{G}\mathbf{K})^{-1} \\ (\mathbf{I}+\mathbf{K}\mathbf{G})^{-1}\mathbf{K} & -(\mathbf{I}+\mathbf{K}\mathbf{G})^{-1}\mathbf{K} \end{pmatrix} \begin{pmatrix} \mathbf{y}_{ref} \\ \mathbf{v} \end{pmatrix} \triangleq \begin{pmatrix} \mathbf{J} & \mathbf{H} \\ \mathbf{N} & \mathbf{M} \end{pmatrix} \begin{pmatrix} \mathbf{y}_{ref} \\ \mathbf{v} \end{pmatrix} . \tag{27.4}$$

Applying the input-output relation of the uncertainty

$$\mathbf{y} = \mathbf{J}\mathbf{y}_{ref} + \mathbf{H}\mathbf{v} \tag{27.5}$$

$$\mathbf{u} = \mathbf{N}\mathbf{y}_{ref} + \mathbf{M}\mathbf{v} = \mathbf{N}\mathbf{y}_{ref} + \mathbf{M}\boldsymbol{\Delta}\,\mathbf{u} \tag{27.6}$$

or

$$\mathbf{y} = [\mathbf{J} + \mathbf{H}\boldsymbol{\Delta}(\mathbf{I} - \mathbf{M}\boldsymbol{\Delta})^{-1}\mathbf{N}]\mathbf{y}_{ref} \triangleq \mathbf{T}\mathbf{y}_{ref} . \tag{27.7}$$

Performance robustness is considered given if

$$\|\mathbf{T}\|_\infty = \|\mathbf{J} + \mathbf{H}\boldsymbol{\Delta}(\mathbf{I} - \mathbf{M}\boldsymbol{\Delta})^{-1}\mathbf{N}\|_\infty < 1 . \tag{27.8}$$

Stability robustness is satisfied if

$$\sup_\omega \mu_D[\mathbf{M}] < 1 . \tag{27.9}$$

27.1.3 Performance Robustness Theorem

Both inequalities mentioned above can be combined by establishing a fictitious $\boldsymbol{\Delta}_o$ from the output $\mathbf{y}$ to the reference $\mathbf{y}_{ref}$ where $\|\boldsymbol{\Delta}_o\|_\infty \leq 1$, see[1] Fig. 27.2. Postulating

$$\sup_\omega \mu_D[\begin{pmatrix} \mathbf{J} & \mathbf{H} \\ \mathbf{N} & \mathbf{M} \end{pmatrix}] < 1 \tag{27.10}$$

is a sufficient condition for both performance and stability condition if $\boldsymbol{\Delta}$ is a block-diagonal matrix $\in \mathcal{X}_1$, see Eq.(26.15) (*Doyle, J.C., et al. 1982*).

Proof: The structured singular value satisfies

$$\mu_D[\begin{pmatrix} \mathbf{J} & \mathbf{H} \\ \mathbf{N} & \mathbf{M} \end{pmatrix}] < 1 \iff \det[\mathbf{I} - \begin{pmatrix} \mathbf{J} & \mathbf{H} \\ \mathbf{N} & \mathbf{M} \end{pmatrix} \begin{pmatrix} \boldsymbol{\Delta}_o & 0 \\ 0 & \boldsymbol{\Delta} \end{pmatrix}] > 0 \tag{27.11}$$

where each entry of $\mathbf{J}, \mathbf{H}, \mathbf{N}, \mathbf{M}$ is a function of $s = j\omega$ with $\|\boldsymbol{\Delta}_o\|_\infty < 1$ and $\boldsymbol{\Delta} \in \mathcal{X}_1$, see Eq.(26.15) for $\delta = 1$. Evaluating the right-hand inequality above

$$\det[\mathbf{I} - \begin{pmatrix} \mathbf{J}\boldsymbol{\Delta}_o & \mathbf{H}\boldsymbol{\Delta} \\ \mathbf{N}\boldsymbol{\Delta}_o & \mathbf{M}\boldsymbol{\Delta} \end{pmatrix}] = \det[\begin{pmatrix} \mathbf{I} - \mathbf{J}\boldsymbol{\Delta}_o & -\mathbf{H}\boldsymbol{\Delta} \\ -\mathbf{N}\boldsymbol{\Delta}_o & \mathbf{I} - \mathbf{M}\boldsymbol{\Delta} \end{pmatrix}] > 0 . \tag{27.12}$$

Using Eq.(C.40),

$$\det(\mathbf{I} - \mathbf{M}\boldsymbol{\Delta}) \det[\mathbf{I} - \mathbf{J}\boldsymbol{\Delta}_o - \mathbf{H}\boldsymbol{\Delta}(\mathbf{I} - \mathbf{M}\boldsymbol{\Delta})^{-1}\mathbf{N}\boldsymbol{\Delta}_o] > 0 . \tag{27.13}$$

The inequality above must hold for separately $\boldsymbol{\Delta} = \mathbf{0}$ or $\boldsymbol{\Delta}_o = \mathbf{0}$. Hence, both determinants must be greater zero. The former determinant condition corresponds to $\sup_\omega \mu_D[\mathbf{M}] < 1$. Referring to Eq.(26.11), the latter determinant condition is rewritten to

$$\det[\mathbf{I} - \left(\mathbf{J} + \mathbf{H}\boldsymbol{\Delta}(\mathbf{I} - \mathbf{M}\boldsymbol{\Delta})^{-1}\mathbf{N}\right)\boldsymbol{\Delta}_o] > 0 \quad \leadsto \quad \|\mathbf{J} + \mathbf{H}\boldsymbol{\Delta}(\mathbf{I} - \mathbf{M}\boldsymbol{\Delta})^{-1}\mathbf{N}\|_\infty < 1 . \tag{27.14}$$

End of the Proof

[1]For comparison purpose, only: In Eq.(26.1) scaling matrices $\mathbf{L}^{-1}$ and $\mathbf{R}$ are used in order to let $\|\boldsymbol{\Theta}\|_\infty \leq 1$. In this section $\boldsymbol{\Delta}_o$ itself is normalized $\|\boldsymbol{\Delta}_o\|_\infty \leq 1$.

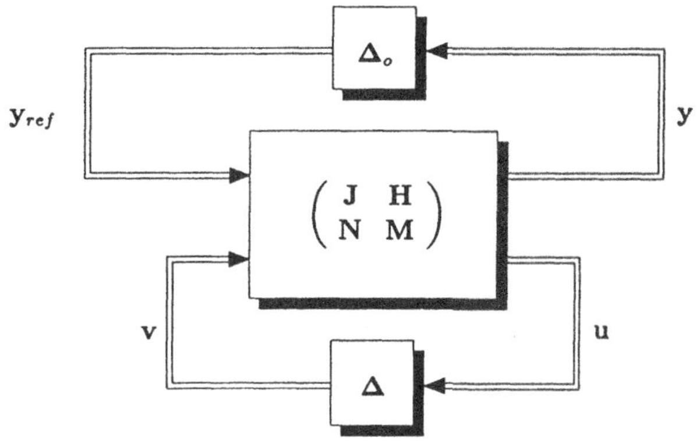

Figure 27.2: Additional feedback by a fictitious $\mathbf{\Delta}_o$

27.2 Design Based on the Index of Performance

27.2.1 Ordinary Linear System

A system is said performance-bounded if the index of performance I_p , taken of the perturbed system, does not exceed the nominal index I_n by more than a factor $k_I > 1$, i.e., if

$$I_p \triangleq \int_0^{\infty} \mathbf{x}_p^T(t)\mathbf{Q}\mathbf{x}_p(t)dt < k_I I_n \triangleq k_I \int_0^{\infty} \mathbf{x}^T(t)\mathbf{Q}\mathbf{x}(t)dt \tag{27.15}$$

where $\dot{\mathbf{x}}_p(t) = [\mathbf{A}(t) + \Delta\mathbf{A}(t)]\mathbf{x}_p(t)$ and $\dot{\mathbf{x}}(t) = \mathbf{A}\mathbf{x}(t)$. From

$$\mathbf{PA} + \mathbf{A}^T\mathbf{P} + 2\mathbf{Q} = 0 \qquad \mathbf{Q} = \mathbf{Q}^T > 0 \tag{27.16}$$

for spectral-norm-bounded $\Delta\mathbf{A}$ the performance robustness criterion is

$$\sigma_{\max}[\Delta\mathbf{A}(t)] \le \frac{k_I - 1}{k_I}\frac{\sigma_{\min}[\mathbf{Q}]}{\sigma_{\max}[\mathbf{P}]} . \tag{27.17}$$

For structured uncertainty one has (*Yedavalli, R.K., 1985c*)

$$\Delta\mathbf{A} \le_e \mathbf{E}(t) \qquad e_{\max} = \max_{i,j} E_{ij}(t) \qquad U_{ei,j} = (\mathbf{U}_e)_{i,j} \triangleq \frac{E_{ij}(t)}{e_{\max}} \tag{27.18}$$

$$e_{\max} \le \frac{k_I - 1}{k_I}\frac{\sigma_{\min}[\mathbf{Q}]}{\sigma_{\max}[\,(|\mathbf{P}|\mathbf{U}_e)_s\,]} . \tag{27.19}$$

27.2.2 Optimal Linear System

The objective is to design a control system, both optimal with respect to a quadratic index of performance and robust with regard to a specified index of performance. The linear time-invariant system

$$\dot{\mathbf{x}}(t) = \mathbf{A}\mathbf{x}(t) + \mathbf{B}\mathbf{u}(t) \qquad \mathbf{x} \in \mathcal{R}^n, \mathbf{u} \in \mathcal{R}^m \tag{27.20}$$

$$\mathbf{y}(t) = \mathbf{C}\mathbf{x}(t) \qquad \mathbf{y} \in \mathcal{R}^r \tag{27.21}$$

under the optimality condition

$$I = \int_0^\infty [\mathbf{y}^T(t)\mathbf{Q}\mathbf{y}(t) + \rho_c \mathbf{u}^T(t)\mathbf{R}\mathbf{u}(t)]dt \rightarrow \min \tag{27.22}$$

yields $\mathbf{K} = -\mathbf{R}^{-1}\mathbf{B}^T\mathbf{P}_o/\rho_c$ where $\mathbf{P}_o$ is given by

$$\mathbf{P}_o\mathbf{A} + \mathbf{A}^T\mathbf{P}_o - \mathbf{P}_o\mathbf{B}\mathbf{R}^{-1}\mathbf{B}^T\mathbf{P}_o/\rho_c + \mathbf{C}^T\mathbf{Q}\mathbf{C} = \mathbf{0} \ . \tag{27.23}$$

The closed-loop nominal behaviour is obtained from $\dot{\mathbf{x}} = (\mathbf{A}+\mathbf{B}\mathbf{K})\mathbf{x} \triangleq \mathbf{F}\mathbf{x}$. The corresponding maximum modulus matrices of the structured uncertainty deviations $\Delta\mathbf{A}$, $\Delta\mathbf{B}$, $\Delta\mathbf{K}$ of $\mathbf{A}, \mathbf{B}$ and $\mathbf{K}$ are termed $\mathbf{E}, \mathbf{E}_B, \mathbf{E}_K$, respectively. Then, the performance robustness is stated by the condition

$$I_p \triangleq \int_0^\infty \mathbf{x}_p^T(t)\mathbf{C}^T\mathbf{Q}\mathbf{C}\mathbf{x}_p(t)dt \le k_I I_n \triangleq k_I \int_0^\infty \mathbf{x}^T(t)\mathbf{C}^T\mathbf{Q}\mathbf{C}\mathbf{x}(t)dt \tag{27.24}$$

where $\mathbf{x}_p$ is given by $\dot{\mathbf{x}}_p(t) = [(\mathbf{A}+\Delta\mathbf{A}) + (\mathbf{B}+\Delta\mathbf{B})(\mathbf{K}+\Delta\mathbf{K})]\mathbf{x}_p(t) \triangleq \mathbf{F}_p\mathbf{x}_p(t)$ and I_n is the performance in the unperturbed case. Defining

$$e_{\max\ CL} \triangleq \max_{i,j}[\ \mathbf{E} + \mathbf{E}_B|\mathbf{K}| + |\mathbf{B}+\mathbf{E}_B|\ \mathbf{K}\ ]_{ij} \ , \tag{27.25}$$

the condition Eq.(27.24) is satisfied if

$$e_{\max\ CL} < \frac{\sigma_{\min}[\mathbf{C}^T\mathbf{Q}\mathbf{C}]}{\sigma_{\max}[\ (|\mathbf{P}|\mathbf{U}_e)_s\ ]} \tag{27.26}$$

where $\mathbf{P}$ results from $\mathbf{P}\mathbf{F} + \mathbf{F}^T\mathbf{P} + 2\mathbf{C}^T\mathbf{Q}\mathbf{C} = \mathbf{0}$ (*Yedavalli, R.K., 1985c*). The parameter ρ_c is considered as a design factor globally weighting the control effort. It seems suitable to determine the factor k_I by equating an index of performance $I_{p\max}$ in the *worst* case of the perturbed system, i.e., $I_{p\max} = k_I I_n(\rho_c)$ where I_n is defined by Eq.(27.24). Hence, it results $k_I(\rho_c) = I_{p\max}/I_n(\rho_c)$, and $I_p \le I_{p\max} = k_I I_n$ is satisfied for all admissible perturbations.

Chapter 28

Robust Controllers Via Spectral Radius Technique

Consider a multivariable system as depicted in Fig. 28.1. The controller is decomposed into three transfer matrices $\mathbf{K}_r(s)$, $\mathbf{K}_f(s)$ and $\mathbf{K}_y(s)$. The plant $\mathbf{G}(s)$ is characterized by an uncertainty in parallel $\Delta\mathbf{G}(s)$. The uncertainty is caused by unknown plant structure and parameters or by reduced-order model approximation of the plant.

28.1 Control System Without Uncertainty

If the plant uncertainty $\Delta\mathbf{G}(s) = \mathbf{0}$, the internal signals are denoted with an overline bar $\bar{\mathbf{u}}$, $\bar{\mathbf{y}}$ and the following definitions and relations in the s-domain are stated. The first and second subscript symbols denote reaction and source signal, respectively.

Return-ratio matrix (break point S):

$$\mathbf{F}_o \triangleq \mathbf{G}\mathbf{K}_f\mathbf{K}_y \tag{28.1}$$

Controller:

$$\mathbf{K}\begin{pmatrix}\mathbf{y}_{ref}\\ \mathbf{y}\end{pmatrix} = \mathbf{K}_f(\mathbf{K}_r\mathbf{y}_{ref} - \mathbf{K}_y\mathbf{y}) \tag{28.2}$$

$\mathbf{y}_{ref} = \mathbf{0}$:

$$\mathbf{F}_{yw} \triangleq (\mathbf{I} + \mathbf{G}\mathbf{K}_f\mathbf{K}_y)^{-1} = (\mathbf{I} + \mathbf{F}_o)^{-1} \quad \text{and} \quad \bar{\mathbf{y}} = \mathbf{F}_{yw}\mathbf{w} \tag{28.3}$$

$$\mathbf{F}_{uw} \triangleq \mathbf{K}_f\mathbf{K}_y(\mathbf{I} + \mathbf{G}\mathbf{K}_f\mathbf{K}_y)^{-1} \quad \text{and} \quad \bar{\mathbf{u}} = -\mathbf{F}_{uw}\mathbf{w} \tag{28.4}$$

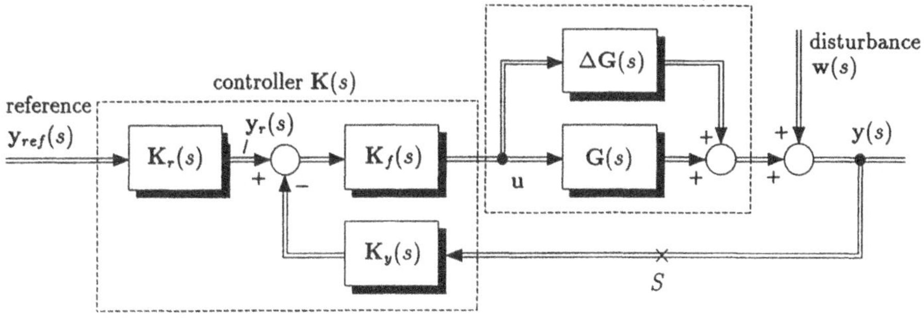

Figure 28.1: Multivariable controller $\mathbf{K}(s)$ and plant $\mathbf{G}(s)$ with uncertainty $\Delta\mathbf{G}(s)$

$\mathbf{w} = \mathbf{0}$:

$$\mathbf{F}_{yy_r} \triangleq (\mathbf{I} + \mathbf{G}\mathbf{K}_f\mathbf{K}_y)^{-1}\mathbf{G}\mathbf{K}_f \quad \text{and} \quad \bar{\mathbf{y}} = \mathbf{F}_{yy_r}\mathbf{y}_r \qquad (28.5)$$

$$\mathbf{F}_{uy_r} \triangleq \mathbf{K}_f(\mathbf{I} + \mathbf{K}_y\mathbf{G}\mathbf{K}_f)^{-1} \quad \text{and} \quad \bar{\mathbf{u}} = \mathbf{F}_{uy_r}\mathbf{y}_r \qquad (28.6)$$

$\mathbf{w} \neq \mathbf{0},\ \mathbf{y}_{ref} \neq \mathbf{0}$:

$$\bar{\mathbf{y}} = \mathbf{F}_{yy_r}\mathbf{y}_r + \mathbf{F}_{yw}\mathbf{w} = \mathbf{F}_{yy_r}\mathbf{y}_r + (\mathbf{I} + \mathbf{G}\mathbf{K}_f\mathbf{K}_y)^{-1}\mathbf{w} = \mathbf{F}_{yy_r}\mathbf{y}_r + (\mathbf{I} - \mathbf{F}_{yy_r}\mathbf{K}_y)\mathbf{w} \qquad (28.7)$$

$$\bar{\mathbf{u}} = -\mathbf{F}_{uw}\mathbf{w} + \mathbf{K}_f(\mathbf{I} + \mathbf{K}_y\mathbf{G}\mathbf{K}_f)^{-1}\mathbf{y}_r = \mathbf{K}_f(\mathbf{I} + \mathbf{K}_y\mathbf{G}\mathbf{K}_f)^{-1}(\mathbf{y}_r - \mathbf{K}_y\mathbf{w}) \qquad (28.8)$$

$$= \mathbf{F}_{uy_r}(\mathbf{y}_r - \mathbf{K}_y\mathbf{w})\ . \qquad (28.9)$$

Note that $\mathbf{I} - \mathbf{F}_{yy_r}\bar{\mathbf{y}} = \mathbf{I} - (\mathbf{I} + \mathbf{G}\mathbf{K}_f\mathbf{K}_y)^{-1}\mathbf{G}\mathbf{K}_f\mathbf{K}_y = \ldots = (\mathbf{I} + \mathbf{G}\mathbf{K}_f\mathbf{K}_y)^{-1}$.

28.2 Control System Including Uncertainty $\Delta\mathbf{G}(s)$

If the uncertainty $\Delta\mathbf{G}(s) \neq \mathbf{0}$, the controlling signal and the output are denoted without an overbar: $\mathbf{u}, \mathbf{y}$. By simple calculation, Eq.(30.50), it results

$$\mathbf{y} = [\mathbf{I} + (\mathbf{G} + \Delta\mathbf{G})\mathbf{K}_f\mathbf{K}_y]^{-1}[(\mathbf{G} + \Delta\mathbf{G})\mathbf{K}_f\mathbf{K}_r\mathbf{y}_{ref} + \mathbf{w}]\ . \qquad (28.10)$$

Using the definition

$$\mathbf{F}_d \triangleq \mathbf{F}_{yw}(\mathbf{I} + \Delta\mathbf{G}\ \mathbf{F}_{uw})^{-1}\Delta\mathbf{G}\ , \qquad (28.11)$$

the following identities can easily be proved

$$(\mathbf{F}_{yy_r} + \mathbf{F}_d\mathbf{F}_{uy_r}) = [\mathbf{I} + (\mathbf{G} + \Delta\mathbf{G})\mathbf{K}_f\mathbf{K}_y]^{-1}(\mathbf{G} + \Delta\mathbf{G})\mathbf{K}_f \qquad (28.12)$$

$$\mathbf{I} - (\mathbf{F}_{yy_r} + \mathbf{F}_d\mathbf{F}_{uy_r})\mathbf{F}_y = [\mathbf{I} + (\mathbf{G} + \Delta\mathbf{G})\mathbf{K}_f\mathbf{K}_y]^{-1}\ . \qquad (28.13)$$

Hence, Eq.(28.10) is rewritten to

$$\mathbf{y} = (\mathbf{F}_{yy_r} + \mathbf{F}_d\mathbf{F}_{uy_r})\mathbf{K}_r\mathbf{y}_{ref} + [\mathbf{I} - (\mathbf{F}_{yy_r} + \mathbf{F}_d\mathbf{F}_{uy_r})\mathbf{K}_y]\mathbf{w}\ . \qquad (28.14)$$

Setting $\Delta\mathbf{G}(s) = \mathbf{0}$ yields $\mathbf{F}_d = \mathbf{0}$ and the result $\bar{\mathbf{y}}$ as given in Eq.(28.7). Substraction Eq.(28.14) minus Eq.(28.7) yields

$$\mathbf{y} - \bar{\mathbf{y}} = \mathbf{F}_d\mathbf{F}_{uy_r}\mathbf{K}_r\mathbf{y}_{ref} - \mathbf{F}_d\mathbf{F}_{uy_r}\mathbf{K}_y\mathbf{w} = \mathbf{F}_d\mathbf{F}_{uy_r}(\mathbf{K}_r\mathbf{y}_{ref} - \mathbf{K}_y\mathbf{w})\ . \qquad (28.15)$$

Invoking Eq.(28.9) and substituting the definition $\mathbf{F}_d$,

$$\mathbf{y}(s) - \bar{\mathbf{y}}(s) = \mathbf{F}_{yw}(s)[\ \mathbf{I} + \Delta\mathbf{G}(s)\ \mathbf{F}_{uw}(s)\ ]^{-1}\Delta\mathbf{G}(s)\bar{\mathbf{u}}(s) \qquad (28.16)$$

which is an important relation in the s-domain, analyzing the output difference (output uncertainty) $\mathbf{y} - \bar{\mathbf{y}}$ as determined by the uncertainty $\Delta\mathbf{G}$ (see Fig. 28.2). The well-known transfer matrices $\mathbf{F}_{yw}$ and $\mathbf{F}_{uw}$ act as sensitivity gain matrices.

28.3 Modified Spectral Radius

Referring to the spectral radius of a square matrix **X**, i.e. $\rho_s[\mathbf{X}] = \max_i |\lambda_i[\mathbf{X}]\,|$, a modified spectral radius $q_s[\mathbf{X}]$ is defined

$$q_{sX} \triangleq q_s[\mathbf{X}] \triangleq \begin{cases} \rho_s[\mathbf{X}] & \text{if } \rho_s[\mathbf{X}] \neq 0 \\ \|\mathbf{X}\| & \text{if } \rho_s[\mathbf{X}] = 0 \end{cases} \tag{28.17}$$

where $\|\mathbf{X}\|$ is any consistent norm. The modified spectral radius q_{sX} is introduced because

$$\rho_s[\mathbf{AB}] \leq q_s[\mathbf{A}]q_s[\mathbf{B}] = q_{sA}q_{sB} \tag{28.18}$$

is always true (*Carlucci, D., and Vallauri, M., 1981*) but $\rho_s[\mathbf{AB}] \leq \rho_s[\mathbf{A}]\rho_s[\mathbf{B}]$ is not, e.g.,

$$a > d\,, \qquad \mathbf{A} = \begin{pmatrix} 0 & 1 \\ 0 & 0 \end{pmatrix}, \quad \mathbf{B} = \begin{pmatrix} a & 0 \\ c & d \end{pmatrix}, \qquad c \not\leq 0 \times a = 0\,. \tag{28.19}$$

If the matrix **X** is a function of s then $q_s[\mathbf{X}]$ also depends on s.

28.4 Multivariable Control and Spectral Radius Properties

Premultiplying Eq.(28.4) by **G** and applying Eq.(28.3),

$$\begin{aligned} \mathbf{GF}_{uw} &= \mathbf{GK}_f\mathbf{K}_y(\mathbf{I}+\mathbf{GK}_f\mathbf{K}_y)^{-1} + (\mathbf{I}+\mathbf{GK}_f\mathbf{K}_y)^{-1} - (\mathbf{I}+\mathbf{GK}_f\mathbf{K}_y)^{-1} && (28.20)\\ &= (\mathbf{GK}_f\mathbf{K}_y + \mathbf{I})(\mathbf{I}+\mathbf{GK}_f\mathbf{K}_y)^{-1} - \mathbf{F}_{yw} = \mathbf{I} - \mathbf{F}_{yw} && (28.21)\end{aligned}$$

Using Eq.(28.18),

$$q_s[\mathbf{G}]q_s[\mathbf{F}_{uw}] \geq \rho_s[\mathbf{GF}_{uw}] = \max_i |\lambda_i[\mathbf{GF}_{uw}]\,| = \max_i |\lambda_i[\mathbf{I} - \mathbf{F}_{yw}]\,| \tag{28.22}$$

$$q_s[\mathbf{G}](s)q_s[\mathbf{F}_{uw}](s) = \max_i |1 - \lambda_i[\mathbf{F}_{yw}]\,| \geq |1 - q_s[\mathbf{F}_{yw}](s)|\,. \tag{28.23}$$

Similarily,

$$|\lambda_{\min}[\mathbf{G}]\,|(s)\; q_s[\mathbf{F}_{uw}](s) \leq 1 + q_s[\mathbf{F}_{yw}](s)\,. \tag{28.24}$$

From Eq.(28.3) it results

$$\mathbf{F}_{yw}^{-1} = \mathbf{I} + \mathbf{F}_o \qquad \rightsquigarrow \qquad \lambda[\mathbf{F}_{yw}^{-1}] = 1/\lambda[\mathbf{F}_{yw}] = 1 + \lambda[\mathbf{F}_o]\,. \tag{28.25}$$

Since $|1+\mu| \leq 1 + |\mu|$,

$$1/q_s[\mathbf{F}_{yw}] \leq 1 + q_s[\mathbf{F}_o] \qquad \text{or} \qquad q_s[\mathbf{F}_{yw}](s) \geq [1 + q_s[\mathbf{F}_o](s)]^{-1}\,. \tag{28.26}$$

Similarily, one can prove (*Carlucci, D., and Vallauri, M., 1981*)

$$q_s[\mathbf{F}_{yw}](s) = q_s[(\mathbf{I}+\mathbf{F}_o)^{-1}](s) \leq \{1 - q_s[\mathbf{F}_o](s)\}^{-1} \quad \text{if} \quad q_s[\mathbf{F}_o] < 1 \tag{28.27}$$

$$q_s[\mathbf{F}_{yw}](s) \leq |\,1 - \min_i |\lambda_i[\mathbf{F}_o]\,|(s)\,|^{-1} \quad \text{if} \quad \min_i |\lambda_i[\mathbf{F}_o]\,| < 1\,. \tag{28.28}$$

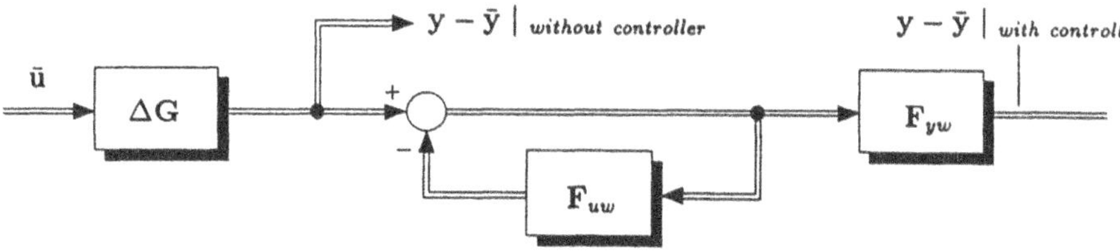

Figure 28.2: Auxiliary control loop, sufficient for stability of the perturbed system

28.5 Sufficient Stability Conditions

Emphasis is put on the stability of the system behaviour if uncertainty $\Delta\mathbf{G}$ occurs. Let all the components $\mathbf{K}_r, \mathbf{K}_y, \mathbf{F}_{yw}, \mathbf{F}_{yy_r}, \mathbf{F}_{uy_r}, \mathbf{F}_{uw}$ and $\Delta\mathbf{G}$ be stable. Sufficient stability conditions for $\mathbf{F}_d$ and Eq.(28.16) will be derived from Fig. 28.2.

Assume that $q_s[\Delta\mathbf{G}]q_s[\mathbf{F}_{uw}] < 1$. Referring to Eq.(28.18),

$$\rho_s[\Delta\mathbf{G}\ \mathbf{F}_{uw}](s) \le q_s[\Delta\mathbf{G}](s).q_s[\mathbf{F}_{uw}](s) < 1\ . \tag{28.29}$$

Applying the generalized Nyquist theorem, the map of the spectral radius of the return ratio matrix $\Delta\mathbf{G}\ \mathbf{F}_{uw}$, i.e. $\rho_s[\Delta\mathbf{G}\ \mathbf{F}_{uw}]$, may not encircle the Nyquist point $(-1+j0)$ in the s-plane.

28.6 Low Frequency Reducibility of the Output Uncertainty

In what follows, consider norms compatible with Eq.(28.17), specialize $s = j\omega$, take the forcing phasor $\bar{\mathbf{u}}(j\omega)$ and let $\mathbf{w} = \mathbf{0}$. Then, from Eq.(28.16) the tracking accuracy is given by

$$\|\mathbf{y}-\bar{\mathbf{y}}\| \le q_s[\mathbf{F}_{yw}]\ q_s[(\mathbf{I}+\Delta\mathbf{G}\ \mathbf{F}_{uw})^{-1}]\ q_s[\Delta\mathbf{G}]\ \|\bar{\mathbf{u}}\|\ . \tag{28.30}$$

Applying Eqs.(28.27) and (28.24) succeedingly yields

$$\|\mathbf{y}-\bar{\mathbf{y}}\| \le \underbrace{q_s[\mathbf{F}_{yw}]\{1-q_s[\Delta\mathbf{G}]q_s[\mathbf{F}_{uw}]\}^{-1}}_{r_u} q_s[\Delta\mathbf{G}]\ \|\bar{\mathbf{u}}\| \tag{28.31}$$

$$\|\mathbf{y}(j\omega)-\bar{\mathbf{y}}(j\omega)\| \le \underbrace{q_s[\mathbf{F}_{yw}]\ \{1-q_s[\Delta\mathbf{G}]\ |\lambda_{\min}[\mathbf{G}]\ |^{-1}\left(1+q_s[\mathbf{F}_{yw}](s)\right)\}^{-1}}_{r_u}\ q_s[\Delta\mathbf{G}]\ \|\bar{\mathbf{u}}\|\ . \tag{28.32}$$

The underbraced factor r_u is an upper bound for a scalar measure to which extent (< 1) the output uncertainty norm can usually be reduced by means of feedback. The factor provides the relation to the open-loop output uncertainty $q_s[\Delta\mathbf{G}]\ \|\bar{\mathbf{u}}\|$.

28.7 Increase of the Output Uncertainty at High Frequencies

Under certain conditions the output uncertainty may grow to an unlimited amount. It will be shown that this can occur if

$$\min_i |\lambda_i[\mathbf{F}_o]\ | \ge 2\ . \tag{28.33}$$

In view of this condition, the following results are derived from the former equations. From Eq.(28.28)

$$q_s[\mathbf{F}_{yw}] \leq |\, 1 - \min_i |\lambda_i[\mathbf{F}_o]\,|\;|^{-1} = (\min_i |\lambda_i[\mathbf{F}_o]\,| - 1)^{-1} \leq 1 \,. \tag{28.34}$$

Inserting the condition above into Eq.(28.23) yields

$$q_s[\mathbf{F}_{uw}] \geq \frac{1}{q_s[\mathbf{G}]}|1 - q_s[\mathbf{F}_{yw}]\,| = \frac{1}{q_s[\mathbf{G}]}[1 - (\min_i |\lambda_i[\mathbf{F}_o]\,| - 1)^{-1}] \,. \tag{28.35}$$

Now, let the expression above and Eq.(28.26) be combined with the definition of r_u , as given in Eq.(28.31),

$$r_u \triangleq q_s[\mathbf{F}_{yw}]\{1 - q_s[\Delta\mathbf{G}]\; q_s[\mathbf{F}_{uw}]\}^{-1} \tag{28.36}$$

in order to find a lower bound for r_u. Note that smaller values both for $q_s[\mathbf{F}_{yw}]$ and for $q_s[\mathbf{F}_{uw}]$ are substituted into the right-hand side of Eq.(28.36), the latter with regard to the double inverse sense in the function $\{1 - (\cdot)\}^{-1}$. Thus, substituting $q_s[\mathbf{F}_{yw}]$, first, from Eq.(28.26) and, second, from Eq.(28.35) into Eq.(28.36), the inequality achieved is

$$\begin{aligned}
r_u(j\omega) &\geq (1 + q_s[\mathbf{F}_o])^{-1}\{1 - q_s[\Delta\mathbf{G}]\frac{1}{q_s[\mathbf{G}]}[1 - (\min_i |\lambda_i[\mathbf{F}_o]\,| - 1)^{-1}]\}^{-1} && (28.37)\\
&= (1 + q_s[\mathbf{F}_o])^{-1}\{1 - \frac{q_s[\Delta\mathbf{G}]}{q_s[\mathbf{G}]}\,[\frac{(\min_i |\lambda_i[\mathbf{F}_o]\,| - 1) - 1}{\min_i |\lambda_i[\mathbf{F}_o]\,| - 1}]\,\}^{-1} && (28.38)\\
&= \frac{\min_i |\lambda_i[\mathbf{F}_o]\,| - 1}{1 + q_s[\mathbf{F}_o]}\{\min_i |\lambda_i[\mathbf{F}_o]\,| - 1 - \frac{q_s[\Delta\mathbf{G}]}{q_s[\mathbf{G}]}(\min_i |\lambda_i[\mathbf{F}_o]\,| - 2)\}^{-1} &&\\
r_u(j\omega) &\geq \frac{\min_i |\lambda_i[\mathbf{F}_o]\,| - 1}{1 + q_s[\mathbf{F}_o]}\{1 + (\min_i |\lambda_i[\mathbf{F}_o]\,| - 2)(1 - \frac{q_s[\Delta\mathbf{G}]}{q_s[\mathbf{G}]})\}^{-1} \,. && (28.39)
\end{aligned}$$

The brace becomes zero and r_u grows to infinity if

$$\min |\lambda_i[\mathbf{F}_o]| = (\frac{q_s[\Delta\mathbf{G}]}{q_s[\mathbf{G}]} - 1)^{-1} + 2 \geq 2 \,. \tag{28.40}$$

The inequality holds if $q_s[\Delta\mathbf{G}] > q_s[\mathbf{G}]$.

28.8 Weak Increasing of the Output Uncertainty at High Frequencies

Assume the condition

$$q_s[\mathbf{K}_f]q_s[\mathbf{K}_y](q_s[\mathbf{G}] + q_s[\Delta\mathbf{G}]) < 1 \tag{28.41}$$

is satisfied. From the definition of the return-ratio matrix $\mathbf{F}_o$ and Eq.(28.18) one has

$$q_s[\mathbf{F}_o] \leq q_s[\mathbf{G}]\; q_s[\mathbf{K}_f]\; q_s[\mathbf{K}_y] < 1 \,, \tag{28.42}$$

the last inequality above with regard to Eq.(28.41). Referring to Eq.(28.4),

$$q_s[\mathbf{F}_{uw}] \leq q_s[\mathbf{K}_f]q_s[\mathbf{K}_y]q_s[\mathbf{F}_{yw}] \leq q_s[\mathbf{K}_f]q_s[\mathbf{K}_y](1 - q_s[\mathbf{F}_o])^{-1} \,. \tag{28.43}$$

The last term was obtained using Eq.(28.27). Now, starting from Eq.(28.36) and substituting Eq.(28.27) again yields

$$\begin{aligned} r_u(j\omega) &= q_s[\mathbf{F}_{yw}]\{1 - q_s[\Delta\mathbf{G}]\ q_s[\mathbf{F}_{uw}])^{-1} && (28.44)\\ &< \frac{1}{1-q_s[\mathbf{F}_o]}\{1 - q_s[\Delta\mathbf{G}]q_s[\mathbf{K}_f]q_s[\mathbf{K}_y]\frac{1}{1-q_s[\mathbf{F}_o]}\}^{-1} && (28.45)\\ &< \frac{1}{1-q_s[\mathbf{F}_o]}\{\frac{1-q_s[\mathbf{F}_o] - q_s[\Delta\mathbf{G}]q_s[\mathbf{K}_f]q_s[\mathbf{K}_y]}{1-q_s[\mathbf{F}_o]}\}^{-1} && (28.46)\\ &< \{1 - q_s[\mathbf{G}]q_s[\mathbf{K}_f]q_s[\mathbf{K}_y] - q_s[\Delta\mathbf{G}]q_s[\mathbf{K}_f]q_s[\mathbf{K}_y]\}^{-1} && (28.47)\\ r_u(j\omega) &< \{1 - q_s[\mathbf{K}_f]q_s[\mathbf{K}_y](q_s[\mathbf{G}] + q_s[\Delta\mathbf{G}])\}^{-1}\ . && (28.48) \end{aligned}$$

Since Eq.(28.41) one has $r_u(j\omega) > 1$ which means that the output uncertainty is worse in feedback case than in open-loop case where $q_s[\mathbf{K}_f]$ or $q_s[\mathbf{K}_y] = 0$ and $r_u(j\omega) = 1$.

28.9 Design Considerations Via Observer-Based Controller

Multivariable proportional plus integral controllers provide satisfactory tracking response and disturbance rejection. The uncertainty behaviour of such a control system leads to the following approach. The derivations in this section are to be augmented by the uncertainty considerations just as in the previous sections of this chapter. For the sake of abbreviation, the uncertainty is not explicitly included in the following calculations. Consider Fig. 28.3a and a corresponding state space description

$$\begin{aligned} \text{Plant:}\quad \dot{\mathbf{x}}(t) &= \mathbf{A}\mathbf{x}(t) + \mathbf{B}\bar{\mathbf{u}}(t) && (28.49)\\ \bar{\mathbf{y}}(t) &= \mathbf{C}\mathbf{x}(t) + \mathbf{w}(t) && (28.50) \end{aligned}$$

$$\begin{aligned} \text{Controller:}\quad \bar{\mathbf{u}}(t) &= \mathbf{K}_P\mathbf{e}(t) + \mathbf{K}_I\mathbf{e}_I(t) && (28.51)\\ \dot{\mathbf{e}}_I(t) &= \mathbf{e}(t) = \mathbf{y}_{ref}(t) - \bar{\mathbf{y}}(t)\ . && (28.52) \end{aligned}$$

Comparison of Figs. 28.3a and 28.1 yields $\mathbf{K}_r(s) = \mathbf{I}$ and $\mathbf{K}_y(s) = \mathbf{I}$. It is observed that there is no degree of freedom to satisfy any of the requirements stated in the previous sections, i.e., there are no design parameters available to take the output uncertainty relations into account as given, e.g., by Eq.(28.31) or Eq.(28.39).

Next, consider an observer-based controller and an observer in the time domain and in the s-domain representation, respectively (see Fig. 28.3b).

$$\begin{aligned} \text{Observer-based controller:}\quad \bar{\mathbf{u}}(t) &= \mathbf{K}_P\hat{\mathbf{e}}(t) + \mathbf{K}_I\hat{\mathbf{e}}_I(t) && (28.53)\\ \dot{\hat{\mathbf{e}}}_I(t) &= \hat{\mathbf{e}}(t) = \mathbf{y}_{ref}(t) - \hat{\mathbf{y}}(t) && (28.54) \end{aligned}$$

$$\begin{aligned} \text{Observer:}\quad \hat{\mathbf{x}}(s) &= \mathbf{L}_y(s)\bar{\mathbf{y}}(s) + \mathbf{L}_u(s)\bar{\mathbf{u}}(s) && (28.55)\\ \hat{\mathbf{y}}(s) &= \mathbf{C}\hat{\mathbf{x}}(s)\ . && (28.56) \end{aligned}$$

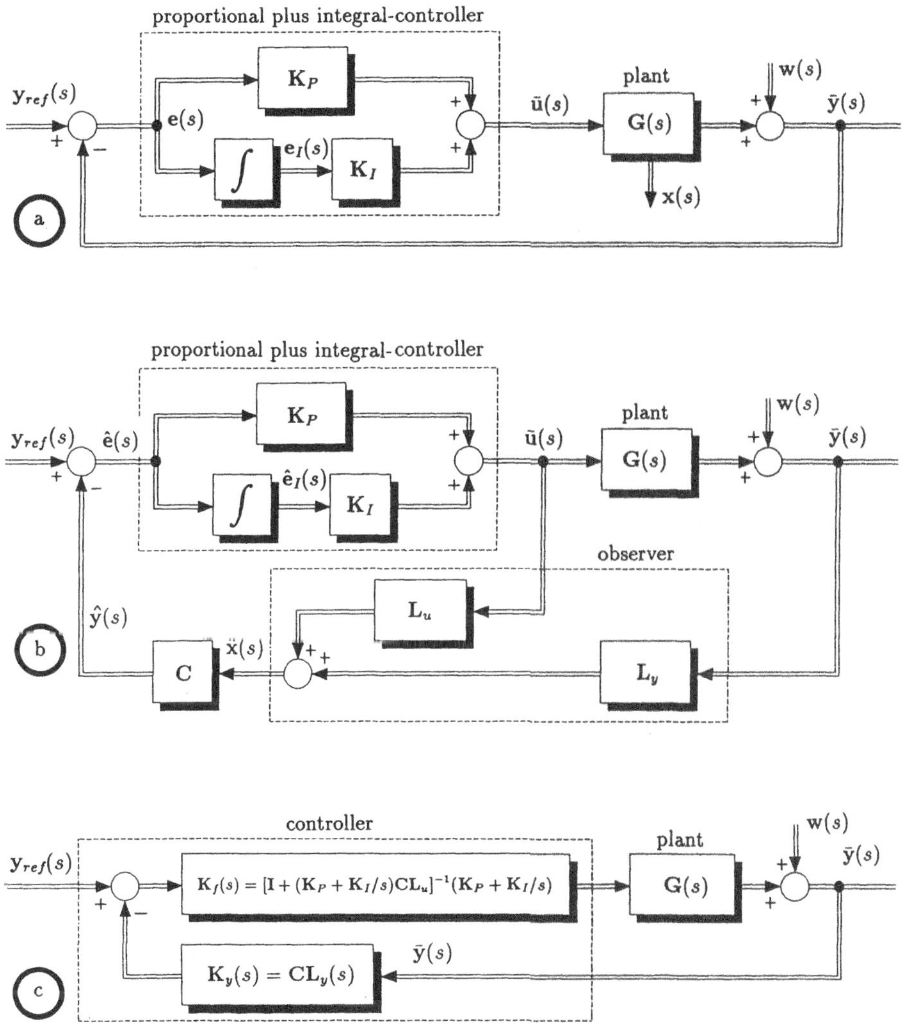

Figure 28.3: Control system with simple proportional plus integral-controller (*a*). Control system with observer-based proportional plus integral-controller (*b*). Control system with observer-based proportional plus integral-controller redrawn according to the structure chosen for the purpose of output uncertainty estimation (*c*)

From observer theory it is known that $\hat{\mathbf{y}}(s)$ converges to $\bar{\mathbf{y}}(s)$. Hence, $\hat{\mathbf{y}}(s)$ is replaced by $\bar{\mathbf{y}}(s)$. Combining Eqs.(28.53) through (28.56) yields

$$\bar{\mathbf{u}}(s) = (\mathbf{K}_P + \mathbf{K}_I/s)[\mathbf{y}_{ref}(s) - \bar{\mathbf{y}}(s)] = (\mathbf{K}_P + \mathbf{K}_I/s)[\mathbf{y}_{ref}(s) - \mathbf{CL}_y(s)\bar{\mathbf{y}}(s) - \mathbf{CL}_u(s)\bar{\mathbf{u}}(s)] \tag{28.57}$$

$$\bar{\mathbf{u}}(s) = \underbrace{[1 + (\mathbf{K}_P + \mathbf{K}_I/s)\mathbf{CL}_u(s)]^{-1}(\mathbf{K}_P + \mathbf{K}_I/s)}_{\mathbf{K}_f(s)}[\mathbf{y}_{ref}(s) - \underbrace{\mathbf{CL}_y(s)}_{\mathbf{K}_y(s)}\bar{\mathbf{y}}(s)] \,. \tag{28.58}$$

Comparison with Eq.(28.2) or Fig. 28.1 yields the relations as underbraced above and $\mathbf{K}_r = \mathbf{I}$. The results are depicted in Fig. 28.3c.

In view of this comparison, note that the overall transfer matrix is given by

$$\bar{\mathbf{y}}(s) = [\mathbf{I} + \mathbf{G}(s)(\mathbf{K}_P + \mathbf{K}_I/s)]^{-1}\mathbf{G}(s)(\mathbf{K}_P + \mathbf{K}_I/s)\mathbf{y}_{ref}(s) \tag{28.59}$$

and is predetermined by the parameters in $\mathbf{K}_P$ and $\mathbf{K}_I$.

Choosing $\mathbf{L}_y(s)$, both observer poles, observer gains and $\mathbf{K}_y(s) = \mathbf{CL}_y(s)$ can be determined. The uncertainty reducibility can be governed by $\mathbf{K}_y(s)$. Separatedly, the input-output behaviour is given by Eq.(28.59) (preassuming that the observer transients have already decayed). The input-output dynamic behaviour is designed by $\mathbf{K}_P$ and $\mathbf{K}_I$, the uncertainty reducibility is governed by $\mathbf{L}_y(s)$ in a separate way.

The following design approach proves advantageous as presented by *Carlucci, D., and Vallauri, M., 1981*. In a first stage, low-frequency values are considered where the open-loop uncertainty is still small. The proportional plus integral-controller is designed to reduce the output uncertainty via high gain feedback. The result are small values $q_s[\mathbf{F}_{yw}]$, see Eq.(28.31). In a second stage, high-frequency values with large open-loop uncertainty are treated and observer poles and gains are used in order to keep $q_s[\mathbf{K}_y]$ small and to obtain $r_u(j\omega)$ near unity, see Eq.(28.48).

Chou, H.H., et al. 1990a presented a similar procedure for the design of a robust controller with the additional facility of placing the closed-loop poles into a desired region.

Part V

Coprime Factorization and Minimax Frequency Optimization

Chapter 29

Robustness Based on the Internal Model Principle

In this chapter conditions are derived such that asymptotic tracking and output regulation (or disturbance rejection) occurs in multivariable systems. The system in considered robust with respect to plant parameter perturbations as long as they do not affect system stability and the asymptotic behaviour mentioned above is guaranteed. The reference input and the disturbance are assumed to satisfy a certain differential equation. This problem is also known as the robust servomechanism problem. A servo-compensator is incorporated which consists of r unstable compensators with identical dynamics, corresponding to the class of reference/disturbance. The integer r is the dimension of the output signal $\mathbf{y}(t)$.

The subject of this chapter is treated in the spectral domain, using polynomial matrices as well as the state space.

29.1 Internal Model Principle Using Polynomial Matrices

29.1.1 Disturbance Rejection Problem via Internal Model Principle

Consider the disturbance rejection (or output regulation) problem as depicted in Fig. 29.1a. The system generally consists of plant, sensor, compensator and a disturbance model. The transfer matrices are stated in relative left-prime factorization. For comparison: lt the state space representation were to be used (as done in a later section) the disturbance model and plant would obey the state equations

$$\dot{\mathbf{x}} = \mathbf{A}\mathbf{x} + \mathbf{B}_1\mathbf{u} + \mathbf{B}_2\mathbf{w} \quad \text{and} \quad \mathbf{y} = \mathbf{C}_1\mathbf{x} + \mathbf{D}_1\mathbf{u} + \mathbf{D}_2\mathbf{w} \;, \tag{29.1}$$

see Eq.(29.27). The zeros of $\det \mathbf{N}$ are considered to be located throughout in the right s-plane, i.e., the model of the disturbance is completely unstable, even in the case of stable or zero signal $\mathbf{v}_w(s)$. The sensor is assumed stable. Comparator, plant and sensor are assumed to be stabilizable, i.e., all the inner loop poles or zeros of $\mathbf{Q}, \mathbf{P}, \mathbf{H}$ are in the open left-half s-plane. Finally, it is assumed that unstable plant modes can be observed and controlled by the compensator.

Collecting all the component equations, the differential operator form in the s-domain is

$$\begin{pmatrix} -\mathbf{R} & 0 & \mathbf{P} & -\mathbf{I} \\ 0 & \mathbf{K} & -\mathbf{J} & 0 \\ \mathbf{Q} & -\mathbf{H} & 0 & 0 \\ 0 & 0 & 0 & \mathbf{N} \end{pmatrix} \begin{pmatrix} \mathbf{u} \\ \mathbf{g} \\ \mathbf{y} \\ \mathbf{w} \end{pmatrix} = \begin{pmatrix} 0 \\ 0 \\ 0 \\ \mathbf{I} \end{pmatrix} \mathbf{v}_w, \ \mathbf{W} \triangleq \begin{pmatrix} -\mathbf{R} & 0 & \mathbf{P} & -\mathbf{I} \\ 0 & \mathbf{K} & -\mathbf{J} & 0 \\ \mathbf{Q} & -\mathbf{H} & 0 & 0 \\ 0 & 0 & 0 & \mathbf{N} \end{pmatrix}, \mathbf{F} \triangleq \begin{pmatrix} 0 \\ 0 \\ 0 \\ \mathbf{I} \end{pmatrix} \tag{29.2}$$

$$\mathbf{y} = (\mathbf{0\,0\,I\,0}) \begin{pmatrix} \mathbf{u} \\ \mathbf{g} \\ \mathbf{y} \\ \mathbf{w} \end{pmatrix}, \qquad \mathbf{C} \triangleq (\mathbf{0\,0\,I\,0})\,. \tag{29.3}$$

Combining the above definitions (*Wolovich, W.A., and Ferreira, P., 1979*), it results

$$\mathbf{y} = \mathbf{C}\mathbf{W}^{-1}\mathbf{F}\mathbf{v}_w \tag{29.4}$$

with the closed-loop transfer matrix $\mathbf{CW}^{-1}\mathbf{F}$. To begin with, the theory of polynomial matrices is introduced to present the internal model principle.

29.1.2 Unobservable Disturbance Modes Associated with the Closed-Loop Transfer Matrix

The poles of the closed-loop system are given by the zeros of $\det \mathbf{W}$. Developing $\mathbf{W}$ with respect to the last row of $\mathbf{W}$ in Eq.(29.2),

$$\det \mathbf{W} = (\det \mathbf{N})\ \det \begin{pmatrix} -\mathbf{R} & \mathbf{0} & \mathbf{P} \\ \mathbf{0} & \mathbf{K} & -\mathbf{J} \\ \mathbf{Q} & -\mathbf{H} & \mathbf{0} \end{pmatrix} = 0\,. \tag{29.5}$$

In view of Eq.(29.5), the unstable poles of the disturbance $\mathbf{w}(s)$ cause the closed-loop system to be unstable. The internal model principle aims at finding a matrix $\mathbf{W}$ such that $\mathbf{C}$ and $\mathbf{W}$ are *not* relatively left-prime, i.e., in such a way that $\mathbf{C}$ and $\mathbf{W}$ contain a common right-divisior $\mathbf{D}_{cr}$. If $\partial[\det \mathbf{D}_{cr}]$ equals the number of unstable poles of $\mathbf{N}^{-1}$ and $\det \mathbf{D}_{cr} = \det \mathbf{N}$, it can be achieved that the poles of $\mathbf{N}^{-1}$ become unobservable modes associated with $\mathbf{CW}^{-1}\mathbf{F}$. The divisor can be found as follows

$$\begin{pmatrix} \mathbf{0} & \mathbf{0} & \mathbf{I} & \mathbf{0} \\ \mathbf{0} & \mathbf{K} & \mathbf{0} & -\mathbf{J} \\ \bar{\mathbf{Q}} & -\mathbf{H} & \mathbf{0} & \mathbf{0} \\ \bar{\mathbf{R}} & \mathbf{0} & \mathbf{N} & \mathbf{NP} \end{pmatrix} \begin{pmatrix} \bar{\mathbf{N}} & \mathbf{0} & \mathbf{0} & \mathbf{0} \\ \mathbf{0} & \mathbf{I} & \mathbf{0} & \mathbf{0} \\ -\mathbf{R} & \mathbf{0} & \mathbf{P} & -\mathbf{I} \\ \mathbf{0} & \mathbf{0} & \mathbf{I} & \mathbf{0} \end{pmatrix} = \begin{pmatrix} -\mathbf{R} & \mathbf{0} & \mathbf{P} & -\mathbf{I} \\ \mathbf{0} & \mathbf{K} & -\mathbf{J} & \mathbf{0} \\ \mathbf{Q} & -\mathbf{H} & \mathbf{0} & \mathbf{0} \\ \mathbf{0} & \mathbf{0} & \mathbf{0} & \mathbf{N} \end{pmatrix} \quad \text{or } \hat{\mathbf{W}}\mathbf{D}_{cr} = \mathbf{W} \tag{29.6}$$

$$(\mathbf{0\,0\,0\,I}) \begin{pmatrix} \bar{\mathbf{N}} & \mathbf{0} & \mathbf{0} & \mathbf{0} \\ \mathbf{0} & \mathbf{I} & \mathbf{0} & \mathbf{0} \\ -\mathbf{R} & \mathbf{0} & \mathbf{P} & -\mathbf{I} \\ \mathbf{0} & \mathbf{0} & \mathbf{I} & \mathbf{0} \end{pmatrix} = (\mathbf{0\,0\,I\,0}) \quad \text{or} \quad \hat{\mathbf{C}}\mathbf{D}_{cr} = \mathbf{C}\,. \tag{29.7}$$

Combining yields, see Eq.(30.6), from $\mathbf{y} = \mathbf{CW}^{-1}\mathbf{F}\mathbf{v}_w$

$$\mathbf{y} = \hat{\mathbf{C}}\mathbf{D}_{cr}(\hat{\mathbf{W}}\mathbf{D}_{cr})^{-1}\mathbf{F}\mathbf{v}_w = \hat{\mathbf{C}}\hat{\mathbf{W}}^{-1}\mathbf{F}\mathbf{v}_w = (\mathbf{0\,0\,0\,I}) \begin{pmatrix} \mathbf{0} & \mathbf{0} & \mathbf{I} & \mathbf{0} \\ \mathbf{0} & \mathbf{K} & \mathbf{0} & -\mathbf{J} \\ \bar{\mathbf{Q}} & -\mathbf{H} & \mathbf{0} & \mathbf{0} \\ \bar{\mathbf{R}} & \mathbf{0} & \mathbf{N} & \mathbf{NP} \end{pmatrix}^{-1} \begin{pmatrix} \mathbf{0} \\ \mathbf{0} \\ \mathbf{0} \\ \mathbf{I} \end{pmatrix} \mathbf{v}_w\,. \tag{29.8}$$

The calculation of the result above can be replaced by an easier way avoiding the inverse of the large partitioned matrix $\hat{\mathbf{W}}$. This is done in the sections to follow. Eq.(29.19) is considered as the solution. In order to satisfy the products of submatrices in Eq.(29.6), the equivalences below must hold

$$\bar{\mathbf{Q}}\bar{\mathbf{N}} = \mathbf{Q} \tag{29.9}$$

$$\bar{\mathbf{R}}\bar{\mathbf{N}} - \mathbf{N}\mathbf{R} = \mathbf{0} , \tag{29.10}$$

i.e., **R** and **N** must be externally skew-prime, see Eq.(30.7). Referring to the definition of $\mathbf{D}_{cr}$ in Eq.(29.6), one has det $\mathbf{D}_{cr}$ = det $\bar{\mathbf{N}}$ and the former relation det $\mathbf{D}_{cr}$ = det **N** requires det **N** = det $\bar{\mathbf{N}}$.

29.1.3 Inner Loop Stability (Internal Stability)

Assume that $\mathbf{K}^{-1}\mathbf{J}\mathbf{P}^{-1}\mathbf{R}$ is closed-loop stabilizable. That is, matrices **Q** and **H** of the compensator can be found such that the second right-hand determinant (of the inner loop) in Eq.(29.5) is a polynomial with zeros completely in $\mathcal{C}^-$. Replacing **Q** by $\bar{\mathbf{Q}}\bar{\mathbf{N}}$ it must be proved whether $\mathbf{K}^{-1}\mathbf{J}\mathbf{P}^{-1}\mathbf{R}\bar{\mathbf{N}}$ is stabilizable by the matrices $\bar{\mathbf{Q}}$ and **H**. Defining $\bar{\mathbf{P}}^{-1}\bar{\mathbf{J}}$ as any relatively left-prime factorization of $\mathbf{J}\mathbf{P}^{-1}$,

$$\mathbf{K}^{-1}\mathbf{J}\mathbf{P}^{-1}\mathbf{R}\bar{\mathbf{N}} = \mathbf{K}^{-1}\bar{\mathbf{P}}^{-1}\bar{\mathbf{J}}\mathbf{R}\bar{\mathbf{N}} . \tag{29.11}$$

Applying the externally skew-primeness of **R** and **N**,

$$\mathbf{K}^{-1}\bar{\mathbf{P}}^{-1}\bar{\mathbf{J}}\mathbf{R}\bar{\mathbf{N}} = \mathbf{K}^{-1}\bar{\mathbf{P}}^{-1}\bar{\mathbf{J}}\mathbf{N}^{-1}\bar{\mathbf{R}} . \tag{29.12}$$

In order to stabilize the unstable modes of $\mathbf{N}^{-1}$ in the inner loop, each series connection of matrices has to be relatively prime. Otherwise the unstable modes could not be observed or controlled, and stabilization would be impossible. The only link in the chain of matrices not yet being inspected re primeness is $\bar{\mathbf{J}}\mathbf{N}^{-1}$. Hence, one has to demand that $\bar{\mathbf{J}}$ and **N** must be relatively right-prime.

29.1.4 Practical Interpretations

The matricial operations stated above can be discussed from a practical view point. Fig. 29.1 gives a graphic interpretation of the matrix substitutions.

After having substituted $\mathbf{Q} = \bar{\mathbf{Q}}\bar{\mathbf{N}}$ and $\mathbf{N}\mathbf{R} = \bar{\mathbf{R}}\bar{\mathbf{N}}$ into corresponding block elements in Fig. 29.1a, b and c, it can be observed that the internal loop contains $\mathbf{N}^{-1}$ as a duplicate of the exogenous system $\mathbf{N}^{-1}$ at the input of the summation junction. This is the key statement of the internal model principle.

The property of **R** and **N** being externally skew-prime yields a pseudo-commutativity property: **NR** is substituted by $\bar{\mathbf{R}}\bar{\mathbf{N}}$. Originally, $\bar{\mathbf{N}}$ is introduced by replacing **Q** with $\bar{\mathbf{Q}}\bar{\mathbf{N}}$, then, the series connection of $\bar{\mathbf{N}}^{-1}$ and **R** is replaced by $\bar{\mathbf{R}}^{-1}$ and $\mathbf{N}^{-1}$ invoking the pseudo-commutativity property. Shifting $\bar{\mathbf{N}}$ through the loop leads to $\mathbf{N}^{-1}$ in front of the summation junction.

Transposing $\mathbf{N}^{-1}$ in front of the summation junction enables $\mathbf{v}_w(s)$ to be used as a signal input directly to the system (Fig. 29.1e). Generating the disturbance the matrix $\mathbf{N}^{-1}$ was considered unstable which is sufficient for increasing disturbance even in the case of $\mathbf{v}_w = \mathbf{0}$. If the signal $\mathbf{v}_w$ can be used to excite the control system (Fig. 29.1e), there is no excitation and reaction at all. The only slight difference may occur from initial condition differences within the $\mathbf{N}^{-1}$-elements of the disturbance path and the internal model in the loop (Fig. 29.1d).

The chain of block elements, beginning with $\mathbf{P}^{-1}$, ending at $\bar{\mathbf{R}}$, can be rewritten in relative left-prime factorization

$$\bar{\mathbf{R}}\bar{\mathbf{Q}}^{-1}\mathbf{H}\mathbf{K}^{-1}\mathbf{J}\mathbf{P}^{-1} \triangleq \mathbf{A}^{-1}\mathbf{B} . \tag{29.13}$$

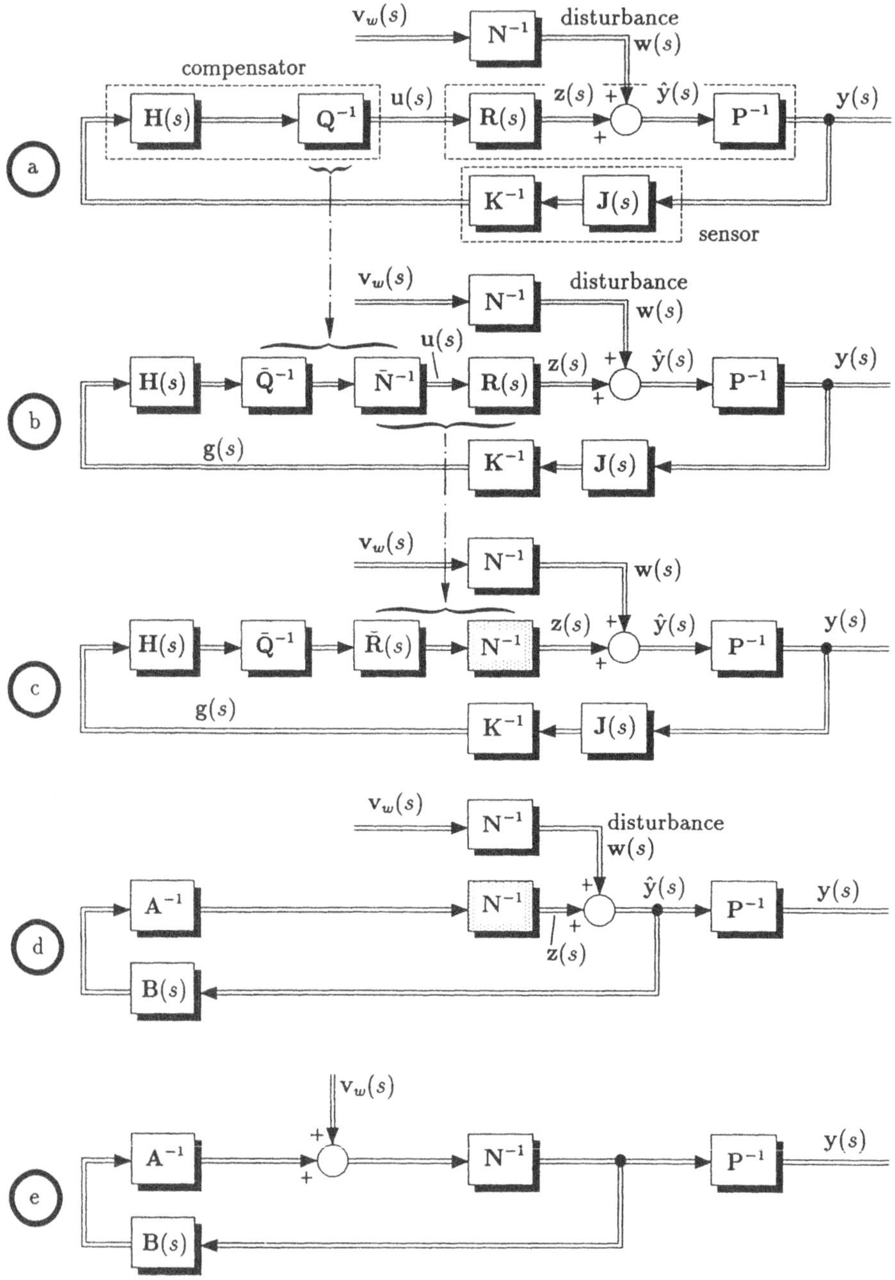

Figure 29.1: Output regulation and equivalent representations leading to the internal model principle and resulting in unobservable disturbance models

Using these definitions, the system is redrawn to Fig. 29.1d and the differential operator form rewritten to

$$\mathbf{N}^{-1}\bar{\mathbf{R}}\bar{\mathbf{Q}}^{-1}\mathbf{H}\mathbf{K}^{-1}\mathbf{J}\mathbf{P}^{-1}(\mathbf{z}+\mathbf{w}) = \mathbf{z} \tag{29.14}$$

$$\mathbf{N}^{-1}\mathbf{A}^{-1}\mathbf{B}(\mathbf{z}+\mathbf{w}) = \mathbf{z} \tag{29.15}$$

$$\mathbf{B}\mathbf{z}+\mathbf{B}\mathbf{w} = \mathbf{A}\mathbf{N}\mathbf{z} \quad \leadsto \quad (\mathbf{A}\mathbf{N}-\mathbf{B})\mathbf{z}-\mathbf{B}\mathbf{w} = \mathbf{0}\ . \tag{29.16}$$

Combining with $\mathbf{N}^{-1}\mathbf{v}_w = \mathbf{w}$ yields[1]

$$\begin{pmatrix} \mathbf{AN}-\mathbf{B} & -\mathbf{B} \\ \mathbf{0} & \mathbf{N} \end{pmatrix} \begin{pmatrix} \mathbf{z} \\ \mathbf{w} \end{pmatrix} = \begin{pmatrix} \mathbf{0} \\ \mathbf{I} \end{pmatrix} \mathbf{v}_w \tag{29.17}$$

$$\hat{\mathbf{y}} = (\mathbf{I} \quad \mathbf{I}) \begin{pmatrix} \mathbf{z} \\ \mathbf{w} \end{pmatrix} . \tag{29.18}$$

The overall transfer relation is

$$\hat{\mathbf{y}} = (\mathbf{I} \quad \mathbf{I}) \begin{pmatrix} \mathbf{AN}-\mathbf{B} & -\mathbf{B} \\ \mathbf{0} & \mathbf{N} \end{pmatrix}^{-1} \begin{pmatrix} \mathbf{0} \\ \mathbf{I} \end{pmatrix} \mathbf{v}_w = (\mathbf{AN}-\mathbf{B})^{-1}\mathbf{A}\mathbf{v}_w\ . \tag{29.19}$$

Calculating the first inverse in Eq.(29.19) yields

$$\begin{pmatrix} (\mathbf{AN}-\mathbf{B})^{-1} & (\mathbf{AN}-\mathbf{B})^{-1}\mathbf{B}\mathbf{N}^{-1} \\ \mathbf{0} & \mathbf{N}^{-1} \end{pmatrix} \tag{29.20}$$

but premultiplication with $(\mathbf{I} \quad \mathbf{I})$ and postmultiplication with $\binom{\mathbf{0}}{\mathbf{I}}$ cancels $\mathbf{N}^{-1}$.

The poles of the resulting system are given by $\det(\mathbf{AN}-\mathbf{B}) = 0$. Comparing with Eq.(29.5) makes clear that the unstable modes of $\mathbf{N}$ are no more of influence to the poles of the resulting system. Finally, it can be found out that the common right-divisior $\mathbf{D}_{cr}$ exists

$$(\mathbf{I} \quad \mathbf{I}) = (\mathbf{0} \quad \mathbf{I}) \begin{pmatrix} \mathbf{N} & \mathbf{0} \\ \mathbf{I} & \mathbf{I} \end{pmatrix} = (\mathbf{0} \quad \mathbf{I})\ \mathbf{D}_{cr} \qquad \text{and} \tag{29.21}$$

$$\begin{pmatrix} \mathbf{AN}-\mathbf{B} & -\mathbf{B} \\ \mathbf{0} & \mathbf{N} \end{pmatrix} = \begin{pmatrix} \mathbf{A} & -\mathbf{B} \\ -\mathbf{I} & \mathbf{N} \end{pmatrix} \begin{pmatrix} \mathbf{N} & \mathbf{0} \\ \mathbf{I} & \mathbf{I} \end{pmatrix} = \begin{pmatrix} \mathbf{A} & -\mathbf{B} \\ -\mathbf{I} & \mathbf{N} \end{pmatrix} \mathbf{D}_{cr} \tag{29.22}$$

$$\text{where} \quad \mathbf{D}_{cr} = \begin{pmatrix} \mathbf{N} & \mathbf{0} \\ \mathbf{I} & \mathbf{I} \end{pmatrix}, \qquad \det \mathbf{D}_{cr} = \det \mathbf{N}, \qquad \partial[\det \mathbf{D}_{cr}] = \partial[\det \mathbf{N}]\ . \tag{29.23}$$

In view of these results, the left-hand side of Eq.(29.19) is put into the form

$$\hat{\mathbf{y}} = (\mathbf{0} \quad \mathbf{I})\mathbf{D}_{cr}\mathbf{D}_{cr}^{-1} \begin{pmatrix} \mathbf{A} & -\mathbf{B} \\ -\mathbf{I} & \mathbf{N} \end{pmatrix}^{-1} \begin{pmatrix} \mathbf{0} \\ \mathbf{I} \end{pmatrix} \mathbf{v}_w \tag{29.24}$$

which is equal to the right-hand side of Eq.(29.19). The integer $\partial[\det \mathbf{N}]$ represents the number of unobservable modes. Hence, with regard to $\mathbf{NR} = \bar{\mathbf{R}}\bar{\mathbf{N}}$ and $\det \mathbf{N} = \det \bar{\mathbf{N}}$, all the modes of the exogenous system $\mathbf{N}^{-1}$ remain unobservable with respect to the output $\hat{\mathbf{y}}$ although the disturbance is completely observable by the sensor.

[1]The hat symbol is used only for the purpose of distinction. There is no relation to estimates.

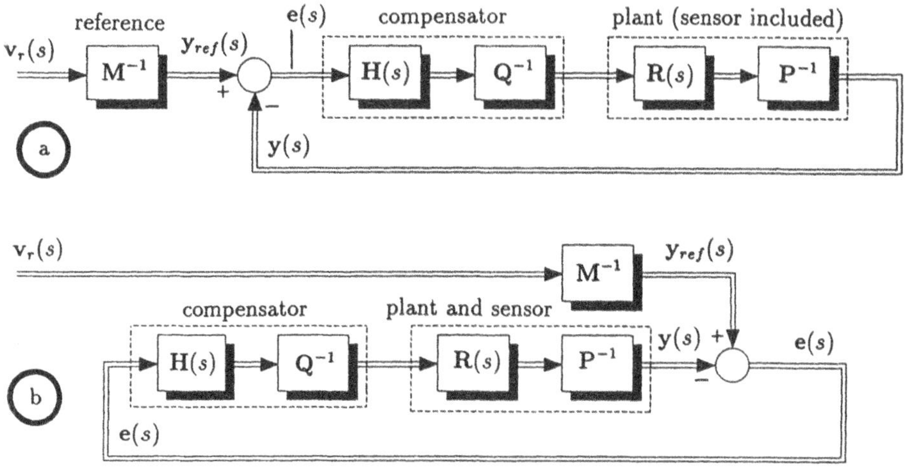

Figure 29.2: Scheme of tracking problem and an adjusted representation in order to fit the facilities of the output regulator problem with the internal model principle

29.1.5 Tracking Problem Via Internal Model Principle

The tracking facility of a control loop is demonstrated in Fig. 29.2a. The structure can be easily redrawn to the Fig. 29.2b. Comparison of the output regulation problem (Fig. 29.1a) with the tracking problem (Fig. 29.2b) shows the similarity between $\lim_{t\to\infty} \mathbf{y}(t) = \mathbf{0}$ and $\lim_{t\to\infty} \mathbf{e}(t) = \mathbf{0}$. Slight differences can be stated and overcome as follows: In the tracking problem $\mathbf{P}^{-1}, \mathbf{R}$ has to include the sensor and $\mathbf{y}_{ref}$ is added to the output of $\mathbf{P}^{-1}$ while $\mathbf{w}$ acts on the input of $\mathbf{P}^{-1}$. The separate sensor in the output regulation problem $\mathbf{K}^{-1}\mathbf{J}$ has to be replaced by $-\mathbf{I}$.

29.1.6 Robustness of Disturbance Rejection

The output regulator problem has to guarantee, first, $\lim_{t\to\infty} \mathbf{y}(t) = \mathbf{0}$ irrespective of the disturbance and, second, asymptotic stability of the closed loop. If the internal model principle is realized the disturbance has no influence on the output. Following Eq.(29.5) $\mathbf{H}$ and $\mathbf{Q}^{-1}$ can be chosen in order to agree with the transient and stability requirements.

29.2 Internal Model Principle Using State-Space Approach

The design of a servo-compensator via the internal model principle, presented in the preceding section and known as the robust servomechanism problem, has also been solved by a state-space approach (*Davison, E.J., 1976; Cheok, K.C., et al. 1988*). Following these investigations, Fig. 29.3 summarizes the disturbance and reference models, the plant, servo-compensator and a servo-controller. Consider the following state equations

$$\text{Disturbance:} \quad \dot{\mathbf{x}}_w = \mathbf{A}_w\mathbf{x}_w + \mathbf{v}_w \quad \text{and} \quad \mathbf{w} = \mathbf{C}_w\mathbf{x}_w \tag{29.25}$$

$$\begin{aligned} \text{Reference:}\quad \dot{\mathbf{x}}_r &= \mathbf{A}_r\mathbf{x}_r + \mathbf{v}_r \quad \text{and} \quad \mathbf{y}_{ref} = \mathbf{C}_r\mathbf{x}_r & (29.26)\\ \text{Plant:}\quad \dot{\mathbf{x}}_p &= \mathbf{A}_p\mathbf{x}_p + \mathbf{B}_p\mathbf{u} + \mathbf{G}_p\mathbf{w} & (29.27)\\ \mathbf{y} &= \mathbf{C}_p\mathbf{x}_p + \mathbf{D}_p\mathbf{u} + \mathbf{H}_p\mathbf{w} & (29.28)\\ \mathbf{e} &= \mathbf{y}_{ref} - \mathbf{y}\,. & (29.29) \end{aligned}$$

The error-driven servo-compensator is a high-order system. The order is at least rl (*Desoer, C.A., and Wang, Y.T., 1978*) where r is the number of outputs y_i, $l \triangleq \partial[\phi]$ and ψ_i are defined in Eq.(29.37).

$$\text{Servo-compensator:}\quad \dot{\mathbf{x}}_c = \mathbf{A}_c\mathbf{x}_c + \mathbf{B}_c\mathbf{e} \quad \text{and} \quad \mathbf{u}_c = \mathbf{F}_c\mathbf{x}_c \tag{29.30}$$

$$\mathbf{A}_c = \begin{pmatrix} -\psi_1\mathbf{I}_r & -\psi_2\mathbf{I}_r & -\psi_3\mathbf{I}_r & \dots & -\psi_l\mathbf{I}_r \\ \mathbf{I}_r & 0 & 0 & \dots & 0 \\ 0 & \mathbf{I}_r & 0 & \dots & 0 \\ 0 & 0 & \ddots & 0 & 0 \\ 0 & \dots & 0 & \mathbf{I}_r & 0 \end{pmatrix}\mathbf{x}_c, \qquad \mathbf{x}_c \triangleq \begin{pmatrix} \mathbf{x}_{c1} \\ \mathbf{x}_{c2} \\ \mathbf{x}_{c3} \\ \vdots \\ \mathbf{x}_{cl} \end{pmatrix} \tag{29.31}$$

$$\mathbf{B}_c = \begin{pmatrix} \mathbf{I}_r \\ 0 \\ 0 \\ \vdots \\ 0 \end{pmatrix} \qquad \mathbf{F}_c = (\mathbf{F}_{c1}\ \ \mathbf{F}_{c2}\ \ \mathbf{F}_{c3}\ \dots\ \mathbf{F}_{cl})\,. \tag{29.32}$$

The reason for this structure can be seen from the derivation of Eq.(29.55). The servo-controller assembles the servo-compensator output $\mathbf{u}_c$, a plant state feedback $\mathbf{K}_p\mathbf{x}_p$ and the feedforward $\mathbf{K}_w\mathbf{x}_w$ and $\mathbf{K}_r\mathbf{x}_r$. The disturbance state feedforward $\mathbf{K}_w\mathbf{x}_w$ and reference state feedforward $\mathbf{K}_r\mathbf{x}_r$ provide anticipative responsive control by influence onto the closed loop (*Weihrich, G., 1977*)

$$\mathbf{u} = \mathbf{u}_c + \mathbf{K}_p\mathbf{x}_p + \mathbf{K}_w\mathbf{x}_w + \mathbf{K}_r\mathbf{x}_r\,. \tag{29.33}$$

Combining Eqs.(29.25) through (29.33) yields

$$\underbrace{\begin{pmatrix} \dot{\mathbf{x}}_c \\ \dot{\mathbf{x}}_p \end{pmatrix}}_{\dot{\mathbf{x}}_{cp}} = \underbrace{\begin{pmatrix} \mathbf{A}_c & -\mathbf{B}_c\mathbf{C}_p \\ 0 & \mathbf{A}_p \end{pmatrix}}_{\mathbf{A}_{cp}} \underbrace{\begin{pmatrix} \mathbf{x}_c \\ \mathbf{x}_p \end{pmatrix}}_{\mathbf{x}_{cp}} + \underbrace{\begin{pmatrix} -\mathbf{B}_c\mathbf{H}_p\mathbf{C}_w \\ \mathbf{G}_p\mathbf{C}_w \end{pmatrix}}_{\mathbf{G}_{cp}} \mathbf{x}_w + \underbrace{\begin{pmatrix} \mathbf{B}_c\mathbf{C}_r \\ 0 \end{pmatrix}}_{\mathbf{E}_{cp}} \mathbf{x}_r + \underbrace{\begin{pmatrix} -\mathbf{B}_c\mathbf{D}_p \\ \mathbf{B}_p \end{pmatrix}}_{\mathbf{B}_{cp}} \mathbf{u} \tag{29.34}$$

$$\mathbf{u} = \underbrace{(\mathbf{K}_c\ \ \mathbf{K}_p)}_{\mathbf{K}_{cp}} \begin{pmatrix} \mathbf{x}_c \\ \mathbf{x}_p \end{pmatrix} + \mathbf{K}_w\,\mathbf{x}_w + \mathbf{K}_r\mathbf{x}_r \tag{29.35}$$

$$\mathbf{e} = \underbrace{(0\ \ -\mathbf{C}_p)}_{\mathbf{C}_{cp}} \begin{pmatrix} \mathbf{x}_c \\ \mathbf{x}_p \end{pmatrix} - \mathbf{H}_p\mathbf{C}_w\mathbf{x}_w + \mathbf{C}_r\mathbf{x}_r - \mathbf{D}_p\mathbf{u}\,. \tag{29.36}$$

29.2.1 ϕ-Transformation

Referring to Fig. 29.3, a combination vector $\mathbf{x}_{cp} = (\mathbf{x}_c^T\mathbf{x}_p^T)^T$ is defined, describing both the servo-compensator $\mathbf{x}_c$ and the plant $\mathbf{x}_p$. The minimal polynomials of $\mathbf{A}_w$ and $\mathbf{A}_r$ are termed $\phi_w(s)$ and $\phi_r(s)$, respectively. The roots are assumed on the imaginary axis or in the right-half complex s-plane. The dominant dynamic behaviour of the disturbance $\mathbf{w}$ and the reference $\mathbf{y}_{ref}$ are given by the matrices $\mathbf{A}_w$ and $\mathbf{A}_r$, preassuming that $\mathbf{v}_w$ and $\mathbf{v}_r$ act as shaping functions. The minimum polynomial is applied because it is the annihilating polynomial of lowest degree. Let the polynomial $\phi(s)$ be the least common multiple of $\phi_w(s)$ and $\phi_r(s)$. By definition, the signal $\mathbf{x}(t)$ premultiplied by a polynomial $\phi(s)$ is

$$\bar{\mathbf{x}}(t) \triangleq \phi(s)\mathbf{x}(t) = \phi\mathbf{x}(t) = (s^l + \psi_1 s^{l-1} + \ldots + \psi_l)\mathbf{x}(t) \triangleq \mathbf{x}^{(l)}(t) + \psi_1\mathbf{x}^{(l-1)}(t) + \ldots + \psi_l\mathbf{x}(t) \tag{29.37}$$

where s is considered the time-domain differential operator d/dt. Hence, $\phi\mathbf{x}(t)$ is a weighted sum of $\mathbf{x}(t)$ and derivatives of $\mathbf{x}(t)$ up to $\mathbf{x}^{(l)}(t)$. This operation is referred to as ϕ-transformation (*Cheok, K.C., et al. 1988*).

29.2.2 Servo-Compensator and ϕ-Transformation

Combining Eqs.(29.30) and (29.31) yields

$$\begin{pmatrix} \dot{\mathbf{x}}_{c1} \\ \dot{\mathbf{x}}_{c2} \\ \dot{\mathbf{x}}_{c3} \\ \vdots \\ \dot{\mathbf{x}}_{cl} \end{pmatrix} = \begin{pmatrix} -\psi_1\mathbf{I}_r & -\psi_2\mathbf{I}_r & \ldots & \ldots & -\psi_l\mathbf{I}_r \\ \mathbf{I}_r & 0 & \ldots & \ldots & 0 \\ 0 & \mathbf{I}_r & 0 & \ldots & 0 \\ 0 & 0 & \ddots & 0 & 0 \\ 0 & 0 & 0 & \mathbf{I}_r & 0 \end{pmatrix} \begin{pmatrix} \mathbf{x}_{c1} \\ \mathbf{x}_{c2} \\ \mathbf{x}_{c3} \\ \vdots \\ \mathbf{x}_{cl} \end{pmatrix} + \begin{pmatrix} \mathbf{I}_r \\ 0 \\ 0 \\ \vdots \\ 0 \end{pmatrix} \mathbf{e} \tag{29.38}$$

or in equivalent vector notation

$$\dot{\mathbf{x}}_{c1} = -\psi_1\mathbf{x}_{c1} - \psi_2\mathbf{x}_{c2} - \ldots - \psi_l\mathbf{x}_{cl} + \mathbf{e} \tag{29.39}$$

$$\dot{\mathbf{x}}_{c2} = \mathbf{x}_{c1} \tag{29.40}$$

$$\dot{\mathbf{x}}_{c3} = \mathbf{x}_{c2} \tag{29.41}$$

$$\vdots$$

$$\dot{\mathbf{x}}_{cl} = \mathbf{x}_{c,l-1}\ . \tag{29.42}$$

Rewriting the preceeding $l-1$ equations yields

$$\mathbf{x}_{c1} = \dot{\mathbf{x}}_{c2} = \mathbf{x}_{c3}^{(2)} = \mathbf{x}_{c4}^{(3)} = \ldots = \mathbf{x}_{cl}^{(l-1)} \tag{29.43}$$

$$\mathbf{x}_{c2} = \dot{\mathbf{x}}_{c3} = \mathbf{x}_{c4}^{(2)} = \mathbf{x}_{c5}^{(3)} = \ldots = \mathbf{x}_{cl}^{(l-2)} \tag{29.44}$$

$$\vdots$$

$$\mathbf{x}_{c,l-1} = \dot{\mathbf{x}}_{cl}\ . \tag{29.45}$$

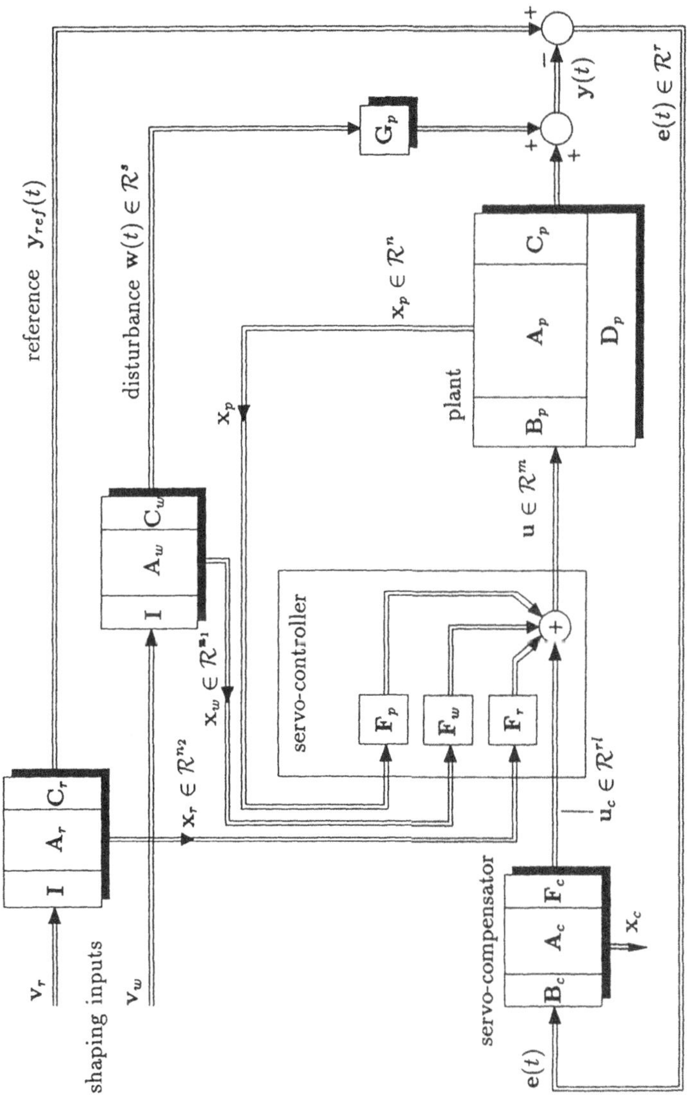

Figure 29.3: Servo-compensator using the internal model principle in the state space

Combining the preceding $l-1$ equations and sufficing $\mathbf{x}_{cl}$ exposes the partitioned vector $\mathbf{x}_c$

$$\mathbf{x}_c = \begin{pmatrix} \mathbf{x}_{c1} \\ \mathbf{x}_{c2} \\ \mathbf{x}_{c3} \\ \vdots \\ \mathbf{x}_{cl} \end{pmatrix} = \begin{pmatrix} \mathbf{x}_{cl}^{(l-1)} \\ \mathbf{x}_{cl}^{(l-2)} \\ \mathbf{x}_{cl}^{(l-3)} \\ \vdots \\ \mathbf{x}_{cl} \end{pmatrix} . \tag{29.46}$$

From Eq.(29.39) and from $\dot{\mathbf{x}}_{c1} = \mathbf{x}_{cl}^{(l)}$

$$\mathbf{x}_{cl}^{(l)} = -\psi_1 \mathbf{x}_{cl}^{(l-1)} - \psi_2 \mathbf{x}_{cl}^{(l-2)} - \ \ldots \ - \psi_l \mathbf{x}_{cl} + \mathbf{e} \tag{29.47}$$

$$\text{or} \quad \mathbf{x}_{cl}^{(l)} + \psi_1 \mathbf{x}_{cl}^{(l-1)} + \ldots + \psi_l \mathbf{x}_{cl} = \mathbf{e} \tag{29.48}$$

$$\phi \mathbf{x}_{cl} \triangleq \bar{\mathbf{x}}_{cl} = \mathbf{e} . \tag{29.49}$$

Repeating this procedure by differentiating Eq.(29.48) once with respect to time and replacing $\dot{\mathbf{x}}_{cl} = \mathbf{x}_{c,l-1}$ through $\mathbf{x}_{cl}^{(l+1)} = \mathbf{x}_{c,l-1}^{(l)}$ yields

$$\mathbf{x}_{cl}^{(l+1)} + \psi_1 \mathbf{x}_{cl}^{(l)} + \ldots + \psi_l \dot{\mathbf{x}}_{cl} = \dot{\mathbf{e}} \tag{29.50}$$

$$\mathbf{x}_{c,l-1}^{(l)} + \psi_1 \mathbf{x}_{c,l-1}^{(l-1)} + \ldots + \psi_l \mathbf{x}_{c,l-1} = \dot{\mathbf{e}} \tag{29.51}$$

$$\phi \mathbf{x}_{c,l-1} = \dot{\mathbf{e}} . \tag{29.52}$$

By repetition, the results are

$$\phi \mathbf{x}_{c,l-\mu} = \mathbf{e}^{(\mu)} \qquad \forall \mu = 0 \ldots l-1 \tag{29.53}$$

$$\phi \mathbf{x}_{c1} = \mathbf{e}^{(l-1)} . \tag{29.54}$$

Summarizing the statements of Eq. (29.53), the key result is obtained (*Cheok, K.C., et al. 1988*)

$$\bar{\mathbf{x}}_c = \phi \mathbf{x}_c = \phi \begin{pmatrix} \mathbf{x}_{c1} \\ \mathbf{x}_{c2} \\ \mathbf{x}_{c3} \\ \vdots \\ \mathbf{x}_{cl} \end{pmatrix} = \begin{pmatrix} \mathbf{e}^{(l-1)} \\ \mathbf{e}^{(l-2)} \\ \mathbf{e}^{(l-3)} \\ \vdots \\ \mathbf{e} \end{pmatrix} . \tag{29.55}$$

In order to obtain a servo-compensator, the outputs after ϕ-transformation correspond to the system error and its derivatives. The servo-compensator has to be modelled on the spectral richness of both the disturbance and the reference signal models (*Davison, E.J., 1976*). This is the main result of applying ϕ-transformation to the signal $\mathbf{x}(t)$ and structuring $\mathbf{A}_c$ as given in Eq.(29.31).

29.2.3 ϕ-Transformed Closed-Loop System

Let the ϕ-transformation be applied to the state space variables

$$\bar{\mathbf{x}}_c(t) \triangleq \phi \mathbf{x}_c(t), \quad \bar{\mathbf{x}}_p(t) \triangleq \phi \mathbf{x}_p(t), \quad \bar{\mathbf{x}}_w(t) \triangleq \phi \mathbf{x}_w(t), \quad \bar{\mathbf{x}}_r(t) \triangleq \phi \mathbf{x}_r(t) \tag{29.56}$$

$$\bar{\mathbf{x}}_{cp}(t) \triangleq \phi \mathbf{x}_{cp}(t), \quad \bar{\mathbf{w}}(t) = \mathbf{C}_w \bar{\mathbf{x}}_w(t), \quad \bar{\mathbf{r}}(t) \triangleq \mathbf{C}_r \bar{\mathbf{x}}_r(t), \quad \bar{\mathbf{y}} \triangleq \mathbf{C}_p \bar{\mathbf{x}}_p + \mathbf{D}_p \bar{\mathbf{u}} + \mathbf{H}_p \bar{\mathbf{w}} \tag{29.57}$$

and to the overall closed-loop system Eqs.(29.34) and (29.35)

$$\dot{\bar{\mathbf{x}}}_{cp} = \mathbf{A}_{cp}\bar{\mathbf{x}}_{cp} + \mathbf{B}_{cp}\bar{\mathbf{u}} + \mathbf{G}_{cp}\bar{\mathbf{x}}_w + \mathbf{E}_{cp}\bar{\mathbf{x}}_r \tag{29.58}$$

$$\bar{\mathbf{u}} = \mathbf{K}_{cp}\bar{\mathbf{x}}_{cp} + \mathbf{K}_w\bar{\mathbf{x}}_w + \mathbf{K}_r\bar{\mathbf{x}}_r \ . \tag{29.59}$$

The third system equation Eq.(29.36) can be replaced by the key property Eq.(29.55). With the definition of $\bar{\mathbf{C}}_{cp}$ Eq.(29.55) is rewritten to

$$\mathbf{e} = \bar{\mathbf{C}}_{cp}\bar{\mathbf{x}}_{cp} \qquad \text{where} \qquad \bar{\mathbf{C}}_{cp} \triangleq [\mathbf{0}^{[r\times(l-1)r]} \ \vdots \ \mathbf{I}_r^{(r\times r)} \ \vdots \ \mathbf{0}^{(r\times n)}] \ . \tag{29.60}$$

29.2.4 Unexcited Case

The unexcited case is defined by zero shaping signals $\mathbf{v}_w(t)$ or $\mathbf{v}_r(t)$. Consider the disturbance and its minimal polynomial $\phi_w(s)$ associated with $\mathbf{A}_w$

$$\phi_w(s) = s^{l_1} + \psi_{w_1}s^{l_1-1} + \ldots + \psi_{wl_1} \ . \tag{29.61}$$

The minimal polynomial is that annihilating polynomial of the square matrix $\mathbf{A}$, i.e $f(\mathbf{A})$, with minimal degree $\partial[f(\mathbf{A})]$. The factor of the highest power is considered 1. With regard to the annihilating property of the minimal polynomial the matrix $\mathbf{A}_w$ satisfies the condition of the minimal polynomial being zero: $\phi_w(\mathbf{A}_w) = \mathbf{0}$, i.e.

$$\phi_w(\mathbf{A}_w) = \mathbf{A}_w^{l_1} + \psi_{w_1}\mathbf{A}_w^{l_1-1} + \ldots + \psi_{wl_1}\mathbf{I} = \mathbf{0} \ . \tag{29.62}$$

From Eq.(29.25) it follows $\dot{\mathbf{x}}_w = \mathbf{A}_w\mathbf{x}_w$ or

$$s^i\mathbf{x}_w(t) = \mathbf{x}_w^{(i)}(t) = \mathbf{A}_w^i\mathbf{x}_w(t) \ . \tag{29.63}$$

Applying the operator $\phi_w(s)$ to the signal $\mathbf{x}_w(t)$ yields

$$\phi_w(s)\mathbf{x}_w(t) = s^{l_1}\mathbf{x}_w + \psi_{w_1}s^{l_1-1}\mathbf{x}_w + \ldots + \psi_{wl_1}\mathbf{x}_w \ . \tag{29.64}$$

Referring to Eq.(29.63), first, and to Eq.(29.62), second, it results

$$\mathbf{A}_w^{l_1}\mathbf{x}_w + \psi_{w_1}\mathbf{A}_w^{l_1-1}\mathbf{x}_w + \ldots + \psi_{wl_1}\mathbf{x}_w = (\mathbf{A}_w^{l_1} + \psi_{w_1}\mathbf{A}_w^{l_1-1} + \ldots + \psi_{wl_1})\mathbf{x}_w = \mathbf{0} \times \mathbf{x}_w = \mathbf{0} \ . \tag{29.65}$$

Finally, the ϕ-transformed signal $\bar{\mathbf{x}}_w$ is zero in the unexcited case

$$\bar{\mathbf{x}}_w = \phi\mathbf{x}_w = \phi_1\phi_w\mathbf{x}_w = \phi_1\mathbf{0} = \mathbf{0} \ . \tag{29.66}$$

With regard to instability of $\mathbf{A}_w$ the signal $\mathbf{x}_w$ is not zero but the ϕ-transformed $\bar{\mathbf{x}}_w$ is.

The derivation for the reference signal $\mathbf{x}_r(t)$ is very similar and omitted for abbreviation.

From Eqs.(29.58) and (29.59), using arbitrary initial conditions $\mathbf{x}_{cpo}$,

$$\dot{\bar{\mathbf{x}}}_{cp} = (\mathbf{A}_{cp} + \mathbf{B}_{cp}\mathbf{F}_{cp})\,\bar{\mathbf{x}}_{cp} \qquad \mathbf{x}_{cp}(0) = \mathbf{x}_{cpo} \tag{29.67}$$

$$\bar{\mathbf{u}} = \mathbf{F}_{cp}\bar{\mathbf{x}}_{cp} = \mathbf{F}_c\bar{\mathbf{x}}_c + \mathbf{F}_p\bar{\mathbf{x}}_p \ . \tag{29.68}$$

From any initial condition, $\mathbf{x}_{cp} \to \mathbf{0}$ if the pair $\mathbf{A}_{cp}, \mathbf{B}_{cp}$ is completely stabilizable. In order to obtain satisfactory transient behaviour, the pair $\mathbf{A}_{cp}, \mathbf{B}_{cp}$ has to be completely reachable. Eq.(29.60) implies $\mathbf{e}(t) \to \mathbf{0}$ when $\bar{\mathbf{x}}_{cp} \to \mathbf{0}$.

Robustness (weak robustness) is given because $\mathbf{e}(t) \to \mathbf{0}$ is achieved, in spite of plant parameter variations, preassuming that the stability of Eq.(29.67) is not severely affected by the parameter variations.

29.2.5 Limited-Time Exciting Case

The limited-time exciting case is given by the shaping signal $\mathbf{v}_w(t)$

$$\mathbf{v}_w(t) = \begin{cases} \neq 0 & t \leq t_1 \\ = 0 & t > t_1 \,. \end{cases} \tag{29.69}$$

The disturbance state signal $\mathbf{x}_w$ causes excitation until t_1 and, then, the system retires to the unexcited case after t_1 .

If the control system is excited, there is $\bar{\mathbf{x}}_w(t) = \mathbf{0}$ but not $\mathbf{x}_w(t) = \mathbf{0}$ for $t > t_1$. Stability and robustness hold for $t > t_1$, as in the unexcited case. In order to obtain optimal behaviour also for $t < t_1$, the controller is structured not only as Eq. (29.68) but also as Eq.(29.59), i.e., the controller is augmented by $\mathbf{x}_w$ and $\mathbf{x}_r$

$$\mathbf{u}(t) = \mathbf{K}_c\mathbf{x}_c(t) + \mathbf{K}_p\mathbf{x}_p(t) + \mathbf{K}_w\mathbf{x}_w(t) + \mathbf{K}_r\mathbf{x}_r(t). \tag{29.70}$$

29.2.6 Robust Servo-Compensator Conditions

Summarizing the servo-compensator conditions, it is necessary that the pair $\mathbf{A}_{cp}, \mathbf{B}_{cp}$ is completely reachable. This condition requires

- $\mathbf{A}_p, \mathbf{B}_p$ completely reachable
- $\mathbf{A}_p, \mathbf{C}_p$ completely observable
- rank $\mathbf{B}_p = m >$ rank $\mathbf{C}_p = r$
- The transmission zeros of the matrices $\mathbf{A}_p, \mathbf{B}_p, \mathbf{C}_p, \mathbf{D}_p$ must not be equal to the roots of $\phi(s) = 0$. The transmission zeros of a system (*Fallside, F.,(Ed.) 1977*)

$$\dot{\mathbf{x}} = \mathbf{A}_p\mathbf{x} + \mathbf{B}_p\mathbf{u} \qquad\qquad \mathbf{y} = \mathbf{C}_p\mathbf{x} + \mathbf{D}_p\mathbf{u} \tag{29.71}$$

  are defined to be the set of complex λ where

$$\text{rank} \begin{pmatrix} \mathbf{A}_p - \lambda\mathbf{I}_n & \mathbf{B}_p \\ \mathbf{C}_p & \mathbf{D}_p \end{pmatrix} < n + \min(r, m) \tag{29.72}$$

  and $\mathbf{x} \in \mathcal{R}^n$, $\mathbf{u} \in \mathcal{R}^m$, $\mathbf{y} \in \mathcal{R}^r$. In this application $\min(r, m) = r$.

29.2.7 Servo-Compensator and Constraints

The eigenvalues of the closed-loop coefficient matrix in Eq.(29.67) is a continuous function of the elements of $\mathbf{K}_{cp}$. If $\mathbf{A}_{cp}$ satisfies a certain constraint, then there exists an $\varepsilon_o > 0$ so that for any given ε, $0 < \varepsilon < \varepsilon_o$, the gain matrix exists, with the property $\|\mathbf{K}_{cp}\| < \varepsilon$, such that the closed-loop system remains stable.

The constraint mentioned above can be given as a damping factor constraint, controller gain constraint or integrity constraint. Integrity is given if the stability of the system is not affected in the case of single or multiple sensor or actuator failure where the signals y_i or u_i remain constant during failure (*Davison, E.J., and Ferguson, I.J., 1981*).

29.3 Optimal Design of the Servo-Controller

The robust servo-compensator is defined such that the output of the controlled plant both achieves asymptotic reference tracking and disturbance rejection, in spite of plant parameter variations. Moreover, the design procedure can minimize a certain performance measure. An extension to decentralized servo-compensators for large-scale interconnected systems is given by *Vaz, A.F., and Davison, E.J., 1989.*

29.3.1 Performance Measure on ϕ-Transformed Signals

A generalized control performance measure I is stated, using $\bar{\mathbf{x}}_{cp}(t) = \phi(s)\mathbf{x}_{cp}(t)$ and $\bar{\mathbf{u}}(t) = \phi(s)\mathbf{u}(t)$

$$I = \int_0^{t_f} [\bar{\mathbf{x}}_{cp}^T(t)\mathbf{Q}_{cp}\bar{\mathbf{x}}_{cp}(t) + 2\bar{\mathbf{u}}^T(t)\mathbf{S}_{cp}\bar{\mathbf{x}}_{cp}(t) + \bar{\mathbf{u}}^T(t)\mathbf{R}_{cp}\bar{\mathbf{u}}(t)]dt \tag{29.73}$$

where the weighting matrices have to satisfy the conditions

$$\mathbf{Q}_{cp} \geq 0\,, \qquad \mathbf{R}_{cp} > 0\,, \qquad \mathbf{Q}_{cp} - \mathbf{S}_{cp}^T\mathbf{R}_{cp}^{-1}\mathbf{S}_{cp} \geq 0\,. \tag{29.74}$$

The transformation function $\phi(s) = \phi(d/dt)$ penalizes the state vector and control vector motions. If the control input frequencies are near the zeros of $\phi(s)$, the amplitude $\mathbf{u}(t)$ causes less performance contribution. Removing penalty near the roots of $\phi(s)$ complies with the necessity of nonpenalizing the control input effort in the vicinity of the exciting signal frequencies. Motions far away from the zeros of $\phi(s)$ are penalized more. This consideration can be repeated as far as the reference signal $\mathbf{x}_r$ excites the system or the compensator and plant motions $\mathbf{x}_{cp}$ are both concerned.

29.3.2 Optimal Robust Servo-Compensator

Defining a combination vector $\bar{\mathbf{x}}$ containing the compensator, plant, disturbance and reference after ϕ-transformation

$$\bar{\mathbf{x}} = \begin{pmatrix} \bar{\mathbf{x}}_c \\ \bar{\mathbf{x}}_p \\ \bar{\mathbf{x}}_w \\ \bar{\mathbf{x}}_r \end{pmatrix} = \begin{pmatrix} \bar{\mathbf{x}}_{cp} \\ \bar{\mathbf{x}}_w \\ \bar{\mathbf{x}}_r \end{pmatrix} \qquad \bar{\mathbf{x}} \in \mathcal{R}^{(rl+n+n_w+n_r)}\,, \tag{29.75}$$

the compensator, plant, disturbance and reference, Eqs.(29.34) and (29.25), can be described by a unified equation in the unexcited case

$$\dot{\bar{\mathbf{x}}}(t) = \bar{\mathbf{A}}\bar{\mathbf{x}}(t) + \bar{\mathbf{B}}\bar{\mathbf{u}}(t) \tag{29.76}$$

$$\text{where} \qquad \bar{\mathbf{A}} = \begin{pmatrix} \mathbf{A}_{cp} & \mathbf{G}_{cp} & \mathbf{E}_{cp} \\ \mathbf{0} & \mathbf{A}_w & \mathbf{0} \\ \mathbf{0} & \mathbf{0} & \mathbf{A}_r \end{pmatrix} \qquad \bar{\mathbf{B}} = \begin{pmatrix} \mathbf{B}_{cp}^{(rl+n)\times m} \\ \mathbf{0} \\ \mathbf{0} \end{pmatrix}. \tag{29.77}$$

The performance index can be generalized once more

$$I = \int_0^{t_f} [\bar{\mathbf{x}}^T(t)\bar{\mathbf{Q}}\bar{\mathbf{x}}(t) + 2\bar{\mathbf{u}}^T(t)\bar{\mathbf{S}}\bar{\mathbf{x}}(t) + \bar{\mathbf{u}}^T\bar{\mathbf{R}}\bar{\mathbf{u}}]dt \tag{29.78}$$

where $$\bar{\mathbf{Q}} = \begin{pmatrix} \mathbf{Q}_{cp}^{(rl+n)\times(rl+n)} & \vdots & \mathbf{0}^{(rl+n)\times(n_w+n_r)} \\ \mathbf{0}^{(n_w+n_r)\times(rl+n)} & \vdots & \mathbf{0}^{(n_w+n_r)\times(n_w+n_r)} \end{pmatrix}, \tag{29.79}$$

$$\bar{\mathbf{S}} = \left(\mathbf{S}_{cp}^{m\times(rl+n)} \;\vdots\; \mathbf{0}^{m\times(n_w+n_r)}\right) \quad \text{and} \quad \bar{\mathbf{R}} = \mathbf{R}_{cp}\,. \tag{29.80}$$

The standard Riccati regulator is then given by (*Cheok, K.C., et al. 1988a*)

$$\bar{\mathbf{u}}(t) = -\bar{\mathbf{K}}(t)\bar{\mathbf{x}}(t) \quad \text{where} \quad \bar{\mathbf{K}}(t) = \bar{\mathbf{R}}^{-1}[\bar{\mathbf{B}}^{-1}\bar{\mathbf{P}}(t) + \bar{\mathbf{S}}] \tag{29.81}$$

$$\dot{\bar{\mathbf{P}}}(t) = \bar{\mathbf{A}}^T\bar{\mathbf{P}}(t) + \bar{\mathbf{P}}(t)\bar{\mathbf{A}} + \bar{\mathbf{Q}} - [\bar{\mathbf{P}}(t)\bar{\mathbf{B}} + \bar{\mathbf{S}}^T]\bar{\mathbf{R}}^{-1}[\bar{\mathbf{B}}^T\bar{\mathbf{P}}(t) + \bar{\mathbf{S}}]\,. \tag{29.82}$$

Partitioning the symmetric Riccati matrix $\bar{\mathbf{P}}(t)$ and the controller matrix $\bar{\mathbf{K}}(t)$ according to

$$\bar{\mathbf{P}}(t) = \begin{pmatrix} \mathbf{P}_{cp} & \mathbf{P}_w & \mathbf{P}_r \\ \mathbf{P}_w^T & \mathbf{P}_{ww} & \mathbf{P}_{wr} \\ \mathbf{P}_r^T & \mathbf{P}_{wr}^T & \mathbf{P}_{rr} \end{pmatrix} \qquad \bar{\mathbf{K}}(t) = \left(\mathbf{K}_{cp} \;\vdots\; \mathbf{K}_w \;\vdots\; \mathbf{K}_r\right), \tag{29.83}$$

and setting $\dot{\bar{\mathbf{P}}}(t) = \mathbf{0}$ for small values of t yields

$$\mathbf{P}_{cp}\mathbf{A}_{cp} + \mathbf{A}_{cp}^T\mathbf{P}_{cp} + \mathbf{Q}_{cp} - (\mathbf{P}_{cp}\mathbf{B}_{cp} + \mathbf{S}_{cp}^T)\mathbf{R}_{cp}^{-1}(\mathbf{B}_{cp}^T\mathbf{P}_{cp} + \mathbf{S}_{cp}) = \mathbf{0} \tag{29.84}$$

$$\mathbf{P}_w\mathbf{A}_w + (\mathbf{A}_{cp} + \mathbf{B}_{cp}\mathbf{K}_{cp})^T\mathbf{P}_w + \mathbf{P}_{cp}\mathbf{G}_{cp} = \mathbf{0} \tag{29.85}$$

$$\mathbf{P}_r\mathbf{A}_r + (\mathbf{A}_{cp} + \mathbf{B}_{cp}\mathbf{K}_{cp})^T\mathbf{P}_r + \mathbf{P}_{cp}\mathbf{E}_{cp} = \mathbf{0} \tag{29.86}$$

$$\mathbf{K}_{cp} = (\mathbf{K}_c \;\vdots\; \mathbf{K}_p) = -\mathbf{R}_{cp}^{-1}(\mathbf{B}_{cp}^T\mathbf{P}_{cp} + \mathbf{S}_{cp}) \tag{29.87}$$

$$\mathbf{K}_w = -\mathbf{R}_{cp}^{-1}\mathbf{B}_{cp}^T\mathbf{P}_w \tag{29.88}$$

$$\mathbf{K}_r = -\mathbf{R}_{cp}^{-1}\mathbf{B}_{cp}^T\mathbf{P}_r \tag{29.89}$$

to be applied in $$\mathbf{u}(t) = -\mathbf{K}_c\mathbf{x}_c(t) - \mathbf{K}_p\mathbf{x}_p(t) - \mathbf{K}_w\mathbf{x}_w(t) - \mathbf{K}_r\mathbf{x}_r(t)\,. \tag{29.90}$$

29.3.3 Observer-Based Servo-Compensator

If the state variables $\mathbf{x}_p, \mathbf{x}_w, \mathbf{x}_r$ are unavailable, they can be replaced by their estimates $\hat{\mathbf{x}}_p$, $\hat{\mathbf{x}}_w$, $\hat{\mathbf{x}}_r$. Using the output $\mathbf{y}$ and the control input $\mathbf{u}$ to excite the observer, and combining the plant and the disturbance equation, Eqs. (29.27) and (29.25), then equations of the type of Eqs.(18.15) and (18.16) are obtained. The estimation error only vanishes if the disturbance or reference is limited-time exciting, e.g. a step function. If a measured output $\mathbf{y}(t) = \mathbf{C}_p\mathbf{x}_p + \mathbf{H}_p\mathbf{w}$ is available and the plant, disturbance model and output equation are combined it results

$$\dot{\mathbf{x}}_{pw}(t) = \mathbf{A}_{pw}\mathbf{x}_{pw}(t) + \mathbf{B}_{pw}\mathbf{u}(t) + \mathbf{N}_{pw}\mathbf{v}_w(t) \tag{29.91}$$

$$\mathbf{y}(t) = \mathbf{C}_{pw}\mathbf{x}_{pw} \quad \text{where} \quad \mathbf{x}_{pw} = \begin{pmatrix} \mathbf{x}_p \\ \mathbf{x}_w \end{pmatrix}, \tag{29.92}$$

$$\mathbf{A}_{pw} = \begin{pmatrix} \mathbf{A}_p & \mathbf{G}_p\mathbf{C}_w \\ \mathbf{0} & \mathbf{A}_w \end{pmatrix}, \quad \mathbf{B}_{pw} = \begin{pmatrix} \mathbf{B}_p \\ \mathbf{0} \end{pmatrix}, \quad \mathbf{N}_{pw} = \begin{pmatrix} \mathbf{0} \\ \mathbf{I}_{n_1} \end{pmatrix}, \quad \mathbf{C}_{pw} = (\mathbf{C}_p \;\vdots\; \mathbf{H}_p\mathbf{C}_w)\,. \tag{29.93}$$

Then, the observer which estimates both the plant and the disturbance state obeys the Eqs.(18.1) and (18.13)

$$\dot{\mathbf{z}}(t) = \mathbf{F}\mathbf{z}(t) + \mathbf{N}\mathbf{y}(t) + \mathbf{G}\mathbf{u}(t) \tag{29.94}$$

$$\hat{\mathbf{x}}(t) = \mathbf{D}\mathbf{y}(t) + \mathbf{E}\mathbf{z}(t)\ . \tag{29.95}$$

The reduced-order observer matrices are chosen to satisfy Eqs.(18.8) and (18.9), respectively,

$$\mathbf{F}\mathbf{T}_z = \mathbf{T}_z\mathbf{A}_{pw} - \mathbf{N}\mathbf{C}_{pw} \tag{29.96}$$

$$\mathbf{G} = \mathbf{T}_z\mathbf{B}_{pw} \tag{29.97}$$

$$\mathbf{D}\mathbf{C}_{pw} + \mathbf{E}\mathbf{T}_z = \mathbf{I}\ . \tag{29.98}$$

29.4 Suggestions for Further Reading

Occasionally, the frequency dependent structure of the index of performance seems suitable. Parseval's theorem is used for linking up with time-domain control performance measure (*Cheok, K.C., et al. 1988a*).

A two-level stategy, comprising a fast first-level analog and a slow second-level digital-analog robust servomechanism, guaranteeing asymptotic tracking and good disturbance rejection, is presented by *Doraiswami, R., and Bordry, F., 1987*.

The optimal decentralized robust servomechanism problem using local servo-compensators and a square cost functional is presented by *Iftar, A., and Özgüner, Ü., 1989*. Rewriting the performance index into the frequency domain shows that the results have the property of low control effort, even at high frequencies. Furthermore, the system is free of high-frequency oscillations.

Chapter 30

Parametrization and Factorization of Systems

The factorization of rational matrix-valued functions of the complex variable s is frequently employed in designing robust controllers. Various factorization algorithms developed, e.g., by *Youla, D.C., 1961; Youla, D.C., et al. 1976; Francis, B.A., 1987; Safonov, M.G., and Verma, M.S., 1985; Vidyasagar, M., and Kimura, H., 1986* are presented in this chapter.

30.1 Polynomial Matrices and Control

Rational matrices $\mathbf{G}(s)$ can be decomposed into polynomial matrices and inverse polynomial matrices easily, e.g.,

$$\mathbf{G} = \begin{pmatrix} N_{11}/D_{11} & N_{12}/D_{12} \\ N_{21}/D_{21} & N_{22}/D_{22} \end{pmatrix} = \begin{pmatrix} N_{11}D_{21} & N_{12}D_{22} \\ N_{21}D_{11} & N_{22}D_{12} \end{pmatrix} \begin{pmatrix} D_{11}D_{21} & 0 \\ 0 & D_{12}D_{22} \end{pmatrix}^{-1} \quad (30.1)$$

$$\mathbf{G} \triangleq \mathbf{C}\mathbf{W}^{-1} = \mathbf{C}\mathbf{W}^{-1}\mathbf{I}\,. \quad (30.2)$$

30.1.1 Relatively Left-Prime and Right-Prime Matrices

Consider a multivariable system with transfer matrix $\mathbf{G}(s) = \mathbf{C}(s)\mathbf{W}^{-1}(s)\mathbf{F}(s)$ as depicted in Fig. 30.1. The Laplace transformed differential operator form is

$$\mathbf{W}(s)\mathbf{x}(s) = \mathbf{F}(s)\mathbf{u}(s) \quad (30.3)$$

$$\mathbf{y}(s) = \mathbf{C}(s)\mathbf{x}(s) \quad (30.4)$$

where $\mathbf{W}, \mathbf{F}$ and $\mathbf{C}$ are polynomial matrices in Laplace operator s (*Wolovich, W.A., 1974*). Suppose $\mathbf{W}$ nonsingular, then $\det \mathbf{W}(s)$ is a polynomial whose zeros correspond to the poles of the multivariable system. Replacing s by d/dt yields the time-domain differential operator form.

The matrix $\mathbf{D}_l$ is a left-divisor of $\mathbf{W}$ if $\mathbf{D}_l\hat{\mathbf{W}} = \mathbf{W}$ is true. The matrix $\mathbf{D}_r$ is a right-divisor of $\mathbf{C}$ if $\hat{\mathbf{C}}\mathbf{D}_r = \mathbf{C}$ is true. The matrices $\mathbf{W}, \mathbf{D}_l, \hat{\mathbf{W}}, \mathbf{C}, \mathbf{D}_r, \hat{\mathbf{C}}$ all are polynomial matrices.

Looking ahead of $\mathbf{W}^{-1}\mathbf{F}$, the matrices $\mathbf{W}$ and $\mathbf{F}$ are said to be relatively left-prime if any greatest common left-divisor $\mathbf{D}_{cl}$ of both matrices is unimodular

$$\mathbf{W} = \mathbf{D}_{cl}\hat{\mathbf{W}} \qquad \mathbf{F} = \mathbf{D}_{cl}\hat{\mathbf{F}} \quad \leadsto \quad \mathbf{W}^{-1}\mathbf{F} = \hat{\mathbf{W}}^{-1}\mathbf{D}_{cl}^{-1}\mathbf{D}_{cl}\hat{\mathbf{F}} = \hat{\mathbf{W}}^{-1}\hat{\mathbf{F}}\,. \quad (30.5)$$

The poles of $\mathbf{W}^{-1}\mathbf{F}$ and $\hat{\mathbf{W}}^{-1}\hat{\mathbf{F}}$ are given by $\det \mathbf{W} = 0$ or $\det \hat{\mathbf{W}} = 0$. If the number of poles of $\hat{\mathbf{W}}^{-1}$ should equal the number of poles of $\mathbf{W}^{-1}$, in view of $\det \mathbf{W} = \det \hat{\mathbf{W}} \det \mathbf{D}_{cl}$,

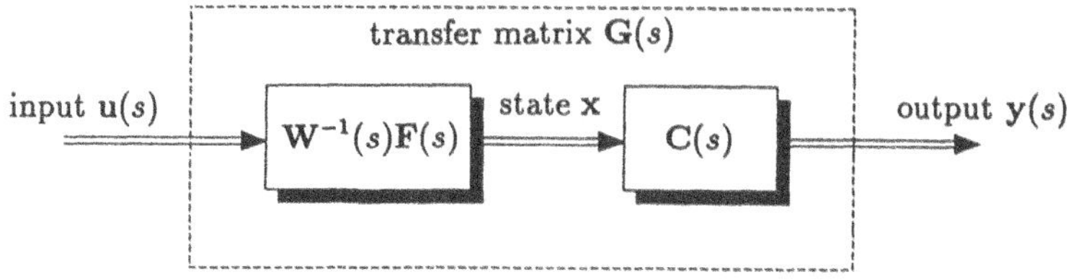

Figure 30.1: Multivariable system structured by polynomial matrices

the degree $\partial[\det \mathbf{D}_{cl}]$ is required zero to guarantee $\partial[\det \mathbf{W}] = \partial[\det \hat{\mathbf{W}}]$. With regard to the unimodular property, i.e. the degree of $\det \mathbf{D}_{cl} = 0$, no pole zero cancellation occurs in the transfer matrix $\mathbf{G}$ caused by $\mathbf{W}$ and $\mathbf{F}$. The corresponding state-space representation is completely state controllable.

Looking ahead of $\mathbf{CW}^{-1}$, the matrices $\mathbf{C}$ and$\mathbf{W}$ are said to be relatively right-prime if any greatest common right-divisor $\mathbf{D}_{cr}$ of both matrices is unimodular

$$\mathbf{C} = \hat{\mathbf{C}}\mathbf{D}_{cr} \quad \mathbf{W} = \hat{\mathbf{W}}\mathbf{D}_{cr} \quad \rightsquigarrow \quad \mathbf{CW}^{-1} = \hat{\mathbf{C}}\mathbf{D}_{cr}\mathbf{D}_{cr}^{-1}\hat{\mathbf{W}}^{-1} = \hat{\mathbf{C}}\hat{\mathbf{W}}^{-1} \,. \tag{30.6}$$

With regard to the unimodular property, i.e. the degree of $\det \mathbf{D}_{cr} = 0$, no pole zero cancellation occurs in the transfer matrix $\mathbf{G}$ caused by $\mathbf{C}$ and $\mathbf{W}$. The corresponding state-space representation is completely state observable.

The degree of $\det \mathbf{D}_{cl}$ corresponds to the number of uncontrollable modes and $\partial[\det \mathbf{D}_{cr}]$ gives the number of unobservable modes.

30.1.2 Externally Skew-Prime Matrices

Consider two polynomial matrices $\mathbf{R}(s)$ and $\mathbf{N}(s)$ where $\mathbf{R} \in \mathcal{R}^{r\times m}, \mathbf{N} \in \mathcal{R}^{r\times r}$ and r and m correspond to the dimension of $\mathbf{y}$ and $\mathbf{u}$, respectively (Fig. 30.1). The matrices $\mathbf{R}$ and $\mathbf{N}$ are externally skew-prime if polynomial matrices $\mathbf{X}$ and $\mathbf{Y}$ of appropriate dimension can be found such that

$$\mathbf{RX} + \mathbf{YN} = \mathbf{I}_p \,. \tag{30.7}$$

The matrices $\mathbf{R}$ and $\mathbf{N}$ possess the following properties

- $\mathbf{NR} = \bar{\mathbf{R}}\bar{\mathbf{N}}$ is true (also defining $\bar{\mathbf{R}}$ and $\bar{\mathbf{N}}$).
- Having in mind $\mathbf{R}\bar{\mathbf{N}}^{-1} = \mathbf{N}^{-1}\bar{\mathbf{R}} = \mathbf{T}_{RN}$, the pair $\mathbf{R}, \bar{\mathbf{N}}$ is relatively right-prime and the pair $\mathbf{N}, \bar{\mathbf{R}}$ is relatively left-prime. The matrix $\mathbf{T}_{RN}$ is a transfer matrix in dual prime factorization.
- $\mathbf{N}$ or $\bar{\mathbf{N}}$ can be considered a denominator matrix, $\mathbf{R}$ and $\bar{\mathbf{R}}$ a numerator matrix of $\mathbf{T}_{RN}$.

30.1.3 Coprime Polynomials

Two polynomials $a(s)$ and $b(s)$ are said to be coprime if they have no common zeros or their greatest common divisor is unity. Then, there exist polynomials $x(s)$ and $y(s)$ such that

$$a(s)x(s) + b(s)y(s) = 1 \qquad \text{(Bezout identity)} \,. \tag{30.8}$$

Two polynomials in RH_∞ are coprime if $x(s)$ and $y(s)$ in RH_∞ exist such that the Bezout identity holds.

30.1.4 Left-Coprime Matrices

Let the transfer matrix $\mathbf{G}(s)$ be represented fractionally, i.e., $\mathbf{G}(s) = \mathbf{A}^{-1}(s)\mathbf{B}(s)$ where $\mathbf{A}(s)$ and $\mathbf{B}(s)$ are left-coprime polynomial matrices. Then, further polynomial matrices $\mathbf{X}$ and $\mathbf{Y}$ can be found such that

$$\mathbf{AX} + \mathbf{BY} = \mathbf{I}_r \quad \text{or} \quad (\mathbf{A} \vdots \mathbf{B})\begin{pmatrix}\mathbf{X}\\ \mathbf{Y}\end{pmatrix} = \mathbf{I}_r \tag{30.9}$$

$$\text{where} \quad \mathbf{A}, \mathbf{X} \in \mathcal{C}^{r\times r}, \quad \mathbf{G}, \mathbf{B} = \mathcal{C}^{r\times m}, \quad \mathbf{Y} \in \mathcal{C}^{m\times r} . \tag{30.10}$$

The matrices $\mathbf{A}(s)$ and $\mathbf{B}(s)$ have equal number r of rows. The matrix $(\mathbf{A} \vdots \mathbf{B})$ is right-invertible.

30.1.5 Right-Coprime Matrices

Let the transfer matrix $\mathbf{G}(s)$ be represented fractionally, i.e., $\mathbf{G}(s) = \mathbf{B}_1(s)\mathbf{A}_1^{-1}(s)$ where $\mathbf{B}_1(s)$ and $\mathbf{A}_1(s)$ are right-coprime polynomial matrices. Then, further polynomial matrices $\mathbf{X}_1$ and $\mathbf{Y}_1$ can be found such that

$$\mathbf{X}_1\mathbf{A}_1 + \mathbf{Y}_1\mathbf{B}_1 = \mathbf{I}_m \quad \text{or} \quad (\mathbf{X}_1 \vdots \mathbf{Y}_1)\begin{pmatrix}\mathbf{A}_1\\ \mathbf{B}_1\end{pmatrix} = \mathbf{I}_m \tag{30.11}$$

$$\text{where} \quad \mathbf{A}_1, \mathbf{X}_1 \in \mathcal{C}^{m\times m}, \quad \mathbf{G}, \mathbf{B}_1 \in \mathcal{C}^{r\times m}, \quad \mathbf{Y}_1 \in \mathcal{C}^{m\times r} . \tag{30.12}$$

The matrices $\mathbf{B}_1$ and $\mathbf{A}_1$ have equal number m of columns. The matrix $\binom{\mathbf{A}_1}{\mathbf{B}_1}$ is left-invertible.

See *Qianhua, W., and Zhongjun, Z., 1987; Francis, B.A., 1987; Desoer, C.A., and Gustafson, C.L., 1984* for more details.

30.2 Parametrizing the Scalar-Valued Compensator Transfer Function

The scalar-valued rational transfer function $G(s)$ is exponentially stable if $G(s)$ is proper rational and has poles with negative real parts, only. The function $G(s)$ is represented as a fraction of rational functions $N(s)$, $D(s)$ (not polynomials)

$$G(s) = \frac{N(s)}{D(s)} = \frac{\frac{n(s)}{h(s)}}{\frac{d(s)}{h(s)}} . \tag{30.13}$$

If $n(s)$ and $d(s)$ are coprime polynomials and all the zeros of the polynomial $h(s)$ are in $\mathcal{C}^-$, then $N(s)$ and $D(s)$ are coprime. There exist exponentially stable rational functions $U(s)$ and $V(s)$ such that

$$U(s)N(s) + V(s)D(s) = 1 . \tag{30.14}$$

Similarly, let the compensator or controller $K(s)$ in Fig. 30.2 be

$$K(s) = \frac{X(s)}{Y(s)} . \tag{30.15}$$

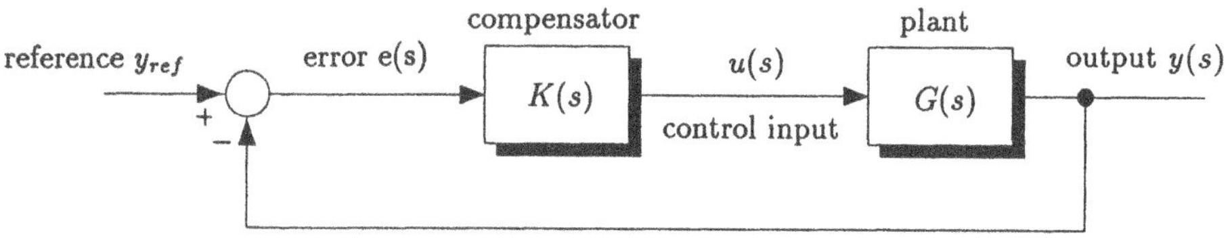

Figure 30.2: Single-input single-output loop

The input-output transfer function from the reference $y_{ref}(s)$ to the output $y(s)$ is given[1] by

$$\frac{y}{y_{ref}} = N(YD+XN)^{-1}X \triangleq NZ^{-1}X \quad \text{where} \quad Z = YD + XN \, . \tag{30.16}$$

The function $Z(s)$ must not contain any zero in C^+ . Thus, $Z(s)$ has to be a minimum-phase function. Solving Eq.(30.16), X and Y are decomposed into the homogeneous and particular solution. Let R be an exponentially stable function. Substituting the homogeneous solution $Y = Y^h = RN$ and $X = X^h = -RD$ into the homogeneous equation associated with Eq.(30.16),

$$Z = YD + XN \quad \leadsto \quad 0 = Y^hD + X^hN \, . \tag{30.17}$$

Multiplying Eq.(30.14) with Z yields

$$ZUN + ZVD = Z \, . \tag{30.18}$$

Rewriting Eq.(30.16),

$$XN + YD = Z \, . \tag{30.19}$$

Comparing Eqs.(30.18) and (30.19) makes clear that the factors of N and D, i.e.,

$$ZU = X^p \quad \text{and} \quad ZV = Y^p \, , \tag{30.20}$$

are the particular solutions. Adding both solutions (*Desoer, C.A., Liu, R.W. et al. 1980*),

$$X = X^h + X^p = -RD + ZU \qquad Y = Y^h + Y^p = RN + ZV \tag{30.21}$$

$$K(s) = \frac{X(s)}{Y(s)} = \frac{-R(s)D(s) + Z(s)U(s)}{R(s)N(s) + Z(s)V(s)} = \frac{-W(s)D(s) + U(s)}{W(s)N(s) + V(s)} \quad \text{where} \quad W(s) \triangleq \frac{R(s)}{Z(s)} \, . \tag{30.22}$$

Eq.(30.22) is a complete parametrization of the compensator $K(s)$ stabilizing the control loop of Fig. 30.2. The scalar-valued functions $X(s), Y(s), U(s), V(s), R(s)$ vary over the set of exponentially stable rational functions. The function $Z(s)$ belongs to the set of minimum-phase functions with an exponentially stable inverse. Using Eq.(30.22),

$$\frac{y}{y_{ref}} = \frac{KG}{1+KG} = \frac{XN}{YD+XN} = \frac{1}{1+\frac{1}{KG}} \tag{30.23}$$

$$= \frac{1}{1+\frac{RN+VZ}{-RD+UZ}\frac{D}{N}} = \frac{-RDN+UZN}{-RDN+UZN+RND+VZD} \tag{30.24}$$

$$\frac{y}{y_{ref}} = \frac{(-RD+UZ)N}{(UN+VD)Z} = \frac{(-RD+UZ)N}{1 \times Z} = (-WD+U)N \tag{30.25}$$

[1]Function argument s omitted for convenience.

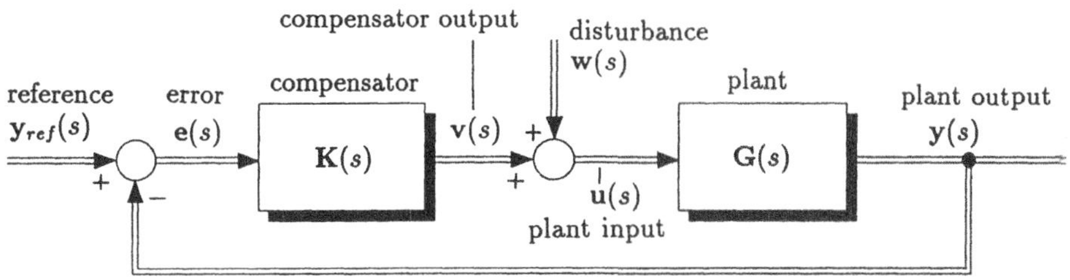

Figure 30.3: Multivariable system with multi-inputs **w** and **r**

which is a parametrization of the closed-loop transfer function and at the same time a feasibility representation of the closed-loop result.

30.3 Parametrizing Compensator Transfer Matrices

30.3.1 Transfer Matrix Representations

The pair of rational transfer matrices $(\mathbf{N}_r, \mathbf{D}_r)$ is named right-coprime if two matrices with rational function entries $\mathbf{U}_r$ and $\mathbf{V}_r$ exist such that

$$\mathbf{U}_r\mathbf{N}_r + \mathbf{V}_r\mathbf{D}_r = \mathbf{I}\,. \tag{30.26}$$

The transfer matrix $\mathbf{G} = \mathbf{N}_r\mathbf{D}_r^{-1}$ is said to be a right-fractional representation. The right-fractional representation $\mathbf{N}_r\mathbf{D}_r^{-1}$ is said to be right-coprime if the pair $(\mathbf{N}_r, \mathbf{D}_r)$ is right-coprime. If the pair $(\mathbf{N}_r, \mathbf{D}_r)$ has a common right-divisor (factor), i.e. $\mathbf{N}_r = \hat{\mathbf{N}}_r\mathbf{R}_r$ and $\mathbf{D}_r = \hat{\mathbf{D}}_r\mathbf{R}$, then

$$\mathbf{U}_r\mathbf{N}_r + \mathbf{V}_r\mathbf{D}_r = \mathbf{U}_r\hat{\mathbf{N}}_r\mathbf{R} + \mathbf{V}_r\hat{\mathbf{D}}_r\mathbf{R} = (\mathbf{U}_r\hat{\mathbf{N}}_r + \mathbf{V}_r\hat{\mathbf{D}}_r)\mathbf{R} = \mathbf{I} \tag{30.27}$$

and $\mathbf{U}_r\hat{\mathbf{N}}_r + \mathbf{V}_r\hat{\mathbf{D}}_r$ is a left-inverse $\mathbf{R}^{LI}$ of $\mathbf{R}$.

The statements outlined above can be repeated analogously as far as left-coprime fractional representations are concerned (*Vidyasagar, M., 1975; 1985*).

30.3.2 Multivariable System Representations (Youla Parametrization)

Consider the multivariable system as depicted in Fig. 30.3. The vector $\mathbf{y}_{ref}$ is the reference signal. It can also be used to represent measurement noise or controller initial conditions while plant initial conditions are adjoined to the disturbance signal $\mathbf{w}$. Defining

$$\begin{pmatrix}\mathbf{y}_{ref}\\ \mathbf{w}\end{pmatrix},\quad \begin{pmatrix}\mathbf{e}\\ \mathbf{u}\end{pmatrix}\quad \text{and}\quad \begin{pmatrix}\mathbf{v}\\ \mathbf{y}\end{pmatrix}\qquad \mathbf{y}_{ref}, \mathbf{w}, \mathbf{e}, \mathbf{v}, \mathbf{u}, \mathbf{y} \in \mathcal{C}^m \tag{30.28}$$

as the combined, error and output vector, respectively, the following equivalences in s-domain can be obtained by simple algebraic operations and by Eq.(C.33)

$$\begin{pmatrix}\mathbf{e}\\ \mathbf{u}\end{pmatrix} \triangleq \mathbf{F}\begin{pmatrix}\mathbf{y}_{ref}\\ \mathbf{w}\end{pmatrix} \qquad \begin{pmatrix}\mathbf{v}\\ \mathbf{y}\end{pmatrix} \triangleq \mathbf{H}\begin{pmatrix}\mathbf{y}_{ref}\\ \mathbf{w}\end{pmatrix} = \begin{pmatrix}\mathbf{K} & 0\\ 0 & \mathbf{G}\end{pmatrix}\mathbf{F}\begin{pmatrix}\mathbf{y}_{ref}\\ \mathbf{w}\end{pmatrix} \tag{30.29}$$

$$\mathbf{F}=\begin{pmatrix} \mathbf{I} & \mathbf{G} \\ -\mathbf{K} & \mathbf{I} \end{pmatrix}^{-1}=\begin{pmatrix} (\mathbf{I}+\mathbf{GK})^{-1} & -\mathbf{G}(\mathbf{I}+\mathbf{KG})^{-1} \\ \mathbf{K}(\mathbf{I}+\mathbf{GK})^{-1} & (\mathbf{I}+\mathbf{KG})^{-1} \end{pmatrix} \qquad \mathbf{K},\mathbf{G},\mathbf{I}\in \mathcal{C}^{m\times m} \quad (30.30)$$

$$\mathbf{H}=\begin{pmatrix} \mathbf{K}(\mathbf{I}+\mathbf{GK})^{-1} & -\mathbf{KG}(\mathbf{I}+\mathbf{KG})^{-1} \\ \mathbf{GK}(\mathbf{I}+\mathbf{GK})^{-1} & \mathbf{G}(\mathbf{I}+\mathbf{KG})^{-1} \end{pmatrix} \qquad \mathbf{F},\mathbf{H}\in \mathcal{C}^{2m\times 2m}\,. \quad (30.31)$$

When deriving the matrices above simple identities have been taken into account

$$\mathbf{I}-\mathbf{G}(\mathbf{I}+\mathbf{KG})^{-1}\mathbf{K}\equiv(\mathbf{I}+\mathbf{GK})^{-1} \qquad \mathbf{I}-(\mathbf{I}+\mathbf{GK})^{-1}\equiv\mathbf{GK}(\mathbf{I}+\mathbf{GK})^{-1}\,. \quad (30.32)$$

Using the definition $\mathbf{Q}\triangleq\mathbf{K}(\mathbf{I}+\mathbf{GK})^{-1}$, the well-known sensitivity can be rewritten to

$$\mathbf{S}\triangleq(\mathbf{I}+\mathbf{GK})^{-1}=(\mathbf{I}-\mathbf{GQ})\,. \quad (30.33)$$

Then, from the definition of $\mathbf{Q}$, one has $\mathbf{K}=(\mathbf{I}-\mathbf{QG})^{-1}\mathbf{Q}$ and from Eq.(30.30)

$$\mathbf{F}=\begin{pmatrix} \mathbf{I}-\mathbf{GQ} & (\mathbf{I}-\mathbf{GQ})\mathbf{G} \\ \mathbf{Q} & (\mathbf{I}-\mathbf{QG}) \end{pmatrix}\,. \quad (30.34)$$

The closed-loop system is internally stable if and only if the transfer matrix function $\mathbf{F}$ is stable, i.e., each of the submatrices of $\mathbf{F}$ is stable and $\det(\mathbf{I}+\mathbf{GK})(j\omega)\mid_{\omega=\infty}\neq 0$. Note that each submatrix contains only the plant $\mathbf{G}$ and the matrix $\mathbf{Q}$ as simple products or differences (*Youla, D.C., 1961*). If $\mathbf{G}$ is assumed stable $\mathbf{Q}$ must be stable, too. The matrix $\mathbf{Q}$ must be proper but the restriction to stable $\mathbf{G}$ can be dropped.

Extending the idea of internal stability leads to the method of structural synthesis (*Ohm, D.Y., et al. 1985*).

Using the symplectic matrix and its properties

$$\mathbf{I}^{\oslash}=\begin{pmatrix} 0 & -\mathbf{I} \\ \mathbf{I} & 0 \end{pmatrix}\in\mathcal{C}^{2m\times 2m} \qquad \mathbf{I}^{\oslash -1}=\mathbf{I}^{\oslash T}=-\mathbf{I}^{\oslash}\,, \quad (30.35)$$

the interrelation $\mathbf{F}-\mathbf{I}_{2m}=\mathbf{I}^{\oslash}\mathbf{H}$ is true.

Assume $\mathbf{G}$ is given as a right-fractional representation $\mathbf{G}=\mathbf{N}_r\mathbf{D}_r^{-1}$ and $\mathbf{K}$ can be found as a left-fractional representation $\mathbf{K}=\mathbf{Y}_l^{-1}\mathbf{X}_l$, then $\mathbf{F}$ can be rewritten to the form

$$\mathbf{F}=\begin{pmatrix} \mathbf{I}-\mathbf{N}_r(\mathbf{Y}_l\mathbf{D}_r+\mathbf{X}_l\mathbf{N}_r)^{-1}\mathbf{X}_l & -\mathbf{N}_r(\mathbf{Y}_l\mathbf{D}_r+\mathbf{X}_l\mathbf{N}_r)^{-1}\mathbf{Y}_l \\ \mathbf{D}_r(\mathbf{Y}_l\mathbf{D}_r+\mathbf{X}_l\mathbf{N}_r)^{-1}\mathbf{X}_l & \mathbf{D}_r(\mathbf{Y}_l\mathbf{D}_r+\mathbf{X}_l\mathbf{N}_r)^{-1}\mathbf{Y}_l \end{pmatrix}\,. \quad (30.36)$$

The result of parametrization $\mathbf{K}(s)$ in Eq.(30.41) will confirm the assumption of $\mathbf{K}(s)$ having a left-fractional representation.

30.3.3 H_∞-Optimization Problem

Referring to Eqs.(31.31) and (30.50), consider the norm $\|\mathbf{S}\|_\infty$ or the norm of the weighted $\mathbf{S}$, i.e., $\|\mathbf{WS}\|_\infty$as given by the supremum versus ω of the maximum singular value, $\sup_\omega\|\mathbf{S}\|_s$ or $\sup_\omega\|\mathbf{WS}\|_s$ where $\mathbf{S}$ is the sensitivity function matrix (output control system error, input reference and/or disturbance). The procedure of minimizing $\|\mathbf{S}\|_\infty$ with respect to the plant perturbation and controller transfer matrix is an important control system application of the H_∞-optimization problem. Then, $\mathbf{K}(s)\in H_\infty$ denotes a realizable, asymptotically stable controller, i.e., $\mathbf{K}(s)$ and $\mathbf{G}(s)$ yield an internally stable feedback control system. Referring to Eq.(30.34), $\|\mathbf{WS}\|_\infty \to \min$ is reformulated to the unconstrained problem

$$\min_{\mathbf{Q}} \|\mathbf{W}(\mathbf{I}-\mathbf{GQ})(\omega)\|_\infty \qquad \mathbf{Q} \in H_\infty . \tag{30.37}$$

Youla parametrization approach is used to ensure the control system is internally stable. The H_∞-controller may be put into the form of Youla parametrization. A link with the equivalent LQG robust optimal controller is presented by *Grimble, M.J., 1989.*

The coprime factorization approach can be applied to solve alternative problems of robust system design in the frequency domain representation. E.g., consider a given plant of l different linearized models, each in the vicinity of an operating point. Does a compensator exist which simultaneously stabilizes all these models? What is the smallest number of right-half-plane-poles that any stabilizing compensator must have? If the problem is solved by stable compensators, then strong stabilization is given (*Vidyasagar, M., and Viswanadham, N., 1982*). Simulataneous stabilization of two scalar plants is considered deriving Eq.(30.160).

30.3.4 Complete Parametrization of the Compensator

The complete parametrization of the compensator $\mathbf{K}$ (Fig. 30.3) is given by the following result. Assume that the plant $\mathbf{G}$ is already decomposed both in a right-coprime and a left-coprime fractional representation which is always possible if $\mathbf{G}$ is square. Thus,

$$\mathbf{G}(s) = \mathbf{N}_r(s)\mathbf{D}_r^{-1}(s) = \mathbf{D}_l^{-1}(s)\mathbf{N}_l(s) \tag{30.38}$$

$$\mathbf{U}_r(s)\mathbf{N}_r(s) + \mathbf{V}_r(s)\mathbf{D}_r(s) = \mathbf{I} . \tag{30.39}$$

Choose $\mathbf{W}(s)$ such that

$$\det[\mathbf{W}(s)\mathbf{N}_l(s) + \mathbf{V}_r] \neq 0 \quad \forall s \in C^+ . \tag{30.40}$$

Then,

$$\mathbf{K}(s) = [\mathbf{W}(s)\mathbf{N}_l(s) + \mathbf{V}_r(s)]^{-1}[-\mathbf{W}(s)\mathbf{D}_l(s) + \mathbf{U}_r(s)] . \tag{30.41}$$

(Proof see *Desoer, C.A., Liu, R.W., et al. 1980* and Eq.(30.85).) Using Eq.(30.41), the parametrization of the partition $\mathbf{H}_{21}$ of Eq.(30.31) points out the realizable representations

$$\begin{aligned}
\mathbf{H}_{21} &= \mathbf{GK}(\mathbf{I}+\mathbf{GK})^{-1} = [(\mathbf{GK})^{-1}+\mathbf{I}]^{-1} = [\mathbf{K}^{-1}\mathbf{G}^{-1}+\mathbf{I}]^{-1} && (30.42)\\
&= [(-\mathbf{WD}_l+\mathbf{U}_r)^{-1}(\mathbf{WN}_l+\mathbf{V}_r)\mathbf{G}^{-1}+\mathbf{I}]^{-1} && (30.43)\\
\mathbf{H}_{21} &= [(-\mathbf{WD}_l+\mathbf{U}_r)^{-1}(\mathbf{WN}_l\mathbf{D}_r\mathbf{N}_r^{-1}+\mathbf{V}_r\mathbf{D}_r\mathbf{N}_r^{-1})+\mathbf{I}]^{-1} . && (30.44)
\end{aligned}$$

In the last step above $\mathbf{G} = \mathbf{N}_r\mathbf{D}_r^{-1}$ was used. Substituting $\mathbf{V}_r\mathbf{D}_r := \mathbf{I} - \mathbf{U}_r\mathbf{N}_r$ and $\mathbf{D}_r\mathbf{N}_r^{-1} := \mathbf{N}_l^{-1}\mathbf{D}_l$ it results

$$\begin{aligned}
\mathbf{H}_{21} &= [(-\mathbf{WD}_l+\mathbf{U}_r)^{-1}(\mathbf{WN}_l\mathbf{N}_l^{-1}\mathbf{D}_l+\mathbf{N}_r^{-1}-\mathbf{U}_r)+\mathbf{I}]^{-1} && (30.45)\\
&= [(-\mathbf{WD}_l+\mathbf{U}_r)^{-1}(\mathbf{WD}_l+\mathbf{N}_r^{-1}-\mathbf{U}_r)+(-\mathbf{WD}_l+\mathbf{U}_r)^{-1}(-\mathbf{WD}_l+\mathbf{U}_r)]^{-1} &&\\
&= [(-\mathbf{WD}_l+\mathbf{U}_r)^{-1}(\mathbf{WD}_l+\mathbf{N}_r^{-1}-\mathbf{U}_r-\mathbf{WD}_l+\mathbf{U}_r)]^{-1} && (30.46)\\
\mathbf{H}_{21} &= \mathbf{N}_r(-\mathbf{WD}_l+\mathbf{U}_r) . && (30.47)
\end{aligned}$$

Similarly,

$$\mathbf{H}_{12} = \mathbf{D}_r(\mathbf{WN}_l+\mathbf{V}_r)-\mathbf{I} \qquad \text{and} \qquad \mathbf{H}_{11} = \mathbf{D}_r(-\mathbf{WD}_l+\mathbf{U}_r) . \tag{30.48}$$

30.4 Generalized System

30.4.1 Sensitivity

Define external signals $\mathbf{z}_j$, e.g., disturbance, measurement noise, reference, and internal signals $\mathbf{v}_i$, e.g., control signal, controlled variable, sensor output (see Fig. 30.4). When the signal flow diagram contains some bypasses then additional signal lines and partitioned transfer matrices are established in order to obtain the structure of Fig. 30.4. Alternatively, the bypassed signal may be generated by a separate transfer function, pulled out of the main loop. For an example see Fig. 30.5. Use the following facts:

- There are several sensitivity matrices. The number of sensitivity matrices equals the number of internal signals $\mathbf{v}_i$. The sensitivity matrix is given by the inverse return difference. "Excite" the open loop at break point $\mathbf{v}_i$ by "unity matrix" $\mathbf{I}$ and determine the output $\mathbf{J}_{v_i}\mathbf{I}$, passing the loop in direction D . The return difference is $\mathbf{I}-\mathbf{J}_{v_i}\mathbf{I}=\mathbf{I}-\mathbf{J}_{v_i}$. Signs at the summing points must be valuated. Then, the nominal sensitivity with respect to the internal signal $\mathbf{v}_i$ is $(\mathbf{I}-\mathbf{J}_{v_i})^{-1}$.
- From the external signal $\mathbf{z}_j$ to the breaking point at signal $\mathbf{v}_i$ there results a transfer matrix $\mathbf{L}_{v_i\ z_j}$ where signs at summing points must be taken into consideration.
- Then, $\mathbf{\Omega}_{ij}=(\mathbf{I}-\mathbf{J}_{v_i})^{-1}\mathbf{L}_{v_i\ z_j}$ is the transfer matrix from the input or external signal $\mathbf{z}_j$ to the internal signal $\mathbf{v}_i$, e.g., for Fig. 30.4

$$\mathbf{z}=(\mathbf{w}^T\ \mathbf{n}^T\ \mathbf{y}_{ref}^T)^T,\quad \mathbf{v}=(\mathbf{u}^T\ \mathbf{y}^T\ \mathbf{m}^T)^T,\quad \mathbf{\Omega}_{12}=(\mathbf{I}+\mathbf{C}_2\mathbf{C}_3\mathbf{H}_2\mathbf{H}_3\mathbf{P}_2\mathbf{P}_3)^{-1}(-\mathbf{C}_2\mathbf{C}_3\mathbf{H}_2\mathbf{H}_1)\,.\Box \tag{30.49}$$

  Additionally, relations of the type Eq.(30.76) are multiply applicable.
- The entire internal signal $\mathbf{v}$ results from

$$\mathbf{v}\triangleq\mathrm{vec}[\mathbf{v}_i]=\mathbf{\Omega}\mathbf{z}\triangleq\mathrm{matrix}[\mathbf{\Omega}_{ij}]\ \mathrm{vec}[\mathbf{z}_j]\,. \tag{30.50}$$

30.4.2 Generalized Signals

Consider a generalized system $\mathbf{\Gamma}(s)$ and a controller $\mathbf{K}(s)$ as depicted in Fig. 30.6. The input, output and error signals are generalized signals. Their meaning depends on the chosen application. Partitioning yields

$$\mathbf{\Gamma}=\begin{pmatrix}\mathbf{\Gamma}_{11} & \mathbf{\Gamma}_{12}\\ \mathbf{\Gamma}_{21} & \mathbf{\Gamma}_{22}\end{pmatrix}\quad\text{or}\quad \begin{aligned}\mathbf{e}&=\mathbf{\Gamma}_{11}\mathbf{y}_{in}+\mathbf{\Gamma}_{12}\mathbf{u}\\ \mathbf{y}&=\mathbf{\Gamma}_{21}\mathbf{y}_{in}+\mathbf{\Gamma}_{22}\mathbf{u}\,.\end{aligned} \tag{30.51}$$

When $\mathbf{u}=\mathbf{K}\mathbf{y}$ then $\mathbf{y}=\mathbf{\Gamma}_{21}\mathbf{y}_{in}+\mathbf{\Gamma}_{22}\mathbf{K}\mathbf{y}$ and

$$\mathbf{e}=[\mathbf{\Gamma}_{11}+\mathbf{\Gamma}_{12}\mathbf{K}(\mathbf{I}-\mathbf{\Gamma}_{22}\mathbf{K})^{-1}\mathbf{\Gamma}_{21}]\mathbf{y}_{in}\,. \tag{30.52}$$

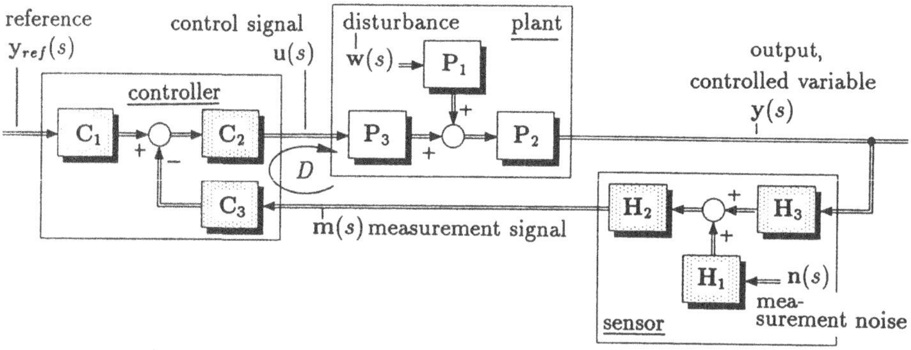

Figure 30.4: Generalized multivariable system in frequency domain

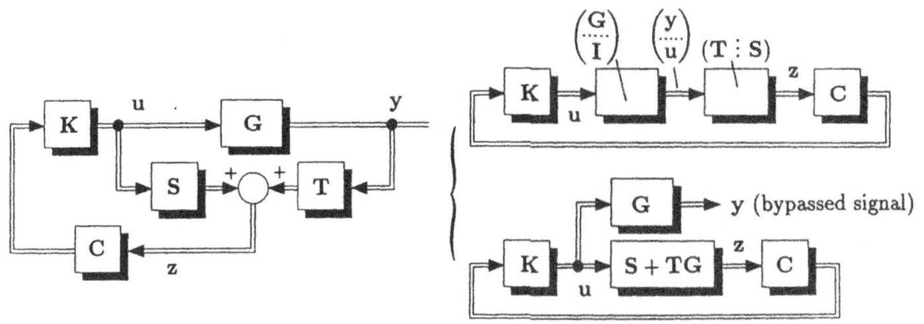

Figure 30.5: Equivalent single-loop multivariable systems

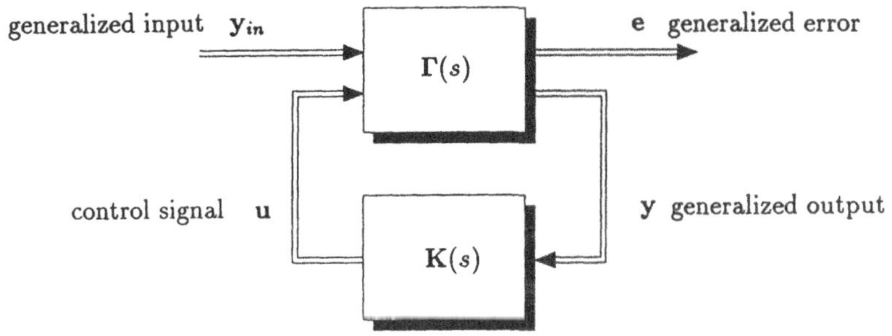

Figure 30.6: Generalized system and generalized input/output

30.4.3 Sensitivity Minimization

Application to sensitivity minimization $\|\mathbf{WS}\|_\infty \to \min$ and equating $\mathbf{y}_{in} = \mathbf{y}_{ref}$ yields

$$\mathbf{WS} = \mathbf{W}(\mathbf{I}+\mathbf{GK})^{-1} \equiv \mathbf{W}[\mathbf{I}-\mathbf{GK}(\mathbf{I}+\mathbf{GK})^{-1}] \tag{30.53}$$

and, by comparison, $\quad \mathbf{\Gamma}_{11} = \mathbf{W} \quad \mathbf{\Gamma}_{12} = -\mathbf{WG} \quad \mathbf{\Gamma}_{21} = \mathbf{I} \quad \mathbf{\Gamma}_{22} = -\mathbf{G}\,.$ (30.54)

30.4.4 Both Optimizing Disturbance Rejection and Minimizing Sensitivity

Minimizing both $\mathbf{S}$ and $(\mathbf{I}-\mathbf{S})$ guarantees good disturbance rejection and good stability robustness in the face of unstructured multiplicative output perturbations, respectively. See also Eqs.(24.1) and (31.56). Hence,

$$\left\|\begin{pmatrix}\mathbf{W}_1\mathbf{S}\\ \mathbf{W}_2(\mathbf{I}-\mathbf{S})\end{pmatrix}\right\|_\infty = \left\|\begin{pmatrix}\mathbf{W}_1(\mathbf{I}+\mathbf{GK})^{-1}\\ \mathbf{W}_2[\mathbf{I}-(\mathbf{I}+\mathbf{GK})^{-1}]\end{pmatrix}\right\|_\infty = \left\|\begin{pmatrix}\mathbf{W}_1(\mathbf{I}+\mathbf{GK})^{-1}\\ \mathbf{W}_2\mathbf{GK}(\mathbf{I}+\mathbf{GK})^{-1}\end{pmatrix}\right\|_\infty \to \min \tag{30.55}$$

where $\mathbf{W}_1$ and $\mathbf{W}_2$ are frequency-dependent weighting matrices. Choosing

$$\mathbf{\Gamma}_{11} = \begin{pmatrix}\mathbf{W}_1\\ \mathbf{0}\end{pmatrix} \qquad \mathbf{\Gamma}_{12} = \begin{pmatrix}-\mathbf{W}_1\mathbf{G}\\ \mathbf{W}_2\mathbf{G}\end{pmatrix} \qquad \mathbf{\Gamma}_{21} = \mathbf{I} \qquad \mathbf{\Gamma}_{22} = -\mathbf{G} \tag{30.56}$$

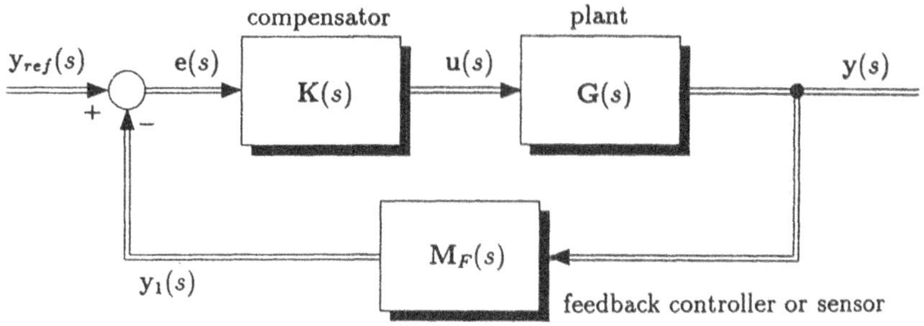

Figure 30.7: Three-element multivariable control system

complies with the objective given in Eq.(30.55) because

$$\begin{pmatrix} \mathbf{e}_1 \\ \mathbf{e}_2 \end{pmatrix} = [\begin{pmatrix} \mathbf{W}_1 \\ 0 \end{pmatrix} + \begin{pmatrix} -\mathbf{W}_1\mathbf{G} \\ \mathbf{W}_2\mathbf{G} \end{pmatrix} \mathbf{K}(\mathbf{I}+\mathbf{GK})^{-1}\mathbf{I}\,]\mathbf{y}_{in} \tag{30.57}$$

$$\begin{pmatrix} \mathbf{W}_1 - \mathbf{W}_1\mathbf{GK}(\mathbf{I}+\mathbf{GK})^{-1} \\ \mathbf{W}_2\mathbf{GK}(\mathbf{I}+\mathbf{GK})^{-1} \end{pmatrix} \mathbf{y}_{in} = \begin{pmatrix} \mathbf{W}_1(\mathbf{I}+\mathbf{GK})^{-1} \\ \mathbf{W}_2\mathbf{GK}(\mathbf{I}+\mathbf{GK})^{-1} \end{pmatrix} \begin{pmatrix} \mathbf{w} \\ \mathbf{y}_{ref} \end{pmatrix} . \tag{30.58}$$

30.5 Parametrizing Compensator Matrices Using Polynomial Matrices

Consider the multivariable system of Fig. 30.7 with (rational) transfer matrices $\mathbf{G}(s)$ and $\mathbf{M}_F(s)$ of plant and feedback controller (or sensor). The compensator transfer matrix is termed $\mathbf{K}(s)$. The sensitivity matrix $\mathbf{S}(s)$ at e-node is given by

$$\mathbf{S}(s) = [\mathbf{I}_r + \mathbf{M}_F(s)\mathbf{G}(s)\mathbf{K}(s)]^{-1} \tag{30.59}$$

$$\text{where} \quad \mathbf{u} \in \mathcal{C}^m, \quad \mathbf{y}_{ref}, \mathbf{y}, \mathbf{e} \in \mathcal{C}^r , \quad \mathbf{G} \in \mathcal{C}^{r\times m}, \quad \mathbf{S}, \mathbf{M}_F \in \mathcal{C}^{r\times r}, \quad \mathbf{Z}, \mathbf{K} \in \mathcal{C}^{m\times r} . \tag{30.60}$$

Polynomial matrices of the following dimensions are used

$$\mathbf{A}, \mathbf{X} \in \mathcal{C}^{r\times r}, \quad \mathbf{B}, \mathbf{B}_1 \in \mathcal{C}^{r\times m}, \quad \mathbf{A}_1, \mathbf{X}_1 \in \mathcal{C}^{m\times m}, \quad \mathbf{Y}, \mathbf{Y}_1 \in \mathcal{C}^{m\times r} . \tag{30.61}$$

Suppose $\mathbf{A}, \mathbf{B}$ and $\mathbf{A}_1, \mathbf{B}_1$ have the same number of rows and columns, respectively. Let the non-square transfer matrix of plant and feedback be rewritten to the fractional representation

$$\mathbf{M}_F(s)\mathbf{G}(s) = \mathbf{A}^{-1}(s)\mathbf{B}(s) = \mathbf{B}_1(s)\mathbf{A}_1^{-1}(s) \tag{30.62}$$

where $\mathbf{A}, \mathbf{B}, \mathbf{A}_1, \mathbf{B}_1$ are polynomial matrices. The pairs $\mathbf{A}, \mathbf{B}$ and $\mathbf{B}_1, \mathbf{A}_1$ are any left- and right-coprime decompositions, respectively. With regard to the left-coprimeness of $\mathbf{A}$ and $\mathbf{B}$ the polynomial matrices $\mathbf{X}$ and $\mathbf{Y}$ can be chosen such that

$$\mathbf{AX} + \mathbf{BY} = \mathbf{I}_r . \tag{30.63}$$

The matrices $\mathbf{A}_1$ and $\mathbf{B}_1$ are right-coprime if and only if there exist matrices $\mathbf{X}_1$ and $\mathbf{Y}_1$ such that

$$\mathbf{X}_1\mathbf{A}_1 + \mathbf{Y}_1\mathbf{B}_1 = \mathbf{I}_m \tag{30.64}$$

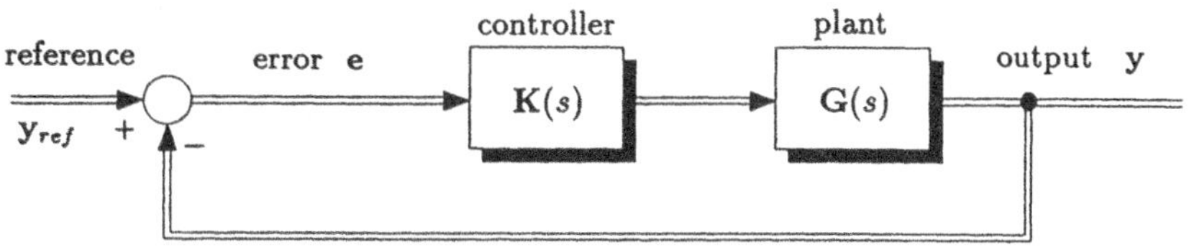

Figure 30.8: Two-element multivariable control system

(right Bezout identity, right Diophantine identity). The closed-loop system of Fig. 30.7 is asymptotically stable if and only if **KS** equals[2]

$$\mathbf{KS} = (\mathbf{Y} + \mathbf{A}_1\mathbf{Z})\mathbf{A} \tag{30.65}$$

where $\mathbf{Z}(s)$ is any rational (m, n)-matrix analytic in C^+ satisfying $\det(\mathbf{X} - \mathbf{B}_1\mathbf{Z}) \neq 0$. Then, the stabilizing compensator $\mathbf{K}(s)$ is given by (*Youla, D.C., et al. 1976*)

$$\mathbf{K}(s) = (\mathbf{Y} + \mathbf{A}_1\mathbf{Z})(\mathbf{X} - \mathbf{B}_1\mathbf{Z})^{-1} , \tag{30.66}$$

proof see Eq.(30.85). To obtain an overview the transfer matrices are listed as follows

$$\mathbf{S} = (\mathbf{X} - \mathbf{B}_1\mathbf{Z})\mathbf{A} \qquad \mathbf{e} = \mathbf{S}\,\mathbf{y}_{ref} \tag{30.67}$$
$$\mathbf{KS} = (\mathbf{Y} + \mathbf{A}_1\mathbf{Z})\mathbf{A} \qquad \mathbf{u} = \mathbf{KS}\,\mathbf{y}_{ref} \tag{30.68}$$
$$\mathbf{GKS} = \mathbf{G}(\mathbf{Y} + \mathbf{A}_1\mathbf{Z})\mathbf{A} \qquad \mathbf{y} = \mathbf{GKS}\,\mathbf{y}_{ref} \tag{30.69}$$
$$\mathbf{M}_F\mathbf{GKS} = \mathbf{I} - \mathbf{S} \qquad \mathbf{y}_1 = \mathbf{M}_F\mathbf{GKS}\,\mathbf{y}_{ref} \; . \tag{30.70}$$

Note that the following relation holds

$$\mathbf{A}(\mathbf{X} - \mathbf{B}_1\mathbf{Z}) + \mathbf{B}(\mathbf{Y} + \mathbf{A}_1\mathbf{Z}) = \mathbf{I} \; . \tag{30.71}$$

Considering the simplified case of Fig. 30.8 where $\mathbf{M}_F = \mathbf{I}$, and defining the matrix $\mathbf{S}_o \triangleq (\mathbf{I} + \mathbf{KG})^{-1}$ yields

$$\mathbf{I} - \mathbf{S}_o = \mathbf{KSG} = (\mathbf{Y} + \mathbf{A}_1\mathbf{K})\mathbf{B} \; . \tag{30.72}$$

Proof that $\mathbf{I} - \mathbf{S}_o = \mathbf{KSG}$ is true:

$$\mathbf{I} - \mathbf{S}_o = \mathbf{I} - (\mathbf{I} - \mathbf{KG})^{-1} = \mathbf{KSG} = \mathbf{K}(\mathbf{I} + \mathbf{GK})^{-1}\mathbf{G} \quad | \times (\mathbf{I} + \mathbf{KG}) \tag{30.73}$$
$$\mathbf{I} + \mathbf{KG} - \mathbf{I} = \mathbf{K}(\mathbf{I} + \mathbf{GK})^{-1}\mathbf{G}(\mathbf{I} + \mathbf{KG}) \tag{30.74}$$
$$(\mathbf{I} + \mathbf{GK}) \times | \quad \mathbf{G} = (\mathbf{I} + \mathbf{GK})^{-1}\mathbf{G}(\mathbf{I} + \mathbf{KG}) \tag{30.75}$$
$$(\mathbf{I} + \mathbf{GK})\mathbf{G} = \mathbf{G}(\mathbf{I} + \mathbf{KG}) \quad \text{q.e.d.} \tag{30.76}$$

A parametrization of output feedback controllers re eigenvalue assignment to a set of distinct eigenvalues is presented by *Roppenecker, G., and O'Reilly, J., 1989*. Parametrization concepts of decentralized feedback compensators are documented by *Gündes, A.N., and Desoer, C.A., 1990*.

[2]Function arguments (s) are partly omitted for the sake of abbreviation.

30.5.1 Parametrization Using RH_∞-Standards

The result in Eq.(30.66) is verified by a derivation using the RH_∞-notation. First, a lemma is stated.

- Lemma: Any matrix $\mathbf{F} \in RH_\infty$ can be factorized according to

$$\mathbf{F} = \mathbf{G}_1 \begin{pmatrix} \mathbf{L} & \mathbf{0} \\ \mathbf{0} & \mathbf{0} \end{pmatrix} \mathbf{G}_2 \tag{30.77}$$

where $\mathbf{G}_1, \mathbf{G}_2, \mathbf{L} \in RH_\infty$; $\mathbf{G}_1, \mathbf{G}_2$ are invertible in RH_∞ and $\mathbf{L}$ is nonsingular. □

Using this lemma the proof of Eq.(30.66) is given (*Francis, B.A., 1987*). Assume $\mathbf{K}$ real-rational and proper. Let $\mathbf{M}_F = -\mathbf{I}_r$ and the plant transfer matrix $\mathbf{G}$ be doubly-coprime factorized, i.e.,

$$\mathbf{G} = -\mathbf{A}^{-1}\mathbf{B} = -\mathbf{B}_1\mathbf{A}_1^{-1} \in \mathcal{C}^{r\times m} \tag{30.78}$$

and

$$\begin{pmatrix} \tilde{\mathbf{X}} & -\tilde{\mathbf{Y}} \\ \mathbf{B} & \mathbf{A} \end{pmatrix} \begin{pmatrix} -\mathbf{A}_1 & \mathbf{Y} \\ \mathbf{B}_1 & \mathbf{X} \end{pmatrix} = \mathbf{I}_{(m+r)} \qquad \tilde{\mathbf{X}} \in \mathcal{C}^{m\times m}, \quad \tilde{\mathbf{Y}} \in \mathcal{C}^{m\times r} . \tag{30.79}$$

Let $\mathbf{Z}$ be any real-rational function in $\Re e\ s > 0$, i.e. $\mathbf{Z} \in RH_\infty$, then Eq.(30.66) holds, as will be shown in what follows. Combining the identity

$$\begin{pmatrix} \mathbf{I}_m & \mathbf{Z} \\ \mathbf{0}_{r,m} & \mathbf{I}_r \end{pmatrix} \begin{pmatrix} \mathbf{I}_m & -\mathbf{Z} \\ \mathbf{0}_{r,m} & \mathbf{I}_r \end{pmatrix} \equiv \mathbf{I}_{(m+r)} \tag{30.80}$$

with Eq.(30.79) yields

$$\begin{pmatrix} \mathbf{I}_m & \mathbf{Z} \\ \mathbf{0}_{r,m} & \mathbf{I} \end{pmatrix} \begin{pmatrix} \tilde{\mathbf{X}} & -\tilde{\mathbf{Y}} \\ \mathbf{B} & \mathbf{A} \end{pmatrix} \begin{pmatrix} -\mathbf{A}_1 & \mathbf{Y} \\ \mathbf{B}_1 & \mathbf{X} \end{pmatrix} \begin{pmatrix} \mathbf{I}_m & -\mathbf{Z} \\ \mathbf{0}_{r,m} & \mathbf{I}_r \end{pmatrix} = \mathbf{I}_{(m+r)} \tag{30.81}$$

$$\begin{pmatrix} \tilde{\mathbf{X}} + \mathbf{Z}\mathbf{B} & -\tilde{\mathbf{Y}} + \mathbf{Z}\mathbf{A} \\ \mathbf{B} & \mathbf{A} \end{pmatrix} \begin{pmatrix} -\mathbf{A}_1 & \mathbf{Y} + \mathbf{A}_1\mathbf{Z} \\ \mathbf{B}_1 & \mathbf{X} - \mathbf{B}_1\mathbf{Z} \end{pmatrix} = \mathbf{I} . \tag{30.82}$$

By calculating the matrix product above, statements are achieved as already derived or defined, except the position in row 1 and column 2 . That is,

$$(\tilde{\mathbf{X}} + \mathbf{Z}\mathbf{B})(\mathbf{Y} + \mathbf{A}_1\mathbf{Z}) - (\tilde{\mathbf{Y}} - \mathbf{Z}\mathbf{A})(\mathbf{X} - \mathbf{B}_1\mathbf{Z}) = 0 \tag{30.83}$$

$$(\tilde{\mathbf{X}} + \mathbf{Z}\mathbf{B})(\mathbf{Y} + \mathbf{A}_1\mathbf{Z}) = (\tilde{\mathbf{Y}} - \mathbf{Z}\mathbf{A})(\mathbf{X} - \mathbf{B}_1\mathbf{Z}) . \tag{30.84}$$

Comparison with Eq.(30.66) yields

$$\mathbf{K} = (\mathbf{Y} + \mathbf{A}_1\mathbf{Z})(\mathbf{X} - \mathbf{B}_1\mathbf{Z})^{-1} = (\tilde{\mathbf{X}} + \mathbf{Z}\mathbf{B})^{-1}(\tilde{\mathbf{Y}} - \mathbf{Z}\mathbf{A}) . \tag{30.85}$$

Eq.(30.82) can be rewritten by defining $\mathbf{U} \in \mathcal{C}^{m\times r}$, $\mathbf{V} \in \mathcal{C}^{r\times r}$, $\tilde{\mathbf{U}} \in \mathcal{C}^{m\times r}$, $\tilde{\mathbf{V}} \in \mathcal{C}^{m\times m}$

$$\mathbf{U} \triangleq \mathbf{Y} + \mathbf{A}_1\mathbf{Z}, \quad \mathbf{V} \triangleq \mathbf{X} - \mathbf{B}_1\mathbf{Z}, \quad \tilde{\mathbf{U}} \triangleq \tilde{\mathbf{Y}} - \mathbf{Z}\mathbf{A}, \quad \tilde{\mathbf{V}} \triangleq \tilde{\mathbf{X}} + \mathbf{Z}\mathbf{B} \tag{30.86}$$

$$\begin{pmatrix} \tilde{\mathbf{V}} & -\tilde{\mathbf{U}} \\ \mathbf{B} & \mathbf{A} \end{pmatrix} \begin{pmatrix} -\mathbf{A}_1 & \mathbf{U} \\ \mathbf{B}_1 & \mathbf{V} \end{pmatrix} = \begin{pmatrix} -\tilde{\mathbf{V}}\mathbf{A}_1 - \tilde{\mathbf{U}}\mathbf{B}_1 & \tilde{\mathbf{V}}\mathbf{U} - \tilde{\mathbf{U}}\mathbf{V} \\ -\mathbf{B}\mathbf{A}_1 + \mathbf{A}\mathbf{B}_1 & \mathbf{B}\mathbf{U} + \mathbf{A}\mathbf{V} \end{pmatrix} = \mathbf{I} \tag{30.87}$$

$$\tilde{\mathbf{V}}\mathbf{U} - \tilde{\mathbf{U}}\mathbf{V} = 0 \ \rightsquigarrow\ \tilde{\mathbf{V}}\mathbf{U} = \tilde{\mathbf{U}}\mathbf{V} \ \rightsquigarrow\ \mathbf{U}\mathbf{V}^{-1} = \tilde{\mathbf{V}}^{-1}\tilde{\mathbf{U}} . \tag{30.88}$$

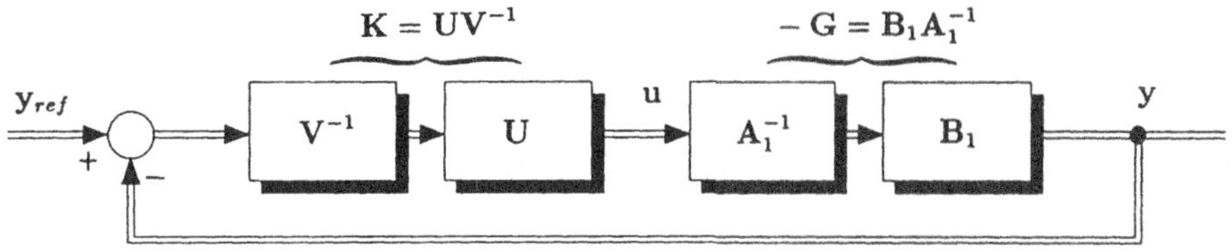

Figure 30.9: Factorization of the plant and the controller

Hence, $\mathbf{U}, \mathbf{V}$ are right-coprime matrices, $\tilde{\mathbf{V}}, \tilde{\mathbf{U}}$ are left-coprime. Invoking Eq.(30.85), the matrix $\mathbf{K}$ is factorized

$$\mathbf{K} = \mathbf{U}\mathbf{V}^{-1} = \tilde{\mathbf{V}}^{-1}\tilde{\mathbf{U}} \; . \tag{30.89}$$

This result is depicted in Fig. 30.9. By Eq.(30.87), invertability in RH_∞ is obtained

$$\begin{pmatrix} \tilde{\mathbf{V}} & -\tilde{\mathbf{U}} \\ \mathbf{B} & \mathbf{A} \end{pmatrix}^{-1} \in RH_\infty, \qquad \begin{pmatrix} -\mathbf{A}_1 & \mathbf{U} \\ \mathbf{B}_1 & \mathbf{V} \end{pmatrix}^{-1} \in RH_\infty \; . \tag{30.90}$$

From the invertability in RH_∞ there follows that the closed-loop system has no poles in $\Re e\; s > 0$, i.e., the system is stable. Hence, the controller $\mathbf{K}$ stabilizes the plant $\mathbf{G}$.

Now, the arbitrary product of partitioned matrices $\mathbf{\Gamma}$ and $\mathbf{\Omega}$ is considered

$$\mathbf{\Gamma\Omega} = \begin{pmatrix} \tilde{\mathbf{X}} & -\tilde{\mathbf{Y}} \\ \mathbf{B} & \mathbf{A} \end{pmatrix} \begin{pmatrix} -\mathbf{A}_1 & \mathbf{U} \\ \mathbf{B}_1 & \mathbf{V} \end{pmatrix} = \begin{pmatrix} -\tilde{\mathbf{X}}\mathbf{A}_1 - \tilde{\mathbf{Y}}\mathbf{B}_1 & \tilde{\mathbf{X}}\mathbf{U} - \tilde{\mathbf{Y}}\mathbf{V} \\ -\mathbf{B}\mathbf{A}_1 + \mathbf{A}\mathbf{B}_1 & \mathbf{B}\mathbf{U} + \mathbf{A}\mathbf{V} \end{pmatrix} = \begin{pmatrix} \mathbf{I} & \tilde{\mathbf{X}}\mathbf{U} - \tilde{\mathbf{Y}}\mathbf{V} \\ \mathbf{0} & \mathbf{D} \end{pmatrix} . \tag{30.91}$$

In Eq.(30.91) the statement $-\tilde{\mathbf{X}}\mathbf{A}_1 - \tilde{\mathbf{Y}}\mathbf{B}_1 = \mathbf{I}$ is true since $\mathbf{B}_1, \mathbf{A}_1$ is a right-coprime factorization of $\mathbf{K}$. The (r,r)-matrix $\mathbf{D}$ is defined for abbreviation $\mathbf{D} \triangleq \mathbf{B}\mathbf{U} + \mathbf{A}\mathbf{V}$. Since, by Eq.(30.77), $\mathbf{\Gamma}$ and $\mathbf{\Omega}$ have inverses in RH_∞ it results

$$\mathbf{D}^{-1} \in RH_\infty \quad \text{and} \quad \tilde{\mathbf{X}}\mathbf{U} - \tilde{\mathbf{Y}}\mathbf{V} \in RH_\infty \; . \tag{30.92}$$

Defining $\mathbf{Z} \triangleq -(\tilde{\mathbf{X}}\mathbf{U} - \tilde{\mathbf{Y}}\mathbf{V})\mathbf{D}^{-1}$, Eq.(30.91) is rewritten

$$\begin{pmatrix} \tilde{\mathbf{X}} & -\tilde{\mathbf{Y}} \\ \mathbf{B} & \mathbf{A} \end{pmatrix} \begin{pmatrix} -\mathbf{A}_1 & \mathbf{U} \\ \mathbf{B}_1 & \mathbf{V} \end{pmatrix} = \begin{pmatrix} \mathbf{I} & -\mathbf{Z}\mathbf{D} \\ \mathbf{0} & \mathbf{D} \end{pmatrix} . \tag{30.93}$$

Premultiplying by $\begin{pmatrix} -\mathbf{A}_1 & \mathbf{Y} \\ \mathbf{B}_1 & \mathbf{X} \end{pmatrix}$ yields

$$\begin{pmatrix} -\mathbf{A}_1 & \mathbf{Y} \\ \mathbf{B}_1 & \mathbf{X} \end{pmatrix} \begin{pmatrix} \tilde{\mathbf{X}} & -\tilde{\mathbf{Y}} \\ \mathbf{B} & \mathbf{A} \end{pmatrix} \begin{pmatrix} -\mathbf{A}_1 & \mathbf{U} \\ \mathbf{B}_1 & \mathbf{V} \end{pmatrix} = \begin{pmatrix} -\mathbf{A}_1 & \mathbf{Y} \\ \mathbf{B}_1 & \mathbf{X} \end{pmatrix} \begin{pmatrix} \mathbf{I} & -\mathbf{Z}\mathbf{D} \\ \mathbf{0} & \mathbf{D} \end{pmatrix} . \tag{30.94}$$

Applying Eq.(30.79), the product of the first and second matrix above yields $\mathbf{I}$ and

$$\begin{pmatrix} -\mathbf{A}_1 & \mathbf{U} \\ \mathbf{B}_1 & \mathbf{V} \end{pmatrix} = \begin{pmatrix} -\mathbf{A}_1 & \mathbf{Y}\mathbf{D} + \mathbf{A}_1\mathbf{Z}\mathbf{D} \\ \mathbf{B}_1 & \mathbf{X}\mathbf{D} - \mathbf{B}_1\mathbf{Z}\mathbf{D} \end{pmatrix} . \tag{30.95}$$

By comparison of the partitions $\mathbf{U} = (\mathbf{Y} + \mathbf{A}_1\mathbf{Z})\mathbf{D}$ and $\mathbf{V} = (\mathbf{X} - \mathbf{B}_1\mathbf{Z})\mathbf{D}$,

$$\mathbf{K} = \mathbf{U}\mathbf{V}^{-1} = (\mathbf{Y} + \mathbf{A}_1\mathbf{Z})(\mathbf{X} - \mathbf{B}_1\mathbf{Z})^{-1} \; . \tag{30.96}$$

30.5.2 Specialization to Stable Plants

If $\mathbf{G}$ is stable the simplifications

$$\mathbf{G} = -\mathbf{B} = \mathbf{B}_1 \qquad \mathbf{X} = \tilde{\mathbf{X}} = \mathbf{A} = -\mathbf{A}_1 = \mathbf{I} \qquad \mathbf{Y} = \tilde{\mathbf{Y}} = 0 \tag{30.97}$$

are admissible. Then, Eq.(30.66) is given by

$$\mathbf{K} = -\mathbf{Z}(\mathbf{I} - \mathbf{G}\mathbf{Z})^{-1} = -(\mathbf{I} - \mathbf{Z}\mathbf{G})^{-1}\mathbf{Z} \,. \tag{30.98}$$

In this case the transfer matrix from the reference $\mathbf{y}_{ref}$ to the control signal $\mathbf{u}$ turns out as $-\mathbf{Z}$ which can be verified easily (*Francis, B.A., 1987*).

30.6 Parametrization of an Ordinary Compensator

The nominal system is given by $\mathbf{G} = \mathbf{N}\mathbf{M}^{-1}$. Now, consider the perturbed system $\mathbf{G}_p(s) = (\mathbf{N} + \Delta\mathbf{N})(\mathbf{M} + \Delta\mathbf{M})^{-1}$, the bounded perturbation $\|\binom{\Delta \mathbf{N}}{\Delta \mathbf{M}}\|_\infty < \varepsilon_f$ and a controller $\mathbf{K}$ as given by Fig. 30.10. Assume that the unperturbed system, i.e., $\mathbf{G}$ and $\mathbf{K}$, is internally stable, see Eq.(30.30).

The perturbed closed-loop system is robustly stable if and only if the control system of $\mathbf{G}_p$ and $\mathbf{K}$ is internally stable, for all $\Delta\mathbf{N}$, $\Delta\mathbf{M}$ within the given bound. If a solution $\mathbf{K}$ exists, $\mathbf{G}_p$ is referred to as robustly stabilizable. The largest valued $\varepsilon_{f\max}$ is known as the maximum stability margin. Defining

$$\mathbf{P} = \begin{pmatrix} \mathbf{P}_{11} & \mathbf{P}_{12} \\ \mathbf{P}_{21} & \mathbf{P}_{22} \end{pmatrix} \triangleq \begin{pmatrix} (0 \,\vdots\, \mathbf{M}^{-1}) & \mathbf{M}^{-1} \\ (\mathbf{I} \,\vdots\, \mathbf{G}) & \mathbf{G} \end{pmatrix} , \tag{30.99}$$

$$\text{then,} \quad \mathbf{G}_p = (\mathbf{N}+\Delta\mathbf{N})(\mathbf{M}+\Delta\mathbf{M})^{-1} = \mathbf{P}_{22}+\mathbf{P}_{21}\begin{pmatrix}\Delta\mathbf{N}\\ \Delta\mathbf{M}\end{pmatrix}[\mathbf{I}-\mathbf{P}_{11}\begin{pmatrix}\Delta\mathbf{N}\\ \Delta\mathbf{M}\end{pmatrix}]^{-1}\mathbf{P}_{12} \tag{30.100}$$

$$\text{and} \quad \mathbf{M}^{-1}(\mathbf{I} + \mathbf{K}\mathbf{G})^{-1}(-\mathbf{K} \,\vdots\, \mathbf{I}) = \mathbf{P}_{11} - \mathbf{P}_{12}\mathbf{K}(\mathbf{I} + \mathbf{P}_{22}\mathbf{K})^{-1}\mathbf{P}_{21} \,. \tag{30.101}$$

The problem of finding maximum stability margin is resolved by

$$\varepsilon_{f\max} = [\ \inf_{\mathbf{K}} \|\mathbf{M}^{-1}(\mathbf{I} + \mathbf{K}\mathbf{G})^{-1}(-\mathbf{K} \,\vdots\, \mathbf{I})\|_\infty\ ]^{-1} \tag{30.102}$$

where the infimum is selected from all stabilizing controllers $\mathbf{K}$.

Obviously, there exists an infinite number of factorizations given by Eq.(30.122). Consider, additionally, for $\mathbf{Q}$, $\mathbf{R} > 0$ that $\mathbf{N}^R\mathbf{Q}\mathbf{N} + \mathbf{M}^R\mathbf{R}\mathbf{M} = \mathbf{I} \quad \forall\ s$ is satisfied, then, the number of factorizations $\mathbf{G}(s) = \mathbf{N}(s)\mathbf{M}^{-1}(s)$ is reduced. For $\mathbf{Q} = \mathbf{R} = \mathbf{I}$ the right-coprime factorization $\mathbf{G}(s) = \mathbf{N}(s)\mathbf{M}^{-1}(s)$ is referred to as normalized, that is $\mathbf{N}^R\mathbf{N} + \mathbf{M}^R\mathbf{M} = \mathbf{I}$ or the matrix $\binom{\mathbf{N}}{\mathbf{M}}$ is inner.

For this restriction, it can be shown that $\mathbf{M}^{-1} = (\mathbf{R} + \mathbf{G}^R\mathbf{Q}\mathbf{G})^{-1}$. Defining $\mathbf{G}_S \triangleq \mathbf{Q}^{1/2}\mathbf{G}\mathbf{R}^{-1/2}$ and $\mathbf{K}_S \triangleq \mathbf{R}^{1/2}\mathbf{K}\mathbf{Q}^{-1/2}$, the result is (*Glover, K., et al. 1990*)

$$\varepsilon_{f\max} = [\ \inf_{\mathbf{K}_S} \|\mathbf{M}_S^{-1}(\mathbf{I} + \mathbf{K}_S\mathbf{G}_S)^{-1}(-\mathbf{K}_S \,\vdots\, \mathbf{I})\begin{pmatrix} \mathbf{Q}^{1/2} & 0 \\ 0 & \mathbf{R}^{1/2} \end{pmatrix}\|_\infty\ ]^{-1} \tag{30.103}$$

where $\mathbf{M}_S$ is derived from the normalized right-coprime factorization $\mathbf{G}_S = \mathbf{N}_S\mathbf{M}_S^{-1}$. Note that the result is obtained for a shaped system.

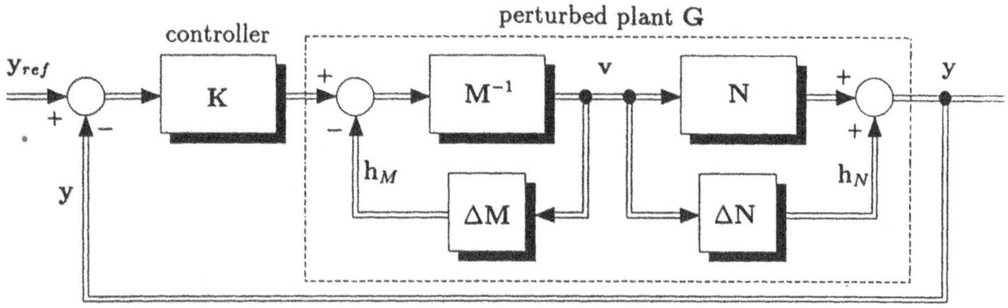

Figure 30.10: Control system using fractional perturbation

30.7 Radius of Robust Stabilizability

Consider the control system as depicted in Fig. 30.10. In the unperturbed case $\Delta\mathbf{M} = \mathbf{0}$ and $\Delta\mathbf{N} = \mathbf{0}$. The state-space representation of the transfer matrix $\mathbf{G}$ is abbreviated as minimal realization $\mathbf{G} = [\mathbf{A}, \mathbf{B}, \mathbf{C}, \mathbf{D}]$. If $\mathbf{D} = \mathbf{0}$ the system is strictly proper. The transfer function matrix of the unperturbed plant is factorized

$$\mathbf{G} = [\mathbf{A}, \mathbf{B}, \mathbf{C}, \mathbf{D}] = \tilde{\mathbf{M}}^{-1}\tilde{\mathbf{N}} = \mathbf{N}\mathbf{M}^{-1} \tag{30.104}$$

where $\tilde{\mathbf{M}}^{-1}\tilde{\mathbf{N}}$ and $\mathbf{N}\mathbf{M}^{-1}$ are, respectively, the left-coprime and right-coprime factorization obeying the associated diophantine equation

$$\begin{pmatrix} \tilde{\mathbf{X}} & -\tilde{\mathbf{Y}} \\ -\tilde{\mathbf{N}} & \tilde{\mathbf{M}} \end{pmatrix} \begin{pmatrix} \mathbf{M} & \mathbf{Y} \\ \mathbf{N} & \mathbf{X} \end{pmatrix} = \begin{pmatrix} \mathbf{I} & \mathbf{0} \\ \mathbf{0} & \mathbf{I} \end{pmatrix} . \tag{30.105}$$

For a normalized right-coprime factorization the maximum radius of stabilizability will be derived.

It is assumed that $(\mathbf{A}, \mathbf{B})$ is stabilizable and $(\mathbf{A}, \mathbf{C})$ detectable. Let stable perturbation transfer function matrices be $\Delta\mathbf{N}$ and $\Delta\mathbf{M}$. Then, the perturbed plant transfer function is

$$\mathbf{G}_p = (\mathbf{N} + \Delta\mathbf{N})(\mathbf{M} + \Delta\mathbf{M})^{-1} . \tag{30.106}$$

Introducing auxiliary signals $\mathbf{h}_N$ and $\mathbf{h}_M$ and $\mathbf{h} \triangleq (\mathbf{h}_N^T \,\vdots\, \mathbf{h}_M^T)^T$, it results

$$\begin{pmatrix} \mathbf{v} \\ \mathbf{y} \end{pmatrix} = \begin{pmatrix} -(\mathbf{M}+\mathbf{K}\mathbf{N})^{-1}[\mathbf{K} \,\vdots\, \mathbf{I}] & (\mathbf{M}+\mathbf{K}\mathbf{N})^{-1}\mathbf{K} \\ (\mathbf{I}+\mathbf{G}\mathbf{K})^{-1}[-\mathbf{I} \,\vdots\, \mathbf{G}] & (\mathbf{I}+\mathbf{G}\mathbf{K})^{-1}\mathbf{G}\mathbf{K} \end{pmatrix} \begin{pmatrix} \mathbf{h} \\ \mathbf{y}_{ref} \end{pmatrix} \triangleq \begin{pmatrix} \mathbf{S}_{11} & \mathbf{S}_{12} \\ \mathbf{S}_{21} & \mathbf{S}_{22} \end{pmatrix} \begin{pmatrix} \mathbf{h} \\ \mathbf{y}_{ref} \end{pmatrix} . \tag{30.107}$$

Defining

$$\boldsymbol{\Delta}_{res} \triangleq \begin{pmatrix} \Delta\mathbf{N} \\ \Delta\mathbf{M} \end{pmatrix} , \quad \text{it results} \quad \mathbf{h} = \boldsymbol{\Delta}_{res}\mathbf{v} = \boldsymbol{\Delta}_{res}(\mathbf{S}_{11}\mathbf{h} + \mathbf{S}_{12}\mathbf{y}_{ref}) \tag{30.108}$$

and Eq.(30.107) is reduced to

$$\mathbf{y} = \mathbf{S}_{21}\mathbf{h} + \mathbf{S}_{22}\mathbf{y}_{ref} \tag{30.109}$$

$$\mathbf{y} = [\mathbf{S}_{22} + \mathbf{S}_{21}\boldsymbol{\Delta}_{res}(\mathbf{I} - \mathbf{S}_{11}\boldsymbol{\Delta}_{res})^{-1}\mathbf{S}_{12}]\mathbf{y}_{ref} . \tag{30.110}$$

Using Youla parametrization, the set of controllers[3] that guarantee internal stability in the unperturbed system are given either by

$$\mathbf{K} = (\mathbf{MZ} - \mathbf{Y})(\mathbf{X} - \mathbf{NZ})^{-1} \quad \text{or by} \quad = (\mathbf{Z}\tilde{\mathbf{N}} - \tilde{\mathbf{X}})^{-1}(\tilde{\mathbf{Y}} - \mathbf{Z}\tilde{\mathbf{M}}) . \tag{30.111}$$

From Small gain theorem Eqs.(20.33) and (26.11), internal stability of the perturbed system is guaranteed if for stable $\mathbf{\Delta}_{res}$

$$\|\mathbf{S}_{11}\mathbf{\Delta}_{res}\|_\infty < 1 . \tag{30.112}$$

Loosely speaking, larger uncertainty $\mathbf{\Delta}_{res}$ can be tolerated if $\mathbf{S}_{11}$ is reduced. Note that $\mathbf{S}_{11}$ depends on $\mathbf{K}$. Optimal robustness is achieved by selecting $\mathbf{K}$ such that $\|\mathbf{S}_{11}\|_\infty$ is minimized. This minimization is recast to the minimum distance problem as given in Eqs.(31.118) and (31.123). Choosing $\mathbf{K}$ according to the right-hand side of Eq.(30.111),

$$\mathbf{S}_{11} = -(\mathbf{M} + \mathbf{KN})^{-1}[\mathbf{K} \vdots \mathbf{I}] = [\tilde{\mathbf{Y}} \vdots -\mathbf{X}] + \mathbf{\Theta}[-\tilde{\mathbf{M}} \vdots \tilde{\mathbf{N}}] \tag{30.113}$$

leads to

$$\|\mathbf{S}_{11}\|_\infty = \| \; (\mathbf{R}_{11} + \mathbf{\Theta}) \; \vdots \; -\mathbf{I} \; \|_\infty . \tag{30.114}$$

Using all-pass extensions, Eq.(30.166), and the all-pass transfer function matrix property of preserving the H_∞-norm, Eq.(30.168), the minimum is obtained for

$$\gamma_{opt}^2 = \min_{\mathbf{\Theta}} \|\mathbf{S}_{11}\|_\infty^2 = 1 + \|\mathbf{R}_{11}\|_H^2 \tag{30.115}$$

where $\|\cdot\|_H$ denotes the Hankel norm. The reciprocated value of γ_{opt} denotes the maximum radius of robust stabilizability with respect to unconstrained perturbations, and is given by

$$\rho_{opt} = \frac{1}{\gamma_{opt}} = \sqrt{1 - \| \begin{pmatrix} \mathbf{N} \\ \mathbf{M} \end{pmatrix} \|_H^2} . \tag{30.116}$$

See *Walker, D.J., 1990* for the proof. Any $\mathbf{K}$ in any parametrization which achieves γ_{opt} is an optimal stabilizing controller. The tool of Nehari extensions of $\mathbf{R}_{11}$ yields the set of all such stabilizing controllers (*Glover, K., and McFarlane, D., 1989; McFarlane, D.C., and Glover, K., 1990*). For practical calculations of γ_{opt} the Hankel norm $\|\cdot\|_H$ is given by Eq.(2.124).

30.8 Plant Uncertainty Principle

The Hankel singular value is frequently used to indicate for what classes of control problems nonlinear time-varying (adaptive) feedback is superior to linear time-invariant feedback (*Glover, K., and McFarlane, D., 1989; Poolla, K., 1990*). If the plant is particularly characterized by more *parametric uncertainty* than *dynamic uncertainty*, then nonlinear time-varying feedback is more useful than time-invariant feedback. If the plant mainly contains *dynamic uncertainty*, time-invariant feedback is sufficient in most cases (*Plant uncertainty principle, Poolla, K., 1990*).

Parametric uncertainty is characterized, e.g., by the linear time-invariant plant models

[3]Observe the corresponding notation $\mathbf{M}$, $\mathbf{N}$, $\tilde{\mathbf{M}}$, $\tilde{\mathbf{N}}$ and $-\mathbf{A}_1$, $\mathbf{B}_1$, $\mathbf{A}$, $-\mathbf{B}$ in Eqs.(30.105) and (30.79), respectively. Hence the opposite sign of $\mathbf{K}$ results in Eqs.(30.111) and (30.85).

- $\{v\mathbf{G}_o(z) \; : \; v \in \Omega \subseteq \mathcal{R}^+ \}$ (gain and phase margin problem) and
- $\{\mathbf{G}_i(z) \; : \; i = 1 \ldots N \}$

Dynamic modelling uncertainty is given, e.g., by the set

$$\{\mathbf{G}(z) = \mathbf{G}_o(z) + \mathbf{\Delta}(z)\mathbf{W}(z) \; : \; \|\mathbf{\Delta}(z)\|_\infty \leq \delta \} \tag{30.117}$$

where $\mathbf{W}(z)$ is a weighting function. A linear time-invariant controller robustly stabilizes the set of plants cited above if

$$\delta < \sigma_{H\,\min}[\mathbf{G}_o(z)\mathbf{W}^{-1}(z)] \tag{30.118}$$

where $\sigma_{H\,\min}[\cdot]$ is the smallest nonzero Hankel singular value of the unstable part of $\mathbf{G}_o(z)\mathbf{W}^{-1}(z)$. In Eqs.(30.112) and (30.116) another family of uncertain plant models is presented, yielding sufficient results based on a linear time-invariant feedback.

The reason for the Plant uncertainty principle is as follows: Too much *dynamic uncertainty* makes it impossible to *accurately and further* identify the dynamic details of the plant model and, hence, the merit of adaptive (or nonlinear time-varying) control is small.

30.9 Parametrization of Observer-Based Compensator Matrices

Consider a perturbed plant $\mathbf{G}_p(s)$ including uncertainty and an observer-based compensator as depicted in Fig. 30.11. The control variable $\mathbf{u}(s)$ is given by

$$\mathbf{u}(s) = \mathbf{y}_{ref}(s) - \mathbf{R}_c^{-1}(s)[\mathbf{S}_c(s)\mathbf{u}(s) + \mathbf{T}_c(s)\mathbf{y}(s)] \; . \tag{30.119}$$

Assume $\mathbf{R}_c, \mathbf{S}_c, \mathbf{T}_c$ rational, proper and stable, $\mathbf{R}_c$, $\mathbf{S}_c \in \mathcal{C}^{m\times m}$, $\mathbf{T}_c \in \mathcal{C}^{m\times r}$, $\mathbf{y} \in \mathcal{R}^r$, $\mathbf{u} \in \mathcal{R}^m$. Define robust stability as L_2-stability, i.e., the function norms satisfy

$$\|\mathbf{u}(t)\|_2 \leq \beta_1 \|\mathbf{y}_{ref}(t)\|_2 + d_1 \quad \text{and} \quad \|\mathbf{y}(t)\|_2 \leq \alpha_1 \|\mathbf{y}_{ref}(t)\|_2 + c_1 \; . \tag{30.120}$$

The constants $\alpha_1, \beta_1, c_1, d_1$ are positive numbers. Moreover, the relation

$$[\mathbf{R}_c(s) + \mathbf{S}_c(s)]\mathbf{M}(s) + \mathbf{T}_c(s)\mathbf{N}(s) = \mathbf{I} \tag{30.121}$$

is established. This relation is arbitrary but proves very useful when designing the robust compensator. The matrices $\mathbf{M}(s)$ and $\mathbf{N}(s)$ are defined by the factorization of the unperturbed plant.

Assume the right-coprime and left-coprime factorizations of the matrix transfer function $\mathbf{G}(s) = \mathbf{N}(s)\mathbf{M}^{-1}(s) = \bar{\mathbf{M}}^{-1}(s)\bar{\mathbf{N}}(s)$. Use the Bezout identity

$$\mathbf{X}(s)\mathbf{M}(s) + \mathbf{Y}(s)\mathbf{N}(s) = \mathbf{I} \tag{30.122}$$

$$\bar{\mathbf{M}}(s)\bar{\mathbf{X}}(s) + \bar{\mathbf{N}}(s)\bar{\mathbf{Y}}(s) = \mathbf{I} \tag{30.123}$$

where $\mathbf{X}(s), \mathbf{Y}(s), \bar{\mathbf{X}}(s), \bar{\mathbf{Y}}(s), \bar{\mathbf{M}}(s)$ and $\bar{\mathbf{N}}(s)$ are rational, stable and proper. Comparing $\bar{\mathbf{N}}(s)\mathbf{M}(s) - \bar{\mathbf{M}}(s)\mathbf{N}(s) = \mathbf{0}$, premultiplied by $\mathbf{Z}(s)$, with Eq.(30.121) in homogeneous form

$$\mathbf{Z}(s)\bar{\mathbf{N}}(s)\mathbf{M}(s) - \mathbf{Z}(s)\bar{\mathbf{M}}(s)\mathbf{N}(s) = \mathbf{0} \qquad [\mathbf{R}_c(s) + \mathbf{S}_c(s)]\mathbf{M}(s) + \mathbf{T}_c(s)\mathbf{N}(s) = \mathbf{0} \tag{30.124}$$

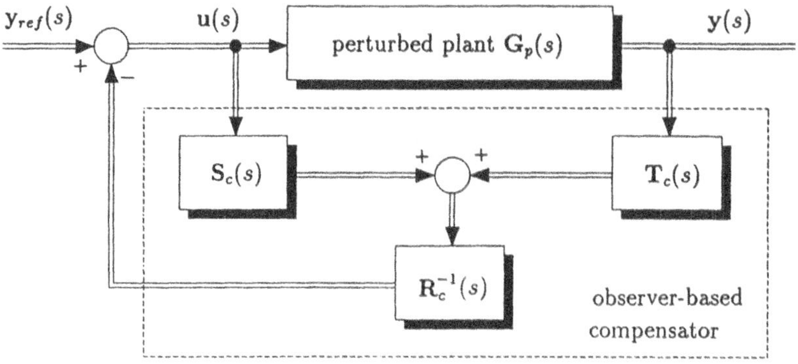

Figure 30.11: Perturbed plant and observer-based compensator

yields

$$\mathbf{R}_c(s) + \mathbf{S}_c(s) = \mathbf{Z}(s)\tilde{\mathbf{N}}(s) \qquad \mathbf{T}_c(s) = -\mathbf{Z}(s)\tilde{\mathbf{M}}(s) \tag{30.125}$$

for any rational stable and proper $\mathbf{Z}(s)$. Comparison between Eq.(30.121) and (30.122) yields the particular solution

$$\mathbf{R}_c(s) + \mathbf{S}_c(s) = \mathbf{X}(s) \qquad \mathbf{T}_c(s) = \mathbf{Y}(s) \ . \tag{30.126}$$

By superposition of particular and homogeneous solution, the parametrization is

$$\mathbf{R}_c(s) + \mathbf{S}_c(s) = \mathbf{X}(s) + \mathbf{Z}(s)\tilde{\mathbf{N}}(s) \tag{30.127}$$
$$\mathbf{T}_c(s) = \mathbf{Y}(s) - \mathbf{Z}(s)\tilde{\mathbf{M}}(s) \ . \tag{30.128}$$

30.9.1 Additive Uncertainty

Suppose $\mathbf{G}_p(s) = \mathbf{G}(s) + \Delta\mathbf{G}(s)$ where the additive uncertainty $\Delta\mathbf{G}(s)$ is free of poles and zeros in $\Re e\ s \geq 0$ and $\|\Delta\mathbf{G}(j\omega)\|_s \leq |l_a(j\omega)|$. Then, robust stability in the L_2-sense is given if and only if (*Chen, B.S., and Lo, C.H., 1989*)

$$\|l_a(j\omega)\mathbf{M}(j\omega)\mathbf{T}_c(j\omega)\|_\infty < 1 \ . \tag{30.129}$$

Combining Eqs.(30.128) and (30.129), the system is robustly stabilizable if and only if

$$\min_{\mathbf{Z}(s)} \|l_a(j\omega)\mathbf{M}(j\omega)[\mathbf{Y}(j\omega) - \mathbf{Z}(j\omega)\tilde{\mathbf{M}}(j\omega)]\ \|_\infty < 1 \ . \tag{30.130}$$

Note that the robust compensator always exists if $\mathbf{G}(s)$ is stable (which means that $\mathbf{M}^{-1}$ and $\tilde{\mathbf{M}}^{-1}$ are rational stable proper) and if

$$\mathbf{Z}(s) = [\mathbf{Y}(s) - \mathbf{M}^{-1}(s)\frac{\varepsilon_1}{l_a(s)}]\tilde{\mathbf{M}}^{-1}(s) \tag{30.131}$$

where ε_1 is a scalar with $|\varepsilon_1| < 1$. This can easily be verified by substitution.

30.9.2 Multiplicative Uncertainty

In the case of multiplicative input associated uncertainty, $\mathbf{G}_p(s) = \mathbf{G}(s)[\mathbf{I}+\Delta\mathbf{L}_i(s)]$ where $\Delta\mathbf{L}_i(s)$ is bounded $\|\Delta\mathbf{L}_i(s)\|_s \leq |l_m(j\omega)|$. Then, following *Chen, B.S., and Lo, C.H., 1989*, the robust stability and stabilizability conditions are, respectively,

$$\|l_m(j\omega)\{\mathbf{I} - \mathbf{M}(j\omega)[\mathbf{R}_c(j\omega) + \mathbf{S}_c(j\omega)]\}\|_\infty < 1 \tag{30.132}$$

$$\min_{\mathbf{Z}(s)} \|l_m(j\omega)\{\mathbf{I} - \mathbf{M}(j\omega)[\mathbf{X}(j\omega) + \mathbf{Z}(j\omega)\tilde{\mathbf{N}}(j\omega)]\}\|_\infty < 1\ . \tag{30.133}$$

If $\mathbf{G}(s)$ is free of poles and zeros in $\Re e\ s \geq 0$ and if $\mathbf{Z}(s)$ is chosen as

$$\mathbf{Z}(s) = [\mathbf{M}^{-1}(s)(1 - \frac{\varepsilon_2}{l_m(s)}) - \mathbf{X}(s)]\tilde{\mathbf{N}}^{-1}(s) \tag{30.134}$$

where ε_2 is a scalar with $|\varepsilon_2| < 1$, then the robust stability problem using an observer-based compensator can always be solved.

Large-scale systems using decentralized compensators are presented by *Chen, B.S., and Wang, W.J., 1990*.

30.9.3 Fractional Uncertainty

The case of fractional uncertainty is given by

$$\mathbf{G}(s) = \mathbf{N}(s)\mathbf{M}^{-1}(s) \qquad \in \mathcal{C}^{r\times m} \tag{30.135}$$

$$\mathbf{G}_p(s) = [\mathbf{N}(s) + \Delta\mathbf{N}(s)][\mathbf{M}(s) + \Delta\mathbf{M}(s)]^{-1} \tag{30.136}$$

where $\mathbf{M}(s), \mathbf{N}(s), \Delta\mathbf{M}(s), \Delta\mathbf{N}(s)$ are rational, stable and proper. The system is robustly L_2-stable if and only if

$$\|l_f(j\omega)[\mathbf{R}_c(j\omega) + \mathbf{S}_c(j\omega) \ \vdots\ \mathbf{T}_c(j\omega)]\ \|_\infty < 1 \tag{30.137}$$

where $l_f(j\omega)$ is a stable rational function free of zeros in $\Re e\ s \geq 0$ bounding $\Delta\mathbf{M}(s)$ and $\Delta\mathbf{N}(s)$

$$\|\begin{pmatrix}\Delta\mathbf{M}(j\omega)\\ \Delta\mathbf{N}(j\omega)\end{pmatrix}\|_s \leq |l_f(j\omega)|\ . \tag{30.138}$$

L_2-Stabilizability requires

$$\min_{\mathbf{Z}(s)}\ \|l_f(j\omega)[\mathbf{X}(j\omega) + \mathbf{Z}(j\omega)\tilde{\mathbf{N}}(j\omega) \ \vdots\ \mathbf{Y}(j\omega) - \mathbf{Z}(j\omega)\tilde{\mathbf{M}}(j\omega)]\ \|_\infty < 1\ . \tag{30.139}$$

Proof: **Sufficiency:** Combining Eqs.(30.119) and (30.136)

$$\mathbf{u}(s) = \mathbf{y}_{ref}(s) - \mathbf{R}_c^{-1}[\mathbf{S}_c\mathbf{u}(s) + \mathbf{T}_c(\mathbf{N}+\Delta\mathbf{N})(\mathbf{M}+\Delta\mathbf{M})^{-1}\mathbf{u}(s)] \tag{30.140}$$

$$[(\mathbf{R}_c + \mathbf{S}_c) + \mathbf{T}_c(\mathbf{N}+\Delta\mathbf{N})(\mathbf{M}+\Delta\mathbf{M})^{-1}]\mathbf{u}(s) = \mathbf{R}_c\mathbf{y}_{ref}(s) \tag{30.141}$$

$$\begin{aligned}
\mathbf{u}(s) &= (\mathbf{M}+\Delta\mathbf{M})[(\mathbf{R}_c+\mathbf{S}_c)(\mathbf{M}+\Delta\mathbf{M}) + \mathbf{T}_c(\mathbf{N}+\Delta\mathbf{N})]^{-1}\mathbf{R}_c\mathbf{y}_{ref}(s) && (30.142)\\
&= (\mathbf{M}+\Delta\mathbf{M})[\underbrace{(\mathbf{R}_c+\mathbf{S}_c)\mathbf{M} + \mathbf{T}_c\mathbf{N}}_{\mathbf{I}} + (\mathbf{R}_c+\mathbf{S}_c \ \vdots\ \mathbf{T}_c)\begin{pmatrix}\Delta\mathbf{M}\\ \Delta\mathbf{N}\end{pmatrix}]^{-1}\mathbf{R}_c\mathbf{y}_{ref}(s) && (30.143)\\
&= (\mathbf{M}+\Delta\mathbf{M})[\mathbf{I} + (\mathbf{R}_c+\mathbf{S}_c \ \vdots\ \mathbf{T}_c)\begin{pmatrix}\Delta\mathbf{M}\\ \Delta\mathbf{N}\end{pmatrix}]^{-1}\mathbf{R}_c\mathbf{y}_{ref}(s)\ . && (30.144)
\end{aligned}$$

Referring to Eqs.(20.101), for the linear causal stable operator $\mathbf{H}$ in the time domain the following equality exists

$$||\mathbf{H}||_2 \triangleq \sup_{\mathbf{x}(t)\in L^n} \frac{||\mathbf{Hx}(t)||_2}{||\mathbf{x}(t)||_2} = ||\mathbf{H}(j\omega)||_\infty \triangleq \sup_\omega ||\mathbf{H}(j\omega)||, \tag{30.145}$$

where $\mathbf{H}(s)$ and $\mathbf{H}(j\omega)$ denote the Laplace and Fourier transform of the corresponding operator $\mathbf{H}$, respectively. Then, the second part in the bracket using the bound $l_f(j\omega)$ is

$$||(\mathbf{R}_c + \mathbf{S}_c \vdots \mathbf{T}_c)\begin{pmatrix}\Delta\mathbf{M}\\ \Delta\mathbf{N}\end{pmatrix}||_2 = ||(\mathbf{R}_c + \mathbf{S}_c \vdots \mathbf{T}_c)\begin{pmatrix}\Delta\mathbf{M}\\ \Delta\mathbf{N}\end{pmatrix}||_\infty \le ||l_f(j\omega)\{\mathbf{R}_c(j\omega) + \mathbf{S}_c(j\omega) \vdots \mathbf{T}_c(j\omega)\}||_\infty < 1\,. \tag{30.146}$$

Hence, the inverse $[\;]^{-1}$ exists and

$$||\mathbf{u}_T(t)||_2 \le ||\mathbf{M}+\Delta\mathbf{M}||_2\; ||\,[\mathbf{I}+(\mathbf{R}_c+\mathbf{S}_c \vdots \mathbf{T}_c)\begin{pmatrix}\Delta\mathbf{M}\\ \Delta\mathbf{N}\end{pmatrix}]^{-1}||_2||\mathbf{R}_c||_2||\mathbf{y}_{refT}(t)||_2 \triangleq \beta_1||\mathbf{y}_{refT}(t)||_2\,. \tag{30.147}$$

$$\text{As } T\to\infty \qquad ||\mathbf{u}(t)||_2 \le \beta_1||\mathbf{y}_{ref}(t)||_2 \qquad \beta_1\in\mathcal{R}^+\,. \tag{30.148}$$

Similarly, from $\mathbf{y}(s) = (\mathbf{N}+\Delta\mathbf{N})(\mathbf{M}+\Delta\mathbf{M})^{-1}\mathbf{u}(s) \rightsquigarrow ||\mathbf{y}(t)||_2 \le \alpha_1||\mathbf{y}_{ref}(t)||_2$ and the system is L_2-stable.

Necessity: The condition Eq.(30.137) is a necessary condition if *any* assumption opposite to Eq.(30.137) yields an inadmissible statement which does not provide a compensator stabilizing the system. Replacing <1 in Eq.(30.137) by ≥ 1, a frequency ω_o is postulated such that

$$|l_f(j\omega_o)|\; ||\;[\mathbf{R}_c(j\omega_o) + \mathbf{S}_c(j\omega_o) \vdots \mathbf{T}_c(j\omega_o)]\;||_\infty \ge 1\,. \tag{30.149}$$

Applying singular-value decomposition and unitary matrices $\mathbf{U}$ and $\mathbf{V}$,

$$[\mathbf{R}_c(j\omega_o) + \mathbf{S}_c(j\omega_o) \vdots \mathbf{T}_c(j\omega_o)] = \mathbf{U\Sigma V} \tag{30.150}$$

$$\text{where } \sigma_1\ge\sigma_2\ge\ldots\ldots\ge\sigma_m \quad\text{and}\quad \Sigma = \begin{pmatrix}\sigma_1 & 0 & \ldots & & 0\\ 0 & \sigma_2 & & & 0\\ \vdots & & \ddots & & \vdots\\ 0 & & & \sigma_m & 0\end{pmatrix}. \tag{30.151}$$

Since the relation ≥ 1 in Eq.(30.149) it follows $\sigma_1|l_f(j\omega_o)|\ge 1$. An additional and arbitrary assumption in order to obtain a contradiction is

$$\begin{pmatrix}\Delta\mathbf{M}(s)\\ \Delta\mathbf{N}(s)\end{pmatrix} = -\mathbf{V}^{-1}\begin{pmatrix}\mathrm{diag}_m\{1/\sigma_1\}\\ 0_{r\times m}\end{pmatrix}\mathbf{U}^{-1}\frac{l_f(s)}{l_f(j\omega_o)} \tag{30.152}$$

which is permissible since

$$||\begin{pmatrix}\Delta\mathbf{M}(s)\\ \Delta\mathbf{N}(s)\end{pmatrix}||_s = \frac{1}{\sigma_1}|\frac{l_f(j\omega)}{l_f(j\omega_o)}| < |l_f(j\omega)|\,. \tag{30.153}$$

The inequality in the right-hand side of the expression above results from replacing the term $\sigma_1|l_f(j\omega_o)|$ by 1 although $\sigma_1|l_f(j\omega_o)|$ has been derived ≥ 1. Finally,

$$||\begin{pmatrix}\Delta\mathbf{M}(j\omega_o)\\ \Delta\mathbf{N}(j\omega_o)\end{pmatrix}||_s = \frac{1}{\sigma_1} \le |l_f(j\omega_o)| \tag{30.154}$$

which agrees with the definition Eq.(30.138). Thus, the assumption Eq.(30.152) is a permissible perturbation. Calculating $[\;]^{-1}$ of Eq.(30.144) at $\omega=\omega_o$ and using the assumptions Eq.(30.151) and (30.152) yields

$$\mathbf{I}+(\mathbf{R}_c \;+\; \mathbf{S}_c \;\vdots\; \mathbf{T}_c)\begin{pmatrix}\Delta\mathbf{M}\\ \Delta\mathbf{N}\end{pmatrix} = \mathbf{I}+[\mathbf{R}_c(j\omega_o) + \mathbf{S}_c(j\omega_o) \vdots \mathbf{T}_c(j\omega_o)]\begin{pmatrix}\Delta\mathbf{M}(j\omega_o)\\ \Delta\mathbf{N}(j\omega_o)\end{pmatrix} \tag{30.155}$$

$$= \mathbf{I}+\mathbf{U}\begin{pmatrix}-1 & 0 & 0\\ 0 & -\sigma_2/\sigma_1 & 0\\ 0 & 0 & \ddots\end{pmatrix}\mathbf{U}^{-1} = \mathbf{U}\begin{pmatrix}0 & 0 & 0\\ 0 & 1-\sigma_2/\sigma_1 & 0\\ 0 & 0 & \ddots\end{pmatrix}\mathbf{U}^{-1}\,. \tag{30.156}$$

Since the matrix above is not invertible $\mathbf{u}(j\omega_o)$ is infinite and does not stabilize the system. Hence, the necessity is proved.

End of the Proof

30.10 Stability Robustness and Interpolation Algorithm

The robust stabilization problem and the problem of its existence can be solved by finding a unit in H_∞ which interpolates to given points in the right-half plane. In Eqs.(30.131) and (30.134) it has been stated that the robust observer-based compensator can always be found for additive and multiplicative perturbation if the nominal plant is stable. In the unstable case of nominal plant the existence of robust observer-based compensators is solved by using Nevanlinna-Pick theory, see below. Before doing so, the problem of strong and simultaneous stabilization is presented to give a basic introduction to the efficiency of interpolation methods.

30.10.1 Strong Stabilization

Consider the scalar plant $G(s) = N(s)/D(s)$ where $N(s)$ and $D(s)$ are relatively prime functions in H_∞ . Then, the controller $K(s)$ is stable and stabilizes the feedback system if

$$K(s) = \frac{U(s) - D(s)}{N(s)}, \qquad U(s) \in H_\infty \quad \text{where} \quad U(z_i) = D(z_i) \tag{30.157}$$

and z_i are the (distinct) zeros of $N(s)$ in $\Re e\ s \geq 0$ including infinity. The function $U(s)$ interpolates to $U(z_i) = D(z_i)$. Thus, the strong-stabilization problem is reduced to the interpolation with a unit in H_∞.

$$\text{Proof}: \qquad KG = (U - D)/D \quad \rightsquigarrow \quad KG/(1 + KG) = 1 - D/U \ . \ \square \tag{30.158}$$

30.10.2 Simultaneous Stabilization

Consider two scalar plants $G_o(s)$ and $G_1(s)$ where $G_o(s)$ is stable and $G_1(s)$ arbitrary. Then, the fixed function $K(s)$ is a robust controller or compensator

$$K(s) = [I - R(s)G_o(s)]^{-1}R(s) \tag{30.159}$$

$$R(s) = \frac{U(s) - D_A(s)}{N_A(s)} \qquad \frac{N_A}{D_A} \triangleq G_1 - G_o \qquad N_A, D_A \in H_\infty \tag{30.160}$$

$$U(s) \in H_\infty \quad \text{where} \quad U(z_i) = D_A(z_i), \quad N_A(z_i) = 0, \quad \Re e\ z_i \geq 0\ . \tag{30.161}$$

Proof: Calculate the closed-loop transfer function $T(s) = K(s)G(s)/[1 + K(s)G(s)]$
(i) $T\,|_{G=G_o} = \frac{U-D_A}{N_A}G_o$. Since $U(z_i) = D_A(z_i)$ at z_i where $N_A(z_i) = 0$ the closed loop is stable.
(ii) $T\,|_{G=G_1} = (1 - \frac{D_A}{U})(\frac{D_A}{N_A}G_o + 1)$: Unstable poles of T are compensated by $1 - D_A/U$.
Comparison of Eq.(30.157) and (30.160) shows that a necessary and sufficient condition for the existence of a robust controller is that the difference plant $G_1 - G_o$ must be strongly stabilizable (*Dorato, P., et al. 1989; Vidyasagar, M., 1985*). □

Ghosh, B.K., 1986 investigates the existence of a single compensator which simultaneously renders a given set of plants internally stable. A p-tuple of plants $\in C^{r \times m}$ is stabilized via simultaneous partial pole placement provided that $p \leq \max\{r, m\}$. The problem of stabilizing l different models is presented by *Vidyasagar, M., and Viswanadham, N., 1982*.

30.10.3 Nevanlinna-Pick Theory

Several interpolation methods and algorithms can be used to obtain robust controllers. However, the algorithm of finding interpolation functions often yields very high degree interpolation units. Hence, interpolation algorithms requiring the lowest possible degree are very important. See, e.g., *Dorato, P., et al. 1989; Kimura, H., 1984; Chen, B.S., and Lo, C.H., 1989; Chang, B.C., and Pearson, J.B.Jr., 1984* and Nevanlinna-Pick theory (*Delsarte, P., et al. 1981*).

The Nevanlinna-Pick theory serves to solve the following interpolation problem. Assume n complex numbers $s_1, s_2 \dots s_m$ satisfying $\Re e\ s_i > 0$ and $\mathbf{F}_1, \mathbf{F}_2 \dots \mathbf{F}_m$ complex matrices where $\|\mathbf{F}_i\|_s < 1 \ \forall i$. Then, there exists a rational stable proper matrix $\mathbf{F}(s)$ such that $\|\mathbf{F}(s)\|_\infty < 1$ and such that $\mathbf{F}(s_i) = \mathbf{F}_i \ \forall i$ if and only if the Pick matrix $\mathbf{P}$ is positive definite

$$\mathbf{P} = \begin{pmatrix} \mathbf{P}_{11} & \mathbf{P}_{12} & \dots & \mathbf{P}_{1m} \\ \vdots & \ddots & & \\ \mathbf{P}_{m1} & \dots & \dots & \mathbf{P}_{mm} \end{pmatrix} > 0 \quad \text{where} \quad \mathbf{P}_{ij} = \frac{\mathbf{I} - \mathbf{F}_i^H \mathbf{F}_j}{s_i^* + s_j} . \tag{30.162}$$

An alternative approach of multivariable interpolation using all-pass transfer function matrices is presented by *Shaked, U., 1990*.

30.11 Properties Using Inner and Outer Matrices

30.11.1 Continuous-Time Systems

A rational Hardy space matrix $\mathbf{H}_i(s)$ with proper stable elements is all-pass if

$$\mathbf{H}_i^H(j\omega)\mathbf{H}_i(j\omega) = \mathbf{I} \quad \text{or} \quad \mathbf{H}_i(j\omega)\mathbf{H}_i^H(j\omega) = \mathbf{I} \quad \forall \omega . \tag{30.163}$$

The matrix is inner if it is also stable. A corresponding definition is: A rational nonsquare matrix $\mathbf{H}_i \in \mathcal{C}^{m \times n}$, $\mathbf{H}_i \in H_\infty$, is inner if for $m > n$ or $m < n$, respectively,

$$\mathbf{H}_i^R(s)\mathbf{H}_i(s) = \mathbf{I}_n \quad \text{or} \quad \mathbf{H}_i(s)\mathbf{H}_i^R(s) = \mathbf{I}_m \quad \forall s \tag{30.164}$$

$$\mathbf{H}_i^T(-s)\mathbf{H}_i(s) = \mathbf{I}_n \quad \text{or} \quad \mathbf{H}_i(s)\mathbf{H}_i^T(-s) = \mathbf{I}_m \quad \forall s . \tag{30.165}$$

For $m \neq n$ the matrix $\mathbf{H}_i(s)$ has a complementary inner

$$\mathbf{H}_i^\perp(s) \quad \text{such that} \quad \big(\mathbf{H}_i(s) \ \vdots \ \mathbf{H}_i^\perp(s)\big) \quad \text{or} \quad \begin{pmatrix} \mathbf{H}_i(s) \\ \mathbf{H}_i^\perp(s) \end{pmatrix} \tag{30.166}$$

is square and inner. The matrix $\mathbf{H}_i^\perp(s)$ is named all-pass extension.

A matrix $\mathbf{H}_o(s)$ is outer (or minimum-phase) if $\mathbf{H}_o(s)$ has no right half-plane zeros or if $\mathbf{H}_o(s)$ has full row rank at all s for $\Re e\ s > 0$. A rational stable proper matrix is outer if it has no right-half-plane zeros. A square transfer matrix is outer if it is proper stable and has a proper stable inverse. If a matrix $\mathbf{H}(s)$ is rational proper stable and square, then there exist the following factorizations, using an inner and an outer matrix,

$$\mathbf{H}(s) = \mathbf{H}_i(s)\mathbf{H}_o(s) = \tilde{\mathbf{H}}_o(s)\tilde{\mathbf{H}}_i(s) \tag{30.167}$$

where $\mathbf{H}_i, \tilde{\mathbf{H}}_i$ are square inner and $\mathbf{H}_o, \tilde{\mathbf{H}}_o$ are square outer.

If $\mathbf{H}(s)$ is square and inner, then $\det\mathbf{H}(s)$ and the adjoint matrix $\mathrm{adj}\mathbf{H}(s)$ also are inner.

For $\mathbf{G}(s)$ rational proper and stable

$$\|\mathbf{H}_i(j\omega)\mathbf{G}(j\omega)\|_\infty = \|\mathbf{G}(j\omega)\|_\infty \quad \text{or} \quad \|\mathbf{G}(j\omega)\mathbf{H}_i(j\omega)\|_\infty = \|\mathbf{G}(j\omega)\|_\infty , \tag{30.168}$$

i.e., the H_∞-norm is invariant with respect to multiplication with an inner matrix since

$$\|\mathbf{H}_i(j\omega)\mathbf{G}(j\omega)\|_\infty = \sup_\omega \sqrt{\lambda_{\max}[\mathbf{G}^H\mathbf{H}_i^H\mathbf{H}_i\mathbf{G}]} = \sup_\omega \sqrt{\lambda_{\max}[\mathbf{G}^H\mathbf{H}_i^R\mathbf{H}_i\mathbf{G}]} = \sup_\omega \sqrt{\lambda_{\max}[\mathbf{G}^H\mathbf{G}]} = \|\mathbf{G}(s)\|_\infty . \tag{30.169}$$

(*Safonov, M.G., and Verma, M.S., 1985; Vidyasagar, M., and Kimura, H., 1986*).

Replacing a pole $s = s_P$ by its conjugate $s = -s_P$ in a transfer function or multiplying a transfer function by an all-pass function $(s-s_P)/(s+s_P)$ does not influence the modulus of a transfer function. Inner-outer factorization is a special class of conjugation (*Kimura, H., 1989*).

30.11.2 All-Pass and Inner Matrices for Discrete-Time Systems

The matrix $\mathbf{R}(z^{-1})$ is inner (all-pass) if it is square and stable and if it obeys the equation $\mathbf{R}^R(z^{-1})\mathbf{R}(z^{-1}) = \mathbf{R}^T(z)\mathbf{R}(z^{-1}) = \mathbf{I}$ or $\mathbf{R}^R(z)\mathbf{R}(z) = \mathbf{I}$ where $\mathbf{R}^R(z^{-1}) = \mathbf{R}^T(z)$.

30.12 Existence of an Observer-Based Compensator for Additive Uncertainty

In the case of unstable additive plant uncertainty, the existence of a robust observer-based compensator is solved using Nevanlinna-Pick theory. Starting with Eqs.(30.130) and (30.129), factorization of $\mathbf{M}(s)$ and $\bar{\mathbf{M}}(s)$ is applied and Eq.(30.168) is employed. Hence,

$$\mathbf{M}(s) = \mathbf{M}_i(s)\mathbf{M}_o(s) \qquad \bar{\mathbf{M}}(s) = \bar{\mathbf{M}}_o(s)\bar{\mathbf{M}}_i(s) \tag{30.170}$$

$$\begin{aligned} \|l_a(j\omega)\mathbf{M}_i(j\omega)\mathbf{M}_o(j\omega)[\mathbf{Y}(j\omega) &- \mathbf{Z}(j\omega)\mathbf{M}(j\omega)]\ \|_\infty \\ &= \|l_a(j\omega)\mathbf{M}_o(j\omega)[\mathbf{Y}(j\omega) - \mathbf{Z}(j\omega)\bar{\mathbf{M}}_o(j\omega)\bar{\mathbf{M}}_i(j\omega)]\ \|_\infty . \end{aligned} \tag{30.171}$$

Since $\mathrm{adj}\bar{\mathbf{M}}_i(j\omega)$ is inner it can be postmultiplied to the argument of the $\|\cdot\|_\infty$-norm, i.e.,

$$= \|l_a(j\omega)\mathbf{M}_o(j\omega)\mathbf{Y}(j\omega)\ \mathrm{adj}\bar{\mathbf{M}}_i(j\omega) - l_a(j\omega)\mathbf{M}_o(j\omega)\mathbf{Z}(j\omega)\bar{\mathbf{M}}_o(j\omega)\bar{\mathbf{M}}_i(j\omega)\ \mathrm{adj}\bar{\mathbf{M}}_i(j\omega)\|_\infty \tag{30.172}$$

$$= \|l_a(j\omega)\mathbf{M}_o(j\omega)\mathbf{Y}(j\omega)\ \mathrm{adj}\bar{\mathbf{M}}_i(j\omega) - l_a(j\omega)\mathbf{M}_o(j\omega)\mathbf{Z}(j\omega)\bar{\mathbf{M}}_o(j\omega)\ \det\bar{\mathbf{M}}_i(j\omega)\|_\infty \triangleq \|\mathbf{\Phi}(s)|_{s=j\omega}\|_\infty . \tag{30.173}$$

Assume that the zeros of $\det\bar{\mathbf{M}}_i(s)$ in $\Re e\ s > 0$ are calculated as $s_1, s_2 \dots s_m$, then

$$\mathbf{\Phi}(s_i) = l_a(s_i)\mathbf{M}_o(s_i)\mathbf{Y}(s_i)\ \mathrm{adj}\bar{\mathbf{M}}_i(s_i) + \mathbf{0} \qquad \forall i = 1 \dots m . \tag{30.174}$$

If $\mathbf{\Phi}(s_i) = \mathbf{F}_i$ satisfies $\|\mathbf{F}_i\|_s = \|\mathbf{\Phi}(s_i)\|_s < 1$, as required by Nevanlinna-Pick theory, the robust stability inequality of Eq.(30.130) can be solved if the Pick matrix of Eq.(30.162) is positive definite (*Chen, B.S., and Lo, C.H., 1989*).

An extension to an unstable pole at $s = 0$, i.e. an integrator, is given by *Kimura, H., 1984*.

Chapter 31

Hardy Space Robust Design

31.1 Introduction to the Min-Max-Problem of Optimum Robustness

In Fig. 31.1 a single-input single-output system is given. A robust controller $K(s)$ should be designed to guarantee sufficient closed-loop performance even in the presence of plant uncertainties. The actual plant transfer function is $G_p = \{1 + \Delta N[u(t), t]\}G$ where G denotes the nominal plant and $\Delta N[u(t), t]$ is a real-valued nonlinear time-varying function of the input signal $u(t)$. The uncertainty ΔN is considered bounded by the sector gain γ_N

$$\Delta N[u(t), t] \leq \gamma_N |u(t)| \quad \forall u(t) \quad \text{where} \quad \Delta N[0, t] = 0 \quad \forall t\ . \tag{31.1}$$

The nonlinearity is not essential in this chapter but the derivations can easily be extended to several types of nonlinear uncertainty.

31.1.1 Internal Stability for Scalar Systems

The internal stability of a single-input single-output system, as depicted in Fig. 31.1, is defined in the nominal case with omitted nonlinearity ($\Delta N = 0$). The transfer function of the plant $G(s)$ is decomposed into a numerator and a denominator polynomial $b(s)$ and $a(s)$, respectively, i.e., $G(s) = b(s)/a(s)$. The sensitivity function $S(s)$ and the complementary sensitivity function $1 - S(s)$ of the closed loop are given by

$$S(s) \triangleq \frac{1}{1 + G(s)K(s)} = \frac{a(s)}{a(s) + b(s)K(s)} \tag{31.2}$$

$$1 - S(s) = \frac{G(s)K(s)}{1 + G(s)K(s)} = \frac{b(s)K(s)}{a(s) + b(s)K(s)}\ . \tag{31.3}$$

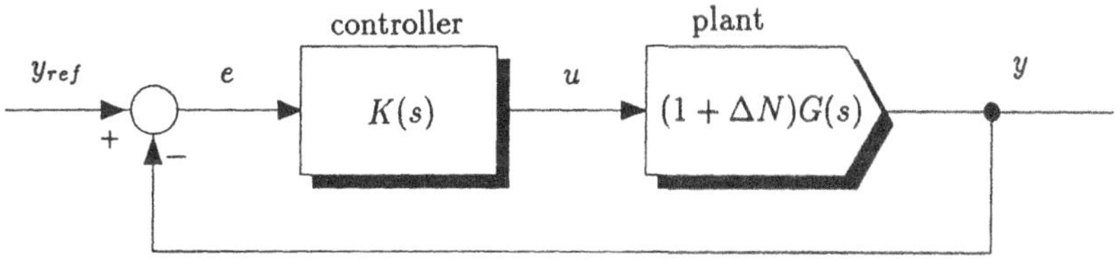

Figure 31.1: Single-input single-output system containing a time-varying nonlinear plant

For a given $S(s)$ and $G(s)$ the controller is

$$K(s) = \frac{[1 - S(s)]}{S(s)} \frac{a(s)}{b(s)} = \frac{1 - S(s)}{S(s)G(s)} . \tag{31.4}$$

The sensitivity function $S(s)$ is said to be internally stable (or realizable) if the system is asymptotically stable for some classes of the controller $K(s)$ which is true if three conditions hold (*Youla, D.C., et al. 1976*)

(i)

$$S(s) \text{ is analytic in } \Re e\ s \geq 0 , \tag{31.5}$$

(ii)

$$\frac{S(s)}{a(s)} \text{ is analytic in } \Re e\ s \geq 0 , \tag{31.6}$$

i.e., every zero of $a(s)$ in $\Re e\ s \geq 0$ is a zero of $S(s)$ of at least the same multiplicity,

(iii)

$$\frac{1 - S(s)}{b(s)} \text{ is analytic in } \Re e\ s \geq 0 , \tag{31.7}$$

i.e., every zero of $b(s)$ in $\Re e\ s \geq 0$ is a zero of $1 - S(s)$ of at least the same multiplicity.

31.1.2 Robust Stability of the Nonlinear Time-Varying System

By decomposing the plant $(1 + \Delta N)G(s)$ and by adjoining an element $G'(s)$ twice, as depicted in Fig. 31.2a, the system dynamics are not changed. If $G'(s) := G(s)$ the elements $G'(s)$ and $G(s)$ in parallel to $\Delta N\ G(s)$ can be cancelled and Fig. 31.2b results. Now, the circle criterion as given in Eq.(20.58) is applied. Replacing $G(s)$ in Eq.(20.58) by $GK/(1 + GK)$ of Eq.(31.3), it follows from Eq.(20.58)

$$\sup_{\omega} \left|\frac{GK}{1 + GK}\right| < \frac{1}{\gamma_N} \quad \rightsquigarrow \quad \gamma_N \sup_{\omega} |1 - S(s)| = \gamma_N \|1 - S(s)\|_\infty < 1 . \tag{31.8}$$

The norm $\|\cdot\|_\infty$ is the *function norm* in L_∞ , applied to the scalar-valued spectral function $1 - S(s)$ in Eq.(31.8). This norm $\|\cdot\|_\infty$ corresponds to the worst-case frequency-domain norm given by $\|\mathbf{H}\|_\infty = \sup_\omega \sigma_{\max}[\mathbf{H}(j\omega)]$ in the case of matrix-valued $\mathbf{H}(j\omega)$.

31.1.3 Optimal Robustness Based on Internal Stability

Maximum robustness is obtained when the sector gain γ_N is maximized with respect to the controller transfer function $K(s)$. From Eq.(31.8), maximizing γ_N corresponds to minimizing the H_∞-norm $\|1 - S(s)\|_\infty$ with respect to $K(s)$, i.e.,

$$\min_{K(s)} \sup_{\omega} |1 - S(j\omega)| . \tag{31.9}$$

For mathematical aspects see *Rogosinski, W.W., and Shapiro, H.S., 1953*. Following *Lin, C.M., and Chou, W.D., 1989*, the minimum of the H_∞-norm is of all-pass form

$$\min_{K(s)} \sup_{\omega} |1 - S(j\omega)| = \rho \frac{m(-s)}{m(s)} \tag{31.10}$$

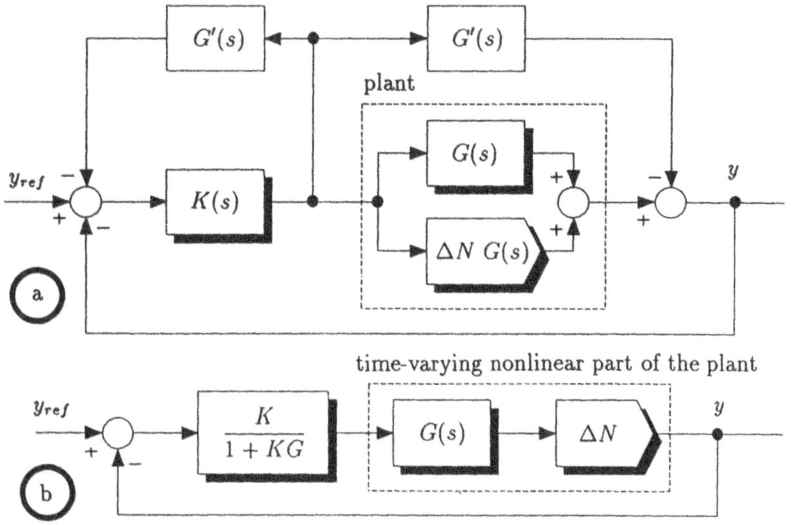

Figure 31.2: Separating the nonlinearity from the nominal loop

where $m(s)$ is a Hurwitz polynomial (a polynomial with all roots $\Re e\ s < 0$) and ρ is a real-valued constant. The result is given by

$$\min_{K(s)} \sup_{\omega} |1 - S(j\omega)| = \|1 - S^{\star}(j\omega)\|_{\infty} = |\rho|\ \|\frac{m(-s)}{m(s)}\|_{\infty} = |\rho| \times 1 - |\rho| \,. \tag{31.11}$$

The superscript star * denotes the optimum with respect to $K(s)$. Thus, referring to Eq.(31.8), the maximum sector gain γ_N is

$$\gamma_{N\,\max} = 1/|\rho| \,. \tag{31.12}$$

Let the polynomials $a(s)$ and $b(s)$ be decomposed into polynomial factors with roots in $\Re e\ s \geq 0$ and in $\Re e\ s < 0$, i.e. $a^-(s), a^+(s), b^-(s), b^+(s)$, respectively. The meaning of the superscript $^+$ is *free of zeros in the closed right-half s-plane*. Decomposition yields

$$a(s) = a^+(s)a^-(s) \qquad b(s) = b^+(s)b^-(s) \qquad G(s) = \frac{b(s)}{a(s)} = \frac{b^+(s)b^-(s)}{a^+(s)a^-(s)} \,. \tag{31.13}$$

In order to satisfy the condition Eq.(31.7) the optimal function $1 - S^{\star}(s)$ must contain $b^-(s)$ as a factor and, furthermore, the all-pass behaviour must be satisfied

$$1 - S^{\star}(s) = \rho\frac{m(s)}{m(-s)} \triangleq \rho\frac{l(s)b^-(s)}{l(-s)b^-(-s)} \,. \tag{31.14}$$

In Eq.(31.14) $l(-s)$ is an unknown Hurwitz polynomial which has to provide stable closed-loop transients. Rearranging yields

$$S^{\star}(s) = \frac{l(-s)b^-(-s) - \rho\ l(s)b^-(s)}{l(-s)b^-(-s)} \,. \tag{31.15}$$

The condition Eq.(31.6) is satisfied if the numerator of $S^\star(s)$, as given in the above result, contains $a^-(s)$ as a factor, i.e.,

$$l(-s)b^-(-s) - \rho\ l(s)b^-(s) = a^-(s)f(s)\ . \tag{31.16}$$

The polynomials $l(s)$ and $f(s)$ and the constant parameter ρ are determined from Eq.(31.16) by comparing the coefficients of s^i. Combining Eqs.(31.2) and (31.14) through (31.16) yields

$$K(s) = \frac{\rho\ l(s)a^+(s)}{b^+(s)f(s)}\ . \tag{31.17}$$

Example: Consider the plant $G(s) = (s-10)/(s-2)$, then

$$a^-(s) = s-2, \qquad a^+(s) = 1, \qquad b^-(s) = s-10, \qquad b^+(s) = 1\ . \tag{31.18}$$

From Eq.(31.16),

$$l(-s)(-s-10) - \rho\ l(s)(s-10) = (s-2)f(s)\ , \tag{31.19}$$

it can be seen that $l(s) = 1$ is a solution

$$-s-10-\rho(s-10) = (s-2)f(s)\ . \tag{31.20}$$

From $s = 2$ follows $\rho = 1.5$. Further comparison yields $f(s) = -2.5$. Finally, $K^\star(s) = -0.6$ and $\gamma_{N\,\max} = 1/|\rho| = 0.67$. □

31.2 Theory of Hardy Space on the Open Unit Disc

The Hardy space (H_p-space) consists of all analytic functions f on the open unit disc in the complex plane if the following property is satisfied

$$||f||_p = [\frac{1}{2\pi}\int_0^{2\pi} |f(e^{j\theta})|^p d\theta]^{1/p} < \infty \qquad \forall p = [1,\infty) \tag{31.21}$$

$$||f||_\infty = \text{ess.}\sup_{\theta\in[0,2\pi]} |f(e^{j\theta})| < \infty\ . \tag{31.22}$$

For each $p \in [1,\infty]$ the space H_p is a Banach space (*Sz.-Nagy, B., and Foias, C., 1970*). The space H_2 is a Hilbert space with the inner product

$$\langle f,g\rangle = \frac{1}{2\pi}\int_0^{2\pi} f^*(e^{j\theta})g(e^{j\theta})d\theta\ . \tag{31.23}$$

31.3 Theory of Hardy Space on the Open Right Half-Plane

For any function f analytic in the open right half-plane the H_p-space is defined by the Banach space of complex-valued functions $f(s)$ of a complex variable $s = \sigma + j\omega$, $f \in H_p$ if

$$||f||_p \triangleq \sup_{\sigma>0}[\frac{1}{2\pi}\int_{-\infty}^{\infty} |f(\sigma+j\omega)|^p d\omega]^{1/p} < \infty \qquad p = [1,\infty) \tag{31.24}$$

$$||f||_\infty \triangleq \sup_{\sigma>0}\sup_{\omega} |f(\sigma+j\omega)| < \infty\ . \tag{31.25}$$

If a scalar function $f(s)$ is analytic in the open right-half of the s-plane then max $|f(s)|$ is reached at $s = j\omega$. Furthermore, $f(s)$ must not be identically constant. This property is referred to as *maximum-modulus principle*. Hence, $\sup_{\sigma=\Re e\ s>0}$ can be replaced by $\sup_\omega$

or the open right half-plane can be replaced by the $j\omega$-axis or the $j\omega$-axis can be included and

$$||f||_p \triangleq [\frac{1}{2\pi}\int_{-\infty}^{\infty}|f(j\omega)|^p d\omega]^{1/p} < \infty \qquad p = [1,\infty) \tag{31.26}$$

$$||f||_\infty \triangleq \text{ess.}\sup_{\omega}|f(j\omega)| \qquad \omega \in \mathcal{R}\,. \tag{31.27}$$

The imaginary axis $j\omega$ in this section corresponds to the unit circle in the previous section.

The subset of rational functions with real coefficients is termed RH_p . The function $f(s)$ is denoted proper if $f(\infty)$ if finite and strictly proper if $f(\infty) = 0$. The function is denoted stable if $f(s)$ does not possess poles in the closed right-half s-plane, i.e., $\Re e\ s \geq 0$. When $f(s)$ is real rational then $f(s) \in RH_\infty$ if and only if $f(s)$ is proper and stable. When $f(s)$ is real rational then $f(s) \in RH_2$ if and only if $f(s)$ is *strictly* proper and stable.

31.4 Hardy Space and Lebesque Space on Matrix-Valued Functions

Hardy spaces (H_2, H_∞) require that each entry of $\mathbf{G}(s)$ is analytic off the $j\omega$-axis. Lebesque spaces (L_2, L_∞) only require that $\mathbf{G}(s)$ is analytic on the $j\omega$-axis.

The set of all matrix-valued functions that are square integrable on the imaginary axis in the sense of the inner product

$$\langle \mathbf{F},\mathbf{G}\rangle \triangleq \frac{1}{2\pi}\int_{-\infty}^{\infty} \text{tr}\ [\mathbf{F}^R(j\omega)\mathbf{G}(j\omega)]d\omega \tag{31.28}$$

are elements of the Hilbert space L_2. Matrix-valued functions $\mathbf{G}(s)$ on L_2 that are analytic in the open right half-plane and are satisfying

$$\sup_{\sigma>0}\frac{1}{2\pi}\int_{-\infty}^{\infty} \text{tr}\ [\mathbf{G}^R(\sigma+j\omega)\mathbf{G}(\sigma+j\omega)]d\omega < \infty \tag{31.29}$$

belong to H_2-space.

Using maximum modulus theorem,

$$||\mathbf{G}(s)||_2 = \sqrt{\frac{1}{2\pi}\int_{-\infty}^{\infty} \text{tr}\ \mathbf{G}^H(j\omega)\mathbf{G}(j\omega)d\omega}\ . \tag{31.30}$$

The Hardy H_∞-norm of an operator $\mathbf{G}(j\omega)$ is given by

$$||\mathbf{G}(s)||_\infty = ||\mathbf{G}(j\omega)||_\infty \;=\; \sup_{\omega}||\mathbf{G}(j\omega)||_s = \sup_{\omega}\sigma_{\max}[\mathbf{G}(j\omega)] \tag{31.31}$$

$$= \sup_{||\mathbf{x}||_2\neq 0}\frac{||\mathbf{G}(j\omega)\mathbf{x}||_2}{||\mathbf{x}||_2} \ \text{ if }\ \mathbf{G}(s)\in RH_\infty \tag{31.32}$$

where the supremum above is the induced L_2^n- norm of the operator $\mathbf{G}(j\omega)$ and where $||\mathbf{x}||_2$ is the L_2^n-norm. The norm $||\mathbf{G}(s)||_\infty$ is denoted L_∞-norm if $\mathbf{G} \in L_\infty$ on the imaginary axis. Note that

$$||\mathbf{G}(s)||_{L\infty} \triangleq \text{ess.}\ \sup_{\omega}\ \sigma_{\max}[\mathbf{G}(j\omega)] \tag{31.33}$$

where $\mathbf{G}(s) \in L_\infty$ is a complex function essentially uniformly bounded on the $j\omega$-axis ($\in L_\infty$). If the matrix $\mathbf{G}(s) \in H_\infty$, i.e. $\in L_\infty$ with a bounded analytic continuation in the right-half s-plane, then (*Sandberg, I.W., 1964*)

$$||\mathbf{G}(s)||_{L\infty} = ||\mathbf{G}(s)||_\infty \triangleq \sup_{s,\Re e\ s>0}\ \sigma_{\max}[\mathbf{G}(s)]\ . \tag{31.34}$$

The set of all proper transfer functions $\mathbf{G}(s)$ is denoted by the Lebesgue space L_∞ if $||\mathbf{G}||_\infty < \infty$. If, additionally, the matrices $\mathbf{G}(s)$ are exponentially stable, those transfer functions are denoted as the Hardy space H_∞. In other words, the Hardy space H_∞ is the space of bounded-input bounded-output stable transfer functions, i.e., functions in Laplace domain of the variable s which are proper and analytic for $\Re e\ s \geq 0$.

If the plant $\mathbf{G}$ has poles on the $j\omega$-axis, then the method described above can be generalized if the technique of $j\omega$-axis shifting is introduced. Let $v \triangleq s+d, \quad d>0$, the poles at $s=j\omega$ in the s-plane are shifted to poles at $d+j\omega$ in the v-plane (*Xu, J.H., and Mansour, M., 1988*).

A matrix $\mathbf{G}(s)$ is denoted ***bounded real*** if $\mathbf{G}(s)$ is analytic for $\Re e\ s>0$, $\mathbf{G}(s)$ real for s real and $\|\mathbf{G}(s)\|_\infty \leq 1$.

The spectral factorization of a matrix $\mathbf{\Phi}(s)$ is defined by

$$\mathbf{\Phi}(s)=\mathbf{G}^T(-s)\mathbf{G}(s)=\mathbf{G}^R(s)\mathbf{G}(s) \tag{31.35}$$

where $\mathbf{G}(s)$ is analytic and has constant rank in the right half-plane. If a spectral factorization exists then $\mathbf{\Phi}^T(-s)=\mathbf{\Phi}(s)$ and $\mathbf{\Phi}(j\omega)\geq 0 \ \forall\, \omega$ (*Youla, D.C., 1961*). Spectral factorization is frequently applied in H_2-optimization theory.

31.5 Application on Control Theory

Considering the system error $\mathbf{e}(t)$ or $\mathbf{e}(j\omega)$, or the transfer matrix $\mathbf{G}(s)$, the expressions

$$\|\mathbf{e}(s)\|_2^2=\int_{-\infty}^{\infty}\|\mathbf{e}(j\omega)\|_F^2 d\omega\ , \qquad \|\mathbf{G}(s)\|_2^2=\frac{1}{2\pi}\int_{-\infty}^{\infty}\mathrm{tr}\ \mathbf{G}^H(j\omega)\mathbf{G}(j\omega)d\omega, \tag{31.36}$$

$$\|\mathbf{e}(s)\|_\infty=\sup_\omega\|\mathbf{e}(j\omega)\|_F\ , \qquad \|\mathbf{G}(s)\|_\infty=\sup_\omega\|\mathbf{G}(j\omega)\|_s \tag{31.37}$$

are known as H_2-norm and H_∞-norm, respectively. Occasionally, a weighting matrix $\mathbf{W}(s)$ is premultiplied. These matrices correspond to bounded-input bounded-output stable systems, abbreviated BIBO-systems. In the time domain, one has, e.g.,

$$\|\mathbf{h}(t)\|_2=[\int_0^{\infty}\|\mathbf{h}(t)\|_F^2 dt]^{1/2} \tag{31.38}$$

Linear quadratic Gaussian theory uses the norm $\|\mathbf{G}(s)\|_2$. Optimization problems in most cases are solved by minimizing $\|\mathbf{G}(s)\|_2$. The problem of finding the best possible sensitivity function $\mathbf{S}(s)=[\mathbf{I}+\mathbf{M}_F(s)\mathbf{G}(s)\mathbf{K}(s)]^{-1}$ is performed by minimizing $\|\mathbf{S}(s)\|_\infty$. The sensitivity function $\mathbf{S}(s)$ relates the error vector $\mathbf{e}(s)$ to the reference vector $\mathbf{y}_{ref}(s)$, i.e., $\mathbf{e}(s)=\mathbf{S}(s)\mathbf{y}_{ref}(s)$.

31.5.1 Energy Gain Factor

Let an output signal be $\mathbf{y}(s)=\mathbf{G}(s)\mathbf{u}(s)$ and consider the input signal $\mathbf{u}$ has finite energy

$$\|\mathbf{u}(t)\|_2=\sqrt{\int_{-\infty}^{\infty}\mathbf{u}^T(t)\mathbf{u}(t)dt}\ . \tag{31.39}$$

Then, with regard to the definition of the H_∞-norm belonging to a stable $\mathbf{G}(s)$, see Eq.(20.90),

$$\sup_\omega\|\mathbf{G}(j\omega)\|_s=\|\mathbf{G}(s)\|_\infty=\sup_{\mathbf{u}}\frac{\|\mathbf{y}(t)\|_2}{\|\mathbf{u}(t)\|_2}\ , \tag{31.40}$$

the number $\|\mathbf{G}\|_\infty$ bounds the energy gain factor between the energy associated with the input $\mathbf{u}(s)$ and the energy of the output $\mathbf{y}(s)$ of the multivariable system $\mathbf{G}(s)$.

31.5.2 H_∞-Norm (Operator Norm) $\|\mathbf{G}(j\omega)\|_\infty$ and Stochastic Signals

For single-input single-output systems, $y(s) = G(s)u(s)$, with input power spectral density $S_{uu}(\omega)$

$$E\{y^2(t)\} = \frac{1}{2\pi}\int_{-\infty}^{\infty} S_{yy}(\omega)d\omega = \frac{1}{2\pi}\int_{-\infty}^{\infty} |G(j\omega)|^2 S_{uu}(\omega)d\omega \,. \tag{31.41}$$

For multivariable systems with stable $\mathbf{G}(s)$ and transfer property $\mathbf{y}(s) = \mathbf{G}(s)\mathbf{u}(s)$ and stochastic input signals with power spectral density

$$\mathbf{S}_{uu}(\omega) = \mathcal{F}\{\mathbf{R}_{uu}(\tau)\} = \mathcal{F}E\{\mathbf{u}(t)\mathbf{u}^T(t+\tau)\} \,, \tag{31.42}$$

the stochastic transmission property is

$$\mathbf{S}_{yy}(\omega) = \mathbf{G}(j\omega)\mathbf{S}_{uu}(\omega)\mathbf{G}^H(j\omega) \,. \tag{31.43}$$

When a weighting matrix $\mathbf{Q} \triangleq \mathbf{V}^T\mathbf{V}$ is applied then the output power is

$$E\{\mathbf{y}^T(t)\mathbf{Q}\mathbf{y}(t)\} = \frac{1}{2\pi}\int_{-\infty}^{\infty} \mathrm{tr}[\mathbf{V}^T\mathbf{V}\mathbf{S}_{yy}(\omega)]d\omega = \frac{1}{2\pi}\int_{-\infty}^{\infty} \mathrm{tr}[\mathbf{V}\mathbf{G}(j\omega)\mathbf{S}_{uu}(\omega)\mathbf{G}^H(j\omega)\mathbf{V}^T]d\omega \,. \tag{31.44}$$

The same result could be obtained omitting the weighting matrix $\mathbf{V}^T\mathbf{V}$ but rewriting $\mathbf{y}(s) = \mathbf{G}(s)\mathbf{u}(s)$ to $\mathbf{y}(s) = \mathbf{V}\mathbf{G}(s)\mathbf{u}(s)$. The output signal power using singular values σ_i is given (*Postlethwaite, I., et al. 1981*) by

$$E\{\mathbf{y}^T\mathbf{y}\} = \frac{1}{2\pi}\int_{-\infty}^{\infty} \sum_i \sigma_i^2[\mathbf{S}_{uu}^{1/2}\mathbf{G}(j\omega)]d\omega \tag{31.45}$$

$$\sigma_{\min}^2[\mathbf{G}(j\omega)] \le \frac{\mathrm{tr}\ \mathbf{S}_{yy}(\omega)}{\mathrm{tr}\ \mathbf{S}_{uu}(\omega)} \le \sigma_{\max}^2[\mathbf{G}(j\omega)] \,. \tag{31.46}$$

From Eq.(31.44) when $\mathbf{Q} = \mathbf{I}$, $\mathbf{S}_{uu} = \mathbf{I}$ (white stochastic process),

$$E\{\mathbf{y}^T\mathbf{y}\} = \frac{1}{2\pi}\int_{-\infty}^{\infty} \mathrm{tr}[\mathbf{G}(j\omega)\mathbf{G}^H(j\omega)]d\omega = \|\mathbf{G}(j\omega)\|_2^2 \tag{31.47}$$

which explains the importance of the operator norm $\|\mathbf{G}(j\omega)\|_2$ from a physical point of view.

31.5.3 Input-Output Relations

The following properties in Hardy space H_p are very important for valuating the relation between the input space (set of input signals) and the performance of a control unit, particularly given by the output space (*Liu, G.P., 1990*). The output is preferably termed $y(t)$. Furthermore, scalar-valued functions $P(s)$, $G(s)$ are used, $G(s)$ and $K(s)$ are the transfer functions of the plant and the controller, respectively.

If $|y(t)| \le D_o$, $t > 0$ and $y(t) = 0$, $t \le 0$ and $\mathcal{L}\{y(t)\}(s) \in H_1$ then

$$y(t) = \frac{1}{2\pi}\int_{-\infty}^{\infty} \mathcal{L}\{y(t)\}(j\omega)e^{j\omega t}d\omega \,. \ \square \tag{31.48}$$

$$\text{If}\quad P(s) \in RH_p \quad (1 \le p \le \infty),\quad Q(s) \in H_1\,,\quad \text{then}\quad P(s)Q(s) \in H_1 \,. \ \square \tag{31.49}$$

$$\text{If}\quad P(s) \in RH_p \quad (1 \le p < \infty)\quad \text{then}\quad \lim_{p\to\infty} \|P(j\omega)\|_p = \|P(j\omega)\|_\infty \,. \ \square \tag{31.50}$$

Let $y(t) = \mathcal{L}^{-1}\{P(s)Q(s)\}$ where $P(s) \in RH_p$ $(1 \leq p \leq \infty)$ and $\|Q(j\omega)\|_1 \leq D_o$ then

$$\sup\{ \, |y(t)| \; : \; t > 0 \, \} \; \leq D_o \; \|P(j\omega)\|_\infty \; . \; \square \tag{31.51}$$

This theorem is applied to feedback control systems, replacing (i) $Q(s)$ by the input signal, (ii) $P(s) = (1 + GK)^{-1}L$ by the transfer function of a closed-loop system where L is a certain feedforward transfer function and (iii) $P(s)Q(s)$ by the output $y(s)$.

31.5.4 General Results of the H_∞-Sensitivity Design

The main objective of H_∞-control theory is the design of controllers which minimize the H_∞-norm of the appropriatedly weighted sensitivity function.

The optimally weighted sensitivity is all-pass. H_∞-norm optimization technique has the advantage that stability margins and other robustness measures can be taken into account in the performance criterion. Sensitivity of multivariable systems without right-half-plane poles can be made arbitrarily small. In the presence of right-half-plane zeros, the reduction of the sensitivity is considerably restricted. Furthermore, plant uncertainty distinctly diminishes the ability of feedback to reduce sensitivity (*Zames, G., 1981*). Hence, sensitivity optimal solution may have poor robust stabilization properties. Furthermore, the optimal compensator is unstable if there arises more than one right-half s-plane *zero*.

31.6 Robust SISO-Stability on H_∞-Space

Let a single-input single-output system be given as depicted in Fig. 31.3a. The nominal plant transfer function is $G(s)$, the additive perturbation of the plant is termed $\Delta G(s)$. Suppose $K, G, \Delta G$ scalar, real-rational, ΔG and G strictly proper and stable, K proper. Stability of the system for $\Delta G \equiv 0$ is preassumed. Hence, $K(1 + GK)^{-1} \in \mathcal{R}H_\infty$. The H_∞-norm simply is given by the maximum modulus achieved with respect to ω, i.e. $\|G\|_\infty = \sup_\omega |G(j\omega)|$. Let $\Delta G(j\omega)$ be bounded by the inequality using the real-valued function $g_o(\omega)$

$$|\Delta G(j\omega)| < g_o(\omega) \qquad \forall\omega = [0, \infty) \; . \tag{31.52}$$

Applying simple block transformations the perturbation $\Delta G(s)$ can be separated, see Fig 31.3d. Invoking Eq.(3.11) yields a sufficient condition for the upper bound g_o of the perturbation in order to guarantee stability

$$\|K(1 + GK)^{-1}\Delta G\|_\infty < 1 \tag{31.53}$$

$$\|\Delta G\|_\infty \|K(1 + GK)^{-1}\|_\infty < \|g_o(\omega)\|_\infty \|K(1 + GK)^{-1}\|_\infty < 1 \; . \tag{31.54}$$

31.7 Disturbance Attenuation Via H_∞-Norm

Consider a control system as depicted in Fig. 31.4. Suppose the disturbance $w \in RH_2$ generated by the transfer system $W(j\omega)$ and the input w' where $\|w'\|_2 = 1$. The signal energy is given by the H_2-norm, see Eq.(3.7). Then, with regard to Eqs.(20.56) and (31.40) the H_∞-norm $\|w\|_\infty$ is an upper bound for the signal energy in the worst case $w' \in RH_2$.

The transfer function from the input w' to the output y is $W(s)S(s)$ where the function $S(s) = [1 + G(s)K(s)]^{-1}$ is the sensitivity function. Minimizing the signal energy of y in the worst case of the disturbance results in minimizing the H_∞-norm $\|W(s)S(s)\|_\infty$.

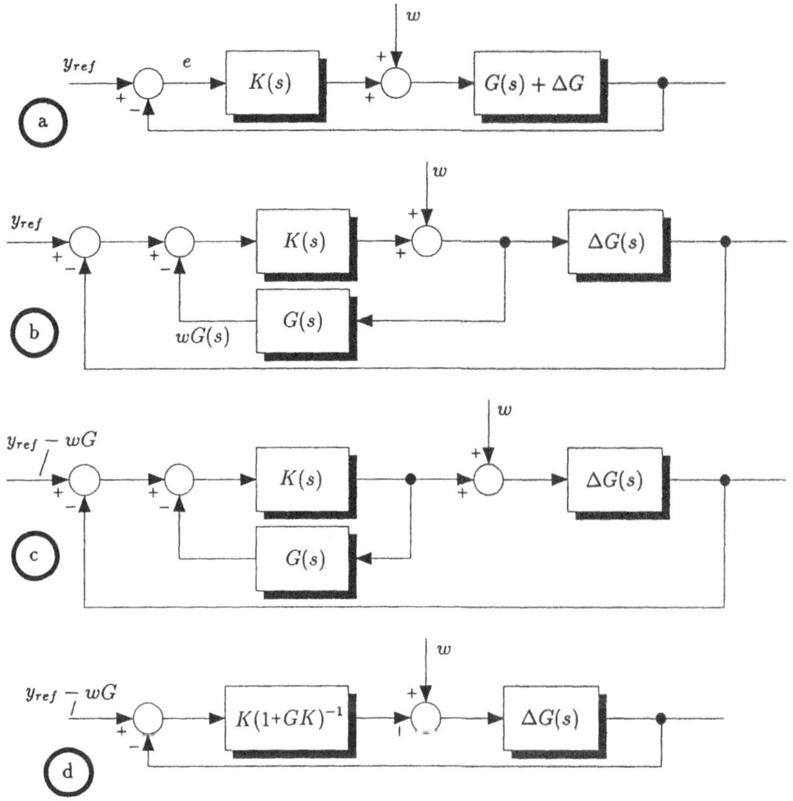

Figure 31.3: Single-input single-output loop with block transformations

For a given $W(s)$ and $G(s)$, the controller $K(s)$ can be chosen in order to minimize $\|W(1+GK)\|_\infty$.

Bhattacharyya, S.P., et al. 1983 have investigated the problem of disturbance rejection under state feedback based on making the transfer function zero, namely from disturbance input to control output. The class of state-space parameter variations is usually too rich for practical purpose and engineering concern, in comparison with transfer function variations. The problem of minimizing both the error signal and the control effort under worst frequency-domain condition for arbitrary reference is solved by *Verma, M., and Jonckheere, E., 1984*, using the results of a broadband matching problem (*Helton, J.W., 1983*).

Additionally, minimizing the expression $\sup_\omega |W_1(j\omega)S(j\omega)|^2 + |W_2(j\omega)T(j\omega)|^2$ by choosing a suitable controller K has the advantage that via $T(j\omega)$ the inverse of the stability margin is taken into account

$$T(j\omega) = 1 - S(j\omega) = [1 + (GK)^{-1}]^{-1} = GK/(1+GK) , \tag{31.55}$$

see also next section. The same minimax problem is solved via polynomial equation by *Kwakernaak, H., 1985a*.

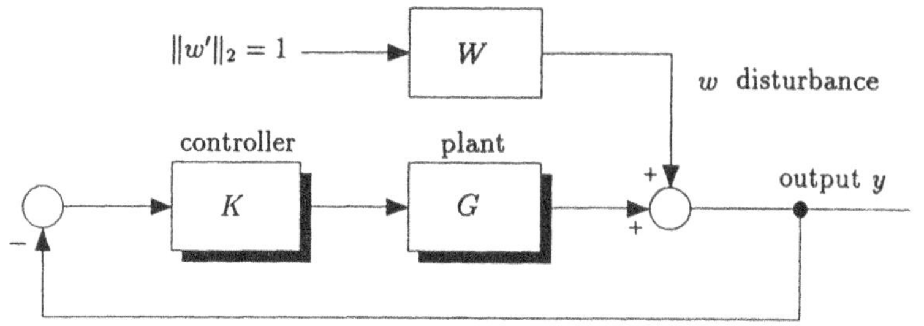

Figure 31.4: Control system and disturbance rejection

31.8 Sensitivity Reduction, Noise Suppression, Stability Margin

Using the open-loop transfer matrix $\mathbf{L}$, it is a well-known fact that the return difference matrix $\mathbf{I}+\mathbf{L} \triangleq \mathbf{S}^{-1}$ is responsible for the disturbance attenuation, i.e., the sensitivity function of the output with respect to the disturbance. Furthermore, the complementary sensitivity function $\mathbf{T} = \mathbf{L}(\mathbf{I}+\mathbf{L})^{-1}$ determines the noise suppression. Since $\mathbf{S}+\mathbf{T} = \mathbf{I}$, a tradeoff is necessary between sensitivity reduction and noise suppression. Finally, the inverse complementary sensitivity function $\mathbf{T}^{-1}$ is a measure for the stability margin (*Safonov, M.G., et al. 1981*).

The stability margin of the system is considered as the size of the smallest multiplicative perturbation $\Delta\mathbf{L}$ in the expression $\mathbf{L}_p = (\mathbf{I}+\Delta\mathbf{L})\mathbf{L}$ that destabilizes the closed-loop sensitivity function $\mathbf{S} = (\mathbf{I}+\mathbf{L}_p)^{-1}$, i.e., that $\mathbf{I}+\mathbf{L}_p$ has zeros in the closed right half-plane, or, in other words, that for some s_o with $\Re e\ s_o > 0$ there exists a solution $[\mathbf{I}+\mathbf{L}_p(s_o)]\mathbf{x}_o e^{s_o t} = \mathbf{0}\ \ \forall t\ ,\ \forall \mathbf{x}_o \neq \mathbf{0}$. Under the assumption of a stable unperturbed system the expression $[\mathbf{I}+\mathbf{L}(s_o)]\mathbf{x}_o$ does not equal $\mathbf{0}$, but equals say $\mathbf{x}_1$. Combining yields in abbreviated notation

$$\mathbf{I}+\mathbf{L}_p = \mathbf{I}+(\mathbf{I}+\Delta\mathbf{L})\mathbf{L} = \mathbf{0} \quad \text{and} \quad \mathbf{I}+\mathbf{L} \neq \mathbf{0} \quad \rightsquigarrow \quad \Delta\mathbf{L}(s_o) = -(\mathbf{I}+\mathbf{L})\mathbf{L}^{-1} = -(\mathbf{I}+\mathbf{L}^{-1}) = -\mathbf{T}^{-1}(s_o)\,. \tag{31.56}$$

This is the maximum tolerable uncertainty $\Delta\mathbf{L}$. Hence, in any norm, $\mathbf{T}$ should be small, at least at those frequencies where the plant perturbation $\Delta\mathbf{L}$ is large.

As far as the unperturbed system is considered, it is proven (*Kalman, R.E., 1964*) that the scalar optimal full-state feedback LQ-controller $u = \mathbf{Kx}$ and the plant $(s\mathbf{I}-\mathbf{A})^{-1}\mathbf{b}$ yield a return difference $1+\mathbf{K}(s\mathbf{I}-\mathbf{A})^{-1}\mathbf{b} \triangleq 1+L(s)$. The loop gain $L(s)$ at u-node (break point at u) is such that $|1+L(j\omega)| > 1\ \ \forall\omega$. Thus, the scalar sensitivity function $S(s) = [1+L(s)]^{-1}$ obeys $|S(j\omega)| < 1\ \ \forall\omega$. This property does not hold for sampled-data LQG-optimal systems and also fails in the case of output feedback.

31.9 Generalized Plant and Controller

Consider the multivariable system as given by the block diagram in Fig. 31.5. The generalized vector-valued input signal $\mathbf{y}_{in}$ contains the vector-valued components set-point, measurement noise and disturbances. The signal $\mathbf{e}$ is considered the deviation, tracking error etc., $\mathbf{y}$ is the output. The feedback controller matrix is $\mathbf{K}$. The matrix $\mathbf{\Gamma}$ is partitioned according to

$$\begin{pmatrix}\mathbf{e}\\ \mathbf{y}\end{pmatrix} \triangleq \begin{pmatrix}\mathbf{\Gamma}_{11} & \mathbf{\Gamma}_{12}\\ \mathbf{\Gamma}_{21} & \mathbf{\Gamma}_{22}\end{pmatrix}\begin{pmatrix}\mathbf{y}_{in}\\ \mathbf{u}\end{pmatrix} = \mathbf{\Gamma}\begin{pmatrix}\mathbf{y}_{in}\\ \mathbf{u}\end{pmatrix}. \tag{31.57}$$

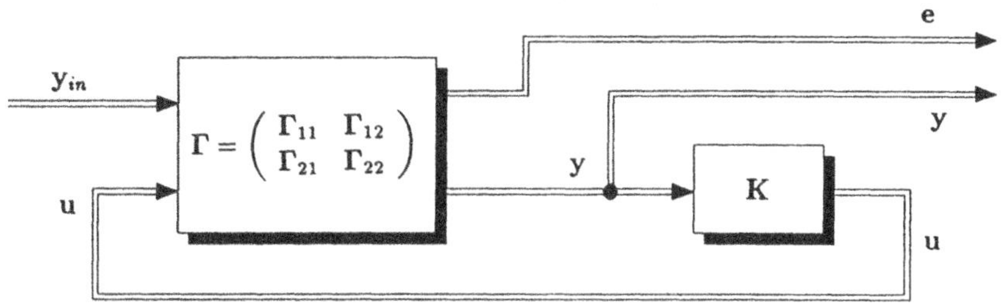

Figure 31.5: Generalized plant and controller

The controller output (control signal) results from $\mathbf{u} = \mathbf{K}\mathbf{y}$. Minimizing the energy of the error, i.e. $\|\mathbf{e}\|_2^2$, for the worst input $\mathbf{y}_{in}$ leads to minimizing the H_∞-norm of the transfer matrix $\mathbf{F}$ from the input $\mathbf{y}_{in}$ to the error $\mathbf{e}$

$$\|\mathbf{F}\|_\infty = \|\mathbf{\Gamma}_{11} + \mathbf{\Gamma}_{12}\mathbf{K}(\mathbf{I} - \mathbf{\Gamma}_{22}\mathbf{K})^{-1}\mathbf{\Gamma}_{21}\|_\infty \quad \rightarrow \quad \min \ . \tag{31.58}$$

Relaxing the minimum of Eq.(31.58) to an upper bound by installing a non-minimal controller $\mathbf{K}'$,

$$\|\mathbf{F}'\|_\infty = \|\mathbf{\Gamma}_{11} + \mathbf{\Gamma}_{12}\mathbf{K}'(\mathbf{I} - \mathbf{\Gamma}_{22}\mathbf{K}')^{-1}\mathbf{\Gamma}_{21}\|_\infty < \gamma > \|\mathbf{F}\|_\infty \tag{31.59}$$

and there results a class of controllers $\mathbf{K}'$. A unique solution is obtained by specifying that an entropy function I_E of the closed-loop transfer function is minimized with respect to $\mathbf{K}'$. The entropy function is defined by

$$I_E(\mathbf{F}';\ \gamma;\ s_o) \triangleq -\frac{\gamma^2}{2\pi}\int_{-\infty}^{\infty} \ln|\det[\mathbf{I} - \frac{\mathbf{F}'^R(j\omega)\mathbf{F}'(j\omega)}{\gamma^2}]\,| \times \frac{(\Re e\ s_o)^2}{|s_o - j\omega|^2}d\omega \tag{31.60}$$

where s_o is any point in the open right-half s-plane (*Mustafa, D., et al. 1991*).

31.10 Optimal Model-Matching in H_∞-Space

Consider the diagram in Fig. 31.6 with parallel blocks $\mathbf{T}_1$ and $\mathbf{T}_2\mathbf{K}\mathbf{T}_3$ where the matrices $\mathbf{T}_1$, $\mathbf{T}_2$, $\mathbf{T}_3$ are given. Preassuming that $\mathbf{w} \in H_2$ and $\|\mathbf{w}\|_2 \leq 1$ the matrix $\mathbf{K}$ has to be chosen in order to minimize $\|\mathbf{e}\|_2$ for the worst input $\mathbf{w}$. The result is obtained by minimizing the norm $\|\mathbf{T}_1 - \mathbf{T}_2\mathbf{K}\mathbf{T}_3\|_\infty$ with respect to the matrix $\mathbf{K}$. The model-matching result can be substituted by the generalized result in Eq.(31.58) (*Kimura, H., 1989*). *Francis, B.A., et al. 1984* present a family of improper optimal feedback controllers which minimize the H_∞-norm of a weighted sensitivity matrix based on the operator theory by *Ball, J.A., and Helton, J.W., 1983*. For certain conditions on the weighting matrix, proper feedback transfer elements are used to approximate the optimal improper ones.

Robust stability conditions for model matching systems under external disturbances are derived by *Hsiao, F.B., and Lin, C.L., 1991*.

31.11 Optimal Tracking in H_∞-Space

In Fig. 31.7 the feedback controller matrix $\mathbf{K}$ and the compensator $\mathbf{C}$ will be designed with the aim of optimal tracking. The reference is $\mathbf{y}_{ref}$ and the output $\mathbf{y}$. The matrices

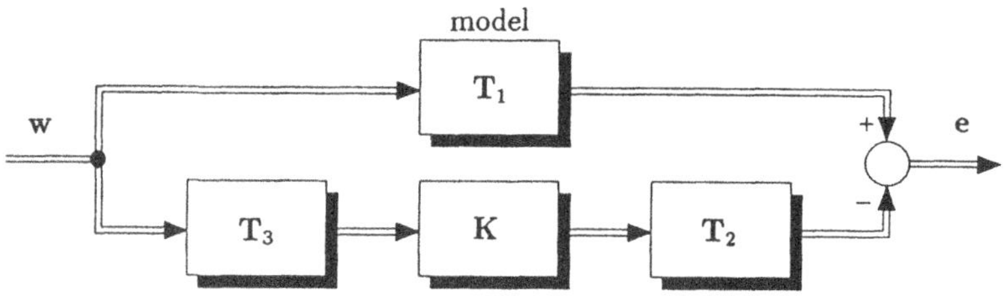

Figure 31.6: Optimal model-matching

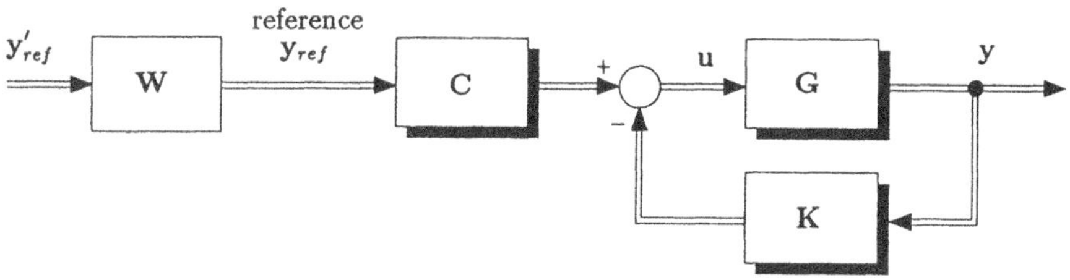

Figure 31.7: Control system and tracking facility

G and **W** are given. Let the reference signal be energy bounded

$$\mathbf{y}_{ref} = \mathbf{W}\mathbf{y}'_{ref} \quad \text{where} \quad \mathbf{y}'_{ref} \in H_2 \quad \text{and} \quad \|\mathbf{y}'_{ref}\|_2 \leq 1 \ . \tag{31.61}$$

The tracking error is $\mathbf{y}_{ref} - \mathbf{y}$ and its energy $(\|\mathbf{y}_{ref} - \mathbf{y}\|_2)^2$. The control effort is $\|\mathbf{u}\|_2^2$. A combined performance norm is stated using the factor q^2

$$I = \sqrt{(\|\mathbf{y}_{ref} - \mathbf{y}\|_2)^2 + q^2(\|\mathbf{u}\|_2)^2} \triangleq \|\bar{\mathbf{e}}\|_2 \ . \tag{31.62}$$

The resulting performance is given by

$$\sup \|\bar{\mathbf{e}}\|_2 \quad \text{where} \quad \mathbf{y}'_{ref} \in H_2, \quad \|\mathbf{y}'_{ref}\|_2 \leq 1, \quad \bar{\mathbf{e}} = \begin{pmatrix} \mathbf{y}_{ref} - \mathbf{y} \\ q\mathbf{u} \end{pmatrix} \ . \tag{31.63}$$

From Fig. 31.7 the output **y** and control variable **u** are

$$\mathbf{y} = (\mathbf{I} + \mathbf{G}\mathbf{K})^{-1}\mathbf{G}\mathbf{C}\mathbf{W}\mathbf{y}'_{ref} \qquad \mathbf{u} = (\mathbf{I} + \mathbf{K}\mathbf{G})^{-1}\mathbf{C}\mathbf{W}\mathbf{y}'_{ref} \ . \tag{31.64}$$

Combining yields

$$\bar{\mathbf{e}} = \begin{pmatrix} \mathbf{y}_{ref} - \mathbf{y} \\ q\mathbf{u} \end{pmatrix} = \begin{pmatrix} \mathbf{I} - (\mathbf{I} + \mathbf{G}\mathbf{K})^{-1}\mathbf{G}\mathbf{C}\mathbf{W} \\ q[(\mathbf{I} + \mathbf{K}\mathbf{G})^{-1}\mathbf{C}\mathbf{W}] \end{pmatrix} \mathbf{y}'_{ref} \triangleq \mathbf{F}_e \, \mathbf{y}'_{ref} \ . \tag{31.65}$$

Optimal tracking for the worst input $\mathbf{y}'_{ref}$ is obtained by $\|\mathbf{F}_e\|_\infty \to \min$ with respect to **C** and **K** where q, **W** and **G** are given.

As outlined above, the polynomial approach gives good theoretical insight. Using state-space models, fully developed numerical algorithms are available. Combination of the polynomial approach to the H_∞-optimal control with the state-space description is presented by *Grimble, M.J., 1988*. The flexibility of model matching control can be utilized to minimize the effect of external disturbances (*Chen, B.S., et al. 1989*).

In any case of application, the selection of weighting functions is an important tool to satisfy the design specifications such as tracking properties, disturbance rejection, minimum sensitivity and stability robustness (*Postlethwaite, I., et al. 1990*).

31.12 Minimizing Sensitivity in H_2-Norm

An algorithm is introduced to design an optimal controller $\mathbf{K}(s)$ for minimum H_2-norm of the weighted sensitivity function $\mathbf{S}(s)$. The sensitivity function of a multivariable system is defined as follows. Referring to Fig. 31.8, by premultiplying the reference vector $\mathbf{y}_{ref}(s)$ by $\mathbf{S}(s)$ the error vector $\mathbf{e}(s)$ is achieved: $\mathbf{S}(s)\mathbf{y}_{ref} = \mathbf{e}(s)$. Thus, the sensitivity matrix relates the closed-loop error to the closed-loop input. High performance I or low error signal is given in the case of small sensitivity $\mathbf{S}(s)$

$$\mathbf{S}(s) \triangleq [\mathbf{I} + \mathbf{M}_F(s)\mathbf{G}(s)\mathbf{K}(s)]^{-1} \qquad \mathbf{S}(s) \in \mathcal{C}^{n\times n} . \tag{31.66}$$

The weighting matrix $\mathbf{W}(s)$ is chosen such that the norm $\|\mathbf{W}(s)\|$ is nearly unity in the frequency range of interest and is nearly zero outside. Then, using Eq.(31.44),

$$I_2 = \|\mathbf{W}(s)\mathbf{S}(s)\|_2 \to \min \quad \text{or} \quad I_2^* = \min_{\mathbf{K}(s)} \|\mathbf{W}(s)\mathbf{S}(s)\|_2 \tag{31.67}$$

where the H_2-norm of the matrix-valued function $\mathbf{A}(j\omega) \in RH_\infty$ is given by

$$\|\mathbf{A}(s)\|_2 = \|\mathbf{A}(j\omega)\|_2 \triangleq \sqrt{\text{tr}\,\frac{1}{2\pi}\int_{-\infty}^{\infty} \|\mathbf{A}(j\omega)\|_F^2 d\omega} = \sqrt{\frac{1}{2\pi}\int_{-\infty}^{\infty} \text{tr}\ \mathbf{A}^R(j\omega)\mathbf{A}(j\omega) d\omega}\ , \tag{31.68}$$

the matrix $\mathbf{A}(s)$ being a proper transfer function with no poles on the imaginary axis. The superscript ^{R} denotes the transposed matrix with reflected argument or para-Hermitian conjugate transpose, i.e., $\mathbf{A}^R(s) = \mathbf{A}^T(-s)$. When the argument is merely imaginary then $\mathbf{A}^R(j\omega) = \mathbf{A}^H(j\omega)$. If $\mathbf{A} \in L_2$ on the imaginary axis, then the norm above is denoted L_2-norm. Both $\|\mathbf{A}\|_2$ and $\|\mathbf{A}\|_\infty$ satisfy the norm properties. Additionally, the infinity operator norm $\|\mathbf{A}\|_\infty$ is consistent, i.e., $\|\mathbf{GH}\|_\infty \le \|\mathbf{G}\|_\infty \|\mathbf{H}\|_\infty$. Weighting the input additionally with $\mathbf{W}_2(s)$, the spectral density of the output, weighted by $\mathbf{W}_1(s)$, and the energy of the process are, respectively,

$$\|\mathbf{W}_1(s)\mathbf{S}(s)\mathbf{W}_2(s)\|_2^2 \qquad \frac{1}{2\pi}\int_{-\infty}^{\infty} \|\mathbf{W}_1(s)\mathbf{S}(s)\mathbf{W}_2(s)\|_2^2 d\omega\ . \tag{31.69}$$

Plant and sensor, as depicted in Fig. 31.8, are decomposed such that $\mathbf{A}(s), \mathbf{B}(s)$ and $\mathbf{A}_1(s), \mathbf{B}_1(s)$ constitute some left and right-coprime matrices in $\Re e\ s \ge 0$

$$\mathbf{M}_F\mathbf{G} = \mathbf{A}^{-1}\mathbf{B} = \mathbf{B}_1\mathbf{A}_1^{-1}\ . \tag{31.70}$$

Assume $\mathbf{A}, \mathbf{B}, \mathbf{A}_1, \mathbf{B}_1$ given. Then, the matrices $\mathbf{U}_1(s), \mathbf{U}_2(s), \mathbf{V}_1(s), \mathbf{V}_2(s)$ result from the partitioned matrix identity[1] (Bezout identity)(*Chen, B.S., 1984*)

$$\begin{pmatrix} \mathbf{B}_1 & \mathbf{V}_2 \\ \mathbf{A}_1 & -\mathbf{U}_2 \end{pmatrix} \begin{pmatrix} \mathbf{U}_1 & \mathbf{V}_1 \\ \mathbf{A} & -\mathbf{B} \end{pmatrix} = \mathbf{I} = \begin{pmatrix} \mathbf{U}_1 & \mathbf{V}_1 \\ \mathbf{A} & -\mathbf{B} \end{pmatrix} \begin{pmatrix} \mathbf{B}_1 & \mathbf{V}_2 \\ \mathbf{A}_1 & -\mathbf{U}_2 \end{pmatrix} \tag{31.71}$$

[1]There are several possibilities of notations. Observe the corresponding lists of symbols in Eqs.(30.79), (30.85) $\mathbf{X}$, $\mathbf{Y}$, $\tilde{\mathbf{X}}$, $\tilde{\mathbf{Y}}$, $\mathbf{A}_1$, $\mathbf{B}_1$, $\mathbf{A}$, $\mathbf{B}$ and Eqs.(31.71), (31.73) $\mathbf{V}_2$, $\mathbf{U}_2$, $\mathbf{V}_1$, $\mathbf{U}_1$, $-\mathbf{A}_1$, $-\mathbf{B}_1$, $\mathbf{A}$, $\mathbf{B}$.

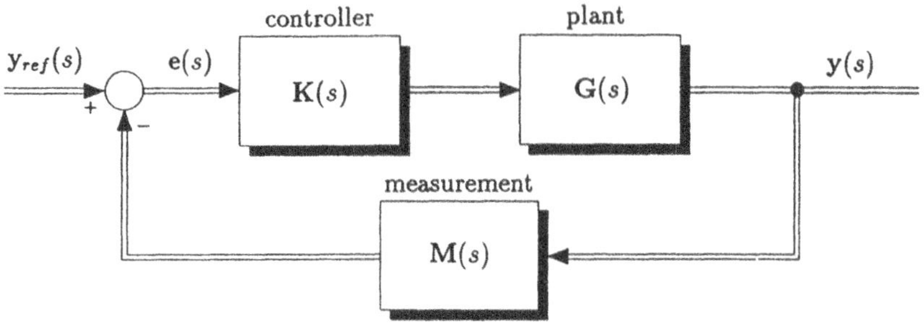

Figure 31.8: Multivariable control system

where the argument (s) was omitted for abbreviation. The closed-loop system is said internally stable (or realizable) if $\mathbf{S}(s)$ is given by

$$\mathbf{S}(s) = \mathbf{I} - \mathbf{B}_1(\mathbf{U}_1 - \mathbf{Z}\mathbf{A}) \tag{31.72}$$

where $\mathbf{Z} = \mathbf{Z}(s)$ is any rational matrix analytic in $\Re e\ s \geq 0$.

The controller $\mathbf{K}(s)$ is given by

$$\mathbf{K} = (\mathbf{V}_1 + \mathbf{Z}\mathbf{B})^{-1}(\mathbf{U}_1 - \mathbf{Z}\mathbf{A}) = (\mathbf{U}_2 - \mathbf{A}_1\mathbf{Z})(\mathbf{V}_2 + \mathbf{B}_1\mathbf{Z})^{-1} \tag{31.73}$$

which can be proved by substitution into Eq.(31.66) and by combining with (31.72). From Eq.(31.71) it results

$$\mathbf{B}_1\mathbf{U}_1 + \mathbf{V}_2\mathbf{A} = \mathbf{I}\,. \tag{31.74}$$

Combination with Eq.(31.72) yields

$$\mathbf{S} = (\mathbf{V}_2 + \mathbf{B}_1\mathbf{Z})\mathbf{A}\,. \tag{31.75}$$

With the definition Eq.(31.68), the Eq.(31.67) is rewritten

$$\begin{aligned} I_2^2 = \|\mathbf{W}(s)\mathbf{S}(s)\|_2^2 &= \operatorname{tr}\frac{1}{2\pi}\int_{-\infty}^{\infty}\mathbf{S}^R\mathbf{W}^R\mathbf{W}\mathbf{S}d\omega \\ &= \frac{1}{2\pi}\operatorname{tr}\int_{-\infty}^{\infty}\mathbf{A}^R(\mathbf{V}_2^R + \mathbf{Z}^R\mathbf{B}_1^R)\mathbf{W}^R\mathbf{W}(\mathbf{V}_2 + \mathbf{B}_1\mathbf{Z})\mathbf{A}d\omega\,. \end{aligned} \tag{31.76}$$

31.12.1 Variational Calculus

Applying the standard variational calculus, let

$$\mathbf{Z}(s) = \mathbf{Z}^*(s) + \varepsilon\mathbf{\Upsilon}(s) \tag{31.77}$$

where $\mathbf{\Upsilon}(s)$ is an arbitrary physically realizable rational transfer matrix free of poles and zeros in $\Re e\ s \geq 0$. The optimal I_2, denoted I_2^*, should be achieved at $\mathbf{Z}^*$ characterized by the necessary condition

$$\frac{\partial I_2^2}{\partial \varepsilon}\mid_{\varepsilon=0} = 0\,. \tag{31.78}$$

Combination with Eq.(31.76) yields

$$I_2^2 = \frac{1}{2\pi}\operatorname{tr}\int_{-\infty}^{\infty}\mathbf{A}^R(\mathbf{V}_2^R + \mathbf{Z}^{*R}\mathbf{B}_1^R + \varepsilon\mathbf{\Upsilon}^R\mathbf{B}_1^R)\mathbf{W}^R\mathbf{W}(\mathbf{V}_2 + \mathbf{B}_1\mathbf{Z}^* + \varepsilon\mathbf{B}_1\mathbf{\Upsilon})\mathbf{A}d\omega\,. \tag{31.79}$$

The integrand is a quadratic function in the parameter ε : $\mathbf{C}_2\varepsilon^2 + \mathbf{C}_1\varepsilon + \mathbf{C}_o$. The derivative with respect to ε is $2\mathbf{C}_2\varepsilon + \mathbf{C}_1$. Putting ε to zero yields $\mathbf{C}_1$. Thus, simplifying and deriving with respect to ε

$$\frac{\partial I_2^2}{\partial \varepsilon}|_{\varepsilon=0} = \frac{1}{2\pi}\text{tr}\int_{-\infty}^{\infty} [\mathbf{A}^R\mathbf{\Upsilon}^R\mathbf{B}_1\mathbf{W}^R\mathbf{W}(\mathbf{V}_2+\mathbf{B}_1\mathbf{Z}^\star)\mathbf{A}+\mathbf{A}^R(\mathbf{V}_2^R+\mathbf{Z}^{\star R}\mathbf{B}_1^R)\mathbf{W}^R\mathbf{W}\mathbf{B}_1\mathbf{\Upsilon}\mathbf{A}]\, d\omega = 0 . \tag{31.80}$$

Using the cyclic property of the trace yields

$$\frac{1}{2\pi}\text{tr}\int_{-\infty}^{\infty} [\mathbf{\Upsilon}^R\mathbf{B}_1\mathbf{W}^R\mathbf{W}(\mathbf{V}_2 + \mathbf{B}_1\mathbf{Z}^\star)\mathbf{A}\mathbf{A}^R + \mathbf{A}\mathbf{A}^R(\mathbf{V}_2^R + \mathbf{Z}^{\star R}\mathbf{B}_1^R)\mathbf{W}^R\mathbf{W}\mathbf{B}_1\mathbf{\Upsilon}]\, d\omega = 0 \tag{31.81}$$

$$\frac{1}{2\pi}\text{tr}\int_{-\infty}^{\infty} \mathbf{\Upsilon}^R(s)\mathbf{H}(s)+[\mathbf{\Upsilon}^T\mathbf{H}(-s)]^T d\omega = \frac{1}{2\pi}\text{tr}\int_{-\infty}^{\infty} \mathbf{\Upsilon}^R(s)\mathbf{H}(s)+[\mathbf{\Upsilon}^R(-s)\mathbf{H}(-s)]^T d\omega = 0 \tag{31.82}$$

$$\text{where} \quad \mathbf{H}(s) \triangleq \mathbf{B}_1^R(s)\mathbf{W}^R(s)\mathbf{W}(s)[\mathbf{V}_2(s) + \mathbf{B}_1(s)\mathbf{Z}^\star(s)]\mathbf{A}(s)\mathbf{A}^R(s) \quad \text{or} \tag{31.83}$$

$$\mathbf{H}(-s) = \mathbf{B}_1^T(s)\mathbf{W}^T(s)\mathbf{W}(-s)[\mathbf{V}_2(-s) + \mathbf{B}_1(-s)\mathbf{Z}^\star(-s)]\mathbf{A}(-s)\mathbf{A}^T(s) . \tag{31.84}$$

Since the integration path is the imaginary axis and the bounds are symmetric $-\infty$ and ∞, the minimum is given by

$$\frac{1}{\pi}\text{tr}\int_{-j\infty}^{j\infty} \mathbf{\Upsilon}^R(s)\mathbf{H}(s)ds = 0 . \tag{31.85}$$

The matrix $\mathbf{\Upsilon}(s)$ is assumed analytic in $\Re e\ s \geq 0$, i.e., all its poles are located in $\Re e\ s < 0$. Hence, $\mathbf{\Upsilon}^R(s)$ is analytic in $\Re e\ s \leq 0$. In order to obtain

$$\text{tr}\int_{-j\infty}^{j\infty} \mathbf{\Upsilon}^R(s)\mathbf{H}(s)ds = 0 \tag{31.86}$$

not only $\mathbf{\Upsilon}^R(s)$ must be analytic in $\Re e\ s \leq 0$ but also $\mathbf{H}(s)$ (*Bongiorno, J.J., 1969*).

Proving this, assume that each component of the trace in Eq.(31.86) approaches zero as $s \to \infty$. Then the integral along the imaginary axis, Eq.(31.86), may be substituted by an integral along the contour C_L in the complex s-plane consisting of the imaginary axis *and* the semicircle with infinity radius around the left s-plane. By Cauchy's residue theorem,

$$\oint_{C_L} P(s)ds = 2\pi j \sum_k \text{residues of } P(s) \text{ at } s_k , \tag{31.87}$$

the integral results from the sum of the residues. The residues of a function $P(s)$ at the pole of the multiplicity ν is

$$\text{residues of } P(s) \text{ at } s_k = \lim_{s\to s_k} \frac{1}{(\nu-1)!}\frac{d^{\nu-1}}{ds^{\nu-1}}[(s-s_k)^\nu P(s)] . \tag{31.88}$$

If the number of enclosed poles is zero then the sum of residues vanishes.

Since $\mathbf{\Upsilon}^R(s)$ is analytic in $\Re e\ s \leq 0$ the integral Eq.(31.86) is zero along a contour C_L around the left half plane if there is no pole of $\mathbf{H}(s)$ in the left half plane. If $\mathbf{\Upsilon}^R(s)$ is analytic in $\Re e\ s \leq 0$ then $\mathbf{H}(s)$ must also be analytic in $\Re e\ s \leq 0$. Otherwise it may be arbitrary.

31.12.2 Spectral Factorization

Applying spectral factorization (*Shaked, U., 1976; Bongiorno, J.J., 1969*),

$$\mathbf{B}_1^R\mathbf{W}^R\mathbf{W}\mathbf{B}_1 = (\mathbf{W}\mathbf{B}_1)^R\mathbf{W}\mathbf{B}_1 = \mathbf{\Delta}_1^R\mathbf{\Delta}_1 \quad \text{and} \tag{31.89}$$

$$\mathbf{A}\mathbf{A}^R = \mathbf{\Delta}_2\mathbf{\Delta}^R \tag{31.90}$$

where $\mathbf{\Delta}_1$ and $\mathbf{\Delta}_2$ are free of poles and zeros in $\Re e\ s \geq 0$. Then, from Eq.(31.83),

$$\mathbf{H} = \mathbf{\Delta}_1^R\mathbf{\Delta}_1\mathbf{B}_1^{-1}[\mathbf{V}_2 + \mathbf{B}_1\mathbf{Z}^\star]\mathbf{\Delta}_2\mathbf{\Delta}_2^R \tag{31.91}$$

$$\mathbf{H} = \mathbf{\Delta}_1^R\mathbf{\Delta}_1\mathbf{B}_1^{-1}\mathbf{V}_2\mathbf{\Delta}_2\mathbf{\Delta}_2^R + \mathbf{\Delta}_1^R\mathbf{\Delta}_1\mathbf{Z}^\star\mathbf{\Delta}_2\mathbf{\Delta}_2^R \tag{31.92}$$

$$\mathbf{\Delta}_1^{R,-1}\mathbf{H}\mathbf{\Delta}_2^{R,-1} = \mathbf{\Delta}_1\mathbf{B}_1^{-1}\mathbf{V}_2\mathbf{\Delta}_2 + \mathbf{\Delta}_1\mathbf{Z}^\star\mathbf{\Delta}_2\,. \tag{31.93}$$

Now, fractional expansion $\mathbf{A}(s) = \mathbf{A}^-(s) + \mathbf{A}^+(s)$ is introduced. The matrix $\mathbf{A}^+(s)$ with superscript $^+$ denotes that part of the expansion associated with all poles in $\Re e\ s < 0$ and, thus, is analytic in the right-half s-plane, i.e. $\Re e\ s \geq 0$. Fractional expansion and separation yields

$$\mathbf{\Delta}_1^{R,-1}\mathbf{H}\mathbf{\Delta}_2^{R,-1} - [\mathbf{\Delta}_1\mathbf{B}_1^{-1}\mathbf{V}_2\mathbf{\Delta}_2]^+ = [\mathbf{\Delta}_1\mathbf{B}_1^{-1}\mathbf{V}_2\mathbf{\Delta}_2]^- + \mathbf{\Delta}_1\mathbf{Z}^\star\mathbf{\Delta}_2\,. \tag{31.94}$$

Since $\mathbf{Z}^\star$ is considered analytical in $\Re e\ s \geq 0$ and $\mathbf{\Delta}_1$, $\mathbf{\Delta}_2$ are also analytic in $\Re e\ s > 0$, the right-hand side of the equation above is analytic in $\Re e\ s \geq 0$. The left-hand side of the equation above is analytic in $\Re e\ s \leq 0$. Hence, both sides of the equation must be zero and

$$\mathbf{Z}^\star = -\mathbf{\Delta}_1^{-1}[\mathbf{\Delta}_1\mathbf{B}_1^{-1}\mathbf{V}_2\mathbf{\Delta}_2]^-\mathbf{\Delta}_2^{-1}\,. \tag{31.95}$$

The resulting optimum H_2-norm controller $\mathbf{K}^\star$ is given by Eq.(31.73), invoking $\mathbf{Z}^\star$ from Eq.(31.95).

In combination with spectral factorization, this analytic concept can also be used as a Wiener-like design technique minimizing the closed-loop performance for given statistics of system inputs in terms of power-spectral-density matrices (*Weston, J.E., and Bongiorno, J.J., 1972*).

The minimum weighted H_2-norm of the sensitivity matrix is given from Eq.(31.76) by

$$I_2^\star = \min_{\mathbf{K}(s)} \|\mathbf{W}(s)\mathbf{S}(s)\|_2 = \sqrt{\mathrm{tr}\frac{1}{2\pi j}\int_{-j\infty}^{j\infty} \mathbf{S}^{\star R}\mathbf{W}^R\mathbf{W}\mathbf{S}^\star ds} \tag{31.96}$$

$$= \sqrt{\mathrm{tr}\frac{1}{2\pi j}\int_{-j\infty}^{j\infty} \mathbf{A}^R(\mathbf{V}_2^R + \mathbf{Z}^{\star R}\mathbf{B}_1^R)\mathbf{W}^R\mathbf{W}(\mathbf{V}_2 + \mathbf{B}_1\mathbf{Z}^\star)\mathbf{A}ds}\,. \tag{31.97}$$

(31.98)

31.12.3 Special Case

In the special case that $\mathbf{B}_1^{-1}$ is free of poles and zeros in $\Re e\ s \geq 0$, i.e., the matrix product $\mathbf{M}_F\mathbf{G}$ is free of zeros in $\Re e\ s \geq 0$, then from Eq.(31.95) it results

$$\mathbf{Z}^\star = -\mathbf{B}_1^{-1}\mathbf{V}_2 \tag{31.99}$$

and the controller $\mathbf{K}^\star(s)$ can make $I_2^\star$ vanish or arbitrarily small

$$I_2^\star = \min_{\mathbf{K}(s)} \|\mathbf{W}(s)\mathbf{S}(s)\|_2 = 0\,. \tag{31.100}$$

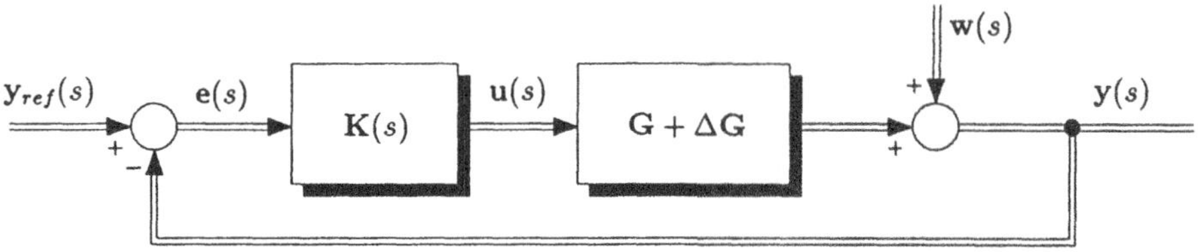

Figure 31.9: Multivariable control system with perturbed plant

The H_2-optimal controller with the restriction to stable controller behaviour was investigated by *Ganesh, C., and Pearson, J.B., 1989*. State-feedback controllers minimizing the H_2-norm of a closed-loop transfer matrix and, additionally, satisfying a prescribed H_∞-norm bound on some other closed-loop transfer matrix are designed by *Rotea, M.A., and Khargonekar, P.P., 1991*.

31.13 Minimizing Weighted Sensitivity Function

Let the system be given as depicted in Fig. 31.9. The plant uncertainty $\Delta\mathbf{G}$ is of additive type, bounded by

$$\Delta\mathbf{G} \triangleq \mathbf{L}^{-1}(s)\mathbf{\Theta}(s)\mathbf{R}(s) \quad \text{where} \quad \|\mathbf{\Theta}\|_{L\infty} \leq 1 \tag{31.101}$$

and the definition Eq.(31.33) was used. The matrices $\mathbf{L}^{-1}(s)$ and $\mathbf{R}(s)$ are given constant transfer matrices, $\mathbf{\Theta}(s)$ is a stable matrix. Then, referring to Eqs.(24.68) and (26.1), if and only if

$$\frac{1}{k_m(\mathbf{K})} \triangleq \|\mathbf{RK}(\mathbf{I}+\mathbf{GK})^{-1}\mathbf{L}^{-1}\|_\infty < 1 \tag{31.102}$$

$$\mathbf{G},\ \mathbf{\Theta}\ \in \mathcal{C}^{r\times m}, \quad \mathbf{K} \in \mathcal{C}^{m\times r}, \quad \mathbf{L} \in \mathcal{C}^{r\times r}, \quad \mathbf{R} \in \mathcal{C}^{m\times m} \tag{31.103}$$

the closed-loop system is internally stable under all perturbations given by Eq.(31.101) (*Doyle, J.C., et al. 1982*). The reciprocal value of the left-hand side of Eq.(31.102) is denoted excess stability margin $k_m(\mathbf{K})$.

31.14 Perturbed Plant Parametrization

Assume $\mathbf{G} \in RH_\infty$ (real, rational in H_∞, i.e. stable and proper with real coefficients) and $\mathbf{G}$ has no poles on the imaginary axis. Using the factorization $\mathbf{G}(s) = \mathbf{A}^{-1}\mathbf{B} = \mathbf{B}_1\mathbf{A}_1^{-1}$ where $\mathbf{A}$, $\mathbf{B}$, $\mathbf{A}_1$, $\mathbf{B}_1 \in RH_\infty$ and coprime. Additionally, $\mathbf{U}_i$, $\mathbf{V}_i$ $\forall i = 1,2$ exist with appropriate dimension such that Eq.(31.71) holds. Then, the set of matrices $\mathbf{K}(s)$ (real, rational, proper) stabilizing $\mathbf{G}(s)$ can be parametrized according to Eq.(31.73). Defining the sensitivity function $\mathbf{S}_o$ including the additional scaling matrices

$$\mathbf{S}_o \triangleq \mathbf{RK}(\mathbf{I}+\mathbf{GK})^{-1}\mathbf{L}^{-1} \tag{31.104}$$

and combining with Eq.(31.75) yields

$$\mathbf{S}_o = \mathbf{R}(\mathbf{V}_2 + \mathbf{B}_1\mathbf{Z})\mathbf{A}\mathbf{L}^{-1} \qquad \mathbf{Z} \in \mathcal{C}^{m\times r}\ . \tag{31.105}$$

Maximizing the excess stability margin $k_m(\mathbf{K})$ is equivalent to the problem of minimizing $\|\mathbf{S}_o\|_\infty$ with respect to $\mathbf{Z}$, $\|\mathbf{S}_o\|_\infty \to \min$. The function to be minimized can be rewritten

$$\min_{\mathbf{Z}} \|\mathbf{S}_o\|_\infty = \min_{\mathbf{Z}}\{\|\mathbf{T}_{11} - \mathbf{T}_{12}\mathbf{Z}\mathbf{T}_{21}\|_\infty\} \tag{31.106}$$

$$\text{where} \quad \mathbf{T}_{11} = \mathbf{R}\mathbf{V}_2\mathbf{A}\mathbf{L}^{-1}, \quad \mathbf{T}_{12} = -\mathbf{R}\mathbf{B}_1, \quad \mathbf{T}_{21} = \mathbf{A}\mathbf{L}^{-1} \tag{31.107}$$

which shows the similarity to the model-matching problem.

This method is only applicable if asymptotic regulation is not required.

31.15 Minimum Sensitivity in H_∞-Norm

The question is how to design a controller $\mathbf{K}(s)$ such that the minimum function infinity norm or H_∞-norm of the weighted sensitivity function

$$I_\infty^\star = \min_{\mathbf{K}(s)} \|\mathbf{W}(s)\mathbf{S}(s)\|_\infty \tag{31.108}$$

is obtained. Using Eq.(31.75) and defining a matrix $\mathbf{L}(s)$,

$$I_\infty^\star = \min_{\mathbf{K}(s)} \|\mathbf{W}(s)(\mathbf{V}_2 + \mathbf{B}_1\mathbf{Z})\mathbf{A}\|_\infty \triangleq \min_{\mathbf{K}(s)} \|\mathbf{L}(s)\|_\infty = \|\mathbf{L}^\star(s)\|_\infty \tag{31.109}$$

where $\mathbf{L}^\star(s)$ is the minimum with respect to $\mathbf{K}(s)$. As derived by *Adamjan, V.M., et al. 1978, p. 147,* the minimal $\mathbf{L}^\star(s)$ is an all-pass matrix form

$$\mathbf{L}^\star(s) = \rho\, \mathbf{Y}^-(s)\mathbf{X}^{+,-1}(s) \tag{31.110}$$

where the matrices satisfy

$$(\mathbf{X}^+)^R\mathbf{X}^+ = (\mathbf{Y}^-)^R\mathbf{Y}^- . \tag{31.111}$$

Hence, the infinity norm $\|\mathbf{L}^\star(s)\|_\infty = |\rho|$. Premultiplying with $(\mathbf{X}^+)^{-1,R}$ and postmultiplying with $(\mathbf{X}^+)^{-1}$ yields the identity matrix

$$\mathbf{I} = [\mathbf{Y}^-(\mathbf{X}^+)^{-1}]^R\mathbf{Y}^-(\mathbf{X}^+)^{-1} \triangleq \Xi^R\Xi . \tag{31.112}$$

The structure of the matrix Ξ is

$$\Xi = \mathbf{Y}^-(\mathbf{X}^+)^{-1} = \mathbf{U}\begin{pmatrix} q_1^R(s)/q_1(s) & \dots & 0 \\ \vdots & \ddots & 0 \\ 0 & 0 & q_n^R(s)/q_n(s) \end{pmatrix}\mathbf{U}^{-1} \tag{31.113}$$

where $q_i(s)$ $\forall i = [1,n]$ are Hurwitz polynomials and $\mathbf{U}$ is an unitary (n,n)-matrix. Then, the controller $\mathbf{K}(s)$ for minimal H_∞-norm is given by

$$\mathbf{K}(s) = (\mathbf{V}_1 + \mathbf{Z}^\star\mathbf{B}_1)^{-1}(\mathbf{U}_1 - \mathbf{Z}^\star\mathbf{A}) = (\mathbf{U}_2 - \mathbf{A}_1\mathbf{Z}^\star)(\mathbf{V}_2 + \mathbf{B}_1\mathbf{Z}^\star)^{-1} \tag{31.114}$$

where $\mathbf{W}(s) = w(s)\mathbf{I}$ and

$$\mathbf{Z}^\star = \mathbf{B}_1^{-1}\Big(\frac{\rho}{w(s)}\Xi\mathbf{A}^{-1} - \mathbf{V}_2\Big) . \tag{31.115}$$

The minimum achievable H_∞-norm is

$$I_\infty^\star = \min_{\mathbf{K}(s)} \|\mathbf{W}(s)\mathbf{S}(s)\|_\infty = \min_{\mathbf{K}(s)} \|\mathbf{W}[\mathbf{V}_2 + \mathbf{B}_1\mathbf{Z}^\star]\mathbf{A}\|_\infty = |\rho| . \tag{31.116}$$

Special Case: If $\mathbf{B}_1^{-1}$ or $\mathbf{M}_F\mathbf{G}$ is free of poles in $\Re e\ s \geq 0$ then a matrix Ξ always exists such that $\Xi\mathbf{A}^{-1}$ is analytic in $\Re e\ s \geq 0$ and that $\rho = 0$, i.e.,

$$\mathbf{Z}^* = -\mathbf{B}_1^{-1}\mathbf{V}_2 \qquad \text{and} \qquad I_\infty^* = \min_{\mathbf{K}(s)} \|\mathbf{W}(s)\mathbf{S}(s)\|_\infty = |\rho| = 0\ . \tag{31.117}$$

Substituting $\mathbf{Z}^*$ of Eq.(31.117) into Eq.(31.114), $\mathbf{K}^* \to \infty$ results. Such a controller is not realizable. In order to obtain a defined controller this controller is a non-minimum one. For any chosen $\rho \neq 0$, a non-minimum $\mathbf{Z}$ is attained by Eq.(31.115).

To sum up, the weighted sensitivity matrix both in H_2-norm and H_∞-norm can be made arbitrarily small if the plant and measurement have no zero in $\Re e\ s \geq 0$. Otherwise lower limits exist.

31.16 Distance Problem in H_∞ (1-Block-Problem)

For $\mathbf{N}, \mathbf{T}_{ij} \in H_\infty$ and $\mathbf{T}_{12}, \mathbf{T}_{21}$ square, $\mathbf{T}_{12}\mathbf{T}_{12}^R = \mathbf{I}$, $\mathbf{T}_{21}^R\mathbf{T}_{21} = \mathbf{I}$ solve

$$\begin{aligned}\|\mathbf{T}_{11} + \mathbf{T}_{12}\mathbf{N}\mathbf{T}_{21}\|_\infty &= \|(\mathbf{T}_{12}\mathbf{T}_{12}^R)\mathbf{T}_{11}(\mathbf{T}_{21}^R\mathbf{T}_{21}) + \mathbf{T}_{12}\mathbf{N}\mathbf{T}_{21}\|_\infty = \|\mathbf{T}_{12}(\mathbf{T}_{12}^R\mathbf{T}_{11}\mathbf{T}_{21}^R + \mathbf{N})\mathbf{T}_{21}\|_\infty \\ &= \|\mathbf{T}_{12}^R\mathbf{T}_{11}\mathbf{T}_{21}^R + \mathbf{N}\|_\infty = \|\mathbf{T}_{12}\mathbf{T}_{11}^R\mathbf{T}_{21} + \mathbf{N}^R\|_\infty \triangleq \|\mathbf{R} + \mathbf{N}^R\|_\infty \end{aligned} \tag{31.118}$$

Since $\mathbf{T}_{12}$ and $\mathbf{T}_{21}$ are all-pass they were omitted when calculating the H_∞-norm. With regard to $\mathbf{Q} \in H_\infty$, the matrix $\mathbf{N}^R$ has only unstable poles but $\mathbf{N} \triangleq \mathbf{T}_{12}\mathbf{T}_{11}^R\mathbf{T}_{21}$ has only stable poles (*Limebeer, D.J.N., and Hung, Y.S., 1987*). Since $\mathbf{R}$ is stable, the model matching problem can be considered as a problem of approximating a stable transfer matrix $\mathbf{R}$ by an unstable matrix $-\mathbf{N}^R$, in the sense of minimum norm $\|\cdot\|_\infty$. This problem is solved by the so-called Hankel approximation (*Safonov, M.G., and Verma, M.S., 1985*).

Reducing factorizations to solving an algebraic Riccati equation and using state-space methods for performing factorizations is a key idea given by *Doyle, J.C., 1983*. Computations required for H_∞-sensitivity optimization are reduced to the solution of the standard LQG-optimal control problem, to Riccati equation and to model reduction using Hankel norm (*Doyle, J.C., 1983*). The polynomial solution approach is presented by *Kwakernaak, H., 1990*, for the standard H_2-optimal problem by *Grimble, M.J., 1991*.

31.17 General Distance Problem in H_∞ (4-Block-Problem)

Referring to Eq.(31.106), the sensitivity $\|\mathbf{S}_o\|_p$ again has to be minimized for $p = 2$ or ∞.

For combined performance robustness and stability robustness problems, the matrices $\mathbf{T}_{12}$ and $\mathbf{T}_{21}$ are not square matrices. Fortunately, it happens that it is always possible to choose $\mathbf{T}_{12}$ and $\mathbf{T}_{21}$ all-pass. The assumptions $\mathbf{T}_{12}\mathbf{T}_{12}^R = \mathbf{I}$ and $\mathbf{T}_{21}^R\mathbf{T}_{21} = \mathbf{I}$ can be easily satisfied if $\mathbf{T}_{12}$ and $\mathbf{T}_{21}$ are row-like[2] and column-like matrices, respectively. Otherwise, if $\mathbf{T}_{12}$ is column-like, $\mathbf{T}_{12\perp}$ is chosen such that $(\mathbf{T}_{12}\vdots\mathbf{T}_{12\perp})$ is square and all-pass. Then, one has

$$\begin{pmatrix}\mathbf{T}_{12}^R \\ \mathbf{T}_{12\perp}^R\end{pmatrix}(\mathbf{T}_{12}\vdots\mathbf{T}_{12\perp}) = \begin{pmatrix}\mathbf{T}_{12}^R\mathbf{T}_{12} & \mathbf{T}_{12}^R\mathbf{T}_{12\perp} \\ \mathbf{T}_{12\perp}^R\mathbf{T}_{12} & \mathbf{T}_{12\perp}^R\mathbf{T}_{12\perp}\end{pmatrix} = \begin{pmatrix}\mathbf{I} & 0 \\ 0 & \mathbf{I}\end{pmatrix} = \mathbf{I}\ . \tag{31.119}$$

Similarly,

$$\begin{pmatrix}\mathbf{T}_{21} \\ \mathbf{T}_{21\perp}\end{pmatrix}(\mathbf{T}_{21}^R\vdots\mathbf{T}_{21\perp}^R) = \begin{pmatrix}\mathbf{T}_{21}\mathbf{T}_{21}^R & \mathbf{T}_{21}\mathbf{T}_{21\perp}^R \\ \mathbf{T}_{21\perp}\mathbf{T}_{21}^R & \mathbf{T}_{21\perp}\mathbf{T}_{21\perp}^R\end{pmatrix} = \begin{pmatrix}\mathbf{I} & 0 \\ 0 & \mathbf{I}\end{pmatrix} = \mathbf{I}\ . \tag{31.120}$$

[2] that is, more columns than rows

Then,

$$\begin{aligned}
\|\mathbf{T}_{11} + \mathbf{T}_{12}\mathbf{N}\mathbf{T}_{21}\|_\infty &= \| \underbrace{\begin{pmatrix} \mathbf{T}_{12}^R \\ \mathbf{T}_{12\perp}^R \end{pmatrix} (\mathbf{T}_{12} \vdots \mathbf{T}_{12\perp})}_{\mathbf{I}} (\mathbf{T}_{11} + \mathbf{T}_{12}\mathbf{N}\mathbf{T}_{21}) \underbrace{\begin{pmatrix} \mathbf{T}_{21} \\ \mathbf{T}_{21\perp} \end{pmatrix} (\mathbf{T}_{21}^R \vdots \mathbf{T}_{21\perp}^R)}_{\mathbf{I}} \|_\infty \\
&= \| \underbrace{(\mathbf{T}_{12} \vdots \mathbf{T}_{12\perp})}_{\mathbf{X}} \begin{pmatrix} \mathbf{T}_{12}^R \\ \mathbf{T}_{12\perp}^R \end{pmatrix} (\mathbf{T}_{11} + \mathbf{T}_{12}\mathbf{N}\mathbf{T}_{21}) (\mathbf{T}_{21}^R \vdots \mathbf{T}_{21\perp}^R) \underbrace{\begin{pmatrix} \mathbf{T}_{21} \\ \mathbf{T}_{21\perp} \end{pmatrix}}_{\mathbf{Y}} \|_\infty \\
&= \| \begin{pmatrix} \mathbf{T}_{12}^R \\ \mathbf{T}_{12\perp}^R \end{pmatrix} (\mathbf{T}_{11} + \mathbf{T}_{12}\mathbf{N}\mathbf{T}_{21}) (\mathbf{T}_{21}^R \vdots \mathbf{T}_{21\perp}^R) \|_\infty \qquad (31.121) \\
&= \| \begin{pmatrix} \mathbf{T}_{12}^R \\ \mathbf{T}_{12\perp}^R \end{pmatrix} \mathbf{T}_{11} (\mathbf{T}_{21}^R \vdots \mathbf{T}_{21\perp}^R) + \begin{pmatrix} \mathbf{N} & \mathbf{0} \\ \mathbf{0} & \mathbf{0} \end{pmatrix} \|_\infty \qquad (31.122) \\
&\triangleq \| \begin{pmatrix} \mathbf{R}_{11} + \mathbf{N} & \mathbf{R}_{12} \\ \mathbf{R}_{21} & \mathbf{R}_{22} \end{pmatrix} \|_\infty \qquad (31.123)
\end{aligned}$$

(*Chu, C.C., et al. 1986*) where $\mathbf{X}$ and $\mathbf{Y}$ could have been omitted since they are all-pass and the H_∞-norm is invariant with respect to the multiplication of its argument with an all-pass matrix. Minimizing $\|\mathbf{S}_c\|_\infty$ with respect to $\mathbf{N}$

$$\min_{\mathbf{N} \in RH_\infty} \| \begin{pmatrix} \mathbf{R}_{11} + \mathbf{N} & \mathbf{R}_{12} \\ \mathbf{R}_{21} & \mathbf{R}_{22} \end{pmatrix} \|_p = \gamma_o \qquad (31.124)$$

for $p = \infty$ yields

$$\gamma_o = \min_{\mathbf{N} \in H_\infty} \| \begin{pmatrix} \mathbf{R}_{11} + \mathbf{N} & \mathbf{R}_{12} \\ \mathbf{R}_{21} & \mathbf{R}_{22} \end{pmatrix} \|_\infty = \|\mathbf{\Gamma}_R\|_s = \| \begin{pmatrix} \mathbf{P}_{H_2^\perp} & \mathbf{0} \\ \mathbf{0} & \mathbf{I} \end{pmatrix} \begin{pmatrix} \mathbf{R}_{11} & \mathbf{R}_{12} \\ \mathbf{R}_{21} & \mathbf{R}_{22} \end{pmatrix} \|_s \qquad (31.125)$$

where the orthogonal projection from L_2 to $H_2^\perp$ is termed $\mathbf{P}_{H_2^\perp}$ (*Feintuch, A., and Francis, B.A., 1985*).

Several robustness problems do not need the minimization but only require a check to see whether the H_∞-norm is smaller than a predetermined number.

31.18 H_∞-Norm of a Column-Like Partitioned Matrix

If $\mathbf{A}$, $\mathbf{B} \in L_\infty$ then

$$\| \begin{pmatrix} \mathbf{A} \\ \mathbf{B} \end{pmatrix} \|_\infty = \sup_\omega \sigma_{\max}[\begin{pmatrix} \mathbf{A} \\ \mathbf{B} \end{pmatrix}] = \sup_\omega \sqrt{\lambda_{\max}[\mathbf{A}^H\mathbf{A} + \mathbf{B}^H\mathbf{B}]} \le \sup_\omega \sqrt{\|\mathbf{A}\|_s^2 + \|\mathbf{B}\|_s^2} = \| \begin{pmatrix} \|\mathbf{A}\|_\infty \\ \|\mathbf{B}\|_\infty \end{pmatrix} \|_F . \qquad (31.126)$$

Since

$$\|\mathbf{A}\|_\infty \le \| \begin{pmatrix} \mathbf{A} \\ \mathbf{B} \end{pmatrix} \|_\infty , \qquad (31.127)$$

$$\| \begin{pmatrix} \|\mathbf{A}\|_\infty \\ \|\mathbf{B}\|_\infty \end{pmatrix} \|_F \frac{1}{\| \begin{pmatrix} \mathbf{A} \\ \mathbf{B} \end{pmatrix} \|_\infty} \le \| \begin{pmatrix} \| \begin{pmatrix} \mathbf{A} \\ \mathbf{B} \end{pmatrix} \|_\infty \\ \| \begin{pmatrix} \mathbf{A} \\ \mathbf{B} \end{pmatrix} \|_\infty \end{pmatrix} \|_F \frac{1}{\| \begin{pmatrix} \mathbf{A} \\ \mathbf{B} \end{pmatrix} \|_\infty} = \| \begin{pmatrix} 1 \\ 1 \end{pmatrix} \|_F = \sqrt{2} . \qquad (31.128)$$

Hence,

$$\| \begin{pmatrix} \mathbf{A} \\ \mathbf{B} \end{pmatrix} \|_\infty \le \| \begin{pmatrix} \|\mathbf{A}\|_\infty \\ \|\mathbf{B}\|_\infty \end{pmatrix} \|_F \le \sqrt{2} \, \| \begin{pmatrix} \mathbf{A} \\ \mathbf{B} \end{pmatrix} \|_\infty . \qquad (31.129)$$

31.19 H_∞-Norm of a Partitioned Matrix

Similarly to Eq.(31.129)(*Chu, C.C., 1986*),

$$\left\| \begin{pmatrix} \mathbf{A} & \mathbf{B} \\ \mathbf{C} & \mathbf{D} \end{pmatrix} \right\|_\infty \le \left\| \begin{pmatrix} \|\mathbf{A}\|_\infty & \|\mathbf{B}\|_\infty \\ \|\mathbf{C}\|_\infty & \|\mathbf{D}\|_\infty \end{pmatrix} \right\|_F \le 2 \left\| \begin{pmatrix} \mathbf{A} & \mathbf{B} \\ \mathbf{C} & \mathbf{D} \end{pmatrix} \right\|_\infty . \tag{31.130}$$

31.20 Hankel Approximation

Given $\mathbf{G} \in H_\infty$, the H_∞-norm taken of the sum of $\mathbf{G}$ and the reflected matrix $\mathbf{N}^R$ of any unknown matrix $\mathbf{N} \in H_\infty$ (1-block-problem) has a lower bound $\|\mathbf{G}\|_H$, i.e.,

$$\|\mathbf{G} + \mathbf{N}^R\|_\infty \ge \|\mathbf{G}\|_H \tag{31.131}$$

and

$$\min_{\mathbf{N}} \|\mathbf{G} + \mathbf{N}^R\|_\infty = \|\mathbf{G} + \mathbf{N}^R_{\min}\|_\infty = \|\mathbf{G}\|_H \tag{31.132}$$

(*Glover, K., 1984*). For the 4-block-problem solving the minimization of Eq.(31.118) and Eq.(31.123) see *Safonov, M.G., et al. 1987; Glover, K., and Doyle, J.C., 1988; Doyle, J.C., et al. 1988; 1989.*

31.21 Graph Metric Perturbations

The uncertainty of a plant is denoted graph metric perturbation if it is characterized by a graph metric distance. The right-coprime factorization $(\mathbf{N}, \mathbf{D})$ of a plant $\mathbf{P}$ is said to be normalized if

$$\left(\mathbf{D}^T(-s) \,\vdots\, \mathbf{N}^T(-s)\right) \begin{pmatrix} \mathbf{D}(s) \\ \mathbf{N}(s) \end{pmatrix} = \mathbf{I} \qquad \forall\, s . \tag{31.133}$$

Then, $\mathbf{A} = (\mathbf{D}^T \,\vdots\, \mathbf{N}^T)^T = \binom{\mathbf{D}}{\mathbf{N}}$ is inner. The graph metric distance $d_G(\mathbf{G}_1, \mathbf{G}_2)$ between two plants $\mathbf{G}_1$ and $\mathbf{G}_2$ is (*Vidyasagar, M., 1984*)

$$d_G(\mathbf{G}_1, \mathbf{G}_2) = \max\{\delta(\mathbf{G}_1, \mathbf{G}_2),\ \delta(\mathbf{G}_2, \mathbf{G}_1)\} \quad \text{where} \quad \delta(\mathbf{G}_1, \mathbf{G}_2) = \inf_{\mathbf{U}} \|\mathbf{G}_1 - \mathbf{G}_2\mathbf{U}\|_\infty \tag{31.134}$$

and the matrices $\mathbf{G}_1$, $\mathbf{G}_2$ are given by the normalized right-coprime factorization of the plant $\mathbf{G}_1, \mathbf{G}_2$, respectively. The matrix $\mathbf{U}$ belongs to the set of symmetric rational functions in H_∞ and is bounded by $\|\mathbf{U}\|_\infty \le 1$.

Consider the unperturbed plant $\mathbf{G}$ and controller $\mathbf{C}$ a stable pair $(\mathbf{G}, \mathbf{C})$. The perturbed plant and controller is $(\mathbf{G}_p, \mathbf{C}_p)$. Then, the pair $(\mathbf{G}_p, \mathbf{C}_p)$ is stable if

$$d_G(\mathbf{G}_p, \mathbf{G})\ \|\mathbf{T}(\mathbf{G}, \mathbf{C})\|_\infty + d_G(\mathbf{C}_p, \mathbf{C})\ \|\mathbf{T}(\mathbf{C}, \mathbf{G})\|_\infty < 1 \tag{31.135}$$

where

$$\mathbf{T}(\mathbf{G}, \mathbf{C}) = \mathbf{H}(\mathbf{G}, \mathbf{C}) - \begin{pmatrix} \mathbf{I} & 0 \\ 0 & 0 \end{pmatrix} \tag{31.136}$$

$$\mathbf{H}(\mathbf{G}, \mathbf{C}) = \begin{pmatrix} (\mathbf{I} + \mathbf{G}\mathbf{C})^{-1} & -\mathbf{G}(\mathbf{I} + \mathbf{C}\mathbf{G})^{-1} \\ \mathbf{C}(\mathbf{I} + \mathbf{G}\mathbf{C})^{-1} & (\mathbf{I} + \mathbf{C}\mathbf{G})^{-1} \end{pmatrix} . \tag{31.137}$$

Note that the graph metric perturbation is "very" unstructured since there is no restriction on the number of right-half plane poles of perturbed and unperturbed plant being the same. The perturbations are not expressed in terms of a particular coprime factorization of the plant. The plant and controller uncertainty effects enter additively in Eq.(31.135). If only the plant is uncertain then

$$d_G(\mathbf{C}_p, \mathbf{C}) = 0 \quad \text{and} \quad d_G(\mathbf{G}_p, \mathbf{G}) < \|\mathbf{T}(\mathbf{G}, \mathbf{C})\|_\infty^{-1} . \tag{31.138}$$

(Proof see *Vidyasagar, M., and Kimura, H., 1986.*)

31.22 Robustness Using Gap Metric

Consider a nominal plant $\mathbf{G}(s) = \mathbf{N}(s)\mathbf{M}^{-1}(s) = \tilde{\mathbf{M}}^{-1}(s)\tilde{\mathbf{N}}(s)$ where $\mathbf{M}, \tilde{\mathbf{M}}, \mathbf{N}, \tilde{\mathbf{N}} \in RH_\infty$. The perturbed plant is $\mathbf{G}_p(s) = (\mathbf{N}+\Delta\mathbf{N})(\mathbf{M}+\Delta\mathbf{M})^{-1}$. The directed gap between two systems $\mathbf{G}_1$ and $\mathbf{G}_2$ or $\mathbf{G}$ and $\mathbf{G}_p$ is, respectively, defined as

$$\vec{\delta}(\mathbf{G}_1, \mathbf{G}_2) = \inf_{\mathbf{Z}} \| \begin{pmatrix} \mathbf{M}_1 \\ \mathbf{N}_1 \end{pmatrix} - \begin{pmatrix} \mathbf{M}_2 \\ \mathbf{N}_2 \end{pmatrix} \mathbf{Z} \|_\infty \quad \text{or} \quad \vec{\delta}(\mathbf{G}, \mathbf{G}_p) = \inf_{\mathbf{Z}} \| \begin{pmatrix} \mathbf{M} \\ \mathbf{N} \end{pmatrix} - \begin{pmatrix} \mathbf{M}+\Delta\mathbf{M} \\ \mathbf{N}+\Delta\mathbf{N} \end{pmatrix} \mathbf{Z} \|_\infty \tag{31.139}$$

where $\mathbf{Z} \in H_\infty$. In view of this, the computation of the gap metric results from a "two-block" H_∞-optimization.

Now, dissolving the robustness problem, consider a normalized coprime fraction plant $\mathbf{G}$ as cited above and a stabilizing controller $\mathbf{K}$. Then, the system $\mathbf{G}_p, \mathbf{K}$ is stable for all $\mathbf{G}_p$ with $\vec{\delta}(\mathbf{G}, \mathbf{G}_p) < b$ and for the controller

$$\mathbf{K} = (\tilde{\mathbf{X}} + \mathbf{Z}\tilde{\mathbf{N}})^{-1}(\tilde{\mathbf{Y}} + \mathbf{Z}\tilde{\mathbf{M}}) \quad \text{for some} \quad \mathbf{Z} \in H_\infty \tag{31.140}$$

where

$$\| \begin{pmatrix} \Delta\mathbf{M} \\ \Delta\mathbf{N} \end{pmatrix} \|_\infty < b , \quad \begin{pmatrix} \tilde{\mathbf{X}} & -\tilde{\mathbf{Y}} \\ -\tilde{\mathbf{N}} & \tilde{\mathbf{M}} \end{pmatrix} \begin{pmatrix} \mathbf{M} & \mathbf{Y} \\ \mathbf{N} & \mathbf{X} \end{pmatrix} = \begin{pmatrix} \mathbf{I} & \mathbf{0} \\ \mathbf{0} & \mathbf{I} \end{pmatrix} , \tag{31.141}$$

$\mathbf{X}, \mathbf{Y}, \tilde{\mathbf{X}}, \tilde{\mathbf{Y}} \in RH_\infty$ and where $0 < b \leq 1$ (*Georgiou, T.T., and Smith, M.C., 1990*).

31.23 Robust Controller For a Dithered Nonlinear System

Saturating actuators in linear perturbed dither-free systems were frequently investigated in the time domain (*Su, T.J., et al. 1988; 1990; 1990a*). Consider, now, the control system as depicted in Fig. 31.10. A dither is introduced in order to smooth the nonlinearity $\mathrm{N} = \mathrm{diag}\{N\}$ and to improve the performance. The objective is to determine conditions for robust control, in terms of dither and system parameters.

If both the dither has sufficiently large amplitude, the frequency of the dither is high enough and the linearized feedback system has a sufficient high-frequency attenuation, the actual overall system behaviour and the behaviour of the system with smoothed nonlinearity can be made as close as desired (*Mossaheb, S., 1983*).

Consider a nonlinearity with an input signal $f + h(t)$ where f is a slow motion and $h(t)$ a high-frequency signal known as dither. Let the dither be a saw-tooth or triangle signal with period T and amplitude distribution function $F(\eta)$ (see Fig. 31.11)

$$F(\eta) = \frac{1}{T} \sum E_T(\eta) \quad \text{where} \quad E_T = \{t : 0 \leq t \leq T, \; h(t) \leq \eta\} . \tag{31.142}$$

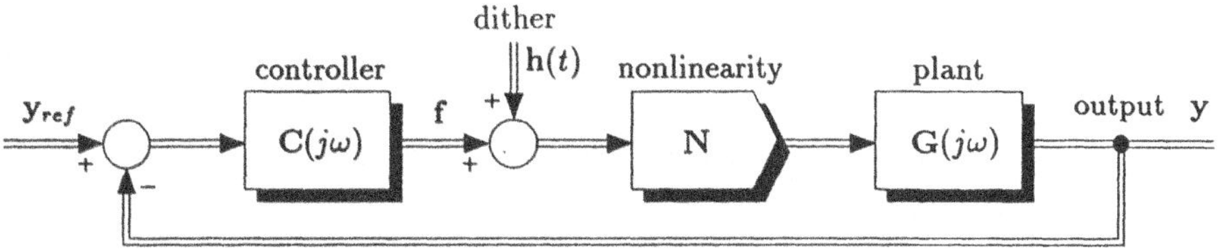

Figure 31.10: Dithered nonlinear system

The distribution density function is given by $dF(\eta)/d\eta$. Let $q = 4$ and $N(\eta)$ be a saturation as depicted in Fig. 31.11 (*Zames, G., and Shneydor, N.A., 1976; 1977*). Then, the smoothed nonlinearity N_s results from the integral of the product of the distribution density and the nonlinearity displaced by f

$$\begin{aligned} N_s(f) &= \int_{-\infty}^{\infty} N(\eta - f)dF(\eta) = \int_{-\infty}^{\infty} N(\eta - f)\frac{dF(\eta)}{d\eta}d\eta = \int_{-q+f}^{q+f} N(\eta)\frac{1}{2q}d\eta \quad (31.143) \\ &= [-\frac{3+(3+\frac{3-f}{3})}{2}(3-f) + \frac{3+(3+\frac{3+f}{3})}{2}(3+f)]\frac{1}{8} = f \quad \forall f \in [0,3] \,. \end{aligned}$$

In a wide range of f it follows
$$N_s(f) = \begin{cases} -\frac{1}{6}f^2 + 2f - \frac{3}{2} & \forall f \in [3,5] \\ \frac{8}{3} + \frac{f}{3} & \forall f \geq 5 \\ -N_s(-f) & \forall f \leq 0 \,. \end{cases} \quad (31.144)$$

The smoothed nonlinearity $N_s(f)$ is depicted in Fig. 31.11. Note that the dither amplitude affects N_s but dither frequency does not. The smoothed nonlinearity lies inside a sector $\{\alpha_s, \beta_s\}$. The quantity $\theta_s = (\beta_s + \alpha_s)/2$ is denoted centre of the sector, and the quantity $\delta_s = (\beta_s - \alpha_s)/2$ the radius of the sector. In the case of Fig. 31.11 $\alpha_s = 0,\ \beta_s = 1$.

31.24 Conditions for Sobolev-Boundedness

The Sobolev-boundedness is chosen to additionally check the high-frequency attenuation properties of the plant. The overall closed-loop system is (Sobolev) S_2-bounded, i.e.,

$$\|\mathbf{y}(t)\|_{S_2} \leq k_1\|\mathbf{y}_{ref}(t)\|_2 \qquad \|\mathbf{u}(t)\|_{S_2} \leq k_2\|\mathbf{y}_{ref}(t)\|_2 \,, \quad (31.145)$$

if the following five conditions are satisfied:

(i) $(\mathbf{I} + \theta_s\mathbf{GK})^{-1}$ exists and is causal.

(ii) The smoothed nonlinearity $\mathbf{N}_s$ lies inside a sector $\{\alpha_s, \beta_s\}$.

(iii) Asymptotic stability of the linearized feedback system exists and is obtained if the nonlinear system is replaced by $\theta_s\mathbf{I}$.

(iv) The L_2-S_2-gain is finite, i.e.,

$$\sup_\omega |(1+j\omega)\sigma_{\max}[\mathbf{I}-\mathbf{S}_{os}(j\omega)]\,| < \infty \quad \text{where} \quad \mathbf{S}_{os}(s) \triangleq [\mathbf{I}+\theta_s\mathbf{G}(s)\mathbf{C}(s)]^{-1} \,. \quad (31.146)$$

(v) The quantity $\rho_2 < 1$, where $\|\cdot\|_\infty$ denotes the Hardy H_∞-norm (infinity norm)

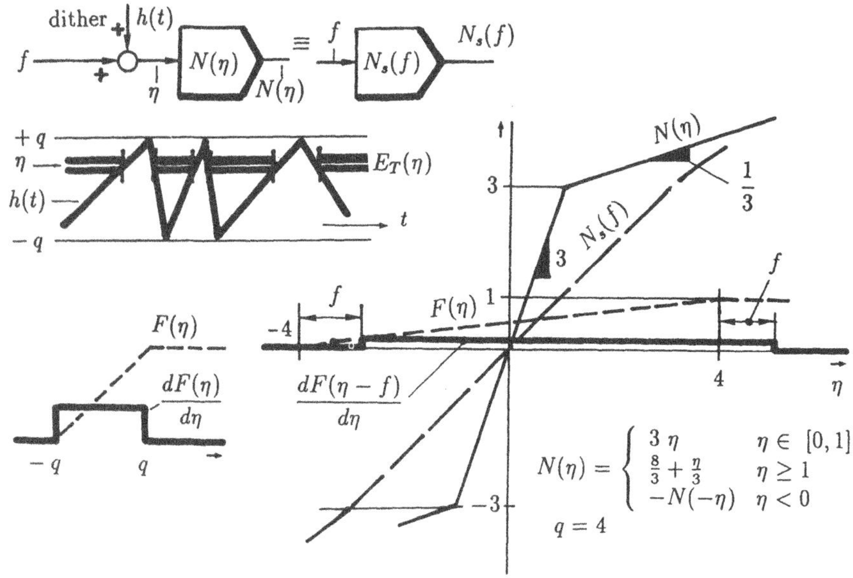

Figure 31.11: Nonlinearity with dithered input. Original and smoothed nonlinearity

$$\rho_2 \triangleq \|\frac{\delta_s}{\theta_s}[\mathbf{I}-\mathbf{S}_{1s}(j\omega)]\,\|_\infty = \|\frac{\delta_s}{\theta_s}[\mathbf{Y}(j\omega)+\mathbf{A}_1(j\omega)\mathbf{Z}(j\omega)]\mathbf{B}(j\omega)\|_\infty \tag{31.147}$$

$$\mathbf{S}_{1s}(s) = [\mathbf{I}+\theta_s\mathbf{C}(s)\mathbf{G}(s)]^{-1} \quad \text{where} \quad \theta_s\mathbf{G}(s) = \mathbf{A}^{-1}(s)\mathbf{B}(s) = \mathbf{B}_1(s)\mathbf{A}_1^{-1}(s) \tag{31.148}$$

$$\begin{pmatrix} -\mathbf{Y}_1(s) & \mathbf{X}_1(s) \\ \mathbf{A}(s) & \mathbf{B}(s) \end{pmatrix}\begin{pmatrix} -\mathbf{B}_1(s) & \mathbf{X}(s) \\ \mathbf{A}_1(s) & \mathbf{Y}(s) \end{pmatrix} = \begin{pmatrix} \mathbf{I} & \mathbf{0} \\ \mathbf{0} & \mathbf{I} \end{pmatrix}. \tag{31.149}$$

Solving $\rho_2 < 1$ for all $\omega \geq 0$ yields

$$\mathbf{I} - \frac{\delta_s}{\theta_s}[\mathbf{I}-\mathbf{S}_{1s}(j\omega)]^H[\mathbf{I}-\mathbf{S}_{1s}(j\omega)]\frac{\delta_s}{\theta_s} > 0 \tag{31.150}$$

and conditions for the matrix $\mathbf{Z}(j\omega)$. Alternatively, conditions for minimizing the H_∞-norm (*Hsieh, J.G., and Hsiao, F.H., 1989*) can be obtained. If $\mathbf{I}-\mathbf{S}_{os}(s)$ is chosen strictly proper this is necessary for adequate high-frequency attenuation of the plant.

The objective of this section could have been reformulated: Find a lower bound on the dither amplitude to stabilize the dithered nonlinear system.

Part VI

Robustness Via Approximative Models

Chapter 32

Robust Hyperplane Design in Variable Structure Control

32.1 Sliding Mode

Sliding motions in a variable structure control system yield a design method which is applicable to uncertain dynamical systems. A variable structure control system has the central feature of switching on one or more manifolds in the state space. When acting in the sliding motion, the system state continuously intersects the switching hyperplane in different directions. If a system switches on all the switching manifolds together, the system is said to be in the sliding mode (*Dorling, C.M., and Zinober, A.S.I., 1986*). The sliding mode is obtained by a discontinuous control action. The system changes its structure when crossing the switching surface. Systems operating in sliding mode are therefore denoted variable structure systems.

When the system has to act along these switching manifolds the motion is constrained to a certain subspace of the full state space. This subspace corresponds to a system of lower order, denoted as equivalent system (*Utkin, V.I., 1977*). The system usually moves towards the sliding mode in a rapid stage of motion ($t < t_s$). After having arrived at the sliding mode ($t \geq t_s$) the motion normally becomes slower. This two-stage version helps to find a good compromise between dynamic and static accuracy and to obtain a robust controller when an uncertain plant has to be taken into account. The nominal plant is given by

$$\dot{\mathbf{x}}(t) = \mathbf{A}\mathbf{x}(t) + \mathbf{B}\mathbf{u}(t) \qquad \mathbf{x}(t) \in \mathcal{R}^n, \quad \mathbf{u}(t) \in \mathcal{R}^m . \tag{32.1}$$

Regarding a state-feedback controller, the components u_j of $\mathbf{u}$ are assumed state dependent $u_j(\mathbf{x})$. On the jth switching surface S_j the control variable u_j is switching. This surface S_j is a hyperplane comprising the origin with respect to the steady-state behaviour. The sliding mode is given in the intersection S of all hyperplanes S_j by

$$S = \{\mathbf{x} \ : \ \mathbf{C}_s\mathbf{x} \equiv \mathbf{0}\} . \tag{32.2}$$

The intersection S is a subspace and equals the null space or kernel of $\mathbf{C}_s$. There are at most m intersecting hyperplanes S_i . The dimension of the intersection S is $n - m$. The sliding mode dynamic behaviour ($t \geq t_s$) is given by combining Eqs.(32.1) and (32.2). If $\mathbf{C}_s$ is chosen to a certain matrix, then this combination yields solutions $\mathbf{u}(t)$ such that both the plant equation and the condition $s_w \triangleq \mathbf{C}_s\mathbf{x} \equiv \mathbf{0}$ are satisfied

$$\mathbf{C}_s\dot{\mathbf{x}} = \mathbf{C}_s\mathbf{A}\mathbf{x} + \mathbf{C}_s\mathbf{B}\mathbf{u} \equiv \mathbf{0} . \tag{32.3}$$

Defining the equivalent controller $\mathbf{K}_{eq}$ in the sliding mode as $\mathbf{u}_{eq} = \mathbf{K}_{eq}\mathbf{x}$, it results

$$\mathbf{C}_s\mathbf{A}\mathbf{x} + \mathbf{C}_s\mathbf{B}\mathbf{K}_{eq}\mathbf{x} = \mathbf{C}_s(\mathbf{A} + \mathbf{B}\mathbf{K}_{eq})\mathbf{x} \triangleq \mathbf{C}_s\mathbf{F}_{eq}\mathbf{x} \equiv \mathbf{0} \tag{32.4}$$

$$\mathbf{K}_{eq} = -(\mathbf{C}_s\mathbf{B})^{-1}\mathbf{C}_s\mathbf{A} \tag{32.5}$$

$$\dot{\mathbf{x}} = \mathbf{A}\mathbf{x} + \mathbf{B}\mathbf{u} = \mathbf{A}\mathbf{x} + \mathbf{B}\mathbf{K}_{eq}\mathbf{x} = \mathbf{F}_{eq}\mathbf{x} = [\mathbf{I}_n - \mathbf{B}(\mathbf{C}_s\mathbf{B})^{-1}\mathbf{C}_s]\mathbf{A}\mathbf{x} \,. \tag{32.6}$$

The invertability of $\mathbf{C}_s\mathbf{B}$ requires $\mathbf{C}_s \in \mathcal{R}^{m\times n}$. The sliding motion is determined by Eq.(32.6) and depends only on the choice of $\mathbf{C}_s$, irrespective of the actual values of u_j. The non-singularity of $\mathbf{C}_s\mathbf{B}$ implies that $[\mathbf{I} - \mathbf{B}(\mathbf{C}_s\mathbf{B})^{-1}\mathbf{C}_s]$ can be considered a projector of $\mathbf{A}$ into a subspace (*Knobloch, H.W., und Kwakernaak, H., 1985*). Note that $[\mathbf{I} - \mathbf{B}(\mathbf{C}_s\mathbf{B})^{-1}\mathbf{C}_s]$ is an idempotent matrix.

Example: One dimensional sliding mode $n = 3, \ m = 2$: Consider

$$\dot{\mathbf{x}} = \mathbf{A}\mathbf{x} + \mathbf{B}\mathbf{u}\,, \qquad \mathbf{A} = \begin{pmatrix} 0 & 1 & 0 \\ 0 & 0 & 1 \\ -6 & -11 & -6 \end{pmatrix}, \qquad \mathbf{B} = \begin{pmatrix} 0 & 0 \\ 1 & 0 \\ 0 & 2 \end{pmatrix}. \tag{32.7}$$

Anticipating Eq.(32.82),

$$\mathbf{u} = \mathbf{L}\mathbf{x} + \rho \begin{pmatrix} \text{sign}\,(x_1 + x_2) \\ \text{sign}\, x_3 \end{pmatrix}, \qquad \mathbf{L} = \begin{pmatrix} -1 & -2 & -1 \\ 3 & 5.5 & 2.5 \end{pmatrix}, \qquad \rho = 5 \tag{32.8}$$

$$\mathbf{C}_s = \begin{pmatrix} 1 & 1 & 0 \\ 0 & 0 & 1 \end{pmatrix}, \qquad \mathcal{N}(\mathbf{C}_s) \text{ represented by } x_1 + x_2 = 0 \text{ and } x_3 = 0 \tag{32.9}$$

$$\mathbf{B}(\mathbf{C}_s\mathbf{B})^{-1}\mathbf{C}_s = \begin{pmatrix} 0 & 0 & 0 \\ 1 & 1 & 0 \\ 0 & 0 & 1 \end{pmatrix}, \qquad \mathbf{I} - \mathbf{B}(\mathbf{C}_s\mathbf{B})^{-1}\mathbf{C}_s = \begin{pmatrix} 1 & 0 & 0 \\ -1 & 0 & 0 \\ 0 & 0 & 0 \end{pmatrix} \tag{32.10}$$

$$\mathbf{F}_{eq} = [\mathbf{I} - \mathbf{B}(\mathbf{C}_s\mathbf{B})^{-1}\mathbf{C}_s]\mathbf{A} = \begin{pmatrix} 0 & 1 & 0 \\ 0 & -1 & 0 \\ 0 & 0 & 0 \end{pmatrix}. \tag{32.11}$$

End of Example

Consider a specific choice of the matrices $\mathbf{K} = -(\mathbf{C}_s\mathbf{B})^{-1}\mathbf{C}_s$ and $\mathbf{H} = \mathbf{B}(\mathbf{C}_s\mathbf{B})^{-1}$ in Eq.(A.62). Since $\mathcal{R}[\mathbf{B}\mathbf{K}] = \mathcal{R}[\mathbf{B}]$ and $\mathcal{N}[\mathbf{H}\mathbf{C}_s] = \mathcal{N}[\mathbf{C}_s]$,

$$\mathcal{R}[\mathbf{B}(\mathbf{C}_s\mathbf{B})^{-1}\mathbf{C}_s] = \mathcal{R}[\mathbf{B}] \quad \text{and} \quad \mathcal{N}[\mathbf{B}(\mathbf{C}_s\mathbf{B})^{-1}\mathbf{C}_s] = \mathcal{N}[\mathbf{C}_s] \,. \tag{32.12}$$

Summarizing, the projector

$$\mathbf{P}_r = \mathbf{B}(\mathbf{C}_s\mathbf{B})^{-1}\mathbf{C}_s \quad \text{projects } \mathcal{R}^n \text{ on } \mathcal{R}[\mathbf{B}] \text{ along } \mathcal{N}[\mathbf{C}_s] \tag{32.13}$$

$$\text{and } \mathbf{I} - \mathbf{B}(\mathbf{C}_s\mathbf{B})^{-1}\mathbf{C}_s \quad \text{projects } \mathcal{R}^n \text{ on } \mathcal{N}[\mathbf{C}_s] \text{ along } \mathcal{R}[\mathbf{B}] \,. \tag{32.14}$$

The preceding expression in Eq.(32.14) is very important as far as the application to Eq.(32.6) is concerned. Premultiplication of $\mathbf{A}$ with $\mathbf{I}_n - \mathbf{P}_r$ means that each column of $\mathbf{A}$, say e.g. the ith $\mathbf{A}_{.i}$, is projected on $\mathcal{N}[\mathbf{C}_s]$. This corresponds to an order reduction from n to $n - m$. The matrix product $[\mathbf{I}_n - \mathbf{P}_r]\mathbf{A}$ lies in a subspace of dimension $n - m$.

The control signals act in the range space $\mathcal{R}[\mathbf{B}]$ of $\mathbf{B}$. The sliding mode lies completely within the null space $\mathcal{N}[\mathbf{C}_s]$ of $\mathbf{C}_s$. With regard to the fact that the range space of $\mathbf{B}$ and the null space of $\mathbf{C}_s$ are complementary (and, hence, possess intersection $\{\emptyset\}$), sliding motion and control action do not interfere. The function of the controlling variable $\mathbf{u}$ is to drive the motion into S and to keep it within S.

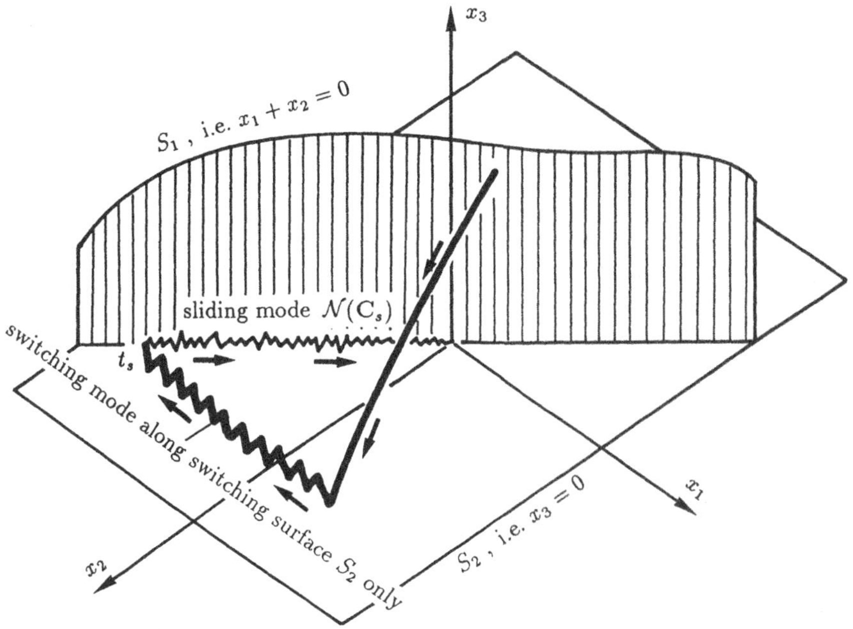

Figure 32.1: Three dimensional state-space and one dimensional sliding mode

32.2 Disturbance Rejection

Consider the system as given in Eq.(32.1), augmented by a disturbance $\mathbf{Gw}$

$$\dot{\mathbf{x}} = \mathbf{Ax} + \mathbf{Bu} + \mathbf{Gw} \ . \tag{32.15}$$

The operation of the projector $\mathbf{P}_r = \mathbf{B}(\mathbf{C}_s\mathbf{B})^{-1}\mathbf{C}_s$ during sliding mode is governed similarly to Eq.(32.6)

$$\dot{\mathbf{x}} = [\mathbf{I}_n - \mathbf{P}_r]\mathbf{Ax} + [\mathbf{I}_n - \mathbf{P}_r]\mathbf{Gw} \ . \tag{32.16}$$

If $[\mathbf{I}_n - \mathbf{P}_r]\mathbf{G} = \mathbf{0}$ is satisfied, robustness with respect to $\mathbf{w}$ is achieved. Referring to Eqs.(A.54) and (A.55), the relations

$$(\mathbf{I}_n - \mathbf{P}_r)\mathbf{x} = \mathbf{0} \quad \rightsquigarrow \quad \mathbf{P}_r\mathbf{x} = \mathbf{x} \qquad \mathbf{x} \in \mathcal{R}[\mathbf{P}_r] \tag{32.17}$$

make clear that all columns $\mathbf{G}_{.i}$ must lie in range space $\mathcal{R}[\mathbf{P}_r]$, i.e. $\mathbf{G}_{.i} \in \mathcal{R}[\mathbf{P}_r]$.

32.3 Robustness to Plant Parameter Variations

Since Eq.(32.6) one has $\dot{\mathbf{x}} = (\mathbf{I}_n - \mathbf{P}_r)\mathbf{Ax}$. Similarly to the statements in the preceding section,

$$\mathbf{A}_{.i} \in \mathcal{R}[\mathbf{P}_r] = \mathcal{R}[\mathbf{B}] \quad \rightsquigarrow \quad \dot{\mathbf{x}} = \mathbf{0} \ . \tag{32.18}$$

If the matrices $\mathbf{A}, \mathbf{B}$ are such that all the variable parameters of $\mathbf{A}$ are completely located in $\mathcal{R}[\mathbf{B}]$, then $\mathbf{x}(t)$ becomes independent of the plant parameters.

Example: Robustness to parameter changes of a single-input plant during sliding mode:

$$n = 3, \quad m = 1, \quad \mathbf{B} = \mathbf{b} = (0\ \ 0\ \ 1)^T \tag{32.19}$$

$$\text{Given } \mathbf{P}_r = \begin{pmatrix} 0 & 0 & 0 \\ 0 & 0 & 0 \\ 4 & 2 & 1 \end{pmatrix}, \quad \mathbf{I}_n - \mathbf{P}_r = \begin{pmatrix} 1 & 0 & 0 \\ 0 & 1 & 0 \\ -4 & -2 & 0 \end{pmatrix}, \tag{32.20}$$

$$\mathcal{R}[\mathbf{P}_r] = \mathcal{R}[\mathbf{B}] = \mathcal{N}[\mathbf{I}_n - \mathbf{P}_r] : \ \mathrm{x}_3\text{-axis} . \tag{32.21}$$

$$\mathcal{R}[\mathbf{I}_n - \mathbf{P}_r] = \mathcal{N}[\mathbf{P}_r] = \mathcal{N}[\mathbf{C}_s] : \ \text{plane } s_w = 4x_1 + 2x_2 + x_3 = 0 . \tag{32.22}$$

$$\text{The matrix } \mathbf{I}_n - \mathbf{P}_r \text{ projects } \mathcal{R}^n \text{ on } \mathcal{N}[\mathbf{C}_s], \text{ i.e., the plane } s_w = 4x_1 + 2x_2 + x_3 = 0 , \tag{32.23}$$

see Fig. A.2.

$$\text{Given } \mathbf{A} = \begin{pmatrix} 0 & 1 & 0 \\ -0.36 & -2 & 0 \\ -a_1 & -a_2 & -a_3 \end{pmatrix} \rightsquigarrow (\mathbf{I}_n - \mathbf{P}_r)\mathbf{A} = \mathbf{F}_{eq} = \begin{pmatrix} 0 & 1 & 0 \\ -0.36 & -2 & 0 \\ 0.72 & 0 & 0 \end{pmatrix} \tag{32.24}$$

$$\lambda[\mathbf{F}_{eq}] = -1.8 \ ; \ \ -0.2 \ ; \ \ 0 . \tag{32.25}$$

$$\text{Right eigenvectors of } \mathbf{F}_{eq} : \quad \mathbf{f}_i = \begin{pmatrix} 10 \\ -2 \\ -36 \end{pmatrix}, \begin{pmatrix} 10 \\ -18 \\ -4 \end{pmatrix}, \begin{pmatrix} 0 \\ 0 \\ 0 \end{pmatrix} \tag{32.26}$$

$$\mathbf{T}_f \triangleq (\mathbf{f}_1 \vdots \mathbf{f}_2) = \begin{pmatrix} 10 & 10 \\ -2 & -18 \\ -36 & -4 \end{pmatrix} . \tag{32.27}$$

All columns $(\mathbf{F}_{eq})_{.i} \in \mathcal{N}[\mathbf{C}_s]$ or $\mathbf{P}_r\mathbf{F}_{eq} = \mathbf{0}$ can be checked. Furthermore, all eigenvectors $\mathbf{f}_i \in \mathcal{N}[\mathbf{C}_s]$ or $\mathbf{P}_r\mathbf{T}_f = \mathbf{0}$ can also be checked. A matrix $\mathbf{C}_s = (4\ \ 2\ \ 1)$ is required, in order to obtain the given $\mathbf{P}_r$

$$\mathbf{P}_r = \mathbf{B}(\mathbf{C}_s\mathbf{B})^{-1}\mathbf{C}_s = \begin{pmatrix} 0 \\ 0 \\ 1 \end{pmatrix} [(4\ \ 2\ \ 1) \begin{pmatrix} 0 \\ 0 \\ 1 \end{pmatrix}]^{-1}(4\ \ 2\ \ 1) = \begin{pmatrix} 0 & 0 & 0 \\ 0 & 0 & 0 \\ 4 & 2 & 1 \end{pmatrix} . \tag{32.28}$$

Anticipating the assertion in arriving at Eq.(32.52), the control input is

$$u = \mathbf{d}^T\mathbf{x} - k \text{ sign } s_w \tag{32.29}$$

and from Eq.(32.52) with $\mathbf{C}_s = \mathbf{c}_s^T = (c_1\ \ c_2\ \ c_3)^T = (4\ \ 2\ \ 1)^T, \ \ b = 1$

$$\underbrace{s_w[(-0.72 - a_1 + d_1)x_1 - (a_2 - d_2)x_2 - (a_3 - d_3)x_3]}_{\rightarrow\, <0} - k|s_w| < 0 . \tag{32.30}$$

During sliding mode:

$$\left.\begin{aligned} \dot{x}_1 &= x_2 \\ \dot{x}_2 &= -0.36\ x_1 - 2\ x_2 \end{aligned}\right\} \text{ independent of } x_3 \quad \text{where} \quad \dot{x}_3 = 0.72\ x_1 \tag{32.31}$$

$$u_{eq} = -(\mathbf{C}_s\mathbf{B})^{-1}\mathbf{C}_s\mathbf{A}\mathbf{x} = (0.72 + a_1 \vdots a_2 \vdots a_3)\mathbf{x} . \tag{32.32}$$

End of Example

32.4 Calculating $\mathbf{T}_f$ from λ_i During Sliding Mode

Let certain eigenvalues λ_i be assigned to the system during sliding. When the eigenvalues are distinct

$$(\mathbf{A}+\mathbf{B}\mathbf{K}_{eq})\mathbf{T}_f = \mathbf{T}_f \text{ diag } \lambda_i \quad \text{where} \quad \lambda_i = \lambda_i[\mathbf{A}+\mathbf{B}\mathbf{K}_{eq}] \tag{32.33}$$

and $\mathbf{T}_f$ is the $(n, n-m)$-matrix of the non-zero right eigenvectors of $\mathbf{F}_{eq} = \mathbf{A}+\mathbf{B}\mathbf{K}_{eq}$. If $\mathbf{K}_{eq}$ has full rank by Eq.(32.33), it results

$$\mathcal{R}[\mathbf{B}] = \mathcal{R}[\mathbf{B}\mathbf{K}_{eq}\mathbf{T}_f] = \mathcal{R}[-\mathbf{A}\mathbf{T}_f + \mathbf{T}_f \text{ diag } \lambda_i] = \mathcal{R}[\mathbf{B}\mathbf{L}_m] \tag{32.34}$$

where $\mathbf{L}_m$ is an arbitrary non-singular (m,m)-matrix. Assume $\mathbf{A}$ and λ_i given, then Eq.(32.34) provides the result

$$-\mathbf{A}\mathbf{T}_f + \mathbf{T}_f \text{ diag } \lambda_i = \mathbf{B}\mathbf{L}_m , \tag{32.35}$$

i.e., $n(n-m)$ linear algebraic equations in the entries of $\mathbf{T}_f$, parametrized on $\mathbf{L}_m$. In closed form,

$$\text{col } \mathbf{T}_f = [-\mathbf{I}_{n-m} \otimes \mathbf{A} + (\text{diag } \lambda_i) \otimes \mathbf{I}_n]^{-1} \text{ col } \mathbf{B}\mathbf{L}_m . \tag{32.36}$$

32.5 Properties of A, B, $\mathbf{T}_f$, $\mathbf{P}_r$ and $\mathbf{F}_{eq}$

32.5.1 Range Space of $\mathbf{T}_f$ and B do not intersect

Since $\mathbf{B}_{.i} \notin \mathcal{N}[\mathbf{C}_s]$ and $(\mathbf{T}_f)_{.i} \in \mathcal{N}[\mathbf{C}_s]$, it results (*El-Ghezawi, O.M.E., et al. 1983*)

$$\mathcal{R}[\mathbf{T}_f] \cap \mathcal{R}[\mathbf{B}] = \{\emptyset\} . \tag{32.37}$$

32.5.2 Projectors and Left-Inverses $\mathbf{B}^L$ and $\mathbf{T}_f^L$

Since $\mathcal{R}[\mathbf{B}] = \mathcal{R}[\mathbf{P}_r]$ and $(\mathbf{T}_f)_{.i} \in \mathcal{N}[\mathbf{C}_s]$

$$\mathbf{P}_r \times (\mathbf{B} \vdots \mathbf{T}_f) = \left(\mathbf{B}(\mathbf{C}_s\mathbf{B})^{-1}\mathbf{C}_s\mathbf{B} \vdots \mathbf{P}_r\mathbf{T}_f\right) = (\mathbf{B} \vdots \mathbf{0}) \tag{32.38}$$

$$\mathbf{P}_r = (\mathbf{B} \vdots \mathbf{0})(\mathbf{B} \vdots \mathbf{T}_f)^{-1} . \tag{32.39}$$

The assertion

$$(\mathbf{B} \vdots \mathbf{T}_f)^{-1} = \begin{pmatrix} \mathbf{B}^L \\ \mathbf{T}_f^L \end{pmatrix} \tag{32.40}$$

with regard to

$$(\mathbf{B} \vdots \mathbf{T}_f)^{-1}(\mathbf{B} \vdots \mathbf{T}_f) = \begin{pmatrix} \mathbf{B}^L \\ \mathbf{T}_f^L \end{pmatrix}(\mathbf{B} \vdots \mathbf{T}_f) = \begin{pmatrix} \mathbf{B}^L\mathbf{B} & \mathbf{B}^L\mathbf{T}_f \\ \mathbf{T}_f^L\mathbf{B} & \mathbf{T}_f^L\mathbf{T}_f \end{pmatrix} = \begin{pmatrix} \mathbf{I}_m & \mathbf{0} \\ \mathbf{0} & \mathbf{I}_{n-m} \end{pmatrix} \tag{32.41}$$

is only true if $\mathbf{B}^L\mathbf{T}_f = \mathbf{0}$ and $\mathbf{T}_f^L\mathbf{B} = \mathbf{0}$.

32.5.3 General Solution of $\mathbf{P}_r$

From Eq.(32.39) the relationship between $\mathbf{P}_r$ and $\mathbf{B}^L$ is

$$\mathbf{P}_r = (\mathbf{B} \,\vdots\, \mathbf{0}) \begin{pmatrix} \mathbf{B}^L \\ \mathbf{T}_f^L \end{pmatrix} = \mathbf{B}\mathbf{B}^L \, . \tag{32.42}$$

From $(\mathbf{T}_f)_{.i} \in \mathcal{N}[\mathbf{C}_s]$ the relation between $\mathbf{P}_r$ and $\mathbf{T}_f^L$ is

$$\mathbf{P}_r\mathbf{T}_f = \mathbf{0} \;\rightsquigarrow\; \mathbf{P}_r = \mathbf{H}(\mathbf{I}_n - \mathbf{T}_f\mathbf{T}_f^L) \tag{32.43}$$

where $\mathbf{H}$ is any nonsingular (n,n)-matrix. Proof by substitution.

32.5.4 Relation between $\mathbf{P}_r$ and $\mathbf{F}_{eq}$

From $\mathbf{F}_{eq} = (\mathbf{I}_n - \mathbf{P}_r)\mathbf{A}$ in Eq.(32.6) one has

$$(\mathbf{I}_n - \mathbf{P}_r)\mathbf{F}_{eq} = (\mathbf{I}_n - \mathbf{P}_r)^2\mathbf{A} = \mathbf{F}_{eq} \, , \tag{32.44}$$

$$\text{i.e.,} \quad (\mathbf{I}_n - \mathbf{P}_r)\mathbf{F}_{eq} = \mathbf{F}_{eq} \quad \text{or} \quad (\mathbf{F}_{eq})_{.i} \in \mathcal{R}[\mathbf{I}_n - \mathbf{P}_r] \tag{32.45}$$

$$\mathbf{P}_r\mathbf{F}_{eq} = \mathbf{0} \quad \text{or} \quad (\mathbf{F}_{eq})_{.i} \in \mathcal{N}[\mathbf{P}_r] = \mathcal{N}[\mathbf{C}_s] \, . \tag{32.46}$$

Premultiplying Eq.(32.6) with $\mathbf{B}^L$ yields

$$\mathbf{B}^L\mathbf{F}_{eq} = \mathbf{B}^L(\mathbf{I}_n - \mathbf{P}_r)\mathbf{A} = \mathbf{B}^L\mathbf{A} - \mathbf{B}^L\mathbf{P}_r\mathbf{A} = \mathbf{B}^L\mathbf{A} - \mathbf{B}^L(\mathbf{B}\mathbf{B}^L)\mathbf{A} = \mathbf{0} \, . \tag{32.47}$$

32.6 Design of $\mathbf{C}_s$ Using the Annihilator of $\mathbf{T}_f$

Since $\mathbf{P}_r\mathbf{T}_f = \mathbf{0}$ one has $\mathbf{B}(\mathbf{C}_s\mathbf{B})^{-1}\mathbf{C}_s\mathbf{T}_f = \mathbf{0}$. If the matrix $\mathbf{C}_s\mathbf{B} = \mathbf{S}$ is a nonsingular (m,m)-matrix the aforementioned equation must be satisfied by $\mathbf{C}_s\mathbf{T}_f = \mathbf{0}$. The matrix $\mathbf{C}_s = \mathbf{L}_o\mathbf{T}_f^{\perp}$ is a solution where $\mathbf{T}_f^{\perp}$ is the annihilator of $\mathbf{T}_f$ and $\mathbf{L}_o$ is any nonsingular matrix. Postmultiplying with $\mathbf{B}$ yields

$$\mathbf{C}_s\mathbf{B} = \mathbf{L}_o\mathbf{T}_f^{\perp}\mathbf{B} \;\rightsquigarrow\; \mathbf{S} = \mathbf{L}_o\mathbf{T}_f^{\perp}\mathbf{B} \;\rightsquigarrow\; \mathbf{L}_o = \mathbf{S}(\mathbf{T}_f^{\perp}\mathbf{B})^{-1} \;\rightsquigarrow\; \mathbf{C}_s = \mathbf{S}(\mathbf{T}_f^{\perp}\mathbf{B})^{-1}\mathbf{T}_f^{\perp} \, . \tag{32.48}$$

32.7 Design of $\mathbf{C}_s$ Using Left-Inverse of B

For a given $\mathbf{T}_f$ the matrix has to satisfy both $\mathbf{C}_s\mathbf{T}_f = \mathbf{0}$ and $\mathbf{C}_s\mathbf{B} = \mathbf{S}$ where $\mathbf{S}$ is an arbitrary nonsingular (m,m)-matrix. The last equation can be rewritten to $\mathbf{C}_s = \mathbf{S}\mathbf{B}^L$.

32.8 Control Variable During Sliding Mode

The hyperplane motion does not correspond to a natural motion as given by $\mathbf{A}$. Only by variable structure control $\mathbf{u}$, the system can be forced in a motion given by the hyperplane s_w in the subspace S . The hyperplane $s_w \triangleq \mathbf{C}_s\mathbf{x} = \mathbf{0}$ is defined by Eq.(32.2). In order to keep the sliding motion within the subspace S it is necessary to provide negative derivative $\dot{s}_w$ when s_w is positive and positive derivative $\dot{s}_w$ when s_w is negative. Hence,

$$s_w\dot{s}_w < 0 \quad \text{or} \quad \frac{1}{2}\frac{ds_w^2}{dt} < 0 \quad \text{or} \quad s_w\mathbf{C}_s\dot{\mathbf{x}} < 0 \tag{32.49}$$

$$s_w(\mathbf{C}_s\mathbf{A}\mathbf{x} + \mathbf{C}_s\mathbf{B}\mathbf{u}) < 0 \ . \tag{32.50}$$

If, e.g., $m = 1$, $\mathbf{C}_s = \mathbf{c}_s^T = (c_1 \ \ c_2 \ \ c_3)^T$, $\mathbf{B} = \mathbf{b} = (0 \ \ 0 \ \ b)^T$, then $s_w(\mathbf{c}_s^T\mathbf{A}\mathbf{x} + c_3bu) < 0$. By assertion of $u = \mathbf{d}^T\mathbf{x} - k \text{ sign } s_w$, it results

$$s_w\mathbf{c}_s^T\mathbf{A}\mathbf{x} + s_wc_3b[\mathbf{d}^T\mathbf{x} - k \text{ sign } s_w] < 0 \tag{32.51}$$

$$s_w(\mathbf{c}_s^T\mathbf{A} + c_3b\mathbf{d}^T)\mathbf{x} - c_3bk|s_w| < 0 \ . \tag{32.52}$$

If $c_3b > 0$ and $k > 0$, suitable conditions (switching operations) in the entries of $\mathbf{d}$ must be chosen in order to satisfy

$$s_w(\mathbf{c}_s^T\mathbf{A} + c_3b\mathbf{d}^T)\mathbf{x} < 0 \tag{32.53}$$

(*Slotine, J.J., and Sastry, S.S., 1983; Asada, H., and Slotine, J.J., 1986*). Then, the control input u can be realized by the assertion above.

32.9 Canonical Form and Sliding Mode

Applying a transformation $\mathbf{T}$ to the plant equation, it is intended to decompose the plant into two parts: The first part being unexcited by $\mathbf{u}$, the second one being influenced by the control input $\mathbf{u}$. For the sake of computational simplicity when taking the inverse, an orthogonal transformation $\mathbf{T}$ to the plant equation is employed

$$\mathbf{T}^T = \mathbf{T}^{-1}, \qquad \mathbf{T} \in \mathcal{R}^{n \times n}, \qquad \mathbf{v} = \mathbf{T}\mathbf{x} \ , \qquad \mathbf{x} = \mathbf{T}^T\mathbf{v} \ . \tag{32.54}$$

The matrix $\mathbf{B}$ is assumed of full rank and $\mathbf{T}$ is selected to obtain

$$\mathbf{T}\mathbf{B} \triangleq \begin{pmatrix} \mathbf{0} \\ \mathbf{B}_2 \end{pmatrix} \qquad \mathbf{B}_2 \in \mathcal{R}^{m \times m} \ , \tag{32.55}$$

see Fig. 32.2. The zero component in the transformed input matrix provides a transformed system being partially independent of $\mathbf{u}$. From Eqs.(32.1), (32.2) and (32.54)

$$\dot{\mathbf{v}} = \mathbf{T}\mathbf{A}\mathbf{T}^T\mathbf{v} + \mathbf{T}\mathbf{B}\mathbf{u} \qquad \mathbf{C}_s\mathbf{T}^T\mathbf{v} \equiv \mathbf{0} \ . \tag{32.56}$$

Defining the following partitioned vectors and matrices

$$\mathbf{v} \triangleq \begin{pmatrix} \mathbf{v}_1 \\ \mathbf{v}_2 \end{pmatrix} \qquad \mathbf{v}_1 \in \mathcal{R}^{n-m}, \quad \mathbf{v}_2 \in \mathcal{R}^m \tag{32.57}$$

$$\mathbf{T}\mathbf{A}\mathbf{T}^T \triangleq \begin{pmatrix} \mathbf{A}_{11} & \mathbf{A}_{12} \\ \mathbf{A}_{21} & \mathbf{A}_{22} \end{pmatrix} \qquad \mathbf{A}_{11} \in \mathcal{R}^{(n-m)\times(n-m)} \ , \ \mathbf{A}_{22} \in \mathcal{R}^{m \times m} \text{ etc.} \tag{32.58}$$

$$\mathbf{C}_s\mathbf{T}^T = (\mathbf{C}_{s1} \ \vdots \ \mathbf{C}_{s2}) \quad \mathbf{C}_{s1} \in \mathcal{R}^{m\times(n-m)} \ , \ \mathbf{C}_{s2} \in \mathcal{R}^{m \times m} \ , \ \mathbf{C}_s \in \mathcal{R}^{m \times n} \ , \tag{32.59}$$

the sliding mode condition is

$$\mathbf{C}_s\mathbf{T}^T\mathbf{v} \equiv \mathbf{0} \ \rightsquigarrow \ (\mathbf{C}_{s1} \ \vdots \ \mathbf{C}_{s2})\begin{pmatrix} \mathbf{v}_1 \\ \mathbf{v}_2 \end{pmatrix} \equiv \mathbf{0} \ \rightsquigarrow \ \mathbf{v}_2 = -\mathbf{C}_{s2}^{-1}\mathbf{C}_{s1}\mathbf{v}_1 \triangleq \mathbf{K}_v\mathbf{v}_1 \ . \tag{32.60}$$

The transformed system equations turn out as

$$\dot{\mathbf{v}}_1 = \mathbf{A}_{11}\mathbf{v}_1 + \mathbf{A}_{12}\mathbf{v}_2 \quad \text{(transformed plant independent of } \mathbf{u}) \tag{32.61}$$

$$\dot{\mathbf{v}}_2 = \mathbf{A}_{21}\mathbf{v}_1 + \mathbf{A}_{22}\mathbf{v}_2 + \mathbf{B}_2\mathbf{u} \tag{32.62}$$

$$\mathbf{v}_2 = \mathbf{K}_v\mathbf{v}_1 \quad \text{(transformed equivalent controller).} \tag{32.63}$$

The sliding condition $\mathbf{v}_2 = \mathbf{K}_v\mathbf{v}_1$ acts as an equivalent state feedback controller equation using $\mathbf{K}_v \in \mathcal{R}^{m\times(n-m)}$. Combining Eqs. (32.61) and (32.63), the canonical form during sliding mode is

$$\dot{\mathbf{v}}_1 = \mathbf{A}_{11}\mathbf{v}_1 + \mathbf{A}_{12}(\mathbf{K}_v\mathbf{v}_1) = (\mathbf{A}_{11} + \mathbf{A}_{12}\mathbf{K}_v)\mathbf{v}_1 \,. \tag{32.64}$$

The controller design can be achieved by minimizing an index of performance or by pole assignment, outlined in the following sections.

32.10 Hyperplane Riccati Controller

Let a quadratic performance criterion be minimized with symmetric $\mathbf{Q}$

$$I = \int_{t_s}^{\infty} \mathbf{x}^T(t)\mathbf{Q}\mathbf{x}(t)dt \rightarrow \min, \qquad \mathbf{TQT}^T \triangleq \begin{pmatrix} \mathbf{Q}_{11} & \mathbf{Q}_{12} \\ \mathbf{Q}_{21} & \mathbf{Q}_{22} \end{pmatrix} \tag{32.65}$$

where $\mathbf{TQT}^T$ is symmetric, too, and $\mathbf{Q}_{11} = \mathbf{Q}_{11}^T$, $\mathbf{Q}_{22} = \mathbf{Q}_{22}^T$, $\mathbf{Q}_{12}^T = \mathbf{Q}_{21}$, $\mathbf{Q}_{21}^T = \mathbf{Q}_{12}^T$. Supposing $\mathbf{Q} > 0$, then $\mathbf{Q}_{22} > 0$ and $\mathbf{Q}_{22}^{-1}$ exists. Defining

$$\mathbf{u}' \triangleq \mathbf{v}_2 + \mathbf{Q}_{22}^{-1}\mathbf{Q}_{21}\mathbf{v}_1 \tag{32.66}$$
$$\mathbf{Q}' \triangleq \mathbf{Q}_{11} - \mathbf{Q}_{12}\mathbf{Q}_{22}^{-1}\mathbf{Q}_{21} \tag{32.67}$$
$$\mathbf{A}' \triangleq \mathbf{A}_{11} - \mathbf{A}_{12}\mathbf{Q}_{22}^{-1}\mathbf{Q}_{21} \,, \tag{32.68}$$

it can be easily checked that an identical formulation of I is given by

$$I = \int_{t_s}^{\infty} \mathbf{v}_1^T\mathbf{Q}'\mathbf{v}_1 + \mathbf{u}'^T\mathbf{Q}_{22}\mathbf{u}'dt \,. \tag{32.69}$$

Therefore, the optimal $\mathbf{u}'$ is

$$\mathbf{u}'^{\star} = -\mathbf{Q}_{22}^{-1}\mathbf{A}_{12}^T\mathbf{P}\ \mathbf{v}_1 \,. \tag{32.70}$$

The matrix $\mathbf{P}$ is calculated from the Riccati equation

$$\mathbf{A}'^T\mathbf{P} + \mathbf{P}\mathbf{A}' - \mathbf{P}\mathbf{A}_{12}\mathbf{Q}_{22}^{-1}\mathbf{A}_{12}^T + \mathbf{Q}' = \mathbf{0} \,. \tag{32.71}$$

Substituting $\mathbf{u}'^{\star}$ from above into Eq.(32.66), the optimal gain matrix $\mathbf{K}_v^{\star}$ is obtained

$$\mathbf{v}_2 = -\mathbf{Q}_{22}^{-1}(\mathbf{A}_{12}^T\mathbf{P} + \mathbf{Q}_{21})\mathbf{v}_1 = \mathbf{K}_v^{\star}\mathbf{v}_1 \quad \leadsto \quad \mathbf{K}_v^{\star} = -\mathbf{Q}_{22}^{-1}(\mathbf{A}_{12}^T\mathbf{P} + \mathbf{Q}_{21}) \,. \tag{32.72}$$

32.11 Hyperplane Eigenvalue Assignment

During sliding mode the system is governed by $\dot{\mathbf{x}}(t) = (\mathbf{A} + \mathbf{BK}_{eq})\mathbf{x}(t) = \mathbf{F}_{eq}\mathbf{x}(t)$. Then, the matrix $\mathbf{K}_{eq}$ is designed by choosing the (m,n)-matrix $\mathbf{C}_s = (\mathbf{C}_{s1} \vdots \mathbf{C}_{s2})\mathbf{T}$ according to $\mathbf{K}_{eq} = -(\mathbf{C}_s\mathbf{B})^{-1}\mathbf{C}_s\mathbf{A}$, see Eq.(32.5). For simpler inversion the matrix product $\mathbf{C}_s\mathbf{B}$ is preferably chosen diagonal or diagonally dominant. Then, regarding Eqs.(32.55) and (32.59), it results

$$(\mathbf{C}_s\mathbf{B}) = \mathbf{C}_s\mathbf{T}^T\mathbf{T}\mathbf{B} = (\mathbf{C}_{s1} \vdots \mathbf{C}_{s2})\begin{pmatrix} \mathbf{0} \\ \mathbf{B}_2 \end{pmatrix} = \mathbf{C}_{s2}\mathbf{B}_2 \quad \leadsto \quad \mathbf{C}_{s2} = (\mathbf{C}_s\mathbf{B})\mathbf{B}_2^{-1} \,. \tag{32.73}$$

From Eqs.(32.60) and (32.59), $\mathbf{K}_v = \mathbf{C}_{s2}^{-1}\mathbf{C}_{s1} \in \mathcal{R}^{m\times(n-m)}$ and $\mathbf{C}_s = (\mathbf{C}_{s1} \vdots \mathbf{C}_{s2})\mathbf{T}$ and

$$\mathbf{C}_s = [\mathbf{C}_{s2}\mathbf{K}_v \vdots (\mathbf{C}_s\mathbf{B})\mathbf{B}_2^{-1}]\mathbf{T} = (\mathbf{C}_s\mathbf{B})\mathbf{B}_2^{-1}[\mathbf{K}_v \vdots \mathbf{I}_m]\mathbf{T} \,. \tag{32.74}$$

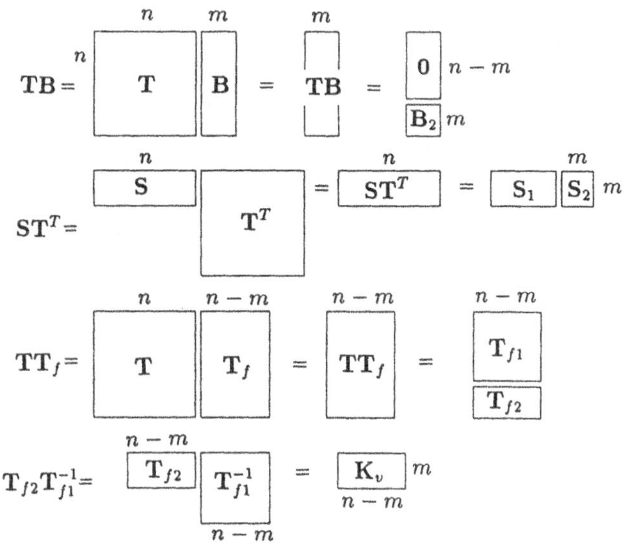

Figure 32.2: Transformation matrices

Designing $\mathbf{C}_s$ by $\mathbf{K}_v$, proceed as follows: Sliding mode operation requires $\mathbf{C}_s\mathbf{F}_{eq} = \mathbf{0}$ from Eq.(32.4). Using the right eigenvectors $\mathbf{f}_i$ of $\mathbf{F}_{eq}$ as given by $\mathbf{F}_{eq}\mathbf{f}_i = \lambda_i[\mathbf{F}_{eq}]\mathbf{f}_i$, and combination with $\mathbf{C}_s\mathbf{F}_{eq} = \mathbf{0}$ yields

$$\mathbf{C}_s\mathbf{F}_{eq} = \mathbf{0} \quad \rightsquigarrow \quad \mathbf{C}_s\mathbf{F}_{eq}\mathbf{f}_i = \mathbf{C}_s\lambda_i\mathbf{f}_i = \lambda_i\mathbf{C}_s\mathbf{f}_i = \mathbf{0}\ . \tag{32.75}$$

This implies that either $\lambda_i = 0$ or $\mathbf{C}_s\mathbf{f}_i = \mathbf{0}$, i.e., $\mathbf{f}_i$ is an element of the null space of $\mathbf{C}_s$. In order to achieve sliding mode as given in Eq. (32.4), m eigenvalues λ_i must be zero and $n-m$ eigenvalues non-zero. Hence, $n-m$ eigenvectors span the null space of $\mathbf{C}_s$. These eigenvectors are referred to as $\mathbf{f}_i\ \forall i = 1\ldots(n-m)$. Then,

$$\mathbf{T}_f \triangleq (\mathbf{f}_1 \ \vdots\ \mathbf{f}_2\ \ldots\ \mathbf{f}_{n-m}) \tag{32.76}$$

and $\mathbf{C}_s\mathbf{f}_i = \mathbf{0}$ can be written $\mathbf{C}_s\mathbf{T}_f = \mathbf{0}$. Partitioning the matrix product $\mathbf{T}\mathbf{T}_f$, see Fig. 32.2, finally yields

$$\mathbf{C}_s\mathbf{T}_f = \mathbf{0} \quad \rightsquigarrow \quad \mathbf{C}_s\mathbf{T}^T\mathbf{T}\mathbf{T}_f = (\mathbf{C}_{s1} \ \vdots\ \mathbf{C}_{s2})\begin{pmatrix}\mathbf{T}_{f1}\\ \mathbf{T}_{f2}\end{pmatrix} = (\mathbf{C}_{s2}\mathbf{K}_v \ \vdots\ \mathbf{C}_{s2})\begin{pmatrix}\mathbf{T}_{f1}\\ \mathbf{T}_{f2}\end{pmatrix} = \mathbf{0} \tag{32.77}$$

$$\mathbf{C}_{s2}(\mathbf{K}_v \ \vdots\ \mathbf{I}_m)\begin{pmatrix}\mathbf{T}_{f1}\\ \mathbf{T}_{f2}\end{pmatrix} = 0 \quad \rightsquigarrow \quad \mathbf{K}_v = -\mathbf{T}_{f2}\mathbf{T}_{f1}^{-1}\ . \tag{32.78}$$

The sliding mode system is governed by the matrix

$$\mathbf{A} + \mathbf{B}\mathbf{K}_{eq} = \mathbf{A} - \mathbf{B}(\mathbf{C}_s\mathbf{B})^{-1}\mathbf{C}_s\mathbf{A} = [\mathbf{I}_n - \mathbf{B}(\mathbf{C}_s\mathbf{B})^{-1}\mathbf{C}_s]\mathbf{A}\ . \tag{32.79}$$

Eigenvalues of this matrix are zero, namely of number m which results from

$$\det\ \{s\mathbf{I}_n - [\mathbf{I}_n - \mathbf{B}(\mathbf{C}_s\mathbf{B})^{-1}\mathbf{C}_s]\mathbf{A}\}\ \big|_{s=0,\ m\ times} = 0 \tag{32.80}$$

$$\det[\mathbf{I}_n - \mathbf{B}(\mathbf{C}_s\mathbf{B})^{-1}\mathbf{C}_s]\mathbf{A} = 0 \quad \rightsquigarrow \quad \det[\mathbf{I}_n - \mathbf{B}(\mathbf{C}_s\mathbf{B})^{-1}\mathbf{C}_s]\det\mathbf{A} = 0 \; . \quad (32.81)$$

The matrix $\mathbf{I}_n - \mathbf{B}(\mathbf{C}_s\mathbf{B})^{-1}\mathbf{C}_s$ has eigenvalues zero m times which can be seen from Eq.(A.64): $\text{rank}[(\mathbf{I}_n - \mathbf{P}_r)\mathbf{A}] \leq n - m$. With increasing number m of inputs or increasing range space of $\mathbf{B}$, the design is facilitated by choosing the eigenvectors associated with the $n - m$ eigenvalues.

32.12 Unit Vector Control

In the previous sections, the conditions for the existence of a sliding mode are outlined. This section presents a condition to reach the sliding mode. In order to reach the hyperplane as quickly as possible, remember the time-optimal performance assumption and on-off control. The on-off control using signum function can be generalized to the unit vector control. Realizing quick approach to switching manifold ($t < t_s$) , this part of system behaviour becomes a fast mode, in comparison with the slow sliding mode. The control structure can be given by combined linear and unit vector control

$$\mathbf{u}(\mathbf{x}) = \mathbf{L}\mathbf{x} + \rho\frac{\mathbf{N}\mathbf{x}}{\|\mathbf{M}\mathbf{x}\|_F} = \mathbf{u}_L + \mathbf{u}_N \qquad \rho > 0 \; . \quad (32.82)$$

The null spaces of $\mathbf{N}, \mathbf{M}$ and $\mathbf{C}_s$ coincide. Recall the vector components $\mathbf{v}_2$ and $\mathbf{v}_1$ being linearly dependent, Eq.(32.63), $\mathbf{v}_2 = \mathbf{K}_v\mathbf{v}_1$, thus $-\mathbf{K}_v\mathbf{v}_1 + \mathbf{v}_2 = \mathbf{0}$. Defining a variable $\mathbf{z}_2 \triangleq -\mathbf{K}_v\mathbf{v}_1 + \mathbf{v}_2$, the controlling variable $\mathbf{u}$ has to guarantee $\mathbf{z}_2 = \mathbf{0}$ (and $\dot{\mathbf{z}}_2 = \mathbf{0}$). Hence, an additional transformation is defined

$$\left.\begin{array}{l}\mathbf{z}_1 = \mathbf{v}_1\\ \mathbf{z}_2 = -\mathbf{K}_v\mathbf{v}_1 + \mathbf{v}_2\end{array}\right\} \qquad \mathbf{z} \triangleq \begin{pmatrix}\mathbf{z}_1\\ \mathbf{z}_2\end{pmatrix} = \mathbf{T}_2\mathbf{v} = \mathbf{T}_2\begin{pmatrix}\mathbf{v}_1\\ \mathbf{v}_2\end{pmatrix} \qquad \mathbf{T}_2 \triangleq \begin{pmatrix}\mathbf{I}_{n-m} & \mathbf{0}\\ -\mathbf{K}_v & \mathbf{I}_m\end{pmatrix} . \quad (32.83)$$

The $\mathbf{v}$-system, Eqs.(32.61) to (32.63), is transformed into the $\mathbf{z}$-system

$$\dot{\mathbf{z}} = \begin{pmatrix}\mathbf{A}_{11} + \mathbf{A}_{12}\mathbf{K}_v & \mathbf{A}_{12}\\ -\mathbf{K}_v\mathbf{A}_{11} - \mathbf{K}_v\mathbf{A}_{12}\mathbf{K}_v + \mathbf{A}_{22}\mathbf{K}_v + \mathbf{A}_{21} \; \vdots & -\mathbf{K}_v\mathbf{A}_{12} + \mathbf{A}_{22}\end{pmatrix}\mathbf{z} + \begin{pmatrix}\mathbf{0}\\ \mathbf{B}_2\end{pmatrix}\mathbf{u} \; . \quad (32.84)$$

Substituting $\dot{\mathbf{z}}_2 = \mathbf{A}_z\mathbf{z}_2$ with a stable (m, m)-matrix $\mathbf{A}_z$, the controlling variable can be written

$$\mathbf{u} = \mathbf{B}_2^{-1}(\mathbf{K}_v\mathbf{A}_{11} + \mathbf{K}_v\mathbf{A}_{12}\mathbf{K}_v - \mathbf{A}_{22}\mathbf{K}_v - \mathbf{A}_{21} \; \vdots \; \mathbf{K}_v\mathbf{A}_{12} - \mathbf{A}_{22} + \mathbf{A}_z)\mathbf{z}_2 \; . \quad (32.85)$$

Using $\mathbf{z}_2 = \mathbf{T}_2\mathbf{v}$ and $\mathbf{v} = \mathbf{T}\mathbf{x}$, the linear component $\mathbf{L}$ in the controller matrix is

$$\mathbf{L} = \mathbf{B}_2^{-1}(\mathbf{K}_v\mathbf{A}_{11} + \mathbf{K}_v\mathbf{A}_{12}\mathbf{K}_v - \mathbf{A}_{22}\mathbf{K}_v - \mathbf{A}_{21} \; \vdots \; \mathbf{K}_v\mathbf{A}_{12} - \mathbf{A}_{22} + \mathbf{A}_z)\mathbf{T}_2\mathbf{T} \; . \quad (32.86)$$

Referring to Eq.(13.16) ($\mathbf{A} := \mathbf{A}_z$, $\mathbf{P} := \mathbf{P}_2$, $\mathbf{Q} := \mathbf{I}_m$), the positive definite solution $\mathbf{P}_2$ of the Lyapunov equation

$$\mathbf{A}_z^T\mathbf{P}_2 + \mathbf{P}_2\mathbf{A}_z = -\mathbf{I}_m \quad (32.87)$$

complies with the condition $\mathbf{P}_2\mathbf{z}_2 = \mathbf{0}$ if and only if $\mathbf{z}_2 = \mathbf{0}$. Since the nonlinear part of the control has to be discontinuous when $\mathbf{z}_2$ passes zero, the nonlinear part of the controlling variable can thus be defined as

$$-\rho\frac{\mathbf{B}_2^{-1}\mathbf{P}_2\mathbf{z}_2}{\|\mathbf{P}_2\mathbf{z}_2\|_F} \qquad \text{when} \qquad \mathbf{z}_2 \neq \mathbf{0} \; . \quad (32.88)$$

Observing the relations

$$-\mathbf{B}_2^{-1}\mathbf{P}_2\mathbf{z}_2 = -\mathbf{B}_2^{-1}\mathbf{P}_2\mathbf{T}_2\mathbf{v}_2 = -\mathbf{B}_2^{-1}\mathbf{P}_2\mathbf{T}_2\mathbf{T}\mathbf{T}^T\mathbf{v}_2$$

$$= -\mathbf{B}_2^{-1}(\mathbf{0} \vdots \mathbf{P}_2)\mathbf{T}_2\mathbf{T}\mathbf{T}^T\begin{pmatrix}\mathbf{v}_1\\ \mathbf{v}_2\end{pmatrix} = -\mathbf{B}_2^{-1}(\mathbf{0} \vdots \mathbf{P}_2)\mathbf{T}_2\mathbf{T}\mathbf{x} \triangleq \mathbf{N}\mathbf{x}\,, \tag{32.89}$$

the matrices $\mathbf{N}$ and $\mathbf{M}$ of Eq.(32.82) are determined by

$$\mathbf{N} = -\mathbf{B}_2^{-1}(\mathbf{0} \ \vdots \ \mathbf{P}_2)\mathbf{T}_2\mathbf{T} \qquad \text{and} \qquad \mathbf{M} = (\mathbf{0} \ \vdots \ \mathbf{P}_2)\mathbf{T}_2\mathbf{T}\,. \tag{32.90}$$

The linear part $\mathbf{L}$ forces $\mathbf{z}_2$ to zero, asymptotically. The nonlinear part $\mathbf{N}$ switches whenever $\mathbf{z}_2$ is crossing zero, thus driving the system into the sliding mode quickly. Whenever $\mathbf{z}_2 = \mathbf{0}$, any assumption $\|\mathbf{u}_N\|_F \leq \rho$ is a solution.

In practice, the ideal sliding mode controller as derived above yields high frequency switching, known as chattering. This is an undesirable effect and must be smoothed, by introducing amplifiers with limited switching speed or by using elements with small hysteresis or by using a boundary layer neighboring the sliding surface (*Leung, T.P., et al. 1991*).

Reachability conditions for scalar discrete-time systems to reach the switching function from anywhere in the state space and a simplified design algorithm are presented by *Aly, G.M., and Ali, W.G., 1990.*

Asymptotic tracking facilities using a slightly modified control law were investigated by *Fu, L.C., and Liao, T.L., 1990.*

The application of the sliding mode control to the decentralized tracking problem is presented by *Matthews, G.P., and DeCarlo, R.A., 1988.*

32.13 Insensitivity of Hyperplane Design

In the preceding sections stating the sliding mode system behaviour, see Eq.(32.64), it has been outlined that the matrices $\mathbf{A}_{21}$ and $\mathbf{A}_{22}$ can be left out of consideration. The system is robust with respect to the dynamics incorporated by these matrices.

Insensitivity within the remaining matrices $\mathbf{A}_{11}, \mathbf{A}_{12}$ and $\mathbf{K}_v$ can be achieved by carrying out the following general design principle, using the eigenvalue differential sensitivity $\partial\lambda_i[\mathbf{F}_r]/\partial\mathbf{M}$. The matrix $\mathbf{M}$ equals either $\mathbf{A}_{11}$, $\mathbf{A}_{12}$ or $\mathbf{K}_v$. The matrix $\mathbf{F}_r$ is defined as $\mathbf{F}_r = \mathbf{A}_{11} + \mathbf{A}_{12}\mathbf{K}_v$. The sensitivity is inversely proportional to the cosine $\cos\alpha_i$ of the angle α_i between $\mathbf{f}_{ri}$ and $\mathbf{f}_{ri}^{\triangleleft}$ (*Wilkinson, J.H., 1965*)

$$\frac{\partial\lambda_i[\mathbf{F}_r]}{\partial\mathbf{M}} \sim \frac{1}{\cos\alpha_i} = \frac{\|\mathbf{f}_{ri}\|_F\ \|\mathbf{f}_{ri}^{\triangleleft}\|_F}{\mathbf{f}_{ri}^T\mathbf{f}_{ri}^{\triangleleft}}\,. \tag{32.91}$$

Thus, $\mathbf{f}_{ri}$ should be chosen to maximize $\cos\alpha_i$ for each α_i .

Another general principle observes minimizing the spectral condition number $\kappa_s[\mathbf{T}_{fr}]$

$$\kappa_s[\mathbf{T}_{fr}] = \frac{\sigma_{\max}\,[\mathbf{T}_{fr}]}{\sigma_{\min}\,[\mathbf{T}_{fr}]} \qquad \text{where } \mathbf{T}_{fr} = (\mathbf{f}_{r1}\ \ \mathbf{f}_{r2}\ \ldots\ \mathbf{f}_{r,n-m})\,. \tag{32.92}$$

Minimizing $\kappa_s[\mathbf{T}_{fr}]$ guarantees that the differential sensitivity of all the assigned eigenvalues is bounded (*Dorling, C.M., and Zinober, A.S.I., 1988*).

Chapter 33

Singular Perturbations. Unmodelled High-Frequency Dynamics

Models of dynamic systems are frequently lacking in sufficient knowledge of several high-frequency parameters, is spite of strength in identification. Hence, those parts of the system incorporating inaccessible high-frequency dynamics are referred to as parasitic high-frequency dynamics. Neglecting high-frequency dynamics seems comfortable on the one hand since the system is of reduced order but on the other hand, the resulting closed-loop control may have poor stability properties.

Classical control system design is based on algorithms compensating for the slow modes and retaining the fast modes, with known or unknown parameters. Design methods have to be carefully selected in order to tolerate fast and possibly unknown modes.

An alternative design proposal for robust control systems is given by cancelling the poles with the fast mode reverse to the classical principle of cancelling the slow mode plant poles (*Aida, K., and Kitamori, T., 1990*).

33.1 High-Frequency Parasitic Dynamics

The actual plant model is structured using state-space differential equations

$$\dot{\mathbf{x}}(t) = \mathbf{A}_{11}\mathbf{x}(t) + \mathbf{A}_{12}\mathbf{z}(t) + \mathbf{B}_1\mathbf{u}(t) \qquad \mathbf{x}(t_o) = \mathbf{x}_o \qquad \mathbf{x}(t) \in \mathcal{R}^n \tag{33.1}$$

$$\varepsilon\dot{\mathbf{z}}(t) = \mathbf{A}_{21}\mathbf{x}(t) + \mathbf{A}_{22}\mathbf{z}(t) + \mathbf{B}_2\mathbf{u}(t) \qquad \mathbf{z}(t_o) = \mathbf{z}_o \qquad \mathbf{z}(t) \in \mathcal{R}^{n_z} \tag{33.2}$$

$$\mathbf{y}(t) = \mathbf{C}_1\mathbf{x}(t) + \mathbf{C}_2\mathbf{z}(t) \qquad \mathbf{y}(t) \in \mathcal{R}^r, \quad \mathbf{u}(t) \in \mathcal{R}^m \,. \tag{33.3}$$

The vector $\mathbf{x}(t)$ denotes the dominant state vector, $\mathbf{z}(t)$ the vector of parasitic dynamics, $\mathbf{y}(t)$ the output, $\mathbf{u}(t)$ the input and ε a small positive scalar representing and comprising the effect of parasitic elements.

Fig. 33.1 shows the block diagram of the actual plant and anticipates the feedback controller as discussed later. The idea given by Eqs. (33.1) through (33.3) is named singular perturbation method analysis. Specifying $\mathbf{B}_2 = \mathbf{0}$ means weak control, and $\mathbf{C}_2 = \mathbf{0}$ weak observation of parasitics.

If the parasitic elements are neglected, then $\varepsilon = 0$ and a reduced-order or slow model is achieved where $\mathbf{x}(t)|_{\varepsilon=0} \triangleq \mathbf{x}_s(t)$, $\mathbf{u}(t)|_{\varepsilon=0} \triangleq \mathbf{u}_s(t)$, $\mathbf{z}(t)|_{\varepsilon=0} \triangleq \mathbf{z}_s(t)$

$$\dot{\mathbf{x}}_s = \mathbf{A}_{11}\mathbf{x}_s + \mathbf{A}_{12}\mathbf{z}_s + \mathbf{B}_1\mathbf{u}_s \tag{33.4}$$

$$\mathbf{0} = \mathbf{A}_{21}\mathbf{x}_s + \mathbf{A}_{22}\mathbf{z}_s + \mathbf{B}_2\mathbf{u}_s \quad \rightsquigarrow \quad \mathbf{z}_s = -\mathbf{A}_{22}^{-1}(\mathbf{A}_{21}\mathbf{x}_s + \mathbf{B}_2\mathbf{u}_s) \tag{33.5}$$

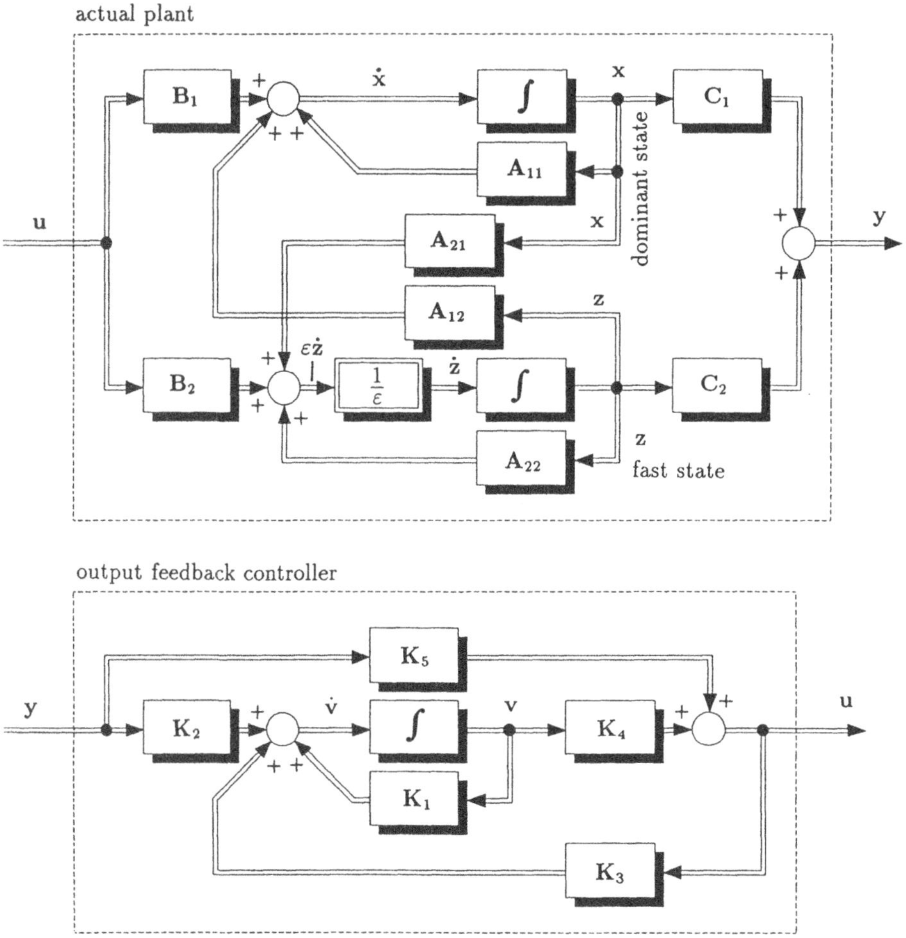

Figure 33.1: Actual plant and output feedback controller

$$\text{Hence,}\quad \dot{\mathbf{x}}_s = (\mathbf{A}_{11} - \mathbf{A}_{12}\mathbf{A}_{22}^{-1}\mathbf{A}_{21})\mathbf{x}_s + (\mathbf{B}_1 - \mathbf{A}_{12}\mathbf{A}_{22}^{-1}\mathbf{B}_2)\mathbf{u}_s \triangleq \mathbf{A}_o\mathbf{x}_s + \mathbf{B}_o\mathbf{u}_s \quad (33.6)$$

$$\mathbf{y}_s = (\mathbf{C}_1 - \mathbf{C}_2\mathbf{A}_{22}^{-1}\mathbf{A}_{21})\mathbf{x}_s - \mathbf{C}_2\mathbf{A}_{22}^{-1}\mathbf{B}_2\mathbf{u}_s \triangleq \mathbf{C}_o\mathbf{x}_s + \mathbf{D}_o\mathbf{u}_s\,. \quad (33.7)$$

The robustness investigation aims at a controller design using the plant model $(\mathbf{A}_o, \mathbf{B}_o, \mathbf{C}_o, \mathbf{D}_o)$ shown in Fig. 33.2 but ensuring that the resulting controller will be stable and will have enough stability margin even when operating with the actual plant system $(\mathbf{A}_{ij}, \mathbf{B}_i, \mathbf{C}_i\ \forall i,j = 1,2)$ (*O'Reilly, J., 1986; Kokotović, P.V., et al. 1986; Gajic, Z., et al. 1990*).

The actual plant is said to be of the standard form if $\mathbf{A}_{22}$ is nonsingular at $\varepsilon = 0$. If $\varepsilon \to 0$ then also $\mathbf{A}_{21}\mathbf{x}_s + \mathbf{A}_{22}\mathbf{z}_s + \mathbf{B}_2\mathbf{u}_s \to 0$ in order to obtain a finite $\mathbf{z}$-integrator input.

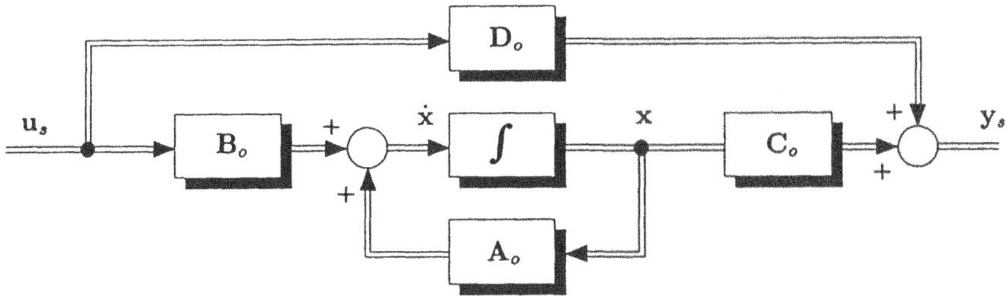

Figure 33.2: Plant model neglecting parasitic high-frequency dynamics

33.2 Two Time-Scale System Behaviour

33.2.1 Time-Scale t

As ε tends to zero, the fast variable $\mathbf{z}(t)$ becomes very rapid. The variable $\mathbf{z}(t)$ seems to be instantaneous from the t-time-scale point of view, i.e., in relation to the slow motion $\mathbf{x}(t)$. Of course, it can be assumed that $\dot{\mathbf{z}}(t)$ does not grow to infinity when control engineering problems are considered. Thus, $\mathbf{z}(t)$ is a motion that has passed instantaneously and has nearly arrived at a quasi-steady-state, at the beginning of the slow motion $\mathbf{x}(t)$.

33.2.2 Fast Time-Scale τ

The dynamic behaviour can be treated by introducing an additional time-scale τ

$$\tau = \frac{t - t_o}{\varepsilon}, \qquad \varepsilon d\tau = dt, \qquad \frac{d}{d\tau} = \varepsilon\frac{d}{dt} \, . \tag{33.8}$$

Recognize that τ serves as a fast time scale. As ε tends to small values, the fast time-scale variable τ grows rapidly, even within small intervals of t. Proceeding both with t and τ, a two-time-scale is given (*O'Reilly, J., 1980*).

The original system of dimension $n+n_z$ is decomposed into two lower-order subsystems. These subsystems are of dimension n and n_z, respectively, and operate in separate time-scales t and τ. Applying the fast time-scale, the original system meets the equations

$$\frac{d\mathbf{x}}{d\tau} = \varepsilon(\mathbf{A}_{11}\mathbf{x} + \mathbf{A}_{12}\mathbf{z} + \mathbf{B}_1\mathbf{u}) \quad \leadsto \quad \frac{d\mathbf{x}}{d\tau} \mid_{\varepsilon=0} = \mathbf{0} \tag{33.9}$$

$$\frac{d\mathbf{z}}{d\tau} = \mathbf{A}_{21}\mathbf{x} + \mathbf{A}_{22}\mathbf{z} + \mathbf{B}_2\mathbf{u} \, . \tag{33.10}$$

Note that in fast time-scale the variable $\mathbf{z}(\tau)$ is the solution of a normal differential equation whereas the dynamic behaviour of the slow state $\mathbf{x}$ Eq.(33.29) turns out almost constant because $d\mathbf{x}/d\tau \to \mathbf{0}$ if $\varepsilon \to 0$. The slow variable $\mathbf{x}$ plays the role of a constant parameter in Eq.(33.10). Treating the dominant variable $\mathbf{x}$ this quasi-steady-state is taken into account via initial conditions $\mathbf{x}(t_o)$. The singularly perturbed system in original (slow) time scale t, Eqs.(33.1) through (33.3), is converted into a regularly perturbed system in fast time-scale τ, Eqs.(33.9) and (33.10). Note that $\mathbf{x}_s(t)|_{t=t_o} = \mathbf{x}_s(t_o) = \mathbf{x}_o$. Invoking Eq.(33.5), $\mathbf{z}_s(t_o)$ and $\mathbf{x}_o$ are interrelated by

$$\mathbf{z}_s(t_o) = \mathbf{z}_{so} = \mathbf{A}_{22}^{-1}[\mathbf{A}_{21}\mathbf{x}_s(t_o) + \mathbf{B}_2\mathbf{u}_s(t_o)] = -\mathbf{A}_{22}^{-1}(\mathbf{A}_{21}\mathbf{x}_o + \mathbf{B}_2\mathbf{u}_{so}) \, . \tag{33.11}$$

Defining, first, a purely fast variable $\mathbf{z}_f(\tau) \triangleq \mathbf{z} - \mathbf{z}_s$ and a fast controlling variable $\mathbf{u}_f \triangleq \mathbf{u} - \mathbf{u}_s$ and, second, taking into consideration that $\mathbf{x}(t)$ and $\mathbf{z}_s(t)$ behaves constant during fast time motion, from Eq.(33.2) it results (*Kokotović, P.V., et al. 1976*)

$$\frac{d\mathbf{z}_f(\tau)}{d\tau} = \mathbf{A}_{21}\mathbf{x} + \mathbf{A}_{22}\mathbf{z}(\tau) + \mathbf{B}_2\mathbf{u} \qquad (33.12)$$

$$= \mathbf{A}_{21}\mathbf{x}_o + \mathbf{A}_{22}\mathbf{z}_f(\tau) + \mathbf{A}_{22}\mathbf{z}_{so} + \mathbf{B}_2\mathbf{u}_{so} + \mathbf{B}_2\mathbf{u}_f \qquad (33.13)$$

$$= \mathbf{A}_{21}\mathbf{x}_o + \mathbf{A}_{22}\mathbf{z}_f(\tau) - \mathbf{A}_{21}\mathbf{x}_o - \mathbf{B}_2\mathbf{u}_{so} + \mathbf{B}_2\mathbf{u}_{so} + \mathbf{B}_2\mathbf{u}_f \qquad (33.14)$$

$$\frac{d\mathbf{z}_f(\tau)}{d\tau} = \mathbf{A}_{22}\mathbf{z}_f(\tau) + \mathbf{B}_2\mathbf{u}_f \quad \text{and} \quad \mathbf{z}_f(\tau = 0) = \mathbf{z}(t_o) - \mathbf{z}_s(t_o) \,. \qquad (33.15)$$

Multi-time-scale concepts for singular perturbation modelling of large-scale systems are presented by *Kando, H., et al. 1988*.

33.3 Dynamic Output Feedback Controller

Introducing an l-vector $\mathbf{v}(t)$, a dynamic feedback controller can be stated as

$$\dot{\mathbf{v}}(t) = \mathbf{K}_1\mathbf{v}(t) + \mathbf{K}_2\mathbf{y}(t) + \mathbf{K}_3\mathbf{u}(t) \qquad (33.16)$$

$$\mathbf{u}(t) = \mathbf{K}_4\mathbf{v}(t) + \mathbf{K}_5\mathbf{y}(t) \qquad (33.17)$$

where $0 \le l \le n$. Combining the reduced-order model, Eq.(33.6) and (33.7), and the controller, Eq.(33.16) and (33.17), by eliminating $\mathbf{u}(t)$ and $\mathbf{y}(t)$ the closed-loop behaviour of the reduced-order model is

$$\begin{pmatrix} \dot{\mathbf{x}} \\ \dot{\mathbf{v}} \end{pmatrix} = \begin{pmatrix} \mathbf{F}_{11} & \mathbf{F}_{12} \\ \mathbf{F}_{21} & \mathbf{F}_{22} \end{pmatrix} \begin{pmatrix} \mathbf{x} \\ \mathbf{v} \end{pmatrix} \qquad (33.18)$$

where

$$\mathbf{F}_{11} \triangleq \mathbf{A}_o + \mathbf{B}_o(\mathbf{I} - \mathbf{K}_5\mathbf{D}_o)^{-1}\mathbf{K}_5\mathbf{C}_o \equiv \mathbf{A}_o + \mathbf{B}_o\mathbf{K}_5(\mathbf{I} - \mathbf{D}_o\mathbf{K}_5)^{-1}\mathbf{C}_o$$

$$\mathbf{F}_{12} \triangleq \mathbf{B}_o(\mathbf{I} - \mathbf{K}_5\mathbf{D}_o)^{-1}\mathbf{K}_4 \qquad (33.19)$$

$$\mathbf{F}_{21} \triangleq [\mathbf{K}_2 + (\mathbf{K}_3 + \mathbf{K}_2\mathbf{D}_o)(\mathbf{I} - \mathbf{K}_5\mathbf{D}_o)^{-1}\mathbf{K}_5]\mathbf{C}_o \qquad (33.20)$$

$$\mathbf{F}_{22} \triangleq \mathbf{K}_1 + (\mathbf{K}_3 + \mathbf{K}_2\mathbf{D}_o)(\mathbf{I} - \mathbf{K}_5\mathbf{D}_o)^{-1}\mathbf{K}_4 \,. \qquad (33.21)$$

The matrices $\mathbf{A}_o, \mathbf{B}_o, \mathbf{C}_o, \mathbf{D}_o$ are given by Eqs.(33.6) and (33.7).

Combining the actual plant and the controller, Eqs.(33.1), (33.2), (33.3), (33.16) and (33.17) the actual closed-loop transients are obtained

$$\begin{pmatrix} \dot{\mathbf{x}} \\ \dot{\mathbf{v}} \\ \cdots \\ \varepsilon\dot{\mathbf{z}} \end{pmatrix} = \left(\begin{array}{cc:c} \mathbf{A}_{11} + \mathbf{B}_1\mathbf{K}_5\mathbf{C}_1 & \mathbf{B}_1\mathbf{K}_4 & \mathbf{A}_{12} + \mathbf{B}_1\mathbf{K}_5\mathbf{C}_2 \\ (\mathbf{K}_2 + \mathbf{K}_3\mathbf{K}_5)\mathbf{C}_1 & \mathbf{K}_1 + \mathbf{K}_3\mathbf{K}_4 & (\mathbf{K}_2 + \mathbf{K}_3\mathbf{K}_5)\mathbf{C}_2 \\ \cdots & \cdots & \cdots \\ \mathbf{A}_{21} + \mathbf{B}_2\mathbf{K}_5\mathbf{C}_1 & \mathbf{B}_2\mathbf{K}_4 & \mathbf{A}_{22} + \mathbf{B}_2\mathbf{K}_5\mathbf{C}_2 \end{array}\right) \begin{pmatrix} \mathbf{x} \\ \mathbf{v} \\ \cdots \\ \mathbf{z} \end{pmatrix}$$

$$\triangleq \left(\begin{array}{c:c} \mathbf{\Gamma}_1 & \mathbf{\Gamma}_2 \\ \cdots\;\cdots\;\cdots & \cdots \\ \mathbf{\Gamma}_3 & \mathbf{\Gamma}_4 \end{array}\right) \begin{pmatrix} \mathbf{x} \\ \mathbf{v} \\ \cdots \\ \mathbf{z} \end{pmatrix} \triangleq \mathbf{\Gamma} \begin{pmatrix} \mathbf{x} \\ \mathbf{v} \\ \cdots \\ \mathbf{z} \end{pmatrix} . \qquad (33.22)$$

The coefficient matrix is partitioned by dotted lines, in order to prepare the similarity transformation and the two-time-scale introduced in Eqs. (33.29) and (33.34).

33.4 Nonlinear Singularly Perturbed System. Basic Relations

Linear singularly perturbed systems are outlined in the preceding sections. An extension to nonlinear systems is given in this section. The actual plant is governed by the nonlinear relations (*Sharkey, P.M., and O'Reilly, J., 1988*)

$$\begin{aligned} \dot{\mathbf{x}}(t) &= \mathbf{f}(\mathbf{x},\mathbf{z},\mathbf{u},\varepsilon) \qquad \mathbf{x}(t_o)=\mathbf{x}_o \qquad \mathbf{x}(t)\in\mathcal{R}^n & (33.23)\\ \varepsilon\dot{\mathbf{z}}(t) &= \mathbf{g}(\mathbf{x},\mathbf{z},\mathbf{u},\varepsilon) \qquad \mathbf{z}(t_o)=\mathbf{z}_o \qquad \mathbf{z}(t)\in\mathcal{R}^{n_z}, \quad \mathbf{u}\in\mathcal{R}^m\,. & (33.24)\end{aligned}$$

The variable $\mathbf{x}(t)$ is the slow state, $\mathbf{z}(t)$ the fast state, ε represents a small positive scalar. The controlling variable $\mathbf{u}$ is partitioned into a slow and a fast component, both established as a state feedback

$$\mathbf{u} = \mathbf{u}_s(\mathbf{x},\varepsilon) + \mathbf{u}_f(\mathbf{x},\mathbf{z},\varepsilon)\,. \qquad (33.25)$$

Combining Eqs.(33.23) through (33.25), a system of order $n+n_z$ is obtained. In the case $\varepsilon = 0$

$$\begin{aligned} \dot{\mathbf{x}}_s(t) &= \mathbf{f}\big(\mathbf{x}_s,\mathbf{z}_s,\mathbf{u}_s(\mathbf{x}_s,0)+\mathbf{u}_f\{\mathbf{x}_s,\mathbf{z}_s(0),0\},0\big) \qquad \mathbf{x}(t_o)=\mathbf{x}_o & (33.26)\\ 0 &= \mathbf{g}[\mathbf{x}_s,\mathbf{z}_s,\mathbf{u}_s(\mathbf{x}_s,0)+\mathbf{u}_f\{\mathbf{x}_s,\mathbf{z}_s(0),0\}]\,. & (33.27)\end{aligned}$$

These equations represent a dynamical system of order n modelling the slow dynamics. From the algebraic Eq.(33.27) the solution $\mathbf{z}_s = \mathbf{h}[\mathbf{x},\mathbf{u}_s(\mathbf{x}_s,0),0]$ can be derived and substituted into Eq.(33.26), yielding the reduced-order (nth order) slow motion $\mathbf{x}_s$

$$\dot{\mathbf{x}}_s(t) = \mathbf{f}\big(\mathbf{x}_s,\mathbf{h}[\mathbf{x}_s,\mathbf{u}_s(\mathbf{x}_s,0),0],\mathbf{u}_s(\mathbf{x}_s,0)+\mathbf{u}_f\{\mathbf{x}_s,\mathbf{h}[\mathbf{x}_s,\mathbf{u}_s(\mathbf{x}_s,0),0],0\},0\big)\,. \qquad (33.28)$$

If the original system is considered with ε non-zero but small, the motion of this singularly perturbed system is characterized by a rapid motion $\mathbf{z}(t)$ followed by a slow motion $\mathbf{x}_s(t)$.

The original system in the fast time-scale τ obeys the equations

$$\frac{d\mathbf{x}(\tau)}{d\tau} = \varepsilon\,\mathbf{f}(\mathbf{x},\mathbf{z},\mathbf{u},\varepsilon) \qquad \mathbf{x}(\tau=0)=\mathbf{x}_o \qquad (33.29)$$

$$\frac{d\mathbf{z}(\tau)}{d\tau} = \mathbf{g}(\mathbf{x},\mathbf{z},\mathbf{u},\varepsilon) \qquad \mathbf{z}(\tau=0)=\mathbf{z}_o\,. \qquad (33.30)$$

If ε is zero

$$\begin{aligned} \frac{d\mathbf{x}(\tau)}{d\tau} &= 0 \qquad \mathbf{x}(\tau=0)=\mathbf{x}_o & (33.31)\\ \frac{d\mathbf{z}(\tau)}{d\tau} &= \mathbf{g}[\mathbf{x}_o,\mathbf{z},\ \mathbf{u}_s(\mathbf{x}_o,0)+\mathbf{u}_f(\mathbf{x}_o,\mathbf{z},0)\ ,\ 0] & (33.32)\end{aligned}$$

where $\mathbf{z}(\tau) = \mathbf{z}_s + \mathbf{z}_f(\tau)$ and $\mathbf{z}(\tau=0) = \mathbf{z}_o$. This is a process of n_zth order. The property Eq.(33.32) corresponds with Eqs.(33.10) and (33.15) in linear theory. The *boundary layer* correction $\mathbf{z}_f(\tau)$ has the same structure in the nonlinear case as in the linear case. The boundary layer is known as the interval that passes from t_o until the quasi-steady state of $\mathbf{z}_f(\tau)$.

33.5 Two-Time-Scale Decomposition

Starting with a combined vector $(\mathbf{x}^T \mathbf{v}^T \mathbf{z}^T)^T$, a similarity transformation employing a new vector $(\mathbf{x}^{\bullet T}\ \mathbf{v}^{\bullet T} \mathbf{z}^{\bullet T})^T$ is defined. The vector components designed with a superscript bullet are introduced as an image of $\mathbf{x}$, $\mathbf{v}$ and $\mathbf{z}$. Thus, the dimensions are the same. The similarity transformation is introduced by the following matrix and its inverse[1]

$$\begin{pmatrix} \mathbf{x} \\ \mathbf{v} \\ \cdots \\ \mathbf{z} \end{pmatrix} = \begin{pmatrix} \mathbf{I}_{n+l} & \vdots & \varepsilon\mathbf{H} \\ \cdots\ \cdots\ \cdots & & \cdots \\ -\mathbf{L} & \vdots & \mathbf{I}_{n_z} - \varepsilon\mathbf{L}\mathbf{H} \end{pmatrix} \begin{pmatrix} \mathbf{x}^\bullet \\ \mathbf{v}^\bullet \\ \cdots \\ \mathbf{z}^\bullet \end{pmatrix} \qquad \begin{matrix} \mathbf{H} \in \mathcal{R}^{(n+l)\times n_z}, \\ \mathbf{L} \in \mathcal{R}^{n_z \times (n+l)} \end{matrix} \tag{33.34}$$

$$\begin{pmatrix} \mathbf{x}^\bullet \\ \mathbf{v}^\bullet \\ \cdots \\ \mathbf{z}^\bullet \end{pmatrix} = \begin{pmatrix} \mathbf{I}_{n+l} - \varepsilon\mathbf{H}\mathbf{L} & \vdots & -\varepsilon\mathbf{H} \\ \cdots\ \cdots\ \cdots & & \cdots \\ \mathbf{L} & \vdots & \mathbf{I}_{n_z} \end{pmatrix} \begin{pmatrix} \mathbf{x} \\ \mathbf{v} \\ \cdots \\ \mathbf{z} \end{pmatrix} = \begin{pmatrix} \mathbf{I}_{n+l} & \vdots & \varepsilon\mathbf{H} \\ \cdots\ \cdots\ \cdots & & \cdots \\ -\mathbf{L} & \vdots & \mathbf{I}_{n_z} - \varepsilon\mathbf{L}\mathbf{H} \end{pmatrix}^{-1} \begin{pmatrix} \mathbf{x} \\ \mathbf{v} \\ \cdots \\ \mathbf{z} \end{pmatrix} \tag{33.35}$$

The matrices $\mathbf{H}$ and $\mathbf{L}$ will be chosen in order to achieve a transformed singularly perturbed system that satisfies the condition of decoupled form. Treating vector equations with vector components $\dot{\mathbf{x}}, \dot{\mathbf{v}}$ and $\varepsilon\dot{\mathbf{z}}$ (i.e., $\varepsilon\dot{\mathbf{z}}$ instead of $\dot{\mathbf{z}}$), the transformation must be slightly modified

$$\begin{pmatrix} \dot{\mathbf{x}} \\ \dot{\mathbf{v}} \\ \cdots \\ \varepsilon\dot{\mathbf{z}} \end{pmatrix} = \begin{pmatrix} \mathbf{I}_{n+l} & \vdots & \mathbf{H} \\ \cdots\ \cdots\ \cdots & & \cdots \\ -\varepsilon\mathbf{L} & \vdots & \mathbf{I}_{n_z} - \varepsilon\mathbf{L}\mathbf{H} \end{pmatrix} \begin{pmatrix} \dot{\mathbf{x}}^\bullet \\ \dot{\mathbf{v}}^\bullet \\ \cdots \\ \varepsilon\dot{\mathbf{z}}^\bullet \end{pmatrix} . \tag{33.36}$$

Now, consider the given singularly perturbed system of Eq.(33.22)

$$\begin{pmatrix} \dot{\mathbf{x}} \\ \dot{\mathbf{v}} \\ \cdots \\ \varepsilon\dot{\mathbf{z}} \end{pmatrix} = \begin{pmatrix} \mathbf{\Gamma}_1 & \vdots & \mathbf{\Gamma}_2 \\ \cdots\ \cdots\ \cdots & & \cdots \\ \mathbf{\Gamma}_3 & \vdots & \mathbf{\Gamma}_4 \end{pmatrix} \begin{pmatrix} \mathbf{x} \\ \mathbf{v} \\ \cdots \\ \mathbf{z} \end{pmatrix} \triangleq \mathbf{\Gamma} \begin{pmatrix} \mathbf{x} \\ \mathbf{v} \\ \cdots \\ \mathbf{z} \end{pmatrix} . \tag{33.37}$$

---

[1]The simple rule

$$\mathbf{A}^{-1} = \begin{pmatrix} a & b \\ c & d \end{pmatrix}^{-1} = \begin{pmatrix} d & -b \\ -c & a \end{pmatrix} / \det \begin{pmatrix} a & b \\ c & d \end{pmatrix} \tag{33.33}$$

when inverting a 2×2-matrix with scalar entries must not be used in the case of partitioned matrices.

Applying the transformations Eq.(33.36) to the left-hand side, and Eq.(33.35) on the right-hand side, it results

$$\begin{pmatrix} \mathbf{I}_{n+l} & \vdots & \mathbf{H} \\ \cdots & \cdots & \cdots \\ -\varepsilon\mathbf{L} & \vdots & \mathbf{I}_{n_s} - \varepsilon\mathbf{HL} \end{pmatrix} \begin{pmatrix} \dot{\mathbf{x}}^\bullet \\ \dot{\mathbf{v}}^\bullet \\ \cdots \\ \dot{\mathbf{z}}^\bullet \end{pmatrix} = \mathbf{\Gamma} \begin{pmatrix} \mathbf{I}_{n+l} & \vdots & \varepsilon\mathbf{H} \\ \cdots & \cdots & \cdots \\ -\mathbf{L} & \vdots & \mathbf{I}_{n_s} - \varepsilon\mathbf{HL} \end{pmatrix} \begin{pmatrix} \mathbf{x}^\bullet \\ \mathbf{v}^\bullet \\ \cdots \\ \mathbf{z}^\bullet \end{pmatrix} . \tag{33.38}$$

Premultiplication with the inverse of the left-hand side coefficient matrix yields

$$\begin{pmatrix} \dot{\mathbf{x}}^\bullet \\ \dot{\mathbf{v}}^\bullet \\ \cdots \\ \dot{\mathbf{z}}^\bullet \end{pmatrix} = \begin{pmatrix} \mathbf{I}_{n+l} - \varepsilon\mathbf{HL} & \vdots & -\mathbf{H} \\ \cdots & \cdots & \cdots \\ \varepsilon\mathbf{L} & \vdots & \mathbf{I}_{n_s} \end{pmatrix} \mathbf{\Gamma} \begin{pmatrix} \mathbf{I}_{n+l} & \vdots & \varepsilon\mathbf{H} \\ \cdots & \cdots & \cdots \\ -\mathbf{L} & \vdots & \mathbf{I}_{n_s} - \varepsilon\mathbf{HL} \end{pmatrix} \begin{pmatrix} \mathbf{x}^\bullet \\ \mathbf{v}^\bullet \\ \cdots \\ \mathbf{z}^\bullet \end{pmatrix} . \tag{33.39}$$

In order to obtain a coefficient matrix with block diagonal structure in the right-hand side of Eq.(33.39), it can easily be verified, dissolving $\mathbf{\Gamma}$ into its partitions $\mathbf{\Gamma}_1$ through $\mathbf{\Gamma}_4$, that $\mathbf{H}$ and $\mathbf{L}$ must obey the conditions

$$\varepsilon(\mathbf{L\Gamma}_1 - \mathbf{L\Gamma}_2\mathbf{L}) + \mathbf{\Gamma}_3 - \mathbf{\Gamma}_4\mathbf{L} = 0 \tag{33.40}$$

$$\varepsilon(\mathbf{\Gamma}_1\mathbf{H} - \mathbf{\Gamma}_2\mathbf{LH} - \mathbf{HL\Gamma}_2) + \mathbf{\Gamma}_2 - \mathbf{H\Gamma}_4 = 0 \ . \tag{33.41}$$

Provided that $\mathbf{\Gamma}_4$ is nonsingular,

$$\mathbf{L} = \mathbf{\Gamma}_4^{-1}\mathbf{\Gamma}_3 + \varepsilon\mathbf{\Gamma}_4^{-1}(\mathbf{L\Gamma}_1 - \mathbf{L\Gamma}_2\mathbf{L}) = \mathbf{\Gamma}_4^{-1}\mathbf{\Gamma}_3 + 0_L(\varepsilon) \tag{33.42}$$

$$\mathbf{H} = \mathbf{\Gamma}_2\mathbf{\Gamma}_4^{-1} + \varepsilon(\mathbf{\Gamma}_1\mathbf{H} - \mathbf{\Gamma}_2\mathbf{LH} - \mathbf{HL\Gamma}_2)\mathbf{\Gamma}_4^{-1} = \mathbf{\Gamma}_2\mathbf{\Gamma}_4^{-1} + 0_H(\varepsilon) \ . \tag{33.43}$$

Combining these results with Eq.(33.39) yields

$$\begin{pmatrix} \dot{\mathbf{x}}^\bullet \\ \dot{\mathbf{v}}^\bullet \\ \cdots \\ \varepsilon\dot{\mathbf{z}}^\bullet \end{pmatrix} = \begin{pmatrix} \mathbf{\Gamma}_1 - \mathbf{\Gamma}_2\mathbf{\Gamma}_4^{-1}\mathbf{\Gamma}_3 + 0_1(\varepsilon) & \vdots & 0 \\ \cdots & \cdots & \cdots \\ 0 & \vdots & \mathbf{\Gamma}_4 + 0_2(\varepsilon) \end{pmatrix} \begin{pmatrix} \mathbf{x}^\bullet \\ \mathbf{v}^\bullet \\ \cdots \\ \mathbf{z}^\bullet \end{pmatrix} . \tag{33.44}$$

In Eq.(33.44) substitute $\mathbf{\Gamma}_1$ through $\mathbf{\Gamma}_4$ as defined in Eq.(33.22): $\mathbf{\Gamma}_4 = \mathbf{A}_{22} + \mathbf{B}_2\mathbf{K}_5\mathbf{C}_2$ and

$$\mathbf{\Gamma}_1 - \mathbf{\Gamma}_2\mathbf{\Gamma}_4^{-1}\mathbf{\Gamma}_3 = \begin{pmatrix} \mathbf{A}_{11} + \mathbf{B}_1\mathbf{K}_5\mathbf{C}_1 & \mathbf{B}_1\mathbf{K}_4 \\ (\mathbf{K}_2 + \mathbf{K}_3\mathbf{K}_5)\mathbf{C}_1 & \mathbf{K}_1 + \mathbf{K}_3\mathbf{K}_4 \end{pmatrix}$$

$$- \begin{pmatrix} \mathbf{A}_{12} + \mathbf{B}_1\mathbf{K}_5\mathbf{C}_2 \\ (\mathbf{K}_2 + \mathbf{K}_3\mathbf{K}_5)\mathbf{C}_2 \end{pmatrix} (\mathbf{A}_{22} + \mathbf{B}_2\mathbf{K}_5\mathbf{C}_2)^{-1} (\mathbf{A}_{21} + \mathbf{B}_2\mathbf{K}_5\mathbf{C}_1 \ \vdots \ \mathbf{B}_2\mathbf{K}_4) \ . \tag{33.45}$$

Omitting lengthy calculations and referring to Eq.(33.18) (*O'Reilly, J., 1986*), the short result is

$$\mathbf{\Gamma}_1 - \mathbf{\Gamma}_2\mathbf{\Gamma}_4^{-1}\mathbf{\Gamma}_3 \triangleq \begin{pmatrix} \mathbf{F}_{11} & \mathbf{F}_{12} \\ \mathbf{F}_{21} & \mathbf{F}_{22} \end{pmatrix} + 0(\varepsilon) \tag{33.46}$$

$$\begin{pmatrix} \dot{\mathbf{x}}^\bullet \\ \dot{\mathbf{v}}^\bullet \\ \varepsilon\dot{\mathbf{z}}^\bullet \end{pmatrix} = \begin{pmatrix} \mathbf{F}_{11} & \mathbf{F}_{12} & \mathbf{0} \\ \mathbf{F}_{21} & \mathbf{F}_{22} & \mathbf{0} \\ \mathbf{0} & \mathbf{0} & \mathbf{A}_{22} + \mathbf{B}_2\mathbf{K}_5\mathbf{C}_2 \end{pmatrix} \begin{pmatrix} \mathbf{x}^\bullet \\ \mathbf{v}^\bullet \\ \mathbf{z}^\bullet \end{pmatrix} . \tag{33.47}$$

High-order terms $\mathbf{O}(\varepsilon)$ have been omitted for abbreviation. Note that the vector components $\mathbf{x}^\bullet, \mathbf{v}^\bullet$ represent the dominant state and $\mathbf{z}^\bullet$ is the fast state. With regard to the block-diagonal structure, both vector partitions are decoupled and can be treated as independent systems in the transformed space. The main result is: The rate $\dot{\mathbf{z}}^\bullet$ is premultiplied with ε and yields a two-time-scale decomposition of the actual control, similarly to the actual plant in Eqs. (33.1) through (33.3), but the vector partitions are decoupled.

33.6 Robust State Controller Design

An important question is: In which cases is the term $\mathbf{B}_2\mathbf{K}_5\mathbf{C}_2$ in Eq.(33.47) of no influence to the eigenvalues ? The answer is

- (i) if the product $\mathbf{B}_2\mathbf{K}_5\mathbf{C}_2$ vanishes,
- (ii) if $(\mathbf{A}_{22}, \mathbf{B}_2)$ is weakly controllable or
- (iii) if $(\mathbf{C}_2, \mathbf{A}_{22})$ is weakly observable. Besides this, if $(\mathbf{A}_{22}, \mathbf{B}_2)$-controllability or $(\mathbf{C}_2, \mathbf{A}_{22})$-observability is weak, the high frequency parasitics can only be weakly influenced by the controller Eqs.(33.16) and (33.17).
- (iv) No influence can be ensured if $\mathbf{K}_5 = \mathbf{0}$ is chosen, i.e., the controller provides no direct static signal from $\mathbf{y}$ to $\mathbf{u}$. Then, only the eigenvalues $\lambda[\mathbf{A}_{22}]$ are relevant to the dynamics of the signals $\mathbf{z}^\bullet$. From Eq.(33.47) it results that the actual control loop embodies the same matrices $\mathbf{F}_{ij}$ as the reduced-order control loop of Eq.(33.18). Hence, $\mathbf{K}_5 = 0$ is sufficient for robustness. According to the similarity transformation the controller of Eqs. (33.16) and (33.17) is stable operating with the actual plant.
- (v) If $\mathbf{C}_1 = \mathbf{I}$, $\mathbf{C}_2 = \mathbf{0}$ and $\mathbf{K}_5 = \mathbf{K}$ are implemented this constellation corresponds to the classical feedback state controller $\mathbf{u} = \mathbf{K}\mathbf{x}$ which is also stable because of $\mathbf{C}_2 = \mathbf{0}$. The full-order observer-based state controller also has the property of being robust to unmodelled high-frequency parasitics.

Investigating robustness, *singular-value based* stability bounds on the frequency response representation both for the high-frequency part and the reduced-order system (with $\varepsilon = 0$) were presented by *Chen, B.S., and Lin, C.L., 1990*. Sliding-mode controllers for singularly perturbed systems were designed by *Heck, B.S., 1991*.

33.7 Two-Frequency-Scale Decomposition

Using the complex frequency domain, the system is described by separate transfer functions of the plant and of the controller.

33.7.1 Plant

To obtain decoupled coordinates, again a transformation is applied. Similarly to Eqs.(33.34) and (33.22), a new plant vector with lower dimension $(\mathbf{x}'^T, \mathbf{z}'^T)^T$ is stated. Use the following transformation matrices

$$\begin{pmatrix}\mathbf{x}\\ \mathbf{z}\end{pmatrix} = \begin{pmatrix}\mathbf{I}_n & \varepsilon\mathbf{H}'\\ -\mathbf{L}' & \mathbf{I}_{n_z} - \varepsilon\mathbf{L}'\mathbf{H}'\end{pmatrix}\begin{pmatrix}\mathbf{x}'\\ \mathbf{z}'\end{pmatrix}, \qquad \begin{pmatrix}\mathbf{x}'\\ \mathbf{z}'\end{pmatrix} = \begin{pmatrix}\mathbf{I}_n - \varepsilon\mathbf{H}'\mathbf{L}' & -\varepsilon\mathbf{H}'\\ \mathbf{L}' & \mathbf{I}_{n_z}\end{pmatrix}\begin{pmatrix}\mathbf{x}\\ \mathbf{z}\end{pmatrix} \tag{33.48}$$

$$\begin{pmatrix}\dot{\mathbf{x}}\\ \varepsilon\dot{\mathbf{z}}\end{pmatrix} = \begin{pmatrix}\mathbf{I}_n & \mathbf{H}'\\ -\varepsilon\mathbf{L}' & \mathbf{I}_{n_z} - \varepsilon\mathbf{L}'\mathbf{H}'\end{pmatrix}\begin{pmatrix}\dot{\mathbf{x}}'\\ \varepsilon\dot{\mathbf{z}}'\end{pmatrix}. \tag{33.49}$$

In order to obtain a block diagonal plant matrix, $\mathbf{H}' = \mathbf{A}_{12}\mathbf{A}_{22}^{-1}$ and $\mathbf{L}' = \mathbf{A}_{22}^{-1}\mathbf{A}_{21}$ must be chosen, see Eqs.(33.42) ,(33.43) and (33.22). Then, the transformed plant equation in state space turns out as

$$\begin{aligned}\begin{pmatrix}\dot{\mathbf{x}}'\\ \varepsilon\dot{\mathbf{z}}'\end{pmatrix} &= \begin{pmatrix}\mathbf{A}_o & \mathbf{0}\\ \mathbf{0} & \mathbf{A}_{22}\end{pmatrix}\begin{pmatrix}\mathbf{x}'\\ \mathbf{z}'\end{pmatrix} + \begin{pmatrix}\mathbf{B}_o\\ \mathbf{B}_2\end{pmatrix}\mathbf{u}\\ &= \begin{pmatrix}\mathbf{A}_{11} - \mathbf{A}_{12}\mathbf{A}_{22}^{-1}\mathbf{A}_{21} & \mathbf{0}\\ \mathbf{0} & \mathbf{A}_{22}\end{pmatrix}\begin{pmatrix}\mathbf{x}'\\ \mathbf{z}'\end{pmatrix} + \begin{pmatrix}\mathbf{B}_1 - \mathbf{A}_{12}\mathbf{A}_{22}^{-1}\mathbf{B}_2\\ \mathbf{B}_2\end{pmatrix}\mathbf{u}\end{aligned} \tag{33.50}$$

$$\mathbf{y} = (\mathbf{C}_o \;\; \mathbf{C}_2)\begin{pmatrix}\mathbf{x}'\\ \mathbf{z}'\end{pmatrix}. \tag{33.51}$$

Note that each matrix component is deteriorated with high-order terms in ε but these terms were omitted for the sake of abbreviation.

In the frequency domain the plant transfer matrix is

$$\mathbf{G}(s) = \mathbf{C}_o(s\mathbf{I}_n - \mathbf{A}_o)^{-1}\mathbf{B}_o + \mathbf{C}_2(\varepsilon s\mathbf{I}_{n_z} - \mathbf{A}_{22})^{-1}\mathbf{B}_2 + \mathbf{0}(\varepsilon). \tag{33.52}$$

When the Laplace operator s is multiplied by ε, the consequences are well known

$$G(s) = \mathcal{L}g(t) \qquad \varepsilon G(\varepsilon s) = \mathcal{L}g(\frac{t}{\varepsilon}). \tag{33.53}$$

As ε approaches zero the corresponding time function becomes faster and faster.

33.7.2 Controller

With reference to Eqs.(33.16) and (33.17), the controller can be written

$$\dot{\mathbf{v}} = \mathbf{K}_1\mathbf{v}+\mathbf{K}_2\mathbf{y}+\mathbf{K}_3\mathbf{u} = \mathbf{K}_1\mathbf{v}+\mathbf{K}_2\mathbf{y}+\mathbf{K}_3(\mathbf{K}_4\mathbf{v}+\mathbf{K}_5\mathbf{y}) = (\mathbf{K}_1+\mathbf{K}_3\mathbf{K}_4)\mathbf{v}+(\mathbf{K}_2+\mathbf{K}_3\mathbf{K}_5)\mathbf{y}. \tag{33.54}$$

This equation, in combination with Eq.(33.17), $\mathbf{u} = \mathbf{K}_4\mathbf{v} + \mathbf{K}_5\mathbf{y}$, yields the output $\mathbf{u}$ of the controller forced by the input $\mathbf{y}$, in the time domain. The transfer matrix of the controller in the frequency domain, from the input $\mathbf{y}$ to the output $\mathbf{u}$, is

$$\mathbf{K}(s) = \mathbf{K}_4(s\mathbf{I}_n - \mathbf{K}_1 - \mathbf{K}_3\mathbf{K}_4)^{-1}(\mathbf{K}_2 + \mathbf{K}_3\mathbf{K}_5) + \mathbf{K}_5. \tag{33.55}$$

Both transfer matrices are depicted in Fig. 33.3.

The condition $\mathbf{K}_5 = \mathbf{0}$, known from two-time-scale robust controller design, requires the transfer matrix $\mathbf{K}(s)$ not to possess a static (dynamic free) component. The matrix $\mathbf{K}(s)$ must be strictly proper[2].

[2] $\mathbf{G}(s)$ is proper if and only if $\mathbf{G}(\infty)$ is bounded. $\mathbf{G}(s)$ is strictly proper if $\mathbf{G}(\infty) = \mathbf{0}$.

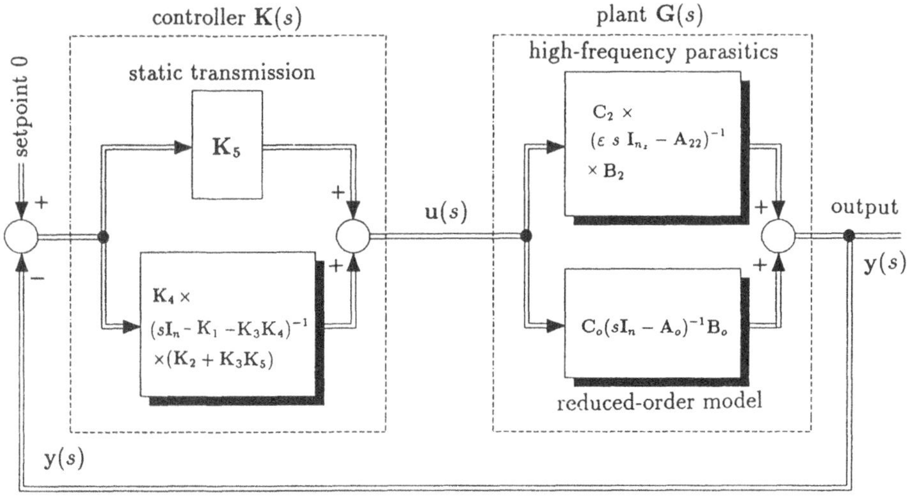

Figure 33.3: Frequency domain representation of robust controller design with singularly perturbed plant

33.8 Observer-Based Control for Singularly Perturbed Systems

Singularly perturbed linear systems can be controlled by an output feedback controller or by an observer-based state feedback controller when the system output is accessible, only. Applying full-order observers to time-varying systems with inaccessible state variables, both slow and fast dynamic state may be used for the state observers. In this section a brief presentation is given of how the observer in a singularly perturbed system can be recast to an ordinary one (*O'Reilly, J., 1980*). The singularly perturbed system, Eqs.(33.1) to (33.3), is rewritten to an ordinary state-space formulation, by defining the following vectors and matrices

$$\mathbf{x}^{\circ}(t) \triangleq \begin{pmatrix} \mathbf{x}(t) \\ \mathbf{z}(t) \end{pmatrix} \qquad \bar{\mathbf{I}} \triangleq \begin{pmatrix} \mathbf{I}_n & 0 \\ 0 & \frac{1}{\varepsilon}\mathbf{I}_{n_z} \end{pmatrix} . \tag{33.56}$$

Then, unified system equations are

$$\dot{\mathbf{x}}^{\circ}(t) = \begin{pmatrix} \dot{\mathbf{x}}(t) \\ \dot{\mathbf{z}}(t) \end{pmatrix} = \bar{\mathbf{I}} \begin{pmatrix} \mathbf{A}_{11} & \mathbf{A}_{12} \\ \mathbf{A}_{21} & \mathbf{A}_{22} \end{pmatrix} \mathbf{x}^{\circ}(t) + \bar{\mathbf{I}} \begin{pmatrix} \mathbf{B}_1 \\ \mathbf{B}_2 \end{pmatrix} \mathbf{u}(t) = \mathbf{A}^{\circ}\mathbf{x}^{\circ} + \mathbf{B}^{\circ}\mathbf{u} \tag{33.57}$$

$$\mathbf{y}(t) = (\mathbf{C}_1 \,\vdots\, \mathbf{C}_2)\mathbf{x}^{\circ}(t) = \mathbf{C}^{\circ}\mathbf{x}^{\circ} . \tag{33.58}$$

Referring to Eqs.(33.56) through (33.58), a full-order observer is stated by Eq.(18.22). Combining state and error equation results in a well-known overall matrix of block-diagonal structure. The observer may be separated into a slow and a fast part. The slow part in real time t, the fast one in fast time τ . A composite controller for slow and fast system can be built up using separate subsystem matrices. Since the fast observer is usually dominated by the slow one, the latter is neglected in many cases.

Chapter 34

Control Using Aggregation Models

34.1 Aggregation Matrix

Modelling of large-scale dynamic systems requires differential equations of high dimension. Control algorithms can often only be derived and implemented for models of reduced dimension. In view of this, the concept of aggregation provides simplified representation and is suitable for approximate synthesis. The relation $\mathbf{z} = \mathbf{\Gamma x}$ with constant matrix $\mathbf{\Gamma}$ is used to transform the original model or plant state vector $\mathbf{x}$ into the aggregated system with the aggregated state vector $\mathbf{z}$. The objective is that $\mathbf{z}$ contains the significant portion of the plant dynamics, see Fig. 34.1. Then, one has

$$\begin{aligned} \text{original plant:} \quad \dot{\mathbf{x}} &= \mathbf{Ax} + \mathbf{Bu} \qquad \mathbf{x} \in \mathcal{R}^n, \quad \mathbf{u} \in \mathcal{R}^m && (34.1)\\ \text{aggregation:} \quad \mathbf{z} &= \mathbf{\Gamma x}, \qquad \mathbf{\Gamma} \in \mathcal{R}^{l \times n} && (34.2)\\ \text{aggregated model:} \quad \dot{\mathbf{z}} &= \mathbf{Fz} + \mathbf{Gu} \qquad \mathbf{z} \in \mathcal{R}^l . && (34.3) \end{aligned}$$

The differential equations Eqs.(34.1) and (34.3) are equivalent if the relations hold that

$$\mathbf{F\Gamma} = \mathbf{\Gamma A} \quad \text{and} \tag{34.4}$$

$$\mathbf{G} = \mathbf{\Gamma B} . \tag{34.5}$$

If all state variables are controllable and observable, the plant model is irreducible without cancelling information. Hence, exact aggregation is impossible. However, approximate aggregation is of great practical interest and is realized by omitting non-dominant state variables and eigenvalues.

34.2 Aggregation Matrix Γ and Dominant Eigenvalues

Defining the eigenvalues $\lambda_i[\mathbf{A}]$ and the associated normalized eigenvectors of $\mathbf{A}$ as $\mathbf{a}_i$, and supposing that $\mathbf{A}$ and $\mathbf{\Gamma}$ satisfy Eq.(34.4),

$$\mathbf{\Gamma A a}_i = \mathbf{\Gamma}(\mathbf{A a}_i) = \mathbf{\Gamma}(\lambda_i \mathbf{a}_i) = \lambda_i \mathbf{\Gamma a}_i \qquad \text{and} \qquad \mathbf{F\Gamma a}_i = \mathbf{\Gamma A a}_i = \lambda_i \mathbf{\Gamma a}_i . \tag{34.6}$$

Eq.(34.6) shows that $\mathbf{\Gamma a}_i$ is also an eigenvector of $\mathbf{F}$ (if it is nonzero) and $\mathbf{F}$ contains the same eigenvalue λ_i. Choosing $\mathbf{\Gamma}$ such that

$$\mathbf{\Gamma a}_i \neq \mathbf{0} \qquad \forall\, i = 1 \ldots l \qquad \text{and} \qquad \mathbf{\Gamma a}_i = \mathbf{0} \qquad \forall\, i = l+1, \ldots n \tag{34.7}$$

one makes the matrix $\mathbf{F}$ inherit the eigenvalues $\lambda_1 \ldots \lambda_l$. In most cases, these eigenvalues should be the dominant ones. The aggregated matrix $\mathbf{F}$, in this way, retains the dominant eigenvalues. The remaining eigenvalues $\lambda_{l+1} \ldots \lambda_n$ loose any significance in the aggregated model $\mathbf{F}$.

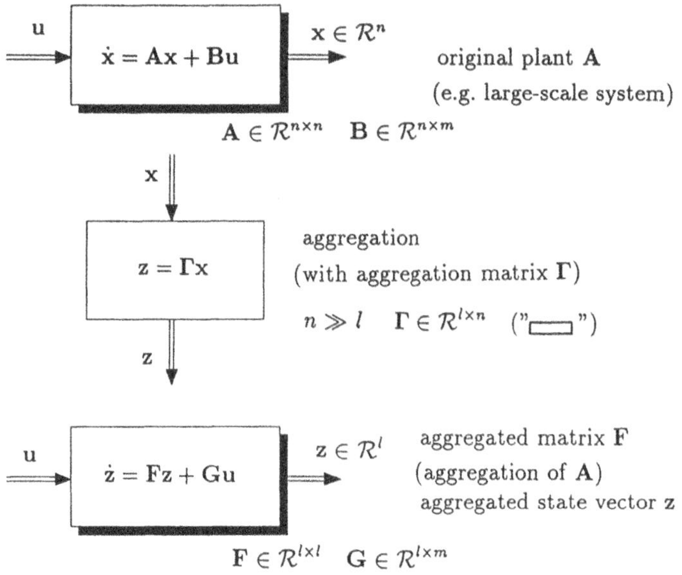

Figure 34.1: Original plant and aggregated system

34.3 Designing Γ via Controllability Matrices

In order to design the aggregation matrix $\mathbf{\Gamma}$, the relation between Kalman controllability matrices $\mathbf{W}_A$ and $\mathbf{W}_F$ can be utilized. By definition, the controllability matrices for the original and for the aggregated system are as follows

$$\mathbf{W}_A \triangleq (\mathbf{B} \vdots \mathbf{AB} \vdots \dots \vdots \mathbf{A}^{n-1}\mathbf{B}) \qquad \mathbf{W}_A \in \mathcal{R}^{n\times nm} \tag{34.8}$$

$$\mathbf{W}_F \triangleq (\mathbf{G} \vdots \mathbf{FG} \vdots \dots \vdots \mathbf{F}^{l-1}\mathbf{G}, \dots \mathbf{F}^{n-1}\mathbf{G}) \qquad \mathbf{W}_F \in \mathcal{R}^{l\times nm} . \tag{34.9}$$

The matrix $\mathbf{W}_F$ has more elements than the well-known Kalman controllability matrices. Since $\mathbf{B}, \mathbf{G}, \mathbf{A}$ and $\mathbf{F}$ are given by Eqs.(34.1) and (34.3), the extended controllability matrix $\mathbf{W}_F$ can be deduced by premultiplying $\mathbf{W}_A$ by $\mathbf{\Gamma}$

$$\mathbf{\Gamma W}_A = \mathbf{W}_F \qquad \mathbf{\Gamma} \in \mathcal{R}^{l\times n} . \tag{34.10}$$

In this equation, $\mathbf{\Gamma}$ is overdetermined if $m > 1$. Eq.(34.10) is a matrix expression for lmn scalar equations with ln unknown elements of matrix $\mathbf{\Gamma}$. Transposing and applying the left-inverse $\mathbf{M}^{\sharp L}$ for $\mathbf{M} := \mathbf{W}_A^T$ yields

$$\mathbf{W}_A^T \mathbf{\Gamma}^T = \mathbf{M\Gamma}^T = \mathbf{W}_F^T \tag{34.11}$$

$$\mathbf{\Gamma}^T = \mathbf{M}^{\sharp L}\mathbf{W}_F^T \tag{34.12}$$

$$\mathbf{\Gamma} = \mathbf{\Gamma}^* = \mathbf{W}_F(\mathbf{M}^{\sharp L})^T = \mathbf{W}_F[(\mathbf{M}^T\mathbf{M})^{-1}\mathbf{M}^T]^T = \mathbf{W}_F\mathbf{W}_A^T(\mathbf{W}_A\mathbf{W}_A^T)^{-1} . \tag{34.13}$$

For given matrices **A** and **F** the aforementioned $\mathbf{\Gamma}^*$ yields difference vectors (if $\mathbf{W}_A^T\mathbf{\Gamma}^T - \mathbf{W}_F^T$ is decomposed into columns) which are minimized in norm.

$$\mathbf{Y} = \mathbf{MP} + \mathbf{E} \quad \rightsquigarrow \quad \text{col }\mathbf{Y} = (\mathbf{I}\otimes\mathbf{M})\text{col }\mathbf{P} + \text{col }\mathbf{E} \;\hat{=}\; \mathbf{y} = (\mathbf{I}\otimes\mathbf{M})\mathbf{p} + \boldsymbol{\varepsilon}\,. \tag{34.14}$$

In order to obtain $\boldsymbol{\varepsilon}^T\boldsymbol{\varepsilon} \to \min$ or $\text{tr}\mathbf{E}^T\mathbf{E} \to \min$, use Eqs.(D.32) and (4.10)

$$\mathbf{p} = [(\mathbf{I}\otimes\mathbf{M})^T(\mathbf{I}\otimes\mathbf{M})]^{-1}(\mathbf{I}\otimes\mathbf{M})^T\mathbf{y} = [\mathbf{I}\otimes\mathbf{M}^T\mathbf{M}]^{-1}(\mathbf{I}\otimes\mathbf{M})^T\mathbf{y} = [\mathbf{I}\otimes(\mathbf{M}^T\mathbf{M})^{-1}](\mathbf{I}\otimes\mathbf{M})^T\mathbf{y} \tag{34.15}$$

$$\text{col }\mathbf{P} = \mathbf{p} = \mathbf{I}\otimes(\mathbf{M}^T\mathbf{M})^{-1}\mathbf{M}^T\text{col }\mathbf{Y} \quad \rightsquigarrow \quad \mathbf{P} = (\mathbf{M}^T\mathbf{M})^{-1}\mathbf{M}^T\mathbf{Y} = \mathbf{M}^{\sharp L}\mathbf{Y}\,. \tag{34.16}$$

Substituting $\mathbf{P} := \mathbf{\Gamma}^T$, $\mathbf{Y} := \mathbf{W}_F^T$ and $\mathbf{M} := \mathbf{W}_A^T$ yields the result of Eq.(34.12). Since, by assumption, $(\mathbf{A},\mathbf{B})$ is observable, $\mathbf{W}_A$ is of rank n and $(\mathbf{W}_A\mathbf{W}_A^T)$ is invertible.

34.4 Choosing the Aggregated Matrix F

Eq.(34.4) is overdetermined in **F**: ln equations for l^2 unknowns. Hence, an approximate aggregated matrix **F** has to be found, referred to as approximate aggregation of **A** . To obtain the optimal components in **F**, transpose Eq.(34.4), find a scalar representative for

$$\mathbf{\Gamma}^T\mathbf{F}^T - \mathbf{A}^T\mathbf{\Gamma}^T \to \min, \tag{34.17}$$

partition $\mathbf{F}^T$ into columns and solve the problem of linear regression. The optimum $\mathbf{F}^T$ is given, see Eq.(D.32), by premultiplying $\mathbf{A}^T\mathbf{\Gamma}^T$ by the left-pseudo-inverse $(\mathbf{\Gamma}^T)^{\sharp L}$ of $\mathbf{\Gamma}^T$

$$\mathbf{F}^T = (\mathbf{\Gamma}^T)^{\sharp L}\mathbf{A}^T\mathbf{\Gamma}^T \tag{34.18}$$

$$\mathbf{F} = \mathbf{F}^* = \mathbf{\Gamma A}[(\mathbf{\Gamma}^T)^{\sharp L}]^T = \mathbf{\Gamma A}[(\mathbf{\Gamma\Gamma}^T)^{-1}\mathbf{\Gamma}]^T = \mathbf{\Gamma A\Gamma}^T(\mathbf{\Gamma\Gamma}^T)^{-1}\,. \tag{34.19}$$

Example: Minimizing the norm of $||\mathbf{F\Gamma} - \mathbf{\Gamma A}||_F$: Directly minimizing the Frobenius norm

$$||\mathbf{F\Gamma} - \mathbf{\Gamma A}||_F^2 = \text{tr}\;[\;(\mathbf{F\Gamma} - \mathbf{\Gamma A})^T(\mathbf{F\Gamma} - \mathbf{\Gamma A})\;] \tag{34.20}$$

with respect to **F** yields $\frac{\partial}{\partial\mathbf{F}}$ tr $[\mathbf{\Gamma}^T\mathbf{F}^T\mathbf{F\Gamma} - \mathbf{A}^T\mathbf{\Gamma}^T\mathbf{F\Gamma} - \mathbf{\Gamma}^T\mathbf{F}^T\mathbf{\Gamma A} + \mathbf{A}^T\mathbf{\Gamma\Gamma A}] = \mathbf{0}$. Using $\frac{\partial}{\partial\mathbf{M}}\text{tr}\mathbf{BM}^T = \mathbf{B}$ and $\frac{\partial}{\partial\mathbf{M}}$ tr $\mathbf{BM}^T\mathbf{M} = \mathbf{M}(\mathbf{B} + \mathbf{B}^T)$, the same result $\mathbf{F}^*$ is achieved. □

34.5 Suboptimal Control Via the Aggregated Model

The design of an optimal controller for the original system $(\mathbf{A},\mathbf{B})$ requires high effort, on account of the high dimension of **A** and **B** . If the lower-order aggregated system $(\mathbf{F},\mathbf{G})$ is selected as a basis, computational effort is reduced. Then, the control result can only be regarded as a suboptimal one. Preferably, it is applicable in cases of weakly coupled subsystems within the system **A**. Considering the original system Eq.(34.1) with the criterion

$$I = \int_0^\infty (\mathbf{x}^T\mathbf{Qx} + \mathbf{u}^T\mathbf{Ru})\;dt\;, \tag{34.21}$$

the optimal linear control law is

$$\mathbf{u}^* = \mathbf{K}^*\mathbf{x} = -\mathbf{R}^{-1}\mathbf{B}^T\mathbf{P}^*\mathbf{x} \tag{34.22}$$

where $\mathbf{P}^*$ obeys the Riccati equation

$$\mathbf{A}^T\mathbf{P}^* + \mathbf{P}^*\mathbf{A} - \mathbf{P}^*\mathbf{BR}^{-1}\mathbf{B}^T\mathbf{P}^* + \mathbf{Q} = \mathbf{0}\,. \tag{34.23}$$

The control policy based on the aggregated model Eq.(34.3) is given by $\mathbf{u} = \mathbf{Kz}$. Using a criterion I_z yields a suboptimal $\mathbf{K}, \mathbf{P}$ and $\mathbf{u}$ as follows

$$I_z = \int_0^\infty (\mathbf{z}^T \mathbf{Q}_z \mathbf{z} + \mathbf{u}^T \mathbf{R} \mathbf{u})\, dt \tag{34.24}$$

$$\mathbf{u} = \mathbf{Kz} = -\mathbf{R}^{-1}\mathbf{G}^T\mathbf{Pz} \tag{34.25}$$

$$\mathbf{F}^T\mathbf{P} + \mathbf{PF} - \mathbf{PGR}^{-1}\mathbf{G}^T\mathbf{P} + \mathbf{Q}_z = \mathbf{0}\,. \tag{34.26}$$

Supposing Eq.(34.19) satisfied, the associated optimal expression for $\mathbf{\Gamma A}$ is

$$\mathbf{F\Gamma} = \mathbf{\Gamma A \Gamma}^T(\mathbf{\Gamma\Gamma}^T)^{-1}\mathbf{\Gamma}\,. \tag{34.27}$$

Premultiplying and postmultiplying Eq.(34.26) by $\mathbf{\Gamma}^T$ and $\mathbf{\Gamma}$, respectively, yields

$$\mathbf{\Gamma}^T\mathbf{F}^T\mathbf{P\Gamma} + \mathbf{\Gamma}^T\mathbf{PF\Gamma} - \mathbf{\Gamma}^T\mathbf{PGR}^{-1}\mathbf{G}^T\mathbf{P\Gamma} + \mathbf{\Gamma}^T\mathbf{Q}_z\mathbf{\Gamma} = \mathbf{0}\,. \tag{34.28}$$

Substituting $\mathbf{\Gamma A}$ for $\mathbf{F\Gamma}$, as aforementioned, and $\mathbf{\Gamma B}$ for $\mathbf{G}$, one has

$$\mathbf{A}^T\mathbf{\Gamma}^T\mathbf{P\Gamma} + \mathbf{\Gamma}^T\mathbf{P\Gamma A} - \mathbf{\Gamma}^T\mathbf{P\Gamma BR}^{-1}\mathbf{B}^T\mathbf{\Gamma}^T\mathbf{P\Gamma} + \mathbf{\Gamma}^T\mathbf{Q}_z\mathbf{\Gamma} = \mathbf{0}\,. \tag{34.29}$$

Comparing Eqs.(34.29) and (34.23) shows that $\mathbf{P}^*$ corresponds to $\mathbf{\Gamma}^T\mathbf{P\Gamma}$ if $\mathbf{Q}$ is made to correspond with $\mathbf{\Gamma}^T\mathbf{Q}_z\mathbf{\Gamma}$ optimally. Invoking Eq.(34.19), this leads to

$$\mathbf{Q}_z = (\mathbf{\Gamma\Gamma}^T)^{-1}\mathbf{\Gamma Q \Gamma}^T(\mathbf{\Gamma\Gamma}^T)^{-1}. \tag{34.30}$$

Considering the residual difference remaining after approximation, the following result can be found by analogy to Eq.(16.50) (see *Aoki, M., 1968*)

$$(\mathbf{A} + \mathbf{BK\Gamma})^T\mathbf{P}^o + \mathbf{P}^o(\mathbf{A} + \mathbf{BK\Gamma}) + \mathbf{\Gamma}^T\mathbf{K}^T\mathbf{RK\Gamma} + \mathbf{Q} = \mathbf{0}\,. \tag{34.31}$$

The matrix $\mathbf{P}^o$ is an upper bound and $\mathbf{\Gamma}^T\mathbf{P\Gamma}$ a lower bound for $\mathbf{P}^*$

$$\mathbf{\Gamma}^T\mathbf{P\Gamma} \le \mathbf{P}^* \le \mathbf{P}^o. \tag{34.32}$$

Eq.(34.31) can be solved much easier than the Riccati equation in $\mathbf{P}^*$.

34.6 Iterative Algorithm Approximating the Aggregation Matrix

If the system or system component is controllable as well as observable, then exact aggregation, i.e., exact order reduction is impossible. *Feliachi, A., and Bhurtun, C., 1987* developed an iterative algorithm for the aggregation matrix by preserving some dominant key state variables and by compensating for the deleted ones. Given the plant Eq.(34.1), the problem is to reduce the model equation to dimension l optimally. Thus, the vector is partitioned

$$\mathbf{x} = \begin{pmatrix} \mathbf{x}_1 \\ x_2 \\ \mathbf{x}_3 \end{pmatrix} \quad \text{where} \quad \begin{cases} \mathbf{x}_1 \ldots \text{state vector of key variables,} \;\; \mathbf{x}_1 \in \mathcal{R}^{l-1} \\ x_2 \ldots \text{scalar } l\text{th state variable,} \;\; x_2 \in \mathcal{R} \\ \mathbf{x}_3 \ldots \text{state vector to be deleted,} \;\; \mathbf{x}_3 \in \mathcal{R}^{n-l}\,. \end{cases} \tag{34.33}$$

The aggregated vector $\mathbf{z}$ is given by $\mathbf{z} = (\mathbf{x}_1 \;\; z_m)^T$ where z_m is the scalar state variable compensating for x_2 and $\mathbf{x}_3$. If exact aggregation is possible then $z_m = x_2$. The system matrix **A** and control matrix **B** are partitioned, introducing the first step of an iteration algorithm

$$\dot{\mathbf{x}} = \begin{pmatrix} \mathbf{A}_{11} & \mathbf{A}_{12} & \mathbf{A}_{13} \\ \mathbf{A}_{21} & A_{22} & \mathbf{A}_{23} \\ \mathbf{A}_{31} & \mathbf{A}_{32} & \mathbf{A}_{33} \end{pmatrix} \begin{pmatrix} \mathbf{x}_1 \\ x_2 \\ \mathbf{x}_3 \end{pmatrix} + \begin{pmatrix} \mathbf{B}_1 \\ \mathbf{B}_2 \\ \mathbf{B}_3 \end{pmatrix} \mathbf{u} \triangleq \begin{pmatrix} \mathbf{F}_1 & \mathbf{H}_1 \\ \mathbf{E}_1 & \mathbf{K}_1 \end{pmatrix} \mathbf{x} + \begin{pmatrix} \mathbf{G}_1 \\ \mathbf{R}_1 \end{pmatrix} \mathbf{u} \,. \tag{34.34}$$

Note that $\mathbf{A}_{12}$ and $\mathbf{A}_{32}$ are columns and $\mathbf{A}_{21}$ and $\mathbf{A}_{23}$ are row matrices of the dimension $l-1$ and $n-l$, respectively, and A_{22} is a scalar. For abbreviation, use the substitutions $\mathbf{F}_1 := \mathbf{A}_{11}$, $\mathbf{H}_1 := (\mathbf{A}_{12} \vdots \mathbf{A}_{13})$, $\mathbf{G}_1 := \mathbf{B}_1$ and

$$\mathbf{E}_1 := \begin{pmatrix} \mathbf{A}_{21} \\ \mathbf{A}_{31} \end{pmatrix} \qquad \mathbf{K}_1 := \begin{pmatrix} A_{22} & \mathbf{A}_{23} \\ \mathbf{A}_{32} & \mathbf{A}_{33} \end{pmatrix} \qquad \mathbf{R}_1 := \begin{pmatrix} \mathbf{B}_2 \\ \mathbf{B}_3 \end{pmatrix} . \tag{34.35}$$

If $\mathbf{K}_1^{-1}$ exists, the matrix $\mathbf{K}_1$ is decomposed into a product of an orthogonal matrix $\mathbf{J}_1$ and a lower triangular matrix $\mathbf{L}_1$

$$\mathbf{K}_1 = \mathbf{J}_1\mathbf{L}_1 \qquad \mathbf{J}_1^T = \mathbf{J}_1^{-1} \,. \tag{34.36}$$

The matrices $\mathbf{J}_1$ and $\mathbf{L}_1$ are easily computed, on account of the triangular structure of $\mathbf{L}_1$ and the orthogonality of $\mathbf{J}_1$. Multiplying both sides of the aforementioned equation by its transpose from the left

$$\mathbf{K}_1^T\mathbf{K}_1 = (\mathbf{J}_1\mathbf{L}_1)^T\mathbf{J}_1\mathbf{L}_1 = \mathbf{L}_1^T\mathbf{J}_1^T\mathbf{J}_1\mathbf{L}_1 = \mathbf{L}_1^T\mathbf{I}\,\mathbf{L}_1 = \mathbf{L}_1^T\mathbf{L}_1 \,. \tag{34.37}$$

With regard to the triangular structure the entries of the product $\mathbf{L}_1^T\mathbf{L}_1$ are simple bilinear functions of the entries $\mathbf{L}_1$. Eq.(34.37) can be separated into scalar equations with only one unknown each, and can be solved successively. The inverse of the triangular matrix $\mathbf{L}_1$ also needs little computational effort.

Next, similarity transformation $\mathbf{S}_1$ and $\mathbf{S}_k$ are defined in the first and in the higher steps, respectively:

$$\mathbf{S}_1 \triangleq \text{block diag}\,\{\mathbf{I}_{l-1},\ \mathbf{L}_1\} \ \in \mathcal{R}^{n\times n} \,, \qquad \mathbf{L}_1 \in \mathcal{R}^{(n-l-1)\times(n-l-1)} \ . \tag{34.38}$$

Excited by the input **x**, the output $\mathbf{x}^{(1)}$ of the first iteration step is obtained by premultiplying **x** with $\mathbf{S}_1$

$$\mathbf{x}^{(1)} = \mathbf{S}_1\mathbf{x} \qquad \text{or} \qquad \mathbf{x} = \mathbf{S}_1^{-1}\mathbf{x}^{(1)} \,. \tag{34.39}$$

Combining with Eq.(34.34) yields

$$\dot{\mathbf{x}}^{(1)} = \mathbf{S}_1\dot{\mathbf{x}} = \begin{pmatrix} \mathbf{I}_{l-1} & 0 \\ 0 & \mathbf{L}_1 \end{pmatrix} \begin{pmatrix} \mathbf{F}_1 & \mathbf{H}_1 \\ \mathbf{E}_1 & \mathbf{K}_1 \end{pmatrix} \begin{pmatrix} \mathbf{I}_{l-1} & 0 \\ 0 & \mathbf{L}_1^{-1} \end{pmatrix} \mathbf{x}^{(1)} + \begin{pmatrix} \mathbf{I}_{l-1} & 0 \\ 0 & \mathbf{L}_1 \end{pmatrix} \begin{pmatrix} \mathbf{G}_1 \\ \mathbf{R}_1 \end{pmatrix} \mathbf{u} \tag{34.40}$$

$$\dot{\mathbf{x}}^{(1)} = \begin{pmatrix} \mathbf{F}_1 & \mathbf{H}_1\mathbf{L}_1^{-1} \\ \mathbf{L}_1\mathbf{E}_1 & \mathbf{L}_1\mathbf{K}_1\mathbf{L}_1^{-1} \end{pmatrix} \mathbf{x}^{(1)} + \begin{pmatrix} \mathbf{G}_1 \\ \mathbf{L}_1\mathbf{R}_1 \end{pmatrix} \mathbf{u} = \begin{pmatrix} \mathbf{F}_1 & \mathbf{H}_2 \\ \mathbf{E}_2 & \mathbf{K}_2 \end{pmatrix} \mathbf{x}^{(1)} + \begin{pmatrix} \mathbf{G}_1 \\ \mathbf{R}_2 \end{pmatrix} \mathbf{u} \,. \tag{34.41}$$

The second step is initiated by the decomposition of

$$\mathbf{K}_2 = \mathbf{L}_1\mathbf{K}_1\mathbf{L}_1^{-1} = \mathbf{L}_1\mathbf{J}_1 \triangleq \mathbf{J}_2\mathbf{L}_2 \tag{34.42}$$

etc. The complete algorithm obeys the following equations

$$\begin{aligned}
&\textit{Step 1}: && \mathbf{K}_k=\mathbf{K}_1,\ \mathbf{R}_k=\mathbf{R}_1,\ \mathbf{H}_k=\mathbf{H}_1,\ \mathbf{E}_k=\mathbf{E}_1 && (34.43)\\
&\textit{Step 2}: && \mathbf{K}_k=\mathbf{J}_k\mathbf{L}_k\ ;\ \mathbf{J}_k\mathbf{J}_k^T=\mathbf{I} && (34.44)\\
&\textit{Step 3}: && \mathbf{H}_{k+1}=\mathbf{H}_k\mathbf{L}_k^{-1} && (34.45)\\
& && \mathbf{E}_{k+1}=\mathbf{L}_k\mathbf{E}_k && (34.46)\\
& && \mathbf{R}_{k+1}=\mathbf{L}_k\mathbf{R}_k && (34.47)\\
& && \mathbf{K}_{k+1}=\mathbf{L}_k\mathbf{K}_k\mathbf{L}_k^{-1} && (34.48)
\end{aligned}$$

Step 4 : Return to step 2 until convergence is satisfactory, i.e., until the most dominant eigenvalue of $\mathbf{K}_k$ remains as the first entry of $\mathbf{K}_k$.

Finally, the result of the approximate aggregation is

$$\mathbf{S}=\mathbf{S}_k\mathbf{S}_{k-1}\ldots\mathbf{S}_1=\ \text{block diag}\ \{\mathbf{I}_{l-1},\mathbf{L}_k\mathbf{L}_{k-1}\ldots\mathbf{L}_1\} \tag{34.49}$$

$$\mathbf{x}^{(k)}=\mathbf{S}\mathbf{x}=\begin{pmatrix}\mathbf{x}_1\\ x_2^{(k)}\\ \mathbf{x}_3^{(k)}\end{pmatrix}=\begin{pmatrix}\mathbf{x}_1\\ z_m\\ \mathbf{x}_3^{(k)}\end{pmatrix} \tag{34.50}$$

$$\boldsymbol{\Gamma}_R=\begin{pmatrix}\mathbf{I}_{l-1} & \mathbf{0}_{l-1,1} & \mathbf{0}_{l-1,n-l}\\ \mathbf{0}_{1,l-1} & 1 & \mathbf{0}_{1,n-l}\end{pmatrix} \tag{34.51}$$

$$\mathbf{z}=\boldsymbol{\Gamma}_R\,\mathbf{x}^{(k)} \tag{34.52}$$

$$\dot{\mathbf{x}}^{(k)}=\begin{pmatrix}\mathbf{F}_1 & \mathbf{H}_{k+1}\\ \mathbf{E}_{k+1} & \mathbf{K}_{k+1}\end{pmatrix}\mathbf{x}^{(k)}+\begin{pmatrix}\mathbf{G}_1\\ \mathbf{LR}_1\end{pmatrix}\mathbf{u} \qquad \mathbf{L}=\mathbf{L}_k\ \ldots\ \mathbf{L}_2\mathbf{L}_1 \tag{34.53}$$

$$\dot{\mathbf{z}}=\boldsymbol{\Gamma}_R\begin{pmatrix}\mathbf{F}_1 & \mathbf{H}_1\mathbf{L}^{-1}\\ \mathbf{LE}_1 & \mathbf{LK}_1\mathbf{L}^{-1}\end{pmatrix}\boldsymbol{\Gamma}_R^T\mathbf{z}+\boldsymbol{\Gamma}_R\begin{pmatrix}\mathbf{G}_1\\ \mathbf{LR}_1\end{pmatrix}\mathbf{u}\ . \tag{34.54}$$

The aggregated matrices are

$$\mathbf{F}=\boldsymbol{\Gamma}_R\begin{pmatrix}\mathbf{F}_1 & \mathbf{H}_1\mathbf{L}^{-1}\\ \mathbf{LE}_1 & \mathbf{LK}_1\mathbf{L}^{-1}\end{pmatrix}\boldsymbol{\Gamma}_R^T \quad \text{and} \quad \mathbf{G}=\boldsymbol{\Gamma}_R\begin{pmatrix}\mathbf{G}_1\\ \mathbf{LR}_1\end{pmatrix}\ . \tag{34.55}$$

34.7 Aggregation Algorithm and Optimal System Performance

An aggregation method guided by an optimal index of performance was documented by *Zheng, J., and Lu, Y., 1988*. This method does not preserve all slow-decaying modes of the system but uses an index of performance. Based on this idea, the parameters of the reduced-order system of arbitrary order are determined optimally.

Consider the single-input system

$$\dot{\mathbf{x}}=\mathbf{A}\mathbf{x}+\mathbf{b}u=\begin{pmatrix}\mathbf{A}_1 & 0\\ 0 & \mathbf{A}_2\end{pmatrix}\mathbf{x}+\begin{pmatrix}\mathbf{b}_1\\ \mathbf{b}_2\end{pmatrix}u \tag{34.56}$$

where $\mathbf{A}\in\mathcal{R}^{n\times n}$, $\mathbf{A}_1\in\mathcal{R}^{l\times l}$ and $\mathbf{A}_1$ contains all dominant modes to be preserved. The index of performance with given $\mathbf{Q}\geq 0$ and $r>0$ is

$$I=\int_0^\infty(\mathbf{x}^T\mathbf{Q}\mathbf{x}+ru^2)dt \tag{34.57}$$

and has to be minimized. The original state $\mathbf{x}(t)$ is approximated by a reduced-order state $\mathbf{z}(t)$ premultiplied by $\mathbf{T}$, i.e. $\mathbf{Tz}(t) \rightarrow \mathbf{x}(t)$. The reduced-order state $\mathbf{z}(t)$ is governed by

$$\dot{\mathbf{z}}(t) = \mathbf{A}_z\mathbf{z}(t) + \mathbf{b}_z u = \begin{pmatrix} \mathbf{A}_1 & \mathbf{0} \\ \mathbf{0} & \mathbf{A}_{z2} \end{pmatrix} \mathbf{z}(t) + \begin{pmatrix} \mathbf{b}_1 \\ \mathbf{b}_{z2} \end{pmatrix} u \tag{34.58}$$

and has to minimize

$$I_z = \int_0^\infty (\mathbf{z}^T\mathbf{Q}_z\mathbf{z} + ru^2)dt \quad \text{where} \quad \mathbf{Q}_z = \mathbf{T}^T\mathbf{Q}\mathbf{T}, \quad \mathbf{T} = \begin{pmatrix} \mathbf{I} & \mathbf{0} \\ \mathbf{0} & \mathbf{L} \end{pmatrix} \tag{34.59}$$

and $\mathbf{L}$ is a non-square matrix.

Now, an algorithm is suggested as follows: A first step with a regular aggregation is chosen as outlined in the preceding section. Then, I_z is minimized yielding a suboptimal control u_1 and an optimal system in the sense of Riccati with eigenvalues λ_{i1}. Using λ_{i1} the matrix $\mathbf{L}$ is modified and denoted $\mathbf{L}_1$. As a result

$$\mathbf{T}_1 = \begin{pmatrix} \mathbf{I} & \mathbf{0} \\ \mathbf{0} & \mathbf{L}_1 \end{pmatrix} \tag{34.60}$$

is achieved. Repeating this algorithm and supposing convergence (*Zheng, J., and Lu, Y., 1988*) the optimal solution, in the above sense, is $\lim \mathbf{z} = \mathbf{z}^\star$ and

$$\lim \mathbf{T}_i = \mathbf{T}^\star, \quad \lim u = u^\star = \mathbf{k}^\star e^{(\mathbf{A}_z + \mathbf{b}_z\mathbf{k}^\star)t}\ \mathbf{z}_o\ . \tag{34.61}$$

Finally, the optimum reduced-order strategy is given by

$$u^\star = \mathbf{k}^\star\mathbf{T}^{\star\dagger T}\ \mathbf{x} \qquad \text{or} \qquad \mathbf{k}^\star\mathbf{\Gamma}\mathbf{x} \tag{34.62}$$

where $\mathbf{\Gamma}\mathbf{x}$ is the reduced-order state vector.

34.8 Component Connection Model

A component of an interconnected dynamic system is defined as

$$\dot{\mathbf{x}}_i = \mathbf{A}_i\mathbf{x}_i + \mathbf{B}_i\mathbf{u}_i \qquad \mathbf{x}_i \in \mathcal{R}^{n_i} \quad \mathbf{u}_i \in \mathcal{R}^{m_i} \tag{34.63}$$

$$\mathbf{y}_i = \mathbf{C}_i\mathbf{x}_i + \mathbf{D}_i\mathbf{u}_i \qquad \mathbf{y}_i \in \mathcal{R}^{r_i} \tag{34.64}$$

The number of components is L. With the abbreviations

$$\mathbf{x} = (\mathbf{x}_1^T \ldots \mathbf{x}_i^T \ldots \mathbf{x}_L^T)^T \quad \mathbf{u} = (\mathbf{u}_1^T \ldots \mathbf{u}_i^T \ldots \mathbf{u}_L^T)^T \quad \mathbf{y} = (\mathbf{y}_1^T \ldots \mathbf{y}_i^T \ldots \mathbf{y}_L^T)^T \tag{34.65}$$

$$\mathbf{A} = \text{block diag } (\mathbf{A}_1, \ldots \mathbf{A}_L) \qquad \mathbf{B} = \text{block diag } (\mathbf{B}_1, \ldots \mathbf{B}_L) \tag{34.66}$$

$$\mathbf{C} = \text{block diag } (\mathbf{C}_1, \ldots \mathbf{C}_L) \qquad \mathbf{D} = \text{block diag } (\mathbf{D}_1, \ldots \mathbf{D}_L) \tag{34.67}$$

the system containing L components can be written

$$\dot{\mathbf{x}} = \mathbf{A}\mathbf{x} + \mathbf{B}\mathbf{u} \qquad \mathbf{A} \in \mathcal{R}^{n\times n}, \quad \mathbf{B} \in \mathcal{R}^{n\times m} \tag{34.68}$$

$$\mathbf{y} = \mathbf{C}\mathbf{x} + \mathbf{D}\mathbf{u} \qquad \mathbf{C} \in \mathcal{R}^{r\times n}, \quad \mathbf{D} \in \mathcal{R}^{r\times m} \tag{34.69}$$

$$\sum_{i=1}^{L} m_i = m, \quad \sum_{i=1}^{L} n_i = n, \quad \sum_{i=1}^{L} r_i = r, \quad \mathbf{x} \in \mathcal{R}^n, \quad \mathbf{u} \in \mathcal{R}^m, \quad \mathbf{u}_c \in \mathcal{R}^{m_c}, \quad \mathbf{y} \in \mathcal{R}^r, \quad \mathbf{y}_c \in \mathcal{R}^{r_c}.$$

The interconnection properties are given by the algebraic equations

$$\mathbf{u} = \mathbf{M}_y \mathbf{y} + \mathbf{M}_u \mathbf{u}_c \qquad \mathbf{M}_y \in \mathcal{R}^{m \times r}, \quad \mathbf{M}_u \in \mathcal{R}^{m \times m_c} \tag{34.70}$$

$$\mathbf{y}_c = \mathbf{N}_y \mathbf{y} + \mathbf{N}_u \mathbf{u}_c \qquad \mathbf{N}_y \in \mathcal{R}^{r_c \times r}, \quad \mathbf{N}_u \in \mathcal{R}^{r_c \times m_c}, \tag{34.71}$$

the subscript $_c$ indicating the composite system. The overall state-space description of the interconnected system in the component connection model framework is

$$\dot{\mathbf{x}} = [\mathbf{A} + \mathbf{B}(\mathbf{I} - \mathbf{M}_y\mathbf{D})^{-1}\mathbf{M}_y\mathbf{C}]\mathbf{x} + \mathbf{B}(\mathbf{I} - \mathbf{M}_y\mathbf{D})^{-1}\mathbf{M}_u\mathbf{u}_c \tag{34.72}$$

$$\mathbf{y}_c = \mathbf{N}_y(\mathbf{I} - \mathbf{D}\mathbf{M}_y)^{-1}\mathbf{C}\mathbf{x} + [\mathbf{N}_y(\mathbf{I} - \mathbf{D}\mathbf{M}_y)^{-1}\mathbf{D}\mathbf{M}_u + \mathbf{N}_u]\mathbf{u}_c. \tag{34.73}$$

In Eq.(34.73), the identities

$$\mathbf{D}(\mathbf{I} - \mathbf{M}_y\mathbf{D})^{-1} \equiv (\mathbf{I} - \mathbf{D}\mathbf{M}_y)^{-1}\mathbf{D} \qquad \mathbf{I} + \mathbf{D}(\mathbf{I} - \mathbf{M}_y\mathbf{D})^{-1}\mathbf{M}_y \equiv (\mathbf{I} - \mathbf{D}\mathbf{M}_y)^{-1} \tag{34.74}$$

were applied.

34.9 Interconnected System and Component Connection Model

If a large-scale interconnected system is given in the component connection model framework (Fig. 34.2) (*De Carlo, R.A., and Saeks, R., 1981*), it can be shown that the model reduction at the component level is equivalent to the model reduction at the level of the interconnected system level (*Feliachi, A., and Bhurtun, C., 1987*). An interconnected system is exactly aggregable if it comprises at least one component i which is exactly aggregable. This theorem can be proved as follows.

Component i is pretended to be exactly aggregable

$$\dot{\mathbf{z}}_i = \mathbf{F}_i\mathbf{z}_i + \mathbf{G}_i\mathbf{u}_i \tag{34.75}$$

$$\mathbf{y}_i = \mathbf{H}_i\mathbf{z}_i + \mathbf{T}_i\mathbf{u}_i \tag{34.76}$$

with the aggregation conditions

$$\mathbf{F}_i\mathbf{\Gamma}_i = \mathbf{\Gamma}_i\mathbf{A}_i \qquad \mathbf{G}_i = \mathbf{\Gamma}_i\mathbf{B}_i \qquad \mathbf{H}_i\mathbf{\Gamma}_i = \mathbf{C}_i \qquad \mathbf{T}_i = \mathbf{D}_i. \tag{34.77}$$

Since the algebraic input output relation $\mathbf{y}_i$ over $\mathbf{u}_i$ is identical in the original and in the aggregated model, this requires $\mathbf{T}_i = \mathbf{D}_i$. The aggregated vector $\mathbf{z}$ contains all component vectors $\mathbf{x}_k$ except $\mathbf{x}_i$ which is replaced by $\mathbf{z}_i$

$$\mathbf{z} = (\mathbf{x}_1^T \; \mathbf{x}_2^T \; \dots \; \mathbf{x}_{i-1}^T \; \mathbf{z}_i^T \; \mathbf{x}_{i+1}^T \; \dots \; \mathbf{x}^T)^T = \mathbf{\Gamma}\mathbf{x} \tag{34.78}$$

$$\mathbf{\Gamma} = \text{block diag}\,\{\mathbf{I}_1, \mathbf{I}_2, \dots \mathbf{I}_{i-1}, \mathbf{\Gamma}_i, \mathbf{I}_{i+1}, \dots \mathbf{I}_L\} \qquad \mathbf{I}_k \triangleq \mathbf{I}_{n_k, n_k} \tag{34.79}$$

$$\mathbf{F} = \text{block diag}\,\{\mathbf{A}_1, \mathbf{A}_2, \dots \mathbf{A}_{i-1}, \mathbf{F}_i, \mathbf{A}_{i+1}, \dots \mathbf{A}_L\} \tag{34.80}$$

$$\mathbf{H} = \text{block diag}\,\{\mathbf{C}_1, \mathbf{C}_2, \dots \mathbf{C}_{i-1}, \mathbf{H}_i, \mathbf{C}_{i+1}, \dots \mathbf{C}_L\} \tag{34.81}$$

$$\mathbf{T} = \text{block diag}\,\{\mathbf{D}_1, \mathbf{D}_2, \dots \mathbf{D}_{i-1}, \mathbf{T}_i, \mathbf{D}_{i+1}, \dots \mathbf{D}_L\} \qquad \mathbf{T}_i = \mathbf{D}_i. \tag{34.82}$$

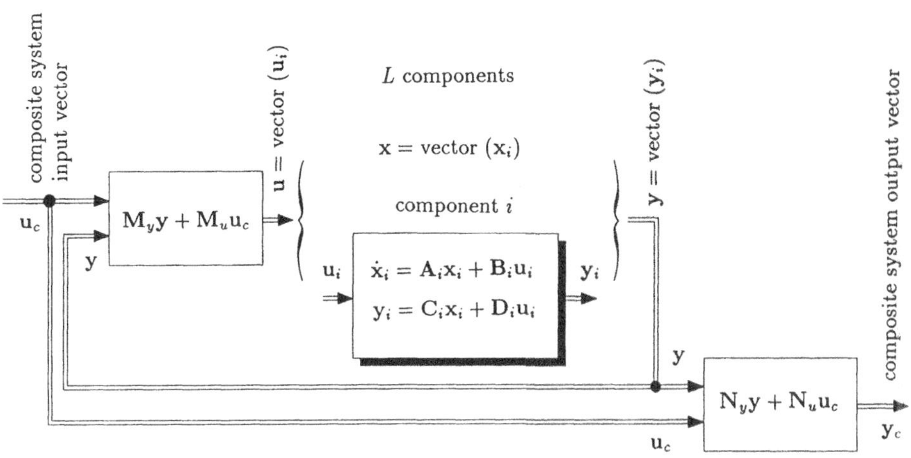

Figure 34.2: Component connection model framework

If the component i is replaced by its aggregated model, the state-space description of the component connection model can immediately be transferred from Eqs.(34.72) and (34.73), substituting $\mathbf{x}, \mathbf{A}, \mathbf{B}, \mathbf{C}, \mathbf{D}$ by $\mathbf{z}, \mathbf{F}, \mathbf{G}, \mathbf{H}, \mathbf{T} = \mathbf{D}$, respectively,

$$\dot{\mathbf{z}} = [\mathbf{F} + \mathbf{G}(\mathbf{I} - \mathbf{M}_y\mathbf{D})^{-1}\mathbf{M}_y\mathbf{H}]\mathbf{z} + \mathbf{G}(\mathbf{I} - \mathbf{M}_y\mathbf{D})^{-1}\mathbf{M}_u\mathbf{u}_c \tag{34.83}$$

$$\mathbf{y}_c = \mathbf{N}_y(\mathbf{I} - \mathbf{D}\mathbf{M}_y)^{-1}\mathbf{H}\mathbf{z} + [\mathbf{N}_y(\mathbf{I} - \mathbf{D}\mathbf{M}_y)^{-1}\mathbf{D}\mathbf{M}_y + \mathbf{N}_u]\mathbf{u}_c\ . \tag{34.84}$$

Comparing Eqs.(34.83) and (34.84) with (34.72) and (34.73), respectively, the aggregation conditions are

$$\mathbf{F}\mathbf{\Gamma} = \mathbf{\Gamma}\mathbf{A} \tag{34.85}$$

$$\mathbf{G}(\mathbf{I} - \mathbf{M}_y\mathbf{D})^{-1}\mathbf{M}_y\mathbf{H}\mathbf{\Gamma} = \mathbf{\Gamma}\mathbf{B}(\mathbf{I} - \mathbf{M}_y\mathbf{D})^{-1}\mathbf{M}_y\mathbf{C} \tag{34.86}$$

$$\mathbf{G} = \mathbf{\Gamma}\mathbf{B}, \qquad \mathbf{H}\mathbf{\Gamma} = \mathbf{C} \quad \text{and} \quad \mathbf{T} = \mathbf{D}\ . \tag{34.87}$$

If the exact aggregation has to be replaced by the approximate one, then one can follow the iterative approach of Eq.(34.54). Thc component related reduction is treated much more easily because it is evident that the size of a component is much smaller than that of the overall plant. The results derived in the iterative algorithm are simply inserted into the corresponding component block in Fig. 34.2.

Chapter 35

Optimum Control of Approximate and Nonlinear Systems

35.1 Taylor Series Models

Consider the matrix set $\mathcal{X} = \{\mathbf{X}_o \ \ \mathbf{X}_1 \ \ \mathbf{X}_2 \ ... \ \mathbf{X}_{q-1}\}$ and a mapping T_t , given by the matrix sum weighted with powers of time t

$$T_t\mathcal{X} \triangleq \sum_{k=0}^{q-1} \mathbf{X}_k t^k \ . \tag{35.1}$$

The matrix set $\mathcal{X}$ may be specialized to a vector set or scalar set. The calligraphic symbol denotes the set corresponding to the elements in boldface symbols. The integral of $T_t\mathcal{X}$ with respect to time t

$$\int_0^t T_\tau \mathcal{X} d\tau = \sum_{k=0}^{q-1} \frac{1}{k+1}\mathbf{X}_k t^{k+1} = T_t(L\mathcal{X}) \tag{35.2}$$

can be replaced by the mapping T_t applied to $L\mathcal{X}$ if a transformer L is used which is defined by replacing

$$\mathcal{X} = \{\mathbf{X}_o \ \ \mathbf{X}_1 \ \ \mathbf{X}_2 \ \ \mathbf{X}_3 \ .... \ \mathbf{X}_{q-1}\} \quad \text{by} \quad L\mathcal{X} = \{0 \ \ \frac{\mathbf{X}_o}{1} \ \ \frac{\mathbf{X}_1}{2} \ \ \frac{\mathbf{X}_2}{3} \ .... \ \frac{\mathbf{X}_{q-2}}{q-1}\} \ . \tag{35.3}$$

The Taylor series expansion and truncated series is expressed using T_t and $\mathcal{X}$

$$\mathbf{X}(t) = \sum_{k=0}^{\infty} \mathbf{X}_k t^k \doteq \sum_{k=0}^{q-1} \mathbf{X}_k t^k = T_t\mathcal{X} \qquad \text{where} \qquad \mathbf{X}_k = \frac{1}{k!}\frac{d^k\mathbf{X}(t)}{dt^k}\Big|_{t=0} \ . \tag{35.4}$$

Supposing that the matrices $\mathbf{X}_i$ and $\mathbf{Y}_i$ possess appropriate dimension, the matrix product $\mathbf{X}(t)\mathbf{Y}(t)$ in power series expansion is

$$\mathbf{X}(t)\mathbf{Y}(t) \doteq \sum_{k=0}^{q-1} \Big(\sum_{i+j=k} \mathbf{X}_i\mathbf{Y}_j \Big) t^k \qquad i,j = 0,1,2...q-1 \tag{35.5}$$

and can be rewritten to

$$\mathbf{X}(t)\mathbf{Y}(t) = T_t(\mathcal{X} \odot \mathcal{Y}) \tag{35.6}$$

where $\odot$ denotes the generalized product corresponding to a polynomial product

$$\mathcal{X} \odot \mathcal{Y} = \mathcal{Z} \qquad \mathbf{Z}_k = \sum_{i+j=k} \mathbf{X}_i\mathbf{Y}_j \qquad i,j = 0,1,2...q-1 \ . \tag{35.7}$$

Then, a linear time-varying system

$$\dot{\mathbf{x}}(t) = \mathbf{A}(t)\mathbf{x}(t)+\mathbf{B}(t)\mathbf{u}(t), \quad \mathbf{x}(0) = \mathbf{x}_o \quad \rightsquigarrow \quad \mathbf{x}(t)-\mathbf{x}_o = \int_0^t \mathbf{A}(\tau)\mathbf{x}(\tau)d\tau+\int_0^t \mathbf{B}(\tau)\mathbf{u}(\tau)d\tau \tag{35.8}$$

is reformulated to

$$T_t\mathcal{X} - T_t\mathcal{X}_o = \int_0^t (T_\tau\mathcal{A})(T_\tau\mathcal{X})d\tau + \int_0^t (T_\tau\mathcal{B})(T_\tau\mathcal{U})d\tau = T_t[L(\mathcal{A}\odot\mathcal{X})]+T_t[L(\mathcal{B}\odot\mathcal{U})] \tag{35.9}$$

$$\mathcal{X} - \mathcal{X}_o = L(\mathcal{A}\odot\mathcal{X}) + L(\mathcal{B}\odot\mathcal{U}) . \tag{35.10}$$

This result is a suitable form for computation (*Wang, W.P., 1990*). The time-varying matrix $\mathbf{A}(t)$ is taken into account by the matrix set $\mathcal{A}$.

The analysis can be extended easily to approximate bilinear systems and optimal time-varying control systems with quadratic performance. Let $\phi = \{\mathbf{\Phi}_o \ \ \mathbf{\Phi}_1 \ \ldots . \ \mathbf{\Phi}_{q-1}\}$ denote the set of transition matrices corresponding to $\mathbf{M}_R$, see Eq.(9.19), and let $\mathcal{I}, \mathcal{M}_R$ denote the matrix sets corresponding to $\mathbf{I}, \mathbf{M}_R$, respectively, then, for the optimal control a recursive algorithm is given by

$$\phi - \mathcal{I} + L(\phi\odot\mathcal{M}_R) = 0 \qquad \text{starting with} \quad \mathbf{\Phi} = \mathbf{I} . \tag{35.11}$$

35.2 Robustness Assessment Via Taylor Expansion

Consider the vector-valued signal $\mathbf{x}(t) \in \mathcal{R}^n$ and its Taylor expansion

$$\mathbf{x}(t) = \mathbf{x}_o + \mathbf{x}_1 t + \mathbf{x}_2 t^2 + \ \ldots . \ + \mathbf{x}_{q-2}t^{q-2} + \mathbf{x}_{q-1}t^{q-1} \tag{35.12}$$

where $\mathbf{x}_k$ is the Taylor coefficient associated with the power t^k . The derivative with respect to time is

$$\dot{\mathbf{x}}(t) = \dot{\mathbf{x}}_o+\dot{\mathbf{x}}_1 t+\dot{\mathbf{x}}_2 t^2+ \ \ldots . \ +\dot{\mathbf{x}}_{q-2}t^{q-2}+\dot{\mathbf{x}}_{q-1}t^{q-1} = \mathbf{x}_1+2\mathbf{x}_2 t+ \ \ldots . \ +(q-2)\mathbf{x}_{q-2}t^{q-3}+(q-1)\mathbf{x}_{q-1}t^{q-2} . \tag{35.13}$$

Utilizing the vector definitions, the additional Taylor coefficients $\dot{\mathbf{x}}_k$ and the matrix $\dot{\mathbf{X}}$ of expansion coefficients $\dot{\mathbf{x}}_k$

$$\mathbf{X} \triangleq \begin{pmatrix} \mathbf{x}_o^T \\ \mathbf{x}_1^T \\ \vdots \\ \mathbf{x}_{q-2}^T \\ \mathbf{x}_{q-1}^T \end{pmatrix}, \qquad \dot{\mathbf{X}} \triangleq \begin{pmatrix} \dot{\mathbf{x}}_o^T \\ \dot{\mathbf{x}}_1^T \\ \vdots \\ \dot{\mathbf{x}}_{q-2}^T \\ \dot{\mathbf{x}}_{q-1}^T \end{pmatrix} = \begin{pmatrix} \mathbf{x}_1^T \\ 2\,\mathbf{x}_2^T \\ \vdots \\ (q-1)\mathbf{x}_{q-1}^T \\ \mathbf{0}^T \end{pmatrix}, \qquad \mathbf{t} \triangleq \begin{pmatrix} 1 \\ t \\ \vdots \\ t^{q-2} \\ t^{q-1} \end{pmatrix} \tag{35.14}$$

the integral with respect to time is replaced by an operational matrix $\mathbf{P}$ (*Wang, W.P., 1990*)

$$\mathbf{X} = \mathbf{P}\dot{\mathbf{X}} + \mathbf{X}_o \triangleq \begin{pmatrix} 0 & 0 & 0 & \ldots & 0 & 0 \\ 1 & 0 & 0 & \ldots & 0 & 0 \\ 0 & 1/2 & 0 & \ldots & 0 & 0 \\ 0 & 0 & 1/3 & \ldots & 0 & 0 \\ \vdots & \vdots & \vdots & \ddots & \vdots & 0 \\ 0 & 0 & 0 & \ldots & 0 & 0 \end{pmatrix} \dot{\mathbf{X}} + \begin{pmatrix} \mathbf{x}_o^T \\ \mathbf{0}^T \\ \mathbf{0}^T \\ \mathbf{0}^T \\ \vdots \\ \mathbf{0}^T \end{pmatrix} . \tag{35.15}$$

Then, the vector-valued signal $\mathbf{x}(t)$ is rewritten

$$\mathbf{x}(t) = \mathbf{X}^T\mathbf{t} = (\mathbf{x}_o \vdots \mathbf{x}_1 \ldots \mathbf{x}_{q-2} \vdots \mathbf{x}_{q-1}) \begin{pmatrix} 1 \\ t \\ \vdots \\ t^{q-2} \\ t^{q-1} \end{pmatrix} = \sum_{i=0}^{q-1} \mathbf{x}_i t^i \ . \tag{35.16}$$

The solution $\mathbf{x}(t)$ of the unperturbed system $\dot{\mathbf{x}}(t) = \mathbf{A}\mathbf{x}(t)$ is given by any classical method using, e.g., matrix exponential. The perturbed system is given by

$$\dot{\mathbf{x}}(t)+\Delta\dot{\mathbf{x}}(t) = (\mathbf{A}+\Delta\mathbf{A})[\mathbf{x}(t)+\Delta\mathbf{x}(t)] \quad \leadsto \quad \Delta\dot{\mathbf{x}}(t) = \mathbf{A}\Delta\mathbf{x}(t)+\Delta\mathbf{A}\ \mathbf{x}(t)+\Delta\mathbf{A}\ \Delta\mathbf{x}(t) \tag{35.17}$$

where $\Delta\mathbf{x}(t)$ denotes the deviation from the unperturbed signal $\mathbf{x}(t)$. Referring to $\Delta\dot{\mathbf{x}}(t) \triangleq \Delta\dot{\mathbf{X}}^T\mathbf{t}$ and $\Delta\mathbf{X} \triangleq \mathbf{P}\Delta\dot{\mathbf{X}} + \mathbf{0}$ one has

$$\Delta\dot{\mathbf{X}}^T = \mathbf{A}\ \Delta\mathbf{X}^T + \Delta\mathbf{A}\ \mathbf{X}^T + \Delta\mathbf{A}\ \Delta\mathbf{X}^T \ . \tag{35.18}$$

Transposing and premultiplying by $\mathbf{P}$ yields

$$\Delta\mathbf{X} = \mathbf{P}\ \Delta\dot{\mathbf{X}} = \mathbf{P}\ \Delta\mathbf{X}\ \mathbf{A}^T + \mathbf{P}\mathbf{X}\ \Delta\mathbf{A}^T + \mathbf{P}\ \Delta\mathbf{X}\ \Delta\mathbf{A}^T \tag{35.19}$$

$$(\mathbf{I} - \mathbf{A}\otimes\mathbf{P} - \Delta\mathbf{A}\otimes\mathbf{P})\text{col}\Delta\mathbf{X} = \text{col}(\mathbf{P}\mathbf{X}\ \Delta\mathbf{A}^T) = (\Delta\mathbf{A}\otimes\mathbf{P})\text{col}\mathbf{X} \tag{35.20}$$

$$\text{col}\Delta\mathbf{X} = (\mathbf{I} - \mathbf{A}\otimes\mathbf{P} - \Delta\mathbf{A}\otimes\mathbf{P})^{-1} = (\Delta\mathbf{A}\otimes\mathbf{P})\text{col}\mathbf{X} \ . \tag{35.21}$$

Now, employ the Frobenius norm $\|\text{col}\Delta\mathbf{X}\|_F = \|\Delta\mathbf{X}\|_F$ and the spectral norm property of the Kronecker product, i.e., $\|\mathbf{M}\otimes\mathbf{N}\|_s = \|\mathbf{M}\|_s\|\mathbf{N}\|_s$. Then, for any $t \in [0 \ \ t_f]$

$$\|\mathbf{x}(t)\|_F \le \|\mathbf{X}^T\|_s\|\mathbf{t}\|_F \ . \tag{35.22}$$

In view of this, an upper bound for $\|\Delta\mathbf{x}(t)\|_F$ is (*Weinmann, A., 1991*)

$$\|\Delta\mathbf{x}(t)\|_F \ \le \ \|\Delta\mathbf{X}\|_s\|\mathbf{t}\|_F \le \|\Delta\mathbf{X}\|_F\|\mathbf{t}\|_F \tag{35.23}$$

$$\|\Delta\mathbf{x}(t)\|_F \ \le \ \|\text{col}\Delta\mathbf{X}\|_F\|\mathbf{t}\|_F \le \frac{\|\Delta\mathbf{A}\|_s\|\mathbf{P}\|_s\|\mathbf{X}\|_F\|\mathbf{t}\|_F}{\sigma_{\min}[\mathbf{I} - \mathbf{A}\otimes\mathbf{P} - \Delta\mathbf{A}\otimes\mathbf{P}]} \ . \tag{35.24}$$

35.3 Kronecker Power Models

Using the Kronecker notation Eq.(38.94), the nonlinear and time-varying system is stated

$$\dot{\mathbf{x}}(t) = \mathbf{A}[\mathbf{x}(t)] + \sum_{i=1}^{m} \mathbf{B}_i[\mathbf{x}(t)]u_i(t) \qquad \text{where} \tag{35.25}$$

$$\mathbf{A}[\mathbf{x}(t)] = \sum_{j=1}^{\infty} \mathbf{A}_j\mathbf{x}^{[j]}; \quad \mathbf{B}_i[\mathbf{x}(t)] = \sum_{j=0}^{\infty} \mathbf{B}_{ij}\mathbf{x}^{[j]}; \quad \mathbf{B}(\mathbf{x}) = \sum_{k=1}^{m}\sum_{j=0}^{\infty} \mathbf{B}_{kj}\mathbf{x}^{[j]} \tag{35.26}$$

$$\mathbf{y}(t) = \mathbf{c}(\mathbf{x}) = \sum_{j=1}^{\infty} \mathbf{C}_j\mathbf{x}^{[j]} \ . \tag{35.27}$$

Several nonlinear systems can be treated is this way[1]. The problem of finding the optimal state feedback controller minimizing the performance index

$$\mathbf{I} = \int_0^{\infty} [\mathbf{y}^T(t)\mathbf{Q}\mathbf{y}(t) + \mathbf{u}^T(t)\mathbf{R}\mathbf{u}(t)]dt \rightarrow \min \tag{35.28}$$

is given by

$$\mathbf{u}^*(t) = -\mathbf{R}^{-1}\mathbf{B}^T(\mathbf{x})\mathbf{P}(\mathbf{x}) \tag{35.29}$$

$$\mathbf{c}^T(\mathbf{x})\mathbf{Q}\mathbf{c}(\mathbf{x}) + \mathbf{P}^T(\mathbf{x})\mathbf{A}(\mathbf{x}) + \mathbf{A}^T(\mathbf{x})\mathbf{P}(\mathbf{x}) - \mathbf{P}^T(\mathbf{x})\mathbf{B}(\mathbf{x})\mathbf{R}^{-1}\mathbf{B}^T(\mathbf{x})\mathbf{P}(\mathbf{x}) = 0\,. \tag{35.30}$$

This result is very similar to the structure of the Riccati controller (*Rotella, F., and Dauphin-Tanguy, G., 1988*). Solving Eq.(35.30), use is made of the following decomposition

$$\mathbf{P}(\mathbf{x}) = \sum_{j=1}^{\infty} \mathbf{P}_j \mathbf{x}^{[j]}\,. \tag{35.31}$$

$$\mathbf{x}(t),\ \mathbf{A}[\mathbf{x}(t)],\ \mathbf{B}_i[\mathbf{x}(t)],\ \mathbf{P}(\mathbf{x}) \in \mathcal{R}^n \quad \mathbf{x}^{[j]} \in \mathcal{R}^{(n^j)} \quad \mathbf{y}(t),\ \mathbf{c}(\mathbf{x}) \in \mathcal{R}^r \quad \mathbf{u}(t) \in \mathcal{R}^m \quad \mathbf{A}_j,\ \mathbf{B}_{ij} \in \mathcal{R}^{n\times(n^j)}\,. \tag{35.32}$$

The following lemmas prove helpful when reducing Eq.(35.30): The scalar expression $\mathbf{y}^T\mathbf{A}\mathbf{x}$ is - though scalar - subject to the *col*-operation, then formally Eq. (4.40) can be applied and finally, with regard to the scalar dimension, the transpose is admissible

$$\mathbf{x}^T\mathbf{A}\mathbf{y} = \mathrm{col}\ (\mathbf{x}^T\mathbf{A}\mathbf{y}) = (\mathbf{y}^T \otimes \mathbf{x}^T)\mathrm{col}\ \mathbf{A} = (\mathrm{col}^T\mathbf{A})(\mathbf{y}\otimes\mathbf{x}). \tag{35.33}$$

If $\mathbf{y} = \mathbf{x}$, it results $\mathbf{x}^T\mathbf{A}\mathbf{x} = (\mathrm{col}^T\mathbf{A})\mathbf{x}^{[2]}$. Substituting $\mathbf{x}^{[j]}$ and $\mathbf{x}^{[i]}$ for $\mathbf{x}$ and $\mathbf{y}$

$$\mathbf{x}^{[j]T}\mathbf{A}\mathbf{x}^{[j]} = (\mathrm{col}^T\mathbf{A})\mathbf{x}^{[i+j]}\,. \tag{35.34}$$

Combining Eqs.(35.31), (35.27), (35.26) and defining $\mathbf{R}^{-1}$ as a matrix with elements r_{kl}, from Eq.(35.30)

$$\Big(\sum_{j=1}^{\infty}\mathbf{x}^{[j]T}\mathbf{C}_j^T\Big)\mathbf{Q}\Big(\sum_{j=1}^{\infty}\mathbf{C}_j^T\mathbf{x}^{[j]}\Big) + \Big(\sum_{j=1}^{\infty}\mathbf{x}^{[j]T}\mathbf{P}_j^T\Big)\Big(\sum_{j=1}^{\infty}\mathbf{A}_j\mathbf{x}^{[j]}\Big) + \Big(\sum_{j=1}^{\infty}\mathbf{x}^{[j]T}\mathbf{A}_j^T\Big)\Big(\sum_{j=1}^{\infty}\mathbf{P}_j\mathbf{x}^{[j]}\Big)$$

$$-\sum_{k=1}^{m}\sum_{l=1}^{m}\Big(\sum_{i=1}^{\infty}\mathbf{x}^{[i]T}\mathbf{P}_i^T\Big)\Big(\sum_{j=0}^{\infty}\mathbf{B}_{kj}\mathbf{x}^{[j]}\Big)r_{kl}\Big(\sum_{j'=0}^{\infty}\mathbf{x}^{[j']T}\mathbf{B}_{lj'}^T\Big)\Big(\sum_{i'=1}^{\infty}\mathbf{P}_{i'}\mathbf{x}^{[i']}\Big) = 0 \tag{35.35}$$

is obtained. Applying Eq.(35.34),

$$\sum_{i=1}^{\infty}\sum_{j=1}^{\infty}\mathrm{col}^T\big(\mathbf{C}_i^T\mathbf{Q}\mathbf{C}_j + \mathbf{P}_i^T\mathbf{A}_j + \mathbf{A}_i^T\mathbf{P}_j\big)\mathbf{x}^{[i+j]} - \sum_{k=1}^{m}\sum_{l=1}^{m}\Omega_k\ r_{kl}\ \bar{\Omega}_l = 0 \tag{35.36}$$

$$\text{where}\quad \Omega_k = \sum_{i=1}^{\infty}\sum_{j=0}^{\infty}\mathbf{x}^{[i+j]T}\ \mathrm{col}\ (\mathbf{P}_i^T\mathbf{B}_{kj}) \quad\text{and}\quad \bar{\Omega}_l = \sum_{j'=0}^{\infty}\sum_{i'=1}^{\infty}\mathrm{col}^T(\mathbf{P}_{i'}^T\mathbf{B}_{lj'})\mathbf{x}^{[i'+j']} = 0\,. \tag{35.37}$$

Using

$$\mathbf{V}_{ij} \triangleq \big(\mathrm{col}\ \mathbf{P}_i^T\mathbf{B}_{1j} \ \vdots\ \mathrm{col}\ \mathbf{P}_i^T\mathbf{B}_{2j} \ \vdots \ldots \mathrm{col}\ \mathbf{P}_i^T\mathbf{B}_{mj}\big) \tag{35.38}$$

[1]Continuing the original sense of the coefficient matrix, input matrix and matrix variable of the Riccati equation the terms $\mathbf{A}, \mathbf{B}_i, \mathbf{P}$ are respectively written as capital letters although they represent vector-valued functions.

and omitting simple calculation yields

$$\sum_{i=1}^{\infty}\sum_{j=1}^{\infty} \mathrm{col}^T(\mathbf{C}_i^T\mathbf{Q}\mathbf{C}_j + \mathbf{P}_i^T\mathbf{A}_j + \mathbf{A}_i^T\mathbf{P}_j)\mathbf{x}^{[i+j]}$$

$$- \sum_{k=1}^{\infty}\sum_{i'=1}^{\infty}\sum_{l=0}^{\infty}\sum_{j'=0}^{\infty} \mathrm{col}^T(\mathbf{V}_{kl}\mathbf{R}^{-1}\mathbf{V}_{i'j'})\mathbf{x}^{[k+l+i'+j']} = \mathbf{0} \,. \quad (35.39)$$

The unknown matrices $\mathbf{P}_i$ are achieved by comparing the coefficients of $\mathbf{x}^{[i+1]}$ (*Rotella, F., and Dauphin-Tanguy, G., 1988*), that is,
the matrix $\mathbf{P}_1$ from a classical algebraic Riccati equation

$$\mathbf{C}_1^T\mathbf{Q}\mathbf{C}_1 + \mathbf{P}_1^T\mathbf{A}_1 + \mathbf{A}_1^T\mathbf{P}_1 - \mathbf{P}_1^T\mathbf{B}_0\mathbf{R}^{-1}\mathbf{B}_0^T\mathbf{P}_1 = \mathbf{0} \quad (35.40)$$

and the matrix $\mathbf{P}_2$

$$\mathbf{P}_2 = (-\mathbf{A}_1^T - \mathbf{P}_1^T\mathbf{B}_0\mathbf{R}^{-1}\mathbf{B}_0^T)^{-1}(\mathbf{C}_1^T\mathbf{Q}\mathbf{C}_2 + \mathbf{P}_1^T\mathbf{A}_2 - \mathbf{V}_{10}\mathbf{R}^{-1}\mathbf{V}_{11}^T) \,. \quad (35.41)$$

In order to calculate $\mathbf{P}_i$ $\forall i \geq 2$, a slightly modified notation must be used to avoid singularity. A non-redundant "power" $\check{\mathbf{x}}^{|j|}$ is defined. This expression partly replaces the former Kronecker power expression

$$\check{\mathbf{x}}^{|j|} \triangleq \Big(x_1^j \,\vdots\, x_1^{j-1}x_2 \,\vdots\, x_1^{j-1}x_3 \,\vdots \ldots x_1^{j-1}x_n \,\vdots \quad (35.42)$$

$$x_1^{j-2}x_2^2 \,\vdots\, x_1^{j-2}x_2x_3 \,\vdots \ldots x_1^{j-2}x_2x_n \,\vdots \ldots x_1^{j-2}x_n^2 \,\vdots \quad (35.43)$$

$$x_1^{j-3}x_2^3 \,\vdots\, x_1^{j-3}x_2^2x_3 \,\vdots \ldots \ldots \ldots x_q^j\Big)^T \quad (35.44)$$

$$\mathbf{x}^{[j]} \triangleq \mathbf{T}_j\check{\mathbf{x}}^{|j|} \text{ (defining } \mathbf{T}_j) \quad \text{and} \quad \check{\mathbf{x}}^{|j|} = \mathbf{T}_j^{\sharp L}\mathbf{x}^{[j]}, \quad \mathbf{T}_j \in \mathcal{R}^{n^j \times \alpha_j}, \quad n^j > \alpha_j \quad (35.45)$$

$$\mathbf{A}(\mathbf{x}) = \sum_{j=0}^{\infty}\mathbf{A}_j\mathbf{x}^{[j]} = \sum_{j=0}^{\infty}\check{\mathbf{A}}\check{\mathbf{x}}^{|j|} \quad \text{where} \quad \check{\mathbf{A}}_j = \mathbf{A}_j\mathbf{T}_j \quad \text{and} \quad \mathbf{A}_j^\star = \check{\mathbf{A}}_j\mathbf{T}^{\sharp L} \,. \quad (35.46)$$

The expression $\mathbf{x}^{[j]} = \mathbf{T}_j\check{\mathbf{x}}^{|j|}$ corresponds to $\mathbf{y}_m = \mathbf{M}\mathbf{p}$, well-known from linear regression, $\check{\mathbf{x}}^{|j|} = \mathbf{T}_j^{\sharp L}\mathbf{x}^{[j]}$ is equivalent to $\mathbf{p}^\star = \mathbf{M}^{\sharp L}\mathbf{y}_m$.

Given the algebraic system of linear equations

$$\mathbf{M}\mathbf{p} = \mathbf{y}_m, \quad \mathbf{p} \in \mathcal{R}^{n_p}, \quad \mathbf{y}_m \in \mathcal{R}^{n_m}, \quad n_p > n_m \,. \quad (35.47)$$

There are more unknown parameters p_i than measured values y_{mi} and an infinite variety of solutions $\mathbf{p}$. Within this set of solutions a particular $\mathbf{p}$ is selected, characterized by $||\mathbf{p}||_F \to$ min and termed as $\mathbf{p}^\star$. From regression techniques, $\mathbf{p}^\star = \mathbf{M}^{\sharp R}\mathbf{y}_m$. Substituting $\mathbf{p}^\star$ into $\mathbf{M}\mathbf{p} = \mathbf{y}_m$, the identity can be proved where $\mathbf{p}^\star$ is a special solution of $\mathbf{p}$, of course. Note that this substitution yields n_m scalar equivalences.

A mistake might happen if one would try to substitute $\mathbf{y}_m = \mathbf{M}\mathbf{p}$ into $\mathbf{p}^\star = \mathbf{M}^{\sharp R}\mathbf{y}_m \stackrel{?}{=} \mathbf{M}^{\sharp R}\mathbf{M}\mathbf{p}$ which is obviously incorrect: $\mathbf{M}^{\sharp R}\mathbf{M} \neq \mathbf{I}$. The last equivalence, labelled with a question mark, pretends n_p scalar equivalences, in consequence of n_m given equations. Such an increase must be invalid. Finally, $\mathrm{rank}[\mathbf{M}^{\sharp R}\mathbf{M}] = n_m$ leads to the same statement.

From Eq.(35.26), one has

$$\mathbf{A}_j\mathbf{x}^{[j]} = \check{\mathbf{A}}_j\check{\mathbf{x}}^{|j|} \quad \leadsto \quad \mathbf{A}_j\mathbf{T}_j\check{\mathbf{x}}^{|j|} = \check{\mathbf{A}}_j\check{x}^{|j|} \quad \leadsto \quad \check{\mathbf{A}}_j = \mathbf{A}_j\mathbf{T}_j \,. \quad (35.48)$$

By this equation, redundant information $\mathbf{A}_j$ is reduced to $\check{\mathbf{A}}_j$ by postmultiplying $\mathbf{A}_j$ by $\mathbf{T}_j$. The inverse operation, i.e., the operation to find redundant $\mathbf{A}_j$ from non-redundant $\check{\mathbf{A}}_j$ is given by $\mathbf{A}_j^\star = \check{\mathbf{A}}_j\mathbf{T}^{\sharp L}$.

Proof: Substituting $\mathbf{A}_j^\star = \check{\mathbf{A}}_j \mathbf{T}^{\sharp L}$ as a special result of the equation $\check{\mathbf{A}}_j = \mathbf{A}_j \mathbf{T}_j$ (underdetermined in $\mathbf{A}_j$) yields

$$\check{\mathbf{A}}_j = \mathbf{A}_j^\star \mathbf{T}_j = \check{\mathbf{A}}_j \mathbf{T}_j^{\sharp L} \mathbf{T}_j = \check{\mathbf{A}}_j \mathbf{T}_j = \check{\mathbf{A}}_j \ . \tag{35.49}$$

Comparing with the demarcation procedure $\mathbf{p}^\star = \mathbf{M}^{\sharp R}\mathbf{y}_m$ mentioned above, note that Eq.(35.49) corresponds to the same number of scalar equations given by the non-redundant elements of $\check{\mathbf{A}}_j$. (Again a mistake might arise if $\check{\mathbf{A}}_j = \mathbf{A}_j \mathbf{T}_j$ were to be substituted to $\mathbf{A}_j = \check{\mathbf{A}}_j \mathbf{T}_j^{\sharp L}$.)

In order to derive the result a great many calculations are required. Omitting this procedure, the result in vector notation col $\check{\mathbf{P}}_j$ is

$$\text{col}\ \check{\mathbf{P}}_j = -[\mathbf{I}_{\alpha_j} \otimes (\mathbf{A}_1^T - \mathbf{P}_1 \mathbf{B}_0 \mathbf{R}^{-1} \mathbf{B}_0^T)]^{-1} \mathbf{W}_j^{\sharp R} \mathbf{T}_{j+1}^T \mathbf{g}_j \tag{35.50}$$

where the abbreviations

$$\alpha_j = \binom{n+j-1}{n} \qquad \mathbf{P}(\mathbf{x}) = \sum_{j=1}^{\infty} \mathbf{P}_j \mathbf{x}^{[j]} \triangleq \sum_{j=1}^{\infty} \check{\mathbf{P}}_j \check{\mathbf{x}}^{[j]} \qquad \mathbf{W}_j^{\sharp R} = \mathbf{Z}_j^T (\mathbf{Z}_j \mathbf{Z}_j^T)^{-1} \tag{35.51}$$

$$\mathbf{Z}_j = \mathbf{T}_{j+1}^T (\mathbf{I}_{(n^{j+1})} + \mathbf{U}_{n,(n^j)})(\mathbf{T}_j^{\sharp L} \otimes \mathbf{I}_n) \qquad \mathbf{T}_j^{\sharp L} = (\mathbf{T}_j^T \mathbf{T}_j)^{-1} \mathbf{T}_j^T \tag{35.52}$$

$$\begin{aligned}\mathbf{g}_j = (\mathbf{I}_{(n^{j+1})} + \mathbf{U}_{n,(n^j)})\ \text{col}\ (\mathbf{C}_1^T \mathbf{Q} \mathbf{C}_j + \mathbf{P}_1^T \mathbf{A}_j) + \sum_{i=2}^{j-1} \text{col}\ (\mathbf{C}_i \mathbf{Q} \mathbf{C}_{j-i+1} + \mathbf{P}_i^T \mathbf{A}_{j-i+1} + \mathbf{A}_i^T \mathbf{P}_{j-i+1}) \\ - \sum_{k=1,i'=1}^{j-1} \sum_{l=0,j'=0}^{j-1} \text{col}\ (\mathbf{V}_{kl} \mathbf{R}^{-1} \mathbf{V}_{i'j'}^T)|_{k+l+i'+j'=j+1}\end{aligned} \tag{35.53}$$

are used and

$$\check{\mathbf{x}}^{[j]} \in \mathcal{R}^{\alpha_j},\ \ \mathbf{T}_j \in \mathcal{R}^{n^j \times \alpha_j},\ \ \mathbf{Z}_j \in \mathcal{R}^{\alpha_{j+1} \times n\alpha_j},\ \ \mathbf{g}_j \in \mathcal{R}^{n^{j+1}},\ \ \mathbf{P}_j \in \mathcal{R}^{n \times n^j},\ \ \mathbf{P}_i^T \mathbf{A}_{j-i+1} \in \mathcal{R}^{n^i \times n^{j-i+1}} . \tag{35.54}$$

The bracket term in Eq.(35.50) is easily invertible with regard to its block diagonal structure. Finally, the state-feedback controller is given by

$$\mathbf{u}^\star(\mathbf{x}) = -\mathbf{R}^{-1}\mathbf{B}^T\mathbf{P} = -\mathbf{R}^{-1} \begin{pmatrix} \sum_{j=0}^{\infty} \mathbf{x}^{[j]T} \mathbf{B}_{1j}^T \\ \vdots \\ \sum_{j=0}^{\infty} \mathbf{x}^{[j]T} \mathbf{B}_{mj}^T \end{pmatrix} \sum_{j=1}^{\infty} \mathbf{P}_j \mathbf{x}^{[j]} \ . \tag{35.55}$$

When the approximation of the Riccati equation is truncated at the first term, the bilinear process model and control is achieved. Even in this case, the control result is better than that obtained by merely linear model. Irrespective of the order of the nonlinear model, the conditions of existence (stabilizability) are already fixed by the linearized system expansion.

35.4 Continuously Operating Variable Structure Mode

Consider the linear plant $\dot{\mathbf{x}} = \mathbf{A}\mathbf{x} + \mathbf{B}\mathbf{u}$, $\mathbf{y} = \mathbf{C}\mathbf{x}$ and the nonlinear controller

$$\mathbf{u}(t) = \mathbf{K}\mathbf{y}(t) + \sum_{i=1}^{r} p_i(t) \mathbf{N}_i \mathbf{y}(t) \ , \tag{35.56}$$

the continuous vector-valued variable $\mathbf{p}$ being generated by the nonlinear equation

$$\dot{\mathbf{p}} = \text{vec}\ \dot{p}_i = \mathbf{f}(\mathbf{x}, \mathbf{p}) \ . \tag{35.57}$$

This operating mode opens a wide area of design possibilities. Undesired high-frequency switching behaviour or sliding modes are excluded. The Lyapunov approach allows for general design criteria. Control variable bounds can be entirely exhausted (*Franke, D., 1982a,b*).

35.5 Models Approximating Continuous Nonlinearities

Extending the convolution integral of the continuous-time plant

$$y(t) = \int_0^t g(\tau)u(t-\tau)d\tau \ , \tag{35.58}$$

the Volterra series approximation (*Peyton, J.C., and Billings, S.A., 1991*)

$$y(t) = g_o + \int_0^t g_1(\tau)u(t-\tau)d\tau + \int_0^t \int_0^t g_2(\tau,\eta)u(t-\tau)u(t-\eta)d\tau d\eta + \ \dots \tag{35.59}$$

is the basis of a *nonparametric* model. For discrete-time systems there are corresponding relations using multiple sums

$$y(k) = g_o + \sum_{i=0}^{n} g_1(i)u(k-d-i) + \sum_{i=0}^{n}\sum_{j=0}^{n} g_2(i,j)u(k-d-i)u(k-d-j) + \ \dots \tag{35.60}$$

where d denotes a positive number resulting from the time delay divided by the sampling period. Bounding the series by the order n, nonlinear *parametric* models are achieved where the generalized weighting functions g_l are replaced by suitable approximate polynomials (*Lachmann,K.H., 1985; Banks, S.P., 1988*).

Hammerstein models are obtained if $\eta = \tau$ etc. and if the nonlinearity is of the type

$$v(k) = \sum_{i=0}^{n} r_i u^i(k) \qquad y(k) = \frac{b(z^{-1})z^{-d}}{a(z^{-1})}v(k) \ . \tag{35.61}$$

Wiener models are defined by a structure where the linear part is associated with the input and the nonlinearity is connected to the output.

Choosing a structure with nonlinearities in the feedback part of the transfer structure, Lachmann models are obtained.

Chapter 36

System Analysis via Orthogonal Functions

Classical orthogonal polynomials, generalized and shifted orthogonal polynomials serve as an excellent tool for modelling plants and processes in an approximative way. Moreover, solutions via orthogonal polynomials give a straightforward algorithm in reducing various dynamic problems to the solution of a set of algebraic linear equations. The integral of a function with respect to time is replaced by premultiplying the vector of the decomposed signal by an operational matrix. Furthermore, the product of functions and even time-varying coefficients and nonlinear relations are reduced to algebraic manipulations. Among these problems are system analysis, variational calculus, optimal control, identification and model reduction. Orthogonal functions have been applied in control measurement and correlation techniques for a long period, see e.g. *Kitamori, T., 1960; Douce, J.L., and Roberts, P.D., 1964*. The classical Fourier series expansion, truncated to a certain amount of terms, can also be applied to solve various dynamic problems (*Paraskevopoulos, P.N., et al. 1985; Cheok, K.C., et al. 1989*).

36.1 General Orthogonal Polynomials $\varphi_i(\zeta)$

A polynomial $\varphi_i(\zeta)$ is regarded as an orthogonal one if it satisfies the following conditions $\forall i = 0, 1, \ldots k-1$

$$\int_a^b w(\zeta)\varphi_i(\zeta)\varphi_j(\zeta)d\zeta = \begin{cases} \rho_i & i = j \\ 0 & i \neq j. \end{cases} \tag{36.1}$$

These polynomials φ_i, φ_j are orthogonal with respect to the weighting function $w(\zeta)$ and orthogonal over an interval $[a, b]$. Orthogonal polynomials (*Szegö, G., 1975; Sansone, G., 1959*) of various kind have been applied to control problems. In particular, Jacobi, ultraspherical, Chebyshev, Legendre, Laguerre and Hermite polynomials are used. The latter, Laguerre and Hermite polynomials, usually have less convergence properties than Jacobi, ultraspherical, Chebyshev and Legendre polynomials. Note that all these polynomials (in ζ) can be derived from the same recurrence relation, only the coefficients are different. These general orthogonal polynomials are $\varphi_o(\zeta) = 1$, $\varphi_1(\zeta) = a_o\zeta + b_o$ and

$$\varphi_{i+1}(\zeta) = (a_i\zeta + b_i)\,\varphi_i(\zeta) - c_i\varphi_{i-1}(\zeta). \tag{36.2}$$

The values of the recurrence coefficients a_i, b_i, c_i are listed in Table 36.3 (*Chang, Y.F., and Lee, T.T., 1986*). The variable ζ is defined in the interval $a \leq \zeta \leq b$, only. The subscript $_i$ is the degree of the polynomial. Chebyshev polynomials are widely discussed by *Paraskevopoulos, P.N., 1983* and compared with Walsh, block-pulse and Laguerre functions.

Table 36.1: General correspondence between interval boundaries in ζ-domain and time domain

ζ-domain	t-domain
$[a,b]$	a,b,p,q given; $a' = \frac{a-q}{p}; b' = \frac{b-q}{p}$: $[\min(a',b'), \max(a',b')]$
$[a,b]$	a,b,a',b' given; $p = \frac{b-a}{b'-a'}$; $q = \frac{ab'-a'b}{b'-a'}$: $[a',b']$
$[-1,1]$	$[0,t_f]$ $p = \frac{2}{t_f}$; $q = -1$

36.2 General Shifted Orthogonal Polynomials $h_i(t)$

For control purposes the interval $a \le \zeta \le b$ is not suitable. Selecting an appropriate transformation

$$\zeta = pt + q \,, \tag{36.3}$$

general shifted orthogonal polynomials in time domain t are obtained. The numbers p and q are constant. Setting, e.g.,

$$p = \frac{2}{t_f}, \qquad q = -1 \qquad \rightsquigarrow \qquad \zeta = -1 + \frac{2}{t_f}t \,, \tag{36.4}$$

the interval $[-1,1]$ is transformed to the interval $[0,t_f]$. For general transformation parameters see Table 36.1. Using Eq.(36.3), the shifted general orthogonal polynomials $h_i(t)$ become

$$h_o(t) = 1 \tag{36.5}$$

$$h_1(t) = a_o pt + (b_o + a_o q) = a'_o t + b'_o \tag{36.6}$$

$$h_{i+1}(t) = (a'_i t + b'_i)h_i(t) - c'_i h_{i-1}(t) \tag{36.7}$$

$$\text{where} \quad a'_i = a_i p; \quad b'_i = b_i + a_i q; \quad c'_i = c_i \quad \forall i = 0,1,2\dots \,. \tag{36.8}$$

The orthogonality condition in time domain which is satisfied by the general shifted orthogonal polynomials (or orthogonal functions) is given by

$$\int_{a'}^{b'} w'(t)h_i(t)h_j(t)dt = \begin{cases} \rho'_i & i = j \\ 0 & i \neq j \end{cases} \qquad \forall i = 0,1,\dots k-1 \,. \tag{36.9}$$

The weighting functions are listed in Table 36.2.

The Jacobi polynomial coefficients and recurrence relations coefficients are listed below where $\alpha > -1$, $\beta > -1$ and $a = -1$, $b = 1$

$$a_i = \frac{(\alpha+\beta+2i+1)(\alpha+\beta+2i+2)}{2(i+1)(\alpha+\beta+i+1)} \tag{36.10}$$

$$b_i = \frac{(\alpha+\beta+2i+1)(\alpha^2-\beta^2)}{2(i+1)(\alpha+\beta+i+1)(\alpha+\beta+2i)} \tag{36.11}$$

$$c_i = \frac{(\alpha+i)(\beta+i)(\alpha+\beta+2i+2)}{(i+1)(\alpha+\beta+i+1)(\alpha+\beta+2i)} \tag{36.12}$$

Table 36.2: Shifted interval and weighting function $w'(t)$ with shifting operation $\zeta = \frac{2t}{t_f} - 1$

	a'	b'	$w'(t)$
Shifted Jacobi	0	t_f	$\left(1-\frac{t}{t_f}\right)^{\alpha}\left(\frac{t}{t_f}\right)^{\beta}$
Shifted ultraspherical	0	t_f	$\left(\frac{t}{t_f}-\frac{t^2}{t_f^2}\right)^{\lambda-0.5}$
Shifted Chebyshev 1st	0	t_f	$\left(\frac{t}{t_f}-\frac{t^2}{t_f^2}\right)^{-0.5}$
Shifted Chebyshev 2nd	0	t_f	$\left(\frac{t}{t_f}-\frac{t^2}{t_f^2}\right)^{0.5}$
Shifted Legendre	0	t_f	1
Modified Laguerre	0	∞	$t^{\alpha}e^{-\eta t}$
Modified Hermite	$-\infty$	∞	$e^{-\eta^2 t^2}$

Table 36.3: Orthogonal polynomials recurrence coefficients. Note that $\lambda > -0.5$.

	a	b	a_i	b_i	c_i	A_i	B_i	C_i
Ultraspherical	-1	1	$\frac{2(i+\lambda)}{i+1}$	0	$\frac{i+2\lambda-1}{i+1}$	$\frac{1}{2(\lambda+i)}$	0	$\frac{-1}{2(\lambda+i)}$
Chebyshev $1^{st}(a_o=1)$	-1	1	2	0	1	$\frac{1}{2i+2}$	0	$\frac{-1}{2i-2}$
Chebyshev 2nd	-1	1	2	0	1	$\frac{1}{2i+2}$	0	$\frac{-1}{2i+2}$
Legendre	-1	1	$\frac{2i+1}{i+1}$	0	$\frac{i}{i+1}$	$\frac{1}{2i+1}$	0	$\frac{-1}{2i+1}$
Laguerre	0	∞	$-\frac{1}{n+1}$	$\frac{\alpha+2i+1}{i+1}$	$\frac{\alpha+i}{i+1}$	-1	1	0
Hermite	$-\infty$	∞	2	0	$2i$	$\frac{1}{2(i+1)}$	0	0

$$A_i = \frac{2(\alpha+\beta+i+1)}{(\alpha+\beta+2i+2)(\alpha+\beta+2i+1)} \tag{36.13}$$

$$B_i = \frac{2(\alpha-\beta)}{(\alpha+\beta+2i)(\alpha+\beta+2i+2)} \tag{36.14}$$

$$C_i = \frac{-2(\alpha+i)(\beta+i)}{(\alpha+\beta+2i)(\alpha+\beta+i)(\alpha+\beta+2i+1)}. \tag{36.15}$$

The further orthogonal polynomial recurrence coefficients are given in Table 36.3.

36.3 Decomposition into Orthogonal Functions

An arbitrary function $x(t) \in L_{2e}$ can be approximated by a weighted sum of k terms of generalized shifted orthogonal polynomials in t

$$x(t) \doteq \sum_{i=0}^{k-1} x_i^{\neg} h_i(t) = \mathbf{x}^{\neg T}\mathbf{h}(t) = \mathbf{h}^T(t)\mathbf{x}^{\neg} \tag{36.16}$$

where $\mathbf{x}^{\neg}$ is the general orthogonal coefficient vector and $\mathbf{h}(t)$ the general orthogonal polynomial (function) vector with the elements $h_i(t)$.

The unknown general orthogonal coefficient vector $\mathbf{x}^{\neg}$ with elements $x_i^{\neg}$ is determined by minimizing the integral of weighted squared deviation

$$I(\mathbf{x}^{\neg}) = \int_{a'}^{b'} w'(t)[x(t) - \sum_{j=0}^{k-1} x_j^{\neg} h_j(t)]^2 dt \tag{36.17}$$

$$\frac{\partial I(\mathbf{x}^{\neg})}{\partial \mathbf{x}^{\neg}} = 2\int_{a'}^{b'} w'(t)[x(t) - \sum_{j=0}^{k-1} x_j^{\neg} h_j(t)]\big(h_1(t) \;\; h_2(t) \ldots h_{k-1}(t)\big)^T dt = \mathbf{0}\,. \tag{36.18}$$

This gradient vector has the component i

$$2\int_{a'}^{b'} [w'(t)x(t)h_i(t) - w'(t)x_i^{\neg} h_i^2]dt - 2\int_{a'}^{b'} w'(t) \sum_{j=0\neq i}^{k-1} x_j^{\neg} h_i h_j dt = 0\,. \tag{36.19}$$

With regard to the orthogonality condition Eq.(36.9), the last term comprising the sum vanishes. Thus,

$$\int_{a'}^{b'} w'(t)x(t)h_i(t)dt - x_i^{\neg}\rho_i' = 0 \quad \rightsquigarrow \quad x_i^{\neg} = \frac{1}{\rho_i'}\int_{a'}^{b'} w'(t)x(t)h_i(t)dt\,. \tag{36.20}$$

The decoupling property and the remarkable simplification using an orthogonal basis is shown by repeating the derivation in vector notation. For convenience, $w'(t) = 1$ and orthonormality ($\rho_i' = 1$) are assumed, first,

$$\frac{\partial}{\partial \mathbf{x}^{\neg}}\int_{a'}^{b'} [x(t) - \mathbf{x}^{\neg T}\mathbf{h}(t)]^2 dt = 2\int_{a'}^{b'} [x(t) - \mathbf{x}^{\neg T}\mathbf{h}(t)]\mathbf{h}(t)dt = \mathbf{0} \tag{36.21}$$

$$\int_{a'}^{b'} x(t)\mathbf{h}(t)dt - \int_{a'}^{b'} \mathbf{x}^{\neg T}\mathbf{h}(t)\mathbf{h}(t)dt = \int_{a'}^{b'} x(t)\mathbf{h}(t)dt - [\int_{a'}^{b'} \mathbf{h}(t)\mathbf{h}^T(t)dt]\mathbf{x}^{\neg} = \mathbf{0} \tag{36.22}$$

$$\int_{a'}^{b'} x(t)\mathbf{h}(t)dt - (\int_{a'}^{b'} \text{diag}\, h_i^2 dt)\mathbf{x}^{\neg} = \int_{a'}^{b'} x(t)\mathbf{h}(t)dt - \mathbf{I}\mathbf{x}^{\neg} \quad \rightsquigarrow \quad \mathbf{x}^{\neg} = \int_{a'}^{b'} x(t)\mathbf{h}(t)dt\,. \tag{36.23}$$

Including $w'(t)$ and ρ_i', it results

$$\mathbf{x}^{\neg} = [\text{diag}\,(\rho_i')^{-1}]\int_{a'}^{b'} w'(t)x(t)\mathbf{h}(t)dt\,. \tag{36.24}$$

If an orthogonal basis is used the expression $\int_{a'}^{b'} w'(t)\mathbf{h}(t)\mathbf{h}^T(t)dt$ is considerably simplified to $\mathbf{I}$ or diag ρ_i'. The decoupling property is obvious from the diagonalization of the premultiplication matrix.

This derivation shows the numerical advantage due to orthogonality condition Eq.(36.9): To obtain the element $x_i^{\neg}$ of the coefficient vector $\mathbf{x}^{\neg}$ the quantities x, h_i, w must be taken into consideration, only. The remaining $x_j^{\neg}$ $(j \neq i)$ are of no influence to $x_i^{\neg}$, regardless whether they have already been computed or not. The individual computations of various $x_i^{\neg}$ are decoupled from each other.

Computational difficulties calculating $x_i^{\neg}$ by numerical integration may arise from the singularities at the integration boundaries. This difficulty can be overcome by Gauss-quadrature relation (*Chang, Y.F., and Lee, T.T., 1986*).

An alternative notational convenience is

$$\langle f, g\rangle = \int_{a'}^{b'} w' f(t)g(t)dt, \qquad \langle h_i, h_j\rangle = \rho_i'\delta_{ij}\,, \qquad x_i^{\neg} = \frac{\langle x, h_i\rangle}{\langle h_i, h_i\rangle}\,. \tag{36.25}$$

36.4 Operational Matrix of Integration in ζ-Domain

In the previous section an introduction to orthogonal functions recurrence relations is given. Another important property of orthogonal polynomials is the differential recurrence relation

$$\varphi_i(\zeta) = A_i\dot{\varphi}_{i+1}(\zeta) + B_i\dot{\varphi}_i(\zeta) + C_i\dot{\varphi}_{i-1}(\zeta) \,. \tag{36.26}$$

The values of the coefficients A_i, B_i, C_i are summarized in Table 36.3. Taking the integral with respect to ζ,

$$\int_a^\zeta \varphi_i(\zeta)d\zeta = A_i\varphi_{i+1}(\zeta) + B_i\varphi_i(\zeta) + C_i\varphi_{i-1}(\zeta) \underbrace{- A_i\varphi_{i+1}(a) - B_i\varphi_i(a) - C_i\varphi_{i-1}(a)}_{D_i} \,. \tag{36.27}$$

A coefficient D_i is defined equal to the underbraced term which is a constant, resulting from the amount of the integral at the lower bound a. With regard to $\varphi_o(\zeta) = 1$ this constant can be declared as a coefficient of $\varphi_o(\zeta)$ and the underbraced term is substituted by $D_i\varphi_o(\zeta)$. Starting with $\varphi_o(\zeta)$ and letting the coefficients be A_o and $B_o - a$,

$$\int_a^\zeta \varphi_o(\zeta)d(\zeta) = A_o\varphi_1(\zeta) + (B_o - a)\varphi_o(\zeta) = \int_a^\zeta 1 \; d\zeta = \zeta - a \,. \tag{36.28}$$

From $\varphi_1 = a_o\zeta + b_o$ the coefficients are

$$A_o = \frac{1}{a_o} \quad \text{and} \quad B_o = -\frac{b_o}{a_o} \,. \tag{36.29}$$

Continuing with $\varphi_1(\zeta), \varphi_2(\zeta)$ and so on, one can express the integral of the ith orthogonal function by the orthogonal functions with degree $i+1, i, i-1$ and 0. The function $\varphi_o(\zeta)$ is utilized to include the integration constant at the lower bound. Finally, the result in matrix form is

$$\int_a^\zeta \varphi(\zeta)d\zeta = \mathbf{P}_{M\zeta}\, \varphi(\zeta) \tag{36.30}$$

where $\mathbf{P}_{M\zeta}$ is the so-called operational matrix of integration

$$\mathbf{P}_{M\zeta} = \begin{pmatrix} B_o - a & A_o & 0 & 0 & 0 & \dots & 0 & 0 & 0 \\ C_1 + D_1 & B_1 & A_1 & 0 & 0 & \dots & 0 & 0 & 0 \\ D_2 & C_2 & B_2 & A_2 & 0 & \dots & 0 & 0 & 0 \\ D_3 & 0 & C_3 & B_3 & A_3 & \dots & 0 & 0 & 0 \\ \vdots & \vdots & \vdots & \vdots & \vdots & \ddots & \vdots & \vdots & \vdots \\ D_{m-2} & 0 & 0 & 0 & 0 & \dots & C_{m-2} & B_{m-2} & A_{m-2} \\ D_{m-1} & 0 & 0 & 0 & 0 & \dots & 0 & C_{m-1} & B_{m-1} \end{pmatrix} \,. \tag{36.31}$$

Within the matrix above in parallel to the main diagonal there are A_i, B_i, C_i sequences with increasing subscript. The operational matrix of integration is the best orthogonal function approximation of the integral operation.

36.5 Operational Matrix $\mathbf{P}_M$ of Integration in Time-Domain

The derivation known from ζ-domain is repeated in time-domain using the shifting transformation $\zeta = pt + q$ and $d\zeta = p\;dt$. For $\varphi_o(\zeta), h_o(t), \varphi_i(\zeta)$ and $h_i(t)$

$$\int_a^\zeta \varphi_o(\zeta)d\zeta = \int_{a'}^t \varphi_o(pt+q)p\;dt = p(t - a') \tag{36.32}$$

$$\int_{a'}^{t} h_i(t)dt = \frac{1}{p}\int_{a}^{\zeta} \varphi_i(\zeta)d\zeta. \tag{36.33}$$

Therefore, the operational matrix $\mathbf{P}_{Mt}$ for integration with respect to time follows from $\mathbf{P}_{M\zeta}$ using slight modifications, only,

$$\int_{a'}^{t} \mathbf{h}(t)dt = \mathbf{P}_{Mt}\ \mathbf{h}(t) \triangleq \mathbf{P}_M\ \mathbf{h}(t) \quad \text{where} \tag{36.34}$$

$$\mathbf{P}_M = \frac{1}{p}\begin{pmatrix} B_o - a'p - q & A_o & 0 & 0 & 0 & \dots & 0 & 0 & 0 \\ C_1 + D_1 & B_1 & A_1 & 0 & 0 & \dots & 0 & 0 & 0 \\ D_2 & C_2 & B_2 & A_2 & 0 & \dots & 0 & 0 & 0 \\ D_3 & 0 & C_3 & B_3 & A_3 & \dots & 0 & 0 & 0 \\ \vdots & \vdots & \vdots & \vdots & \vdots & \ddots & \vdots & \vdots & \vdots \\ D_{m-2} & 0 & 0 & 0 & 0 & \dots & C_{m-2} & B_{m-2} & A_{m-2} \\ D_{m-1} & 0 & 0 & 0 & 0 & \dots & 0 & C_{m-1} & B_{m-1} \end{pmatrix}. \tag{36.35}$$

If $a' = 0$ (and $t > a'$), then $\mathbf{P}_M$ is the operational matrix for forward integration. When $a' = t_f$ (and $t < a'$), the operational matrix for backward integration is obtained.

Specializing for various individual polynomials yields
using Chebyshev polynomials

$$\mathbf{P}_M = t_f\begin{pmatrix} .5 & -.5 & 0 & 0 & 0 & \dots & 0 & 0 & 0 \\ .125 & 0 & -.125 & 0 & 0 & \dots & 0 & 0 & 0 \\ -.16\dot{6} & .25 & 0 & -.08\dot{3} & 0 & \dots & 0 & 0 & 0 \\ -.063 & 0 & .125 & 0 & -.063 & \dots & 0 & 0 & 0 \\ \vdots & \vdots & \vdots & \vdots & \vdots & \ddots & \vdots & \vdots & \vdots \\ \frac{-.5}{(k-1)(k-3)} & 0 & 0 & 0 & 0 & \dots & \frac{.25}{k-3} & 0 & \frac{-.25}{k-1} \\ \frac{-.5}{k(k-2)} & 0 & 0 & 0 & 0 & \dots & 0 & \frac{.25}{k-2} & 0 \end{pmatrix}, \tag{36.36}$$

using Legendre polynomials

$$\mathbf{P}_M = t_f\begin{pmatrix} .5 & .5 & 0 & 0 & 0 & \dots & 0 & 0 & 0 \\ -.16\dot{6} & 0 & .16\dot{6} & 0 & 0 & \dots & 0 & 0 & 0 \\ 0 & -.1 & 0 & .1 & 0 & \dots & 0 & 0 & 0 \\ 0 & 0 & -.071 & 0 & .071 & \dots & 0 & 0 & 0 \\ \vdots & \vdots & \vdots & \vdots & \vdots & \ddots & \vdots & \vdots & \vdots \\ 0 & 0 & 0 & 0 & 0 & \dots & \frac{-1}{4k-6} & 0 & \frac{-1}{4k-6} \\ 0 & 0 & 0 & 0 & 0 & \dots & 0 & \frac{-1}{4k-2} & 0 \end{pmatrix}, \tag{36.37}$$

using harmonic (sinusoidal) functions in the interval T (*Cheok, K.C., 1989*)

$$\mathbf{h}(t) = \Big(1,\quad \cos\frac{2\pi i}{T}t,\ \dots.\ \cos\frac{2\pi r}{T}t,\ \sin\frac{2\pi i}{T}t,\ \dots.\ \sin\frac{2\pi r}{T}t\Big)^T \tag{36.38}$$

$$\mathbf{P}_M = \frac{T}{2\pi}\begin{pmatrix} \pi & \mathbf{0}^{[1\times r]} & -2\mathbf{e}_r^T \\ \mathbf{0}^{[r\times 1]} & \mathbf{0}_r^{[r\times r]} & \mathbf{I}_d \\ \mathbf{e}_r & -\mathbf{I}_d & \mathbf{0}_r^{[r\times r]} \end{pmatrix} \in \mathcal{R}^{(2r+1)\times(2r+1)} \tag{36.39}$$

$$\mathbf{e}_r^T = (1\ \ \frac{1}{2}\ \ \frac{1}{3}\dots\dots\frac{1}{r})^T \qquad \mathbf{I}_d = \mathrm{diag}\{1\ \ \frac{1}{2}\ \ \frac{1}{3}\dots\dots\frac{1}{r}\}\,. \tag{36.40}$$

36.6 Power Series Expansion of Orthogonal Functions

In the interval $[0, t_f]$ the generalized shifted orthogonal function $h_i(t)$ can be expanded into a power series in t

$$h_i(t) = \sum_{j=0}^{i} f_{ij}\, t^j \quad \rightsquigarrow \quad \mathbf{h}(t) = [h_o(t) \;\; h_1(t) \dots h_{k-1}(t)]^T = \mathbf{F}\; \mathbf{t}(t) \tag{36.41}$$

where $\mathbf{t}(t)$ is a power series vector and $\mathbf{F}$ a lower triangular matrix of the expansion coefficients f_{ij}

$$\mathbf{t}(t) \triangleq (1 \;\; t \;\; t^2 \;\dots\; t^{k-1})^T \tag{36.42}$$

$$\mathbf{F} = \begin{pmatrix} f_{00} & 0 & \dots & 0 \\ f_{10} & f_{11} & \dots & 0 \\ \vdots & \vdots & \ddots & \vdots \\ f_{k-1,0} & f_{k-1,1} & \dots & f_{k-1,k-1} \end{pmatrix} . \tag{36.43}$$

Substituting Eq.(36.41) into Eq.(36.7) yields

$$\sum_{j=0}^{i+1} f_{i+1,j}\; t^j = a'_i \sum_{j=0}^{i} f_{ij} t^{j+1} + b'_i \sum_{j=0}^{i} f_{ij} t^j - c'_i \sum_{j=0}^{i-1} f_{i-1,j}\; t^j . \tag{36.44}$$

Comparing the coefficients of each power t^j, a recurrence formula is obtained

$$f_{i+1,j} = a'_i f_{i,j-1} + b'_i f_{ij} - c'_i f_{i-1,j} \quad \text{where} \qquad f_{i,-1} = 0, \;\; f_{ij} \mid_{j<i} = 0, \;\; f_{00} = 1 \; . \tag{36.45}$$

The ith power of t can also be expressed in terms of $h_i(t)$, i.e., $t^i = \sum_{j=0}^{i} \vartheta_{ij} h_j(t)$ or

$$\mathbf{t}(t) = \mathbf{\Theta h}(t) = \mathbf{\Theta F t}(t) \qquad \mathbf{\Theta} \triangleq \begin{pmatrix} \vartheta_{00} & 0 & \dots & 0 \\ \vartheta_{10} & \vartheta_{11} & \dots & 0 \\ \vdots & \vdots & \ddots & \vdots \\ \vartheta_{k-1,0} & \vartheta_{k-1,1} & \dots & \vartheta_{k-1,k-1} \end{pmatrix} \tag{36.46}$$

where $\mathbf{\Theta F} = \mathbf{I}_k$ and $\vartheta_{ii} = 1/f_{ii}$, $\vartheta_{ij} \mid_{j>i} = 0$.

Integrating Eq.(36.41) and substituting Eq.(36.46) yields

$$\int_o^t h_i(t) dt = \sum_{j=0}^{i} f_{ij} \frac{t^{j+1}}{j+1} = \sum_{j=0}^{i} \sum_{\mu=0}^{j+1} \frac{f_{ij}}{j+1} \vartheta_{j+1,\mu}\; h_\mu(t) = \sum_{\mu=0}^{i+1} \vartheta^\circ_{i\mu} h_\mu(t) \tag{36.47}$$

$$\text{where} \qquad \vartheta^\circ_{i\mu} \triangleq \sum_{j=\xi}^{i} \frac{f_{ij}\; \vartheta_{j+1,\mu}}{j+1} \qquad \xi \mid_{\mu=0} = 0, \qquad \xi \mid_{\mu\geq 1} = \mu - 1 \; . \tag{36.48}$$

Thus, the operational matrix of integration is formulated

$$\mathbf{P}_M = \begin{pmatrix} \vartheta^\circ_{00} & \vartheta^\circ_{01} & 0 & \dots & 0 \\ \vartheta^\circ_{10} & \vartheta^\circ_{11} & \vartheta^\circ_{12} & \dots & 0 \\ \vdots & \vdots & \vdots & \ddots & 0 \\ \vartheta^\circ_{k-1,0} & \dots & \dots & \dots & \vartheta^\circ_{k-1,k-1} \end{pmatrix} . \tag{36.49}$$

This agrees with Eq.(36.34), without making use of the differential recurrence relation (see *Chang, R.Y., et al. 1987*).

36.7 Orthogonal Expansion of the Product of Two Functions

The product of two orthogonal polynomials h_i and h_j in t is a polynomial in t, too. Decomposing this product $h_i(t)h_j(t)$ into orthogonal polynomials yields the definition of the vector $\mathbf{l}_{ij}$ with entries $l_{ij\mu}$

$$h_i(t)h_j(t) \doteq \mathbf{l}_{ij}^T\mathbf{h}(t) \qquad \forall i,j = 1,2\ldots k-1. \tag{36.50}$$

The product of two functions $e(t)g(t)$ can be expressed in terms of general orthogonal polynomials

$$e(t)g(t) = \mathbf{e}^{\neg T}\mathbf{h}\mathbf{g}^{\neg T}\mathbf{h} = \mathbf{e}^{\neg T}\mathbf{h}\mathbf{h}^{\neg T}\mathbf{g} = \mathbf{e}^{\neg T}\{\mathbf{h}\mathbf{h}^T\}\mathbf{g}^{\neg} \tag{36.51}$$

$$e(t)g(t) = \mathbf{e}^{\neg T}\{\text{matrix}[\mathbf{l}_{ij}^T\mathbf{h}(t)]\}\mathbf{g}^{\neg} = \mathbf{e}^{\neg T}\begin{pmatrix} \mathbf{l}_{00}^T\mathbf{h} & \mathbf{l}_{01}^T\mathbf{h} & \ldots \\ \vdots & \vdots & \ddots \\ \mathbf{l}_{k-1,0}\,\mathbf{h} & \ldots & \mathbf{l}_{k-1,k-1}\,\mathbf{h} \end{pmatrix}\mathbf{g}^{\neg}$$

$$e(t)g(t) = \Big(\sum_{i=0}^{k-1}\sum_{j=0}^{k-1} e_i g_i \mathbf{l}_{ij}^T\Big)\mathbf{h}(t) = (\mathbf{e}^{\neg}\mathbf{g}^{\neg})^T\mathbf{h}(t)\,. \tag{36.52}$$

The expression $(\mathbf{e}^{\neg}\mathbf{g}^{\neg})$ is used as an abbreviation for the general orthogonal polynomial coefficient vector of the product $e(t)g(t)$. This vector is designated as the operational product vector. The dyadic product $\mathbf{h}\mathbf{h}^T$ itself can be detailed as follows

$$\mathbf{h}\mathbf{h}^T = \text{matrix}[\mathbf{l}_{ij}^T\mathbf{h}] = (\mathbf{L}_o\mathbf{h} \vdots \mathbf{L}_1\mathbf{h}\ldots\mathbf{L}_{k-1}\mathbf{h}) \quad \text{where } \mathbf{L}_j = \begin{pmatrix} \mathbf{l}_{oj}^T \\ \mathbf{l}_{1j}^T \\ \vdots \\ \mathbf{l}_{k-1,j}^T \end{pmatrix} \in \mathcal{R}^{k\times k}\,. \tag{36.53}$$

Note that for any vector $\mathbf{u}$ (*Rotella, F., and Dauphin-Tanguy, G., 1988*) it results

$$\mathbf{h}\mathbf{h}^T\mathbf{u} \doteq (\mathbf{L}_0^T\mathbf{u} \vdots \mathbf{L}_1^T\mathbf{u}\ldots\mathbf{L}_{k-1}^T\mathbf{u})\mathbf{h}\,. \tag{36.54}$$

36.8 Power Series Expansion of the Integral of the Product of Two Functions

In optimal control theory, the integral of squared functions or product of two functions $e(t), g(t)$ is applied frequently

$$\int_o^{t_f} e(t)g(t)dt = \int_o^{t_f} \mathbf{e}^{\neg T}\mathbf{h}(t)\;\mathbf{g}^{\neg T}\mathbf{h}(t)dt \tag{36.55}$$

$$\int_o^{t_f} e(t)g(t)dt = \int_o^{t_f}(\mathbf{e}^{\neg T}\mathbf{h})\;(\mathbf{h}^T\mathbf{g}^{\neg})dt = \mathbf{e}^{\neg T}\Big(\int_o^{t_f}\mathbf{h}\mathbf{h}^T dt\Big)\mathbf{g}^{\neg} = \mathbf{e}^{\neg T}\mathbf{H}\mathbf{g}^{\neg} \tag{36.56}$$

$$\mathbf{H} \triangleq \int_o^{t_f}\mathbf{F}\mathbf{t}\mathbf{t}^T\mathbf{F}^T dt = \mathbf{F}\Big(\int_o^{t_f}\mathbf{t}\mathbf{t}^T dt\Big)\mathbf{F}^T = \mathbf{F}\mathbf{T}_f(t_f)\mathbf{F}^T \tag{36.57}$$

$$\mathbf{T}_t(t_f) = \begin{pmatrix} t_f & t_f^2/2 & \ldots & t_f^k/k \\ t_f^2/2 & t_f^3/3 & \ldots & t_f^{k+1}/(k+1) \\ \vdots & \vdots & \ddots & \vdots \\ t_f^k/k & t_f^{k+1}/(k+1) & \ldots & t_f^{2k-1}/(2k-1) \end{pmatrix}. \tag{36.58}$$

Another version of the matrix $\mathbf{H}$ is

$$\mathbf{H} = \text{matrix}[\sum_{l=0}^{j} f_{jl}\sum_{k=0}^{i} f_{ik}\frac{t_f^{k+l+1}}{k+l+1}]\,. \tag{36.59}$$

36.9 Orthogonally Based Linear State Feedback Control

Consider the system equation

$$\dot{\mathbf{x}}(t) = \mathbf{A}\mathbf{x}(t) + \mathbf{B}\mathbf{u}(t) \qquad \mathbf{x}(0) = \mathbf{x}_o \,. \tag{36.60}$$

Supposing that the input signal $\mathbf{u}(t)$ and state variable $\mathbf{x}(t)$ can be decomposed by generalized shifted orthogonal functions,

$$\mathbf{x}(t) \doteq \sum_{i=0}^{k-1} \mathbf{x}_i^{\neg} h_i(t) = \mathbf{X}^{\neg T}\mathbf{h}(t) \qquad \mathbf{u}(t) \doteq \sum_{i=0}^{k-1} \mathbf{u}_i^{\neg} h_i(t) = \mathbf{U}^{\neg T}\mathbf{h}(t) \tag{36.61}$$

where the matrices $\mathbf{X}^{\neg T}$ and $\mathbf{U}^{\neg T}$ are regarded as the orthogonal spectrum (e.g. Chebyshev spectrum). Applying Eq.(36.34) and setting $a' = 0$,

$$\mathbf{x}(t) - \mathbf{x}_o = \int_o^t [\mathbf{A}\mathbf{x}(t) + \mathbf{B}\mathbf{u}(t)]dt \tag{36.62}$$

$$\mathbf{X}^{\neg T}\mathbf{h}(t) - \mathbf{X}_o^{\neg T}\mathbf{h}(t) = \mathbf{A}\mathbf{X}^{\neg T}\mathbf{P}_M\mathbf{h}(t) + \mathbf{B}\mathbf{U}^{\neg T}\mathbf{P}_M\mathbf{h}(t) \tag{36.63}$$

$$\mathbf{X}^{\neg T} - \mathbf{A}\mathbf{X}^{\neg T}\mathbf{P}_M = \mathbf{X}_o^{\neg T} + \mathbf{B}\mathbf{U}^{\neg T}\mathbf{P}_M \quad \text{where} \quad \mathbf{X}_o^{\neg T} = (\mathbf{x}_o\, \mathbf{0}\, \mathbf{0} \ldots \mathbf{0}) \in \mathcal{R}^{n \times k} \quad \text{and} \quad h_o(t) = 1\,. \tag{36.64}$$

Using the vector-valued column function col $\mathbf{X}^{\neg T} = (\mathbf{X}_{1.}^{\neg} \;\; \mathbf{X}_{2.}^{\neg} \ldots \mathbf{X}_{k.}^{\neg})$ and applying the Kronecker matrix product, Eq.(4.40), it follows from Eq.(36.64)

$$\text{col}\, \mathbf{X}^{\neg T} - (\mathbf{P}_M^T \otimes \mathbf{A})\, \text{col}\, \mathbf{X}^{\neg T} = \text{col}\, \mathbf{X}_o^{\neg T} + (\mathbf{P}_M^T \otimes \mathbf{B})\, \text{col}\, \mathbf{U}^{\neg T} \tag{36.65}$$

$$\text{col}\, \mathbf{X}^{\neg T} = (\mathbf{I}_{kn} - \mathbf{P}_M^T \otimes \mathbf{A})^{-1}\, [\text{col}\, \mathbf{X}_o^{\neg T} + (\mathbf{P}_M^T \otimes \mathbf{B})\text{col}\, \mathbf{U}^{\neg T}] \tag{36.66}$$

$$\mathbf{x}(t) \in \mathcal{R}^n, \;\; \mathbf{u}(t) \in \mathcal{R}^m, \;\; \mathbf{A} \in \mathcal{R}^{n\times n}, \;\; \mathbf{B} \in \mathcal{R}^{n\times m}, \;\; \mathbf{K} \in \mathcal{R}^{m\times n}, \;\; \mathbf{P}_M \in \mathcal{R}^{k\times k}, \;\; \mathbf{X}^{\neg T} \in \mathcal{R}^{n\times k}, \tag{36.67}$$

$$\mathbf{P}_M^T \otimes \mathbf{A} \in \mathcal{R}^{kn\times kn}, \;\; \mathbf{P}_M^T \otimes \mathbf{B} \in \mathcal{R}^{kn\times km}, \;\; \mathbf{I}_{kn} \in \mathcal{R}^{kn\times kn}, \;\; \mathbf{I}\otimes\mathbf{K} \in \mathcal{R}^{km\times kn}, \;\; \text{col}\, \mathbf{X}^{\neg T} \in \mathcal{R}^{kn}, \;\; \text{col}\, \mathbf{U}^{\neg T} \in \mathcal{R}^{km}\,. \tag{36.68}$$

Introducing a state feedback controller,

$$\mathbf{u} = \mathbf{K}\mathbf{x} \quad \rightsquigarrow \quad \mathbf{U}^{\neg T}\mathbf{h}(t) = \mathbf{K}\mathbf{X}^{\neg T}\mathbf{h}(t) \tag{36.69}$$

$$\text{col}\, \mathbf{U}^{\neg T} = \text{col}\, (\mathbf{K}\mathbf{X}^{\neg T}\mathbf{I}_k) = (\mathbf{I}_k \otimes \mathbf{K})\text{col}\, \mathbf{X}^{\neg T} \tag{36.70}$$

$$\text{col}\, \mathbf{X}^{\neg T} = [\mathbf{I}_{kn} - (\mathbf{P}_M^T \otimes \mathbf{A}) - (\mathbf{P}_M^T \otimes \mathbf{B})(\mathbf{I}_k \otimes \mathbf{K})]^{-1}\, \text{col}\, \mathbf{X}_o^{\neg T}. \tag{36.71}$$

The calculation is repeated with the expansion $\dot{\mathbf{x}}(t) = \mathbf{C}^{\neg}\mathbf{h}(t)$, namely for the sake of comparison with Laplace domain $\mathcal{L}$

$$\text{col}\, \mathbf{C}^{\neg} = (\mathbf{I}_{kn} - \mathbf{P}_M^T \otimes \mathbf{A})^{-1}\, \text{col}\, (\mathbf{A}\mathbf{X}_o^{\neg} + \mathbf{B}\mathbf{U}^{\neg T}) \tag{36.72}$$

$$\text{col}\, \mathbf{X}^{\neg} = (\mathbf{I}_{kn} - \mathbf{P}_M^T \otimes \mathbf{A})^{-1}\, \text{col}\, (\mathbf{X}_o^{\neg} + \mathbf{B}\mathbf{U}^{\neg T}\mathbf{P}_M) \tag{36.73}$$

$$\mathcal{L}\dot{\mathbf{x}}(t) = \left(\mathbf{I}_n - \frac{1}{s}\mathbf{A}\right)^{-1}\left(\mathbf{x}_o + \mathbf{B}\mathbf{U}\right). \tag{36.74}$$

$$\mathcal{L}\mathbf{x}(t) = \left(\mathbf{I}_n - \frac{1}{s}\mathbf{A}\right)^{-1}\left(\frac{1}{s}\mathbf{x}_o + \mathbf{B}\mathbf{U}\frac{1}{s}\right). \tag{36.75}$$

Differences in the aforementioned results arise from differences embodied in the basic properties

$$\mathcal{L}\dot{\mathbf{x}}(t) = [\mathcal{L}\mathbf{x}(t)]s - \mathbf{x}_o \quad \text{and} \quad \mathbf{C}^{\neg} = \mathbf{X}^{\neg}\mathbf{P}_M^{-1} - \mathbf{X}_o\mathbf{P}_M^{-1}. \tag{36.76}$$

36.10 State Feedback Control Including Dead-Time

36.10.1 Operational Matrix for Delay

First, the operational matrix for differentiation

$$\frac{d}{dt}\,\mathbf{h}(t) \doteq \mathbf{D}_M\mathbf{h}(t) \tag{36.77}$$

is introduced. Note that the operational matrices depend on the polynomial-type and on the shifting operation details, see e.g. Eq.(36.35). To derive the operational matrix $\mathbf{D}_d$ for dead-time delay, the operational matrix $\mathbf{D}_M$ for differentiation is required. The matrix $\mathbf{D}_M$ is derived in a similar way as the operational matrix $\mathbf{P}_M$ for integration. Consider a scalar function $v(t)$, first. Orthogonal expanding yields

$$v(t) \doteq \sum_{i=0}^{k-1} v_i h_i(t) = \mathbf{v}^{\neg}\mathbf{h}(t). \tag{36.78}$$

Backward extrapolation of the function $v(t)$ from t back to $t - T_d$ can be achieved by Taylor expansion

$$v(t-T_d) = \sum_{j=0}^{\infty} \frac{(-T_d)^j}{j!} v^{(j)}(t) = \mathbf{v}^{\neg T}(t-T_d)\mathbf{h}(t) \tag{36.79}$$

$$v(t-T_d) = \sum_{j=0}^{\infty} \frac{(-T_d)^j}{j!} \mathbf{v}^{\neg T}\mathbf{D}_M^j\mathbf{h}(t) = \mathbf{v}^{\neg T}\sum_{j=0}^{\infty} \frac{(-T_d)^j}{j!}\mathbf{D}_M^j\mathbf{h}(t)\,. \tag{36.80}$$

Defining the operational matrix $\mathbf{D}_d$ for dead-time delay, the matrix $\mathbf{D}_d$ results

$$\mathbf{v}^{\neg T}(t-T_d) \triangleq \mathbf{v}^{\neg T}\mathbf{D}_d \qquad \mathbf{D}_d = \sum_{j=0}^{\infty}\frac{(-T_d\mathbf{D}_M)^j}{j!} = \exp\left(-T_d\mathbf{D}_M\right) \qquad \mathbf{D}_d,\ \mathbf{D}_M \in \mathcal{R}^{k\times k}\,. \tag{36.81}$$

The similarity to the simple transfer function of a dead-time delay $\exp(-sT_d)$ is obvious.

36.10.2 Feedback Control Equation Including Dead-Time Delay

Let the linear time-invariant system including dead-time delay be

$$\dot{\mathbf{x}}(t) = \mathbf{A}\mathbf{x}(t) + \mathbf{A}_d\mathbf{x}(t-T_d) + \mathbf{B}\mathbf{u}(t) \tag{36.82}$$

$$\mathbf{x}(t)\,|_{-T_d\le t<0} = \mathbf{x}_{in}(t), \qquad \mathbf{x}_o = \mathbf{x}(0) = \mathbf{x}_{in}(0)\,. \tag{36.83}$$

Decomposing into orthogonal components yields

$$\mathbf{x}(t) \doteq \mathbf{X}^{\neg T}\mathbf{h}(t), \quad \mathbf{u}(t) \doteq \mathbf{U}^{\neg T}\mathbf{h}(t), \quad \mathbf{x}_{in}(t) = \mathbf{X}_{in}^{\neg T}\mathbf{h}(t) \tag{36.84}$$

$$\mathbf{x}(t-T_d) \doteq \mathbf{X}^{\neg T}\mathbf{h}(t-T_d) = \mathbf{X}^{\neg T}\mathbf{D}_d\mathbf{h}(t)\,. \tag{36.85}$$

Applying the integration and delay operational formulas to Eq.(36.82), one has to distinguish between the first interval $0 \le t \le T_d$ where the prehistory $\mathbf{x}_{in}(t)$ has influence and a second interval $T_d \le t \le t_f$ in which the delay $\mathbf{A}_d$ comes into action forced by $\mathbf{x}(t)$.

In the time range $0 \le t \le T_d$:

$$\mathbf{X}^{\neg T} - \mathbf{X}_o^{\neg T} = \mathbf{A}\mathbf{X}^{\neg T}\mathbf{P}_M + \mathbf{A}_d\mathbf{X}_{in}^{\neg T}\mathbf{P}_M + \mathbf{B}\mathbf{U}^T\mathbf{P}_M \tag{36.86}$$

$$\text{col } \mathbf{X}^{\neg T} = [\mathbf{I}_{kn} - (\mathbf{P}_M^T \otimes \mathbf{A})]^{-1} (\text{col } \mathbf{X}_o^{\neg} + \text{col } \mathbf{A}_d \mathbf{X}_{in}^{\neg T} \mathbf{P}_M + \text{col } \mathbf{B}\mathbf{U}^T \mathbf{P}_M) \quad (36.87)$$

and in the time range $T_d \leq t \leq t_f$:

$$\mathbf{X}^{\neg T} - \mathbf{X}_{T_d}^{\neg T} = \mathbf{A}\mathbf{X}^{\neg T}\mathbf{P}_M + \mathbf{A}_d\mathbf{X}^{\neg T}\mathbf{D}_d\mathbf{P}_M + \mathbf{B}\mathbf{U}^{\neg T}\mathbf{P}_M \quad (36.88)$$

$$\text{col } \mathbf{X}^{\neg T} = [\mathbf{I}_{kn} - (\mathbf{P}_M^T \otimes \mathbf{A}) - (\mathbf{D}_d\mathbf{P}_M)^T \otimes \mathbf{A}_d]^{-1} \left[\text{col } \mathbf{X}_{T_d}^{\neg T} + \text{col } \mathbf{B}\mathbf{U}^{\neg T}\mathbf{P}_M\right] . \quad (36.89)$$

If a state feedback controller is implemented, the following results can be achieved, similarly to Eq.(36.71):

In the time range $0 \leq t \leq T_d$:

$$\text{col } \mathbf{X}^{\neg T} = [\mathbf{I}_{kn} - (\mathbf{P}_M^T \otimes \mathbf{A}) + (\mathbf{P}_M^T \otimes \mathbf{B})(\mathbf{I}_k \otimes \mathbf{K})]^{-1} \left[\text{col } \mathbf{X}_o^{\neg T} + \text{col}\mathbf{A}_d\mathbf{X}_{in}^{\neg T}\mathbf{P}_M\right] \quad (36.90)$$

and in the time range $T_d \leq t \leq t_f$:

$$\text{col } \mathbf{X}^{\neg T} = [\mathbf{I}_{kn} - \mathbf{P}_M^T \otimes \mathbf{A} - (\mathbf{D}_d\mathbf{P}_M)^T \otimes \mathbf{A}_d + (\mathbf{P}_M^T \otimes \mathbf{B})(\mathbf{I}_k \otimes \mathbf{K})]^{-1} \text{ col } \mathbf{A}_d\mathbf{X}_{in}^{\neg T}\mathbf{P}_M . \quad (36.91)$$

Problems probably arising from initial value transfer at $t = T_d$ are left out of consideration. A method to reduce the effort involved with the computation of the inverse is given by *Hadzer, C.M., et al. 1989*.

36.11 Time-Varying System. State-Space Approach

36.11.1 System Analysis

Rewriting Eq.(38.50) into the n-dimensional state space and applying Kronecker calculus,

$$\dot{\mathbf{x}}(t) = \sum_{i=0}^{j} \mathbf{A}_i t^i \mathbf{x}(t) + \sum_{i=0}^{l} \mathbf{B}_i t^i \mathbf{u}(t) \qquad \mathbf{x}(0) = \mathbf{x}_o . \quad (36.92)$$

Use an operational matrix $\mathbf{L}$ of time powers multiplication, i.e., $t^i\mathbf{h}(t) = \mathbf{L}^i\mathbf{h}(t)$, and the operational matrix $\mathbf{P}_M$ of intergration $\int_a^t \mathbf{h}(t)dt = \mathbf{P}_M\mathbf{h}$ and the expansion $\mathbf{x} = \mathbf{X}^{\neg T}\mathbf{h}$. Then, the key properties solving Eq.(36.92) are

$$\mathbf{A}_i t^i \mathbf{x}(t) = \mathbf{A}_i t^i \mathbf{X}^{\neg T}\mathbf{h}(t) = \mathbf{A}_i \mathbf{X}^{\neg T} t^i \mathbf{h}(t) = \mathbf{A}_i \mathbf{X}^{\neg T}\mathbf{L}^i\mathbf{h}(t) \quad (36.93)$$

$$\int_a^t \mathbf{A}_i t^i \mathbf{x}(t)dt = \mathbf{A}_i\mathbf{X}^{\neg T}\mathbf{L}^i \int_a^t \mathbf{h}(t)dt = \mathbf{A}_i\mathbf{X}^{\neg T}\mathbf{L}^i\mathbf{P}_M\mathbf{h} . \quad (36.94)$$

Integrating Eq.(36.92) once with respect to time and substituting the properties mentioned above yields, similarly to Eq.(36.64),

$$\mathbf{x}(t) = \mathbf{X}^{\neg T}\mathbf{h} = \sum_{i=0}^{j} \mathbf{A}_i\mathbf{X}^{\neg T}\mathbf{L}^i\mathbf{P}_M\mathbf{h} + \sum_{i=0}^{l} \mathbf{B}_i\mathbf{U}^{\neg T}\mathbf{L}^i\mathbf{P}_M\mathbf{h} + (\mathbf{x}_o \vdots \mathbf{0} \,....\, \mathbf{0})^T \quad (36.95)$$

$$\begin{aligned}
\text{col } \mathbf{X}^{\neg T} &= \sum_{i=0}^{j}(\mathbf{L}^i\mathbf{P}_M)^T \otimes \mathbf{A}_i \text{col } \mathbf{X}^{\neg T} + \sum_{i=0}^{l}(\mathbf{L}^i\mathbf{P}_M)^T \otimes \mathbf{B}_i \text{col } \mathbf{U}^{\neg T} + \mathbf{X}_o^{\neg T} && (36.96)\\
&= [\, \mathbf{I}_{kn} - \sum_{i=0}^{j}(\mathbf{L}^i\mathbf{P}_M)^T \otimes \mathbf{A}_i \,]^{-1}\{[\sum_{i=0}^{l}(\mathbf{L}^i\mathbf{P}_M)^T \otimes \mathbf{B}_i]\text{col } \mathbf{U}^{\neg T} + \mathbf{X}_o^{\neg T}\} . && (36.97)
\end{aligned}$$

36.11.2 Time-Varying System. Parameter Estimation

Let the input and state vector be known at sufficiently many instants. For a given $\mathbf{X}_o^{\neg T}$ the matrices $\mathbf{A}_i$ and $\mathbf{B}_i$ can be determined. From Eq.(36.95) it follows (*Lee, T.T., and Tsay, S.C., 1989*)

$$\mathbf{X}^{\neg T} = \mathbf{A}_o\mathbf{X}^{\neg T}\mathbf{P}_M + \mathbf{A}_1\mathbf{X}^{\neg T}\mathbf{L}\mathbf{P}_M + \dots\dots + \mathbf{B}_l\mathbf{U}^{\neg T}\mathbf{L}^l\mathbf{P}_M + \mathbf{X}^{\neg T} \tag{36.98}$$

$$\mathrm{col}(\mathbf{X}^{\neg T}-\mathbf{X}_o^{\neg T}) = [(\mathbf{X}^{\neg T}\mathbf{P}_M)^T\otimes\mathbf{I}_n]\mathrm{col}\mathbf{A}_o+[(\mathbf{X}^{\neg T}\mathbf{L}\mathbf{P}_M)^T\otimes\mathbf{I}_n]\mathrm{col}\mathbf{A}_1+\dots.[(\mathbf{U}^{\neg T}\mathbf{L}^l\mathbf{P}_M)^T\otimes\mathbf{I}_n]\mathrm{col}\mathbf{B}_l \tag{36.99}$$

$$\begin{pmatrix} \mathrm{col}\mathbf{A}_o \\ \vdots \\ \mathrm{col}\mathbf{A}_j \\ \mathrm{col}\mathbf{B}_o \\ \vdots \\ \mathrm{col}\mathbf{B}_l \end{pmatrix} = [\,(\mathbf{X}^{\neg T}\mathbf{P}_M)^T\otimes\mathbf{I}_n \vdots (\mathbf{X}^{\neg T}\mathbf{L}\mathbf{P}_M)^T\otimes\mathbf{I}_n \vdots \dots\ (\mathbf{U}^{\neg T}\mathbf{L}^l\mathbf{P}_M)^T\otimes\mathbf{I}_n\,]^{\sharp L}\mathrm{col}(\mathbf{X}^{\neg T}-\mathbf{X}_o^{\neg T}) \tag{36.100}$$

36.11.3 Optimal Design

Following the optimal design procedure as given by Eq.(9.19) where $\mathbf{A} = \mathbf{A}(t)$, the transition matrix $\mathbf{\Theta}(t_f, t)$ is calculated. By Taylor expansion of $\mathbf{M}_R = \sum_{i=1}^{l} \mathbf{M}^i t^i$, the matrix differential equation $\dot{\mathbf{\Theta}}(t_f, t) = -\mathbf{\Theta}(t_f, t)\mathbf{M}_R(t)$ can be solved via orthogonal expansion of the matrix $\mathbf{\Theta}(t_f, t)$, i.e.,

$$\mathbf{\Theta}(t_f, t) = \mathbf{\Theta}^{\neg T}(\mathrm{h} \otimes \mathbf{I}_{2n}) \qquad \mathbf{\Theta}^{\neg T} \in \mathcal{R}^{2n\times 2kn} \ . \tag{36.101}$$

36.12 Orthogonal Polynomial Solution of Functional Differential Equations

36.12.1 Operational Matrix of Integration for $\mathbf{t}(t)$

Integrating $\mathbf{t}(t)$ as defined in Eq.(36.42) and using Eqs.(36.46) and (36.42) yields

$$\int_0^t \mathbf{t}(t)dt = \mathbf{\Theta}\int_0^t \mathrm{h}(t) = \mathbf{\Theta}\mathbf{P}_M\mathrm{h}(t) = \mathbf{\Theta}\mathbf{P}_M\mathbf{F}\mathbf{t}(t) = \mathbf{P}_{Mp}\mathbf{t}(t) \ . \tag{36.102}$$

The matrix $\mathbf{P}_{Mp}$ is the operational matrix of integration, associated with the power series expansion vector $\mathbf{t}(t)$

$$\mathbf{P}_{Mp} \triangleq \begin{pmatrix} 0 & 1 & 0 & 0 & \dots & 0 \\ 0 & 0 & 1/2 & 0 & \dots & 0 \\ 0 & 0 & 0 & 1/3 & \dots & 0 \\ \vdots & \vdots & \vdots & \vdots & \ddots & 0 \\ 0 & 0 & 0 & 0 & \dots & 1/(k-1) \\ \chi_0 & \chi_1 & \chi_2 & \chi_3 & \cdots & \chi_{k-1} \end{pmatrix} \qquad \begin{array}{l} \chi_i = -\dfrac{f_{ki}}{k f_{kk}} \\ \forall i = 0,1,2\dots k-1 \\ \mathbf{P}_{Mp} \in \mathcal{R}^{k\times k} \end{array} \tag{36.103}$$

where f_{ki} are the expansion coefficients of $h_k(t)$ by powers of t, see Eq. (36.41). The operational matrix $\mathbf{P}_{Mp}$ is defined for any type of orthogonal polynomial.

36.12.2 Stretched Operational Matrix of Integration

Considering the stretched time scale λt of a separate model

$$\mathbf{t}(\lambda t) = (1 \quad \lambda t \quad \lambda^2 t^2 \ \ldots \ \lambda^{k-1} t^{k-1})^T \tag{36.104}$$

and comparing with $\mathbf{t}(t)$ in real time t of Eq.(36.42) yields

$$\mathbf{t}(\lambda t) = (\text{diag } \lambda^i)\mathbf{t}(t) \qquad i = 0, 1 \ldots k-1 \,. \tag{36.105}$$

The diagonal (k,k)-matrix serves as a premultiplication matrix. Integrating $\mathbf{t}(t)$ from 0 to λt yields

$$\int_0^{\lambda t} \mathbf{t}(\tau)d\tau \mid_{\tau=\lambda\tau'} = \int_0^t \mathbf{t}(\lambda\tau')\lambda d\tau' = \lambda(\text{diag } \lambda^i) \int_0^t \mathbf{t}(t)dt = \mathbf{P}_{Ms}\,\mathbf{t}(t) \tag{36.106}$$

$$\mathbf{P}_{Ms} \triangleq \lambda(\text{diag } \lambda^i)\mathbf{P}_{Mp} = (\text{diag } \lambda^{i+1})\mathbf{P}_{Mp} \,. \tag{36.107}$$

The matrix $\mathbf{P}_{Ms}$ is named stretched operational matrix of integration.

36.12.3 Power Series Expansion of the Solution

Consider the n-dimensional functional ordinary differential equation with time-invariant coefficient matrices $\mathbf{A}$ and $\mathbf{B}$, occurring, e.g., in particulate processes (*Hwang, C., and Shih, Y.P., 1982*) or in real-time processes where the control variable $\mathbf{u}(t)$ is generated by an asynchronous model $\mathbf{u}(t) = \mathbf{x}(\lambda t)$. Then,

$$\dot{\mathbf{x}}(t) = \mathbf{A}\mathbf{x}(t) + \mathbf{B}\mathbf{x}(\lambda t) \qquad \mathbf{x}(0) = \mathbf{x}_o \tag{36.108}$$

with initial conditon $\mathbf{x}_o$. The yet unknown $\dot{\mathbf{x}}(t)$ is expressed by power series up to power t^{k-1}

$$\dot{\mathbf{x}}(t) = \mathbf{C}\mathbf{t}(t) \qquad \mathbf{C} = (\mathbf{c}_o\ \mathbf{c}_1 \ldots \mathbf{c}_{k-1}) \qquad \mathbf{C} \in \mathcal{R}^{n \times k} \,. \tag{36.109}$$

The *rows* of $\mathbf{C}$ are the unknown power series expansion coefficients. However, $\mathbf{C}$ is decomposed into *columns* $\mathbf{c}_i$, for convenient calculation purposes. Integrating Eq.(36.109) yields

$$\mathbf{x}(t) = \mathbf{x}_o + \int_0^t \dot{\mathbf{x}}(t)dt = \mathbf{x}_o + \mathbf{C}\mathbf{P}_{Mp}\mathbf{t}(t) = [(\mathbf{x}_o \ \ \mathbf{0} \ \ldots \ \mathbf{0}) + \mathbf{C}\mathbf{P}_{Mp}]\mathbf{t}(t) \,. \tag{36.110}$$

The first term in the brackets is an (n,k)-matrix, with one column $\mathbf{x}_o$ and zero columns $\mathbf{0}$ otherwise. Considering the scaled argument λt,

$$\mathbf{x}(\lambda t) = \mathbf{x}_o + \int_0^{\lambda t} \dot{x}(t)dt = \mathbf{x}_o + \mathbf{C}\int_0^{\lambda t} \mathbf{t}(t)dt = [(\mathbf{x}_o \ \ \mathbf{0} \ \ldots \ \mathbf{0}) + \mathbf{C}\mathbf{P}_{Ms}]\mathbf{t}(t) \,. \tag{36.111}$$

Substituting Eqs.(36.109), (36.110) and (36.111) into Eq.(36.108) yields

$$\mathbf{C}\mathbf{t}(t) = \mathbf{A}[(\mathbf{x}_o \ \ \mathbf{0} \ \ldots \ \mathbf{0}) + \mathbf{C}\mathbf{P}_{Mp}]\mathbf{t}(t) + \mathbf{B}[(\mathbf{x}_o \ \ \mathbf{0} \ \ldots \ \mathbf{0}) + \mathbf{C}\mathbf{P}_{Ms}]\,\mathbf{t}(t) \tag{36.112}$$

$$\mathbf{C} - \mathbf{A}\mathbf{C}\mathbf{P}_{Mp} - \mathbf{B}\mathbf{C}\mathbf{P}_{Ms} = [(\mathbf{A}+\mathbf{B})\mathbf{x}_o \ \ \mathbf{0} \ldots \mathbf{0}] \,. \tag{36.113}$$

Eq.(36.113) is a linear algebraic equation in the unknown (n,k)-matrix $\mathbf{C}$. Using the $(kn,1)$-vector col$\mathbf{C}$ and the initial vector $\mathbf{x}'_o = [\mathbf{x}_o^T(\mathbf{A}^T + \mathbf{B}^T) \ \ \mathbf{0}\ \mathbf{0} \ldots \mathbf{0}]^T$, Eq.(36.113) is rewritten to

$$\text{col } \mathbf{C} - (\mathbf{P}_{Mp}^T \otimes \mathbf{A}) \text{ col } \mathbf{C} - (\mathbf{P}_{Ms}^T \otimes \mathbf{B}) \text{ col } \mathbf{C} = \mathbf{x}'_o \tag{36.114}$$

$$\text{col } \mathbf{C} = (\mathbf{I}_{kn} - \mathbf{P}_{Mp}^T \otimes \mathbf{A} - \mathbf{P}_{Ms}^T \otimes \mathbf{B})^{-1}\mathbf{x}'_o \,. \tag{36.115}$$

36.12.4 Recursive Algorithm Solving Functional Differential Equations

The conventional method shown in the previous section requires the inverse of a (kn, kn)-matrix containing the stretched operational matrix $\mathbf{P}_{M_s}$. *Chang, R.Y., et al. 1987* proposed a simple and straightforward algorithm using (n, n)-matrix inversion, only. This computational algorithm expands Eq.(36.113) into vector components

$$\mathbf{c}_o = (\mathbf{A} + \mathbf{B})\mathbf{x}_o + \chi_o(\mathbf{A} + \lambda^k\mathbf{B})\mathbf{c}_{k-1} \tag{36.116}$$

$$\mathbf{c}_i = \frac{1}{i}(\mathbf{A} + \lambda^i\mathbf{B})\mathbf{c}_{i-1} + \chi_i(\mathbf{A} + \lambda^k\mathbf{B})\mathbf{c}_{k-1} \quad \forall i = 1, 2 \ldots k-1 \ . \tag{36.117}$$

In order to obtain a recursive algorithm, the definitions

$$\mathbf{c}_i = \mathbf{p}_i + \mathbf{q}_i \ \mathbf{c}_{k-1} \tag{36.118}$$

$$\mathbf{c}_{i-1} = \mathbf{p}_{i-1} + \mathbf{q}_{i-1} \ \mathbf{c}_{k-1} \ . \tag{36.119}$$

are useful. Substituting Eq.(36.119) into Eq.(36.117) yields

$$\mathbf{c}_i = [\frac{1}{i}(\mathbf{A} + \lambda^i\mathbf{B})\mathbf{q}_{i-1} + \chi_i(\mathbf{A} + \lambda^k\mathbf{B})] \ \mathbf{c}_{k-1} + \frac{1}{i}(\mathbf{A} + \lambda^i\mathbf{B}) \ \mathbf{p}_{i-1} \ . \tag{36.120}$$

Comparing with Eq.(36.118) gives the recurrence $\forall i = 1, 2 \ldots k-1$

$$\mathbf{p}_i = \frac{1}{i}(\mathbf{A} + \lambda^i\mathbf{B})\mathbf{p}_{i-1} \qquad \mathbf{q}_i = \frac{1}{i}(\mathbf{A} + \lambda^i\mathbf{B})\mathbf{q}_{i-1} + \chi_i(\mathbf{A} + \lambda^k\mathbf{B}) \tag{36.121}$$

where the initial vector components $\mathbf{p}_o$ and $\mathbf{q}_o$ are already given by Eq.(36.116)

$$\mathbf{p}_o = (\mathbf{A} + \mathbf{B})\mathbf{x}_o \qquad \mathbf{q}_o = \chi_o(\mathbf{A} + \lambda^k\mathbf{B}). \tag{36.122}$$

The last column $\mathbf{c}_{k-1}$ can be calculated from Eq.(36.119) with $i = k$

$$\mathbf{c}_{k-1} = (\mathbf{I} - \mathbf{q}_{k-1})^{-1}\mathbf{p}_{k-1} \ . \tag{36.123}$$

Chapter 37

System Analysis Via Pulse Functions and Piecewise Linear Functions

For obvious reasons, approximating signals by piecewise constant or piecewise linear signals is a very simple approach. Walsh functions or block-pulse functions are applied in order to obtain concise results. Piecewise constant signals provide simple algorithms for approximating signals and processes.

37.1 Control Design Via Walsh Functions

The approach of piecewise constant gains using Walsh functions was proposed by *Chen, C.F., and Hsiao, C.H., 1975*. This approximation method proves very powerful especially in the design of optimal control. Algorithms can be implemented easily.

Walsh functions $h_i(t)$ are a set of square waves with orthonormal property. They are depicted in Fig. 37.1 up to $h_7(t)$ (*Harmuth, H.H., 1969*). Expanding an arbitrary but periodic and absolutely integrable function

$$x(t) \doteq \sum_{i=0}^{k-1} x_i^w h_i(t) , \tag{37.1}$$

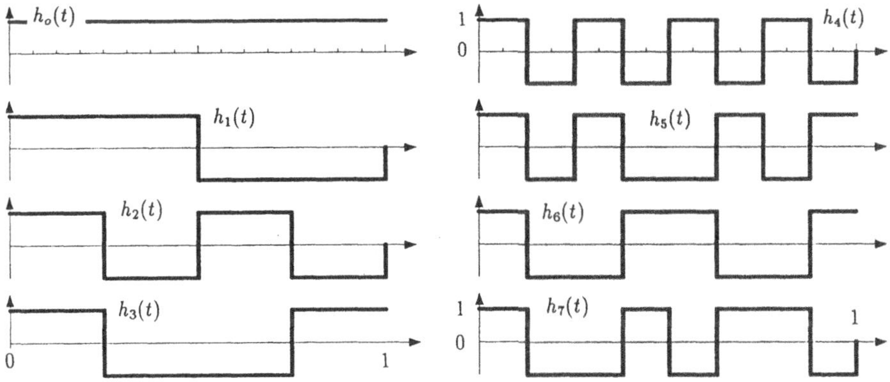

Figure 37.1: Walsh functions $h_i(t)$

the coefficients $x_i^w \; \forall i = 0, 1, 2 \dots k-1$ are determined such that

$$\int_0^1 \|x(t) - \sum_{i=0}^{k-1} x_i^w h_i(t)\|_F^2 dt \to \min \quad \rightsquigarrow \quad x_i^w = \int_0^1 x(t) h_i(t) dt \; . \tag{37.2}$$

It is convenient to define $\mathbf{h}(t) \triangleq [h_o(t) \; h_1(t) \dots h_{k-1}(t)]^T$. To obtain simpler results, k is preferably chosen $k = 2^\alpha$ where α is an integer. In order to replace the integration by premultiplication $\int_0^1 \mathbf{h}(t)dt \doteq \mathbf{P}_{Mw}\mathbf{h}(t)$ the operational matrix for integration $\mathbf{P}_{Mw}$ is defined. When $\alpha = 3$ then $\mathbf{P}_{Mw}$ is of dimension 8×8

$$\mathbf{P}_{Mw} = \begin{pmatrix} 0.5 & -0.25 & -0.125 & 0 & -0.0625 & 0 & 0 & 0 \\ 0.25 & 0 & 0 & -0.125 & 0 & -0.0625 & 0 & 0 \\ 0.125 & 0 & 0 & 0 & 0 & 0 & -0.0625 & 0 \\ 0 & 0.125 & 0 & 0 & 0 & 0 & 0 & -0.0625 \\ 0.0625 & 0 & 0 & 0 & 0 & 0 & 0 & 0 \\ 0 & 0.0625 & 0 & 0 & 0 & 0 & 0 & 0 \\ 0 & 0 & 0.0625 & 0 & 0 & 0 & 0 & 0 \\ 0 & 0 & 0 & 0.0625 & 0 & 0 & 0 & 0 \end{pmatrix} . \tag{37.3}$$

If $k = 2^\alpha$ a general formula can be found yielding the (k,k)-matrix $\mathbf{P}_{Mw(k \times k)}$ partitioned into matrices of half dimension

$$\mathbf{P}_{Mw(k \times k)} = \begin{pmatrix} \mathbf{P}_{Mw(\frac{k}{2} \times \frac{k}{2})} & \frac{-1}{2k}\mathbf{I}_{(\frac{k}{2})} \\ \frac{1}{2k}\mathbf{I}_{(\frac{k}{2})} & \mathbf{0}_{(\frac{k}{2})} \end{pmatrix} \qquad \int_0^1 \mathbf{h}(t)dt \doteq \mathbf{P}_{Mw}\mathbf{h}(t) \; . \tag{37.4}$$

The entry in the upper left corner always is $\frac{1}{2}$.

Decomposing the rate variable $\dot{\mathbf{x}}(t)$ into a weighted sum of Walsh functions yields

$$\dot{\mathbf{x}}(t) = \mathbf{C}\mathbf{h}(t) \qquad \int_0^t \dot{\mathbf{x}}(t)dt = \mathbf{C}\int_0^t \mathbf{h}(t)dt = \mathbf{C}\mathbf{P}_{Mw}\mathbf{h}(t) = \mathbf{x}(t) - \mathbf{x}_o \tag{37.5}$$

$$\mathbf{x}(t) = \mathbf{C}\mathbf{P}_{Mw}\mathbf{h}(t) + \mathbf{x}_o = \mathbf{C}\mathbf{P}_{Mw}\mathbf{h}(t) + \mathbf{x}_o h_o = \mathbf{C}\mathbf{P}_{Mw}\mathbf{h}(t) + (\mathbf{x}_o \; \mathbf{0} \dots \mathbf{0})\mathbf{h}(t) \tag{37.6}$$

where $h_o(t) = 1$. Using the state variable expansion within the interval $0 \le t \le 1$

$$\mathbf{x}(t) \triangleq \mathbf{X}^w\mathbf{h}(t) \qquad \mathbf{C}\mathbf{P}_{Mw} + (\mathbf{x}_o \; \mathbf{0} \dots \mathbf{0}) = \mathbf{C}\mathbf{P}_{Mw} + \mathbf{X}_o = \mathbf{X}^w \; . \tag{37.7}$$

37.1.1 System Analysis

Solving the system state equation

$$\dot{\mathbf{x}}(t) = \mathbf{A}\mathbf{x}(t) + \mathbf{B}\mathbf{u}(t) \qquad \mathbf{x}(0) = \mathbf{x}_o \tag{37.8}$$

by using $\mathbf{u}(t) = \mathbf{U}^w\mathbf{h}(t)$

$$\mathbf{C} = \mathbf{A}\mathbf{X}^w + \mathbf{B}\mathbf{U}^w = \mathbf{A}\mathbf{C}\mathbf{P}_{Mw} + \mathbf{A}\mathbf{X}_o + \mathbf{B}\mathbf{U}^w \tag{37.9}$$

$$\text{col } \mathbf{C} = (\mathbf{P}_{Mw}^T \otimes \mathbf{A}) \text{ col } \mathbf{C} + \text{ col } (\mathbf{A}\mathbf{X}_o + \mathbf{B}\mathbf{U}^w) \tag{37.10}$$

$$\text{col } \mathbf{C} = (\mathbf{I}_{kn} - \mathbf{P}_{Mw}^T \otimes \mathbf{A})^{-1} \text{ col } (\mathbf{A}\mathbf{X}_o + \mathbf{B}\mathbf{U}^w) \tag{37.11}$$

is obtained. The solution cited above may be considered as a solution of a generalized Lyapunov equation. *Bilinear* systems and bilinear differential equations are transformed into *linear algebraic* equations by applying Walsh functions (*Lewis, F.L., et al. 1990; Lewis, F.L., and Fountain, D.W., 1991*).

37.1.2 Optimal Control

Applying Walsh series approximation, the optimal control problem is solved. Starting with an initial $\mathbf{x}(0)$, a combined vector of state and costate variable is used, see Eq.(9.19),

$$\begin{pmatrix}\dot{\mathbf{x}}(\mu)\\ \dot{\boldsymbol{\lambda}}(\mu)\end{pmatrix} = -t_f \mathbf{M}_R \begin{pmatrix}\mathbf{x}(\mu)\\ \boldsymbol{\lambda}(\mu)\end{pmatrix}, \quad 0 \le \mu \le 1 \qquad \mathbf{x},\ \boldsymbol{\lambda} \in \mathcal{R}^n, \qquad \mathbf{h} \in \mathcal{R}^k\,. \tag{37.12}$$

The normalized time scale variable μ serves as a variable, running from final to initial state

$$\frac{t_f - t}{t_f} = 1 - \frac{t}{t_f} = \mu, \qquad \frac{d}{d\mu} = -\frac{d}{t_f dt}\,. \tag{37.13}$$

The factor t_f in Eq.(37.12) is determined by the relation between dt and $d\mu$. Defining a decomposition matrix $\mathbf{C}$ of dimension $2n \times k$

$$\begin{pmatrix}\dot{\mathbf{x}}(\mu)\\ \dot{\boldsymbol{\lambda}}(\mu)\end{pmatrix} = \mathbf{C}\mathbf{h}(\mu) \qquad \begin{pmatrix}\mathbf{x}(\mu)\\ \boldsymbol{\lambda}(\mu)\end{pmatrix} = \mathbf{C}\mathbf{P}_{Mw}\mathbf{h}(\mu) + \begin{pmatrix}\mathbf{x}(\mu=0)\\ \mathbf{0}_n\end{pmatrix} \tag{37.14}$$

$$\begin{pmatrix}\dot{\mathbf{x}}(\mu)\\ \dot{\boldsymbol{\lambda}}(\mu)\end{pmatrix} = \mathbf{C}\mathbf{h}(\mu) = -t_f\mathbf{M}_R[\mathbf{C}\mathbf{P}_{Mw} + \begin{pmatrix}\mathbf{x}(\mu=0)\\ \mathbf{0}_n\end{pmatrix} \vdots\ \mathbf{0}_{2n}\ldots\mathbf{0}_{2n})]\mathbf{h}(\mu) \tag{37.15}$$

$$\mathbf{C} = -t_f\begin{pmatrix}\mathbf{A} & \mathbf{B}\mathbf{R}^{-1}\mathbf{B}^T\\ \mathbf{Q} & -\mathbf{A}^T\end{pmatrix}[\mathbf{C}\mathbf{P}_{Mw} + \begin{pmatrix}\mathbf{x}(\mu=0)\\ \mathbf{0}_n\end{pmatrix} \vdots\ \mathbf{0}_{2n}\ldots\mathbf{0}_{2n})] \tag{37.16}$$

$$\mathbf{C} = -t_f\begin{pmatrix}\mathbf{A} & \mathbf{B}\mathbf{R}^{-1}\mathbf{B}^T\\ \mathbf{Q} & -\mathbf{A}^T\end{pmatrix}\mathbf{C}\mathbf{P}_{Mw} - t_f\begin{pmatrix}\mathbf{A}\mathbf{x}(\mu=0)\\ \mathbf{Q}\mathbf{x}(\mu=0)\end{pmatrix} \vdots\ \mathbf{0}_{2n}\ldots\mathbf{0}_{2n}) \tag{37.17}$$

$$\text{col}\ \mathbf{C} = -t_f(\mathbf{P}_{Mw}^T \otimes \mathbf{M}_R)\ \text{col}\ \mathbf{C} - t_f\begin{pmatrix}\mathbf{A}\mathbf{x}(\mu=0)\\ \mathbf{Q}\mathbf{x}(\mu=0)\\ \mathbf{0}_{2n}\\ \vdots\end{pmatrix} \tag{37.18}$$

$$\text{col}\ \mathbf{C} = -t_f(\mathbf{I} + t_f\mathbf{P}_{Mw}^T \otimes \mathbf{M}_R)^{-1}\begin{pmatrix}\mathbf{A}\mathbf{x}(\mu=0)\\ \mathbf{Q}\mathbf{x}(\mu=0)\\ \mathbf{0}_{2n}\\ \vdots\end{pmatrix}. \tag{37.19}$$

The calculation of the matrix inverse $(\mathbf{I} + t_f\mathbf{P}_{Mw}^T \otimes \mathbf{M}_R)^{-1}$ may be very tedious, with regard to the high dimension. This difficulty can be overcome by putting the equations into recursive form (see *Chen, C.F., and Hsiao, C.H., 1975*). Based on the decomposition-coordination approach and on hierarchically recursive algorithms, large-scale problems are solved by *Zhu, J.M., 1988*.

37.2 System Analysis and Identification Via Block-Pulse Functions

Piecewise constant solutions often are used to approximate the exact motion of a dynamic system. To obtain piecewise constant solutions, the method of block-pulse functions can be applied. Piecewise constant control signals can easily be implemented by computer control. This method, presented by *Sannuti, P., 1977*, is computationally simple and has good numerical accuracy.

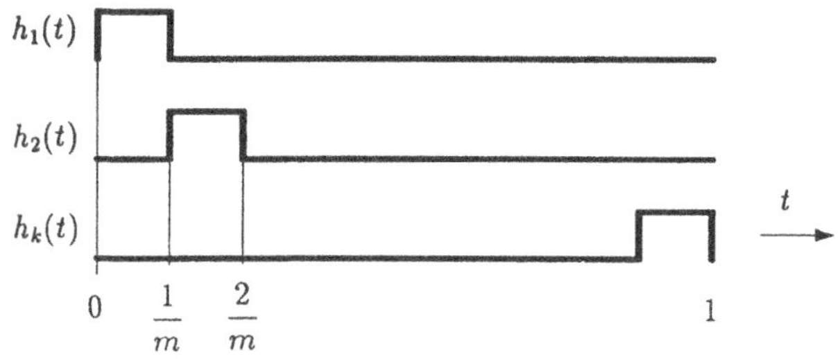

Figure 37.2: Oscillogram of block-pulse functions

37.2.1 Block-Pulse Functions

On a unit interval $(0,1]$ block-pulse functions are defined $\forall i = 1,2\ldots k$ by

$$h_i(t) = \begin{cases} 1 & \text{if } \frac{i-1}{k} < t \leq \frac{i}{k} \\ 0 & \text{otherwise} . \end{cases} \tag{37.20}$$

As to be seen from Fig. 37.2, the functions $h_i(t)$ are obviously orthogonal functions from an engineering point of view, see orthogonality condition Eq.(36.9).

If the interval $(0,1]$ is divided into k parts, k block-pulse functions are defined. Approximating the scalar signal $x(t)$ by a weighted sum of block-pulse functions

$$x(t) \doteq \sum_{i=1}^{k} x_i^{\sqcap} h_i(t) = \mathbf{x}^{\sqcap T}\mathbf{h}(t) \qquad \text{where} \qquad x_i^{\sqcap} = k \int_{\frac{i-1}{k}}^{\frac{i}{k}} x(t)dt \; . \tag{37.21}$$

The coefficient vector $\mathbf{x}^{\sqcap}$ can be determined by minimizing the integral of squared error. Since $h_i(t)$ can only take on level 1 or 0 the entries of the coefficient vector are obtained by the simple integral inside the bounds i/k and $(i-1)/k$. This value corresponds to the linear average value of $x(t)$ over the interval in which the block-pulse function is unity. Considering

$$\int_0^t \mathbf{h}(t)dt \doteq \mathbf{P}_{Mb}\, \mathbf{h}(t) \tag{37.22}$$

the integral is decomposed into a linear set of block-pulse functions h_i. This is equivalent to an operational matrix $\mathbf{P}_{Mb}$ approximating the integral optimally. The entries of $\mathbf{P}_{Mb}$ are determined by

$$\frac{\partial}{\partial \mathbf{P}_{Mb}} \int_0^1 \| \int_0^t \mathbf{h}(t)dt - \mathbf{P}_{Mb}\mathbf{h}(t) \|_F^2 \, dt = \mathbf{0} \, . \tag{37.23}$$

It results

$$\mathbf{P}_{Mb} = \frac{1}{k} \begin{pmatrix} \frac{1}{2} & 1 & 1 & \ldots & 1 \\ 0 & \frac{1}{2} & 1 & \ldots & 1 \\ 0 & 0 & \frac{1}{2} & \ldots & 1 \\ 0 & 0 & 0 & \ddots & \vdots \\ 0 & 0 & 0 & \ldots & \frac{1}{2} \end{pmatrix} = \frac{1}{k}\,(\text{trig } \mathbf{I}^{\swarrow} \mathbf{p}_b)\, \mathbf{I}^{\swarrow} \qquad \mathbf{p}_b = \left(\frac{1}{2} \;\; 1 \;\; 1 \ldots 1\right)^T . \tag{37.24}$$

37.2.2 Block-Pulse Spectra of $\dot{\mathbf{x}}(t)$ and $\mathbf{x}(t)$

Considering vector state signals $\mathbf{x}(t)$ and $\dot{\mathbf{x}}(t)$ and expanding both into a block-pulse spectrum

$$\mathbf{x}(t) = \mathbf{X}^{\sqcap}\mathbf{h}(t) \qquad \dot{\mathbf{x}}(t) = \dot{\mathbf{X}}^{\sqcap}\mathbf{h}(t) = \mathbf{C}\mathbf{h}(t) \qquad \dot{\mathbf{X}}^{\sqcap} \equiv \mathbf{C}\,. \tag{37.25}$$

For the sake of simplicity, the block-pulse spectrum is not designated as $\dot{\mathbf{X}}^{\sqcap T}$ but as $\dot{\mathbf{X}}^{\sqcap}$ although some other chapters did so, e.g. $\dot{\mathbf{X}}^{\neg T}$. Simplification becomes important in this chapter with respect to the intended recurrence equations. The block-pulse spectrum of $\mathbf{x}(t)$ results from

$$\mathbf{x}(t) = \mathbf{X}^{\sqcap}\mathbf{h}(t) = \begin{pmatrix} X_{11}^{\sqcap} & X_{12}^{\sqcap} & \dots & X_{1k}^{\sqcap} \\ X_{21}^{\sqcap} & X_{22}^{\sqcap} & \dots & \vdots \\ \vdots & \vdots & \ddots & \\ X_{n1}^{\sqcap} & & & X_{nk}^{\sqcap} \end{pmatrix} \begin{pmatrix} h_1(t) \\ h_2(t) \\ \vdots \\ h_k(t) \end{pmatrix} = \begin{pmatrix} X_{11}^{\sqcap}h_1(t) + X_{12}^{\sqcap}h_2(t) + \ldots + X_{1k}^{\sqcap}h_k(t) \\ X_{21}^{\sqcap}h_1(t) + X_{22}^{\sqcap}h_2(t) + \ldots + X_{2k}^{\sqcap}h_k(t) \\ \vdots \\ X_{n1}^{\sqcap}h_1(t) + X_{n2}^{\sqcap}h_2(t) + \ldots + X_{nk}^{\sqcap}h_k(t) \end{pmatrix} \tag{37.26}$$

The expansion above illustrates $X_{ji}^{\sqcap}$: The entry $X_{ji}^{\sqcap}$ stands for the approximation of the jth component $x_j(t)$ of state vector $\mathbf{x}(t)$ during time interval i. Utilizing these notations and Eq.(37.22) and expanding the derivative $\dot{\mathbf{x}}(t)$

$$\int_0^t \dot{\mathbf{x}}(t)dt \;=\; \mathbf{x}(t) - \mathbf{x}_o \tag{37.27}$$

$$\int_0^t \dot{\mathbf{x}}(t)dt = \dot{\mathbf{X}}^{\sqcap}\int_0^t \mathbf{h}(t)dt = \dot{\mathbf{X}}^{\sqcap}\mathbf{P}_{Mb}\mathbf{h}(t) \;=\; \mathbf{X}^{\sqcap}\mathbf{h}(t) - \mathbf{X}_o^{\sqcap}\mathbf{h}(t) \tag{37.28}$$

$$\dot{\mathbf{X}}^{\sqcap}\mathbf{P}_{Mb} \;=\; \mathbf{X}^{\sqcap} - \mathbf{X}_o^{\sqcap}\,. \tag{37.29}$$

With regard to the special properties of the block-pulse functions, $\mathbf{X}_o^{\sqcap}$ turns out as a matrix with identic columns $\mathbf{x}_o$, i.e., $\mathbf{X}_o^{\sqcap} = (\mathbf{x}_o\ \mathbf{x}_o \ldots \mathbf{x}_o)$. Partitioning $\dot{\mathbf{X}}^{\sqcap}\mathbf{P}_{Mb}$ and $\mathbf{X}^{\sqcap}$ into columns yields

$$\dot{\mathbf{X}}^{\sqcap}\mathbf{P}_{Mb} = \left((\dot{\mathbf{X}}^{\sqcap}\mathbf{P}_{Mb})_{.1} \vdots (\dot{\mathbf{X}}^{\sqcap}\mathbf{P}_{Mb})_{.2} \ldots (\dot{\mathbf{X}}^{\sqcap}\mathbf{P}_{Mb})_{.k}\right) \quad \text{and} \quad \mathbf{X}^{\sqcap} = (\mathbf{X}_{.1}^{\sqcap}\ \mathbf{X}_{.2}^{\sqcap} \ldots \mathbf{X}_{.k}^{\sqcap})\,. \tag{37.30}$$

Studying the matrix product $\dot{\mathbf{X}}^{\sqcap}\mathbf{P}_{Mb}$ in detail and employing the abbreviation $\mathbf{C} \triangleq \dot{\mathbf{X}}^{\sqcap}$

$$\dot{\mathbf{X}}^{\sqcap}\mathbf{P}_{Mb} = \begin{pmatrix} C_{11} & C_{12} & C_{13} & \dots & C_{1n} \\ C_{21} & C_{22} & C_{23} & \dots & C_{2n} \\ C_{31} & \dots & & & \\ \vdots & & & & \end{pmatrix} \frac{1}{k} \begin{pmatrix} \frac{1}{2} & 1 & 1 & \dots & 1 \\ 0 & \frac{1}{2} & 1 & \dots & 1 \\ 0 & 0 & \frac{1}{2} & \dots & 1 \\ \vdots & & & & \end{pmatrix} \tag{37.31}$$

$$= \frac{1}{k} \begin{pmatrix} \frac{1}{2}C_{11} & C_{11} + \frac{1}{2}C_{12} & C_{11} + C_{12} + \frac{1}{2}C_{13} & \dots \\ \frac{1}{2}C_{21} & C_{21} + \frac{1}{2}C_{22} & C_{21} + C_{22} + \frac{1}{2}C_{23} & \\ \frac{1}{2}C_{31} & C_{31} + \frac{1}{2}C_{32} & \vdots & \\ \vdots & & & \end{pmatrix}. \tag{37.32}$$

By inspection, realize that the ith column of $\mathbf{C}\mathbf{P}_{Mb}$ can easily be expressed as a function of $\mathbf{C}_{.i}$. The interdependence between the columns $\mathbf{C}_{.i}$ of the matrix $\mathbf{C}$ and the block-pulse spectrum $\dot{\mathbf{X}}^{\sqcap}$ is

$$(\mathbf{C}\mathbf{P}_{Mb})_{.i} = (\dot{\mathbf{X}}^{\sqcap}\mathbf{P}_{Mb})_{.i} = \frac{1}{k}\sum_{j=1}^{i-1} C_{.j} + \frac{1}{2k}C_{.i}\,. \tag{37.33}$$

37.2.3 Recurrence Equation

Eq.(37.33) can be put into recurrence equation form. This is easy to be seen by rewriting some columns in detail

$$(\mathbf{CP}_{Mb})_{.1} = \frac{1}{2k}C_{.1} \tag{37.34}$$

$$(\mathbf{CP}_{Mb})_{.2} = \frac{1}{k}C_{.1} + \frac{1}{2k}C_{.2} \tag{37.35}$$

$$(\mathbf{CP}_{Mb})_{.3} = \frac{1}{k}C_{.1} + \frac{1}{k}C_{.2} + \frac{1}{2k}C_{.3} = \frac{1}{k}C_{.1} + \frac{1}{2k}C_{.2} + \frac{1}{2k}C_{.2} + \frac{1}{2k}C_{.3} \,. \tag{37.36}$$

Finally,

$$(\mathbf{CP}_{Mb})_{.(i+1)} = (\mathbf{CP}_{Mb})_{.i} + \frac{1}{2k}(\mathbf{C}_{.\,i} + C_{.(i+1)}) \,. \tag{37.37}$$

Considering the $(i+1)$th column of the matrices in Eq.(37.29) and introducing the original notation yields the relation

$$\mathbf{X}^{\sqcap}_{.(i+1)} = (\dot{\mathbf{X}}^{\sqcap}\mathbf{P}_{Mb})_{.(i+1)} + \mathbf{x}_o \,. \tag{37.38}$$

Combining Eqs.(37.37) and (37.38), one has $\forall i = 1 \ldots k$

$$\mathbf{X}^{\sqcap}_{.1} = \frac{1}{2k}\dot{\mathbf{X}}^{\sqcap}_{.1} + \mathbf{x}_o \tag{37.39}$$

$$\mathbf{X}^{\sqcap}_{.(i+1)} = \mathbf{X}^{\sqcap}_{.i} + \frac{1}{2k}(\dot{\mathbf{X}}^{\sqcap}_{.i} + \dot{\mathbf{X}}^{\sqcap}_{.(i+1)}) \,. \tag{37.40}$$

Eqs.(37.39) and (37.40) show the interdependence between the block-pulse spectra of $\mathbf{x}(t)$ and $\dot{\mathbf{x}}(t)$.

37.2.4 Linear System Analysis Via Block-Pulse Functions

Considering a linear time-invariant process, possibly after having shifted real-time scale $(0, t_f)$ to time scale t in the 0 to 1 interval,

$$\dot{\mathbf{x}}(t) = \mathbf{A}\mathbf{x}(t) + \mathbf{B}\mathbf{u}(t), \qquad \mathbf{x}(0) = \mathbf{x}_o \tag{37.41}$$

and applying block-pulse expansion yields the ith column $\forall i = 1 \ldots k$

$$\dot{\mathbf{X}}^{\sqcap}_{.i} = \mathbf{A}\mathbf{X}^{\sqcap}_{.i} + (\mathbf{B}\mathbf{U}^{\sqcap})_{.i} \,. \tag{37.42}$$

Substituting $\dot{\mathbf{X}}^{\sqcap}_{.1}$ from Eq.(37.42) into Eq.(37.39),

$$\mathbf{X}^{\sqcap}_{.1} = \frac{1}{2k}\mathbf{A}\mathbf{X}^{\sqcap}_{.1} + \frac{1}{2k}(\mathbf{B}\mathbf{U}^{\sqcap})_{.1} + \mathbf{x}_o \tag{37.43}$$

$$\mathbf{X}^{\sqcap}_{.1} = (\mathbf{I} - \frac{1}{2k}\mathbf{A})^{-1}\mathbf{x}_o + \frac{1}{2k}(\mathbf{I} - \frac{1}{2k}\mathbf{A})^{-1}(\mathbf{B}\mathbf{U}^{\sqcap})_{.1} \,. \tag{37.44}$$

Combining Eqs.(37.42) and (37.40), finally,

$$\mathbf{X}^{\sqcap}_{.(i+1)} = (\mathbf{I} - \frac{1}{2k}\mathbf{A})^{-1}(\mathbf{I} + \frac{1}{2k}\mathbf{A})\mathbf{X}^{\sqcap}_{.i} + \frac{1}{2k}(\mathbf{I} - \frac{1}{2k}\mathbf{A})^{-1}[(\mathbf{B}\mathbf{U}^{\sqcap})_{.i} + (\mathbf{B}\mathbf{U}^{\sqcap})_{.(i+1)}\,] \tag{37.45}$$

is achieved. Since $\lambda[\mathbf{A} - 2k\mathbf{I}] = \lambda[\mathbf{A}] - 2k$ and $\lambda[\alpha\mathbf{A}] = \alpha\lambda[\mathbf{A}]$

$$\lambda[\mathbf{I} - \frac{1}{2k}\mathbf{A}] = -\frac{1}{2k}(\lambda[\mathbf{A}] - 2k) = 1 - \frac{1}{2k}\lambda[\mathbf{A}] \,. \tag{37.46}$$

The expression $\det[\mathbf{I} - \frac{1}{2k}\mathbf{A}]$ and the inverse $(\mathbf{I} - \frac{1}{2k}\mathbf{A})^{-1}$ always exist except if $\lambda[\mathbf{I} - \frac{1}{2k}\mathbf{A}] = 0$, $\lambda[\mathbf{A}] = 2k$, $\mathbf{A}$ unstable.

Considering a state-space controller $\mathbf{u} = \mathbf{K}\mathbf{x}$, the aforementioned equations can be used by substituting $\mathbf{A} := \mathbf{A} + \mathbf{B}\mathbf{K}$. A similar approach concerning suboptimal controller techniques is shown in Eq.(16.143).

37.2.5 Nonlinear and Time-Varying System Analysis Via Block-Pulse Functions

On account of the particular framework of the block-pulse functions, the recurrence method can be extended to time-varying and nonlinear systems $\dot{\mathbf{x}}(t) = f(\mathbf{x}, u, t)$

$$\dot{\mathbf{x}}(t) = \dot{\mathbf{X}}^{\sqcap}\mathbf{h}(t) = \sum_{i=1}^{k} \dot{\mathbf{X}}^{\sqcap}_{.i} h_i(t) = f\Big(\sum_{i=1}^{k} \mathbf{X}^{\sqcap}_{.i} h_i,\ \sum_{i=1}^{k} u^{\sqcap}_i h_i,\ \sum_{i=1}^{k} t^{\sqcap}_i h_i\Big) \,. \tag{37.47}$$

The scalars $u^{\sqcap}_i$ and $t^{\sqcap}_i$ denote the coefficients of expansion of the functions of time $u(t)$ and t. In spite of the nonlinear function, the block-pulse decomposition can be separated with respect to columns $\dot{\mathbf{X}}^{\sqcap}_{.i}$, $\mathbf{X}^{\sqcap}_{.i}$ etc.

$$\dot{\mathbf{X}}^{\sqcap}_{.i} = f(\mathbf{X}^{\sqcap}_{.i},\ u^{\sqcap}_i,\ t^{\sqcap}_i) \qquad \forall i = 1 \ldots k \,. \tag{37.48}$$

Combining with the basic relations in $\dot{\mathbf{X}}^{\sqcap}_{.i}$ and $\mathbf{X}^{\sqcap}_{.i}$, given by Eqs.(37.39) and (37.40), it results $\forall i = 1 \ldots k$

$$\mathbf{X}^{\sqcap}_{.1} = \frac{1}{2k} f(\mathbf{X}^{\sqcap}_{.1}, u^{\sqcap}_1,\ t^{\sqcap}_1) + \mathbf{x}_o \tag{37.49}$$

$$\mathbf{X}^{\sqcap}_{.(i+1)} = \mathbf{X}^{\sqcap}_{.i} + \frac{1}{2k}[f(\mathbf{X}^{\sqcap}_{.i},\ u^{\sqcap}_i,\ t^{\sqcap}_i) + f(\mathbf{X}^{\sqcap}_{.(i+1)}\ ,\ u^{\sqcap}_{i+1}\ ,\ t^{\sqcap}_{i+1}\ )] \,. \tag{37.50}$$

37.2.6 System Identification Via Block-Pulse Functions

System identification using orthogonal functions and the basic algorithms are outlined in Eqs.(38.58) (38.62) etc.. Two modifications are presented below, using block-pulse functions and taking into account the unknown initial states. A recursive method identifying a system with constant parameters yields the same results as known from Eq.(D.88). Let the system be of order n

$$\sum_{i=0}^{n} a_i y^{(i)}(t) - \sum_{j=0}^{n} b_j u^{(j)}(t) = u(t), \qquad b_o = 0 \,. \tag{37.51}$$

Defining

$$\begin{array}{ll} y(t) \doteq \mathbf{y}^{\sqcap T}\mathbf{h}(t) = \mathbf{h}^T(t)\mathbf{y}^{\sqcap}, & \mathbf{y}^{\sqcap} \triangleq (y_1\ y_2 \ldots y_k), \\ u(t) \doteq \mathbf{u}^{\sqcap T}\mathbf{h}(t) = \mathbf{h}^T(t)\mathbf{u}^{\sqcap}, & \mathbf{u}^{\sqcap} \triangleq (u_1\ u_2 \ldots u_k), \end{array} \qquad \int_0^t \mathbf{h}(t)dt = \mathbf{P}_{Mb}\mathbf{h}(t) \,. \tag{37.52}$$

The vector-valued variable $\mathbf{y}_1$ is used for approximating the first integral of y . Similarly, $\mathbf{y}_i$ is defined as the ith integral

$$\int_0^t y(t)dt \left\{ \begin{array}{l} = \mathbf{y}^{\sqcap T} \int_0^t \mathbf{h}(t)dt = \mathbf{y}^{\sqcap T}\mathbf{P}_{Mb}\mathbf{h}(t) \\ \triangleq \mathbf{y}_1^{\sqcap T}\mathbf{h}(t) \end{array} \right\} \rightsquigarrow \mathbf{y}^{\sqcap}_1 = \mathbf{P}^T_{Mb}\mathbf{y}^{\sqcap} \rightsquigarrow \mathbf{y}^{\sqcap}_i = (\mathbf{P}^i_{Mb})^T\mathbf{y}^{\sqcap} \,. \tag{37.53}$$

Integrating Eq.(37.51) n times

$$\sum_{i=0}^{n} a_i \mathbf{y}_{n-i}^{\sqcap} - \sum_{j=1}^{n} b_j \mathbf{u}_{n-j}^{\sqcap} = \mathbf{u}_n^{\sqcap} \tag{37.54}$$

(see *Palanisamy, K.R., and Bhattacharya, D.K., 1981; Cheng, B., and Hsu, N.S., 1982*). Abbreviating yields

$$\mathbf{M}\mathbf{p} = \mathbf{u}_n^{\sqcap} \quad \text{where} \quad \mathbf{p} \triangleq (a_o\ a_1 \dots a_n\ b_1\ b_2 \dots b_n)^T \tag{37.55}$$

$$\text{and} \quad \mathbf{M} \triangleq (\mathbf{y}_n^{\sqcap} \vdots\ \mathbf{y}_{n-1}^{\sqcap} \cdots \mathbf{y}_o^{\sqcap} \vdots\ -\mathbf{u}_{n-1}^{\sqcap} \vdots\ -\mathbf{u}_{n-2}^{\sqcap} \cdots -\mathbf{u}_o^{\sqcap}) \,. \tag{37.56}$$

Applying the weighted least squares regression method and using the weighting matrix of powers of forgetting factor α , i.e., $\mathbf{W} = \text{diag}\ \{\alpha^{k-1}, \alpha^{k-2} \dots \alpha,\ 1\}$, from Eq.(D.30) it results

$$\mathbf{p}^\star = (\mathbf{M}^T\mathbf{W}\mathbf{M})^{-1}\mathbf{M}^T\mathbf{W}\mathbf{u}_n^{\sqcap} \,. \tag{37.57}$$

As shown in Eq.(A.80), the initials $y^{(i)}(0)$ may not be chosen arbitrarily in the case of transfer function zeros. The initials are predetermined and depend on the zeros and the initials of $u^{(j)}(0)$. Determining, thus, appropriate and compatible initials,

$$\sum_{i=0}^{n} a_i \mathbf{y}_{n-i}^{\sqcap} - \sum_{j=1}^{n} b_j \mathbf{u}_{n-j}^{\sqcap} + \sum_{k=0}^{n-1} w_k \mathbf{z}_k^{\sqcap} = \mathbf{u}_n^{\sqcap} \tag{37.58}$$

$$\text{where} \quad w_k = \sum_{\nu=0}^{k} a_{n-k+\nu}\ y^{(\nu)}(0) \qquad \mathbf{z}_k^{\sqcap} = \frac{1}{k!}\int_0^1 t^k \mathbf{h}(t)dt \tag{37.59}$$

and defining $\mathbf{p} \triangleq (a_o\ a_1 \dots a_n\ b_1\ b_2 \dots b_n\ w_o\ w_1 \dots w_{n-1})^T$ and

$$\mathbf{M} \triangleq (\mathbf{y}_n^{\sqcap} \vdots\ \mathbf{y}_{n-1}^{\sqcap} \cdots \mathbf{y}_o^{\sqcap} \vdots\ -\mathbf{u}_{n-1}^{\sqcap} \vdots\ -\mathbf{u}_{n-2}^{\sqcap} \cdots -\mathbf{u}_o^{\sqcap} \vdots\ \mathbf{z}_o^{\sqcap} \dots \mathbf{z}_{n-1}^{\sqcap}) \tag{37.60}$$

the result is given by Eq.(37.57). The result includes the parameters as well as the initials, provided that sufficient measurements exist. Partitioning the matrix $\mathbf{M}$ into a matrix $\mathbf{M}_l$ containing the first l columns, only, and separating $\mathbf{u}_{nl}^{\sqcap}$ as a vector with the first l elements of $\mathbf{u}_n^{\sqcap}$, then $\mathbf{p}_l^\star$ is an estimate based on the first measurements, only (*Cheng, B., and Hsu, N.S., 1982*)

$$\mathbf{p}_{l+1}^\star = \mathbf{p}_l^\star + \gamma_{l+1}\mathbf{P}_l\mathbf{M}_{.(l+1)}\ (u_{n,l+1}^{\sqcap} - \mathbf{M}_{.(l+1)}^T\ \mathbf{p}_l^\star) \tag{37.61}$$

$$\gamma_{l+1} = \frac{1}{\alpha + \mathbf{M}_{.(l+1)}^T\ \mathbf{P}_l\mathbf{M}_{.(l+1)}} \qquad \mathbf{P}_{l+1} = \frac{1}{\alpha}[\mathbf{P}_l - \gamma_{l+1}\mathbf{P}_l\mathbf{M}_{.(l+1)}^T\ \mathbf{M}_{.(l+1)}\ \mathbf{P}_l] \,. \tag{37.62}$$

The vector $\mathbf{M}_{.(l+1)}$ is the $l+1$ column of $\mathbf{M}$. Substituting the elements of $\mathbf{M}$ and $\mathbf{u}_n^{\sqcap}$ by recent measurements, a recursion is achieved as outlined in Eqs.(D.86) to (D.88).

37.2.7 Generalized Block-Pulse Functions

If generalized block-pulse functions are introduced, multi-delay state equations occurring with many control problems can be solved in an attractive way. The time interval of calculation and the number of expansion coefficients can be adjusted to the special problem. Referring to Fig. 37.2, the instants $\frac{i}{k}$ (equally spaced within the unit interval) are replaced by arbitrarily spaced instants t_i. The advantage of applying generalized block-pulse functions is better approximation and less computational effort (*Wang, M.L., et al. 1987*).

The results based on block-pulse function approach can also be obtained by the modulating function or the differential operator approach as presented by *Kraus, F., and Schaufelberger, W., 1990*.

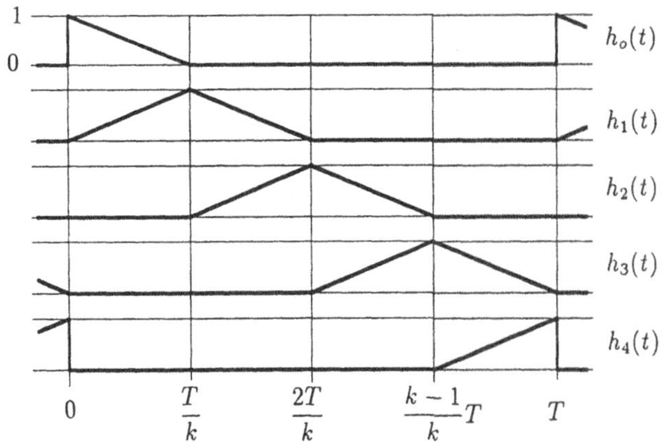

Figure 37.3: Piecewise linear polynomial function with $k = 4$

37.3 Collocation Point Approximation With Piecewise Linear Functions

Selecting piecewise linear polynomial functions as an approximating basis and considering collocation point approximation yields an advantageous approach solving linear time-varying systems.

37.3.1 Piecewise Linear Polynomial Functions

A set of $k+1$ piecewise linear polynomial functions is defined by the following equations $\forall i = 1 \dots k-1$

$$h_o(t) = \begin{cases} 1 - \frac{k}{T}t & 0 \le t \le \frac{T}{k} \\ 0 & \text{otherwise} \end{cases} \tag{37.63}$$

$$\vdots$$

$$h_i(t) = \begin{cases} (1-i) + \frac{k}{T}t & (i-1)\frac{T}{k} \le t \le i\frac{T}{k} \\ (1+i) - \frac{k}{T}t & i\frac{T}{k} \le t \le (i+1)\frac{T}{k} \end{cases} \tag{37.64}$$

$$\vdots$$

$$h_k(t) = \begin{cases} 0 & \text{otherwise} \\ (1-k) + \frac{k}{T}t & (k-1)\frac{T}{k} \le t \le T \; . \end{cases} \tag{37.65}$$

Fig. 37.3 shows these piecewise linear functions with $k = 4$. The interval T is partitioned into k equal parts. At the break points $h_i(i\frac{T}{k}) = 1$ and $h_j(i\frac{T}{k})|_{j \ne i} = 0$. The functions $h_i(t)$ are linearly independent but not orthogonal.

37.3.2 Expansion of a Scalar Function

Functions only within the interval $(0,T)$ are of interest. An arbitrary function $f(t)$ is expanded into k piecewise linear polynomial functions

$$f(t) \doteq \sum_{i=0}^{k} f_i^{\wedge} h_i(t) \,. \tag{37.66}$$

At the break points $t = i\frac{T}{k}$, equality between $f(t)$ and the approximating function can be achieved

$$f(i\frac{T}{k}) = \ldots 0 + f_i^{\wedge} h_i(i\frac{T}{k}) + 0 \ldots = f_i^{\wedge} \times 1 = f_i^{\wedge} \,. \tag{37.67}$$

There is no deviation at the break (collocation) points. Hence, this method is denoted as collocation approximation. Defining

$$\mathbf{h}(t) \triangleq \left(h_o(t) \;\; h_1(t) \ldots h_k(t)\right)^T \qquad \text{and} \qquad \mathbf{f}^{\wedge} \triangleq (f_o \;\; f_1 \ldots f_k)^T \tag{37.68}$$

it results

$$f(t) \;\doteq\; \mathbf{f}^{\wedge T}\mathbf{h}(t) \,. \tag{37.69}$$

37.3.3 Integration of Scalar Functions

The integration of the piecewise linear polynomial function can be reduced to the multiplication by an operational matrix of integration $\mathbf{P}_{Ml}$

$$\int_0^t f(t)dt = \int_0^t \mathbf{f}^{\wedge T}\mathbf{h}(t)dt = \mathbf{f}^{\wedge T}\int_0^t \mathbf{h}(t)dt = \mathbf{f}^{\wedge T}\mathbf{P}_{Ml}\mathbf{h}(t) \tag{37.70}$$

where (*Liou, C.T., and Chou, Y.S., 1987*)

$$\mathbf{P}_{Ml} = \frac{T}{k}\begin{pmatrix} 0 & \frac{1}{2} & \frac{1}{2} & \frac{1}{2} & \ldots & \frac{1}{2} \\ 0 & \frac{1}{2} & 1 & 1 & \ldots & 1 \\ 0 & 0 & \frac{1}{2} & 1 & \ldots & 1 \\ \vdots & \vdots & \vdots & \vdots & & \\ 0 & 0 & 0 & 0 & \ldots & \frac{1}{2} \end{pmatrix} . \tag{37.71}$$

37.3.4 Expansion and Integration of Vector-Valued Functions

Expanding a vector function $\mathbf{f}(t)$ into piecewise linear functions yields

$$\mathbf{f}(t) \doteq \mathbf{F}^{\wedge}\; \mathbf{h}(t) \,. \tag{37.72}$$

Each row of $\mathbf{F}^{\wedge}$ corresponds to a row with elements coinciding with the values of an entry of $\mathbf{f}(t)$ at consecutive instants. Each column of $\mathbf{F}^{\wedge}$ equals to the vector $\mathbf{f}(i\frac{T}{k})$ at the ith instant. Integration yields

$$\int_0^t f(t)dt \doteq \mathbf{F}^{\wedge}\int_0^t \mathbf{h}(t)dt = \mathbf{F}^{\wedge}\;\mathbf{P}_{Ml}\mathbf{h}(t) \,. \tag{37.73}$$

37.3.5 Expansion of the Product of Two Functions

Let $f(t)$ be the product of $g(t)y(t)$. Then,

$$\int_0^t f(t)dt = \int_0^t g(t)y(t)dt = \mathbf{f}^{\wedge T}\mathbf{P}_{Ml}\mathbf{h}(t) \tag{37.74}$$

$$f(t) \doteq \sum_{i=0}^{k} f_i h_i(t) = \sum_{i=0}^{k} f(i\frac{T}{k})\ h_i(t) = \sum_{i=0}^{k} g(i\frac{T}{k})\ y(i\frac{T}{k})\ h_i(t) \tag{37.75}$$

$$f_i = f(i\frac{T}{k}) = g(i\frac{T}{k})\ y(i\frac{T}{k}) = g_i y_i\ . \tag{37.76}$$

Considering the integral of the vector $\mathbf{f}(t)$ and generating $\mathbf{f}(t)$ by the product of two time-varying terms, namely the matrix $\mathbf{A}(t)$ and the vector $\mathbf{x}(t)$, then the matrix $\mathbf{F}^{\wedge}$ in Eq.(37.72) becomes

$$\mathbf{F}^{\wedge} = \left(\mathbf{A}(0\frac{T}{k})\mathbf{x}(0\frac{T}{k}) \vdots \mathbf{A}(1\frac{T}{k})\mathbf{x}(1\frac{T}{k}) \vdots \ldots\right) = (\mathbf{A}_o\mathbf{x}_o \vdots \mathbf{A}_1\mathbf{x}_1 \vdots \mathbf{A}_2\mathbf{x}_2 \ldots \mathbf{A}_k\mathbf{x}_k)\ . \tag{37.77}$$

37.3.6 Time-Varying System Analysis

Consider the time-varying linear system forced by $\mathbf{u}(t)$. Performing the decomposition via piecewise linear functions,

$$\dot{\mathbf{x}}(t) = \mathbf{A}(t)\mathbf{x}(t) + \mathbf{B}(t)\mathbf{u}(t) \qquad \mathbf{x}(0) = \mathbf{x}_o \tag{37.78}$$

$$\int_0^t \dot{\mathbf{x}}(t)dt = \mathbf{x}(t) - \mathbf{x}_o = \int_0^t \mathbf{A}(t)\mathbf{x}(t)dt + \int_0^t \mathbf{B}(t)\mathbf{u}(t)dt \tag{37.79}$$

$$(\mathbf{x}_o \vdots \mathbf{x}_1 \vdots \ldots \mathbf{x}_k)\mathbf{h}(t) - (\mathbf{x}_o \vdots \mathbf{x}_o \vdots \ldots \mathbf{x}_o)\mathbf{h}(t) = \left(\mathbf{A}_o\mathbf{x}_o + \mathbf{B}_o\mathbf{u}_o \vdots \mathbf{A}_1\mathbf{x}_1 + \mathbf{B}_1\mathbf{u}_1 \vdots \ldots\right)\mathbf{P}_{Ml}\mathbf{h}(t) \tag{37.80}$$

$$(\mathbf{x}_o \vdots \mathbf{x}_1 \vdots \ldots \mathbf{x}_k) - (\mathbf{x}_o \vdots \mathbf{x}_o \vdots \ldots \mathbf{x}_o) = \left(\mathbf{A}_o\mathbf{x}_o + \mathbf{B}_o\mathbf{u}_o \vdots \mathbf{A}_1\mathbf{x}_1 + \mathbf{B}_1\mathbf{u}_1 \vdots \ldots\right)\mathbf{P}_{Ml}\ . \tag{37.81}$$

Applying the matrix product formula reduced to the columns of the product

$$\mathbf{AB} = \mathbf{C} \quad \leadsto \quad \mathbf{C}_{.j} = \sum_{i=1}^{k} \mathbf{A}_{.j} B_{ij}\ , \tag{37.82}$$

taking care of the subscripts (the first column $\mathbf{A}_o\mathbf{x}_o + \mathbf{B}_o\mathbf{u}_o$ has subscripts $_o$) and referring to Eq.(37.71) yields

$$\mathbf{x}_o - \mathbf{x}_o = 0 \tag{37.83}$$

$$\mathbf{x}_1 - \mathbf{x}_o = (\mathbf{A}_o\mathbf{x}_o + \mathbf{B}_o\mathbf{u}_o)\frac{T}{2k} + (\mathbf{A}_1\mathbf{x}_1 + \mathbf{B}_1\mathbf{u}_1)\frac{T}{2k} + 0 + 0 \ldots \tag{37.84}$$

$$\mathbf{x}_2 - \mathbf{x}_o = (\mathbf{A}_o\mathbf{x}_o + \mathbf{B}_o\mathbf{u}_o)\frac{T}{2k} + (\mathbf{A}_1\mathbf{x}_1 + \mathbf{B}_1\mathbf{u}_1)\frac{T}{k} + (\mathbf{A}_2\mathbf{x}_2 + \mathbf{B}_2\mathbf{u}_2)\frac{T}{2k} + 0 + \ldots \tag{37.85}$$

or

$$\mathbf{x}_2 = \mathbf{x}_1 + (\mathbf{A}_1\mathbf{x}_1 + \mathbf{B}_1\mathbf{u}_1 + \mathbf{A}_2\mathbf{x}_2 + \mathbf{u}_2)\frac{T}{2k} \tag{37.86}$$

$$\vdots$$

$$\mathbf{x}_i = \mathbf{x}_{i-1} + \frac{T}{2k}(\mathbf{A}_{i-1}\mathbf{x}_{i-1} + \mathbf{B}_{i-1}\mathbf{u}_{i-1} + \mathbf{A}_i\mathbf{x}_i + \mathbf{B}_i\mathbf{u}_i)\ . \tag{37.87}$$

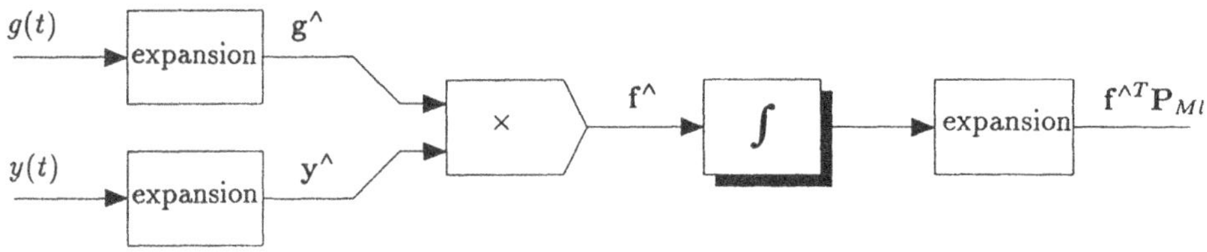

Figure 37.4: Linear piecewise expansion integrating a product of two functions

The expression of $\mathbf{x}_2$ and $\mathbf{x}_i$ are recurrence equations. The vectors $\mathbf{x}_i$ can be calculated successively and recursively by linear algebraic operations saving calculation time and memory storage. Referring to Eqs.(37.67) and (37.72), the result is

$$\mathbf{x}_i = \mathbf{x}(i\frac{T}{k}) \quad \rightsquigarrow \quad \mathbf{X}^{\wedge} = (\mathbf{x}_o \vdots \mathbf{x}_1 \vdots \ldots \mathbf{x}_k) \quad \rightsquigarrow \quad \mathbf{x}(t) \doteq \mathbf{X}^{\wedge}\mathbf{h}(t) \ . \qquad (37.88)$$

Fig. 37.4 gives an outline of the integration and expansion procedure listed above. The expansion of the integral succeeds the multiplication and the previous expansion. The expansion before multiplication could be omitted by directly applying the integration to a product of two time functions. By that means another method is defined which is more complex but more accurate.

Chapter 38

Orthogonal Decomposition Applications

38.1 Linear Optimal Control Via Generalized Orthogonal Functions

38.1.1 Expansion of Various Functions

The orthogonal component coefficient vector $\mathbf{x}^{\neg}$ is given by Eq.(36.16). Thus, $x(t)$ is decomposed into k terms. Starting now with the orthogonal expansion coefficient vector $\dot{\mathbf{x}}^{\neg}$ associated with $\dot{x}(t)$ and using $h_o(t) = 1$, the original function $x(t)$ can be expressed as follows

$$x(t) = \int_o^t \dot{x}(t)dt + x(0) = \int_o^t \dot{\mathbf{x}}^{\neg T}\mathbf{h}(t)dt + x(0) = \dot{\mathbf{x}}^{\neg T}\int_o^t \mathbf{h}(t)dt + x(0) \tag{38.1}$$

$$x(t) = \dot{\mathbf{x}}^{\neg T}\mathbf{P}_M\mathbf{h}(t) + x(0)h_o(t) = \dot{\mathbf{x}}^{\neg T}\mathbf{P}_M\mathbf{h} + \mathbf{x}_o^T\mathbf{h}; \quad \mathbf{x}_o \triangleq [x(0)\ 0\ 0\ldots]^T. \tag{38.2}$$

If one needs $x(t)$ at a fixed t, only, e.g. $t = 1$, the following expression is chosen assuming $x(0) = 0$ and, applying Eq.(36.41),

$$x(1) = \int_o^1 \dot{x}(t)dt + x(0) = \int_o^1 \dot{\mathbf{x}}^{\neg T}\mathbf{h}dt + 0 = \dot{\mathbf{x}}^{\neg T}\int_o^1 \mathbf{h}dt \tag{38.3}$$

$$x(1) = \dot{\mathbf{x}}^{\neg T}\mathbf{F}\int_o^1 \mathbf{t}(t)dt = \dot{\mathbf{x}}^{\neg T}\mathbf{F}\Big(1\ \frac{1}{2}\ \frac{1}{3}\cdots\frac{1}{k}\Big)^T = \dot{\mathbf{x}}^{\neg T}\mathbf{v}_o; \quad \mathbf{v}_o \triangleq \mathbf{F}\Big(1\ \frac{1}{2}\ \frac{1}{3}\cdots\frac{1}{k}\Big)^T. \tag{38.4}$$

The variable t also is expressed in terms of orthogonal functions

$$t = -\frac{b_o'}{a_o'} + \frac{b_o'}{a_o'} + \frac{a_o't}{a_o'} = -\frac{b_o'}{a_o'} + \frac{a_o't + b_o'}{a_o'} = -\frac{b_o'}{a_o'}h_o + \frac{1}{a_o'}h_1 \tag{38.5}$$

$$t = \Big(-\frac{b_o'}{a_o'}\ \frac{1}{a_o'}\ 0\ 0\ldots\Big)^T\mathbf{h} = \mathbf{a}_o^T\mathbf{h} \qquad \text{where} \qquad \mathbf{a}_o \triangleq \Big(-\frac{b_o'}{a_o'}\ \frac{1}{a_o'}\ 0\ 0\ldots\Big)^T. \tag{38.6}$$

Referring to Eq.(36.56), the integral of $\dot{x}^2$ or x^2 can be exposed as follows. For the sake of simplicity, the upper bound is $t_f = 1$, $x(0) = 0$ and $\mathbf{x}_o = \mathbf{0}$

$$\int_o^1 \dot{x}^2(t)dt = \dot{\mathbf{x}}^{\neg T}\mathbf{H}(1)\dot{\mathbf{x}}^{\neg} \tag{38.7}$$

$$\int_o^1 x^2(t)dt = \int_o^1(\dot{\mathbf{x}}^{\neg T}\mathbf{P}_M\mathbf{h})(\dot{\mathbf{x}}^{\neg T}\mathbf{P}_M\mathbf{h})dt = \int_o^1 \dot{\mathbf{x}}^{\neg T}\mathbf{P}_M\mathbf{h}(\mathbf{P}_M\mathbf{h})^T\dot{\mathbf{x}}^{\neg}dt \tag{38.8}$$

$$= \dot{\mathbf{x}}^{\neg T}\mathbf{P}_M\Big(\int_o^1 \mathbf{h}\mathbf{h}^T dt\Big)\mathbf{P}_M^T\dot{\mathbf{x}}^{\neg} = \dot{\mathbf{x}}^{\neg T}\mathbf{P}_M\mathbf{H}(1)\mathbf{P}_M^T\dot{\mathbf{x}}^{\neg}. \tag{38.9}$$

Finally, the time weighted integral of $\dot{x}(t)$ is

$$\int_o^1 t\dot{x}(t)dt = \int_o^1 \mathbf{a}_o^T \mathbf{h} \dot{\mathbf{x}}^{\neg T} \mathbf{h}\, dt = \int_o^1 \mathbf{a}_o^T \mathbf{h}\mathbf{h}^T \dot{\mathbf{x}}^{\neg} dt = \mathbf{a}_o^T \mathbf{H}(1)\dot{\mathbf{x}}^{\neg} = \dot{\mathbf{x}}^{\neg T}\mathbf{H}(1)\mathbf{a}_o \,. \quad (38.10)$$

38.1.2 Calculus of Variations

Based on the operational matrix of integration and on the power series expansion of the general orthogonal functions, *Tsay, S.C., and Lee, T.T., 1988* proposed a straightforward algorithm to manage variational problems or optimal control problems by means of algebraic equations. Consider a scalar two-point boundary value problem

$$I = \int_o^{t_f} f[x(t),\ \dot{x}(t), t]dt \qquad x(0) = x_o \qquad x(t_f) = x_f \quad (38.11)$$

without being subject to any condition e.g. given by plant differential equation. In most cases, the expansions given in Eqs.(38.2) through (38.10) are enough to decompose the performance I. For the sake of abbreviation, a simple example of variational calculus is

$$I = \int_o^1 [\dot{x}^2(t) + t\dot{x}(t) + x^2(t)]dt \qquad x_o = 0 \quad x_f = 0.25 \,. \quad (38.12)$$

Applying Eqs.(38.7), (38.9) and (38.10),

$$I = \dot{\mathbf{x}}^{\neg T}\mathbf{H}(1)\dot{\mathbf{x}}^{\neg} + \dot{\mathbf{x}}^{\neg T}\mathbf{H}(1)\ \mathbf{a}_o + \dot{\mathbf{x}}^{\neg T}\mathbf{P}_M\mathbf{H}(1)\mathbf{P}_M^T\dot{\mathbf{x}}^{\neg} \,. \quad (38.13)$$

The upper boundary condition, using Eq.(38.4), can be appended to I by means of a Lagrange multiplier λ

$$I' = I + \lambda(\dot{\mathbf{x}}^{\neg T}\mathbf{v}_o - 0.25) \,. \quad (38.14)$$

Calculating the gradient $\partial I'/\partial\dot{\mathbf{x}}^{\neg} = 0$ yields

$$2\mathbf{H}(1)\dot{\mathbf{x}}^{\neg} + \mathbf{H}(1)\ \mathbf{a}_o + 2\mathbf{P}_M\mathbf{H}(1)\mathbf{P}_M^T\dot{\mathbf{x}}^{\neg} + \lambda\mathbf{v}_o = \mathbf{0} \,. \quad (38.15)$$

By defining a resulting vector $(\dot{\mathbf{x}}^{\neg T}\ \ \lambda)^T$ and combining with the boundary condition $\dot{\mathbf{x}}^{\neg T}\mathbf{v}_o = 0.25$ yields

$$\begin{pmatrix} \dot{\mathbf{x}}^{\neg} \\ \lambda \end{pmatrix} = \begin{pmatrix} 2\mathbf{H}(1) + 2\mathbf{P}_M\mathbf{H}(1)\mathbf{P}_M^T & \vdots\ \mathbf{v}_o \\ \mathbf{v}_o^T & \vdots\ 0 \end{pmatrix}^{-1} \begin{pmatrix} -\mathbf{H}(1)\ \mathbf{a}_o \\ 0.25 \end{pmatrix} , \quad (38.16)$$

i.e., $k+1$ linear algebraic equations with as many unknowns. The result is an orthogonal expansion of $\dot{x}(t) \doteq \sum_0^{k-1} \dot{\mathbf{x}}^{\neg T}\mathbf{h}(t)$. The unknown function $x(t)$ is calculated applying Eq.(38.2).

38.1.3 Approximating the Riccati Controller by Orthogonal Functions

Consider the time-varying linear system $\dot{\mathbf{x}}(t) = \mathbf{A}(t)\mathbf{x}(t) + \mathbf{B}(t)\mathbf{u}(t)$, $\mathbf{x}(0) = \mathbf{x}_o$, with linear state feedback controller $\mathbf{u}(t) = \mathbf{K}(t)\mathbf{x}(t)$, minimizing the quadratic performance index

$$I = \mathbf{x}^T(t_f)\mathbf{F}_f\mathbf{x}(t_f) + \int_o^{t_f} [\mathbf{x}^T(t)\mathbf{Q}(t)\mathbf{x}(t) + \mathbf{u}^T(t)\mathbf{R}(t)\mathbf{u}(t)]dt \quad (38.17)$$

$$\mathbf{x} \in \mathcal{R}^n, \quad \mathbf{B} \in \mathcal{R}^{n \times m}, \quad \mathbf{A}, \mathbf{P}, \mathbf{Q} \in \mathcal{R}^{n \times n}, \quad \mathbf{R} \in \mathcal{R}^{m \times m}, \ \mathbf{Q} \geq 0, \mathbf{R} > 0, \mathbf{F}_f > 0 \,. \quad (38.18)$$

The optimal time-varying control law is well known as linear controller of the Riccati type

$$\mathbf{K}(t) = -\mathbf{R}^{-1}(t)\mathbf{B}^T(t)\mathbf{P}(t) \tag{38.19}$$

where $\mathbf{P}(t)$ is satisfying the Riccati differential equation

$$\dot{\mathbf{P}}(t) = -\mathbf{P}(t)\mathbf{A}(t) - \mathbf{A}^T(t)\mathbf{P}(t) - \mathbf{Q}(t) + \mathbf{P}(t)\mathbf{B}(t)\mathbf{R}^{-1}(t)\mathbf{B}^T(t)\mathbf{P}(t)\ ; \quad \mathbf{P}(t_f) = \mathbf{F}_f\ . \tag{38.20}$$

Given $\mathbf{A}(t), \mathbf{B}(t), \mathbf{Q}(t), \mathbf{R}(t)$ as arbitrary time-varying matrices, the (nonlinear) Riccati differential equation can only be solved by approximation (*Chang, Y.F., and Lee, T.T., 1986*).

38.1.4 Approximating Matrices by Orthogonal Functions $h_i(t)$

Let the matrix Υ be assembled by $k-1$ submatrices, each of them built up as a Kronecker product $\mathbf{I}_n \otimes h_i(t)$

$$\Upsilon \triangleq \begin{pmatrix} \mathbf{I}_n \otimes h_o(t) \\ \mathbf{I}_n \otimes h_1(t) \\ \vdots \\ \mathbf{I}_n \otimes h_{k-1}(t) \end{pmatrix} = \begin{pmatrix} h_o & 0 & \dots & 0 \\ 0 & h_o & \dots & 0 \\ \vdots & \vdots & \ddots & \vdots \\ 0 & 0 & \dots & h_o \\ h_1 & 0 & \dots & 0 \\ 0 & h_1 & \dots & 0 \\ \vdots & \vdots & \ddots & \vdots \\ 0 & 0 & \dots & h_1 \\ \vdots & \vdots & \vdots & \vdots \\ \vdots & \vdots & \vdots & \vdots \\ h_{k-1} & 0 & \dots & 0 \\ 0 & h_{k-1} & \dots & 0 \\ \vdots & \vdots & \ddots & \vdots \\ 0 & 0 & \dots & h_{k-1} \end{pmatrix} \qquad \Upsilon \in \mathcal{R}^{kn\times n}\ . \tag{38.21}$$

Writing down in detail the product of a row-like partitioned (n, kn)-matrix with (n, n)-submatrices $\mathbf{A}_i^{\neg}\ \forall i = 0, 1, 2 \dots k-1$, i.e., $(\mathbf{A}_o^{\neg} \vdots \mathbf{A}_1^{\neg} \dots \mathbf{A}_{k-1}^{\neg})\Upsilon$, and calculating the product with transposed matrices, the following identity can be easily proved

$$(\mathbf{A}_o^{\neg} \vdots \mathbf{A}_1^{\neg} \dots \mathbf{A}_{k-1}^{\neg})\Upsilon \equiv \Upsilon^T \begin{pmatrix} \mathbf{A}_o^{\neg} \\ \mathbf{A}_1^{\neg} \\ \vdots \\ \mathbf{A}_{k-1}^{\neg} \end{pmatrix}. \tag{38.22}$$

Note that the column-like matrix in Eq.(38.22) contains submatrices $\mathbf{A}_i^{\neg}$ that have *not* been transposed; the column-like matrix above is not the transpose of the row-like matrix.

Generalizing the orthogonal expansion of time-varying scalar functions $x(t)$ in Eq.(36.16), the time-varying matrix $\mathbf{P}(t)$ being approximated by the sum of orthogonal functions $h_i(t)$ with matrix coefficients $\mathbf{P}_i^{\neg}$ is

$$\mathbf{P}(t) \doteq \sum_{i=0}^{k-1} \mathbf{P}_i^{\neg} h_i(t) = (\mathbf{P}_o^{\neg}\ \ \mathbf{P}_1^{\neg} \dots \mathbf{P}_{k-1}^{\neg})\Upsilon(t)\ . \tag{38.23}$$

In a similar way, matrices $\mathbf{A}_i^{\neg}, \mathbf{Q}_i^{\neg}$ and $\mathbf{W}_i^{\neg}$ are defined where $\mathbf{W}_i^{\neg}$ is the matrix coefficient associated with $\mathbf{W}(t) \triangleq \mathbf{B}(t)\mathbf{R}^{-1}(t)\mathbf{B}^T(t)$.

38.1.5 Approximating Matrix Products by Orthogonal Functions

As a result of the identity Eq.(38.22) the matrix product can be developed

$$\mathbf{P}(t)\mathbf{A}(t) = (\mathbf{P}_o^{\neg} \vdots \mathbf{P}_1^{\neg} \dots \mathbf{P}_{k-1}^{\neg})\, \boldsymbol{\Upsilon}(t)\boldsymbol{\Upsilon}^T(t) \begin{pmatrix} \mathbf{A}_o^{\neg} \\ \mathbf{A}_1^{\neg} \\ \vdots \\ \mathbf{A}_{k-1}^{\neg} \end{pmatrix} \tag{38.24}$$

$$\mathbf{P}(t)\mathbf{A}(t) = (\mathbf{P}_o^{\neg} \vdots \mathbf{P}_1^{\neg} \dots \mathbf{P}_{k-1}^{\neg}) \begin{pmatrix} \mathbf{I}_n \otimes h_o(t)h_o(t)\vdots & \dots & \mathbf{I}_n \otimes h_o(t)h_{k-1}(t) \\ \mathbf{I}_n \otimes h_1(t)h_o(t)\vdots & \dots & \vdots \\ \vdots & \ddots & \vdots \\ \mathbf{I}_n \otimes h_{k-1}(t)h_o(t)\vdots & \dots & \mathbf{I}_n \otimes h_{k-1}(t)h_{k-1}(t) \end{pmatrix} \begin{pmatrix} \mathbf{A}_o^{\neg} \\ \mathbf{A}_1^{\neg} \\ \vdots \\ \mathbf{A}_{k-1}^{\neg} \end{pmatrix} \tag{38.25}$$

$$\mathbf{P}(t)\mathbf{A}(t) = (\mathbf{P}_o^{\neg} \vdots \mathbf{P}_1^{\neg} \dots \mathbf{P}_{k-1}^{\neg}) \begin{pmatrix} \mathbf{A}_o^{\neg} \otimes h_o h_o + \mathbf{A}_1^{\neg} \otimes h_o h_1 + \dots + \mathbf{A}_{k-1}^{\neg} \otimes h_o h_{k-1} \\ \mathbf{A}_o^{\neg} \otimes h_1 h_o + \mathbf{A}_1^{\neg} \otimes h_1 h_1 + \dots + \mathbf{A}_{k-1}^{\neg} \otimes h_1 h_{k-1} \\ \vdots \\ \mathbf{A}_o^{\neg} \otimes h_{k-1} h_o + \dots \quad + \dots + \mathbf{A}_{k-1}^{\neg} \otimes h_{k-1} h_{k-1} \end{pmatrix} \tag{38.26}$$

$$\mathbf{P}(t)\mathbf{A}(t) = \sum_{i=0}^{k-1}\sum_{j=0}^{k-1} \mathbf{P}_i^{\neg}\mathbf{A}_j^{\neg} \otimes h_i(t)h_j(t) = [\sum_{i=0}^{k-1}\sum_{j=0}^{k-1} \mathbf{P}_i^{\neg}\mathbf{A}_j^{\neg}\mathbf{L}_{ij}]\boldsymbol{\Upsilon}(t) \tag{38.27}$$

where the (n, kn)-matrix $\mathbf{L}_{ij} = (l_{ijo}\mathbf{I}_n \vdots l_{ij1}\mathbf{I}_n \dots l_{ij,k-1}\mathbf{I}_n)$ corresponds to $l_{ij\mu}$ in Eq.(36.50).

38.1.6 Operational Matrix $\mathbf{P}_{Mn}$ of Integration

Before integrating the Riccati differential equation, the operational matrix $\mathbf{P}_{Mn}$ of integration belonging to the matrix $\boldsymbol{\Upsilon}$ for forward and backward integration, respectively, is calculated

$$\int_0^{t_f} \boldsymbol{\Upsilon}(t)dt = \mathbf{P}_{Mn}\boldsymbol{\Upsilon}(t) \qquad a = o,\ t > a \tag{38.28}$$

$$\int_{t_f}^{t} \boldsymbol{\Upsilon}(t)dt = \mathbf{P}_{Mn}\boldsymbol{\Upsilon}(t) \qquad a = t_f,\ t < a \tag{38.29}$$

$$\mathbf{P}_{Mn} = \frac{1}{p} \begin{pmatrix} (B_o - a'p - q)\mathbf{I}_n & A_o\mathbf{I}_n & 0 & 0 & 0 & \dots & 0 & 0 & 0 \\ (C_1 + D_1)\mathbf{I}_n & B_1\mathbf{I}_n & A_1\mathbf{I}_n & 0 & 0 & \dots & 0 & 0 & 0 \\ D_2\mathbf{I}_n & C_2\mathbf{I}_n & B_2\mathbf{I}_n & A_2\mathbf{I}_n & 0 & \dots & 0 & 0 & 0 \\ D_3\mathbf{I}_n & 0 & C_3\mathbf{I}_n & B_3\mathbf{I}_n & A_3\mathbf{I}_n & \dots & 0 & 0 & 0 \\ \vdots & \vdots & \vdots & \vdots & \vdots & \ddots & \vdots & \vdots & \vdots \\ D_{m-2}\mathbf{I}_n & 0 & 0 & 0 & 0 & \dots & C_{m-2}\mathbf{I}_n & B_{m-2}\mathbf{I}_n & A_{m-2}\mathbf{I}_n \\ D_{m-1}\mathbf{I}_n & 0 & 0 & 0 & 0 & \dots & 0 & C_{m-1}\mathbf{I}_n & B_{m-1}\mathbf{I}_n \end{pmatrix} . \tag{38.30}$$

With regard to the block diagonal structure of $\boldsymbol{\Upsilon}$, the operational matrix $\mathbf{P}_{Mn} \in \mathcal{R}^{kn \times kn}$ can be derived from $\mathbf{P}_M$ by appending the identity matrix $\mathbf{I}_n$ to each entry.

38.1.7 Approximating the Riccati Equation by Orthogonal Functions

Taking the backward integration form of the operational matrix

$$\int_{t_f}^{t} \dot{\mathbf{P}}(t)dt = \mathbf{P}(t) - \mathbf{F}_f = \int_{t_f}^{t} [-\mathbf{P}(t)\mathbf{A}(t) + \mathbf{A}^T(t)\mathbf{P}(t) - \mathbf{P}(t)\mathbf{W}(t)\mathbf{P}(t) + \mathbf{Q}(t)]dt \,. \quad (38.31)$$

Substituting Eq.(38.23) into the left-hand side of Eq.(38.31) yields

$$(\mathbf{P}_o^{\neg} - \mathbf{F}_f \vdots \mathbf{P}_1^{\neg} \vdots \ldots \mathbf{P}_{k-1}^{\neg})\Upsilon(t), \quad (38.32)$$

substituting Eq.(38.27) into the right-hand side of Eq. (38.31)

$$-[\sum_{i=0}^{k-1}\sum_{j=0}^{k-1}(\mathbf{P}_i^{\neg}\mathbf{A}_j^{\neg} + \mathbf{A}_i^{\neg T}\mathbf{P}_j^{\neg} + \mathbf{P}_i^{\neg}\mathbf{V}_j)\mathbf{L}_{ij} + (\mathbf{Q}_o^{\neg} \vdots \mathbf{Q}_1^{\neg} \ldots \mathbf{Q}_{k-1}^{\neg})]\mathbf{P}_{Mn}\Upsilon(t) \quad (38.33)$$

where $\mathbf{V}_j$ results from keeping $\mathbf{W}(t)\mathbf{P}(t)$ in coherence similarly to Eq.(38.27)

$$(\mathbf{V}_o \ \ \mathbf{V}_1 \ldots \mathbf{V}_{k-1}) = \sum_{i=0}^{k-1}\sum_{j=0}^{k-1} \mathbf{W}_i^{\neg}\mathbf{P}_j^{\neg}\mathbf{L}_{ij} \,. \quad (38.34)$$

Equating both sides yields an algebraic equation with kn^2 unknowns, represented by the elements of k expansion matrices $\mathbf{P}_i^{\neg}$

$$(\mathbf{P}_o^{\neg} - \mathbf{F}_f \vdots \mathbf{P}_1^{\neg} \ldots \mathbf{P}_{k-1}^{\neg}) = -[\sum_{i=0}^{k-1}\sum_{j=0}^{k-1}(\mathbf{P}_i^{\neg}\mathbf{A}_j^{\neg} + \mathbf{A}_i^{\neg T}\mathbf{P}_j^{\neg} + \mathbf{P}_i^{\neg}\mathbf{V}_j)\mathbf{L}_{ij} + (\mathbf{Q}_o^{\neg} \vdots \mathbf{Q}_1^{\neg} \ldots \mathbf{Q}_{k-1}^{\neg}]\mathbf{P}_{Mn} \,. \quad (38.35)$$

Note that by application of the orthogonal polynomial expansion, the two-point boundary value constraints could be reduced to simple initial conditions. *Tsay, S.C., and al. 1988* presented an approach with a combined state and costate vector to solve the two-point boundary value problem with an initial condition method.

38.2 Parameter Estimation in Linear Systems

The method of generalized orthogonal function expansion and the operational matrix approach replacing the integration is extended to the identification of linear processes with lumped parameters (linear ordinary differential equations), even with time-varying coefficients. Expanding the single-input and single-output of a process into generalized orthogonal polynomials yields a set of linear algebraic equations. If the number k of orthogonal polynomials is sufficient, a set of overdetermined algebraic equations is obtained. By regression method, the parameters and their time-dependence can be evaluated (see *Chang, R.Y., et al. 1988; Hwang, C., and Guo, T.Y., 1984*).

38.2.1 Expansions and Operational Matrices

Use is made again of decomposing the orthogonal polynomial function $\mathbf{h}(t)$ by a power series $\mathbf{F}\,\mathbf{t}(t)$ as shown in Eq.(36.41). The operational matrix $\mathbf{P}_{Mp}$ for the single integration of $\mathbf{t}(t)$ is taken from Eq.(36.102). For an α-tuples integral one has

$$\int_o^t \overset{\alpha}{\cdots} \int_o^t \mathbf{t}(t)\ (dt)^\alpha = \mathbf{P}_{Mp}^\alpha\ \mathbf{t}(t) \,. \quad (38.36)$$

Starting with Eq.(36.46), the power of t can be expressed

$$t^i = \sum_{j=0}^{i} \vartheta_{ij} h_j(t) = \sum_{j=0}^{i} \vartheta_{ij} \Big(\sum_{l=0}^{j} f_{jl} t^l \Big) = \sum_{j=0}^{i} \sum_{l=0}^{j} \vartheta_{ij} f_{il} t^l . \tag{38.37}$$

Rearranging yields

$$t^i = \sum_{j=0}^{i} \vartheta_{ij} (f_{j0} t^0 + f_{j1} t^1 + \ldots + f_{jj} t^j) \tag{38.38}$$

$$t^i = \vartheta_{i0} f_{j0} + \vartheta_{i1}(f_{10} + f_{11} t) + \vartheta_{i2}(f_{20} + f_{21} t + f_{22} t^2) + \ldots + \vartheta_{ii}(f_{i0} + f_{i1} t + \ldots f_{ii} t^i) \tag{38.39}$$

$$t^i = \sum_{\nu=0}^{i} \sum_{\mu=\nu}^{i} \vartheta_{i\mu} f_{\mu\nu} t^\nu \doteq \sum_{\nu=0}^{k-1} \sum_{\mu=\nu}^{k-1} \vartheta_{i\mu} f_{\mu\nu} t^\nu = \sum_{\nu=0}^{k-1} \Big(\sum_{\mu=\nu}^{k-1} \vartheta_{i\mu} f_{\mu\nu} \Big) t^\nu \tag{38.40}$$

$$t^{k+1} \doteq \sum_{\nu=0}^{k-1} \Big(\sum_{\mu=\nu}^{k-1} \vartheta_{k+i,\mu}\, f_{\mu\nu} \Big) t^\nu = \sum_{\nu=0}^{k-1} r_{i\mu} t^\nu \quad \text{where} \quad r_{i\mu} = \sum_{\mu=\nu}^{m-1} \vartheta_{k+i,\mu}\, f_{\mu\nu} \, . \tag{38.41}$$

The following properties are very helpful $\forall i = 1, 2...k$

$$t^i \mathbf{t}(t) = (t^i \;\; t^{i+1} \ldots t^{k-1+i}\,)^T = \mathbf{R}_i \mathbf{t}(t) \tag{38.42}$$

$$\mathbf{R}_i \triangleq \begin{pmatrix} 0 & 0 & 0 & \ldots & 1 & 0 & 0 & \ldots & 0 \\ 0 & 0 & 0 & \ldots & 0 & 1 & 0 & \ldots & 0 \\ 0 & 0 & 0 & \ldots & 0 & 0 & 1 & \ldots & 0 \\ \vdots & \vdots & \vdots & & \vdots & \vdots & \vdots & \ddots & \vdots \\ 0 & 0 & 0 & \ldots & 0 & 0 & 0 & \ldots & 1 \\ r_{oo} & r_{o1} & r_{o2} & \ldots & & & & & r_{o,k-1} \\ \vdots & \vdots & \vdots & & & & & & \vdots \\ r_{i-1,o} & r_{i-1,1} & r_{i-1,2} & \ldots & & & & & r_{i-1,k-1} \end{pmatrix} \in \mathcal{R}^{k \times k} \tag{38.43}$$

$$\breve{\mathbf{I}}_i \triangleq (\mathbf{I}_i \;\vdots\; \mathbf{0}_{i,k-i}) = \begin{pmatrix} 1 & 0 & 0 & \ldots & 0 & 0 & 0 & \ldots & 0 \\ 0 & 1 & 0 & & 0 & 0 & 0 & \ldots & 0 \\ 0 & 0 & 1 & & 0 & 0 & 0 & \ldots & 0 \\ \vdots & & & \ddots & & \vdots & \vdots & & \vdots \\ 0 & & & & 1 & 0 & 0 & \ldots & 0 \end{pmatrix} \tag{38.44}$$

$$\mathbf{R}_i = \begin{pmatrix} \mathbf{0}_{k-i,i} \;\vdots\; \mathbf{I}_{k-i} \\ \breve{\mathbf{I}}_i \mathbf{R}_k \end{pmatrix} \qquad \breve{\mathbf{I}}_i \in \mathcal{R}^{i \times k} \tag{38.45}$$

where $\mathbf{R}_k = \mathbf{R}_i \,|_{i=k}$. By letting $\chi_i = 0$ in $\mathbf{P}_{M_p}$ from Eq.(36.102), $\mathbf{R}_k = 0$ is achieved. The generalized polynomial approximation is then reduced to the Taylor series approximation.

Let a new vector $\grave{\mathbf{t}}(t)$ similar to $\mathbf{t}(t)$ be of dimension $\eta + 1$

$$\grave{\mathbf{t}} \triangleq (1 \quad t \quad t^2 \;\ldots\; t^\eta)^T \qquad \grave{\mathbf{t}} \in \mathcal{R}^{(\eta+1)} \, . \tag{38.46}$$

Differentiating the vector $\grave{\mathbf{t}}$ with respect to time t yields

$$\frac{d\grave{\mathbf{t}}}{dt} = (0 \quad 1 \quad 2t \quad 3t^2 \ldots \eta t^{\eta-1})^T \, . \tag{38.47}$$

This result is also obtained from $\grave{\mathbf{t}}$ itself by shifting to the right and by multiplying with natural numbers. Defining an operational matrix $\mathbf{D}$

$$\mathbf{D} \triangleq \begin{pmatrix} 0 & 1 & 0 & \dots & 0 \\ 0 & 0 & 2 & \dots & 0 \\ 0 & 0 & 0 & \dots & 0 \\ \vdots & \vdots & \vdots & \ddots & \eta \\ 0 & 0 & 0 & 0 & 0 \end{pmatrix} \in \mathcal{R}^{(\eta+1)\times(\eta+1)} \tag{38.48}$$

to be applied to the transpose $\grave{\mathbf{t}}^T$ yields

$$\frac{d\grave{\mathbf{t}}^T}{dt} = \grave{\mathbf{t}}^T\mathbf{D} \quad \text{and} \quad \frac{d^j\grave{\mathbf{t}}^T}{dt^j} = \grave{\mathbf{t}}^T\mathbf{D}^j . \tag{38.49}$$

In addition to operational matrices for integration and differentiation, a time delay (dead time) operational matrix was presented by *Mohan, B.M., and Datta, K.B., 1988*.

38.2.2 Time-Varying System. Differential Equation of nth Order

Consider a time-varying system with the time-varying differential equation

$$\sum_{l=0}^{n} a_l(t)y^{(l)}(t) = \sum_{l=0}^{n_1} b_l(t)u^{(l)}(t) . \tag{38.50}$$

Usually $n_1 < n$. If $b_l\,|_{l>n_1} = 0$ only one summation boundary $n_1 = n$ has to be considered. The coefficient $a_l(t)$ is assumed a polynomial with terms up to $a_{l\eta}t^\eta$. The subcoefficients $a_{l\nu}$ are constant.

Term j of Time-Varying System Equation

The coefficient $a_l(t)$ can be described as follows

$$a_l(t) = a_{lo} + a_{l1}t + \ldots + a_{l\eta}t^\eta = \mathbf{a}_l^T\grave{\mathbf{t}} = \grave{\mathbf{t}}^T\mathbf{a}_l \qquad \text{where} \quad \mathbf{a}_l \triangleq (a_{lo} \;\; a_{l1} \ldots a_{l\eta})^T \in \mathcal{R}^{(\eta+1)} \tag{38.51}$$

$$a_l^{(j)} = \frac{d^j a_l(t)}{dt^j} = \frac{d\grave{\mathbf{t}}^T}{dt}\mathbf{a}_l = \grave{\mathbf{t}}^T\mathbf{D}^j\mathbf{a}_l . \tag{38.52}$$

Decomposing the output signal $y(t)$ into a power series of t with coefficients $\breve{y}_i$

$$y(t) = \sum_{i=0}^{k-1} \breve{y}_i t^i = \breve{\mathbf{y}}^T\mathbf{t} \qquad \breve{\mathbf{y}} \in \mathcal{R}^k \tag{38.53}$$

yields

$$\begin{aligned} a_l^{(j)}(t)y(t) &= y(t)a_l^{(j)}(t) = \breve{\mathbf{y}}^T\mathbf{t}\,\grave{\mathbf{t}}^T\mathbf{D}^j\mathbf{a}_l = \breve{\mathbf{y}}^T\big(\mathbf{t} \;\; t\mathbf{t} \;\; t^2\mathbf{t}\ldots t^\eta\mathbf{t}\big)_{[k\times(\eta+1)]}\mathbf{D}^j\mathbf{a}_l & (38.54)\\ &= \breve{\mathbf{y}}^T\big(\mathbf{t} \;\; \mathbf{R}_1\mathbf{t} \;\; \mathbf{R}_2\mathbf{t}\ldots\mathbf{R}_\eta\mathbf{t}\big)\mathbf{D}^j\mathbf{a}_l = \big(\mathbf{t} \;\; \mathbf{R}_1\mathbf{t}\ldots\mathbf{R}_\eta\mathbf{t}\big)^T\breve{\mathbf{y}}\mathbf{D}^j\mathbf{a}_l & (38.55)\\ &= \big(\mathbf{t}^T\breve{\mathbf{y}} \vdots \mathbf{t}^T\mathbf{R}_1^T\breve{\mathbf{y}} \vdots \ldots\mathbf{t}^T\mathbf{R}_\eta^T\breve{\mathbf{y}}\big)\mathbf{D}^j\mathbf{a}_l = \mathbf{t}^T(t)\big(\breve{\mathbf{y}} \vdots \mathbf{R}_1^T\breve{\mathbf{y}}\ldots\mathbf{R}_\eta^T\breve{\mathbf{y}}\big)_{[k\times(\eta+1)]}\mathbf{D}^j\mathbf{a}_l . \end{aligned}$$

n-tuples Integral

Integrating both sides α times and applying Eq.(38.36)

$$\int_o^t \overset{\alpha}{\cdots} \int_o^t a_l^{(j)}(t)y(t)(dt)^\alpha = \mathbf{t}^T(t)(\mathbf{P}_{Mp}^\alpha)^T\left(\check{\mathbf{y}} \vdots \mathbf{R}_1^T\check{\mathbf{y}} \vdots \ldots \mathbf{R}_\eta^T\check{\mathbf{y}}\right)\mathbf{D}^j\mathbf{a}_l \ . \tag{38.56}$$

Integrating Eq.(38.50) n times and making use of the method of integration by parts yields

$$\begin{aligned}\int_o^t \overset{n}{\cdots} \int_o^t \sum_{l=0}^n a_l(t)y^{(l)}(t)(dt)^n &= \sum_{l=0}^n\sum_{j=0}^l(-1)^j\binom{l}{j}\int_o^t \overset{(n-l+j)}{\cdots\cdots} \int_o^t a_l^{(j)}(t)y(t)(dt)^{n-l+j} \\ &+ \sum_{l=0}^n\sum_{i=1}^l\sum_{j=0}^{l-i}(-1)^{j+1}\binom{j+i-1}{i-1}a_l^{(j)}(0)y^{(l-i-j)}(0)\frac{t^{n-i}}{(n-i)!}\end{aligned} \tag{38.57}$$

and

$$\sum_{l=0}^n\sum_{j=0}^l(-1)^j\binom{l}{j}\int_o^t \overset{n-l+j}{\cdots\cdots} \int_o^t \left(a_l^{(j)}(t)y(t) - b_l^{(j)}(t)u(t)\right)(dt)^{n-l+j} - \sum_{i=0}^{n-1} v_i\frac{t^i}{i!} = 0 \tag{38.58}$$

where

$$v_i = \sum_{l=n-i}^n\sum_{j=0}^{l-n+i}(-1)^j\binom{n+j-i-1}{n-i-1}\left(a_l^{(j)}(0)y^{(l-n+i-j)}(0) - b_l^{(j)}(0)u^{(l-n+i-j)}(0)\right) \tag{38.59}$$

$$\frac{t^i}{i!} = (0\ 0\ldots 0\ \frac{1}{i!}\ 0\ldots 0)\mathbf{t}(t) = \frac{\mathbf{e}_{i+1}^T\mathbf{t}(t)}{i!} \qquad \forall i = 0,1,2\ldots n-1, \quad \mathbf{e}_{i+1} \in \mathcal{R}^k \ . \tag{38.60}$$

In the first term of Eq.(38.58) the first sum with index l is partitioned

$$\sum_{l=0}^n Z_l = Z_n + \sum_{l=0}^{n-1} Z_l \ . \tag{38.61}$$

Specializing $a_{no} = 1$

$$\begin{aligned} y(t) &+ \sum_{j=0}^n(-1)^j\binom{n}{j}\int_o^t \overset{j}{\cdots} \int_o^t \bar{a}_n^{(j)}y(t)(dt)^j + \sum_{l=0}^{n-1}\sum_{j=0}^l(-1)^j\binom{l}{j}\int_o^t \overset{n-l+j}{\cdots\cdots} \int_o^t a_l^{(j)}y(t)(dt)^{n-l+j} \\ &- \sum_{l=0}^n\sum_{j=0}^l(-1)^j\binom{l}{j}\int_o^t \overset{n-l+j}{\cdots\cdots} \int_o^t b_l^{(j)}u(t)(dt)^{n-l+j} - \sum_{i=0}^{n-1}\frac{v_i\mathbf{e}_{i+1}^T\mathbf{t}(t)}{i!} = 0 \ . \end{aligned} \tag{38.62}$$

Making use of $\Theta^T\check{\mathbf{y}} = \mathbf{y}^\neg$, $\check{\mathbf{y}} = \mathbf{F}^\neg\mathbf{y}^\neg$ and referring to Eq.(36.46),

$$y(t) = \mathbf{t}^T\check{\mathbf{y}} = \mathbf{h}^T(t)\Theta^T\check{\mathbf{y}} = \mathbf{h}^T\mathbf{y}^\neg = [\Theta^{-1}\mathbf{t}(t)]^T\mathbf{y}^\neg = \mathbf{t}^T(t)\Theta^{-1,T}\mathbf{y}^\neg = \mathbf{t}^T(t)\mathbf{F}^T\mathbf{y}^\neg \tag{38.63}$$

the following abbreviations are stated

$$\mathbf{Y}_l = \sum_{j=0}^l(-1)^j\binom{l}{j}(\mathbf{P}_{Mp}^{n-l+j}\ )^T(\check{\mathbf{y}}\vdots\mathbf{R}_1^T\check{\mathbf{y}}\ldots\mathbf{R}_\eta^T\check{\mathbf{y}})\mathbf{D}^j \tag{38.64}$$

$$\mathbf{U}_l = \sum_{j=0}^l(-1)^j\binom{l}{j}(\mathbf{P}_{Mp}^{n-l+j}\ )^T(\check{\mathbf{u}}\vdots\mathbf{R}_1^T\check{\mathbf{u}}\ldots\mathbf{R}_\eta^T\check{\mathbf{u}})\mathbf{D}^j \tag{38.65}$$

$$\bar{\mathbf{Y}}_n = \mathbf{Y}_l|_{l=n \textit{ and first column removed}} \tag{38.66}$$

$$\bar{\mathbf{a}}_n = (a_{n1}\ \ a_{n2}\ldots a_{n\eta})^T \qquad \mathbf{a}_l = (a_{lo}\ \ a_{l1}\ldots a_{l\eta})^T \qquad \mathbf{b}_l = (b_{lo}\ \ b_{l1}\ldots b_{l\eta})^T \tag{38.67}$$

$$\mathbf{I}_{mod} = (\mathbf{e}_1\ \frac{\mathbf{e}_2}{2!}\ \frac{\mathbf{e}_3}{3!}\cdots\frac{\mathbf{e}_\eta}{\eta!}) = \text{diag}\ \frac{1}{i!} \qquad \mathbf{v} = (v_o\ \ v_1\ldots v_{\eta-1})^T \ . \tag{38.68}$$

The dimensions of the vector and matrix variables are:

$$
\begin{array}{lll}
\mathbf{t}, \check{\mathbf{y}}, \mathbf{y}^{\urcorner} & k \times 1 \\
\dot{\mathbf{t}}, \mathbf{a}_l, \mathbf{b}_l, \mathbf{v} & (\eta+1) \times 1 \\
\mathbf{p} & [(2n+3)(\eta+1)-1] \times 1
\end{array}
\qquad
\begin{array}{ll}
\mathbf{R}_i, \Theta, \mathbf{F} & k \times k \\
\mathbf{Y}_l & k \times (\eta+1) \\
\mathbf{D}, \mathbf{I}_{mod} & (\eta+1)(\eta+1) \\
\boldsymbol{\Omega} & k \times [(2n+3)(\eta+1)-1] \,.
\end{array}
\tag{38.69}
$$

Utilizing these abbreviations, the original differential equation can be presented in three domains, i.e. signal domain $y(t)$, power series domain $\check{\mathbf{y}}$ and orthogonal expansion domain $\mathbf{y}^{\urcorner}$.

Signal domain $y(t)$ from Eq.(38.62):

$$
\mathbf{t}^T(t)\Big(\check{\mathbf{y}} + \bar{\mathbf{Y}}_n \bar{\mathbf{a}}_n + \sum_{l=0}^{n-1} \mathbf{Y}_l \mathbf{a}_l - \sum_{l=0}^{n} \mathbf{U}_l \mathbf{b}_l - \mathbf{I}_{mod}\mathbf{v}\Big) = 0 \,. \tag{38.70}
$$

Removing $\mathbf{t}^T(t)$, the embraced term is the result in power series domain $\check{\mathbf{y}}$

$$
\Big(\check{\mathbf{y}} + \bar{\mathbf{Y}}_n \bar{\mathbf{a}}_n + \sum_{l=0}^{n-1} \mathbf{Y}_l \mathbf{a}_l - \sum_{l=0}^{n} \mathbf{U}_l \mathbf{b}_l - \mathbf{I}_{mod}\mathbf{v}\Big) = 0. \tag{38.71}
$$

Finally, if removing only $\mathbf{h}^T(t)$, the orthogonal expansion domain $\mathbf{y}^{\urcorner}$ is

$$
\Theta^T\Big(\check{\mathbf{y}} + \bar{\mathbf{Y}}_n \bar{\mathbf{a}}_n + \sum_{l=0}^{n-1} \mathbf{Y}_l \mathbf{a}_l - \sum_{l=0}^{n} \mathbf{U}_l \mathbf{b}_l - \mathbf{I}_{mod}\mathbf{v}\Big) = 0 \,. \tag{38.72}
$$

Defining the parameter vector $\mathbf{p}$ and the orthogonal expression domain matrix $\boldsymbol{\Omega}$

$$
\mathbf{p} = (\mathbf{a}_o^T \vdots \mathbf{a}_1^T \ldots \mathbf{a}_{n-1}^T \vdots \bar{\mathbf{a}}_n^T \vdots \mathbf{b}_o^T \ldots \mathbf{b}_n^T \vdots \mathbf{v}^T)^T \tag{38.73}
$$

$$
\boldsymbol{\Omega} = \Big(-\Theta \mathbf{Y}_o \vdots \ldots -\Theta^T \mathbf{Y}_{n-1} \vdots -\Theta^T \bar{\mathbf{Y}}_n \vdots \Theta^T \mathbf{U}_o \ldots \Theta^T \mathbf{U}_n \vdots \mathbf{I}_{mod}\Big) \tag{38.74}
$$

from Eq.(38.72)

$$
\boldsymbol{\Omega}\mathbf{p} = \Theta^T \check{\mathbf{y}} \,. \tag{38.75}
$$

If the number of k is greater than the dimension of the parameter vector $\mathbf{p}$, i.e., $k > (2n+3)(\eta+1)-1$, the optimum least-squares approximation $\mathbf{p}^\star$ is obtained by linear regression technique

$$
\mathbf{p}^\star = \boldsymbol{\Omega}^{\sharp L}\, \Theta^T\, \check{\mathbf{y}} = \boldsymbol{\Omega}^{\sharp L}\, \mathbf{y}^{\urcorner} \,. \tag{38.76}
$$

Using Laguerre polynomials, a similar identification was presented by *Hwang, C., and Shih, Y.P., 1982*, identifying the system with zero and non-zero initial conditions and under the influence of output signal noise. Identification via generalized orthogonal polynomials applying state space characterization of the system equations was carried out by *Wang, M.L., Chang, R.Y., and Yang, S.Y., 1987*. Time-delay systems identification was investigated by *Wang, M.L., Jan, Y.J., and Chang, R.Y., 1987*. A Fourier-series approach was given by *Chung, H.Y., 1987*.

38.3 Parameter Estimation in Nonlinear Systems

38.3.1 Elementary Differential Equations

In this subsection, the identification strategies are not restricted to elementary cases. It seems suitable from an illustration point of view to outline the essential steps in the case of rather simple identification problems (*Horng, I.R., and Chou, J.H., 1987*).

Second-Order Model. Zero Initial Conditions

Consider a nonlinear time-invariant second-order differential equation

$$a_2 \frac{d^2}{dt^2} y^2(t) + a_1 \frac{d}{dt} y(t) = u(t) \ , \tag{38.77}$$

first. In spite of the nonlinearity, the objective of identification in this case proves rather simple because the equation is still linear in the parameters a_2 and a_1. Integrating twice under zero initial conditions

$$a_2 y^2(t) + a_1 \int_0^t y(t)dt = \int_0^t \int_0^t u(t)dt \ dt \ . \tag{38.78}$$

With regard to the fact that $y^2(t)$ and $y(t)$ are known by measured values of the output it is possible to provide separate approximations for each. Referring to Eq.(36.16),

$$y(t) = \mathbf{y}^{\neg T}\mathbf{h}(t), \quad y^2(t) = \mathbf{z}^{\neg T}\mathbf{h}(t), \quad u(t) = \mathbf{u}^{\neg T}\mathbf{h}(t) \ . \tag{38.79}$$

Applying the operational matrix of integration $\mathbf{P}_M$ as defined in Eq.(36.34),

$$a_2 \mathbf{z}^{\neg T}\mathbf{h}(t) + a_1 \int_0^t \mathbf{y}^{\neg T}\mathbf{h}(t)dt = \int_0^t \int_0^t \mathbf{u}^{\neg T}\mathbf{h}(t)dt \ dt \tag{38.80}$$

$$a_2 \mathbf{z}^{\neg T}\mathbf{h}(t) + a_1 \mathbf{y}^{\neg T} \int_0^t \mathbf{h}(t)dt = \mathbf{u}^{\neg T} \int_0^t \int_0^t \mathbf{h}(t)dt \ dt \tag{38.81}$$

$$a_2 \mathbf{z}^{\neg T}\mathbf{h}(t) + a_1 \mathbf{y}^{\neg T}\mathbf{P}_M\mathbf{h}(t) = \mathbf{u}^{\neg T}\mathbf{P}_M^2\mathbf{h}(t) \ . \tag{38.82}$$

Omitting $\mathbf{h}(t)$ and defining $\mathbf{p} = (a_1 \ \ a_2)^T$ yields

$$(\mathbf{y}^{\neg T}\mathbf{P}_M \vdots \mathbf{z}^{\neg T})\mathbf{p} \triangleq \mathbf{M}\mathbf{p} = \mathbf{u}^{\neg T}\mathbf{P}_M^2 \quad \rightsquigarrow \quad \mathbf{p}^* = \mathbf{M}^{\sharp L}\mathbf{u}^{\neg T}\mathbf{P}_M^2 \ . \tag{38.83}$$

First-Order Model. Non-Zero Initial Conditions

Non-zero initial conditions can be taken into consideration as explained with an example of first order

$$\frac{d}{dt}[a_2 y^2(t) + a_1 y(t)] = u(t) \ . \tag{38.84}$$

Applying

$$\int_0^t a_1 \frac{d}{dt} y(t)dt = a_1 y(t) - a_1 y(0) \tag{38.85}$$

and a corresponding expression for $y^2(t)$ yields

$$a_2 \mathbf{z}^{\neg T}\mathbf{h}(t) - a_2 y^2(0) + a_1 \mathbf{y}^{\neg T}\mathbf{h}(t) - a_1 y(0) = \mathbf{u}^{\neg T}\mathbf{P}_M\mathbf{h}(t) \ . \tag{38.86}$$

Referring to the unit vector $\mathbf{e}_1^{(k\times 1)} = (1 \ 0 \ 0 \ldots)^T$ and to $h_o(t) = 1$ as outlined in Eq.(36.7), the constant terms $-a_2 y^2(0)$ and $-a_1 y(0)$ can be regarded as a coefficient of $\mathbf{e}_1^T\mathbf{h}(t) \equiv 1$. Thus,

$$a_2 \mathbf{z}^{\neg T}\mathbf{h}(t) - a_2 y^2(0)\mathbf{e}_1^T\mathbf{h}(t) + a_1 \mathbf{y}^{\neg T}\mathbf{h}(t) - a_1 y(0)\mathbf{e}_1^T\mathbf{h}(t) = \mathbf{u}^{\neg T}\mathbf{P}_M\mathbf{h}(t) \tag{38.87}$$

$$a_2[\mathbf{z}^{\neg T} - y^2(0)\mathbf{e}_1^T] + a_1[\mathbf{y}^{\neg T} - y(0)\mathbf{e}_1^T] = \mathbf{u}^{\neg T}\mathbf{P}_M \tag{38.88}$$

$$[\mathbf{y}^{\neg T} - y(0)\mathbf{e}_1^T \vdots \mathbf{z}^{\neg T} - y^2(0)\mathbf{e}_1^T]\begin{pmatrix} a_1 \\ a_2 \end{pmatrix} \triangleq \mathbf{M}\mathbf{p} = \mathbf{u}^{\neg T}\mathbf{P}_M \quad \rightsquigarrow \quad \mathbf{p}^* = \mathbf{M}^{\sharp L}\mathbf{u}^{\neg T}\mathbf{P}_M \ . \tag{38.89}$$

38.3.2 Nonlinear Model Using Kronecker Powers

Referring to the formulation of linear systems $\dot{\mathbf{x}}(t) = \mathbf{A}(t)\mathbf{x}(t) + \mathbf{B}(t)\mathbf{u}(t), \quad \mathbf{x}(0) = \mathbf{x}_o$, nonlinear systems can be described in the following notation[1]

$$\dot{\mathbf{x}}(t) = \mathbf{A}[\mathbf{x}(t)] + \sum_{i=1}^{m} \mathbf{B}_i[\mathbf{x}(t)]u_i(t) \qquad \mathbf{x} \in \mathcal{R}^n, \quad \mathbf{u} = (u_1\ u_2\ \ldots u_m)^T \in \mathcal{R}^m . \tag{38.90}$$

This is only a particular form of a nonlinear system but it gives an excellent model for physical and ecological systems.

Following *Rotella, F., and Dauphin-Tanguy, G., 1988*, the matrices $\mathbf{A}$ and $\mathbf{B}_i$ are decomposed using the ith Kronecker power $\mathbf{x}^{[i]}(t)$ of the vector $\mathbf{x}(t)$

$$\mathbf{A}[\mathbf{x}(t)] = \mathbf{A}(\mathbf{x}) = \sum_{j=1}^{\infty} \mathbf{A}_j \mathbf{x}^{[j]} \doteq \sum_{j=1}^{r_A} \mathbf{A}_j \mathbf{x}^{[j]} \qquad \mathbf{x}^{[j]} \in \mathcal{R}^{n^j}, \tag{38.91}$$

$$\mathbf{B}_i[\mathbf{x}(t)] = \mathbf{B}_i(\mathbf{x}) = \sum_{j=0}^{\infty} \mathbf{B}_{ij} \mathbf{x}^{[j]} \doteq \sum_{j=0}^{r_B} \mathbf{B}_{ij} \mathbf{x}^{[j]} \qquad \forall i = 1 \ldots m \qquad \mathbf{A}_j,\ \mathbf{B}_{ij} \in \mathcal{R}^{n \times (n^j)} . \tag{38.92}$$

The approximation is truncated by highest powers r_A and r_B. In order to illustrate the nonlinear expressions including the fact of redundancy, a few tensor powers of a two-dimensional vector $\mathbf{x}$ are presented below

$$\mathbf{x}^{[0]} = 1 \qquad \mathbf{x}^{[1]} = \begin{pmatrix} x_1 \\ x_2 \end{pmatrix} \qquad \mathbf{x}^{[2]} = \begin{pmatrix} x_1\mathbf{x} \\ x_2\mathbf{x} \end{pmatrix} = \begin{pmatrix} x_1x_1 \\ x_1x_2 \\ x_2x_1 \\ x_2x_2 \end{pmatrix} = \begin{pmatrix} x_1^2 \\ x_1x_2 \\ x_1x_2 \\ x_2^2 \end{pmatrix} \tag{38.93}$$

$$\mathbf{x}^{[3]} = \mathbf{x} \otimes \mathbf{x}^{[2]} = \begin{pmatrix} x_1\mathbf{x}^{[2]} \\ x_2\mathbf{x}^{[2]} \end{pmatrix} = \begin{pmatrix} x_1^3 \\ x_1^2x_2 \\ x_1^2x_2 \\ x_1x_2^2 \\ x_2x_1^2 \\ x_1x_2^2 \\ x_1x_2^2 \\ x_2^3 \end{pmatrix} . \tag{38.94}$$

38.3.3 Decomposing the Kronecker Product to Orthogonal Functions

Consider the Kronecker product of two vector-valued functions $\mathbf{x}(t)$ and $\mathbf{y}(t)$ on the interval $t \in [0, t_f]$. Using the mixed product rule Eq.(4.10),

$$\mathbf{x}(t)\otimes\mathbf{y}(t) = \left(\mathbf{X}^{\neg T}\mathbf{h}(t)\right)\otimes\left(\mathbf{Y}^{\neg T}\mathbf{h}(t)\right) = (\mathbf{X}^{\neg T}\otimes\mathbf{Y}^{\neg T})(\mathbf{h}\otimes\mathbf{h}) = (\mathbf{X}^{\neg T}\otimes\mathbf{Y}^{\neg T})\mathbf{h}^{[2]}. \tag{38.95}$$

[1]Continuing the original sense of the coefficient matrix and input matrix, the terms $\mathbf{A}, \mathbf{B}_i$ are written as capital letters although they represent vector-valued functions.

Regarding the vector $\mathbf{l}_{ij}$ defined in Eq.(36.50) for the second tensor power, it results

$$\mathbf{h}^{[2]} = \mathbf{h} \otimes \mathbf{h} = \begin{pmatrix} h_o\mathbf{h} \\ h_1\mathbf{h} \\ \vdots \\ h_{k-1}\mathbf{h} \end{pmatrix} = \begin{pmatrix} h_o h_o \\ h_o h_1 \\ h_o h_2 \\ \vdots \\ h_o h_{k-1} \\ \cdots \\ h_1 h_o \\ \vdots \end{pmatrix} = \begin{pmatrix} \mathbf{l}_{00}^T\mathbf{h} \\ \mathbf{l}_{01}^T\mathbf{h} \\ \mathbf{l}_{02}^T\mathbf{h} \\ \vdots \\ \mathbf{l}_{0,k-1}^T\mathbf{h} \\ \cdots \\ \mathbf{l}_{10}^T\mathbf{h} \\ \vdots \end{pmatrix} = \begin{pmatrix} \mathbf{K}_o\mathbf{h} \\ \cdots \\ \mathbf{K}_1\mathbf{h} \\ \vdots \end{pmatrix} = \mathbf{K}\mathbf{h} \tag{38.96}$$

$$\text{where } \mathbf{K}_j = \begin{pmatrix} \mathbf{l}_{j0}^T \\ \mathbf{l}_{j1}^T \\ \vdots \\ l_{j,k-1}^T \end{pmatrix} \quad \text{and} \quad \mathbf{K} = \begin{pmatrix} \mathbf{K}_o \\ \mathbf{K}_1 \\ \vdots \\ \mathbf{K}_{k-1} \end{pmatrix} . \tag{38.97}$$

38.3.4 Decomposing the Kronecker Power to Orthogonal Functions

If the νth Kronecker power of $\mathbf{x}$ has already been decomposed to orthogonal functions

$$\mathbf{x}^{[\nu]} = \mathbf{X}^{\nu\neg T}\mathbf{h} \tag{38.98}$$

a recurrence respecting $\mathbf{X}^{\nu\neg T}$ and $\mathbf{X}^{\nu+1,\neg T}$ is

$$\begin{aligned} \mathbf{x}^{[\nu+1]} &= \mathbf{x}^{[\nu]} \otimes \mathbf{x} = (\mathbf{X}^{\nu\neg T}\mathbf{h}) \otimes (\mathbf{X}^{\neg T}\mathbf{h}) = (\mathbf{X}^{\nu\neg T} \otimes \mathbf{X}^{\neg T})\mathbf{h}^{[2]} && (38.99) \\ &= (\mathbf{X}^{\nu\neg T} \otimes \mathbf{X}^{\neg T})\mathbf{K}\mathbf{h} = \mathbf{X}^{\nu+1,\neg T}\, \mathbf{h}(t) && (38.100) \end{aligned}$$

$$\mathbf{X}^{\nu+1,\neg T} = (\mathbf{X}^{\nu\neg T} \otimes \mathbf{X}^{\neg T})\mathbf{K} \tag{38.101}$$

$$\text{where} \quad \mathbf{X}^{0\neg T} = (1 \;\; 0 \;\; 0 \ldots 0) , \qquad \mathbf{X}^{0\neg T} \in \mathcal{R}^{1\times k}, \qquad \mathbf{X}^{\nu\neg T} \in \mathcal{R}^{n^\nu \times k} . \tag{38.102}$$

38.3.5 Identification of the State Space Model $\mathbf{A}(\mathbf{x})$ and $\mathbf{B}_i(\mathbf{x})$

The identification process based on the model of Eqs. (38.90) to (38.92) starts with the integration of Eq.(38.90)

$$\mathbf{x}(t) - \mathbf{x}_o = \sum_{j=1}^{r_A} \mathbf{A}_j \int_0^t \mathbf{x}^{[j]}(\tau)d\tau + \sum_{i=1}^{m}\sum_{j=0}^{r_B} \mathbf{B}_{ij} \int_0^t \mathbf{x}^{[j]}(\tau)u_i(\tau)d\tau . \tag{38.103}$$

Denoting the orthogonal function approximation of order k

$$\mathbf{x}(t) \doteq \mathbf{X}^{\neg T}\mathbf{h}(t) \quad \text{and} \quad \int_0^t \mathbf{x}^{[\nu]}(\tau)d\tau = \mathbf{X}^{\nu\neg T}\mathbf{P}_M\mathbf{h}(t), \tag{38.104}$$

$$u_i(t) \doteq \mathbf{u}_i^{\neg T}\mathbf{h}(t) = \mathbf{h}^T(t)\mathbf{u}_i^{\neg T} \tag{38.105}$$

Eqs. (36.54) and (38.98) yield

$$\begin{aligned} \int_0^t \mathbf{x}^{[j]}(\tau)u_i(\tau)d\tau &= \mathbf{X}^{j\neg T}\int_0^t \mathbf{h}(\tau)\mathbf{h}^T(\tau)\mathbf{u}_i^{\neg T}d\tau && (38.106) \\ &= \mathbf{X}^{j\neg T}(\mathbf{L}_o^T\mathbf{u}_i \,\vdots\, \mathbf{L}_1^T\mathbf{u}_i \ldots \mathbf{L}_{k-1}\mathbf{u}_i)\mathbf{P}_M h(t) = \mathbf{X}^{j\neg T}\mathbf{L}'\mathbf{P}_M\mathbf{h}(t) . \end{aligned}$$

Combining with Eq.(38.103) and defining $\mathbf{x}_o = (\mathbf{x}_o \ \ \mathbf{0} \ \ \mathbf{0} \ldots \mathbf{0})\mathbf{h}(t) \triangleq \mathbf{X}_0^{\neg}\mathbf{h}(t)$

$$\mathbf{X}^{\neg T} - \mathbf{X}_0^{\neg} = \sum_{j=1}^{r_A} \mathbf{A}_j \mathbf{X}^{j\neg T}\mathbf{P}_M + \sum_{i=1}^{m}\sum_{j=0}^{r_B} \mathbf{B}_{ij}\mathbf{X}^{j\neg T}\mathbf{L}'\mathbf{P}_M . \tag{38.107}$$

The matrices $\mathbf{X}^{\neg T}, \mathbf{X}_0^{\neg}, \mathbf{X}^{j\neg T}$ and $\mathbf{L}'$ are known from measurements, the unknown parameters in Eq. (38.107) are the coefficients in $\mathbf{A}_j$ and $\mathbf{B}_{ij}$. The number of unknowns is

$$\sum_{j=1}^{r_A} n\ n^j + m \sum_{j=0}^{r_B} n\ n^j \triangleq n\beta_c \tag{38.108}$$

and the number of equations is nk. If $k \geq \beta_c$ a solution exists. If k highly exceeds the aforementioned amount, a good result can be expected by linear regression. Obviously, a high number k (orthogonal function approximations) is required, small quantities n (order of process) and small r_A and r_B (nonlinear decomposition terms). The matrix equation Eq.(38.107) can be rewritten to a vector equation using Eq.(4.40)

$$\text{col}\ (\mathbf{A}^{(n\times n^j)}\ \mathbf{D}^{(n^j\times k)}) = \text{col}\ (\mathbf{I}_n\mathbf{A}\mathbf{D}) = (\mathbf{D}^T \otimes \mathbf{I}_n)\text{col}\ \mathbf{A} \tag{38.109}$$

where $\mathbf{A}$ and $\mathbf{D}$ correspond with $\mathbf{A}_j$ and $\mathbf{X}^{j\neg T}\mathbf{P}_M$ in Eq.(38.107), respectively. Thus,

$$\text{col}\ (\mathbf{X}^{\neg T} - \mathbf{X}_0^{\neg}) = \sum_{j=1}^{r_A}[(\mathbf{P}_M^T\mathbf{X}^{j\neg})\otimes\mathbf{I}_n]\text{col}\ \mathbf{A}_j + \sum_{i=1}^{m}\sum_{j=0}^{r_B}[(\mathbf{P}_M^T\mathbf{L}'^T\mathbf{X}^{j\neg})\otimes\mathbf{I}_n]\text{col}\ \mathbf{B}_{ij}\ . \tag{38.110}$$

Using the abbreviations

$$\mathbf{M} = [(\mathbf{P}_M^T\mathbf{X}^{1\neg})\otimes\mathbf{I}_n \vdots \ldots \vdots (\mathbf{P}_M^T\mathbf{X}^{r_A\neg})\otimes\mathbf{I}_n \vdots (\mathbf{P}_M^T\mathbf{L}'^T\mathbf{M}^{0\neg})\otimes\mathbf{I}_n \vdots \ldots \vdots (\mathbf{P}_M^T\mathbf{L}'^T\mathbf{X}^{r_B\neg})\otimes\mathbf{I}_n] \tag{38.111}$$

$$\mathbf{p} = \begin{pmatrix} \text{col}\ \mathbf{A}_1 \\ \text{col}\ \mathbf{A}_2 \\ \vdots \\ \text{col}\ \mathbf{A}_{r_A} \\ \text{col}\ \mathbf{B}_{10} \\ \vdots \\ \text{col}\ \mathbf{B}_{1r_B} \\ \text{col}\ \mathbf{B}_{20} \\ \vdots \\ \text{col}\ \mathbf{B}_{mr_B} \end{pmatrix} \qquad \text{and} \qquad \mathbf{x}_c = \text{col}\ (\mathbf{X}^{\neg T} - \mathbf{X}_o^{\neg}) \tag{38.112}$$

Eq.(38.110) can be put into the form $\mathbf{M}\mathbf{p} = \mathbf{x}_c$ or $\mathbf{M}\mathbf{p} = \mathbf{x}_c - \boldsymbol{\varepsilon}$. By linear regression method Eq.(D.32), the solution is

$$\mathbf{p}^* = \mathbf{M}^{\sharp L}\mathbf{x}_c\ . \tag{38.113}$$

In the previous expressions the following dimensions occur: $\mathbf{p} \in \mathcal{R}^{\beta_c}$,

$$\mathbf{X}^{\neg T} \in \mathcal{R}^{n\times k}, \quad \mathbf{X}^{j\neg T} \in \mathcal{R}^{n^j\times k}, \quad \mathbf{A}_j \in \mathcal{R}^{n\times n^j}, \quad \mathbf{P}_M \in \mathcal{R}^{k\times k}, \quad \mathbf{x}_c,\ \text{col}\ \mathbf{X}^{\neg T} \in \mathcal{R}^{kn\times 1}, \tag{38.114}$$

$$\mathbf{L}',\mathbf{L}_j \in \mathcal{R}^{k\times k}, \quad \dim \mathbf{M} = kn\times(n^2+n^3+\ldots+n^{r_A+1}+mn+mn^2+\ldots+mn^{r_B+1}) = kn\times\beta_c\ . \tag{38.115}$$

38.3.6 Identification of the Output Matrix C(x)

Replacing the linear representation

$$\mathbf{y}(t) = \mathbf{C}\mathbf{x}(t) \qquad \mathbf{y} \in \mathcal{R}^r \qquad \mathbf{x}(t) \in \mathcal{R}^n \tag{38.116}$$

by the nonlinear expression with truncation r_C, one has

$$\mathbf{y}(t) = \mathbf{c}[\mathbf{x}(t)] = \sum_{j=1}^{\infty} \mathbf{C}_j \mathbf{x}^{[j]} \doteq \sum_{j=1}^{r_C} \mathbf{C}_j \mathbf{x}^{[j]} \doteq \mathbf{Y}^{\neg T}\mathbf{h}(t) = \sum_{j=1}^{r_C} \mathbf{C}_j \mathbf{X}^{j\neg T}\mathbf{h}(t) \tag{38.117}$$

$$\text{col } \mathbf{Y}^{\neg T} = (\mathbf{X}^{1\neg} \otimes \mathbf{I}_r \vdots \mathbf{X}^{2\neg} \otimes \mathbf{I}_r \ldots \mathbf{X}^{r_C\neg} \otimes \mathbf{I}_r) \begin{pmatrix} \text{col } \mathbf{C}_1 \\ \text{col } \mathbf{C}_2 \\ \vdots \\ \text{col } \mathbf{C}_{r_C} \end{pmatrix} \tag{38.118}$$

$$\mathbf{y} \in \mathcal{R}^r, \quad \mathbf{C}_j \in \mathcal{R}^{r\times(n^j)}, \quad \mathbf{h}(t) \in \mathcal{R}^k, \quad \mathbf{Y}^{\neg T} \in \mathcal{R}^{r\times k}, \quad \mathbf{X}^{1\neg T} \in \mathcal{R}^{n\times k}, \quad \mathbf{X}^{j\neg T} \in \mathcal{R}^{n^j\times k} \tag{38.119}$$

$$\dim\,(\mathbf{X}^{1\neg} \otimes \mathbf{I}_r \vdots \mathbf{X}^{2\neg} \otimes \mathbf{I}_r \ldots \mathbf{X}^{r_C\neg} \otimes \mathbf{I}_r) = kr \times r\sum_{i=1}^{r_C} n^i . \tag{38.120}$$

If $\mathbf{x}(t)$ and $\mathbf{y}(t)$ are known by measurements of adequate number $k > \sum_{i=1}^{r_C} n^i$, the solution can be achieved by regression method as outlined before.

Identification algorithms based on the linear least squares methods of Cholesky decomposition, singular-value decomposition etc. are applied to nonlinear systems by *Chen, S., et al. 1989*. Severely nonlinear systems are presented by *Billings, S.A., and Chen, S., 1989*.

38.4 Distributed Linear Time-Varying Systems and Laguerre Polynomials

Expanding linear partial differential equations in the two-variable function $y(x,t)$, double Laguerre series play a key role, i.e.,

$$y(x,t) = \sum_{i=0}^{\infty}\sum_{j=0}^{\infty} y_{ij} v_j(x) h_i(t) \tag{38.121}$$

where $v_j(x)$ and $h_i(t)$ form an orthogonal basis over space $x \in [0,\infty)$ and time $t \in [0,\infty)$, respectively. The coefficients y_{ij} of the double Laguerre expansion are calculated as

$$y_{ij} = \int_0^{\infty}\int_0^{\infty} y(x,t) v_j(x) h_i(t) dx\, dt\ . \tag{38.122}$$

Defining the truncated vectors $\mathbf{v}(x), \mathbf{h}(t)$ and the matrix $\mathbf{Y}$ of the expansion coefficients y_{ij} , the approximation is given by

$$\mathbf{v}(x) = \Big(v_o(x) \;\; v_1(x) \ldots v_{l-1}(x)\Big)^T \qquad \mathbf{h}(t) = \Big(h_o(t) \;\; h_1(t) \ldots h_{k-1}(t)\Big)^T \tag{38.123}$$

$$\mathbf{Y} = \begin{pmatrix} y_{00} & y_{01} & \cdots & y_{0,k-1} \\ y_{10} & \ddots & \cdots & \vdots \\ \vdots & \vdots & \ddots & \vdots \\ y_{l-1,0} & y_{l-1,1} & \cdots & y_{l-1,k-1} \end{pmatrix} \qquad y(x,t) \doteq \mathbf{v}^T(x)\mathbf{Y}\mathbf{h}(t)\ . \tag{38.124}$$

Integrating p times with respect to x and q times with respect to t yields (*Ranganathan, V., 1984 and 1986*)

$$\int_o^x \overset{p}{\cdots} \int_o^x \int_o^t \overset{q}{\cdots} \int_o^t \mathbf{v}^T(x)\mathbf{Y}\mathbf{h}(t)\, dt \ldots dt\, dx \ldots dx \doteq \mathbf{v}^T(\mathbf{P}_M^T)^p\, \mathbf{Y}\, (\mathbf{P}_M)^q\, \mathbf{h}\,. \tag{38.125}$$

Example: Laguerre polynomials bi-diagonal operational matrix. Laguerre polynomials h_i are given by $h_i(t) = \sum_{j=0}^{i} \frac{(-1)^j}{j!}\binom{i}{j} t^j$. Hence, for $k = 4$

$$\mathbf{h}(t) = \begin{pmatrix} h_0(t) \\ h_1(t) \\ h_2(t) \\ h_3(t) \end{pmatrix} = \begin{pmatrix} 1 \\ 1-t \\ 1-2t+0.5t^2 \\ 1-3t+1.5t^2-0.167t^3 \end{pmatrix} \qquad \mathbf{P}_M = \begin{pmatrix} 1 & -1 & 0 & 0 \\ 0 & 1 & -1 & 0 \\ 0 & 0 & 1 & -1 \\ 0 & 0 & 0 & 1 \end{pmatrix} \quad \text{(bi-diagonal)} \tag{38.126}$$

$$\int_0^t \mathbf{h}(t)dt = \begin{pmatrix} t \\ t-0.5t^2 \\ t-t^2+0.167t^3 \\ t-1.5t^2+0.5t^3-0.041t^4 \end{pmatrix} \doteq \mathbf{P}_M\mathbf{h}(t) = \begin{pmatrix} t \\ t-0.5t^2 \\ t-t^2-0.167t^3 \\ 1-3t+1.5t^2-0.167t^3 \end{pmatrix} . \;\square \tag{38.127}$$

38.4.1 Analysis

Consider the first-order distributed parameter system with time-varying coefficients

$$\frac{\partial}{\partial t}y(x,t) + A_n(t)\frac{\partial}{\partial x}y(x,t) + B_n(t)y(x,t) = C_n(t)u(x,t) \tag{38.128}$$

$$A_n(t) = \sum_0^n a_i t^i, \quad B_n(t) = \sum_0^n b_i t^i, \quad C_n(t) = \sum_0^n c_i t^i\,. \tag{38.129}$$

For a given input $u(x,t)$, the output $y(x,t)$ is obtained by applying the expansions outlined in the preceding subsection

$$\frac{\partial}{\partial t}\mathbf{v}^T\mathbf{Y}\mathbf{h} + A_n\frac{\partial}{\partial x}\mathbf{v}^T\mathbf{Y}\mathbf{h} + B_n\mathbf{v}^T\mathbf{Y}\mathbf{h} = C_n\mathbf{v}^T\mathbf{U}\mathbf{h}\,. \tag{38.130}$$

Now, use is made of the definitions $\frac{\partial \mathbf{v}^T}{\partial x} \triangleq \mathbf{v}^T\mathbf{D}_x$ and $\frac{\partial \mathbf{h}}{\partial t} \triangleq \mathbf{D}_t\mathbf{h}$, see Eq.(38.49), comprising operational matrices $\mathbf{D}_x$ and $\mathbf{D}_t$ for differentiation. Referring to Eq.(4.40), one has

$$\mathbf{v}^T\mathbf{Y}\mathbf{D}_t\mathbf{h} + \mathbf{v}^T A_n\mathbf{D}_x\mathbf{Y}\mathbf{h} + \mathbf{v}^T B_n\mathbf{Y}\mathbf{h} = \mathbf{v}^T C_n\mathbf{U}\mathbf{h} \tag{38.131}$$

$$\mathbf{Y}\mathbf{D}_t + A_n\mathbf{D}_x\mathbf{Y} + B_n\mathbf{Y} = C_n\mathbf{U} \tag{38.132}$$

$$\text{col}\,\mathbf{Y} = [\mathbf{D}_t^T \otimes \mathbf{I} + \mathbf{I} \otimes (A_n\mathbf{D}_x + B_n\mathbf{I})]^{-1}\text{col}(C_n\mathbf{U})\,. \tag{38.133}$$

38.4.2 Identification

Considering Eq.(38.128) again, the problem is to estimate the parameters a_i, b_i, c_i from measurements of the input and output. The estimation of the initial and boundary conditions can be included in this calculation. Before integrating Eq.(38.128) once with respect to t and once with respect to x, consider the term $A_n(t)\frac{\partial}{\partial x}y(x,t)$ and its integral with respect to x and t. Use is also made of the integration by parts. Then,

$$\int_0^t \int_0^x A_n(t)\frac{\partial}{\partial x}y(x,t)dx\,dt = \int_0^t A_n(t)y(x,t)dt = A_n(t)y(x,t) - \int_0^t y(x,t)\frac{dA_n(t)}{dt}dt \tag{38.134}$$

Integrating Eq.(38.128) as a whole

$$\int_0^x y(x,t)dx + A_n(t)y(x,t) - \int_0^t y(x,t)\frac{dA_n(t)}{dt}dt + \int_0^x\int_0^t B_n(t)y(x,t)dt\,dx = \int_0^x\int_0^t C_n(t)u(x,t)dtdx \,. \tag{38.135}$$

In this equation the terms can be further developed as follows.

First term: $\int_0^x y(x,t)dx = \mathbf{v}^T\mathbf{P}_M^T\mathbf{Y}\mathbf{h}$.

Second term comprising the following subterms:

$$a_o y(x,t) = a_o \mathbf{v}^T\mathbf{Y}_o\mathbf{h} \qquad \mathbf{Y}_o \equiv \mathbf{Y}, \qquad a_i t^i y(x,t) = a_i\mathbf{v}^T\mathbf{Y}_i\mathbf{h} \tag{38.136}$$

$$A_n(t)y(x,t) = \sum_0^n a_i\mathbf{v}^T\mathbf{Y}_i\mathbf{h} = \mathbf{v}^T(\sum_0^n a_i\mathbf{Y}_i)\mathbf{h} \,. \tag{38.137}$$

Third term:

$$\int_0^t y(x,t)\frac{dA_n(t)}{dt}dt = \int_0^t (0a_o + a_1 + 2ta_2 + \ldots + nt^{n-1}a_n)y(x,t)dt \tag{38.138}$$

$$= \int_0^t \sum_0^n ia_it^{i-1}y(x,t)dt = \sum_0^n ia_i\int_0^t t^{i-1}y(x,t)dt \tag{38.139}$$

$$= \sum_0^n ia_i\mathbf{v}^T\mathbf{Y}_{i-1}\mathbf{P}_M\mathbf{h} = \mathbf{v}^T(\sum_0^n a_i i\mathbf{Y}_{i-1}\mathbf{P}_M)\mathbf{h} \,. \tag{38.140}$$

Fourth term: $\int_0^x\int_0^t B_n(t)y(x,t)dt\,dx = \mathbf{v}^T\mathbf{P}_M^T\sum_0^n b_i\mathbf{Y}_i\mathbf{P}_M\mathbf{h}$.

Combining the second and third term yields $\mathbf{v}^T[\sum_0^n a_i(\mathbf{Y}_i - i\mathbf{Y}_{i-1}\mathbf{P}_M)]\mathbf{h}$.

Applying the col operation of Eq.(4.40) to the bracket yields

$$\text{col}\sum_0^n a_i(\mathbf{Y}_i - i\mathbf{Y}_{i-1}\mathbf{P}_M) = \left(\text{col}\,\mathbf{Y}_o \,\vdots\, \text{col}(\mathbf{Y}_1 - \mathbf{Y}_o\mathbf{P}_M) \,\vdots\, \text{col}(\mathbf{Y}_2 - 2\mathbf{Y}_1\mathbf{P}_M) \ldots \vdots\, \text{col}(\mathbf{Y}_n - n\mathbf{Y}_{n-1}\mathbf{P}_M)\right)\mathbf{a} \triangleq -\mathbf{M}_a a \,. \tag{38.141}$$

Finally, from Eq.(38.135) an equivalence in the parentheses and bracket terms is given

$$\text{col}(\mathbf{P}_M^T\mathbf{Y}) = \mathbf{M}_a\mathbf{a} + \mathbf{M}_b\mathbf{b} + \mathbf{M}_c\mathbf{c} \tag{38.142}$$

where
$$-\mathbf{M}_b \triangleq \left(\text{col}(\mathbf{P}_M^T\mathbf{Y}_o\mathbf{P}_M) \,\vdots\, \text{col}(\mathbf{P}_M^T\mathbf{Y}_1\mathbf{P}_M) \,\vdots\, \ldots \,\vdots\, \text{col}(\mathbf{P}_M^T\mathbf{Y}_n\mathbf{P}_M)\right) \tag{38.143}$$

$$\mathbf{M}_c \triangleq \left(\text{col}(\mathbf{P}_M^T\mathbf{U}_o\mathbf{P}_M) \,\vdots\, \text{col}(\mathbf{P}_M^T\mathbf{U}_1\mathbf{P}_M) \,\vdots\, \ldots \,\vdots\, \text{col}(\mathbf{P}_M^T\mathbf{U}_n\mathbf{P}_M)\right) \tag{38.144}$$

$$\mathbf{a} = (a_o\; a_1 \ldots a_n)^T, \qquad \mathbf{b} = (b_o\; b_1 \ldots b_n)^T, \qquad \mathbf{c} = (c_o\; c_1 \ldots c_n)^T \,. \tag{38.145}$$

Hence, the identification result is

$$(\mathbf{M}_a \,\vdots\, \mathbf{M}_b \,\vdots\, \mathbf{M}_c)\begin{pmatrix}\mathbf{a}\\ \mathbf{b}\\ \mathbf{c}\end{pmatrix} = \text{col}(\mathbf{P}_M^T\mathbf{Y}) \tag{38.146}$$

$$\begin{pmatrix}\mathbf{a}\\ \mathbf{b}\\ \mathbf{c}\end{pmatrix}^* = (\mathbf{M}_a \,\vdots\, \mathbf{M}_b \,\vdots\, \mathbf{M}_c)^{\sharp L}\,\text{col}(\mathbf{P}_M^T\mathbf{Y}) \,. \tag{38.147}$$

Owing to the bi-diagonal properties of the Laguerre polynomials, the operational matrix for integration (*Hwang, C., and Shih, Y.P., 1982a*) can easily be derived by recursive relations, see *Ranganathan, V., et al. 1987*. The method can also be extended to higher order systems.

38.5 Component Detection Via Robust Observation

Based on the so-called Beard-Jones filters (*Beard, R.V., 1971*), robust observation can be applied to the detection of specified component failures.

Ge, W., and Fang, C.Z., 1988 suggest a bank of detection observers each of which is sensitive to only one component failure $d_i(t)$ of the failure signal $\mathbf{d}(t) \in \mathcal{R}^q$ but insensitive to all others. *Ding, X., and Frank, P.M., 1990* presented a scheme for localization and identification of component failures $\mathbf{d}(t)$, based on the orthogonal function series representation.

Consider the multivariable system where $\mathbf{d}(t)$ represents a system failure signal. An observer with observer gain matrix $\mathbf{L}$ is implemented to reconstruct the state vector given by the estimate $\mathbf{f}(t)$. The output error is termed $\mathbf{e}(t)$. The plant is given by

$$\dot{\mathbf{x}}(t) = \mathbf{A}\mathbf{x}(t) + \mathbf{B}\mathbf{u}(t) + \mathbf{D}_d\mathbf{d}(t), \quad \mathbf{B} \in \mathcal{R}^{n\times m}, \quad \mathbf{y}(t) = \mathbf{C}\mathbf{x}(t)\ . \tag{38.148}$$

Using a bank of detection observers and combining with Eq.(38.148) yields

$$\dot{\mathbf{z}}_l(t) = \mathbf{F}_l\mathbf{z}_l(t) + \mathbf{G}_l\mathbf{u}(t) + \mathbf{L}_l\mathbf{y}(t) \tag{38.149}$$

$$\mathbf{w}_l(t) = \mathbf{K}_l\mathbf{z}(t) + \mathbf{S}_l\mathbf{y}(t)\ . \tag{38.150}$$

With the help of the relation $\mathbf{K}_l\mathbf{T}_l + \mathbf{S}_l\mathbf{C} = \mathbf{H}_l\mathbf{C}$, the state estimation error $\mathbf{f}_l(t)$ and the output estimation error $\mathbf{e}_l(t)$ are given by

$$\dot{\mathbf{f}}_l(t) = \mathbf{F}_l\mathbf{f}_l(t) + \mathbf{T}_l\mathbf{D}_d\mathbf{d}(t) \qquad \mathbf{e}_l(t) = \mathbf{K}_l\mathbf{f}_l(t)\ . \tag{38.151}$$

If the error $\mathbf{e}_l(t)$ of the lth observer is only sensitive to the lth component of $\mathbf{d}(t)$ and robust (insensitive) to all the other components, then the component failures are isolated.

Applying orthogonal decomposition and reducing the observer equations yields an expression of the following structure

$$\text{col}\ \mathbf{E}^{\cdot T} = \mathbf{\Omega}[(\mathbf{I} \otimes \mathbf{d}_{d1})\mathbf{R}_1 + \ \dots\ + (\mathbf{I} \otimes \mathbf{d}_{dq})\mathbf{R}_q] \tag{38.152}$$

where $\mathbf{d}_{dl}$ is the lth column of $\mathbf{D}_d$. The error $\mathbf{e}_l$ can be localized if a filter is constructed using premultiplication by a vector $\mathbf{v}_i$ such that for

$$i \neq l \qquad \mathbf{v}_i\mathbf{\Omega}(\mathbf{I} \otimes \mathbf{d}_{dl})\mathbf{R}_l = \mathbf{0} \qquad i,l = 1,2....q\ . \tag{38.153}$$

The vector $\mathbf{v}_i$ is chosen such that it is an element both of the kernel of the matrix expression $\sum_{l\neq i} \mathbf{\Omega}(\mathbf{I} \otimes \mathbf{d}_{dl})\mathbf{R}_l$ and of the range space of $\mathbf{\Omega}(\mathbf{I} \otimes \mathbf{d}_{dl})\mathbf{R}_l$. Finally, as shown by *Ding, X., and Frank, P.M., 1990* the design of the filter does *not* depend on $\mathbf{L}$. Hence, the design of the filter, as given by the premultiplication with $\mathbf{v}_i$, and the observer design, i.e. determining $\mathbf{L}$, are separable.

A procedure similar to the algorithm outlined above is given by *Noisser, R., 1976*; *Müller, P.C., and Lückel, J., 1977* maintaining disturbance rejection including estimation and counteraction.

38.6 Robustness Assessment Using Orthogonal Decomposition

Consider the autonomous unperturbed and perturbed system, respectively, namely $\dot{\mathbf{x}}(t) = \mathbf{A}\mathbf{x}(t)$ and $\dot{\mathbf{x}}_p(t) = \mathbf{A}_p\mathbf{x}_p(t)$ where $\mathbf{A}_p = \mathbf{A} + \Delta\mathbf{A}$. Referring to Eq.(36.64),

$$\mathbf{X}^{\cdot T} - \mathbf{A}\mathbf{X}^{\cdot T}\mathbf{P}_M = \mathbf{X}_o^{\cdot T} \quad \text{or} \quad \text{col}\ \mathbf{X}^{\cdot T} = (\mathbf{I} - \mathbf{P}_M^T \otimes \mathbf{A})^{-1}\text{col}\ \mathbf{X}_o^{\cdot T} \tag{38.154}$$

$$\mathbf{X}_p^{\cdot T} - (\mathbf{A} + \Delta\mathbf{A})\mathbf{X}_p^{\cdot T}\mathbf{P}_M = \mathbf{X}_o^{\cdot T}\ . \tag{38.155}$$

Defining $\mathbf{X}_p^{\neg T} \triangleq \mathbf{X}^{\neg T} + \Delta\mathbf{X}^{\neg T}$ and subtracting both equations above, i.e., $\mathbf{X}_p^{\neg T} - \mathbf{X}^{\neg T}$ yields

$$\Delta\mathbf{X}^{\neg T} - \mathbf{A}\Delta\mathbf{X}^{\neg T}\mathbf{P}_m - \Delta\mathbf{A}\ \mathbf{X}^{\neg T}\mathbf{P}_M - \Delta\mathbf{A}\ \Delta\mathbf{X}^{\neg T}\mathbf{P}_M = \mathbf{0} \tag{38.156}$$

$$\mathrm{col}\Delta\mathbf{X}^{\neg T} - (\mathbf{P}_M^T \otimes \mathbf{A})\mathrm{col}\Delta\mathbf{X}^{\neg T} - (\mathbf{P}_M^T \otimes \Delta\mathbf{A})\mathrm{col}\ \mathbf{X}^{\neg T} - (\mathbf{P}_M^T \otimes \Delta\mathbf{A})\mathrm{col}\ \mathbf{X}^{\neg T} = \mathbf{0} \tag{38.157}$$

$$[\mathbf{I} - (\mathbf{P}_M^T \otimes \mathbf{A}) - (\mathbf{P}_M^T \otimes \Delta\mathbf{A})]\mathrm{col}\Delta\mathbf{X}^{\neg T} = (\mathbf{P}_M^T \otimes \Delta\mathbf{A})(\mathbf{I} - \mathbf{P}_M^T \otimes \mathbf{A})^{-1}\mathrm{col}\mathbf{X}_o^{\neg T}\ . \tag{38.158}$$

Using Eqs.(24.11) and (24.16), to avoid singularity

$$\det\{(\mathbf{I} - \mathbf{P}_M^T \otimes \mathbf{A})[\mathbf{I} - (\mathbf{I} - \mathbf{P}_M^T \otimes \mathbf{A})^{-1}(\mathbf{P}_M^T \otimes \Delta\mathbf{A})]\} \neq 0\ . \tag{38.159}$$

Since $\mathbf{A}$ is considered stable, the determinant of the second term must not vanish in order to guarantee stability of the perturbed system, i.e.,

$$\sigma_{\min}[\mathbf{I} - (\mathbf{I} - \mathbf{P}_M^T \otimes \mathbf{A})^{-1}(\mathbf{P}_M^T \otimes \Delta\mathbf{A})] > 0 \tag{38.160}$$

$$1 - \sigma_{\max}[(\mathbf{I} - \mathbf{P}_M^T \otimes \mathbf{A})^{-1}(\mathbf{P}_M^T \otimes \Delta\mathbf{A})] > 0 \tag{38.161}$$

$$1 > \sigma_{\max}[(\mathbf{I} - \mathbf{P}_M^T \otimes \mathbf{A})^{-1}]\sigma_{\max}[(\mathbf{P}_M^T \otimes \Delta\mathbf{A})] \tag{38.162}$$

$$\sigma_{\max}[(\mathbf{P}_M^T \otimes \Delta\mathbf{A})] < \sigma_{\min}[\mathbf{I} - \mathbf{P}_M^T \otimes \mathbf{A}]\ . \tag{38.163}$$

Referring to Eqs.(4.21) and (22.69) (*Weinmann, A., 1991*),

$$\sigma_{\max}[\Delta\mathbf{A}] < \frac{\sigma_{\min}[\mathbf{I} - \mathbf{P}_M^T \otimes \mathbf{A}]}{\sigma_{\max}[\mathbf{P}_M]}\ . \tag{38.164}$$

Appendix A

Matrix Algebra and Control

Boldface lower case letters, e.g., **a** or **b**, denote vectors, boldface capital letters, e.g., **A**,**M**, denote matrices. A vector is a column matrix. Containing m elements (entries) it is referred to as an m-vector. The number of rows and columns of a matrix **A** is n and m, respectively. Then, **A** is an (n,m)-matrix or $n \times m$-matrix (dimension $n \times m$).

The matrix **A** is called positive or non-negative if $\mathbf{A} >_e \mathbf{0}$ or $\mathbf{A} \geq_e \mathbf{0}$, respectively, i.e., if the elements are real, positive and non-negative, respectively.

A.1 Matrix Multiplication

Two matrices **A** and **B** may only be multiplied, $\mathbf{C} = \mathbf{AB}$, if they are conformable. **A** has size $n \times m$, **B** $m \times r$, **C** $n \times r$. Two matrices are conformable for multiplication if the number m of columns of the first matrix **A** equals the number m of rows of the second matrix **B**. Kronecker matrix products do not require conformable multiplicands.

The elements or entries of the matrices are related as follows $C_{ij} = \sum_{\nu=1}^{m} A_{i\nu} B_{\nu j} \quad \forall i = 1 \ldots n,\ j = 1 \ldots r$. The jth column vector $\mathbf{C}_{.j}$ of the matrix **C** as denoted in Eq.(A.13) can be calculated from the columns $\mathbf{A}_{.\nu}$ and the entries $B_{\nu j}$ by the following relation; the jth row $\mathbf{C}_{j.}$ from rows $\mathbf{B}_{\nu.}$ and $A_{j\nu}$:

$$\text{column } \mathbf{C}_{.j} = \sum_{\nu=1}^{m} \mathbf{A}_{.\nu} B_{\nu j} \ , \qquad \text{row } \mathbf{C}_{j.} = (\mathbf{C}^T)_{.j} = \sum_{\nu=1}^{m} A_{j\nu} \mathbf{B}_{\nu.} \ . \tag{A.1}$$

A matrix product, e.g., $\mathbf{AB} = \begin{pmatrix} a & a \\ ca & ca \end{pmatrix} \begin{pmatrix} b & -b \\ -b & b \end{pmatrix} = \mathbf{O}$, may be zero although neither multiplicand **A** nor multiplicator **B** is zero. Without **A** or **B** being the nullmatrix the product **AB** only vanishes if both **A** and **B** are singular. The matrix $\mathbf{B} = \mathbf{A}^{\perp}$ is the (right) annihilator of **A**, i.e., $\mathbf{AA}^{\perp} = 0$.

A.2 Properties of Matrix Operations

Distributivity: $\mathbf{A}(\mathbf{B}+\mathbf{C}) = \mathbf{AB} + \mathbf{AC}$. Associativity of addition $(\mathbf{A}+\mathbf{B})+\mathbf{C} = \mathbf{A}+(\mathbf{B}+\mathbf{C})$ and multiplication $(\mathbf{AB})\mathbf{C} = \mathbf{A}(\mathbf{BC})$. Commutativity of addition, non-commutativity of multiplication and raise to higher powers:

$$\mathbf{A}+\mathbf{B} = \mathbf{B}+\mathbf{A} \ , \qquad \mathbf{AB} \neq \mathbf{BA}, \qquad (\mathbf{AB})^i \neq \mathbf{A}^i\mathbf{B}^i \ . \tag{A.2}$$

Exceptions: Consider, first, a multivariable control with transfer matrix $\mathbf{G}(s)$ in the forward path and unity feedback and, second, **H** in the forward path and **F** in the feedback where $\mathbf{G} = \mathbf{FH}$. The overall transfer matrix is given by

$$(\mathbf{I}+\mathbf{G})^{-1}\mathbf{G} = \mathbf{G}(\mathbf{I}+\mathbf{G})^{-1} \quad \text{and} \quad (\mathbf{I}+\mathbf{FH})^{-1}\mathbf{FH} = \mathbf{FH}(\mathbf{I}+\mathbf{FH})^{-1} \ . \tag{A.3}$$

The inverse of the return-difference matrix and **G** commute unexpectedly. Another exceptional case is the product **A** exp $(\mathbf{A}t)$, i.e. the coefficient matrix and the state transition matrix. Finally, suppose **A** and **B** nonsingular. Then, **A** and **B** commute if their product is the identity matrix: Both $\mathbf{AB} = \mathbf{I}$ and $\mathbf{BA} = \mathbf{I}$ yield $\mathbf{A} = \mathbf{B}^{-1}$. Generally, the matrices **A** and **B** commute with respect to multiplication if **B** is a function of **A** , e.g. as given by a matrix polynomial or by the decomposition in Eq.(A.45).

Note that $\mathbf{F}(\mathbf{I}+\mathbf{HF})^{-1} = (\mathbf{I}+\mathbf{FH})^{-1}\mathbf{F}$ and $\mathbf{F}(\mathbf{I}+\mathbf{HF})^{-1}\mathbf{H} = (\mathbf{I}+\mathbf{FH})^{-1}\mathbf{FH}$, particularly observe the change of order within the parentheses. □

Properties when transposing or inverting a matrix product:

$$(\mathbf{Ab})^T = \mathbf{b}^T\mathbf{A}^T , \quad (\mathbf{AB})^T = \mathbf{B}^T\mathbf{A}^T , \quad (\mathbf{AB})^{-1} = \mathbf{B}^{-1}\mathbf{A}^{-1} . \tag{A.4}$$

Inverse and transpose operations (symbols) may be permuted: $(\mathbf{A}^T)^{-1} = (\mathbf{A}^{-1})^T$. If $\mathbf{A}^{-1} = \mathbf{A}^T$ is true then $\mathbf{A}$ is referred to as an orthogonal matrix.

An idempotent matrix $\mathbf{A}$ has the property $\mathbf{A}^2 = \mathbf{A}$. This result can be observed in least squares, estimation and sliding mode theory, e.g.,

$$\mathbf{A} = \mathbf{I} - \mathbf{M}(\mathbf{M}^T\mathbf{M})^{-1}\mathbf{M}^T \quad \text{or} \quad \mathbf{A} = \mathbf{I} - \mathbf{B}(\mathbf{C}_s\mathbf{B})^{-1}\mathbf{C}_s . \tag{A.5}$$

A matrix $\mathbf{A}$ is nilpotent if $\mathbf{A}^k = \mathbf{0}$ for some k. Such a matrix appears in the case of the state-space representation of a k-tuples integrator.

A.3 Diagonal Matrices

A diagonal matrix $\mathbf{A}$ is a square matrix with non-zero entries A_{ii} in the main diagonal, only, e.g.,

$$\mathrm{diag}_n A_i = \begin{pmatrix} A_1 & 0 & 0 & \dots \\ 0 & A_2 & 0 & \dots \\ 0 & 0 & A_3 & \dots \\ \vdots & \vdots & \vdots & \ddots \end{pmatrix} . \tag{A.6}$$

If these entries A_{ii} are equal to each other $\mathbf{A}$ is a scalar matrix. The identity matrix $\mathbf{I}_n$ is a scalar matrix with elements 1 and dimension $n \times n$: $\mathbf{I}_n \triangleq \mathrm{diag}_n(1,1,1,\dots,1)$, $\mathbf{I}_n \in \mathcal{R}^{n\times n}$.

Given a rectangular (n,r)-matrix $\mathbf{B}$, premultiplying $\mathbf{B}$ by the identity matrix $\mathbf{I}_n$ or postmultiplying $\mathbf{B}$ by $\mathbf{I}_r$ yields $\mathbf{I}_n\mathbf{B} = \mathbf{B}$ or $\mathbf{B}\mathbf{I}_r = \mathbf{B}$.

Premultiplying [postmultiplying] a matrix $\mathbf{A}$ by a diagonal matrix yields

$$(\mathrm{diag}\ d_i)\mathbf{A} = \begin{pmatrix} d_1A_{11} & d_1A_{12} & \dots \\ d_2A_{21} & d_2A_{22} & \dots \\ \vdots & \vdots & \ddots \end{pmatrix} \qquad \mathbf{A}(\mathrm{diag}\ d_i) = \begin{pmatrix} d_1A_{11} & d_2A_{12} & \dots \\ d_1A_{21} & d_2A_{22} & \dots \\ \vdots & \vdots & \ddots \end{pmatrix} , \tag{A.7}$$

i.e., a new matrix the rows [columns] of which are successively multiplied (scaled) by d_i (i.e., i-th row [column] with d_i).

A.4 Triangular Matrices

A lower triangular matrix is a square matrix having all elements zero above the main diagonal, an upper triangular matrix only contains zero elements below the main diagonal. The product of two triangular matrices produces a triangular one again.

If $\mathbf{A}$ is given as a diagonal matrix or an upper or lower triangular matrix, the eigenvalues $\lambda[\mathbf{A}]$ are already given by the entries in the main diagonal A_{ii} $\forall i = 1...n$.

A.5 Column Matrices (Vectors) and Row Matrices

The unit m-vector with k-th component 1 is termed $\mathbf{e}_k$. Defining this m-vector $\mathbf{e}_k$ and the n-vector $\mathbf{e}_i$, the elementary matrix or Kronecker matrix $\mathbf{E}_{ik}$ is given by the dyadic product

$$\mathbf{e}_k = (0,0,\dots 1,\dots,0,0)^T = \mathbf{e}_k^{(m\times 1)} = (\mathbf{I}_m)_{.k} \qquad \mathbf{E}_{ik} = \mathbf{e}_i\mathbf{e}_k^T = \mathbf{E}_{ik}^{(n\times m)} \qquad \mathbf{E}_{ik} \in \mathcal{R}^{n\times m} \tag{A.8}$$

as an (n,m)-matrix with entry 1 only in the i,k-element and zero elsewhere. Thus, the (n,m)-matrix $\mathbf{A}$ can be established element by element: $\mathbf{A} = \sum_{i=1}^n \sum_{k=1}^m A_{ik}\mathbf{E}_{ik}$ where $\mathbf{A} \in \mathcal{R}^{n\times m}$ or $\dim \mathbf{A} = n\times m$.

The identity matrix can be achieved by the sum $\mathbf{I}_n = \sum_{i=1}^n \mathbf{e}_i\mathbf{e}_i^T = \sum_{i=1}^n \mathbf{E}_{ii}$. The sum vector with elements throughout unity $\mathbf{1} = (1,1,\dots,1)^T$ serves as a summation operator for an m-vector: $\mathbf{1}^T\mathbf{a} = \sum_1^m a_i$.

The inner product of two vectors **a** and **b** is a scalar $\mathbf{a}^T\mathbf{b} = \mathbf{b}^T\mathbf{a}$. Orthogonal vectors have zero inner product.

Assume a vector output signal **y** given by the linear combination of a vector input signal **x** governed by the transfer equation

$$\mathbf{y} = \mathbf{C}\mathbf{x} \quad \text{where} \quad \mathbf{y} \in \mathcal{R}^r, \quad \mathbf{x} \in \mathcal{R}^n, \quad \mathbf{C} \in \mathcal{R}^{r \times n} . \tag{A.9}$$

The entry C_{ij} of **C** is considered as an operational factor from the input component x_j to the output component y_i. Note that the output subscript i (effect) is written first and j (source) is written second.

A partitioned vector **u** is denoted by the "vec" symbol

$$\mathbf{u} = \begin{pmatrix} \mathbf{u}_1 \\ \mathbf{u}_2 \\ \vdots \\ \mathbf{u}_N \end{pmatrix} = \mathrm{vec}\,(\mathbf{u}_1, \mathbf{u}_2, \ldots \mathbf{u}_N) = \mathrm{vec}\,\mathbf{u}_i = (\mathbf{u}_1^T, \mathbf{u}_2^T, \ldots \mathbf{u}_N^T)^T \quad \mathbf{u}_i \in \mathcal{R}^{m_i}, \quad \mathbf{u} \in \mathcal{R}^m, \quad m = \sum_i m_i . \tag{A.10}$$

The norm or length of **a** is the distance to the nullvector (origin) and is defined by the Frobenius norm $||\mathbf{a}||_F = \sqrt{\mathbf{a}^T\mathbf{a}}$. The inner product is always smaller than the product of the norms of the multiplicators (Schwartz inequality): $|\mathbf{a}^T\mathbf{b}| \leq ||\mathbf{a}||_F||\mathbf{b}||_F$.

Triangle inequality (for any kind of norm): $||\mathbf{a} + \mathbf{b}|| \leq ||\mathbf{a}|| + ||\mathbf{b}||$.

The angle θ between two vectors **u** and **v** is defined by $\cos\theta = \mathbf{u}^T\mathbf{v}/(||\mathbf{u}||_F||\mathbf{v}||_F)$.

Mapping a matrix **A**

$$\mathbf{A} = \begin{pmatrix} A_{11} & \ldots & A_{1n} \\ \vdots & \ddots & \vdots \\ A_{m1} & \ldots & A_{mn} \end{pmatrix} = \mathrm{matrix}[A_{ij}] \qquad \mathbf{A} \in \mathcal{R}^{m \times n}, \qquad \mathbf{a} \in \mathcal{R}^{mn} \tag{A.11}$$

to a vector **a** is provided by the operator "col" (or by the operator "vec")

$$\mathrm{col}\mathbf{A} = \mathrm{vec}\mathbf{A} = \mathbf{a} = (A_{11} \ \ldots \ A_{m1} \ \vdots \ A_{12} \ \ldots \ A_{m2} \ \ldots \ A_{mn})^T \tag{A.12}$$

The operator col lists the entries of **A** column-wise.

Separating the ith column of **A** will be termed by $(\mathbf{A})_{\cdot i}$. For abbreviation also $\mathbf{A}_{\cdot i}$ is used although, usually, only matrices are denoted by upper case boldface letters, irrespective of the subscript. $\mathbf{A}_{\cdot i}$ is defined as a column matrix (vector). The operator col (column string) can also be written as

$$\mathrm{col}\ \mathbf{A} = (\mathbf{A}_{\cdot 1}^T \ \vdots \ \mathbf{A}_{\cdot 2}^T \ \ldots \ \mathbf{A}_{\cdot n}^T)^T . \tag{A.13}$$

The inner product of two real matrices **A** and **B** of equal dimension $m \times n$ is a scalar and coincides with the trace of the matrix product: $(\mathbf{A}, \mathbf{B}) = (\mathrm{col}\mathbf{A})^T\mathrm{col}\mathbf{B} = \mathrm{tr}\mathbf{A}^T\mathbf{B} = \sum_{i=1}^m \sum_{j=1}^n A_{ij}B_{ij} \leq ||\mathbf{A}||_F||\mathbf{B}||_F$.

The Frobenius or Euclidian norm of a real matrix **A** is given by

$$||\mathbf{A}||_F = ||\mathrm{col}\mathbf{A}||_F = \sqrt{(\mathrm{col}\mathbf{A})^T\mathrm{col}\mathbf{A}} = \sqrt{\mathrm{tr}\mathbf{A}^T\mathbf{A}} . \tag{A.14}$$

A.6 Reduced Matrix, Minor, Cofactor, Adjoint

Given the (n,n)-matrix **A** , the reduced matrix $\mathbf{A}_{red\ ik}$ of the size $(n-1) \times (n-1)$ is obtained by cancelling row i and column k. Repeating for $i\forall 1 \ldots n, k\forall 1 \ldots n$ yields n^2 different matrices. The minor on the i,k-component of **A** is defined as the determinant $\det \mathbf{A}_{red\ ik}$. The cofactor of the i,k-element of **A** is obtained by permuting the sign of the minor, precisely by multiplying the minor with $(-1)^{i+k}$, that is, $\mathrm{cof}_{ik}\mathbf{A} = (-1)^{i+k} \det \mathbf{A}_{red\ ik}$. Given the (n,n)-matrix **A** on the elements A_{ik}, then $\mathbf{A} = (A_{ik}) = \mathrm{matrix}[A_{ik}]$ and the adjoint is given as the transposed matrix of cofactors: $\mathrm{adj}\mathbf{A} = [\mathrm{matrix}(\mathrm{cof}_{ik}\mathbf{A})]^T$.

The determinant can be decomposed with respect to the row i , or with respect to the kth column, that is,

$$\det \mathbf{A} = \sum_{k=1}^n A_{ik}\mathrm{cof}_{ik}\mathbf{A} \qquad \forall i = 1 \ldots n \quad \text{or} \quad \det \mathbf{A} = \sum_{i=1}^n A_{ik}\mathrm{cof}_{ik}\mathbf{A} \qquad \forall k = 1 \ldots n . \tag{A.15}$$

Interpretation of the equations above as a matrix multiplication yields

$$\mathbf{I}\ \det \mathbf{A} = \mathbf{A}\ \mathrm{adj}\mathbf{A} = (\mathrm{adj}\mathbf{A})\ \mathbf{A} \qquad \rightsquigarrow \qquad \mathbf{A}^{-1} = \frac{\mathrm{adj}\mathbf{A}}{\det \mathbf{A}} . \tag{A.16}$$

A.7 Similar Matrices

Matrices $\mathbf{A}$ and Λ are similar, i.e., $\mathbf{A} \sim \Lambda$, if $\mathbf{A} = \mathbf{T}\Lambda\mathbf{T}^{-1}$. The preceding equation is named similarity transformation. Similar matrices $\mathbf{A}$ and Λ are characterized by the property det $\mathbf{A} =$ det Λ, by the same eigenvalues, eigenvalue multiplicities and eigenvalue indices (and by the same number of generalized eigenvectors), see Eq.(B.107). Examples of similar matrices are $\mathbf{A}$ and $\Lambda = \text{diag}\ \lambda_i[\mathbf{A}] = \mathbf{T}^{-1}\mathbf{A}\mathbf{T}$ in Eq.(B.13).

A.8 Some Properties of Determinants

Multiplying an (n,n)-matrix $\mathbf{A}$ by a scalar μ yields a (n,n)-matrix $\mathbf{\Gamma} = \mu\mathbf{A} = (\mu A_{ik}) = \text{matrix}[\mu A_{ik}]$. The determinants of $\mathbf{A}$ and $\mathbf{\Gamma}$ comply with the relation $\det\mathbf{\Gamma} = \det\mu\mathbf{A} = \mu^n \det\mathbf{A}$. Determinant of the transpose: $\det\mathbf{A}^T = \det\mathbf{A}$. Determinant of the product $\mathbf{AB}$ if both $\mathbf{A}$ and $\mathbf{B}$ are square: $\det(\mathbf{AB}) = (\det\mathbf{A})(\det\mathbf{B})$. Determinant of the inverse: $\det(\mathbf{A}^{-1}) = (\det\mathbf{A})^{-1} = 1/\det\mathbf{A}$.

In the case of an orthogonal matrix $\mathbf{R}$ the relation $\mathbf{R}^T = \mathbf{R}^{-1}$ holds and $\det\mathbf{R} = \pm 1$.

The eigenvalues $\lambda[\mathbf{A}]$ of an (n,n)-matrix $\mathbf{A}$ are related with the determinant and with the trace as follows

$$\det\mathbf{A} = \prod_1^n \lambda_i[\mathbf{A}] \qquad \text{tr}\mathbf{A} = \sum_1^n A_{ii} = \sum_1^n \lambda_i[\mathbf{A}] \,. \tag{A.17}$$

Selecting two conformable matrices, the (m,n)-matrix $\mathbf{A}$ and the (n,m)-matrix $\mathbf{B}$, the equivalence holds

$$\det(\mathbf{I}_m + \mathbf{AB}) = \det(\mathbf{I}_n + \mathbf{BA}). \tag{A.18}$$

A.9 Singularity

The matrix $\mathbf{U}$ is singular and no inverse $\mathbf{U}^{-1}$ exists if $\det\mathbf{U} = 0$. The matrix $\mathbf{U}$ is nonsingular if its determinant does not vanish. A nonsquare matrix always is singular.

A.10 System of Linear Equations

Solving a system of n linear algebraic equations $\mathbf{Ax} = \mathbf{b}$ yields

$$\mathbf{x} = \mathbf{A}^{-1}\mathbf{b} = \frac{\text{adj}\mathbf{A}}{\det\mathbf{A}}\mathbf{b} \quad \text{or} \quad x_k = \frac{1}{\det\mathbf{A}}\sum_i(\text{cof}_{ik}\mathbf{A})b_i \tag{A.19}$$

where the transposed matrix of cofactors replaces the adjoint.

A.11 Stable Matrices

Consider the set $\Lambda(\mathbf{A})$ of eigenvalues $\lambda_i[\mathbf{A}]$ of the (square) matrix $\mathbf{A}$

$$\Lambda[\mathbf{A}] = \{\lambda_i[\mathbf{A}] \in \mathcal{C} \quad : \quad \det(\mathbf{A} - \lambda_i\mathbf{I}) = 0\ \} \,. \tag{A.20}$$

If and only if Λ is a subset of $\mathcal{C}^-$ (complex numbers with negative real part), i.e., $\Lambda[\mathbf{A}] \subseteq \mathcal{C}^-$, the matrix $\mathbf{A}$ is said to be asymptotically stable.

A.12 Range Space. Rank. Null Space

The range (or image) of a linear operator (map) $\mathcal{X} \rightarrow \mathcal{Y}$ is the set of all linear combinations of the columns of the matrix $\mathbf{A}$ of this operator and is termed range space $\mathcal{R}[\mathbf{A}]$, see Figs. A.1 and A.2. The range space of $\mathbf{A}$ is the set of all vectors $\mathbf{Ax}$ where $\mathbf{x}$ ranges over the set $\mathcal{R}^n$.

$$\mathcal{R}[\mathbf{A}] \triangleq \{\mathbf{y} : \mathbf{y} \in \mathcal{Y},\ \mathbf{x} \in \mathcal{X},\ \mathbf{y} = \mathbf{Ax}\} \subset \mathcal{Y} \qquad \mathbf{y} \in \mathcal{Y} \in \mathcal{R}^m, \qquad \mathbf{x} \in \mathcal{X} \in \mathcal{R}^n, \qquad \mathbf{A} \in \mathcal{R}^{m \times n} \,. \tag{A.21}$$

Example: Consider $m > n$, $m = 3$, $n = 2$, $\mathbf{A}$ column-like

$$\mathbf{y} = \mathbf{A}\mathbf{x} = (\mathbf{A}_{\cdot 1} \vdots \mathbf{A}_{\cdot 2})\begin{pmatrix} x_1 \\ x_2 \end{pmatrix} = x_1\mathbf{A}_{\cdot 1} + x_2\mathbf{A}_{\cdot 2} \ . \tag{A.22}$$

Within the three-dimensional Euclidian space $\mathcal{Y}$ the range space of $\mathbf{A}$ is a plane through the origin spanned by both columns of $\mathbf{A}$. In a case $m < n$ the columns of $\mathbf{A}$ are linearly dependent which yields redundancy but is not in contradiction to the definition Eq.(A.21). □

The dimension of the range space is the rank of $\mathbf{A}$: dim $\mathcal{R}[\mathbf{A}]$ = rank $\mathbf{A}$, i.e., the number of linearly independent columns of $\mathbf{A}$. In any square or rectangular matrix the maximum number of linearly independent columns is equal to the maximum number of linearly independent rows. The rank can be checked either from rows or columns of a matrix. Given the (m,n)-matrix $\mathbf{A}$, if $m \leq n$ and rank $\mathbf{A} = m$ the matrix is of full row rank. If $m \geq n$ and rank $\mathbf{A} = n$ the matrix is of full column rank. In the case $m = n =$ rank $\mathbf{A}$ the matrix $\mathbf{A}$ is nonsingular.

An (m,n)-matrix $\mathbf{A}$ is said to be of full rank if rank $\mathbf{A}$ is n or m whichever is less, thus

$$\text{rank } \mathbf{A} = \min \{n, m\}. \tag{A.23}$$

A square matrix of full rank is nonsingular, a nonsingular square matrix is of full rank. If a matrix is nonsingular, the inverse of the matrix exists.

Since the rank of a matrix $\mathbf{A}$ is the maximum number of linear independent rows *or* columns of $\mathbf{A}$ the rank may be considered the order of its largest nonsingular square submatrix (non-vanishing minor).

Two matrices $\mathbf{A}$ and $\mathbf{B}$, $\mathbf{B} = \mathbf{\Gamma}\mathbf{A}\mathbf{\Omega}$, are equivalent if rank $\mathbf{A}$ = rank $\mathbf{B}$ and both $\mathbf{\Gamma}$ and $\mathbf{\Omega}$ are nonsingular. The rank of a square matrix $\mathbf{A}$ equals the number of eigenvalues $\lambda[\mathbf{A}]$ that are nonzero,

$$\text{e. g.,} \quad \mathbf{A} = \begin{pmatrix} 2 & 5 \\ 4 & 10 \end{pmatrix} \quad \rightsquigarrow \quad \text{rank } \mathbf{A} = 1 \quad \rightsquigarrow \quad \lambda[\mathbf{A}] = 0 \ ; \ -8 \ . \ \square \tag{A.24}$$

Usually, the input matrix $\mathbf{B} \in \mathcal{R}^{n \times m}$ in state-space representation is full rank: rank$\mathbf{B} = m$. If not, the deficiency of rank$\mathbf{B}$ corresponds to redundant input signals u_i .

Additionally, rank $(\mathbf{B}\mathbf{B}^{IL} - \mathbf{I}_n) = n -$ rank $\mathbf{B}$.

$$\text{Rank of the dyadic (outer) product:} \quad \text{rank } (\mathbf{a}\mathbf{b}^T) = \begin{cases} 1 & \text{if} \quad \mathbf{a} \neq 0 \quad \text{and} \quad \mathbf{b} \neq 0 \\ 0 & \text{if} \quad \mathbf{a} = 0 \quad \text{or} \quad \mathbf{b} = 0 \, . \end{cases} \tag{A.25}$$

Some more properties:

$$\text{rank } \mathbf{a} = 1 \quad \text{if} \quad \mathbf{a} \neq 0 \, , \quad \text{rank } (\mathbf{A}+\mathbf{B}) \leq \text{rank } \mathbf{A} + \text{rank } \mathbf{B} \, , \quad \text{rank } (\mathbf{A}\mathbf{B}) \leq \min(\text{rank } \mathbf{A}, \text{rank } \mathbf{B}) \, . \tag{A.26}$$

The null space (or kernel) of a matrix $\mathbf{A}$ is termed $\mathcal{N}[\mathbf{A}]$ and is defined as a subspace

$$\mathcal{N}[\mathbf{A}] \triangleq \{\mathbf{x} : \mathbf{x} \in \mathcal{X} \ \text{ where } \ \mathbf{A}\mathbf{x} = 0\} \subset \mathcal{X} \tag{A.27}$$

In other words: If $\mathbf{A}\mathbf{x} = \mathbf{0}$, then $\mathbf{x}$ belongs to the null space of $\mathbf{A}$, i.e., $\mathbf{x} \in \mathcal{N}[\mathbf{A}]$, see Figs. A.1 and A.2.

The range space $\mathcal{R}[\mathbf{A}^T]$ and the null space $\mathcal{N}[\mathbf{A}]$ are orthogonal since $\mathbf{y}_1 = \mathbf{A}^T\mathbf{x}$ and $\mathbf{A}\mathbf{y}_2 = \mathbf{0}$ yields $\mathbf{y}_1^T\mathbf{y}_2 = \mathbf{x}^T\mathbf{A}\mathbf{y}_2 = 0$. The dimension of the null space plus dimension of column space of the (n,m)-matrix $\mathbf{B}$ is

$$\dim \mathcal{N}[\mathbf{B}] + \dim \mathcal{R}[\mathbf{B}] = n \ . \tag{A.28}$$

Example: Consider a multivariable system

$$\dot{\mathbf{x}}(t) = \mathbf{A}\mathbf{x}(t) + \mathbf{B}\mathbf{u}(t) \ , \tag{A.29}$$

its Kalman controllability matrix $\mathbf{S}_K$

$$\mathbf{S}_K = (\mathbf{B} \vdots \mathbf{A}\mathbf{B} \vdots \ \ldots \ \vdots \mathbf{A}^{n-1}\mathbf{B}) \qquad \in \mathcal{R}^{n \times nm} \tag{A.30}$$

and a subspace $\mathcal{S}_K$ spanned by the columns of $\mathbf{S}_K$. Then $\mathcal{S}_K$ is the controllable subspace of the system given in Eq.(A.29). Only intitial conditions $\mathbf{x}(0) \in \mathcal{S}_K$ can be controlled (*Kalman, R.E., 1963*). Within a limited time interval any system state in $\mathcal{S}_K$ can be reached. If any vector $\mathbf{z}$ satisfies $\mathbf{z} \in \mathcal{R}[\mathbf{S}_K]$ then also $\mathbf{A}\mathbf{z} \in \mathcal{R}[\mathbf{S}_K]$. Consider a special case $(m = 2, n = 2)$

$$\mathbf{A} = \begin{pmatrix} 0 & -1 \\ -5 & 4 \end{pmatrix} \qquad \mathbf{B} = \mathbf{b} = \begin{pmatrix} 1 \\ 1 \end{pmatrix} \sim \mathbf{a}_1^\triangleright = \begin{pmatrix} 0.707 \\ 0.707 \end{pmatrix} \tag{A.31}$$

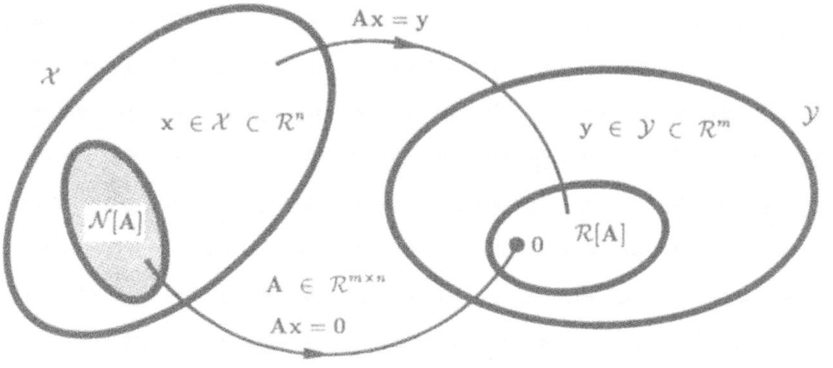

Figure A.1: Range space (or image or column space) and null space (or kernel)

$$\mathbf{S}_K = \begin{pmatrix} 1 & -1 \\ 1 & -1 \end{pmatrix} \quad \text{rank } \mathbf{S}_K = 1 \neq 2, \text{ hence } (\mathbf{A},\mathbf{B}) \text{ not completely controllable .} \tag{A.32}$$

$$\mathcal{S}_K = \mathcal{R}[\mathbf{A}] = \{\mathbf{y} \ : \ \mathbf{y} = \alpha \begin{pmatrix} 1 \\ 1 \end{pmatrix}, \ \alpha \in (-\infty,\infty)\} \ , \tag{A.33}$$

i.e., only the line $\alpha(1 \vdots 1)$ is the controllable subspace of the $\mathbf{x}$-plane. In this example with distinct eigenvalues

$$\mathbf{T} = (\mathbf{a}_1^\triangleright \vdots \mathbf{a}_2^\triangleright) = \begin{pmatrix} 0.707 & -0.196 \\ 0.707 & 0.981 \end{pmatrix}, \quad \mathbf{T}^{-1} = \begin{pmatrix} 1.178 & 0.236 \\ -0.850 & 0.850 \end{pmatrix} \quad \text{and} \quad \mathbf{T}^{-1}\mathbf{B} = \mathbf{T}^{-1}\mathbf{b} = \begin{pmatrix} 1.304 \\ 0 \end{pmatrix}. \tag{A.34}$$

As far as controllability is concerned, Eq.(B.20) shows that x_2^{mo} cannot be influenced by u since the second entry of $\mathbf{T}^{-1}\mathbf{b}$ is zero and the system equations in $\mathbf{x}^{mo}$ are decoupled. There is a transient motion in x_2^{mo} , only, starting from the initial conditions, and decaying to zero. The first variable x_1^{mo} is generated from $u(t)$ via PT_1-delay, conventionally. Invoking $\mathbf{x} = \mathbf{T}\mathbf{x}^{mo}$ and omitting, for brevity, the homogeneous part of the solution, yields

$$\mathbf{x} = \begin{pmatrix} x_1 \\ x_2 \end{pmatrix} = \mathbf{T}\begin{pmatrix} x_1^{mo} \\ x_2^{mo} \end{pmatrix} = \mathbf{T}\begin{pmatrix} x_1^{mo} \\ 0 \end{pmatrix} = \begin{pmatrix} 0.707 \\ 0.707 \end{pmatrix} x_1^{mo} \tag{A.35}$$

which corresponds to the line $\alpha(1 \vdots 1)$. If $\mathbf{B} = \mathbf{b}$ coincides with an eigenvector $\mathbf{b} = \mathbf{a}_1^\triangleright$ then the controllable subspace is reduced to this eigenvector. □

A.13 Trace

The trace of a square matrix is defined as the sum of the elements A_{ii} in the main diagonal. Moreover, the trace is identical to the sum of the eigenvalues. For comparison: The product of the eigenvalues yields the determinant. The trace of the dyadic product of two vectors gives the inner product: $\text{tr } \mathbf{a}\mathbf{b}^T = \mathbf{a}^T\mathbf{b}$.

Taking the trace of a matrix product, the matrix factors are commutative $\text{tr } \mathbf{AB} = \text{tr } \mathbf{BA} = \sum_{i=1}^n \sum_{k=1}^p A_{ik}B_{ki}$ although $\mathbf{AB}$ or $\mathbf{BA}$ and their dimensions are strongly different, $\mathbf{A}$ is an (n,p)-matrix and $\mathbf{B}$ a (p,n)-matrix. This cyclic property for a product matrix in a trace argument is very convenient when matrix derivative operations with respect to matrices are studied. For applications see e.g. sensitivity theory. Further properties of the trace are

$$\text{tr}(\mathbf{A}+\mathbf{B}) = \text{tr}\mathbf{A} + \text{tr}\mathbf{B}, \qquad \text{tr}(\mathbf{A}\mathbf{a}\mathbf{a}^T) = \mathbf{a}^T\mathbf{A}\mathbf{a}, \qquad \text{tr } \mathbf{I}_n = n \ . \tag{A.36}$$

A.14 Matrix Functions

Consider a function $f(\lambda)$ and assume that its Taylor expansion

$$f(\lambda) = \sum_{j=0}^{\infty} \frac{1}{j!}\left(\frac{\partial^j f(\lambda)}{\partial \lambda^j}\right)_{\lambda=\lambda_i} (\lambda - \lambda_i)^j \tag{A.37}$$

exists in the vicinity of λ_i where $\lambda_i = \lambda_i[\mathbf{A}]$ is one of the p eigenvalues of $\mathbf{A}$ of multiplicity m_i and degeneracy q_i. Inverting the characteristic polynomial $c(\lambda) = \det(\lambda \mathbf{I} - \mathbf{A})$ and expanding into partial fractions, using a numerator polynomial $b_i(\lambda)$

$$\frac{1}{c(\lambda)} = \sum_{i=1}^{p} \frac{b_i(\lambda)}{(\lambda - \lambda_i)^{m_i}} \, . \tag{A.38}$$

The definition of $a_i(\lambda)$ is

$$a_i(\lambda) \triangleq \frac{b_i(\lambda) c(\lambda)}{(\lambda - \lambda_i)^{m_i}} \quad \text{or} \quad (\lambda - \lambda_i)^{m_i} a_i(\lambda) = b_i(\lambda) c(\lambda) \, . \tag{A.39}$$

Multiplying Eq.(A.38) with $c(\lambda)$ and invoking Eq.(A.39) yields

$$1 = \sum_{i=1}^{p} a_i(\lambda) = c(\lambda) \sum_{i=1}^{p} \frac{b_i(\lambda)}{(\lambda - \lambda_i)^{m_i}} \, . \tag{A.40}$$

The matrix function $f(\mathbf{A})$ associated with the scalar function $f(\lambda)$ is obtained by substituting λ by $\mathbf{A}$. From Eqs.(A.39) ,(A.40) and using Cayley-Hamiltion theorem $c(\mathbf{A}) = 0$

$$(\mathbf{A} - \lambda_i \mathbf{I})^{m_i} a_i(\mathbf{A}) = b_i(\mathbf{A}) c(\mathbf{A}) = \mathbf{0} \tag{A.41}$$

and $\sum_{i=1}^{p} a_i(\mathbf{A}) = \mathbf{I}$. Multiplying $f(\mathbf{A})$ with $a_i(\mathbf{A})$ yields

$$f(\mathbf{A}) a_i(\mathbf{A}) = \sum_{j=0}^{\infty} \frac{f^{(j)}(\lambda_i)}{j!} (\mathbf{A} - \lambda_i \mathbf{I})^j a_i(\mathbf{A}) \, . \tag{A.42}$$

The upper bound in the sum above can be replaced by m_i. This results since one has $(\mathbf{A} - \lambda_i \mathbf{I})^j|_{j>m_i-1} = \mathbf{0}$ in Eq.(A.41). If the multiplicity m_{mi} of λ_i in the minimal polynomial is $m_{mi} < m_i$ then the upper bound can be replaced by m_{mi} . Thus, taking the sum with respect to i yields

$$\sum_{i=1}^{p} f(\mathbf{A}) a_i(\mathbf{A}) = f(\mathbf{A}) = \sum_{i=1}^{p} \sum_{j=0}^{m_i-1} \frac{f^{(j)}(\lambda_i)}{j!} (\mathbf{A} - \lambda_i \mathbf{I})^j a_i(\mathbf{A}) \, . \tag{A.43}$$

With the abbreviation

$$\mathbf{D}_{i,j+1} \triangleq (\mathbf{A} - \lambda_i \mathbf{I})^j a_i(\mathbf{A}) = j! \; \mathbf{Z}_{ij} \tag{A.44}$$

where $\mathbf{Z}_{ij}$ denotes the interpolating matrix polynomial or *component* of $\mathbf{A}$, one has

$$f(\mathbf{A}) = \sum_{i=1}^{p} \sum_{j=0}^{m_i-1} \frac{f^{(j)}(\lambda_i)}{j!} \mathbf{D}_{i,j+1} \, . \tag{A.45}$$

Note that the matrices $\mathbf{D}_{i,j+1}$ or $\mathbf{Z}_{ij}$ do not depend on the function f but only on $\mathbf{A}$. Hence, any f can be used to determine $\mathbf{D}_{i,j+1}$ or $\mathbf{Z}_{ij}$ for a given $\mathbf{A}$. The derivatives are very simple if f is taken the rth power $f(\mathbf{A}) = \mathbf{A}^r$

$$\mathbf{A}^r = \sum_{i=1}^{p} \sum_{j=0}^{m_i-1} \binom{r}{j} \lambda_i^{r-j} \, \mathbf{D}_{i,j+1} \, . \tag{A.46}$$

The last equation can be used to determine $\mathbf{D}_{i,j+1}$. Now, defining column-like partitioned matrices $\mathbf{D}_c = (\mathbf{D}_{11}^T \vdots \mathbf{D}_{12}^T \vdots \ldots \mathbf{D}_{pm_p}^T)^T$ and $\mathbf{A}_c = (\mathbf{I} \vdots \mathbf{A}^T \vdots \ldots \mathbf{A}^{Tn})^T$, Eq.(A.46) is rewritten to $\mathbf{V}_{dmg} \, \mathbf{D}_c = \mathbf{A}_c$ where $\mathbf{V}_{dmg}$ is the generalized Vandermonde matrix. In the case $m_{mi} < m_i$ one has $\mathbf{D}_{i,m_{mi}+2} = \mathbf{0}$ through $\mathbf{D}_{i,m_i+1} = \mathbf{0}$ (*Frame, J.S., 1964; Gantmacher, F.R., 1986*).

A.15 Metzler Matrices

A Metzler matrix $\mathbf{A} \in \mathcal{R}^{n \times n}$ is characterized $\forall i, j = [1, n]$ by

$$a_{ij} = \begin{cases} < 0 & i = j \quad \text{main diagonal} \\ \geq 0 & i \neq j \quad \text{elsewhere} . \end{cases} \tag{A.47}$$

A Metzler matrix is stable if and only if all *leading* principal minors $\forall k = [1, n]$ premultiplied by $(-1)^k$

$$(-1)^k \det \begin{pmatrix} a_{11} & a_{12} & \dots & a_{1k} \\ a_{21} & a_{22} & \dots & a_{2k} \\ \vdots & \vdots & \ddots & \vdots \\ a_{k1} & \dots & \dots & a_{kk} \end{pmatrix} > 0 . \tag{A.48}$$

Remark: To check the stability of an ordinary matrix one has to test *all* the principal minors.

A Metzler matrix is stable if and only if it is quasidominant negativ diagonal. A matrix $\mathbf{A}$ is said quasidominant negative dominant if all $a_{ii} < 0$ and $\forall i, j = [1, n]$ there exist positive numbers d_i such that

$$\text{either} \quad d_i |a_{ii}| > \sum_{j=1,\, j \neq i}^{n} d_j |a_{ij}| \quad \text{or} \quad d_j |a_{jj}| > \sum_{i=1,\, i \neq j}^{n} d_i |a_{ij}| \tag{A.49}$$

is satisfied (*Metzler, L.A., 1950; Šiljak, D.D., 1978; Mansour, M., 1987; Xin, L.X., 1987*).

Example:

$$\mathbf{A} = \begin{pmatrix} -1 & 0.5 \\ 0.5 & -3 \end{pmatrix} \qquad \lambda_{1,2}[\mathbf{A}] = -3.118; \ -0.882 \tag{A.50}$$

$$\text{leading principal minors} \times (-1)^k : \quad (-1)^1(-1) > 0 ; \quad (-1)^2 \det \begin{pmatrix} -1 & 0.5 \\ 0.5 & -3 \end{pmatrix} = 2.75 > 0 . \, \square \tag{A.51}$$

There exists another definition of M-matrices which is explained using the notation $\mathbf{P}$. The matrix $\mathbf{P}$ is an M-matrix if $p_{ij} \leq 0 \ \forall i \neq j$ and $\Re e \ \lambda[\mathbf{P}] > 0$. Then, Eq.(A.48) has to be rewritten with $a_{ij} := p_{ij}$ and $(-1)^k$ must be omitted. For any non-negative matrix $\mathbf{B}$ the matrix $\mathbf{P} = \nu \mathbf{I} - \mathbf{B}$ is an M-matrix if and only if $\nu > \pi[\mathbf{B}]$ where $\pi[\cdot]$ denotes the Perron root. Finally, if $\mathbf{P}$ and $\mathbf{D}$ are square and non-negative and $\pi[\mathbf{P}] < 1$ then all the matrices

$$\mathbf{I} - \mathbf{P} - \mathbf{D} ; \quad \mathbf{I} - \mathbf{D}(\mathbf{I} - \mathbf{P})^{-1} ; \quad \mathbf{I} - (\mathbf{I} - \mathbf{P})^{-1}\mathbf{D} \tag{A.52}$$

are M-matrices and the Perron root of all the following matrices is less than unity

$$\mathbf{P} + \mathbf{D} ; \quad \mathbf{D}(\mathbf{I} - \mathbf{P})^{-1} ; \quad (\mathbf{I} - \mathbf{P})^{-1}\mathbf{D} . \tag{A.53}$$

A.16 Projectors

Consider the n-dimensional space decomposed into two subspaces namely the range space and null space. The matrix $\mathbf{P}_r$ is a projector if it projects $\mathcal{R}^n$ on the range space $\mathcal{R}[\mathbf{P}_r]$ of $\mathbf{P}_r$ along the null space $\mathcal{N}[\mathbf{P}_r]$ of $\mathbf{P}_r$. A matrix acts as a projector if it is idempotent, i.e. $\mathbf{P}_r^2 = \mathbf{P}_r$. The projection operator function and the range space details can be seen by premultiplying a vector $\mathbf{x}$ of $\mathcal{R}^n$ by $\mathbf{P}_r$

$$\mathbf{P}_r \mathbf{x} = \mathbf{x} . \tag{A.54}$$

The above projection is essentially augmented by another projection: If the same vector $\mathbf{x}$ is considered to be projected by the projector $\mathbf{I}_n - \mathbf{P}_r$ the projection result

$$(\mathbf{I} - \mathbf{P}_r)\mathbf{x} = 0 \tag{A.55}$$

is the null space, in other words, $\mathbf{I}_n - \mathbf{P}_r$ projects $\mathbf{x}$ of $\mathcal{R}^n$ on the null space $\mathcal{N}[\mathbf{P}_r]$ along the range space $\mathcal{R}[\mathbf{P}_r]$. Summarizing,

$$\mathbf{P}_r \ \text{projects} \ \mathcal{R}^n \ \text{on} \ \mathcal{R}[\mathbf{P}_r] \ \text{along} \ \mathcal{N}[\mathbf{P}_r] \tag{A.56}$$

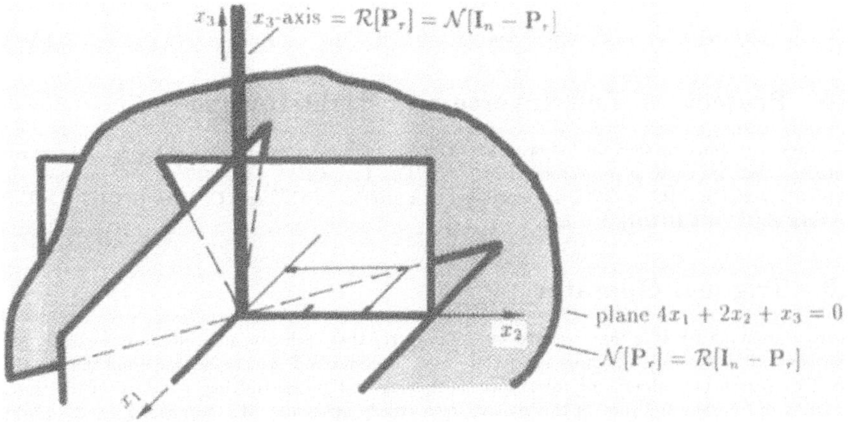

Figure A.2: Range space and null space of a projector

$$\mathbf{I}-\mathbf{P}_r \text{ projects } \mathcal{R}^n \text{ on } \mathcal{N}[\mathbf{P}_r] \text{ along } \mathcal{R}[\mathbf{P}_r]\,. \tag{A.57}$$

Hence, $\mathcal{R}[\mathbf{P}_r] = \mathcal{N}[\mathbf{I}-\mathbf{P}_r]$ and $\mathcal{N}[\mathbf{P}_r] = \mathcal{R}[\mathbf{I}-\mathbf{P}_r]$.

Example: Range space and null space: Consider $n = 3$ and

$$\mathbf{P}_r = \begin{pmatrix} 0 & 0 & 0 \\ 0 & 0 & 0 \\ 4 & 2 & 1 \end{pmatrix}; \quad \mathbf{P}_r\mathbf{x} = \mathbf{x} \quad \leadsto \begin{pmatrix} 0 & 0 & 0 \\ 0 & 0 & 0 \\ 4 & 2 & 1 \end{pmatrix} \begin{pmatrix} x_1 \\ x_2 \\ x_3 \end{pmatrix} = \begin{pmatrix} x_1 \\ x_2 \\ x_3 \end{pmatrix}. \tag{A.58}$$

Hence, $x_1 = 0$, $x_2 = 0$, x_3 undetermined, i.e., the range space $\mathcal{R}[\mathbf{P}_r]$ is given by the x_3-axis in this example (see Fig. A.2). Check

$$\mathbf{I}-\mathbf{P}_r = \begin{pmatrix} 1 & 0 & 0 \\ 0 & 1 & 0 \\ -4 & -2 & 0 \end{pmatrix}; \quad (\mathbf{I}-\mathbf{P}_r)\mathbf{x} = \mathbf{0} \quad \leadsto \begin{pmatrix} 1 & 0 & 0 \\ 0 & 1 & 0 \\ -4 & -2 & 0 \end{pmatrix} \begin{pmatrix} 0 \\ 0 \\ x_3 \end{pmatrix} = \mathbf{0}\,. \tag{A.59}$$

The null space $\mathcal{N}[\mathbf{P}_r]$ is obtained from $\mathbf{P}_r\mathbf{x}' = \mathbf{0}$ or from the range space of $\mathbf{I}-\mathbf{P}_r$, i.e.,

$$\mathcal{R}[\mathbf{I}-\mathbf{P}_r]: \quad (\mathbf{I}-\mathbf{P}_r)\mathbf{x}' = \mathbf{x}' \tag{A.60}$$

$$(\mathbf{I}-\mathbf{P}_r)\mathbf{x}' = \mathbf{x}' \quad \leadsto \begin{pmatrix} 1 & 0 & 0 \\ 0 & 1 & 0 \\ -4 & -2 & 0 \end{pmatrix} \begin{pmatrix} x_1' \\ x_2' \\ x_3' \end{pmatrix} = \begin{pmatrix} x_1' \\ x_2' \\ x_3' \end{pmatrix}. \tag{A.61}$$

Thus, $x_1' = x_1'$; $x_2' = x_2'$; $-4x_1' - 2x_2' = x_3'$, i.e. the null space $\mathcal{N}[\mathbf{P}_r]$ is the plane $4x_1 + 2x_2 + x_3 = 0$ (see Fig. A.2). The matrix $\mathbf{P_r}$ is a projector on the x_3-axis along the plane. The matrix $(\mathbf{I}-\mathbf{P}_r)$ is a projector on the plane along the x_3-axis and rank $\mathbf{P}_r = 1$, rank $(\mathbf{I}-\mathbf{P}_r) = 2$, tr $\mathbf{P}_r = 1$. □

A.17 Projectors and Rank

Note that rank $\mathbf{P}_r = \text{tr}\,\mathbf{P}_r$ and rank $(\mathbf{I}_n - \mathbf{P}_r) = n - \text{rank}\,\mathbf{P}_r$.

If the matrices $\mathbf{B}, \mathbf{K}, \mathbf{H}, \mathbf{C}$ have full rank, the following relations are true

$$\mathcal{R}[\mathbf{BK}] = \mathcal{R}[\mathbf{B}] \quad \text{and} \quad \mathcal{N}[\mathbf{HC}] = \mathcal{N}[\mathbf{C}] \tag{A.62}$$

which is often used when treating the matter of sliding mode (*El-Ghezawi, O.M.E., et al. 1983*). Summarizing: The rank of a projector matrix equals the dimension of its range space.

$$\text{If } \mathbf{P}_r = \mathbf{B}(\mathbf{CB})^{-1}\mathbf{C} \quad \leadsto \text{rank}\,\mathbf{P}_r = \text{rank}\,\mathbf{B} = m \quad \text{and} \quad \text{rank}\,(\mathbf{I}_n - \mathbf{P}_r) = n - \text{rank}\,\mathbf{P}_r = n - m\,. \tag{A.63}$$

Premultiplying any square matrix by $\mathbf{I}_n - \mathbf{P}_r$ will cause the product to have the rank of $n-m$ (or less):

$$\text{rank}\ (\mathbf{I}_n - \mathbf{P}_r)\mathbf{A} = n - m\ . \tag{A.64}$$

A.18 Projectors. Left-Inverse and Right-Inverse

Let $\mathbf{C}$ be any (m,n)-matrix then the following list of idempotent matrices, subdivided into left-inverse and right-inverse matrices, serve as projectors: $\mathbf{CC}^{LI}$, $\mathbf{C}^{RI}\mathbf{C}$, $\mathbf{I}_m - \mathbf{CC}^{LI}$, $\mathbf{I}_n - \mathbf{C}^{RI}\mathbf{C}$. For instance, the projector $\mathbf{P}_r = \mathbf{B}(\mathbf{CB})^{-1}\mathbf{C} := \mathbf{C}^{RI}\mathbf{C}$ is considered as a product of $\mathbf{C}^{RI}$ and $\mathbf{C}$. Then $\mathbf{B}(\mathbf{CB})^{-1} = \mathbf{C}^{RI}$ is true since $\mathbf{CC}^{RI} = \mathbf{CB}(\mathbf{CB})^{-1} = \mathbf{I}_m$.

A.19 Trigonal Operator

The permutation matrix $\mathbf{P}$ is characterized by the property that each row and column has exactly one element equal to unity while the other entries are zero. The matrix $\mathbf{P}$ can be derived from the identity matrix $\mathbf{I}$ by permuting columns or rows. Postmultiplying and premultiplying a matrix by $\mathbf{P}$ causes permutation of columns and rows of this matrix, respectively (*Qianhua, W., and Zhongjun, Z., 1987*). Finally $\det\mathbf{P} = 1$.

As a special case of the permutation matrix the rotation matrix is defined as the unity matrix with unity entries in the secondary diagonal, only,

$$\mathbf{I}^{\swarrow} = \begin{pmatrix} 0 & 0 & 1 \\ 0 & 1 & 0 \\ 1 & 0 & 0 \end{pmatrix} . \quad \text{For comparison,} \quad \mathbf{I} = \mathbf{I}^{\searrow} = \text{diag}\{1,1,\ldots,1\}\ . \tag{A.65}$$

*Post*multiplication of a matrix $\mathbf{A}$ with $\mathbf{I}^{\swarrow}$ yields a reflection of $\mathbf{A}$ across the *vert*ical symmetric axis

$$\begin{pmatrix} a_1 & a_2 & a_3 \\ a_4 & a_5 & a_6 \\ a_7 & a_8 & a_9 \end{pmatrix} \mathbf{I}^{\swarrow} = \begin{pmatrix} a_1 & a_2 & a_3 \\ a_4 & a_5 & a_6 \\ a_7 & a_8 & a_9 \end{pmatrix} \begin{pmatrix} 0 & 0 & 1 \\ 0 & 1 & 0 \\ 1 & 0 & 0 \end{pmatrix} = \begin{pmatrix} a_3 & a_2 & a_1 \\ a_6 & a_5 & a_4 \\ a_9 & a_8 & a_7 \end{pmatrix} . \tag{A.66}$$

*Pre*multiplication determines a reflection across *hor*izontal symmetric axis

$$\begin{pmatrix} 0 & 0 & 1 \\ 0 & 1 & 0 \\ 1 & 0 & 0 \end{pmatrix} \begin{pmatrix} a_1 & a_2 & a_3 \\ a_4 & a_5 & a_6 \\ a_7 & a_8 & a_9 \end{pmatrix} = \begin{pmatrix} a_7 & a_8 & a_9 \\ a_4 & a_5 & a_6 \\ a_1 & a_2 & a_3 \end{pmatrix} . \tag{A.67}$$

The reflection operation holds for every kind of vector or matrix.

The secondary diagonal matrix possesses the following properties: $(\mathbf{I}^{\swarrow})^{-1} = \mathbf{I}^{\swarrow}$, $(\mathbf{I}^{\swarrow})^2 = \mathbf{I}$, $\mathbf{I}^{\swarrow} = \sqrt{\mathbf{I}}$, $(\mathbf{I}^{\swarrow}\mathbf{C})^T = \mathbf{C}^T\mathbf{I}^{\swarrow T} = \mathbf{C}^T\mathbf{I}^{\swarrow}$. With regard to the symmetry of $\mathbf{I}^{\swarrow}$ and trigonal matrices

$$(\mathbf{I}^{\swarrow}\text{trig}\ \mathbf{b})^T = (\text{trig}\ \mathbf{b})^T\mathbf{I}^{\swarrow T} = (\text{trig}\ \mathbf{b})\mathbf{I}^{\swarrow}\ , \quad \text{trig}\ \mathbf{a} = \text{trig}(a_i) = \text{trig}[\mathbf{I}^{\swarrow} \times (a_{n+1-i})]. \tag{A.68}$$

For a given vector $\mathbf{b}$ the trigonal matrix is defined as trig $\mathbf{b}$

$$\mathbf{b} = \begin{pmatrix} b_1 \\ b_2 \\ b_3 \\ b_4 \end{pmatrix} , \qquad \text{trig}\ \mathbf{b} \triangleq \begin{pmatrix} b_1 & b_2 & b_3 & b_4 \\ b_2 & b_3 & b_4 & 0 \\ b_3 & b_4 & 0 & 0 \\ b_4 & 0 & 0 & 0 \end{pmatrix} \ldots \text{trigonal matrix} \tag{A.69}$$

The secondary diagonal of the original trigonal matrix is filled with b_n, i.e., the entry b_i of highest index n. Reflecting the vector $\mathbf{b}$

$$\text{trig}\ (\mathbf{I}^{\swarrow}\mathbf{b}) = \begin{pmatrix} b_4 & b_3 & b_2 & b_1 \\ b_3 & b_2 & b_1 & 0 \\ b_2 & b_1 & 0 & 0 \\ b_1 & 0 & 0 & 0 \end{pmatrix} . \tag{A.70}$$

The elements of the secondary diagonal of the result are troughout b_1.

Further properties:

$$\text{Upper triangular matrix:} \quad (\text{trig } \mathbf{b})\mathbf{I}^{\swarrow} = \begin{pmatrix} b_4 & b_3 & b_2 & b_1 \\ 0 & b_4 & b_3 & b_2 \\ 0 & 0 & b_4 & b_3 \\ 0 & 0 & 0 & b_4 \end{pmatrix} \tag{A.71}$$

$\updownarrow$(transpose)

$$\text{Lower triangular matrix:} \quad \mathbf{I}^{\swarrow}(\text{trig } \mathbf{b}) = \begin{pmatrix} b_4 & 0 & 0 & 0 \\ b_3 & b_4 & 0 & 0 \\ b_2 & b_3 & b_4 & 0 \\ b_1 & b_2 & b_3 & b_4 \end{pmatrix} \tag{A.72}$$

$$(\text{trig } \mathbf{I}^{\swarrow}\mathbf{b})\mathbf{I}^{\swarrow} = \begin{pmatrix} b_1 & b_2 & b_3 & b_4 \\ 0 & b_1 & b_2 & b_3 \\ 0 & 0 & b_1 & b_2 \\ 0 & 0 & 0 & b_1 \end{pmatrix} \tag{A.73}$$

$\updownarrow$(transpose)

$$\mathbf{I}^{\swarrow}\text{trig } (\mathbf{I}^{\swarrow}\mathbf{b}) = \begin{pmatrix} b_1 & 0 & 0 & 0 \\ b_2 & b_1 & 0 & 0 \\ b_3 & b_2 & b_1 & 0 \\ b_4 & b_3 & b_2 & b_1 \end{pmatrix} . \tag{A.74}$$

The trigonal matrices in Eqs.(A.71) through (A.74) are persymmetric and Toeplitz matrices because the elements along and parallel to the main diagonal are identical (*Grenander, U., and Szegö, G., 1958, Makhoul, J., 1981; Brillinger, D.R., 1981*). The remaining trigonal matrices are of Toeplitz type with respect to the secondary diagonal.

A.20 Transfer Function Zeros and Initial Step Transients

Consider the single-input single-output system with the transfer function $G(s)$

$$\frac{Y(s)}{U(s)} = G(s) = \frac{\sum_o^n f_k s^k}{\sum_o^n a_k s^k} = \frac{Z(s)}{N(s)} \tag{A.75}$$

and with input signal $u(t)$ ($\neq 0 \; t \geq 0; = 0 \; t < 0$). For the sake of abbreviation, the following vectors of constants, signal derivatives and powers of the Laplace operator s are defined

$$\mathbf{a} = (a_1 \; a_2 \ldots a_n)^T \qquad \mathbf{a}_r = (a_o \; a_1 \ldots a_{n-1})^T \qquad \mathbf{f} = (f_1 \; f_2 \ldots f_n)^T \qquad \mathbf{f}_r = (f_o \; f_1 \ldots f_{n-1})^T \tag{A.76}$$

$$\mathbf{u}(t) = (u(t) \; \dot{u}(t) \ldots u^{(n-1)}(t))^T \qquad \mathbf{y}(t) = (y(t) \; \dot{y}(t) \ldots y^{(n-1)}(t))^T \qquad \mathbf{s} = (1 \; s \; s^2 \ldots s^{n-1})^T \; . \tag{A.77}$$

If $G(s)$ has zeros, i.e., coefficients f_i exist ($i > 0$), the output signal y changes suddenly between 0^- and 0^+ (*Föllinger, O., 1961; Wunsch, G., 1971*). The initial conditions of y obey the relation

$$\mathbf{y}(0^+) = \mathbf{y}(0^-) + \mathbf{L}^{\natural}\mathbf{u}(0^+) \qquad \text{where} \qquad \mathbf{L}^{\natural} = (\text{trig } \mathbf{a})^{-1}(\text{trig } \mathbf{f}) \tag{A.78}$$

is a matrix, relating the initial condition step transients in a general view of $\mathbf{a}$ and $\mathbf{f}$ (*Weinmann, A., 1988*).

Note that the Laplace transform of the differential equation corresponding to Eq.(A.75) ***does not*** depend on $\mathbf{y}(0^+)$ but only on $\mathbf{y}(0^-)$ (*Wunsch, G., 1971*)

$$Y(s)N(s) - \mathbf{y}^T(0^-)(\text{trig } \mathbf{a}) \; \mathbf{s} = U(s)Z(s). \tag{A.79}$$

The nth derivatives $u^{(n)}(0^+)$ and $y^{(n)}(0^+)$ comply with the following relation

$$y^{(n)}(0^+) = \frac{(\mathbf{f}_r^T - \mathbf{a}_r^T\mathbf{L}^{\natural})\mathbf{u}(0^+) - f_n u^{(n)}(0^+)}{a_n} \; . \tag{A.80}$$

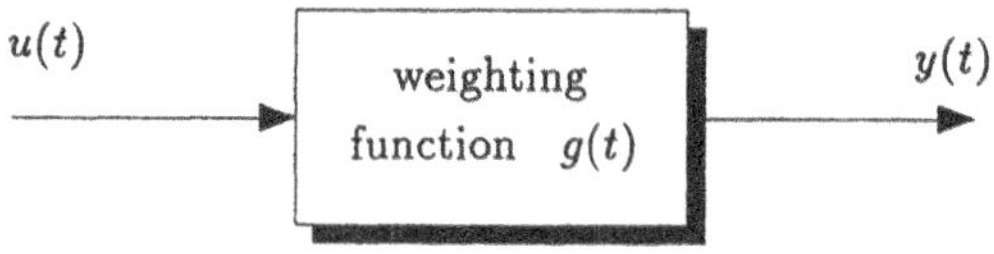

Figure A.3: System output analysis

Even in the case of non-existing zeros of $G(s)$, i.e., $f_i|_{i\neq o} = 0$ and $\mathbf{L}^{\natural} = \mathbf{0}$, the nth derivative $y^{(n)}(0^+)$ has the non-zero amount

$$y^{(n)}(0^+) = \frac{f_o}{a_n} u(0^+). \tag{A.81}$$

Using the state-space representation for the single-input single-output system of Eq.(A.75)

$$\dot{\mathbf{x}}(t) = \mathbf{A}\mathbf{x}(t) + \mathbf{b}u(t) \qquad y(t) = \mathbf{c}^T\mathbf{x}(t) + du(t) \tag{A.82}$$

and observing the definitions of boldface $\mathbf{u}(t)$ and $\mathbf{y}(t)$ in Eq.(A.77) which are different from the scalars $u(t)$ and $y(t)$, it results

$$\mathbf{y}(0^+) = (\mathbf{c} \quad \mathbf{A}^T\mathbf{c} \quad \mathbf{A}^{2,T}\mathbf{c} \ \ldots \ \mathbf{A}^{n-1,T}\mathbf{c})^T\mathbf{x}(0^+) + \mathbf{L}^{\natural}\mathbf{u}(0^+) \tag{A.83}$$

where $\mathbf{L}^{\natural} = \mathbf{I}^{\checkmark}\ \text{trig}[\mathbf{I}^{\checkmark}(d \quad \mathbf{c}^T\mathbf{b} \quad \mathbf{c}^T\mathbf{A}\mathbf{b} \ \ldots\ )^T]$ is equivalent to Eq. (A.78).

A.21 Convolution Sum and Trigonal Operator

Definitions:

$$\mathbf{u} = (u_i) \triangleq \begin{pmatrix} u(\Delta t) \\ u(2\Delta t) \\ \vdots \\ u(i\Delta t) \\ \vdots \\ u(N\Delta t) \end{pmatrix}, \quad \mathbf{y} = (y_i) \triangleq \begin{pmatrix} y(\Delta t) \\ y(2\Delta t) \\ \vdots \\ y(i\Delta t) \\ \vdots \\ y(N\Delta t) \end{pmatrix}, \quad \mathbf{g} \triangleq \begin{pmatrix} g(o) \\ g(\Delta t) \\ \vdots \\ g(i\Delta t) \\ \vdots \\ g[(N-1)\Delta t] \end{pmatrix}. \tag{A.84}$$

Convolution integral and convolution sum (see Fig. A.3):

$$y(t) = \int_0^t g(\tau)u(t-\tau)d\tau\,, \quad y(k\Delta t) = \sum_{i=o}^{k-1} g(i\Delta t)u(k\Delta t - i\Delta t)\Delta t = \sum_{i=o}^{k-1} u(i\Delta t)g(k\Delta t - i\Delta t)\Delta t \tag{A.85}$$

System analysis:

$$\mathbf{y} = \mathbf{I}^{\checkmark}\ [\text{trig}\ (\mathbf{I}^{\checkmark}\mathbf{u})]\ \mathbf{g}\ \Delta t \quad \text{or} \quad \mathbf{y} = \mathbf{I}^{\checkmark}\ [\text{trig}\ (\mathbf{I}^{\checkmark}\mathbf{g})]\ \mathbf{u}\ \Delta t \tag{A.86}$$

System identification:

$$\mathbf{g} = \frac{1}{\Delta t}(\text{trig}\ \mathbf{I}^{\checkmark}\mathbf{u})\ \mathbf{I}^{\checkmark}\ \mathbf{y} \tag{A.87}$$

Input synthesis:

$$\mathbf{u} = \frac{1}{\Delta t}(\text{trig}\ \mathbf{I}^{\checkmark}\mathbf{g})^{-1}\ \mathbf{I}^{\checkmark}\ \mathbf{y} \tag{A.88}$$

Compensator design (see Fig. A.4):

$$\mathbf{c} = \frac{1}{\Delta t^2}(\text{trig}\ \mathbf{I}^{\checkmark}\mathbf{r})^{-1}\ \mathbf{I}^{\checkmark}(\text{trig}\ \mathbf{I}^{\checkmark}\mathbf{g})^{-1}\ \mathbf{I}^{\checkmark}\mathbf{y}\,. \tag{A.89}$$

Considering $k \to \infty$ the equation above corresponds to infinite matrix equations in expanded form. For stability considerations and norm definitions of multivariable infinite matrices etc. see *Makhlouf, M.A., 1972*.

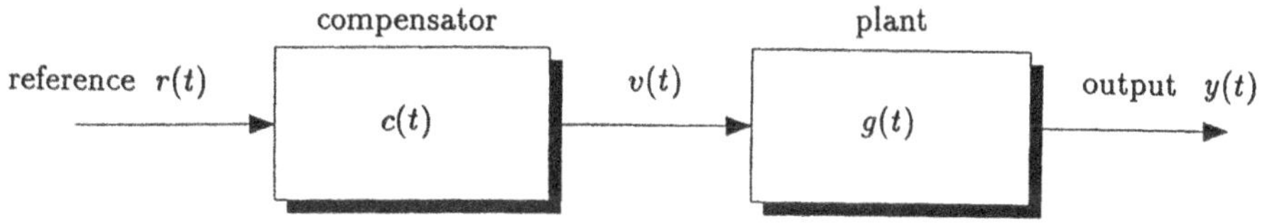

Figure A.4: Compensator design

Appendix B

Eigenvalues and Eigenvectors

Assume a stable (n, n)-matrix $\mathbf{A}$ with the following definitions:

Characteristic matrix of $\mathbf{A}$: $\lambda \mathbf{I} - \mathbf{A}$.

Characteristic function (polynomial) in λ : $c(\lambda) = \det(\lambda \mathbf{I} - \mathbf{A})$.

Characteristic equation of $\mathbf{A}$: $c(\lambda) = \det(\lambda \mathbf{I} - \mathbf{A}) = (\lambda - \lambda_1)(\lambda - \lambda_2)\ldots(\lambda - \lambda_n) = 0$. Roots of the characteristic equation yield the eigenvalues $\lambda[\mathbf{A}]$. The coefficients of the characteristic polynomial

$$c(\lambda) = \lambda^n + c_{n-1}\,\lambda^{n-1} + \ldots c_0 \tag{B.1}$$

can be written as

$$c_{n-i} = (-1)^i \operatorname{tr}_{(i)} \mathbf{A} \qquad \text{where} \quad \operatorname{tr}_{(i)} \mathbf{A} = \sum_{j=1}^{K} \det\,(\text{principal minors of order } r), \qquad K = \binom{n}{r}. \tag{B.2}$$

Note that $\operatorname{tr}_{(n)} \mathbf{A} = \det \mathbf{A}$ and $\operatorname{tr}_{(1)} \mathbf{A} = \operatorname{tr} \mathbf{A}$. The polynomial $c(\lambda)$ is a monic polynomial, i.e., a polynomial with coefficients one of the highest power in λ . The coefficients c_{n-i} of the characteristic polynomial can be calculated from the trace of $\mathbf{A}$ up to the power $n - i$ (starting with $c_n = 1$) according to the following formulas

$$c_{n-1} = -\operatorname{tr} \mathbf{A} \qquad \text{and } \forall i = [1, 2 \ldots n] \tag{B.3}$$

$$c_{n-i} = -\frac{1}{i}(c_{n-i+1} \operatorname{tr} \mathbf{A} + c_{n-i+2} \operatorname{tr} \mathbf{A}^2 + \ldots + c_{n-1} \operatorname{tr} \mathbf{A}^{i-1} + \operatorname{tr} \mathbf{A}^i) . \tag{B.4}$$

The inverse matrix $(\lambda \mathbf{I} - \mathbf{A})^{-1} \triangleq \mathbf{\Phi}(\lambda)$ is known as resolvent matrix for $\mathbf{A}$ and is identical to the fundamental or transition matrix $\mathbf{\Phi}(\lambda)$.

Cayley-Hamilton Theorem: In the characteristic equation λ may be substituted by $\mathbf{A}$: $c(\mathbf{A}) = \mathbf{0}$. The matrix $\mathbf{A}$ satisfies its own characteristic equation.

Examples: $\lambda[\mathbf{I}_n] = 1 (n \text{ times})$; $\lambda[\operatorname{diag}_n a_i] = a_1, a_2, \ldots a_n$; $\lambda[\alpha \mathbf{A}] = \alpha \lambda[\mathbf{A}]$. If any eigenvalue $\lambda[\mathbf{A}] = 0$ then $\det(\mathbf{A} - \lambda \mathbf{I}) = \det \mathbf{A} = 0$ and $\mathbf{A}^{-1}$ does not exist. See Eq.(A.16). □

B.1 Right-Eigenvectors

The eigenvector $\mathbf{a}_i$ is derived from

$$\mathbf{A}\mathbf{a}_i = \lambda_i \mathbf{a}_i = \lambda_i[\mathbf{A}]\,\mathbf{a}_i . \tag{B.5}$$

These eigenvectors are denoted as right-eigenvectors. The transpose $\mathbf{A}^T$ possesses the same eigenvalues λ_i. The right-eigenvectors $\mathbf{p}_i$ of $\mathbf{A}^T$ are defined by

$$\mathbf{A}^T \mathbf{p}_i = \lambda_i \mathbf{p}_i . \tag{B.6}$$

B.2 Left-Eigenvectors

If eigenvectors $\mathbf{a}_i$ are taken into consideration, they are interpreted as right-eigenvectors $\mathbf{a}_i = \mathbf{a}_i^{\triangleright} = \mathbf{a}_i^{\triangleright}[\mathbf{A}]$. Left-eigenvectors $\mathbf{a}_i^{\triangleleft}$ are defined as $\mathbf{a}_i^{\triangleleft T} \mathbf{A} = \lambda_i \mathbf{a}_i^{\triangleleft T}$. Transposing and comparing with Eq. B.6 shows the conformity with the right-eigenvector of $\mathbf{A}^T$. The left-eigenvector of $\mathbf{A}$ is identical to the right-eigenvector of $\mathbf{A}^T$

$$\mathbf{a}_i^{\triangleleft} = \mathbf{a}_i^{\triangleleft}[\mathbf{A}] = \mathbf{a}_i^{\triangleright}[\mathbf{A}^T] = \mathbf{a}_i[\mathbf{A}^T] = \mathbf{p}_i . \tag{B.7}$$

B.3 Complex-Conjugate Eigenvalues

If the eigenvalues $\lambda_i[\mathbf{A}]$ of the real matrix $\mathbf{A}$ are complex-conjugate note the following properties

$$\begin{aligned} \mathbf{A}\mathbf{a}_i &= \lambda_i \mathbf{a}_i & \rightsquigarrow \text{ right-eigenvectors } \mathbf{a}_i^{\triangleright} = \mathbf{a}_i \text{ of } \mathbf{A} \,, \qquad (B.8) \\ \mathbf{a}_i^{\triangleleft T}\mathbf{A} &= \lambda_i^* \mathbf{a}_i^{\triangleleft T} & \rightsquigarrow \text{ left-eigenvectors } \mathbf{a}_i^{\triangleleft} \text{ of } \mathbf{A} \,, \qquad (B.9) \\ \mathbf{A}^T\mathbf{a}_i^{\triangleleft} &= \lambda_i^* \mathbf{a}_i^{\triangleleft} & \rightsquigarrow \text{ right-eigenvectors } \mathbf{a}_i^{\triangleleft} \text{ of } \mathbf{A}^T \,, \qquad (B.10) \end{aligned}$$

$$\mathbf{a}_i^{*T}\mathbf{a}_i^{\triangleleft} = 1 \quad \text{and} \quad \mathbf{a}_i^{*T}\mathbf{a}_k^{\triangleleft} = 0 \;\text{ or }\; \mathbf{a}_i^{*T}\mathbf{a}_k^{\triangleleft} = \delta_{ik} \,. \qquad (B.11)$$

The asterix superscript denotes the complex conjugation. It does not matter if these properties are applied in the case of real eigenvalues but not vice versa. The pair $(\lambda_i, \mathbf{a}_i)$ is denoted eigensolution of $\mathbf{A}$. Then, the corresponding eigensolution of the transpose $\mathbf{A}^T$ is $(\lambda_i^*, \mathbf{a}_i^{\triangleleft})$. See also Eq.(6.70).

Example:

$$\mathbf{A} = \begin{pmatrix} 0 & -1 \\ 2 & -2 \end{pmatrix} \qquad j = \sqrt{-1} \qquad (B.12)$$

i	λ_i	$\mathbf{a}_i$	$\mathbf{a}_i^{\triangleleft}$
1	$-1+j$	$\begin{pmatrix} 1+j \\ 2 \end{pmatrix}$	$\begin{pmatrix} .50j \\ .25-.25j \end{pmatrix}$
2	$-1-j$	$\begin{pmatrix} 1-j \\ 2 \end{pmatrix}$	$\begin{pmatrix} -.50j \\ .25+.25j \end{pmatrix}$

End of Example

Perturbation theory of eigenvalues see separate chapter. Numerical solutions see *Wilkinson, J.H., 1965*.

B.4 Modal Matrix of Eigenvectors

Assume all eigenvalues $\lambda_i[\mathbf{A}]$ distinct. Then, the modal matrix $\mathbf{T}$ of a given square matrix $\mathbf{A}$ is built up by the right-eigenvectors $\mathbf{T} = (\mathbf{a}_1, \mathbf{a}_2, \ldots, \mathbf{a}_n)$. Multiplying $\mathbf{T}$ by $\mathbf{A}$ from the left and substituting Eqs.(B.8) and (A.7), right-hand side,

$$\mathbf{A}\mathbf{T} = \mathbf{A}(\mathbf{a}_1, \mathbf{a}_2, \ldots, \mathbf{a}_n) = \mathbf{T}\,\text{diag}\lambda_i \quad \rightsquigarrow \quad \mathbf{T}^{-1}\mathbf{A}\mathbf{T} = \text{diag}\lambda_i = \mathbf{\Lambda} \,. \qquad (B.13)$$

In this expression $\mathbf{\Lambda} = \mathbf{\Lambda}[\mathbf{A}] = \text{diag}\ \lambda_i[\mathbf{A}]$ is the diagonal matrix of the eigenvalues λ_i of $\mathbf{A}$ in the main diagonal.

The eigenvector matrix associated with $\mathbf{A}^T$ is named $\mathbf{P}$, i.e., $\mathbf{P} = (\mathbf{p}_1, \mathbf{p}_2, \ldots, \mathbf{p}_n)$. Combining the modal matrices $\mathbf{T}$ and $\mathbf{P}$ and referring to Eq.(B.13),

$$\mathbf{T}^{-1}\mathbf{A}\mathbf{T} = \mathbf{P}^{-1}\mathbf{A}^T\mathbf{P} = \text{diag}\lambda_i \,. \qquad (B.14)$$

Taking the transpose and comparing with the left-hand side of the equation above yields

$$\mathbf{T}^T\mathbf{A}^T\mathbf{T}^{-1T} = \mathbf{P}^T\mathbf{A}\mathbf{P}^{-1T} = \text{diag}\lambda_i \quad \rightsquigarrow \quad \mathbf{P}^T = \mathbf{T}^{-1} \,. \qquad (B.15)$$

In other words: The modal matrices $\mathbf{T}$ and $\mathbf{P}$ associated with $\mathbf{A}$ and $\mathbf{A}^T$ are orthogonal. Hence, the corresponding eigenvectors are also orthogonal:

$$\mathbf{p}_i^T\mathbf{a}_k = \delta_{ik} \,. \qquad (B.16)$$

Table B.1 presents an overview of the matrices cited above where double lines denote identities.

If a matrix $\mathbf{A}$ is symmetric the eigenvectors associated with $\mathbf{A}$ and $\mathbf{A}^T$ are identical. The right and left-eigenvectors are identical, too. With regard to $\mathbf{P} = \mathbf{T}$ and Eq.(B.15)

$$\mathbf{T}^T = \mathbf{T}^{-1} \qquad \mathbf{T} = \mathbf{T}[\mathbf{A}] \quad \text{if} \quad \mathbf{A} = \mathbf{A}^T. \qquad (B.17)$$

Although the eigenvectors are normalized to unity the eigenvectors are only fixed with the exception of the sign. See also Eq.(6.70).

Table B.1: Modal matrices $\mathbf{T} = \mathbf{T}[\mathbf{A}]$ and $\mathbf{P} = \mathbf{T}[\mathbf{A}^T]$

$\mathbf{T}^{-1T}$	$\mathbf{T}$	$\mathbf{T}^{-1}$	$\mathbf{T}^T$
$\mathbf{P}^{-1T}$	$\mathbf{P}$	$\mathbf{P}^{-1}$	$\mathbf{P}^T$

In the case of real and complex conjugate eigenvalues and eigenvectors, the following relations are used. The definition of the modal matrix $\mathbf{P} \triangleq (\mathbf{a}_1^q,\ \mathbf{a}_2^q \ldots \mathbf{a}_n^q)$ associated with the transposed matrix $\mathbf{A}^T$ corresponds to $\mathbf{A}^T\mathbf{a}_i^q = \lambda_i^*\mathbf{a}_i^q$. Then, since $\mathbf{a}_k^H\mathbf{a}_i^q = \delta_{ik}$

$$\mathbf{P}^H\mathbf{T} = \begin{pmatrix} \mathbf{a}_1^{qH} \\ \mathbf{a}_2^{qH} \\ \vdots \end{pmatrix} (\mathbf{a}_1, \mathbf{a}_2 \ldots) = (\mathbf{a}_1, \mathbf{a}_2 \ldots) \begin{pmatrix} \mathbf{a}_1^{qH} \\ \mathbf{a}_2^{qH} \\ \vdots \end{pmatrix} = \mathbf{I}\,. \tag{B.18}$$

B.5 Complex Matrices

If $\mathbf{A} \in \mathcal{C}^{n\times n}$ the following notation is used: $\mathbf{A}\mathbf{a}_i \triangleq \lambda_i\mathbf{a}_i$ and $\mathbf{A}^H\mathbf{a}_i^q \triangleq \lambda_i^*\mathbf{a}_i^q$ where $\mathbf{a}_k^T\mathbf{a}_i^{q*} = \delta_{ki}$ or $\mathbf{a}_k^H\mathbf{a}_i^q = \delta_{ki}$.

B.6 Modal Decomposition

Transformation of the original system

$$\dot{\mathbf{x}}(t) = \mathbf{A}\mathbf{x}(t) + \mathbf{B}\mathbf{u}(t), \quad \mathbf{x}(0) = \mathbf{x}_o \qquad \mathbf{y}(t) = \mathbf{C}\mathbf{x}(t) \tag{B.19}$$

into the diagonalized modal system by defining the modal state variable $\mathbf{x}^{mo}$ is achieved by substituting $\mathbf{x} = \mathbf{T}\mathbf{x}^{mo}$. Thus,

$$\dot{\mathbf{x}}^{mo} = \mathbf{T}^{-1}\mathbf{A}\mathbf{T}\,\mathbf{x}^{mo}(t) + \mathbf{T}^{-1}\mathbf{B}\mathbf{u}(t) = (\text{diag}\,\lambda_i[\mathbf{A}])\,\mathbf{x}^{mo}(t) + \mathbf{T}^{-1}\mathbf{B}\mathbf{u}(t) \qquad \mathbf{y}(t) = \mathbf{C}\mathbf{T}\mathbf{x}^{mo}\,. \tag{B.20}$$

If the sign of eigenvector i is altered, the sign of the i-th component of the modal vector $\mathbf{x}^{mo}$ is merely changed.

B.7 Linear Differential Equations and Modal Transformations

The interrelations between original and modal domain are listed in Table B.2. An eigenvector $\mathbf{a}_i$, in any case, is determined except a constant factor α_i. To verify that the modal matrices $\mathbf{T} = (\mathbf{a}_1 \vdots \mathbf{a}_2 \ldots \mathbf{a}_n)$ and $\bar{\mathbf{T}} = (\alpha_1\mathbf{a}_1 \vdots \alpha_2\mathbf{a}_2 \ldots \alpha_n\mathbf{a}_n) = \mathbf{T}$ diag α_i possess the same modal decomposition, $\mathbf{T}$ is substituted by $\bar{\mathbf{T}}$ in the expression $\mathbf{A} = \mathbf{T}\Lambda\mathbf{T}^{-1}$

$$\mathbf{A} = \bar{\mathbf{T}}\Lambda\bar{\mathbf{T}}^{-1} = \mathbf{T}\,(\text{diag}\,\alpha_i)\Lambda(\text{diag}\,\alpha_i)^{-1}\mathbf{T}^{-1} = \mathbf{T}\Lambda\,(\text{diag}\,\alpha_i)(\text{diag}\,\alpha_i)^{-1}\mathbf{T}^{-1} = \mathbf{T}\Lambda\mathbf{T}^{-1}\,. \tag{B.21}$$

Example:

$$\mathbf{A} = \begin{pmatrix} -3 & -1 \\ 2 & 0 \end{pmatrix}, \quad \mathbf{T} = \begin{pmatrix} 1/\sqrt{5} & 1/\sqrt{2} \\ -2/\sqrt{2} & -1/\sqrt{2} \end{pmatrix}, \quad \mathbf{T}^{-1} = \begin{pmatrix} -\sqrt{5} & -\sqrt{5} \\ 2\sqrt{2} & \sqrt{2} \end{pmatrix}, \quad \lambda_{1,2} = -1; -2 \tag{B.22}$$

$$\mathbf{\Phi}(s) = \mathbf{T}\begin{pmatrix} \frac{1}{s+1} & 0 \\ 0 & \frac{1}{s+2} \end{pmatrix}\mathbf{T}^{-1} = \begin{pmatrix} -\frac{1}{s+1} + \frac{2}{s+2} & \vdots & -\frac{1}{s+1} + \frac{1}{s+2} \\ \frac{2}{s+1} - \frac{2}{s+2} & \vdots & \frac{2}{s+1} - \frac{1}{s+2} \end{pmatrix}.\ \square \tag{B.23}$$

Table B.2: Transformation into the modal domain

Original domain	Transformation	Modal domain
$\dot{\mathbf{x}} = \mathbf{A}\mathbf{x}$	$\mathbf{x} = \mathbf{T}\mathbf{x}^{mo}$	$\dot{\mathbf{x}}^{mo} = \mathbf{\Lambda}\mathbf{x}^{mo}$
$\mathbf{A}$	$\mathbf{T}^{-1}\mathbf{A}\mathbf{T} = \mathbf{\Lambda}$ or $\mathbf{A} = \mathbf{T}\mathbf{\Lambda}\mathbf{T}^{-1}$	$\mathbf{\Lambda}$
$\mathbf{x}(t) = e^{\mathbf{A}t}\,\mathbf{x}(0^+)$		$\mathbf{x}^{mo}(t) = e^{\mathbf{\Lambda}t}\,\mathbf{x}^{mo}(0^+)$
$\mathbf{\Phi}(t) = e^{\mathbf{A}t}$	$e^{\mathbf{A}t} = \mathbf{T}e^{\mathbf{\Lambda}t}\,\mathbf{T}^{-1}$	$e^{\mathbf{\Lambda}t} =$ diag $\{\exp(\lambda_i t)\}$
$\mathbf{\Phi}(s) = (s\mathbf{I}-\mathbf{A})^{-1}$	$(s\mathbf{I}-\mathbf{A})^{-1} = \mathbf{T}(s\mathbf{I}-\mathbf{\Lambda})^{-1}\mathbf{T}^{-1}$	$(s\mathbf{I}-\mathbf{\Lambda})^{-1} =$ diag $\frac{1}{s-\lambda_i}$

B.8 Eigenvalue Assignment

Defining a polynomial

$$p_f(\lambda) \triangleq p_{fo} + p_{f1}\lambda + \ldots\ldots + p_{f,n-1}\lambda^{n-1} + \lambda^n = (\mathbf{p}_f^T \,\vdots\, 1)\boldsymbol{\lambda} \tag{B.24}$$

$$\text{where } \mathbf{p}_f \triangleq (p_{fo}\;\; p_{f1}\cdots p_{f,n-1})^T \;\text{ and }\; \boldsymbol{\lambda} \triangleq (1\;\; \lambda\; \ldots\; \lambda^{n-1}\;\; \lambda^n)^T \tag{B.25}$$

consider an (n,n)-matrix $\mathbf{A}$ and an n-vector $\mathbf{b}$ and let $(\mathbf{A},\mathbf{b})$ be an observable pair, i.e. det $\mathbf{R} \neq 0$ where $\mathbf{R} = (\mathbf{b}, \mathbf{A}\mathbf{b}, \mathbf{A}^2\mathbf{b}, \ldots \mathbf{A}^{n-1}\mathbf{b})$. Then, the zeros of $p_f(\lambda)$ are assigned to the eigenvalues of $\mathbf{A} + \mathbf{b}\mathbf{k}^T$ if

$$\det\,(\lambda\mathbf{I} - \mathbf{A} - \mathbf{b}\mathbf{k}^T) = p_f(\lambda) = (\mathbf{p}_f^T \,\vdots\, 1)\boldsymbol{\lambda}\,. \tag{B.26}$$

The unique solution, given by Rissanen (*Ackermann, J., 1980*), is

$$\mathbf{k}^T = (\mathbf{p}_f^T \,\vdots\, 1)\begin{pmatrix} (\mathbf{R}^{-1})_{n\cdot} \\ (\mathbf{R}^{-1})_{n\cdot}\;\mathbf{A} \\ \vdots \\ (\mathbf{R}^{-1})_{n\cdot}\;\mathbf{A}^n \end{pmatrix} \triangleq (\mathbf{p}_f^T \,\vdots\, 1)\mathbf{W}^{-1} \tag{B.27}$$

the vector $(\mathbf{R}^{-1})_{n\cdot}$ being the last row of $\mathbf{R}^{-1}$. It can be rewritten

$$(\mathbf{R}^{-1})_{n\cdot} = (0\;\;0\;\ldots\;1)\mathbf{R}^{-1} = \mathbf{e}_n^T\mathbf{R}^{-1} \tag{B.28}$$

where $\mathbf{e}_n$ is the unit n-vector $\mathbf{e}_k$ with $k = n$. Note that the matrix $\mathbf{W}$ has the property of transforming the system $(\mathbf{A},\mathbf{b})$ to the control canonical form, i.e., $\mathbf{W}^{-1}\mathbf{A}\mathbf{W}$ is the companion matrix and $\mathbf{W}^{-1}\mathbf{b} = \mathbf{e}_n$.

B.9 Eigensystem Assignment

The closed-loop system given by the plant and the state feedback, respectively,

$$\dot{\mathbf{x}}(t) = \mathbf{A}\mathbf{x}(t) + \mathbf{B}\mathbf{u}(t), \qquad \mathbf{x}(0) = \mathbf{x}_o, \qquad \mathbf{u}(t) = \mathbf{K}\mathbf{x}(t) \tag{B.29}$$

can be rewritten to

$$\dot{\mathbf{x}}(t) = (\mathbf{A} + \mathbf{B}\mathbf{K})\mathbf{x}(t) \triangleq \mathbf{F}\mathbf{x}(t)\,. \tag{B.30}$$

Using the diagonal matrix of distinct eigenvalues or the Jordan canonical form $\mathbf{J}$, and the modal matrix of eigenvectors $\mathbf{T}$, one has

$$\mathbf{F} = \mathbf{A} + \mathbf{B}\mathbf{K} = \mathbf{T}\mathbf{J}\mathbf{T}^{-1} \quad\rightsquigarrow\quad \mathbf{A}\mathbf{T} - \mathbf{T}\mathbf{J} = -\mathbf{B}\mathbf{K}\mathbf{T} \tag{B.31}$$

and a Lyapunov-type of closed-loop system equation is obtained.

Let m_α denote a non-negative integer, $m_\alpha \geq 0$ and $\alpha > \|\mathbf{A}\|_F$ a positive constant. Then, it will be proved in what follows that choosing $\mathbf{T}$ according with the Lyapunov equation

$$-(\mathbf{A} + m_\alpha\alpha\mathbf{I}_n)\mathbf{T} + \mathbf{T}[-(\mathbf{A} + m_\alpha\alpha\mathbf{I}_n)]^T = -(m_\alpha + 1)\mathbf{B}\mathbf{R}^{-1}\mathbf{B}^T \quad\text{where}\quad \mathbf{R} > 0, \mathbf{T} = \mathbf{T}^T > 0 \tag{B.32}$$

makes all the closed-loop eigenvalues $\lambda[\mathbf{F}]$ lie to the left of $-m_\alpha\alpha$ in the complex s-plane, i.e., the controller matrix $\mathbf{K}$ is

$$\mathbf{K} = -(m_\alpha + 1)\mathbf{R}^{-1}\mathbf{B}^T\mathbf{T}^{-1} \quad \rightsquigarrow \quad \lambda[\mathbf{F}] < -m_\alpha\alpha \ . \tag{B.33}$$

Proof: From Eq.(B.32)

$$-\mathbf{AT} - \mathbf{TA}^T - 2m_\alpha\alpha\mathbf{T} = -(m_\alpha + 1)\mathbf{BR}^{-1}\mathbf{B}^T \tag{B.34}$$

$$\mathbf{AT} + \mathbf{T}(\mathbf{A}^T + 2m_\alpha\alpha\mathbf{I}_n) - m_\alpha\mathbf{BR}^{-1}\mathbf{B}^T\mathbf{T}^{-1}\mathbf{T} = \mathbf{BR}^{-1}\mathbf{B}^T \tag{B.35}$$

$$\underbrace{(\mathbf{A} - m_\alpha\mathbf{BR}^{-1}\mathbf{B}^T\mathbf{T}^{-1})}_{\mathbf{A}_1}T + \mathbf{T}\underbrace{(\mathbf{A}^T + 2m_\alpha\alpha\mathbf{I}_n)}_{-\mathbf{J}} = \mathbf{BR}^{-1}\mathbf{B}^T \ . \tag{B.36}$$

From Eq.(B.36) the following statements can be derived: First,

$$\mathbf{A}_1 = \mathbf{A} - m_\alpha\mathbf{BR}^{-1}\mathbf{B}^T\mathbf{T}^{-1} \triangleq \mathbf{A} + \mathbf{BK}_1 \quad \rightsquigarrow \quad \mathbf{K}_1 = -m_\alpha\mathbf{R}^{-1}\mathbf{B}^T\mathbf{T}^{-1} \tag{B.37}$$

where $\mathbf{K}_1$ is a feedback gain to stabilize $\mathbf{A}$. Second, comparing Eq.(B.36) with Eq.(B.31),

$$\mathbf{BKT} = -\mathbf{BR}^{-1}\mathbf{B}^T \quad \rightsquigarrow \quad \mathbf{K}_2 = -\mathbf{R}^{-1}\mathbf{B}^T\mathbf{T}^{-1} \tag{B.38}$$

where $\mathbf{K}_2$ is an additional feedback to shift $\lambda[\mathbf{F}]$ to the left of $-\alpha$. Hence

$$\mathbf{F} = \mathbf{A}_1 + \mathbf{BK}_2 = \mathbf{A} + \mathbf{B}(\mathbf{K}_1 + \mathbf{K}_2) = \mathbf{A} + \mathbf{BK} \tag{B.39}$$

$$\mathbf{K} = \mathbf{K}_1 + \mathbf{K}_2 = -(m_\alpha + 1)\mathbf{R}^{-1}\mathbf{B}^T\mathbf{T}^{-1} \ . \tag{B.40}$$

Third, in order to obtain all the eigenvalues of $\mathbf{J} = -(\mathbf{A}^T + 2m_\alpha\alpha\mathbf{I}_n)$ in the left of $-m_\alpha\alpha$ in the complex plane, the decay factor α must satisfy $\alpha > ||\mathbf{A}||_F$ since

$$\lambda[\mathbf{J}] = \lambda[-\mathbf{A}^T - 2m_\alpha\alpha\mathbf{I}] = \lambda[-\mathbf{A}^T] - 2m_\alpha\alpha < -m_\alpha\alpha \tag{B.41}$$

$$-\lambda[\mathbf{A}] < m_\alpha\alpha \quad \rightsquigarrow \quad \alpha > -\lambda[\mathbf{A}]/m_\alpha \quad \Leftarrow \quad \alpha > ||\mathbf{A}||_F \ . \tag{B.42}$$

End of the Proof

Example:

$$\mathbf{A} = \begin{pmatrix} 0 & 1 \\ -2 & -3 \end{pmatrix} \qquad \lambda_i[A] = -1; -2 \qquad \mathbf{B} = \begin{pmatrix} 0 \\ 1 \end{pmatrix} \qquad \mathbf{R} = \mathbf{I} \tag{B.43}$$

$$m_\alpha = 1 \qquad ||\mathbf{A}||_F = \sqrt{14} = 3.7417 \qquad \sigma_{\max}[\mathbf{A}] = 3.7025 \ . \tag{B.44}$$

$$\alpha > \sigma_{\max}[\mathbf{A}] \rightsquigarrow \alpha = 4 \ . \text{ From Lyapunov equation } \mathbf{T} = \frac{1}{30}\begin{pmatrix} 1 & -4 \\ -4 & 22 \end{pmatrix}, \quad \mathbf{T}^{-1} = \begin{pmatrix} 110 & 20 \\ 20 & 5 \end{pmatrix} \tag{B.45}$$

$$\mathbf{K} = -(m_\alpha + 1)\mathbf{R}^{-1}\mathbf{B}^T\mathbf{T}^{-1} = -2(0 \ \ 1)\begin{pmatrix} 110 & 20 \\ 20 & 5 \end{pmatrix} = (-40 \ \ -10) \tag{B.46}$$

$$\mathbf{F} = \mathbf{A} + \mathbf{BK} = \begin{pmatrix} 0 & 1 \\ -2 & -3 \end{pmatrix} + \begin{pmatrix} 0 \\ 1 \end{pmatrix}(-40 \ \ -10) = \begin{pmatrix} 0 & 0 \\ -42 & -13 \end{pmatrix} \quad \rightsquigarrow \quad \lambda_i[\mathbf{F}] = -6; -7 \ . \tag{B.47}$$

Check $\lambda[-\mathbf{A}] - 2m_\alpha\alpha < -4$.

End of Example

B.10 Complete Modal Synthesis

Consider the nth order open-loop system and its matrices $\mathbf{A}, \mathbf{B} \in \mathcal{R}^{n\times m}$, the closed-loop matrix with state feedback $\mathbf{F} = \mathbf{A} + \mathbf{BK}$ and the eigenvectors $\mathbf{f}_i$ associated with $\mathbf{F}$. Then,

$$(\mathbf{A} + \mathbf{BK})\mathbf{f}_i = \mathbf{Ff}_i = \lambda_i[\mathbf{F}]\ \mathbf{f}_i \quad \rightsquigarrow \quad (\mathbf{A} - \lambda_i[\mathbf{F}]\ \mathbf{I})\mathbf{f}_i = -\mathbf{BKf}_i \triangleq -\mathbf{Bp}_i \ . \tag{B.48}$$

With the aforementioned definition $\mathbf{p}_i \triangleq \mathbf{Kf}_i$ of n parameter vectors $\mathbf{p}_i \in \mathcal{R}^m$ and a parameter matrix $\mathbf{P} \in \mathcal{R}^{m\times n}$ it results, using the modal matrix $\mathbf{T}[\mathbf{F}]$ of the closed loop (*Roppenecker, G., and Lohmann, B., 1989 ; Roppenecker, G., 1987 and 1988*),

$$\mathbf{f}_i = -(\mathbf{A} - \lambda_i[\mathbf{F}]\mathbf{I}_n)^{-1}\mathbf{Bp}_i \qquad \mathbf{K} = \mathbf{P}\ \mathbf{T}^{-1}[\mathbf{F}] \tag{B.49}$$

$$\mathbf{P} = (\mathbf{p}_1 \vdots \mathbf{p}_2 \ \dots \ \mathbf{p}_n) = (\mathbf{Kf}_1 \vdots \mathbf{Kf}_2 \ \dots \ \mathbf{Kf}_n) = \mathbf{K}(\mathbf{f}_1 \vdots \mathbf{f}_2 \ \dots \ \mathbf{f}_n) = \mathbf{K}\ \mathbf{T}[\mathbf{F}] \ . \tag{B.50}$$

B.11 Vandermonde Matrix

Consider a system with (n,n)-matrix $\mathbf{A}$ in companion form (regulator form, controllable canonical form) and distinct eigenvalues,

$$\mathbf{A} = \begin{pmatrix} 0 & 1 & 0 & \dots & 0 \\ 0 & 0 & 1 & \dots & 0 \\ \vdots & \vdots & & \ddots & \\ 0 & 0 & 0 & & 1 \\ -a_o & -a_1 & \dots & & -a_{n-1} \end{pmatrix}, \qquad \mathbf{b}_1 = \begin{pmatrix} 0 \\ 0 \\ \vdots \\ 0 \\ 1 \end{pmatrix}. \tag{B.51}$$

Then, the eigenvectors (the eigenvector direction) and the modal matrix are given by the Vandermonde matrix $\mathbf{V}_{dm}$

$$\mathbf{T} = \mathbf{V}_{dm} = \begin{pmatrix} 1 & 1 & \dots & \dots & 1 \\ \lambda_1 & \lambda_2 & \dots & \dots & \lambda_n \\ \lambda_1^2 & \lambda_2^2 & \dots & \dots & \lambda_n^2 \\ \vdots & \vdots & \dots & \dots & \vdots \\ \lambda_1^{n-1} & \lambda_2^{n-1} & \dots & \dots & \lambda_n^{n-1} \end{pmatrix} \tag{B.52}$$

which simplifies computation to a high extent. Note the helpful relation that the determimant of the Vandermonde matrix is equal to $\prod_{i,j=1;i<j}^{n} (\lambda_j - \lambda_i)$. If $\mathbf{A}$ is given as a companion matrix and the modal decomposition is to be calculated, note for computational purpose: $\mathbf{T}[\mathbf{A}]$ can be calculated simply via Vandermonde matrix but Vandermonde matrix cannot be applied to $\mathbf{T}^{-1} = \mathbf{P}^T = \mathbf{T}^T[\mathbf{A}^T]$ (Table B.1) since $\mathbf{A}^T$ is not a companion matrix.

Consider another canonical form (*Isermann, R., 1981*) with the system matrix $\mathbf{A}'$

$$\mathbf{A}' \triangleq \mathbf{I}^{\checkmark}\mathbf{A}^T\mathbf{I}^{\checkmark} = \mathbf{A} = \begin{pmatrix} -a_{n-1} & 1 & 0 & \dots & 0 \\ -a_{n-2} & 0 & 1 & \dots & 0 \\ \vdots & \vdots & & \ddots & \\ -a_1 & 0 & 0 & & 1 \\ -a_o & 0 & 0 & \dots & 0 \end{pmatrix} \tag{B.53}$$

where $\mathbf{A}'$ is reflected from $\mathbf{A}$ with respect to the secondary axis and $\mathbf{A}$ is taken from the aforementioned $\mathbf{A}$ in companion form. Then, the corresponding modal matrix is denoted $\mathbf{T}' = \mathbf{T}[\mathbf{A}']$, the eigenvalues are the same $\lambda_i[\mathbf{A}] = \lambda_i[\mathbf{A}']$. Combining $\mathbf{T}\Lambda\mathbf{T}^{-1} = \mathbf{A}$ and $\mathbf{T}'\Lambda(\mathbf{T}')^{-1} = \mathbf{A}' = \mathbf{I}^{\checkmark}\mathbf{A}^T\mathbf{I}^{\checkmark}$ in order to eliminate $\mathbf{A}$, and referring to $\mathbf{I}^{\checkmark} = (\mathbf{I}^{\checkmark})^{-1}$

$$\mathbf{T}'\Lambda(\mathbf{T}')^{-1} = \mathbf{I}^{\checkmark}(\mathbf{T}^{-1})^T\Lambda\mathbf{T}^T\mathbf{I}^{\checkmark} = (\mathbf{T}^T\mathbf{I}^{\checkmark})^{-1}\Lambda(\mathbf{T}^T\mathbf{I}^{\checkmark}) \quad \rightsquigarrow \quad (\mathbf{T}')^{-1} = \mathbf{T}^T\mathbf{I}^{\checkmark} = \mathbf{V}_{md}^T\mathbf{I}^{\checkmark} . \tag{B.54}$$

Example:

$$\mathbf{A} = \begin{pmatrix} 0 & 1 \\ 5 & 4 \end{pmatrix}; \quad \lambda_i[\mathbf{A}] = -1,\ 5; \quad \mathbf{a}_i = \begin{pmatrix} 1 \\ -1 \end{pmatrix}, \begin{pmatrix} 1 \\ 5 \end{pmatrix} \quad \mathbf{T} = \mathbf{V}_{dm} = \begin{pmatrix} 1 & 1 \\ -1 & 5 \end{pmatrix} \tag{B.55}$$

$$\mathbf{A} = \mathbf{T}\ \mathrm{diag}\lambda_i[\mathbf{A}]\ \mathbf{T}^{-1} = \begin{pmatrix} 1 & 1 \\ -1 & 5 \end{pmatrix}\begin{pmatrix} -1 & 0 \\ 0 & 5 \end{pmatrix}\begin{pmatrix} 5 & -1 \\ 1 & 1 \end{pmatrix}/6 . \tag{B.56}$$

Eq.(B.54) $\rightsquigarrow$

$$(\mathbf{T}')^{-1} = \begin{pmatrix} 1 & 1 \\ -1 & 5 \end{pmatrix}^T \begin{pmatrix} 0 & 1 \\ 1 & 0 \end{pmatrix} = \begin{pmatrix} -1 & 1 \\ 5 & 1 \end{pmatrix} \quad \rightsquigarrow \quad \mathbf{T}' = \begin{pmatrix} -1 & 1 \\ 5 & 1 \end{pmatrix}/6 \tag{B.57}$$

which is equivalent to $\mathbf{T}'$ or $\tilde{\mathbf{T}}'$, calculated directly from

$$\mathbf{A}' = \begin{pmatrix} 4 & 1 \\ 5 & 0 \end{pmatrix} \quad \text{yielding, e.g.,} \quad \tilde{\mathbf{T}}' = \begin{pmatrix} 1 & 1 \\ -5 & 1 \end{pmatrix} . \ \square \tag{B.58}$$

Consider the nth order system as given in Eq.(B.20) but single-input ($\mathbf{u} := u$) and distinct eigenvalues, only. The Vandermonde matrix $\mathbf{V}_{dm}$ also plays an important role in decomposing the Kalman controllability matrix $\mathbf{S}_k$ as mentioned in Eq.(A.30).

$$\mathbf{S}_k = \text{diag}\ \{(\mathbf{T}^{-1}\mathbf{b})_i\}\mathbf{V}_{dm}^T \ . \tag{B.59}$$

Then, rank $\mathbf{S}_k < n$ occurs if one or more elements $(\mathbf{T}^{-1}\mathbf{b})_i$ should vanish and controllability is no longer given. Furthermore, rank$\mathbf{S}_k$ is the dimension of the controllable subspace (*Ackermann, J., 1988*).

B.12 Decompostion into Eigenvectors

In this section the solution of a system of linear algebraic equations is decomposed into a linear combination of the eigenvectors of the system matrix. Consider the nonhomogenous equation $\mathbf{Ax} = \mathbf{b}$ with the (n,n)-matrix $\mathbf{A}$ real and symmetric. The eigenvalues $\lambda[\mathbf{A}]$ are assumed to be distinct. The (right) eigenvectors $\mathbf{a}_i$ of $\mathbf{A}$ are orthogonal. Moreover, $\mathbf{a}_i$ are normalized to unity. Expanding the unknown solution $\mathbf{x}$ into vector components according with $\mathbf{a}_i$

$$\mathbf{x} = \sum_{i=1}^{n} \alpha_i \mathbf{a}_i \ , \qquad \mathbf{Ax} = \mathbf{A}\sum_{i=1}^{n} \alpha_i \mathbf{a}_i = \sum_{i=1}^{n} \alpha_i \mathbf{A}\mathbf{a}_i = \sum_{i=1}^{n} \alpha_i\ \lambda_i[\mathbf{A}]\ \mathbf{a}_i = \mathbf{b} \ . \tag{B.60}$$

Premultiplying this equation with $\mathbf{a}_j^T$ yields

$$\alpha_j \lambda_j[\mathbf{A}]\mathbf{a}_j^T\mathbf{a}_j + \sum_{i=1;i\neq j}^{n} \alpha_i \lambda_i[A]\mathbf{a}_j^T\mathbf{a}_i = \alpha_j\lambda_j[\mathbf{A}]\times 1 + 0 = \mathbf{a}_j^T\mathbf{b} \ \leadsto \ \alpha_j = \frac{\mathbf{a}_j^T\mathbf{b}}{\lambda_j[\mathbf{A}]} \quad \text{and} \quad \mathbf{x} = \sum_{j=1}^{n} \frac{\mathbf{a}_j^T\mathbf{b}}{\lambda_j[\mathbf{A}]}\mathbf{a}_j \ . \tag{B.61}$$

B.13 Properties of Eigenvalues

B.13.1 Smallest and Largest Eigenvalue of Symmetric Matrices

Consider $\lambda_{\min}$ and $\lambda_{\max}$, the smallest and largest eigenvalues of a symmetric (n,n)-matrix $\mathbf{A}$, respectively. Then, the following relations hold for any vector $\mathbf{x}$

$$|\lambda_{\min}|\ ||\mathbf{x}||_F \le ||\mathbf{Ax}||_F \le |\lambda_{\max}|\ ||\mathbf{x}||_F \qquad |\lambda_{\min}|\ ||\mathbf{x}||_F^2 \le |\mathbf{x}^T\mathbf{Ax}| \le |\lambda_{\max}|\ ||\mathbf{x}||_F^2 \ . \tag{B.62}$$

If the matrix $\mathbf{A}$ is positive definite, the notation for the modulus can be omitted.

All eigenvalues $\lambda[\mathbf{A}]$ are real and positive if $\mathbf{A}$ is positive definite. Assume all eigenvalues distinct. Thus, all eigenvectors $\mathbf{a}$ are distinct, too. With regard to the symmetry $\mathbf{A} = \mathbf{A}^T$ the eigenvectors are orthogonal, see Eq.(B.16), and form an orthogonal basis in n-dimensional space. Any vector variable $\mathbf{x}$ can be assumed of the form

$$\mathbf{x} = \sum_{i=1}^{n} \alpha_i \mathbf{a}_i . \tag{B.63}$$

The quadratic forms $\mathbf{x}^T\mathbf{Ax}$ and $\mathbf{x}^T\mathbf{x}$ using Eq.(B.63) can be deduced as follows

$$\mathbf{Ax} = \sum_{i=1}^{n} \alpha_i \mathbf{A}\mathbf{a}_i = \sum_{i=1}^{n} \alpha_i\lambda_i[\mathbf{A}]\ \mathbf{a}_i \tag{B.64}$$

$$\mathbf{x}^T\mathbf{Ax} = \Big(\sum_{j=1}^{n} \alpha_j \mathbf{a}_j^T\Big) \sum_{i=1}^{n} \alpha_i\lambda_i[\mathbf{A}]\ \mathbf{a}_i = \sum_j\sum_i \alpha_i\alpha_j\lambda_i\mathbf{a}_j^T\mathbf{a}_i = \sum_i \alpha_i^2\lambda_i\mathbf{a}_i^T\mathbf{a}_i \ . \tag{B.65}$$

Using $\lambda_{\max}$ leads to $\mathbf{x}^T\mathbf{Ax} = \sum_i \alpha_i^2\lambda_i\mathbf{a}_i^T\mathbf{a}_i \le \sum_i \alpha_i^2\lambda_{\max}\mathbf{a}_i^T\mathbf{a}_i = \lambda_{\max}\mathbf{x}^T\mathbf{x}$. Finally, an upper and a lower bound for the quadratic (Lyapunov) form is

$$\lambda_{\min}[\mathbf{A}]\ \mathbf{x}^T\mathbf{x} \le \mathbf{x}^T\mathbf{Ax} \le \lambda_{\max}[\mathbf{A}]\ \mathbf{x}^T\mathbf{x} \ . \tag{B.66}$$

From Eq.(B.64) one can derive

$$||\mathbf{Ax}||_F \ \le \ ||\sum_i \alpha_i\lambda_{\max}\mathbf{a}_i||_F = \lambda_{\max}||\sum_i \alpha_i\mathbf{a}_i||_F = \lambda_{\max}\ ||\mathbf{x}||_F \ . \ \square \tag{B.67}$$

If **A** and **B** are symmetric matrices and $\mathbf{A} > 0$ then a nonsingular matrix **S** exists such that

$$\mathbf{S}^T(\mathbf{A}+\mathbf{B})\mathbf{S} = \mathbf{I} + \mathrm{diag}\{\lambda_i[\mathbf{A}^{-1}\mathbf{B}]\} \tag{B.68}$$

(*Thrall, R.M., and Tornheim, L., 1957, p.188; Petkovski, D.B., 1985*).

For **A**, **B**, $\mathbf{C} \in \mathcal{R}^{n\times n}$ symmetric and $\mathbf{A}, \mathbf{B} \geq 0, \mathbf{C} > 0$ one has

$$\lambda_{\max}[\mathbf{AB}] \leq \lambda_{\max}[\mathbf{AC}]\ \lambda_{\max}[\mathbf{C}^{-1}\mathbf{B}]\ . \tag{B.69}$$

Proof: Setting $\mathbf{C} = \mathbf{T}^T\mathbf{T}$ and using $\lambda_i[\mathbf{M}] = \lambda_i[\mathbf{TMT}^{-1}]$ and Eq.(22.25)

$$\begin{aligned}\lambda_{\max}[\mathbf{AC}]\lambda_{\max}[\mathbf{C}^{-1}\mathbf{B}] &= \lambda_{\max}[\mathbf{AT}^T\mathbf{T}]\ \lambda_{\max}[\mathbf{T}^{-1}\mathbf{T}^{-1,T}\mathbf{B}] = \lambda_{\max}[\mathbf{TAT}^T]\ \lambda_{\max}[\mathbf{T}^{-1,T}\mathbf{BT}^{-1}] \\ &= \|\mathbf{TAT}^T\|_s\|\mathbf{T}^{-1,T}\mathbf{BT}^{-1}\|_s \geq \|\mathbf{TABT}^{-1}\|_s = \lambda_{\max}[\mathbf{TABT}^{-1}] = \lambda_{\max}[\mathbf{AB}]\end{aligned} \tag{B.70}$$

(*Corless, M., and Da, D., 1989*) . □

B.13.2 Eigenvalues and Trace

A similar property is given as follows. If $\mathbf{G}, \mathbf{H} \in \mathcal{R}^{n\times n}$ where $\mathbf{H} \geq 0$ and **G** such that either the relations hold $\mathbf{G}, \mathbf{H} \geq 0$ (*Kleinman, D.L., and Athans, M., 1968*) or $\mathbf{H} = \mathbf{H}^T$ (*Wang, S.D., et al. 1986*), then,

$$\lambda_{\min}[\mathbf{G}]\ \mathrm{tr}\ \mathbf{H} \leq \mathrm{tr}\ [\mathbf{GH}] \leq \lambda_{\max}[\mathbf{G}]\ \mathrm{tr}\ \mathbf{H}\ . \tag{B.71}$$

Calculating bounds of the Riccati and Lyapunov matrix equation, these conditions are useful (*Kwon, B.H., et al. 1985; Mori, T., et al. 1987*).

B.13.3 Maximum Real Part of an Eigenvalue

For $\mathbf{A} \in \mathcal{C}^{n\times n}$ consider $\mathbf{Aa}_i = \lambda[\mathbf{A}]\mathbf{a}_i$, the expression $\mathbf{a}_i^H\mathbf{Aa}_i = \lambda_i[\mathbf{A}]\mathbf{a}_i^H\mathbf{a}_i$ and its conjugate transpose $\mathbf{a}_i^H\mathbf{A}^H\mathbf{a}_i = \lambda_i^*[\mathbf{A}]\mathbf{a}_i^H\mathbf{a}_i$. Adding both expressions above,

$$\mathbf{a}_i^H(\mathbf{A}+\mathbf{A}^H)\mathbf{a}_i = (\lambda_i[\mathbf{A}] + \lambda_i^*[\mathbf{A}])\mathbf{a}_i^H\mathbf{a}_i = 2(\Re e\ \lambda_i[\mathbf{A}])\mathbf{a}_i^H\mathbf{a}_i\ . \tag{B.72}$$

Since the sum $\mathbf{A}+\mathbf{A}^H$ is Hermitian, from Rayleigh's theorem it is known that

$$\max_{\mathbf{a}_i} \frac{\mathbf{a}_i^H\frac{\mathbf{A}+\mathbf{A}^H}{2}\mathbf{a}_i}{\mathbf{a}_i^H\mathbf{a}_i} = \lambda_{\max}[\frac{\mathbf{A}+\mathbf{A}^H}{2}]\ . \tag{B.73}$$

Hence,

$$\lambda_{\max}[\frac{\mathbf{A}+\mathbf{A}^H}{2}] = \max_i \Re e\ \lambda_i[\mathbf{A}] \geq \Re e\ \lambda_i[\mathbf{A}]\ . \tag{B.74}$$

If $\mathbf{A} \in \mathcal{R}^{n\times n}$ then the symmetric part $\mathbf{A}_s$ is used to evaluate $\Re e\ \lambda_i[\mathbf{A}]$

$$\lambda_{\max}[\frac{\mathbf{A}+\mathbf{A}^T}{2}] = \lambda_{\max}[\mathbf{A}_s] = \max_i \Re e\ \lambda_i[\mathbf{A}] \geq \Re e\ \lambda_i[\mathbf{A}]\ .\ \square \tag{B.75}$$

Another property is

$$\lambda_{\min}[\mathbf{A}_s] + \lambda_{\min}[\mathbf{B}_s] \leq \Re e\ \lambda[\mathbf{A}+\mathbf{B}] \leq \lambda_{\max}[\mathbf{A}_s] + \lambda_{\max}[\mathbf{B}_s] \tag{B.76}$$

where $\mathbf{A}_s \triangleq (\mathbf{A}+\mathbf{A}^T)/2$, $\mathbf{A}, \mathbf{B} \in \mathcal{R}^{n\times n}$ □ (*Jiang, C.L., 1987*).

B.13.4 Definiteness of $\mathbf{A}_s$ and Stability of A

The symmetric part of a matrix **A** is $\mathbf{A}_s \triangleq (\mathbf{A}+\mathbf{A}^T)/2$. If the matrix $\mathbf{A}_s$ is negative definite ($\mathbf{x}^T\mathbf{A}_s\mathbf{x} < 0$ for any **x**) since Eq.(B.73) and since $\mathbf{a}_i \in \{\mathbf{x}\} \rightsquigarrow \lambda_{\max}[\mathbf{A}_s] < 0$. Since Eq.(B.75) $\lambda_{\max}[\mathbf{A}_s] > \Re e\ \lambda_i[\mathbf{A}]$. Hence,

$$\mathbf{A}_s < 0 \quad \rightsquigarrow \quad \Re e\ \lambda_i[\mathbf{A}] < 0 \quad \text{and } \mathbf{A} \text{ is stable}\ . \tag{B.77}$$

Further properties concerning norms of a matrix see separate chapter and *Heitzinger, W., et al. 1985*.

B.13.5 Adding the Identity Matrix

Adding $\mu\mathbf{I}$ (μ positive) to a matrix $\mathbf{A}$ causes increased eigenvalues

$$\lambda[\mu\mathbf{I} + \mathbf{A}] = \mu + \lambda[\mathbf{A}]. \tag{B.78}$$

The eigenvalues are uniformly shifted to the right. The amount of shifting is μ. Frequently $\mathbf{A}$ serves as a system coefficient matrix in state-space approach. Then, the displacement follows towards the imaginary axis, towards instability. Adding the μ-scaled identity matrix to $\mathbf{A}^{-1}$ results in $\lambda[\mu\mathbf{I}+\mathbf{A}^{-1}] = \mu+1/\lambda[\mathbf{A}]$.

Example: Inverting $(s\mathbf{I}_n - \mathbf{A})$, the determinant $\det(s\mathbf{I}_n - \mathbf{A})$ is needed. Note that the s-dependent eigenvalues of the combined matrix $(s\mathbf{I}_n - \mathbf{A})$ are related to $\lambda_i[\mathbf{A}]$ as follows $\lambda_i[s\mathbf{I}_n - \mathbf{A}] = -\lambda_i[\mathbf{A}] + s$. For the sake of completeness,

$$\lambda_i[-\mathbf{A}] = -\lambda_i[\mathbf{A}] \qquad \det(-\mathbf{A}) = (-1)^n \det \mathbf{A} = (-1)^n \prod_{i=1}^{n} \lambda_i[\mathbf{A}]. \tag{B.79}$$

Characteristic equation and the sum and product of eigenvalues (Vieta's rule):

$$\det(s\mathbf{I}_n - \mathbf{A}) = 0 \qquad s^n + s^{n-1}\sum_{i=1}^{n}\{-\lambda_i[\mathbf{A}]\} + \ldots + \prod_{i=1}^{n}\{-\lambda_i[\mathbf{A}]\} = 0\,. \tag{B.80}$$

End of Example

B.13.6 Eigenvalues of Matrix Products

Note the property

$$\lambda[\mathbf{AB}] = \lambda[\mathbf{BA}] \qquad \text{if } \mathbf{A} \text{ and } \mathbf{B} \text{ are square}\,. \tag{B.81}$$

B.13.7 Eigenvalue of a Matrix Polynomial

For a matrix polynomial $f(\mathbf{A})$ (or a function which can be represented by a matrix polynomial) there exists the relation between eigenvectors

$$\lambda[f(\mathbf{A})] = f(\lambda[\mathbf{A}])\,. \tag{B.82}$$

B.13.8 Weyl Inequality

Let $\mathbf{A}$ and $\mathbf{E}$ be Hermitian matrices and let a perturbed matrix be defined as $\mathbf{A}_p = \mathbf{A} + \mathbf{E}$ where the largest eigenvalue is denoted by the subscript $_n$ etc.

$$\lambda_{\min}[\mathbf{A}] = \lambda_1[\mathbf{A}] \le \lambda_2[\mathbf{A}] \le \ldots \le \lambda_n[\mathbf{A}] = \lambda_{\max}[\mathbf{A}]\,. \tag{B.83}$$

Then,

$$\lambda_i[\mathbf{A}] + \lambda_{\max}[\mathbf{E}] \ge \lambda_i[\mathbf{A}_p] \ge \lambda_i[\mathbf{A}] + \lambda_{\min}[\mathbf{E}] \qquad \forall i = [1, n]\,. \tag{B.84}$$

If $\mathbf{E} > 0$ then $\lambda_i[\mathbf{A}_p] > \lambda_i[\mathbf{A}] \quad \forall i = [1, n]$ and

$$|\ \lambda_i[\mathbf{A}_p] - \lambda_i[\mathbf{A}]\ | \le ||\mathbf{E}||_s \qquad \forall i = [1, n] \tag{B.85}$$

(*Franklin, J.N., 1968, p.157*). For comparison see Eq.(15.43).

B.14 Rayleigh's Theorem

Consider $\mathbf{G} = \mathbf{G}^H \in \mathcal{C}^{n\times n}$ and let $\mathbf{G}_{n-r} \in \mathcal{C}^{(n-r)\times(n-r)}$ be a principal submatrix of $\mathbf{G}$ $\forall r = 1 \ \ldots \ n-1$.

$$\begin{array}{ll} \text{If the eigenvalues of } \mathbf{G}_{n-r} \text{ are} & \lambda_1[\mathbf{G}_{n-r}] \le \lambda_2[\mathbf{G}_{n-r}] \le \ldots \le \lambda_{n-r}[\mathbf{G}_{n-r}] \\ \text{and the eigenvalues of } \mathbf{G} \text{ are} & \lambda_1[\mathbf{G}] \le \lambda_2[\mathbf{G}] \le \ldots \le \lambda_n[\mathbf{G}] \end{array} \tag{B.86}$$

then (*Lancaster, P., and Tismenetsky, M., 1985, p. 294*)

$$\lambda_i[\mathbf{G}] \le \lambda_i[\mathbf{G}_{n-r}] \le \lambda_{i+r}[\mathbf{G}] \ \ \forall i = 1 \ \ldots \ n-r\,. \tag{B.87}$$

B.15 Eigenvalues and Eigenvectors of the Inverse

Let an eigenvalue of $\mathbf{A}$ and $\mathbf{A}^{-1}$ be λ and μ, respectively,

$$\det(\lambda\mathbf{I}-\mathbf{A})=0 \qquad \det(\mu\mathbf{I}-\mathbf{A}^{-1})=0\ . \tag{B.88}$$

Multiplying by $\det\mathbf{A}$ yields

$$\det\mathbf{A}\,\det(\mu\mathbf{I}-\mathbf{A}^{-1})=\det(\mu\mathbf{A}-\mathbf{I})=0\ . \tag{B.89}$$

Comparing with Eq.(B.88) yields $\mu=1/\lambda$. If $\lambda[\cdot]$ is used as an operator for determining the eigenvalue, then generally, $\lambda[\mathbf{A}^{-1}]=\lambda^{-1}[\mathbf{A}]$. Denoting the eigenvectors of $\mathbf{A}^{-1}$ as $\mathbf{a}^{\bullet}$ then $\mathbf{A}^{-1}\mathbf{a}^{\bullet}=\lambda^{-1}\mathbf{a}^{\bullet}$. Multiplying by $\mathbf{A}$ from the left shows that $\mathbf{a}^{\bullet}=\mathbf{a}$ is the solution. If a square matrix $\mathbf{A}$ has eigenvalues λ_i and eigenvectors $\mathbf{a}_i$, then, the inverted matrix $\mathbf{A}^{-1}$ has inverted eigenvalues $1/\lambda_i$ but identical eigenvectors $\mathbf{a}_i^{\bullet}=\mathbf{a}_i$. Eigenvalues are reciprocals, the eigenvectors are shared.

Example:

$$\mathbf{A}=\begin{pmatrix}0 & -1\\ -5 & 4\end{pmatrix} \quad\leadsto\quad \det\mathbf{A}=-5 \tag{B.90}$$

$$\mathbf{A}^{-1}=\begin{pmatrix}-0.8 & -0.2\\ -1 & 0\end{pmatrix} \quad\leadsto\quad \det\mathbf{A}^{-1}=-0.2 \tag{B.91}$$

$$\lambda_{1,2}=-1;\ 5 \qquad \mu_{1,2}=-1;\ 0.2 \qquad \mathbf{a}_{1,2}^{\bullet}=\begin{pmatrix}0.7071\\0.7071\end{pmatrix},\begin{pmatrix}-0.1961\\0.9806\end{pmatrix}. \tag{B.92}$$

End of Example

B.16 Dyadic Decomposition (Spectral Representation)

Given the m-vector $\mathbf{a}$ and the n-vector $\mathbf{b}$, the (m,n)-matrix $\mathbf{ab}^T$ is the dyadic product (or outer product). Using the modal matrices $\mathbf{T}[\mathbf{A}]$ and $\mathbf{P}=\mathbf{T}[\mathbf{A}^T]$ of $\mathbf{A}$ and $\mathbf{A}^T$, respectively, decomposed to columns (eigenvectors)

$$\mathbf{T}=\mathbf{T}[\mathbf{A}]=(\mathbf{a}_1,\mathbf{a}_2\ldots\mathbf{a}_n) \qquad \mathbf{P}=\mathbf{T}[\mathbf{A}^T]=(\mathbf{p}_1,\mathbf{p}_2\cdots\mathbf{p}_n)\ , \tag{B.93}$$

one can decompose the (n,n)-matrix $\mathbf{A}$

$$\mathbf{A}=\mathbf{T}(\text{diag}\ \lambda_i)\mathbf{T}^{-1}=\mathbf{T}(\text{diag}\ \lambda_i)\mathbf{P}^T \tag{B.94}$$

$$\mathbf{A}=\sum_{i=1}^{n}\lambda_i(\mathbf{a}_i\mathbf{p}_i^T)=\sum_{i=1}^{n}(\lambda_i\mathbf{a}_i)\mathbf{p}_i^T=\sum_{i=1}^{n}(\mathbf{A}\mathbf{a}_i)\mathbf{p}_i^T=\mathbf{A}\sum_{i=1}^{n}\mathbf{a}_i\mathbf{p}_i^T=\mathbf{AI}=\mathbf{A}\ . \tag{B.95}$$

In this equation use is made of the following result

$$\sum_{i=1}^{n}\mathbf{a}_i\mathbf{p}_i^T=\text{matrix}[\sum_i a_{ik}p_{il}]=\mathbf{TP}^T=\mathbf{TT}^{-1}=\mathbf{I} \tag{B.96}$$

caused by the orthonormality of $\mathbf{T}$ and $\mathbf{P}^T$. Both expressions of $\mathbf{A}$, mentioned in the right-hand side of Eqs. (B.94) and (B.95), are equal to a square matrix the (k,l)-element of which is $\sum_{i=1}^{n}\lambda_i\,a_{ik}\,p_{il}$.

Example:

$$\mathbf{A}=\begin{pmatrix}0 & -1\\ -5 & 4\end{pmatrix} \qquad \lambda=-1;\ 5 \qquad \mathbf{T}=\begin{pmatrix}0.707 & -0.196\\ 0.707 & 0.981\end{pmatrix} \qquad \mathbf{P}=(\mathbf{T}^{-1})^T=\begin{pmatrix}1.178 & -0.850\\ 0.236 & 0.850\end{pmatrix} \tag{B.97}$$

$$\mathbf{A}=(-1)\begin{pmatrix}0.707\\0.707\end{pmatrix}(1.178\quad 0.236)+(5)\begin{pmatrix}-0.196\\0.981\end{pmatrix}(-0.850\quad 0.850) \tag{B.98}$$

$$\mathbf{A}=(-1)\begin{pmatrix}0.833 & 0.167\\ 0.833 & 0.167\end{pmatrix}+(5)\begin{pmatrix}0.167 & -0.167\\ -0.834 & 0.834\end{pmatrix}. \tag{B.99}$$

End of Example

Starting once more from Eq.(B.94), the decomposition of $\mathbf{A}$ is $\mathbf{A} = \mathbf{T}(\text{diag }\lambda_i)\mathbf{T}^{-1}$. Inverting yields

$$\mathbf{A}^{-1} = \mathbf{T}(\text{diag }\lambda_i)^{-1}\mathbf{T}^{-1} = \mathbf{T}(\text{diag }\lambda_i^{-1})\mathbf{T}^{-1}. \tag{B.100}$$

Comparison with Eq.(B.94) gives an expression corresponding with Eq.(B.95)

$$\mathbf{A}^{-1} = \sum_{i=1}^{n} \lambda_i^{-1}(\mathbf{a}_i\mathbf{p}_i^T) \,. \tag{B.101}$$

The dyadic product matrices $(\mathbf{a}_i\mathbf{p}_i^T)$ are the same no matter if $\mathbf{A}$ or $\mathbf{A}^{-1}$ is decomposed. The factors of these decomposition matrices are λ_i and λ_i^{-1}, respectively.

With respect to the quadratic structure of the dyadic decomposition in the variables $\mathbf{a}_i$ and $\mathbf{p}_i$, the notation ***spectral representation*** is also of common use.

B.17 Spectral Representation of the Exponential Matrix

Suppose $\mathbf{A} \in \mathcal{R}^{n\times n}, \lambda_i[\mathbf{A}]$ distinct, the eigenvectors $\mathbf{a}_i$ and $\mathbf{a}_i^q$ linear independent and, finally, $\mathbf{a}_i^q$ normalized: $\mathbf{a}_i^{qT}\mathbf{a}_k = \delta_{ik}$. Decomposing $\mathbf{A} = \sum_{i=1}^{n} \mathbf{a}_i\mathbf{a}_i^{qT}\lambda_i[\mathbf{A}]$ and applying the well-known Taylor expansion for $e^{\mathbf{A}}$,

$$e^{\mathbf{A}} = \sum_{i=1}^{n} \mathbf{a}_i\mathbf{a}_i^{qT}e^{\lambda_i[\mathbf{A}]} \quad \leadsto \quad \lambda_i[e^{\mathbf{A}}] = e^{\lambda_i[\mathbf{A}]} \tag{B.102}$$

$$\mathbf{\Phi}(t) = e^{\mathbf{A}t} = \sum_{i=1}^{n} \mathbf{a}_i\mathbf{a}_i^{qT}e^{\lambda_i[\mathbf{A}]t} \quad \leadsto \quad \lambda_i[e^{\mathbf{A}t}] = \lambda_i[\mathbf{\Phi}(t)] = e^{\lambda_i[\mathbf{A}]t}. \tag{B.103}$$

B.18 Perron-Frobenius Theorem

A non-negative (n,n)-matrix $\mathbf{A}$ always has a non-negative eigenvalue $\pi[\mathbf{A}]$, named Perron root or Perron eigenvalue or Perron-Frobenius radius of $\mathbf{A}$, such that $|\lambda_i[\mathbf{A}]\,| \leq \pi[\mathbf{A}]\ \forall i = [1,n]$. If $\mathbf{A}$ is irreducible then $\pi[\mathbf{A}]$ is positive and simple and the corresponding eigenvector can be chosen as a positive vector.

A matrix is reducible if by identical row and column transpositions the matrix can be brought to an upper trigonal block matrix $\begin{pmatrix} \mathbf{A}_{11} & \mathbf{A}_{12} \\ \mathbf{0} & \mathbf{A}_{22} \end{pmatrix}$ where $\mathbf{A}_{11}$ and $\mathbf{A}_{22}$ are square.

B.19 Multiple Eigenvalues. Generalized Eigenvectors

Consider $\mathbf{A} \in \mathcal{C}^{n\times n}$. The n eigenvalues $\lambda_i[\mathbf{A}]\ \forall i = 1 \ldots n$ are the zeros of the characteristic polynomial $\det(\lambda_i[\mathbf{A}]\,\mathbf{I}_n - \mathbf{A})$ with multiplicities counted. Multiple eigenvalues $\lambda_i[\mathbf{A}]$ of (algebraic) multiplicity m_i correspond to an m_i-order zero. If there are p distinct eigenvalues $\sum_{i=1}^{p} m_i = n$. In the case $m_i > 1$ a full set of n linearly independent eigenvectors need not exist. Assume q_i linearly independent eigenvectors associated with λ_i. The number q_i is the degeneracy of $\lambda_i[\mathbf{A}]\,\mathbf{I}_n - \mathbf{A}$

$$q_i \triangleq n - \text{rank }(\lambda_i[\mathbf{A}]\,\mathbf{I}_n - \mathbf{A}) \quad \text{where} \quad 1 \leq q_i \leq m_i \ . \tag{B.104}$$

In addition to the q_i distinct eigenvectors, $m_i - q_i$ generalized eigenvectors (principal vectors) associated with $\lambda_i[\mathbf{A}]$ are needed.

There is the ambiguity which of the q_i linearly independent eigenvectors the generalized eigenvector chain is associated with. Associated with an eigenvector means that the generalized eigenvector $\mathbf{a}_i^1$ equals $\mathbf{a}_i$, see Eq. (B.107) and Fig. B.1.

These $m_i - q_i$ generalized eigenvectors appear in q_i chains (see e.g. Fig. B.1). The generalized eigenvectors associated with $\lambda_i[\mathbf{A}]$ are given as follows: Calculating $(\lambda_i[\mathbf{A}]\,\mathbf{I}_n - \mathbf{A})^{\mu}$ for $\mu = 1,2\ldots$ until for some $\mu = k$ the relation

$$\text{rank }(\lambda_i[\mathbf{A}]\,\mathbf{I}_n - \mathbf{A})^k = \text{rank }(\lambda_i[\mathbf{A}]\,\mathbf{I}_n - \mathbf{A})^{k+1} \tag{B.105}$$

is obtained. Then, it can be verified that the highest generalized eigenvector $\mathbf{a}_{ij}^k$ of rank k is given by

$$(\lambda_i[\mathbf{A}]\,\mathbf{I}_n - \mathbf{A})^k\mathbf{a}_{ij}^k = 0 \quad \text{where} \quad (\lambda_i[\mathbf{A}]\,\mathbf{I}_n - \mathbf{A})^{k-1}\mathbf{a}_{ij}^k \neq 0 \,, \tag{B.106}$$

see $\mathbf{a}_{i1}^{3}$ in Fig. B.1. The superscript k is used to distinguish between various generalized eigenvectors within a chain. This superscript, of course, cannot be mistaken with an exponent. If $k \neq m_i$ there are two or more chains of generalized eigenvectors. The entire chain of generalized eigenvectors of rank $\mu < k$ is given by

$$\mathbf{a}_{ij}^{\mu} = (\lambda_i[\mathbf{A}]\,\mathbf{I}_n - \mathbf{A})^{k-\mu}\,\mathbf{a}_{ij}^{k} \qquad \forall\,\mu = 1\ldots(k-1)\,. \tag{B.107}$$

Totally, there are k generalized eigenvectors $\mathbf{a}_{ij}^{\mu}$ $\forall\mu = 1\ldots k$ within this chain. This manifold k is denoted $k = s_{ij}$. The generalized eigenvectors are linearly independent, as can be easily verified.

Each eigenvector or each chain of eigenvectors is labelled by an additional subscript j. The number of eigenvectors belonging to a chain is denoted s_{ij}. Following this procedure, j is affixed to every $\mathbf{a}_{ij}$ in the following Eqs. (B.108) up to (B.111). Writing $\mathbf{a}_{ij}$, the index j varies from 1 to q_i. The overall sum of the manifold eigenvectors is $\sum_{j=1}^{j=q_i} s_{ij} = m_i$. Note that Eq.(B.107) can be rewritten to

$$\mathbf{a}_{ij}^{k-1} = (\lambda_i[\mathbf{A}]\,\mathbf{I}_n - \mathbf{A})\mathbf{a}_{ij}^{k} \quad \text{or} \quad \mathbf{A}\mathbf{a}_{ij}^{k} = \mathbf{a}_{ij}^{k-1} + \lambda_i[\mathbf{A}]\mathbf{a}_{ij}^{k} \tag{B.108}$$

$$\mathbf{a}_{ij}^{k-2} = (\lambda_i[\mathbf{A}]\,\mathbf{I}_n - \mathbf{A})^2\mathbf{a}_{ij}^{k} = (\lambda_i[\mathbf{A}]\,\mathbf{I}_n - A)\mathbf{a}_{ij}^{k-1} \quad \text{or} \quad \mathbf{A}\mathbf{a}_{ij}^{k-1} = \mathbf{a}_{ij}^{k-2} + \lambda_i[\mathbf{A}]\mathbf{a}_{i}^{k-1} \tag{B.109}$$

$$\vdots \qquad \vdots$$

$$\mathbf{a}_{ij}^{1} = (\lambda_i[\mathbf{A}]\,\mathbf{I}_n - \mathbf{A})^{k-1}\mathbf{a}_{ij}^{k} = (\lambda_i[\mathbf{A}]\,\mathbf{I}_n - \mathbf{A})\mathbf{a}_{ij}^{2} \quad \text{or} \quad \mathbf{A}\mathbf{a}_{ij}^{2} = \mathbf{a}_{ij}^{1} + \lambda_i[\mathbf{A}]\mathbf{a}_{ij}^{2}\,. \tag{B.110}$$

Premultiplication of the last row by $\lambda_i[\mathbf{A}]\,\mathbf{I}_n - \mathbf{A}$ and referring to Eq.(B.106) yields

$$(\lambda_i[\mathbf{A}]\,\mathbf{I}_n - \mathbf{A})\mathbf{a}_{ij}^{1} = (\lambda_i[\mathbf{A}]\,\mathbf{I}_n - \mathbf{A})^{k}\mathbf{a}_{ij}^{k} = \mathbf{0} \tag{B.111}$$

which reveals that $\mathbf{a}_{ij}^{1}$ (within the chain of generalized eigenvectors) corresponds to the ordinary eigenvector $\mathbf{a}_{ij}$.

Synopsis of symbols:

m_i ... algebraic multiplicity of the eigenvalue $\lambda_i[\mathbf{A}]$

q_i .... geometric multiplicity, degeneracy, number of linear independent (distinct) eigenvectors associated with $\lambda_i[\mathbf{A}]$, dimension of the space spanned by the eigenvectors, number of chains

$m_i - q_i$ ... number of generalized eigenvectors associated with $\lambda_i[\mathbf{A}]$ (total number for all j associated with λ_i)

m_{mi}.. index of the eigenvalue λ_i in $\mathbf{A}$ given by the largest order of the Jordan blocks associated with λ_i , i.e., $m_{mi} = \max_j\ s_{ij}$

s_{ij} ... length of the chain containing the eigenvector $\mathbf{a}_{ij}$ and the generalized eigenvectors $\mathbf{a}_{ij}^{2}$ through $\mathbf{a}_{ij}^{s_{ij}}$ associated with λ_i, $s_{ij} \in [s_{i1}, s_{i2}\ldots s_{iq_i}]$.

B.20 Jordan Canonical Form and Jordan Blocks

In order to obtain a canonical representation the following definitions are used:

i) The Jordan canonical form $\mathbf{J}$, i.e., a block diagonal matrix containing q_i Jordan blocks for each eigenvalue $\lambda_i[\mathbf{A}]$ (one Jordan block for each distinct eigenvector $\mathbf{a}_{ij}$) .

ii) The Jordan block $\mathbf{J}_{ij}$ is given by

$$\mathbf{J}_{ij} \triangleq \begin{pmatrix} \lambda_i & 1 & 0 & 0 & 0 & \ldots & 0 \\ 0 & \lambda_i & 1 & 0 & 0 & \ldots & 0 \\ 0 & 0 & \lambda_i & 1 & 0 & \ldots & 0 \\ 0 & 0 & 0 & \lambda_i & 1 & \ldots & 0 \\ \vdots & \vdots & \vdots & \vdots & \ddots & \ddots & \vdots \\ \vdots & \vdots & \vdots & \vdots & \vdots & \ddots & 1 \\ 0 & 0 & 0 & 0 & 0 & & \lambda_i \end{pmatrix} \in \mathcal{C}^{s_{ij}\times s_{ij}} \tag{B.112}$$

i.e., a matrix with the same $\lambda_i[\mathbf{A}]$ for all main diagonal elements and ones on the diagonal just above the main diagonal. The size of the Jordan blocks within the Jordan canonical form is $s_{ij} \times s_{ij}$ and is not given merely by m_i or q_i. (In the example of Fig. B.1 there are three Jordan blocks of dimension

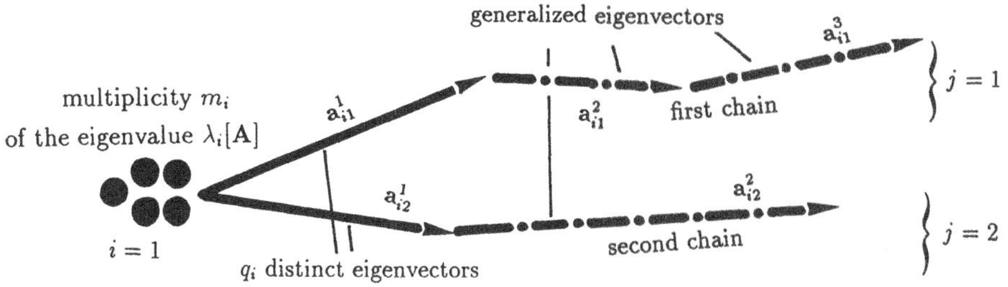

Figure B.1: Brief sketch to obtain orientation about eigenvectors and generalized eigenvectors (associated with $\lambda_i = 2,\ i = 1$) and their multiplicity. Two chains ($q_i = 2$), length of eigenvector chains is 3 and 2 vectors ($s_{i1} = 3, s_{i2} = 2$), $m_i = 5$

$3 \times 3, 2 \times 2, 1 \times 1$ associated with λ_i.) The order of the Jordan blocks within the Jordan canonical form is given by choosing $s_{i1} > s_{i2} > \ldots > s_{iq_i} > 1$. Of course, ordering the eigenvectors within the modal matrix $\mathbf{T}$, see Eq.(B.117), must correspond to the Jordan canonical form $\mathbf{J}$.

The Jordan canonical form $\mathbf{J}$ is given by

$$\mathbf{J} \triangleq \text{block diag}\,(\mathbf{J}_{11},\ \mathbf{J}_{12}, \ldots, \mathbf{J}_{1q_1},\ \mathbf{J}_{21} \ldots \mathbf{J}_{ij} \ldots \mathbf{J}_{2q_2}, \ldots \mathbf{J}_{pq_p}) \in \mathcal{C}^{n\times n}\ . \tag{B.113}$$

Rewriting the right-hand equation block of Eqs.(B.108) to (B.110), one has

$$\mathbf{A}\mathbf{a}_{ij}^1 = \lambda_i \mathbf{a}_{ij}^1 \tag{B.114}$$

$$\mathbf{A}\mathbf{a}_{ij}^2 = \mathbf{a}_{ij}^1 + \lambda_i \mathbf{a}_{ij}^2 \tag{B.115}$$

$$\vdots$$

$$\mathbf{A}\mathbf{a}_{ij}^k = \mathbf{a}_{ij}^{k-1} + \lambda_i \mathbf{a}_{ij}^k \tag{B.116}$$

which can be easily abbreviated to and identified as a special part of $\mathbf{AT} = \mathbf{TJ}$ given in Eq.(B.117). The nonsingular modal matrix $\mathbf{T} \in \mathcal{C}^{n\times n}$ is composed by the eigenvectors and generalized eigenvectors. With the help of Jordan canonical form $\mathbf{J}$ and the modal matrix $\mathbf{T}$ the matrix $\mathbf{A}$ can be decomposed as follows

$$\mathbf{A} = \mathbf{TJT}^{-1}\ . \tag{B.117}$$

Example (corresponding to Fig. B.1 if the single-input case is selected), $n = 6, p = 2$

$$\mathbf{A} = \begin{pmatrix} 2 & 0 & -1 & 0 & -1 & 0 \\ 0 & 0 & 0 & 4 & 0 & 0 \\ 0 & 0 & 1 & 0 & 1 & 0 \\ 1 & -1 & 0 & 4 & 0 & 1 \\ 0 & 0 & 1 & 0 & 1 & 0 \\ 0 & 0 & 1 & 0 & 1 & 2 \end{pmatrix} \qquad \mathbf{T}^{-1}\mathbf{AT} = \mathbf{J} \tag{B.118}$$

generalized eigenvectors associated with λ_1

$$\mathbf{T} = (\mathbf{a}_{11}\ \ \mathbf{a}_{11}^2\ \ \mathbf{a}_{11}^3\ \ \mathbf{a}_{12}\ \ \mathbf{a}_{12}^2\ \ \mathbf{a}_2) \tag{B.119}$$

eigenvectors associated with λ_1 — eigenvector associated with λ_2

$$\mathbf{T} = \begin{pmatrix} 0 & 0 & 0 & 0.876 & 0 & 0 \\ 1 & 0 & 0 & 1 & 0 & 0 \\ 0 & 0 & 0 & 0 & -0.438 & 1 \\ 0.5 & 0.25 & 0 & 0.5 & 0.25 & 0 \\ 0 & 0 & 0 & 0 & -0.438 & -1 \\ 0 & 0 & 0.25 & -0.876 & 0 & 0 \end{pmatrix},\quad \mathbf{T}^{-1} = \begin{pmatrix} -1.142 & 1 & 0 & 0 & 0 & 0 \\ 0 & -2 & 1.142 & 4 & 1.142 & 0 \\ 4 & 0 & 0 & 0 & 0 & 4 \\ 1.142 & 0 & 0 & 0 & 0 & 0 \\ 0 & 0 & -1.142 & 0 & -1.142 & 0 \\ 0 & 0 & 0.5 & 0 & -0.5 & 0 \end{pmatrix} \tag{B.120}$$

$$q_1 = n - \text{rank}\,(\mathbf{A} - \lambda_1\mathbf{I}) = 6 - 4 = 2 \tag{B.121}$$

i	$\lambda_i[\mathbf{A}]$	q_i	m_i	m_{mi}
1	2	2	5	3
2	0	1	1	1

$$c(\lambda) = (\lambda - 2)^5\lambda = \prod_{i=1}^{p}(\lambda - \lambda_i)^{m_i} \quad \text{characteristic polynomial} \tag{B.122}$$

$$m(\lambda) = (\lambda - 2)^3\lambda = \prod_{i=1}^{p}(\lambda - \lambda_i)^{m_{mi}} \quad \text{minimal polynomial} \tag{B.123}$$

$$\mathbf{J} = \begin{pmatrix} 2 & 1 & 0 & 0 & 0 & 0 \\ 0 & 2 & 1 & 0 & 0 & 0 \\ 0 & 0 & 2 & 0 & 0 & 0 \\ 0 & 0 & 0 & 2 & 1 & 0 \\ 0 & 0 & 0 & 0 & 2 & 0 \\ 0 & 0 & 0 & 0 & 0 & 0 \end{pmatrix} = \text{block diag}\left[\begin{pmatrix} 2 & 1 & 0 \\ 0 & 2 & 1 \\ 0 & 0 & 2 \end{pmatrix}, \begin{pmatrix} 2 & 1 \\ 0 & 2 \end{pmatrix}, 0\right]. \tag{B.124}$$

B.21 Special Cases

In the special case $q_i = m_i$ (full degeneracy), there are m_i separate eigenvectors and $s_{ij} = 1\ \forall j = 1 \ldots m_i$. The Jordan blocks become 1×1 matrices, identical to the scalar λ_i, the *diagonalization* of $\mathbf{A}$ by similarity transformation is still available if $q_i = m_i\ \forall i$, see Eq. (B.117). Whenever $q_i \neq m_i$ for any i, diagonalization is substituted by the Jordan canonical form.

In the special case of $q_i = 1$ (simple degeneracy), one has $s_{ij}|_{j=1} = s_{i1} = m_i$ and $\mathbf{J}_{ij} \in \mathcal{C}^{m_i \times m_i}$, i.e. only one Jordan block for $\lambda_i[\mathbf{A}]$. If this should be true for each λ_i then $\mathbf{J} = \text{diag}\,(\mathbf{J}_1, \mathbf{J}_2 \ldots \mathbf{J}_p)$. This case arises, e.g., if the (n,n)-matrix $\mathbf{A}$ is given in companion form, irrespective of the multiplicity of the roots λ_i since rank $(\lambda_i\mathbf{I}_n - \mathbf{A}) = n - 1$ is always true. Various examples are given by *Chen, C.T., 1970*.

B.22 Fundamental Matrix

Calculating the fundamental matrix (state transition matrix) yields

$$e^{\mathbf{A}t} = \sum_{i=0}^{\infty}\frac{1}{i!}\mathbf{A}^i t^i = \sum_{i=0}^{\infty}\frac{1}{i!}(\mathbf{TJT}^{-1})^i t^i = \sum_{i=0}^{\infty}\frac{1}{i!}\mathbf{TJ}^i\mathbf{T}^{-1}t^i = \mathbf{T}\sum_{i=0}^{\infty}\frac{1}{i!}\mathbf{J}^i t^i\mathbf{T}^{-1} = \mathbf{T}e^{\mathbf{J}t}\mathbf{T}^{-1}. \tag{B.125}$$

The matrix exponential of the Jordan canonical form can be calculated block by block, thus

$$e^{\mathbf{J}t} = \text{block diag}\,(\mathbf{S}_{11} \ldots \mathbf{S}_{ij} \ldots \mathbf{S}_{pq_p})\, e^{\text{block diag}\,(\mathbf{J}_{11}, \ldots \mathbf{J}_{ij} \ldots \mathbf{J}_{pq_p})} \tag{B.126}$$

where the trigonal matrices as given below are used

$$\mathbf{S}_{ij} = \begin{pmatrix} 1 & t & t^2/2! & t^3/3! & \ldots & t^{s_{ij}-1}/(s_{ij}-1)! \\ 0 & 1 & t & t^2/2! & \ldots & \ldots \\ 0 & 0 & 1 & t & \ldots & \ldots \\ \vdots & \vdots & \vdots & \ddots & \ddots & \\ 0 & 0 & 0 & \ldots & \ldots & 1 \end{pmatrix} \in \mathcal{C}^{s_{ij} \times s_{ij}}. \tag{B.127}$$

For comparison, in the case of distinct eigenvalues there is $\mathbf{S}_{ij} \equiv 1$ and the simple version

$$\exp(\text{diag}\,\lambda_i t) = \text{diag}\,e^{\lambda_i t}. \tag{B.128}$$

Table B.3: Multiple eigenvalues and degeneracy

Only distinct eigenvalues	Multiple eigenvalues λ_i with multiplicity m_i		
	simple degeneracy	degeneracy, general case	full degeneracy
$m_i = 1$ $q_i = 1$	$q_i = 1$ $s_{ij} = s_{i1} = m_i(> 1)$	$q_i = n - \text{rank}(\mathbf{A} - \lambda_i \mathbf{I}_n)$ $s_{ij} \in [s_{i1}, \dots s_{iq_i}]$	$q_i = m_i$ $s_{ij} = 1$
$m_{mi} = 1$ $m(\lambda) = c(\lambda)$	$m_{mi} = m_i$ $m(\lambda) = c(\lambda)$	$m_{mi} = \max_j \; s_{ij}$ $m(\lambda) = \prod_1^p (\lambda - \lambda_i)^{m_{m_i}}$	$m_{mi} = 1$ $m(\lambda) = \prod_1^p (\lambda - \lambda_i)$
diagonalization feasible where $\mathbf{J} = \text{diag}\ \lambda_i$	diagonalization impossible	diagonalization impossible	diagonalization feasible only if $q_i = m_i \ \forall i$
one eigenvector per eigenvalue	one eigenvector irrespective of $m_i(> 1)$	q_i eigenvectors	q_i eigenvectors
	$m_i - 1$ generalized eigenvectors	$m_i - q_i$ generalized eigenvectors	no generalized eigenvectors
	only one Jordan block $m_i \times m_i$ per eigenvalue	q_i Jordan blocks $s_{ij} \times s_{ij}$ per eigenvalue	m_i Jordan blocks 1×1 identical to the scalars λ_i
	e.g., $\mathbf{A}$ in companion form regardless of m_i		e.g., identity matrix $\mathbf{I}_n$

B.23 Eigenvector Assignment

B.23.1 Assignable Subspaces. Parametrization of Controllers

Applying state feedback by (m, n)-matrix $\mathbf{K}$ in multivariable systems, the degree of freedom is mn in order to assign n closed-loop eigenvalues. The remaining degree of freedom $(m - 1)n$ can be utilized to assign eigenvectors. Assigning eigenvectors in addition to eigenvalues is an important matter because the eigenvectors are responsible for the relation between the state variables and the modal state variables (see Table B.2) and thus the eigenvectors determine the strength of eigenvalue dominance in the state variables. For a given control system in state space representation, i.e., $(\mathbf{A}, \mathbf{B})$ and state feedback $\mathbf{K}$ with the closed-loop behaviour $(\mathbf{A} + \mathbf{BK}) = \mathbf{F}$, there exists a subspace within which the (right) eigenvectors $\mathbf{f}_i$ of the closed-loop system are located and can be assigned arbitrarily.

B.23.2 Single Real or Complex-Conjugate Eigenvalues

From the eigenvector definition

$$(\mathbf{A} + \mathbf{BK})\mathbf{f} = \mathbf{Ff} = \lambda[\mathbf{F}]\,\mathbf{f} \tag{B.129}$$

$$\mathbf{BKf} = (\lambda[\mathbf{F}]\,\mathbf{I}_n - \mathbf{A})\,\mathbf{f}\ . \tag{B.130}$$

Referring to Eq.(C.55), by substitution of $\mathbf{M} := \mathbf{K},\ \mathbf{P} := \mathbf{B},\ \mathbf{Q} := \mathbf{f},\ \mathbf{L} := (\lambda[\mathbf{F}]\,\mathbf{I}_n - \mathbf{A})\mathbf{f}$

$$\mathbf{BB}^{!}(\lambda[\mathbf{F}]\,\mathbf{I}_n - \mathbf{A})\,\mathbf{f}\,\mathbf{f}^{!}\mathbf{f} = (\lambda[\mathbf{F}]\,\mathbf{I}_n - \mathbf{A})\mathbf{f}\ . \tag{B.131}$$

Noting that the pseudoinverse $\mathbf{B}^{!}$ of the column-like matrix $\mathbf{B}$ is $\mathbf{B}^{!L}$ and that $\mathbf{f}\,\mathbf{f}^{!}\mathbf{f} = \mathbf{f}$ a solution $\mathbf{K}$ exists if *(Sinswat, V., and Fallside, F., 1977)*

$$(\mathbf{BB}^{!L} - \mathbf{I}_n)(\lambda[\mathbf{F}]\,\mathbf{I}_n - \mathbf{A})\mathbf{f} = \mathbf{0}\ , \tag{B.132}$$

i.e., the vector $[(\lambda[\mathbf{F}]\,\mathbf{I}_n - \mathbf{A})\mathbf{f}]$ is in the null space of $(\mathbf{BB}^{!L} - \mathbf{I}_n)$. Since

$$\mathbf{BB}^{!L}\mathbf{B} - \mathbf{B} = \mathbf{0} \ \rightsquigarrow\ (\mathbf{BB}^{!L} - \mathbf{I}_n)\mathbf{B} = \mathbf{0} \tag{B.133}$$

and the column space of $\mathbf{B}$ equals the null space of $(\mathbf{B}\mathbf{B}^{!L} - \mathbf{I}_n)$. Hence, the eigenvector $\mathbf{f}$ to be assigned premultiplied by a modified characteristic matrix lies within the column space $\mathcal{R}[\mathbf{B}]$

$$(\lambda[\mathbf{F}]\ \mathbf{I}_n - \mathbf{A})\mathbf{f} \quad \in \mathcal{R}[\mathbf{B}] \ . \tag{B.134}$$

Note that the modified characteristic matrix uses the eigenvalues of the closed loop and, thus, differs from the ordinary characteristic matrix $(\lambda[\mathbf{A}]\ \mathbf{I}_n - \mathbf{A})$.

B.23.3 Multiple Eigenvalues and Linearly Independent Eigenvectors

Find a controller gain matrix $\mathbf{K}$ to a given system $\mathbf{A}, \mathbf{B}$ where a number of k eigenvalues and k linearly independent eigenvectors should be assigned. The numbers $\lambda_1[\mathbf{F}] \ldots \lambda_k[\mathbf{F}]$ are multiple real or conjugate complex closed-loop eigenvalues. The vectors $\mathbf{f}_i$ must be linearly independent ($q_i = m_i$, $s_{ij} = 1$). If $\lambda_1 \ldots \lambda_k$ and $\mathbf{f}_1 \ldots \mathbf{f}_k$ belong to an assignable set

$$(\mathbf{A} + \mathbf{B}\mathbf{K})(\mathbf{f}_1 \vdots \mathbf{f}_2 \ldots \mathbf{f}_k) = (\mathbf{f}_1 \vdots \mathbf{f}_2 \ldots \mathbf{f}_k) \text{ diag } \lambda_i[\mathbf{F}] \quad \text{or} \quad (\mathbf{A} + \mathbf{B}\mathbf{K})\ \mathbf{T}_f = \mathbf{T}_f \text{ diag } \lambda_i[\mathbf{F}] \tag{B.135}$$

$$\mathbf{B}\mathbf{K}\mathbf{T}_f = \mathbf{T}_f \text{ diag } \lambda_i[\mathbf{F}] - \mathbf{A}\mathbf{T}_f \ . \tag{B.136}$$

The matrix diag $\lambda_i[\mathbf{F}]$ is a (k, k)-diagonal matrix with $\lambda_i[\mathbf{F}]$ in the main diagonal. Referring to Eq.(C.55) and applying the following substitutions yields

$$\mathbf{M} := \mathbf{K},\ \mathbf{P} := \mathbf{B},\ \mathbf{Q} := \mathbf{T}_f,\ \mathbf{L} := \mathbf{T}_f \text{ diag } \lambda_i - \mathbf{A}\mathbf{T}_f \tag{B.137}$$

$$\mathbf{B}\mathbf{B}^{!L}(\mathbf{T}_f \text{ diag } \lambda_i[\mathbf{F}] - \mathbf{A}\mathbf{T}_f)\mathbf{T}_f^{!}\mathbf{T}_f = \mathbf{T}_f \text{ diag } \lambda_i[\mathbf{F}] - \mathbf{A}\mathbf{T}_f \ . \tag{B.138}$$

Since $\mathbf{T}_f^{!}\mathbf{T}_f = \mathbf{I}_k$

$$(\mathbf{B}\mathbf{B}^{!L} - \mathbf{I}_n)(\mathbf{T}_f \text{ diag } \lambda_i[\mathbf{F}] - \mathbf{A}\mathbf{T}_f) = \mathbf{0} \ . \tag{B.139}$$

Resubstituting the definition $\mathbf{T}_f$, one has

$$(\mathbf{B}\mathbf{B}^{!L} - \mathbf{I}_n)[(\mathbf{f}_1 \vdots \mathbf{f}_2 \ldots \mathbf{f}_k) \text{ diag } \lambda_i[\mathbf{F}] - \mathbf{A}(\mathbf{f}_1 \vdots \mathbf{f}_2 \ldots \mathbf{f}_k)] = \mathbf{0} \ . \tag{B.140}$$

Postmultiplication by a diagonal matrix yields column-wise multiplication with λ_i

$$(\mathbf{B}\mathbf{B}^{!L} - \mathbf{I}_n)[(\mathbf{f}_1\lambda_1 \vdots \mathbf{f}_2\lambda_2 \ldots \mathbf{f}_k\lambda_k) - \mathbf{A}(\mathbf{f}_1 \vdots \mathbf{f}_2 \ldots \mathbf{f}_k)] = \mathbf{0} \tag{B.141}$$

$$(\mathbf{B}\mathbf{B}^{!L} - \mathbf{I}_n)[(\lambda_1\mathbf{I}_n - \mathbf{A})\mathbf{f}_1 \vdots (\lambda_2\mathbf{I}_n - \mathbf{A})\mathbf{f}_2 \vdots \ldots (\lambda_k\mathbf{I}_n - \mathbf{A})\mathbf{f}_k] = \mathbf{0} \ . \tag{B.142}$$

From the above equation, a necessary and sufficient condition is obtained for $\mathbf{f}_i$ being in the nullspace of $(\mathbf{B}\mathbf{B}^{!L} - \mathbf{I}_n)(\lambda_i\mathbf{I}_n - \mathbf{A})$

$$\mathbf{f}_i \in \mathcal{N}[(\mathbf{B}\mathbf{B}^{!L} - \mathbf{I}_n)(\lambda_i[\mathbf{F}]\ \mathbf{I}_n - \mathbf{A})] \ . \tag{B.143}$$

Referring to Eq.(C.56) and applying the same substitutions as mentioned above, the solution $\mathbf{K}$ of Eq.(B.136) is given by

$$\mathbf{K} = \mathbf{B}^{!L}(\mathbf{T}_f \text{ diag } \lambda_i - \mathbf{A}\mathbf{T}_f)\mathbf{T}_f^{!L} + \mathbf{Z} - \mathbf{Z}\mathbf{T}_f\mathbf{T}_f^{!L} \tag{B.144}$$

where $\mathbf{Z}$ is any (m, n)-matrix. Eq.(B.144) points out the parametrization of the result.

In the case of complex conjugate eigenvalues $\lambda_i[\mathbf{F}]$, the matrices $\mathbf{T}_f$ and diag $\lambda_i[\mathbf{F}]$ can be modified in order to obtain a real-valued controller gain matrix[1]. Consider λ_1 and λ_2 complex-conjugate and the remaining $k - 2$ eigenvalues real,

$$\lambda_1 \triangleq \lambda_{1re} + j\lambda_{1im}, \qquad \lambda_2 \triangleq \lambda_{1re} - j\lambda_{1im} \tag{B.145}$$

$$\mathbf{f}_1 \triangleq \mathbf{f}_{1re} + j\mathbf{f}_{1im}, \qquad \mathbf{f}_2 \triangleq \mathbf{f}_{1re} - j\mathbf{f}_{1im} \ . \tag{B.146}$$

It can be easily verified that the result as given by Eq.(B.144) is obtained by the modified real matrices $\mathbf{T}_f$ and diag λ_i (*Korn, U., and Wilfert, H.H., 1982; Schwarz, H., 1971*)

$$\mathbf{T}_f := (\mathbf{f}_{1re} \vdots \mathbf{f}_{1im} \vdots \mathbf{f}_3 \ldots \mathbf{f}_k) \tag{B.147}$$

$$\text{diag } \lambda_i[\mathbf{F}] := \text{ block diag } [\begin{pmatrix} \lambda_{1re} & \lambda_{1im} \\ -\lambda_{1im} & \lambda_{1re} \end{pmatrix}, \lambda_3 \ldots \lambda_k] \ . \tag{B.148}$$

[1]The modification can be applied but need not be applied if complex controller gain $\mathbf{K}$ is overcome by any other programming facility.

B.23.4 Multiple Eigenvalues and Generalized Eigenvectors

First, the conditions for the existence of **K** are derived. The desired set of eigenvalues $\lambda_i[\mathbf{F}]$ of the closed-loop system should comprise p distinct eigenvalues, with multiplicity m_i each. The design of **K** should provide q_i linear independent eigenvectors $\mathbf{f}_{ij}$ with manifold s_{ij} each (see Fig. B.1 but **a** replaced by **f**). The overall number of linear independent eigenvectors is d. Thus,

$$\sum_{j=1}^{q_i} s_{ij} = m_i , \qquad \sum_{i=1}^{p} m_i = n , \qquad \sum_{i=1}^{p} q_i = d . \tag{B.149}$$

If the set of eigenvalues $\lambda_i[\mathbf{F}]$ and eigenvectors $\mathbf{f}_i$ is preassumed assignable, then, referring to Eq.(B.117), the closed-loop system matrix **F** can be decomposed using the nonsingular modal matrix $\mathbf{T}_f$ and the Jordan canonical form $\mathbf{J}_f$

$$\mathbf{F} = (\mathbf{A}+\mathbf{BK}) = \mathbf{T}\mathbf{J}_f\mathbf{T}^{-1} \qquad \mathbf{J}_f, \mathbf{T} \in \mathcal{C}^{n\times n} \tag{B.150}$$

where the Jordan canonical form is defined by Eq.(B.113)

$$\mathbf{J}_f = \text{block diag}\,(\mathbf{J}_{11},\, \mathbf{J}_{12}\ldots\mathbf{J}_{1q_1},\, \mathbf{J}_{21}\ldots\mathbf{J}_{2q_2},\ldots\mathbf{J}_{ij}\ldots,\mathbf{J}_{pq_p}) \in \mathcal{C}^{n\times n} . \tag{B.151}$$

The Jordan blocks are given by Eq.(B.112) with $\lambda_i = \lambda_i[\mathbf{F}]$. The modal matrix **T** is

$$\mathbf{T} = (\mathbf{T}_{11} \vdots \mathbf{T}_{12}\ldots\mathbf{T}_{1q_1}\, \mathbf{T}_{21}\ldots\mathbf{T}_{2q_2}\ldots\mathbf{T}_{rq_r}) \quad \in \mathcal{C}^{n\times n} \quad \text{where} \quad \mathbf{T}_{ij} = (\mathbf{f}_{ij}^1 \vdots f_{ij}^2 \vdots \ldots\mathbf{f}_{ij}^{s_{ij}}) \quad \in \mathcal{C}^{n\times s_{ij}} . \tag{B.152}$$

Note that the vectors $\mathbf{f}_{ij}^{\mu}$ are linearly independent eigenvectors and generalized eigenvectors, to be assigned to $\mathbf{F} = \mathbf{A}+\mathbf{BK}$ satisfying

$$(\mathbf{A}+\mathbf{BK})\mathbf{f}_{ij}^1 = \lambda_i[\mathbf{F}]\,\mathbf{f}_{ij} \tag{B.153}$$

$$(\mathbf{A}+\mathbf{BK})\mathbf{f}_{ij}^{\mu} = \lambda_i[\mathbf{F}]\,\mathbf{f}_{ij}^{\mu} + \mathbf{f}_{ij}^{\mu-1} \qquad \forall \mu = 2\ldots s_{ij} . \tag{B.154}$$

Applying the condition Eq.(C.55) to Eqs.(B.153) and (B.154), there result d conditions

$$(\mathbf{B}\mathbf{B}^{\dagger L} - \mathbf{I}_n)(\lambda_i\mathbf{I}_n - \mathbf{A})\mathbf{f}_{ij}^1 = 0 \qquad \forall j = 1,2\ldots q_i\ ; \quad i = 1,2\ldots p \tag{B.155}$$

associated with the linearly independent eigenvectors $\mathbf{f}_{ij} \equiv \mathbf{f}_{ij}^1$, and $n-d$ conditions associated with the generalized eigenvectors

$$(\mathbf{B}\mathbf{B}^{\dagger L} - \mathbf{I}_n)\,[(\lambda_i\mathbf{I}_n - \mathbf{A})\mathbf{f}_{ij}^k - \mathbf{f}_{ij}^{k-1}] = 0 \qquad \forall k = 2,3\ldots s_{ij} \tag{B.156}$$

(*Sinswat, V., and Fallside, F., 1977*). In the case of a given system $(\mathbf{A},\mathbf{B})$ and k predetermined eigenvalues and eigenvectors $(k < n)$

$$(\mathbf{A}+\mathbf{BK})\mathbf{T}_f = \mathbf{T}_f\mathbf{J}_f \tag{B.157}$$

is used. Referring to Eq.(C.56),

$$\mathbf{K} = \mathbf{B}^{\dagger L}(\mathbf{T}_f\mathbf{J}_f - \mathbf{A}\mathbf{T}_f)\mathbf{T}_f^{\dagger L} + \mathbf{Z}(\mathbf{I} - \mathbf{T}_f\mathbf{T}_f^{\dagger L}\ ) . \tag{B.158}$$

In the case $k = n$,

$$\mathbf{K} = \mathbf{B}^{\dagger}(\mathbf{T}\mathbf{J}_f\mathbf{T}^{-1} - \mathbf{A}) + \mathbf{Z} - \mathbf{Z} = \mathbf{B}^{\dagger L}(\mathbf{T}\mathbf{J}_f\mathbf{T}^{-1} - \mathbf{A}) \tag{B.159}$$

all degrees of freedom are required and the solution **K** is unique.

B.23.5 Assignable Subspace. Concluding Remarks

In order to determine n eigenvalues and n eigenvectors, the degree of freedom must be n for the eigenvalues and $n-1$ for each given eigenvector since the direction of the eigenvectors is sufficient. This leads to the (m,n)-matrix **K** of the controller. One has to satisfy the condition

$$n + (n-1)k \le\ nm . \tag{B.160}$$

The designer is free to predetermine a complete set of eigenvalues and/or an incomplete set of additionally given eigenvectors. Determining the null space $\mathcal{N}[\mathbf{H}] = \mathcal{N}[(\mathbf{B}\mathbf{B}^{\dagger L} - \mathbf{I}_n)(\lambda_i\mathbf{I}_n - \mathbf{A})]$ for the given $\mathbf{A},\mathbf{B}$

and the desired λ_i , a matrix can be established whose columns span a subspace (or whose column space is) identical to the assignable eigenvector space. It has to be checked if the desired set of eigenvalues and eigenvectors belongs to the assignable eigenvalues and the assignable eigenvector subspace.

If the open-loop system has an uncontrollable eigenvalue, then the closed-loop system must comprise this eigenvalue. Regardless of the fact that this eigenvalue cannot be moved the eigenvector associated with this uncontrollable eigenvalue can be arbitrarily assigned. The only condition is that the eigenvalue to be assigned lies within the assignable subspace $\mathcal{N}[\mathbf{H}]$ associated with the uncontrollable eigenvalue.

The eigenvector assignment (or eigenstructure method) is carried out by finding the right and left-eigenvectors for the closed loop such that the eigenvalues meet the expected closed-loop values. The eigenvector assignment has the disadvantage that the eigenvalue sensitivity with respect to eigenvector errors may be very high, resulting in a non-robust approach. In the case of state feedback, the sensitivity can be reduced by choosing the eigenvectors as orthogonal to each other as possible.

Prespecifying the closed-loop eigenvectors associated with unchanged open-loop eigenvalues is denoted partial eigenstructure assignment (*Jin Lu et al., 1991*).

The output feedback case is investigated by *Ho, W.C., and Fletcher, L.R., 1988*. Norm bounds are given on the closed-loop eigenvalues, caused by the perturbation of the assigned closed-loop eigenvectors. Moreover, the influence of the plant and controller matrices $\mathbf{A}, \mathbf{B}, \mathbf{C}$ and $\mathbf{K}$ is estimated by using their spectral norm.

Appendix C

Matrix Inversion

In addition to the simple inversion as given by Eq.(A.16), the control engineer is involved in many problems associated with matrix inversion such as the right and left-inverse, the pseudo-inverse and the inverse of partitioned matrices. Furthermore, in many applications the inversion of matrices is related to the operations of conditioning, scaling and orthogonalizing.

C.1 Matrix Inversion Using Cayley-Hamilton Theorem

The Cayley-Hamilton theorem gives a tool to calculate the matrix inverse via powers of the same matrix. In detail, a matrix polynomial up to the power $\mathbf{A}^{n-1}$ yields the result. Defining the characteristic polynomial $c(\lambda) = \lambda^n + c_{n-1}\lambda^{n-1} + \ldots + c_o$ where $c_o = \det \mathbf{A}$ the Cayley-Hamilton theorem is given by

$$c(\mathbf{A}) = \mathbf{A}^n + c_{n-1}\mathbf{A}^{n-1} + \ldots + c_o\mathbf{I}_n = \mathbf{0} \; . \tag{C.1}$$

Postmultiplying by $\mathbf{A}^{-1}$ (assuming $\mathbf{A}^{-1}$ exists) yields

$$\mathbf{A}^{n-1} + c_{n-1}\mathbf{A}^{n-2} + \ldots + c_1\mathbf{I}_n + c_o\mathbf{A}^{-1} = 0 \tag{C.2}$$

$$\mathbf{A}^{-1} = -(\mathbf{A}^{n-1} + c_{n-1}\mathbf{A}^{n-2} + \ldots + c_1\mathbf{I}_n)/\det \mathbf{A} \; . \tag{C.3}$$

$$\text{E.g.} \quad \mathbf{A} = \begin{pmatrix} 0 & 1 \\ 5 & 4 \end{pmatrix}, \quad n = 2, \quad c(\lambda) = \lambda^2 - 4\lambda - 5 \; ; \quad c_1 = -4, \quad c_o = -5 = \det \mathbf{A} \tag{C.4}$$

$$\mathbf{A}^{-1} = -[\begin{pmatrix} 0 & 1 \\ 5 & 4 \end{pmatrix} - 4\begin{pmatrix} 1 & 0 \\ 0 & 1 \end{pmatrix}]/(-5) = \begin{pmatrix} 4 & -1 \\ -5 & 0 \end{pmatrix}/(-5) = \begin{pmatrix} -0.8 & 0.2 \\ 1 & 0 \end{pmatrix} \; . \; \square \tag{C.5}$$

C.2 Matrix Inversion Lemma

The matrix inversion lemma proves very helpful in inverting complicated matrices and economizing computational effort. Take into consideration: The quadratic (n, n)-matrix $\mathbf{A}$, its inverse $\mathbf{A}^{-1}$ and two rectangular matrices, the (n, r)-matrix $\mathbf{B}$ and the (r, n)-matrix $\mathbf{C}$. These matrices are given. The dimensions are related as $r < n$ or $r \ll n$. The inverse of $\mathbf{A} + \mathbf{BC}$ is to be calculated, i.e.,

$$\mathbf{\Gamma} \triangleq (\mathbf{A} + \mathbf{BC})^{-1} \qquad (\mathbf{BC}) \in \mathcal{R}^{n \times n} \; , \quad (\mathbf{CB}) \in \mathcal{R}^{r \times r} \tag{C.6}$$

$$\begin{aligned}
\mathbf{\Gamma} \times | \qquad \mathbf{\Gamma}^{-1} &= \mathbf{A} + \mathbf{BC} && \text{(C.7)} \\
\mathbf{I}_n &= \mathbf{\Gamma A} + \mathbf{\Gamma BC} \qquad | \times \mathbf{A}^{-1} && \text{(C.8)} \\
\mathbf{A}^{-1} &= \mathbf{\Gamma} + \mathbf{\Gamma BCA}^{-1} \qquad | \times \mathbf{B} && \text{(C.9)} \\
\mathbf{A}^{-1}\mathbf{B} &= \mathbf{\Gamma B} + \mathbf{\Gamma BCA}^{-1}\mathbf{B} = \mathbf{\Gamma B}(\mathbf{I}_r + \mathbf{CA}^{-1}\mathbf{B}) && \text{(C.10)} \\
\mathbf{A}^{-1}\mathbf{B}(\mathbf{I}_r + \mathbf{CA}^{-1}\mathbf{B})^{-1} &= \mathbf{\Gamma B} \; . && \text{(C.11)}
\end{aligned}$$

Postmultiplying by $(-\mathbf{CA}^{-1})$ yields $-\mathbf{A}^{-1}\mathbf{B}(\mathbf{I}_r + \mathbf{CA}^{-1}\mathbf{B})^{-1}\mathbf{CA}^{-1} = -\mathbf{\Gamma BCA}^{-1}$. Adding $\mathbf{A}^{-1}$ on both sides,

$$\mathbf{A}^{-1} - \mathbf{A}^{-1}\mathbf{B}(\mathbf{I}_r + \mathbf{CA}^{-1}\mathbf{B})^{-1}\mathbf{CA}^{-1} = \mathbf{A}^{-1} - \mathbf{\Gamma BCA}^{-1} = \mathbf{\Gamma} \; . \tag{C.12}$$

The last equating operation follows directly from Eq.(C.9). So one has the result

$$\Gamma = (\mathbf{A}+\mathbf{BC})^{-1} = \mathbf{A}^{-1} - \mathbf{A}^{-1}\mathbf{B}(\mathbf{I}_r + \mathbf{CA}^{-1}\mathbf{B})^{-1}\mathbf{CA}^{-1} . \tag{C.13}$$

If the inverse $\mathbf{A}^{-1}$ is already given, then the matrix inversion lemma Eq.C.13 requires the inversion of an $(r \times r)$-matrix $(\mathbf{I}_r + \mathbf{CA}^{-1}\mathbf{B})$, only. Eq.(C.13) can easily be proved: Examining $\Gamma(\mathbf{A}+\mathbf{BC})$ results in the identity matrix $\mathbf{I}_n$.

Example: If $||\mathbf{F}|| < ||\mathbf{A}||$, from Eq.(C.13) it results

$$\begin{aligned}(\mathbf{A}+\mathbf{F})^{-1} &= \mathbf{A}^{-1} - \mathbf{A}^{-1}\mathbf{F}[\mathbf{I}+\mathbf{A}^{-1}\mathbf{F}]^{-1}\mathbf{A}^{-1} = \mathbf{A}^{-1} - \mathbf{A}^{-1}\mathbf{F}[\mathbf{I} - \mathbf{IA}^{-1}(\mathbf{I}+\mathbf{FA}^{-1})^{-1}\mathbf{F}]\mathbf{A}^{-1} \\ &= \mathbf{A}^{-1} - \mathbf{A}^{-1}\mathbf{FA}^{-1} + \mathbf{A}^{-1}\mathbf{FA}^{-1}(\mathbf{I}+\mathbf{FA}^{-1})^{-1}\mathbf{FA}^{-1} = \text{etc.}\end{aligned} \tag{C.14}$$

Using $(\mathbf{I}+\mathbf{B})^{-1} = \mathbf{I} - \mathbf{B} + \mathbf{B}^2 - \mathbf{B}^3 + \ldots$ where $||\mathbf{B}|| < 1$

$$(\mathbf{A}+\mathbf{F})^{-1} = [\mathbf{A}(\mathbf{I}+\mathbf{A}^{-1}\mathbf{F})]^{-1} = [\mathbf{I} - \mathbf{A}^{-1}\mathbf{F} + (\mathbf{A}^{-1}\mathbf{F})^2 - (\mathbf{A}^{-1}\mathbf{F})^3 + \ldots]\mathbf{A}^{-1} \tag{C.15}$$

$$= \mathbf{A}^{-1} - \mathbf{A}^{-1}\mathbf{FA}^{-1} + \mathbf{A}^{-1}\mathbf{FA}^{-1}\mathbf{FA}^{-1} - \mathbf{A}^{-1}\mathbf{FA}^{-1}\mathbf{FA}^{-1}\mathbf{FA}^{-1} + \ldots \tag{C.16}$$

End of Example

C.3 Simplified Version of the Matrix Inversion Lemma

Given the (n,r)-matrix $\mathbf{B}$ and the (r,n)-matrix $\mathbf{C}$ where $r < n$ the inverse $(\mathbf{I}_n + \mathbf{BC})^{-1}$ is derived as follows

$$\mathbf{B} + \mathbf{BCB} = (\mathbf{I}_n + \mathbf{BC})\mathbf{B} = \mathbf{B}(\mathbf{I}_r + \mathbf{CB}) \qquad | \times (\mathbf{I}_r + \mathbf{CB})^{-1} \tag{C.17}$$

$$(\mathbf{I}_n + \mathbf{BC})\mathbf{B}(\mathbf{I}_r + \mathbf{CB})^{-1} = \mathbf{B} \qquad | \times \mathbf{C} \tag{C.18}$$

$$(\mathbf{I}_n + \mathbf{BC})\mathbf{B}(I_r + \mathbf{CB})^{-1}\mathbf{C} = \mathbf{BC} \qquad | + \mathbf{I}_n \tag{C.19}$$

$$(\mathbf{I}_n + \mathbf{BC})^{-1} \times | \qquad \mathbf{I}_n + (\mathbf{I}_n + \mathbf{BC})\mathbf{B}(\mathbf{I}_r + \mathbf{CB})^{-1}\mathbf{C} = \mathbf{I}_n + \mathbf{BC} \tag{C.20}$$

$$(\mathbf{I}_n + \mathbf{BC})^{-1} + \mathbf{B}(\mathbf{I}_r + \mathbf{CB})^{-1}\mathbf{C} = \mathbf{I}_n \quad \leadsto \quad (\mathbf{I}_n + \mathbf{BC})^{-1} = \mathbf{I}_n - \mathbf{B}(\mathbf{I}_r + \mathbf{CB})^{-1}\mathbf{C} . \tag{C.21}$$

C.4 Matrices in Partitioned Form

C.4.1 Algebraic Properties

The transpose of a partitioned matrix is given by

$$\mathbf{M}^T = \begin{pmatrix} \mathbf{A} & \mathbf{B} \\ \mathbf{C} & \mathbf{D} \end{pmatrix}^T = \begin{pmatrix} \mathbf{A}^T & \mathbf{C}^T \\ \mathbf{B}^T & \mathbf{D}^T \end{pmatrix} . \tag{C.22}$$

Partial matrices or partial vectors within matrices and vectors (in partitioned form) may be multiplied or added as though the submatrices were scalar elements (numerical or scalar functions) provided the submatrices or subvectors are conformable. The matrices can be treated as follows

$$\begin{pmatrix} \mathbf{A} & \mathbf{B} \\ \mathbf{C} & \mathbf{D} \end{pmatrix} \begin{pmatrix} \mathbf{u} \\ \mathbf{v} \end{pmatrix} = \begin{pmatrix} \mathbf{Au} + \mathbf{Bv} \\ \mathbf{Cu} + \mathbf{Dv} \end{pmatrix} \tag{C.23}$$

$$\begin{pmatrix} \mathbf{A} & \mathbf{B} \\ \mathbf{C} & \mathbf{D} \end{pmatrix} \begin{pmatrix} \mathbf{E} & \mathbf{F} \\ \mathbf{G} & \mathbf{H} \end{pmatrix} = \begin{pmatrix} \mathbf{AE} + \mathbf{BG} & \mathbf{AF} + \mathbf{BH} \\ \mathbf{CE} + \mathbf{DG} & \mathbf{CF} + \mathbf{DH} \end{pmatrix} . \tag{C.24}$$

C.4.2 Inversion of a Partitioned Matrix

Preassuming the matrix **M** in partitioned form is a square $(n+q, n+q)$-matrix containing the (n,n)-submatrix **A**, the (q,q)-submatrix **D** and corresponding rectangular (n,q)-matrix **B** and (q,n)-matrix **C**, then, find $\mathbf{M}^{-1}$. On condition det $\mathbf{A} \neq 0$ multiply the first submatrix row of **M** with $-\mathbf{CA}^{-1}$ and add the result to the second row in order to get a block triangular matrix. These operations can be combined by premultiplying **M** with a certain matrix as follows

$$\begin{pmatrix} \mathbf{I} & 0 \\ -\mathbf{CA}^{-1} & \mathbf{I} \end{pmatrix} \mathbf{M} = \begin{pmatrix} \mathbf{I} & 0 \\ -\mathbf{CA}^{-1} & \mathbf{I} \end{pmatrix} \begin{pmatrix} \mathbf{A} & \mathbf{B} \\ \mathbf{C} & \mathbf{D} \end{pmatrix} = \begin{pmatrix} \mathbf{A} & \mathbf{B} \\ 0 & \mathbf{D}-\mathbf{CA}^{-1}\mathbf{B} \end{pmatrix} . \tag{C.25}$$

Inverting this equation,

$$\mathbf{M}^{-1} \begin{pmatrix} \mathbf{I} & 0 \\ -\mathbf{CA}^{-1} & \mathbf{I} \end{pmatrix}^{-1} = \begin{pmatrix} \mathbf{A} & \mathbf{B} \\ 0 & \mathbf{D}-\mathbf{CA}^{-1}\mathbf{B} \end{pmatrix}^{-1} . \tag{C.26}$$

With regard to the upper trigonal property in the right-hand side of Eq.(C.26), it is easy to postulate the structure of the inverse containing a specific submatrix **S** and the inverse of the submatrices

$$\begin{pmatrix} \mathbf{A} & \mathbf{B} \\ 0 & \mathbf{D}-\mathbf{CA}^{-1}\mathbf{B} \end{pmatrix}^{-1} = \begin{pmatrix} \mathbf{A}^{-1} & \mathbf{S} \\ 0 & (\mathbf{D}-\mathbf{CA}^{-1}\mathbf{B})^{-1} \end{pmatrix} . \tag{C.27}$$

In order to obtain $\mathbf{I}_{(n+q)}$ by transferring the left-hand side of Eq.(C.27) to the right side, i.e., to get zero matrices outside the main diagonal, the condition is

$$\mathbf{AS} + \mathbf{B}(\mathbf{D}-\mathbf{CA}^{-1}\mathbf{B})^{-1} = 0 \quad \rightsquigarrow \quad \mathbf{S} = -\mathbf{A}^{-1}\mathbf{B}(\mathbf{D}-\mathbf{CA}^{-1}\mathbf{B})^{-1} . \tag{C.28}$$

Combining this result with Eqs.(C.26) and (C.27),

$$\mathbf{M}^{-1} = \begin{pmatrix} \mathbf{A}^{-1} & -\mathbf{A}^{-1}\mathbf{B}(\mathbf{D}-\mathbf{CA}^{-1}\mathbf{B})^{-1} \\ 0 & (\mathbf{D}-\mathbf{CA}^{-1}\mathbf{B})^{-1} \end{pmatrix} \begin{pmatrix} \mathbf{I} & 0 \\ -\mathbf{CA}^{-1} & \mathbf{I} \end{pmatrix} \tag{C.29}$$

$$\mathbf{M}^{-1} = \begin{pmatrix} \mathbf{A}^{-1} + \mathbf{A}^{-1}\mathbf{B}(\mathbf{D}-\mathbf{CA}^{-1}\mathbf{B})^{-1}\mathbf{CA}^{-1} & -\mathbf{A}^{-1}\mathbf{B}(\mathbf{D}-\mathbf{CA}^{-1}\mathbf{B})^{-1} \\ -(\mathbf{D}-\mathbf{CA}^{-1}\mathbf{B})^{-1}\mathbf{CA}^{-1} & (\mathbf{D}-\mathbf{CA}^{-1}\mathbf{B})^{-1} \end{pmatrix} \tag{C.30}$$

This result is the Frobenius formula for inverting a partitioned matrix. Eq.(C.30) can be rewritten to

$$\mathbf{M}^{-1} = \begin{pmatrix} \mathbf{I}_n & -\mathbf{A}^{-1}\mathbf{B} \\ 0_{q\times n} & \mathbf{I}_n \end{pmatrix} \begin{pmatrix} \mathbf{A}^{-1} & 0_{n\times q} \\ 0_{q\times n} & (\mathbf{D}-\mathbf{CA}^{-1}\mathbf{B})^{-1} \end{pmatrix} \begin{pmatrix} \mathbf{I}_n & 0_{n\times q} \\ -\mathbf{CA}^{-1} & \mathbf{I}_n \end{pmatrix} . \tag{C.31}$$

If det $\mathbf{D} \neq 0$ can be preassumed instead of det $\mathbf{A} \neq 0$

$$\mathbf{M}^{-1} = \begin{pmatrix} (\mathbf{A}-\mathbf{BD}^{-1}\mathbf{C})^{-1} & -(\mathbf{A}-\mathbf{BD}^{-1}\mathbf{C})^{-1}\mathbf{BD}^{-1} \\ -\mathbf{D}^{-1}\mathbf{C}(\mathbf{A}-\mathbf{BD}^{-1}\mathbf{C})^{-1} & \mathbf{D}^{-1} + \mathbf{D}^{-1}\mathbf{C}(\mathbf{A}-\mathbf{BD}^{-1}\mathbf{C})^{-1}\mathbf{BD}^{-1} \end{pmatrix} . \tag{C.32}$$

C.4.3 Inversion of a Partitioned Matrix. Nonsingular Submatrices

If both det $\mathbf{A} \neq 0$ and det $\mathbf{D} \neq 0$ can be presupposed, then

$$\mathbf{M}^{-1} = \begin{pmatrix} (\mathbf{A}-\mathbf{BD}^{-1}\mathbf{C})^{-1} & -\mathbf{A}^{-1}\mathbf{B}(\mathbf{D}-\mathbf{CA}^{-1}\mathbf{B})^{-1} \\ -\mathbf{D}^{-1}\mathbf{C}(\mathbf{A}-\mathbf{BD}^{-1}\mathbf{C})^{-1} & (\mathbf{D}-\mathbf{CA}^{-1}\mathbf{B})^{-1} \end{pmatrix} . \tag{C.33}$$

The result can be confirmed by applying the matrix inversion lemma to the expression $(\mathbf{D}-\mathbf{CA}^{-1}\mathbf{B})^{-1}$ in the left column of the matrix in Eq.(C.30). Deriving the above result by constituting

$$\mathbf{M}\begin{pmatrix} \mathbf{x} \\ \mathbf{y} \end{pmatrix} = \begin{pmatrix} \mathbf{A} & \mathbf{B} \\ \mathbf{C} & \mathbf{D} \end{pmatrix} \begin{pmatrix} \mathbf{x} \\ \mathbf{y} \end{pmatrix} = \begin{pmatrix} \mathbf{u} \\ \mathbf{v} \end{pmatrix} \quad \text{or} \quad \mathbf{Ax} + \mathbf{By} = \mathbf{u} \quad \text{and} \quad \mathbf{Cx} + \mathbf{Dy} = \mathbf{v} , \tag{C.34}$$

it follows

$$\mathbf{x} = (\mathbf{A}-\mathbf{BD}^{-1}\mathbf{C})^{-1} (\mathbf{u}-\mathbf{BD}^{-1}\mathbf{v}) \tag{C.35}$$

$$\mathbf{y} = (\mathbf{D}-\mathbf{CA}^{-1}\mathbf{B})^{-1} (\mathbf{v}-\mathbf{CA}^{-1}\mathbf{u}) . \tag{C.36}$$

Rearranging yields

$$\begin{pmatrix}\mathbf{x}\\ \mathbf{y}\end{pmatrix} = \begin{pmatrix} (\mathbf{A}-\mathbf{B}\mathbf{D}^{-1}\mathbf{C})^{-1} & -(\mathbf{A}-\mathbf{B}\mathbf{D}^{-1}\mathbf{C})^{-1}\mathbf{B}\mathbf{D}^{-1} \\ -(\mathbf{D}-\mathbf{C}\mathbf{A}^{-1}\mathbf{B})^{-1}\mathbf{C}\mathbf{A}^{-1} & (\mathbf{D}-\mathbf{C}\mathbf{A}^{-1}\mathbf{B})^{-1}\end{pmatrix}\begin{pmatrix}\mathbf{u}\\ \mathbf{v}\end{pmatrix} = \mathbf{M}^{-1}\begin{pmatrix}\mathbf{u}\\ \mathbf{v}\end{pmatrix} . \quad \text{(C.37)}$$

The elements in the secondary diagonal of Eq.(C.33) can simply be checked as equal to those in Eq.(C.37), e.g., $-\mathbf{A}^{-1}\mathbf{B}(\mathbf{D}-\mathbf{C}\mathbf{A}^{-1}\mathbf{B})^{-1} \equiv -(\mathbf{A}-\mathbf{B}\mathbf{D}^{-1}\mathbf{C})^{-1}\mathbf{B}\mathbf{D}^{-1}$. Applying the Frobenius formula in any version, the inversion of a $(n+q, n+q)$-matrix can be reduced to the inversion of (n,n)-matrices and (q,q)-matrices. To the former class $\mathbf{A}, \mathbf{A}^{-1}$ and $\mathbf{B}\mathbf{D}^{-1}\mathbf{C}$ belong, to the latter $\mathbf{D}, \mathbf{D}^{-1}$ and $\mathbf{C}\mathbf{A}^{-1}\mathbf{B}$.

C.4.4 Inversion of a Block-Diagonal Matrix

$$\bar{\mathbf{M}} = \begin{pmatrix}\mathbf{A} & \mathbf{0}\\ \mathbf{0} & \mathbf{D}\end{pmatrix} = \text{block diag}\,(\mathbf{A},\mathbf{D}) \quad \rightsquigarrow \quad \bar{\mathbf{M}}^{-1} = \begin{pmatrix}\mathbf{A}^{-1} & \mathbf{0}\\ \mathbf{0} & \mathbf{D}^{-1}\end{pmatrix} = \text{block diag}\,(\mathbf{A}^{-1},\mathbf{D}^{-1}) . \quad \text{(C.38)}$$

C.4.5 Determinants of Matrices in Partitioned Form

A determinant remains unchanged if any row (column), multiplied by a certain constant factor, is added to another row (column). Thus, presuming $\mathbf{A}$ square and det $\mathbf{A} \neq 0$, premultiply the first submatrix with $-\mathbf{C}\mathbf{A}^{-1}$ and add the result to the second submatrix row

$$\det\ \mathbf{M} = \det\begin{pmatrix}\mathbf{A} & \mathbf{B}\\ \mathbf{C} & \mathbf{D}\end{pmatrix} = \det\begin{pmatrix}\mathbf{A} & \mathbf{B}\\ \mathbf{0} & \mathbf{D}-\mathbf{C}\mathbf{A}^{-1}\mathbf{B}\end{pmatrix} = [\det \mathbf{A}][\det(\mathbf{D}-\mathbf{C}\mathbf{A}^{-1}\mathbf{B})] \quad \text{(C.39)}$$

since submatrices in triangle determinants may be treated as scalar numbers. Similarly, if $\mathbf{D}$ is square and det $\mathbf{D} \neq 0$

$$\det\begin{pmatrix}\mathbf{A} & \mathbf{B}\\ \mathbf{C} & \mathbf{D}\end{pmatrix} = \det\begin{pmatrix}\mathbf{A}-\mathbf{B}\mathbf{D}^{-1}\mathbf{C} & \mathbf{0}\\ \mathbf{C} & \mathbf{D}\end{pmatrix} = [\det \mathbf{D}][\det(\mathbf{A}-\mathbf{B}\mathbf{D}^{-1}\mathbf{C})] . \quad \text{(C.40)}$$

When a partitioned matrix with submatrices $\mathbf{A}$ through $\mathbf{D}$ of equal dimensions is considered, the result $\det\mathbf{M} = \det(\mathbf{A}\mathbf{D}-\mathbf{A}\mathbf{C}\mathbf{A}^{-1}\mathbf{B})$ is obtained. If the matrices $\mathbf{A}$ and $\mathbf{C}$ commute (e.g., if $\mathbf{A}$ or $\mathbf{C}$ is the identity matrix) the result is $\det(\mathbf{A}\mathbf{D}-\mathbf{C}\mathbf{B})$, regardless if det $\mathbf{A} = 0$. Considering Eq.(C.39) in the special case $\mathbf{C} = \mathbf{0}$ and $\mathbf{A}$, $\mathbf{D}$ square of any order, the result is $[\det \mathbf{A}][\det \mathbf{D}]$.

C.4.6 Reducible Matrix

If a partitioned decomposition $\begin{pmatrix}\mathbf{A} & \mathbf{B}\\ \mathbf{0} & \mathbf{D}\end{pmatrix} = \mathbf{P}\mathbf{H}\mathbf{P}^{-1}$ with $\mathbf{C} = \mathbf{0}$, $\mathbf{A}$ and $\mathbf{D}$ square can be deduced from a matrix $\mathbf{H}$ where $\mathbf{P}$ is the permutation matrix then the matrix $\mathbf{H}$ is denoted reducible.

C.5 Right-Inverse

Define a rectangular matrix $\mathbf{W}^{RI}$ as the right-inverse of the nonsquare matrix $\mathbf{W}$

$$\mathbf{W}\mathbf{W}^{RI} = \mathbf{I} . \quad \text{(C.41)}$$

The right-pseudo-inverse

$$\mathbf{W}^{\dagger R} = \mathbf{W}^T(\mathbf{W}\mathbf{W}^T)^{-1} \quad \text{(C.42)}$$

satisfies Eq.(C.41) $\mathbf{W}\mathbf{W}^{RI} = \mathbf{I}$ and is, therefore, a possible solution $\mathbf{W}^{RI} = \mathbf{W}^{\dagger R}$ but not the only one.

Given an n_p-vector $\mathbf{p}$, an n_m-vector $\mathbf{y}$ and an (n_m, n_p)-matrix $\mathbf{M}$ in the relation $n_p > n_m$ the equation $\mathbf{M}\mathbf{p} = \mathbf{y}$ has more unknowns than knowns. There exist solutions in a large variety $\mathbf{p} = \mathbf{M}^{RI}\mathbf{y}$. The matrix $\mathbf{M}^{RI}$ is an (n_p, n_m)-matrix and the right-inverse (inverse to the right). The matrix $\mathbf{M}^{RI}$ exists if rank $\mathbf{M} \geq n_m$ and obeys $\mathbf{M}\mathbf{M}^{RI} = \mathbf{I}_{n_m}$. The right-pseudo-inverse $\mathbf{M}^{\dagger R}$ is the minimum right-inverse (*Morari, M., 1989*), i.e.,

$$\mathbf{M}^{\dagger R} = \mathbf{M}^{RI} \mid_{\|\mathbf{p}\|_F = \min} = \mathbf{M}^T(\mathbf{M}\mathbf{M}^T)^{-1} . \quad \text{(C.43)}$$

C.6 Left-Inverse

Given an n_p-vector $\mathbf{p}$, an n_m-vector $\mathbf{y}$ and an (n_m, n_p)-matrix $\mathbf{M}$ the formula $\mathbf{Mp} = \mathbf{y}$ is inadmissible if the relation $n_m > n_p$ holds, if more knowns exist than unknowns, if the number of scalar equations exceeds the number of unknown variables in $\mathbf{p}$.

The left-inverse (or inverse form the left) $\mathbf{M}^{LI}$ does *not* supply the solution. Although a variety of $\mathbf{M}^{LI}$ exists, none of the associated products $\mathbf{p}^? = \mathbf{M}^{LI}\mathbf{y}$ may be declared as a solution that is compatible to the overdetermined system, completely.

Example:

$$\mathbf{M} = \begin{pmatrix} 0 & 0.5 \\ -0.5 & 1 \\ -1.1506 & 0 \end{pmatrix} \qquad \mathbf{y} = \begin{pmatrix} 0.6 \\ 1.3 \\ 0.5 \end{pmatrix} \tag{C.44}$$

$$\mathbf{M}^{LI} = \begin{pmatrix} 4 + 4.6024\,\alpha & -2 - 2.3012\,\alpha & \alpha \\ 2 + 4.6024\,\beta & -2.3012\,\beta & \beta \end{pmatrix} \qquad \mathbf{M}^{LI}\mathbf{M} = \mathbf{I}_2 \quad \forall \alpha, \beta\,. \tag{C.45}$$

Selecting, e.g., $\alpha = \beta = 0$ then $\mathbf{M}^{LI}\mathbf{y} = (-0.2 \;\; 1.2)^T = \mathbf{p}^?$. Computing $\mathbf{Mp}^? = (0.6 \;\; 1.3 \;\; 0.230)^T$, this result is not consistent with $\mathbf{y}$ in Eq.(C.44). □

The formulation by addition (subtraction) of $\boldsymbol{\varepsilon}$, i.e., $\mathbf{Mp} = \mathbf{y} - \boldsymbol{\varepsilon}$, is permissible even in the case $n_m > n_p$. Within the infinite variety of $\mathbf{M}^{LI}$ the minimum left-inverse is defined

$$\mathbf{M}^{!L} = \mathbf{M}^{LI} \mid_{\|\boldsymbol{\varepsilon}\|_F = \min} = (\mathbf{M}^T\mathbf{M})^{-1}\mathbf{M}^T\,. \tag{C.46}$$

C.7 Pseudo-Inverse

C.7.1 General Pseudo-Inverse

Let the (n, m)-matrix $\mathbf{M}$ with rank$\mathbf{M} = r$ contain only real elements. With regard to rank r, the product form can be attained

$$\mathbf{M} = \mathbf{VW} = \begin{pmatrix} \leftarrow & r & \rightarrow \\ \uparrow & & \\ \vdots & & \\ n & \mathbf{V} & \\ \vdots & & \\ \downarrow & & \end{pmatrix} \begin{pmatrix} \leftarrow & \cdots & m & \cdots & \rightarrow \\ \uparrow & & & & \\ r & & \mathbf{W} & & \\ \downarrow & & & & \end{pmatrix} \qquad \begin{array}{l} \text{rank } \mathbf{V} = r \\ \text{rank } \mathbf{W} = r \\ r < n, \quad r < m\,. \end{array} \tag{C.47}$$

The general pseudo-inverse $\mathbf{M}^!$ is defined by the following four equations

$$\mathbf{MM}^!\mathbf{M} = \mathbf{M} \qquad \mathbf{M}^!\mathbf{MM}^! = \mathbf{M}^! \tag{C.48}$$

$$(\mathbf{MM}^!)^T = \mathbf{MM}^! \qquad (\mathbf{M}^!\mathbf{M})^T = \mathbf{M}^!\mathbf{M}\,. \tag{C.49}$$

Explicitly, $\mathbf{M}^!$ is given by

$$\mathbf{M}^! = \mathbf{W}^T(\mathbf{WW}^T)^{-1}(\mathbf{V}^T\mathbf{V})^{-1}\mathbf{V}^T\,. \tag{C.50}$$

The pseudo-inverse $\mathbf{M}^!$ is an (m, n)-matrix.

Inserting Eq.(C.50) into Eq. (C.48), i.e., $\mathbf{VWW}^T(\mathbf{WW}^T)^{-1}(\mathbf{V}^T\mathbf{V})^{-1}\mathbf{V}^T\mathbf{VW} = \mathbf{VW}$ makes evident that it is not permissible to omit $\mathbf{M}$ neither in the left nor in the right half of Eq.(C.48).

Some more properties for the pseudoinverse:

$$(\mathbf{M}^!)^! = \mathbf{M} \qquad (\mathbf{M}^!)^T = (\mathbf{M}^T)^! \qquad (\beta\mathbf{M})^! = (1/\beta)\mathbf{M}^! \quad (\text{if } \;\beta \neq 0) \tag{C.51}$$

$$(\mathbf{M}^T\mathbf{M})^! = \mathbf{M}^!(\mathbf{M}^!)^T \qquad (\mathbf{M}^!\mathbf{M})^! = \mathbf{M}^!\mathbf{M} \qquad (\mathbf{MM}^!)^! = \mathbf{MM}^! \qquad (\mathbf{MB})^! \neq \mathbf{B}^!\mathbf{M}^! \tag{C.52}$$

$$\text{rank } \mathbf{M} = \text{rank } \mathbf{M}^! = \text{rank } \mathbf{M}^T\mathbf{M} = \text{rank } \mathbf{M}^!\mathbf{M} \qquad \text{tr } \mathbf{M}^!\mathbf{M} = \text{rank } \mathbf{M}\,. \tag{C.53}$$

C.7.2 General Pseudo-Inverse and a General Matrix Equation

Given the matrix equation with known matrices $\mathbf{P}, \mathbf{Q}, \mathbf{L}$ and unknown $\mathbf{M}$

$$\mathbf{PMQ} = \mathbf{L} \; . \tag{C.54}$$

All matrices are assumed of adequate dimensions. If the condition

$$\mathbf{PP}^{!}\mathbf{LQ}^{!}\mathbf{Q} = \mathbf{L} \tag{C.55}$$

is satisfied, the solution of Eq.(C.54) is (*Rao, C.R., and Mitra, S.K., 1971*)

$$\mathbf{M} = \mathbf{P}^{!}\mathbf{LQ}^{!} + \mathbf{Z} - \mathbf{P}^{!}\mathbf{PZQQ}^{!} \tag{C.56}$$

for any matrix $\mathbf{Z}$. This result can easily be proved by substitution.

The condition Eq.(C.55) is necessary and sufficient for Eq.(C.54) to have a solution.

(i) Necessity: It is unimaginable that the equation to be solved, Eq.(C.54), and the condition Eq.(C.55) for the existence of the solution are incompatible. Hence, it is a necessary condition. Consistency of Eq.(C.54) and (C.55) requires that $\mathbf{L}$ of Eq.(C.54) substituted into the left-hand side of Eq.(C.55) gives a true statement

$$\mathbf{PP}^{!}(\mathbf{PMQ})\mathbf{Q}^{!}\mathbf{Q} = \mathbf{PP}^{!}\mathbf{PMQQ}^{!}\mathbf{Q} = \mathbf{PMQ} = \mathbf{L} \; . \tag{C.57}$$

(ii) Sufficiency: Substituting the solution Eq.(C.56) into Eq.(C.54) yields the expression Eq.(C.55) irrespective of the value of $\mathbf{Z}$. Thus, Eq.(C.55) is sufficient.

C.7.3 Right-Pseudo-Inverse

If $n = r$ and $\mathbf{V} = \mathbf{I}_r$ then $\mathbf{M} = \mathbf{W}$ and the right-pseudo-inverse is

$$\mathbf{M}^{!R} = \mathbf{M}^T(\mathbf{MM}^T)^{-1} \quad \text{and simply} \quad \mathbf{MM}^{!} = \mathbf{I}_r \; . \tag{C.58}$$

Example: Specializing Eqs.(C.54) and (C.56), namely $\mathbf{Mp} = \mathbf{y}$, $\mathbf{M} \in \mathcal{R}^{n_m \times n_p}$, $n_p > n_m$. If $\mathbf{MM}^{!}\mathbf{y} = \mathbf{y}$ exists ($\mathbf{M}^{!} = \mathbf{M}^{!R}$) then the solution $\mathbf{p}$ is

$$\mathbf{p} = \mathbf{M}^{!}\mathbf{y} + (\mathbf{I}_{n_p} - \mathbf{M}^{!}\mathbf{M})\mathbf{z} = \mathbf{M}^{!R}\mathbf{y} + (\mathbf{I}_{n_p} - \mathbf{M}^{!R}\mathbf{M})\mathbf{z} \tag{C.59}$$

where $\mathbf{z}$ is any n_p-vector. This can easily be proved by substituting the solution into $\mathbf{Mp} = \mathbf{y}$ and using Eq.(C.48)

$$\mathbf{Mp} = \mathbf{MM}^{!}\mathbf{y} + \mathbf{M}(\mathbf{I}_{n_p} - \mathbf{M}^{!}\mathbf{M})\mathbf{z} = \mathbf{y} + \mathbf{Mz} - \mathbf{MM}^{!}\mathbf{Mz} = \mathbf{y} \; . \tag{C.60}$$

Differentiating $\mathbf{p}^T\mathbf{p}$ with respect to $\mathbf{z}$ yields $\mathbf{z} = \mathbf{0}$, i.e. the special case $\mathbf{p} = \mathbf{M}^{!}\mathbf{y}$ has minimal norm and $\mathbf{M}^{!} = \mathbf{M}^{!R}$. The solution Eq.(C.59) can be generalized to solve Eq.(C.54) by rewriting Eq.(C.54) into vector form using Eq.(4.40)

$$(\mathbf{Q}^T \otimes \mathbf{P})\text{col } \mathbf{M} = \text{col } \mathbf{L} \tag{C.61}$$

$$\text{col } \mathbf{M} = (\mathbf{Q}^T \otimes \mathbf{P})^{!R} \text{ col } \mathbf{L} + [\mathbf{I} - (\mathbf{Q}^T \otimes \mathbf{P})^{!R} \, (\mathbf{Q}^T \otimes \mathbf{P})]\text{col } \mathbf{Z} \tag{C.62}$$

$$\text{col } \mathbf{M} = (\mathbf{Q}^T \otimes \mathbf{P})^{!R} \text{ col } \mathbf{L} + \text{col } \mathbf{Z} - (\mathbf{Q}^{T!R} \otimes \mathbf{P}^{!R})(\mathbf{Q}^T \otimes \mathbf{P})\text{col } \mathbf{Z} \tag{C.63}$$

$$\mathbf{M} = \mathbf{P}^{!}\mathbf{LQ}^{!} + \mathbf{Z} - \mathbf{P}^{!}\mathbf{PZQQ}^{!} \; . \tag{C.64}$$

End of Example

C.7.4 Left-Pseudo-Inverse

Considering $m = r$ and $\mathbf{W} = \mathbf{I}_r$ then $\mathbf{M} = \mathbf{V}$, the left-pseudo-inverse is

$$\mathbf{M}^{\dagger L} = (\mathbf{M}^T\mathbf{M})^{-1}\mathbf{M}^T \tag{C.65}$$

and $\mathbf{M}^{\dagger L}\mathbf{M} = \mathbf{I}_r$. The left-pseudo-inverse is frequently applied in linear regression. Applying the left-pseudo-inverse, only, the superscript ^{L} is often omitted.

Example: Specializing the general pseudo-inverse to the left-pseudo-inverse,

$$\mathbf{Q} = \mathbf{B} \text{ (input matrix)}, \quad \mathbf{P} = \mathbf{I}_m, \quad \mathbf{B} \in \mathcal{C}^{n\times m}, \quad \text{then } \mathbf{L} = \mathbf{I}_m \ . \tag{C.66}$$

Eq.(C.55) yields $\mathbf{B}^{\dagger}\mathbf{B} = \mathbf{I}_m$ or $\mathbf{B}^{\dagger} = \mathbf{B}^{\dagger L}$. From Eq.(C.56) one has

$$\mathbf{M} = \mathbf{B}^{\dagger} + \mathbf{Z}(\mathbf{I}_n - \mathbf{B}\mathbf{B}^{\dagger}) = \mathbf{B}^{\dagger L} + \mathbf{Z}(\mathbf{I}_n - \mathbf{B}\mathbf{B}^{\dagger L}) \ . \ \square \tag{C.67}$$

Note that for $\mathbf{M} \in \mathcal{R}^{n\times m}$, $n > m$,

$$\det(s\mathbf{I}_n - \mathbf{M}\mathbf{M}^{\dagger}) = s^n \det(\mathbf{I}_n - \frac{1}{s}\mathbf{M}\mathbf{M}^{\dagger}) = s^n \det(\mathbf{I}_m - \frac{1}{s}\mathbf{M}^{\dagger}\mathbf{M}) \tag{C.68}$$

$$= s^{n-m} \det(s\mathbf{I}_m - \mathbf{I}_m) = s^{n-m}(s-1)^m \ . \tag{C.69}$$

C.7.5 Projector Properties of $\mathbf{M}\mathbf{M}^{\dagger}$ and $\mathbf{M}^{\dagger R}\mathbf{M}$

The matrix product, in this order, possesses the property of a projector matrix, in both cases left-pseudo-inverse and right-pseudo-inverse.

C.8 General System Inverse

Consider the system

$$\dot{\mathbf{x}}(t) = \mathbf{A}\mathbf{x}(t) + \mathbf{B}\mathbf{u}(t) \qquad \mathbf{x}(t) \in \mathcal{R}^n, \quad \mathbf{u}(t) \in \mathcal{R}^m \tag{C.70}$$

$$\mathbf{y}(t) = \mathbf{C}\mathbf{x}(t) + \mathbf{D}\mathbf{u}(t) \qquad \mathbf{y}(t) \in \mathcal{R}^r, \quad \text{rank } \mathbf{D} = \min(r, m) \ . \tag{C.71}$$

C.8.1 Case $r < m$

Eq.(C.71) is solved with respect to the input vector using right-pseudo-inverse and an arbitrary m-vector $\mathbf{z}$ as shown in Eq.(C.59)

$$\mathbf{u} = \mathbf{D}^{\dagger R}(\mathbf{y} - \mathbf{C}\mathbf{x}) + (\mathbf{I}_m - \mathbf{D}^{\dagger R}\mathbf{D})\mathbf{z} \tag{C.72}$$

$$\text{where} \quad \mathbf{D}^{\dagger R} = \mathbf{D}^T(\mathbf{D}\mathbf{D}^T)^{-1} \qquad \mathbf{D} \in \mathcal{R}^{r\times m}, \ \mathbf{D}^{\dagger R} \in \mathcal{R}^{m\times r} \ . \tag{C.73}$$

Deriving $\mathbf{z}$ from $\mathbf{x}$ by defining an arbitrary (m, n)-matrix $\mathbf{N}$, i.e., $\mathbf{z} = \mathbf{N}\mathbf{x}$, from Eqs.(C.70) and (C.72), it results

$$\dot{\mathbf{x}} = \mathbf{A}\mathbf{x} + \mathbf{B}[\mathbf{D}^{\dagger R}(\mathbf{y} - \mathbf{C}\mathbf{x}) + (\mathbf{I}_m - \mathbf{D}^{\dagger R}\mathbf{D})\mathbf{N}\mathbf{x}] \tag{C.74}$$

$$= [\mathbf{A} - \mathbf{B}\mathbf{D}^{\dagger R}\mathbf{C} + \mathbf{B}(\mathbf{I}_m - \mathbf{D}^{\dagger R}\mathbf{D})\mathbf{N}]\mathbf{x} + \mathbf{B}\mathbf{D}^{\dagger R}\mathbf{y} \ . \tag{C.75}$$

Eqs.(C.75) and (C.72) as a system with left-hand vectors $(\dot{\mathbf{x}}, \mathbf{u})$ is called general system inverse of the system with left-hand vectors $(\dot{\mathbf{x}}, \mathbf{y})$.

C.8.2 Case $r = m$

Equating $\mathbf{D}^{\dagger R} = \mathbf{D}^{-1}$ yields a unique inverse

$$\mathbf{u} = \mathbf{D}^{-1}\mathbf{y} - \mathbf{D}^{-1}\mathbf{C}\mathbf{x} \tag{C.76}$$

$$\dot{\mathbf{x}} = (\mathbf{A} - \mathbf{B}\mathbf{D}^{-1}\mathbf{C})\mathbf{x} + \mathbf{B}\mathbf{D}^{-1}\mathbf{y} \ . \tag{C.77}$$

C.8.3 Case $r > m$

With regard to $r > m$, Eq.(C.71) is overdetermined and incompatible in this version. Considering a dual system (*Sinswat, V., 1976*)

$$\dot{\mathbf{x}}' = \mathbf{A}^T\mathbf{x}' + \mathbf{C}^T\mathbf{u}' \qquad \mathbf{y}' = \mathbf{B}^T\mathbf{x}' + \mathbf{D}^T\mathbf{u}' \tag{C.78}$$

with an r-vector $\mathbf{y}'$ as an input and an m-vector $\mathbf{u}'$ as an output, the aforementioned calculus can be applied again. Using the right-pseudo-inverse of $\mathbf{D}^T$, i.e.,

$$(\mathbf{D}^T)^{!R} = \mathbf{D}(\mathbf{D}^T\mathbf{D})^{-1} = (\mathbf{D}^{!L})^T , \tag{C.79}$$

the transpose of the left-pseudo-inverse comes into action, corresponding to the fact that the system matrices are defined as transposed matrices in the case of Eq.(C.78).

C.9 Pseudo-Inverse and Singular-Value Decomposition

If the singular-value decomposition of a complex matrix $\mathbf{G}$ is given by $\mathbf{G} = \mathbf{U\Sigma V}^H$, the pseudo-inverse amounts to

$$\mathbf{G}^! = \mathbf{V}\begin{pmatrix} \mathbf{\Sigma}_r^{-1} & 0 \\ 0 & 0 \end{pmatrix}\mathbf{U}^H \quad \text{where} \quad \mathbf{\Sigma} = \begin{pmatrix} \mathbf{\Sigma}_r & 0 \\ 0 & 0 \end{pmatrix}, \quad r = \text{rank}\mathbf{G} \tag{C.80}$$

since the basic properties of the pseudo-inverse are satisfied by the decomposition cited above.

C.10 Pseudo-Inverse of a Matrix Partitioned into Submatrices

If a matrix $\mathbf{M}$ is partitioned into matrices $\mathbf{A}$ through $\mathbf{D}$

$$\mathbf{M} = \begin{pmatrix} \mathbf{A} & \mathbf{B} \\ \mathbf{C} & \mathbf{D} \end{pmatrix} = \begin{pmatrix} \mathbf{A} \\ \mathbf{C} \end{pmatrix}\mathbf{A}^{-1}(\mathbf{A} \vdots \mathbf{B}) = \begin{pmatrix} \mathbf{A} \\ \mathbf{C} \end{pmatrix}(\mathbf{I} \vdots \mathbf{A}^{-1}\mathbf{B}) = \begin{pmatrix} \mathbf{A} & \mathbf{B} \\ \mathbf{C} & \mathbf{CA}^{-1}\mathbf{B} \end{pmatrix} \tag{C.81}$$

and the submatrix rows in $\mathbf{M}$ are linearly dependent, i.e., the second row results from the first one by multiplying with $\mathbf{CA}^{-1}$ ($\mathbf{A}$ square and nonsingular) then $\mathbf{D} = \mathbf{CA}^{-1}\mathbf{B}$, and rank$\mathbf{M}$ = rank$\mathbf{A}$. In this case, the pseudo-inverse can be calculated in partitioned form

$$\mathbf{M}^! = \begin{pmatrix} \mathbf{A}^! \\ \mathbf{B}^! \end{pmatrix}(\mathbf{AA}^! + \mathbf{BB}^!)^{-1}\mathbf{A}(\mathbf{A}^!\mathbf{A} + \mathbf{C}^!\mathbf{C})^{-1}(\mathbf{A}^! \vdots \mathbf{C}^!) . \tag{C.82}$$

C.11 Pseudo-Inverse of a Matrix Partitioned into Columns

Using Greville's method, the pseudo-inverse can be calculated in a recursive way. The matrix $\mathbf{M}$ is disassembled into columns $\mathbf{M} = (\mathbf{m}_{c1}, \mathbf{m}_{c2}, ..., \mathbf{m}_{cn})$. On the other hand, the matrix $\mathbf{M}$ is recursively composed by defining

$$\mathbf{M}_1 = \mathbf{m}_{c1} \qquad \mathbf{M}_k = (\mathbf{M}_{k-1}, \mathbf{m}_{ck}) \quad \forall k = 2, ...n \qquad \mathbf{M}_k|_{k=n} = \mathbf{M}_n = \mathbf{M} . \tag{C.83}$$

With the definitions

$$\mathbf{M}_1^! \triangleq \mathbf{a}_1^! = \begin{cases} \mathbf{a}_1^T/(\mathbf{a}_1^T\mathbf{a}_1) & \text{if } \mathbf{a}_1 \neq 0 \\ 0^T & \text{if } \mathbf{a}_1 = 0 \end{cases} \quad \text{and} \quad \mathbf{d}_k \triangleq \mathbf{M}_{k-1}^!\mathbf{a}_k , \tag{C.84}$$

$$\mathbf{b}_k^T = \begin{cases} (\mathbf{a}_k - \mathbf{M}_{k-1}\mathbf{d}_k)^! & \text{if } \mathbf{a}_k - \mathbf{M}_{k-1}\mathbf{d}_k \neq 0 \\ (1 + \mathbf{d}_k^T\mathbf{d}_k)^{-1}\mathbf{d}_k^T\mathbf{M}_{k-1}^! & \text{if } \mathbf{a}_k - \mathbf{M}_{k-1}\mathbf{d}_k = 0 . \end{cases} \tag{C.85}$$

Taking these abbreviations, the pseudo-inverse associated with $\mathbf{M}_k \ \forall \ k = 2, \ldots, n$ is obtained

$$\mathbf{M}_k^{!R} = \begin{pmatrix} \mathbf{M}_{k-1}^! - \mathbf{d}_k\mathbf{b}_k^T \\ \mathbf{b}_k^T \end{pmatrix} . \tag{C.86}$$

C.12 Successive Application of Right and Left-Pseudo-Inverse Operator

According to the relation n_p and n_m , the following identities hold

$$n_p > n_m : \quad (\mathbf{M}^{!R})^{!L} = \mathbf{M} \,, \qquad n_m > n_p : \quad (\mathbf{M}^{!L})^{!R} = \mathbf{M} \,. \tag{C.87}$$

Checking, e.g., Eq.(C.87)

$$(\mathbf{M}^{!L})^{!R} = \mathbf{M}^{!LT}\left(\mathbf{M}^{!L}\mathbf{M}^{!LT}\right)^{-1} \tag{C.88}$$

$$(\mathbf{M}^{!L})^{!R} = [(\mathbf{M}^T\mathbf{M})^{-1}\mathbf{M}^T]^T\left([(\mathbf{M}^T\mathbf{M})^{-1}\mathbf{M}^T][(\mathbf{M}^T\mathbf{M})^{-1}\mathbf{M}^T]^T\right)^{-1} \tag{C.89}$$

$$(\mathbf{M}^{!L})^{!R} = \mathbf{M}(\mathbf{M}^T\mathbf{M})^{-1}\left((\mathbf{M}^T\mathbf{M})^{-1}\mathbf{M}^T\mathbf{M}(\mathbf{M}^T\mathbf{M})^{-1}\right)^{-1} = \mathbf{M} \tag{C.90}$$

(*Barnett, S., 1971; Rao, C.R., and Mitra, S.K., 1971; Lancaster, P., and Tismenetsky, M., 1985*) .

C.13 Conditioning and Scaling

C.13.1 Condition Number of a Matrix

The condition number $\kappa_s[\mathbf{A}]$ of a matrix $\mathbf{A}$ is defined for any unit vector $\mathbf{u}$ and $\mathbf{v}$

$$\kappa_s[\mathbf{A}] \triangleq \max_{\|\mathbf{u}\|_F=1,\ \|\mathbf{v}\|_F=1} \frac{\|\mathbf{Au}\|_F}{\|\mathbf{Av}\|_F} = \frac{\max_{\|\mathbf{u}\|_F=1} \|\mathbf{Au}\|_F}{\min_{\|\mathbf{v}\|_F=1} \|\mathbf{Av}\|_F} = \frac{\sigma_{\max}[\mathbf{A}]}{\sigma_{\min}[\mathbf{A}]} \tag{C.91}$$

$$\kappa_s[\mathbf{A}] = \frac{\|\mathbf{A}\|_s}{\min_{\|\mathbf{A}^{-1}\mathbf{Av}\|_F=1} \|\mathbf{Av}\|_F} = \frac{\|\mathbf{A}\|_s}{\|\mathbf{A}^{-1}\|_s^{-1}} = \|\mathbf{A}\|_s\|\mathbf{A}^{-1}\|_s \geq 1 \,. \tag{C.92}$$

Using the relation between norm and maximum/minimum modulus of eigenvalues,

$$\kappa_s[\mathbf{A}] = \frac{\|\mathbf{A}\|_s}{\dfrac{1}{\|\mathbf{A}^{-1}\|_s}} \geq \frac{\max_i\ |\lambda_i[\mathbf{A}]\ |}{\min_i\ |\lambda_i[\mathbf{A}]\ |} \,. \tag{C.93}$$

Solving, e.g., $\mathbf{Ay} = \mathbf{b}$ where $\mathbf{A}$ is nonsingular, $\mathbf{b}$ and $\mathbf{A}$ are only known to be in the limit of a tolerance $\delta\mathbf{b}$ and $\delta\mathbf{A}$, an estimate of the error $\delta\mathbf{y}$ of the solution $\mathbf{y}$ is (*Franklin, J.N., 1968*)

$$\frac{\|\delta\mathbf{y}\|_F}{\|\mathbf{y}\|_F} \leq \frac{\kappa_s[\mathbf{A}]}{1 - \kappa_s[\mathbf{A}]\ \|\delta\mathbf{A}\|_s/\|\mathbf{A}\|_s}\left(\frac{\|\delta\mathbf{b}\|_F}{\|\mathbf{b}\|_F} + \frac{\|\delta\mathbf{A}\|_s}{\|\mathbf{A}\|_s}\right) \,. \tag{C.94}$$

A matrix $\mathbf{A}$ is ill-conditioned if $\kappa_s[\mathbf{A}]$ is large. Then, the solution of the equation $\mathbf{Ay} = \mathbf{b}$ or the computation of the inverse of $\mathbf{A}$ or of the eigenvalues $\lambda[\mathbf{A}]$ becomes inaccurate, especially as far as the smaller eigenvalues are concerned. Conditioning and scaling methods help to overcome this problem.

A general approach transforming a given (n,n)-matrix $\mathbf{A}$ into a decomposed version $\mathbf{D}$ is considered as $\mathbf{D} = \mathcal{T}[\mathbf{A}]$. Some transformations $\mathcal{T}$ are listed in Table C.1. With regard to their quadratic structure, they are referred to as spectral decompositions.

C.13.2 General Spectral Decomposition

In the general case of spectral decomposition $\mathbf{A} = \mathbf{EDE}^T$, the matrices $\mathbf{D}$ and $\mathbf{A}$ are congruent matrices. Substituting $\mathbf{x}^* = \mathbf{E}^T\mathbf{x}$ to the quadratic form $a(\mathbf{x})$

$$a(\mathbf{x}) = \mathbf{x}^T\mathbf{Ax} = \mathbf{x}^T\mathbf{EDE}^T\mathbf{x} = \mathbf{x}^{*T}\mathbf{Dx}^* \,. \tag{C.95}$$

If $\mathbf{D}$ is desired diagonal, $\mathbf{D} = \operatorname{diag} d_i$, the quadratic form turns out a sum $a(\mathbf{x}) = \sum_1^n d_i x_i^{*2}$.

Table C.1: Various decompositions (matrix transformations)$\mathcal{T}$

$\mathcal{T}$	name, details, main properties
$\mathbf{A} = \mathbf{EDE}^T$	general spectral decomposition (congruent matrix transformation)
$\mathbf{A} = \mathbf{TDT}^{-1}$	eigenvalue decomposition , $\mathbf{D}$... diagonal ($\mathbf{A} = \mathbf{A}^T \rightsquigarrow \mathbf{T}^{-1} = \mathbf{T}^T \rightsquigarrow \mathbf{E} = \mathbf{T}$) (orthogonal matrix transformation)
$\mathbf{A} = \mathbf{EDE}$	scaled decomposition
$\mathbf{A} = \mathbf{D}^2$	square root decomposition
$\mathbf{A} = \mathbf{DD}^T$	Cholesky decomposition

C.13.3 Eigenvalue Decomposition

Assume the matrix $\mathbf{A}$ symmetric, only. The modal matrix $\mathbf{T} = \mathbf{T}[\mathbf{A}]$ associated with $\mathbf{A}$ then becomes $\mathbf{T}^T = \mathbf{T}^{-1}$. The matrix $\mathbf{T}$ has orthogonal property. Using $\mathbf{T}$ for transformation $\mathbf{E} := \mathbf{T}$, $\mathbf{E}^T = \mathbf{E}^{-1}$. Since $\mathbf{A}^T = \mathbf{A} \rightsquigarrow \mathbf{p}_i = \mathbf{a}_i$ and from Eq.(B.95)

$$\mathbf{A} = \mathbf{T}(\text{diag }\lambda_i[\mathbf{A}])\mathbf{T}^T = \sum_{i=1}^{n} \lambda_i(\mathbf{a}_i\mathbf{a}_i^T) \tag{C.96}$$

$$\mathbf{A}^{-1} = (\mathbf{T}^T)^{-1}(\text{diag }\lambda_i^{-1}[\mathbf{A}])\mathbf{T}^{-1} = \mathbf{T}(\text{diag }\lambda_i^{-1})\mathbf{T}^T = \sum_{i=1}^{n} \lambda_i^{-1}(\mathbf{a}_i\mathbf{a}_i^T) \; . \tag{C.97}$$

Since $\mathbf{E} = \mathbf{T}$, $\mathbf{D} = \mathbf{T}^{-1}\mathbf{A}\mathbf{T}$ and $\det \mathbf{D} = \det \mathbf{A}$, the matrices $\mathbf{D}$ and $\mathbf{A}$ are similar, in this special case.

C.13.4 Orthogonal Transformation

Applying an orthogonal transformation $\mathbf{T}$ to second-order polynomials or quadratic forms, it can be arranged that the transformed variables v_i are non-interacting. Consider a quadratic polynomial in the vector-valued variable $\mathbf{x}$

$$f(\mathbf{x}) \triangleq \mathbf{x}^T\mathbf{Q}\mathbf{x} + 2\mathbf{b}^T\mathbf{x} + c \qquad \text{where} \qquad \mathbf{Q} = \mathbf{Q}^T \in \mathcal{R}^{n\times n} \; . \tag{C.98}$$

By the linear transformation $\mathbf{x} = \mathbf{T}^T\mathbf{v}$, the scalar $f(\mathbf{x})$ is transformed to

$$f(\mathbf{v}) = \mathbf{v}^T\mathbf{T}\mathbf{Q}\mathbf{T}^T\mathbf{v} + 2\mathbf{b}^T\mathbf{T}^T\mathbf{v} + c \; . \tag{C.99}$$

The transformation matrix $\mathbf{T}$ is defined as the modal matrix associated with $\mathbf{Q}$. Invoking Eq.(B.16),

$$\mathbf{T}^T = \mathbf{T}^{-1} \;\; \text{and} \;\; \mathbf{T}\mathbf{Q}\mathbf{T}^T = \mathbf{T}\mathbf{Q}\mathbf{T}^{-1} = \text{diag }\lambda_i[\mathbf{Q}] = \mathbf{\Lambda} \tag{C.100}$$

$$f(\mathbf{v}) = \mathbf{v}^T\mathbf{\Lambda}\mathbf{v} + 2\mathbf{b}^T\mathbf{T}^T\mathbf{v} + c \; . \tag{C.101}$$

The gradient of $f(\mathbf{v})$ with respect to $\mathbf{v}$ is $\partial f(\mathbf{v})/\partial\mathbf{v} = 2\mathbf{\Lambda}\mathbf{v} + 2\mathbf{T}\mathbf{b}$. Since $\mathbf{\Lambda}$ is diagonal, the gradient component $[\partial f(\mathbf{v})/\partial\mathbf{v}]_i$ only depends on the component v_i. In order to find the optimum of $f(\mathbf{v})$, the variables v_i may be adjusted independently of each other. Moreover, the optimum is given by (*Roberts, P.D., 1967*)

$$\mathbf{v}^\star = -\mathbf{\Lambda}^{-1}\mathbf{T}\mathbf{b}, \qquad \mathbf{x}^\star = -\mathbf{T}^T\mathbf{\Lambda}^{-1}\mathbf{T}\mathbf{b} = -\mathbf{Q}^{-1}\mathbf{b}, \qquad \mathbf{f}^\star = -\mathbf{b}^T\mathbf{Q}^{-1}\mathbf{b} + c \; . \tag{C.102}$$

Important applications are given in the field of parameter identification and optimal system design.

C.13.5 Scaled Decomposition

Consider $\mathbf{A}$ symmetric, only, and $\mathbf{E}$ diagonal, $\mathbf{E} := \text{diag } e_i$, $\mathbf{E} = \mathbf{E}^T$, then

$$\mathbf{A} = \mathbf{EDE} \; . \tag{C.103}$$

The matrix $\mathbf{D}$ is the scaled version of $\mathbf{A}$. The elements e_i are assigned to

$$e_i = \begin{cases} \sqrt{|A_{ii}|} & \text{if} \quad A_{ii} \neq 0 \\ 1 & \text{if} \quad A_{ii} = 0 \quad \text{(initial data).} \end{cases} \tag{C.104}$$

In the following paragraphs, A_{ii} has to be replaced by 1 if the initial data A_{ii} are zero. Substituting Eq.(C.104) into Eq.(C.103) and rearranging yields

$$\mathbf{D} = (\text{diag } e_i^{-1})\mathbf{A}(\text{diag } e_i^{-1}) \quad \text{or} \quad D_{ij} = A_{ij}/\sqrt{|A_{ii}|\,|A_{jj}|} \quad \text{and} \quad D_{ii} = 1. \tag{C.105}$$

The matrix $\mathbf{D}$ is symmetric since $\mathbf{A}$ is symmetric. The modal matrix associated with $\mathbf{D}$ is $\mathbf{T} = \mathbf{T}[\mathbf{D}]$. With regard to the symmetry of $\mathbf{D}$ it results $\mathbf{T}^T = \mathbf{T}^{-1}$ and $\mathbf{D} = \mathbf{T}(\text{diag } \lambda_i[\mathbf{D}])\mathbf{T}^T$. Note that $\mathbf{T}[\mathbf{D}]$ is used, *not* $\mathbf{T}[\mathbf{A}]$. Finally, the scaled decomposition of $\mathbf{A}$ is

$$\mathbf{A} = \mathbf{EDE} = \mathbf{ET}(\text{diag } \lambda_i[\mathbf{D}])\mathbf{T}^T\mathbf{E} = \mathbf{ET}(\text{diag } \lambda_i[\mathbf{D}])(\mathbf{ET})^T \; . \tag{C.106}$$

Inverting Eq.(C.106), the scaled decomposition of $\mathbf{A}^{-1}$ is achieved

$$\mathbf{A}^{-1} = \mathbf{E}^{-1}\mathbf{T}(\text{diag } \lambda_i^{-1})(\mathbf{E}^{-1}\mathbf{T})^T \; . \tag{C.107}$$

Partitioning the matrices $\mathbf{ET}$ and $\mathbf{E}^{-1}\mathbf{T}$ into columns $(\mathbf{ET})_{\cdot i}$ and $(\mathbf{E}^{-1}\mathbf{T})_{\cdot i}$, respectively, and denoting the kth element with the additional subscript k, the relations can be proved

$$(\mathbf{ET})_{\cdot ik} = (\mathbf{d}_i^{\triangleright})_k\sqrt{|A_{kk}|}, \qquad (\mathbf{E}^{-1}\mathbf{T})_{\cdot ik} = (\mathbf{d}_i^{\triangleright})_k\frac{1}{\sqrt{|A_{kk}|}} \tag{C.108}$$

where $\mathbf{d}_i^{\triangleright}$ is the eigenvector of $\mathbf{D}$. The scaled decomposition of $\mathbf{A}$ and $\mathbf{A}^{-1}$ can be written as the sum of λ_i-weighted dyadic products

$$\mathbf{A} = \sum_{i=1}^{n} \lambda_i[\mathbf{D}]\,(\mathbf{ET})_{\cdot i}(\mathbf{ET})_{\cdot i}^T \qquad \mathbf{A}^{-1} = \sum_{i=1}^{n} \lambda_i^{-1}[\mathbf{D}]\,(\mathbf{E}^{-1}\mathbf{T})_{\cdot i}\,(\mathbf{E}^{-1}\mathbf{T})_{\cdot i}^T \; . \tag{C.109}$$

Example:

$$\mathbf{A} = \begin{pmatrix} 10000 & 10 \\ 10 & 1 \end{pmatrix} \quad \mathbf{E} = \begin{pmatrix} 100 & 0 \\ 0 & 1 \end{pmatrix} \quad n_c(\mathbf{A}) = |\,\lambda[\mathbf{A}]\,|_{\max}/|\,\lambda[\mathbf{A}]\,|_{\min} = 10000.01 \,/\, 0.99 = 10101 \tag{C.110}$$

$$\text{Scaled version of } \mathbf{A}: \; \mathbf{D} = \begin{pmatrix} 0.01 & 0 \\ 0 & 1 \end{pmatrix}\begin{pmatrix} 10000 & 10 \\ 10 & 1 \end{pmatrix}\begin{pmatrix} 0.01 & 0 \\ 0 & 1 \end{pmatrix} = \begin{pmatrix} 1 & 0.1 \\ 0.1 & 1 \end{pmatrix} \tag{C.111}$$

$$\text{Eigenvalues } \lambda[\mathbf{D}]: \; \lambda_1 = 1.10, \; \lambda_2 = 0.90 \; ; \quad n_c(\mathbf{D}) = 1.10 \,/\, 0.90 = 1.22 \tag{C.112}$$

$$\text{Modal matrix: } \mathbf{T} = \begin{pmatrix} 0.7071 & 0.7071 \\ 0.7071 & -0.7071 \end{pmatrix} = (\mathbf{d}_1^{\triangleright} \; \mathbf{d}_2^{\triangleright}) \tag{C.113}$$

$$\mathbf{ET} = \begin{pmatrix} 70,71 & 70,71 \\ 0.71 & -0.71 \end{pmatrix} = \left((\mathbf{ET})_{\cdot 1} \;\vdots\; (\mathbf{ET})_{\cdot 2}\right) \tag{C.114}$$

$$\text{Matrix decomposed: } \mathbf{A} = \begin{pmatrix} 70,71 & 70,71 \\ 0.71 & -0.71 \end{pmatrix}\begin{pmatrix} 1.1 & 0 \\ 0 & 0.9 \end{pmatrix}\begin{pmatrix} 70,71 & 0.71 \\ 70,71 & -0.71 \end{pmatrix} \tag{C.115}$$

$$\mathbf{E}^{-1}\mathbf{T} = \begin{pmatrix} 0.0071 & 0.0071 \\ 0.7071 & -0.7071 \end{pmatrix} = \left((\mathbf{E}^{-1}\mathbf{T})_{\cdot 1} \;\vdots\; (\mathbf{E}^{-1}\mathbf{T})_{\cdot 2}\right) \tag{C.116}$$

$$\text{Inverse decomposed: } \mathbf{A}^{-1} = \begin{pmatrix} 0.0071 & 0.0071 \\ 0.7071 & -0.7071 \end{pmatrix}\begin{pmatrix} 0.9091 & 0 \\ 0 & 1.1111 \end{pmatrix}\begin{pmatrix} 0.0071 & 0.7071 \\ 0.0071 & -0.7071 \end{pmatrix} . \tag{C.117}$$

End of Example

C.13.6 Square Root Decomposition

Considering nonsingular $\mathbf{A} = \mathbf{D}\mathbf{D}^T$ and symmetric $\mathbf{D}$, the square root decomposition is obtained. The matrix $\mathbf{D}$ is named the square root of $\mathbf{A}$. That is, $\mathbf{A} = \mathbf{D}^2$ and $\mathbf{D} = \sqrt{\mathbf{A}}$. In view of the eigenvalue decomposition Eq.(C.96),

$$\mathbf{A} = \mathbf{T}(\text{diag } \lambda_i)\mathbf{T}^T = \mathbf{T} \sqrt{\text{diag } \lambda_i}\ \mathbf{T}^T\mathbf{T} \sqrt{\text{diag } \lambda_i}\ \mathbf{T}^T . \tag{C.118}$$

The term $\mathbf{T}^T\mathbf{T}$ in Eq.(C.118) is $\mathbf{I}$. Thus, it results $\mathbf{D} = \mathbf{T} \sqrt{\text{diag } \lambda_i}\ \mathbf{T}^T$.

C.13.7 Cholesky Decomposition

Consider nonsingular $\mathbf{A}$, only, and a lower triangular matrix $\mathbf{L}$ with elements $L_{ij} = 0 \quad \forall i < j$. The decomposition

$$\mathbf{A} = \mathbf{L}\mathbf{L}^T \quad \text{or} \quad A_{ij} = \sum_{v=1}^{n} L_{iv}L_{iv} \quad \forall i,j = 1 \ldots n \tag{C.119}$$

is chosen. Then, L_{ij} can be found out by a simple recursion formula.

If a new variable $\mathbf{y}$ is defined when solving $\mathbf{Ax} = \mathbf{b}$ then

$$\mathbf{L}\mathbf{L}^T\mathbf{x} = \mathbf{b} \text{ and } \mathbf{L}^T\mathbf{x} = \mathbf{y} \quad \rightsquigarrow \quad \mathbf{Ly} = \mathbf{b} . \tag{C.120}$$

Both equations for $\mathbf{y}$ and $\mathbf{x}$ can easily be solved on account of the triangular nature of $\mathbf{L}$ and its simple inverse.

C.14 Orthogonalizing

A given set of data $\mathbf{z}_i$, i.e., a set of n-vectors, possibly contains linear dependent information. Composing these data $\mathbf{z}_i$ to a matrix $\mathbf{Z} = (\mathbf{z}_i, \mathbf{z}_2, \ldots)$, the information of $\mathbf{Z}$ should be transferred into the column information of a new matrix $\mathbf{X}$. The aim is to receive orthogonal columns $\mathbf{x}_i$, only. The matrix $\mathbf{X}_{i-1}$ is the submatrix of $\mathbf{X}$, containing columns $\mathbf{x}_j$ $(j \leq i-1)$ already transformed by previous operations

$$\mathbf{X}_i = (\mathbf{x}_1, \mathbf{x}_2 \ldots \mathbf{x}_i) \qquad \mathbf{X} = (\mathbf{x}_1, \mathbf{x}_2 \ldots \mathbf{x}_q) \qquad \mathbf{X}_i \in \mathcal{R}^{n \times i}, \quad \mathbf{X} \in \mathcal{R}^{n \times q}, \quad \mathbf{x}_i \in \mathcal{R}^n , \quad i < q \leq n . \tag{C.121}$$

Remember the orthogonality properties treated when deriving Eq.(D.37). Thus, the operator matrix (premultiplication matrix) to be applied to $\mathbf{z}_i$, in order to obtain $\mathbf{x}_i$ orthogonal to each column of $\mathbf{X}_{i-1}$, is $(\mathbf{I} - \mathbf{X}_{i-1}\mathbf{X}_{i-1}^{!L})$, i.e.,

$$\mathbf{x}_i = (\mathbf{I} - \mathbf{X}_{i-1}\mathbf{X}_{i-1}^{!L})\mathbf{z}_i . \tag{C.122}$$

Multiplying $\mathbf{x}_i^T$ with $\mathbf{X}_{i-1}$ yields the null row $\mathbf{0}^T$

$$\mathbf{z}_i^T(\mathbf{I} - \mathbf{X}_{i-1}\mathbf{X}_{i-1}^{!L})^T\mathbf{X}_{i-1} = \mathbf{z}_i^T\left(\mathbf{X}_{i-1} - \mathbf{X}_{i-1}(\mathbf{X}_{i-1}^T\mathbf{X}_{i-1})^{-1}\mathbf{X}_{i-1}^T\mathbf{X}_{i-1}\right) = \mathbf{z}_i^T\mathbf{0} = \mathbf{0}^T . \tag{C.123}$$

An alternative way to achieve $\mathbf{x}_i$ is to apply the pseudo-left-inverse $\mathbf{X}_{i-1}^{!L}$ to the vector $\mathbf{z}_i$

$$\mathbf{p}_i = (\mathbf{X}_{i-1}^T\ \mathbf{X}_{i-1})^{-1}\mathbf{X}_{i-1}^T\mathbf{z}_i . \tag{C.124}$$

The parameter vector $\mathbf{p}_i$ fits $\mathbf{z}_i$ optimally into the $\mathbf{X}_{i-1}$-space. The vector $\mathbf{X}_{i-1}\mathbf{p}_i$ is the optimal least square regression of $\mathbf{z}_i$. The residual $\mathbf{z}_i - \mathbf{X}_{i-1}\mathbf{p}_i$ is orthogonal to all previous vectors from $\mathbf{x}_1$ until $\mathbf{x}_{i-1}$. This residual is declared as $\mathbf{x}_i$

$$\mathbf{x}_i = \mathbf{z}_i - \mathbf{X}_{i-1}(\mathbf{X}_{i-1}^T\mathbf{X}_{i-1})^{-1}\mathbf{X}_{i-1}^T\mathbf{z}_i \tag{C.125}$$

and is chosen to complete $\mathbf{X}_{i-1}$ to $\mathbf{X}_i$.

The algorithm is started with $\mathbf{x}_1 = \mathbf{z}_1$, without any regression manipulation. After having regressed all input vectors $\mathbf{z}_i$ and having computed the residuals, all the vectors $\mathbf{x}_i$ are orthogonal.

Example:

$$\mathbf{z}_1 = \begin{pmatrix} 3 \\ 4 \\ 5 \end{pmatrix} = \mathbf{x}_1 \qquad \mathbf{z}_2 = \begin{pmatrix} 1 \\ 1 \\ 2 \end{pmatrix} \tag{C.126}$$

$$\mathbf{x}_2 = \begin{pmatrix} 1 \\ 1 \\ 2 \end{pmatrix} - \begin{pmatrix} 3 \\ 4 \\ 5 \end{pmatrix} (50)^{-1}(3 \quad 4 \quad 5) \begin{pmatrix} 1 \\ 1 \\ 2 \end{pmatrix} = \begin{pmatrix} -0.02 \\ -0.36 \\ 0.30 \end{pmatrix} \perp \mathbf{x}_1 \,. \tag{C.127}$$

End of Example

If a zero column $\mathbf{x}_i$ appears, the linear dependence is obvious. This datum $\mathbf{x}_i = \mathbf{0}$ has to be cancelled to avoid singularity of $\mathbf{X}_{i-1}^T\mathbf{X}_{i-1}$. Linear dependence occurs if at least one column of $\mathbf{X}_{i-1}$ is a linear combination of the other ones. Then, $\mathrm{rank}(\mathbf{X}_{i-1}^T\mathbf{X}_{i-1}) < n$, $\det\mathbf{X}_{i-1}^T\mathbf{X}_{i-1} = 0$ and $\mathbf{X}_{i-1}^T\mathbf{X}_{i-1}$ is singular.

Example:

$$\mathbf{z}_1 = \begin{pmatrix} 3 \\ 4 \\ 5 \end{pmatrix} = \mathbf{x}_1; \quad \mathbf{z}_2 = \begin{pmatrix} 0 \\ 0 \\ 0 \end{pmatrix} \rightsquigarrow \mathbf{x}_2 = \begin{pmatrix} 0 \\ 0 \\ 0 \end{pmatrix}, \quad \mathbf{X}_2 = \begin{pmatrix} 3 & 0 \\ 4 & 0 \\ 5 & 0 \end{pmatrix}, \quad \mathbf{X}_2^T\mathbf{X}_2 = \begin{pmatrix} 50 & 0 \\ 0 & 0 \end{pmatrix}. \tag{C.128}$$

End of Example

Appendix D

Linear Regression and Estimation

Linear regression is considered as the deterministic problem of minimizing the error between measurement sequences and a deterministic model. The aim is to find the optimum model parameters based on least squares. Linear models are taken into account, only. The model is defined as

$$\mathbf{M}\mathbf{p} = \mathbf{y} \qquad \mathbf{M} \in \mathcal{R}^{n_m \times n_p}, \quad \mathbf{p} \in \mathcal{R}^{n_p}, \quad \mathbf{y} \in \mathcal{R}^{n_m} . \tag{D.1}$$

The matrix $\mathbf{M}$ is an (n_m, n_p)-matrix with fixed elements, $\mathbf{p}$ (usually) a vector with n_p elements of unknown parameters and $\mathbf{y}$ an observation vector with dimension n_m (the number of measurements or observations).

D.1 Parameter Demarcation

When measurements are available to state n_m linear equations with n_p parameters and if $n_m < n_p$, the system is underdetermined and a solution does not exist. The shape of the matrix $\mathbf{M}$ is row-like. Although there is no solution the matrix $\mathbf{M}$ defines a demarcation which relation between the elements of $\mathbf{p}$ exists.

Within the scope given by $\mathbf{M}$ and $\mathbf{y}$, find $\mathbf{p}$ such that the least square $\mathbf{p}^T\mathbf{p}$ or the Frobenius norm $||\mathbf{p}||_F^2$ is minimum. This investigation can be formulated with the help of a vector-Lagrange-multiplier $\boldsymbol{\lambda}$

$$\mathbf{p}^T\mathbf{p} + \boldsymbol{\lambda}^T(\mathbf{M}\mathbf{p} - \mathbf{y}) \rightarrow \min \tag{D.2}$$

$$\frac{\partial}{\partial \mathbf{p}} [\mathbf{p}^T\mathbf{p} + \boldsymbol{\lambda}^T(\mathbf{M}\mathbf{p} - \mathbf{y})] = 0 \quad \leadsto \quad 2\mathbf{p} + (\boldsymbol{\lambda}^T\mathbf{M})^T = 0 \quad \leadsto \quad \mathbf{p} = -0.5\ \mathbf{M}^T\boldsymbol{\lambda} . \tag{D.3}$$

In view of Eq.(D.1) and (D.3),

$$\mathbf{M}(-0.5\ \mathbf{M}^T\boldsymbol{\lambda}) = \mathbf{y} \quad \text{and} \quad \boldsymbol{\lambda} = -2(\mathbf{M}\mathbf{M}^T)^{-1}\mathbf{y} . \tag{D.4}$$

The matrix $\mathbf{M}\mathbf{M}^T$ is preassumed nonsingular. Combining Eq.(D.3) and (D.4), the result is

$$\mathbf{p} = \mathbf{p}^\star = \mathbf{M}^T(\mathbf{M}\mathbf{M}^T)^{-1}\mathbf{y} = \mathbf{M}^{!R}\mathbf{y}. \tag{D.5}$$

The starred parameter $\mathbf{p}^\star$ is the optimal (minimal least square) parameter vector, complying with Eq. (D.1).

Eq.(D.5) shows the linear relationship between $\mathbf{y}$ and $\mathbf{p}^\star$. The matrix operator to be applied to $\mathbf{y}$ is the right-pseudo-inverse $\mathbf{M}^{!R}$. It is a right-inverse of $\mathbf{M}$ because $\mathbf{M}$ has to be multiplied by $\mathbf{M}^{!R}$ from the right to yield the identity matrix $\mathbf{I}_{n_m}$.

Example 1: Suppose that $n_m = 2$ measurements are available and $n_p = 3$ parameters are to be determined, one has two linear equations with three variables p_1, p_2 and p_3. Each equation corresponds to a plane in Fig. D.1.

$$\mathbf{M} = \begin{pmatrix} 0 & -0.5 & -1.1506 \\ 0.5 & 1 & 0 \end{pmatrix} \qquad (\mathbf{M}\mathbf{M}^T)^{-1} = \begin{pmatrix} 0.7279 & 0.2911 \\ 0.2911 & 0.9165 \end{pmatrix} \tag{D.6}$$

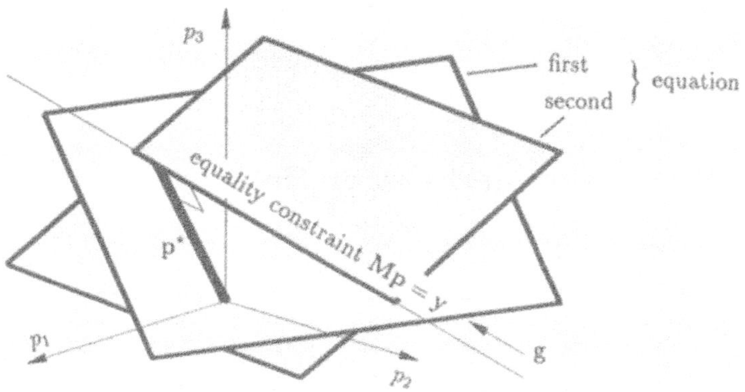

Figure D.1: Illustration of two linear equations with three variables

$$\mathbf{M}^{IR} = \mathbf{M}^T(\mathbf{M}\mathbf{M}^T)^{-1} = \begin{pmatrix} 0.1456 & 0.4583 \\ -0.0729 & 0.7710 \\ -0.8375 & -0.3349 \end{pmatrix}, \quad \mathbf{y} = \begin{pmatrix} 0.5 \\ 1.1506 \end{pmatrix} \rightsquigarrow \mathbf{p}^\star = \mathbf{M}^{IR}\mathbf{y} = \begin{pmatrix} 0.6001 \\ 0.8507 \\ -0.8041 \end{pmatrix}. \tag{D.7}$$

Vectors perpendicular to the two planes given by $\mathbf{M}$:

$$\begin{pmatrix} 0 \\ -0.5 \\ -1.1506 \end{pmatrix} \quad \text{and} \quad \begin{pmatrix} 0.5 \\ 1 \\ 0 \end{pmatrix}. \tag{D.8}$$

Vector $\mathbf{g}$ in the direction of the intersection of both planes:

$$\mathbf{g} = \begin{pmatrix} -0.5 \times 0 + 1.1506 \times 1 \\ -1.1506 \times 0.5 - 0 \times 0 \\ 0 \times 1 + 0.5 \times 0.5 \end{pmatrix} = \begin{pmatrix} 1.1506 \\ -0.5753 \\ 0.25 \end{pmatrix}. \tag{D.9}$$

Checking if $\mathbf{g}$ is perpendicular to $\mathbf{p}^\star$:

$$\mathbf{g}^T\mathbf{p}^\star = 0.6001 \times 1.1506 + 0.8507(-0.5753) - 0.8041 \times 0.25 = 4.23 \times 10^{-5} \doteq 0\ . \ \square \tag{D.10}$$

Example 2: Optimal initial conditions for observers.

The initial condition is usually chosen $\hat{\mathbf{x}}(0) = \mathbf{0}$. This default choice does not reduce $\tilde{\mathbf{x}}(t)$ as much as possible and is not compatible with the initial output $\mathbf{y}(0)$. The optimal choice is given by $||\hat{\mathbf{x}}(0)||_F \rightarrow \min$, subject to $\mathbf{C}\hat{\mathbf{x}}(0) = \mathbf{y}(0)$ (*Johnson, C.D., 1988*) which is exactly the problem posed in this section. Substituting $\mathbf{M} := \mathbf{C}$, $\mathbf{y} := \mathbf{y}(0)$, $\mathbf{p} := \hat{\mathbf{x}}(0)$, the optimal initial condition is

$$\hat{\mathbf{x}}^\star(0) = \mathbf{C}^{IR}\mathbf{y}(0) = \mathbf{C}^T(\mathbf{C}\mathbf{C}^T)^{-1}\mathbf{y}(0)\ . \tag{D.11}$$

End of Examples

D.2 Interpolation

Consider the case $n_m = n_p$. The number of measurements is supposed equal to the number of parameters to fit the model of Eq. (D.1). The measurements are preassumed linearly independent. Eq.(D.1) with square matrix $\mathbf{M}$ can be solved

$$\mathbf{p}^\star = \mathbf{M}^{-1}\mathbf{y}. \tag{D.12}$$

The inverse exists with regard to the independence of the measurements. The unique solution in the parameter vector $\mathbf{p}^\star$ provides a unique interpolation of the observation $\mathbf{y}$.

D.3 Weighted Least Squares Approximation

In the case $n_m > n_p$, more measurement information is given than one needs at minimum to solve Eq.(D.1), see Fig. D.3. The system of equations of this model is overdetermined. Hence, Eq.(D.1) must be rewritten to

$$\mathbf{Mp} = \mathbf{y} - \boldsymbol{\varepsilon} \qquad \mathbf{M} \in \mathcal{R}^{n_m \times n_p}\,, \qquad n_m > n_p \tag{D.13}$$

in order to have a consistent system of equations.

The solution of Eq.(D.13) yields an optimal approximation of the measurements without rejecting any information. Optimal approximation is gained by postulating $\mathbf{p}^\star$ in such a way that an index of performance is minimized. In most cases this index of performance C is defined as the sum of squares ε_i^2 produced by a special $\mathbf{p}$

$$\boldsymbol{\varepsilon} = \mathbf{y} - \mathbf{Mp} = (\varepsilon_1 \ldots \varepsilon_{n_m})^T = \boldsymbol{\varepsilon}(\mathbf{p}) \quad \rightsquigarrow \quad C = \sum_{i=1}^{n_m} \varepsilon_i^2 = \boldsymbol{\varepsilon}^T\boldsymbol{\varepsilon} \;\rightarrow\; \min\,. \tag{D.14}$$

This case is shown in a subsequent section in detail. The function C can also be established in a more general way, using different weighting factors for ε_i^2 and $\varepsilon_i\varepsilon_j$ thus forming a weighting matrix $\mathbf{W}$

$$C = \boldsymbol{\varepsilon}^T\mathbf{W}\boldsymbol{\varepsilon} \;\rightarrow\; \min \tag{D.15}$$

where $\mathbf{W}$ denotes an arbitrary positive definite matrix, normally symmetric.

Following the necessary condition for the minimum of C with respect to $\mathbf{p}$ yields

$$\frac{\partial C}{\partial \mathbf{p}} = \frac{\partial}{\partial \mathbf{p}}[(-\mathbf{Mp}+\mathbf{y})^T\mathbf{W}(-\mathbf{Mp}+\mathbf{y})] = \frac{\partial}{\partial \mathbf{p}}[\mathbf{p}^T\mathbf{M}^T\mathbf{WMp} - \mathbf{y}^T\mathbf{WMp} - \mathbf{p}^T\mathbf{M}^T\mathbf{Wy} + \mathbf{y}^T\mathbf{Wy}] = 0 \tag{D.16}$$

$$[\mathbf{M}^T\mathbf{WM} + (\mathbf{M}^T\mathbf{WM})^T]\mathbf{p} - (\mathbf{y}^T\mathbf{WM})^T - \mathbf{M}^T\mathbf{Wy} = 0 \tag{D.17}$$

$$\mathbf{p} = \mathbf{p}^\star = (\mathbf{M}^T\mathbf{WM})^{-1}\mathbf{M}^T\mathbf{Wy}\,. \tag{D.18}$$

Calculating the second derivative of C with respect to $\mathbf{p}$, this derivative matrix has to be positive definite. If $\mathbf{W} = \mathbf{W}^T$

$$\frac{\partial^2 C}{\partial \mathbf{p}^T \partial \mathbf{p}} = \mathbf{M}^T\mathbf{WM} > 0\,. \tag{D.19}$$

Both Eqs.(D.18) and (D.19) together are a necessary and sufficient condition for achieving a minimum of C with respect to $\mathbf{p}$. The condition Eq.(D.19) is satisfied if $\mathbf{W}$ is a positive definite matrix.

If the matrix $\mathbf{H} = \mathbf{M}^T\mathbf{WM}$ is singular, linear independent information must be cancelled. If cancelling leads to $n_m < n_p$ then Eq.(D.5) has to be employed. After cancelling, the matrix $\mathbf{H}_{red}$ plays the same role as $\mathbf{M}$ in Eq.(D.1). Using the right-pseudo-inverse $\mathbf{H}_{red}^{IR}$ yields the optimal solution as in Eq.(D.5).

A necessary condition to uniquely estimate $\mathbf{p}$ in the model $\mathbf{Mp} = \mathbf{y}$ when applying the least square method is the nonsingularity of $\mathbf{M}^T\mathbf{M}$ or $\det \mathbf{M}^T\mathbf{M} \neq 0$. It is a necessary condition even in the weighted least square case. This can easily be realized by inverting the assertion: If one has $\det\mathbf{M}^T\mathbf{M} = 0$ then also $\det \mathbf{M}^T\mathbf{WM} = 0$. It is not a sufficient condition: If $\det \mathbf{M}^T\mathbf{M} \neq 0$ then $\det \mathbf{M}^T\mathbf{WM} \neq 0$ or $= 0$ is possible, depending on $\mathbf{W}$. Thus, for the model $\mathbf{Mp} = \mathbf{y}$ the statement $\det \mathbf{M}^T\mathbf{M} \neq 0$ is denoted ***identifiability condition***.

Example: Four simple cases are chosen to illustrate various assumptions on n_m, n_p and rank $\mathbf{M}$. To obtain definite solutions rank $\mathbf{M} \geq n_p$ must be satisfied.

Case a) $n_m = 3,\ n_p = 2$:

$$\mathbf{M} = \begin{pmatrix} 3 & 4 \\ 2 & 1 \\ 1 & 2 \end{pmatrix} \qquad \mathbf{y} = \begin{pmatrix} 11 \\ 4 \\ 5 \end{pmatrix} \tag{D.20}$$

Third equation (row) linear dependent on the first and second one: rank $\mathbf{M} = 2$

$$\mathbf{M}^T\mathbf{M} = \begin{pmatrix} 14 & 16 \\ 16 & 21 \end{pmatrix}, \qquad \det \mathbf{M}^T\mathbf{M} \neq 0\,, \qquad \hat{\mathbf{p}} = (1 \quad 2)^T\,. \tag{D.21}$$

Case b) $n_m = 2,\ n_p = 2$:

$$\mathbf{M} = \begin{pmatrix} 3 & 4 \\ 6 & 8 \end{pmatrix} \tag{D.22}$$

Second equation linear dependent on the first: rank $\mathbf{M} = 1$

$$\mathbf{M}^T\mathbf{M} = \begin{pmatrix} 45 & 60 \\ 60 & 80 \end{pmatrix}, \qquad \det \mathbf{M}^T\mathbf{M} = 0\ . \tag{D.23}$$

Case c) $n_m = 3,\ n_p = 2$:

$$\mathbf{M} = \begin{pmatrix} 3 & 4 \\ 6 & 8 \\ -3 & -4 \end{pmatrix} \tag{D.24}$$

Second and third row linear dependent on the first one: rank $\mathbf{M} = 1$

$$\mathbf{H} = \mathbf{M}^T\mathbf{M} = \begin{pmatrix} 54 & 72 \\ 72 & 96 \end{pmatrix}, \qquad \det \mathbf{M}^T\mathbf{M} = 0\ . \tag{D.25}$$

Case d) : Case c) continued:
The equations

$$\mathbf{H}\hat{\mathbf{p}} = \mathbf{M}^T\mathbf{y} = \begin{pmatrix} 198 \\ 264 \end{pmatrix} \tag{D.26}$$

are linear dependent just as Eq.(D.24). The system contains the same information as the first equation in Eq. (D.24) or any other one : $(3 \quad 4)\hat{\mathbf{p}} = 11$. After cancelling useless information in Eq.(D.26),

$$\mathbf{H}_{red} = (54 \quad 72), \qquad (\mathbf{M}^T\mathbf{y})_{red} = 198 \quad \leadsto \quad \hat{\mathbf{p}} = \mathbf{H}_{red}^T(\mathbf{H}_{red}\,\mathbf{H}_{red}^T)^{-1}(\mathbf{M}^T\mathbf{y})_{red} = \mathbf{H}_{red}^{IR}\,(\mathbf{M}^T\mathbf{y})_{red} \tag{D.27}$$

$$\hat{\mathbf{p}} = \begin{pmatrix} 54 \\ 72 \end{pmatrix}[(54 \quad 72)\begin{pmatrix} 54 \\ 72 \end{pmatrix}]^{-1} \times 198 = \begin{pmatrix} 0.0067 \\ 0.0089 \end{pmatrix} 198 = \begin{pmatrix} 1.32 \\ 1.76 \end{pmatrix}\ . \tag{D.28}$$

End of Example

The error vector $\boldsymbol{\varepsilon}$ at the minimum $\mathbf{p}^\star$, $\boldsymbol{\varepsilon}(\mathbf{p}^\star) = \boldsymbol{\varepsilon}^\star = \mathbf{e}$ is named residual $\mathbf{e}$. In statistic estimation theory the expectation $E[\mathbf{p}^\star]$ turns out to be the optimal estimation $\hat{\mathbf{p}}$. Anticipating this fact, the superscript * is frequently omitted and the optimal parameter $\mathbf{p}^\star$ is equated with $\hat{\mathbf{p}}$, i.e. $\mathbf{p}^\star = \hat{\mathbf{p}}$. The quantity $\mathbf{M}\hat{\mathbf{p}} = \hat{\mathbf{y}}$ is the regression value of $\mathbf{y}$. Fig. D.2 gives an impression in three dimensional sample space.

The weighted squared error index of performance $C = \boldsymbol{\varepsilon}^T\mathbf{W}\boldsymbol{\varepsilon} = C(\mathbf{p})$ depends on $\mathbf{p}$. After having minimized C, there remains the residual sum of squares $C^\star = \boldsymbol{\varepsilon}^{\star T}\mathbf{W}\boldsymbol{\varepsilon}^\star = C(\hat{\mathbf{p}})$. In view of this, substituting $\mathbf{e} = \boldsymbol{\varepsilon}^\star = -\mathbf{M}\hat{\mathbf{p}} + \mathbf{y}$ and $\hat{\mathbf{p}}$ into C yields

$$\begin{aligned}
C^\star_{WLS} &= (-\mathbf{M}\hat{\mathbf{p}} + \mathbf{y})^T\mathbf{W}(-\mathbf{M}\hat{\mathbf{p}} + \mathbf{y}) = \hat{\mathbf{p}}^T\mathbf{M}^T\mathbf{W}\mathbf{M}\hat{\mathbf{p}} - \mathbf{y}^T\mathbf{W}\mathbf{M}\hat{\mathbf{p}} - \hat{\mathbf{p}}^T\mathbf{M}^T\mathbf{W}\mathbf{y} + \mathbf{y}^T\mathbf{W}\mathbf{y} \\
C^\star_{WLS} &= \mathbf{y}^T\mathbf{W}^T\mathbf{M}(\mathbf{M}^T\mathbf{W}\mathbf{M})^{-1,T}\mathbf{M}^T\mathbf{W}\mathbf{M}(\mathbf{M}^T\mathbf{W}\mathbf{M})^{-1}\mathbf{M}^T\mathbf{W}\mathbf{y} \\
&\quad - \mathbf{y}^T\mathbf{W}\mathbf{M}(\mathbf{M}^T\mathbf{W}\mathbf{M})^{-1}\mathbf{M}^T\mathbf{W}\mathbf{y} - \mathbf{y}^T\mathbf{W}^T\mathbf{M}(\mathbf{M}^T\mathbf{W}\mathbf{M})^{-1,T}\mathbf{M}^T\mathbf{W}\mathbf{y} + \mathbf{y}^T\mathbf{W}\mathbf{y}\ .
\end{aligned} \tag{D.29}$$

The first and second term above can be cancelled. Supposing $\mathbf{W}$ symmetric, simplifying yields

$$C^\star_{WLS} = \mathbf{y}^T\mathbf{W}[\mathbf{I} - \mathbf{M}(\mathbf{M}^T\mathbf{W}\mathbf{M})^{-1}\mathbf{M}^T\mathbf{W}]\mathbf{y}. \tag{D.30}$$

By substituting $\mathbf{y} = \mathbf{M}\hat{\mathbf{p}} + \mathbf{e}$, the expression Eq.(D.30) is rewritten by elementary algebraic operations

$$\begin{aligned}
C^\star_{WLS} &= (\mathbf{M}\hat{\mathbf{p}} + \mathbf{e})^T\mathbf{W}[\mathbf{I} - \mathbf{M}(\mathbf{M}^T\mathbf{W}\mathbf{M})^{-1}\mathbf{M}^T\mathbf{W}](\mathbf{M}\hat{\mathbf{p}} + \mathbf{e}) \\
&= (\hat{\mathbf{p}}^T\mathbf{M}^T + \mathbf{e}^T)[\mathbf{W} - \mathbf{W}\mathbf{M}(\mathbf{M}^T\mathbf{W}\mathbf{M})^{-1}\mathbf{M}^T\mathbf{W}](\mathbf{M}\hat{\mathbf{p}} + \mathbf{e}) \\
&= [\hat{\mathbf{p}}^T\mathbf{M}^T\mathbf{W} + \mathbf{e}^T\mathbf{W} - \hat{\mathbf{p}}^T\mathbf{M}^T\mathbf{W}\mathbf{M}(\mathbf{M}^T\mathbf{W}\mathbf{M})^{-1}\mathbf{M}^T\mathbf{W} \\
&\qquad -\mathbf{e}^T\mathbf{W}\mathbf{M}(\mathbf{M}^T\mathbf{W}\mathbf{M})^{-1}\mathbf{M}^T\mathbf{W}](\mathbf{M}\hat{\mathbf{p}} + \mathbf{e}) \\
&= \mathbf{e}^T\mathbf{W}\mathbf{M}\hat{\mathbf{p}} - \mathbf{e}^T\mathbf{W}\mathbf{M}(\mathbf{M}^T\mathbf{W}\mathbf{M})^{-1}\mathbf{M}^T\mathbf{W}\mathbf{M}\hat{\mathbf{p}} \\
&\qquad +\mathbf{e}^T\mathbf{W}\mathbf{e} - \mathbf{e}^T\mathbf{W}\mathbf{M}(\mathbf{M}^T\mathbf{W}\mathbf{M})^{-1}\mathbf{M}^T\mathbf{W}\mathbf{e} \\
C^\star_{WLS} &= \mathbf{e}^T\mathbf{W}[\mathbf{I} - \mathbf{M}(\mathbf{M}^T\mathbf{W}\mathbf{M})^{-1}\mathbf{M}^T\mathbf{W}]\mathbf{e}\ .
\end{aligned} \tag{D.31}$$

The matrix $\mathbf{W}[\mathbf{I} - \mathbf{M}(\mathbf{M}^T\mathbf{W}\mathbf{M})^{-1}\mathbf{M}^T\mathbf{W}]$ weights the squares of $\mathbf{y}$ and $\mathbf{e}$ in such a way that equal results $C^\star_{WLS}$ are obtained (*Rosen, J.B., 1960*).

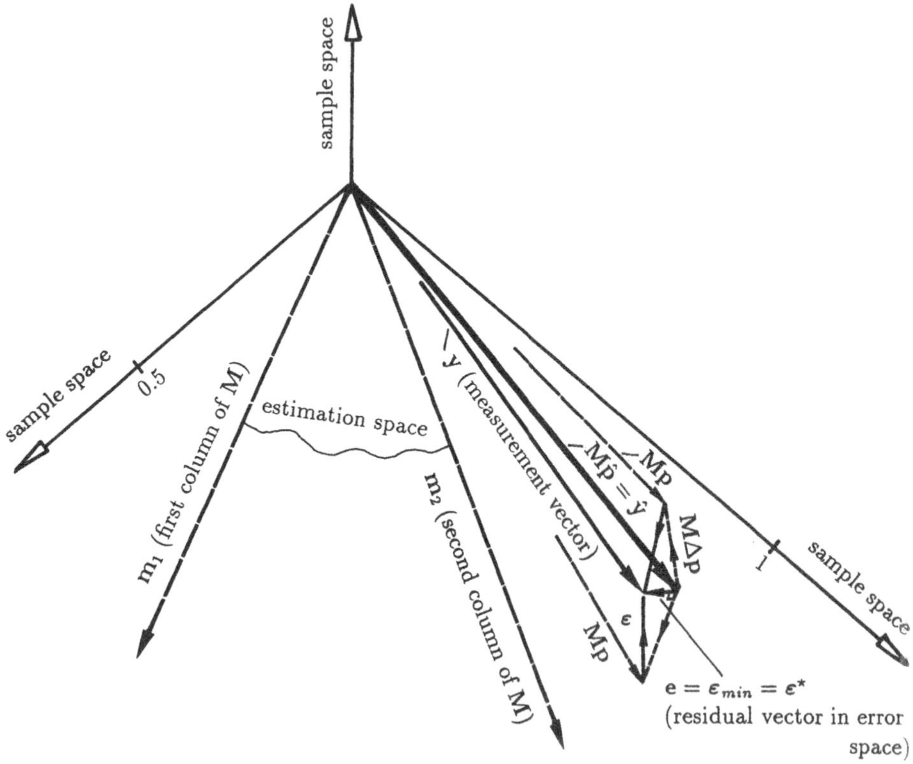

Figure D.2: Sample space, estimation space and error space in the case of $n_m = 3$ measurements and $n_p = 2$ parameters

D.4 Ordinary Least Squares Approximation

Setting $\mathbf{W} = \mathbf{I}$, the weighted least squares solution is specialized to the ordinary least squares one. The ordinary least squares solution is

$$\mathbf{p} = \mathbf{p}^{\star} = (\mathbf{M}^T\mathbf{M})^{-1}\mathbf{M}^T\mathbf{y} = \mathbf{M}^{!L}\mathbf{y}. \tag{D.32}$$

The matrix $\mathbf{M}^{!L}$ is the left-pseudo-inverse of the matrix $\mathbf{M}$ and is of dimension $n_p \times n_m$ (vice versa to the dimension $n_m \times n_p$ of the model matrix $\mathbf{M}$).

For the sake of comparison, previous results are repeated: If solutions of $\mathbf{Mp} = \mathbf{y}$ exist then either $\mathbf{M}^{-1}\mathbf{y}$ is the unique solution or $\hat{\mathbf{p}} = \mathbf{M}^{!R}\mathbf{y}$ is the solution of minimum norm (length) $||\mathbf{p}||_F$. If solutions of $\mathbf{Mp} = \mathbf{y}$ do not exist then the sum of squares $\boldsymbol{\varepsilon}^T\boldsymbol{\varepsilon}$ of deviations $\boldsymbol{\varepsilon} = -\mathbf{Mp} + \mathbf{y}$ is minimized by $\hat{\mathbf{p}} = \mathbf{M}^{!L}\mathbf{y}$.

For automatic control purposes the left-pseudo-inverse $\mathbf{M}^{!L}$ plays a dominant role. If only the left-pseudo-inverse $\mathbf{M}^{!L}$ is used within a section the superscript ^{L} is omitted in $\mathbf{M}^{!}$, for simplicity. In the ordinary least squares case the resulting residual least squares becomes more simple

$$C^{\star}_{LS} = \mathbf{y}^T(\mathbf{I} - \mathbf{M}\mathbf{M}^{!})\mathbf{y} = \mathbf{e}^T(\mathbf{I} - \mathbf{M}\mathbf{M}^{!})\mathbf{e}\ . \tag{D.33}$$

Corresponding with

$$\mathbf{Mp} = \mathbf{y} - \boldsymbol{\varepsilon} \quad \text{(arbitrary } \mathbf{p}) \tag{D.34}$$

and with

$$\hat{\mathbf{y}} = \mathbf{M}\hat{\mathbf{p}} = \mathbf{y} - \mathbf{e} \quad \text{(optimal parameter } \hat{\mathbf{p}}) \tag{D.35}$$

Table D.1: Vectors and sum of squares

vector	squared amount of the vector (distance)
$\boldsymbol{\varepsilon}$	$C(\mathbf{p}) = \boldsymbol{\varepsilon}^T\boldsymbol{\varepsilon}$ (sum of squares in sample space representing the squared distance from any point $\mathbf{Mp}$ in estimation space to point $\mathbf{y}$)
$\mathbf{M}\Delta\mathbf{p} = \mathbf{M}\tilde{\mathbf{p}}$	$C(\mathbf{p}) - C(\hat{\mathbf{p}})$
$\mathbf{M}\hat{\mathbf{p}} = \hat{\mathbf{y}}$	regression sum of squares
$\mathbf{e}(= \tilde{\mathbf{y}})$	residual sum of squares $C_{min} = C^{\star}_{LS} = C(\hat{\mathbf{p}}) = \mathbf{e}^T\mathbf{e}$

the squared amount of the distances $\|\hat{\mathbf{y}}\|_F^2, \|\mathbf{y}\|_F^2$ and $\|\mathbf{e}\|_F^2$ is investigated. With regard to the fact that $\mathbf{e}$ is minimum in quadratic sense, $\hat{\mathbf{y}} = \mathbf{M}\hat{\mathbf{p}}$ and $\mathbf{e} = \boldsymbol{\varepsilon}(\hat{\mathbf{p}})$ are vertical. Applying Eq. (D.32),

$$\hat{\mathbf{y}}^T\mathbf{e} = (\mathbf{M}\hat{\mathbf{p}})^T(\mathbf{y} - \mathbf{M}\hat{\mathbf{p}}) = \hat{\mathbf{p}}^T(\mathbf{M}^T\mathbf{y} - \mathbf{M}^T\mathbf{M}\hat{\mathbf{p}}) = \hat{\mathbf{p}}^T \times 0 = 0. \tag{D.36}$$

Hence, the above-mentioned distances $\hat{\mathbf{y}}, \mathbf{y}$ and $\mathbf{e}$ satisfy $\|\mathbf{y}\|_F^2 = \|\hat{\mathbf{y}}\|_F^2 + \|\mathbf{e}\|_F^2$ (see Fig. D.2). The regression sum of squares is defined by the Frobenius norm $\|\hat{\mathbf{y}}\|_F^2 = \hat{\mathbf{y}}^T\hat{\mathbf{y}}$. In terms of the observation vector $\mathbf{y}$ the regression sum can be expressed as

$$\|\hat{\mathbf{y}}\|_F^2 = \hat{\mathbf{y}}^T\hat{\mathbf{y}} = (\mathbf{M}\hat{\mathbf{p}})^T\mathbf{M}\hat{\mathbf{p}} = (\mathbf{M}\mathbf{M}^I\mathbf{y})^T\mathbf{M}\mathbf{M}^I\mathbf{y} = \mathbf{y}^T\mathbf{M}\mathbf{M}^I\mathbf{y} \,. \tag{D.37}$$

Referring to the orthogonality between $\hat{\mathbf{y}}$ and $\mathbf{e}$, the same result is obtained from $(\mathbf{M}\hat{\mathbf{p}})^T\mathbf{y}$ or $(\mathbf{M}\hat{\mathbf{p}})^T\hat{\mathbf{y}}$. The projection of the vector $\mathbf{y}$ on to the plane spanned by $\mathbf{m}_1$ and $\mathbf{m}_2$ in Fig. D.2 is identical to $\hat{\mathbf{y}}$. The squared length of the vector $\mathbf{y}$ is given by $\mathbf{y}^T\mathbf{y}$, the optimal C by $C^{\star}_{LS}$ in Eq.(D.33). Combining Eqs. (D.37) and (D.33),

$$C^{\star}_{LS} = \mathbf{y}^T[\mathbf{I} - \mathbf{M}\mathbf{M}^I]\mathbf{y} = \mathbf{y}^T\mathbf{y} - \hat{\mathbf{y}}^T\hat{\mathbf{y}} \,. \tag{D.38}$$

This expression $C^{\star}_{LS}$ is named the residual sum of squares. Thus, the residual sum $C^{\star}_{LS}$ plus the regression sum $\|\hat{\mathbf{y}}\|_F^2$ yields the squared sum of the observation vector $\|\mathbf{y}\|_F^2$

$$C^{\star}_{LS} + \|\hat{\mathbf{y}}\|_F^2 = \|\mathbf{y}\|_F^2 \,. \tag{D.39}$$

The vector quantity $\mathbf{y} = \hat{\mathbf{y}} + \mathbf{e} = \hat{\mathbf{y}} + \tilde{\mathbf{y}}$ is used to define $\tilde{\mathbf{y}} = \mathbf{e}$. The residual $\mathbf{e}$ frequently is declared as the minimum estimation error $\tilde{\mathbf{y}}$ (deterministic equation error vector), see Table D.1.

Combining Eq.(D.18) and $\hat{\mathbf{y}} = \mathbf{M}\mathbf{p}^{\star}$,

$$\hat{\mathbf{y}} = \mathbf{M}\mathbf{p}^{\star} = \mathbf{M}(\mathbf{M}^T\mathbf{W}\mathbf{M})^{-1}\mathbf{M}^T\mathbf{W}\mathbf{y} \triangleq \mathbf{T}_M\mathbf{y} \tag{D.40}$$

where $\mathbf{T}_M$ is an idempotent transformation matrix. Applying $\mathbf{T}_M$ twice, three times etc., the same result is to be expected. This can easily be verified in Fig. D.2, by inspection.

In Fig. D.2 one has to distinguish between three spaces (*Draper, N.R., and Smith, H., 1966*). An additional space is sketched in Fig. D.4.

(i) The n_m-dimensional *sample space* containing the observation vector $\mathbf{y}$ and all columns of $\mathbf{M}$. Each of the n_p columns of $\mathbf{M}$ and the vector $\mathbf{y}$ is represented by a single point in the sample space.

(ii) The n_p-dimensional subspace named *estimation space* defined by the n_p column vectors of $\mathbf{M}$.

(iii) The *error space* assembled by the error vectors $\boldsymbol{\varepsilon}$ and the residual vector $\mathbf{e}$. The dimension of the error space is $n_m - n_p$. In Fig. D.2 it is a cone with the apex pointed out by $\mathbf{y}$.

(iv) The *parameter space* assembled by the components of $\mathbf{p}$.

Example: Linear regression of two parameters on the basis of three observations $n_m = 3, n_p = 2$.

a) Non-faulty observations:

$$\mathbf{M} = \begin{pmatrix} 0 & 0.5 \\ -0.5 & 1 \\ -1.1506 & 0 \end{pmatrix} \qquad \mathbf{y} = \begin{pmatrix} 0.5 \\ 1.1506 \\ 0.3466 \end{pmatrix} \tag{D.41}$$

$$\mathbf{M}^T\mathbf{M} = \begin{pmatrix} 0.7279 & 0.2911 \\ 0.2911 & 0.9165 \end{pmatrix} \qquad \mathbf{M}^{!L} = (\mathbf{M}^T\mathbf{M})^{-1}\mathbf{M}^T = \begin{pmatrix} 0.1456 & -0.0728 & -0.8375 \\ 0.4582 & 0.7709 & -0.3350 \end{pmatrix} \tag{D.42}$$

$$\mathbf{p}^* = \mathbf{M}^{!L}\mathbf{y} = \begin{pmatrix} -0.3012 \\ 1.0000 \end{pmatrix} . \tag{D.43}$$

b) Faulty observations:

$$\mathbf{y} = \begin{pmatrix} 0.6 \\ 1.3 \\ 0.5 \end{pmatrix} , \quad \hat{\mathbf{p}} = \mathbf{M}^{!L}\mathbf{y} = \begin{pmatrix} -0.426 \\ 1.1096 \end{pmatrix} , \quad \hat{\mathbf{y}} = \mathbf{M}\hat{\mathbf{p}} = \begin{pmatrix} 0.5548 \\ 1.3226 \\ 0.4902 \end{pmatrix} , \quad \mathbf{e} = \mathbf{y} - \mathbf{M}\hat{\mathbf{p}} = \begin{pmatrix} 0.0452 \\ -0.0226 \\ 0.0098 \end{pmatrix} \tag{D.44}$$

Check if $\mathbf{e}$ and sample space are orthogonal: $\mathbf{e}^T\mathbf{M} \doteq (0 \;\; 0)$.
Residual sum of squares: $\mathbf{e}^T\mathbf{e} = C^*_{LS} = C(\hat{\mathbf{p}}) = 0.0026, \quad ||\mathbf{e}||_F = 0.0515$.
Regression sum of squares: $\hat{\mathbf{y}}^T\hat{\mathbf{y}} = 2.2974, \quad ||\hat{\mathbf{y}}||_F = \sqrt{2.2974} = 1.5157$.
Norm of observation: $\mathbf{y}^T\mathbf{y} = 2.3000$, $\quad ||\mathbf{y}||_F = 1.5166$.
Check: $\mathbf{e}^T\mathbf{e} + \hat{\mathbf{y}}^T\hat{\mathbf{y}} = \,? \, = \mathbf{y}^T\mathbf{y} \qquad 0.0026 + 2.2974 = 2.3000$.
Parameter space (see subsequent section):

$$\Delta C = C(\mathbf{p}) - C(\hat{\mathbf{p}}) = \tilde{\mathbf{p}}^T\mathbf{M}^T\mathbf{M}\tilde{\mathbf{p}} \qquad \Delta C = 0.7279\,\tilde{p}_1^2 + 0.5822\,\tilde{p}_1\tilde{p}_2 + 0.9165\,\tilde{p}_2^2 . \tag{D.45}$$

End of Example

D.5 Left Inverse and Right Inverse. Mnemonic Aid

Considering $\mathbf{Mp} = \mathbf{y}$, $\mathbf{M} \in \mathcal{R}^{n_m \times n_p}$ and comparing (i) the least squares approximation $n_m > n_p$ with (ii) the parameter demarcation $n_m < n_p$, note the following facts when solving $\mathbf{Mp} = \mathbf{y}$, i.e. when "separating" $\mathbf{p}$ as a result.

(i) Least squares approximation: $\hat{\mathbf{p}} = \mathbf{M}^{!L}\mathbf{y}$ can be achieved by "multiplying" $\mathbf{Mp} = \mathbf{y}$ by $\mathbf{M}^{!L}$ from the left although *none* of the scalar equations $\mathbf{Mp} = \mathbf{y}$ is true (ϵ omitted) but the premultiplication by $\mathbf{M}^{!L}$ is a good mnemonic aid.

(ii) Parameter demarcation: $\mathbf{p} = \mathbf{M}^{!R}\mathbf{y}$ cannot be obtained by a multiplication operation from the statement $\mathbf{Mp} = \mathbf{y}$ although the result can easily be proved by multiplying the result $\mathbf{p} = \mathbf{M}^{!R}\mathbf{y}$ by $\mathbf{M}$ from the left. Note that each scalar equation of $\mathbf{Mp} = \mathbf{y}$ is true but multiplication of $\mathbf{Mp} = \mathbf{y}$ by $\mathbf{M}^{!L}$ from the left is inadmissible although one is induced to do this: $\mathbf{M}^T\mathbf{M}$ in the case $n_m < n_p$ has dimension $n_p \times n_p$, $\text{rank}\mathbf{M}^T\mathbf{M} = n_m$. Hence, the inverse $(\mathbf{M}^T\mathbf{M})^{-1}$ and $\mathbf{M}^{!L}$ do not exist.

D.6 Complex Matrix M

If the matrices involved are complex but only a real parameter vector $\mathbf{p}$ is admissible the result

$$\mathbf{p}^* = \arg\min_{\mathbf{p}} C_{WLS} \quad \text{subject to} \quad \Im m\, \mathbf{p} = 0 \tag{D.46}$$

is given by

$$\mathbf{p}^* = (\Re e\,\{\mathbf{M}^H\mathbf{M}\})^{-1}\Re e\,\{\mathbf{M}^H\mathbf{y}\} = (\Re e\,\{\mathbf{M}^H\mathbf{W}\mathbf{M}\})^{-1}\Re e\,\{\mathbf{M}^H\mathbf{W}\mathbf{y}\}, \qquad \mathbf{W} > 0 . \tag{D.47}$$

D.7 Sum of Errors and Residual Sum in Parameter Space

The parameter space is defined by the n_p-dimensional parameter vector $\mathbf{p}$. The sum of squared errors $\boldsymbol{\epsilon}^T\boldsymbol{\epsilon}$ is

$$C(\mathbf{p}) = \boldsymbol{\epsilon}^T\boldsymbol{\epsilon} = (-\mathbf{Mp} + \mathbf{y})^T(-\mathbf{Mp} + \mathbf{y}) = \mathbf{p}^T\mathbf{M}^T\mathbf{Mp} - \mathbf{y}^T\mathbf{Mp} - \mathbf{p}^T\mathbf{M}^T\mathbf{y} + \mathbf{y}^T\mathbf{y} . \tag{D.48}$$

The smallest value $C^*_{LS} = C(\hat{\mathbf{p}}) = \mathbf{y}^T\mathbf{y} - \mathbf{y}^T\mathbf{M}\mathbf{M}^!\mathbf{y}$ is subtracted from $C(\mathbf{p})$, thus obtaining

$$\Delta C = C(\mathbf{p}) - C(\hat{\mathbf{p}}) = \mathbf{p}^T\mathbf{M}^T\mathbf{Mp} - \mathbf{y}^T\mathbf{Mp} - \mathbf{p}^T\mathbf{M}^T\mathbf{y} + \mathbf{y}^T\mathbf{M}\mathbf{M}^!\mathbf{y} . \tag{D.49}$$

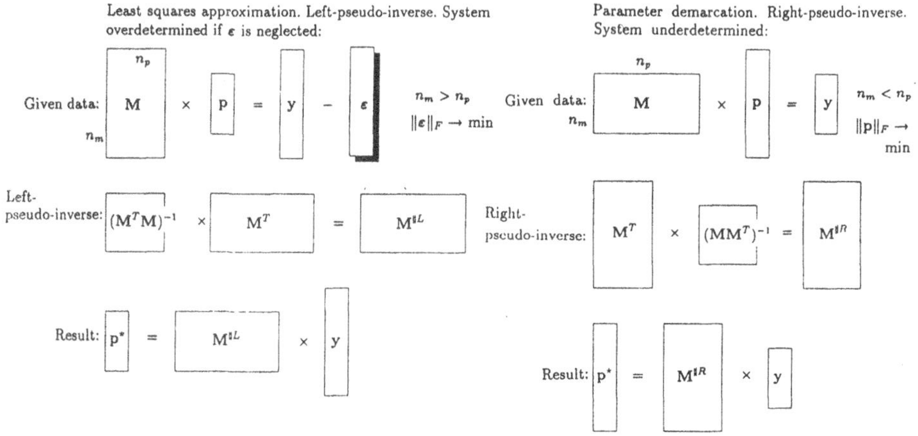

Figure D.3: Ordinary least squares approximation and parameter demarcation. Graphic interpretation: left-pseudo-inverse and right-pseudo-inverse

It can easily be checked that the formulations

$$\Delta C = (\mathbf{p}-\hat{\mathbf{p}})^T(\mathbf{M}^T\mathbf{M})(\mathbf{p}-\hat{\mathbf{p}}) = [\mathbf{M}(\mathbf{p}-\hat{\mathbf{p}})]^T[\mathbf{M}(\mathbf{p}-\hat{\mathbf{p}})] = (\mathbf{M}\Delta\mathbf{p})^T(\mathbf{M}\Delta\mathbf{p}) = (\mathbf{M}\tilde{\mathbf{p}})^T(\mathbf{M}\tilde{\mathbf{p}}) \quad (D.50)$$

are identical to the previous version. The vector $\Delta\mathbf{p} = \tilde{\mathbf{p}}$ is the parameter difference. The error sum difference to the residual sum expressed by ΔC is equal to the squared sum of $\Delta\mathbf{p}$ weighted by the matrix $\mathbf{M}^T\mathbf{M}$, as known from identifiability condition. The gradient of ΔC in the parameter space reads as follows

$$\frac{\partial \Delta C}{\partial \Delta\mathbf{p}} = 2(\mathbf{M}^T\mathbf{M})\ \Delta\mathbf{p}\ . \quad (D.51)$$

If the columns $\mathbf{m}_{ci}$ of the matrix $\mathbf{M} = (\mathbf{m}_{c1},\ \mathbf{m}_{c2}\ldots\mathbf{m}_{cn_p})$ are orthogonal $\forall\mu,\nu = 1\ldots n_p$ then

$$\mathbf{m}_{c\nu}^T\mathbf{m}_{c\mu} = 0 \quad \text{if} \quad \mu \neq \nu \quad \text{and} \quad \mathbf{m}_{c\nu}^T\mathbf{m}_{c\nu} = \alpha_\nu \quad \text{if} \quad \mu = \nu\ . \quad (D.52)$$

The matrix $\mathbf{M}^T\mathbf{M}$ becomes diagonal, $\mathbf{M}^T\mathbf{M} = \text{diag}\ \alpha_\nu$. The increment ΔC and its gradient turn out as

$$\Delta C = \Delta\mathbf{p}^T(\mathbf{M}^T\mathbf{M})\Delta\mathbf{p} = \Delta\mathbf{p}^T(\text{diag}\ \alpha_\nu)\Delta\mathbf{p} = \sum_{\nu=1}^{n_p}\alpha_\nu(\Delta p_\nu)^2 \quad \rightsquigarrow \quad \frac{\partial \Delta C}{\partial \Delta\mathbf{p}} = 2(\text{diag}\ \alpha_\nu)\ \Delta\mathbf{p}\ . \quad (D.53)$$

The shape ΔC does not contain cross products. Hence, the shape is symmetrical to the axes Δp_ν. The component ν of the gradient only depends on Δp_ν.

If $\mathbf{M}^T\mathbf{M}$ is diagonal recognize the following fact: Fixing p_2 at an arbitrary value and minimizing ΔC with regard to p_1 , the minimum $p_{1\,\min} = \hat{p}_1$ is always the same, no matter which arbitrary p_2 was initially chosen (see Fig. D.4).

D.8 Successive Estimation in Large-Scale Systems

In the field of large-scale systems, the parameter estimation is based on a decomposition technique. For small-scale systems, the model $\mathbf{y} = \mathbf{M}\mathbf{p}+\boldsymbol{\varepsilon}$ is applied as given in Eq.(D.13). For a large-scale system, with regard to computational difficulties, it is troublesome to apply Eq.(D.13). Decomposing into L submodels yields

$$\mathbf{y}_i = \sum_{j=1}^{L}\mathbf{M}_{ij}\mathbf{p}_j + \boldsymbol{\varepsilon}_i \quad \mathbf{y}_i \in \mathcal{R}^{n_{mi}},\quad \mathbf{p}_i, \boldsymbol{\varepsilon}_i \in \mathcal{R}^{n_{pi}},\quad \mathbf{M}_{ij} \in \mathcal{R}^{n_{mi} \times n_{pi}}\ ,\quad \sum_{j=1}^{L} n_{p_j} = n_p\ ,\quad \sum_{j=1}^{L} n_{m_j} = n_m\ . \quad (D.54)$$

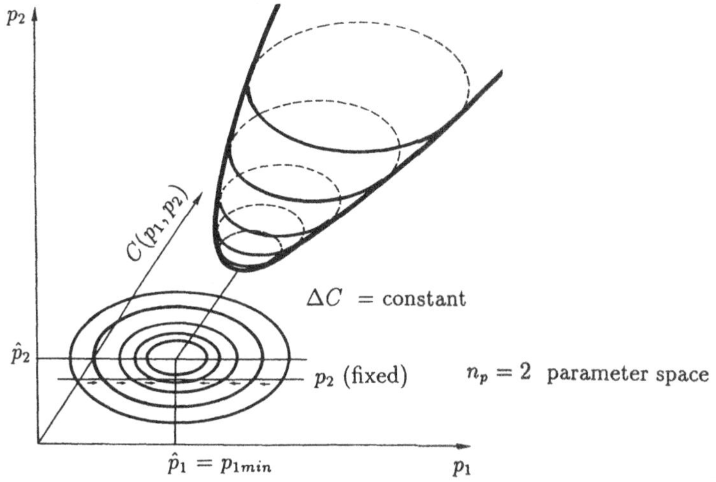

Figure D.4: Shape $C(\mathbf{p})$ and contour lines of constant amount ΔC in parameter space

The parameter vector $\mathbf{p}$ in decomposed form is $\mathbf{p}^T = (\mathbf{p}_1^T \; \mathbf{p}_2^T \ldots \mathbf{p}_L^T)$. The least squares estimation is obtained by minimizing

$$C = \sum_{i=1}^{L} \boldsymbol{\epsilon}_i^T \boldsymbol{\epsilon}_i \tag{D.55}$$

with respect to $\mathbf{p}_i$. Furthermore, try to find $\mathbf{p}_i$ in such a way that the interaction between the L submodels is minimized. Thus, an interaction variable

$$\mathbf{w}_i = \sum_{j=1, j\neq i}^{L} \mathbf{M}_{ij}\mathbf{p}_j \tag{D.56}$$

is defined, providing the influence from all subsystems $j \neq i$ to the subsystem i. Following Eq.(D.55),

$$C = \sum_{i=1}^{L} (\mathbf{y}_i - \mathbf{M}_{ii}\mathbf{p}_i - \mathbf{w}_i)^T (\mathbf{y}_i - \mathbf{M}_{ii}\mathbf{p}_i - \mathbf{w}_i) \tag{D.57}$$

has to be minimized with respect to $\mathbf{p}_i$ and $\mathbf{w}_i$ within the constraint of Eq.(D.56). By application of the method of the Lagrange multiplier, the new criterion function I is

$$I = \sum_{i=1}^{L} (\mathbf{y}_i - \mathbf{M}_{ii}\mathbf{p}_i - \mathbf{w}_i)^T (\mathbf{y}_i - \mathbf{M}_{ii}\mathbf{p}_i - \mathbf{w}_i) + \boldsymbol{\lambda}_i^T (\mathbf{w}_i - \sum_{j=1, j\neq i}^{L} \mathbf{M}_{ij}\mathbf{p}_j) \, . \tag{D.58}$$

Differentiating with respect to $\mathbf{p}_i, \boldsymbol{\lambda}_i, \mathbf{w}_i$ yields

$$\hat{\mathbf{p}}_i = (\mathbf{M}_{ii}^T \mathbf{M}_{ii})^{-1} [\mathbf{M}_{ii}^T (\mathbf{y}_i - \hat{\mathbf{w}}_i) + \sum_{j=1, j\neq i}^{L} \mathbf{M}_{ji}^T \boldsymbol{\lambda}_j] \tag{D.59}$$

$$\hat{\mathbf{w}}_i = \sum_{j=1, j\neq i}^{L} \mathbf{M}_{ij}\hat{\mathbf{p}}_j \qquad \boldsymbol{\lambda}_i = \mathbf{y}_i - \mathbf{M}_{ii}\hat{\mathbf{p}}_i - \hat{\mathbf{w}}_i , \tag{D.60}$$

respectively, $\forall i = 1 \ldots L$. The previous equations can only be solved by iteration using appropriate starting values for $\hat{\mathbf{w}}_i^{(0)}$ and $\boldsymbol{\lambda}_i^{(0)}$ in Eq.(D.59). Eq.(D.59) gives $\hat{\mathbf{p}}_i^{(0)}$. Taking this value, Eqs. (D.60) yield $\hat{\mathbf{w}}_i^{(1)}$ and $\boldsymbol{\lambda}_i^{(1)}$, respectively. This interaction-prediction algorithm is carried out until some preset stopping condition in $\boldsymbol{\lambda}_i^{(k)}$ and $\hat{\mathbf{w}}_i^{(k)}$ is satisfied, e.g., in the norm of the difference between consecutive quantities.

The above algorithm can be modified by substituting Eq.(D.60) into Eq.(D.59). In this way, a successive algorithm in $\hat{\mathbf{p}}_i^{(k)}$ is formulated. In order to process on-line measurements, a recursive version of this algorithm also was developed (see *Sultan, M.A., et al. 1988*), corresponding with the recursive algorithm given in the next section.

D.9 Recursive Least-Squares Estimation

In practice, measurements are recorded successively with time. An estimator version updating the latest parameter state is of interest. By processing recent data this recursive estimator provides an incremental complement to the hitherto existing parameter result. Up to the kth sample assume to process the measurement matrix $\mathbf{M}(k)$ and the measurement vector $\mathbf{y}(k)$. Advancing from sample instant k to $k+1$, the measurement matrix $\mathbf{M}(k)$ is supplemented by a new row $\mathbf{m}^T(k+1)$ and $\mathbf{M}(k)$ turns out as the new measurement matrix $\mathbf{M}(k+1)$.

Assume the model specified by the difference equation of nth order

$$y(k) + a_1 y(k-1) + \ldots + a_n y(k-n) = b_1 u(k-1) + \ldots + b_n u(k-n) \tag{D.61}$$

and define the parameter vector $\mathbf{p} \triangleq (a_1 \ldots a_n \; b_1 \ldots b_n)^T$. Let $L+1$ measurements be available from sampling instant $k-L$ through k. Without measurement errors this equation is rewritten, applying the definitions $\mathbf{m}(k)$, $\mathbf{M}(k)$ and $\mathbf{y}(k)$

$$y(k-L) - \mathbf{m}^T(k-L)\mathbf{p} = 0, \qquad y(k-L+1) - \mathbf{m}^T(k-L+1)\mathbf{p} = 0, \quad \ldots\ldots \quad y(k) - \mathbf{m}^T(k)\mathbf{p} = 0\,. \tag{D.62}$$

Taking into account some measurement error $\boldsymbol{\varepsilon}$,

$$\mathbf{y}(k) - \mathbf{M}(k)\mathbf{p} = 0 \qquad \leadsto \qquad \mathbf{y}(k) - \mathbf{M}(k)\mathbf{p} = \boldsymbol{\varepsilon} \tag{D.63}$$

$$\mathbf{m}(k) = \begin{pmatrix} -y(k-1) \\ \vdots \\ -y(k-n) \\ ----- \\ u(k-1) \\ \vdots \\ u(k-n) \end{pmatrix} \in \mathcal{R}^{2n}, \qquad \mathbf{y}(k) = \begin{pmatrix} y(k-L) \\ \vdots \\ y(k) \end{pmatrix} \in \mathcal{R}^{(L+1)} \tag{D.64}$$

$$\mathbf{M}(k) = \begin{pmatrix} -y(k-L-1) & -y(k-L-2).. & -y(k-L-n) & u(k-L-1).. & u(k-L-n) \\ -y(k-L) & -y(k-L-1).. & & & \\ -y(k-L+1) & \vdots & & & \\ \vdots & & & & \\ -y(k-1) & \ldots & \ldots & u(k-1) & u(k-n) \end{pmatrix} \tag{D.65}$$

$$\mathbf{M}(k) = [\,\mathbf{m}(k-L) \;\vdots\; \mathbf{m}(k-L+1) \;\vdots\; \ldots \vdots\; \mathbf{m}(k)\,]^T \qquad \mathbf{M}(k) \in \mathcal{R}^{(L+1)\times(2n)}\,. \tag{D.66}$$

Starting the recursive algorithm, care must be taken of negative arguments and $k \geq L+n$ must be provided. Advancing from sample instant k to $k+1$ and assuming enlarged measurement data from $L+1$ to $L+2$, the dimension of $\mathbf{y}(k)$ to $\mathbf{y}(k+1)$ is increased by one

$$\mathbf{y}(k) = \begin{pmatrix} y(k-L) \\ \vdots \\ y(k) \end{pmatrix} \in \mathcal{R}^{(L+1)}, \qquad \mathbf{y}(k+1) = \begin{pmatrix} y(k-L) \\ \vdots \\ y(k) \\ y(k+1) \end{pmatrix} = \begin{pmatrix} \mathbf{y}(k) \\ y(k+1) \end{pmatrix} \in \mathcal{R}^{(L+2)}\,. \tag{D.67}$$

Within the vectors $\mathbf{m}(k)$ and $\mathbf{m}(k+1)$ only a shift operation is carried out. Thus, the dimension of $\mathbf{m}(k)$

does not increase when proceeding to $\mathbf{m}(k+1)$.

$$\text{From }\ \mathbf{m}(k) = \begin{pmatrix} -y(k-1) \\ \vdots \\ -y(k-n) \\ ----- \\ u(k-1) \\ \vdots \\ u(k-n) \end{pmatrix} \ \text{ get }\ \mathbf{m}(k+1) = \begin{pmatrix} -y(k) \\ \vdots \\ -y(k-n+1) \\ ------- \\ u(k) \\ \vdots \\ u(k-n+1) \end{pmatrix} \qquad \begin{array}{l} \mathbf{m}(k) \in \mathcal{R}^{2n}, \\ \mathbf{m}(k+1) \in \mathcal{R}^{2n}. \end{array} \tag{D.68}$$

From $\mathbf{M}(k)$ to $\mathbf{M}(k+1)$ proceed by enclosing a row $\mathbf{m}^T(k+1)$. Thus,

$$\mathbf{M}(k+1) = \begin{pmatrix} \mathbf{M}(k) \\ \mathbf{m}^T(k+1) \end{pmatrix} \qquad \mathbf{M}(k+1) \in \mathcal{R}^{(L+2)\times 2n}. \tag{D.69}$$

The least-squares estimator $\hat{\mathbf{p}}(k)$ and $\hat{\mathbf{p}}(k+1)$ resulting from $(L+1)$ or $(L+2)$ measurements between samples k and $k+L$ or $k+L+1$ is obtained by applying Eq.(D.32)

$$\hat{\mathbf{p}}(k) = [\mathbf{M}^T(k)\mathbf{M}(k)]^{-1}\mathbf{M}^T(k)\mathbf{y}(k) \quad \text{and} \quad \hat{\mathbf{p}}(k+1) = [\mathbf{M}^T(k+1)\mathbf{M}(k+1)]^{-1}\mathbf{M}^T(k+1)\mathbf{y}(k+1), \tag{D.70}$$

respectively. Making use of Eq.(C.23) and substituting

$$\mathbf{A} := \mathbf{M}^T(k), \quad \mathbf{B} := \mathbf{m}(k+1), \quad \mathbf{C} = \mathbf{D} := 0, \quad \mathbf{u} := \mathbf{y}(k), \quad \mathbf{v} := y(k+1), \tag{D.71}$$

one has

$$\mathbf{M}^T(k+1)\mathbf{y}(k+1) = \Big(\underbrace{\mathbf{m}(k-L)\ \ \mathbf{m}(k-L+1)\ldots\mathbf{m}(k)}_{\mathbf{M}^T(k)}\ \ \mathbf{m}(k+1)\Big)\begin{pmatrix} \mathbf{y}(k) \\ y(k+1) \end{pmatrix} \tag{D.72}$$

$$\mathbf{M}^T(k+1)\mathbf{y}(k+1) = \mathbf{M}^T(k)\mathbf{y}(k) + \mathbf{m}(k+1)y(k+1). \tag{D.73}$$

Invoking Eq.(C.24) and substituting $\mathbf{C} = \mathbf{D} = \mathbf{F} = \mathbf{H} := 0$

$$\mathbf{A} := \mathbf{M}^T(k), \quad \mathbf{B} := \mathbf{m}(k+1), \quad \mathbf{E} := \mathbf{M}(k), \quad \mathbf{G} := \mathbf{m}^T(k+1), \tag{D.74}$$

$$\mathbf{M}^T(k+1)\mathbf{M}(k+1) = \begin{pmatrix} \mathbf{M}(k) \\ \mathbf{m}^T(k+1) \end{pmatrix}^T \begin{pmatrix} \mathbf{M}(k) \\ \mathbf{m}^T(k+1) \end{pmatrix} = \Big(\mathbf{M}^T(k) \ \vdots\ \mathbf{m}(k+1)\Big)\begin{pmatrix} \mathbf{M}(k) \\ \mathbf{m}^T(k+1) \end{pmatrix} \tag{D.75}$$

$$\mathbf{M}^T(k+1)\mathbf{M}(k+1) = \mathbf{M}^T(k)\mathbf{M}(k) + \mathbf{m}(k+1)\mathbf{m}^T(k+1) \tag{D.76}$$

is achieved. Taking Eqs.(D.73), (D.70) and (D.76) into consideration,

$$\hat{\mathbf{p}}(k+1) = [\mathbf{M}^T(k+1)\mathbf{M}(k+1)]^{-1}[\mathbf{M}^T(k)\mathbf{y}(k) + \mathbf{m}(k+1)y(k+1)] \tag{D.77}$$

$$\hat{\mathbf{p}}(k+1) = [\mathbf{M}^T(k+1)\mathbf{M}(k+1)]^{-1}[\mathbf{M}^T(k)\mathbf{M}(k)\hat{\mathbf{p}}(k) + \mathbf{m}(k+1)y(k+1)] \tag{D.78}$$

$$\begin{aligned}\hat{\mathbf{p}}(k+1) = {} & [\mathbf{M}^T(k+1)\mathbf{M}(k+1)]^{-1}[\mathbf{M}^T(k+1)\mathbf{M}(k+1)\hat{\mathbf{p}}(k) \\ & -\mathbf{m}(k+1)\mathbf{m}^T(k+1)\hat{\mathbf{p}}(k) + \mathbf{m}(k+1)y(k+1)]\end{aligned} \tag{D.79}$$

$$\underbrace{\hat{\mathbf{p}}(k+1)}_{\text{new estimate including sample } k+1} = \underbrace{\hat{\mathbf{p}}(k)}_{\text{old estimate}} + \underbrace{[\mathbf{M}^T(k+1)\mathbf{M}(k+1)]^{-1}\mathbf{m}(k+1)}_{\text{correcting gain vector (Kalman gain factor) } \gamma(k)}\underbrace{[\underbrace{y(k+1)}_{\text{new measurement}} - \underbrace{\mathbf{m}^T(k+1)\hat{\mathbf{p}}(k)}_{\text{one sample step ahead prediction measurement } \hat{\mathbf{y}}(k+1|k)}]}_{\text{prediction error}}. \tag{D.80}$$

Applying the matrix inversion lemma Eq.(C.13) to the inverse of Eq.(D.76), substituting and defining

$$\mathbf{A} := [\mathbf{M}^T(k)\mathbf{M}(k)]^{-1}, \qquad \mathbf{B} = \mathbf{C}^T := \mathbf{m}(k+1), \qquad \mathbf{P}(k) \triangleq \mathbf{A} = [\mathbf{M}^T(k)\mathbf{M}(k)]^{-1}, \tag{D.81}$$

the result may be considered as a covariance matrix

$$\mathbf{P}(k+1) = \mathbf{P}(k) - \underbrace{\mathbf{P}(k)\mathbf{m}(k+1)[1+\mathbf{m}^T(k+1)\mathbf{P}(k)\mathbf{m}(k+1)]^{-1}}_{\gamma_1}\mathbf{m}^T(k+1)\mathbf{P}(k). \tag{D.82}$$

Note that the bracket expression in the above equation is a scalar. Hence, it can be shifted to any other position within the matrix product

$$\mathbf{P}(k+1) = \frac{\mathbf{P}(k)[1+\mathbf{m}^T(k+1)\mathbf{P}(k)\mathbf{m}(k+1)] - \mathbf{P}(k)\mathbf{m}(k+1)\mathbf{m}^T(k+1)\mathbf{P}(k)}{1+\mathbf{m}^T(k+1)\mathbf{P}(k)\mathbf{m}(k+1)} . \tag{D.83}$$

Postmultiplying the above numerator by $\mathbf{m}(k+1)$ yields

$$\mathbf{P}(k)\mathbf{m}(k+1) + \mathbf{P}(k)\mathbf{m}(k+1)\mathbf{m}^T(k+1)\mathbf{P}(k)\mathbf{m}(k+1) - \mathbf{P}(k)\mathbf{m}(k+1)\mathbf{m}^T(k+1)\mathbf{P}(k)\mathbf{m}(k+1) . \tag{D.84}$$

The second and third term are cancelled. Hence, by Eq.(D.83),

$$\mathbf{P}(k+1)\mathbf{m}(k+1) = \frac{\mathbf{P}(k)\mathbf{m}(k+1)}{1+\mathbf{m}^T(k+1)\mathbf{P}(k)\mathbf{m}(k+1)} . \tag{D.85}$$

The left side above corresponds to the correcting gain vector $\gamma(k)$ in Eq. (D.80), the right side above equals $\gamma_1(k)$, as defined in Eq.(D.82): $\gamma_1(k) = \gamma(k)$. Finally, the resulting recursive algorithm is expressed by the following three equations

$$\gamma(k) = \frac{\mathbf{P}(k)\mathbf{m}(k+1)}{1+\mathbf{m}^T(k+1)\mathbf{P}(k)\mathbf{m}(k+1)} \qquad \gamma(k) \in \mathcal{R}^{2n} \tag{D.86}$$

$$\mathbf{P}(k+1) = \mathbf{P}(k) - \gamma(k)\mathbf{m}^T(k+1)\mathbf{P}(k) \tag{D.87}$$

$$\hat{\mathbf{p}}(k+1) = \hat{\mathbf{p}}(k) + \gamma(k)[y(k+1) - \mathbf{m}^T(k+1)\hat{\mathbf{p}}(k)] . \tag{D.88}$$

Note that the matrix $\mathbf{M}$ increases its dimension from sample to sample but the product $\mathbf{M}^T\mathbf{M}$ does not. The matrix $\mathbf{P}$ as its inverse also keeps the dimension $(2n) \times (2n)$ unchanged. For the sake of comparison with block-pulse function approximation, see Eq.(37.58).

D.10 Recursive Instrumental Variable Method

The recursive (and one-shot) least-squares method only yields an unbiased parameter vector if the error signal $\boldsymbol{\varepsilon}$ is uncorrelated both with the input and output signal of the process u and y, respectively. If uncorrelated error signals cannot be preassumed the method of instrumental variables is used since it also works with correlated $\boldsymbol{\varepsilon}$. The instrumental variable vector $\mathbf{y}_{aux}$ is synthesized as the output of the instrumental model, see Fig. D.5

$$\mathbf{m}_{aux}(k) = \begin{pmatrix} -y_{aux}(k-1) \\ \vdots \\ -y_{aux}(k-n) \\ ------ \\ u(k-1) \\ \vdots \\ u(k-n) \end{pmatrix} \qquad \mathbf{y}_{aux} = \begin{pmatrix} y_{aux}(k-1) \\ \vdots \\ y_{aux}(k-n) \end{pmatrix} . \tag{D.89}$$

The instrumental variable y_{aux} is generated by exciting the instrumental model with $\mathbf{u}$ and by declaration $y_{aux} = \mathbf{m}_{aux}^T\,\hat{\mathbf{p}}_{aux}$ using an auxiliary parameter vector $\hat{\mathbf{p}}_{aux}$. To avoid stochastic dependence between $\mathbf{y}_{aux}$ and $\boldsymbol{\varepsilon}$, the auxiliary parameter $\hat{\mathbf{p}}_{aux}$ is taken from the real $\hat{\mathbf{p}}$ after having passed PT_1-delay. A memory is employed to generate the vector $\mathbf{y}_{aux}$ from the past scalars $y_{aux}(k-i)$. The recursive least-squares algorithm runs as given by Eqs.(D.87), (D.88) and, additionally, by

$$\gamma(k) = \mathbf{P}(k)\mathbf{m}_{aux}(k+1)[1+\mathbf{m}^T(k+1)\mathbf{P}(k)\mathbf{m}_{aux}(k+1)]^{-1} . \tag{D.90}$$

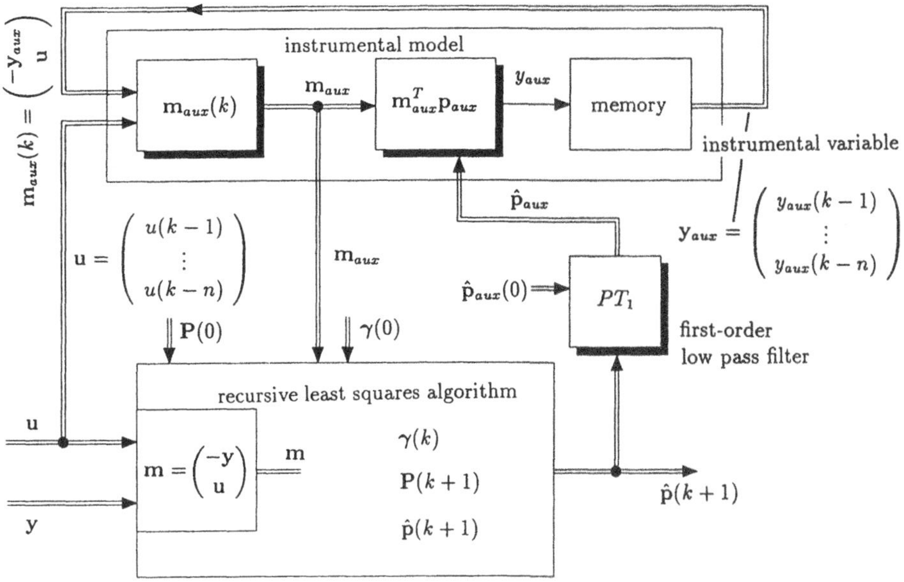

Figure D.5: Recursive instrumental variable method

The matrix $\mathbf{P}(k)$ is a modified covariance $\mathbf{P}(k) = [\mathbf{M}_{aux}^T(k)\mathbf{M}(k)]^{-1}$ and $\mathbf{M}_{aux}(k)$ is the instrumental variable matrix

$$\mathbf{M}_{aux}(k) = \begin{pmatrix} \mathbf{m}_{aux}^T(k) \\ \vdots \\ \mathbf{m}_{aux}^T(k+L) \end{pmatrix} \tag{D.91}$$

(*Jakoby, W., 1985*). The aim is to establish a signal y_{aux} strongly correlated with the undisturbed process output but weakly correlated with measurement noise (*Isermann, R., 1988*).

D.11 Linear Estimation

For $x(k)$ scalar, the set $\{x(k)\}$ with $k = 1...N$ is referred to as a scalar random process. The vector-valued variable $\{\mathbf{x}(k)\}$ denotes a vector process.

Linear expectation (mean): $\bar{x} \triangleq E\{x(k)\} \triangleq \lim_{N\to\infty} \frac{1}{N}\sum_{k=1}^{N} x(k)$.

Variance: $\sigma_x^2 \triangleq E\{[x(k)-\bar{x}]^2\} \triangleq \lim_{N\to\infty} \frac{1}{N}\sum_{k=1}^{N}[x(k)-\bar{x}]^2$.

Autocorrelation function: $R_{xx}(\tau) \triangleq E\{x(k)x(k+\tau)\} \triangleq \lim_{N\to\infty} \frac{1}{N}\sum_{k=1}^{N} x(k)x(k+\tau)$.

Autocovariance function:

$$\text{cov}\ [x(k)x(k+\tau)] \triangleq \text{cov}\ (x,x,\tau) \triangleq E\{[x(k)-\bar{x}][x(k+\tau)-\bar{x}]\} \triangleq Q_{xx}(\tau) \triangleq X(\tau)\ . \tag{D.92}$$

If the signal is Gauss distributed (= normally distributed), then the process is completely determined by $\bar{x}$ and cov $[x(k)x(k+\tau)]$. The process is denoted stationary if $\bar{x}$ and cov $[x(k)x(k+\tau)]$ are time-invariant.

Cross covariance function: cov $[x,y,\tau] \triangleq E\{[x(k)-\bar{x}][y(k)-\bar{y}]\} = R_{xy}(\tau) - \bar{x}\bar{y} \triangleq Q_{xy}(\tau)$.

A random discrete-time process is named white if the signal $x(k)$ has no stochastic interrelation to earlier signals $x(k-\mu)$. The covariance function of a white process (white noise) is

$$\text{cov}\,[x,\tau] = \sigma_x^2 \delta_\tau(\tau) \qquad \delta_\tau(\tau) = \begin{cases} 1 & \text{if } \tau = 0 \\ 0 & \text{if } \tau \neq 0 \end{cases} \tag{D.93}$$

where $\delta_\tau(\tau)$ is the Kronecker delta. The power density function is constant versus frequency $\forall\ |\omega| \leq \pi/T$ where T is the sampling interval.

For comparison: Continuous-time white random processes are characterized by a vanishing stochastic interrelation, even between instants of infinitesimally small time-shift. The power density function is constant for all frequencies up to infinity and signals may grow to infinity. Hence, a white continuous-time process does not exist in reality and is only imaginable as a *limes*. Kronecker delta in Eq.(D.93) is replaced by the dirac function $\delta(\tau)$.

If the stochastic signal contains several components $\mathbf{x}(k) = [x_1(k), x_2(k) \ldots x_n(k)]^T$ a vector process $\{\mathbf{x}(k)\}$ is established and a covariance function matrix is given by

$$\text{cov}\,[\mathbf{x},\tau] \triangleq \text{cov}\,\mathbf{x} \triangleq E\{[\mathbf{x}(k) - \bar{\mathbf{x}}][\mathbf{x}(k+\tau) - \bar{\mathbf{x}}]^T\} \tag{D.94}$$

$$\triangleq \begin{pmatrix} \text{cov}\,[x_1, x_1, \tau] & \text{cov}\,[x_1, x_2, \tau] & \cdots \\ \text{cov}\,[x_2, x_1, \tau] & \text{cov}\,[x_2, x_2, \tau] & \cdots \\ \vdots & \vdots & \ddots \end{pmatrix} \triangleq \mathbf{Q}_{xx}(\tau) \triangleq \mathbf{X}(\tau)\ . \tag{D.95}$$

Note that $\mathbf{Q}_{xx}$ is positive semidefinite symmetric for $\tau = 0$ and that the main diagonal positions of $\mathbf{Q}_{xx}(\tau)$ are given by the autocovariance functions.

D.11.1 Parametric Models. Markov Processes

Markov signal processes (of first order) are defined by the conditioned probability distribution function $p[x(k)|x(k-1), x(k-2) \ldots x(0)] = p[x(k)|x(k-1)]$ depending only on the preceding $x(k-1)$. A signal process obeys the probability distribution above if it is generated by the first order difference equation $x(k+1) = a_M x(k) + b_M v(k+1)$. Higher-order time-invariant Markov signal processes are given by

$$\mathbf{x}(k+1) = \mathbf{A}_M \mathbf{x}(k) + \mathbf{B}_M \mathbf{v}(k)\ . \tag{D.96}$$

If $\mathbf{A}_M(k)$ and/or $\mathbf{B}_M(k)$ are time-dependent the Markov process is time-varying. The process $\{\mathbf{x}(k)\}$ is a white Markov process if $\mathbf{v}(k)$ is white. Defining

$$\text{cov}\,[\mathbf{x}(k+1), \mathbf{x}(k+1), \tau = 0] = \text{cov}\,[\mathbf{x}(k+1)] \triangleq E\{[\mathbf{x}(k+1) - \bar{\mathbf{x}}][x(k+1) - \bar{x}]^T\} \triangleq \mathbf{X}(k+1) \tag{D.97}$$

and substituting Eq.(D.97) into (D.96) yields (calculations omitted) in the case of white noise $\mathbf{v}(k)$

$$\mathbf{X}(k+1) = \mathbf{A}_M \mathbf{X}(k) \mathbf{A}_M^T + \mathbf{B}_M \mathbf{V}(k) \mathbf{B}_M^T\ . \tag{D.98}$$

If $\mathbf{A}_M$ is constant and stable and $\mathbf{B}_M$ is constant for $k \to \infty$ then $\mathbf{X}(k+1) = \mathbf{X}(k)$. Using Eq.(4.40)

$$\text{col}\,\mathbf{X} = (\mathbf{I} - \mathbf{A}_M \otimes \mathbf{A}_M)^{-1} (\mathbf{B}_M \otimes \mathbf{B}_M) \text{col}\,\mathbf{V}\ . \tag{D.99}$$

Assuming a positive semidefinite weighting function matrix $\mathbf{R}$, the Markov signal process $\mathbf{x}(k)$ can be assessed globally by $I = E\{\mathbf{x}^T(k)\mathbf{R}\mathbf{x}(k)\} \rightsquigarrow I = \bar{\mathbf{x}}^T \mathbf{R} \bar{\mathbf{x}} + \text{tr}\,[\mathbf{R}\mathbf{X}]$.

D.11.2 Observation as a Random Process

Techniques of linear regression yield deterministic linear models with weighted least squares $||\varepsilon||^2$ where ε is a vector of distinct errors. If there is a high number of data the error space is of high dimension. With regard to the high amount of data it is suitable to consider ε as a stochastic process.

The observation vector $\mathbf{y}$ now is regarded as the sum of the model function $\mathbf{Mp}$ plus an additive zero mean noise ε where the matrix $\mathbf{M}$ is given by $\mathbf{y} = \mathbf{Mp} + \varepsilon$. The parameter $\mathbf{p}$ should be estimated optimally with respect to a risk function cov $\tilde{\mathbf{p}}$. Supposing a linear relationship $\mathbf{p}^* = \mathbf{\Gamma y}$, in order to derive the optimal estimation $\mathbf{p}^*$ from the observation $\mathbf{y}$, and using the model $\varepsilon = -\mathbf{Mp} + \mathbf{y}$

$$\hat{\mathbf{p}} = E\{\mathbf{p}^*\} = E\{\mathbf{\Gamma}(\mathbf{Mp} + \varepsilon)\} = E\{\mathbf{\Gamma Mp}\} + E\{\mathbf{\Gamma}\varepsilon\}. \tag{D.100}$$

The second term in Eq.(D.100) is named bias $\mathbf{b} = E\{\mathbf{\Gamma}\varepsilon\}$. The bias can be separated into $E\{\mathbf{\Gamma}\}$ and $E\{\varepsilon\}$ if both processes are uncorrelated. Regardless of $E\{\mathbf{\Gamma}\}$, the bias $\mathbf{b}$ vanishes if ε has zero mean as preassumed. The condition of being uncorrelated is met if ε is a white noise (see *Bard, Y., 1974; Isermann, R., 1988*). If $\mathbf{b} = \mathbf{0}$ then only the first term remains in Eq. (D.100) $\hat{\mathbf{p}} = E\{\mathbf{\Gamma Mp}\} = E\{\mathbf{\Gamma M}\}\mathbf{p}$. In order to obtain $\hat{\mathbf{p}} = E\{\mathbf{p}^*\} = \mathbf{p}$, the condition $\mathbf{\Gamma M} = \mathbf{I}$ must be satisfied. Hence, $\mathbf{\Gamma}$ is a left inverse of $\mathbf{M}$.

D.11.3 Minimum Variance Estimator. Gauss-Markov Theorem

Which linear estimator $\hat{\mathbf{p}} = \mathbf{\Gamma y}$ or which $\mathbf{\Gamma}$ matches the minimum of some scalar valuation of the covariance cov $\tilde{\mathbf{p}}$? The variance of the estimation error vector

$$\text{cov } \tilde{\mathbf{p}} = E\{[\tilde{\mathbf{p}} - E\{\tilde{\mathbf{p}}\}][\tilde{\mathbf{p}} - E\{\tilde{\mathbf{p}}\}]^T\} \qquad \text{cov } \tilde{\mathbf{p}} = E\{\tilde{\mathbf{p}}\tilde{\mathbf{p}}^T\} \tag{D.101}$$

could be reduced since $E\{\tilde{\mathbf{p}}\} = \mathbf{0}$ (unbiased estimator). The estimation error $\tilde{\mathbf{p}} = \mathbf{p} - \hat{\mathbf{p}}$, the optimum estimate $\hat{\mathbf{p}} = \mathbf{\Gamma y}$ and the model $\mathbf{y} = \mathbf{Mp} + \boldsymbol{\varepsilon}$ are substituted in Eq.(D.101). Thus,

$$\text{cov } \tilde{\mathbf{p}} = E\{(\mathbf{p}-\hat{\mathbf{p}})(\mathbf{p}-\hat{\mathbf{p}})^T\} = E\{[\mathbf{p}-\mathbf{\Gamma}(\mathbf{Mp}+\boldsymbol{\varepsilon})][\mathbf{p}-\mathbf{\Gamma}(\mathbf{Mp}+\boldsymbol{\varepsilon}]^T\} = E\{(\mathbf{\Gamma}\boldsymbol{\varepsilon})(\mathbf{\Gamma}\boldsymbol{\varepsilon})^T\} = \mathbf{\Gamma}(\text{cov } \boldsymbol{\varepsilon})\mathbf{\Gamma}^T = \mathbf{\Gamma V \Gamma}^T \tag{D.102}$$

where $\mathbf{V}$ is assumed a given function. As scalar valuation of cov $\mathbf{p}$, the determinant is chosen. Hence, the objective is to minimize det cov $\tilde{\mathbf{p}}$ subject to the matrix condition $\hat{\mathbf{p}} = \mathbf{\Gamma y}$ or $\mathbf{\Gamma M} = \mathbf{I}$. Introducing a matrix of Lagrange multipliers $\mathbf{\Lambda}$ and selecting the trace as a scalar measure of the constraint, the resultant form det $(\mathbf{\Gamma V \Gamma}^T) +$ tr $\mathbf{\Lambda}(\mathbf{\Gamma M} - \mathbf{I})$ is achieved. The minimum with respect to $\mathbf{\Gamma}$ is obtained by differentiation

$$\frac{\partial}{\partial \mathbf{\Gamma}}\{\det\,(\mathbf{\Gamma V \Gamma}^T) + \text{tr } \mathbf{\Lambda}(\mathbf{\Gamma M} - \mathbf{I})\} = 0 \quad \rightsquigarrow \quad \mathbf{\Gamma} = (\mathbf{M}^T\mathbf{V}^{-1}\mathbf{M})^{-1}\mathbf{M}^T\mathbf{V}^{-1}. \tag{D.103}$$

Comparing Eq.(D.103) with the optimum weighted least squares estimation in Eq.(D.18) yields the following result. If the weighting matrix $\mathbf{W}$ is defined as $\mathbf{W} = \mathbf{V}^{-1}$ (the inverse of the covariance matrix $\mathbf{V} =$ cov $\boldsymbol{\varepsilon}$), the optimal *linear* estimator $\hat{\mathbf{p}} = \mathbf{\Gamma y}$ in accordance with Eq.(D.103) yields an *unbiased* estimate, additionally a minimal weighted risk function $C = \boldsymbol{\varepsilon}^T\mathbf{W}\boldsymbol{\varepsilon} = \boldsymbol{\varepsilon}^T\mathbf{V}^{-1}\boldsymbol{\varepsilon}$ and, finally, a minimum determinant of the variance of the parameter cov $\tilde{\mathbf{p}} = \mathbf{\Gamma}(\text{cov } \boldsymbol{\varepsilon})\mathbf{\Gamma}^T$. Using *biased* and *nonlinear* estimators better results are available. The linear estimator discussed above is the best linear one and is unbiased but does not supply the least squares estimate at all (*Hoerl, A.E., 1962; Hoerl, A.E., and Kennard, R.W., 1970*).

The minimum variance estimator can only be realized if the covariance matrix of the equation error is known a priori. Otherwise it is known after a recursive process. Substituting Eq.(D.103) into (D.102), the variance in the minimum variance estimation case is $(\text{cov } \tilde{\mathbf{p}})_{MV} = (\mathbf{M}^T\mathbf{V}^{-1}\mathbf{M})^{-1}$. Then, substituting the identity matrix $\mathbf{V} = \mathbf{I}$ in Eqs.(D.103) and (D.102) yields the variance in the least squares estimation case $(\text{cov } \tilde{\mathbf{p}})_{LS} = (\mathbf{M}^T\mathbf{M})^{-1}\mathbf{M}^T(\text{cov } \boldsymbol{\varepsilon})\mathbf{M}(\mathbf{M}^T\mathbf{M})^{-1}$ which cannot be simplified any further for a general matrix $\mathbf{M}$. The covariance $(\text{cov } \tilde{\mathbf{p}})_{MV}$ is smaller than $(\text{cov } \tilde{\mathbf{p}})_{LS}$.

The expectation of the ordinary least squares risk function $C = \boldsymbol{\varepsilon}^T\boldsymbol{\varepsilon}$ is now investigated, provided the standard assumptions are valid (additive zero mean noncorrelated measurement errors with constant variance cov $\boldsymbol{\varepsilon} = \sigma^2\mathbf{I}$; nonrandom parameters; independent variables; no other prior information) (*Beck, J.V., and Arnold, K.J., 1977*). In the ordinary least squares case $\mathbf{\Gamma}$ is the left pseudo-inverse $\mathbf{M}^{!L}$. Thus, the expectation of C turns out

$$E\{\boldsymbol{\varepsilon}^T\boldsymbol{\varepsilon}\} = E\{\boldsymbol{\varepsilon}^T(\mathbf{I}_{n_m} - \mathbf{M}\mathbf{M}^!)\boldsymbol{\varepsilon}\} = \text{tr}\{(\mathbf{I}_{n_m} - \mathbf{M}\mathbf{M}^!)\sigma^2\mathbf{I}\} = (\text{tr } \mathbf{I}_{n_m} - \text{tr } \mathbf{I}_{n_p})\sigma^2 = (n_m - n_p)\sigma^2 \,. \tag{D.104}$$

D.11.4 Estimation Sensitivity

Consider the case that the exact value of the covariance matrix $\mathbf{V} =$ cov $\boldsymbol{\varepsilon}$ is not known but an erroneous (perturbed) symmetric matrix $\mathbf{V}_p = \mathbf{V} + \Delta\mathbf{V}$ is available as a basis for the linear regression formula Eq.(D.103). Then, $\mathbf{\Gamma}_p$ is obtained and the erroneous estimation is $\hat{\mathbf{p}}_p = \mathbf{\Gamma}_p\mathbf{y}$. Using Eq.(D.102) with the true value $\mathbf{V}$ but with the erroneous $\mathbf{\Gamma}_p$ yields cov $\hat{\mathbf{p}}_p = \mathbf{\Gamma}_p\mathbf{V}\mathbf{\Gamma}_p^{-1}$.

A degree of inefficiency η_e can be defined. This efficiency factor is always greater than or equal to unity because $\hat{\mathbf{p}}$ is the minimum variance estimation. An upper bound for η_e can be formulated

$$1 \leq \eta_e \triangleq \frac{\det \text{cov } \hat{\mathbf{p}}_p}{\det \text{cov } \hat{\mathbf{p}}} \leq \frac{(1+\alpha)^2}{4\alpha} \quad \text{where} \quad \alpha = \frac{|\lambda_{\max}[\mathbf{H}]\,|}{|\lambda_{\min}[\mathbf{H}]\,|} \quad \text{and} \quad \mathbf{H} = \sqrt{\mathbf{V}_p}\;\mathbf{V}\;\sqrt{\mathbf{V}_p}\,. \tag{D.105}$$

The condition number α is given by the ratio between largest and smallest eigenvalue of $\mathbf{H}$ with Hermite property, see *Hoerl, A.E., and Kennard, R.W., 1970; Wilkinson, J.H., 1965* .

Appendix E

Notations

E.1 General Conventions

- For detailed definition see boldface page in subject index.
- Boldface lower and capital letters denote vectors and matrices, respectively. Calligraphic letters denote sets.
- Closed set $\leq$ or [ ] , e.g., $1 \leq a \leq 5$ or $a \in [1,5]$ for a real number.
- Open set $<$ or ( ), e.g., $2 < b < 4$ or $b \in (2,4)$ for b real number.
- Set of integers { }, e.g., $i = 1, 2, ...8$ or $i = \{1, 2, ... 8\}$ for i integer.
- □ end of the proof, end of the example, end of discussion.
- Dots stand for undesignated variables.

E.2 Abbreviations and General Symbols

$=$, $\neq$, $\doteq$	equal, is different from, approximately equals, respectively
$\triangleq$	equality by definition
$:=$	equality by simple substitution
$\equiv$	identity
$\angle(\cdot)$	angle corresponding to the argument of the complex number
$>$	positive definite (matrix) ($\mathbf{P} > \mathbf{Q}$ means $\mathbf{P} - \mathbf{Q}$ positive definite)
$>_e \mathbf{0}$ $\geq_e \mathbf{0}$	positive and non-negative (matrix), respectively, element-by-element relation, ratio of size comparison
$\|\cdot\|_p$	p-norm
$\partial\Gamma$	boundary of the region Γ
$\partial[\cdot]$	degree
$\lvert\cdot\rvert$	modulus (absolute value) of a complex number
$\lvert\cdot\rvert$	modulus (matrix), e.g., $\lvert\mathbf{M}\rvert$ =matrix[$\lvert M_{ij}\rvert$]
$\perp$	perpendicular to
$\leadsto$	leads to
$\Re e$, $\Im m$	real part and imaginary part of a complex number, respectively
$\mathcal{C}$	field of complex numbers
$\mathcal{C}^n$	n-dimensional linear vector space over $\mathcal{C}$, $\mathcal{C}^n \equiv \mathcal{C}^{n\times 1}$
$\mathcal{C}^-$	set of complex numbers with negative real part (open complex left-half plane)
$\mathcal{C}^+$	closed right-half complex plane
$\emptyset$	the empty space
$\{a_i\}$	set (with elements a_i)
$\forall$	for all
$\in$	is an element of, belongs to
$\notin$	does not belong to
$\subseteq$	is a subset of (is contained in) or identical to
$\cup$	union (of sets)

$\cap$	intersection (of sets)
$\Leftarrow$, $\Rightarrow$	$A \Rightarrow B$ means: If the statement A (e.g. equation, inequality) is true then B is also true or, equivalently, "A is sufficient for B" or "A implies B", e.g., $\Re e\ \lambda_i[\mathbf{A}] < -4 \Rightarrow \mathbf{A}$ is stable. $N \Leftarrow T$ means "N is implied by T" or, equivalently expressed as a necessary condition, "only if N then T", e.g., the system with state space coefficient matrix $\mathbf{A}$ is stable only if $\mathrm{tr}\mathbf{A} < 0$.
$\Longleftrightarrow$	implies and is implied, if and only if. E.g., $\mathcal{B} \subseteq \mathcal{A} \Longleftrightarrow \forall x\ :\ (x \in \mathcal{B}) \Rightarrow (x \in \mathcal{A})$.
$\langle \cdot, \cdot \rangle$	inner product of two functions or vectors
$\mathcal{F}$, $\mathcal{F}^{-1}$	Fourier transformation and its inverse, respectively. $\mathcal{F}y(t) = y(j\omega)$ or, if necessary, $\hat{y}(j\omega)$
$\mathcal{L}$, $\mathcal{L}^{-1}$	Laplace transformation and its inverse, respectively. $\mathcal{L}y(t) = y(s)$ or, if necessary, $\hat{y}(s)$
$\mathcal{Z}$, $\mathcal{Z}^{-1}$	z-transformation and its inverse, respectively
H_2	Hardy space satisfying H_2-norm, (i) for continuous-time systems: Hardy space of functions analytic in the open right-half plane and square-integrable in the closed right-half plane $f(s) \in H_2 \Longleftrightarrow \sup_{\sigma>0} \int_{-\infty}^{\infty} f^*(\sigma + j\omega) f(\sigma + j\omega) d\omega < \infty$, (ii) for discrete-time systems: Hardy space of functions analytic in the open unit disc ($\lvert z \rvert < 1$) and square-integrable in the closed unit disc $g(z) \in H_2 \Longleftrightarrow \sup_{r<1} \int_0^{2\pi} g^*(r\, e^{j\theta})\, g(r\, e^{j\theta}) d\theta < \infty$.
H_∞	Hardy space of complex-valued functions of a complex variable, (i) for continuous-time systems: Hardy space of functions analytic in the open right-half plane and bounded in the closed right-half plane $f(s) \in H_\infty \Longleftrightarrow \sup_{\sigma>0} \lvert f(\sigma + j\omega) \rvert < \infty$, or functions on L_∞ with bounded analytic continuation in the right-half plane, (ii) for discrete-time systems: Hardy space of functions analytic in the open unit disc and bounded in the closed unit disc $g(z) \in H_\infty \Longleftrightarrow \sup_{r<1} \lvert g(r\, e^{j\theta}) \rvert < \infty$, set of asymptotically stable transfer functions with $\lVert \mathbf{G} \rVert_\infty < \infty$.
RH_∞	subspace of functions in Hardy space H_∞ that are real rational
L_p	space of functions
L_2	Hilbert space of functions square-integrable on $j\mathcal{R}$ $f(s) \in L_2 \Longleftrightarrow \frac{1}{2\pi} \int_{-\infty}^{\infty} f^*(j\omega) f(j\omega) d\omega \triangleq \lVert f(s) \rVert_2^2 < \infty$
L_p^n	space of vector-valued functions
L_{pe}^n	extended space L_p^n
L_∞	Banach space essentially bounded on $j\mathcal{R}$
$\mathcal{N}$	null space (kernel)
$\mathcal{R}[\cdot]$	range space (image)
$\mathcal{R}$	field of real numbers
$\mathcal{R}^n$	n-dimensional linear vector space over $\mathcal{R}$, $\mathcal{R}^n \equiv \mathcal{R}^{n \times 1}$
$\mathcal{R}^+$	field of non-negative real numbers
R	used as a prefix denotes real rational
$\nabla_\mathbf{p}$	Nabla operator (gradient) with respect to $\mathbf{p}$, $\partial(\cdot)/\partial\mathbf{p}$
b^x	exponential of base b
$\log_b x$	logarithm of a real number x to base b
$\odot$	generalized product corresponding to a polynomial product
$\otimes$	Kronecker product
$\oplus$	Kronecker sum
$\star$	convolution
$\mid$	e.g., $M(s) \mid_{s=1}$ means function $M(s)$ selected for $s = 1$
:	$\{a_i \in \mathcal{X}\ :\ f(a_i) = 0\}$ the set of elements of $\mathcal{X}$ having the property $f(a_i) = 0$
adj	adjoint of a matrix
arg	$\arg\min_x(.)$, that is, value of x that minimizes $(\cdot)$
circ	clockwise encirclements

cof	cofactor (of a matrix)
col	column string
conv	convex hull, convex combination
cov	covariance matrix
det	determinant (of a matrix)
$\text{diag}_n\{a_i\}$	diagonal matrix of dimension $n \times n$ with entries a_i
dim	dimension of a vector or a matrix (number of rows $\times$ number of columns)
$E\{x\}$	(expected) mean value of a stochstic process $\{x\}$, expectation operator
$\exp\mathbf{A}t$	matrix exponential function of the matrix $\mathbf{A}t$
$\text{grad}_{\mathbf{p}}$	gradient operator, $\partial(\cdot)/\partial\mathbf{p}$
h.o.t.	higher-order-terms
$\inf_{\mathbf{u}}$	infimum over $\mathbf{u}$, largest lower bound over $\mathbf{u}$
log	logarithm to base 10
$\text{matrix}[a_{ij}]$	arranging a matrix with entries a_{ij}
$\min\{a, b\}$	selecting the minimum of a and b
rank	rank of a matrix
row	row string
sign	signum function (sign $x = \pm 1$ for $x > 0$ and $x < 0$, respectively, and $= 0$ for $x = 0$)
$\sup_{\mathbf{u}}$	supremum over $\mathbf{u}$, least upper bound over $\mathbf{u}$
tr	trace of a matrix
$\text{tr}_{(i)}$	generalized trace
trig	trigonal matrix
vec	stacking operator

E.3 Superscripts

-1	inverse (of a matrix)
$\star$	(star) optimal
$*$	(asterix) complex conjugate
$+$	e.g., 0^+, i.e., $0 + \varepsilon$ where $\varepsilon > 0, \varepsilon \to 0$, e.g., $t = 0^+$ means instant immediately following zero
$+$	$a^+(s)$ is a polynomial free of zeros in the closed right-half s-plane
$-$	$a^-(s)$ is a polyomial free of zeros in the open left-half s-plane
$\perp$	all-pass extension
$\perp$	(right) annihilator (of a matrix)
$\square$	corner matrix
$(n \times m)$	matrix $\mathbf{M}$ of dimension $n \times m$
$\cdot$	(overline dot) first derivative with respect to time
(k)	k-th derivative with respect to time
$[j]$	jth Kronecker power
∂	specified bounds according to Eq.(12.80)
$\flat$	specified bounds according to Eq.(12.66)
$-$	(overline bar) mean value, linear expectation
$-$	(overline bar) *signals* in the case of *unperturbed* plant
$\wedge$	(written as a superscript) piecewise linear (function expansion coefficient)
$\hat{}$	(overline hat) (if necessary) referring to the frequency domain or the Laplace domain
$\hat{}$	(overline hat) estimate
$\tilde{}$	(overline tilde) estimation error
$\tilde{}$	(casually) characterizing a modified signal
$\sqcap$	block-pulse (function expansion coefficient)
$\neg$	general orthogonal (coefficient vector)
$\sharp$	general pseudo-inverse
$\sharp L$, $\sharp R$	left-pseudo-inverse and right-pseudo-inverse, respectively.
	$\mathbf{B}^{\sharp L}$ left-pseudo-inverse where $\mathbf{B} \in \mathcal{C}^{n\times m}$ and rank$\mathbf{B} = m < n$.
	$\mathbf{K}^{\sharp R}$ right-pseudo-inverse where $\mathbf{K} \in \mathcal{C}^{m\times n}$ and rank$\mathbf{K} = m < n$.
$\triangleright$	right (-eigenvalue)
$\triangleleft$	left (-eigenvalue)

D	Hadamard product
h	homogeneous solution
H	conjugate transpose ($\mathbf{A}^H = \mathbf{A}^{*T}$)
LI (L), RI	left-inverse, right-inverse, respectively
mo	modal variable, modal coordinate
op	operator
p	particular solution
R	reciprocated (polynomial)
R	para-Hermitian transpose (of a matrix)
T	transpose
w	Walsh function (expansion coefficient)

E.4 Subscripts

er	e.g., $\mathbf{F}_{er}$ means transfer matrix from the input $\mathbf{r}$ to the output e
1	sum vector norm, largest absolute column sum
∞	vector infinity norm, largest absolute row sum
$(n \times m)$	matrix $\mathbf{M}$ of dimension $n \times m$
$(\,)_{.j}$	jth column of a matrix
$(\,)_{i.}$	ith row of a matrix
c	observer based compensator
c	servo-compensator
cp	combined servo-compensator and plant
CL	closed-loop
d	dead-time delay
D	diagonally weighted or D-weighted Hölder norm
e	element by element
eq	equivalent sliding mode condition
f	final
F	Frobenius norm
H	Hankel (singular value or norm)
i	inner, see Eq.(30.166)
ij	(i, j)-partition of a matrix
I	interval
k	index denoting sampling instant
L	L-step ahead prediction
L	least favourable
LS	least squares
l, r	left-coprime and right-coprime, respectively
L, R	left and right polar decomposition, respectively
m	measurement
M	Markov process
MV	minimum variance
nom	nominal (value) (casually o)
o	initial
o	outer
p	plant
p	perturbed *system*
p	p-norm $\lVert\cdot\rVert_p$, p-measure μ_p
p	$\lVert\cdot\rVert_p$ function norm
p	e.g., $\mathbf{A}_p$ abbreviates $\partial\mathbf{A}/\partial p$
pr	index denoting a system with combined plant state and dynamic controller state
r	dynamic feedback controller
$red\ ij$	reduced (matrix), obtained by cancelling row i and column j
ref	reference, setpoint, particularly $\mathbf{y}_{ref}$
s	symmetric part of a matrix, e.g., $\mathbf{A}_s$

s	spectral norm, e.g., $\|\mathbf{A}\|_s$
s	slow model, neglecting high-frequency dynamics
S_2	(Sobolev) S_2-bounded
T	truncation of a function
α	variable multiplied by $e^{\alpha t}$

E.5 Glossary of Symbols in Alphabetic Order

$a(s)$	unperturbed polynomial in s
$a_e(s)$	extreme polynomial
$a_{ik}(s)$	edge polynomial
$a_p(s)$	perturbed polynomial $a(s)$
$\mathbf{a}_k$	(right) eigenvector of $\mathbf{A}$ associated with $\lambda_k[\mathbf{A}]$, abbreviation for $\mathbf{a}_k^{\triangleright}$
$\mathbf{a}_k^{\triangleleft}$	left-eigenvector of $\mathbf{A}$ associated with $\lambda_k[\mathbf{A}]$
$\mathbf{a}_{nom}$	nominal coefficient vector associated with the nominal polynomial $a(s)$
$\mathbf{A}$	matrix of coefficients of a system in state-space representation, having elements a_{ij} in row i and column j
$\bar{\mathbf{A}}$	average matrix
$\mathbf{A}_b$	bias matrix
$\mathbf{A}_p$	perturbed matrix $\mathbf{A}$
$\mathbf{A}_I$	real interval matrix
$\Delta\mathbf{A}$	perturbation matrix
$\mathbf{B}$	input matrix in state-space representation
C	performance, index of performance
C, $\mathbf{C}$	(matrix) transfer function of the controller
$\mathbf{C}$	output matrix
$\mathbf{C}_s$	matrix expressing sliding mode condition
$\mathcal{C}^{m\times n}$	set of complex matrices with m rows and n columns
db	decibels, e.g., a db means a gain of $10^{(\frac{a}{20})}$
d	path in the s-plane, see Eq.(21.113)
$d[s_1, s_2]$	chordal metric between the points s_1 and s_2
$d_G(\mathbf{G}_1, \mathbf{G}_2)$	graph metric distance
d_{si}	stochastic parameter
D	Nyquist contour enclosing the right half of the s-plane
D_R	Nyquist contour enclosing the right half of the s-plane with indentations of radius $1/R$ along the imaginary axis
$D_o(\omega)$	performance deterioration, see Eq.(24.43)
$\mathbf{D}$	matrix in the dynamic-free part of the observer
$\mathbf{D}$	diagonal scaling matrix
$\mathbf{D}_{cl}$, $\mathbf{D}_{cr}$	common left-divisor and common right-divisor, respectively
$\mathbf{D}_d$	operational matrix for dead-time delay
$\mathbf{D}_l$, $\mathbf{D}_r$	left-divisor and right-divisor, respectively
$\mathbf{D}_M$	operational matrix for differentiation
$\mathbf{D}_p^\star$	optimal p-norm weight
$\mathcal{D}_s$, $\mathcal{D}_u$	set of structured and unstructured perturbations, respectively
e	base of natural logarithm ($e = 2.71828$)
e	perturbation factor ($\Delta\mathbf{A} = e\mathbf{E}$)
$e_{\max}$	maximum perturbation parameter
e_T	tracking error, see Eq.(1.5)
$\mathbf{e}$	residual error
$\mathbf{e}$	error vector ($\mathbf{e} = \mathbf{y}_{ref} - \mathbf{y}$)
$\mathbf{e}_i$	ith standard basis vector (unit i-vector) with 1 only in the ith position
$\mathbf{E}$	matrix in the dynamic-free part of the observer
$\mathbf{E}$	error matrix
$\mathbf{E}$	unidirectional perturbation matrix
$\mathbf{E}$, $\mathbf{E}_d$	upper bound for structured perturbation of continuous-time systems

	and discrete-time systems, respectively
$\mathbf{E}_i$	constant perturbation matrix
$\mathbf{E}_p$	perturbation bound
$\mathbf{E}_{ij}$	Kronecker matrix
$f(z)$	polynomial in z
$\mathbf{f}_i$	eigenvector of $\mathbf{F}$ associated with $\lambda_i[\mathbf{F}]$
$\mathbf{F}$	matrix of closed-loop coefficients in state-space description
$\mathbf{F}$	closed-loop system transfer matrix
$\mathbf{F}$	aggregated matrix
$\mathbf{F}$	power series expansion matrix of orthogonal functions, see Eq.(36.41)
$\mathbf{F}_o$	return-ratio matrix, open-loop transfer matrix
$\mathbf{F}_{\mathbf{uw}}$	transfer matrix of a system with output $\mathbf{u}$ and input $\mathbf{w}$
$\mathbf{F}_z$	system matrix of the observer
$g(t)$	continuous impulse response function, weighting function
$\mathbf{g}$	perturbation vector
$\mathbf{g}(\mathbf{x}, t)$	nonlinear time-varying perturbation vector
G, $\mathbf{G}$	(matrix) transfer function, plant (matrix) transfer function
$\mathbf{G}$	aggregated input matrix
$\mathbf{G}_i$, $\mathbf{G}_n$	invertible and non-invertible part of the matrix $\mathbf{G}$
$\mathbf{G}_N$	nearest normal approximation
$\mathbf{G}_p^{op}$	operator associated with the perturbed plant
h	high-frequency signal, dither
h	degree of stability
h_i	("high") upper limit of a coefficient a_i
h_i, $\mathbf{h}$	general shifted orthogonal polynomial and polynomial function vector, respectively
H	Hamiltonian function
$\mathbf{H}$	Hurwitz testing matrix
$\mathbf{H}$	observer input matrix associated with the control variable
$\mathbf{H}$	upper bound of an interval matrix
$\mathbf{H}$	measurement transfer matrix
i, $\imath$	integer numbers
I	index of performance
$\mathbf{I}$	identity matrix of appropriate dimension
$\mathbf{I}_n$	identity matrix of dimension $n \times n$
$\mathbf{I}^{\oslash}$	symplectic matrix
$\mathbf{I}^{\checkmark}$	rotation matrix
$\mathbf{0}$	null matrix of appropriate dimension
$\mathbf{0}_n$	null matrix of dimension $n \times n$
j, $\jmath$	integer numbers
j	$= \sqrt{-1}$
$\mathbf{J}$	matrix in its Jordan canonical form, Jordan form, Jordan canonical form
$\mathbf{J}_{ij}$	Jordan block
k	integer number
k	number of terms of generalized shifted orthogonal polynomials in t which are taken into consideration
k_i	dimension of $\mathbf{\Delta}_i$, see Eq.(26.15)
k_m	stability margin
k_s	stability margin with respect to block-structured uncertainty
k_u	attenuation factor
$\mathbf{k}$	output feedback controller
K, $\mathbf{K}$	(matrix) transfer function of the controller, matrix gain factor
$\mathbf{K}_K$	steady-state Kalman gain matrix
$\mathbf{K}_v$	equivalent sliding mode state feedback controller
$\mathbf{K}_y$	output feedback controller matrix
l	integer number
l	dimension of the aggregated state vector $\mathbf{z}$
l_i	lower limit of (a coefficient a_i)
$l_o(\omega)$	bound for $\Delta\mathbf{L}_o$

l_q dimension of the parameter vector $\mathbf{q}$
L number of submodels
$\mathbf{L}$ observer input matrix associated with the output variable of the plant
$\mathbf{L}$ lower bound of an interval matrix
$\mathbf{L}$ scaling matrix
$\mathbf{L}_C$, $\mathbf{L}_O$ controllability and observability gramian, respectively
ΔL scalar multiplicative uncertainty
$\Delta\mathbf{L}_o$, $\mathbf{L}_o$, $\Delta\mathbf{L}_i$, $\mathbf{L}_i$ output and input associated uncertainty, respectively
$\mathbf{L}_I$ proportional-integral observer input matrix associated with the output of the plant
$\mathbf{L}_\Lambda$, $\mathbf{L}_V$ matrix-Lagrange-multipliers
m dimension of the input variable $\mathbf{u}$
m scalar function replacing the norm of the perturbation matrix $\mathbf{M}_i$
m_i number of partitions $\mathbf{\Delta}_i$, see Eq.(26.15)
m_i multiplicity of the eigenvalue λ_i
m_m order of the minimal polynomial
m_{mi} index of the eigenvalue λ_i, equivalent to the multiplicity of the eigenvalue in the minimal polynomial
m_M measure of stabilty robustness, see Eq.(23.2)
$\mathbf{m}_{aux}$ auxiliary measurement vector
$\mathbf{M}$ maximal bias matrix, deviation matrix
$\mathbf{M}$ measurement matrix
$\mathbf{M}$ matrix transfer function of the model
$\mathbf{M}$ matrix partition of the unperturbed control system interacting with the uncertainty which is pulled out and is considered as an external part of the control system
$\mathbf{M}_{aux}$ instrumental variable matrix
$\mathbf{M}_F$ feedback controller or sensor transfer matrix
$\mathbf{M}_i$ perturbation matrix associated with the sensor
$\mathbf{M}_R$ partitioned Riccati coefficient matrix
$\mathbf{M}_u$, $\mathbf{M}_y$ component connection model input matrices
n dimension of state variable, order of a dynamical system, state dimension, order of the characteristic polynomial
n_E number of fixed perturbation matrices
n_m dimension of the observation vector $\mathbf{y}$
n_p dimension of the parameter vector $\mathbf{p}$
$\mathbf{n}$ measurement noise
N number of zeros in the closed right half s-plane
N number of corners
N number of local subsystems
N_s smoothed nonlinearity
$\mathbf{N}_o$ resolvent matrix of $\mathbf{L}$
$\mathbf{N}^{op}$ operator associated with a nonlinear system
$\mathbf{N}_u$, $\mathbf{N}_y$ component connection model output matrices
p scalar parameter varying the interconnection
p, $\mathbf{p}$ parameter, parameter vector
$p[x(k)|x(k-1)]$ conditioned probability distribution function
$p(s)$ polynomial in s
$p_o(\omega)$ low-frequency performance
$p_M(j\omega)$ modified polynomial for Mikhailow hodograph
$\mathbf{p}_i$ right-eigenvector of $\mathbf{A}^T$
P number of unstable poles of the open-loop system
$\mathbf{P}$ permutation matrix
$\mathbf{P}$ operational premultiplication matrix
$\mathbf{P}$ solution matrix of the Lyapunov equation
$\mathbf{P}$ solution matrix of the Riccati equation
$\mathbf{P}$ matrix weighting the state vector $\mathbf{x}(t)$
$\mathbf{P}$ Pick matrix
$\mathbf{P}$ covariance matrix
$\mathbf{P}_K$ matrix solution of the Riccati equation determining a Kalman-Bucy filter

$\mathbf{P}_r$ projector matrix
$\mathbf{P}_I$ solution of the Lyapunov equation when $\mathbf{Q} = \mathbf{I}$
$\mathbf{P}_{Mb}$ operational matrix of integration using block-pulse functions
$\mathbf{P}_{Ml}$ operational matrix of integration using piecewise linear polynomial functions
$\mathbf{P}_{M\zeta}$, $\mathbf{P}_{Mt}$, $\mathbf{P}_M$ general operational matrices of integration
$\mathbf{P}_{Mp}$ operational matrix of integration associated with the power series expansion vector, see Eq.(36.102)
$\mathbf{P}_{Ms}$ stretched operational matrix of integration
$\mathbf{P}_{Mw}$ operational matrix of integration using Walsh functions
$\mathbf{P}_{\alpha}^{op}$ operator substituting the multiplication by $e^{\alpha t}$
$\mathbf{P}_{T}^{op}$ truncation operator
$\mathcal{P}_p$ family of perturbed polynomials
$\delta\mathcal{P}_p$ family of perturbation polynomials
q_i degeneracy
$q_s[\cdot]$ modified spectral radius, see Eq.(28.17)
$\mathbf{q}$ real uncertain parameter vector which allows for linear dependent coefficient perturbations
$\mathbf{Q}$ positive definite matrix
$\mathbf{Q}$ weighing matrix, weighting the state vector $\mathbf{x}(t)$
$\mathbf{Q}_\sigma$ weighting matrix, weighting the differential sensitivity vector σ_i
r dimension of the parameter vector $\mathbf{p}$
r dimension of the output vector $\mathbf{y}$
$r_c[\cdot]$, $r_R[\cdot]$ complex and real stability radius, respectively
r_L, r_o uncertainty radius bounding $|\Delta L(j\omega)|$
R region
$R[\mathbf{x}]$ Rayleigh quotient
$\mathbf{R}$ scaling matrix
$\mathbf{R}$ weighting matrix, weighting the control vector $\mathbf{u}(t)$
$\mathbf{R}_T^{op}$ retardation operator
$\mathcal{R}^{(m\times n)}$ set of real matrices with m rows and n columns
s complex Laplace variable
s_{ij} entry of the corner matrix
s_{ij} length of the chain containing the eigenvector and the generalized eigenvectors
s_w sliding mode condition
$\mathbf{s}_o$ output noise of a plant (measurement noise)
$\mathbf{s}_u$ input noise of a plant
S intersection of hyperplanes S_j
S_j hyperplane
$\mathbf{S}(s)$ sensitivity matrix of the nominal system
S_a^F normalized sensitivity function of the closed-loop transfer function $F(s)$ with respect to the characteristic polynomial $a(s)$ of the plant
S_G^F differential sensitivity
t, t_o real time variable, initial time, respectively
t_m discrete sampling instants
t_s sliding mode initial time
T sampling interval
T time constant
T_d time delay
T_I time constant of integration
T_t mapping given by the matrix sum weighted with powers of time
$\mathbf{T}$ matrix relating the plant state variable and the observer state variable
$\mathbf{T}$ transformation matrix (modal matrix)
$\mathbf{T}(s)$ complementary sensitivity matrix function
$\mathbf{T}_F$ modal matrix associated with $\mathbf{F} = \mathbf{A} + \mathbf{BK}$
$\mathcal{T}$ transformation
u, $\mathbf{u}$ control variable, controlling variable (vector), input vector
$\bar{\mathbf{u}}$ input signal in the case of unperturbed plant
$\mathbf{u}_i$ singular vector of $\mathbf{GG}^H$

$\mathbf{U}$	unitary matrix
$\mathbf{U}$	matrix of singular vectors $\mathbf{u}_i$, see Eq.(22.96)
$\mathbf{U}_e$	normalized perturbation bound
$\mathbf{U}_{kl}$	permutation matrix
$\bar{\mathbf{U}}_{kl}$	self-derivative matrix
$\mathbf{U}^{\urcorner}$	orthogonal spectrum of $\mathbf{u}$
$\mathbf{v}$	vector of measurement noise
$\mathbf{v}$	plant state vector applying the orthogonal transformation $\mathbf{T}$, see Eq.(32.54)
$\mathbf{v}$	state vector of the dynamic feedback controller
$\mathbf{v}_i$	singular vector of $\mathbf{G}^H\mathbf{G}$
V	gain constant
V, V_k	Lyapunov function, Lyapunov function at the sampling instant k, respectively
V_N	norm-like Lyapunov function
$V(\delta\mathcal{P}_p, d)$	normalized value set of perturbation polynomials, see Eq.(21.113)
$\mathbf{V}$	matrix of singular vectors $\mathbf{v}_i$, see Eq.(22.97)
$\mathbf{V}$	prefilter matrix
$\mathbf{V}_{dm}$, $\mathbf{V}_{dmg}$	Vandermonde matrix, generalized Vandermonde matrix, respectively
w	weighting function
$\mathbf{w}$	distortion, disturbance vector
$\mathbf{w}$	vector of process noise
$\mathbf{W}$	weighting matrix
$\mathbf{W}_A$	Kalman controllability matrix
$\mathbf{x}, \mathbf{x}_o$	state vector, initial state vector, respectively
$\mathbf{x}^{\urcorner}$	general orthogonal coefficient vector
$\bar{\mathbf{x}}(t)$	signal vector after having multiplied $\mathbf{x}(t)$ by the operator polynomial $\phi(s)$
$\mathbf{x}_{in}$, $\mathbf{x}_{out}$	output and input (*Note the order !*) of the block-diagonal uncertainty, see Fig. 26.3
$\mathbf{x}_r$	state of the dynamic feedback controller
$\mathbf{X}$	covariance matrix
$\mathbf{X}(t)$	state variable matrix
$\mathbf{X}^{\wedge}$	piecewise linear spectrum
$\mathbf{X}^{\urcorner}$	orthogonal spectrum of $\mathbf{x}$
$\mathbf{X}^{\sqcap}$	block-pulse spectrum
$\mathcal{X}$	set of matrices $\mathbf{X}$
$\mathcal{X}_\infty$	set of block-diagonal matrices with no restriction on the norm
$\mathbf{y}$	measurement vector, output vector
$\bar{\mathbf{y}}$	output signal in the case of unperturbed plant
$\tilde{\mathbf{y}}$	residual sum of squares
$\mathbf{y}_{aux}$	instrumental variable vector
$\mathbf{y}_{ref}$	reference, set point
$\mathcal{Y}$	set of matrices $\mathbf{Y}$
z	complex z-transform variable
z	variable characterizing the E-contour
$\mathbf{z}$	observer state vector
$\mathbf{z}$	(fast) state vector of parasitic dynamics
$\mathbf{z}$	aggregated model state vector
$\mathbf{z}$, $\tilde{\mathbf{z}}$	observer state and observer estimation error, respectively
$\mathbf{Z}$	(preferably) any matrix
$\mathbf{Z}_{ij}$	interpolating matrix polynomial, component of a matrix
$\mathcal{Z}$	set of matrices $\mathbf{Z}$
$1(t)$	unit step function at $t = 0$ ($= 1$ for $t \geq 0$, and $= 0$ for $t < 0$)
$\mathbf{1}$	sum vector
α	degree of stability
α	positive constant determining the radius of a disc
α_s	spectral abscissa
β_o	robustness measure
β_G	constant, bounding the plant operator, see Eq.(2.26)

β_N	constant, bounding the nonlinearity, see Eq.(2.26)
γ	sector bound
γ	maximum gain
γ_G	gain term, bounding the plant operator, see Eq.(2.26)
γ_N	gain term, bounding the nonlinearity, see Eq.(2.26)
$\boldsymbol{\gamma}(k)$	correcting gain vector (Kalman gain vector)
Γ	stability region in the s-plane
Γ^c	complement of the region Γ in the s-plane
$\mathbf{\Gamma}$	system matrix of a singularly perturbed system including dynamic feedback controller
$\mathbf{\Gamma}$	aggregation matrix
$\mathbf{\Gamma}$	generalized system transfer matrix
$\mathbf{\Gamma}_\Lambda$, $\mathbf{\Gamma}_V$	matrix constraints
δ	e.g., δx, first-order infinitesimal difference (first variation) of x
$\delta(t)$	unit Dirac delta function, unit impulse occuring at $t = 0$
δ_{ij}	Kronecker delta (= 1 for $i = j$, and = 0 for $i \neq j$)
$\delta_\tau(\tau)$	Kronecker delta (= 1 for $\tau = 0$, and = 0 for $\tau \neq 0$)
$\vec{\delta}(\mathbf{G}_1, \mathbf{G}_2)$	directed gap between two systems $\mathbf{G}_1$ and $\mathbf{G}_2$
Δ	e.g., Δt, (small) increment of t
$\mathbf{\Delta}_{bd}$	block-diagonal uncertainty matrix
$\mathbf{\Delta}_i$	partition of the block-diagonal uncertainty matrix $\mathbf{\Delta}_{bd}$
ε	small quantity
ε	small scalar parameter representing high-frequency parasitic dynamics
ε	quantity preferably moving from 0 to 1
ε	parameter determining the size of the polytope
$\boldsymbol{\varepsilon}$	error vector
$\mathbf{\Theta}$	inverse matrix of power series expansion of orthogonal functions, see Eq.(36.46)
$\mathbf{\Theta}(s)$	norm-bound uncertainty associated with an additive perturbance, see Eq.(26.1)
$\kappa_s[\cdot]$	spectral condition number
λ, $\boldsymbol{\lambda}$	scalar and vector-Lagrange-multiplier, respectively
λ	coefficient of stretched time scale
$\lambda_i[\mathbf{A}]$	ith eigenvalue of the matrix $\mathbf{A}$
λ_p	eigenvalue of the perturbed plant transfer matrix
$\boldsymbol{\lambda}$	costate variable, adjoint variable
Λ	set of eigenvalues
$\mathbf{\Lambda}$	diagonal matrix of the eigenvalues λ_i
$\mathbf{\Lambda}$	matrix-Lagrange-multiplier, costate matrix
μ	integer number
μ_{CN}	upper bound for nonlinear time-varying perturbation, see Eq.(13.75)
$\mu_D[\cdot]$	structured singular value, see Eq.(26.18)
$\mu_p[\cdot]$	matrix measure
μ_P	factor $\lambda_{\min}[\mathbf{Q}]/\lambda_{\max}[\mathbf{P}]$, see Eq.(13.41)
$\mu_s[\cdot]$	spectral matrix measure
ν	integer number
ξ_{1ik} , ξ_{2ik}	multiplicative noise
π	$\pi = 3.14159265$
$\pi[\cdot]$	Perron-Frobenius radius, Perron eigenvalue (root)
ρ	scalar parameter characterizing the ρ-system, see Eq.(25.173)
ρ_H	distance from Hurwitz stability, see Eq.(21.100), radius of largest hypersphere
ρ_{opt}	maximum radius of robust stabilizability, see Eq.(30.116)
ρ_R	numerical radius
ρ_s	spectral radius
$\boldsymbol{\rho}_i$	modified differential sensitivity variable
$\sigma[\cdot]$, $\sigma_i[\cdot]$	singular value
$\sigma_{Hi}[\cdot]$	Hankel singular value
σ_x	standard deviation, square root of the constant variance of observation errors
σ_{yv}, σ_{uv}	maximum plant sensitivity and maximum controller sensitivity, respectively
$\boldsymbol{\sigma}_i$	differential sensitivity vector
Σ_{pj}	region in $\mathbf{k}$-space where Γ-stability is satisfied for a fixed plant parameter vector

$\mathbf{\Sigma}$	(n,m)-matrix of singular values, see Eq.(22.88)
τ	time shift
τ	fast time scale, treating singular perturbation behaviour
$\phi(s)$	least common multiple of the minimal polynomials of $\mathbf{A}_w$ and $\mathbf{A}_r$ where $\mathbf{A}_w$ and $\mathbf{A}_r$ are system matrices associated with the disturbance and reference, respectively
$\varphi_i(\zeta)$	orthogonal polynomial
φ_m	phase margin
$\boldsymbol{\varphi}_i$	right-eigenvector of $\mathbf{\Phi}(T)$
$\mathbf{\Phi}(t,\tau)$	transition matrix for a time-varying system
$\mathbf{\Phi}_{cl}(t)$	transition matrix of the closed-loop system
$\mathbf{\Phi}_{CL}(T)$	closed-loop coefficient matrix in state-space description of a discrete-time system
$\mathbf{\Psi}$	input matrix of a discrete-time system
ω	(real) frequency in radians per second
ω_D, ω_r	crossover frequency
Ω	ellipsoidal set
$\mathbf{\Omega}_{ij}$	transfer matrix from the input $\mathbf{z}_j$ to the internal signal $\mathbf{v}_i$

Appendix F

Author Index

Abdul-Wahab, A.A., 1989 Robustness measure bounds for generalized dynamic output feedback controllers, *Int.J.Systems Sci.* **20**, pp. 2095-2105

Abdul-Wahab, A.A., 1990, Lyapunov-type equations for matrix root-clustering in subregions of the complex plane, *Int.J.Systems Sci.* **21**, pp. 1819-1830

Abdul-Wahab, A.A., 1990a, Lyapunov bounds for root clustering in the presence of system uncertainty, *Int. J. Systems Sci.* **21**, pp. 2603-2611

Abdul-Wahab, A.A., 1990b, Lyapunov stability robustness measures for multivariable, continuous, time-invariant, linear systems, *Int. J. Systems Sci.* **21**, pp.2577-2587

Abdul-Wahab, A.A., 1991, Perturbation bounds for root-clustering of linear continuous-time systems, *Int. J. Systems Sci.* **22**, pp. 921-930

Abdul-Wahab, A.A., and Zohdy, M.A., 1988, Generalized linear transformations on the design of robust dynamic output feedback controllers, *Int.J.Control* **48**, pp. 1241-1266

Abdul-Wahab, A.A., and Zohdy, M.A., 1989, Eigensystem assignment by feedback control, *Int.J.Control* **50**, pp. 1619-1634

Ackermann, J.,1972, Der Entwurf linearer Regelungssysteme im Zustandsraum, ***Regelungstechnik*** **20**, pp. 297-300

Ackermann, J., 1980, Parameter space design of robust control systems, *IEEE-Trans.* **AC-25**, pp. 1058-1072

Ackermann, J., 1984, Robustness against sensor failures, *Automatica* **20**, pp. 211-215

Ackermann, J.,(Ed.) 1985, Uncertainty and Control (Springer, Berlin New York)

Ackermann, J.,1985a, Multi-model approaches to robust control system design, In: *Ackermann, J.,(Ed.) 1985*, Uncertainty and Control (Springer, Berlin New York)

Ackermann, J., 1988, Abtastregelung, 3. Auflage (Springer, Berlin)

Ackermann, J., and Barmish, B.R., 1988, Robust Schur stability of a polytope of polynomials, *IEEE-Trans.* **AC-33**, pp. 984-986

Ackermann, J., Hu, H.Z., and Kaesbauer, D., 1990, Robustness analysis: a case study, *IEEE-Trans.* **AC-35**, pp. 352-356

Ackermann, J., and Hu, H.Z., 1990a, Robustness of sampled-data control systems with uncertain physical parameters, *11th IFAC-Congress Tallinn*, Vol. **5**, pp. 194-199

Ackermann, J., Kaesbauer, D., and Muench, R., 1991, Robust Gamma-stability analysis in a plant parameter space, *Automatica* **27**, pp. 75-85

Adamjan, V.M., Arov, D.Z., and Krein, M.G., 1978, Infinite Hankel block matrices and related extension problems, *Am.Math.Soc.,Transl. Series 2,* **111**, pp. 133-156

Aida, K., and Kitamori, T., 1990, Design of a PI-type state feedback optimal servo system, *Int.J.Control* **52**, pp. 613-625

Aly, G.M., and Ali, W.G., 1990, Digital design of variable structure control systems, *Int.J. Systems Sci.* **21**, pp. 1709-1720

Anagnost, J.J., Desoer, C.A., and Minichelli, R.J., 1989, Generalized Nyquist tests for robust stability: Frequency domain generalizations of Kharitonov's theorem In: *Milanese, M., Tempo, R., and Vicino, A., (Eds.) 1989* Robustness in Identification and Control (Plenum Press, New York London) pp. 79-96

Anderson, B.D.O., 1969, Stability results for optimal systems, *Electronics Letters* **5**, pp. 545.

Anderson, B.D.O., 1973, Exponential data weighting in the Kalman-Bucy filter, *Information Sciences* **5**, pp. 217-230

Anderson, B.D.O., Dasgoupta, S., Khargonekar, P., Kraus, F.J., and Mansour, M., 1989, Robust strict positive realness: characterization and construction *Proc. 28th IEEE-Conference on Decision and Control, Tampa, Florida* pp. 426-430

Anderson, B.D.O., Jury, E.I., and Mansour, M., 1987, On robust Hurwitz polynomials, *IEEE-Trans.* **AC-32**, pp. 909-913

Anderson, B.D.O., and Moore, J.B., 1971, Linear Optimal Control (Prentice-Hall, Englewood Cliffs/New Jersey)

Anderson, B.D.O., and Moore, J.B., 1989, Optimal Control. Linear Quadratic Methods. (Prentice-Hall, Englewood Cliffs)

Aoki, M., 1968, Control of large-scale dynamic systems by aggregation, *IEEE-Trans.* **AC-13**, pp. 246-253

Argoun, M.B., 1986a, Allowable coefficient perturbations with preserved stability of a Hurwitz polynomial, *Int.J.Control* **44**, pp. 927-934

Argoun, M.B., 1986b, On sufficient conditions for the stability of interval matrices, *Int.J.Control* **44**, pp. 1245-1250

Argoun, M.B., 1986c, Allowable coefficient perturbations with preserved stability of a Hurwitz polynomial, *Int.J.Control* **44**, pp. 927-934

Argoun, M.B., 1987, Stability of a Hurwitz polynomial under coefficient perturbations: necessary and sufficient conditions, *Int.J.Control* **45**, pp. 739-744

Arkun, Y., 1987, Dynamic block relative gain and its connection with the performance and stability of decentralized control structures, *Int.J.Control* **46**, pp. 1187-1193

Asada, H., and Slotine, J.J., 1986, Robot Analysis and Control (John Wiley, New York)

Åström, K.J., 1985, Adaptive control - a way to deal with uncertainty, In: Ackermann, J.,(Ed.) 1985, Uncertainty and Control (Springer, Berlin New York)

Athans, M., 1968, The matrix minimum principle, *Information and Control* **11**, pp. 592-606

Ball, J.A., and Helton, J.W., 1983, A Beurling-Lax theorem for the lie group $U(m,n)$ which contains most classical interpolation theory, *J. Operator Theory* **9**, pp. 107-142

Banks, S.P., 1988, Mathematical Theories of Nonlinear Systems (Prentice-Hall, New York London)

Bard, Y., 1974, Nonlinear Parameter Estimation (Academic Press, New York)

Barmish, B.R., 1983, Stabilization of uncertain systems via linear control, *IEEE-Trans.* **AC-28**, pp. 848-850

Barmish, B.R., 1984, Invariance of the strict Hurwitz property for polynomials with perturbed coefficients, *IEEE-Trans.* **AC-29**, pp. 935-936

Barmish, B.R., Corless, M., and Leitmann, G., 1983, A new class of stabilizing controllers for uncertain dynamical systems, *SIAM J. on Control and Optimization* **21**, pp. 246-255

Barmish, B.R., and Hollot, C.V., 1984, Counter-example to a recent result on the stability of interval matrices by S. Bialas, *Int.J.Control* **39**, pp. 1103-1104

Barmish, B.R., and Leitmann, G., 1982, On ultimate boundedness control of uncertain systems in the absence of matching assumptions, *IEEE-Trans.* **AC-27**, pp. 153-158

Barmish, B.R., and Shi, Z., 1990, Robust stability of a class of polynomials with coefficients depending multilinearly on perturbations, *IEEE-Trans.* **AC-35**, pp. 1040-1043

Barmish, B.R., and Tempo, R., 1990, The robust root locus, *Automatica* **26**, pp. 283-292

Barnett, S., 1971, Matrices in Control Theory (Van Nortrand Reinhold, London)

Barnett, S., 1973, Matrix differential equations and Kronecker products, *SIAM J.Appl.Math.* **24**, pp. 1-5

Bartlett, A.C., Hollot, C.V., and Lin, H., 1987, Root locations of an entire polytope of polynomials: it suffices to check the edges, *Proc. American Control Conf.* pp. 1611-1616

Bauer, F.L., 1962, On the field of values subordinate to a norm, *Numer. Math.***4**, pp. 103-113

Bauer, F.L., 1963, Optimally scaled matrices, *Numer.Math.***5**, pp. 73-87

Bauer, F.L., and Fike, C.T., 1960, Norms and exclusion theorems, *Numer. Math.***2**, pp. 137-141

Beale, S., and Shafai, B., 1989, Robust control system design with a proportional integral observer, *Int.J.Control* **50**, pp. 97-111

Beck, J.V., and Arnold, K.J., 1977, Parameter Estimation in Engineering and Science (John Wiley, New York London)

Bell, D.J., (Ed.) 1973, Recent Mathematical Developments in Control. Proceedings of a conference, Bath 1972, (Academic Press, London New York)

Bellman, R., 1964, Perturbation Techniques in Mathematics, Physics and Engineering (Holt, Rinehart and Winston, New York)

Bellman, R., 1967, Introduction to the Mathematical Theory of Control Processes, Vol. 1, Linear Equations and Quadratic Criteria (Academic Press, New York)
Bellman, R., 1970, Introduction to Matrix Analysis, Second Edition, (Mc Graw-Hill, New York London)
Benidir, M., and Picinbono, B., 1988, Comparison between some stability criteria of discrete-time filters, *IEEE-Trans.* **ASSP-36**, pp. 993-1001
Benninger, N.F., 1986, Eine gut konditionierte Transformation auf Sensorkoordinaten, *Automatisierungstechnik* **34** , pp.410-411
Bernussou, J., Peres, P.L.D., and Geromel, J.C., 1990, Stabilizability of uncertain dynamical systems: the continuous and the discrete case, *11th IFAC-Congress Tallinn*, Vol. **5**, pp. 135-140
Bernussou, J., and Titli, A., 1982, Interconnected Dynamical Systems: Stability, Decomposition and Decentralization (North-Holland, Amsterdam New York Oxford)
Bhattacharyya, S.P., 1987, Robust Stabilization Against Structured Perturbations, (Springer, Berlin)
Bhattacharyya, S.P., del Nero Gomes, A.C., and Howze, J.W., 1983, The structure of robust disturbance rejection control, *IEEE-Trans.* **AC-28**, pp. 874-881
Bialas, S., 1983, A necessary and sufficient condition for the stability of interval matrices, *Int.J.Control* **37**, pp.717-722
Bialas, S., and Garloff, J., 1985, Convex combinations of stable polynomials, *J. of the Franklin Institute* **319**, pp. 375-377
Billings, S.A., and Chen, S., 1989, Extended model set, global data and threshold model identification of severely nonlinear systems, *Int.J.Control* **50**, pp. 1897-1923
Birdwell, J.D., and Laub, A.J., 1987, Balanced singular values for LQG/LTR design, *Int.J.Control* **45**, pp. 939-950
Bistritz, Y., 1984, Zero location with respect to the unit circle of discrete-time linear system polynomials, *Proc. IEEE* **72**, pp. 1131-1142
Böcker, J., Hartmann, I., und Zwanzig, Ch., 1986 , Nichtlineare und adaptive Regelungssysteme (Springer, Berlin New York)
Bode, H.W., 1945, Network Analysis and Feedback Amplifier Design, (Van Nostrand, New York)
Bongiorno, J.J.Jr., 1969, Minimum sensitivity design of linear multivariable feedback control systems by matrix spectral factorization, *IEEE-Trans.* **AC-14** pp.665-673
Bose, N.K., 1977, Implementation of a new stability test for two-dimensional filters, *IEEE-Trans.* **ASSP-25**, pp. 117-120
Bose, N.K., 1985, A system-theoretic approach to stability of sets of polynomials, *Contemporary Mathematics* **47**, pp. 25-34
Bose, N.K., and Delansky, J.F., 1989, Boundary implications for interval positive rational functions: preliminaries, In: *Milanese, M., Tempo, R., and Vicino, A., (Eds.) 1989* Robustness in Identification and Control (Plenum Press, New York London) pp. 287-291
Bose, N.K., Jury, E.I., and Zeheb, E., 1988, On robust Hurwitz and Schur polynomials, *IEEE-Trans.* **AC-33**, pp. 1166-1168
Bose, N.K., and Shi, Y.Q., 1987, A simple general proof of Kharitonov's generalized stability criterion, *IEEE-Trans.* **CAS-34**, pp. 1233-1237
Boyd, S., and Yang, Q., 1989, Structured and simultaneous Lyapunov functions for system stability problems, In: *Milanese, M., Tempo, R., and Vicino, A., (Eds.) 1989* Robustness in Identification and Control (Plenum Press, New York London) pp. 243-262
Brammer, K., and Siffling, G., 1975, Kalman-Bucy-Filter (Oldenbourg, München Wien)
Breinl, W., 1980, Entwurf eines parameterunempfindlichen Reglers am Beispiel der Magnetschwebebahn, *Regelungstechnik* **28**, pp. 87-92
Brewer, J.W., 1977, The derivative of the exponential matrix with respect to a matrix, *IEEE-Trans.* **AC-22**, pp. 656-657
Brewer, J.W., 1977a, The derivative of the Riccati matrix with respect to a matrix, *IEEE-Trans.* **AC-22**, pp. 980-983
Brewer, J.W., 1978, Kronecker products and matrix calculus in system theory, *IEEE-Trans.* **CAS-25**, pp. 772-781
Brewer, J.W., 1978a, Matrix calculus and the sensitivity analysis of linear dynamic systems, *IEEE-Trans.* **AC-23**, pp. 748-751
Brillinger, D.R., 1981, Time Series, Data Analysis and Theory (Holden-Day, San Francisco)
Brogan, W.L., 1985, Modern Control Theory, 2nd edition (Prentice-Hall, New York London)

Bühler, H., 1986, Minimierung der Norm des Rückführvektors bei der Polvergabe, *Automatisierungstechnik* **34**, pp.152-155
Byers, R., and Nash, S.G., 1989, Approaches to robust pole assignment, *Int.J.Control* **49**, pp. 97-117
Byrne, P.C., and Burke, M., 1976, Optimization with trajectory sensitivity considerations, *IEEE-Trans.* **AC-21**, pp. 282-283
Cao, C.T., 1989, Entwurf eines robusten prädiktiven Reglers für totzeitbehaftete Regelsysteme, *Automatisierungstechnik* **37**, pp. 104-110
Carlucci, D., and Donati, F., 1975, Control of norm uncertain systems, *IEEE-Trans.* **AC-20**, pp. 792-795
Carlucci, D., and Vallauri, M., 1981, Feedback control of linear multivariable systems with uncertain description in the frequency domain, *Int.J.Control* **33**, pp.903-912
Cavallo, A., Celentano, G., and De Maria, G., 1991, Robust stability analysis of polynomials with linearly dependent coefficient perturbations, *IEEE-Trans.* **AC-36**, pp. 380-384
Chang, B.C., and Pearson, J.B., Jr., 1984, Optimal disturbance reduction in linear multivariable systems, *IEEE-Trans.* **AC-29**, pp. 880-887
Chang, R.Y., Yang, S.Y., and Wang, M.L., 1987, Solution of a scaled system via generalized orthogonal polynomials, *Int.J.Systems Sci.* **18**, pp. 2369-2382
Chang, R.Y., Yang,S,Y., and Wang, M.L., 1988, Parameter identification of time-varying systems via generalized orthogonal polynomials, *Int.J.Systems Sci.* **19**, pp. 471-485
Chang, Y.F., and Lee, T.T., 1986, Application of general orthogonal polynomials to the optimal control of time-varying linear systems, *Int.J.Control* **43**, pp. 1283-1304
Chang, S.S.L., and Peng, T.K.C., 1972, Adaptive guaranteed cost control of systems with uncertain parameters, *IEEE-Trans.* **AC-17**, pp. 474-483
Chapellat, H., and Bhattacharyya, S.P., 1989, A generalization of Kharitonov's theorem: robust stability of interval plants, *IEEE-Trans.* **AC-34**, pp. 306-311
Chapellat, H., and Bhattacharyya, S.P., 1989a, Robust stability and stabilization of interval plants, In: *Milanese, M., Tempo, R., and Vicino, A., (Eds.) 1989* Robustness in Identification and Control (Plenum Press, New York London) pp. 207-229
Chapellat, H., Bhattacharyya, S.P., and Dahleh, M., 1990, Robust stability of a family of disc polynomials, *Int.J.Control* **51**, pp. 1353-1362
Chapellat, H., Dahleh, M., and Bhattacharyya, S.P., 1990a, Robust stability under structured and unstructured perturbations, *IEEE-Trans.* **AC-35**, pp. 1100-1108
Chapellat, H., Dahleh, M., and Bhattacharyya, S.P., 1991, On robust nonlinear stability of interval control systems, *IEEE-Trans.* **AC-36**, pp. 59-67
Chen, B.S., 1984, Controller synthesis of optimal sensitivity: multivariable case, *Proc.IEE Part D* **131**, pp. 47-51
Chen, B.S., and Chou, H.H., 1988, Optimal robustness via constrained controller in SISO discrete-time feedback systems, *Int.J.Control* **48**, pp. 1391-1408
Chen, B.S., and Dong, T.Y., 1988, Robust stability analysis of Kalman-Bucy filter under parametric and noise uncertainties, *Int.J.Control* **48**, pp. 2189-2199
Chen, B.S., and Dong, T.Y., 1989, LQG optimal control system design under plant perturbation and noise uncertainty: a state-space approach, *Automatica* **25**, pp. 431-436
Chen, B.S., and Kung, J.Y., 1988, Robust stability of structured perturbation system in state space model, *Proc. 27th IEEE Conference on Decision and Control, Austin*, pp. 121-122
Chen, B.S., and Lin, C.L., 1990, On the stability bounds of singularly perturbed systems, *IEEE-Trans.* **AC-35**, pp. 1265-1270
Chen, B.S., and Lo, C.H., 1989, Necessary and sufficient conditions for robust stabilization of perturbed observer-based compensating systems, *Int.J.Control* **49**, pp. 937-960
Chen, B.S., Wang, S.S., and Lu, H.C., 1989, Minimal sensitivity perfect model matching control, *IEEE-Trans.* **AC-34**, pp. 1279-1283
Chen, B.S., and Wang, W.J., 1990, Robust stabilization of nonlinearly perturbed large-scale systems by decentralized observer-controller compensators, *Automatica* **26**, pp. 1035-1041
Chen, B.S., and Wong, C.C., 1987, Robust linear controller design: Time domain approach, *IEEE-Trans.* **AC-32**, pp. 161-164
Chen, C.F., and Hsiao, C.H., 1975, Design of piecewise constant gains for optimal control via Walsh functions, *IEEE-Trans.***AC-20**, pp. 596-603
Chen, C.T., 1968, Stability of linear multivariable feedback systems, *Proc. IEEE* **56**, pp. 821-828

Chen, C.T., 1970, Introduction to Linear System Theory (Holt, Rinehart and Winston, New York London)

Chen, K., 1961, Analysis and design of feedback systems with gain and time constant variations, *IRE-Trans.* **AC-6**, pp. 73-79

Chen, K.H., and Chen, W.L., 1989, Robust analysis of multivariable systems with plant perturbation and sensor uncertainty, *Int.J.Systems Sci.* **20**, pp. 2329-2333

Chen, M.J., and Desoer, C.A., 1982, Necessary and sufficient condition for robust stability of linear distributed feedback systems, *Int.J.Control* **35**, pp. 255-267

Chen, S., Billings, S.A., and Luo, W., 1989, Orthogonal least squares methods and their application to nonlinear system identification, *Int.J.Control* **50**, pp. 1873-1896

Chen, Y.H., 1986, On the deterministic performance of uncertain dynamical systems, *Int.J.Control* **43**, pp. 1557-1579

Chen, Y.H., 1987, Robust output feedback controller: direct/indirect design *Int.J.Control* **46**, pp. 1083-1103

Chen, Y.H., 1989, Modified adaptive robust control system design, *Int.J.Control* **49**, pp. 1869-1882

Cheng, B., and Hsu, N.S., 1982, Single-input-single-output system identification via block- pulse functions, *Int.J.Systems Sci.* **13**, pp. 697-702

Cheok, K.C., Hu, H.X., and Loh, N.K., 1988a, Optimal output feedback regulation with frequency-shaped cost functional, *Int.J.Control* **47**, pp. 1665-1681

Cheok, K.C., Hu, H.X., and Loh, N.K., 1989, Representation of dynamic output feedback control system variables via orthogonal functions, *Int.J.Systems Sci.* **20**, pp.1163-1171

Cheok, K.C., Loh, N.K., and Ho, J.B., 1988, Continuous-time optimal robust servo-controller design with internal model principle, *Int.J.Control* **18** , pp. 1993-2010

Cheres, E., Gutman, S., and Palmor, Z.J., 1989, Stabilization of uncertain dynamic systems including state delay, *IEEE-Trans.* **AC-34**, pp. 1199-1203

Cheres, E., Palmor, Z.J., and Gutman, S., 1989a, Quantitative measures of robustness for systems including delayed perturbations, *IEEE-Trans.* **AC-34**, pp. 1203-1204

Chiang, C.C., and Chen, B.S., 1988, Robust stabilization against non-linear time-varying uncertainties, *Int.J.Systems Sci.* **19**, pp. 747-760

Cho, Y.S., and Narendra, K.S., 1968, An off-axis circle criterion for the stability of feedback systems with a monotonic nonlinearity, *IEEE-Trans.* **AC-13**, pp. 413-416

Chou, H.H., Chen, B.S., and Lin, Y.P., 1990, Necessary and sufficient condition for stability robustness under nonlinear time-varying uncertainties, *Int.J.Systems Sci.* **21**, pp. 305-316

Chou, H.H., Chen, B.S., and Lin, Y.P., 1990a, Robust pole placement: a frequency-domain approach, *Int.J.Systems Sci.* **21**, pp. 317-333

Chu, C.C., Doyle, J.C., and Lee, E.B., 1986, The general distance problem in H_∞ optimal control theory, *Int.J.Control* **44**, pp. 565-596

Chung, H.Y., 1987, System identification via Fourier series, *Int.J.Systems Sci.* **18**, pp. 1191-1194

Cieslik, J., 1987, On possibilities of the extension of Kharitonov's stability test for interval polynomials to the discrete-time case, *IEEE-Trans.* **AC-32**, pp. 237-238

Cobb, J.D., 1988, The minimal dimension of stable faces required to guarantee stability of a matrix polytope: D-stability, *Proc. IEEE Conf. on Decision and Control* pp. 119-120

Cobb, J.D., 1990, The minimal dimension of stable faces required to guarantee stability of a matrix polytope: *D*-stability, *IEEE-Trans.* **AC-35**, pp. 469-473

Cobb, J.D., and DeMarco, C.L., 1988, The minimal dimensionality of stable faces required to guarantee stability of a matrix polytope, *American Control Conference* pp. 818-819

Cook, P.A., 1972, Modified multivariable circle theorems. In *Bell, D.J.,(Ed.) 1973*, pp. 367-372

Corless, M., 1985, Stabilization of uncertain discrete-time systems, In: Skelton, R.E. and Owens, D.H. (Eds.) 1985

Corless, M., and Da, D., 1989, New criteria for robust stability In: *Milanese, M., Tempo, R., and Vicino, A., (Eds.) 1989* Robustness in Identification and Control (Plenum Press, New York London) pp. 329-337

Corless, M., Garofalo, F., and Leitmann, G., 1989, Guaranteeing ultimate boundedness of exponential rate of convergence for a class of uncertain systems In: *Milanese, M., Tempo, R., and Vicino, A., (Eds.) 1989* Robustness in Identification and Control (Plenum Press, New York London) pp. 293-301

Corless, M.J., and Leitmann, G., 1981, Continuous state feedback guaranteeing uniform ultimate boundedness for uncertain dynamic systems, *IEEE-Trans.* **AC-26**, pp. 1139-1144

Cruz, I., (Ed.) 1973, System Sensitivity Analysis (Dowden, Hutchinson and Ross, Stroudsburg/Pennsylvania),

Curtain, R.F., and Pritchard, A.J., 1977, Functional Analysis in Modern Applied Mathematics (Academic Press, New York)

Dahleh, M.A., 1990, BIBO stability robustness for coprime factor perturbations, *11th IFAC-Congress Tallinn,* Vol. 5, pp. 201-204

Daniel, R.W., and Kouvaritakis, B., 1985, A new robust stability criterion for linear and nonlinear multivariable feedback systems, *Int.J.Control* **41**, pp. 1349-1379

Darwish, M.G., and Soliman, H.M., 1988, Design of decentralized reliable controllers for large-scale systems, *Int.J.Systems Sci.* **19**, pp. 1529-1538

Dasgupta, S., 1987, A Kharitonov-like theorem for systems under nonlinear passive feedback, *Proc. IEEE-Conf. on Decision and Control, Los Angeles* pp. 2062-2063

Dasgupta, S., 1988, Kharitonov's theorem revisited, *Systems and Control Letters* **11**, pp. 381-384

Dasgupta, S., Parker, P.J., Anderson, B.D.O., Kraus, F.J., and Mansour, M., 1988, Frequency domain conditions for the robust stability of linear and nonlinear dynamical systems, *Proc. American Control Conf., Atlanta* pp.1863-1868

Davision, E.J., 1976, The robust control of a servomechanism problem for linear time-invariant multivariable systems, *IEEE-Trans.* **AC-23**, pp. 25-34

Davison, E.J., and Ferguson, I.J., 1981, The design of controllers for the multivariable robust servomechanism problem using parameter optimization methods, *IEEE-Trans.* **AC-26**, pp. 93-110

Dawson, D.M., Qu, Z., Lewis, F.L., and Dorsey, J.F., 1990, Robust control for the tracking of a robot motion, *Int.J.Control* **52**, pp. 581-595

De Carlo, R.A., and Saeks, R., 1981, Interconnected Dynamical Systems (Marcel Dekker, New York)

de la Sen, M., 1990, Use of Gronwall's Lemma for robust compensation of time-varying linear systems via synthesis of augmented exciting signals, *Int. J. Systems Sci.* **21**, pp. 2317-2335

Delsarte, P., Genin, Y., and Kamp, Y., 1981, On the role of the Nevanlinna-Pick problem in circuit and system theory, *Int. J. on Circuit Theory and Application* **9**, pp. 177-187

Desoer, C.A., and Gustafson, C.L., 1984, Algebraic theory of linear multivariable feedback systems, *IEEE-Trans.* **AC-29**, pp. 909-917

Desoer, C.A., Liu, R.W., Murray, J., and Saeks, R., 1980, Feedback system design: The fractional representation approach to analysis and synthesis, *IEEE-Trans.* **AC-25**, pp. 399-412

Desoer, C.A., and Vidyasagar, M., 1975, Feedback Systems: Input-Output Properties (Academic Press, New York)

Desoer, C.A., and Wang, Y.T., 1978, On the minimum order of a robust servocompensator, *IEEE-Trans.* **AC-23**, pp. 70-73

Dickman, A., and Sivan, R., 1985, On the robustness of multivariable linear feedback systems, *IEEE-Trans.* **AC-30**, pp. 401-404

Diduch, C.P., and Doraiswami, R., 1988, Role of sample period in transients, robustness and sensitivity in optimally designed digital control systems, *Int.J.Systems Sci.* **19**, pp. 921-934

Dieudonné, J., 1969, Foundations of Modern Analysis (Academic Press, New York)

Ding, X., and Frank, P.M., 1990, Komponentenfehlerdetektion mittels Fourier-Analyse im Zustandsraum, *Automatisierungstechnik* **38**, pp.134-143

Djaferis, T.E., 1991, Simple robust stability tests for polynomials with real parameter uncertainty, *Int. J. Control* **53**, pp. 907-927

Doetsch, G., 1967, Guide to the Applications of the Laplace and z-Transforms, (Van Nostrand Reinhold, London)

Doraiswami, R., 1990, Time-domain approach to quantification of overshoot, speed and robustness, *Int.J.Systems Sci.* **21**, pp. 1057-1075

Doraiswami, R., and Bordry, F., 1987, Robust two-time-level control strategy for sampled-data servomechanism problem, *Int.J.Systems Sci.* **18**, pp. 2261-2277

Dorato, P., (Ed.) 1987, Robust Control, (IEEE-Press, New York)

Dorato, P., Park, H.B., and Li, Y., 1989, An algorithm for interpolation with units in H^∞, with applications to feedback stabilization, *Automatica* **25**, pp. 427-430

Dorling, C.M., and Zinober, A.S.I., 1986, Two approaches to hyperplane design in multivariable variable structure control systems, *Int.J.Control* **44**, pp. 65-82

Dorling, C.M., and Zinober, A.S.I., 1988, Robust hyperplane design in multivariable variable structure control systems, *Int.J.Control* **48**, pp. 2043-2054

Douce, J.L., and Roberts, P.D., 1964, Orthogonal functions, *Control* **8**, pp. 457-461

Doyle, J.C., 1982, Analysis of feedback systems with structured uncertainties, *IEE Proc. Part D* **129**, pp. 242-250

Doyle, J.C., 1983, Synthesis of robust controllers and filters, *Proc. 22nd IEEE Conf. on Decision and Control* pp. 109-114

Doyle, J.C., Glover, K., Khargonekar, P., and Francis, B., 1988, State-space solutions to standard H_2 and H_∞ control problems, *Proc. American Control Conference, Atlanta,* pp. 1691-1696

Doyle, J.D., Glover, K., Khargonekar, P.P., and Francis, B.A., 1989, State-space solutions to standard H_2 and H_∞ control, *IEEE-Trans.* **AC-34**, pp.831-847

Doyle, J.C., and Stein, G., 1979, Robustness with observers, *IEEE-Trans.* **AC-24**, pp.607-611

Doyle, J.C., and Stein, G., 1981, Multivariable feedback design: Concepts for a classical/modern synthesis, *IEEE-Trans.* **AC-26**, pp. 4-16

Doyle, J.C., Wall, J.E., and Stein, G., 1982, Performance and robustness analysis for structured uncertainty, *Proc. 21st IEEE Conf. on Decision and Control, Orlando, Florida,* pp. 629-636

Draper, N.R., and Smith, H., 1966, Applied Regression Analysis (John Wiley, New York London)

Dubois, D., and Prade, H., 1980, Fuzzy Sets and Systems: Theory and Applications (Academic Press, New York Toronto London)

Duren, P.L., 1970, Theory of H_p Spaces (Academic Press, New York)

Eitelberg, E., 1986, Deterministic optimal state estimation, *Automatisierungstechnik* **34**, pp. 394-397

El-Ghezawi, O.M.E., Zinober, A.S.I., and Billings, S.A., 1983, Analysis and design of variable structure systems using a geometric approach, *Int.J.Control* **38**, pp. 657-671

Elmetwally, E.E., and Rao, N.D., 1974, Design of low sensitivity optimal regulators for synchronous machines, *Int.J.Control* **19**, pp. 593-607

El-Sakkary, A.K., 1988, Estimating the robust dead time for closed-loop stability, *IEEE-Trans.* **AC-33**, pp. 599-601 (and **AC-34**, p. 1324)

El-Sakkary, A.K., 1989, Estimating robustness on the Riemann sphere, *Int.J.Control* **49**, pp.561-567

Eslami, M., and Russell, D.L., 1980, On stability with large parameter vaiations: stemming from the direct method of Lyapunov, *IEEE-Trans.* **AC-25**, pp. 1231-1234

Evans, R.J., Cantoni, A., and Ahmed, K.M., 1983, Envelope-constrained filters with uncertain input, *Circuits Systems Signal Process* **2**, pp. 131-154

Evans, R.J., and Xianya, X., 1985, Robust regulator design, *Int.J.Control* **41**, pp. 461-476

Eveleigh, V.W., 1967, Adaptive Control and Optimization Techniques, (McGraw-Hill, New York London)

Fallside, F., (Ed.) 1977, Control System Design by Pole-Zero Assignment (Academic Press, London New York)

Fam, A.T., 1989, Stability conditions for polynomials via quadratic inequalities in their coefficients, In: *Milanese, M., Tempo, R., and Vicino, A., (Eds.) 1989* Robustness in Identification and Control (Plenum Press, New York London) pp. 281-286

Fan, M.K.H., and Tits, A.L., 1986, Characterization and efficient computation of the structured singular value, *IEEE-Trans.* **AC-31**, pp. 734-743

Fan, M.K.H., and Tits, A.L., 1988, m-Form numerical range and the computation of the structured singular value, *IEEE-Trans.* **AC-33**, pp. 284-289

Fan, M.K.H., Tits, A.L., and Doyle, J.C., 1991, Robustness in the presence of mixed parametric uncertainty and unmodeled dynamics, *IEEE-Trans.* **AC-36**, pp.25-38

Feintuch, A., and Francis, B.A., 1985 Uniformly optimal control of linear feedback systems, *Automatica* **21**, pp. 563-574

Feliachi, A., 1986, Decentralized stabilization of interconnected systems, *Int.J.Control* **44**, pp. 1499-1505

Feliachi, A., 1988, Linear feedback control with prespecified performance measure, *Int.J.Systems Sci.* **19**, pp. 979-984

Feliachi, A., and Bhurtun, C., 1987, Model reduction of large-scale interconnected systems, *Int.J.Systems Sci.* **18**, pp. 2249-2259

Ficklscherer, P., und Müller, P.C., 1985, Robuste Zustandsbeobachter für lineare Mehrgrößenregelungssysteme mit unbekannten Störeingängen, *Automatisierungstechnik* **33**, pp. 173-179

Fleming, P.J., 1978, Desensitizing constant gain feedback linear regulators, *IEEE-Trans.* **AC-23**, pp. 933-936

Fleming, P.J., and Newmann, M.M., 1977, Design algorithm for a sensitivity constrained suboptimal controller, *Int.J.Control* **25**, pp. 965-978

Fletscher, R., and Reeves, C.M., 1964, Function minimization by conjugate gradients, *The Computer Journal* 7, pp.149-154

Föllinger, O., 1961, Über die Anfangsbedingungen bei linearen Übertragungsgliedern, *Regelungstechnik* 9, pp. 149-153

Föllinger, O., 1976, Entwurf von Regelkreisen durch Transformation der Zustandsvariablen, *Regelungstechnik* **24**, pp. 239-245

Föllinger, O., 1986, Entwurf konstanter Ausgangsrückführungen im Zustandsraum, *Automatisierungstechnik* **34**, pp. 5-15

Foo, Y.K., 1990, Stability and performance robustness analysis with highly structured uncertainties: necessary and sufficient test, *IEEE-Trans.* **AC-35**, pp. 350-352

Foo, Y.K., and Postlethwaite, I., 1988, Extensions of the small-μ test for robust stability, *IEEE Trans.* **AC-33**, pp. 172-176

Foo, Y.K., and Soh, Y.C., 1990, A generalization of strong Kharitonov theorems to polytopes of polynomials, *IEEE-Trans.* **AC-35**, pp. 936-939

Frame, J.S., 1964, Matrix functions and applications, Part IV-Matrix functions and constituent matrices, *IEEE Spectrum* **1**, pp. 123-131

Francis, B.A., 1987, A Course in H_∞ Control Theory, (Springer, Berlin)

Francis, B.A., Helton, J.W., and Zames, G., 1984, H^∞-Optimal feedback controllers for linear multivariable systems, *IEEE-Trans.* **AC-29**, pp. 888-899

Frank, P.M., 1978, Introduction to System Sensitivity Theory, (Academic Press, New York)

Frank, P.M., 1985, Entwurf parameterunempfindlicher und robuster Regelkreise im Zeitbereich - Definitionen, Verfahren und ein Vergleich *Automatisierungstechnik* **33**, pp. 233-240

Franke, D., 1982a, Strukturvariable Regelung ohne Gleitzustände, *Regelungstechnik* **30**, pp. 271-276

Franke, D., 1982b, Ausschöpfen von Stellgrößenbeschränkungen mittels weicher strukturvariabler Regelung, *Regelungstechnik* **30**, pp. 348-355

Franklin, G.F., Powell, J.D., and Emami-Naeini, A., 1988, Feedback Control of Dynamic Systems, (Addison-Wesley, Reading, Mass.)

Franklin, J.N., 1968, Matrix Theory, (Prentice-Hall, Englewood Cliffs)

Freeman, E.A., 1972, Some control system stability and optimality results obtained via functional analysis. In *Bell, D.J., (Ed.) 1973*, pp.45-68

Freudenberg, J.S., 1989, Analysis and design for ill- conditioned plants Part 1. Lower bounds on the structured singular value, *Int.J.Control* **49**, pp. 851-871

Freudenberg, J.S., Looze, D.P., and Cruz, J.B., 1982, Robustness analysis using singular value sensitivities, *Int.J.Control* **35**, pp. 95-116

Freudenberg, J.S., and Looze, D.P., 1985, Right half plane poles and zeros and design tradeoffs in feedback systems, *IEEE-Trans.* **AC-30**, pp. 555-565

Fu, L.C., and Liao, T.L., 1990, Globally stable robust tracking of nonlinear systems using variable structure control and with an application to robotic manipulator, *IEEE-Trans.* **AC-35**, pp. 1345-1350

Fu, M., 1989, Polytopes of polynomials with zeros in a prescribed region: new criteria and algorithms In: *Milanese, M., Tempo, R., and Vicino, A., (Eds.) 1989* Robustness in Identification and Control (Plenum Press, New York London) pp. 125-145

Fu, M., and Barmish, B.R., 1987, Stability of convex and linear combinations of polynomials and matrices arising in robustness problems, *Proc. Conference on Information Science Systems, Baltimore,* pp. 16-21

Fu, M., and Barmish, B.R., 1988, Maximal unidirectional perturbation bounds for stability of polynomials and matrices, *Systems and Control Letters,* **11**, pp. 173-179

Fu, M., and Barmish, B.R., 1988a, Polytopes of polynomials with zeros in a prescribed region, *Proc. American Control Conf., Atlanta,* pp. 2461-2464

Furuta, K., and Kim, S.B., 1987, Pole assignment in a specified disk, *IEEE-Trans.* **AC-32**, pp. 423-427

Gahinet, P.M., Laub, A.J., Kenney, C.S., and Hewer, G.A., 1990, Sensitivity of the stable discrete-time Lyapunov equation, *IEEE-Trans.* **AC-35**, pp. 1209-1217

Gajic, Z., Petkovski, D., and Shen, X., 1990, Singularly Perturbed and Weakly Coupled Linear Control Systems, (Springer, Berlin New York)

Ganesh, C., and Pearson, J.B., 1989, H^2-Optimization with stable controllers, *Automatica* **25**, pp. 629-634

Gantmacher, F.R., 1986, Matrizentheorie (Springer, Berlin)

Garnett, J.B., 1981, Bounded Analytic Functions (Academic Press, New York)
deGaston, R.R.E., and Safonov, M.G., 1986, Calculation of the multiloop stability margin, *Proc. American Control Conference* pp. 761-770
deGaston, R.R.E., and Safonov, M.G., 1988, Exact calculation of the multiloop stability margin, *IEEE-Trans.* **AC-33**, pp. 156-171
Ge, W., and Fang, C.Z., 1988, Detection of faulty components via robust observation, *Int.J.Control* **47**, pp. 581-599
Genesio, R., and Tesi, A., 1988, Feedback of SISO bilinear systems, *Int.J. Control* **48**, pp. 1319-1326
Georgiou, T.T., and Smith, M.C., 1990, Optimal robustness in the gap metric, *IEEE-Trans.* **AC-35**, pp. 673-686, and *11th IFAC-Congress Tallinn*, Vol. **5**, pp. 170-174
Geromel, J.C., and Bernussou, J., 1979, An algorithm for optimal decentralized regulation of linear quadratic interconnected systems, *Automatica* **15**, pp. 489-491 Stability of two-level control schemes subjected to structural perturbations *Int.J.Control* **29**, pp. 313-324
Geromel, J.C., and Bernussou, J., 1982, Optimal decentralized control of dynamic systems, *Automatica* **18**, pp. 545-557
Ghosh, B.K., 1986, Transcendental and interpolation methods in simultaneous stabilization and simultaneous partial pole placement problems, *SIAM J.Control and Optimization* **24**, pp. 1091-1109
Gilbert, E.G., 1984, Conditions for minimizing the norm sensitivity of characteristic roots, *IEEE-Trans.* **AC-29**, pp.658-661
Giri, F., M'Saad, M., Dion, J.M., and Dugard, L., 1991, On the robustness of discrete-time indirect adaptive (linear) controllers, *Automatica* **27**, pp. 153-159
Glad, S.T., 1987, Robustness of nonlinear state feedback - a survey, *Automatica* **23**, pp. 425-435
Glover, K., 1984, All optimal Hankel-norm approximations of linear multivariable systems and their L^{∞}-error bounds, *Int.J.Control* **39**, pp. 1115-1193
Glover, K., and Doyle, J.C., 1988, State-space formulae for all stabilizing controllers that satisfy an H_{∞}-norm bound and relations to risk sensitivity, *Systems and Control Letters* **11**, pp. 167-172
Glover, K., and McFarlane, D., 1989, Robust stabilization of normalized coprime factor plant description with H_{∞}-bounded uncertainty, *IEEE-Trans.* **AC-34**, pp. 821-830
Glover, K., Sefton, J., and McFarlane, D.C., 1990, A tutorial on loop shaping using H-infinity robust stabilization, *11th IFAC-Congress Tallinn*, Vol. **5**, pp. 94-103
Gnanasekaran, R., 1981, A note on the new 1-D and 2-D stability theorems for discrete systems, *IEEE-Trans.* **ASSP-29**, pp. 1211-1212
Golub, G.H., and Van Loan, C.F., 1990, Matrix Computations (John Hopkins University Press, Baltimore)
Graham, A., 1981, Kronecker Products and Matrix Calculus with Applications (Wiley, New York)
Grenander, U., and Szegö, G., 1958, Toeplitz Forms and Their Applications (Univ. of California Press, Berkeley Los Angeles)
Grimble, M.J., 1988, State-space approach to H_{∞} optimal control, *Int. J. Systems Sci.* **19**, pp. 1451-1468
Grimble, M.J., 1989, Extensions to H_{∞} multivariable robust controllers and the relationship to LQG design, *Int.J.Control* **50**, pp. 309-338
Grimble, M.J., 1991, Polynomial matrix solution to the standard H_2 optimal control problem, *Int. J. Systems Sci.* **22**, pp. 793-806
Gu, K., Chen, Y.H., Zohdy, M.A., and Loh, N.K., 1991, Quadratic stabilizability of uncertain systems: a two level optimization setup, *Automatica* **27**, pp. 161-165
Gu, K., Zohdy, M.A., and Loh, N.K., 1990, Necessary and sufficient conditions of quadratic stability of uncertain linear systems, *IEEE-Trans.* **AC-35**, pp. 601-604
Guiver, J.P., and Bose, N.K., 1983, Strictly Hurwitz property invariance of quartics under coefficient perturbation, *IEEE-Trans.* **AC-28**, pp. 106-107
Gündes, A.N., and Desoer, C.A., 1990, Algebraic Theory of Linear Feedback Systems with Full and Decentralized Compensators, (Springer, Berlin New York)
Günther, M., 1986, Zeitdiskrete Steuerungssysteme (Verlag Technik, Berlin)
Gutman, S., and Jury, E.I., 1981, A general theory for matrix root-clustering in subregions of the complex plane, *IEEE-Trans.* **AC-26**, pp. 853-863
Hadzer, C.M., Asmah, N., and Saleh, Y., 1989, Comments on "Discrete pulse orthogonal functions ...", *Int.J.Control* **50**, pp. 2097-2100
Harmuth, H.F., 1969, Transmission of Information by Orthogonal Functions (Springer, Berlin Heidelberg New York)

Heck, B.S., 1991, Sliding-mode for singularly perturbed systems, *Int. J. Control* **53**, pp. 985-1001

Heinen, J.A., 1984, Sufficient conditions for stability of interval matrices, *Int. J. Control* **39**, pp. 1323-1328

Heitzinger, W., Troch, I., und Valentin, G., 1985, Praxis nichtlinearer Gleichungen (Hanser, München)

Helton, J.W., 1983, An H_∞ approach to control, *Proc. IEEE-Conference on Decision and Control, San Antonio, Texas*, pp. 607-611

Hinrichsen, D., Kelb, B., and Linnemann, A., 1989, An algorithm for the computation of the structured complex stability radius, *Automatica* **25**, pp. 771-775

Hinrichsen, D., and Pritchard, A.J., 1986, Stability radii of linear systems, *Systems and Control Letters* **7**, pp. 1-10

Hinrichsen, D., and Pritchard, A.J., 1986a, Stability radius for structured perturbations and the algebraic Riccati equation, *Systems and Control Letters* **8**, pp. 105-113

Hinrichsen, D., and Pritchard, A.J., 1988, New robustness results for linear systems under real perturbations, *Proc.27th IEEE Conference on Decision and Control*, pp. 1375-1379

Hinrichsen, D., and Pritchard, A.J., 1989, An application of state-space methods to obtain explicit formulae for robustness measures of polynomials, In: *Milanese, M., Tempo, R., and Vicino, A., (Eds.) 1989,* Robustness in Identification and Control (Plenum Press, New York London) pp. 183-206

Hinrichsen, D., Pritchard, A.J., and Townley, S.B., 1990, Riccati equation approach to maximizing the complex stability radius by state feedback, *Int. J. Control* **52**, pp. 769-794

Hippe, P., and Wurmthaler, C., 1987, Frequenzbereichsentwurf für optimale Zustandsregelungen, *Automatisierungstechnik* **35**, pp. 317-322

Hmamed, A., 1989, A matrix inequality, *Int.J.Control* **49**,pp. 363-365

Hmamed, A., 1991, Further results on the robust stability of uncertain time-delay systems, *Int. J. Systems Sci.* **22**, pp.605-614

Hmamed, A., 1991a, Sufficient condition for the positive definiteness of symmetric interval matrices, *Int. J. Systems Sci.* **22**, pp.615-619

Ho, W.C., and Fletcher, L.R., 1988, Perturbation theory of output feedback pole assignment, *Int.J.Control* **48**,pp. 1075-1088

Hoerl, A.E., 1962, Application of ridge analysis to regression problems, *Chem. Eng. Progr.* **58**, pp. 54-59

Hoerl, A.E., and Kennard, R.W., 1970, Ridge regression: biased estimation for nonorthogonal problems, *Technometrics* **12**, No.1, pp. 55-67

Hollot, C.V., 1989, Markov's theorem of determinants and the stability of families of polynomials, In: *Milanese, M., Tempo, R., and Vicino, A., (Eds.) 1989* Robustness in Identification and Control (Plenum Press, New York London) pp. 165-181

Hollot, C.V., and Bartlett, A.C., 1986, Some discrete-time counterparts to Kharitonov's stability criterion for uncertain systems, *IEEE-Trans.* **AC-31**, pp. 355-356

Horn, R.A., and Johnson, C.R., 1990, Matrix Analysis (Cambridge University Press, Cambridge London)

Horng, I.R., and Chou, J.H., 1987, Legendre series for the identification of non-linear lumped systems *Int. J. Systems Sci.* **18**, pp.1139-1144

Horowitz, I. M., 1963, Synthesis of Feedback Systems, (Academic Press, New York)

Householder, A.S., 1964, The Theory of Matrices in Numerical Analysis (Dover, New York)

Hsiao, F.B., and Lin, C.L., 1991, Model matching systems: necessary and sufficient conditions for stability under plant perturbations, *Int. J. Systems Sci.* **22**, pp. 905-920

Hsieh, J.G., and Hsiao, F.H., 1989, Robust controller synthesis in nonlinear multivariable systems: using dither as auxiliary, *Int.J.Systems Sci.* **20**, pp. 2515-2537

Hsu, J.C., and Meyer, A.U., 1968, Modern Control Principles and Applications (McGraw-Hill, New York)

Hwang, C., and Guo, T.Y., 1984, Parameter identification of a class of time-varying systems via orthogonal shifted legendre polynomials, *J. Franklin Inst.* **318**, pp. 59-69

Hwang, C., and Shih, Y.P., 1982 Laguerre series solutions of a functional differential equation, *Int.J.Systems Sci.* **13**, pp. 783-788

Hwang, C., and Shih, Y.P., 1982a, Parameter identification via Laguerre polynomials, *Int. J. Systems Sci.* **13**, pp. 209-217

Iglesias, P.A., 1990, On the use of robust regulators in adaptive control *11th IFAC-Congress Tallinn,* Vol. **4**, pp. 214-219

Isermann, R., 1981, Digital Control Systems (Springer, Berlin New York)

Ishihara, T., 1988, Robust stability bounds for a class of discrete-time regulators with computation delays, *Automatica* **24**, pp. 697-700

Jakoby, W., 1985, Diskrete adaptive Regler-Versuch einer Einordnung, *Automatisierungstechnik* **33**, pp. 6-14

Jayasuriya, S., 1987, Robust tracking of a class of uncertain linear systems, *Int. J.Control* **45**, pp. 875-892

Jayasuriya, S., Rabins, M.J., and Barnard, R.D., 1984, Guaranteed tracking behaviour in the sense of input-output spheres for systems with uncertain parameters, *J.Dynamic Syst.Meas.Control* **106**, pp. 273-279

Jiang, C.L., 1987, Sufficient condition for the asymptotic stability of interval matrices, *Int.J.Control* **46**, pp. 1803-1810

Jiang, C.L., 1988, Another sufficient condition for the stability of interval matrices, *Int.J.Control* **47**, pp. 181-186

Jiang, C.L., 1988a, Sufficient and necessary condition for the asymptotic stability of discrete linear interval systems, *Int.J.Control* **47**, pp. 1563-1565

Jin Lu, Chiang, H.D., and Thorp, J.S., 1991, Partial eigenstructure assignment and its application to large-scale systems, *IEEE-Trans.* **AC-36**, pp.340-347

Johnson, C.D., 1988, Optimal initial conditions for full-order observers, *Int.J.Control* **48**, pp.857-864

Johnston, R.D., and Barton, G.W., 1984, Improved process conditioning using internal control loops, *Int.J.Control* **40**, pp. 1051-1063

Johnston, R.D., and Barton, G.W., 1987, Control system performance robustness, *Int.J. Control* **45**, pp. 641-648

Juang, Y.T., 1989, Eigenvalue assignment robustness: the analysis for autonomous system matrices, *Int.J.Control* **49**, pp. 1787-1797

Juang, Y.T., Kuo, T.S., and Hsu, C.F., 1987, Stability robustness analysis of digital control systems in state-space models, *Int.J.Control* **46**, pp. 1547-1556

Juang, Y.T., Kuo, T.S., Hsu, C.F., and Wang, S.D., 1987, Root-locus approach to the stability analysis of interval matrices, *Int.J.Control* **46**, pp. 817-822

Juang, Y.T., and Shao, C.S., 1989, Stability analysis of dynamic interval systems, *Int.J.Control* **49**, pp. 1401-1408

Juang, Y.T., Tung, S.L., and Ho, T.C., 1989, Sufficient condition for asymptotic stability of discret interval systems, *Int.J.Control* **49**, pp. 1799-1803

Jury, E.I., 1990, Robustness of discrete systems: a review, *11th IFAC-Congress Tallinn,* Vol. **5**, pp. 184-189

Jury, E.I., and Pavlidis, T., 1963, Stability and aperiodicity constraints for control system design, *IEEE-Trans.* **CT-10**, pp. 137-141

Jury, E.I., and Tsypkin, Y.Z., 1971, On the theory of discrete systems, *Automatica* **7**, pp. 89-107

Kaesbauer, D., and Ackermann, J., 1990, The distance from stability of Γ-stability boundaries, *11th IFAC-Congress Tallinn,* Vol. **5**, pp. 130-134

Kalman, R.E., 1963, Mathematical description of linear dynamical systems, *SIAM Journal Control Ser.A* **1**, No. 2, pp. 152-192

Kalman, R.E., 1964, When is a linear control system optimal, *Trans.ASME, Ser.D., J.Basic Eng.* **86**, pp. 51-60

Kalman, R.E., and Bertram, J.E., 1960, Control system analysis and design via the "second method" of Lyapunov *Trans. ASME - Series D -J. of Basic Engineering* **82**, pp. 371-400

Kando, H., Iwazumi, T., and Ukai, H., 1988, Singular perturbation modelling of large-scale systems with multi-time-scale property, *Int.J.Control* **48**, pp. 2361-2387

Karl, W.C., Gzeschak, J.P., and Verghese, G.C., 1984, Comments on "A necessary and sufficient condition for the stability of interval matrices", *Int.J.Control* **39**, pp. 849-851

Katbab, A., and Jury, E.I., 1990, Robust Schur stability of control systems with interval plants, *Int.J.Control* **51**, pp. 1343-1352

Katbab, A., and Jury, E.I., 1990a, On the strictly positive realness of Schur interval functions, *IEEE-Trans.* **AC-35**, pp. 1382-1385

Katbab, A., and Jury, E.I., 1991, Generalization and comparison of two recent frequency-domain stability robustness results, *Int. J. Control* **53**, pp. 463-475

Kato, T., 1976, Perturbation Theory of Linear Operators, 2nd edition, (Springer, New York)
Kautsky, J., Nichols, N.K., and Van Dooren, P., 1985, Robust pole assignment in linear state feedback, *Int.J.Control* **41**, pp. 1129-1155
Kay, S.M., 1988, Modern Spectral Estimation: Theory and Application (Prentice Hall, Englewood Cliffs)
Khargonekar, P.P., Petersen, I.R., and Zhou, K., 1990, Robust stabilization of uncertain linear systems: quadratic stabilizability and H^∞, *IEEE-Trans.* **AC-35**, pp. 356-361
Kharitonov, V.L., 1979, Asymptotic stability of an equilibrium position of a family of systems of linear differential equations, *Differential Equations* **14**, pp. 1483-1485
Kiendl, H., 1985, Totale Stabilität von linearen Regelungssystemen bei ungenau bekannten Parametern der Regelstrecke, *Automatisierungstechnik* **33**, pp. 379-386
Kimura, H., 1984, Robust stabilizability for a class of transfer functions, *IEEE-Trans.* **AC-29**, pp. 788-793
Kimura, H., 1989, Conjugation, interpolation and model-matching in H^∞, *Int.J.Control* **49**, pp. 269-307
Kitamori, T., 1960, Applications of orthogonal functions to the determination of process dynamic characteristics and to the construction of self-optimizing control systems, *1st IFAC Congress Moscow*, pp. AB 1-AB 5 (pp.82-86)
Kleinman, D.L., and Athans, M., 1968, The design of suboptimal linear time-varying systems, *IEEE-Trans.* **AC-13**, pp. 150-159
Kleinman, D.L., Fortmann, T., and Athans, M., 1968, On the design of linear systems with piecewise-constant feedback gains, *IEEE-Trans.***AC-13**, pp. 354-361
Klema, V.C., and Laub, A.J., 1980, The singular value decomposition: Its computation and some applications, *IEEE-Trans.* **AC-25**, pp. 164-176
Knobloch, H.W., und Kappel, F., 1974, Gewöhnliche Differentialgleichungen (Teubner, Stuttgart)
Knobloch, H.W., und Kwakernaak, H., 1985, Lineare Kontrolltheorie (Springer, Berlin Heidelberg)
Kofahl, R., 1988, Robuste parameteradaptive Regelungen, (Springer, Berlin Heidelberg)
Kokame, H., and Mori, T., 1991, A root distribution criterion for interval polynomials, *IEEE-Trans.* **AC-36**, pp.362-364
Kokotovics, P.V., O'Malley, R.E.Jr., and Sannuti, P., 1976, Singular perturbations and order reduction in control theory - an overview, *Automatica* **12**, pp. 123-132
Kokotovics, P.V., Khalil, H.K., and O'Reilly, J., 1986, Singular Perturbation Methods in Control: Analysis and Design (Academic Press, London)
Kolla, S.R., and Farison, J.B., 1990, Improved robust stability bounds for discrete-time regulators with computational delays, *Automatica* **26**, pp. 619-621
Kolla, S.R., and Farison, J.B., 1990a, Improved stability robustness bounds using state transformation for linear discrete systems, *Automatica* **26**, pp. 933-935
Kolla, S.R., Yedavalli, R.K., and Farison, J.B., 1989, Robust stability bounds on time-varying perturbations for state-space models of linear discrete-time systems, *Int.J.Control* **50**, pp. 151-159
Konigorski, U., 1987, Entwurf strukturbeschränkter Zustandsregelungen unter besonderer Berücksichtigung des Störverhaltens, *Automatisierungstechnik* **35**, pp. 457-463
Konigorski, U., 1988, Die Nullstellen der Einzelübertragungsfunktionen linearer Mehrgrößensysteme, *Automatisierungstechnik* **36**, pp.353-354
Korn, U., and Wilfert, H.H., 1982, Mehrgrößenregelungen (VEB Verlag Technik, Berlin und Springer, Wien New York)
Kosmidou, O.I., 1990, Robust stability and performance of systems with structured and bounded uncertainties: an extension of the guaranteed cost control approach *Int.J.Control* **52**, pp. 627-640
Kouvaritakis, B., and Latchman, H., 1985, Singular-value and eigenvalue techniques in the analysis of systems with structured perturbations, *Int.J.Control* **41**, pp. 1381-1412
Kouvaritakis, B., and Trimboli, M.S., 1989, Robust multivariable feedback design, *Int.J.Control* **50**, pp. 1327-1377
Kraus, F.J., Anderson, B.D.O., and Mansour, M., 1988, Robust Schur polynomial stability and Kharitonov's theorem, *Int.J.Control* **47**, pp. 1213-1225
Kraus, F.J., Anderson, B.D.O., Jury, E.I. and Mansour, M., 1988a, On the robustness of low-order Schur polynomials, *IEEE-Trans.* **CS-35**, pp. 570-577
Kraus, F.J., Mansour, M., 1987, Robuste Stabilität von zeitkontinuierlichen und zeitdiskreten Systemen, *SGA-Zeitschrift* **7**, pp. 20-25

Kraus, F.J., and Mansour, M., 1990, On robust stability of discrete systems, *Proc. 29th IEEE Conf. on Decision and Control, Honolulu*

Kraus, F.J., Mansour, M., and Jury, E.I., 1990, Robust Schur-stability of interval polynomials

Kraus, F.J., and Schaufelberger, W., 1990, Identification with block pulse functions, modulating functions and differential operators, *Int.J.Control* **51**, pp. 931-942

Kraus, F.J., and Truöl, W., 1991, Robust stability of control systems with polytopical uncertainty: a Nyquist approach, *Int. J. Control* **53**, pp. 967-983

Krause, J.M., Khargonekar, P.P., and Stein, G., 1990, Robust parameter adjustment with nonparametric weighted-ball-in-H^∞ uncertainty, *IEEE-Trans.* **AC-35**, pp. 225-229

Kreindler, E., 1968 On minimization of trajectory sensitivity, *Int.J.Control* **8**, pp. 89-96

Kreisselmeier, G., 1986, A robust indirect adaptive-control approach, *Int.J.Control* **43**, pp. 161-175

Kuhn, U., and Schmidt, G., 1987, Fresh look into the design and computation of optimal output feedback controls for linear multivariable systems, *Int.J.Control* **46**, pp. 75-95

Kwakernaak, H., 1985, Uncertainty models and the design of robust control systems, In: Ackermann, J., (Ed.), 1985, Uncertainty and Control (Springer, Berlin New York)

Kwakernaak, H., 1985a, Minimax frequency domain performance and robustness optimization of linear feedback systems, *IEEE-Trans.* **AC-30**, pp. 994-1004

Kwakernaak, H., 1990, Progress in the polynomial solution of the standard H_∞ optimal control problem, *11th IFAC-Congress Tallinn,* Vol. **5**, pp. 122-129

Kwakernaak, H., and Sivan, R., 1972, Linear Optimal Control Systems (Wiley Interscience, New York)

Kwon, B.H., Youn, M.J., and Bien, Z., 1985, On bounds of the Riccati and Lyapunov matrix equations, *IEEE-Trans.* **AC-30**, pp. 1134-1135

Lachmann, K.H., 1985, Selbsteinstellende nichtlineare Regelalgorithmen für eine bestimmte Klasse nichtlinearer Prozesse, *Automatisierungstechnik* **33**, pp. 210-218

Lancaster, P., and Tismenetsky, M., 1985 The Theory of Matrices, 2nd edition (Academic Press, New York London)

Leal, M.A., and Gibson, J.S., 1990, A first-order Lyapunov robustness method for linear systems with uncertain parameters, *IEEE-Trans.* **AC-35**, pp. 1068-1070

Lee, G.K.F., 1981, Output feedback system design via inverse systems, *Int.J.Control* **34**, pp. 1125-1141

Lee, T.T., and Lee, S.H., 1987, Root clustering in subregions of the complex plane, *Int.J. Systems Sci.* **18**, pp. 117-129

Lee, T.T., and Tsay, S.C., 1989, Analysis, parameter identification and optimal control for time-varying systems via general orthogonal polynomials, *Int.J.Systems Sci.* **20**, pp. 1451-1465

Lee, W.H., 1982, Robustness analysis for state space models, *Alphatech Inc. Report* **TP-151**

Lee, W.H., Gully, S.W., Eterno, J.S, and Sandell, N.R., Jr., 1982, Structured information in robust analysis, *Proc. American Control Conference* pp. 1040-1045

Lefebvre, S., De Carlo, R.A., and Carroll, D.P., 1982, Decentralized eigenvalue assignment in two-terminal HVDC systems, *Int. J. Control* **35**, pp. 427-447

Lehtomaki, N.A., Castanon, D.A., Levy, B.C., Stein, G., Sandell, N.R.,Jr., Athans, M., 1984, Robustness and modelling error characterization, *IEEE-Trans.* **AC-29**, pp. 212-220

Lehtomaki, N.A., Sandell, N.R., Jr., and Athans, M., 1981, Robustness results in linear-quadratic Gaussian based multivariable control designs, *IEEE-Trans.* **AC-26**, pp. 75-92

Leithead, W.E., and O'Reilly, J., 1991, Uncertain SISO systems with fixed stable minimum-phase controllers: relationship of closed-loop systems to plant RHP poles and zeros, *Int. J. Control* **53**, pp. 771-798

Leitmann, G., Ryan, E.P., and Steinberg, A., 1986, Feedback control of uncertain systems: robustness with respect to actuator and sensor dynamics, *Int.J.Control* **43**, pp. 1243-1256

Leondes, C.T., and Novak, L.M., 1972, Optimal minimal-order observers for discrte-time systems - a unified theory, *Automatica* **8**, pp. 379-387

Leung, T.P., Zhou, Q.J., and Su, C.Y., 1991, An adaptive variable structure model following control design for robot manipulators, *IEEE-Trans.* **AC-36**, pp.347-353

Levine, W.S., and Athans, M., 1970 , On the determination of the optimal constant output feedback gains for linear multivariable systems, *IEEE-Trans.***AC-15**, pp. 44-48

Lewis, F.L., and Fountain, D.W., 1991, Walsh function analysis of linear and bilinear discrete-time systems, *Int. J. Control* **53**, pp. 847-853

Lewis, F.L., Mertzios, V.G., Vachtsevanos, G., and Christodoulou, M.A., 1990, Analysis of bilinear systems using Walsh functions, *IEEE-Trans.* **AC-35**, pp. 119-123

Limebeer, D.J.N., 1982, The application of generalized diagonal dominance to linear systems stability theory, *Int.J.Control* **36**, pp. 185-212

Limebeer, D.J.N., and Hung, Y.S., 1987, An analysis of the pole-zero cancellations in H_∞-optimal control problems of the first kind, *SIAM J. on Control and Optimization* **25**, pp. 1457-1493

Lin, C.L., Hsiao, F.B., and Chen, B.S., 1990, Stabilization of large structural systems under mode truncation, parameter perturbations and actuator saturations, *Int.J.Systems Sci.* **21**, pp. 1423-1440

Lin, C.M., and Chou, W.D., 1989, Robust optimization of feedback systems with time-varying nonlinear uncertainties, *Int.J.Systems Sci.* **20**, pp. 1011-1018

Lin, S.H., Fong,I.K., Juang, Y.T., Kuo, T.S., and Hsu, C.F., 1989, Stability of perturbed polynomials based on the argument principle and Nyquist criterion, *Int.J.Control* **50**, pp. 55-63

Lin, S.H., Juang, Y.T., Fong, I.K., Hsu, C.F., and Kuo, T.S., 1988, Dynamic interval systems analysis and design, *Int. J. Control* **48**, pp. 1807-1818

Liou, C.T., and Chou, Y.S., 1987, Operational matrices of piecewise linear polynomial functions with application to linear time-varying systems, *Int.J.Systems Sci* **18**, pp. 1931-1942 and Solution of stiff dynamical systems via piecwiese-linear polynomial functions, *Int.J.Systems Sci.* **18**, pp. 2349-2357, Inverse Laplace transform by piecewise linear polynomial functions, *Int.J.Systems Sci.***18**, pp. 749-754

Litz, L., 1979, Reduktion der Ordnung linearer Zustandsraummodelle mittels modaler Verfahren (Hochschul-Verlag, Stuttgart)

Litz, L., and Preuss, H.P., 1977, Bestimmung einer Ausgangsvektorrückführung mittels Frequenzbereichsmethoden, *Regelungstechnik* **25**, pp.119-126

Liu, G.P., 1990, Frequency-domain approach for critical systems, *Int. J. Control* **52**, pp. 1507-1519

Ljung, L., 1985, System identification, In: *Ackermann, J., (Ed.)*: Uncertainty and Control. Proceedings of the Seminar "Uncertainty and Control", Bonn, (Springer, Berlin New York)

Ljusternik, L.A., and Sobolew, W.I., 1979, Elemente der Funktionalanalysis, 6. Auflage (H.Deutsch, Thun)

Looze, D.P., Poor, H.V., Vastola, K.S., and Darragh, J.C., 1983, Minimax control of linear stochastic systems with noise uncertainty *IEEE-Trans.* **AC-28**, pp. 882-888

Ludyk, G., 1977, Theorie dynamischer Systeme (Elitera, Berlin)

Ludyk, G., 1985, Stability of Time-variant Discrete-time Systems, (Vieweg, Braunschweig)

Ludyk, G., 1990, CAE von Dynamischen Systemen (Springer, Berlin New York)

MacCluer, C.R., and Chait, Y., 1990, Obtaining robust stability operationally, *IEEE-Trans.* **AC-35**, pp. 1350-1351

MacFarlane, A.G.J., and Kouvaritakis, B., 1977, A design technique for linear multivariable feedback systems, *Int.J.Control* **25**, pp. 837-874

MacFarlane, A.G.J., and Postlethwaite, I., 1977, Characteristic frequency functions and characteristic gain functions, *Int.J.Control* **26**, pp. 265-278

Maciejowski, J.M., 1989 Multivariable Feedback Design, (Addison-Wesley, Wokingham)

Mahalanabis, A.K., and Singh, R., 1980, On decentralized feedback stabilization of large-scale interconnected systems, *Int. J. Control* **32**, pp. 115-126

Makhlouf, M.A., 1972, Extension of existing knowledge of the theory of multivariable systems in z domain by using the theory of multivariable infinite matrices. In *Bell, D.J., (Ed.) 1973*, pp.405-420

Makhoul, J., 1981, On the eigenvectors of symmetric Toeplitz matrices, *IEEE-Trans.* **ASSP-29**, pp. 868-872

Mansour, M., 1987, Comment on "Stability of interval matrices", *Int.J.Control* **46**, pp. 1845

Mansour, M., 1988, Sufficient conditions for the asymptotic stability of interval matrices, *Int.J.Control* **47**, pp. 1973-1974

Mansour, M., 1989, Simplified sufficient conditions for the asymtotic stability of interval matrices, *Int.J.Control* **50**, pp. 443-444

Mansour, M., 1989a, Robust stability of interval matrices

Mansour, M., and Kraus, F.J., 1989, Robust stability of systems with nonlinear parameter dependence, *Proc. 28th IEEE Conf. on Decision and Control, Tampa, Florida*, pp. 1931-1933

Mansour, M., Kraus, F.J., and Anderson, B.D.O., 1989, Strong Kharitonov theorem for discrete systems, In: *Milanese, M., Tempo, R., and Vicino, A., (Eds.) 1989* Robustness in Identification and Control (Plenum Press, New York London) pp.

Martin, J.M., 1987, State-space measures for stability robustness, *IEEE-Trans.* **AC-32**, pp. 509-512

Martin, J.M., and Hewer, G.A., 1987, Smallest destabilizing perturbations for linear systems, *Int.J.Control* **45**, pp. 1495-1504
Martin, R.H., Jr., 1976, Nonlinear Operators and Differential Equations in Banach Spaces (Wiley, New York)
Matthews, G.P., and De Carlo, R.A., 1988, Decentralized tracking for a class of interconnected nonlinear systems using variable structure control, *Automatica* **24**, pp. 187-193
Mayer, G., 1984, On the convergence of powers of interval matrices, *Linear Algebra and its Applications* **58**, pp. 201-216
McFarlane, D.C., and Glover, K., 1990, Robust Control Design Using Normalized Coprime Factor Plant Description (Springer, Berlin New York)
Mendel, J.M., 1973, Discrete Techniques of Parameter Estimation (Marcel Dekker, New York)
Metzler, L.A., 1950, A multiple-region theory of income and trade, *Econometrica* **18**, pp. 329-354
Milanese, M., and Negro, A., 1973, Uniform approximation of systems: A Banach space approach, *J. Optimization Theory Applications* **12**, pp. 203-217
Milanese, M., Tempo, R., and Vicino, A., (Eds.) 1989, Robustness in Identification and Control, (Plenum Press, New York London)
Mita, T., 1985, Optimal digital feedback control systems counting computation time of control laws, *IEEE-Trans.* **AC-30**, pp. 542-548
Mohan, B.M., and Datta, K.B., 1988, Analysis of time-delay systems via shifted polynomials of first and second kind, *Int.J.Systems Sci.* **19**, pp. 1843-1851
Mohler, R.R., 1973, Bilinear Control Processes, (Academic Press, New York)
Morari, M., 1983, Flexibility and resiliency of process systems, *Computers & Chemical Engineering* **7**, pp. 423-437
Morari, M., 1983a, Design of resilient processing plants-III, *Chem.Eng.Sci.* **38**, pp. 1881-1891
Mori, T., 1980, *Trans, Inst. Commun. Engng.* **63**, pp. 38
Mori, T., 1989, Estimates for a measure of stability robustness via a Lyapunov matrix equation, *Int.J.Control* **50**, pp. 435-438
Mori, T., Fukuma, N., and Kuwahara, M., 1981, Simple stability criteria for single and composite linear systems with time delays, *Int.J.Control* **34**, pp. 1175-1184
Mori, T., Fukuma, N., and Kuwahara, M., 1987, Bounds in the Lyapunov matrix differential equation, *IEEE-Trans.* **AC-32**, pp. 55 57
Mori, T., and Kokame, H., 1986, A necessary and sufficient condition for stability of linear discrete systems with parameter-variation, *J. of the Franklin Institute* **321**, pp. 135-138
Mori, T., and Kokame, H., 1987, Convergence property of interval matrices and interval polynomials, *Int.J.Control* **45**, pp. 481-484
Mossaheb, S., 1983, Application of a method of averaging to the study of dithers in nonlinear systems, *Int.J.Control* **38**, pp. 557-576
Motscha, M., 1988, Algorithm to compute the complex stability radius, *Int.J. Control* **48**, pp. 2417-2428
Müller, P.C., and Lückel, J., 1977, Zur Theorie der Störgrößenaufschaltung in linearen Mehrgrößenregelsystemen, *Regelungstechnik* **25**, pp. 54-59
Mustafa, D., Glover, K., and Limebeer, D.J.N., 1991, Solutions of the H_∞ general distance problem which minimize an entropy integral, *Automatica* **27**, pp. 193-199
Nagorski, U., and Schmidt, J., 1986, Das "invariante Polprodukt"- ein Hilfsmittel bei der stationär genauen Entkopplung, *Automatisierungstechnik* **34** pp. 498-499
Sz.-Nagy, B., and Foias, C., 1970, Harmonic Analysis of Operators on Hilbert Space (North-Holland, Amsterdam London)
Newmann, M.M., 1970, On attempts to reduce the sensitivity of the optimal linear regulator to a parameter change, *Int.J.Control* **11**, pp. 1079-1084
Niemann, H.H., Sogaard-Andersen, P., and Stroustrup, J., 1991, Loop transfer recovery for general observer architectures, *Int. J. Control* **53**, pp. 1177-1203
Noble, B., and Daniel, J.W., 1988, Applied Linear Algebra, (Prentice-Hall, Englewood Cliffs)
Noisser, R., 1976, Verbesserung des Störungs- und Führungsverhaltens durch eine synthetische Störgrößenaufschaltung, *Regelungstechnik* **24**, pp. 96-101
Noisser, R., 1988, Zusammenhang zwischen Polvorgabe und Stabilitätsreserve bei Mehrgrößenregelungen mittels Kontrollbeobachter, *Automatisierungstechnik* **36**, pp. 339-347
Noldus, E., 1982, Design of robust state feedback laws, *Int.J.Control* **35**, pp. 935-944

Nour Eldin, H.A., 1971, Ein neues Stabilitätskriterium für abgetastete Regelsysteme, *Regelungstechnik* **19**, pp.301-307

Ogata, K., 1970, Modern Control Engineering (Prentice-Hall, Englewood Cliffs London)

Ogata, K., 1987, Discrete-Time Control Systems (Prentice-Hall, Englewood Cliffs)

Ohm, D.Y., Howze, J.W., and Bhattacharyya, S.P., 1985, Structural synthesis of multivariable controllers, *Automatica* **21**, pp. 35-55

Okada, T., Kihara, M., and Furihata, H., 1985, Robust control system with observer, *Int.J.Control* **41**, pp. 1207-1219

Okada, T., Kihara, M., and Nishio, Y., 1988, Reduced conservative singular value analysis for robustness, *Int.J.Control* **48**, pp. 1455-1473

O'Reilly, J., 1980, Dynamical feedback control for a class of singularly perturbed linear systems using a full-order observer, *Int. J. Control* **31**, pp. 1-10

O'Reilly, J., 1986, Robustness of linear feedback control systems to unmodelled high-frequency dynamics, *Int.J.Control* **44**, pp. 1077-1088

Ortega, J.M., 1972, Numerical Analysis (Academic Press, New York)

Osborne, E.E., 1960, On preconditioning of matrices, *J.of ACM* **7**, pp. 338-345

Özgüner, Ü., and Iftar, A., 1989, An optimal control approach to the decentralized robust servomechanism problem, *IEEE-Trans.* **AC-34**, pp. 1268-1271

Pal, S.K., 1991, Fuzzy tools for the management of uncertainty in pattern recognition, image analysis, vision and expert systems, *Int. J. Systems Sci.* **22**, pp.511-549

Palanisamy, K.R., and Bhattacharya, D.K., 1981, System identification via block-pulse functions, *Int. J. Systems Sci.* **12**, pp. 643-647

Palmor, Z.J., and Halevi, Y., 1990, Robustness properties of sampled-data systems with dead time compensators, *Automatica* **26**, pp. 637-640

Paraskevopoulos, P.N., 1983, Chebyshev series approach to system identification, analysis and optimal control, *Journal of the Franklin Institute* **316**, No. 2, pp. 135-157

Paraskevopoulos, P.N., Sparis, P.D., and Mouroutsos, S.G., 1985, The Fourier series operational matrix of integration *Int. J. Systems Sci.* **16**, pp. 171-176

Patel, R.V., 1975, On zeros of multivariable systems, *Int. J. Control* **21**, pp. 599-608

Patel, R.V., and Toda, M., 1980, Quantitative measures of robustness for multivariable systems, *Proceedings of Joint Automatic Control Conference San Francisco* Paper TP 8-A

Patel, R.V., Toda, M., and Sridhar, B., 1977, Robustness of linear quadratic state feedback designs in the presence of system uncertainty, *IEEE-Trans.* **AC-22**, pp. 945-949

Perron, O., 1930, Die Stabilitätsfrage bei Differentialgleichungen, *Mathematische Zeitschrift* **32**, pp. 703-728

Peterka, V., 1986, Control of uncertain process: applied theory and algorithms, *Supplement to the Journal Kybernetika* **22**, pp. 3-102 (Academia, Praha)

Peterka, V., 1989, Adaptation of LQG control design to engineering needs. In: *Warwick, K., (Ed.),* Lecture Notes on Control **158** (1991) (Springer, Berlin)

Petersen, I.R., 1990, A new extension to Kharitonov's theorem, *IEEE-Trans.* **AC-35**, pp. 825-828

Petkovski, D.B., 1985, Robustness of decentralized control systems subject to sensor perturbations, *Proc. IEE* **Part D 132**, pp. 53-60

Petkovski, D.B., 1988, Stability analysis of interval matrices: improved bounds, *Int.J.Control* **48**, pp. 2265-2273

Peyton Jones, J.C., and Billings, S.A., 1991, Describing functions, Volterra series, and the analysis of non-linear systems in the frequency domain, *Int. J. Control* **53**, pp. 871-887

Pierre, C., 1989, Root sensitivity to parameter uncertainties: a statistical approach, *Int.J.Control* **49**, pp. 521-532

Poolla, K., Shamma, J.S., Wise, K.A., 1990, Linear and nonlinear controllers for robust stabilization problems: a survey, *11th IFAC-Congress Tallinn*, Vol. **5**, pp. 176-183

Postlethwaite, I., Edmunds, J.M., and MacFarlane, A.G.J., 1981, Principal gains and principal phases in the analysis of linear multivariable feedback systems, *IEEE-Trans.* **AC-26**, pp. 32-46

Postlethwaite, I., and Foo, Y.K., 1985, Robustness with simultaneous pole and zero movement across the $j\omega$-axis, *Automatica* **21**, pp. 433-443

Postlethwaite, I., Tsai, M.C., and Gu, D.W., 1990, Weighting function selection in H_∞ design, *11th IFAC-Congress Tallinn*, Vol. **5**, pp. 104-109

Prasad, R., Ahson, S.I., and Mahalanabis, A.K., 1988, Decentralized observer for interconnected systems, *Int.J.Systems Sci.* **19**, pp. 1201-1211

Preuss, H.P., 1979, Zustandsregler-Entwurf durch näherungsweise Lösung der multilinearen Bestimmungsgleichungen des Polvorgabeproblems, *Regelungstechnik* **27**, pp. 326-332

Preuss, H.P., 1980, Entwurf stationär genauer Zustandsrückführungen, *Regelungstechnik* **28**, pp. 51-56

Pujara, L.R., 1990, On the stability of uncertain polynomials with dependent coefficients, *IEEE-Trans.* **AC-35**, pp. 756-759

Qianhua, W., and Zhongjun, Z., 1987, Algorithm for oblaining the proper relatively prime matrices of polynomial matrices, *Int.J.Control* **46**, pp. 769-784

Qiu, L., and Davison, E.J., 1986, New perturbation bounds for the robust stability of linear state space models, *Proc. 25th IEEE-Conf. on Decision and Control* pp.751-755

Qiu, L., and Davison, E.J., 1988, A new method for the stability robustness determination of state space models with real perturbations, *Proc.27th IEEE-Conf. on Decision and Control* pp. 538-543

Qiu, L., and Davison, E.J., 1988a, Computation of the stability robustness of large state space models with real perturbations, *Proc. 27th IEEE Conf. on Decision and Control, Austin*, pp. 1380-1385

Rachid, A., 1989, Robustness of discrete systems under structured uncertainties, *Int.J.Control* **50**, pp. 1563-1566

Rachid, A., 1990, Robustness of pole assignment in a specified region for perturbed systems, *Int.J.Systems Sci.* **21**, pp. 579-585

Ranganathan, V., Jha, A.N., and Rajamani, V.S., 1984, Identification of linear distributed systems via Laguerre polynomials, *Int. J. Systems Sci.* **15**, pp. 1101-1106

Ranganathan, V., Jha, A.N., and Rajamani, V.S., 1986, Identification of non-linear distributed systems via a Laguerre-polynomial approach, *Int. J. Systems Sci.* **17**, pp. 241-249

Ranganathan, V., Jha, A.N., and Rajamani, V.S., 1987, Identification of distributed linear time-varying systems via Laguerre polynomials, *Int. J. Systems Sci.* **18**, pp. 681-685

Rao, C.R., and Mitra, S.K., 1971, Generalized Inverse of Matrices and its Applications (John Wiley & Sons, New York London)

Rech, C., 1990, Robustness of interconnected systems to structural disturbances in structural controllability and observability, *Int.J.Control* **51**, pp. 205-217

Reid, J.G., 1983, Linear System Fundamentals, (McGraw-Hill, Auckland)

Rillings, J.H., and Roy, R.J., 1970, Analog sensitivity design of Saturn V launch vehicle, *IEEE-Trans.* **AC-15**, pp 437-442

Roberts, P.D., 1967, Orthogonal transformations applied to control system identification and optimisation (IFAC Symposium on Identification in Automatic Control Systems, Prague, paper 5.12)

Rogosinski, W.W., and Shapiro, H.S., 1953, On certain extremum problems for analytic functions, *Acta math.* **90**, pp. 287-318

Rohrer, R.A., and Sobral, M.Jr., 1965, Sensitivity considerations in optimal system design, *IEEE-Trans.* **AC-10**, pp. 43-48

Rommelfanger, H., 1988, Enscheiden bei Unschärfe, Fuzzy Decision Support-Systeme (Springer, Berlin New York)

Roppenecker, G., 1987, Zeitbereichsentwurf linearer dynamischer Systeme mittels Vollständiger Modaler Synthese, *Automatisieruntstechnik* **35**, pp. 89-95

Roppenecker, G., 1988, Zur Reglerberechnung mittels Vollständiger Modaler Synthese, *Automtisierungstechnik* **36**, pp. 117-119

Roppenecker, G., and Lohmann, B., 1989, Vollständige Modale Synthese und Geometrische Methode: Berührungspunkte, *Automatisierungstechnik* **37**, pp. 160-161

Roppenecker, G., and O'Reilly, J., 1989, Parametric output feedback controller design, *Automatica* **25**, pp. 259-265

Rosen, J.B., 1960, The gradient projection method for nonlinear programming, *J. Soc. Indust.Appl.Math.* **8**, pp. 181-217

Rosenbrock, H.H., 1970, State-space and Multivariable Theory (Nelson, London)

Rosenbrock, H.H., 1972, Multivariable circle theorems. In *Bell, D.J., (Ed.) 1973*, pp. 345-365

Rosenbrock, H.H., 1974, Computer-Aided Control System Design, (Academic Press, New York)

Rotea, M.A., and Khargonekar, P.P., 1991, H^2-optimal control with an H^∞-constraint: the state feedback case, *Automatica* **27**, pp.307-316

Rotella, F., and Dauphin-Tanguy, G., 1988, Non-linear systems: identification and optimal control, *Int.J.Control* **48**, pp. 525-544

Saberi, A., Chen, B.M., and Sannuti, P., 1991, Theory of LTR for non-minimum phase systems, recoverable target loops, and recovery in a subspace. Analysis and Design. *Int. J. Control* **53**, pp.

1067-1160
Saberi, A., and Sannuti, P., 1990, Observer-based control of uncertain systems with nonlinear uncertainties, *Int. J. Control* **52**, pp. 1107-1130
Saberi, A., and Sannuti, P., 1990a, Observer design for loop transfer recovery and for uncertain dynamical systems, *IEEE-Trans.* **AC-35**, pp. 878-897
Sadeghi, T., Hoitsma, D.H., and Schoenberg, L., 1983, A control law for pole/variant zero placement. *Proceedings of the 1983 American Control Conference, San Francisco* **3** , pp. 912-917
Safonov, M.G., 1982, Stability margins of diagonally perturbed multivariable feedback systems, *Proc.IEE Part D* **129**, pp. 251-256
Safonov, M.G., 1990, Future directions in H_∞ robust control theory, *11th IFAC-Congress Tallinn,* Vol. **5**, pp. 147-151
Safonov, M.G., and Athans, M., 1981, A multiloop generalization of the circle criterion for stability margin analysis, *IEEE-Trans.* **AC-26**, pp. 415-422
Safonov, M.G., and Doyle, J.C., 1984, Minimizing conservativeness of robustness singular values. Chapter 11 in Tzafestas, S.G., (Ed.) Multivariable Control, New Concepts and Tools (D. Reidel Publishing Company, Dordrecht Boston Lancaster)
Safonov, M.G., Jonckheere, E.A., Verma, M., and Limebeer, D.J.N., 1987, Synthesis of positive real multivariable feedback systems, *Int.J.Control* **45**, pp. 817-842
Safonov, M.G., Laub, A.J., and Hartmann, G.I., 1981, Feedback properties of multivariable systems: the role and use of the return difference matrix, *IEEE-Trans.* **AC-26**, pp. 47-65
Safonov, M.G., and Verma, M.S., 1985, L^∞ Optimization and Hankel Approximation, *IEEE-Trans.* **AC-30**, pp. 279-280
Sage, A.P., and White, C.C. III, 1977, Optimum Systems Control (Prentice-Hall, Englewood Cliffs)
Sánchez Peña, R.S., and Sideris, A., 1990, Robust with real parametric and structured complex uncertainty, *Int. J. Control* **52**, pp. 753-765
Sandberg, I.W., 1964, A frequency-domain condition for the stability of feedback systems containing a single time-varying nonlinear element, *Bell System Technical J.* **43**, pp. 1601-1608
Sandberg, I.W., 1964, On the L_2-boundedness of solutions of nonlinear functional equations, *The Bell System Technical J.* **43**, pp. 1581-1599
Saniuk, J.M., and Rhodes, I.B., 1987, A matrix inequality associated with bounds on solutions of algebraic Riccati and Lyapunov equations, *IEEE-Trans.* **AC-32**, pp. 739-740
Sannuti, P., 1977, Analysis and synthesis of dynamic systems via block-pulse functions, *Proc. IEE* **124**, pp. 569-571
Sansone, G., 1959, Orthogonal Functions (Interscience Publishers, New York)
Saridis, G.N., 1985, Intelligent control-operating systems in uncertain environments, In: Ackermann, J., (Ed.), 1985, Uncertainty and Control (Springer, Berlin New York)
Saydy, L., Tits, A.L., and Abed, E.H., 1988, Robust stability of linear systems relative to guarded domains, *Proc. 27th IEEE Conf. on Decision and Control, Austin* pp. 544-551
Saydy, L., Tits, A.L., and Abed, E.H., 1990, Guardian maps and the generalized stability of the parametrized families of matrices and polynomials, *Mathematics of Control, Signals, and Systems* **3**, pp. 345-371
Schmidt, J., 1987, Einfache Methoden zur Berechnung des invarianten Polprodukts *Automatisierungstechnik* **35**, pp. 323-326
Schmidt, J., 1988, Spezielle "Robustheitseigenschaften" varianter Nullstellen, *Automatisierungstechnik* **36**, pp. 355-356
Schmitendorf, W.E., 1988, Design of observer-based robust stabilizing controllers, *Automatica* **24**, pp. 693-696
Schüssler, H.W., 1976, A stability theorem for discrete systems, *IEEE-Trans.* **ASSP-24**, pp. 87-89
Schwarz, H., 1971, Mehrfachregelungen, Zweiter Band (Springer, Berlin)
Schweppe, F.C., 1973, Uncertain Dynamic Systems, (Prentice-Hall, Englewood Cliffs)
Seneta, E., 1973, Non-Negative Matrices and Markov Chains, 2nd Edition, (Springer, New York)
Sevaston, G.E., and Longman, R.W., 1985, Gain measures of controllability and observability, *Int.J.Control* **41**, pp.865-893
Sevaston, G.E., and Longman, R.W., 1988, Systematic design of robust control systems using small gain stability concepts, *Int.J.Systems Sci.* **19**, pp. 439-451
Sezer, M.E., and Šiljak, D.D., 1988, Robust stability of discrete systems, *Int. J. Control* **48**, pp. 2055-2063

Sezer, M.E., and Šiljak, D.D., 1989, A note on robust stability bounds, *IEEE-Trans.* **AC-34**, pp. 1212-1215

Shaked, U., 1976, A general transfer-function approach to the steady-state linear quadratic Gaussian stochastic control problem, *Int.J.Control* **24**, pp. 771-800

Shaked, U., 1986, Guaranteed stability margins for the discrete-time linear quadratic optimal regulator *IEEE-Trans.* **AC-31**, pp. 162-165

Shaked, U., 1990, Directional interpolation via all-pass transfer function matrices and its application in Hankel norm approximations, *IEEE-Trans.* **AC-35**, pp. 1167-1170

Shane, B.A., and Barnett, S., 1972, Insensitivity of constant linear systems to finite variations in parameters. In *Bell, D.J., (Ed.) 1973* pp. 393-404

Shane, B.A., and Barnett, S., 1972a, Sensitivity of stable linear systems, *IEEE-Trans.* **AC-17**, pp. 148-150

Shane, B.A., and Barnett, S., 1973, Sensitivity of convergent matrices and polynomials, *J. Mathematical Analysis and Applications* **43**, pp. 114-122

Sharkey, P.M., and O'Reilly, J., 1988, Composite control of non-linear singularly perturbed systems: a geometric approach, *Int. J. Control* **48**, pp. 2491-2506

Shieh, L.S., Tsai, J.S.H., and Coleman, N.P., 1987, Rectangular and polar representations of a complex matrix, *Int.J.Systems Sci.* **18**, pp. 1825-1838

Šiljak, D.D., 1978, Large-Scale Dynamic Systems: Stability and Structure (North-Holland, New York)

Sinswat, V., and Fallside, F., 1977, Eigenvalue/Eigenvector assignment by state feedback, *Int.J.Control* **26**, pp. 389-403

Sinswat, V., Patel, R.V., and Fallside, F., 1976, A method for computing invariant zeros and transmission zeros of invertible systems, *Int. J. Control* **23**, pp. 183-196

Skelton, R.E., and Wagie, D.A., 1984, Minimal root sensitivity in linear systems, *J. Guidance Control Dynam.* **7**, pp. 570-574

Skelton, R.E., and Owens, D.H., (Eds.) 1985, Model Error Concepts and Compensation, *Proc. IFAC Workshop, Boston* (Pergamon Press, Oxford New York)

Skogestad, S., Morari, M., and Doyle, J.C., 1988, Robust control of ill-conditioned plants: high-purity distillation, *IEEE-Trans.* **AC-33**, pp.1092-1105

Slotine, J.J., and Sastry, S.S., 1983, Tracking control of non-linear systems using sliding surfaces, with application to robot manipulators, *Int.J.Control* **38**, pp. 465-492

Sobel, K.M., Banda, S.S., and Yeh, H.H., 1989, Robust control for linear systems with structured state space uncertainty, *Int.J.Control* **50**, pp. 1991-2004

Soh, C.B., 1989, Comments on "Smallest destabilizing perturbations for linear systems, *Int.J.Control* **49**, pp. 1813-1814

Soh, C.B., 1991, Robust stability of discrete-time systems using delta operators, *IEEE-Trans.* **AC-36**, pp.377-380

Soh, C.B., Berger, C.S., and Dabke, K.P., 1985, On the stability properties of polynomials with perturbed coefficients, *IEEE-Trans.* **AC-30**, pp. 1033-1036

Soh, C.B., and Chan, C.K., 1989, Destabilizing perturbations for linear systems, *Int.J.Control* **50**, pp. 2101-2104

Soh, Y.C., and Evans, R.J., 1985, Robust multivariable regulator design - general cases, *Proc. 24th IEEE-Conference on Decision and Control, Lauderdale*, pp. 1323-1327

Soh, Y.C., and Evans, R.J., 1988, Stability analysis of interval matrices - continuous and discrete systems, *Int.J.Control* **47**, pp. 25-32

Soh, Y.C., Evans, R.J., Petersen, I.R., and Betz, R.E., 1987, Robust pole assignment, *Automatica* **23**, pp. 601-610

Soh, Y.C., and Foo, Y.K., 1990, Generalization of strong Kharitonov theorems to the left sector, *IEEE-Trans.* **AC-35**, pp. 1378-1382

Soliman, H.M., Darwish, M., and Fantin, J., 1978, Stabilization of a large-scale power system via a multilevel technique, *Int. J. Systems Sci.* **9**, pp. 1091-1111

Sondergeld, K.P., 1983, A generalization of the Routh-Hurwitz stability criteria and an application to a problem in robust controller design, *IEEE-Trans.* **AC-28**, pp. 965-970

deSouza, E., and Bhattacharyya, S.P., 1981, Controllability, observability and the solution of $AX - XB = C$, *Linear Algebra and Its Applications* **39**, pp. 167-188

Staudte, R.G., and Sheather, S.J., 1990, Robust Estimation and Testing (Wiley, New York)

Stengel, R.F., 1980, Some effects of parameter variations on the lateral-directional stability of aircraft, *J. Guidance Contr.* **3**, pp.124-131

Stengel, R.F., and Ray, L.R., 1991, Stochastic robustness of linear time-invariant control systems, *IEEE-Trans.* **AC-36**, pp.82-86

Stewart, G.W., 1973, Introduction to Matrix Computations, (Academic Press, New York)

Stoer, J., and Witzgall, C., 1962, Transformations by diagonal matrices in a normed space, *Numer.Math.* **4**, pp. 158-171

Strejc, V., 1981, State Space Theory of Discrete Linear Control, (Academica, Prague)

Su, T.J., Fong, I.K., Kuo, T.S., and Sun, Y.Y., 1988, Robust stability for linear time-delay systems with linear parameter perturbations, *Int.J.Systems Sci* **19**, pp. 2123-2129

Su, T.J., Kuo, T.S., and Sun, Y.Y., 1990, Robust stability of linear perturbed discrete systems with saturating actuators, *Int.J.Systems Sci.* **21**, pp. 1273-1279

Su, T.J., Kuo, T.S., and Sun, Y.Y., 1990a, Robust stability of linear perturbed discrete large-scale systems with saturating actuators, *Int. J. Systems Sci.* **21**, pp. 2263-2272

Subbayyan, R., Sarma, V.V.S., and Vaithilingam, M.C., 1978, An approach for sensitivity-reduced design of linear regulators, *Int.J.Systems Sci.* **9**, pp. 65-74

Suh, I.H., Moon, Y.S., and Bien, Z., 1981, Comments on "On decentralized feedback stabilization of large-scale interconnected systems", *Int. J. Control* **34**, pp. 1045-1047

Sultan, M.A., Hassan, M.F., and Calvet, J.L., 1988, Parameter estimation in large-scale systems using sequential decomposition *Int.J.Systems Sci* **19**, pp. 487-496

Szegö, G., 1975, Orthogonal Polynomials, Fourth Edition (American Mathematical Society, New York)

Taussky, O., 1964, On the variation of the characteristic roots of a finite matrix under various changes of its elements In: Recent Advances in Matrix Theory, pp. 125-138 (University of Wisconsin Press, Wisconsin)

Tempo, R., 1990, A dual result to Kharitonov's theorem, *IEEE-Trans.* **AC-35**, pp. 195-198

Tesi, A., and Vicino, A., 1990, Robust stability of state-space models with structured uncertainties, *IEEE-Trans.* **AC-35**, pp. 191-195

Tesi, A., and Vicino, A., 1991, Robustness analysis for linear dynamical systems with linearly correlated parametric uncertainties, *IEEE-Trans.* **AC-35**, pp. 186-191

Tesi, A., and Vicino, A., 1991a, Robust absolute stability of Lur'e control systems in parameter space, *Automatica* **27**, pp. 147-151

Thrall, R.M., and Tornheim, L., 1957, Vector Spaces and Matrices, (Wiley, New York)

Tomović, R., and Vukobratović, M., 1972, General Sensitivity Theory, (Elsevier, New York)

Tsay, S.C., and Lee, T.T., 1988, Solution of variational problems via general orthogonal polynomials, *Int. J. Systems Sci.* **19**, pp. 431-437

Tsay, S.C., Wu I.L., and Lee, T.T., 1988, Optimal control of linear time-delay systems via general orthogonal polynomials, *Int.J.Systems Sci.* **19**, pp. 365-376

Tsui, C.C., 1985, A new algorithm for the design of multifunctional observers, *IEEE-Trans.* **AC-30**, pp. 89-93

Tsui, C.C., 1986, On the order reduction of linear function observers, *IEEE-Trans.* **AC-31**, pp. 447-449

Tsui, C.C., 1988, On robust observer compensator design, *Automatica* **24**, pp. 687-692

Tsui, C.C., 1988a, New approach to robust observer design, *Int.J.Control* **47**, pp. 745-751

Tsypkin, Y. Z., 1985, Optimality in adaptive control systems, In: *Ackermann, J., (Ed.), 1985,* Uncertainty and Control (Springer, Berlin New York)

Tsypkin, Y.Z., and Polyak, B.T., 1991, Frequency-domain criterion for l_p-robust stability of continuous linear systems, *IEEE-Trans.* **AC-36**

Unbehauen, H., 1983, *Regelungstechnik II* (Vieweg, Braunschweig)

Usoro, P.B., Schweppe, F.C., Gould, L.A., and Wormley, D.N., 1982, A Lagrange approach to set-theoretic control synthesis, *IEEE-Trans.* **AC-27**, pp. 393-399

Utkin, V.I., 1977, Variable structure systems with sliding modes, *IEEE-Trans.* **AC-22**, pp. 212-222

Vaz, A.F., and Davison, E.J., 1989, The structured robust decentralized servomechanism problem for interconnected systems, *Automatica* **25**, pp. 267-272

Verde, C., und Frank, P.M., 1987, Vergleich verschiedener Verfahren zum Entwurf parameterunempfindlicher Regelkreise, *Automatisierungstechnik* **35**, pp. 396-401

Verde, C., and Frank, P.M., 1988, Sensitivity reduction of the linear quadratic regulator by matrix modification, *Int.J.Control* **48**, pp. 211-223

Verma, M., and Jonckheere, E., 1984, L^{∞}-compensation with mixed sensitivity as a broadband matching problem, *Systems and Control Letters* **4**, pp. 125-129

Vetter, W.J., 1970, Derivative operations on matrices, *IEEE-Trans.* **AC-15**, pp.241-244 and **AC-16**, p.113

Vetter, W.J., 1971, An extension to gradient matrices, *IEEE-Trans.* **SMC-1**, pp. 184-186

Vetter, W.J., 1973, Matrix calculus operations and Taylor expansions, *SIAM Review* **15**, pp.352-369

Vetter, W.J., Balchen, J.G., and Pohner, F., 1972, Near-optimal control of parameter-dependent linear plants, *Int.J.Control* **15**, pp. 683-691

Vicino, A., 1989, Robustness of pole location in perturbed systems, *Automatica* **25**, pp. 109-113

Vicino, A., and Tesi, A., 1989, Robustness bounds for classes of structured perturbations, In: *Milanese, M., Tempo, R., and Vicino, A., (Eds.) 1989* Robustness in Identification and Control (Plenum Press, New York London) pp. 147-163

Vidyasagar, M., 1975, Coprime factorizations and stability of multivariable distributed feedback systems, *SIAM J.Control* **13**, pp. 1144-1155

Vidyasagar, M., 1978, Nonlinear Systems Analysis (Prentice-Hall, Englewood Cliffs)

Vidyasagar, M., 1984, The graph metric for unstable plants and robustness estimates for feedback stability, *IEEE-Trans.* **AC-29**, pp. 403-417

Vidyasagar, M., 1985, Control System Synthesis: A Factorization Approach, (MIT Press, Cambridge/ Mass. London)

Vidyasagar, M., and Kimura, H., 1986, Robust controllers for uncertain linear multivariable systems, *Automatica* **22**, pp. 85-94

Vidyasagar, M., and Viswanadham, N., 1982, Algebraic design techniques for reliable stabilization, *IEEE-Trans.* **AC-27**, pp. 1085-1095

Walker, D.J., 1990, Robust stabilizability of discrete-time systems with normalized stable factor perturbation, *Int.J.Control* **52**, pp. 441-455

Walton, K., and Marshall, J.E., 1987, Direct method for TDS stability analysis, *Proc. IEE* **Part D 134**, pp. 101-107

Wang, M.L., Chang, R.Y., and Yang, S.Y., 1987, Identification of a single-variable linear time-varying system via generalized orthogonal polynomials, *Int. J. Systems Sci.* **18**, pp. 1659-1671

Wang, M.L., Jan, Y.J., and Chang, R.Y., 1987, Analysis and parameter identification of time-delay linear systems via generalized orthogonal polynomials, *Int. J. Systems Sci.* **18**, pp. 1645-1658

Wang, M.L., Yang, S.Y., and Chang, R.Y., 1987, Analysis of systems with multiple time-varying delays via generalized block-pulse functions, *Int. J. Systems Sci.* **18**, pp. 543-552

Wang, S.D., and Kuo, T.S., 1990, Design of robust linear state feedback laws: ellipsoidal set-theoretic approach, *Automatica* **26**, pp. 303-309

Wang, S.D., Kuo, T.S., and Hsu, C.F., 1986, Trace bounds on the solution of the algebraic Riccati and Lyapunov equation, *IEEE-Trans.* **AC-31**, pp. 654-656

Wang, W.J., Song, C.C., and Kao, C.C., 1991, Robustness bounds for large-scale time-delay systems with structured and unstructured uncertainties, *Int. J. Systems Sci.* **22**, pp. 209-216

Wang, W.P., 1990, New approach to analysis and optimal control of time-varying systems via Taylor series, *Int.J.Systems Sci.* **21**, pp. 1831-1839

Wei, K., 1989, Robust stabilization of linear time-invariant systems via linear control, In: *Milanese, M., Tempo, R., and Vicino, A., (Eds.) 1989* Robustness in Identification and Control (Plenum Press, New York London) pp. 311-319

Weihrich, G., 1973, Optimale Regelungen linearer deterministischer Prozesse (Oldenbourg, München Wien)

Weihrich, G., 1977, Mehrgrößen-Zustandsregelung unter Einwirkung von Stör- und Führungssignalen, *Automatisierungstechnik* **25**, pp. 166-172 and pp. 204-209

Weinmann, A., 1981, Variable-rate and random-rate sampling processes in automatic control, *8th IFAC Congress Kyoto,* Paper 53.5

Weinmann, A., 1985, Die zu erwartende Stabilitätsgrenze von Regelungen mit zufälliger Abtastperiode, *Regelungstechnik* **33**, pp. 257-258

Weinmann, A., 1988, Regelungen, Band 1,2,3; 2. Auflage, (Springer, Wien New York)

Weinmann, A., 1988, Die Anfangswertübergabe in Matrizendarstellung, *Elektrotechnik und Informationstechnik e & i* **105**, pp. 313-314

Weinmann, A., 1991, Robust algorithms based on approximate models, *ELIN-Z.* **43**, pp.40-43

Weston, J.E., and Bongiorno, J.J.Jr., 1972, Extension of analytic design techniques to multivariable feedback control systems, *IEEE-Trans.* **AC-17**, pp. 613-620

Wilkinson, J.H., 1965, The Algebraic Eigenvalue Problem (Clarendon Press, Oxford)

Willsky, A.S., 1976, A survey of design methods for failure detection in dynamic systems, *Automatica*,**12**, pp. 601-611

Wilson, D.A., 1989, Tutorial solution to the linear quadratic regulator problem using H_2 optimal control theory, *Int.J.Control* **49**, pp. 1073-1077

Wolovich, W.A., 1974, Linear Multivariable Systems (Springer, New York)

Wolovich, W.A., and Ferreira, P., 1979, Output regulation and tracking in linear multivariable systems, *IEEE-Trans.* **24**, pp. 460-465

Wong, P.K., Stein, G., and Athans, M., 1978, Structural reliability and robustness properties of optimal linear-quadratic multivariable regulators, 7th IFAC Congress Helsinki, pp. 1797-1805

Wonham, M.W., 1967, Optimal stationary control of a linear system with state-dependent noise, *SIAM-J. on Control* **5**, pp. 486-500

Wu, M.Y., and Desoer, C.A., 1969, L^p-Stability ($1 \leq p \leq \infty$) of nonlinear time-varying feedback systems, *SIAM J.Control* **7**, pp. 356-364

Wu, W.T., Jang, Y.J., and Chu, Y.T., 1991, On-line stability index for robust control, *Int. J. Systems Sci.* **22**, pp.495-505

Wunsch, G., 1971, Systemtheorie der Informationstechnik (Geest and Portig, Leipzig)

Xin, L.X., 1987, Stability of interval matrices, *Int.J.Control* **45**, pp. 203-210

Xinogalas, T.C., Mahmoud, M.S., and Singh, M.G., 1982, Hierarchical computation of decentralized gains for interconnected systems, *Automatica***18**, No.4, pp. 473-478

Xu, J.H., and Mansour, M., 1988, H_∞-optimal robust regulation of MIMO-systems, *Int.J.Control* **48**, pp. 1327-1341

Yager, R.R., Ovchinnikov, S., Tong, R.M., and Nguyen, H.T., (Eds.), 1987, Fuzzy Sets and Applications: Selected Papers by L.A. Zadeh (Wiley, New York)

Yahagi, T., 1973, On the design of optimal output feedback control systems, *Int.J.Control* **18**, pp. 839-848

Yahagi, T., 1977, Optimal output feedback control with reduced performance index sensitivity, *Int.J.Control* **25**, pp. 769-783

Yan, W., and Anderson, B.D.O., 1990, The simultaneous optimization problem for sensitivity and gain margin, *IEEE-Trans.* **AC-35**, pp. 558-563

Yang, W.C., and Tomizuka, M., 1990, Discrete-time robust control via state feedback systems for single-input systems, *IEEE-Trans.* **AC-35**, pp. 590-598

Yaz, E., 1988, Deterministic and stochastic robustness measures for discrete systems, *IEEE-Trans.* **AC-33**, pp. 952-955

Yaz, E., 1990, A generalization of the uncertainty threshold principle, *IEEE-Trans.* **AC-35**, pp.942-944

Yaz, E., and Niu, X., 1989, Stability robustness of linear discrete-time systems in the presence of uncertainty, *Int.J.Control* **50**, pp. 173-182

Yaz, E., and Yildizbayrak, N., 1985, Robustness of feedback-stabilized systems in the presence of nonlinear and random perturbations, *Int.J.Control* **41**, pp. 345-353

Yedavalli, R.K., 1985a, Improved measures of stability robustness for linear state space models, *IEEE-Trans.* **AC-30** , pp. 577-579

Yedavalli, R.K., 1985b, Perturbation bounds for robust stability in linear state space models, *Int.J.Control* **42**, pp. 1507-1517

Yedavalli, R.K., 1985c, Time-domain robust control design for linear quadratic regulators by perturbation bound analysis, In: Skelton, R.E., and Owens, D.H., (Eds.), 1985,

Yedavalli, R.K., 1986, Stability analysis of interval matrices: another sufficient condition, *Int.J.Control* **43**, pp. 767-772

Yedavalli, R.K., 1989, On measures of stability robustness for linear uncertain systems, In: *Milanese, M., Tempo, R., and Vicino, A., (Eds.) 1989* Robustness in Identification and Control (Plenum Press, New York London) pp. 303-310

Yeh, H.H., Banda, S.S., and Ridgely, D.B., 1985, Stability robustness measures utilizing structural information, *Int.J.Control* **41**, pp. 365-387

Yeung, K.S., 1983, Linear system stability under parameter uncertainties, *Int.J.Control* **38**, pp. 459-464

Yosida, K., 1980, Functional Analysis, 6th edition (Springer, Berlin New York)

Youla, D.C., 1961, On the factorization of rational matrices, *IRE-Trans.* **IT-7**, pp. 172-189

Youla, D.C., Bongiorno,J.J.Jr., and Jabr, H.A., 1976, Modern Wiener-Hopf design of optimal controllers - Part I: The single-input-output case, *IEEE-Trans.* **AC-21**, pp. 3-13

Youla, D.C., Jabr, H.A., and Bongiorno, J.J., 1976, Modern Wiener-Hopf design of optimal controllers - Part II: The multivariable case, *IEEE-Trans.* **AC-21**, pp. 319-338

Zadeh, L.A., 1965, Fuzzy sets, *Information and Control* **8**, pp.338-353

Zadeh, L.A., and Desoer, C.A., 1963, Linear System Theory (McGraw-Hill, New York London)

Zafiriou, E., and Morari, M., 1985, Digital controllers for SISO systems: a review and a new algorithm, *Int.J.Control* **42**, pp. 855-876

Zafiriou, E., and Morari, m., 1986, Design of robust digital controllers and sampling-time selection for SISO systems, *Int.J.Control* **44**, pp. 711-735

Zak, S.H., 1990, On the stabilization and observations of nonlinear/uncertain dynamic systems, *IEEE-Trans.* **AC-35**, pp. 604-607

Zakian, V., 1989, Critical systems and tolerable inputs, *Int. J. Control* **49**, pp. 1285-1289

Zames, G., 1966, On the input-output stability of time-varying nonlinear feeback systems, *IEEE-Trans.* **AC-11**, pp. 228-238 and pp. 465-476

Zames, G., 1981, Feedback and optimal sensitivity: model reference transformations, multiplicative seminorms, and approximate inverses, *IEEE-Trans.* **AC-26**, pp. 301-320

Zames, G., and Shneydor, N.A., 1976, Dither in nonlinear systems, *IEEE-Trans.* **AC-21**, pp. 660-667

Zames, G., and Shneydor, N.A., 1977, Structural stabilization and quenching by dither in nonlinear systems, *IEEE-Trans.* **AC-22**, pp. 352-361

Zheng, J., and Lu, Y., 1988, Novel aggregation algorithm for large-scale systems control, *Int.J.Control* **48**, pp. 1883-1895

Zhao, K.Y., and Barmish, B.R., 1990, Stability robustness of plant-controller families, *11th IFAC-Congress Tallinn*, Vol. **5**, pp. 190-193

Zhou, C.S., and Deng, J.L., 1986, The stability of the grey linear system, *Int.J.Control* **43**, pp. 313-320

Zhou, K., and Khargonekar, P.P., 1988, Robust stabilization of linear systems with norm-bounded time-varying uncertainty, *Systems and Control Letters* **10**, pp.17-20

Zhou, L., and Jong, M.T., 1989, Comment on "Estimating the robust dead time for closed-loop stability", *IEEE-Trans.* **AC-34**, pp. 1324

Zhu, G., and Skelton, R., 1991, Mixed L_2 and L_∞ problems by weight selection in quadratic optimal control, *Int. J. Control* **53**, pp. 1161-1176

Zhu, J.M., 1988, Novel approach to hierarchical control via single- term Walsh series method, *Int. J. Systems Sci.* **19**, pp. 1859-1870

Zurmühl, R., and Falk, S., 1984, Matrizen und ihre Anwendungen (Springer, Berlin)

Index

Page references in **boldface type** indicate definitions